Introductory Algebra

9th Edition

O'Rourke

Introductory Algebra

9th EDITION

Marvin L. Bittinger

Indiana University Purdue University Indianapolis

Boston San Francisco New York
London Toronto Sydney Singapore Madrid
Mexico City Munich Paris Cape Town Hong Kong Montreal

Publisher Greg Tobin
Editor in Chief Maureen O'Connor
Acquisitions Editor Jennifer Crum
Project Manager Kari Heen
Associate Editor Lauren Morse
Editorial Assistant Katie Nopper
Managing Editor Ron Hampton
Production Supervisor Kathleen A. Manley
Editorial and Production Services Martha K. Morong/Quadrata, Inc.
Art Editor and Photo Researcher Geri Davis/The Davis Group, Inc.
Chapter Opener Art Director Meredith Nightingale
Marketing Manager Dona Kenly
Marketing Coordinator Lindsay Skay
Illustrators Network Graphics, J. B. Woolsey Associates, Doug Hart, and Gary Torissi
Prepress Supervisor Caroline Fell
Compositor The Beacon Group, Inc.
Cover Designer Dennis Schaefer
Cover Photograph Gary Conner/Index Stock Imagery
Interior Designer Geri Davis/The Davis Group, Inc.
Print Buyer Evelyn Beaton
Supplements Production Sheila Spinney
Media Producers Ruth Berry and Beth Standring
Software Development Janet Wann and Marty Wright

Photo credits appear on page I-9.

Library of Congress Cataloging-in-Publication Data
Bittinger, Marvin L.
Introductory algebra.—9th ed. / Marvin L. Bittinger.
p. cm.
Includes index.
ISBN 0-201-74631-X (SE: pbk.)—ISBN 0-201-79251-6 (AIE: pbk.)
ISBN 0-321-15484-3 (hdbk)
1. Algebra. I. Title.
QA152.3 .B57 2002
512.9—dc21 2001053823

Reprinted with corrections.

3 4 5 6 7 8 9 10—WC—06 05 04 03

Contents

R PREALGEBRA REVIEW

1 INTRODUCTION TO REAL NUMBERS AND ALGEBRAIC EXPRESSIONS

2 SOLVING EQUATIONS AND INEQUALITIES

3 GRAPHS OF LINEAR EQUATIONS

4 POLYNOMIALS: OPERATIONS

5 POLYNOMIALS: FACTORING

6 RATIONAL EXPRESSIONS AND EQUATIONS

7 GRAPHS, SLOPE, AND APPLICATIONS

8 SYSTEMS OF EQUATIONS

9 RADICAL EXPRESSIONS AND EQUATIONS

10 QUADRATIC EQUATIONS

APPENDIXES

Preface

This text is the third in a series of texts that includes the following:

Bittinger: *Basic Mathematics,* Ninth Edition

Bittinger: *Fundamental Mathematics,* Third Edition

Bittinger: *Introductory Algebra,* Ninth Edition

Bittinger: *Intermediate Algebra,* Ninth Edition

Bittinger/Beecher: *Introductory and Intermediate Algebra: A Combined Approach,* Second Edition

Introductory Algebra, Ninth Edition, is a significant revision of the Eighth Edition, particularly with respect to design, art program, pedagogy, features, and supplements package. Its unique approach, which has been developed and refined over nine editions, continues to blend the following elements in order to bring students success:

- **Real data** Real-data applications aid in motivating students by connecting the mathematics to their everyday lives. Extensive research was conducted to find new applications that relate mathematics to the real world.
- **Art program** The art program has been expanded to improve the visualization of mathematical concepts and to enhance the real-data applications.

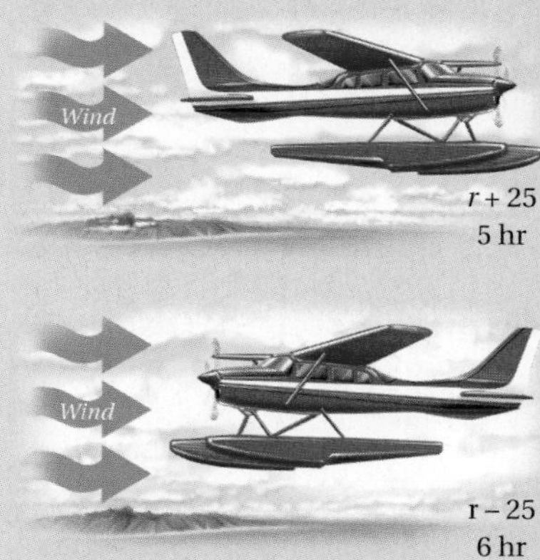

- **Writing style** The author writes in a clear, easy-to-read style that helps students progress from concepts through examples and margin exercises to section exercises.
- **Problem-solving approach** The basis for solving problems and real-data applications is a five-step process (*Familiarize, Translate, Solve, Check,* and *State*) introduced early in the text and used consistently throughout. This problem-solving approach provides students with a consistent framework for solving applications. (See pages 185–186, 231, and 571.)
- **Reviewer feedback** The author solicits feedback from reviewers and students to help fulfill student and instructor needs.
- **Accuracy** The manuscript is subjected to an extensive accuracy-checking process to eliminate errors.
- **Supplements package** All ancillary materials are closely tied with the text and created by members of the author team to provide a complete and consistent package for both students and instructors.

LET'S VISIT THE NINTH EDITION

The style, format, and approach of the Eighth Edition have been strengthened in this new edition in a number of ways.

Updated Applications Extensive research has been done to make the applications in the Ninth Edition even more up-to-date and realistic. A large number of the applications are new to this edition, and many are drawn from the fields of business and economics, life and physical sciences, social sciences, and areas of general interest such as sports and daily life. To encourage students to understand the relevance of mathematics, many applications are enhanced by graphs and drawings similar to those found in today's newspapers and magazines. Many applications are also titled for quick and easy reference, and most real-data applications are authenticated with a source line. (See pages 92, 247–248, 286, 311, and 571.)

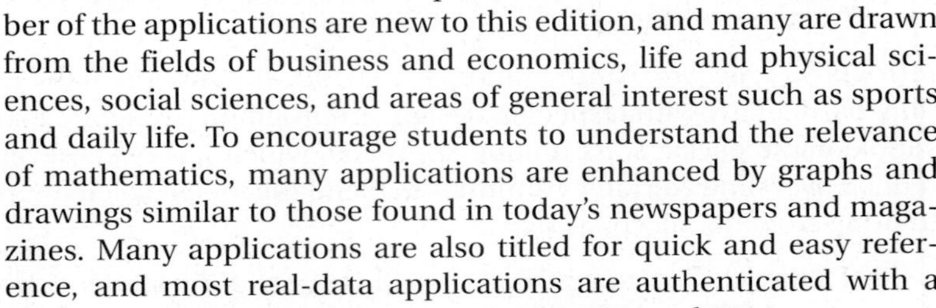

Number of Radio Stations on the Internet

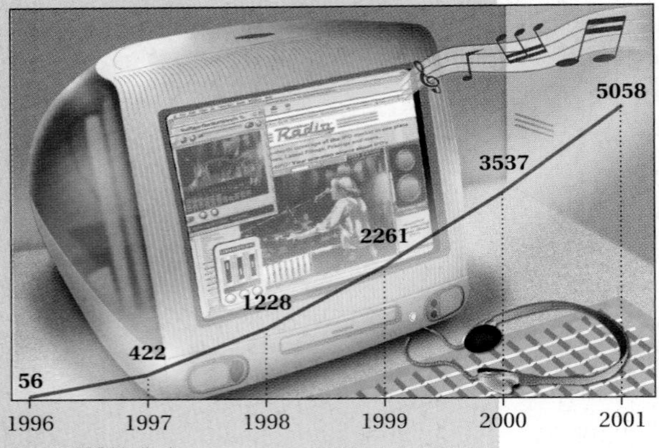

Source: BRS Media, Inc.

Numerous Photographs An application becomes relevant when the connection to the real world is illustrated with a photograph. The Ninth Edition contains approximately 90 photos that immediately spark interest in examples and exercises. (See pages 31, 448, and 640.)

Study Tips Occurring at least twice in every chapter, these mini-lessons provide students with concrete techniques to improve their studying and test-taking skills. These features can be covered in their entirety at the beginning of the course, encouraging good study habits early on (see a complete list of study tips at the back of the book), or they can be used as they occur in the text, allowing students to learn them gradually. These features can also be used in conjunction with Marvin L. Bittinger's "Math Study Skills for Students" videotape, which is free to adopters. Please contact your Addison-Wesley representative for details on how to obtain this videotape. (See pages 154, 368, and 506.)

Calculator Corners Designed specifically for the introductory algebra student, these optional features include graphing-calculator instruction and practice exercises (see pages 131, 250, and 437). Answers to all Calculator Corner exercises appear at the back of the text.

Algebraic–Graphical Connections AG To give students a better visual understanding of algebra, we have included algebraic–graphical connections (see pages 248, 503, and 706). This feature gives the algebra more meaning by connecting it to a graphical interpretation.

New Art To enhance the greater emphasis on real data and applications, we have extensively increased the number of pieces of technical and situational art (see pages 186, 278, 315, 631, and 682).

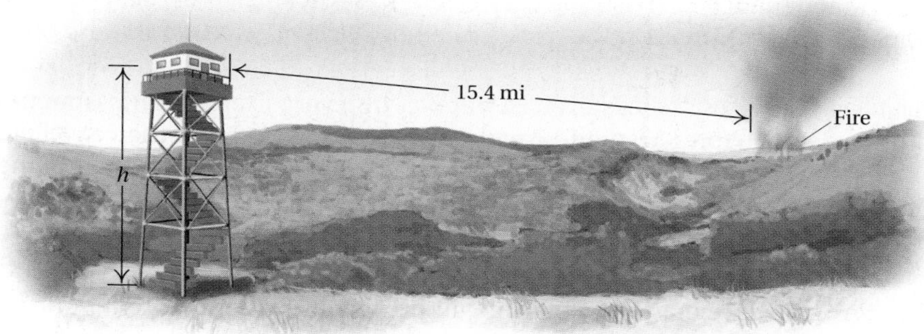

The use of color has been carried out in a methodical and precise manner so that it carries a consistent meaning, which enhances the readability of the text. For example, the use of both red and blue in mathematical art increases understanding of the concepts. When two lines are graphed using the same set of axes, one is usually red and the other blue. Note that equation labels are the same color as the corresponding line to aid in understanding.

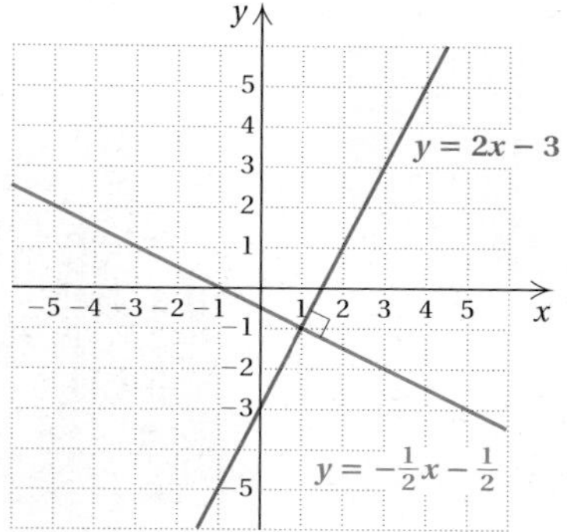

New Design The new design is more open and flexible, allowing for an expanded art and photo package and more prominent headings for the boxed definitions and rules and for the Caution boxes.

Exercises Exercises are paired, meaning that each even-numbered exercise is very much like the odd-numbered one that precedes it. This gives the instructor several options: If an instructor wants the student to have answers available, the odd-numbered exercises are assigned; if an instructor wants the student to practice (perhaps for a test), with no answers available, then the even-numbered exercises are assigned. In this way, each exercise set actually serves as two exercise sets. Answers to all odd-numbered exercises, with the exception of the Discussion and Writing exercises, and *all* Skill Maintenance exercises are provided at the back of the text. If an instructor wants the student to have access to all the answers, a complete answer book is available.

Discussion and Writing Exercises Two Discussion and Writing exercises (denoted by $^{D}{}_{W}$) have been added to every exercise set and Summary and Review. Designed to develop comprehension of critical concepts, these exercises encourage students to both discuss and write about key mathematical ideas in the chapter (see pages 267, 440, and 597).

Skill Maintenance Exercises The Skill Maintenance exercises have been enhanced by the inclusion of 20% more exercises in this edition. These exercises review concepts from other sections of the text in order to prepare students for the Final Examination. Section and objective codes appear next to each Skill Maintenance exercise for easy reference. Answers to all Skill Maintenance exercises appear at the back of the book (see pages 304, 431, and 660).

Synthesis Exercises These exercises appear in every exercise set, Summary and Review, Chapter Test, and Cumulative Review. Synthesis exercises help build critical thinking skills by requiring students to synthesize or combine learning objectives from the section being studied as well as preceding sections in the book. (See pages 113, 239, and 398.)

Content We have made the following improvements to the content of *Introductory Algebra.*

- To provide a review of geometric formulas, Section R.6 ("Geometry") has been added to this edition.

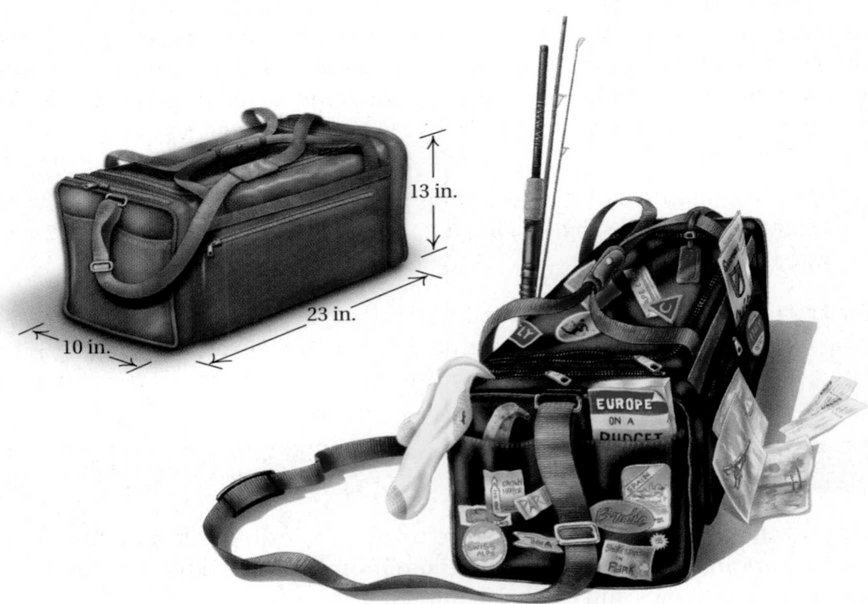

- Emphasis on applications of real numbers has been expanded in Sections 1.3, 1.4, 1.5, and 1.6.
- The concept of slope, formerly located in Chapter 7, is introduced in Section 3.4 ("Slope and Applications") and then reviewed in Section 7.1 ("The Slope–Intercept Equation").
- Section 7.2 ("Graphing Using the Slope and the y-Intercept") is new to the Ninth Edition.
- One new appendix (Appendix D: Mean, Median, and Mode) has been added to the Ninth Edition.

LEARNING AIDS

Interactive Worktext Approach The pedagogy of this text is designed to provide an interactive learning experience between the student and the exposition, annotated examples, art, margin exercises, and exercise sets. This approach provides students with a clear set of learning objectives, involves them with the development of the material, and provides immediate and continual reinforcement and assessment.

Section objectives are keyed by letter not only to section subheadings, but also to exercises in the Pretest, exercise sets, and Summary and Review, as well as to the answers to the Chapter Test and Cumulative Review questions. This enables students to easily find appropriate review material if they are unable to work a particular exercise.

Throughout the text, students are directed to numerous *margin exercises,* which provide immediate reinforcement of the concepts covered in each section.

Objectives

a Given an equation in the form $y = mx + b$, find the slope and the y-intercept; and find an equation of a line when the slope and the y-intercept are given.

b Find an equation of a line when the slope and a point on the line are given.

c Find an equation of a line when two points on the line are given.

Review Material The Ninth Edition of *Introductory Algebra* continues to provide many opportunities for students to prepare for final assessment.

A two-column *Summary and Review* appears at the end of each chapter. The first part is a checklist of some of the Study Tips, as well as a list of important properties and formulas. The second part provides an extensive set of review exercises. Reference codes beside each exercise or direction line preceding it allow the student to easily return to the objective being reviewed (see pages 220, 372, and 453).

Also included at the end of every chapter beginning with Chapter 2 is a *Cumulative Review,* which reviews material from all preceding chapters. At the back of the text are answers to all Cumulative Review exercises, together with section and objective references, so that students know exactly what material to study if they miss a review exercise (see pages 377, 537, and 639).

Both the Summary and Review and the Cumulative Review have been expanded to three pages, allowing for more art and a greater variety of exercises.

For Extra Help Many valuable study aids accompany this text. At the beginning of each exercise set, references to appropriate videos, tutorial software, and other resources make it easy for the student to find the correct support materials.

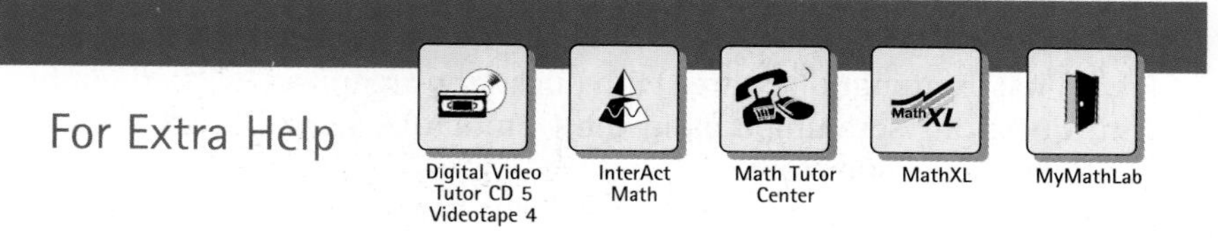

Testing The following assessment opportunities exist in the text.

Chapter Pretests can be used to place students in a specific section of the chapter, allowing them to concentrate on topics with which they have particular difficulty (see pages 144, 542, and 702).

Chapter Tests allow students to review and test comprehension of chapter skills, as well as four objectives from earlier chapters that will be retested (see pages 285, 535, and 583).

In addition, a *Diagnostic Pretest*, found in the *Printed Test Bank/Instructor's Resource Guide* and in MyMathLab, can place students in the appropriate chapter for their skill level by identifying familiar material and specific trouble areas. This may be especially helpful for self-paced courses.

Answers to all Chapter Pretest and Chapter Test questions are found at the back of the book. Section and objective references for Pretest exercises are listed in blue beside each exercise or direction line preceding it. Reference codes for the Chapter Test answers are included with the answers.

SUPPLEMENTS FOR THE INSTRUCTOR

Annotated Instructor's Edition

ISBN 0-201-79251-6

The *Annotated Instructor's Edition* is a specially bound version of the student text with answers to all margin exercises and exercise sets printed in blue near the corresponding exercises.

Instructor's Solutions Manual

ISBN 0-201-79710-0

The *Instructor's Solutions Manual* by Judith A. Penna contains brief worked-out solutions to all even-numbered exercises in the exercise sets and answers to all Discussion and Writing exercises.

Printed Test Bank/Instructor's Resource Guide

by Laurie Hurley
ISBN 0-201-79711-9

The test-bank section of this supplement contains the following:

- A diagnostic test that can place students in the appropriate chapter for their skill level
- Three alternate test forms for each chapter, with questions in the same topic order as the objectives presented in the chapter
- Five alternate test forms for each chapter, modeled after the Chapter Tests in the text
- Three alternate test forms for each chapter, designed for a 50-minute class period
- Two multiple-choice versions of each Chapter Test
- Two cumulative review tests for each chapter, with the exception of Chapter 1
- Eight final examinations: three with questions organized by chapter, three with questions scrambled as in the Cumulative Reviews, and two with multiple-choice questions
- Answers for the Diagnostic Test, Chapter Tests, and Final Examination

The resource-guide section contains the following:

- A conversion guide from the Eighth Edition to the Ninth Edition
- Extra practice exercises (with answers) for 40 of the most difficult topics in the text
- A three-column Summary and Review for each chapter, listing objectives, brief procedures, worked-out examples, multiple-choice problems similar to the example, and the answers to those problems
- Black-line masters of grids and number lines for transparency masters or test preparation
- Indexes to the videotapes and audiotapes that accompany the text

Adjunct Support Manual
ISBN 0-321-12410-3

This manual includes resources designed to help both new and adjunct faculty with course preparation and classroom management, and it also offers helpful teaching tips.

Collaborative Learning Activities Manual
ISBN 0-321-12405-7

The *Collaborative Learning Activities Manual* features group activities that are tied to sections of the text. Instructions for classroom setup are also included in the manual.

Answer Book
ISBN 0-201-79709-7

The *Answer Book* contains answers to all exercises in the exercise sets of the text. Instructors can make quick reference to all answers or have quantities of these booklets made available for sale in the bookstore if they want students to have access to all the answers.

TestGen-EQ/QuizMaster-EQ
ISBN 0-321-10940-6

Available on a dual-platform Windows/Macintosh CD-ROM, this fully networkable software enables instructors to build, edit, print, and administer tests using a computerized test bank of questions organized according to the contents of each chapter. Tests can be printed or saved for on-line testing via a network on the Web, and the software can generate a variety of grading reports for tests and quizzes.

MathXL®: www.mathxl.com
ISBN 0-321-12986-5

The MathXL Web site provides diagnostic testing and tutorial help, all on-line using InterAct Math® tutorial software and TestGen-EQ testing software. Students can take chapter tests correlated to the text, receive individualized study plans based on those test results, work practice problems and receive tutorial instruction for areas in which they need improvement, and take further tests to gauge their progress. Instructors can customize tests and track all student test results, study plans, and practice work. An access card is required.

MyMathLab
MyMathLab is a complete on-line course for Addison-Wesley mathematics textbooks that provides interactive, multimedia instruction correlated to textbook content. MyMathLab is easily customizable to suit the needs of students and instructors and provides a comprehensive and efficient on-line course-management system that allows for diagnosis, assessment, and tracking of students' progress.

MyMathLab features the following:

- Fully interactive multimedia chapter and section folders from the textbook contain a wide range of instructional content, including videos, software tools, audio clips, animations, and electronic supplements.
- Hyperlinks take you directly to on-line testing, diagnosis, tutorials, and gradebooks in MathXL—Addison-Wesley's tutorial and testing system for mathematics and statistics.
- Instructors can create, copy, edit, assign, and track all tests for their course as well as track student tutorial and testing performance.
- With push-button ease, instructors can remove, hide, or annotate Addison-Wesley's preloaded content, add their own course documents, or change the order in which material is presented.
- Using the communication tools found in MyMathLab, instructors can hold on-line office hours, host a discussion board, create communication groups within their class, send e-mails, and maintain a course calendar.
- Print supplements are available on-line, side by side with their textbooks.

For more information, visit our Web site at www.mymathlab.com or contact your Addison-Wesley sales representative for a demonstration.

SUPPLEMENTS FOR THE STUDENT

Student's Solutions Manual
ISBN 0-201-79712-7

The *Student's Solutions Manual* by Judith A. Penna contains fully worked-out solutions with step-by-step annotations for all the odd-numbered exercises in the exercise sets in the text, with the exception of the Discussion and Writing exercises. Students can purchase this manual from Addison-Wesley or their local college bookstore.

Videotapes
ISBN 0-201-88292-2

Digital Video Tutor CD 5 Videotape 4

This videotape series features an engaging team of mathematics instructors who present comprehensive coverage of each section of the text in a student-interactive format. The lecturers' presentations include examples and problems from the text and support an approach that emphasizes visualization and problem solving. A video symbol at the beginning of each exercise set references the appropriate videotape or CD number (see *Digital Video Tutor,* below).

Digital Video Tutor
ISBN 0-321-12384-0, stand-alone

The videotapes for this text are also available on CD-ROM, making it easy and convenient for students to watch video segments from a computer at home or on campus. The complete digitized video set, affordable and portable for students, is ideal for distance learning or supplemental instruction.

"Math Study Skills for Students" Videotape
ISBN 0-321-11739-5

Designed to help students make better use of their math study time, this videotape helps students improve retention of concepts and procedures taught in classes from basic mathematics through intermediate algebra. Through carefully-crafted graphics and comprehensive on-camera explanation, Marvin L. Bittinger helps viewers focus on study skills that are commonly overlooked.

Audiotapes
ISBN 0-201-78619-2

The audiotapes are designed to lead students through the material in each text section. Narrator Bill Saler explains solution steps to examples, cautions students about common errors, and instructs them at certain points to stop the tape and do exercises in the margin. He then reviews the margin-exercise solutions, pointing out potential errors.

InterAct Math® Tutorial CD-ROM
ISBN 0-201-88265-5

InterAct Math

This interactive tutorial software provides algorithmically generated practice exercises that correlate at the objective level to the odd-numbered exercises in the text. Each practice exercise is accompanied by both an example and a guided solution designed to involve students in the solution process. Selected problems also include a video clip that helps students visualize concepts. The software recognizes common student errors and provides appropriate feedback. Instructors can use InterAct Math Plus course management software to create, administer, and track on-line tests and monitor student performance during practice sessions.

MathXL®: www.mathxl.com
Stand-alone ISBN 0-201-72611-4

MathXL

The MathXL Web site provides diagnostic testing and tutorial help, all on-line, using InterAct Math® tutorial software and TestGen-EQ testing software. Students can take chapter tests correlated to the text, receive individualized study plans based on those test results, work practice problems and receive tutorial instruction for areas in which they need improvement, and take further tests to gauge their progress. An access card is required.

New! MyMathLab

Ideal for lecture-based, lab-based, and on-line courses, this state-of-the-art Web site provides students with a centralized point of access to the wide variety of on-line resources available with this text. The pages of the actual book are loaded into MyMathLab, and as students work through a section of the on-line text, they can link directly from the pages to supplementary resources (such as tutorial software, interactive animations, and audio and video clips) that provide instruction, exploration, and practice beyond what is offered in the printed book. MyMathLab generates personalized study plans for students and allows instructors to track all student work on tutorials, quizzes, and tests. Complete course-management capabilities, including a host of communication tools for course participants, are provided to create a user-friendly and interactive on-line learning environment. Contact your Addison-Wesley representative for a demonstration or visit www.mymathlab.com for more information. An access card is required.

AW Math Tutor Center

ISBN 0-201-72170-8, stand-alone

The Addison-Wesley Math Tutor Center is staffed by qualified mathematics instructors who provide students with tutoring on examples and odd-numbered exercises from the textbook. Tutoring is available by toll-free telephone, fax, e-mail, or the Internet. White Board technology allows tutors and students to actually see problems worked while they "talk" in real time over the Internet during tutoring sessions. An access card is required.

Acknowledgments

Many of you have helped to shape the Ninth Edition by reviewing, participating in telephone surveys and focus groups, filling out questionnaires, and spending time with us on your campuses. Our deepest appreciation to all of you and in particular to the following:

Ann Arakawa, *Maui Community College*
Wayne Brown, *Oklahoma State University—Oklahoma City*
Diane Christie, *University of Wisconsin—Stout*
Deirdre Collins, *Glendale Community College*
Karena Curtis, *Labette Community College*
Sharon Edgmon, *Bakersfield College*
Bill Graesser, *Ivy Tech State College*
Joe Jordan, *John Tyler Community College*
Jean-Marie Magnier, *Springfield Technical Community College*
Carol A. Marinas, *Barry University*
Michael Montano, *Riverside Community College—City Campus*
Jane Duncan Nesbit, *Columbia Union College*
Marilyn Platt, *Gaston College*
Susan Santolucito, *Delgado Community College*
Tomesa Smith, *Wallace State Community College*
Diane Trojan, *Kutztown University*
Angela Walters, *Capitol College*
Ray Weaver, *Community College of Allegheny County—Boyce Campus*
Annette Wiesner, *University of Wisconsin—Parkside*

We also wish to recognize the following people who wrote scripts, presented lessons on camera, and checked the accuracy of the videotapes:

Barbara Johnson, *Indiana University Purdue University Indianapolis*
Judith A. Penna, *Indiana University Purdue University Indianapolis*
Patricia Schwarzkopf, *University of Delaware*
Clen Vance, *Houston Community College*

I wish to express my heartfelt appreciation to a number of people who have contributed in special ways to the development of this textbook. My editor, Jennifer Crum, encouraged my vision and provided marketing insight. Kari Heen, the project manager, deserves special recognition for overseeing every phase of the project and keeping it moving. The unwavering support of the Developmental Math group, including Lauren Morse, associate editor, and Kathleen Manley, production supervisor, and the endless hours of hard work by Martha Morong and Geri Davis have led to products of which I am immensely proud.

I also want to thank Judy Beecher, my co-author on many books and my developmental editor on this text. Her steadfast loyalty, vision, and encouragement have been invaluable. In addition to writing the *Student's Solutions Manual,* Judy Penna has continued to provide strong leadership in the preparation of the printed supplements, videotapes, and MyMathLab. Other strong support has come from Laurie Hurley for the *Printed Test Bank*; Bill Saler for the audiotapes; and Barbara Johnson and Judy Penna for their accuracy checking.

M.L.B.

Feature Walkthrough

Chapter Openers

To engage students and prepare them for the upcoming chapter material, two-page gateway chapter openers are designed with exceptional artwork that is tied to a motivating real-world application.

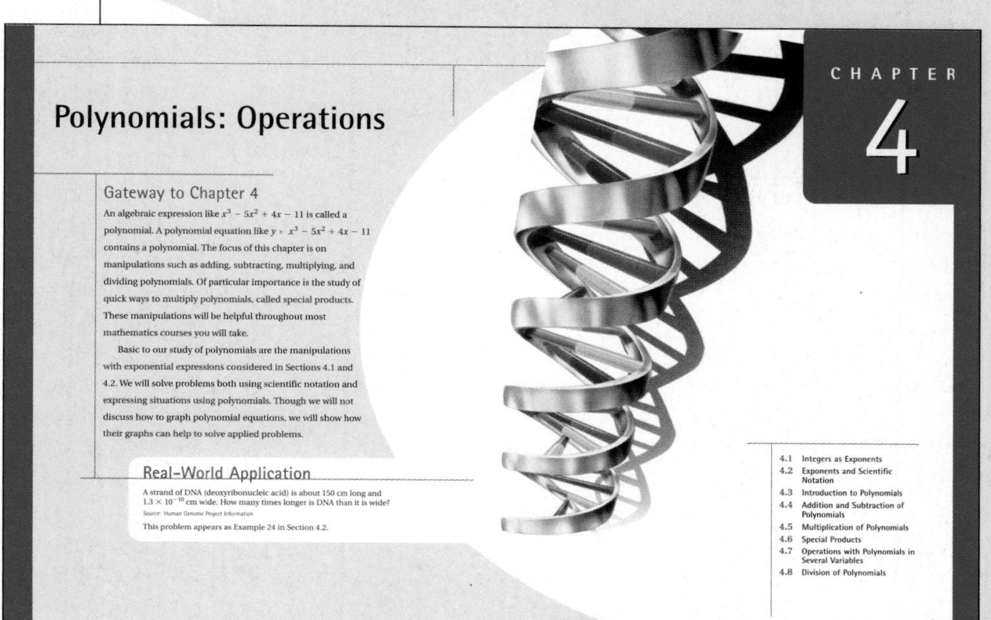

CHAPTER 4

Polynomials: Operations

Gateway to Chapter 4

An algebraic expression like $x^3 - 5x^2 + 4x - 11$ is called a polynomial. A polynomial equation like $y = x^3 - 5x^2 + 4x - 11$ contains a polynomial. The focus of this chapter is on manipulations such as adding, subtracting, multiplying, and dividing polynomials. Of particular importance is the study of quick ways to multiply polynomials, called special products. These manipulations will be helpful throughout most mathematics courses you will take.

Basic to our study of polynomials are the manipulations with exponential expressions considered in Sections 4.1 and 4.2. We will solve problems both using scientific notation and expressing situations using polynomials. Though we will not discuss how to graph polynomial equations, we will show how their graphs can help to solve applied problems.

Real-World Application

A strand of DNA (deoxyribonucleic acid) is about 150 cm long and 1.3×10^{-10} cm wide. How many times longer is DNA than it is wide?

Source: Human Genome Project Information

This problem appears as Example 24 in Section 4.2.

Chapter Pretests

Allowing students to test themselves before beginning each chapter, Chapter Pretests help them to identify material that may be familiar as well as targeting material that may be new or especially challenging. Instructors can use these results to assess student needs.

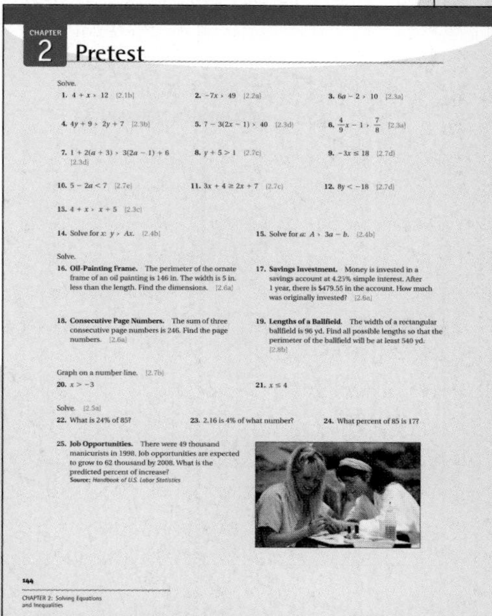

CHAPTER 2 Pretest

Solve.

1. $4 + x = 12$ [2.1b] **2.** $-7x = 49$ [2.2a] **3.** $6a - 2 = 10$ [2.3a]

4. $4y + 9 = 2y + 7$ [2.3b] **5.** $7 - 3(2x - 1) = 40$ [2.3d] **6.** $\frac{4}{9}x - 1 = \frac{7}{8}$ [2.3a]

7. $1 + 2(a + 3) = 3(2a - 1) + 6$ [2.3d] **8.** $y + 5 > 1$ [2.7c] **9.** $-3x \le 18$ [2.7d]

10. $5 - 2a < 7$ [2.7e] **11.** $3x + 4 \ge 2x + 7$ [2.7c] **12.** $8y < -18$ [2.7d]

13. $4 + x = x + 5$ [2.3c]

14. Solve for x: $y = Ax$. [2.4b] **15.** Solve for a: $A = 3a - b$. [2.4b]

Solve.

16. Oil-Painting Frame. The perimeter of the ornate frame of an oil painting is 146 in. The width is 5 in. less than the length. Find the dimensions. [2.6a]

17. Savings Investment. Money is invested in a savings account at 4.25% simple interest. After 1 year, there is $479.55 in the account. How much was originally invested? [2.6a]

18. Consecutive Page Numbers. The sum of three consecutive page numbers is 246. Find the page numbers. [2.6a]

19. Lengths of a Ballfield. The width of a rectangular ballfield is 96 yd. Find all possible lengths so that the perimeter of the ballfield will be at least 540 yd. [2.8b]

Graph on a number line. [2.7b]

20. $x > -3$ **21.** $x \le 4$

Solve. [2.5a]

22. What is 24% of 85? **23.** 2.16 is 4% of what number? **24.** What percent of 85 is 17?

25. Job Opportunities. There were 49 thousand manicurists in 1998. Job opportunities are expected to grow to 62 thousand by 2008. What is the predicted percent of increase?
Source: *Handbook of U.S. Labor Statistics*

144

CHAPTER 2: Solving Equations and Inequalities

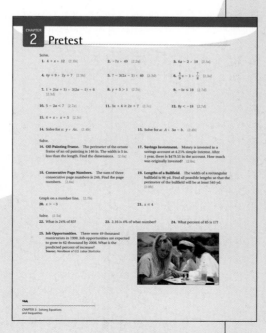

Art Program

Today's students are often visually oriented and their approach to a printed page is no exception. To appeal to students, the situational art in this edition is more dynamic and there are more photographs and art pieces overall. Where possible, mathematics is included in the art pieces to help students visualize the problem at hand.

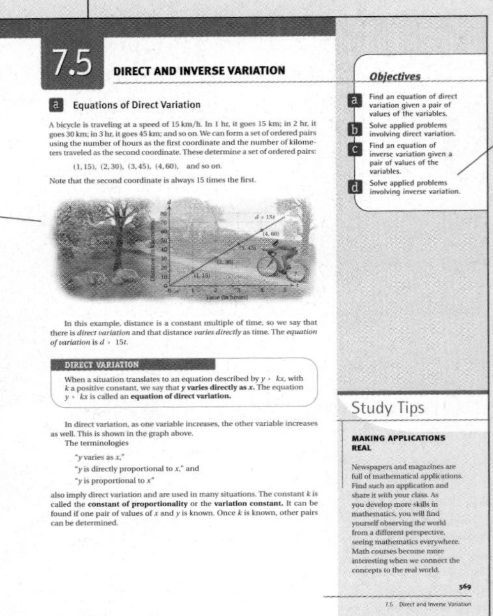

7.5 DIRECT AND INVERSE VARIATION

Objectives

a Find an equation of direct variation given a pair of values of the variables.

b Solve applied problems involving direct variation.

c Find an equation of inverse variation given a pair of values of the variables.

d Solve applied problems involving inverse variation.

a Equations of Direct Variation

A bicycle is traveling at a speed of 15 km/h. In 1 hr, it goes 15 km; in 2 hr, it goes 30 km; in 3 hr, it goes 45 km; and so on. We can form a set of ordered pairs using the number of hours as the first coordinate and the number of kilometers traveled as the second coordinate. These determine a set of ordered pairs:

(1, 15), (2, 30), (3, 45), (4, 60), and so on.

Note that the second coordinate is always 15 times the first.

In this example, distance is a constant multiple of time, so we say that there is *direct variation* and that distance *varies directly* as time. The *equation of variation* is $d = 15t$.

DIRECT VARIATION

When a situation translates to an equation described by $y = kx$, with k a positive constant, we say that **y varies directly as x.** The equation $y = kx$ is called an **equation of direct variation.**

In direct variation, as one variable increases, the other variable increases as well. This is shown in the graph above.

The terminologies

"y varies as x,"

"y is directly proportional to x," and

"y is proportional to x"

also imply direct variation and are used in many situations. The constant k is called the **constant of proportionality** or the **variation constant.** It can be found if one pair of values of x and y is known. Once k is known, other pairs can be determined.

Study Tips

MAKING APPLICATIONS REAL

Newspapers and magazines are full of mathematical applications. Find such an application and share it with your class. As you develop more skills in mathematics, you will find yourself observing the world from a different perspective, seeing mathematics everywhere. Math courses become more interesting when we connect the concepts to the real world.

569

7.5 Direct and Inverse Variation

Objective Boxes

At the beginning of each section, a boxed list of objectives is keyed by letter not only to section subheadings, but also to the exercises in the Pretest, exercise sets, and Summary and Review, as well as answers to the Chapter Test and Cumulative Review questions. This correlation enables students to easily find appropriate review material if they need help with a particular exercise or skill.

Margin Exercises

Throughout the text, students are directed to numerous margin exercises that provide immediate practice and reinforcement of the concepts covered in each section.

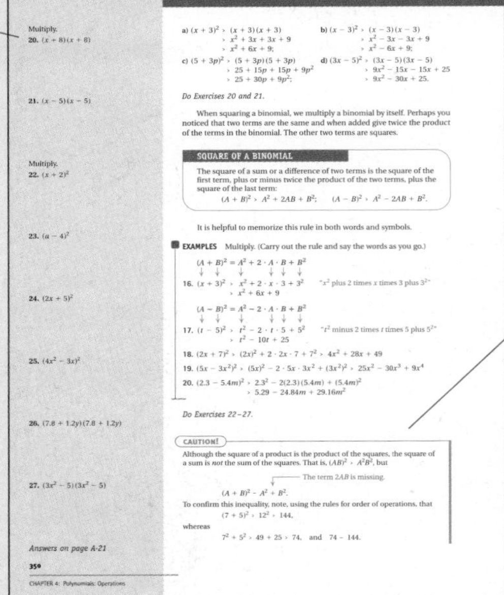

Multiply.

20. $(x + 8)(x + 8)$

21. $(x - 5)(x - 5)$

Multiply.

22. $(x + 2)^2$

23. $(a - 4)^2$

24. $(2x + 5)^2$

25. $(4x^2 - 3x)^2$

26. $(7.8 + 1.2y)(7.8 + 1.2y)$

27. $(3x^2 - 5)(3x^2 - 5)$

Answers on page A-21

350

CHAPTER 4: Polynomials: Operations

a) $(x + 3)^2 = (x + 3)(x + 3) = x^2 + 3x + 3x + 9 = x^2 + 6x + 9$;

b) $(x - 3)^2 = (x - 3)(x - 3) = x^2 - 3x - 3x + 9 = x^2 - 6x + 9$;

c) $(5 + 3p)^2 = (5 + 3p)(5 + 3p) = 25 + 15p + 15p + 9p^2 = 25 + 30p + 9p^2$;

d) $(3x - 5)^2 = (3x - 5)(3x - 5) = 9x^2 - 15x - 15x + 25 = 9x^2 - 30x + 25$.

Do Exercises 20 and 21.

When squaring a binomial, we multiply a binomial by itself. Perhaps you noticed that two terms are the same and when added give twice the product of the terms in the binomial. The other two terms are squares.

SQUARE OF A BINOMIAL

The square of a sum or a difference of two terms is the square of the first term, plus or minus twice the product of the two terms, plus the square of the last term:

$$(A + B)^2 = A^2 + 2AB + B^2; \quad (A - B)^2 = A^2 - 2AB + B^2.$$

It is helpful to memorize this rule in both words and symbols.

EXAMPLES Multiply. (Carry out the rule and say the words as you go.)

$(A + B)^2 = A^2 + 2 \cdot A \cdot B + B^2$

16. $(x + 3)^2 = x^2 + 2 \cdot x \cdot 3 + 3^2 = x^2 + 6x + 9$ "x^2 plus 2 times x times 3 plus 3^2"

$(A - B)^2 = A^2 - 2 \cdot A \cdot B + B^2$

17. $(t - 5)^2 = t^2 - 2 \cdot t \cdot 5 + 5^2 = t^2 - 10t + 25$ "t^2 minus 2 times t times 5 plus 5^2"

18. $(2x + 7)^2 = (2x)^2 + 2 \cdot 2x \cdot 7 + 7^2 = 4x^2 + 28x + 49$

19. $(5x - 3x^2)^2 = (5x)^2 - 2 \cdot 5x \cdot 3x^2 + (3x^2)^2 = 25x^2 - 30x^3 + 9x^4$

20. $(2.3 - 5.4m)^2 = 2.3^2 - 2(2.3)(5.4m) + (5.4m)^2 = 5.29 - 24.84m + 29.16m^2$

Do Exercises 22–27.

CAUTION!

Although the square of a product is the product of the squares, the square of a sum is *not* the sum of the squares. That is, $(AB)^2 = A^2B^2$, but

The term $2AB$ is missing.

$$(A + B)^2 \neq A^2 + B^2.$$

To confirm this inequality, note, using the rules for order of operations, that

$$(7 + 5)^2 = 12^2 = 144,$$

whereas

$$7^2 + 5^2 = 49 + 25 = 74, \text{ and } 74 \neq 144.$$

Caution Boxes

Found at relevant points throughout the text, boxes with the "Caution!" heading warn students of common misconceptions or errors made in performing a particular mathematical operation or skill.

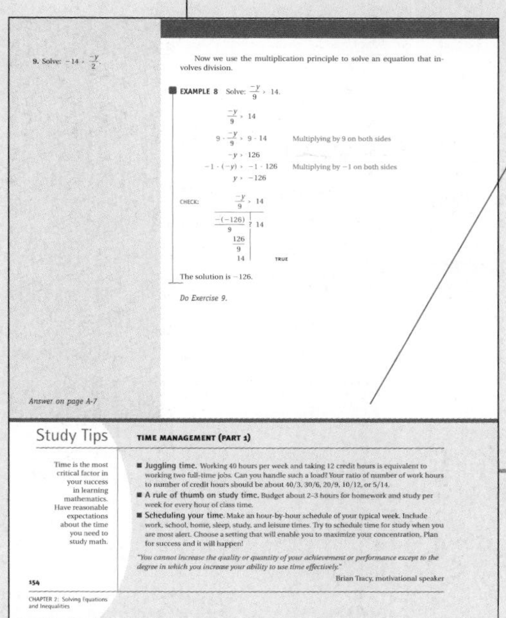

9. Solve: $-14 = \frac{-y}{2}$.

Now we use the multiplication principle to solve an equation that involves division.

EXAMPLE 8 Solve: $\frac{-y}{9} = 14$.

$$\frac{-y}{9} = 14$$
$$9 \cdot \frac{-y}{9} = 9 \cdot 14 \quad \text{Multiplying by 9 on both sides}$$
$$-y = 126$$
$$-1 \cdot (-y) = -1 \cdot 126 \quad \text{Multiplying by } -1 \text{ on both sides}$$
$$y = -126$$

CHECK: $\frac{-y}{9} = 14$; $\frac{-(-126)}{9} \; ? \; 14$; $\frac{126}{9}$; 14 TRUE

The solution is -126.

Do Exercise 9.

Answer on page A-7

Study Tips

TIME MANAGEMENT (PART 1)

Time is the most critical factor in your success in learning mathematics. Have reasonable expectations about the time you need to study math.

- **Juggling time.** Working 40 hours per week and taking 12 credit hours is equivalent to working two full-time jobs. Can you handle such a load? Your ratio of number of work hours to number of credit hours should be about 40/3, 30/6, 20/9, 10/12, or 5/14.
- **A rule of thumb on study time.** Budget about 2–3 hours for homework and study per week for every hour of class time.
- **Scheduling your time.** Make an hour-by-hour schedule of your typical week. Include work, school, home, sleep, study, and leisure times. Try to schedule time for study when you are most alert. Choose a setting that will enable you to maximize your concentration. Plan for success and it will happen!

"You cannot increase the quality or quantity of your achievement or performance except to the degree in which you increase your ability to use time effectively."

Brian Tracy, motivational speaker

154

CHAPTER 2: Solving Equations and Inequalities

Study Tips

Previously called "Improving Your Math Study Skills," a variety of Study Tips throughout the text give students pointers on how to develop good study habits as they progress through the course. At times short snippets and at other times more lengthy discussions, these Study Tips encourage students to input information and get involved in the learning process.

Calculator Corners

Where appropriate throughout the text, students see optional Calculator Corners. Popular in the Eighth Edition, slightly more Calculator Corners have been included in the new edition and the revised content is now more accessible to students.

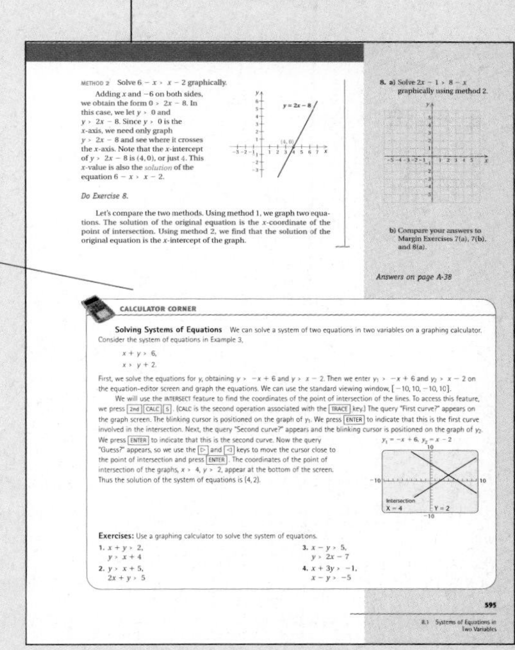

METHOD 2 Solve $6 - x = x - 2$ graphically.

Adding x and -6 on both sides, we obtain the form $0 = 2x - 8$. In this case, we let $y = 0$ and $y = 2x - 8$. Since $y = 0$ is the x-axis, we need only graph $y = 2x - 8$ and see where it crosses the x-axis. Note that the x-intercept of $y = 2x - 8$ is $(4, 0)$, or just 4. This x-value is also the *solution* of the equation $6 - x = x - 2$.

Do Exercise 8.

Let's compare the two methods. Using method 1, we graph two equations. The solution of the original equation is the x-coordinate of the point of intersection. Using method 2, we find that the solution of the original equation is the x-intercept of the graph.

8. a) Solve $2x - 1 = 8 - x$ graphically using method 2.

b) Compare your answers to Margin Exercises 7(a), 7(b), and 8(a).

Answers on page A-38

CALCULATOR CORNER

Solving Systems of Equations We can solve a system of two equations in two variables on a graphing calculator. Consider the system of equations in Example 3,

$x + y = 6,$
$x = y + 2.$

First, we solve the equations for y, obtaining $y = -x + 6$ and $y = x - 2$. Then we enter $y_1 = -x + 6$ and $y_2 = x - 2$ on the equation-editor screen and graph the equations. We can use the standard viewing window, $[-10, 10, -10, 10]$.

We will use the INTERSECT feature to find the coordinates of the point of intersection of the lines. To access this feature, we press 2nd CALC 5. (CALC is the second operation associated with the TRACE key.) The query "First curve?" appears on the graph screen. The blinking cursor is positioned on the graph of y_1. We press ENTER to indicate that this is the first curve involved in the intersection. Next, the query "Second curve?" appears and the blinking cursor is positioned on the graph of y_2. We press ENTER to indicate that this is the second curve. Now the query "Guess?" appears, so we use the ▷ and ◁ keys to move the cursor close to the point of intersection and press ENTER. The coordinates of the point of intersection of the graphs, $x = 4$, $y = 2$, appear at the bottom of the screen. Thus the solution of the system of equations is $(4, 2)$.

Exercises: Use a graphing calculator to solve the system of equations.

1. $x + y = 2,$ $y = x + 4$
2. $y = x + 5,$ $2x + y = 5$
3. $x - y = 5,$ $y = 2x - 7$
4. $x + 3y = -1,$ $x - y = -5$

595

8.1 Systems of Equations in Two Variables

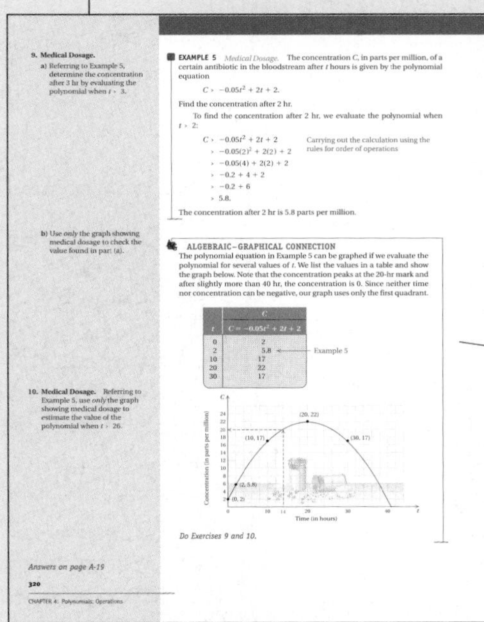

9. **Medical Dosage.**
a) Referring to Example 5, determine the concentration after 3 hr by evaluating the polynomial when $t = 3$.
b) Use *only* the graph showing medical dosage to check the value found in part (a).

10. **Medical Dosage.** Referring to Example 5, use *only* the graph showing medical dosage to estimate the value of the polynomial when $t = 26$.

EXAMPLE 5 *Medical Dosage.* The concentration C, in parts per million, of a certain antibiotic in the bloodstream after t hours is given by the polynomial equation

$$C = -0.05t^2 + 2t + 2.$$

Find the concentration after 2 hr.

To find the concentration after 2 hr, we evaluate the polynomial when $t = 2$:

$$C = -0.05t^2 + 2t + 2 \quad \text{Carrying out the calculation using the rules for order of operations}$$
$$= -0.05(2)^2 + 2(2) + 2$$
$$= -0.05(4) + 2(2) + 2$$
$$= -0.2 + 4 + 2$$
$$= -0.2 + 6$$
$$= 5.8.$$

The concentration after 2 hr is 5.8 parts per million.

ALGEBRAIC–GRAPHICAL CONNECTION

The polynomial equation in Example 5 can be graphed if we evaluate the polynomial for several values of t. We list the values in a table and show the graph below. Note that the concentration peaks at the 20-hr mark and after slightly more than 40 hr, the concentration is 0. Since neither time nor concentration can be negative, our graph uses only the first quadrant.

t	C: $C = -0.05t^2 + 2t + 2$
0	2
2	5.8 ← Example 5
10	17
20	22
30	17

Do Exercises 9 and 10.

Answers on page A-19

320

CHAPTER 4: Polynomials: Operations

Algebraic–Graphical Connections

To provide a visual understanding of algebra, algebraic–graphical connections are included in each chapter beginning with Chapter 3. This feature gives the algebra more meaning by connecting the algebra to a graphical interpretation.

EXERCISE SETS

To give students the opportunity to practice what they have learned, each section is followed by an extensive exercise set designed to reinforce the section concepts. In addition, students also have the opportunity to synthesize the objectives from the current section as well as those from preceding sections.

For Extra Help

Many valuable study aids accompany this text. Located just before each exercise set, "For Extra Help" references list appropriate video, tutorial, and Web resources so students can easily find related support materials.

Exercises

Exercises are keyed by letter to the section objectives for easy review.

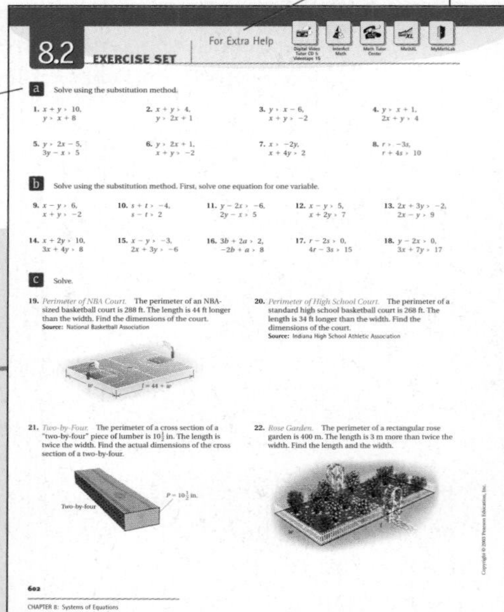

8.2 EXERCISE SET

For Extra Help — Digital Video Tutor CD 5 Videotape 15; InterAct Math; Math Tutor Center; MathXL; MyMathLab

a Solve using the substitution method.

1. $x + y = 10,$ $y = x + 8$
2. $x + y = 4,$ $y = 2x + 1$
3. $y = x - 6,$ $x + y = -2$
4. $y = x + 1,$ $2x + y = 4$
5. $y = 2x - 5,$ $3y - x = 5$
6. $y = 2x + 1,$ $x + y = -2$
7. $x = -2y,$ $x + 4y = 2$
8. $r = -3s,$ $r + 4s = 10$

b Solve using the substitution method. First, solve one equation for one variable.

9. $x - y = 6,$ $x + y = -2$
10. $s + t = -4,$ $s - t = 2$
11. $y - 2x = -6,$ $2y - x = 5$
12. $x - y = 5,$ $x + 2y = 7$
13. $2x + 3y = -2,$ $2x - y = 9$
14. $x + 2y = 10,$ $3x + 4y = 8$
15. $x - y = -3,$ $2x + 3y = -6$
16. $3b + 2a = 2,$ $-2b + a = 8$
17. $r - 2s = 0,$ $4r - 3s = 15$
18. $y - 2x = 0,$ $3x + 7y = 17$

c Solve.

19. *Perimeter of NBA Court.* The perimeter of an NBA-sized basketball court is 288 ft. The length is 44 ft longer than the width. Find the dimensions of the court.
Source: National Basketball Association

$l = 44 + w$

20. *Perimeter of High School Court.* The perimeter of a standard high school basketball court is 268 ft. The length is 34 ft longer than the width. Find the dimensions of the court.
Source: Indiana High School Athletic Association

21. *Two-by-Four.* The perimeter of a cross section of a "two-by-four" piece of lumber is $10\frac{1}{2}$ in. The length is twice the width. Find the actual dimensions of the cross section of a two-by-four.

Two-by-four $P = 10\frac{1}{2}$ in.

22. *Rose Garden.* The perimeter of a rectangular rose garden is 400 m. The length is 3 m more than twice the width. Find the length and the width.

Copyright © 2003 Pearson Education, Inc.

602

CHAPTER 8: Systems of Equations

Discussion and Writing Exercises

Designed to help students develop deeper comprehension of critical concepts, Discussion and Writing exercises (indicated by the $\mathbb{D}_{\mathbb{W}}$ symbol) are suitable for individual or group work. These exercises encourage students to both think and write about key mathematical ideas in the chapter.

Skill Maintenance Exercises

Found in each exercise set, these exercises review concepts from other sections in the text to prepare students for their final examination. Section and objective codes appear next to each Skill Maintenance exercise for easy reference, and in response to user feedback, the overall number of Skill Maintenance exercises has been increased.

Synthesis Exercises

In most exercise sets, Synthesis exercises help build critical-thinking skills by requiring students to synthesize or combine learning objectives from the current section as well as from preceding text sections.

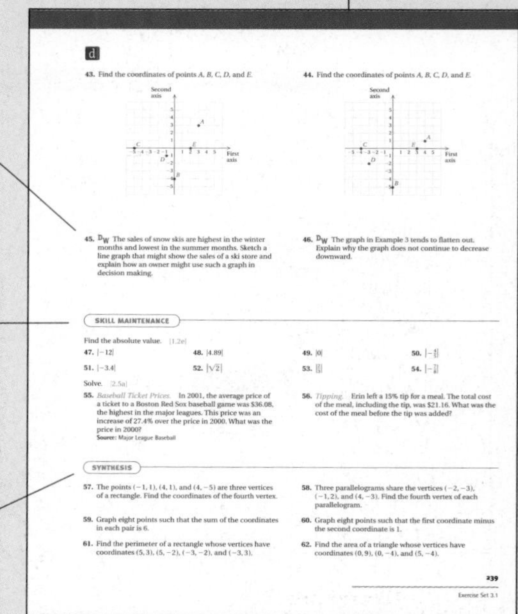

d

43. Find the coordinates of points *A*, *B*, *C*, *D*, and *E*.

44. Find the coordinates of points *A*, *B*, *C*, *D*, and *E*.

45. $\mathbb{D}_{\mathbb{W}}$ The sales of snow skis are highest in the winter months and lowest in the summer months. Sketch a line graph that might show the sales of a ski store and explain how an owner might use such a graph in decision making.

46. $\mathbb{D}_{\mathbb{W}}$ The graph in Example 3 tends to flatten out. Explain why the graph does not continue to decrease downward.

SKILL MAINTENANCE

Find the absolute value. [1.2e]

47. $|-12|$
48. $|4.89|$
49. $|0|$
50. $\left|-\frac{4}{5}\right|$
51. $|-3.4|$
52. $|\sqrt{2}|$
53. $\left|\frac{2}{3}\right|$
54. $\left|-\frac{7}{8}\right|$

Solve. [2.5a]

55. *Baseball Ticket Prices.* In 2001, the average price of a ticket to a Boston Red Sox baseball game was \$36.08, the highest in the major leagues. This price was an increase of 27.4% over the price in 2000. What was the price in 2000?
Source: Major League Baseball

56. *Tipping.* Erin left a 15% tip for a meal. The total cost of the meal, including the tip, was \$21.16. What was the cost of the meal before the tip was added?

SYNTHESIS

57. The points $(-1, 1)$, $(4, 1)$, and $(4, -5)$ are three vertices of a rectangle. Find the coordinates of the fourth vertex.

58. Three parallelograms share the vertices $(-2, -3)$, $(-1, 2)$, and $(4, -3)$. Find the fourth vertex of each parallelogram.

59. Graph eight points such that the sum of the coordinates in each pair is 6.

60. Graph eight points such that the first coordinate minus the second coordinate is 1.

61. Find the perimeter of a rectangle whose vertices have coordinates $(5, 3)$, $(5, -2)$, $(-3, -2)$, and $(-3, 3)$.

62. Find the area of a triangle whose vertices have coordinates $(0, 9)$, $(0, -4)$, and $(5, -4)$.

239

Exercise Set 3.1

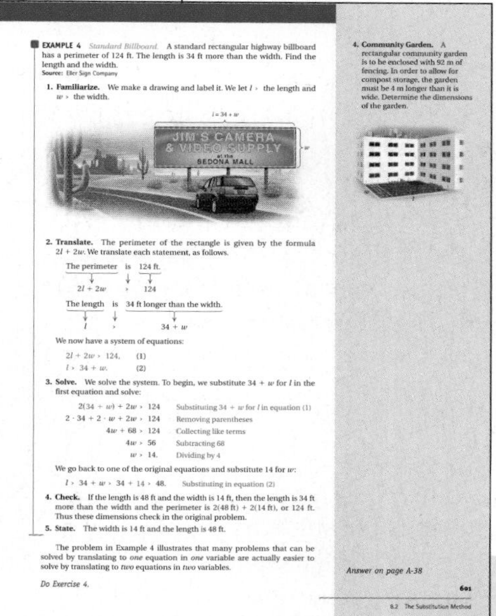

EXAMPLE 4 *Standard Billboard.* A standard rectangular highway billboard has a perimeter of 124 ft. The length is 34 ft more than the width. Find the length and the width.
Source: Eller Sign Company

1. **Familiarize.** We make a drawing and label it. We let l = the length and w = the width.

2. **Translate.** The perimeter of the rectangle is given by the formula $2l + 2w$. We translate each statement, as follows.

The perimeter is 124 ft.
$2l + 2w = 124$

The length is 34 ft longer than the width.
$l = 34 + w$

We now have a system of equations:

$2l + 2w = 124,$ (1)
$l = 34 + w.$ (2)

3. **Solve.** We solve the system. To begin, we substitute $34 + w$ for l in the first equation and solve:

$2(34 + w) + 2w = 124$ Substituting $34 + w$ for l in equation (1)
$2 \cdot 34 + 2 \cdot w + 2w = 124$ Removing parentheses
$4w + 68 = 124$ Collecting like terms
$4w = 56$ Subtracting 68
$w = 14.$ Dividing by 4

We go back to one of the original equations and substitute 14 for w:

$l = 34 + w = 34 + 14 = 48.$ Substituting in equation (2)

4. **Check.** If the length is 48 ft and the width is 14 ft, then the length is 34 ft more than the width and the perimeter is 2(48 ft) + 2(14 ft), or 124 ft. Thus these dimensions check in the original problem.
5. **State.** The width is 14 ft and the length is 48 ft.

The problem in Example 4 illustrates that many problems that can be solved by translating to *one* equation in *one* variable are actually easier to solve by translating to *two* equations in *two* variables.

Do Exercise 4.

4. **Community Garden.** A rectangular community garden is to be enclosed with 92 m of fencing. In order to allow for compost storage, the garden must be 4 m longer than it is wide. Determine the dimensions of the garden.

Answer on page A-38

601

8.2 The Substitution Method

Real-Data Applications

This text encourages students to see and interpret the mathematics that appears every day in the world around them. Throughout the writing process, an energetic search for real-data applications was conducted, and the result is a variety of examples and exercises that connect the mathematical content with the real world. Most of these applications feature source lines and frequently include charts and graphs.

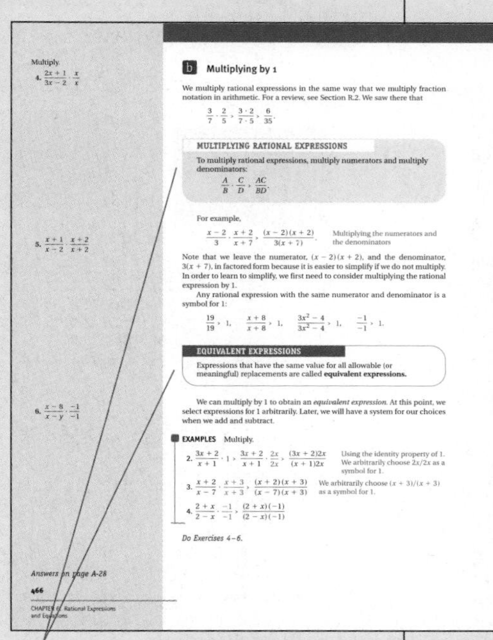

Multiply.

4. $\frac{2x+1}{3x-2} \cdot \frac{x}{x}$

5. $\frac{x+1}{x-2} \cdot \frac{x+2}{x+2}$

6. $\frac{x-8}{x-y} \cdot \frac{-1}{-1}$

b Multiplying by 1

We multiply rational expressions in the same way that we multiply fraction notation in arithmetic. For a review, see Section R.2. We saw there that

$$\frac{3}{7} \cdot \frac{2}{5} = \frac{3 \cdot 2}{7 \cdot 5} = \frac{6}{35}.$$

MULTIPLYING RATIONAL EXPRESSIONS

To multiply rational expressions, multiply numerators and multiply denominators:

$$\frac{A}{B} \cdot \frac{C}{D} = \frac{AC}{BD}.$$

For example,

$$\frac{x-2}{3} \cdot \frac{x+2}{x+7} = \frac{(x-2)(x+2)}{3(x+7)}.$$ Multiplying the numerators and the denominators

Note that we leave the numerator, $(x-2)(x+2)$, and the denominator, $3(x+7)$, in factored form because it is easier to simplify if we do not multiply. In order to learn to simplify, we first need to consider multiplying the rational expression by 1.

Any rational expression with the same numerator and denominator is a symbol for 1:

$$\frac{19}{19} = 1, \quad \frac{x+8}{x+8} = 1, \quad \frac{3x^2-4}{3x^2-4} = 1, \quad \frac{-1}{-1} = 1.$$

EQUIVALENT EXPRESSIONS

Expressions that have the same value for all allowable (or meaningful) replacements are called **equivalent expressions.**

We can multiply by 1 to obtain an *equivalent expression.* At this point, we select expressions for 1 arbitrarily. Later, we will have a system for our choices when we add and subtract.

EXAMPLES Multiply.

2. $\frac{3x+2}{x+1} \cdot 1 = \frac{3x+2}{x+1} \cdot \frac{2x}{2x} = \frac{(3x+2)2x}{(x+1)2x}$ Using the identity property of 1. We arbitrarily choose $2x/2x$ as a symbol for 1.

3. $\frac{x+2}{x-7} \cdot \frac{x+3}{x+3} = \frac{(x+2)(x+3)}{(x-7)(x+3)}$ We arbitrarily choose $(x+3)/(x+3)$ as a symbol for 1.

4. $\frac{2+x}{2-x} \cdot \frac{-1}{-1} = \frac{(2+x)(-1)}{(2-x)(-1)}$

Do Exercises 4–6.

Answers on page A-28

466

CHAPTER 6: Rational Expressions and Equations

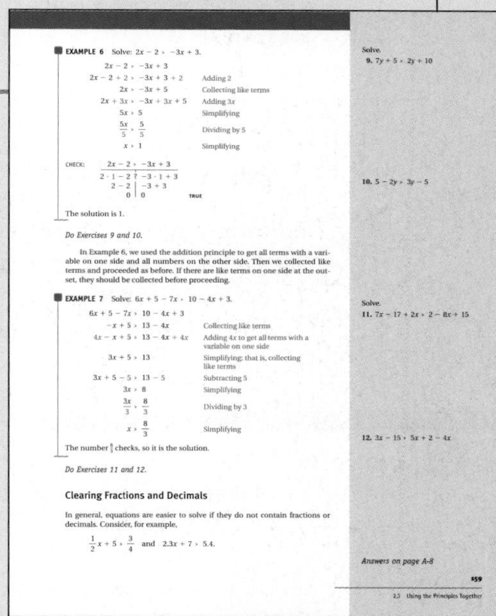

EXAMPLE 6 Solve: $2x - 2 = -3x + 3$.

$2x - 2 = -3x + 3$
$2x - 2 + 2 = -3x + 3 + 2$ Adding 2
$2x = -3x + 5$ Collecting like terms
$2x + 3x = -3x + 3x + 5$ Adding $3x$
$5x = 5$ Simplifying
$\frac{5x}{5} = \frac{5}{5}$ Dividing by 5
$x = 1$ Simplifying

CHECK: $2x - 2 = -3x + 3$
$2 \cdot 1 - 2 \ ? \ -3 \cdot 1 + 3$
$2 - 2 \mid -3 + 3$
$0 \mid 0$ TRUE

The solution is 1.

Do Exercises 9 and 10.

In Example 6, we used the addition principle to get all terms with a variable on one side and all numbers on the other side. Then we collected like terms and proceeded as before. If there are like terms on one side at the outset, they should be collected before proceeding.

EXAMPLE 7 Solve: $6x + 5 - 7x = 10 - 4x + 3$.

$6x + 5 - 7x = 10 - 4x + 3$
$-x + 5 = 13 - 4x$ Collecting like terms
$4x - x + 5 = 13 - 4x + 4x$ Adding $4x$ to get all terms with a variable on one side
$3x + 5 = 13$ Simplifying; that is, collecting like terms
$3x + 5 - 5 = 13 - 5$ Subtracting 5
$3x = 8$ Simplifying
$\frac{3x}{3} = \frac{8}{3}$ Dividing by 3
$x = \frac{8}{3}$ Simplifying

The number $\frac{8}{3}$ checks, so it is the solution.

Do Exercises 11 and 12.

Clearing Fractions and Decimals

In general, equations are easier to solve if they do not contain fractions or decimals. Consider, for example,

$\frac{1}{2}x + 5 = \frac{3}{4}$ and $2.3x + 7 = 5.4.$

Solve.
9. $7y + 5 = 2y + 10$

10. $5 - 2y = 3y - 5$

Solve.
11. $7x - 17 + 2x = 2 - 8x + 15$

12. $3x - 15 = 5x + 2 - 4x$

Answers on page A-8

159

2.3 Using the Principles Together

Highlighted Information

Important definitions, rules, and procedures are highlighted in titled boxes.

Annotated Examples

Detailed annotations and color highlights lead the student through the structured steps of the examples.

END-OF-CHAPTER MATERIAL

At the end of each chapter, students can practice all they have learned as well as tie the current chapter material to material covered in earlier chapters.

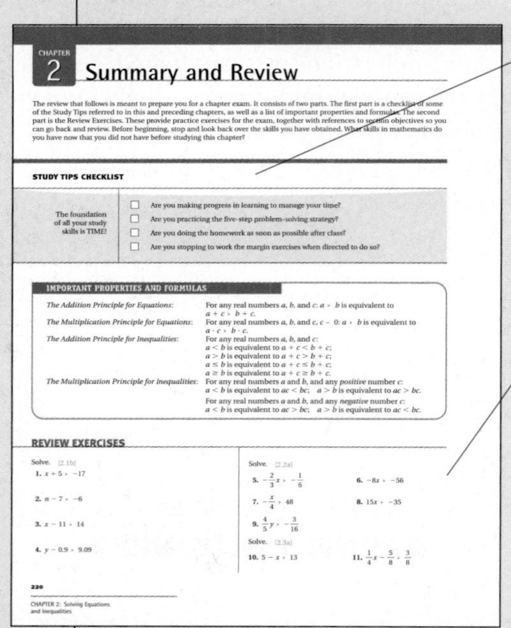

Study Tips Checklist

Each chapter review begins with a Study Tips Checklist that reviews Study Tips introduced in the current and previous chapters, making the use of these Study Tips more interactive.

Review Exercises

At the end of each chapter, students are provided with an extensive set of Review exercises. Reference codes beside each exercise or direction line allow students to easily review the related objective.

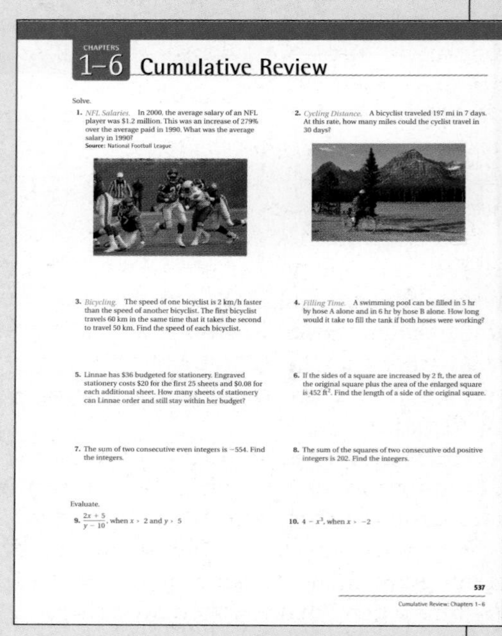

Chapter Test

Following the Review exercises, a sample Chapter Test allows students to review and test comprehension of chapter skills prior to taking an instructor's exam.

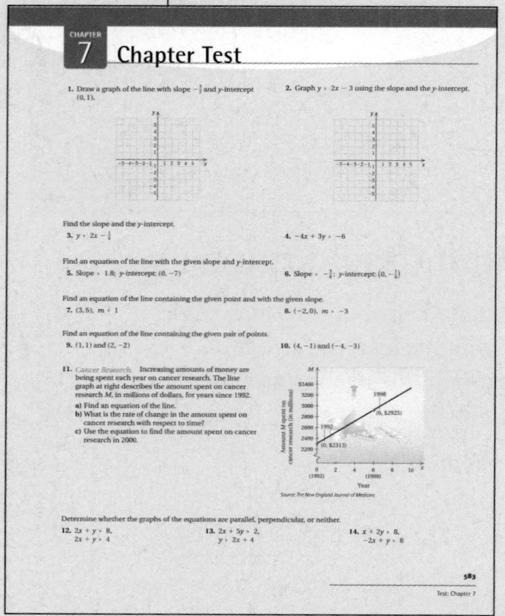

Cumulative Review

Following each chapter (beginning with Chapter 2), students encounter a Cumulative Review. This exercise set reviews skills and concepts from all preceding chapters to help students recall previously learned material and prepare for a final exam.

Introduction to the Student

As your author, I'd like to welcome you to this study of *Introductory Algebra.* Students come to this course with all kinds of backgrounds. Many are recent graduates of high school. Some are returning to college after spending time in the job market or after raising a family.

Whatever your past experiences, I encourage you to look at this mathematics course as a fresh start. Approach your course with a positive attitude about mathematics: This will put you in the right frame of mind to learn. Mathematics is a base for life, for many majors, for personal finances, for most careers, or just for pleasure. There is power in those symbols and equations—believe me!

If you have negative thoughts about mathematics, it is probably because you have had some kind of unpleasant experience in your study of math before now. It is my belief that most people can, and will, be able to learn under the right conditions, but some changes in your approach may be in order.

Let's think about your learning as a team approach. What can "we," meaning your author, your instructor, and, most of all, *you,* do to facilitate real success in your learning? Let's consider the parts of the team.

Your Author: My life has been dedicated to writing mathematics texts for over 30 years. I received my Ph.D. in Mathematics Education at Purdue University in 1968. My first and only teaching position was at Indiana University Purdue University Indianapolis. I sometimes think of myself as a "teacher on paper," or a "global professor," because of the wide use of my textbooks.

I live in Carmel, Indiana, with my wife, Elaine, who is very encouraging and supportive of my writing. We have two grown sons, Lowell and Chris, who are married to Karen and Tricia, respectively. Karen is a wonderful photographer; you will see many of her photographs in this book. Tricia is the mother of our grandchild, Margaret Grace. Needless to say, Maggie occupies much of my time when I'm away from my writing.

Apart from my family, my hobbies include hiking, baseball, golf, and bowling. I'm a terrible baseball player, but one week a year, I go to adult baseball fantasy camp and play ball like a kid. I'm a good bowler, with a 200 average, but a really poor golfer (30 handicap). Golf keeps me humble, and I enjoy learning about the game. I also have an interest in philosophy and theology, in particular, apologetics.

Your Instructor: Clearly, your instructor is at the forefront of your learning. He or she is who you learn from in the classroom. I encourage you to establish a learning relationship with your instructor early on. Feel free to ask questions in and outside of class. Do not wait too long for help or advice! Trivial though it may seem, be sure to get basic information, like his or her name, how he or she can be contacted outside of class, and the location of his or her office.

In addition, learn about your instructor's teaching style and try to adapt your learning to it. Does he or she use an overhead projector or the board? Will there be frequent in-class questions, tests, or quizzes, and so on? How is your grade ultimately determined?

Make use of what your instructor and your college has to offer in the way of help. If your campus has any kind of tutor center or learning lab, be sure to locate it and find out the hours of operation. Too often, students do not avail themselves of help that is there, free for the asking.

Yourself: You are the biggest factor in the success of your learning. This may be the first adjustment you have in college. In earlier experiences, you may have allowed yourself to sit back and let the instructor "pour in" the learning, with little or no follow-up on your part. But now you must take a more assertive and proactive stance. As soon as possible after class, you should thoroughly read the textbook and the supplements and do all you can on your own to learn. In other words, rid yourself of former habits and take responsibility for your own learning. Then, with all the help you have around you, your hard work will lead to success.

A helpful proverb comes to mind here:

"The best way to acquire a virtue is to act as if you already have it."

C. S. Lewis, English scholar and author

If you have never taken an assertive approach to learning mathematics, do so now, and you will soon realize the success of your actions.

One of the most important suggestions I can make is to allow yourself enough *time* to learn. You can have the best book, the best instructor, and the best supplements, but if you do not give yourself time to learn, how can they be of benefit? I usually ask my students the following questions:

- Are you working 40 hours or more per week?
- Are you taking 12 or more hours of classes?
- Do you dislike mathematics or have you had trouble learning in the past?

If you answered "yes" to all three of these questions, you must change one or both of the first two situations listed. You cannot learn without proper time management!

This introduction has contained many points that fall under the category of *Study Tips,* which you will find throughout the book. The following is an example in which I summarize some of the suggestions we have considered in this introduction:

Study Tips

- Establish a learning relationship with your instructor.
- Take more of the responsibility for your learning—do not wait for someone else to provide it for you.
- Use proper time management to allow time to learn.

You may want to study all of the Study Tips before you begin the text or you may decide to wait and encounter them as you go along. Your instructor may have suggestions in this regard.

You probably sense that the purpose of this Introduction is to encourage you, and that is indeed true. This, along with the effort I have made to write the best instructional book I can, is as close as I can come to being your personal instructor.

In closing, I want to wish you well in your new start studying mathematics. I wish I could meet each of you personally, but rest assured I think of you often in the sense that most of my waking moments are spent contemplating textbooks that make your learning more effective. Best wishes!

Marv Bittinger

Prealgebra Review

Gateway to Chapter R

This chapter is a review of skills that are basic to a study of algebra. The Pretest that follows on p. 2 can be used to diagnose your need for this chapter. Consult with your instructor for additional advice.

In this chapter, we review factoring and least common multiples. Then we consider fraction, decimal, percent, and exponential notation, as well as order of operations and basic geometric formulas.

Real-World Application

The largest piece of luggage that you can carry on an airplane measures 23 in. by 10 in. by 13 in. Find the volume of this solid.

This problem appears as Example 14 in Section R.6.

CHAPTER R

EUROPE
ON A
GERMANY
ITALY
SPAIN
CROWN
HARBOR
PARIS
DINO
LAND
SWISS
ALPS
INN
Brodie
Shakespeare
IN the
Park

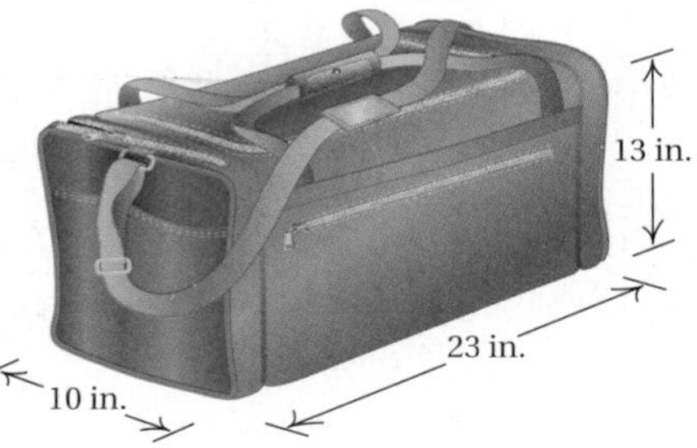
13 in.
23 in.
10 in.

CHAPTER

R Pretest

1. Find the prime factorization of 248.

2. Find the least common multiple of 12, 24, and 42.

3. Write an expression equivalent to $\frac{11}{12}$ with a denominator of 48.

Simplify.

4. $\frac{46}{128}$

5. $\frac{28}{42}$

Compute and simplify.

6. $\frac{3}{5} \div \frac{6}{11}$

7. $\frac{3}{7} - \frac{1}{3}$

8. $\frac{3}{10} + \frac{1}{5}$

9. $\frac{4}{7} \cdot \frac{5}{12}$

10. $8.25 + 91 + 34.7862$

11. $230 - 17.95$

12. 34.78×10.08

13. $78.12 \div 6.3$

14. Convert to fraction notation (do not simplify): 32.17.

15. Convert to decimal notation: $\frac{789}{10{,}000}$.

16. Convert to decimal notation: $\frac{13}{9}$.

17. Round to the nearest hundredth: 345.8395.

18. Round to the nearest tenth: 345.8395.

19. Convert to decimal notation: 11.6%.

20. Convert to fraction notation: 87%.

21. Convert to percent notation: $\frac{7}{8}$.

22. Write exponential notation: $5 \cdot 5 \cdot 5 \cdot 5$.

23. Evaluate: 2^3.

24. Evaluate: $(1.1)^2$.

25. Calculate: $9 \cdot 3 + 24 \div 4 - 5^2 + 10$.

26. Find the area of a square with sides of length 10 ft.

27. Find the perimeter.

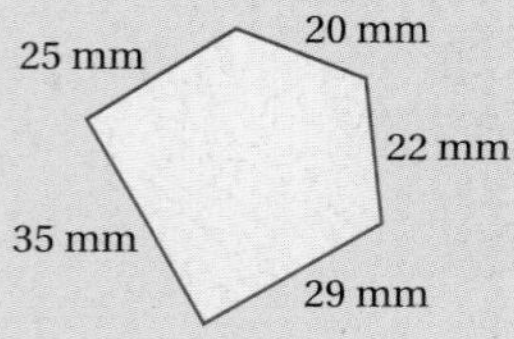

28. Find the volume.

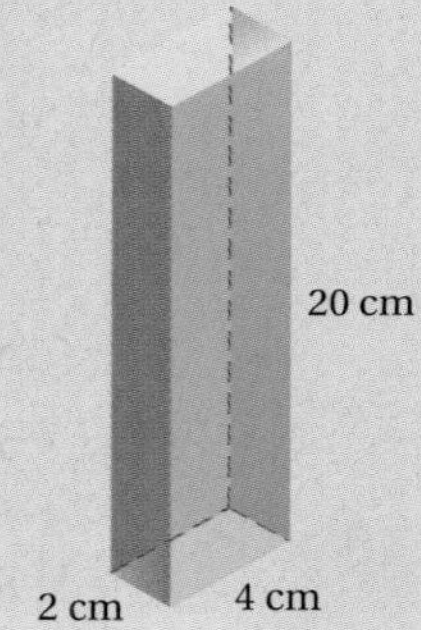

29. Find the length of a diameter of a circle with a radius of 4.8 m.

30. Find the area of the circle in Question 29. Use 3.14 for π.

Find the area.

31.

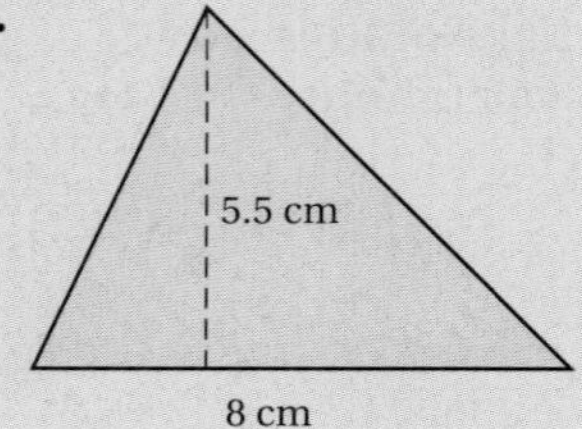

32.

$1\frac{3}{5}$ m

$2\frac{1}{2}$ m

R.1 FACTORING AND LCMS

Objectives

a Find all the factors of numbers and find prime factorizations of numbers.

b Find the LCM of two or more numbers using prime factorizations.

a Factors and Prime Factorizations

We begin our review with *factoring*, which is a necessary skill for addition and subtraction with fraction notation. Factoring is also an important skill in algebra. You will eventually learn to factor algebraic expressions.

The numbers we will be factoring are **natural numbers:**

1, 2, 3, 4, 5, and so on.

To **factor** a number means to express the number as a product. Consider the product $12 = 3 \cdot 4$. We say that 3 and 4 are **factors** of 12 and that $3 \cdot 4$ is a **factorization** of 12. Since $12 = 12 \cdot 1$, we also know that 12 and 1 are factors of 12 and that $12 \cdot 1$ is a factorization of 12.

EXAMPLE 1 Find all the factors of 12.

We first find some factorizations:

$12 = 1 \cdot 12,\quad 12 = 2 \cdot 6,\quad 12 = 3 \cdot 4,\quad 12 = 2 \cdot 2 \cdot 3.$

The factors of 12 are 1, 2, 3, 4, 6, and 12.

EXAMPLE 2 Find all the factors of 150.

We first find some factorizations:

$150 = 1 \cdot 150,\quad 150 = 2 \cdot 75,\quad 150 = 3 \cdot 50,\quad 150 = 5 \cdot 30,$

$150 = 6 \cdot 25,\quad 150 = 10 \cdot 15,\quad 150 = 2 \cdot 5 \cdot 3 \cdot 5.$

The factors of 150 are 1, 2, 3, 5, 6, 10, 15, 25, 30, 50, 75, and 150.

Note that the word "factor" is used both as a noun and as a verb. You **factor** when you express a number as a product. The numbers you multiply together to get the product are **factors.**

Do Exercises 1–4 (in the margin at right).

Find all the factors of the number.

1. 9

2. 16

3. 24

4. 180

Answers on page A-1

PRIME NUMBER

A natural number that has *exactly two different factors,* itself and 1, is called a **prime number.**

EXAMPLE 3 Which of these numbers are prime? 7, 4, 11, 18, 1

7 is prime. It has exactly two different factors, 7 and 1.

4 is not prime. It has three different factors, 1, 2, and 4.

11 is prime. It has exactly two different factors, 11 and 1.

18 is not prime. It has factors 1, 2, 3, 6, 9, and 18.

1 is not prime. It does not have two *different* factors.

A TABLE OF PRIMES

2, 3, 5, 7, 11, 13, 17, 19, 23, 29, 31, 37, 41, 43, 47, 53, 59, 61, 67, 71, 73, 79, 83, 89, 97, 101, 103, 107, 109, 113, 127, 131, 137, 139, 149, 151, 157

5. Which of these numbers are prime?

8, 6, 13, 14, 1

Find the prime factorization.

6. 48

7. 50

8. 770

Answers on page A-1

In the margin at left is a table of the prime numbers from 2 to 157. There are more extensive tables, but these prime numbers will be the most helpful to you in this text.

Do Exercise 5.

If a natural number, other than 1, is not prime, we call it **composite.** Every composite number can be factored into a product of prime numbers. Such a factorization is called a **prime factorization.**

EXAMPLE 4 Find the prime factorization of 36.

We begin by factoring 36 any way we can. One way is like this:

$$36 = 4 \cdot 9.$$

The factors 4 and 9 are not prime, so we factor them:

$$36 = 4 \cdot 9 = 2 \cdot 2 \cdot 3 \cdot 3$$

The factors in the last factorization are all prime, so we now have the *prime factorization* of 36. Note that 1 is *not* part of this factorization because it is not prime.

Another way to find the prime factorization of 36 is like this:

$$36 = 2 \cdot 18 = 2 \cdot 3 \cdot 6 = 2 \cdot 3 \cdot 2 \cdot 3.$$

In effect, we begin factoring any way we can think of and keep factoring until all factors are prime. Using a **factor tree** might also be helpful.

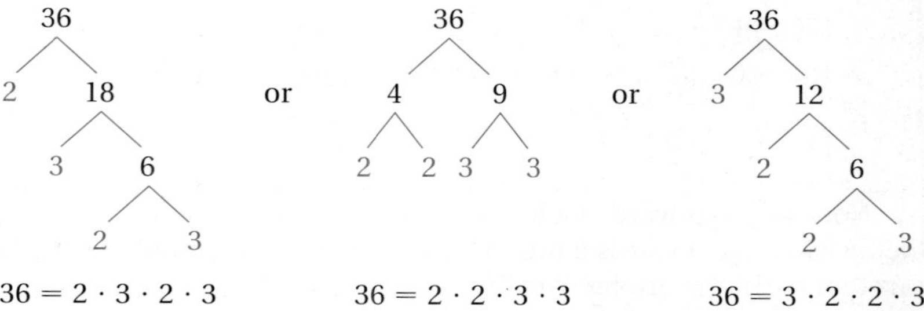

No matter which way we begin, the result is the same: The prime factorization of 36 contains two factors of 2 and two factors of 3. Every composite number has a *unique* prime factorization.

EXAMPLE 5 Find the prime factorization of 60.

This time, we use the list of primes from the table. We go through the table until we find a prime that is a factor of 60. The first such prime is 2.

$$60 = 2 \cdot 30$$

We keep dividing by 2 until it is not possible to do so.

$$60 = 2 \cdot 2 \cdot 15$$

Now we go to the next prime in the table that is a factor of 60. It is 3.

$$60 = 2 \cdot 2 \cdot 3 \cdot 5$$

Each factor in $2 \cdot 2 \cdot 3 \cdot 5$ is a prime. Thus this is the prime factorization.

Do Exercises 6–8.

b Least Common Multiples

Least common multiples are used to add and subtract with fraction notation.

The **multiples** of a number all have that number as a factor. For example, the multiples of 2 are

$$2,\quad 4,\quad 6,\quad 8,\quad 10,\quad 12,\quad 14,\quad 16, \ldots.$$

We could name each of them in such a way as to show 2 as a factor. For example, $14 = 2 \cdot 7$.

The multiples of 3 all have 3 as a factor:

$$3,\quad 6,\quad 9,\quad 12,\quad 15,\quad 18, \ldots.$$

Two or more numbers always have many multiples in common. From lists of multiples, we can find common multiples.

EXAMPLE 6 Find the common multiples of 2 and 3.

We make lists of their multiples and circle the multiples that appear in both lists.

$$2, 4, \textcircled{6}, 8, 10, \textcircled{12}, 14, 16, \textcircled{18}, 20, 22, \textcircled{24}, 26, 28, \textcircled{30}, 32, 34, \textcircled{36}, \ldots;$$
$$3, \textcircled{6}, 9, \textcircled{12}, 15, \textcircled{18}, 21, \textcircled{24}, 27, \textcircled{30}, 33, \textcircled{36}, \ldots.$$

The common multiples of 2 and 3 are

$$6,\quad 12,\quad 18,\quad 24,\quad 30,\quad 36, \ldots.$$

Do Exercises 9 and 10.

In Example 6, we found common multiples of 2 and 3. The *least*, or smallest, of those common multiples is 6. We abbreviate **least common multiple** as **LCM.**

There are several methods that work well for finding the LCM of several numbers. Some of these do not work well in algebra, especially when we consider expressions with variables such as $4ab$ and $12abc$. We now review a method that will work in arithmetic *and in algebra as well.* To see how it works, let's look at the prime factorizations of 9 and 15 in order to find the LCM:

$$9 = 3 \cdot 3, \qquad 15 = 3 \cdot 5.$$

Any multiple of 9 must have *two* 3's as factors. Any multiple of 15 must have *one* 3 and *one* 5 as factors. The smallest multiple of 9 and 5 is

$$3 \cdot 3 \cdot 5 = 45.$$

Two 3's; 9 is a factor (first two factors)

One 3, one 5; 15 is a factor (last two factors)

The LCM must have all the factors of 9 and all the factors of 15, but the factors are not repeated when they are common to both numbers.

To find the LCM of several numbers using prime factorizations:

a) Write the prime factorization of each number.

b) Form the LCM by writing the product of the different factors from step (a), using each factor the greatest number of times that it occurs in any one factorization.

9. Find the common multiples of 3 and 5 by making lists of multiples.

10. Find the common multiples of 9 and 15 by making lists of multiples.

Answers on page A-1

Find the LCM by factoring.

11. 8 and 10

12. 18 and 27

13. Find the LCM of 18, 24, and 30.

Find the LCM.

14. 3, 18

15. 12, 24

Find the LCM.

16. 4, 9

17. 5, 6, 7

Answers on page A-1

EXAMPLE 7 Find the LCM of 40 and 100.

a) We find the prime factorizations:

$$40 = 2 \cdot 2 \cdot 2 \cdot 5,$$
$$100 = 2 \cdot 2 \cdot 5 \cdot 5.$$

b) We write 2 as a factor three times (the greatest number of times that it occurs in any one factorization). We write 5 as a factor two times (the greatest number of times that it occurs in any one factorization).

The LCM is $2 \cdot 2 \cdot 2 \cdot 5 \cdot 5$, or 200.

Do Exercises 11 and 12.

EXAMPLE 8 Find the LCM of 27, 90, and 84.

a) We factor:

$$27 = 3 \cdot 3 \cdot 3,$$
$$90 = 2 \cdot 3 \cdot 3 \cdot 5,$$
$$84 = 2 \cdot 2 \cdot 3 \cdot 7.$$

b) We write 2 as a factor two times, 3 three times, 5 one time, and 7 one time.

The LCM is $2 \cdot 2 \cdot 3 \cdot 3 \cdot 3 \cdot 5 \cdot 7$, or 3780.

Do Exercise 13.

EXAMPLE 9 Find the LCM of 7 and 21.

Since 7 is prime, it has no prime factorization. It still, however, must be a factor of the LCM:

$$7 = 7,$$
$$21 = 3 \cdot 7.$$

The LCM is $7 \cdot 3$, or 21.

If one number is a factor of another, then the LCM is the larger of the two numbers.

Do Exercises 14 and 15.

EXAMPLE 10 Find the LCM of 8 and 9.

We have

$$8 = 2 \cdot 2 \cdot 2,$$
$$9 = 3 \cdot 3.$$

The LCM is $2 \cdot 2 \cdot 2 \cdot 3 \cdot 3$, or 72.

If two or more numbers have no common prime factor, then the LCM is the product of the numbers.

Do Exercises 16 and 17.

Study Tips

Throughout this textbook, you will find a feature called *Study Tips*. We discussed these in the Introduction on pp. xii–xiii of this text. They are intended to help improve your math study skills. An Index of all the Study Tips is found at the back of the book. On the first day of class you should complete this chart.

BASIC INFORMATION ON THE FIRST DAY OF CLASS

Instructor: Name ______________________

Office Hours and Location

____

Phone Number ______________________

Fax Number ______________________

e-mail Address (Instructor) ______________________

e-mail Address (Mathematics Department) ______________________

Find the names of two students whom you could contact for information or study questions:

1. Name ______________________

   Phone Number ______________________

   Fax Number ______________________

   e-mail Address ______________________

2. Name ______________________

   Phone Number ______________________

   Fax Number ______________________

   e-mail Address ______________________

Math lab on Campus:

Location ______________________

Hours ______________________

Phone ______________________

Tutoring:

Campus Location ______________________

Hours ______________________

To order AW Math Tutor Center, call: ______________________

(See the Preface for important information concerning this tutoring.)

Important Supplements:
(See the Preface for a complete list of available supplements.)

Supplements recommended by the instructor

____

____

"I know the price of success: dedication, hard work, and an unremitting devotion to the things you want to see happen."

Frank Lloyd Wright, architect

R.1 EXERCISE SET

For Extra Help

Digital Video Tutor CD 1 Videotape 1

InterAct Math

Math Tutor Center

MathXL

MyMathLab

Always review the objectives before doing an exercise set. See page 3. Note how the objectives are keyed to the exercises.

a Find all the factors of the number.

1. 20 **2.** 36 **3.** 72 **4.** 81

Find the prime factorization of the number.

5. 15 **6.** 14 **7.** 22 **8.** 33 **9.** 9

10. 25 **11.** 49 **12.** 121 **13.** 18 **14.** 24

15. 40 **16.** 56 **17.** 90 **18.** 120 **19.** 210

20. 330 **21.** 91 **22.** 143 **23.** 119 **24.** 221

b Find the prime factorization of the numbers. Then find the LCM.

25. 4, 5 **26.** 18, 40 **27.** 24, 36 **28.** 24, 27 **29.** 3, 15

30. 20, 40 **31.** 30, 40 **32.** 50, 60 **33.** 13, 23 **34.** 12, 18

35. 18, 30 **36.** 45, 72 **37.** 30, 36 **38.** 30, 50 **39.** 24, 30

40. 60, 70 **41.** 17, 29 **42.** 18, 24 **43.** 12, 28 **44.** 35, 45

45. 2, 3, 5 **46.** 3, 5, 7 **47.** 24, 36, 12 **48.** 8, 16, 22

49. 5, 12, 15

50. 12, 18, 40

51. 6, 12, 18

52. 24, 35, 45

Planet Orbits. The earth, Jupiter, Saturn, and Uranus all revolve around the sun. The earth takes 1 yr, Jupiter 12 yr, Saturn 30 yr, and Uranus 84 yr to make a complete revolution. On a certain night, you look at those three distant planets and wonder how many years it will take before they have the same position again. (*Hint*: To find out, you find the LCM of 12, 30, and 84. It will be that number of years.)
Source: *The Handy Science Answer Book*

53. How often will Jupiter and Saturn appear in the same direction in the night sky as seen from the earth?

54. How often will Jupiter and Uranus appear in the same direction in the night sky as seen from the earth?

55. How often will Saturn and Uranus appear in the same direction in the night sky as seen from the earth?

56. How often will Jupiter, Saturn, and Uranus appear in the same direction in the night sky as seen from the earth?

To the student and the instructor: The Discussion and Writing exercises denoted by the symbol **DW** are meant to be answered with one or more sentences. They can be discussed and answered collaboratively by the entire class or by small groups. Because of their open-ended nature, the answers to these exercises do not appear at the back of the book.

57. **DW** Describe the notation of a prime number as though you were talking to a classmate.

58. **DW** Explain a method for finding a composite number that contains exactly two factors other than itself and 1.

SYNTHESIS

To the student and the instructor: The Synthesis exercises found at the end of every exercise set challenge students to combine concepts or skills studied in that section or in preceding parts of the text. Exercises marked with a ▦ symbol are meant to be solved using a calculator.

59. Consider the numbers 8 and 12. Determine whether each of the following is the LCM of 8 and 12. Tell why or why not.

a) $2 \cdot 2 \cdot 3 \cdot 3$ **b)** $2 \cdot 2 \cdot 3$ **c)** $2 \cdot 3 \cdot 3$ **d)** $2 \cdot 2 \cdot 2 \cdot 3$

▦ Use a calculator to find the LCM of the numbers.

60. 288, 324

61. 2700, 7800

Objectives

a Find equivalent fraction expressions by multiplying by 1.

b Simplify fraction notation.

c Add, subtract, multiply, and divide using fraction notation.

R.2 FRACTION NOTATION

We now review fraction notation and its use with addition, subtraction, multiplication, and division of *arithmetic numbers.*

a Equivalent Expressions and Fraction Notation

An example of **fraction notation** for a number is

$$\frac{2}{3} \begin{array}{l} \leftarrow \text{Numerator} \\ \leftarrow \text{Denominator} \end{array}$$

The top number is called the **numerator,** and the bottom number is called the **denominator.**

The **whole numbers** consist of the natural numbers and 0:

$$0, \quad 1, \quad 2, \quad 3, \quad 4, \quad 5, \ldots.$$

The **arithmetic numbers,** also called the **nonnegative rational numbers,** consist of the whole numbers and the fractions such as $\frac{2}{3}$ and $\frac{9}{5}$. The arithmetic numbers can also be described as follows.

ARITHMETIC NUMBERS

The **arithmetic numbers** are the whole numbers and the fractions, such as $\frac{3}{4}$, $\frac{6}{5}$, or 8. All of these numbers can be named with fraction notation $\frac{a}{b}$, where a and b are whole numbers and $b \neq 0$.

Note that all whole numbers can be named with fraction notation. For example, we can name the whole number 8 as $\frac{8}{1}$. We call 8 and $\frac{8}{1}$ **equivalent expressions.**

Being able to find an equivalent expression is critical to a study of algebra. Some simple but powerful properties of numbers that allow us to find equivalent expressions are the identity properties of 0 and 1.

THE IDENTITY PROPERTY OF 0 (ADDITIVE IDENTITY)

For any number a,

$$a + 0 = a.$$

(Adding 0 to any number gives that same number.)

THE IDENTITY PROPERTY OF 1 (MULTIPLICATIVE IDENTITY)

For any number a,

$$a \cdot 1 = a.$$

(Multiplying any number by 1 gives that same number.)

Here are some ways to name the number 1:

$$\frac{5}{5}, \quad \frac{3}{3}, \quad \text{and} \quad \frac{26}{26}.$$

The following property allows us to find equivalent fractional expressions, that is, find other names for arithmetic numbers.

EQUIVALENT EXPRESSIONS FOR 1

For any number a, $a \neq 0$,

$$\frac{a}{a} = 1.$$

We can use the identity property of 1 and the preceding result to find equivalent fractional expressions.

EXAMPLE 1 Write a fractional expression equivalent to $\frac{2}{3}$ with a denominator of 15.

Note that $15 = 3 \cdot 5$. We want fraction notation for $\frac{2}{3}$ that has a denominator of 15, but the denominator 3 is missing a factor of 5. We multiply by 1, using $\frac{5}{5}$ as an equivalent expression for 1. Recall from arithmetic that to multiply with fraction notation, we multiply numerators and denominators:

$$\frac{2}{3} = \frac{2}{3} \cdot 1 \qquad \text{Using the identity property of 1}$$

$$= \frac{2}{3} \cdot \frac{5}{5} \qquad \text{Using } \frac{5}{5} \text{ for 1}$$

$$= \frac{10}{15}. \qquad \text{Multiplying numerators and denominators}$$

Do Exercises 1–3.

b Simplifying Expressions

We know that $\frac{1}{2}, \frac{2}{4}, \frac{4}{8}$, and so on, all name the same number. Any arithmetic number can be named in many ways. The **simplest fraction notation** is the notation that has the smallest numerator and denominator. We call the process of finding the simplest fraction notation **simplifying.** We reverse the process of Example 1 by first factoring the numerator and the denominator. Then we factor the fractional expression and remove a factor of 1 using the identity property of 1.

EXAMPLE 2 Simplify: $\frac{10}{15}$.

$$\frac{10}{15} = \frac{2 \cdot 5}{3 \cdot 5} \qquad \text{Factoring the numerator and the denominator. In this case, each is the prime factorization.}$$

$$= \frac{2}{3} \cdot \frac{5}{5} \qquad \text{Factoring the fractional expression}$$

$$= \frac{2}{3} \cdot 1$$

$$= \frac{2}{3} \qquad \text{Using the identity property of 1 (removing a factor of 1)}$$

1. Write a fractional expression equivalent to $\frac{2}{3}$ with a denominator of 12.

2. Write a fractional expression equivalent to $\frac{3}{4}$ with a denominator of 28.

3. Multiply by 1 to find three different fractional expressions for $\frac{7}{8}$.

Answers on page A-1

Simplify.

4. $\frac{18}{45}$

5. $\frac{38}{18}$

6. $\frac{72}{27}$

Answers on page A-1

EXAMPLE 3 Simplify: $\frac{36}{24}$.

$$\frac{36}{24} = \frac{2 \cdot 3 \cdot 2 \cdot 3}{2 \cdot 2 \cdot 3 \cdot 2} \qquad \text{Factoring the numerator and the denominator}$$

$$= \frac{2 \cdot 3 \cdot 2}{2 \cdot 3 \cdot 2} \cdot \frac{3}{2} \qquad \text{Factoring the fractional expression}$$

$$= 1 \cdot \frac{3}{2}$$

$$= \frac{3}{2} \qquad \text{Removing a factor of 1}$$

It is always a good idea to check at the end to see if you have indeed factored out all the common factors of the numerator and the denominator.

CANCELING

Canceling is a shortcut that you may have used to remove a factor of 1 when working with fraction notation. With *great* concern, we mention it as a possible way to speed up your work. You should use canceling only when removing common factors in numerators and denominators. Each common factor allows us to remove a factor of 1 in a product. **Canceling *cannot* be done when adding.** Our concern is that "canceling" be performed with care and understanding. Example 3 might have been done faster as follows:

$$\frac{36}{24} = \frac{\cancel{2} \cdot \cancel{3} \cdot \cancel{2} \cdot 3}{\cancel{2} \cdot \cancel{2} \cdot \cancel{3} \cdot 2} = \frac{3}{2}, \quad \text{or} \quad \frac{36}{24} = \frac{3 \cdot \cancel{12}}{2 \cdot \cancel{12}} = \frac{3}{2}, \quad \text{or} \quad \frac{\overset{\overset{3}{\cancel{18}}}{\cancel{36}}}{\underset{\underset{2}{\cancel{12}}}{\cancel{24}}} = \frac{3}{2}.$$

CAUTION!

The difficulty with canceling is that it is often applied incorrectly in situations like the following:

$$\frac{\cancel{2} + 3}{\cancel{2}} = 3; \qquad \frac{\cancel{4} + 1}{\cancel{4} + 2} = \frac{1}{2}; \qquad \frac{1\cancel{5}}{\cancel{5}4} = \frac{1}{4}.$$

Wrong! Wrong! Wrong!

The correct answers are

$$\frac{2 + 3}{2} = \frac{5}{2}; \qquad \frac{4 + 1}{4 + 2} = \frac{5}{6}; \qquad \frac{15}{54} = \frac{5}{18}.$$

In each situation, the number canceled was not a factor of 1. Factors are parts of products. For example, in $2 \cdot 3$, 2 and 3 are factors, but in $2 + 3$, 2 and 3 are *not* factors. Canceling may not be done when sums or differences are in numerators or denominators, as shown here.

Do Exercises 4–6.

We can always insert the number 1 as a factor. The identity property of 1 allows us to do that.

EXAMPLE 4 Simplify: $\frac{18}{72}$.

$$\frac{18}{72} = \frac{2 \cdot 9}{8 \cdot 9} = \frac{2}{8} = \frac{2 \cdot 1}{2 \cdot 4} = \frac{1}{4}, \quad \text{or} \quad \frac{18}{72} = \frac{1 \cdot 18}{4 \cdot 18} = \frac{1}{4}$$

EXAMPLE 5 Simplify: $\frac{72}{9}$.

$\frac{72}{9} = \frac{8 \cdot 9}{1 \cdot 9}$ Factoring and inserting a factor of 1 in the denominator

$= \frac{8 \cdot 9}{1 \cdot 9}$ Removing a factor of 1: $\frac{9}{9} = 1$

$= \frac{8}{1} = 8$ Simplifying

Do Exercises 7 and 8.

C Multiplication, Addition, Subtraction, and Division

After we have performed an operation of multiplication, addition, subtraction, or division, the answer may or may not be in simplified form. We simplify, if at all possible.

MULTIPLICATION

To multiply using fraction notation, we multiply the numerators to get the new numerator, and we multiply the denominators to get the new denominator.

MULTIPLICATION USING FRACTION NOTATION

$$\frac{a}{b} \cdot \frac{c}{d} = \frac{a \cdot c}{b \cdot d}$$

EXAMPLE 6 Multiply and simplify: $\frac{5}{6} \cdot \frac{9}{25}$.

$\frac{5}{6} \cdot \frac{9}{25} = \frac{5 \cdot 9}{6 \cdot 25}$ Multiplying numerators and denominators

$= \frac{5 \cdot 3 \cdot 3}{2 \cdot 3 \cdot 5 \cdot 5}$ Factoring the numerator and the denominator

$= \frac{3 \cdot 5 \cdot 3}{3 \cdot 5 \cdot 2 \cdot 5}$ Removing a factor of 1: $\frac{3 \cdot 5}{3 \cdot 5} = 1$

$= \frac{3}{10}$ Simplifying

Do Exercises 9 and 10.

Simplify.

7. $\frac{27}{54}$

8. $\frac{48}{12}$

Multiply and simplify.

9. $\frac{6}{5} \cdot \frac{25}{12}$

10. $\frac{3}{8} \cdot \frac{5}{3} \cdot \frac{7}{2}$

Answers on page A-1

Study Tips

SMALL STEPS LEAD TO GREAT SUCCESS (PART 1)

What is your long-term goal for getting an education? How does math help you to attain that goal? As you begin this course, approach each short-term task, such as going to class, asking questions, using your time wisely, and doing your homework, as part of the framework of your long-term goal.

"What man actually needs is not a tensionless state but rather the struggling and striving for a worthwhile goal, a freely chosen task."

Victor Frankl

ADDITION

When denominators are the same, we can add by adding the numerators and keeping the same denominator.

ADDING FRACTIONS WITH LIKE DENOMINATORS

To add when denominators are the same, add the numerators, keep the same denominator, and simplify, if possible.

$$\frac{a}{c} + \frac{b}{c} = \frac{a + b}{c}$$

EXAMPLE 7 Add and simplify: $\frac{4}{8} + \frac{5}{8}$.

The common denominator is 8. We add the numerators and keep the common denominator:

$$\frac{4}{8} + \frac{5}{8} = \frac{4 + 5}{8} = \frac{9}{8}.$$

In arithmetic, we generally write $\frac{9}{8}$ as $1\frac{1}{8}$. In algebra, you will find that *improper fraction* symbols such as $\frac{9}{8}$ are more useful and are quite *proper* for our purposes.

What do we do when denominators are different? We try to find a common denominator. We can do this by multiplying by 1. Consider adding $\frac{1}{6}$ and $\frac{3}{4}$. There are several common denominators that can be obtained. Let's look at two possibilities.

A. $$\frac{1}{6} + \frac{3}{4} = \frac{1}{6} \cdot 1 + \frac{3}{4} \cdot 1 = \frac{1}{6} \cdot \frac{4}{4} + \frac{3}{4} \cdot \frac{6}{6} = \frac{4}{24} + \frac{18}{24} = \frac{22}{24} = \frac{11}{12}$$

B. $$\frac{1}{6} + \frac{3}{4} = \frac{1}{6} \cdot 1 + \frac{3}{4} \cdot 1 = \frac{1}{6} \cdot \frac{2}{2} + \frac{3}{4} \cdot \frac{3}{3} = \frac{2}{12} + \frac{9}{12} = \frac{11}{12}$$

We had to simplify in (A). We didn't have to simplify in (B). In (B), we used the least common multiple of the denominators, 12. That number is called the **least common denominator,** or **LCD.**

ADDING FRACTIONS WITH DIFFERENT DENOMINATORS

To add when denominators are different:

a) Find the least common multiple of the denominators. That number is the least common denominator, LCD.

b) Multiply by 1, using an appropriate notation, n/n, to express each number in terms of the LCD.

c) Add the numerators, keeping the same denominator.

d) Simplify, if possible.

EXAMPLE 8 Add and simplify: $\frac{3}{8}+\frac{5}{12}$.

The LCM of the denominators, 8 and 12, is 24. Thus the LCD is 24. We multiply each fraction by 1 to obtain the LCD:

$$\frac{3}{8}+\frac{5}{12}=\frac{3}{8}\cdot\frac{3}{3}+\frac{5}{12}\cdot\frac{2}{2}$$

Multiplying by 1. Since $3\cdot 8=24$, we multiply the first number by $\frac{3}{3}$. Since $2\cdot 12=24$, we multiply the second number by $\frac{2}{2}$.

$$=\frac{9}{24}+\frac{10}{24}$$

$$=\frac{9+10}{24}$$

Adding the numerators and keeping the same denominator

$$=\frac{19}{24}.$$

Do Exercises 11–14.

SUBTRACTION

When subtracting, we also multiply by 1 to obtain the LCD. After we have made the denominators the same, we can subtract by subtracting the numerators and keeping the same denominator.

EXAMPLE 9 Subtract and simplify: $\frac{9}{8}-\frac{4}{5}$.

$$\frac{9}{8}-\frac{4}{5}=\frac{9}{8}\cdot\frac{5}{5}-\frac{4}{5}\cdot\frac{8}{8}$$

The LCD is 40.

$$=\frac{45}{40}-\frac{32}{40}=\frac{45-32}{40}$$

$$=\frac{13}{40}$$

EXAMPLE 10 Subtract and simplify: $\frac{7}{10}-\frac{1}{5}$.

$$\frac{7}{10}-\frac{1}{5}=\frac{7}{10}-\frac{1}{5}\cdot\frac{2}{2}$$

The LCD is 10.

$$=\frac{7}{10}-\frac{2}{10}=\frac{7-2}{10}$$

$$=\frac{5}{10}=\frac{1\cdot \cancel{5}}{2\cdot \cancel{5}}=\frac{1}{2}$$

Removing a factor of 1: $\frac{5}{5}=1$

Do Exercises 15 and 16.

Add and simplify.

11. $\frac{4}{5}+\frac{3}{5}$

12. $\frac{5}{6}+\frac{7}{6}$

13. $\frac{5}{6}+\frac{7}{10}$

14. $\frac{1}{4}+\frac{1}{2}$

Subtract and simplify.

15. $\frac{7}{8}-\frac{2}{5}$

16. $\frac{5}{12}-\frac{2}{9}$

Answers on page A-1

Find the reciprocal.

17. $\frac{4}{11}$

18. $\frac{15}{7}$

19. 5

20. $\frac{1}{3}$

21. Divide by multiplying by 1:

$$\frac{\frac{3}{5}}{\frac{4}{7}}.$$

Answers on page A-1

RECIPROCALS

Two numbers whose product is 1 are called **reciprocals,** or **multiplicative inverses,** of each other. All the arithmetic numbers, except zero, have reciprocals.

EXAMPLES

11. The reciprocal of $\frac{2}{3}$ is $\frac{3}{2}$ because $\frac{2}{3} \cdot \frac{3}{2} = \frac{6}{6} = 1$.
12. The reciprocal of 9 is $\frac{1}{9}$ because $9 \cdot \frac{1}{9} = \frac{9}{9} = 1$.
13. The reciprocal of $\frac{1}{4}$ is 4 because $\frac{1}{4} \cdot 4 = \frac{4}{4} = 1$.

Do Exercises 17–20.

RECIPROCALS AND DIVISION

Reciprocals and the number 1 can be used to justify a fast way to divide arithmetic numbers. We multiply by 1, carefully choosing the expression for 1.

EXAMPLE 14 Divide $\frac{2}{3}$ by $\frac{7}{5}$.

This is a symbol for 1.

$$\frac{2}{3} \div \frac{7}{5} = \frac{\frac{2}{3}}{\frac{7}{5}} = \frac{\frac{2}{3}}{\frac{7}{5}} \cdot \frac{\frac{5}{7}}{\frac{5}{7}}$$ Multiplying by $\frac{\frac{5}{7}}{\frac{5}{7}}$. We use $\frac{5}{7}$ because it is the reciprocal of $\frac{7}{5}$.

$$= \frac{\frac{2}{3} \cdot \frac{5}{7}}{\frac{7}{5} \cdot \frac{5}{7}}$$ Multiplying numerators and denominators

$$= \frac{\frac{10}{21}}{\frac{35}{35}} = \frac{\frac{10}{21}}{1}$$ $\frac{35}{35} = 1$

$$= \frac{10}{21}$$ Simplifying

After multiplying, we had a denominator of $\frac{35}{35}$, or 1. That was because we used $\frac{5}{7}$, the reciprocal of the divisor, for both the numerator and the denominator of the symbol for 1.

Do Exercise 21.

When multiplying by 1 to divide, we get a denominator of 1. What do we get in the numerator? In Example 14, we got $\frac{2}{3} \cdot \frac{5}{7}$. This is the product of $\frac{2}{3}$, the dividend, and $\frac{5}{7}$, the reciprocal of the divisor.

DIVISION USING FRACTION NOTATION

To divide, multiply by the reciprocal of the divisor:

$$\frac{a}{b} \div \frac{c}{d} = \frac{a}{b} \cdot \frac{d}{c}.$$

EXAMPLE 15 Divide by multiplying by the reciprocal of the divisor: $\frac{1}{2} \div \frac{3}{5}$.

$$\frac{1}{2} \div \frac{3}{5} = \frac{1}{2} \cdot \frac{5}{3} \qquad \frac{5}{3} \text{ is the reciprocal of } \frac{3}{5}$$
$$= \frac{5}{6} \qquad \text{Multiplying}$$

After dividing, simplification is often possible and should be done.

EXAMPLE 16 Divide and simplify: $\frac{2}{3} \div \frac{4}{9}$.

$$\frac{2}{3} \div \frac{4}{9} = \frac{2}{3} \cdot \frac{9}{4} \qquad \frac{9}{4} \text{ is the reciprocal of } \frac{4}{9}$$
$$= \frac{\not{2} \cdot \not{3} \cdot 3}{\not{3} \cdot \not{2} \cdot 2} \qquad \text{Removing a factor of 1: } \frac{2 \cdot 3}{2 \cdot 3} = 1$$
$$= \frac{3}{2}$$

Do Exercises 22–24.

EXAMPLE 17 Divide and simplify: $\frac{5}{6} \div 30$.

$$\frac{5}{6} \div 30 = \frac{5}{6} \div \frac{30}{1} = \frac{5}{6} \cdot \frac{1}{30} = \frac{5 \cdot 1}{6 \cdot 30} = \underbrace{\frac{\not{5} \cdot 1}{6 \cdot \not{5} \cdot 6} = \frac{1}{6 \cdot 6}} = \frac{1}{36}$$

Removing a factor of 1: $\frac{5}{5} = 1$

EXAMPLE 18 Divide and simplify: $24 \div \frac{3}{8}$.

$$24 \div \frac{3}{8} = \frac{24}{1} \div \frac{3}{8} = \frac{24}{1} \cdot \frac{8}{3} = \frac{24 \cdot 8}{1 \cdot 3} = \underbrace{\frac{\not{3} \cdot 8 \cdot 8}{1 \cdot \not{3}} = \frac{8 \cdot 8}{1}} = 64$$

Removing a factor of 1: $\frac{3}{3} = 1$

Do Exercises 25 and 26.

Divide by multiplying by the reciprocal of the divisor. Then simplify.

22. $\frac{4}{3} \div \frac{7}{2}$

23. $\frac{5}{4} \div \frac{3}{2}$

24. $\frac{\frac{2}{9}}{\frac{5}{12}}$

Divide and simplify.

25. $\frac{7}{8} \div 56$

26. $36 \div \frac{4}{9}$

Answers on page A-1

CALCULATOR CORNER

Graphing Calculators and Operations on Fractions Although a calculator is *not* required for this textbook, the book contains a series of *optional* discussions on using a graphing calculator. The keystrokes for a TI-83 Plus graphing calculator will be shown throughout. For keystrokes for other models of calculators, consult the user's manual for your particular calculator.

Note that there are options above the keys as well as on them. To access the option written on a key, simply press the key. The options written in yellow above the keys are accessed by first pressing the yellow [2nd] key and then pressing the key corresponding to the desired option. The green options are accessed by first pressing the green [ALPHA] key.

To turn the calculator on, press the [ON] key at the bottom left-hand corner of the keypad. You should see a blinking rectangle, or cursor, on the screen. If you do not see the cursor, try adjusting the display contrast. To do this, first press [2nd] and then press and hold [△] to increase the contrast or [▽] to decrease the contrast.

To turn the calculator off, press [2nd] [OFF]. (OFF is the second operation associated with the [ON] key.) The calculator will turn itself off automatically after about five minutes of no activity.

Press [MODE] to display the MODE settings. Initially, you should select the settings on the left side of the display.

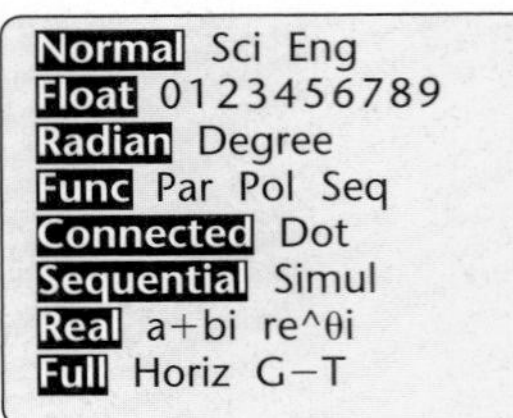

To change a setting on the MODE screen, use [▽] or [△] to move the cursor to the line of that setting. Then use [▷] or [◁] to move the blinking cursor to the desired setting and press [ENTER]. Press [CLEAR] or [2nd] [QUIT] to leave the MODE screen. (QUIT is the second operation associated with the [MODE] key.) Pressing [CLEAR] or [2nd] [QUIT] will take you to the **home screen** where computations are performed.

We can perform operations on fractions on a graphing calculator. To find $\frac{3}{4} + \frac{1}{2}$ and express the result as a fraction, we press [3] [÷] [4] [+] [1] [÷] [2] [MATH] [1] [ENTER]. The keystrokes [MATH] [1] select the ▷FRAC option from the MATH menu, causing the result to be expressed in fraction form. The calculator display is shown below.

```
3/4+1/2►Frac
                5/4
```

Exercises: Perform each calculation. Give the answer in fraction notation.

1. $\frac{1}{3} + \frac{1}{4}$
2. $\frac{5}{6} + \frac{7}{8}$
3. $\frac{7}{5} - \frac{3}{10}$
4. $\frac{31}{16} - \frac{4}{7}$
5. $\frac{15}{4} \cdot \frac{7}{12}$
6. $\frac{3}{2} \cdot \frac{8}{9}$
7. $\frac{4}{5} \div \frac{8}{3}$
8. $\frac{5}{4} \div \frac{3}{7}$

R.2 EXERCISE SET

For Extra Help

Digital Video Tutor CD 1 Videotape 1 | InterAct Math | Math Tutor Center | MathXL | MyMathLab

a Write an equivalent expression for each of the following. Use the indicated name for 1.

1. $\frac{3}{4}\left(\text{Use } \frac{3}{3} \text{ for } 1.\right)$

2. $\frac{5}{6}\left(\text{Use } \frac{10}{10} \text{ for } 1.\right)$

3. $\frac{3}{5}\left(\text{Use } \frac{20}{20} \text{ for } 1.\right)$

4. $\frac{8}{9}\left(\text{Use } \frac{4}{4} \text{ for } 1.\right)$

5. $\frac{13}{20}\left(\text{Use } \frac{8}{8} \text{ for } 1.\right)$

6. $\frac{13}{32}\left(\text{Use } \frac{40}{40} \text{ for } 1.\right)$

Write an equivalent expression with the given denominator.

7. $\frac{7}{8}$ (Denominator: 24)

8. $\frac{5}{6}$ (Denominator: 48)

9. $\frac{5}{4}$ (Denominator: 16)

10. $\frac{2}{9}$ (Denominator: 54)

b Simplify.

11. $\frac{18}{27}$ **12.** $\frac{49}{56}$ **13.** $\frac{56}{14}$ **14.** $\frac{48}{27}$ **15.** $\frac{6}{42}$ **16.** $\frac{13}{104}$

17. $\frac{56}{7}$ **18.** $\frac{132}{11}$ **19.** $\frac{19}{76}$ **20.** $\frac{17}{51}$ **21.** $\frac{100}{20}$ **22.** $\frac{150}{25}$

23. $\frac{425}{525}$ **24.** $\frac{625}{325}$ **25.** $\frac{2600}{1400}$ **26.** $\frac{4800}{1600}$ **27.** $\frac{8 \cdot x}{6 \cdot x}$ **28.** $\frac{13 \cdot v}{39 \cdot v}$

c Compute and simplify.

29. $\frac{1}{3} \cdot \frac{1}{4}$ **30.** $\frac{15}{16} \cdot \frac{8}{5}$ **31.** $\frac{15}{4} \cdot \frac{3}{4}$ **32.** $\frac{10}{11} \cdot \frac{11}{10}$

33. $\frac{1}{3} + \frac{1}{3}$ **34.** $\frac{1}{4} + \frac{1}{3}$ **35.** $\frac{4}{9} + \frac{13}{18}$ **36.** $\frac{4}{5} + \frac{8}{15}$

37. $\frac{3}{10} + \frac{8}{15}$

38. $\frac{9}{8} + \frac{7}{12}$

39. $\frac{5}{4} - \frac{3}{4}$

40. $\frac{12}{5} - \frac{2}{5}$

41. $\frac{11}{12} - \frac{3}{8}$

42. $\frac{15}{16} - \frac{5}{12}$

43. $\frac{11}{12} - \frac{2}{5}$

44. $\frac{15}{16} - \frac{2}{3}$

45. $\frac{7}{6} \div \frac{3}{5}$

46. $\frac{7}{5} \div \frac{3}{4}$

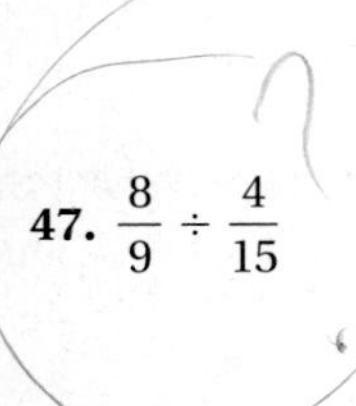

47. $\frac{8}{9} \div \frac{4}{15}$

48. $\frac{3}{4} \div \frac{3}{7}$

49. $\frac{1}{8} \div \frac{1}{4}$

50. $\frac{1}{20} \div \frac{1}{5}$

51. $\frac{\frac{13}{12}}{\frac{39}{5}}$

52. $\frac{\frac{17}{6}}{\frac{3}{8}}$

53. $100 \div \frac{1}{5}$

54. $78 \div \frac{1}{6}$

55. $\frac{3}{4} \div 10$

56. $\frac{5}{6} \div 15$

57. D$_W$ A student incorrectly insists that $\frac{12}{35} \div \frac{4}{5}$ is $\frac{7}{3}$. What mistake is the student likely making?

58. D$_W$ Explain in your own words when it *is* possible to *cancel* and when it is *not* possible to cancel.

SKILL MAINTENANCE

This heading indicates that the exercises that follow are *Skill Maintenance exercises,* which review any skill previously studied in the text. You can expect such exercises in every exercise set. Answers to *all* skill maintenance exercises are found at the back of the book. If you miss an exercise, restudy the objective shown in blue.

Find the prime factorization. [R.1a]

59. 28

60. 56

61. 1000

62. 192

63. 2001

Find the LCM. [R.1b]

64. 18, 63

65. 16, 24

66. 28, 49, 56

67. 48, 64, 96

68. 25, 75, 150

SYNTHESIS

Simplify.

69. $\frac{192}{256}$

70. $\frac{p \cdot q}{r \cdot q}$

71. $\frac{64 \cdot a \cdot b}{16 \cdot a \cdot b}$

72. $\frac{4 \cdot 9 \cdot 24}{2 \cdot 8 \cdot 15}$

73. $\frac{36 \cdot (2 \cdot h)}{8 \cdot (9 \cdot h)}$

R.3 DECIMAL NOTATION

Objectives

a Convert from decimal notation to fraction notation.

b Add, subtract, multiply, and divide using decimal notation.

c Round numbers to a specified decimal place.

Let's say that the cost of a sound system is

$1768.95.

This amount is given in **decimal notation.** The following place-value chart shows the place value of each digit in 1768.95.

PLACE-VALUE CHART								
Ten Thousands	Thousands	Hundreds	Tens	Ones	Ten*ths*	Hundred*ths*	Thousand*ths*	Ten-Thousand*ths*
10,000	1000	100	10	1	$\frac{1}{10}$	$\frac{1}{100}$	$\frac{1}{1000}$	$\frac{1}{10,000}$
	1	7	6	8 .	9	5		

a Converting from Decimal Notation to Fraction Notation

When we multiply by 1, a number is not changed. If we choose the notation $\frac{10}{10}, \frac{100}{100}, \frac{1000}{1000}$, and so on for 1, we can move a decimal point in a numerator to the right to convert from decimal notation to fraction notation.

Look for a pattern in the following products:

$$0.1 = 0.1 \times 1 = 0.1 \times \frac{10}{10} = \frac{0.1 \times 10}{10} = \frac{1}{10};$$

$$0.6875 = 0.6875 \times 1 = 0.6875 \times \frac{10,000}{10,000} = \frac{0.6875 \times 10,000}{10,000} = \frac{6875}{10,000};$$

$$53.47 = 53.47 \times 1 = 53.47 \times \frac{100}{100} = \frac{53.47 \times 100}{100} = \frac{5347}{100}.$$

To convert from decimal notation to fraction notation:

a) Count the number of decimal places. — 4.98 (2 places)

b) Move the decimal point that many places to the right. — 4.98. (Move 2 places.)

c) Write the result over a denominator with that number of zeros. — $\frac{498}{100}$ (2 zeros)

EXAMPLE 1 Convert 0.876 to fraction notation. Do not simplify.

0.876 (3 places) 0.876. (3 places) $0.876 = \frac{876}{1000}$ (3 zeros)

Convert to fraction notation. Do not simplify.

1. 0.568

2. 2.3

3. 89.04

Answers on page A-2

Convert to decimal notation.

4. $\frac{4131}{1000}$

5. $\frac{4131}{10,000}$

6. $\frac{573}{100}$

Add.

7. 69 + 1.785 + 213.67

8. 17.95 + 14.68 + 236

Answers on page A-2

EXAMPLE 2 Convert 1.5018 to fraction notation. Do not simplify.

1.5018 1.5018. (4 places) $1.5018 = \frac{15,018}{10,000}$ (4 zeros)

Do Exercises 1–3 on the preceding page.

To convert from fraction notation to decimal notation when the denominator is a number like 10, 100, or 1000:

a) Count the number of zeros. $\frac{8679}{1000}$ 3 zeros

b) Move the decimal point that number of places to the left. Leave off the denominator. 8.679. Move 3 places.

EXAMPLE 3 Convert to decimal notation: $\frac{123,067}{10,000}$.

$\frac{123,067}{10,000}$ (4 zeros) 12.3067. (4 places) $\frac{123,067}{10,000} = 12.3067$

Do Exercises 4–6.

b Addition, Subtraction, Multiplication, and Division

ADDITION WITH DECIMAL NOTATION

Adding with decimal notation is similar to adding whole numbers. First, line up the decimal points. Then add the thousandths, then the hundredths, and so on, carrying if necessary.

EXAMPLE 4 Add: 74 + 26.46 + 0.998.

$$\begin{array}{r} \overset{1}{7}\overset{1}{4}.\overset{1}{} \\ 26.46 \\ +\ 0.998 \\ \hline 101.458 \end{array}$$

You can place extra zeros to the right of any decimal point so that there are the same number of decimal places in all the addends, but this is not necessary. If you did, the preceding problem would look like this:

$$\begin{array}{r} \overset{1}{7}\overset{1}{4}.\overset{1}{0}00 \\ 26.460 \\ +\ 0.998 \\ \hline 101.458 \end{array}$$

Do Exercises 7 and 8.

SUBTRACTION WITH DECIMAL NOTATION

Subtracting with decimal notation is similar to subtracting whole numbers. First, line up the decimal points. Then subtract the thousandths, then the hundredths, the tenths, and so on, borrowing if necessary. Extra zeros can be added if needed.

EXAMPLES

5. Subtract: 76.14 − 18.953.

```
   15 10 13
 6  5  0  3 10
 7  6. 1  4  0
−1  8. 9  5  3
 5  7. 1  8  7
```

6. Subtract: 200 − 0.68.

```
 1 9 9  9 10
 2 0 0. 0  0
−      0. 6 8
 1 9 9. 3 2
```

Do Exercises 9–12.

Look at this product.

$$\underbrace{5.14}_{\text{2 places}} \times \underbrace{0.8}_{\text{1 place}} = \frac{514}{100} \times \frac{8}{10} = \frac{514 \times 8}{100 \times 10} = \frac{4112}{1000} = \underbrace{4.112}_{\text{3 places}}$$

We can also do this calculation more quickly by multiplying the whole numbers 8 and 514 and then determining the position of the decimal point.

MULTIPLICATION WITH DECIMAL NOTATION

a) Ignore the decimal points and multiply as whole numbers.
b) Place the decimal point in the result of step (a) by adding the number of decimal places in the original factors.

EXAMPLE 7 Multiply: 5.14 × 0.8.

a) Ignore the decimal points and multiply as whole numbers.

```
   1 3
   5.1 4
×    0.8
 4 1 1 2
```

b) Place the decimal point in the result of step (a) by adding the number of decimal places in the original factors.

```
   5.1 4 ←—— 2 decimal places
×    0.8 ←—— 1 decimal place
 4.1 1 2
     ↑——— 3 decimal places
```

Do Exercises 13–15.

Subtract.

9. 29.35 − 1.674

10. 92.375 − 27.692

11. 100 − 0.41

12. 240 − 0.117

Multiply.

13.
```
   6.5 2
×    0.9
```

14.
```
   6.5 2
× 0.0 9
```

15.
```
  5 6.7 6
× 0.9 0 8
```

Answers on page A-2

Divide.

16. $7\overline{)342.3}$

17. $16\overline{)253.12}$

Note that $37.6 \div 8 = 4.7$ because $8 \times 4.7 = 37.6$. If we write this as shown at right, we see how the following method can be used to divide by a whole number.

```
    4.7
8 ) 3 7.6
    3 2
      5 6
      5 6
        0
```

DIVIDING WHEN THE DIVISOR IS A WHOLE NUMBER

a) Place the decimal point in the quotient directly above the decimal point in the dividend.

b) Divide as whole numbers.

EXAMPLE 8 Divide: $216.75 \div 25$.

a)
```
              .
2 5 ) 2 1 6.7 5
           ↑
```
Place the decimal point.

b)
```
          8.6 7
2 5 ) 2 1 6.7 5
      2 0 0
        1 6 7
        1 5 0
          1 7 5
          1 7 5
              0
```
Divide as though dividing whole numbers.

Do Exercises 16 and 17.

Divide.

18. $25\overline{)32}$

19. $38\overline{)682.1}$

Sometimes it is helpful to write extra zeros to the right of the decimal point. Doing so does not change the answer. Remember that the decimal point for a whole number, though not normally written, is to the right of the number.

EXAMPLE 9 Divide: $54 \div 8$.

a)
```
       .
8 ) 5 4.
```

b)
```
    6.7 5
8 ) 5 4.0 0
    4 8
      6 0
      5 6
        4 0
        4 0
          0
```
Extra zeros are written to the right of the decimal point as needed.

Do Exercises 18 and 19.

DIVIDING WHEN THE DIVISOR IS NOT A WHOLE NUMBER

a) Move the decimal point in the divisor as many places to the right as it takes to make it a whole number. Move the decimal point in the dividend the same number of places to the right and place the decimal point in the quotient.

b) Divide as whole numbers, inserting zeros if necessary.

Answers on page A-2

EXAMPLE 10 Divide: 83.79 ÷ 0.098.

a)

$$0.098.\overline{)83.790.}$$

b)

$$
\begin{array}{r}
855. \\
0.098_{\wedge}\overline{)83.790_{\wedge}} \\
\underline{784} \\
539 \\
\underline{490} \\
490 \\
\underline{490} \\
0
\end{array}
$$

Do Exercises 20 and 21.

CONVERTING FROM FRACTION NOTATION TO DECIMAL NOTATION

To convert from fraction notation to decimal notation when the denominator is not a number like 10, 100, or 1000, we divide the numerator by the denominator.

EXAMPLE 11 Convert to decimal notation: $\frac{5}{16}$.

$$
\begin{array}{r}
0.3125 \\
16\overline{)5.0000} \\
\underline{48} \\
20 \\
\underline{16} \\
40 \\
\underline{32} \\
80 \\
\underline{80} \\
0
\end{array}
$$

If we get a remainder of 0, the decimal *terminates*. Decimal notation for $\frac{5}{16}$ is 0.3125.

EXAMPLE 12 Convert to decimal notation: $\frac{7}{12}$.

$$
\begin{array}{r}
0.5833 \\
12\overline{)7.0000} \\
\underline{60} \\
100 \\
\underline{96} \\
40 \\
\underline{36} \\
40 \\
\underline{36} \\
4
\end{array}
$$

The number 4 repeats as a remainder, so the digits will appear in the quotient. Therefore,

$$\frac{7}{12} = 0.583333\ldots.$$

Divide.

20. $0.024\overline{)20.544}$

21. $4.6\overline{)3.91}$

Convert to decimal notation.

22. $\frac{5}{8}$

23. $\frac{2}{3}$

24. $\frac{84}{11}$

Answers on page A-2

Round to the nearest tenth.

25. 2.76

26. 13.85

27. 7.009

Round to the nearest hundredth.

28. 7.834

29. 34.675

30. 0.025

Round to the nearest thousandth.

31. 0.9434

32. 8.0038

33. 43.1119

34. 37.4005

Round 7459.3549 to the nearest:

35. thousandth.

36. hundredth.

37. tenth.

38. one.

39. ten.

Answers on page A-2

Instead of dots, we often put a bar over the repeating part—in this case, only the 3. Thus,

$$\frac{7}{12} = 0.58\overline{3}.$$

Do Exercises 22–24 on the preceding page.

C Rounding

When working with decimal notation in real-life situations, we often shorten notation by **rounding.** Although there are many rules for rounding, we will use the following.

ROUNDING DECIMAL NOTATION

To round to a certain place:

a) Locate the digit in that place.

b) Consider the digit to its right.

c) If the digit to the right is 5 or higher, round up; if the digit to the right is less than 5, round down.

EXAMPLE 13 Round 3872.2459 to the nearest tenth.

a) We locate the digit in the tenths place, 2.

3 8 7 2.2 4 5 9
↑ (under the 2)

b) Then we consider the next digit to the right, 4.

3 8 7 2.2 4 5 9
↑ (under the 4)

c) Since that digit, 4, is less than 5, we round down.

3 8 7 2.2 ⟵ This is the answer.

EXAMPLE 14 Round 3872.2459 to the nearest thousandth, hundredth, tenth, one, ten, hundred, and thousand.

thousandth:	3872.246
hundredth:	3872.25
tenth:	3872.2
one:	3872
ten:	3870
hundred:	3900
thousand:	4000

Do Exercises 25–39.

In rounding, we sometimes use the symbol ≈, which means "is approximately equal to." Thus,

$$46.124 \approx 46.1.$$

Study Tips

USING THIS TEXTBOOK

We began our Study Tips in Section R.1. You will find many of these throughout the book. One of the most important ways to improve your math study skills is to learn the proper use of the textbook. Here we highlight a few points that we consider most helpful.

- **Be sure to note the special symbols [a], [b], [c], and so on, that correspond to the objectives you are to be able to perform.** The first time you see them is in the margin at the beginning of each section; the second time is in the subheadings of each section; and the third time is in the exercise set for the section. You will also find them next to the skill maintenance exercises in each exercise set and in the review exercises at the end of the chapter, as well as in the answers to the chapter tests and the cumulative reviews. These objective symbols allow you to refer to the appropriate place in the text whenever you need to review a topic.
- **Read and study each step of each example.** The examples include important side comments that explain each step. These carefully chosen examples and notes prepare you for success in the exercise set.
- **Stop and do the margin exercises as you study a section.** Doing the margin exercises is one of the most effective ways to enhance your ability to learn mathematics from this text. Don't deprive yourself of its benefits!
- **Note the icons listed at the top of each exercise set.** These refer to the many distinctive multimedia study aids that accompany the book.
- **Odd-numbered exercises.** Usually an instructor assigns some odd-numbered exercises. When you complete these, you can check your answers at the back of the book. If you miss any, check your work in the *Student's Solutions Manual* or ask your instructor for guidance.
- **Even-numbered exercises.** Whether or not your instructor assigns the even-numbered exercises, always do some on your own. Remember, there are no answers given for the chapter tests, so you need to practice doing exercises without answers. Check your answers later with a friend or your instructor.

CALCULATOR CORNER

Operations with Decimal Notation To use a graphing calculator to add, subtract, multiply, and divide with decimal notation, we use the [.], [+], [−], [×], [÷], and [ENTER] keys. To find 62.043 − 48.915, for example, we press [6] [2] [.] [0] [4] [3] [−] [4] [8] [.] [9] [1] [5] [ENTER]. Note that the subtraction operation key [−] must be used rather than the opposite key [(−)] to perform this computation. We will discuss the use of the [(−)] key in Chapter 1.

To find 6.73 × 2.18, we press [6] [.] [7] [3] [×] [2] [.] [1] [8] [ENTER]. The results of these computations are shown below.

```
62.043−48.915
                13.128
6.73*2.18
               14.6714
```

Exercises: Use a calculator to perform each operation.

1. $26 + 13.47 + 0.95$
2. $7.2 - 4.6$
3. $9.03 - 5.7$
4. 0.159×4.36
5. $34.5 \div 4.6$
6. $135.66 \div 57$

R.3 EXERCISE SET

For Extra Help

Digital Video Tutor CD 1 Videotape 1 | InterAct Math | Math Tutor Center | MathXL | MyMathLab

a Convert to fraction notation. Do not simplify.

1. 5.3

2. 2.7

3. 0.67

4. 0.93

5. 2.0007

6. 4.0008

7. 7889.8

8. 1122.3

Convert to decimal notation.

9. $\frac{1}{10}$

10. $\frac{1}{100}$

11. $\frac{1}{10{,}000}$

12. $\frac{1}{1000}$

13. $\frac{9999}{1000}$

14. $\frac{39}{10{,}000}$

15. $\frac{4578}{10{,}000}$

16. $\frac{94}{100{,}000}$

b Add.

17. $\begin{array}{r} 415.78 \\ +\ \ 29.16 \\ \hline \end{array}$

18. $\begin{array}{r} 708.99 \\ +\ \ 75.48 \\ \hline \end{array}$

19. $\begin{array}{r} 234.000 \\ +\ 156.617 \\ \hline \end{array}$

20. $\begin{array}{r} 1345.12 \\ +\ \ 566.98 \\ \hline \end{array}$

21. 85 + 67.95 + 2.774

22. 119 + 43.74 + 18.876

23. 17.95 + 16.99 + 28.85

24. 14.59 + 16.79 + 19.95

Subtract.

25. $\begin{array}{r} 78.110 \\ -\ 45.876 \\ \hline \end{array}$

26. $\begin{array}{r} 14.080 \\ -\ \ 9.199 \\ \hline \end{array}$

27. $\begin{array}{r} 38.7 \\ -\ 11.865 \\ \hline \end{array}$

28. $\begin{array}{r} 300. \\ -\ \ 24.677 \\ \hline \end{array}$

29. 57.86 − 9.95

30. 2.6 − 1.08

31. 3 − 1.0807

32. 5 − 3.4051

Multiply.

33. $\begin{array}{r} 7.34 \\ \times\ \ 1.8 \\ \hline \end{array}$

34. $\begin{array}{r} 6.55 \\ \times\ \ 3.2 \\ \hline \end{array}$

35. $\begin{array}{r} 0.86 \\ \times\ 0.93 \\ \hline \end{array}$

36. $\begin{array}{r} 0.028 \\ \times\ 7.409 \\ \hline \end{array}$

37. $\begin{array}{r} 17.95 \\ \times\ \ \ 10 \\ \hline \end{array}$

38. $\begin{array}{r} 18.94 \\ \times\ \ \ 0.1 \\ \hline \end{array}$

39. $\begin{array}{r} 0.457 \\ \times\ \ 3.08 \\ \hline \end{array}$

40. $\begin{array}{r} 0.0024 \\ \times\ \ 0.015 \\ \hline \end{array}$

41. $\begin{array}{r} 3.642 \\ \times\ \ 0.99 \\ \hline \end{array}$

42. $\begin{array}{r} 287.4 \\ \times\ \ 1.08 \\ \hline \end{array}$

Divide.

43. $72\overline{)165.6}$ **44.** $5.2\overline{)44.2}$ **45.** $8.5\overline{)44.2}$ **46.** $7.8\overline{)72.54}$

47. $9.9\overline{)0.2277}$ **48.** $100\overline{)95}$ **49.** $0.64\overline{)12}$ **50.** $1.6\overline{)75}$

51. $1.05\overline{)693}$ **52.** $25\overline{)4}$ **53.** $8.6\overline{)5.848}$ **54.** $0.47\overline{)0.1222}$

Convert to decimal notation.

55. $\frac{11}{32}$ **56.** $\frac{17}{32}$ **57.** $\frac{13}{11}$ **58.** $\frac{17}{12}$

59. $\frac{5}{9}$ **60.** $\frac{5}{6}$ **61.** $\frac{19}{9}$ **62.** $\frac{9}{11}$

C Round to the nearest hundredth, tenth, one, ten, and hundred.

63. 745.06534 **64.** 317.18565 **65.** 6780.50568 **66.** 840.15493

Round to the nearest cent and to the nearest dollar (nearest one).

67. $17.988 **68.** $20.492 **69.** $346.075 **70.** $4.718

Round to the nearest dollar.

71. $16.95 **72.** $17.50 **73.** $189.50 **74.** $567.24

Divide and round to the nearest ten-thousandth, thousandth, hundredth, tenth, and one.

75. $\frac{1000}{81}$ **76.** $\frac{23}{17}$ **77.** $\frac{23}{39}$ **78.** $\frac{8467}{5603}$

79. ^{D}W A student rounds 536.448 to the nearest one and gets 537. Explain the possible error.

80. ^{D}W A student insists that 5.367 ÷ 0.1 is 0.5367. How could you convince this student that a mistake has been made?

SKILL MAINTENANCE

Calculate. [R.2c]

81. $\frac{7}{8} + \frac{5}{32}$ **82.** $\frac{15}{16} - \frac{11}{12}$ **83.** $\frac{15}{16} \cdot \frac{11}{12}$ **84.** $\frac{15}{32} \div \frac{3}{8}$

Find the prime factorization. [R.1a]

85. 85 **86.** 86 **87.** 87 **88.** 88

89. 208 **90.** 128 **91.** 1250 **92.** 2560

R.4 PERCENT NOTATION

Objectives

a Convert from percent notation to decimal notation.

b Convert from percent notation to fraction notation.

c Convert from decimal notation to percent notation.

d Convert from fraction notation to percent notation.

1. **Suicide.** Suicide accounts for 13.4% of the deaths among people age 15 to 24. Convert 13.4% to decimal notation.

Convert to decimal notation.

2. 100%

3. 66.67%

Answers on page A-2

a Converting to Decimal Notation

Of all the major causes of mortality of people age 15 to 24, homicide accounts for 18%. What does this mean? It means that of every 100 deaths, 18 are from homicide. Thus, 18% is a ratio of 18 to 100.

Major Causes of Mortality Among 15–24 Year Olds

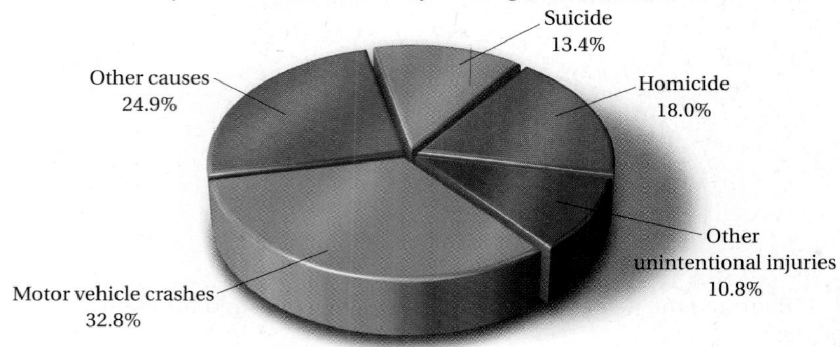

Source: National Center for Health Statistics

The percent symbol % means "per hundred." We can regard the percent symbol as part of a name for a number. For example,

$$28\% \text{ is defined to mean } 28 \times 0.01, \text{ or } 28 \times \frac{1}{100}, \text{ or } \frac{28}{100}.$$

NOTATION FOR $n\%$

$$n\% \text{ means } n \times 0.01, \text{ or } n \times \frac{1}{100}, \text{ or } \frac{n}{100}.$$

EXAMPLE 1 *Motor Vehicle Fatalities.* Motor-vehicle accidents account for 32.8% of the deaths among people age 15 to 24. Convert 32.8% to decimal notation.

$$32.8\% = 32.8 \times 0.01 \quad \text{Replacing \% with } \times 0.01$$
$$= 0.328$$

FROM PERCENT NOTATION TO DECIMAL NOTATION

To convert from percent notation to decimal notation, move the decimal point *two* places to the *left* and drop the percent symbol.

EXAMPLE 2 Convert 43.67% to decimal notation.

43.67%   0.43.67   43.67% = 0.4367

Move the decimal point two places to the left.

Do Exercises 1–3.

b Converting to Fraction Notation

EXAMPLE 3 Convert 88% to fraction notation.

$$88\% = 88 \times \frac{1}{100} \quad \text{Replacing \% with } \times\frac{1}{100}$$
$$= \frac{88}{100} \quad \text{Multiplying. You need not simplify.}$$

EXAMPLE 4 Convert 34.8% to fraction notation.

$$34.8\% = 34.8 \times \frac{1}{100} \quad \text{Replacing \% with } \times\frac{1}{100}$$
$$= \frac{34.8}{100}$$
$$= \frac{34.8}{100} \cdot \frac{10}{10} \quad \text{Multiplying by 1 to get a whole number in the numerator}$$
$$= \frac{348}{1000} \quad \text{You need not simplify.}$$

Do Exercises 4–7.

c Converting from Decimal Notation

By applying the definition of percent in reverse, we can convert from decimal notation to percent notation. We multiply by 1, expressing it as 100×0.01 and replacing $\times 0.01$ with %.

EXAMPLE 5 *Pregnancy Weight Gain.* Of all pregnant women, 0.32 of them gain 21–30 lb. Convert 0.32 to percent notation.
Source: National Vital Statistics Report

$$0.32 = 0.32 \times 1 \quad \text{Identity property of 1}$$
$$= 0.32 \times (100 \times 0.01) \quad \text{Expressing 1 as } 100 \times 0.01$$
$$= (0.32 \times 100) \times 0.01$$
$$= 32 \times 0.01$$
$$= 32\% \quad \text{Replacing } \times 0.01 \text{ with \%}$$

FROM DECIMAL NOTATION TO PERCENT NOTATION

To convert from decimal notation to percent notation, move the decimal point *two* places to the *right* and write the percent symbol.

EXAMPLE 6 Convert 0.082 to percent notation.

0.082 0.08.2 0.082 = 8.2%

Move the decimal point two places to the right.

Do Exercises 8–10.

4. Water in Watermelon. Watermelon is 90% water. Convert 90% to fraction notation.

Convert to fraction notation.

5. 53% **6.** 45.9%

7. 0.23%

8. Foreign Student Enrollment. Of all the foreign students studying in the United States, 0.106 are from China. Convert 0.106 to percent notation.
Source: Institute of International Education

Convert to percent notation.

9. 6.77 **10.** 0.9944

Answers on page A-2

11. **Deaths from Heart Attack.** Of all those who have heart attacks, $\frac{1}{3}$ of them die. Convert $\frac{1}{3}$ to percent notation.
Source: American Heart Association

Convert to percent notation.

12. $\frac{1}{4}$

13. $\frac{7}{8}$

Answers on page A-2

Study Tips

VIDEOTAPES

(ISBN 0-201-88292-2)
Developed and produced especially for this text, the videotapes feature an engaging team of instructors, who present material and concepts by using examples and exercises from every section of the text.

DIGITAL VIDEO TUTOR

(ISBN 0-321-12384-0)
The videotapes for this text are also available on CD-ROM, making it easy and convenient for you to watch video segments from a computer at home or on campus. The complete digitized video set, both affordable and portable, is ideal for distance learning or supplemental instruction.

d Converting from Fraction Notation

We can also convert from fraction notation to percent notation by converting first to decimal notation. Then we move the decimal point two places to the *right* and write a percent symbol.

EXAMPLE 7 Convert $\frac{5}{8}$ to percent notation.

$$\frac{5}{8} = 0.625 = 62.5\%$$

EXAMPLE 8 *Water Content.* The human body is about $\frac{2}{3}$ water. Water is the single most abundant chemical in the body. Convert $\frac{2}{3}$ to percent notation.

a) Find decimal notation for $\frac{2}{3}$ using long division.

$$\begin{array}{r} 0.666 \\ 3\overline{)2.000} \\ \underline{1\,8} \\ 20 \\ \underline{18} \\ 20 \\ \underline{18} \\ 2 \end{array}$$

We get a repeating decimal: $0.66\overline{6}$.

b) Convert the answer to percent notation.

$0.66.\overline{6}$

$$\frac{2}{3} = 66.\overline{6}\%, \text{ or } 66\tfrac{2}{3}\%$$

Do Exercises 11–13.

R.4 EXERCISE SET

For Extra Help

Digital Video Tutor CD 1 Videotape 2 | InterAct Math | Math Tutor Center | MathXL | MyMathLab

a Convert the percent notation in the sentence to decimal notation.

1. *Pluto.* The planet Pluto is 55% water.

2. *Pluto.* The planet Pluto is 45% rock.

3. *CD Rates.* A recent interest rate on a 5-yr certificate of deposit was 5.23%.
Source: Bank Rate Monitor

4. *CD Rates.* A recent interest rate on a $2\frac{1}{2}$-yr certificate of deposit was 4.73%.
Source: Bank Rate Monitor

Convert to decimal notation.

5. 63% **6.** 64% **7.** 94.1% **8.** 34.6%

9. 1% **10.** 100% **11.** 0.61% **12.** 125%

13. 240% **14.** 0.73% **15.** 3.25% **16.** 2.3%

b Convert the percent notation in the sentence to fraction notation.

17. *Left-Handed Golfers.* Of those who golf, 7% are left-handed.
Source: National Association of Left-Handed Golfers

Phil Mickelson, professional golfer

18. *Dehydration.* A 2% drop in water content of the body can affect one's ability to study mathematics.
Source: High Performance Nutrition

19. *Wealth in the Aged.* Those 60 and older own 77% of the nation's financial assets.

20. *Wealth in the Aged.* Those 60 and older make up 66% of the nation's stockholders.

Convert to fraction notation.

21. 60%
22. 40%
23. 28.9%
24. 37.5%
25. 110%
26. 120%
27. 0.042%
28. 0.68%
29. 250%
30. 3.2%
31. 3.47%
32. 12.557%

did only odd nr.

c Convert the decimal notation in the sentence to percent notation.

33. *NASCAR Fans.* Of those who are fans of NASCAR racing, 0.64 of them have attended college or beyond.
Source: NASCAR, Goodyear

34. *NASCAR Fans.* Of those who are fans of NASCAR racing, 0.41 of them earn more than $50,000 per year.
Source: NASCAR, Goodyear

Convert to percent notation.

35. 1
36. 8.56
37. 0.996
38. 0.83
39. 0.0047
40. 2
41. 0.072
42. 1.34
43. 9.2
44. 0.013
45. 0.0068
46. 0.675

d Convert to percent notation.

47. $\frac{1}{6}$
48. $\frac{1}{5}$
49. $\frac{13}{20}$
50. $\frac{14}{25}$
51. $\frac{29}{100}$
52. $\frac{123}{100}$
53. $\frac{8}{10}$
54. $\frac{7}{10}$
55. $\frac{3}{5}$
56. $\frac{17}{50}$
57. $\frac{2}{3}$
58. $\frac{7}{8}$
59. $\frac{7}{4}$
60. $\frac{3}{8}$
61. $\frac{3}{4}$
62. $\frac{99.4}{100}$

Convert the fraction notation in the sentence to percent notation.

63. *Heart-Disease Death Rate.* Of those people living in Florida, 118.6 of every 100,000 will die of heart disease.
Source: "Reforming the Health Care System; State Profiles 1999," AARP

64. *Cancer Death Rate.* Of those people living in Florida, 125.7 of every 100,000 will die of cancer.
Source: "Reforming the Health Care System; State Profiles 1999," AARP

a, b, c, d *Women at Work.* The following table lists the percentages of people in various professions who are women. Fill in the blanks in the table.

WOMEN IN THE WORKPLACE				
	PROFESSION	**DECIMAL NOTATION**	**FRACTION NOTATION**	**PERCENT NOTATION**
65.	Architect			19%
66.	Doctor			22%
67.	Lawyer	0.23		
68.	Engineer	0.09		
69.	Chemist		$\frac{3}{10}$	
70.	College faculty		$\frac{21}{50}$	
71.	Real estate broker (1978)			21%
72.	Real estate broker (mid-90s)			54%

Source: The Indianapolis Star, 9/3/00

73. D_W What would you do to an entry on a calculator in order to get percent notation?

74. D_W Is it always best to convert from fraction notation to percent notation by first finding decimal notation? Why or why not?

SKILL MAINTENANCE

Convert to decimal notation. [R.3b]

75. $\frac{9}{4}$ **76.** $\frac{11}{8}$ **77.** $\frac{17}{12}$ **78.** $\frac{8}{9}$ **79.** $\frac{10}{11}$ **80.** $\frac{17}{11}$

Calculate. [R.3b]

81. 23.458×7.03 **82.** $7.8\overline{)440.154}$ **83.** $809.569 + 86.99$ **84.** $809.569 - 86.99$

SYNTHESIS

Simplify. Express the answer in percent notation.

85. 18% + 14% **86.** 84% − 12% **87.** 1 − 30% **88.** 92% − 10% **89.** 27 × 100%

90. 42% − (1 − 58%) **91.** 3(1 + 15%) **92.** 7(1% + 13%) **93.** $\frac{100\%}{40}$ **94.** $\frac{3}{4} + 20\%$

R.5 EXPONENTIAL NOTATION AND ORDER OF OPERATIONS

Objectives

a Write exponential notation for a product.

b Evaluate exponential expressions.

c Simplify expressions using the rules for order of operations.

a Exponential Notation

Exponents provide a shorter way of writing products. An abbreviation for a product in which the factors are the same is called a **power.** For

$$\underbrace{10 \cdot 10 \cdot 10}_{3 \text{ factors}}, \quad \text{we write} \quad 10^3.$$

This is read "ten to the third power." We call the number 3 an **exponent** and we say that 10 is the **base.** An exponent of 2 or greater tells how many times the base is used as a factor. For example,

$$a \cdot a \cdot a \cdot a = a^4.$$

In this case, the exponent is 4 and the base is a. An expression for a power is called **exponential notation.**

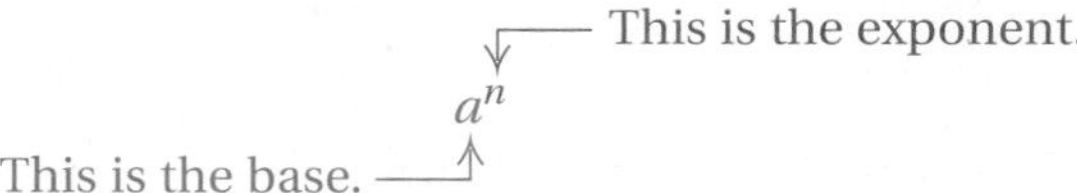

EXAMPLE 1 Write exponential notation for $10 \cdot 10 \cdot 10 \cdot 10 \cdot 10$.

$$10 \cdot 10 \cdot 10 \cdot 10 \cdot 10 = 10^5$$

Do Exercises 1–3.

Write exponential notation.

1. $4 \cdot 4 \cdot 4$

2. $6 \cdot 6 \cdot 6 \cdot 6 \cdot 6$

3. 1.08×1.08

b Evaluating Exponential Expressions

EXAMPLE 2 Evaluate: 5^2.

$$5^2 = 5 \cdot 5 = 25$$

EXAMPLE 3 Evaluate: 3^4.

We have

$$3^4 = 3 \cdot 3 \cdot 3 \cdot 3 = 9 \cdot 9 = 81.$$

We could also carry out the calculation as follows:

$$3^4 = 3 \cdot 3 \cdot 3 \cdot 3 = 9 \cdot 3 \cdot 3 = 27 \cdot 3 = 81.$$

EXPONENTIAL NOTATION

For any natural number n greater than or equal to 2,

$$b^n = \overbrace{b \cdot b \cdot b \cdot b \cdots b}^{n \text{ factors}}.$$

Do Exercises 4–6.

Evaluate.

4. 10^4

5. 8^3

6. $(1.1)^3$

Answers on page A-2

CALCULATOR CORNER

Exponents and Powers We use the [^] key to evaluate exponential notation on a graphing calculator. To find 3^5, for example, we press [3] [^] [5] [ENTER]. To find $\left(\frac{5}{8}\right)^3$ and express the result in fraction notation, we press [(] [5] [÷] [8] [)] [^] [3] [MATH] [1] [ENTER]. Note that the parentheses are necessary in this calculation. If they were not used, we would be calculating $5 \div 8^3$, or $5 \div 512$. The results of these computations are shown on the left below.

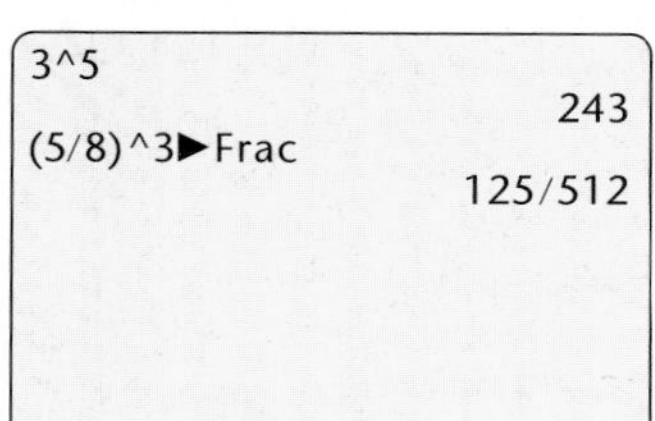

2.4²
5.76
2.4^2
5.76

The calculator has a special [x^2] key that can be used to raise a number to the second power. To find 2.4^2, for example, we press [2] [·] [4] [x^2] [ENTER], as shown on the right above. We could also use the [^] key to do this calculation, pressing [2] [·] [4] [^] [2] [ENTER].

Exercises: Evaluate.

1. 4^5

2. 7^9

3. 19^2

4. 5.718^2

5. 1.8^4

6. 23.4^3

7. $\left(\frac{17}{32}\right)^5$

8. $\left(\frac{2}{3}\right)^9$

C Order of Operations

What does $4 + 5 \times 2$ mean? If we add 4 and 5 and multiply the result by 2, we get 18. If we multiply 5 and 2 and add 4 to the result, we get 14. Since the results are different, we see that the order in which we carry out operations is important. To indicate which operation is to be done first, we use grouping symbols such as parentheses ( ), or brackets [ ], or braces { }. For example, $(3 \times 5) + 6 = 15 + 6 = 21$, but $3 \times (5 + 6) = 3 \times 11 = 33$.

Grouping symbols tell us what to do first. If there are no grouping symbols, we have agreements about the order in which operations should be done.

RULES FOR ORDER OF OPERATIONS

1. Do all calculations within grouping symbols before operations outside.
2. Evaluate all exponential expressions.
3. Do all multiplications and divisions in order from left to right.
4. Do all additions and subtractions in order from left to right.

EXAMPLE 4 Calculate: $15 - 2 \times 5 + 3$.

$$\begin{aligned} 15 - 2 \times 5 + 3 &= 15 - 10 + 3 && \text{Multiplying} \\ &= 5 + 3 && \text{Subtracting} \\ &= 8 && \text{Adding} \end{aligned}$$

Do Exercises 7 and 8.

Calculate.

7. $16 - 3 \times 5 + 4$

8. $4 + 5 \times 2$

Answers on page A-2

Calculate.

9. $18 - 4 \times 3 + 7$

10. $(2 \times 5)^3$

11. 2×5^3

12. $8 + 2 \times 5^3 - 4 \cdot 20$

Calculate.

13. $51.2 \div 0.64 \div 40$

14. $1000 \cdot \frac{1}{10} \div \frac{4}{5}$

Answers on page A-2

Always calculate within parentheses first. When there are exponents and no parentheses, simplify powers before multiplying or dividing.

EXAMPLE 5 Calculate: $(3 \times 4)^2$.

$(3 \times 4)^2 = (12)^2$ Working within parentheses first

$= 144$ Evaluating the exponential expression

EXAMPLE 6 Calculate: 3×4^2.

$3 \times 4^2 = 3 \times 16$ Evaluating the exponential expression

$= 48$ Multiplying

Note that $(3 \times 4)^2 \neq 3 \times 4^2$.

EXAMPLE 7 Calculate: $7 + 3 \times 4^2 - 29$.

$7 + 3 \times 4^2 - 29 = 7 + 3 \times 16 - 29$ There are no parentheses, so we find 4^2 first.

$= 7 + 48 - 29$ Multiplying

$= 55 - 29$ Adding

$= 26$ Subtracting

Do Exercises 9–12.

EXAMPLE 8 Calculate: $2.56 \div 1.6 \div 0.4$.

$2.56 \div 1.6 \div 0.4 = 1.6 \div 0.4$ Doing the divisions in order from left to right

$= 4$ Doing the second division

EXAMPLE 9 Calculate: $1000 \div \frac{1}{10} \cdot \frac{4}{5}$.

$1000 \div \frac{1}{10} \cdot \frac{4}{5} = (1000 \cdot 10) \cdot \frac{4}{5}$ Doing the division first

$= 10{,}000 \cdot \frac{4}{5}$ Multiplying

$= 8000$ Multiplying

Do Exercises 13 and 14.

Sometimes combinations of grouping symbols are used, as in

$5[4 + (8 - 2)]$.

The rules still apply. We begin with the innermost grouping symbols—in this case, the parentheses—and work to the outside.

EXAMPLE 10 Calculate: $5[4 + (8 - 2)]$.

$5[4 + (8 - 2)] = 5[4 + 6]$ Subtracting within the parentheses first

$= 5[10]$ Adding inside the brackets

$= 50$ Multiplying

A fraction bar can play the role of a grouping symbol.

EXAMPLE 11 Calculate: $\dfrac{12(9-7)+4\cdot 5}{3^4+2^3}$.

An equivalent expression with brackets as grouping symbols is

$[12(9-7)+4\cdot 5]\div[3^4+2^3]$.

What this shows, in effect, is that we do the calculations first in the numerator and then in the denominator, and divide the results:

$$\frac{12(9-7)+4\cdot 5}{3^4+2^3}=\frac{12(2)+4\cdot 5}{81+8}=\frac{24+20}{89}=\frac{44}{89}.$$

Do Exercises 15 and 16.

Calculate.

15. $4[(8-3)+7]$

16. $\dfrac{13(10-6)+4\cdot 9}{5^2-3^2}$

Answers on page A-2

CALCULATOR CORNER

Order of Operations Computations are usually entered on a graphing calculator in the same way in which we would write them. To calculate $3+4\cdot 2$, for example, we press [3] [+] [4] [×] [2] [ENTER]. The result is 11.

When an expression contains grouping symbols, we enter them using the [(] and [)] keys. To calculate $7(13-2)-40$, we press [7] [(] [1] [3] [−] [2] [)] [−] [4] [0] [ENTER]. The result is 37.

Since a fraction bar acts as a grouping symbol, we must supply parentheses when entering some fraction expressions. To calculate $\dfrac{38+142}{47-2}$, for example, we think of rewriting it with grouping symbols as $(38+142)\div(47-2)$. We press [(] [3] [8] [+] [1] [4] [2] [)] [÷] [(] [4] [7] [−] [2] [)] [ENTER]. The result is 4.

```
3+4*2
                 11
7(13-2)-40
                 37
(38+142)/(47-2)
                  4
```

Exercises: Calculate.

1. $36\div 2\cdot 3-4\cdot 4$
2. $68-8\div 4+3\cdot 5$
3. $36\div(2\cdot 3-4)\cdot 4$
4. $(15+3)^3+4(12-7)^2$
5. $50.6-8.9\times 3.01+4(5^2-24.7)$
6. $3.2+4.7[159.3-2.1(60.3-59.4)]$
7. $\{(150\cdot 5)\div[(3\cdot 16)\div(8\cdot 3)]\}+25(12\div 4)$
8. $\left(\dfrac{28}{89}+42.8\times 17.01\right)^3\div\left(\dfrac{678}{119}-\dfrac{23.2}{46.08}\right)^2$
9. $\dfrac{178-38}{5+30}$
10. $\dfrac{311-17^2}{13-2}$
11. $785-\dfrac{5^4-285}{17+3\cdot 51}$
12. $12^5-12^4+11^5\div 11^3-10.2^2$
13. What result do you get if you ignore the parentheses when evaluating $(39+141)\div(47-2)$? How does the calculator do the calculation?

R.5 EXERCISE SET

For Extra Help

Digital Video Tutor CD 1 Videotape 2 | InterAct Math | Math Tutor Center | MathXL | MyMathLab

a Write exponential notation.

1. $5 \times 5 \times 5 \times 5$

2. $3 \times 3 \times 3 \times 3 \times 3$

3. $10 \cdot 10 \cdot 10$

4. $1 \cdot 1 \cdot 1$

5. $10 \times 10 \times 10 \times 10 \times 10 \times 10$

6. $18 \cdot 18$

b Evaluate.

7. 7^2

8. 4^3

9. 9^5

10. 12^4

11. 10^2

12. 1^5

13. 1^4

14. $(1.8)^2$

15. $(2.3)^2$

16. $(0.1)^3$

17. $(0.2)^3$

18. $(14.8)^2$

19. $(20.4)^2$

20. $\left(\frac{4}{5}\right)^2$

21. $\left(\frac{3}{8}\right)^2$

22. 2^4

23. 5^3

24. $(1.4)^3$

25. $1000 \times (1.02)^3$

26. $2000 \times (1.06)^2$

c Calculate.

27. $9 + 2 \times 8$

28. $14 + 6 \times 6$

29. $9 \times 8 + 7 \times 6$

30. $30 \times 5 + 2 \times 2$

31. $39 - 4 \times 2 + 2$

32. $14 - 2 \times 6 + 7$

33. $9 \div 3 + 16 \div 8$

34. $32 - 8 \div 4 - 2$

35. $7 + 10 - 10 \div 2$

36. $(5 \cdot 4)^2$

37. $(6 \cdot 3)^2$

38. $3 \cdot 2^3$

39. $4 \cdot 5^2$

40. $(7 + 3)^2$

41. $(8 + 2)^3$

42. $7 + 2^2$

43. $6 + 4^2$

44. $(5 - 2)^2$

45. $(3 - 2)^2$

46. $10 - 3^2$

47. $4^3 \div 8 - 4$

48. $20 + 4^3 \div 8 - 4$

49. $120 - 3^3 \cdot 4 \div 6$

50. $7 \times 3^4 + 18$

51. $6[9 + (3 + 4)]$

52. $8[(13 + 6) - 11]$

53. $8 + (7 + 9)$

54. $(8 + 7) + 9$

55. $15(4 + 2)$

56. $15 \cdot 4 + 15 \cdot 2$

57. $12 - (8 - 4)$

58. $(12 - 8) - 4$

59. $1000 \div 100 \div 10$

60. $256 \div 32 \div 4$

61. $2000 \div \frac{3}{50} \cdot \frac{3}{2}$

62. $400 \times 0.64 \div 3.2$

63. $\frac{80 - 6^2}{9^2 + 3^2}$

64. $\frac{5^2 + 4^3 - 3}{9^2 - 2^2 + 1^5}$

65. $\frac{3(6 + 7) - 5 \cdot 4}{6 \cdot 7 + 8(4 - 1)}$

66. $\frac{20(8 - 3) - 4(10 - 3)}{10(6 + 2) + 2(5 + 2)}$

67. $8 \cdot 2 - (12 - 0) \div 3 - (5 - 2)$

68. $95 - 2^3 \cdot 5 \div (24 - 4)$

69. D_W The expression $(3 \cdot 4)^2$ contains parentheses. Are they necessary? Why or why not?

70. D_W The expression $9 - (4 \cdot 2)$ contains parentheses. Are they necessary? Why or why not?

SKILL MAINTENANCE

Find percent notation. [R.4d]

71. $\frac{5}{16}$

72. $\frac{11}{12}$

73. $\frac{17}{32}$

74. $\frac{11}{6}$

Simplify. [R.2b]

75. $\frac{125}{325}$

76. $\frac{9}{2001}$

77. $\frac{64}{96}$

78. $\frac{2005}{3640}$

79. Find the prime factorization of 48. [R.1a]

80. Find the LCM of 12, 24, and 56. [R.1b]

SYNTHESIS

Write each of the following with a single exponent.

81. $\frac{10^5}{10^3}$

82. $\frac{10^7}{10^2}$

83. $5^4 \cdot 5^2$

84. $\frac{2^8}{8^2}$

85. *Five 5's.* We can use five 5's and grouping symbols to represent the numbers 0 through 10. For example,

$$0 = 5 \cdot 5 \cdot 5(5 - 5), \qquad 1 = \frac{5 + 5}{5} - \frac{5}{5}, \qquad 2 = \frac{5 \cdot 5 - 5}{5 + 5}.$$

Often more than one way to make a representation is possible. Use five 5's to represent the numbers 3 through 10.

R.6 GEOMETRY

Objectives

a Find the perimeter of a polygon.

b Find the area of a rectangle, a square, a parallelogram, and a triangle.

c Find the length of a radius of a circle given the length of a diameter, and find the length of a diameter given the length of a radius; find the circumference and the area of a circle.

d Find the volume of a rectangular solid.

Find the perimeter of the polygon.

1.

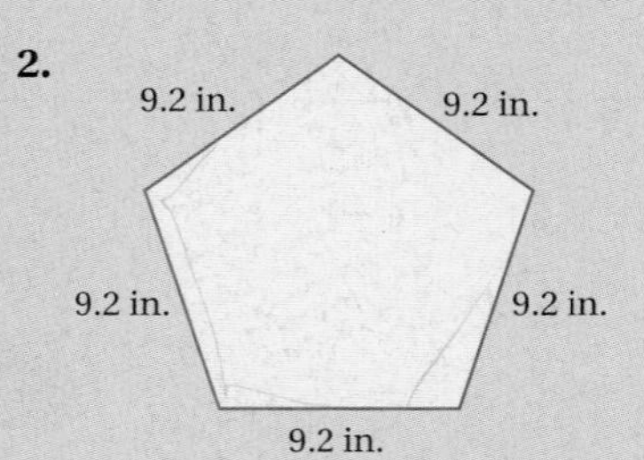

2.

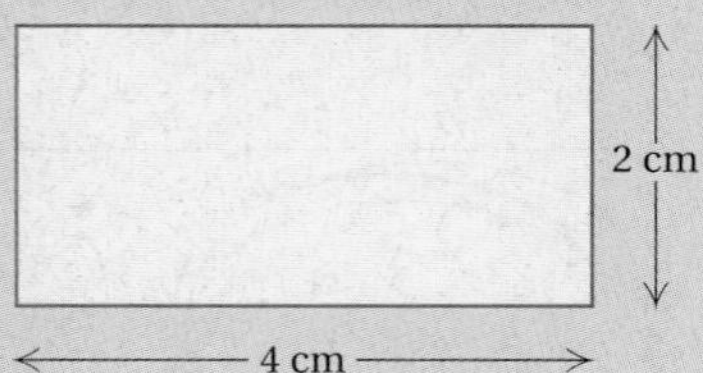

3. Find the perimeter of a rectangle that is 2 cm by 4 cm.

2 cm

4 cm

4. Find the perimeter of a rectangle that is 5.25 yd by 3.5 yd.

Answers on page A-3

a Perimeter

PERIMETER OF A POLYGON

A **polygon** is a geometric figure with three or more sides. The **perimeter of a polygon** is the distance around it, or the sum of the lengths of its sides.

EXAMPLE 1 Find the perimeter of this polygon.

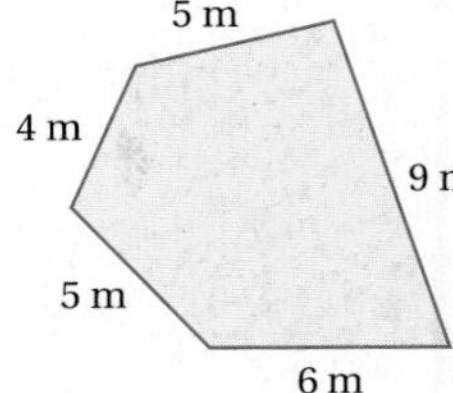

We add the lengths of the sides. Since all the units are the same, we add the numbers, keeping meters (m) as the unit.

$$\begin{aligned}\text{Perimeter} &= 6\text{ m} + 5\text{ m} + 4\text{ m} + 5\text{ m} + 9\text{ m}\\ &= (6 + 5 + 4 + 5 + 9)\text{ m}\\ &= 29\text{ m}\end{aligned}$$

Do Exercises 1 and 2.

A **rectangle** is a figure with four sides and four 90°-angles, like the one shown in Example 2.

PERIMETER OF A RECTANGLE

The **perimeter of a rectangle** is twice the sum of the length and the width, or 2 times the length plus 2 times the width:

$$P = 2 \cdot (l + w), \quad \text{or} \quad P = 2 \cdot l + 2 \cdot w.$$

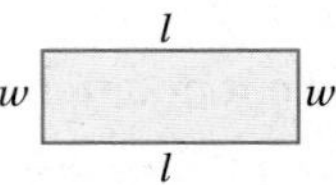

EXAMPLE 2 Find the perimeter of a rectangle that is 7.8 ft by 4.3 ft.

$$\begin{aligned}P = 2 \cdot (l + w) &= 2 \cdot (7.8\text{ ft} + 4.3\text{ ft})\\ &= 2 \cdot (12.1\text{ ft}) = 24.2\text{ ft}\end{aligned}$$

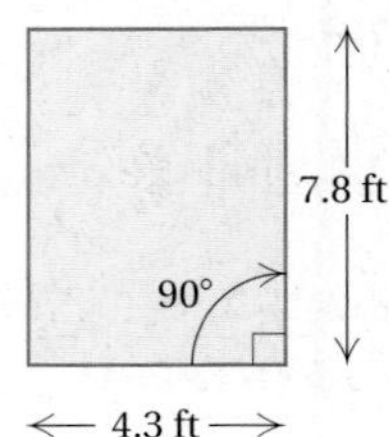

Do Exercises 3 and 4.

A **square** is a rectangle with all sides the same length.

PERIMETER OF A SQUARE

The **perimeter of a square** is four times the length of a side:

$P = 4 \cdot s.$

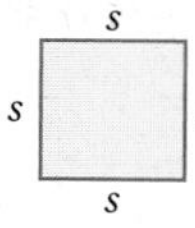

EXAMPLE 3 Find the perimeter of a square whose sides are 9 mm long.

$$P = 4 \cdot s = 4 \cdot (9 \text{ mm}) = 36 \text{ mm}$$

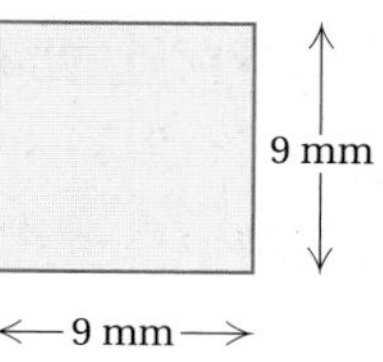

Do Exercises 5 and 6.

Area

RECTANGLES

We can find the area of a rectangle by filling it with square units. Two such units, a *square inch* and a *square centimeter*, are shown below.

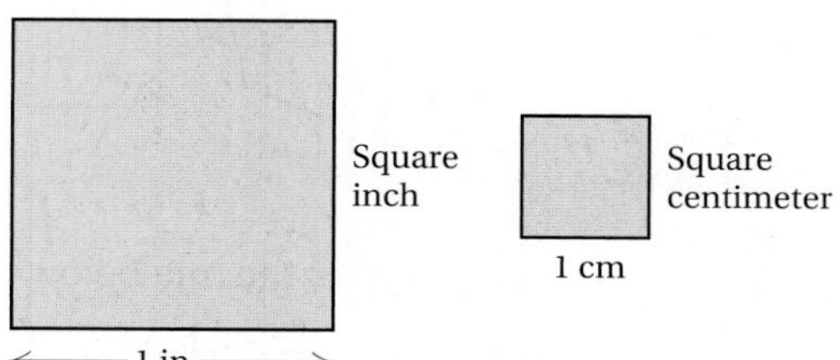

AREA OF A RECTANGLE

The **area of a rectangle** is the product of the length l and the width w:

$A = l \cdot w.$

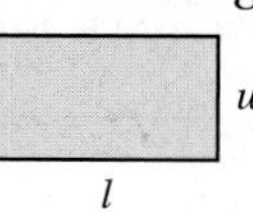

EXAMPLE 4 Find the area of a rectangle that is 7 yd by 4 yd.

$$A = l \cdot w = 7 \text{ yd} \cdot 4 \text{ yd} = 7 \cdot 4 \cdot \text{yd} \cdot \text{yd} = 28 \text{ yd}^2$$

We think of yd · yd as $(\text{yd})^2$ and denote it yd^2. Thus we read "28 yd^2" as "28 square yards."

Do Exercises 7–9.

5. Find the perimeter of a square with sides of length 10 km.

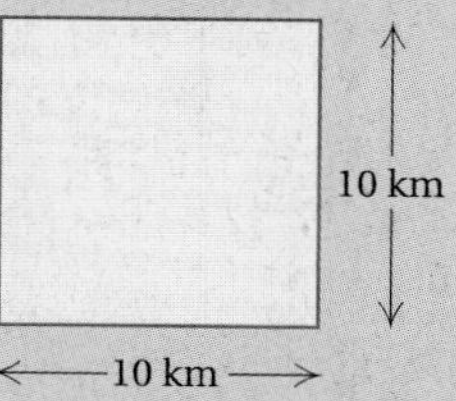

6. Find the perimeter of a square with sides of length $5\frac{1}{4}$ yd.

7. What is the area of this region? Count the number of square centimeters.

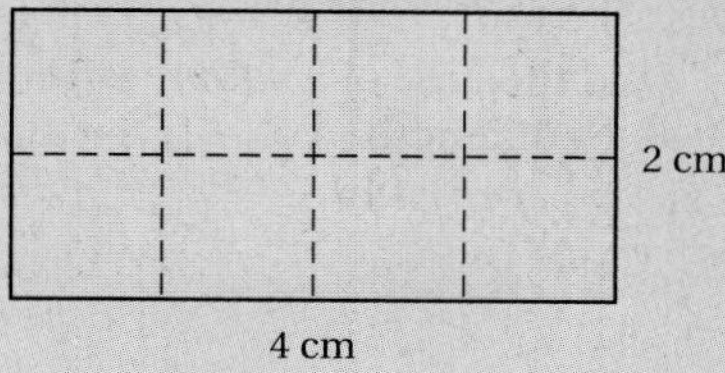

8. Find the area of a rectangle that is 7 km by 8 km.

9. Find the area of a rectangle that is 5.3 yd by 3.2 yd.

Answers on page A-3

10. Find the area of a square with sides of length 10.9 m.

11. Find the area of a square with sides of length $\frac{2}{3}$ yd.

Answers on page A-3

SQUARES

AREA OF A SQUARE

The **area of a square** is the square of the length of a side:

$$A = s \cdot s, \quad \text{or} \quad A = s^2.$$

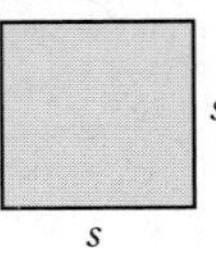

EXAMPLE 5 Find the area of a square with sides of length 20.3 m.

$$A = s \cdot s = 20.3\text{ m} \times 20.3\text{ m}$$
$$= 20.3 \times 20.3 \times \text{m} \times \text{m} = 412.09\text{ m}^2$$

Do Exercises 10 and 11.

PARALLELOGRAMS

A **parallelogram** is a four-sided figure with two pairs of parallel sides, as shown below.

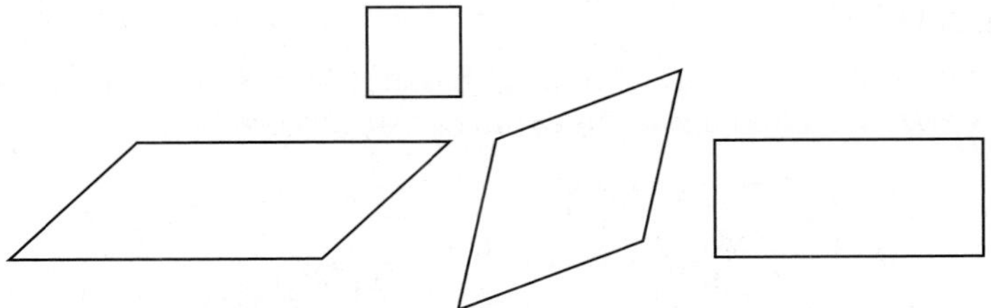

To find the area of a parallelogram, consider the one below.

If we cut off a piece and move it to the other end, we get a rectangle.

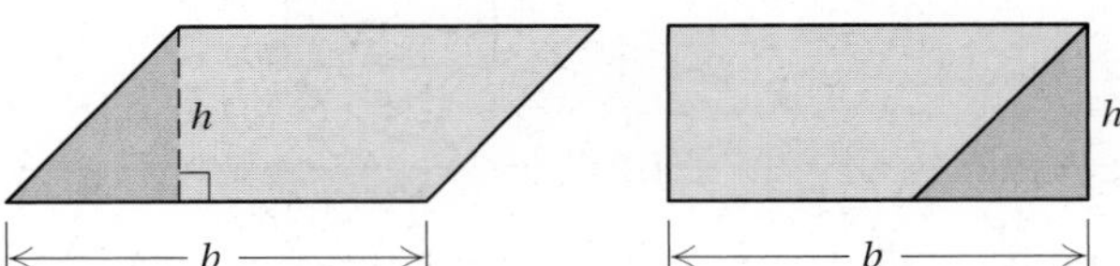

We can find the area by multiplying the length b, called a **base,** by h, called the **height.**

AREA OF A PARALLELOGRAM

The **area of a parallelogram** is the product of a base b and the height h:

$$A = b \cdot h.$$

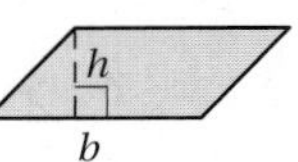

EXAMPLE 6 Find the area of this parallelogram.

$$A = b \cdot h$$
$$= 7\text{ km} \cdot 5\text{ km}$$
$$= 35\text{ km}^2$$

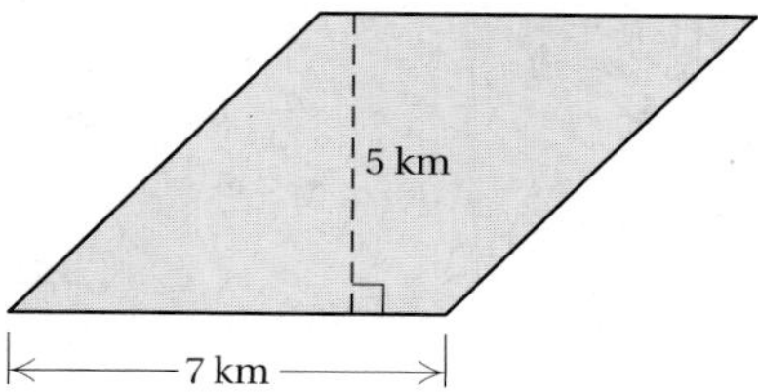

EXAMPLE 7 Find the area of this parallelogram.

$$A = b \cdot h$$
$$= (1.2\text{ m}) \times (6\text{ m})$$
$$= 7.2\text{ m}^2$$

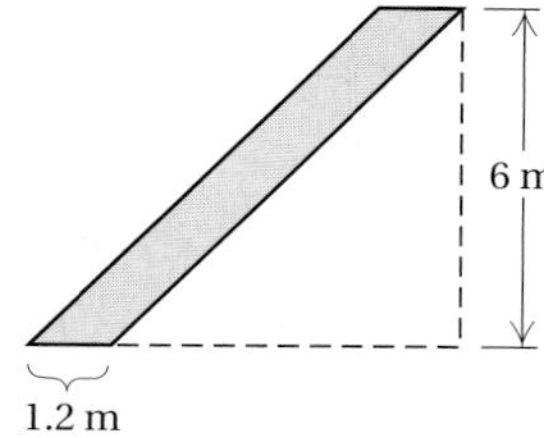

Do Exercises 12 and 13.

TRIANGLES

To find the area of a triangle, think of cutting out another just like it. Then place the second one as shown in the figure on the right.

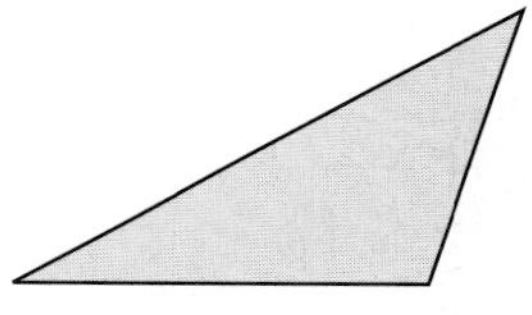

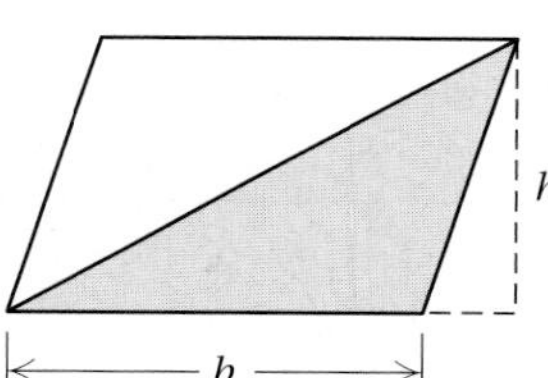

The resulting figure is a parallelogram whose area is $b \cdot h$. The triangle we started with has half the area of the parallelogram, or

$$\frac{1}{2} \cdot b \cdot h.$$

AREA OF A TRIANGLE

The **area of a triangle** is half the length of the base times the height:

$$A = \frac{1}{2} \cdot b \cdot h.$$

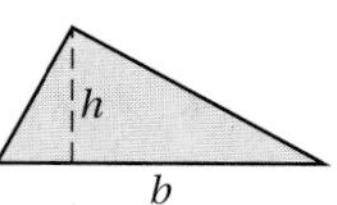

EXAMPLE 8 Find the area of this triangle.

$$A = \frac{1}{2} \cdot b \cdot h$$
$$= \frac{1}{2} \cdot 9\text{ m} \cdot 6\text{ m}$$
$$= \frac{9 \cdot 6}{2}\text{ m}^2$$
$$= 27\text{ m}^2$$

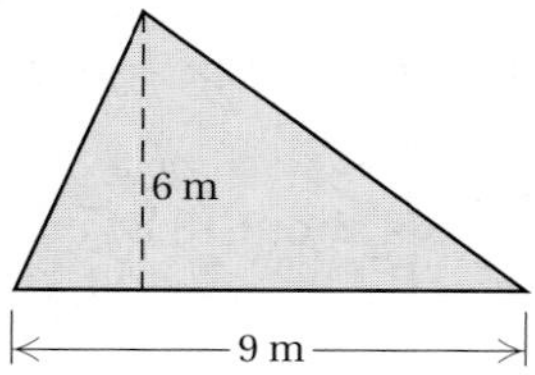

Find the area.

12.

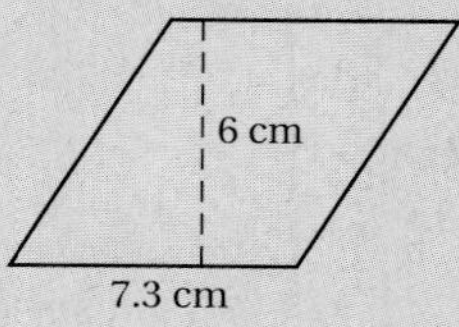

13.

5.5 km

2.25 km

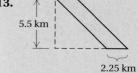

Find the area.

14.

12 m

16 m

15.

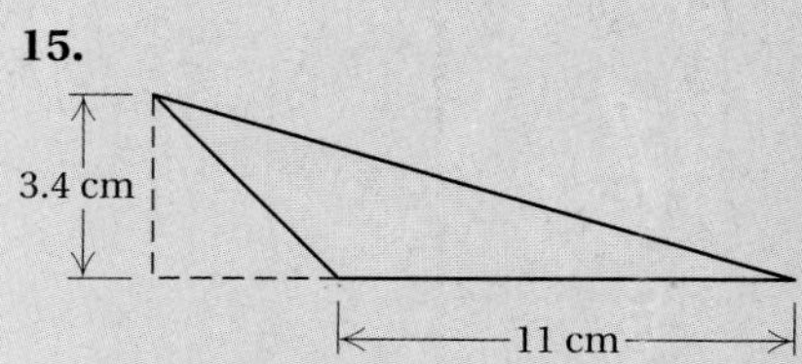

Answers on page A-3

16. Find the length of a radius.

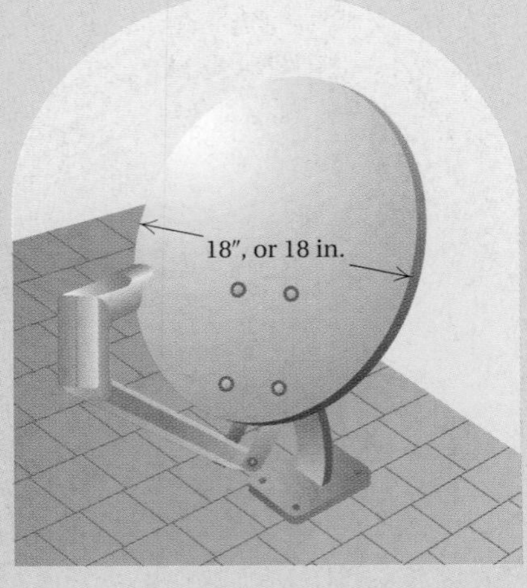

17. Find the length of a diameter.

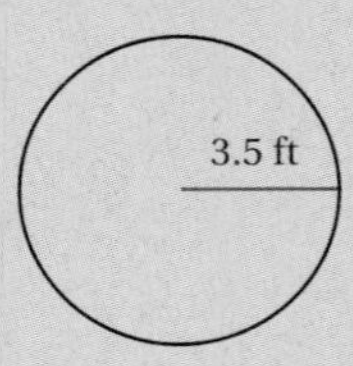

18. Find the circumference of this circle. Use 3.14 for π.

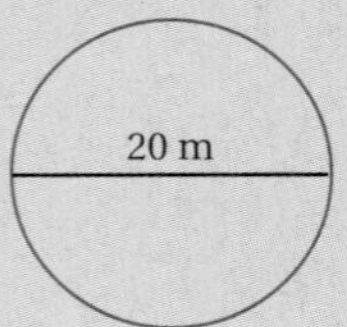

19. Find the circumference of this circle. Use $\frac{22}{7}$ for π.

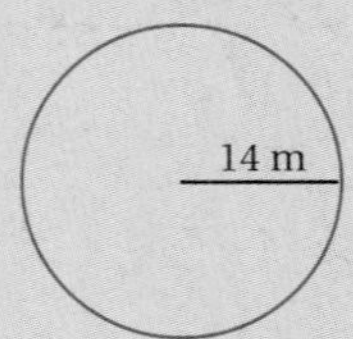

Answers on page A-3

EXAMPLE 9 Find the area of this triangle.

$$A = \frac{1}{2} \cdot b \cdot h$$
$$= \frac{1}{2} \times 6.25 \text{ cm} \times 5.5 \text{ cm}$$
$$= 0.5 \times 6.25 \times 5.5 \text{ cm}^2$$
$$= 17.1875 \text{ cm}^2$$

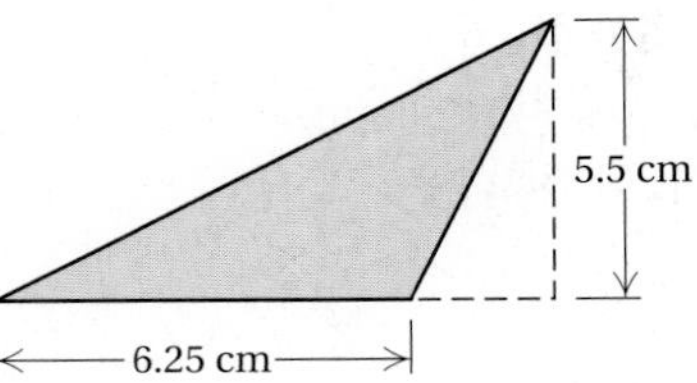

Do Exercises 14 and 15 on the preceding page.

C Circles

DIAMETER AND RADIUS OF A CIRCLE

Shown at right is a circle with center O. Segment $\overline{AC}$ is a *diameter*. A **diameter** is a segment that passes through the center of the circle and has endpoints on the circle. Segment $\overline{OB}$ is called a *radius*. A **radius** is a segment with one endpoint on the center and the other endpoint on the circle.

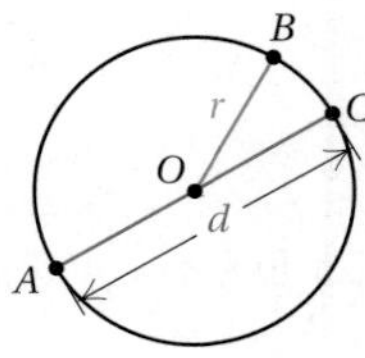

DIAMETER AND RADIUS

Suppose that d is the diameter of a circle and r is the radius. Then

$$d = 2 \cdot r \quad \text{and} \quad r = \frac{d}{2}.$$

Do Exercises 16 and 17.

CIRCUMFERENCE OF A CIRCLE

The **circumference** of a circle is the distance around it. Calculating circumference is similar to finding the perimeter of a polygon. For any circle, if we divide the circumference C by the diameter d, we get the same number. We call this number π (pi).

CIRCUMFERENCE OF A CIRCLE

The **circumference** C of a circle is given by

$$C = \pi \cdot d, \quad \text{or} \quad C = 2 \cdot \pi \cdot r,$$

where d is the diameter and r is the radius. The number π is about 3.14, or about 22/7.

EXAMPLE 10 Find the circumference of this circle. Use 3.14 for π.

$$C = \pi \cdot d$$
$$\approx 3.14 \times 6 \text{ cm}$$
$$= 18.84 \text{ cm}$$

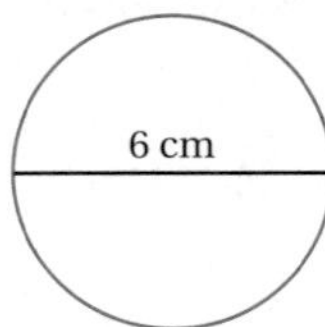

The circumference is about 18.84 cm.

EXAMPLE 11 Find the circumference of this circle. Use $\frac{22}{7}$ for π.

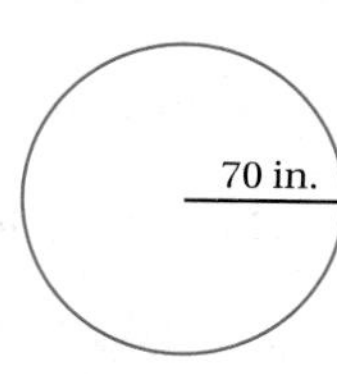

$$\begin{aligned} C &= 2 \cdot \pi \cdot r \\ &\approx 2 \cdot \frac{22}{7} \cdot 70 \text{ in.} \\ &= 2 \cdot 22 \cdot \frac{70}{7} \text{ in.} \\ &= 44 \cdot 10 \text{ in.} \\ &= 440 \text{ in.} \end{aligned}$$

The circumference is about 440 in.

Do Exercises 18 and 19 on the preceding page.

AREA OF A CIRCLE

Now we consider a formula for the area of a circle.

> **AREA OF A CIRCLE**
>
> The **area of a circle** with radius of length r is given by
>
> $A = \pi \cdot r \cdot r$, or $A = \pi \cdot r^2$.
>
> 
>

EXAMPLE 12 Find the area of this circle. Use $\frac{22}{7}$ for π.

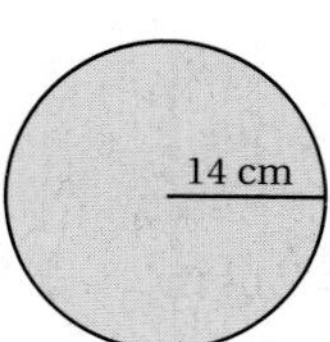

$$\begin{aligned} A &= \pi \cdot r \cdot r \\ &\approx \frac{22}{7} \cdot 14 \text{ cm} \cdot 14 \text{ cm} \\ &= \frac{22}{7} \cdot 196 \text{ cm}^2 \\ &= 616 \text{ cm}^2 \end{aligned}$$

The area is about 616 cm^2.

Do Exercise 20.

EXAMPLE 13 Find the area of this circle. Use 3.14 for π. Round to the nearest hundredth.

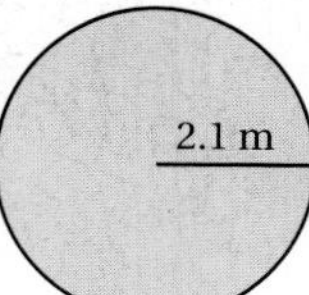

$$\begin{aligned} A &= \pi \cdot r \cdot r \\ &\approx 3.14 \times 2.1 \text{ m} \times 2.1 \text{ m} \\ &= 3.14 \times 4.41 \text{ m}^2 \\ &= 13.8474 \text{ m}^2 \\ &\approx 13.85 \text{ m}^2 \end{aligned}$$

The area is about 13.85 m^2.

Do Exercise 21.

20. Find the area of this circle. Use $\frac{22}{7}$ for π.

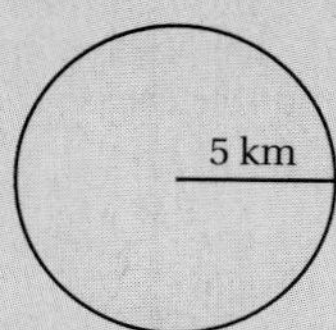

21. Find the area of this circle. Use 3.14 for π. Round to the nearest hundredth.

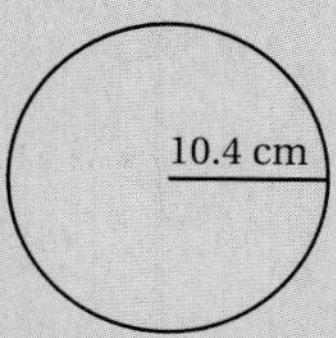

Answers on page A-3

CALCULATOR CORNER

Using the Pi Key On certain calculators, there is a pi key, [π]. You can use a [π] key for most computations instead of stopping to round the value of π. Rounding, if necessary, is done at the end.

Exercises:

1. If you have a [π] key on your calculator, to how many places does this key give the value of π?
2. Find the circumference and the area of a circle with a radius of 225.68 in.
3. Find the area of a circle with a diameter of 46.6 in.
4. Find the area of a large irrigated farming circle with a diameter of 400 ft.

22. **Cord of Wood.** A cord of wood is 4 ft by 4 ft by 8 ft. What is the volume of a cord of wood?
Source: *The American Heritage Dictionary of the English Language*

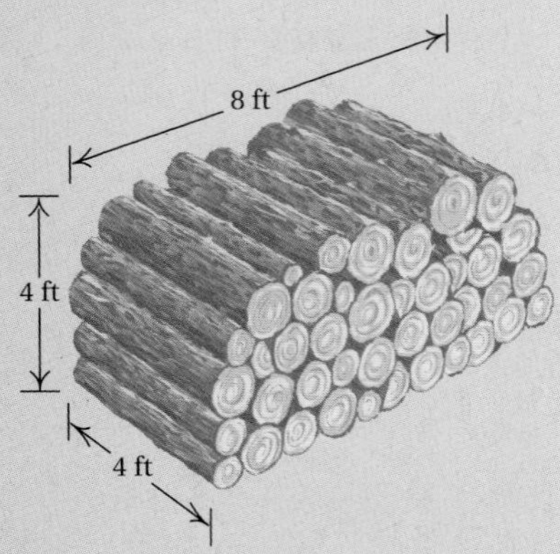

d Volume

The **volume of a rectangular solid** is the number of unit cubes needed to fill it.

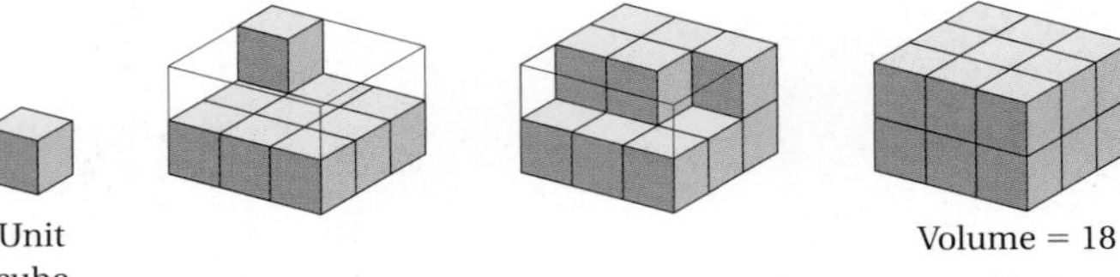

VOLUME OF A RECTANGULAR SOLID

The **volume of a rectangular solid** is found by multiplying length by width by height:

$$V = l \cdot w \cdot h.$$

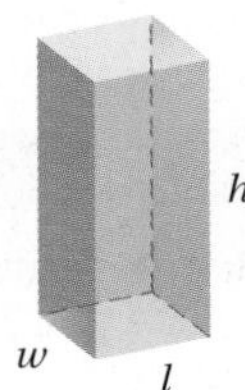

EXAMPLE 14 *Carry-On Luggage.* The largest piece of luggage that you can carry on an airplane measures 23 in. by 10 in. by 13 in. Find the volume of this solid.

$$\begin{aligned} V &= l \cdot w \cdot h \\ &= 23 \text{ in.} \cdot 10 \text{ in.} \cdot 13 \text{ in.} \\ &= 230 \cdot 13 \text{ in}^3 \\ &= 2990 \text{ in}^3 \end{aligned}$$

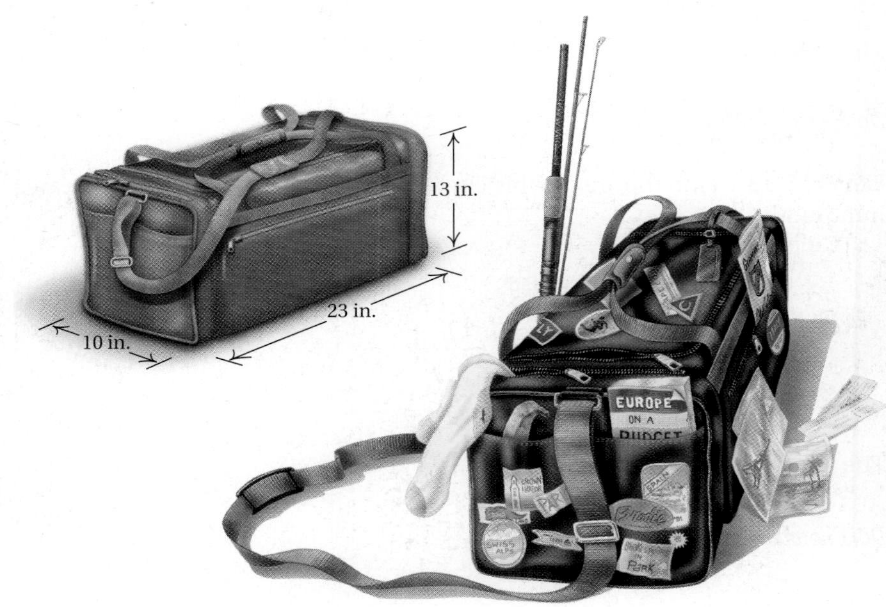

Do Exercise 22.

Answer on page A-3

R.6 EXERCISE SET

For Extra Help

Digital Video Tutor CD 1 Videotape 2 | InterAct Math | Math Tutor Center | MathXL | MyMathLab

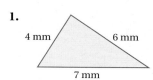

a Find the perimeter of the polygon.

1.

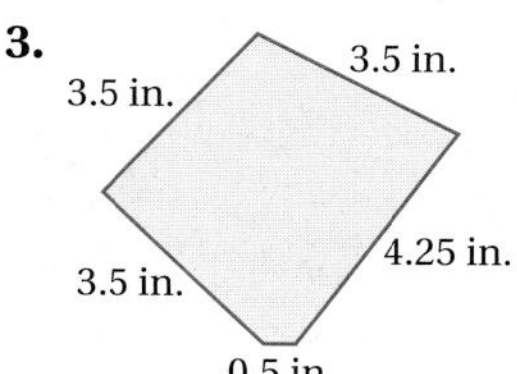

2.

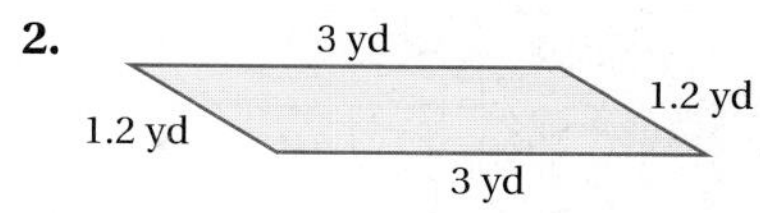

3.

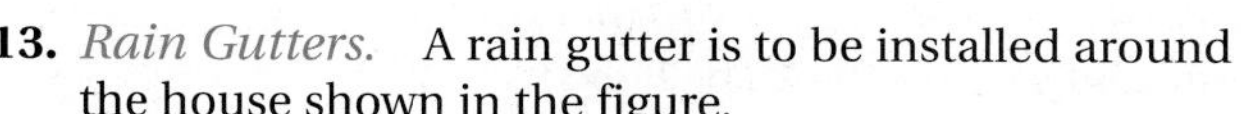

4.

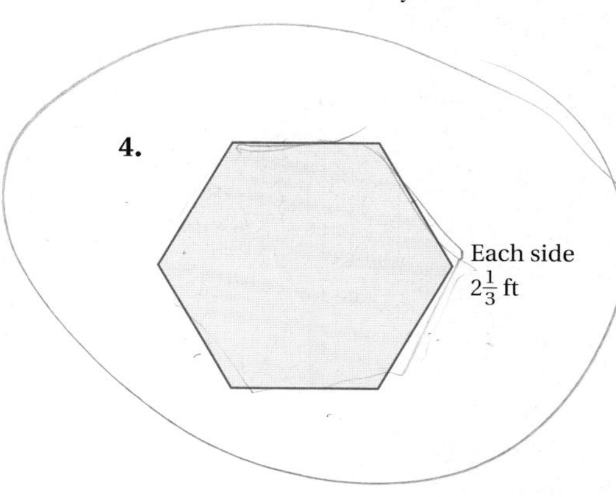

Find the perimeter of the rectangle.

5. 5 ft by 10 ft

6. 2.5 m by 100 m

7. 34.67 cm by 4.9 cm

8. 3.5 yd by 4.5 yd

Find the perimeter of the square.

9. 22 ft on a side

10. 56.9 km on a side

11. 45.5 mm on a side

12. $\frac{1}{8}$ yd on a side

13. *Rain Gutters.* A rain gutter is to be installed around the house shown in the figure.

a) Find the perimeter of the house.

b) If the gutter costs $4.59 per foot, what is the total cost of the gutter?

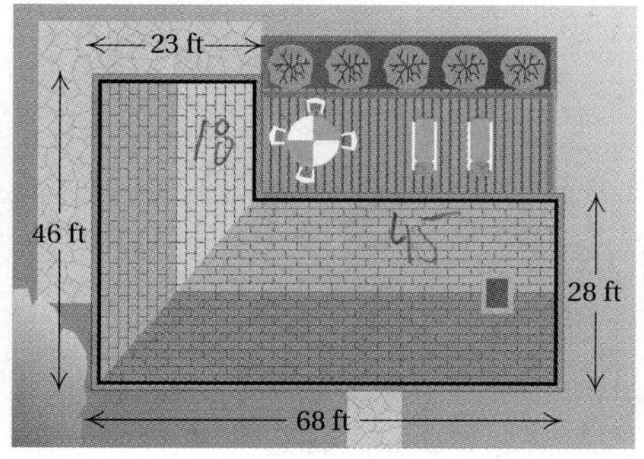

14. *Softball Diamond.* A standard-sized slow-pitch softball diamond is a square with sides of length 65 ft. What is the perimeter of this softball diamond? (This is the distance you would have to run if you hit a home run.)

b Find the area.

15. 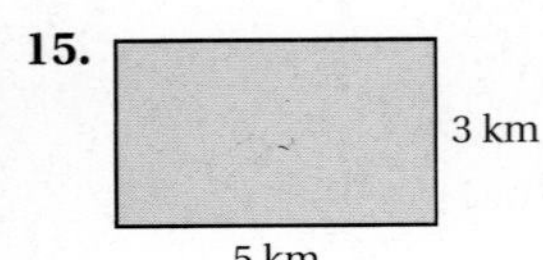

16.  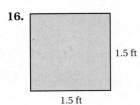

17. 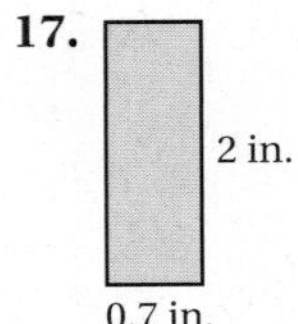

18. 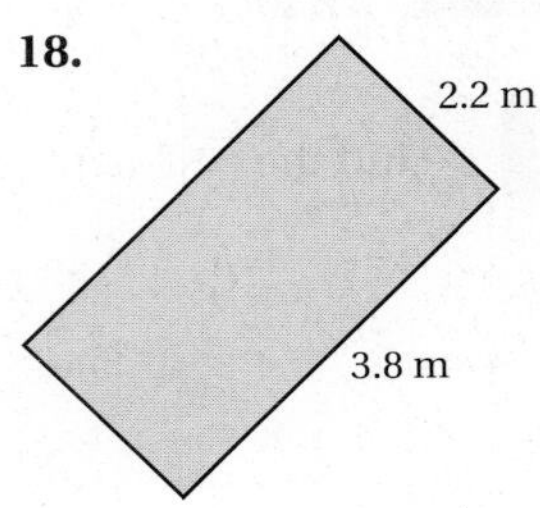

19. 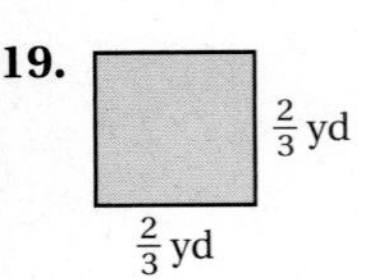

20. 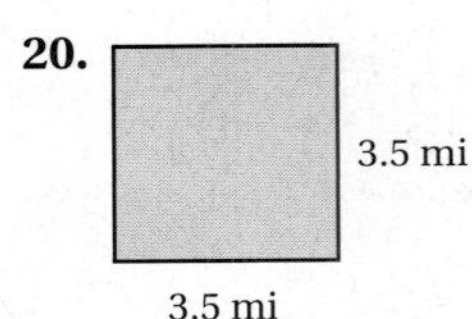

21. 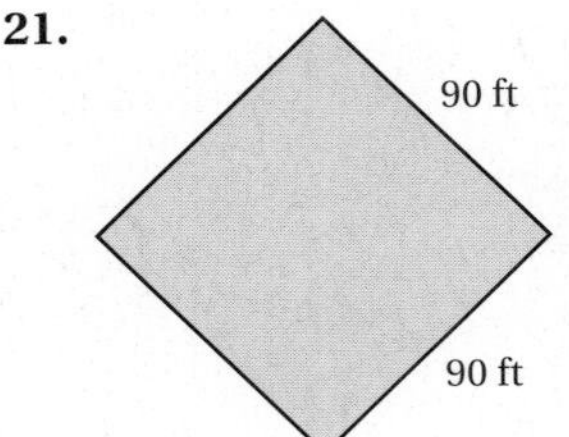

22. 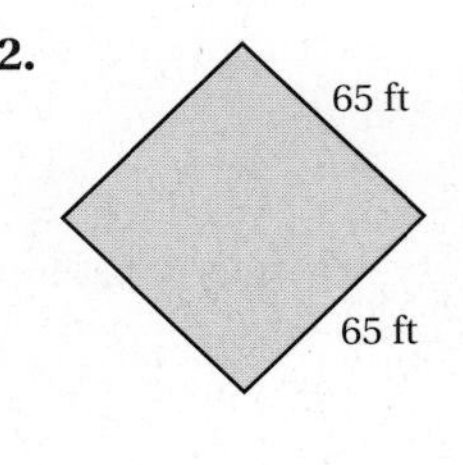

Find the area of the rectangle.

23. 5 ft by 10 ft

24. 14 yd by 8 yd

25. 34.67 cm by 4.9 cm

26. 2.45 km by 100 km

27. $\frac{2}{3}$ in. by $\frac{5}{6}$ in.

28. $\frac{1}{3}$ mi by $\frac{2}{3}$ mi

Find the area of the square.

29. 22 ft on a side

30. 18 yd on a side

31. 56.9 km on a side

32. 45.5 m on a side

33. $\frac{3}{8}$ yd on a side

34. $\frac{1}{10}$ mi on a side

35. *Sidewalk Area.* Franklin Construction Company builds a sidewalk around two sides of the Municipal Trust Bank building, as shown in the figure. What is the area of the sidewalk?

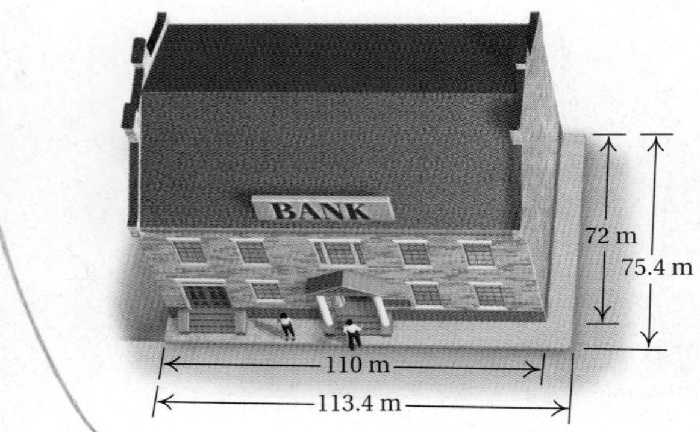

36. *Mowing Expense.* A square sandbox 4.5 ft on a side is placed on a 60-ft by $93\frac{2}{3}$-ft lawn.

a) Find the area of the lawn.

b) It costs $0.03 per square foot to have the lawn mowed. What is the total cost of the mowing?

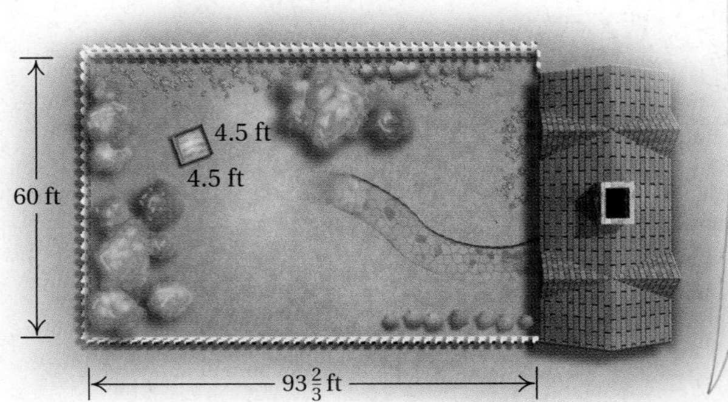

Find the area.

37.

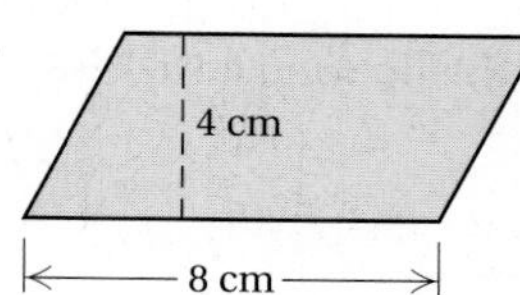

38.

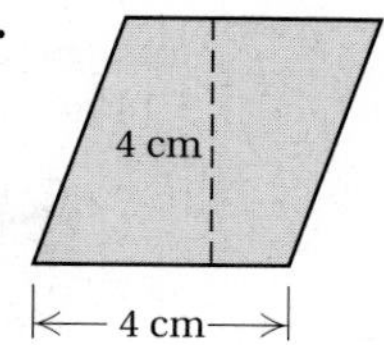

39.

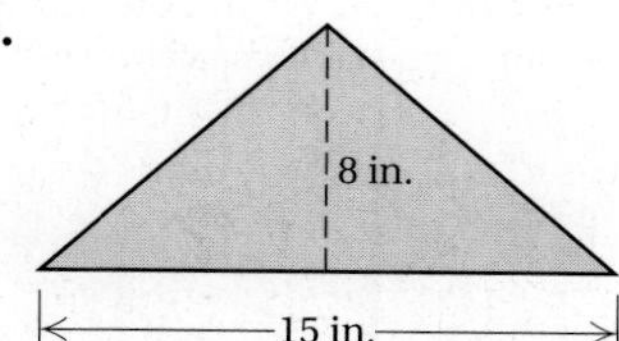

40.

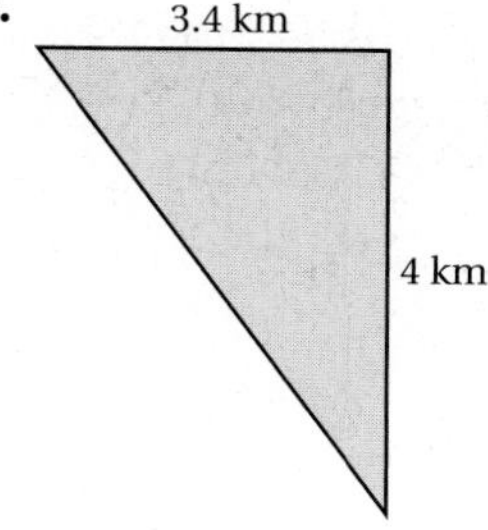

41.

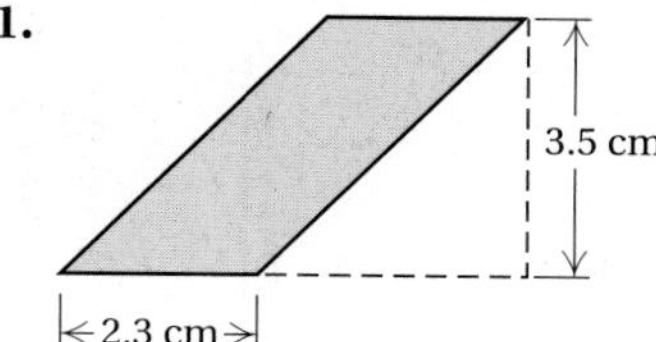

42.

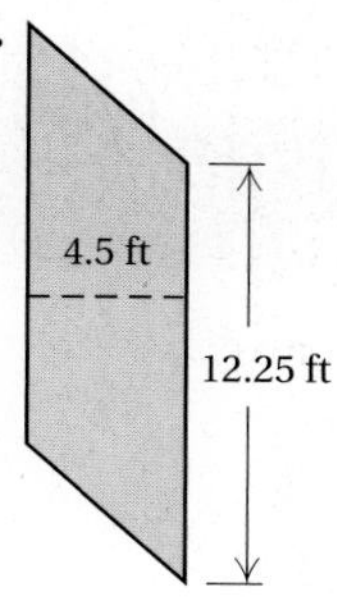

43.

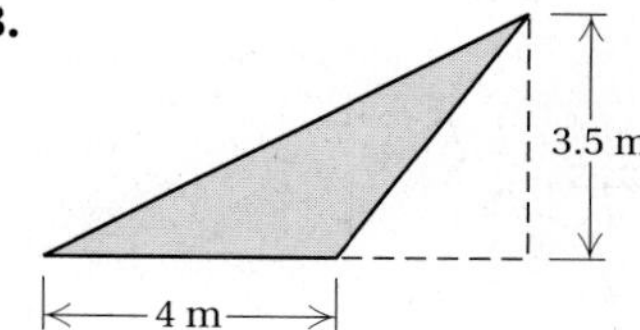

44.

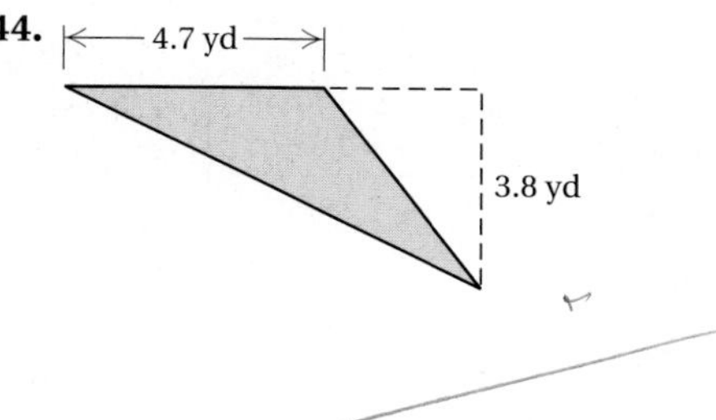

C For each circle, find the length of a diameter, the circumference, and the area. Use $\frac{22}{7}$ for π.

45.

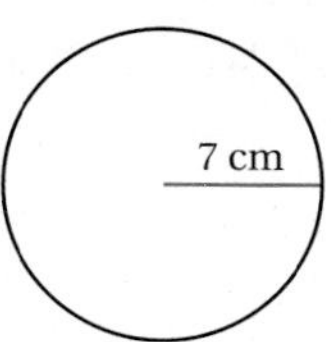

46.

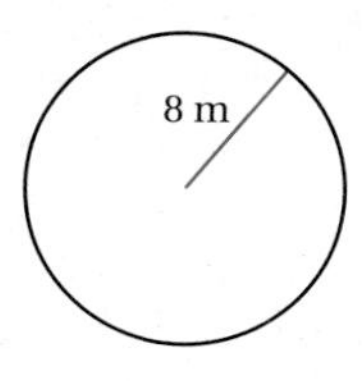

47.

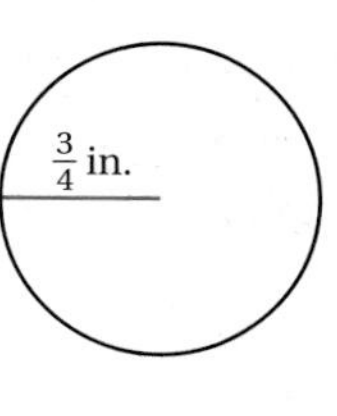

48.

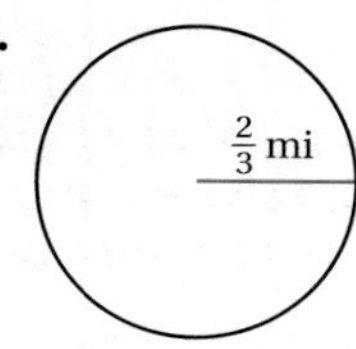

no

For each circle, find the length of a radius, the circumference, and the area. Use 3.14 for π.

49.

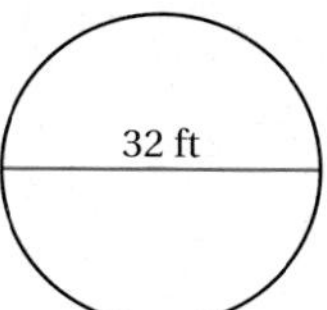

50.

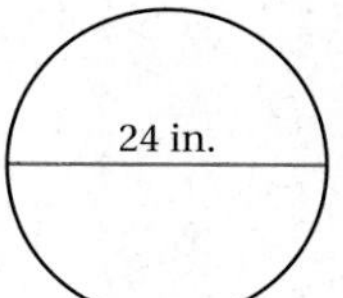

51.

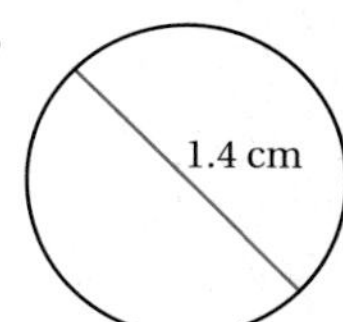

52.

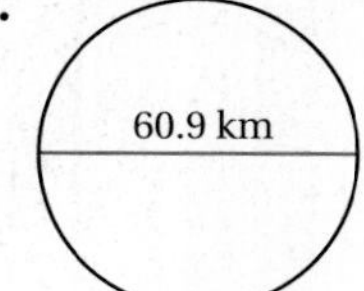

53. *Trampoline.* The standard backyard trampoline has a diameter of 14 ft. What is its area? Use $\frac{22}{7}$ for π.
Source: International Trampoline Industry Association, Inc.

54. *Roller-Rink Floor.* A roller-rink floor is shown below. What is its area? If hardwood flooring costs $10.50 per square meter, how much will the flooring cost? Use 3.14 for π.

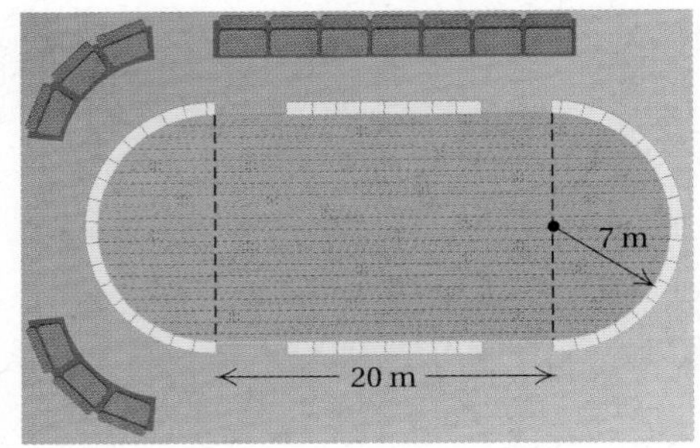

d Find the volume.

55.

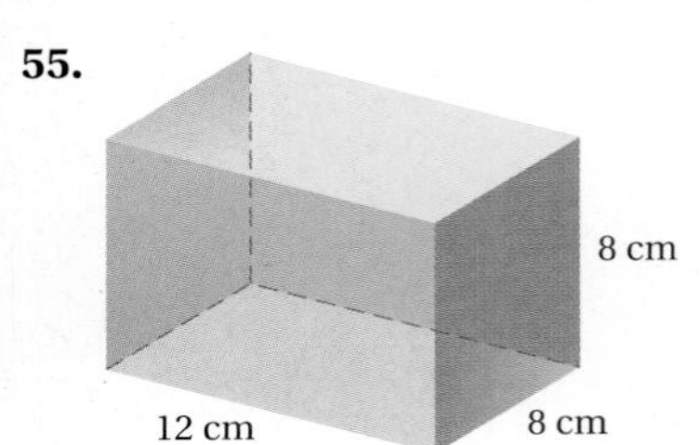

56.

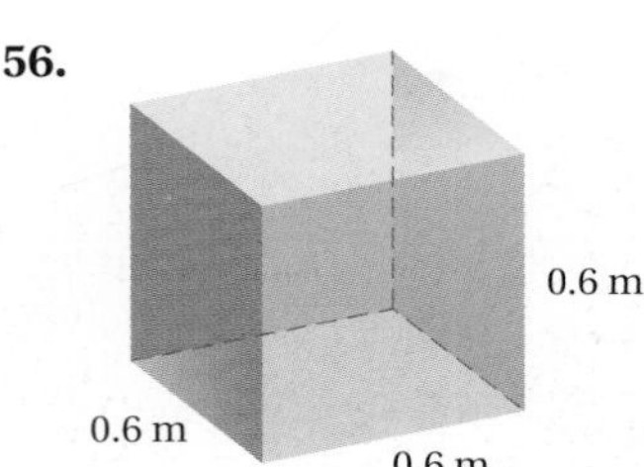

57.

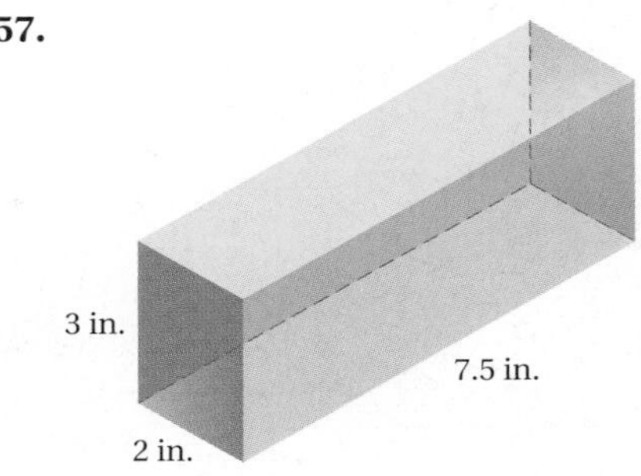

58.

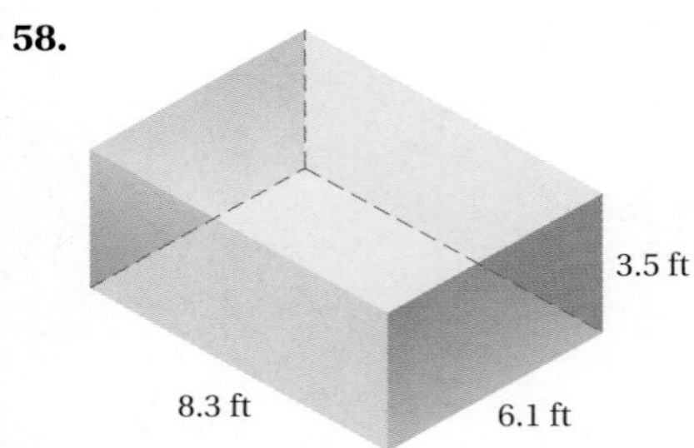

59.

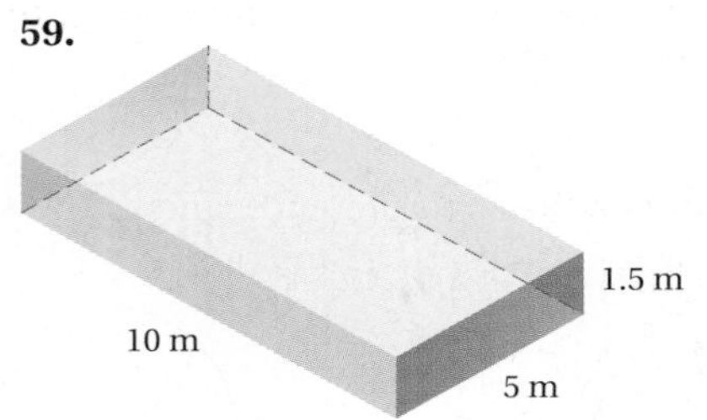

60.

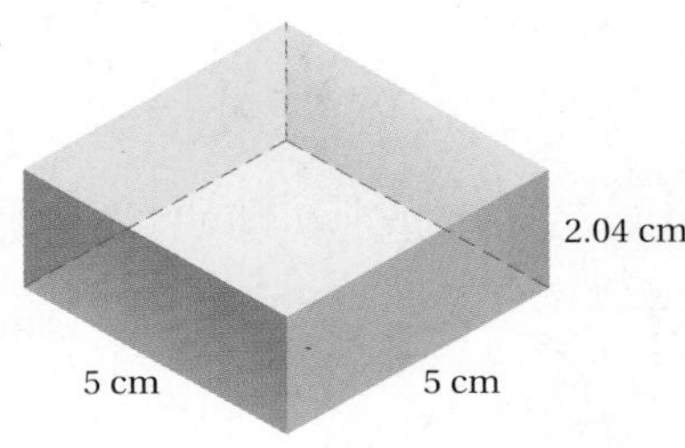

61.

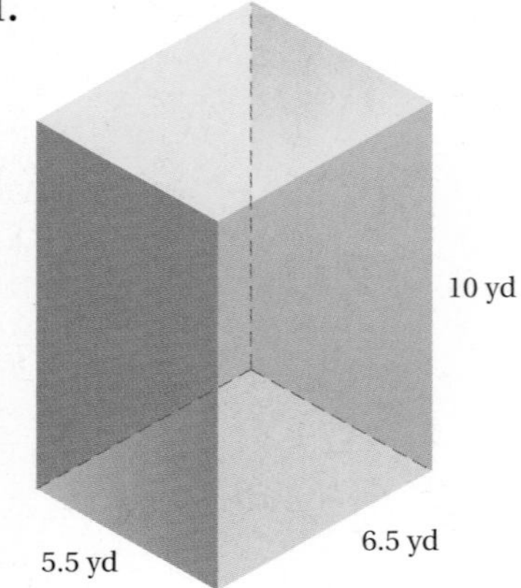

62.

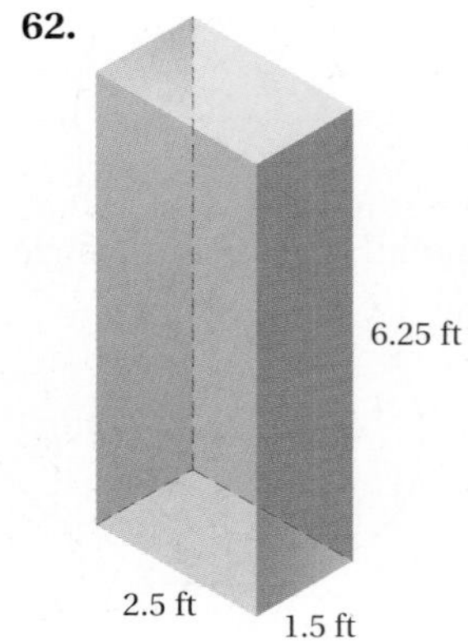

63. $^{D}_{W}$ Explain how the area of a parallelogram can be found by considering the area of a rectangle.

64. $^{D}_{W}$ Explain why a 16-in. diameter pizza that costs \$16.25 is a better buy than a 10-in. diameter pizza that costs \$7.85.

SKILL MAINTENANCE

Convert to percent notation. [R.4c]

65. 0.875 **66.** 0.58 **67.** $0.\overline{6}$ **68.** 0.4361

Convert to percent notation. [R.4d]

69. $\frac{3}{8}$ **70.** $\frac{5}{8}$ **71.** $\frac{2}{3}$ **72.** $\frac{1}{5}$

SYNTHESIS

Circular cylinders have bases of equal area that lie in parallel planes. The bases of circular cylinders are circular regions.

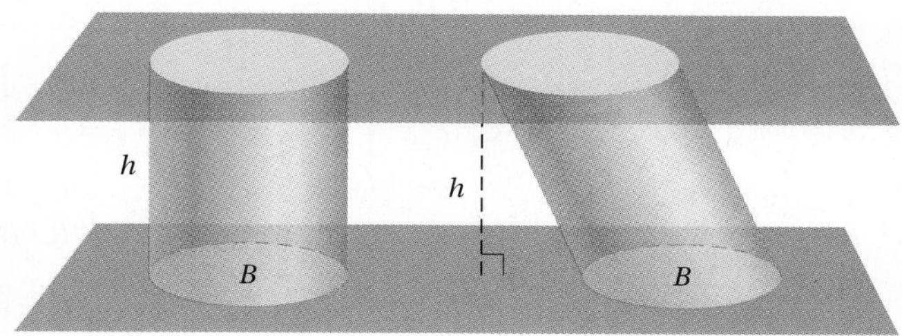

VOLUME OF A CIRCULAR CYLINDER

The **volume of a circular cylinder** is the product of the area of the base B and the height h:

$$V = B \cdot h, \quad \text{or} \quad V = \pi \cdot r^2 \cdot h.$$

Find the volume of the circular cylinder. Use 3.14 for π in Exercises 73–76. Use $\frac{22}{7}$ for π in Exercises 77 and 78.

73.

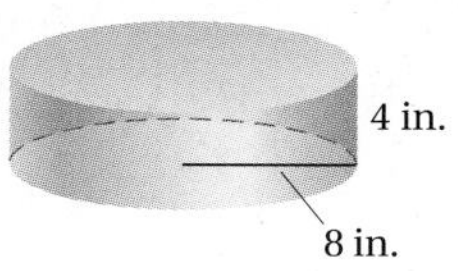

74.

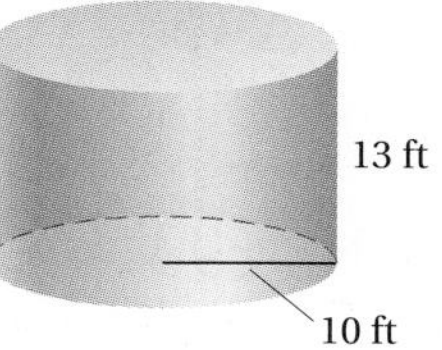

75.

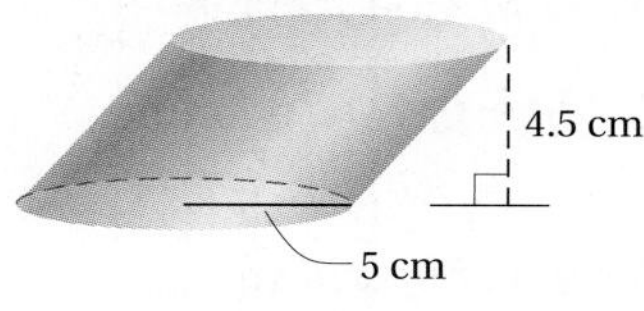

76.

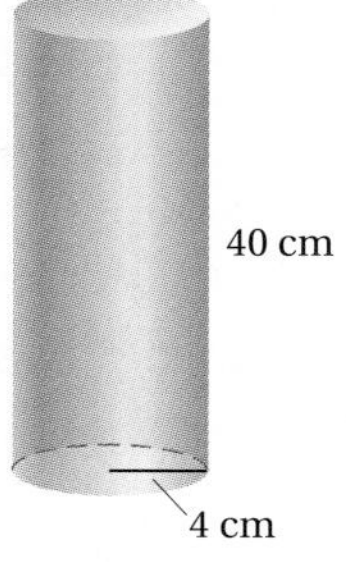

77.

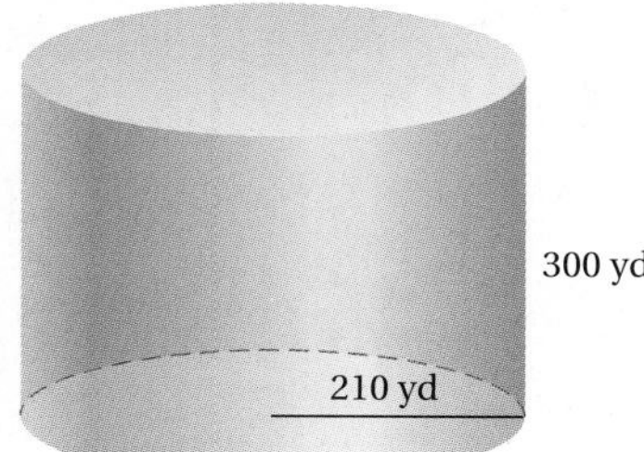

78.

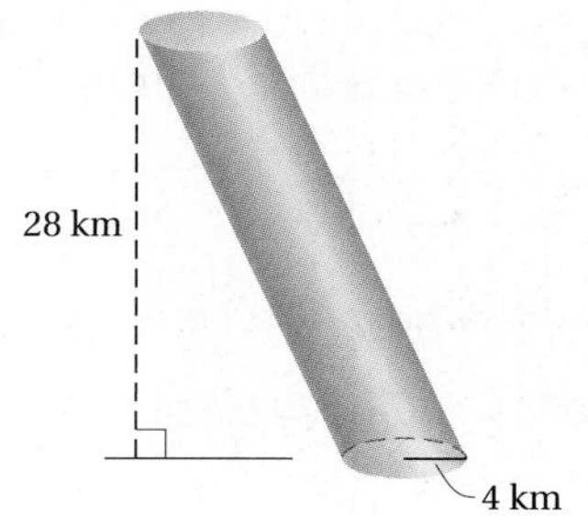

no

CHAPTER

R Summary and Review

The review that follows is meant to prepare you for a chapter exam. It consists of two parts. The first part is a checklist of some of the Study Tips referred to in this chapter, as well as a list of important properties and formulas. The second part is the Review Exercises. These provide practice exercises for the exam, together with references to section objectives so you can go back and review. Before beginning, stop and look back over the skills you have obtained. What skills in mathematics do you have now that you did not have before studying this chapter?

STUDY TIPS CHECKLIST

The foundation of all your study skills is TIME!

- ☐ Have you found adequate time to study?
- ☐ Have you determined the location of the learning resource centers on your campus, such as a math lab, tutor center, and your instructor's office?
- ☐ Are you stopping to work the margin exercises when directed to do so?
- ☐ Are you doing your homework with the attitude that it is a small step that leads to a large success?
- ☐ Are you making use of any of the textbook supplements, such as the videotapes, the Digital Video Tutor, the *Student's Solutions Manual*, and the MyMathLab web site?

IMPORTANT PROPERTIES AND FORMULAS

Identity Property of 0: $a + 0 = a$

Identity Property of 1: $a \cdot 1 = a$

Equivalent Expressions for 1: $\frac{a}{a} = 1, \quad a \neq 0$

$n\% = n \times 0.01 = n \times \frac{1}{100} = \frac{n}{100}$

Exponential Notation: $a^n = \underbrace{a \cdot a \cdot a \cdots a}_{n \text{ factors}}$

Perimeter of a Rectangle: $P = 2 \cdot (l + w)$, or $P = 2 \cdot l + 2 \cdot w$

Perimeter of a Square: $P = 4 \cdot s$

Area of a Rectangle: $A = l \cdot w$

Area of a Square: $A = s \cdot s$, or $A = s^2$

Area of a Parallelogram: $A = b \cdot h$

Area of a Triangle: $A = \frac{1}{2} \cdot b \cdot h$

Radius and Diameter of a Circle: $d = 2 \cdot r$, or $r = \frac{d}{2}$

Circumference of a Circle: $C = \pi \cdot d$, or $C = 2 \cdot \pi \cdot r$

Area of a Circle: $A = \pi \cdot r \cdot r$, or $A = \pi \cdot r^2$

Volume of a Rectangular Solid: $V = l \cdot w \cdot h$

REVIEW EXERCISES

The review exercises that follow are for practice. Answers are at the back of the book. If you miss an exercise, restudy the objective indicated in blue next to the exercise or direction line that precedes it.

Find the prime factorization. [R.1a]

1. 92

2. 1400

Find the LCM. [R.1b]

3. 13, 32

4. 5, 18, 45

Write an equivalent expression using the indicated number for 1. [R.2a]

5. $\frac{2}{5}$ $\left(\text{Use } \frac{6}{6} \text{ for } 1.\right)$

6. $\frac{12}{23}$ $\left(\text{Use } \frac{8}{8} \text{ for } 1.\right)$

Write an equivalent expression with the given denominator. [R.2a]

7. $\frac{5}{8}$ (Denominator: 64)

8. $\frac{13}{12}$ (Denominator: 84)

Simplify. [R.2b]

9. $\frac{20}{48}$

10. $\frac{1020}{1820}$

Compute and simplify. [R.2c]

11. $\frac{4}{9} + \frac{5}{12}$

12. $\frac{3}{4} \div 3$

13. $\frac{2}{3} - \frac{1}{15}$

14. $\frac{9}{10} \cdot \frac{16}{5}$

15. Convert to fraction notation: 17.97. [R.3a]

16. Convert to decimal notation: $\frac{2337}{10{,}000}$. [R.3a]

Add. [R.3b]

17. $\begin{array}{r} 2344.56 \\ +\quad 98.345 \\ \hline \end{array}$

18. 6.04 + 78 + 1.9898

Subtract. [R.3b]

19. 20.4 − 11.058

20. $\begin{array}{r} 789.032 \\ -\ 655.768 \\ \hline \end{array}$

Multiply. [R.3b]

21. $\begin{array}{r} 17.95 \\ \times\quad 24 \\ \hline \end{array}$

22. $\begin{array}{r} 56.95 \\ \times\ 1.94 \\ \hline \end{array}$

Divide. [R3.b]

23. $2.8\,\overline{)\,155.68}$

24. $52\,\overline{)\,23.4}$

25. Convert to decimal notation: $\frac{19}{12}$. [R.3b]

26. Round to the nearest tenth: 34.067. [R.3c]

27. *Homicides.* Homicides account for 18% of the deaths among people age 15 to 24. Convert 18% to decimal notation. [R.4a]
Source: National Center for Health Statistics

28. *Foreign Student Enrollment.* Of all the foreign students studying in the United States, 0.082 are from India. Convert 0.082 to percent notation. [R.4c]
Source: Institute of International Education

29. *Pregnancy Weight Gain.* Of all pregnant women, 22% of them gain less than 20 lb. Convert 22% to fraction notation. [R.4b]
Source: National Vital Statistics Report

30. *Heart-Disease Death Rate.* Of those people living in California, 114.1 of every 100,000 will die of heart disease. Convert $\frac{114.1}{100,000}$ to percent notation. [R.4d]
Source: "Reforming the Health Care System; State Profiles 1999," AARP

Convert to percent notation.

31. $\frac{5}{8}$ [R.4d]

32. $\frac{29}{25}$ [R.4d]

33. Write exponential notation: $6 \cdot 6 \cdot 6$. [R.5a]

34. Evaluate: $(1.06)^2$. [R.5b]

Calculate and compare answers to Exercises 35–37. [R.5c]

35. $120 - 6^2 \div 4 + 8$

36. $(120 - 6^2) \div 4 + 8$

37. $(120 - 6^2) \div (4 + 8)$

38. Calculate: $\frac{4(18 - 8) + 7 \cdot 9}{9^2 - 8^2}$. [R.5c]

Find the perimeter. [R.6a]

39.

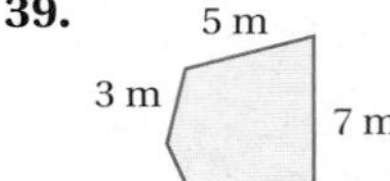

40.

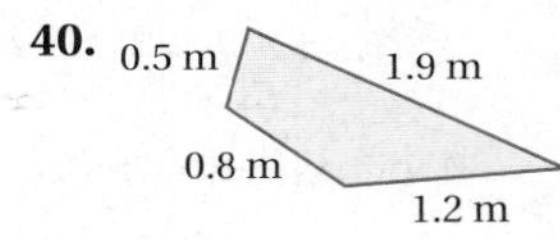

41. *Tennis Court.* The dimensions of a standard-sized tennis court are 78 ft by 36 ft. Find the perimeter and the area of the tennis court. [R.6a, b]

Find the perimeter and the area. [R.6a, b]

42.

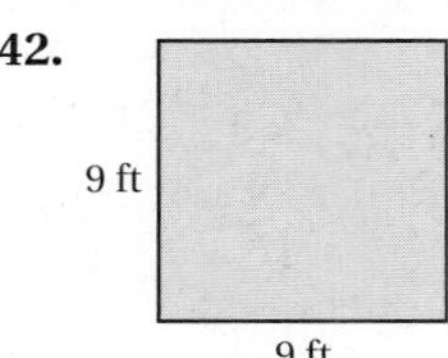

43.

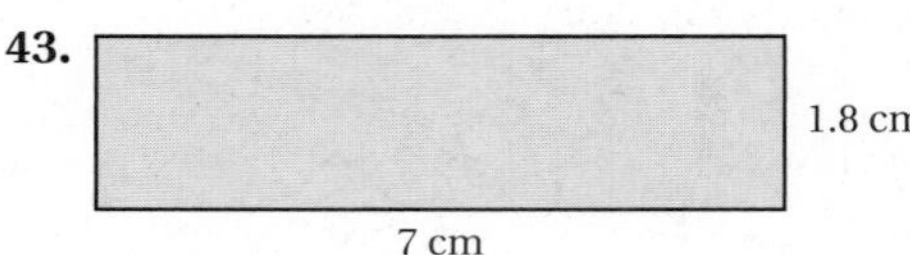

Find the area. [R.6b]

44.

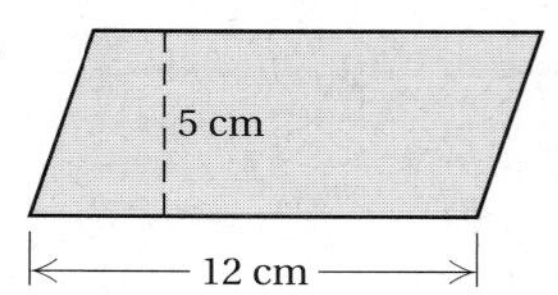

45.

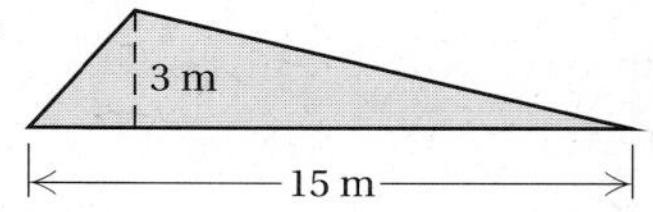

46.

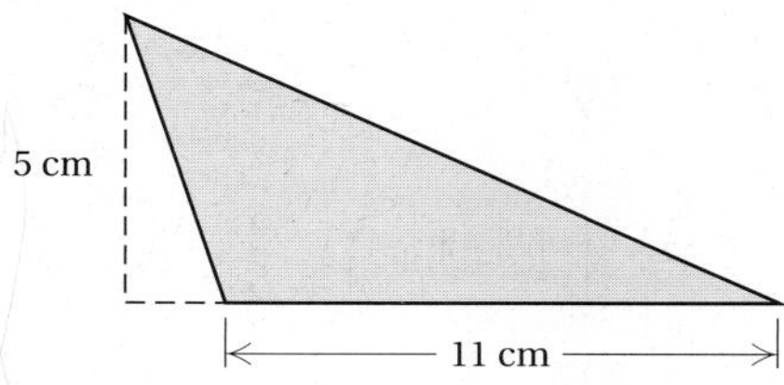

47.

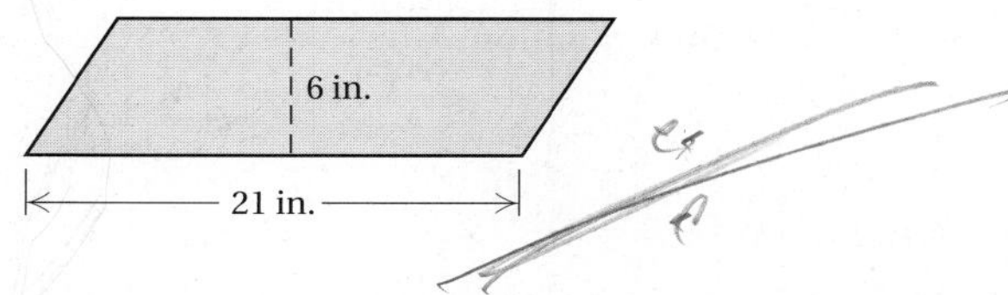

48. *Seeded Area.* A grassy area is to be seeded around three sides of a building and has equal width on the three sides, as shown below. What is the seeded area? [R.6b]

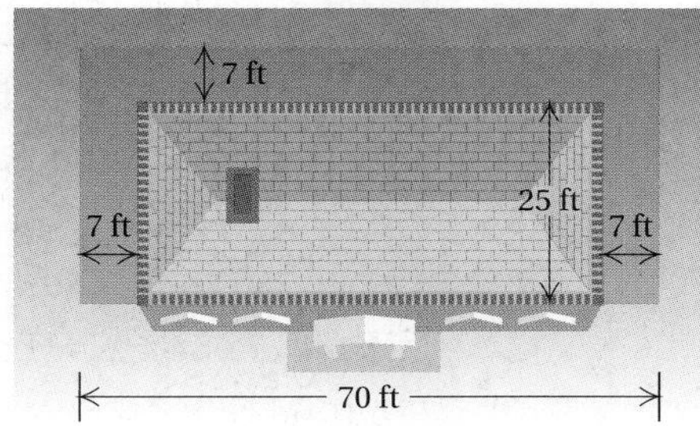

Find the length of a radius of the circle. [R.6c]

49.

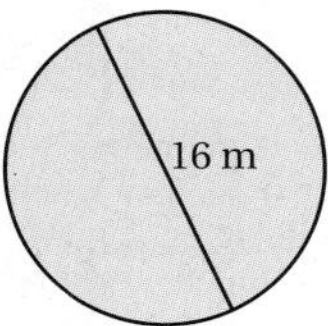

50.

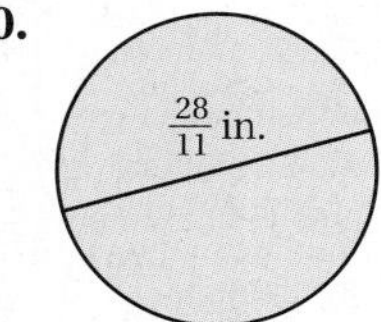

Find the length of a diameter of the circle. [R.6c]

51.

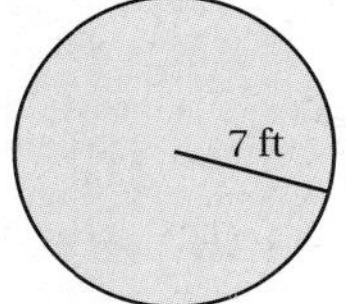

52.

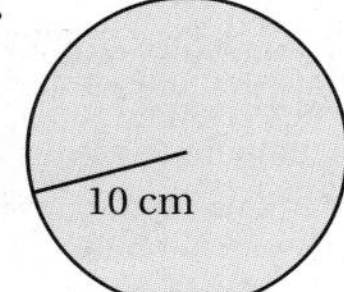

53. Find the circumference of the circle in Exercise 49. Use 3.14 for π. [R.6c]

54. Find the circumference of the circle in Exercise 50. Use $\frac{22}{7}$ for π. [R.6c]

55. Find the area of the circle in Exercise 49. Use 3.14 for π. [R.6c]

56. Find the area of the circle in Exercise 50. Use $\frac{22}{7}$ for π. [R.6c]

Find the volume. [R.6d]

57.

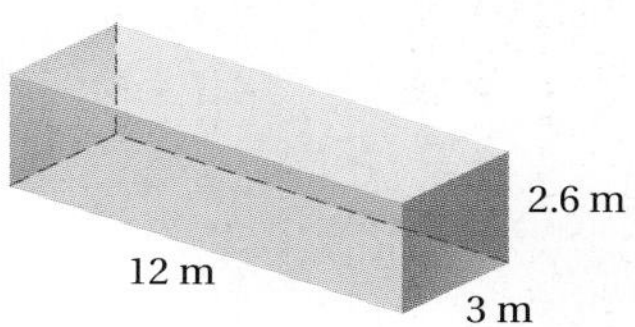

58.

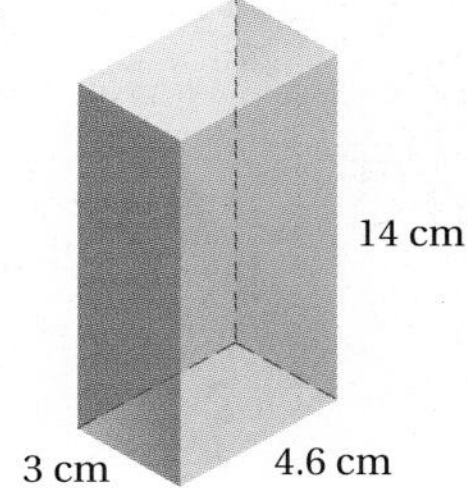

59. $\mathbf{D}_\mathbf{W}$ List and describe all the formulas for area that you have reviewed in this chapter. [R.6b, c]

60. $\mathbf{D}_\mathbf{W}$ What is common to each of the following? Explain. [R.2a, b], [R.3a], [R.4a, b, c, d]

$$\frac{11}{16}, \quad 0.6875, \quad 68.75\%, \quad \frac{55}{80}$$

SYNTHESIS

61. Find the area, in square meters, of the shaded region. [R.6b]

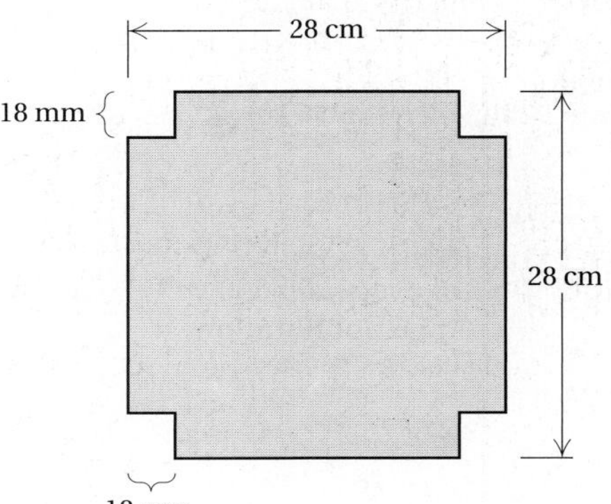

CHAPTER R

Chapter Test

1. Find the prime factorization of 300.

2. Find the LCM of 15, 24, and 60.

3. Write an expression equivalent to $\frac{3}{7}$ using $\frac{7}{7}$ as a name for 1.

4. Write an equivalent expression with the given denominator:

$\frac{11}{16}$. (Denominator: 48)

Simplify.

5. $\frac{16}{24}$

6. $\frac{925}{1525}$

Compute and simplify.

7. $\frac{10}{27} \div \frac{8}{3}$

8. $\frac{9}{10} - \frac{5}{8}$

9. Convert to fraction notation (do not simplify): 6.78.

10. Convert to decimal notation: $\frac{1895}{1000}$.

11. Add: $7.14 + 89 + 2.8787$.

12. Subtract: $1800 - 3.42$.

13. Multiply:

$$\begin{array}{r} 123.6 \\ \times \quad 3.52 \\ \hline \end{array}$$

14. Divide: $7.2 \overline{)\,11.52}$

15. Convert to decimal notation: $\frac{23}{11}$.

16. Round 234.7284 to the nearest tenth.

17. Round 234.7284 to the nearest thousandth.

18. Convert to decimal notation: 0.7%.

19. Convert to fraction notation: 91%.

20. Convert to percent notation: $\frac{11}{25}$.

21. Evaluate: 5^4.

22. Evaluate: $(1.2)^2$.

23. Calculate: $200 - 2^3 + 5 \times 10$.

24. Calculate: $8000 \div 0.16 \div 2.5$.

25. *Heart-Disease Death Rate.* Of those people living in Colorado, 101.4 of every 100,000 will die of cancer. Convert $\frac{101.4}{100{,}000}$ to percent notation.
Source: "Reforming the Health Care System; State Profiles 1999," AARP

26. *CD Rates.* A recent interest rate on a 6-month certificate of deposit was 4.71%. Convert 4.71% to decimal notation.
Source: Bank Rate Monitor

Find the perimeter and the area.

27.

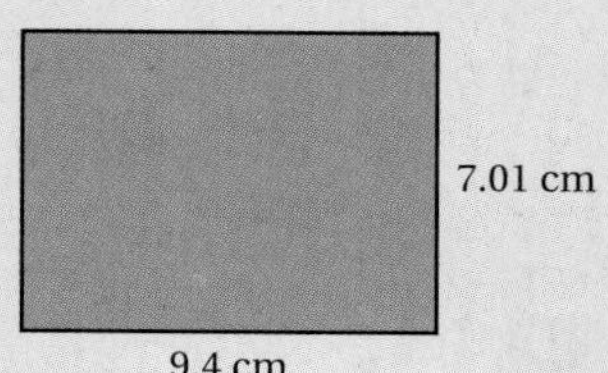

28.

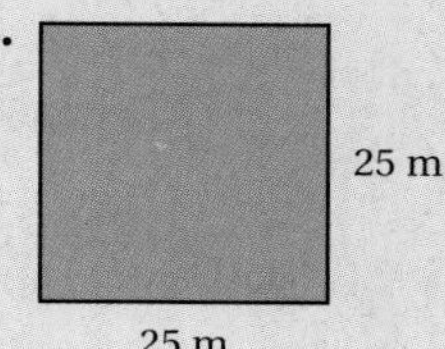

Find the area.

29.

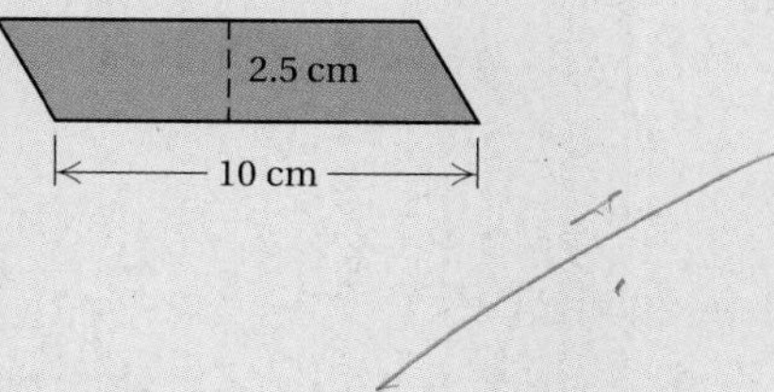

30.

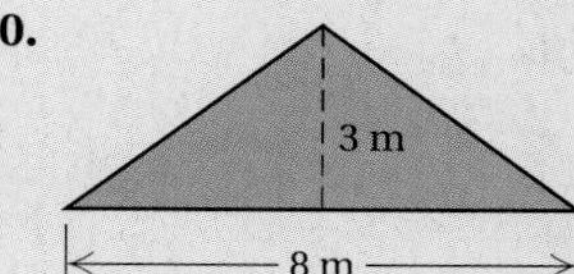

31. Find the volume.

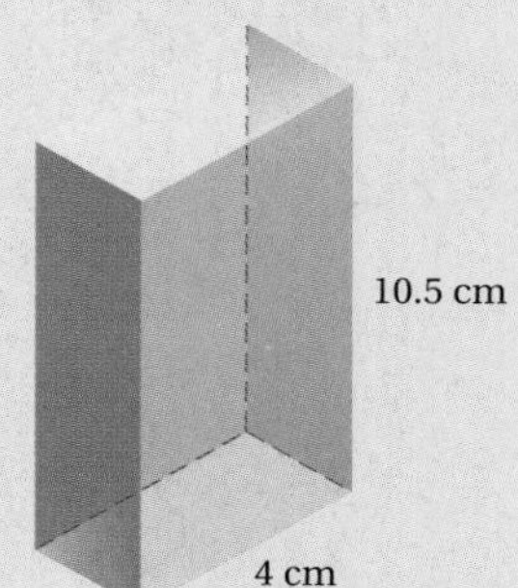

32. Find the length of a diameter, the circumference, and the area of this circle. Use $\frac{22}{7}$ for π.

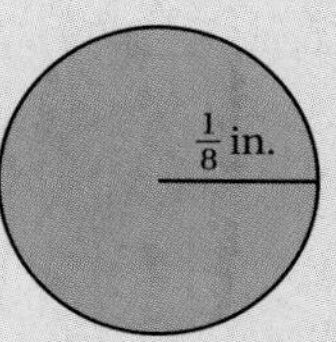

33. Find the length of a radius, the circumference, and the area of this circle. Use 3.14 for π.

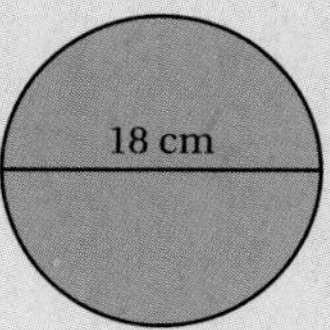

SYNTHESIS

34. A "Norman" window is designed with dimensions as shown. Find its area. Use 3.14 for π.

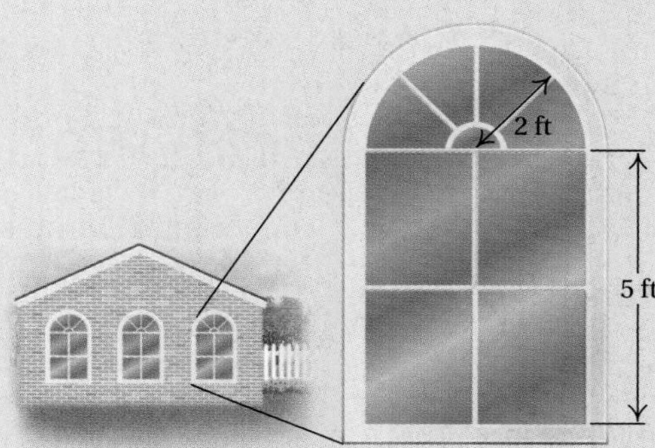

Introduction to Real Numbers and Algebraic Expressions

Gateway to Chapter 1

You are beginning a study of Introductory Algebra. In this chapter, we emphasize skills involving real numbers and algebraic expressions. We will use these skills to solve equations in the next chapter.

Have you read the Preface and the Introduction from the Author to the Student? Be sure to do so before you begin studying the text!

Real-World Application

The Viking 2 Lander spacecraft has determined that temperatures on Mars range from $-125°C$ (Celsius) to $25°C$. Find the difference between the highest value and the lowest value in this temperature range.

Source: The Lunar and Planetary Institute

This problem appears as Example 12 in Section 1.4.

CHAPTER 1

CHAPTER 1

Pretest

1. Evaluate $x/2y$ when $x = 5$ and $y = 8$. [1.1a]

2. Write an algebraic expression: Seventy-eight percent of some number. [1.1b]

3. Find the area of a rectangle when the length is 22.5 ft and the width is 16 ft. [1.1a]

4. Find $-x$ when $x = -12$. [1.3b]

Use either $<$ or $>$ for $\square$ to write a true sentence. [1.2d]

5. $0 \square -5$

6. $10 \square -5$

7. $-35 \square -45$

8. $-\frac{2}{3} \square \frac{4}{5}$

Find the absolute value. [1.2e]

9. $|-12|$

10. $|2.3|$

11. $|0|$

Find the opposite, or additive inverse. [1.3b]

12. 5.4

13. $-\frac{2}{3}$

Find the reciprocal. [1.6b]

14. 10

15. $-\frac{2}{3}$

Compute and simplify.

16. $-9 + (-8)$ [1.3a]

17. $20.2 - (-18.4)$ [1.4a]

18. $-\frac{5}{6} - \frac{3}{10}$ [1.4a]

19. $-11.5 + 6.5$ [1.3a]

20. $-9(-7)$ [1.5a]

21. $\frac{5}{8}\left(-\frac{2}{3}\right)$ [1.5a]

22. $-19.6 \div 0.2$ [1.6c]

23. $-56 \div (-7)$ [1.6a]

24. $12 - (-6) + 14 - 8$ [1.4a]

25. $20 - 10 \div 5 + 2^3$ [1.8d]

Multiply. [1.7c]

26. $9(z - 2)$

27. $-2(2a + b - 5c)$

Factor. [1.7d]

28. $4x - 12$

29. $6y - 9z - 18$

Simplify.

30. $3y - 7 - 2(2y + 3)$ [1.8b]

31. $2[3(y + 1) - 4] - [5(y - 3) - 5]$ [1.8c]

32. Write an inequality with the same meaning as $x > 12$. [1.2d]

33. **Temperature Extremes.** In Churchill, Manitoba, Canada, the average daily low temperature in January is $-31°$C. The average daily low temperature in Key West, Florida, is 19°C. How much higher is the average daily low temperature in Key West, Florida? [1.4b]

1.1 INTRODUCTION TO ALGEBRA

The study of algebra involves the use of equations to solve problems. Equations are constructed from algebraic expressions. The purpose of this section is to introduce you to the types of expressions encountered in algebra.

a Evaluating Algebraic Expressions

In arithmetic, you have worked with expressions such as

$$49 + 75, \quad 8 \times 6.07, \quad 29 - 14, \quad \text{and} \quad \frac{5}{6}.$$

In algebra, we use certain letters for numbers and work with *algebraic expressions* such as

$$x + 75, \quad 8 \times y, \quad 29 - t, \quad \text{and} \quad \frac{a}{b}.$$

Sometimes a letter can represent various numbers. In that case, we call the letter a **variable.** Let a = your age. Then a is a variable since a changes from year to year. Sometimes a letter can stand for just one number. In that case, we call the letter a **constant.** Let b = your date of birth. Then b is a constant.

Where do algebraic expressions occur? Most often we encounter them when we are solving applied problems. For example, consider the bar graph shown at right, one that we might find in a book or magazine. Suppose we want to know how many more moons Saturn has than Jupiter. Using arithmetic, we might simply subtract. But let's see how we might find this out using algebra. We translate the problem into a statement of equality, an equation. It might be done as follows:

Number of moons of Jupiter	plus	How many more	is	Number of moons of Saturn
↓	↓	↓	↓	↓
17	+	x	=	28

Note that we have an algebraic expression, $17 + x$, on the left of the equals sign. To find the number x, we can subtract 17 on both sides of the equation:

$$17 + x = 28$$
$$17 + x - 17 = 28 - 17$$
$$x = 11.$$

The value of x gives the answer, 11 moons.

We call $17 + x$ an *algebraic expression* and $17 + x = 28$ an *algebraic equation.* Note that there is no equals sign, =, in an algebraic expression.

In arithmetic, you probably would do this subtraction without ever considering an equation. *In algebra, more complex problems are difficult to solve without first solving an equation.*

Do Exercise 1.

Objectives

a Evaluate algebraic expressions by substitution.

b Translate phrases to algebraic expressions.

1. Translate this problem to an equation. Use the graph below.

How many more moons does Uranus have than Neptune?

Moons of planets

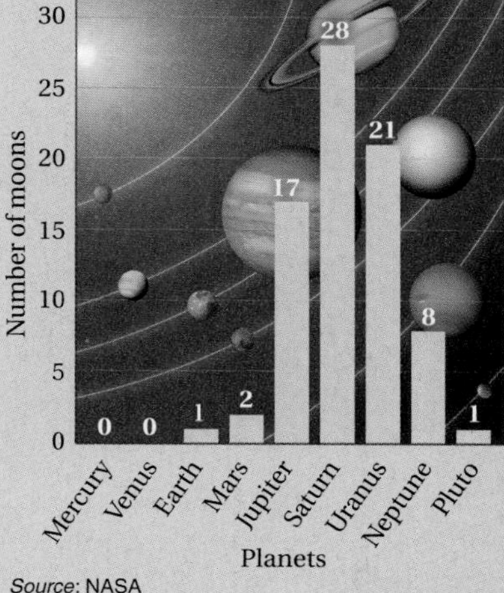

Source: NASA

Answer on page A-3

Study Tips

STUDY TIPS REVIEW

You may or may not have studied parts of Chapter R. In that chapter, we introduced Study Tips. If you have not studied them, go back and review the following:

- Basic Information on the First Day of Class (Section R.1)
- Small Steps Lead to Great Success: Part 1 (Section R.2)
- Using This Textbook (Section R.3)
- Videotapes; Digital Video Tutor (Section R.4)

2. Evaluate $a + b$ when $a = 38$ and $b = 26$.

3. Evaluate $x - y$ when $x = 57$ and $y = 29$.

4. Evaluate $4t$ when $t = 15$.

5. Find the area of a rectangle when l is 24 ft and w is 8 ft.

6. Evaluate a/b when $a = 200$ and $b = 8$.

7. Evaluate $10p/q$ when $p = 40$ and $q = 25$.

Answers on page A-3

An **algebraic expression** consists of variables, constants, numerals, and operation signs. When we replace a variable with a number, we say that we are **substituting** for the variable. This process is called **evaluating the expression.**

EXAMPLE 1 Evaluate $x + y$ when $x = 37$ and $y = 29$.

We substitute 37 for x and 29 for y and carry out the addition:

$$x + y = 37 + 29 = 66.$$

The number 66 is called the **value** of the expression.

Algebraic expressions involving multiplication can be written in several ways. For example, "8 times a" can be written as $8 \times a$, $8 \cdot a$, $8(a)$, or simply $8a$.

Two letters written together without an operation symbol, such as ab, also indicate a multiplication.

EXAMPLE 2 Evaluate $3y$ when $y = 14$.

$$3y = 3(14) = 42$$

Do Exercises 2–4.

EXAMPLE 3 *Area of a Rectangle.* The area A of a rectangle of length l and width w is given by the formula $A = lw$. Find the area when l is 24.5 in. and w is 16 in.

We substitute 24.5 in. for l and 16 in. for w and carry out the multiplication:

$$\begin{aligned} A = lw &= (24.5 \text{ in.})(16 \text{ in.}) \\ &= (24.5)(16)(\text{in.})(\text{in.}) \\ &= 392 \text{ in}^2, \text{ or } 392 \text{ square inches.} \end{aligned}$$

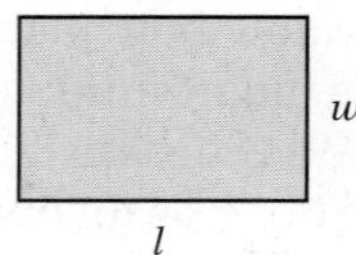

Do Exercise 5.

Algebraic expressions involving division can also be written in several ways. For example, "8 divided by t" can be written as $8 \div t$, $\frac{8}{t}$, $8/t$, or $8 \cdot \frac{1}{t}$, where the fraction bar is a division symbol.

EXAMPLE 4 Evaluate $\frac{a}{b}$ when $a = 63$ and $b = 9$.

We substitute 63 for a and 9 for b and carry out the division:

$$\frac{a}{b} = \frac{63}{9} = 7.$$

EXAMPLE 5 Evaluate $\frac{12m}{n}$ when $m = 8$ and $n = 16$.

$$\frac{12m}{n} = \frac{12 \cdot 8}{16} = \frac{96}{16} = 6$$

Do Exercises 6 and 7.

EXAMPLE 6 *Motorcycle Travel.* Ed takes a trip on his motorcycle. He wants to travel 660 mi on a particular day. The time t, in hours, that it takes to travel 660 mi is given by

$$t = \frac{660}{r},$$

where r is the speed of Ed's motorcycle. Find the time of travel if the speed r is 60 mph.

We substitute 60 for r and carry out the division:

$$t = \frac{660}{r} = \frac{660}{60} = 11 \text{ hr.}$$

Do Exercise 8.

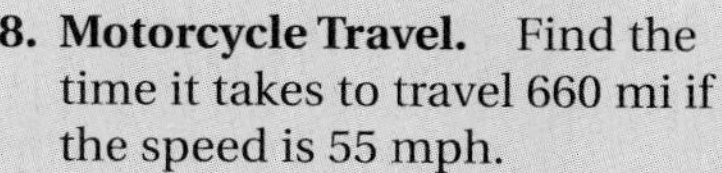

8. Motorcycle Travel. Find the time it takes to travel 660 mi if the speed is 55 mph.

CALCULATOR CORNER

Evaluating Algebraic Expressions *To the student and the instructor*: This book contains a series of *optional* discussions on using a calculator. A calculator is *not* a requirement for this textbook. There are many kinds of calculators and different instructions for their usage. We have included instructions here for the scientific keys on a graphing calculator such as a TI-83 or a TI-83 Plus. Be sure to consult your user's manual as well. Also, check with your instructor about whether you are allowed to use a calculator in the course.

We can evaluate algebraic expressions on a calculator by making the appropriate substitutions, keeping in mind the rules for order of operations, and then carrying out the resulting calculations. To evaluate $12m/n$ when $m = 8$ and $n = 16$, as in Example 5, we enter $12 \cdot 8/16$ by pressing [1] [2] [×] [8] [÷] [1] [6] [ENTER]. The result is 6.

```
12*8/16
                     6
```

Exercises: Evaluate.

1. $\frac{12m}{n}$, when $m = 42$ and $n = 9$
2. $a + b$, when $a = 8.2$ and $b = 3.7$
3. $b - a$, when $a = 7.6$ and $b = 9.4$
4. $27xy$, when $x = 12.7$ and $y = 100.4$
5. $3a + 2b$, when $a = 2.9$ and $b = 5.7$
6. $2a + 3b$, when $a = 7.3$ and $b = 5.1$

b Translating to Algebraic Expressions

In algebra, we translate problems to equations. The different parts of an equation are translations of word phrases to algebraic expressions. It is easier to translate if we know that certain words often translate to certain operation symbols.

Answer on page A-3

Translate to an algebraic expression.

9. Eight less than some number

10. Eight more than some number

11. Four less than some number

12. Half of a number

13. Six more than eight times some number

14. The difference of two numbers

15. Fifty-nine percent of some number

16. Two hundred less than the product of two numbers

17. The sum of two numbers

Answers on page A-3

KEY WORDS, PHRASES, AND CONCEPTS

ADDITION (+)	SUBTRACTION (−)	MULTIPLICATION (·)	DIVISION (÷)
add	subtract	multiply	divide
added to	subtracted from	multiplied by	divided by
sum	difference	product	quotient
total	minus	times	
plus	less than	of	
more than	decreased by		
increased by	take away		

EXAMPLE 7 Translate to an algebraic expression:

Twice (or two times) some number.

Think of some number, say, 8. What number is twice 8? It is 16. How did you get 16? You multiplied by 2. Do the same thing using a variable. We can use any variable we wish, such as x, y, m, or n. Let's use y to stand for some number. If we multiply by 2, we get an expression

$y \times 2,\quad 2 \times y,\quad 2 \cdot y,\quad \text{or}\quad 2y.$

In algebra, $2y$ is the expression generally used.

EXAMPLE 8 Translate to an algebraic expression:

Thirty-eight percent of some number.

The word "of" translates to a multiplication symbol, so we get the following expressions as a translation:

$38\% \cdot n,\quad 0.38 \times n,\quad \text{or}\quad 0.38n.$

EXAMPLE 9 Translate to an algebraic expression:

Seven less than some number.

We let

x represent the number.

Now if the number were 23, then the translation would be "7 subtracted from 23," or $23 - 7$. If we knew the number to be 345, then the translation would be $345 - 7$. If the number is x, then the translation is

$x - 7.$

CAUTION!

Note that $7 - x$ is *not* a correct translation of the expression in Example 9. The expression $7 - x$ is a translation of "seven minus some number" or "some number less than seven."

EXAMPLE 10 Translate to an algebraic expression:

Eighteen more than a number.

We let

$t =$ the number.

Now if the number were 26, then the translation would be $26 + 18$, or $18 + 26$. If we knew the number to be 174, then the translation would be $174 + 18$, or $18 + 174$. If the number is t, then the translation is

$$t + 18, \quad \text{or} \quad 18 + t.$$

EXAMPLE 11 Translate to an algebraic expression:

A number divided by 5.

We let

$m =$ the number.

Now if the number were 76, then the translation would be $76 \div 5$, or $76/5$, or $\frac{76}{5}$. If the number were 213, then the translation would be $213 \div 5$, or $213/5$, or $\frac{213}{5}$. If the number is m, then the translation is

$$m \div 5, \quad m/5, \quad \text{or} \quad \frac{m}{5}.$$

EXAMPLE 12 Translate each phrase to an algebraic expression.

PHRASE	ALGEBRAIC EXPRESSION
Five more than some number	$n + 5$, or $5 + n$
Half of a number	$\frac{1}{2}t$, $\frac{t}{2}$, or $t/2$
Five more than three times some number	$3p + 5$, or $5 + 3p$
The difference of two numbers	$x - y$
Six less than the product of two numbers	$mn - 6$
Seventy-six percent of some number	$76\%z$, or $0.76z$
Four less than twice some number	$2x - 4$

Do Exercises 9–17 on the preceding page.

Study Tips

ATTITUDE AND THE POWER OF YOUR CHOICES

Making the right choices can give you the power to succeed in learning mathematics.

You can choose to improve your attitude and raise the academic goals that you have set for yourself. Projecting a positive attitude toward your study of mathematics and expecting a positive outcome can make it easier for you to learn and to perform well in this course.

Here are some positive choices you can make:

- Choose to allocate the proper amount of time to learn.
- Choose to place the primary responsibility for learning on yourself.
- Choose to establish a learning relationship with your instructor.
- Choose to make a strong commitment to learning.

Well-known American psychologist William James once said, "The one thing that will guarantee the successful conclusion of a doubtful undertaking is faith in the beginning that you can do it." Having a positive attitude and making the right choices will add to your confidence in this course and multiply your successes.

1.1 EXERCISE SET

For Extra Help

Digital Video Tutor CD 2 Videotape 3 | InterAct Math | Math Tutor Center | MathXL | MyMathLab

a Substitute to find values of the expressions in each of the following applied problems.

1. *Enrollment Costs.* At Emmett Community College, it costs \$600 to enroll in the 8 A.M. section of Elementary Algebra. Suppose that the variable n stands for the number of students who enroll. Then $600n$ stands for the total amount of money collected for this course. How much is collected if 34 students enroll? 78 students? 250 students?

2. *Commuting Time.* It takes Erin 24 min less time to commute to work than it does George. Suppose that the variable x stands for the time it takes George to get to work. Then $x - 24$ stands for the time it takes Erin to get to work. How long does it take Erin to get to work if it takes George 56 min? 93 min? 105 min?

3. *Area of a Triangle.* The area A of a triangle with base b and height h is given by $A = \frac{1}{2}bh$. Find the area when $b = 45$ m (meters) and $h = 86$ m.

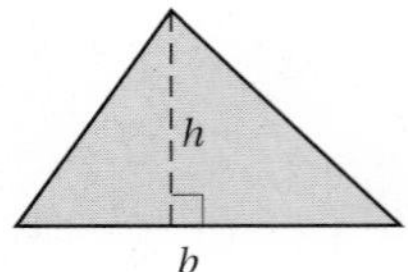

4. *Area of a Parallelogram.* The area A of a parallelogram with base b and height h is given by $A = bh$. Find the area of the parallelogram when the height is 15.4 cm (centimeters) and the base is 6.5 cm.

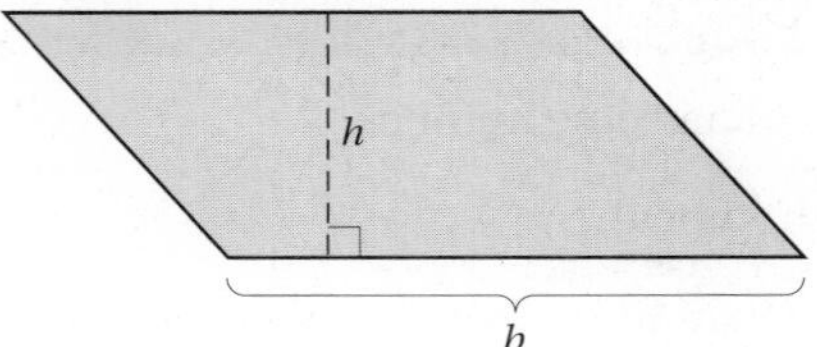

5. *Distance Traveled.* A driver who drives at a speed of r mph for t hr will travel a distance d mi given by $d = rt$ mi. How far will a driver travel at a speed of 65 mph for 4 hr?

6. *Simple Interest.* The simple interest I on a principal of P dollars at interest rate r for time t, in years, is given by $I = Prt$. Find the simple interest on a principal of \$4800 at 9% for 2 yr. (*Hint*: 9% = 0.09.)

7. *Hockey Goal.* The front of a regulation hockey goal is 6 ft wide and 4 ft high. Find its area.
Source: National Hockey League

8. *Zoology.* A great white shark has triangular teeth. Each tooth measures about 5 cm across the base and has a height of 6 cm. Find the surface area of one side of one tooth. (See Exercise 3.)

Evaluate.

9. $8x$, when $x = 7$

10. $6y$, when $y = 7$

11. $\frac{a}{b}$, when $a = 24$ and $b = 3$

12. $\frac{p}{q}$, when $p = 16$ and $q = 2$

13. $\frac{3p}{q}$, when $p = 2$ and $q = 6$

14. $\frac{5y}{z}$, when $y = 15$ and $z = 25$

15. $\frac{x + y}{5}$, when $x = 10$ and $y = 20$

16. $\frac{p + q}{2}$, when $p = 2$ and $q = 16$

17. $\frac{x - y}{8}$, when $x = 20$ and $y = 4$

18. $\frac{m - n}{5}$, when $m = 16$ and $n = 6$

b Translate each phrase to an algebraic expression.

19. Seven more than b

20. Nine more than t

21. Twelve less than c

22. Fourteen less than d

23. Four increased by q

24. Thirteen increased by z

25. b more than a

26. c more than d

27. x divided by y

28. c divided by h

29. x plus w

30. s added to t

31. m subtracted from n

32. p subtracted from q

33. The sum of x and y

34. The sum of a and b

35. Twice z

36. Three times q

37. Three multiplied by m

38. The product of 8 and t

39. The product of 89% and your salary

40. 67% of the women attending

41. Danielle drove at a speed of 65 mph for t hours. How far did Danielle travel?

42. Juan has d dollars before spending \$19.95 on a DVD of the movie *Castaway*. How much did Juan have after the purchase?

43. Lisa had \$50 before spending x dollars on pizza. How much money remains?

44. Dino drove his pickup truck at 55 mph for t hours. How far did he travel?

To the student and the instructor: The Discussion and Writing exercises are meant to be answered with one or more sentences. They can be discussed and answered collaboratively by the entire class or by small groups. Because of their open-ended nature, the answers to these exercises do not appear at the back of the book. They are denoted by the symbol **DW**.

45. **DW** If the length of a rectangle is doubled, does the area double? Why or why not?

46. **DW** If the height and the base of a triangle are doubled, what happens to the area? Explain.

SKILL MAINTENANCE

Find the prime factorization. [R.1a]

47. 54

48. 32

49. 108

50. 192

51. 2001

Find the LCM. [R.1b]

52. 6, 18

53. 6, 24, 32

54. 10, 20, 30

55. 16, 24, 32

56. 18, 36, 44

SYNTHESIS

To the student and the instructor: The Synthesis exercises found at the end of every exercise set challenge students to combine concepts or skills studied in that section or in preceding parts of the text.

Translate to an algebraic expression.

57. Some number x plus three times y

58. Some number a plus 2 plus b

59. A number that is 3 less than twice x

60. Your age in 5 years, if you are a years old now

1.2 THE REAL NUMBERS

A **set** is a collection of objects. (See Appendix C for more on sets.) For our purposes, we will most often be considering sets of numbers. One way to name a set uses what is called **roster notation.** For example, roster notation for the set containing the numbers 0, 2, and 5 is $\{0, 2, 5\}$.

Sets that are part of other sets are called **subsets.** In this section, we become acquainted with the set of *real numbers* and its various subsets.

Two important subsets of the real numbers are listed below using roster notation.

NATURAL NUMBERS

The set of **natural numbers** $= \{1, 2, 3, \ldots\}$. These are the numbers used for counting.

WHOLE NUMBERS

The set of **whole numbers** $= \{0, 1, 2, 3, \ldots\}$. This is the set of natural numbers with 0 included.

We can represent these sets on a number line. The natural numbers are those to the right of zero. The whole numbers are the natural numbers and zero.

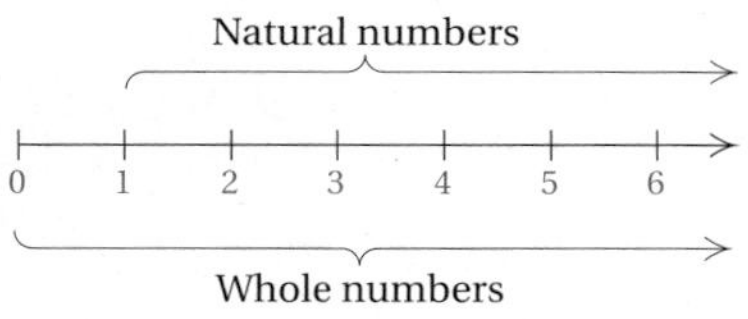

We create a new set, called the *integers,* by starting with the whole numbers, 0, 1, 2, 3, and so on. For each natural number 1, 2, 3, and so on, we obtain a new number to the left of zero on the number line:

For the number 1, there will be an *opposite* number -1 (negative 1).

For the number 2, there will be an *opposite* number -2 (negative 2).

For the number 3, there will be an *opposite* number -3 (negative 3), and so on.

The **integers** consist of the whole numbers and these new numbers.

INTEGERS

The set of **integers** $= \{\ldots, -5, -4, -3, -2, -1, 0, 1, 2, 3, 4, 5, \ldots\}$.

Objectives

- **a** State the integer that corresponds to a real-world situation.
- **b** Graph rational numbers on a number line.
- **c** Convert from fraction notation to decimal notation for a rational number.
- **d** Determine which of two real numbers is greater and indicate which, using < or >; given an inequality like $a > b$, write another inequality with the same meaning. Determine whether an inequality like $-3 \leq 5$ is true or false.
- **e** Find the absolute value of a real number.

Study Tips

THE AW MATH TUTOR CENTER

The AW Math Tutor Center is staffed by highly qualified mathematics instructors who provide students with tutoring on text examples and odd-numbered exercises. Tutoring is provided free to students who have bought a new textbook with a special access card bound with the book. Tutoring is available by telephone (toll-free), fax, and e-mail. White-board technology allows tutors and students to actually see problems worked while they talk live during the tutoring sessions. If you purchased a book without this card, you can purchase an access code through your bookstore using ISBN# 0-201-72170-8. (This is also discussed in the Preface.)

We picture the integers on a number line as follows.

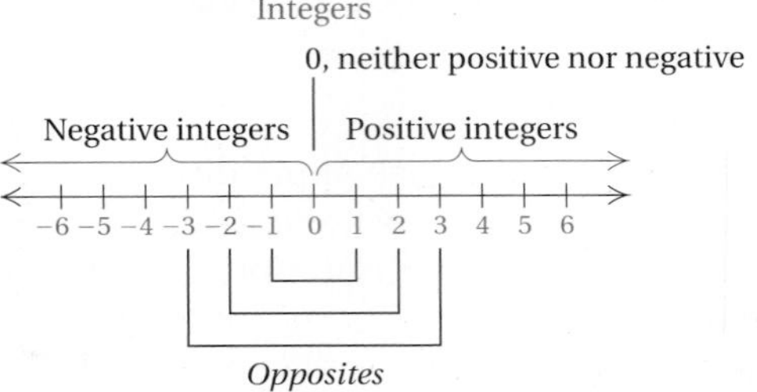

We call these new numbers to the left of 0 **negative integers.** The natural numbers are also called **positive integers.** Zero is neither positive nor negative. We call −1 and 1 **opposites** of each other. Similarly, −2 and 2 are opposites, −3 and 3 are opposites, −100 and 100 are opposites, and 0 is its own opposite. Pairs of opposite numbers like −3 and 3 are the same distance from 0. The integers extend infinitely on the number line to the left and right of zero.

a Integers and the Real World

Integers correspond to many real-world problems and situations. The following examples will help you get ready to translate problem situations that involve integers to mathematical language.

EXAMPLE 1 Tell which integer corresponds to this situation: The temperature is 3 degrees below zero.

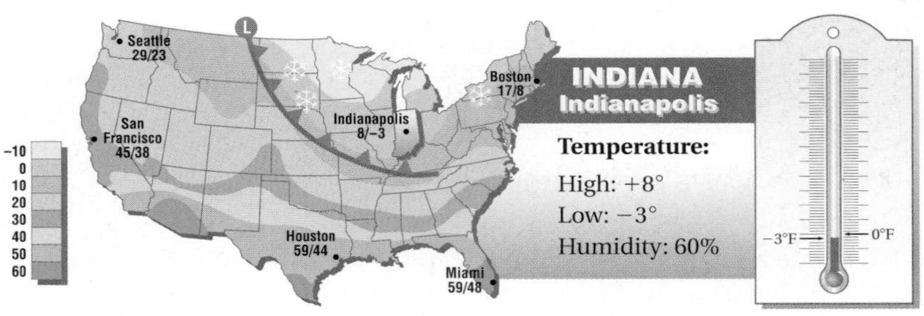

The integer −3 corresponds to the situation. The temperature is −3°.

EXAMPLE 2 *Jeopardy.* Tell which integer corresponds to this situation: A contestant missed a $600 question on the television game show “Jeopardy.”

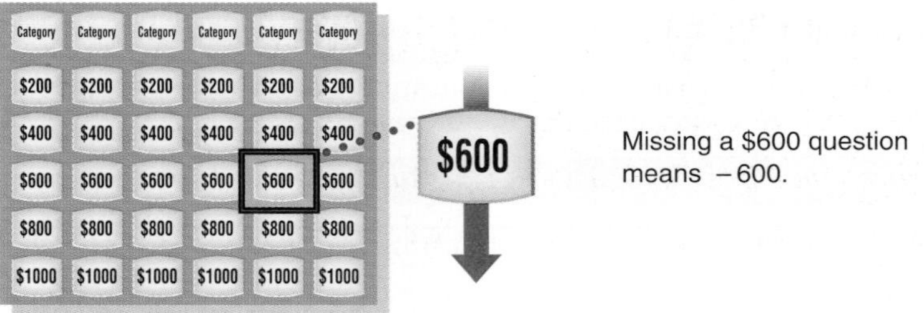

Missing a $600 question causes a $600 loss on the score—that is, the contestant earns −600 dollars.

EXAMPLE 3 *Elevation.* Tell which integer corresponds to this situation: The lowest point in New Orleans is 8 ft below sea level.

The integer −8 corresponds to the situation. The elevation is −8 ft.

EXAMPLE 4 *Stock Price Change.* Tell which integers correspond to this situation: The price of Pearson Education stock decreased from $24 per share to $17 per share over a recent time period. The price of Sherwin Williams Co. stock increased from $21 per share to $25 per share over a recent time period.
Source: The New York Stock Exchange

The integer −7 corresponds to the decrease in the stock value. The integer 4 represents the increase in stock value.

Do Exercises 1–5.

b The Rational Numbers

We created the set of integers by obtaining a negative number for each natural number. To create a larger number system, called the set of **rational numbers,** we consider quotients of integers with nonzero divisors. The following are some examples of rational numbers:

$$\frac{2}{3},\quad -\frac{2}{3},\quad \frac{7}{1},\quad 4,\quad -3,\quad 0,\quad \frac{23}{-8},\quad 2.4,\quad -0.17,\quad 10\frac{1}{2}.$$

The number $-\frac{2}{3}$ (read "negative two-thirds") can also be named $\frac{2}{-3}$ or $\frac{-2}{3}$. The number 2.4 can be named $\frac{24}{10}$ or $\frac{12}{5}$, and −0.17 can be named $-\frac{17}{100}$.

Note that this new set of numbers, the rational numbers, contains the whole numbers, the integers, and the arithmetic numbers (also called the nonnegative rational numbers). We can describe the set of rational numbers as follows.

RATIONAL NUMBERS

The set of **rational numbers** = the set of numbers $\frac{a}{b}$, where a and b are integers and b is not equal to 0 ($b \neq 0$).

State the integers that correspond to the given situation.

1. The halfback gained 8 yd on the first down. The quarterback was sacked for a 5-yd loss on the second down.

2. **Temperature High.** The highest man-made temperature on record is 950,000,000°F. It was created on May 27, 1994, at the Tokamak Fusion Test Reactor at the Princeton Plasma Physics Laboratory in New Jersey.
Source: *The Guinness Book of Records*

3. **Stock Decrease.** The price of Sherwin Williams Co. stock decreased from $25 per share to $19 per share over a recent time period.
Source: The New York Stock Exchange

4. At 10 sec before liftoff, ignition occurs. At 156 sec after liftoff, the first stage is detached from the rocket.

5. A submarine dove 120 ft, rose 50 ft, and then dove 80 ft.

Answers on page A-4

Graph on a number line.

6. $-\frac{7}{2}$

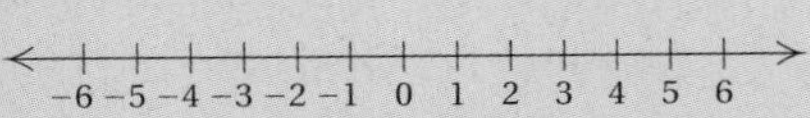

7. -1.4

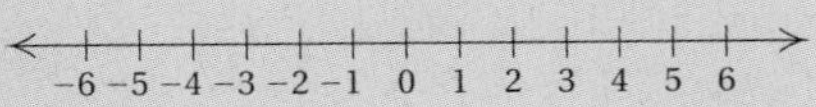

8. $\frac{11}{4}$

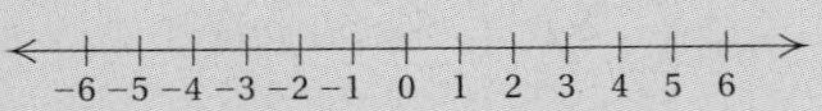

Answers on page A-4

CALCULATOR CORNER

Negative Numbers on a Calculator; Converting to Decimal Notation We use the opposite key [(−)] to enter negative numbers on a graphing calculator. Note that this is different from the [−] key, which is used for the operation of subtraction. To convert $-\frac{5}{8}$ to decimal notation, as in Example 8, we press [(−)] [5] [÷] [8] [ENTER]. The result is −0.625.

```
-5/8
                -.625
```

Exercises: Convert each of the following negative numbers to decimal notation.

1. $-\frac{3}{4}$
2. $-\frac{9}{20}$
3. $-\frac{1}{8}$
4. $-\frac{9}{5}$
5. $-\frac{27}{40}$
6. $-\frac{11}{16}$
7. $-\frac{7}{2}$
8. $-\frac{19}{25}$

We picture the rational numbers on a number line as follows.

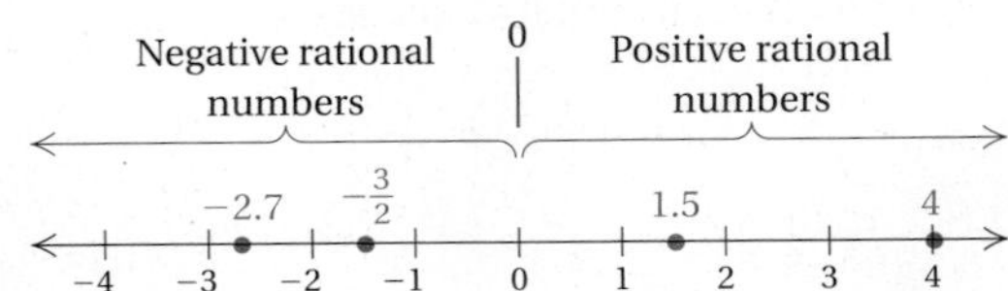

To **graph** a number means to find and mark its point on the number line. Some rational numbers are graphed in the preceding figure.

EXAMPLE 5 Graph: $\frac{5}{2}$.

The number $\frac{5}{2}$ can be named $2\frac{1}{2}$, or 2.5. Its graph is halfway between 2 and 3.

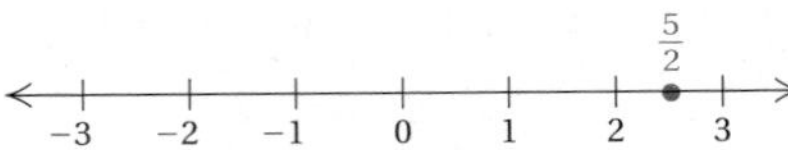

EXAMPLE 6 Graph: −3.2.

The graph of −3.2 is $\frac{2}{10}$ of the way from −3 to −4.

−3.2

−4 −3 −2 −1 0 1 2

EXAMPLE 7 Graph: $\frac{13}{8}$.

The number $\frac{13}{8}$ can be named $1\frac{5}{8}$, or 1.625. The graph is about $\frac{6}{10}$ of the way from 1 to 2.

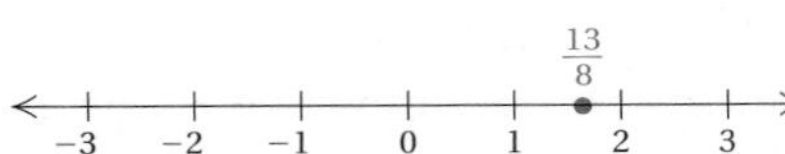

Do Exercises 6–8.

C Notation for Rational Numbers

Each rational number can be named using fraction or decimal notation.

EXAMPLE 8 Convert to decimal notation: $-\frac{5}{8}$.

We first find decimal notation for $\frac{5}{8}$. Since $\frac{5}{8}$ means $5 \div 8$, we divide.

```
     0.6 2 5
8 ) 5.0 0 0
    4 8
      2 0
      1 6
        4 0
        4 0
          0
```

Thus, $\frac{5}{8} = 0.625$, so $-\frac{5}{8} = -0.625$.

Decimal notation for $-\frac{5}{8}$ is -0.625. We consider -0.625 to be a **terminating decimal.** Decimal notation for some numbers repeats.

EXAMPLE 9 Convert to decimal notation: $\frac{7}{11}$.

We divide.

$$\begin{array}{r} 0.6363\ldots \\ 11\overline{)7.0000} \\ \underline{6\,6} \\ 4\,0 \\ \underline{3\,3} \\ 7\,0 \\ \underline{6\,6} \\ 4\,0 \\ \underline{3\,3} \\ 7 \end{array}$$

We can abbreviate repeating decimal notation by writing a bar over the repeating part—in this case, $0.\overline{63}$. Thus, $\frac{7}{11} = 0.\overline{63}$.

The following are other examples to show how each rational number can be named using fraction or decimal notation:

$$0 = \frac{0}{8}, \qquad \frac{27}{100} = 0.27, \qquad -8\frac{3}{4} = -8.75, \qquad -\frac{13}{6} = -2.1\overline{6}.$$

Do Exercises 9–11.

d The Real Numbers and Order

Every rational number has a point on the number line. However, there are some points on the line for which there is no rational number. These points correspond to what are called **irrational numbers.**

What kinds of numbers are irrational? One example is the number π, which is used in finding the area and the circumference of a circle: $A = \pi r^2$ and $C = 2\pi r$.

Another example of an irrational number is the square root of 2, named $\sqrt{2}$.

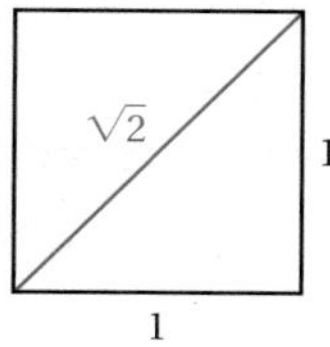

It is the length of the diagonal of a square with sides of length 1. It is also the number that when multiplied by itself gives 2, that is, $\sqrt{2} \cdot \sqrt{2} = 2$. There is no rational number that can be multiplied by itself to get 2. But the following are rational *approximations*:

1.4 is an approximation of $\sqrt{2}$ because $(1.4)^2 = 1.96$;

1.41 is a better approximation because $(1.41)^2 = 1.9881$;

1.4142 is an even better approximation because $(1.4142)^2 = 1.99996164$.

We can find rational approximations for square roots using a calculator.

Convert to decimal notation.

9. $-\frac{3}{8}$

10. $-\frac{6}{11}$

11. $\frac{4}{3}$

Answers on page A-4

CALCULATOR CORNER

Approximating Square Roots and π Square roots are found by pressing [2nd] [√]. (√ is the second operation associated with the [x^2] key.)

To find an approximation for $\sqrt{48}$, we press [2nd] [√] [4] [8] [ENTER]. The approximation 6.92820323 is displayed.

To find $8 \cdot \sqrt{13}$, we press [8] [2nd] [√] [1] [3] [ENTER]. The approximation 28.8444102 is displayed. The number π is used widely enough to have its own key. (π is the second operation associated with the [^] key.)

To approximate π, we press [2nd] [π] [ENTER]. The approximation 3.141592654 is displayed.

Exercises: Approximate.

1. $\sqrt{76}$
2. $\sqrt{317}$
3. $15 \cdot \sqrt{20}$
4. $29 + \sqrt{42}$
5. π
6. $29 \cdot \pi$
7. $\pi \cdot 13^2$
8. $5 \cdot \pi + 8 \cdot \sqrt{237}$

Use either $<$ or $>$ for $\square$ to write a true sentence.

12. $-3 \square 7$

13. $-8 \square -5$

14. $7 \square -10$

15. $3.1 \square -9.5$

16. $-\frac{2}{3} \square -1$

17. $-\frac{11}{8} \square \frac{23}{15}$

18. $-\frac{2}{3} \square -\frac{5}{9}$

19. $-4.78 \square -5.01$

Answers on page A-4

Decimal notation for rational numbers *either* terminates *or* repeats. Decimal notation for irrational numbers *neither* terminates *nor* repeats. Some other examples of irrational numbers are $\sqrt{3}$, $-\sqrt{8}$, $\sqrt{11}$, and 0.121221222122221.... Whenever we take the square root of a number that is not a perfect square, we will get an irrational number.

The rational numbers and the irrational numbers together correspond to all the points on a number line and make up what is called the **real-number system.**

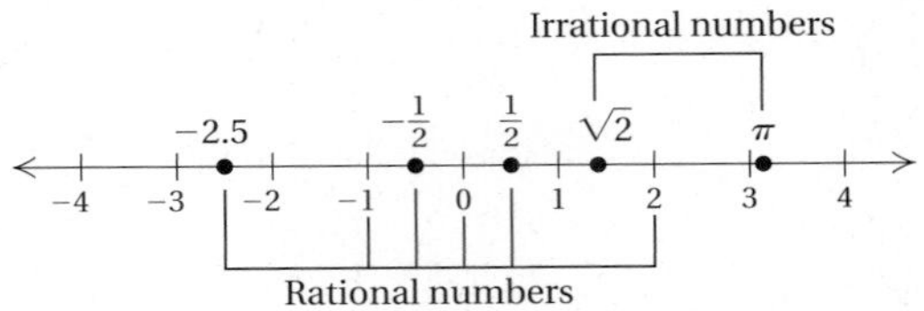

REAL NUMBERS

The set of **real numbers** = The set of all numbers corresponding to points on the number line.

The real numbers consist of the rational numbers and the irrational numbers. The following figure shows the relationships among various kinds of numbers.

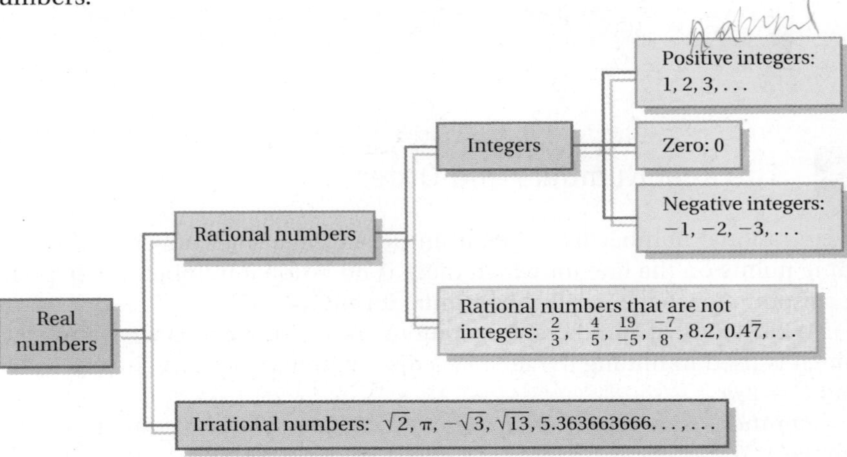

ORDER

Real numbers are named in order on the number line, with larger numbers named farther to the right. For any two numbers on the line, the one to the left is less than the one to the right.

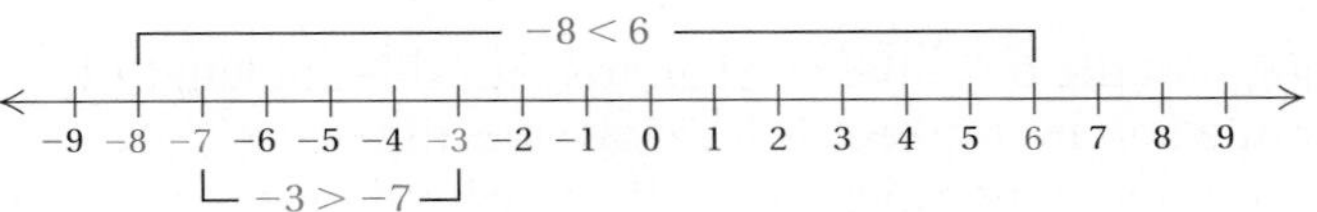

We use the symbol < to mean "**is less than.**" The sentence $-8 < 6$ means "-8 is less than 6." The symbol > means "**is greater than.**" The sentence $-3 > -7$ means "-3 is greater than -7." The sentences $-8 < 6$ and $-3 > -7$ are **inequalities.**

EXAMPLES Use either $<$ or $>$ for $\square$ to write a true sentence.

10. $2 \square 9$ — Since 2 is to the left of 9, 2 is less than 9, so $2 < 9$.
11. $-7 \square 3$ — Since -7 is to the left of 3, we have $-7 < 3$.
12. $6 \square -12$ — Since 6 is to the right of -12, then $6 > -12$.
13. $-18 \square -5$ — Since -18 is to the left of -5, we have $-18 < -5$.
14. $-2.7 \square -\frac{3}{2}$ — The answer is $-2.7 < -\frac{3}{2}$.

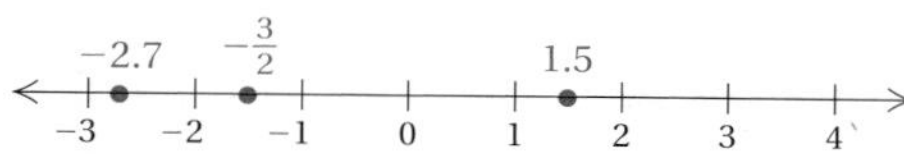

15. $1.5 \square -2.7$ — The answer is $1.5 > -2.7$.
16. $1.38 \square 1.83$ — The answer is $1.38 < 1.83$.
17. $-3.45 \square 1.32$ — The answer is $-3.45 < 1.32$.
18. $-4 \square 0$ — The answer is $-4 < 0$.
19. $5.8 \square 0$ — The answer is $5.8 > 0$.
20. $\frac{5}{8} \square \frac{7}{11}$ — We convert to decimal notation: $\frac{5}{8} = 0.625$ and $\frac{7}{11} = 0.6363\ldots$. Thus, $\frac{5}{8} < \frac{7}{11}$.

Do Exercises 12–19 on the preceding page.

Note that both $-8 < 6$ and $6 > -8$ are true. Every true inequality yields another true inequality when we interchange the numbers or variables and reverse the direction of the inequality sign.

ORDER; >, <

$a < b$ also has the meaning $b > a$.

EXAMPLES Write another inequality with the same meaning.

21. $-3 > -8$ — The inequality $-8 < -3$ has the same meaning.
22. $a < -5$ — The inequality $-5 > a$ has the same meaning.

A helpful mental device is to think of an inequality sign as an "arrow" with the arrow pointing to the smaller number.

Do Exercises 20 and 21.

Note that all positive real numbers are greater than zero and all negative real numbers are less than zero.

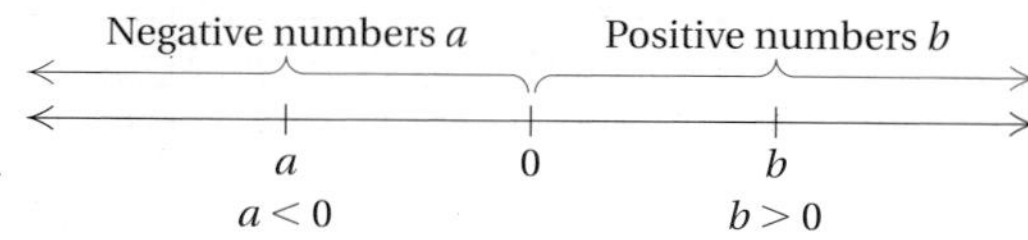

If b is a positive real number, then $b > 0$.
If a is a negative real number, then $a < 0$.

Write another inequality with the same meaning.

20. $-5 < 7$

21. $x > 4$

Write true or false.

22. $-4 \leq -6$

23. $7.8 \geq 7.8$

24. $-2 \leq \frac{3}{8}$

Answers on page A-4

Find the absolute value.

25. $|8|$

26. $|-9|$

27. $\left|-\frac{2}{3}\right|$

28. $|5.6|$

Answers on page A-4

Answers on page A-4

CALCULATOR CORNER

Absolute Value The absolute-value operation is the first item in the Catalog on the TI-83 Plus graphing calculator. To find $|-7|$, as in Example 26, we first press [2nd] [CATALOG] [ENTER] to copy "abs(" to the home screen. (CATALOG is the second operation associated with the [0] numeric key.) Then we press [(−)] [7] [)] [ENTER]. The result is 7. To find $\left|-\frac{1}{2}\right|$ and express the result as a fraction, we press [2nd] [CATALOG] [ENTER] [(−)] [1] [÷] [2] [)] [MATH] [1] [ENTER]. The result is $\frac{1}{2}$.

```
abs(−7)
                    7
abs(−1/2)►Frac
                  1/2
```

Exercises: Find the absolute value.

1. $|-5|$
2. $|17|$
3. $|0|$
4. $|6.48|$
5. $|-12.7|$
6. $|-0.9|$
7. $\left|-\frac{5}{7}\right|$
8. $\left|\frac{4}{3}\right|$

Expressions like $a \le b$ and $b \ge a$ are also inequalities. We read $a \le b$ as "**a is less than or equal to b.**" We read $a \ge b$ as "**a is greater than or equal to b.**"

EXAMPLES Write true or false for each statement.

23. $-3 \le 5.4$ True since $-3 < 5.4$ is true

24. $-3 \le -3$ True since $-3 = -3$ is true

25. $-5 \ge 1\frac{2}{3}$ False since neither $-5 > 1\frac{2}{3}$ nor $-5 = 1\frac{2}{3}$ is true

Do Exercises 22–24 on the preceding page.

e Absolute Value

From the number line, we see that numbers like 4 and -4 are the same distance from zero. Distance is always a nonnegative number. We call the distance of a number from zero on a number line the **absolute value** of the number.

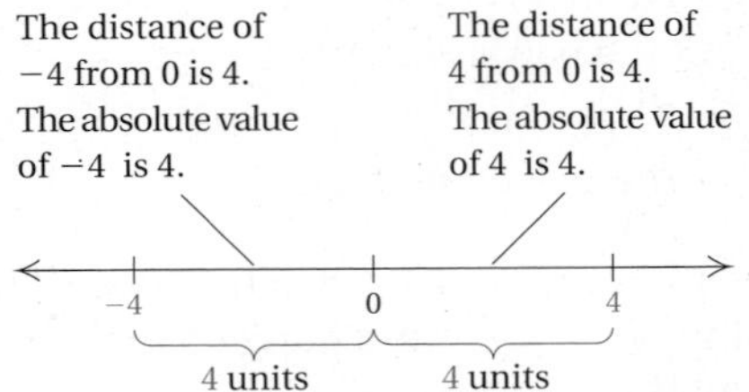
The distance of
−4 from 0 is 4.
The absolute value
of −4 is 4.
The distance of
4 from 0 is 4.
The absolute value
of 4 is 4.
−4
0
4
4 units
4 units

ABSOLUTE VALUE

The **absolute value** of a number is its distance from zero on a number line. We use the symbol $|x|$ to represent the absolute value of a number x.

FINDING ABSOLUTE VALUE

a) If a number is negative, its absolute value is positive.

b) If a number is positive or zero, its absolute value is the same as the number.

EXAMPLES Find the absolute value.

26. $|-7|$ The distance of -7 from 0 is 7, so $|-7| = 7$.

27. $|12|$ The distance of 12 from 0 is 12, so $|12| = 12$.

28. $|0|$ The distance of 0 from 0 is 0, so $|0| = 0$.

29. $\left|\frac{3}{2}\right| = \frac{3}{2}$

30. $|-2.73| = 2.73$

Do Exercises 25–28.

1.2 EXERCISE SET

For Extra Help

Digital Video Tutor CD 2 Videotape 3 | InterAct Math | Math Tutor Center | MathXL | MyMathLab

a State the integers that correspond to the situation.

1. *Elevations.* The Dead Sea, between Jordan and Israel, is 1286 ft below sea level. Mount Rainier in Washington State is 14,410 ft above sea level.
Sources: *The Handy Geography Answer Book; The New York Times Almanac*

2. *Golf Score.* Tiger Woods' score in winning the 2000 PGA Championship was 18 under par.
Source: U.S. Golf Association

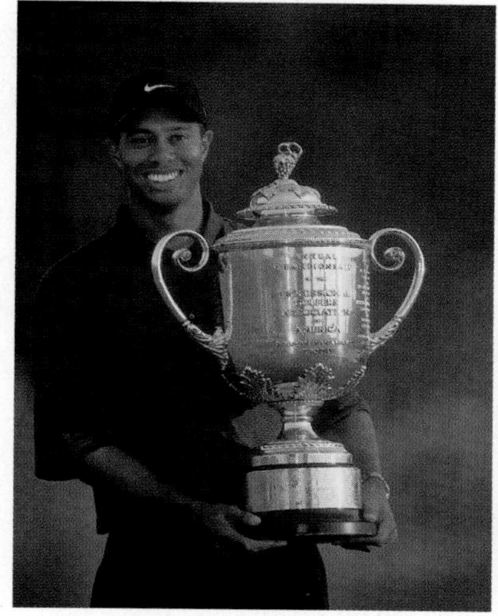

3. On Wednesday, the temperature was 24° above zero. On Thursday, it was 2° below zero.

4. A student deposited her tax refund of $750 in a savings account. Two weeks later, she withdrew $125 to pay sorority fees.

5. *U.S. Public Debt.* Recently, the total public debt of the United States was about $5,600,000,000,000.
Source: U.S. Department of the Treasury

6. *Birth and Death Rates.* Recently, the world birth rate was 270 per ten thousand. The death rate was 97 per ten thousand.
Source: United Nations Population Fund

7. In bowling, the Alley Cats are 34 pins behind the Strikers going into the last frame. Describe the situation of each team.

8. During a video game, Maggie intercepted a missile worth 20 points, lost a starship worth 150 points, and captured a landing base worth 300 points.

b Graph the number on the number line.

9. $\frac{10}{3}$

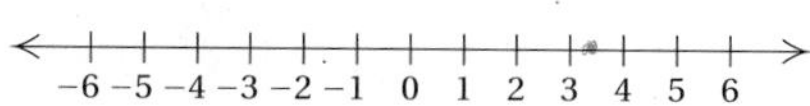

10. $-\frac{17}{4}$

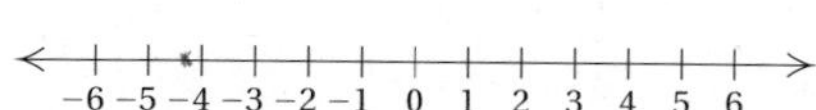

11. -5.2

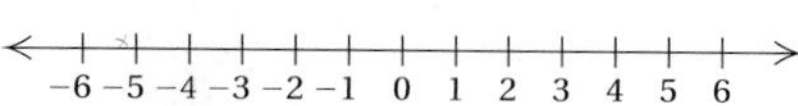

12. 4.78

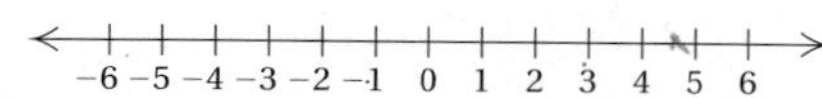

c Convert to decimal notation.

13. $-\frac{7}{8}$

14. $-\frac{1}{8}$

15. $\frac{5}{6}$

16. $\frac{5}{3}$

17. $-\frac{7}{6}$

18. $-\frac{5}{12}$

19. $\frac{2}{3}$

20. $\frac{1}{4}$

21. $-\frac{1}{2}$

22. $\frac{5}{8}$

23. $\frac{1}{10}$

24. $-\frac{7}{20}$

d Use either $<$ or $>$ for $\square$ to write a true sentence.

25. $8 \square 0$

26. $3 \square 0$

27. $-8 \square 3$

28. $6 \square -6$

29. $-8 \square 8$

30. $0 \square -9$

31. $-8 \square -5$

32. $-4 \square -3$

33. $-5 \square -11$

34. $-3 \square -4$

35. $-6 \square -5$

36. $-10 \square -14$

37. $2.14 \square 1.24$

38. $-3.3 \square -2.2$

39. $-14.5 \square 0.011$

40. $17.2 \square -1.67$

41. $-12.88 \square -6.45$

42. $-14.34 \square -17.88$

43. $\frac{5}{12} \square \frac{11}{25}$

44. $-\frac{13}{16} \square -\frac{5}{9}$

Write true or false.

45. $-3 \geq -11$ **46.** $5 \leq -5$ **47.** $0 \geq 8$ **48.** $-5 \leq 7$

Write an inequality with the same meaning.

49. $-6 > x$ **50.** $x < 8$ **51.** $-10 \leq y$ **52.** $12 \geq t$

e Find the absolute value.

53. $|-3|$ **54.** $|-7|$ **55.** $|10|$ **56.** $|11|$ **57.** $|0|$

58. $|-4|$ **59.** $|-24|$ **60.** $|325|$ **61.** $\left|-\frac{2}{3}\right|$ **62.** $\left|-\frac{10}{7}\right|$

63. $\left|\frac{0}{4}\right|$ **64.** $|14.8|$ **65.** $\left|-3\frac{5}{8}\right|$ **66.** $\left|-7\frac{4}{5}\right|$

67. D_W When Jennifer's calculator gives a decimal approximation for $\sqrt{2}$ and that approximation is promptly squared, the result is 2. Yet, when that same approximation is entered by hand and then squared, the result is not exactly 2. Why do you suppose this happens?

68. D_W How many rational numbers are there between 0 and 1? Why?

SKILL MAINTENANCE

Convert to decimal notation. [R.4a]

69. 63% **70.** 8.3% **71.** 110% **72.** 22.76%

Convert to percent notation. [R.4d]

73. $\frac{3}{4}$ **74.** $\frac{5}{8}$ **75.** $\frac{5}{6}$ **76.** $\frac{19}{32}$

SYNTHESIS

List in order from the least to the greatest.

77. $-\frac{2}{3}, \frac{1}{2}, -\frac{3}{4}, -\frac{5}{6}, \frac{3}{8}, \frac{1}{6}$

78. $-8\frac{7}{8}, 7^1, -5, |-6|, 4, |3|, -8\frac{5}{8}, -100, 0, 1^7, \frac{14}{4}, -\frac{67}{8}$

Given that $0.\overline{3} = \frac{1}{3}$ and $0.\overline{6} = \frac{2}{3}$, express each of the following as a quotient or ratio of two integers.

79. $0.\overline{1}$ **80.** $0.\overline{9}$ **81.** $5.\overline{5}$

1.3 ADDITION OF REAL NUMBERS

Objectives

a Add real numbers without using a number line.

b Find the opposite, or additive inverse, of a real number.

c Solve applied problems involving addition of real numbers.

In this section, we consider addition of real numbers. First, to gain an understanding, we add using a number line. Then we consider rules for addition.

ADDITION ON A NUMBER LINE

To do the addition $a + b$ on a number line, we start at 0. Then we move to a and then move according to b.

a) If b is positive, we move to the right.

b) If b is negative, we move to the left.

c) If b is 0, we stay at a.

EXAMPLE 1 Add: $3 + (-5)$.

We start at 0 and move 3 units right since 3 is positive. Then we move 5 units left since -5 is negative.

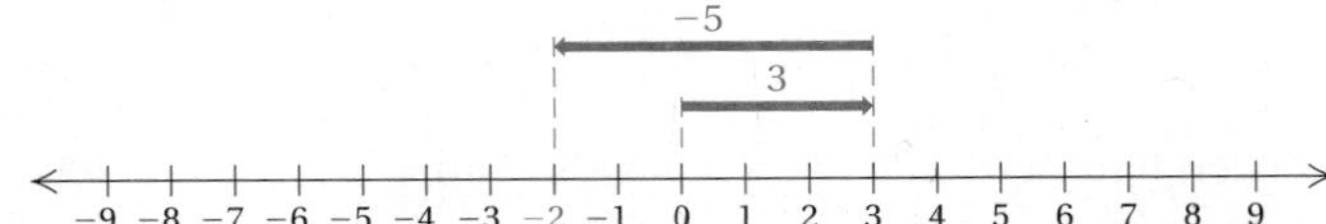

$3 + (-5) = -2$

EXAMPLE 2 Add: $-4 + (-3)$.

We start at 0 and move 4 units left since -4 is negative. Then we move 3 units further left since -3 is negative.

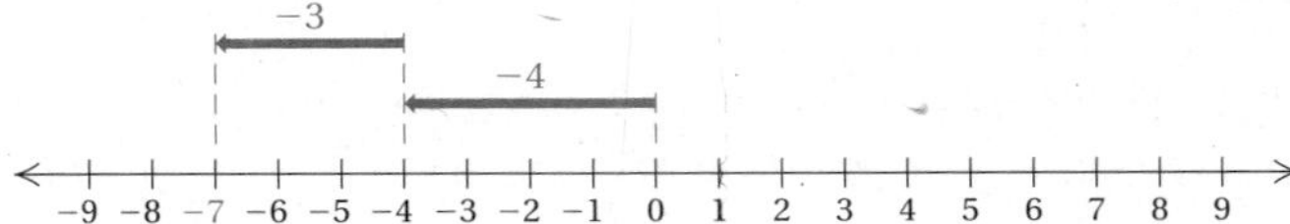

$-4 + (-3) = -7$

EXAMPLE 3 Add: $-4 + 9$.

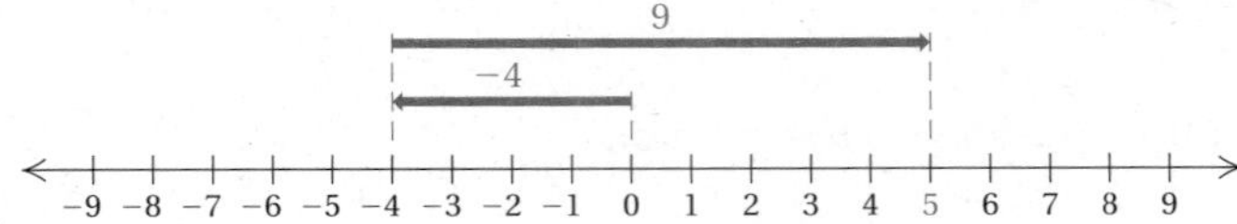

$-4 + 9 = 5$

Add using a number line.

1. $0 + (-3)$

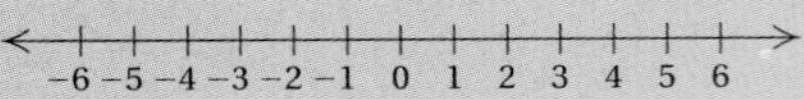

2. $1 + (-4)$

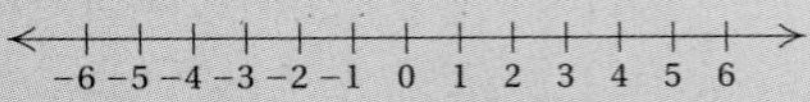

3. $-3 + (-2)$

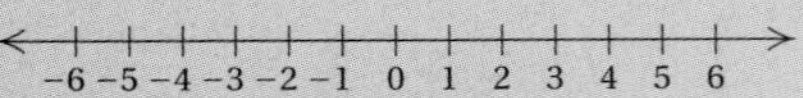

4. $-3 + 7$

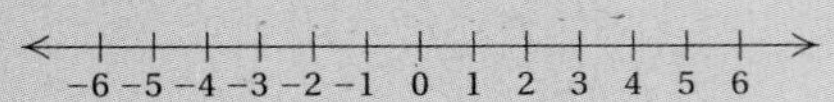

5. $-2.4 + 2.4$

−6 −5 −4 −3 −2 −1 0 1 2 3 4 5 6

6. $-\frac{5}{2} + \frac{1}{2}$

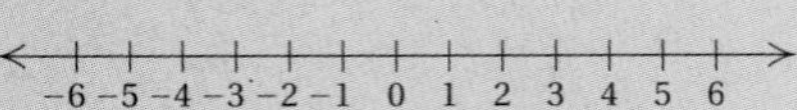

Answers on page A-4

EXAMPLE 4 Add: $-5.2 + 0$.

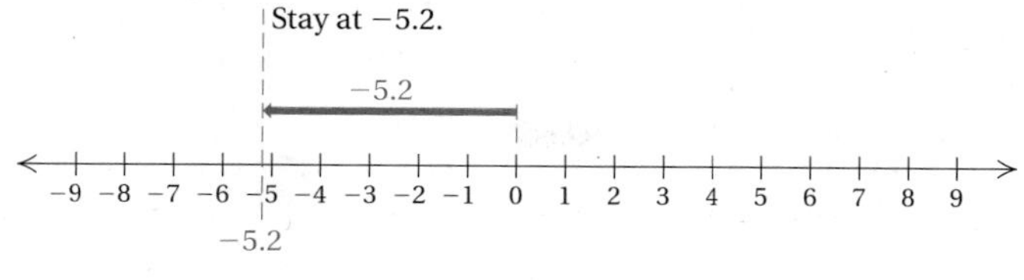

$-5.2 + 0 = -5.2$

Do Exercises 1–6 on the preceding page.

a Adding Without a Number Line

You may have noticed some patterns in the preceding examples. These lead us to rules for adding without using a number line that are more efficient for adding larger numbers.

RULES FOR ADDITION OF REAL NUMBERS

1. *Positive numbers*: Add the same as arithmetic numbers. The answer is positive.
2. *Negative numbers*: Add absolute values. The answer is negative.
3. *A positive and a negative number*: Subtract the smaller absolute value from the larger. Then:
   a) If the positive number has the greater absolute value, the answer is positive.
   b) If the negative number has the greater absolute value, the answer is negative.
   c) If the numbers have the same absolute value, the answer is 0.
4. *One number is zero*: The sum is the other number.

Rule 4 is known as the **identity property of 0.** It says that for any real number a, $a + 0 = a$.

EXAMPLES Add without using a number line.

5. $-12 + (-7) = -19$

Two negatives. Add the absolute values: $|-12| + |-7| = 12 + 7 = 19$. Make the answer *negative*: -19.

6. $-1.4 + 8.5 = 7.1$

One negative, one positive. Find the absolute values: $|-1.4| = 1.4$; $|8.5| = 8.5$. Subtract the smaller absolute value from the larger: $8.5 - 1.4 = 7.1$. The *positive* number, 8.5, has the larger absolute value, so the answer is *positive*, 7.1.

7. $-36 + 21 = -15$

One negative, one positive. Find the absolute values: $|-36| = 36$, $|21| = 21$. Subtract the smaller absolute value from the larger: $36 - 21 = 15$. The negative number, -36, has the larger absolute value, so the answer is *negative*, -15.

Add without using a number line.

7. $-5 + (-6)$

8. $-9 + (-3)$

9. $-4 + 6$

10. $-7 + 3$

11. $5 + (-7)$

12. $-20 + 20$

13. $-11 + (-11)$

14. $10 + (-7)$

15. $-0.17 + 0.7$

16. $-6.4 + 8.7$

17. $-4.5 + (-3.2)$

18. $-8.6 + 2.4$

19. $\frac{5}{9} + \left(-\frac{7}{9}\right)$

20. $-\frac{1}{5} + \left(-\frac{3}{4}\right)$

Answers on page A-4

Add.

21. $(-15) + (-37) + 25 + 42 + (-59) + (-14)$

22. $42 + (-81) + (-28) + 24 + 18 + (-31)$

23. $-2.5 + (-10) + 6 + (-7.5)$

24. $-35 + 17 + 14 + (-27) + 31 + (-12)$

Find the opposite, or additive inverse, of each of the following.

25. -4

26. 8.7

27. -7.74

28. $-\frac{8}{9}$

29. 0

30. 12

Answers on page A-4

8. $1.5 + (-1.5) = 0$ — The numbers have the same absolute value. The sum is 0.

9. $-\frac{7}{8} + 0 = -\frac{7}{8}$ — One number is zero. The sum is $-\frac{7}{8}$.

10. $-9.2 + 3.1 = -6.1$

11. $-\frac{3}{2} + \frac{9}{2} = \frac{6}{2} = 3$

12. $-\frac{2}{3} + \frac{5}{8} = -\frac{16}{24} + \frac{15}{24} = -\frac{1}{24}$

Do Exercises 7–20 on the preceding page.

Suppose we want to add several numbers, some positive and some negative, as follows. How can we proceed?

$$15 + (-2) + 7 + 14 + (-5) + (-12)$$

We can change grouping and order as we please when adding. For instance, we can group the positive numbers together and the negative numbers together and add them separately. Then we add the two results.

EXAMPLE 13 Add: $15 + (-2) + 7 + 14 + (-5) + (-12)$.

a) $15 + 7 + 14 = 36$ — Adding the positive numbers

b) $-2 + (-5) + (-12) = -19$ — Adding the negative numbers

$36 + (-19) = 17$ — Adding (a) and (b)

We can also add the numbers in any other order we wish, say, from left to right as follows:

$$\begin{aligned} 15 + (-2) + 7 + 14 + (-5) + (-12) &= 13 + 7 + 14 + (-5) + (-12) \\ &= 20 + 14 + (-5) + (-12) \\ &= 34 + (-5) + (-12) \\ &= 29 + (-12) \\ &= 17 \end{aligned}$$

Do Exercises 21–24.

b Opposites, or Additive Inverses

Suppose we add two numbers that are **opposites,** such as 6 and -6. The result is 0. When opposites are added, the result is always 0. Such numbers are also called **additive inverses.** Every real number has an opposite, or additive inverse.

OPPOSITES, OR ADDITIVE INVERSES

Two numbers whose sum is 0 are called **opposites,** or **additive inverses,** of each other.

EXAMPLES Find the opposite, or additive inverse, of each number.

14. 34 The opposite of 34 is -34 because $34 + (-34) = 0$.

15. -8 The opposite of -8 is 8 because $-8 + 8 = 0$.

16. 0 The opposite of 0 is 0 because $0 + 0 = 0$.

17. $-\frac{7}{8}$ The opposite of $-\frac{7}{8}$ is $\frac{7}{8}$ because $-\frac{7}{8} + \frac{7}{8} = 0$.

Do Exercises 25–30 on the preceding page.

To name the opposite, we use the symbol $-$, as follows.

SYMBOLIZING OPPOSITES

The opposite, or additive inverse, of a number a can be named $-a$ (read "the opposite of a," or "the additive inverse of a").

Note that if we take a number, say, 8, and find its opposite, -8, and then find the opposite of the result, we will have the original number, 8, again.

THE OPPOSITE OF AN OPPOSITE

The **opposite of the opposite** of a number is the number itself. (The additive inverse of the additive inverse of a number is the number itself.) That is, for any number a,

$$-(-a) = a.$$

EXAMPLE 18 Evaluate $-x$ and $-(-x)$ when $x = 16$.

If $x = 16$, then $-x = -16$. The opposite of 16 is -16.

If $x = 16$, then $-(-x) = -(-16) = 16$. The opposite of the opposite of 16 is 16.

EXAMPLE 19 Evaluate $-x$ and $-(-x)$ when $x = -3$.

If $x = -3$, then $-x = -(-3) = 3$.

If $x = -3$, then $-(-x) = -(-(-3)) = -3$.

Note that in Example 19 we used a second set of parentheses to show that we are substituting the negative number -3 for x. Symbolism like $--x$ is not considered meaningful.

Do Exercises 31–36.

A symbol such as -8 is usually read "negative 8." It could be read "the additive inverse of 8," because the additive inverse of 8 is negative 8. It could also be read "the opposite of 8," because the opposite of 8 is -8. Thus a symbol like -8 can be read in more than one way. It is never correct to read -8 as "minus 8."

CAUTION!

A symbol like $-x$, which has a variable, should be read "the opposite of x" or "the additive inverse of x" and *not* "negative x," because we do not know whether x represents a positive number, a negative number, or 0. You can check this in Examples 18 and 19.

Evaluate $-x$ and $-(-x)$ when:

31. $x = 14$.

32. $x = 1$.

33. $x = -19$.

34. $x = -1.6$.

35. $x = \frac{2}{3}$.

36. $x = -\frac{9}{8}$.

Answers on page A-4

Find the opposite. (Change the sign.)

37. -4

38. -13.4

39. 0

40. $\frac{1}{4}$

41. Class Size. There were 27 students in Eliza's algebra class when the semester began. During the first two weeks, 5 students withdrew, 8 students enrolled in the class, and 6 students were dropped as "no shows." How many were in the class after two weeks?

Answers on page A-4

We can use the symbolism $-a$ to restate the definition of opposite, or additive inverse.

THE SUM OF OPPOSITES

For any real number a, the **opposite,** or **additive inverse,** of a, expressed as $-a$, is such that

$$a + (-a) = (-a) + a = 0.$$

SIGNS OF NUMBERS

A negative number is sometimes said to have a "negative sign." A positive number is said to have a "positive sign." When we replace a number with its opposite, we can say that we have "changed its sign."

EXAMPLES Find the opposite. (Change the sign.)

20. -3 $\quad -(-3) = 3$

21. -10 $\quad -(-10) = 10$

22. 0 $\quad -(0) = 0$

23. 14 $\quad -(14) = -14$

Do Exercises 37–40.

C Applications and Problem Solving

Addition of real numbers occurs in many real-world situations.

EXAMPLE 24 *Lake Level.* In the course of one four-month period, the water level of Lake Champlain went down 2 ft, up 1 ft, down 5 ft, and up 3 ft. How much had the lake level changed at the end of the four months?

We let $T =$ the total change in the level of the lake. Then the problem translates to a sum:

Total change	is	1st change	plus	2nd change	plus	3rd change	plus	4th change.
T	$=$	-2	$+$	1	$+$	(-5)	$+$	3

Adding from left to right, we have

$$\begin{aligned} T = -2 + 1 + (-5) + 3 &= -1 + (-5) + 3 \\ &= -6 + 3 \\ &= -3. \end{aligned}$$

The lake level has dropped 3 ft at the end of the four-month period.

Do Exercise 41.

1.3 EXERCISE SET

For Extra Help

Digital Video Tutor CD 2 Videotape 3 | InterAct Math | Math Tutor Center | MathXL | MyMathLab

a Add. Do not use a number line except as a check.

1. $2 + (-9)$

2. $-5 + 2$

3. $-11 + 5$

4. $4 + (-3)$

5. $-6 + 6$

6. $8 + (-8)$

7. $-3 + (-5)$

8. $-4 + (-6)$

9. $-7 + 0$

10. $-13 + 0$

11. $0 + (-27)$

12. $0 + (-35)$

13. $17 + (-17)$

14. $-15 + 15$

15. $-17 + (-25)$

16. $-24 + (-17)$

17. $18 + (-18)$

18. $-13 + 13$

19. $-28 + 28$

20. $11 + (-11)$

21. $8 + (-5)$

22. $-7 + 8$

23. $-4 + (-5)$

24. $10 + (-12)$

25. $13 + (-6)$

26. $-3 + 14$

27. $-25 + 25$

28. $50 + (-50)$

29. $53 + (-18)$

30. $75 + (-45)$

31. $-8.5 + 4.7$

32. $-4.6 + 1.9$

33. $-2.8 + (-5.3)$

34. $-7.9 + (-6.5)$

35. $-\frac{3}{5} + \frac{2}{5}$

36. $-\frac{4}{3} + \frac{2}{3}$

37. $-\frac{2}{9} + \left(-\frac{5}{9}\right)$

38. $-\frac{4}{7} + \left(-\frac{6}{7}\right)$

39. $-\frac{5}{8} + \frac{1}{4}$

40. $-\frac{5}{6} + \frac{2}{3}$

41. $-\frac{5}{8} + \left(-\frac{1}{6}\right)$

42. $-\frac{5}{6} + \left(-\frac{2}{9}\right)$

43. $-\frac{3}{8} + \frac{5}{12}$

44. $-\frac{7}{16} + \frac{7}{8}$

45. $76 + (-15) + (-18) + (-6)$

46. $29 + (-45) + 18 + 32 + (-96)$

47. $-44 + \left(-\frac{3}{8}\right) + 95 + \left(-\frac{5}{8}\right)$

48. $24 + 3.1 + (-44) + (-8.2) + 63$

49. $98 + (-54) + 113 + (-998) + 44 + (-612)$

50. $-458 + (-124) + 1025 + (-917) + 218$

add to get to zero

b Find the opposite, or additive inverse.

51. 24

52. -64

53. -26.9

54. 48.2

Evaluate $-x$ when:

55. $x = 8$.

56. $x = -27$.

57. $x = -\frac{13}{8}$.

58. $x = \frac{1}{236}$.

Evaluate $-(-x)$ when:

59. $x = -43$.

60. $x = 39$.

61. $x = \frac{4}{3}$.

62. $x = -7.1$.

Find the opposite. (Change the sign.)

63. -24

64. -12.3

65. $-\frac{3}{8}$

66. 10

c Solve.

67. *Tallest Mountain.* The tallest mountain in the world, when measured from base to peak, is Mauna Kea (White Mountain) in Hawaii. From its base 19,684 ft below sea level in the Hawaiian Trough, it rises 33,480 ft. What is the elevation of the peak above sea level?
Source: *The Guinness Book of Records*

68. *Telephone Bills.* Erika's cell-phone bill for July was \$82. She sent a check for \$50 and then made \$37 worth of calls in August. How much did she then owe on her cell-phone bill?

69. *Temperature Changes.* One day the temperature in Lawrence, Kansas, is 32°F at 6:00 A.M. It rises 15° by noon, but falls 50° by midnight when a cold front moves in. What is the final temperature?

70. *Stock Changes.* On a recent day, the price of Quaker Oats stock opened at a value of \$61.38. During the day, it rose \$4.75, dropped \$7.38, and rose \$5.13. Find the value of the stock at the end of the day.
Source: The New York Stock Exchange

71. *Profits and Losses.* A business expresses a profit as a positive number and refers to it as operating "in the black." A loss is expressed as a negative number and is referred to as operating "in the red." The profits and losses of Xponent Corporation over various years are shown in the bar graph below. Find the sum of the profits and losses.

Xponent Corporation

Profits and losses
15000
10000
5000
0
−5000
−10000
−15000
−20000
$10,500
−$16,600
−$12,800
−$9,600
$8,200
2001 2002 2003 2004 2005
Year

72. *Football Yardage.* In a college football game, the quarterback attempted passes with the following results. Find the total gain or loss.

TRY	GAIN OR LOSS
1st	13-yd gain
2nd	12-yd loss
3rd	21-yd gain

73. *Account Balance.* Leah has \$460 in a checking account. She writes a check for \$530, makes a deposit of \$75, and then writes a check for \$90. What is the balance in her account?

74. *Credit Card Bills.* On August 1, Lyle's credit card bill shows that he owes \$470. During the month of August, Lyle sends a check for \$45 to the credit card company, charges another \$160 in merchandise, and then pays off another \$500 of his bill. What is the new balance of Lyle's account at the end of August?

75. $^{\text{D}}{}_{\text{W}}$ Without actually performing the addition, explain why the sum of all integers from -50 to 50 is 0.

76. $^{\text{D}}{}_{\text{W}}$ Explain in your own words why the sum of two negative numbers is always negative.

SKILL MAINTENANCE

Convert to decimal notation. [R.4a]

77. 57%

78. 71.3%

79. $23\frac{4}{5}\%$

80. $92\frac{7}{8}\%$

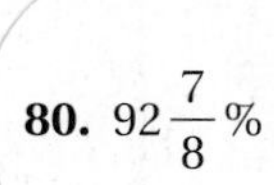

Convert to percent notation. [R.4d]

81. $\frac{5}{4}$

82. $\frac{1}{8}$

83. $\frac{13}{25}$

84. $\frac{13}{32}$

SYNTHESIS

85. For what numbers x is $-x$ negative?

86. For what numbers x is $-x$ positive?

For each of Exercises 87 and 88, choose the correct answer from the selections given.

87. If a is positive and b is negative, then $-a + b$ is:

a) Positive.
b) Negative.
c) 0.
d) Cannot be determined without more information

88. If $a = b$ and a and b are negative, then $-a + (-b)$ is:

a) Positive.
b) Negative.
c) 0.
d) Cannot be determined without more information

1.4 SUBTRACTION OF REAL NUMBERS

Objectives

a Subtract real numbers and simplify combinations of additions and subtractions.

b Solve applied problems involving subtraction of real numbers.

Subtract.

1. $-6 - 4$

Think: What number can be added to 4 to get -6:

$\square + 4 = -6?$

2. $-7 - (-10)$

Think: What number can be added to -10 to get -7:

$\square + (-10) = -7?$

3. $-7 - (-2)$

Think: What number can be added to -2 to get -7:

$\square + (-2) = -7?$

Subtract. Use a number line, doing the "opposite" of addition.

4. $-4 - (-3)$

−6 −5 −4 −3 −2 −1 0 1 2 3 4 5 6

5. $-4 - (-6)$

−6 −5 −4 −3 −2 −1 0 1 2 3 4 5 6

6. $5 - 9$

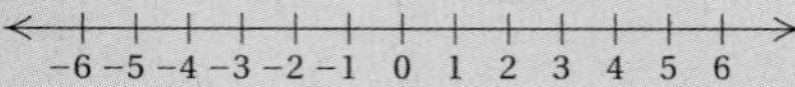

Answers on page A-4

a Subtraction

We now consider subtraction of real numbers.

SUBTRACTION

The difference $a - b$ is the number c for which $a = b + c$.

Consider, for example, $45 - 17$. *Think*: What number can we add to 17 to get 45? Since $45 = 17 + 28$, we know that $45 - 17 = 28$. Let's consider an example whose answer is a negative number.

EXAMPLE 1 Subtract: $3 - 7$.

Think: What number can we add to 7 to get 3? The number must be negative. Since $7 + (-4) = 3$, we know the number is -4: $3 - 7 = -4$. That is, $3 - 7 = -4$ because $7 + (-4) = 3$.

Do Exercises 1–3.

The definition above does not provide the most efficient way to do subtraction. We can develop a faster way to subtract. As a rationale for the faster way, let's compare $3 + 7$ and $3 - 7$ on a number line.

To find $3 + 7$ on a number line, we move 3 units to the right from 0 since 3 is positive. Then we move 7 units farther to the right since 7 is positive.

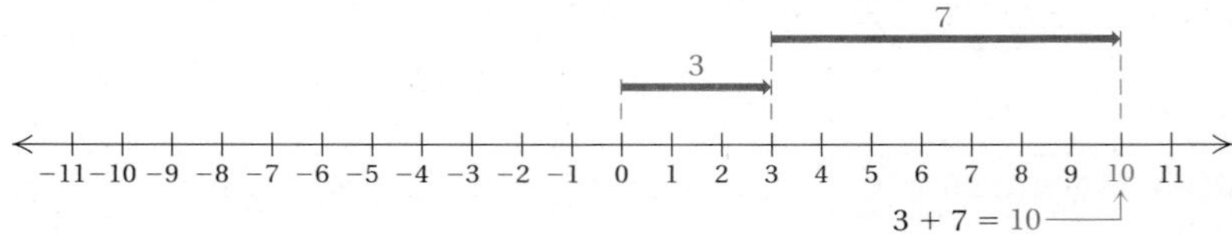

To find $3 - 7$, we do the "opposite" of adding 7: We move 7 units to the *left* to do the subtracting. This is the same as *adding* the opposite of 7, -7, to 3.

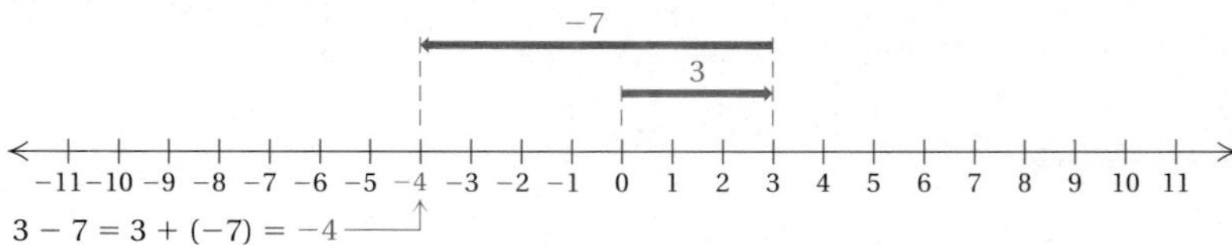

Do Exercises 4–6.

Look for a pattern in the examples shown at right.

SUBTRACTIONS	ADDING AN OPPOSITE
$5 - 8 = -3$	$5 + (-8) = -3$
$-6 - 4 = -10$	$-6 + (-4) = -10$
$-7 - (-2) = -5$	$-7 + 2 = -5$

Do Exercises 7–10.

Perhaps you have noticed that we can subtract by adding the opposite of the number being subtracted. This can always be done.

SUBTRACTING BY ADDING THE OPPOSITE

For any real numbers a and b,

$$a - b = a + (-b).$$

(To subtract, add the opposite, or additive inverse, of the number being subtracted.)

This is the method generally used for quick subtraction of real numbers.

EXAMPLES Subtract.

2. $2 - 6 = 2 + (-6) = -4$ — The opposite of 6 is -6. We change the subtraction to addition and add the opposite. *Check*: $-4 + 6 = 2$.

3. $4 - (-9) = 4 + 9 = 13$ — The opposite of -9 is 9. We change the subtraction to addition and add the opposite. *Check*: $13 + (-9) = 4$.

4. $-4.2 - (-3.6) = -4.2 + 3.6 = -0.6$ — Adding the opposite. *Check*: $-0.6 + (-3.6) = -4.2$.

5. $-\frac{1}{2} - \left(-\frac{3}{4}\right) = -\frac{1}{2} + \frac{3}{4} = \frac{1}{4}$ — Adding the opposite. *Check*: $\frac{1}{4} + \left(-\frac{3}{4}\right) = -\frac{1}{2}$.

Do Exercises 11–16.

EXAMPLES Read each of the following. Then subtract by adding the opposite of the number being subtracted.

6. $3 - 5$ — Read "three minus five is three plus the opposite of five"
   $3 - 5 = 3 + (-5) = -2$

7. $\frac{1}{8} - \frac{7}{8}$ — Read "one-eighth minus seven-eighths is one-eighth plus the opposite of seven-eighths"
   $\frac{1}{8} - \frac{7}{8} = \frac{1}{8} + \left(-\frac{7}{8}\right) = -\frac{6}{8}$, or $-\frac{3}{4}$

8. $-4.6 - (-9.8)$ — Read "negative four point six minus negative nine point eight is negative four point six plus the opposite of negative nine point eight"
   $-4.6 - (-9.8) = -4.6 + 9.8 = 5.2$

9. $-\frac{3}{4} - \frac{7}{5}$ — Read "negative three-fourths minus seven-fifths is negative three-fourths plus the opposite of seven-fifths"
   $-\frac{3}{4} - \frac{7}{5} = -\frac{3}{4} + \left(-\frac{7}{5}\right) = -\frac{15}{20} + \left(-\frac{28}{20}\right) = -\frac{43}{20}$

Do Exercises 17–21 on the following page.

Complete the addition and compare with the subtraction.

7. $4 - 6 = -2$;
   $4 + (-6) =$ ______

8. $-3 - 8 = -11$;
   $-3 + (-8) =$ ______

9. $-5 - (-9) = 4$;
   $-5 + 9 =$ ______

10. $-5 - (-3) = -2$;
    $-5 + 3 =$ ______

Subtract.

11. $2 - 8$

12. $-6 - 10$

13. $12.4 - 5.3$

14. $-8 - (-11)$

15. $-8 - (-8)$

16. $\frac{2}{3} - \left(-\frac{5}{6}\right)$

Answers on page A-4

Read each of the following. Then subtract by adding the opposite of the number being subtracted.

17. $3 - 11$

18. $12 - 5$

19. $-12 - (-9)$

20. $-12.4 - 10.9$

21. $-\frac{4}{5} - \left(-\frac{4}{5}\right)$

Simplify.

22. $-6 - (-2) - (-4) - 12 + 3$

23. $9 - (-6) + 7 - 11 - 14 - (-20)$

24. $-9.6 + 7.4 - (-3.9) - (-11)$

25. Temperature Extremes. The highest temperature ever recorded in the United States is 134°F in Greenland Ranch, California, on July 10, 1913. The lowest temperature ever recorded is −80°F in Prospect Creek, Alaska, on January 23, 1971. How much higher was the temperature in Greenland Ranch than that in Prospect Creek?
Source: National Oceanographic and Atmospheric Administration

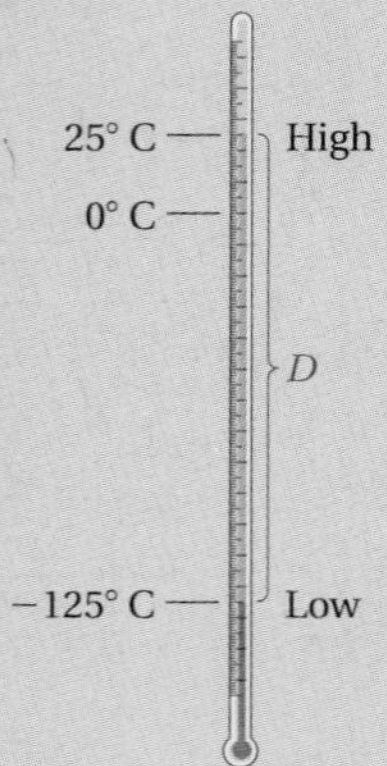

Answers on page A-4

When several additions and subtractions occur together, we can make them all additions.

EXAMPLES Simplify.

10. $8 - (-4) - 2 - (-4) + 2 = 8 + 4 + (-2) + 4 + 2$ Adding the opposite
$= 16$

11. $8.2 - (-6.1) + 2.3 - (-4) = 8.2 + 6.1 + 2.3 + 4 = 20.6$

Do Exercises 22–24.

b Applications and Problem Solving

Let's now see how we can use subtraction of real numbers to solve applied problems.

EXAMPLE 12 *Temperatures on Mars.* The Viking 2 Lander spacecraft has determined that temperatures on Mars range from −125°C (Celsius) to 25°C. Find the difference between the highest value and the lowest value in this temperature range.
Source: The Lunar and Planetary Institute

We let D = the difference in the temperatures. Then the problem translates to the following subtraction:

Difference in temperature	is	Highest temperature	minus	Lowest temperature.
↓	↓	↓	↓	↓
D	$=$	25	$-$	(-125)
D	$= 25 + 125 = 150$			

The difference in the temperatures is 150°C.

Do Exercise 25.

1.4 EXERCISE SET

For Extra Help

Digital Video Tutor CD 2 Videotape 3 | InterAct Math | Math Tutor Center | MathXL | MyMathLab

a Subtract.

1. $2 - 9$
2. $3 - 8$
3. $0 - 4$
4. $0 - 9$
5. $-8 - (-2)$
6. $-6 - (-8)$
7. $-11 - (-11)$
8. $-6 - (-6)$
9. $12 - 16$
10. $14 - 19$
11. $20 - 27$
12. $30 - 4$
13. $-9 - (-3)$
14. $-7 - (-9)$
15. $-40 - (-40)$
16. $-9 - (-9)$
17. $7 - 7$
18. $9 - 9$
19. $7 - (-7)$
20. $4 - (-4)$
21. $8 - (-3)$
22. $-7 - 4$
23. $-6 - 8$
24. $6 - (-10)$
25. $-4 - (-9)$
26. $-14 - 2$
27. $1 - 8$
28. $2 - 8$
29. $-6 - (-5)$
30. $-4 - (-3)$
31. $8 - (-10)$
32. $5 - (-6)$
33. $0 - 10$
34. $0 - 18$
35. $-5 - (-2)$
36. $-3 - (-1)$
37. $-7 - 14$
38. $-9 - 16$
39. $0 - (-5)$
40. $0 - (-1)$

41. $-8 - 0$

42. $-9 - 0$

43. $7 - (-5)$

44. $7 - (-4)$

45. $2 - 25$

46. $18 - 63$

47. $-42 - 26$

48. $-18 - 63$

49. $-71 - 2$

50. $-49 - 3$

51. $24 - (-92)$

52. $48 - (-73)$

53. $-50 - (-50)$

54. $-70 - (-70)$

55. $-\frac{3}{8} - \frac{5}{8}$

56. $\frac{3}{9} - \frac{9}{9}$

57. $\frac{3}{4} - \frac{2}{3}$

58. $\frac{5}{8} - \frac{3}{4}$

59. $-\frac{3}{4} - \frac{2}{3}$

60. $-\frac{5}{8} - \frac{3}{4}$

61. $-\frac{5}{8} - \left(-\frac{3}{4}\right)$

62. $-\frac{3}{4} - \left(-\frac{2}{3}\right)$

63. $6.1 - (-13.8)$

64. $1.5 - (-3.5)$

65. $-2.7 - 5.9$

66. $-3.2 - 5.8$

67. $0.99 - 1$

68. $0.87 - 1$

69. $-79 - 114$

70. $-197 - 216$

71. $0 - (-500)$

72. $500 - (-1000)$

73. $-2.8 - 0$

74. $6.04 - 1.1$

75. $7 - 10.53$

76. $8 - (-9.3)$

77. $\frac{1}{6} - \frac{2}{3}$

78. $-\frac{3}{8} - \left(-\frac{1}{2}\right)$

79. $-\frac{4}{7} - \left(-\frac{10}{7}\right)$

80. $\frac{12}{5} - \frac{12}{5}$

81. $-\frac{7}{10} - \frac{10}{15}$

82. $-\frac{4}{18} - \left(-\frac{2}{9}\right)$

83. $\frac{1}{5} - \frac{1}{3}$

84. $-\frac{1}{7} - \left(-\frac{1}{6}\right)$

Simplify.

85. $18 - (-15) - 3 - (-5) + 2$

86. $22 - (-18) + 7 + (-42) - 27$

87. $-31 + (-28) - (-14) - 17$

88. $-43 - (-19) - (-21) + 25$

89. $-34 - 28 + (-33) - 44$

90. $39 + (-88) - 29 - (-83)$

91. $-93 - (-84) - 41 - (-56)$

92. $84 + (-99) + 44 - (-18) - 43$

93. $-5 - (-30) + 30 + 40 - (-12)$

94. $14 - (-50) + 20 - (-32)$

95. $132 - (-21) + 45 - (-21)$

96. $81 - (-20) - 14 - (-50) + 53$

b Solve.

97. *Ocean Depth.* The deepest point in the Pacific Ocean is the Marianas Trench, with a depth of 11,033 m. The deepest point in the Atlantic Ocean is the Puerto Rico Trench, with a depth of 8648 m. What is the difference in the elevation of the two trenches?
Source: *The Handy Geography Answer Book*

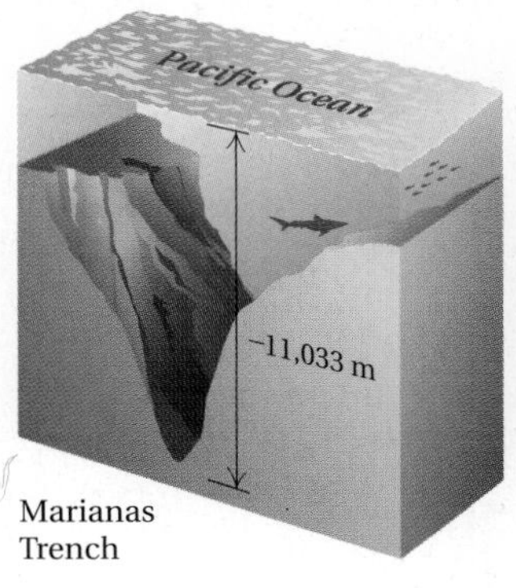

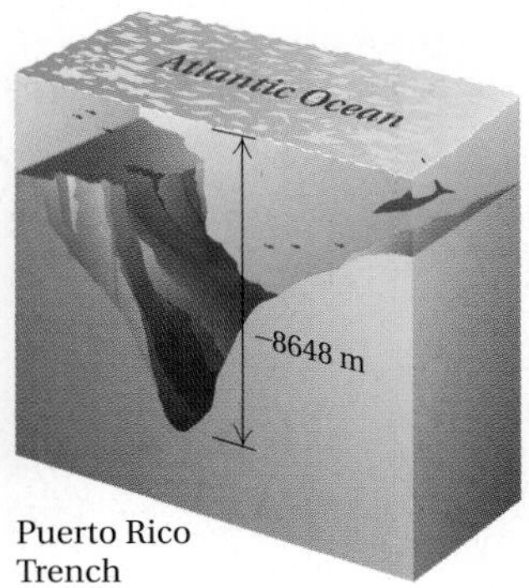

98. *Depth of Offshore Oil Wells.* In 1993, the elevation of the world's deepest offshore oil well was −2860 ft. By 1998, the deepest well was expected to be 360 ft deeper. What was the elevation of the deepest well in 1998?

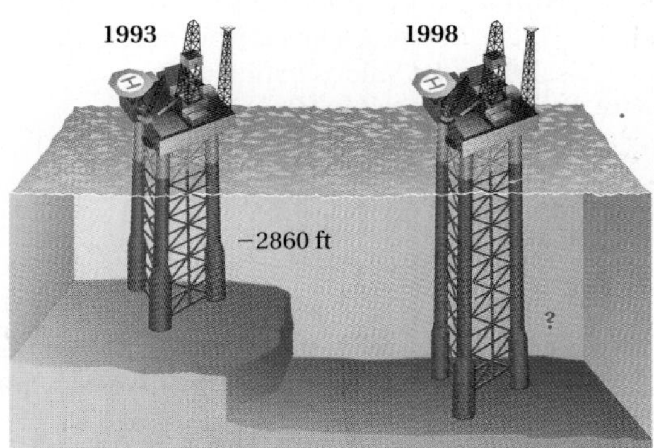

99. Laura has a charge of \$476.89 on her credit card, but she then returns a sweater that cost \$128.95. How much does she now owe on her credit card?

100. Chris has \$720 in a checking account. He writes a check for \$970 to pay for a sound system. What is the balance in his checking account?

101. *Home-Run Differential.* In baseball, the difference between the number of home runs hit by a team's players and the number allowed by its pitchers is called the *home-run differential,* that is,

$$\text{Home run differential} = \begin{array}{c}\text{Number of}\\ \text{home runs hit}\end{array} - \begin{array}{c}\text{Number of home}\\ \text{runs allowed}\end{array}.$$

Teams strive for a positive home-run differential.

a) In a recent year, Atlanta hit 197 home runs and allowed 120. Find its home-run differential.

b) In a recent year, San Francisco hit 153 home runs and allowed 194. Find its home-run differential.

Source: Major League Baseball

102. *Temperature Records.* The greatest recorded temperature change in one day occurred in Browning, Montana, where the temperature fell from 44°F to −56°F. By how much did the temperature drop?

Source: *The Guinness Book of Records*

103. *Low Points on Continents.* The lowest point in Africa is Lake Assal, which is 515 ft below sea level. The lowest point in South America is the Valdes Peninsula, which is 132 ft below sea level. How much lower is Lake Assal than the Valdes Peninsula?

Source: National Geographic Society

104. *Elevation Changes.* The lowest elevation in Asia, the Dead Sea, is 1286 ft below sea level. The highest elevation in Asia, Mount Everest, is 29,028 ft. Find the difference in elevation between the highest point and the lowest.

Source: *The Handy Geography Answer Book*

105. D_W If a negative number is subtracted from a positive number, will the result always be positive? Why or why not?

106. D_W Write a problem for a classmate to solve. Design the problem so that the solution is "The temperature dropped to −9°."

SKILL MAINTENANCE

Evaluate. [R.5b]

107. 5^3

108. 3^5

109. 3^8

110. 10^4

Find the prime factorization. [R.1a]

111. 864

112. 4235

113. Simplify: $256 \div 64 \div 2^3 + 100$. [R.5c]

114. Simplify: $5 \cdot 6 + (7 \cdot 2)^2$. [R.5c]

115. Convert to decimal notation: 58.3%. [R.4a]

116. Simplify: $\frac{164}{256}$. [R.2b]

SYNTHESIS

Subtract.

117. 123,907 − 433,789

118. 23,011 − (−60,432)

Tell whether the statement is true or false for all integers a and b. If false, show why.

119. $a - 0 = 0 - a$

120. $0 - a = a$

121. If $a \neq b$, then $a - b \neq 0$.

122. If $a = -b$, then $a + b = 0$.

123. If $a + b = 0$, then a and b are opposites.

124. If $a - b = 0$, then $a = -b$.

125. *Blackjack Counting System.* The casino game of blackjack makes use of many card-counting systems to give players a winning edge if the count becomes negative. One such system is called *High–Low,* first developed by Harvey Dubner in 1963. Each card counts as −1, 0, or 1 as follows:

2, 3, 4, 5, 6	count as +1;
7, 8, 9	count as 0;
10, J, Q, K, A	count as −1.

Source: Jerry L. Patterson, *Casino Gambling.* New York: Perigee, 1982

a) Find the final count on the sequence of cards

K, A, 2, 4, 5, 10, J, 8, Q, K, 5.

b) Does the player have a winning edge?

Objectives

a. Multiply real numbers.

b. Solve applied problems involving multiplication of real numbers.

1. Complete, as in the example.

$$\begin{aligned} 4 \cdot 10 &= 40 \\ 3 \cdot 10 &= 30 \\ 2 \cdot 10 &= \\ 1 \cdot 10 &= \\ 0 \cdot 10 &= \\ -1 \cdot 10 &= \\ -2 \cdot 10 &= \\ -3 \cdot 10 &= \end{aligned}$$

Multiply.

2. $-3 \cdot 6$

3. $20 \cdot (-5)$

4. $4 \cdot (-20)$

5. $-\frac{2}{3} \cdot \frac{5}{6}$

6. $-4.23(7.1)$

7. $\frac{7}{8}\left(-\frac{4}{5}\right)$

8. Complete, as in the example.

$$\begin{aligned} 3 \cdot (-10) &= -30 \\ 2 \cdot (-10) &= -20 \\ 1 \cdot (-10) &= \\ 0 \cdot (-10) &= \\ -1 \cdot (-10) &= \\ -2 \cdot (-10) &= \\ -3 \cdot (-10) &= \end{aligned}$$

Answers on page A-5

1.5 MULTIPLICATION OF REAL NUMBERS

a Multiplication

Multiplication of real numbers is very much like multiplication of arithmetic numbers. The only difference is that we must determine whether the answer is positive or negative.

MULTIPLICATION OF A POSITIVE NUMBER AND A NEGATIVE NUMBER

To see how to multiply a positive number and a negative number, consider the pattern of the following.

This number decreases by 1 each time. — This number decreases by 5 each time.

$$\begin{aligned} 4 \cdot 5 &= 20 \\ 3 \cdot 5 &= 15 \\ 2 \cdot 5 &= 10 \\ 1 \cdot 5 &= 5 \\ 0 \cdot 5 &= 0 \\ -1 \cdot 5 &= -5 \\ -2 \cdot 5 &= -10 \\ -3 \cdot 5 &= -15 \end{aligned}$$

Do Exercise 1.

According to this pattern, it looks as though the product of a negative number and a positive number is negative. That is the case, and we have the first part of the rule for multiplying numbers.

THE PRODUCT OF A POSITIVE AND A NEGATIVE NUMBER

To multiply a positive number and a negative number, multiply their absolute values. The answer is negative.

EXAMPLES Multiply.

1. $8(-5) = -40$
2. $-\frac{1}{3} \cdot \frac{5}{7} = -\frac{5}{21}$
3. $(-7.2)5 = -36$

Do Exercises 2–7.

MULTIPLICATION OF TWO NEGATIVE NUMBERS

How do we multiply two negative numbers? Again, we look for a pattern.

This number decreases by 1 each time. — This number increases by 5 each time.

$$\begin{aligned} 4 \cdot (-5) &= -20 \\ 3 \cdot (-5) &= -15 \\ 2 \cdot (-5) &= -10 \\ 1 \cdot (-5) &= -5 \\ 0 \cdot (-5) &= 0 \\ -1 \cdot (-5) &= 5 \\ -2 \cdot (-5) &= 10 \\ -3 \cdot (-5) &= 15 \end{aligned}$$

Do Exercise 8 on the preceding page.

According to the pattern, it appears that the product of two negative numbers is positive. That is actually so, and we have the second part of the rule for multiplying real numbers.

THE PRODUCT OF TWO NEGATIVE NUMBERS

To multiply two negative numbers, multiply their absolute values. The answer is positive.

Do Exercises 9–14.

The following is another way to consider the rules we have for multiplication.

To multiply two nonzero real numbers:

a) Multiply the absolute values.

b) If the signs are the same, the answer is positive.

c) If the signs are different, the answer is negative.

MULTIPLICATION BY ZERO

The only case that we have not considered is multiplying by zero. As with other numbers, the product of any real number and 0 is 0.

THE MULTIPLICATION PROPERTY OF ZERO

For any real number a,

$$a \cdot 0 = 0 \cdot a = 0.$$

(The product of 0 and any real number is 0.)

EXAMPLES Multiply.

4. $(-3)(-4) = 12$

5. $-1.6(2) = -3.2$

6. $-19 \cdot 0 = 0$

7. $\left(-\frac{5}{6}\right)\left(-\frac{1}{9}\right) = \frac{5}{54}$

8. $0 \cdot (-452) = 0$

9. $23 \cdot 0 \cdot \left(-8\frac{2}{3}\right) = 0$

Do Exercises 15–20.

Multiply.

9. $-9 \cdot (-3)$

10. $-16 \cdot (-2)$

11. $-7 \cdot (-5)$

12. $-\frac{4}{7}\left(-\frac{5}{9}\right)$

13. $-\frac{3}{2}\left(-\frac{4}{9}\right)$

14. $-3.25(-4.14)$

Multiply.

15. $5(-6)$

16. $(-5)(-6)$

17. $(-3.2) \cdot 0$

18. $\left(-\frac{4}{5}\right)\left(\frac{10}{3}\right)$

19. $0 \cdot (-34.2)$

20. $23 \cdot 0 \cdot \left(-4\frac{2}{3}\right)$

Answers on page A-5

Multiply.

21. $5 \cdot (-3) \cdot 2$

22. $-3 \times (-4.1) \times (-2.5)$

23. $-\frac{1}{2} \cdot \left(-\frac{4}{3}\right) \cdot \left(-\frac{5}{2}\right)$

24. $-2 \cdot (-5) \cdot (-4) \cdot (-3)$

25. $(-4)(-5)(-2)(-3)(-1)$

26. $(-1)(-1)(-2)(-3)(-1)(-1)$

27. Evaluate $(-x)^2$ and $-x^2$ when $x = 2$.

28. Evaluate $(-x)^2$ and $-x^2$ when $x = 3$.

29. Evaluate $3x^2$ when $x = 4$ and when $x = -4$.

Answers on page A-5

MULTIPLYING MORE THAN TWO NUMBERS

When multiplying more than two real numbers, we can choose order and grouping as we please.

EXAMPLES Multiply.

10. $-8 \cdot 2(-3) = -16(-3)$ Multiplying the first two numbers
$= 48$

11. $-8 \cdot 2(-3) = 24 \cdot 2$ Multiplying the negatives. Every pair of negative numbers gives a positive product.
$= 48$

12. $-3(-2)(-5)(4) = 6(-5)(4)$ Multiplying the first two numbers
$= (-30)4$
$= -120$

13. $\left(-\frac{1}{2}\right)(8)\left(-\frac{2}{3}\right)(-6) = (-4)4$ Multiplying the first two numbers and the last two numbers
$= -16$

14. $-5 \cdot (-2) \cdot (-3) \cdot (-6) = 10 \cdot 18$
$= 180$

15. $(-3)(-5)(-2)(-3)(-6) = (-30)(18)$
$= -540$

Considering that the product of a pair of negative numbers is positive, we see the following pattern.

> The product of an even number of negative numbers is positive.
> The product of an odd number of negative numbers is negative.

Do Exercises 21–26.

Let's compare the expressions $(-x)^2$ and $-x^2$.

EXAMPLE 16 Evaluate $(-x)^2$ and $-x^2$ when $x = 5$.

$(-x)^2 = (-5)^2 = (-5)(-5) = 25;$ Substitute 5 for x. Then evaluate the power.

$-x^2 = -(5)^2 = -25$ Substitute 5 for x. Evaluate the power. Then find the opposite.

The expressions $(-x)^2$ and $-x^2$ are *not* equivalent. That is, they do not have the same value for every allowable replacement of the variable by a real number. To find $(-x)^2$, we take the opposite and then square. To find $-x^2$, we find the square and then take the opposite.

Do Exercises 27 and 28.

EXAMPLE 17 Evaluate $2x^2$ when $x = 3$ and $x = -3$.

$2x^2 = 2(3)^2 = 2(9) = 18;$

$2x^2 = 2(-3)^2 = 2(9) = 18$

Do Exercise 29.

b Applications and Problem Solving

We now consider multiplication of real numbers in real-world applications.

EXAMPLE 18 *Chemical Reaction.* During a chemical reaction, the temperature in the beaker decreased by 2°C every minute until 10:23 A.M. If the temperature was 17°C at 10:00 A.M., when the reaction began, what was the temperature at 10:23 A.M.?

This is a multistep problem. We first find the total number of degrees that the temperature dropped, using $-2°$ for each minute. Since it dropped 2° for each of the 23 minutes, we know that the total drop d is given by

$$d = 23 \cdot (-2) = -46.$$

To determine the temperature after this time period, we find the sum of 17 and -46, or

$$T = 17 + (-46) = -29.$$

Thus the temperature at 10:23 A.M. was -29°C.

Do Exercise 30.

30. Chemical Reaction. During a chemical reaction, the temperature in the beaker increased by 3°C every minute until 1:34 P.M. If the temperature was -17°C at 1:10 P.M., when the reaction began, what was the temperature at 1:34 P.M.?

Answer on page A-5

Study Tips

HIGHLIGHTING

Reading and highlighting a section before your instructor lectures on it allows you to maximize your learning and understanding during the lecture.

- **Try to keep one section ahead of your syllabus.** If you study ahead of your lectures, you can concentrate on what is being explained in them, rather than trying to write everything down. You can then take notes only of special points or of questions related to what is happening in class.
- **Highlight important points.** You are probably used to highlighting key points as you study. If that works for you, continue to do so. But you will notice many design features throughout this book that already highlight important points. Thus you may not need to highlight as much as you generally do.
- **Highlight points that you do not understand.** Use a unique mark to indicate trouble spots that can lead to questions to be asked during class, in a tutoring session, or when calling or contacting the AW Math Tutor Center.

1.5 EXERCISE SET

For Extra Help

Digital Video Tutor CD 2 Videotape 4 | InterAct Math | Math Tutor Center | MathXL | MyMathLab

a Multiply.

1. $-4 \cdot 2$

2. $-3 \cdot 5$

3. $-8 \cdot 6$

4. $-5 \cdot 2$

5. $8 \cdot (-3)$

6. $9 \cdot (-5)$

7. $-9 \cdot 8$

8. $-10 \cdot 3$

9. $-8 \cdot (-2)$

10. $-2 \cdot (-5)$

11. $-7 \cdot (-6)$

12. $-9 \cdot (-2)$

13. $15 \cdot (-8)$

14. $-12 \cdot (-10)$

15. $-14 \cdot 17$

16. $-13 \cdot (-15)$

17. $-25 \cdot (-48)$

18. $39 \cdot (-43)$

19. $-3.5 \cdot (-28)$

20. $97 \cdot (-2.1)$

21. $9 \cdot (-8)$

22. $7 \cdot (-9)$

23. $4 \cdot (-3.1)$

24. $3 \cdot (-2.2)$

25. $-5 \cdot (-6)$

26. $-6 \cdot (-4)$

27. $-7 \cdot (-3.1)$

28. $-4 \cdot (-3.2)$

29. $\frac{2}{3} \cdot \left(-\frac{3}{5}\right)$

30. $\frac{5}{7} \cdot \left(-\frac{2}{3}\right)$

31. $-\frac{3}{8} \cdot \left(-\frac{2}{9}\right)$

32. $-\frac{5}{8} \cdot \left(-\frac{2}{5}\right)$

33. -6.3×2.7

34. -4.1×9.5

35. $-\frac{5}{9} \cdot \frac{3}{4}$

36. $-\frac{8}{3} \cdot \frac{9}{4}$

37. $7 \cdot (-4) \cdot (-3) \cdot 5$

38. $9 \cdot (-2) \cdot (-6) \cdot 7$

39. $-\frac{2}{3} \cdot \frac{1}{2} \cdot \left(-\frac{6}{7}\right)$

40. $-\frac{1}{8} \cdot \left(-\frac{1}{4}\right) \cdot \left(-\frac{3}{5}\right)$

41. $-3 \cdot (-4) \cdot (-5)$

42. $-2 \cdot (-5) \cdot (-7)$

43. $-2 \cdot (-5) \cdot (-3) \cdot (-5)$

44. $-3 \cdot (-5) \cdot (-2) \cdot (-1)$

45. $\frac{1}{5}\left(-\frac{2}{9}\right)$

46. $-\frac{3}{5}\left(-\frac{2}{7}\right)$

47. $-7 \cdot (-21) \cdot 13$

48. $-14 \cdot (34) \cdot 12$

49. $-4 \cdot (-1.8) \cdot 7$

50. $-8 \cdot (-1.3) \cdot (-5)$

51. $-\frac{1}{9}\left(-\frac{2}{3}\right)\left(\frac{5}{7}\right)$

52. $-\frac{7}{2}\left(-\frac{5}{7}\right)\left(-\frac{2}{5}\right)$

53. $4 \cdot (-4) \cdot (-5) \cdot (-12)$

54. $-2 \cdot (-3) \cdot (-4) \cdot (-5)$

55. $0.07 \cdot (-7) \cdot 6 \cdot (-6)$

56. $80 \cdot (-0.8) \cdot (-90) \cdot (-0.09)$

57. $\left(-\frac{5}{6}\right)\left(\frac{1}{8}\right)\left(-\frac{3}{7}\right)\left(-\frac{1}{7}\right)$

58. $\left(\frac{4}{5}\right)\left(-\frac{2}{3}\right)\left(-\frac{15}{7}\right)\left(\frac{1}{2}\right)$

59. $(-14) \cdot (-27) \cdot 0$

60. $7 \cdot (-6) \cdot 5 \cdot (-4) \cdot 3 \cdot (-2) \cdot 1 \cdot 0$

61. $(-8)(-9)(-10)$

62. $(-7)(-8)(-9)(-10)$

63. $(-6)(-7)(-8)(-9)(-10)$

64. $(-5)(-6)(-7)(-8)(-9)(-10)$

65. Evaluate $(-3x)^2$ and $-3x^2$ when $x = 7$.

66. Evaluate $(-2x)^2$ and $-2x^2$ when $x = 3$.

67. Evaluate $5x^2$ when $x = 2$ and when $x = -2$.

68. Evaluate $2x^2$ when $x = 5$ and when $x = -5$.

b Solve.

69. *Lost Weight.* Dave lost 2 lb each week for a period of 10 weeks. Express his total weight change as an integer.

70. *Stock Loss.* Michelle lost \$3 each day for a period of 5 days in the value of a stock she owned. Express her total loss as an integer.

71. *Chemical Reaction.* The temperature of a chemical compound was 0°C at 11:00 A.M. During a reaction, it dropped 3°C per minute until 11:18 A.M. What was the temperature at 11:18 A.M.?

72. *Chemical Reaction.* The temperature in a chemical compound was −5°C at 3:20 P.M. During a reaction, it increased 2°C per minute until 3:52 P.M. What was the temperature at 3:52 P.M.?

73. *Stock Price.* The price of ePDQ.com began the day at \$23.75 per share and dropped \$1.38 per hour for 8 hr. What was the price of the stock after 8 hr?

74. *Population Decrease.* The population of a rural town was 12,500. It decreased 380 each year for 4 yr. What was the population of the town after 4 yr?

75. *Diver's Position.* After diving 95 m below the sea level, a diver rises at a rate of 7 meters per minute for 9 min. Where is the diver in relation to the surface?

76. *Checking Account Balance.* Karen had \$68 in her checking account. After writing checks to make seven purchases at \$13 each, what was the balance in her checking account?

77. ^{D}W Multiplication can be thought of as repeated addition. Using this concept and a number line, explain why $3 \cdot (-5) = -15$.

78. ^{D}W What rule have we developed that would tell you the sign of $(-7)^8$ and $(-7)^{11}$ without doing the computations? Explain.

SKILL MAINTENANCE

79. Find the LCM of 36 and 60. [R.1b]

80. Find the prime factorization of 4608. [R.1a]

Simplify. [R.2b]

81. $\frac{26}{39}$

82. $\frac{48}{54}$

83. $\frac{264}{484}$

84. $\frac{1025}{6625}$

85. $\frac{275}{800}$

86. $\frac{111}{201}$

87. $\frac{11}{264}$

88. $\frac{78}{13}$

SYNTHESIS

For each of Exercises 89 and 90, choose the correct answer from the selections given.

89. If a is positive and b is negative, then $-ab$ is:

a) Positive.
b) Negative.
c) 0.
d) Cannot be determined without more information

90. If a is positive and b is negative, then $(-a)(-b)$ is:

a) Positive.
b) Negative.
c) 0.
d) Cannot be determined without more information

91. Below is a number line showing 0 and two positive numbers x and y. Use a compass or ruler and locate as best you can the following:

$2x,\quad 3x,\quad 2y,\quad -x,\quad -y,\quad x+y,\quad x-y,\quad x-2y.$

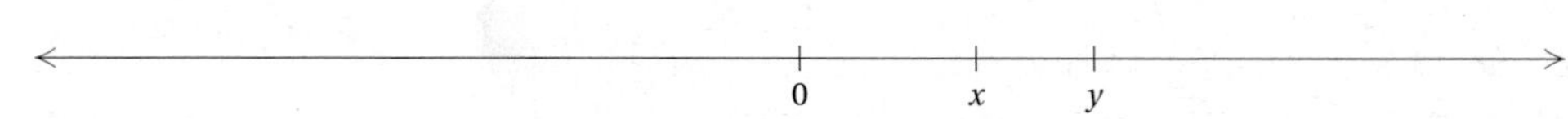

92. Below is a number line showing 0 and two negative numbers x and y. Use a compass or ruler and locate as best you can the following:

$2x,\quad 3x,\quad -x,\quad -y,\quad -3y,\quad x+y,\quad x-y,\quad 2x-y.$

1.6 DIVISION OF REAL NUMBERS

Objectives

a Divide integers.

b Find the reciprocal of a real number.

c Divide real numbers.

d Solve applied problems involving division of real numbers.

We now consider division of real numbers. The definition of division results in rules for division that are the same as those for multiplication.

a Division of Integers

DIVISION

The quotient $a \div b$, or $\frac{a}{b}$, where $b \neq 0$, is that unique real number c for which $a = b \cdot c$.

Let's use the definition to divide integers.

EXAMPLES Divide, if possible. Check your answer.

1. $14 \div (-7) = -2$ — *Think*: What number multiplied by -7 gives 14? That number is -2. *Check*: $(-2)(-7) = 14$.

2. $\frac{-32}{-4} = 8$ — *Think*: What number multiplied by -4 gives -32? That number is 8. *Check*: $8(-4) = -32$.

3. $\frac{-10}{7} = -\frac{10}{7}$ — *Think*: What number multiplied by 7 gives -10? That number is $-\frac{10}{7}$. *Check*: $-\frac{10}{7} \cdot 7 = -10$.

4. $\frac{-17}{0}$ is **not defined.** — *Think*: What number multiplied by 0 gives -17? There is no such number because the product of 0 and *any* number is 0.

The rules for division are the same as those for multiplication.

To multiply or divide two real numbers (where the divisor is nonzero):

a) Multiply or divide the absolute values.
b) If the signs are the same, the answer is positive.
c) If the signs are different, the answer is negative.

Do Exercises 1–6.

EXCLUDING DIVISION BY 0

Example 4 shows why we cannot divide -17 by 0. We can use the same argument to show why we cannot divide any nonzero number b by 0. Consider $b \div 0$. We look for a number that when multiplied by 0 gives b. There is no such number because the product of 0 and any number is 0. Thus we cannot divide a nonzero number b by 0.

On the other hand, if we divide 0 by 0, we look for a number c such that $0 \cdot c = 0$. But $0 \cdot c = 0$ for any number c. Thus it appears that $0 \div 0$ could be any number we choose. Getting any answer we want when we divide 0 by 0 would be very confusing. Thus we agree that division by zero is not defined.

Divide.

1. $6 \div (-3)$

   Think: What number multiplied by -3 gives 6?

2. $\frac{-15}{-3}$

   Think: What number multiplied by -3 gives -15?

3. $-24 \div 8$

   Think: What number multiplied by 8 gives -24?

4. $\frac{-48}{-6}$

5. $\frac{30}{-5}$

6. $\frac{30}{-7}$

Answers on page A-5

Divide, if possible.

7. $\frac{-5}{0}$

8. $\frac{0}{-3}$

Find the reciprocal.

9. $\frac{2}{3}$

10. $-\frac{5}{4}$

11. -3

12. $-\frac{1}{5}$

13. 1.6

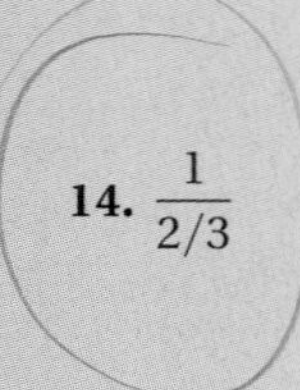

14. $\frac{1}{2/3}$

Answers on page A-5

EXCLUDING DIVISION BY 0

Division by 0 is not defined.

$$a \div 0, \text{ or } \frac{a}{0}, \text{ is not defined for all real numbers } a.$$

DIVIDING 0 BY OTHER NUMBERS

Note that

$$0 \div 8 = 0 \text{ because } 0 = 0 \cdot 8; \qquad \frac{0}{-5} = 0 \text{ because } 0 = 0 \cdot (-5).$$

DIVIDENDS OF 0

Zero divided by any nonzero real number is 0:

$$\frac{0}{a} = 0; \qquad a \neq 0.$$

EXAMPLES Divide.

5. $0 \div (-6) = 0$

6. $\frac{0}{12} = 0$

7. $\frac{-3}{0}$ is not defined.

Do Exercises 7 and 8.

b Reciprocals

When two numbers like $\frac{1}{2}$ and 2 are multiplied, the result is 1. Such numbers are called **reciprocals** of each other. Every nonzero real number has a reciprocal, also called a **multiplicative inverse.**

RECIPROCALS

Two numbers whose product is 1 are called **reciprocals,** or **multiplicative inverses,** of each other.

EXAMPLES Find the reciprocal.

8. $\frac{7}{8}$ The reciprocal of $\frac{7}{8}$ is $\frac{8}{7}$ because $\frac{7}{8} \cdot \frac{8}{7} = 1$.

9. -5 The reciprocal of -5 is $-\frac{1}{5}$ because $-5\left(-\frac{1}{5}\right) = 1$.

10. 3.9 The reciprocal of 3.9 is $\frac{1}{3.9}$ because $3.9\left(\frac{1}{3.9}\right) = 1$.

11. $-\frac{1}{2}$ The reciprocal of $-\frac{1}{2}$ is -2 because $\left(-\frac{1}{2}\right)(-2) = 1$.

12. $-\frac{2}{3}$ The reciprocal of $-\frac{2}{3}$ is $-\frac{3}{2}$ because $\left(-\frac{2}{3}\right)\left(-\frac{3}{2}\right) = 1$.

13. $\frac{1}{3/4}$ The reciprocal of $\frac{1}{3/4}$ is $\frac{3}{4}$ because $\left(\frac{1}{3/4}\right)\left(\frac{3}{4}\right) = 1$.

RECIPROCAL PROPERTIES

For $a \neq 0$, the reciprocal of a can be named $\frac{1}{a}$ and the reciprocal of $\frac{1}{a}$ is a.

The reciprocal of a nonzero number $\frac{a}{b}$ can be named $\frac{b}{a}$.

The number 0 has no reciprocal.

Do Exercises 9–14 on the preceding page.

The reciprocal of a positive number is also a positive number, because their product must be the positive number 1. The reciprocal of a negative number is also a negative number, because their product must be the positive number 1.

THE SIGN OF A RECIPROCAL

The reciprocal of a number has the same sign as the number itself.

CAUTION!

It is important *not* to confuse *opposite* with *reciprocal*. Keep in mind that the opposite, or additive inverse, of a number is what we add to the number to get 0. The reciprocal, or multiplicative inverse, is what we multiply the number by to get 1.

Compare the following.

NUMBER	OPPOSITE (Change the sign.)	RECIPROCAL (Invert but do not change the sign.)
$-\frac{3}{8}$	$\frac{3}{8}$	$-\frac{8}{3}$
19	-19	$\frac{1}{19}$
$\frac{18}{7}$	$-\frac{18}{7}$	$\frac{7}{18}$
-7.9	7.9	$-\frac{1}{7.9}$, or $-\frac{10}{79}$
0	0	Not defined

$\left(-\frac{3}{8}\right)\left(-\frac{8}{3}\right) = 1$

$-\frac{3}{8} + \frac{3}{8} = 0$

Do Exercise 15.

15. Complete the following table.

NUMBER	OPPOSITE	RECIPROCAL
$\frac{2}{3}$		
$-\frac{5}{4}$		
0		
1		
-8		
-4.5		

Answers on page A-5

Study Tips

TAKE THE TIME!

The foundation of all your study skills is *time*! If you invest your time, we will help you achieve success.

"Nine-tenths of wisdom is being wise in time."

Theodore Roosevelt

Rewrite the division as a multiplication.

16. $\frac{4}{7} \div \left(-\frac{3}{5}\right)$

17. $\frac{5}{-8}$

18. $\frac{a-b}{7}$

19. $\frac{-23}{1/a}$

20. $-5 \div 7$

Divide by multiplying by the reciprocal of the divisor.

21. $\frac{4}{7} \div \left(-\frac{3}{5}\right)$

22. $-\frac{8}{5} \div \frac{2}{3}$

23. $-\frac{12}{7} \div \left(-\frac{3}{4}\right)$

24. Divide: $21.7 \div (-3.1)$.

Answers on page A-5

C Division of Real Numbers

We know that we can subtract by adding an opposite. Similarly, we can divide by multiplying by a reciprocal.

RECIPROCALS AND DIVISION

For any real numbers a and b, $b \neq 0$,

$$a \div b = \frac{a}{b} = a \cdot \frac{1}{b}.$$

(To divide, multiply by the reciprocal of the divisor.)

EXAMPLES Rewrite the division as a multiplication.

14. $-4 \div 3$ $\quad$ $-4 \div 3$ is the same as $-4 \cdot \frac{1}{3}$

15. $\frac{6}{-7}$ $\quad$ $\frac{6}{-7} = 6\left(-\frac{1}{7}\right)$

16. $\frac{x+2}{5}$ $\quad$ $\frac{x+2}{5} = (x+2)\frac{1}{5}$ $\quad$ Parentheses are necessary here.

17. $\frac{-17}{1/b}$ $\quad$ $\frac{-17}{1/b} = -17 \cdot b$

18. $\frac{3}{5} \div \left(-\frac{9}{7}\right)$ $\quad$ $\frac{3}{5} \div \left(-\frac{9}{7}\right) = \frac{3}{5}\left(-\frac{7}{9}\right)$

Do Exercises 16–20.

When actually doing division calculations, we sometimes multiply by a reciprocal and we sometimes divide directly. With fraction notation, it is usually better to multiply by a reciprocal. With decimal notation, it is usually better to divide directly.

EXAMPLES Divide by multiplying by the reciprocal of the divisor.

19. $\frac{2}{3} \div \left(-\frac{5}{4}\right) = \frac{2}{3} \cdot \left(-\frac{4}{5}\right) = -\frac{8}{15}$

20. $-\frac{5}{6} \div \left(-\frac{3}{4}\right) = -\frac{5}{6} \cdot \left(-\frac{4}{3}\right) = \frac{20}{18} = \frac{10 \cdot 2}{9 \cdot 2} = \frac{10}{9} \cdot \frac{2}{2} = \frac{10}{9}$

CAUTION!

Be careful not to change the sign when taking a reciprocal!

21. $-\frac{3}{4} \div \frac{3}{10} = -\frac{3}{4} \cdot \left(\frac{10}{3}\right) = -\frac{30}{12} = -\frac{5}{2} \cdot \frac{6}{6} = -\frac{5}{2}$

With decimal notation, it is easier to carry out long division than to multiply by the reciprocal.

EXAMPLES Divide.

22. $-27.9 \div (-3) = \frac{-27.9}{-3} = 9.3$ Do the long division $3\overline{)27.9}$ (quotient 9.3). The answer is positive.

23. $-6.3 \div 2.1 = -3$ Do the long division $2.1\overline{)6.3}$ (quotient 3.). The answer is negative.

Do Exercises 21–24 on the preceding page.

Consider the following:

1. $\frac{2}{3} = \frac{2}{3} \cdot 1 = \frac{2}{3} \cdot \frac{-1}{-1} = \frac{2(-1)}{3(-1)} = \frac{-2}{-3}$. Thus, $\frac{2}{3} = \frac{-2}{-3}$.
   (A negative number divided by a negative number is positive.)

2. $-\frac{2}{3} = -1 \cdot \frac{2}{3} = \frac{-1}{1} \cdot \frac{2}{3} = \frac{-1 \cdot 2}{1 \cdot 3} = \frac{-2}{3}$. Thus, $-\frac{2}{3} = \frac{-2}{3}$.
   (A negative number divided by a positive number is negative.)

3. $\frac{-2}{3} = \frac{-2}{3} \cdot 1 = \frac{-2}{3} \cdot \frac{-1}{-1} = \frac{-2(-1)}{3(-1)} = \frac{2}{-3}$. Thus, $-\frac{2}{3} = \frac{2}{-3}$.
   (A positive number divided by a negative number is negative.)

We can use the following properties to make sign changes in fraction notation.

SIGN CHANGES IN FRACTION NOTATION

For any numbers a and b, $b \neq 0$:

1. $\frac{-a}{-b} = \frac{a}{b}$
   (The opposite of a number a divided by the opposite of another number b is the same as the quotient of the two numbers a and b.)

2. $\frac{-a}{b} = \frac{a}{-b} = -\frac{a}{b}$
   (The opposite of a number a divided by another number b is the same as the number a divided by the opposite of the number b, and both are the same as the opposite of *a divided by b*.)

Do Exercises 25–27.

Find two equal expressions for the number with negative signs in different places.

25. $\frac{-5}{6}$

26. $-\frac{8}{7}$

27. $\frac{10}{-3}$

Answers on page A-5

d Applications and Problem Solving

EXAMPLE 24 *Chemical Reaction.* During a chemical reaction, the temperature in the beaker decreased every minute by the same number of degrees. The temperature was 56°F at 10:10 A.M. By 10:42 A.M., the temperature had dropped to −12°F. By how many degrees did it change each minute?

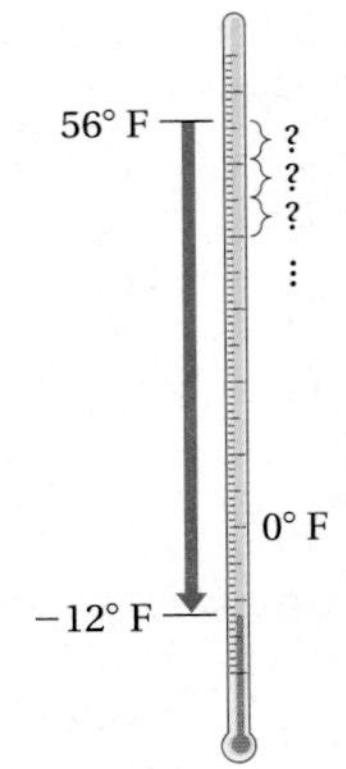

We first determine by how many degrees d the temperature changed altogether. We subtract −12 from 56:

$$d = 56 - (-12) = 56 + 12 = 68.$$

The temperature changed a total of 68°. We can express this as −68° since the temperature dropped.

The amount of time t that passed was 42 − 10, or 32 min. Thus the number of degrees T that the temperature dropped each minute is given by

$$T = \frac{d}{t} = \frac{-68}{32} = -2.125.$$

The change was −2.125°F per minute.

Do Exercise 28.

28. Chemical Reaction. During a chemical reaction, the temperature in the beaker decreased every minute by the same number of degrees. The temperature was 71°F at 2:12 P.M. By 2:37 P.M., the temperature had changed to −14°F. By how many degrees did it change each minute?

Answer on page A-5

CALCULATOR CORNER

Operations on the Real Numbers We can perform operations on the real numbers on a graphing calculator. Recall that negative numbers are entered using the opposite key, [(−)], rather than the subtraction operation key, [−]. Consider the sum −5 + (−3.8). We use parentheses when we write this sum in order to separate the addition symbol and the "opposite of" symbol and thus make the expression more easily read. When we enter this calculation on a graphing calculator, however, the parentheses are not necessary. We can press [(−)] [5] [+] [(−)] [3] [.] [8] [ENTER]. The result is −8.8. Note that it is not incorrect to enter the parentheses. The result will be the same if this is done.

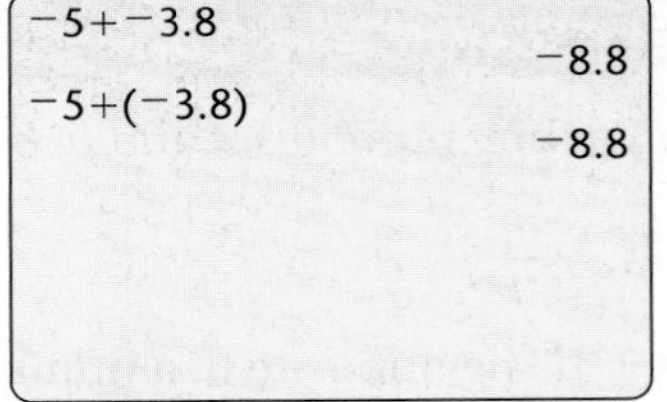

To find the difference 10 − (−17), we press [1] [0] [−] [(−)] [1] [7] [ENTER]. The result is 27. We can also multiply and divide real numbers. To find −5 · (−7), we press [(−)] [5] [×] [(−)] [7] [ENTER], and to find 45 ÷ (−9), we press [4] [5] [÷] [(−)] [9] [ENTER]. Note that it is not necessary to use parentheses in any of these calculations.

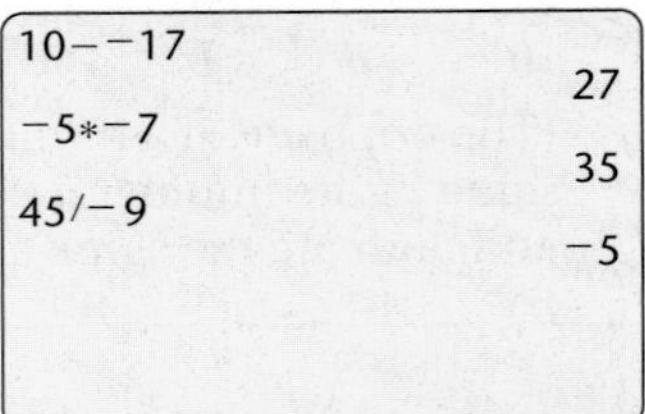

Exercises: Use a calculator to perform the operation.

1. −8 + 4
2. 1.2 + (−1.5)
3. −7 + (−5)
4. −7.6 + (−1.9)
5. −8 − 4
6. 1.2 − (−1.5)
7. −7 − (−5)
8. −7.6 − (−1.9)
9. −8 · 4
10. 1.2 · (−1.5)
11. −7 · (−5)
12. −7.6 · (−1.9)
13. −8 ÷ 4
14. 1.2 ÷ (−1.5)
15. −7 ÷ (−5)
16. −7.6 ÷ (−1.9)

1.6 EXERCISE SET

For Extra Help: Digital Video Tutor CD 2 Videotape 4 · InterAct Math · Math Tutor Center · MathXL · MyMathLab

a Divide, if possible. Check each answer.

1. $48 \div (-6)$

2. $\frac{42}{-7}$

3. $\frac{28}{-2}$

4. $24 \div (-12)$

5. $\frac{-24}{8}$

6. $-18 \div (-2)$

7. $\frac{-36}{-12}$

8. $-72 \div (-9)$

9. $\frac{-72}{9}$

10. $\frac{-50}{25}$

11. $-100 \div (-50)$

12. $\frac{-200}{8}$

13. $-108 \div 9$

14. $\frac{-63}{-7}$

15. $\frac{200}{-25}$

16. $-300 \div (-16)$

17. $\frac{75}{0}$

18. $\frac{0}{-5}$

19. $\frac{-23}{-2}$

20. $\frac{-23}{0}$

b Find the reciprocal.

21. $\frac{15}{7}$

22. $\frac{3}{8}$

23. $-\frac{47}{13}$

24. $-\frac{31}{12}$

25. 13

26. -10

27. 4.3

28. -8.5

29. $\frac{1}{-7.1}$

30. $\frac{1}{-4.9}$

31. $\frac{p}{q}$

32. $\frac{s}{t}$

33. $\frac{1}{4y}$

34. $\frac{-1}{8a}$

35. $\frac{2a}{3b}$

36. $\frac{-4y}{3x}$

why?

C Rewrite the division as a multiplication.

37. $4 \div 17$

38. $5 \div (-8)$

39. $\frac{8}{-13}$

40. $-\frac{13}{47}$

41. $\frac{13.9}{-1.5}$

42. $-\frac{47.3}{21.4}$

43. $\frac{x}{\frac{1}{y}}$

44. $\frac{13}{x}$

45. $\frac{3x + 4}{5}$

46. $\frac{4y - 8}{-7}$

47. $\frac{5a - b}{5a + b}$

48. $\frac{2x + x^2}{x - 5}$

Divide.

49. $\frac{3}{4} \div \left(-\frac{2}{3}\right)$

50. $\frac{7}{8} \div \left(-\frac{1}{2}\right)$

51. $-\frac{5}{4} \div \left(-\frac{3}{4}\right)$

52. $-\frac{5}{9} \div \left(-\frac{5}{6}\right)$

53. $-\frac{2}{7} \div \left(-\frac{4}{9}\right)$

54. $-\frac{3}{5} \div \left(-\frac{5}{8}\right)$

55. $-\frac{3}{8} \div \left(-\frac{8}{3}\right)$

56. $-\frac{5}{8} \div \left(-\frac{6}{5}\right)$

why?

57. $-6.6 \div 3.3$

58. $-44.1 \div (-6.3)$

59. $\frac{-11}{-13}$

60. $\frac{-1.9}{20}$

61. $\frac{48.6}{-3}$

62. $\frac{-17.8}{3.2}$

63. $\frac{-9}{17 - 17}$

64. $\frac{-8}{-5 + 5}$

d *Percent of Increase or Decrease in Employment.* A percent of increase is generally positive and a percent of decrease is generally negative. The following table lists estimates of the number of job opportunities for various occupations in 1998 and 2008. In Exercises 65–68, find the missing numbers.

	OCCUPATION	NUMBER OF JOBS IN 1998 (in thousands)	NUMBER OF JOBS IN 2008 (in thousands)	CHANGE	PERCENT OF INCREASE OR DECREASE
	Court clerk	100	112	12	12%
	Bank teller	560	529	−31	−5.5%
65.	Barber	54	50	−4	
66.	Child-care worker in private household	306	209	−97	
67.	Dental assistant	229	326	97	
68.	Cook (short-order and fast-food)	677	801	124	

Source: Handbook of U.S. Labor Statistics

69. $^{D}_{W}$ Explain how multiplication can be used to justify why a negative number divided by a positive number is negative.

70. $^{D}_{W}$ Explain how multiplication can be used to justify why a negative number divided by a negative number is positive.

SKILL MAINTENANCE

Simplify. [R.5c]

71. $2^3 - 5 \cdot 3 + 8 \cdot 10 \div 2$

72. $16 \cdot 2^3 - 5 \cdot 3 + 80 \div 10 \cdot 2$

73. $1000 \div 100 \div 10$

74. $216 \cdot 6^3 \div 6^2$

75. Simplify: $\frac{264}{468}$. [R.2b]

76. Convert to decimal notation: 47.7%. [R.4a]

77. Convert to percent notation: $\frac{7}{8}$. [R.4d]

78. Simplify: $\frac{40}{60}$. [R.2b]

79. Divide and simplify: $\frac{12}{25} \div \frac{32}{75}$. [R.2c]

80. Multiply and simplify: $\frac{12}{25} \cdot \frac{32}{75}$. [R.2c]

SYNTHESIS

81. Find the reciprocal of -10.5. What happens if you take the reciprocal of the result?

82. Determine those real numbers a for which the opposite of a is the same as the reciprocal of a.

Tell whether the expression represents a positive number or a negative number when a and b are negative.

83. $\frac{-a}{b}$

84. $\frac{-a}{-b}$

85. $-\left(\frac{a}{-b}\right)$

86. $-\left(\frac{-a}{b}\right)$

87. $-\left(\frac{-a}{-b}\right)$

1.7 PROPERTIES OF REAL NUMBERS

Objectives

a Find equivalent fractional expressions and simplify fractional expressions.

b Use the commutative and associative laws to find equivalent expressions.

c Use the distributive laws to multiply expressions like 8 and $x - y$.

d Use the distributive laws to factor expressions like $4x - 12 + 24y$.

e Collect like terms.

Complete the table by evaluating each expression for the given values.

1.

VALUE	$x + x$	$2x$
$x = 3$		
$x = -6$		
$x = 4.8$		

2.

VALUE	$x + 3x$	$5x$
$x = 2$		
$x = -6$		
$x = 4.8$		

Answers on page A-6

a Equivalent Expressions

In solving equations and doing other kinds of work in algebra, we manipulate expressions in various ways. For example, instead of

$$x + x,$$

we might write

$$2x,$$

knowing that the two expressions represent the same number for any allowable replacement of x. In that sense, the expressions $x + x$ and $2x$ are **equivalent,** as are $\frac{3}{x}$ and $\frac{3x}{x^2}$, even though 0 is not an allowable replacement because division by 0 is not defined.

EQUIVALENT EXPRESSIONS

Two expressions that have the same value for all allowable replacements are called **equivalent.**

The expressions $x + 3x$ and $5x$ are *not* equivalent.

Do Exercises 1 and 2.

In this section, we will consider several laws of real numbers that will allow us to find equivalent expressions. The first two laws are the *identity properties of 0 and 1.*

THE IDENTITY PROPERTY OF 0

For any real number a,

$$a + 0 = 0 + a = a.$$

(The number 0 is the *additive identity.*)

THE IDENTITY PROPERTY OF 1

For any real number a,

$$a \cdot 1 = 1 \cdot a = a.$$

(The number 1 is the *multiplicative identity.*)

We often refer to the use of the identity property of 1 as "multiplying by 1." We can use this method to find equivalent fractional expressions. Recall from arithmetic that to multiply with fraction notation, we multiply numerators and denominators. (See also Section R.2.)

EXAMPLE 1 Write a fractional expression equivalent to $\frac{2}{3}$ with a denominator of $3x$.

Note that $3x = 3 \cdot x$. We want fraction notation for $\frac{2}{3}$ that has a denominator of $3x$, but the denominator 3 is missing a factor of x. Thus we multiply by 1, using x/x as an equivalent expression for 1:

$$\frac{2}{3} = \frac{2}{3} \cdot 1 = \frac{2}{3} \cdot \frac{x}{x} = \frac{2x}{3x}.$$

The expressions $2/3$ and $2x/3x$ are equivalent. They have the same value for any allowable replacement. Note that $2x/3x$ is not defined for a replacement of 0, but for all nonzero real numbers, the expressions $2/3$ and $2x/3x$ have the same value.

Do Exercises 3 and 4.

In algebra, we consider an expression like $2/3$ to be "simplified" from $2x/3x$. To find such simplified expressions, we use the identity property of 1 to remove a factor of 1. (See also Section R.2.)

EXAMPLE 2 Simplify: $-\dfrac{20x}{12x}$.

$$-\frac{20x}{12x} = -\frac{5 \cdot 4x}{3 \cdot 4x}$$ We look for the largest factor common to both the numerator and the denominator and factor each.

$$= -\frac{5}{3} \cdot \frac{4x}{4x}$$ Factoring the fractional expression

$$= -\frac{5}{3} \cdot 1$$ $\frac{4x}{4x} = 1$

$$= -\frac{5}{3}$$ Removing a factor of 1 using the identity property of 1

Do Exercises 5 and 6.

b The Commutative and Associative Laws

Let's examine the expressions $x + y$ and $y + x$, as well as xy and yx.

EXAMPLE 3 Evaluate $x + y$ and $y + x$ when $x = 4$ and $y = 3$.

We substitute 4 for x and 3 for y in both expressions:

$$x + y = 4 + 3 = 7; \qquad y + x = 3 + 4 = 7.$$

EXAMPLE 4 Evaluate xy and yx when $x = 23$ and $y = 12$.

We substitute 23 for x and 12 for y in both expressions:

$$xy = 23 \cdot 12 = 276; \qquad yx = 12 \cdot 23 = 276.$$

Do Exercises 7 and 8.

3. Write a fractional expression equivalent to $\frac{3}{4}$ with a denominator of 8.

4. Write a fractional expression equivalent to $\frac{3}{4}$ with a denominator of $4t$.

Simplify.

5. $\dfrac{3y}{4y}$

6. $-\dfrac{16m}{12m}$

7. Evaluate $x + y$ and $y + x$ when $x = -2$ and $y = 3$.

8. Evaluate xy and yx when $x = -2$ and $y = 5$.

Answers on page A-6

Use a commutative law to write an equivalent expression.

9. $x + 9$

10. pq

11. $xy + t$

Answers on page A-6

Note that the expressions

$$x + y \quad \text{and} \quad y + x$$

have the same values no matter what the variables stand for. Thus they are equivalent. Therefore, when we add two numbers, the order in which we add does not matter. Similarly, the expressions xy and yx are equivalent. They also have the same values, no matter what the variables stand for. Therefore, when we multiply two numbers, the order in which we multiply does not matter.

The following are examples of general patterns or laws.

THE COMMUTATIVE LAWS

Addition. For any numbers a and b,

$$a + b = b + a.$$

(We can change the order when adding without affecting the answer.)

Multiplication. For any numbers a and b,

$$ab = ba.$$

(We can change the order when multiplying without affecting the answer.)

Using a commutative law, we know that $x + 2$ and $2 + x$ are equivalent. Similarly, $3x$ and $x(3)$ are equivalent. Thus, in an algebraic expression, we can replace one with the other and the result will be equivalent to the original expression.

EXAMPLE 5 Use the commutative laws to write an expression equivalent to $y + 5$, ab, and $7 + xy$.

An expression equivalent to $y + 5$ is $5 + y$ by the commutative law of addition.

An expression equivalent to ab is ba by the commutative law of multiplication.

An expression equivalent to $7 + xy$ is $xy + 7$ by the commutative law of addition. Another expression equivalent to $7 + xy$ is $7 + yx$ by the commutative law of multiplication.

Do Exercises 9–11.

THE ASSOCIATIVE LAWS

Now let's examine the expressions $a + (b + c)$ and $(a + b) + c$. Note that these expressions involve the use of parentheses as *grouping* symbols, and they also involve three numbers. Calculations within parentheses are to be done first.

EXAMPLE 6 Calculate and compare: $3 + (8 + 5)$ and $(3 + 8) + 5$.

$3 + (8 + 5) = 3 + 13$ Calculating within parentheses first; adding the 8 and 5

$= 16;$

$(3 + 8) + 5 = 11 + 5$ Calculating within parentheses first; adding the 3 and 8

$= 16$

The two expressions in Example 6 name the same number. Moving the parentheses to group the additions differently does not affect the value of the expression.

EXAMPLE 7 Calculate and compare: $3 \cdot (4 \cdot 2)$ and $(3 \cdot 4) \cdot 2$.

$$3 \cdot (4 \cdot 2) = 3 \cdot 8 = 24; \qquad (3 \cdot 4) \cdot 2 = 12 \cdot 2 = 24$$

Do Exercises 12 and 13.

You may have noted that when only addition is involved, parentheses can be placed any way we please without affecting the answer. When only multiplication is involved, parentheses also can be placed any way we please without affecting the answer.

THE ASSOCIATIVE LAWS

Addition. For any numbers a, b, and c,

$$a + (b + c) = (a + b) + c.$$

(Numbers can be grouped in any manner for addition.)

Multiplication. For any numbers a, b, and c,

$$a \cdot (b \cdot c) = (a \cdot b) \cdot c.$$

(Numbers can be grouped in any manner for multiplication.)

EXAMPLE 8 Use an associative law to write an expression equivalent to $(y + z) + 3$ and $8(xy)$.

An expression equivalent to $(y + z) + 3$ is $y + (z + 3)$ by the associative law of addition.

An expression equivalent to $8(xy)$ is $(8x)y$ by the associative law of multiplication.

Do Exercises 14 and 15.

The associative laws say parentheses can be placed any way we please when only additions or only multiplications are involved. Thus we often omit them. For example,

$$x + (y + 2) \quad \text{means} \quad x + y + 2, \qquad \text{and} \qquad (lw)h \quad \text{means} \quad lwh.$$

USING THE COMMUTATIVE AND ASSOCIATIVE LAWS TOGETHER

EXAMPLE 9 Use the commutative and associative laws to write at least three expressions equivalent to $(x + 5) + y$.

a) $(x + 5) + y = x + (5 + y)$ Using the associative law first and then using the commutative law

$= x + (y + 5)$

b) $(x + 5) + y = y + (x + 5)$ Using the commutative law first and then the commutative law again

$= y + (5 + x)$

c) $(x + 5) + y = (5 + x) + y$ Using the commutative law first and then the associative law

$= 5 + (x + y)$

12. Calculate and compare:

$8 + (9 + 2)$ and $(8 + 9) + 2$.

13. Calculate and compare:

$10 \cdot (5 \cdot 3)$ and $(10 \cdot 5) \cdot 3$.

Use an associative law to write an equivalent expression.

14. $r + (s + 7)$

15. $9(ab)$

Answers on page A-6

Use the commutative and associative laws to write at least three equivalent expressions.

16. $4(tu)$

17. $r + (2 + s)$

Compute.

18. a) $7 \cdot (3 + 6)$

b) $(7 \cdot 3) + (7 \cdot 6)$

19. a) $2 \cdot (10 + 30)$

b) $(2 \cdot 10) + (2 \cdot 30)$

20. a) $(2 + 5) \cdot 4$

b) $(2 \cdot 4) + (5 \cdot 4)$

Answers on page A-6

EXAMPLE 10 Use the commutative and associative laws to write at least three expressions equivalent to $(3x)y$.

a) $(3x)y = 3(xy)$ Using the associative law first and then using the commutative law

$= 3(yx)$

b) $(3x)y = y(3x)$ Using the commutative law twice

$= y(x3)$

c) $(3x)y = (x3)y$ Using the commutative law, and then the associative law, and then the commutative law again

$= x(3y)$

$= x(y3)$

Do Exercises 16 and 17.

C The Distributive Laws

The *distributive laws* are the basis of many procedures in both arithmetic and algebra. They are probably the most important laws that we use to manipulate algebraic expressions. The distributive law of multiplication over addition involves two operations: addition and multiplication.

Let's begin by considering a multiplication problem from arithmetic:

$$\begin{array}{r l} 45 & \\ \times\ \ 7 & \\ \hline 35 & \leftarrow \text{This is } 7 \cdot 5. \\ 280 & \leftarrow \text{This is } 7 \cdot 40. \\ \hline 315 & \leftarrow \text{This is the sum } 7 \cdot 40 + 7 \cdot 5. \end{array}$$

To carry out the multiplication, we actually added two products. That is,

$$7 \cdot 45 = 7(40 + 5) = 7 \cdot 40 + 7 \cdot 5.$$

Let's examine this further. If we wish to multiply a sum of several numbers by a factor, we can either add and then multiply, or multiply and then add.

EXAMPLE 11 Compute in two ways: $5 \cdot (4 + 8)$.

a) $5 \cdot (4 + 8)$ Adding within parentheses first, and then multiplying

$= 5 \cdot 12$

$= 60$

b) $(5 \cdot 4) + (5 \cdot 8)$ Distributing the multiplication to terms within parentheses first and then adding

$= 20 + 40$

$= 60$

Do Exercises 18–20.

THE DISTRIBUTIVE LAW OF MULTIPLICATION OVER ADDITION

For any numbers a, b, and c,

$$a(b + c) = ab + ac.$$

In the statement of the distributive law, we know that in an expression such as $ab + ac$, the multiplications are to be done first according to the rules for order of operations. (See Section R.5.) So, instead of writing $(4 \cdot 5) + (4 \cdot 7)$, we can write $4 \cdot 5 + 4 \cdot 7$. However, in $a(b + c)$, we cannot omit the parentheses. If we did, we would have $ab + c$, which means $(ab) + c$. For example, $3(4 + 2) = 18$, but $3 \cdot 4 + 2 = 14$.

There is another distributive law that relates multiplication and subtraction. This law says that to multiply by a difference, we can either subtract and then multiply, or multiply and then subtract.

THE DISTRIBUTIVE LAW OF MULTIPLICATION OVER SUBTRACTION

For any numbers a, b, and c,

$$a(b - c) = ab - ac.$$

We often refer to "*the* distributive law" when we mean *either* or *both* of these laws.

Do Exercises 21–23.

What do we mean by the *terms* of an expression? **Terms** are separated by addition signs. If there are subtraction signs, we can find an equivalent expression that uses addition signs.

EXAMPLE 12 What are the terms of $3x - 4y + 2z$?

We have

$$3x - 4y + 2z = 3x + (-4y) + 2z. \quad \text{Separating parts with + signs}$$

The terms are $3x$, $-4y$, and $2z$.

Do Exercises 24 and 25.

The distributive laws are a basis for a procedure in algebra called **multiplying.** In an expression like $8(a + 2b - 7)$, we multiply each term inside the parentheses by 8:

$$8(a + 2b - 7) = 8 \cdot a + 8 \cdot 2b - 8 \cdot 7 = 8a + 16b - 56.$$

EXAMPLES Multiply.

13. $9(x - 5) = 9x - 9(5)$ Using the distributive law of multiplication over subtraction

$= 9x - 45$

14. $\frac{2}{3}(w + 1) = \frac{2}{3} \cdot w + \frac{2}{3} \cdot 1$ Using the distributive law of multiplication over addition

$= \frac{2}{3}w + \frac{2}{3}$

15. $\frac{4}{3}(s - t + w) = \frac{4}{3}s - \frac{4}{3}t + \frac{4}{3}w$ Using both distributive laws

Do Exercises 26–28.

Calculate.

21. a) $4(5 - 3)$

b) $4 \cdot 5 - 4 \cdot 3$

22. a) $-2 \cdot (5 - 3)$

b) $-2 \cdot 5 - (-2) \cdot 3$

23. a) $5 \cdot (2 - 7)$

b) $5 \cdot 2 - 5 \cdot 7$

What are the terms of the expression?

24. $5x - 8y + 3$

25. $-4y - 2x + 3z$

Multiply.

26. $3(x - 5)$

27. $5(x + 1)$

28. $\frac{3}{5}(p + q - t)$

Answers on page A-6

Multiply.

29. $-2(x - 3)$

30. $5(x - 2y + 4z)$

31. $-5(x - 2y + 4z)$

Answers on page A-6

EXAMPLE 16 Multiply: $-4(x - 2y + 3z)$.

$$\begin{aligned} -4(x - 2y + 3z) &= -4 \cdot x - (-4)(2y) + (-4)(3z) && \text{Using both distributive laws} \\ &= -4x - (-8y) + (-12z) && \text{Multiplying} \\ &= -4x + 8y - 12z \end{aligned}$$

We can also do this problem by first finding an equivalent expression with all plus signs and then multiplying:

$$\begin{aligned} -4(x - 2y + 3z) &= -4[x + (-2y) + 3z] \\ &= -4 \cdot x + (-4)(-2y) + (-4)(3z) \\ &= -4x + 8y - 12z. \end{aligned}$$

Do Exercises 29–31.

d Factoring

Factoring is the reverse of multiplying. To factor, we can use the distributive laws in reverse:

$$ab + ac = a(b + c) \quad \text{and} \quad ab - ac = a(b - c).$$

FACTORING

To **factor** an expression is to find an equivalent expression that is a product.

Look at Example 13. To *factor* $9x - 45$, we find an equivalent expression that is a product, $9(x - 5)$. When all the terms of an expression have a factor in common, we can "factor it out" using the distributive laws. Note the following.

$9x$ has the factors $9, -9, 3, -3, 1, -1, x, -x, 3x, -3x, 9x, -9x$;

-45 has the factors $1, -1, 3, -3, 5, -5, 9, -9, 15, -15, 45, -45$

We generally remove the largest common factor. In this case, that factor is 9. Thus,

$$\begin{aligned} 9x - 45 &= 9 \cdot x - 9 \cdot 5 \\ &= 9(x - 5). \end{aligned}$$

Remember that an expression has been factored when we have found an equivalent expression that is a product.

EXAMPLES Factor.

17. $5x - 10 = 5 \cdot x - 5 \cdot 2$ Try to do this step mentally.

$= 5(x - 2)$ You can check by multiplying.

18. $ax - ay + az = a(x - y + z)$

19. $9x + 27y - 9 = 9 \cdot x + 9 \cdot 3y - 9 \cdot 1 = 9(x + 3y - 1)$

Note in Example 19 that you might, at first, just factor out a 3, as follows:

$$\begin{aligned}9x + 27y - 9 &= 3 \cdot 3x + 3 \cdot 9y - 3 \cdot 3 \\ &= 3(3x + 9y - 3).\end{aligned}$$

At this point, the mathematics is correct, but the answer is not because there is another factor of 3 that can be factored out, as follows:

$$\begin{aligned}3 \cdot 3x + 3 \cdot 9y - 3 \cdot 3 &= 3(3x + 9y - 3) \\ &= 3(3 \cdot x + 3 \cdot 3y - 3 \cdot 1) \\ &= 3 \cdot 3(x + 3y - 1) \\ &= 9(x + 3y - 1).\end{aligned}$$

We now have a correct answer, but it took more work than we did in Example 19. Thus it is better to look for the greatest common factor at the outset.

EXAMPLES Factor. Try to write just the answer, if you can.

20. $5x - 5y = 5(x - y)$

21. $-3x + 6y - 9z = -3(x - 2y + 3z)$

We usually factor out a negative factor when the first term is negative. The way we factor can depend on the situation in which we are working. We might also factor the expression in Example 21 as follows:

$$-3x + 6y - 9z = 3(-x + 2y - 3z).$$

22. $18z - 12x - 24 = 6(3z - 2x - 4)$

23. $\frac{1}{2}x + \frac{3}{2}y - \frac{1}{2} = \frac{1}{2}(x + 3y - 1)$

Remember that you can always check factoring by multiplying. Keep in mind that an expression is factored when it is written as a product.

Do Exercises 32–37.

e Collecting Like Terms

Terms such as $5x$ and $-4x$, whose variable factors are exactly the same, are called **like terms.** Similarly, numbers, such as -7 and 13, are like terms. Also, $3y^2$ and $9y^2$ are like terms because the variables are raised to the same power. Terms such as $4y$ and $5y^2$ are not like terms, and $7x$ and $2y$ are not like terms.

The process of **collecting like terms** is also based on the distributive laws. We can apply the distributive law when a factor is on the right because of the commutative law of multiplication.

Later in this text, terminology like "collecting like terms" and "combining like terms" will also be referred to as "simplifying."

Factor.

32. $6x - 12$

33. $3x - 6y + 9$

34. $bx + by - bz$

35. $16a - 36b + 42$

36. $\frac{3}{8}x - \frac{5}{8}y + \frac{7}{8}$

37. $-12x + 32y - 16z$

Answers on page A-6

Collect like terms.

38. $6x - 3x$

39. $7x - x$

40. $x - 9x$

41. $x - 0.41x$

42. $5x + 4y - 2x - y$

43. $3x - 7x - 11 + 8y + 4 - 13y$

44. $-\frac{2}{3} - \frac{3}{5}x + y + \frac{7}{10}x - \frac{2}{9}y$

Answers on page A-6

EXAMPLES Collect like terms. Try to write just the answer, if you can.

24. $4x + 2x = (4 + 2)x = 6x$ Factoring out the x using a distributive law

25. $2x + 3y - 5x - 2y = 2x - 5x + 3y - 2y$
$= (2 - 5)x + (3 - 2)y = -3x + y$

26. $3x - x = 3x - 1x = (3 - 1)x = 2x$

27. $x - 0.24x = 1 \cdot x - 0.24x = (1 - 0.24)x = 0.76x$

28. $x - 6x = 1 \cdot x - 6 \cdot x = (1 - 6)x = -5x$

29. $4x - 7y + 9x - 5 + 3y - 8 = 13x - 4y - 13$

30. $\frac{2}{3}a - b + \frac{4}{5}a + \frac{1}{4}b - 10 = \frac{2}{3}a - 1 \cdot b + \frac{4}{5}a + \frac{1}{4}b - 10$
$= \left(\frac{2}{3} + \frac{4}{5}\right)a + \left(-1 + \frac{1}{4}\right)b - 10$
$= \left(\frac{10}{15} + \frac{12}{15}\right)a + \left(-\frac{4}{4} + \frac{1}{4}\right)b - 10$
$= \frac{22}{15}a - \frac{3}{4}b - 10$

Do Exercises 38–44.

Study Tips — LEARNING RESOURCES

Are you aware of all the learning resources that exist for this textbook? Many details are given in the Preface.

- The *Student's Solutions Manual* contains worked-out solutions to the odd-numbered exercises in the exercise sets. You can order this through the bookstore or by calling 1-800-282-0693.
- An extensive set of *videotapes* supplements this text. These are available on CD-ROM by calling 1-800-282-0693.
- *Tutorial software* called InterAct Math also accompanies this text. If it is not available in the campus learning center, you can order it by calling 1-800-282-0693.
- The Addison-Wesley *Math Tutor Center* is available for help with the odd-numbered exercises. You can order this service by calling 1-800-824-7799.
- Extensive help is available online via MyMathLab and/or MathXL. Ask your instructor for information about these or visit MyMathLab.com and MathXL.com.

1.7 EXERCISE SET

For Extra Help

Digital Video Tutor CD 2 Videotape 4 | InterAct Math | Math Tutor Center | MathXL | MyMathLab

a Find an equivalent expression with the given denominator.

1. $\frac{3}{5}$; $5y$

2. $\frac{5}{8}$; $8t$

3. $\frac{2}{3}$; $15x$

4. $\frac{6}{7}$; $14y$

Simplify.

5. $-\frac{24a}{16a}$

6. $-\frac{42t}{18t}$

7. $-\frac{42ab}{36ab}$

8. $-\frac{64pq}{48pq}$

b Write an equivalent expression. Use a commutative law.

9. $y + 8$

10. $x + 3$

11. mn

12. ab

13. $9 + xy$

14. $11 + ab$

15. $ab + c$

16. $rs + t$

Write an equivalent expression. Use an associative law.

17. $a + (b + 2)$

18. $3(vw)$

19. $(8x)y$

20. $(y + z) + 7$

21. $(a + b) + 3$

22. $(5 + x) + y$

23. $3(ab)$

24. $(6x)y$

Use the commutative and associative laws to write three equivalent expressions.

25. $(a + b) + 2$

26. $(3 + x) + y$

27. $5 + (v + w)$

28. $6 + (x + y)$

29. $(xy)3$

30. $(ab)5$

31. $7(ab)$

32. $5(xy)$

c Multiply.

33. $2(b + 5)$

34. $4(x + 3)$

35. $7(1 + t)$

36. $4(1 + y)$

37. $6(5x + 2)$

38. $9(6m + 7)$

39. $7(x + 4 + 6y)$

40. $4(5x + 8 + 3p)$

41. $7(x - 3)$

42. $15(y - 6)$

43. $-3(x - 7)$

44. $1.2(x - 2.1)$

45. $\frac{2}{3}(b - 6)$

46. $\frac{5}{8}(y + 16)$

47. $7.3(x - 2)$

48. $5.6(x - 8)$

49. $-\frac{3}{5}(x - y + 10)$

50. $-\frac{2}{3}(a + b - 12)$

51. $-9(-5x - 6y + 8)$

52. $-7(-2x - 5y + 9)$

53. $-4(x - 3y - 2z)$

54. $8(2x - 5y - 8z)$

55. $3.1(-1.2x + 3.2y - 1.1)$

56. $-2.1(-4.2x - 4.3y - 2.2)$

List the terms of the expression.

57. $4x + 3z$

58. $8x - 1.4y$

59. $7x + 8y - 9z$

60. $8a + 10b - 18c$

d Factor. Check by multiplying.

61. $2x + 4$

62. $5y + 20$

63. $30 + 5y$

64. $7x + 28$

65. $14x + 21y$

66. $18a + 24b$

67. $5x + 10 + 15y$

68. $9a + 27b + 81$

69. $8x - 24$

70. $10x - 50$

71. $32 - 4y$

72. $24 - 6m$

73. $8x + 10y - 22$

74. $9a + 6b - 15$

75. $ax - a$

76. $by - 9b$

77. $ax - ay - az$

78. $cx + cy - cz$

79. $18x - 12y + 6$

80. $-14x + 21y + 7$

81. $\frac{2}{3}x - \frac{5}{3}y + \frac{1}{3}$

82. $\frac{3}{5}a + \frac{4}{5}b - \frac{1}{5}$

e Collect like terms.

83. $9a + 10a$

84. $12x + 2x$

85. $10a - a$

86. $-16x + x$

87. $2x + 9z + 6x$

88. $3a - 5b + 7a$

89. $7x + 6y^2 + 9y^2$

90. $12m^2 + 6q + 9m^2$

91. $41a + 90 - 60a - 2$

92. $42x - 6 - 4x + 2$

93. $23 + 5t + 7y - t - y - 27$

94. $45 - 90d - 87 - 9d + 3 + 7d$

95. $\frac{1}{2}b + \frac{1}{2}b$

96. $\frac{2}{3}x + \frac{1}{3}x$

97. $2y + \frac{1}{4}y + y$

98. $\frac{1}{2}a + a + 5a$

99. $11x - 3x$

100. $9t - 17t$

101. $6n - n$

102. $100t - t$

103. $y - 17y$

104. $3m - 9m + 4$

105. $-8 + 11a - 5b + 6a - 7b + 7$

106. $8x - 5x + 6 + 3y - 2y - 4$

107. $9x + 2y - 5x$

108. $8y - 3z + 4y$

109. $11x + 2y - 4x - y$

110. $13a + 9b - 2a - 4b$

111. $2.7x + 2.3y - 1.9x - 1.8y$

112. $6.7a + 4.3b - 4.1a - 2.9b$

113. $\frac{13}{2}a + \frac{9}{5}b - \frac{2}{3}a - \frac{3}{10}b - 42$

114. $\frac{11}{4}x + \frac{2}{3}y - \frac{4}{5}x - \frac{1}{6}y + 12$

115. $^D W$ The distributive law was introduced before the discussion on collecting like terms. Why do you think this was done?

116. $^D W$ Find two different expressions for the total area of the two rectangles shown below. Explain the equivalence of the expressions in terms of the distributive law.

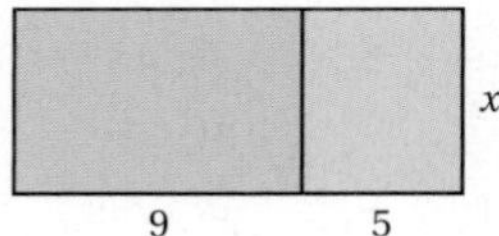

SKILL MAINTENANCE

Find the LCM. [R.1b]

117. 16, 18

118. 18, 24

119. 16, 18, 24

120. 12, 15, 20

121. 16, 32

122. 24, 72

123. 15, 45, 90

124. 18, 54, 108

125. Add and simplify: $\frac{11}{12} + \frac{15}{16}$. [R.2c]

126. Subtract and simplify: $\frac{7}{8} - \frac{2}{3}$. [R.2c]

127. Subtract and simplify: $\frac{1}{8} - \frac{1}{3}$. [R.2c], [1.4a]

128. Convert to percent notation: $\frac{3}{10}$. [R.4d]

SYNTHESIS

Tell whether the expressions are equivalent. Explain.

129. $3t + 5$ and $3 \cdot 5 + t$

130. $4x$ and $x + 4$

131. $5m + 6$ and $6 + 5m$

132. $(x + y) + z$ and $z + (x + y)$

133. Factor: $q + qr + qrs + qrst$.

134. Collect like terms:
$21x + 44xy + 15y - 16x - 8y - 38xy + 2y + xy$.

1.8 SIMPLIFYING EXPRESSIONS; ORDER OF OPERATIONS

Objectives

a Find an equivalent expression for an opposite without parentheses, where an expression has several terms.

b Simplify expressions by removing parentheses and collecting like terms.

c Simplify expressions with parentheses inside parentheses.

d Simplify expressions using rules for order of operations.

We now expand our ability to manipulate expressions by first considering opposites of sums and differences. Then we simplify expressions involving parentheses.

a Opposites of Sums

What happens when we multiply a real number by -1? Consider the following products:

$$-1(7) = -7, \qquad -1(-5) = 5, \qquad -1(0) = 0.$$

From these examples, it appears that when we multiply a number by -1, we get the opposite, or additive inverse, of that number.

THE PROPERTY OF −1

For any real number a,

$$-1 \cdot a = -a.$$

(Negative one times a is the opposite, or additive inverse, of a.)

The property of -1 enables us to find certain expressions equivalent to opposites of sums.

EXAMPLES Find an equivalent expression without parentheses.

1. $-(3 + x) = -1(3 + x)$ — Using the property of -1

   $= -1 \cdot 3 + (-1)x$ — Using a distributive law, multiplying each term by -1

   $= -3 + (-x)$ — Using the property of -1

   $= -3 - x$

2. $-(3x + 2y + 4) = -1(3x + 2y + 4)$ — Using the property of -1

   $= -1(3x) + (-1)(2y) + (-1)4$ — Using a distributive law

   $= -3x - 2y - 4$ — Using the property of -1

Do Exercises 1 and 2.

Find an equivalent expression without parentheses.

1. $-(x + 2)$

2. $-(5x + 2y + 8)$

Answers on page A-6

Suppose we want to remove parentheses in an expression like

$$-(x - 2y + 5).$$

We can first rewrite any subtractions inside the parentheses as additions. Then we take the opposite of each term:

$$-(x - 2y + 5) = -[x + (-2y) + 5]$$
$$= -x + 2y - 5.$$

The most efficient method for removing parentheses is to replace each term in the parentheses with its opposite ("change the sign of every term"). Doing so for $-(x - 2y + 5)$, we obtain $-x + 2y - 5$ as an equivalent expression.

Find an equivalent expression without parentheses. Try to do this in one step.

3. $-(6 - t)$

4. $-(x - y)$

5. $-(-4a + 3t - 10)$

6. $-(18 - m - 2n + 4z)$

Remove parentheses and simplify.

7. $5x - (3x + 9)$

8. $5y - 2 - (2y - 4)$

Remove parentheses and simplify.

9. $6x - (4x + 7)$

10. $8y - 3 - (5y - 6)$

11. $(2a + 3b - c) - (4a - 5b + 2c)$

Answers on page A-6

EXAMPLES Find an equivalent expression without parentheses.

3. $-(5 - y) = -5 + y = y + (-5) = y - 5$ — Changing the sign of each term

4. $-(2a - 7b - 6) = -2a + 7b + 6$

5. $-(-3x + 4y + z - 7w - 23) = 3x - 4y - z + 7w + 23$

Do Exercises 3–6.

b Removing Parentheses and Simplifying

When a sum is added, as in $5x + (2x + 3)$, we can simply remove, or drop, the parentheses and collect like terms because of the associative law of addition:

$$5x + (2x + 3) = 5x + 2x + 3 = 7x + 3.$$

On the other hand, when a sum is subtracted, as in $3x - (4x + 2)$, no "associative" law applies. However, we can subtract by adding an opposite. We then remove parentheses by changing the sign of each term inside the parentheses and collecting like terms.

EXAMPLE 6 Remove parentheses and simplify.

$3x - (4x + 2) = 3x + [-(4x + 2)]$ — Adding the opposite of $(4x + 2)$

$= 3x + (-4x - 2)$ — Changing the sign of each term inside the parentheses

$= 3x - 4x - 2$

$= -x - 2$ — Collecting like terms

Do Exercises 7 and 8.

In practice, the first three steps of Example 6 are usually combined by changing the sign of each term in parentheses and then collecting like terms.

EXAMPLES Remove parentheses and simplify.

7. $5y - (3y + 4) = 5y - 3y - 4$ — Removing parentheses by changing the sign of every term inside the parentheses

$= 2y - 4$ — Collecting like terms

8. $3x - 2 - (5x - 8) = 3x - 2 - 5x + 8$

$= -2x + 6$, or $6 - 2x$

9. $(3a + 4b - 5) - (2a - 7b + 4c - 8)$

$= 3a + 4b - 5 - 2a + 7b - 4c + 8$

$= a + 11b - 4c + 3$

Do Exercises 9–11.

Next, consider subtracting an expression consisting of several terms multiplied by a number other than 1 or -1.

EXAMPLE 10 Remove parentheses and simplify.

$$\begin{aligned} x - 3(x + y) &= x + [-3(x + y)] && \text{Adding the opposite of } 3(x + y) \\ &= x + [-3x - 3y] && \text{Multiplying } x + y \text{ by } -3 \\ &= x - 3x - 3y \\ &= -2x - 3y && \text{Collecting like terms} \end{aligned}$$

EXAMPLES Remove parentheses and simplify.

11. $3y - 2(4y - 5) = 3y - 8y + 10$ Multiplying each term in parentheses by -2

$= -5y + 10$

12. $(2a + 3b - 7) - 4(-5a - 6b + 12)$

$= 2a + 3b - 7 + 20a + 24b - 48$

$= 22a + 27b - 55$

13. $2y - \frac{1}{3}(9y - 12) = 2y - 3y + 4 = -y + 4$

Do Exercises 12–15.

C Parentheses Within Parentheses

In addition to parentheses, some expressions contain other grouping symbols such as brackets [ ] and braces { }.

> When more than one kind of grouping symbol occurs, do the computations in the innermost ones first. Then work from the inside out.

EXAMPLES Simplify.

14. $[3 - (7 + 3)] = [3 - 10]$ Computing $7 + 3$

$= -7$

15. $\{8 - [9 - (12 + 5)]\} = \{8 - [9 - 17]\}$ Computing $12 + 5$

$= \{8 - [-8]\}$ Computing $9 - 17$

$= 8 + 8 = 16$

16. $\left[(-4) \div \left(-\frac{1}{4}\right)\right] \div \frac{1}{4} = [(-4) \cdot (-4)] \div \frac{1}{4}$ Working within the brackets; computing $(-4) \div \left(-\frac{1}{4}\right)$

$= 16 \div \frac{1}{4}$

$= 16 \cdot 4 = 64$

17. $4(2 + 3) - \{7 - [4 - (8 + 5)]\}$

$= 4 \cdot 5 - \{7 - [4 - 13]\}$ Working with the innermost parentheses first

$= 20 - \{7 - [-9]\}$ Computing $4 \cdot 5$ and $4 - 13$

$= 20 - 16$ Computing $7 - [-9]$

$= 4$

Do Exercises 16–19.

Remove parentheses and simplify.

12. $y - 9(x + y)$

13. $5a - 3(7a - 6)$

14. $4a - b - 6(5a - 7b + 8c)$

15. $5x - \dfrac{1}{4}(8x + 28)$

Simplify.

16. $12 - (8 + 2)$

17. $\{9 - [10 - (13 + 6)]\}$

18. $[24 \div (-2)] \div (-2)$

19. $5(3 + 4) - \{8 - [5 - (9 + 6)]\}$

Answers on page A-6

20. Simplify:

$[3(x + 2) + 2x] -$
$[4(y + 2) - 3(y - 2)].$

Simplify.

21. $23 - 42 \cdot 30$

22. $32 \div 8 \cdot 2$

23. $-24 \div 3 - 48 \div (-4)$

Answers on page A-6

EXAMPLE 18 Simplify.

$[5(x + 2) - 3x] - [3(y + 2) - 7(y - 3)]$

$= [5x + 10 - 3x] - [3y + 6 - 7y + 21]$ Working with the innermost parentheses first

$= [2x + 10] - [-4y + 27]$ Collecting like terms within brackets

$= 2x + 10 + 4y - 27$ Removing brackets

$= 2x + 4y - 17$ Collecting like terms

Do Exercise 20.

d Order of Operations

When several operations are to be done in a calculation or a problem, we apply the same rules that we did in Section R.5. We repeat them here for review. (If you did not study that section earlier, you should do so now.)

RULES FOR ORDER OF OPERATIONS

1. Do all calculations within grouping symbols before operations outside.
2. Evaluate all exponential expressions.
3. Do all multiplications and divisions in order from left to right.
4. Do all additions and subtractions in order from left to right.

These rules are consistent with the way in which most computers and scientific calculators perform calculations.

EXAMPLE 19 Simplify: $-34 \cdot 56 - 17$.

There are no parentheses or powers, so we start with the third step.

$-34 \cdot 56 - 17 = -1904 - 17$ Doing all multiplications and divisions in order from left to right

$= -1921$ Doing all additions and subtractions in order from left to right

EXAMPLE 20 Simplify: $25 \div (-5) + 50 \div (-2)$.

There are no calculations inside parentheses or powers. The parentheses with (-5) and (-2) are used only to represent the negative numbers. We begin by doing all multiplications and divisions.

$\underbrace{25 \div (-5)} + \underbrace{50 \div (-2)}$

$= -5 + (-25)$ Doing all multiplications and divisions in order from left to right

$= -30$ Doing all additions and subtractions in order from left to right.

Do Exercises 21–23.

EXAMPLE 21 Simplify: $2^4 + 51 \cdot 4 - (37 + 23 \cdot 2)$.

$2^4 + 51 \cdot 4 - (37 + 23 \cdot 2)$	
$= 2^4 + 51 \cdot 4 - (37 + 46)$	Following the rules for order of operations within the parentheses first
$= 2^4 + 51 \cdot 4 - 83$	Completing the addition inside parentheses
$= 16 + 51 \cdot 4 - 83$	Evaluating exponential expressions
$= 16 + 204 - 83$	Doing all multiplications
$= 220 - 83$	Doing all additions and subtractions in order from left to right
$= 137$	

CALCULATOR CORNER

Order of Operations and Grouping Symbols Parentheses are necessary in some calculations in order to ensure that operations are performed in the desired order. To simplify $-5(3 - 6) - 12$, we press [(−)] [5] [(] [3] [−] [6] [)] [−] [1] [2] [ENTER]. The result is 3. Without parentheses, the computation is $-5 \cdot 3 - 6 - 12$, and the result is -33.

```
-5(3-6)-12
                3
-5*3-6-12
              -33
```

When a negative number is raised to an even power, parentheses must also be used. To find $(-3)^4$, we press [(] [(−)] [3] [)] [∧] [4] [ENTER]. The result is 81. Without parentheses, the computation is $-3^4 = -1 \cdot 3^4 = -1 \cdot 81 = -81$.

```
(-3)^4
               81
-3^4
              -81
```

To simplify an expression like $\dfrac{49 - 104}{7 + 4}$, we must enter it as $(49 - 104) \div (7 + 4)$. We press [(] [4] [9] [−] [1] [0] [4] [)] [÷] [(] [7] [+] [4] [)] [ENTER]. The result is -5.

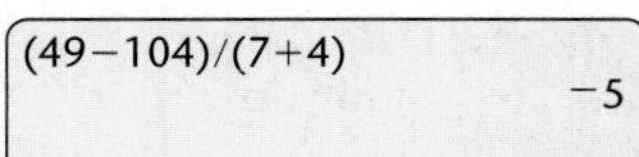

Exercises: Calculate.

1. $-8 + 4(7 - 9) + 5$
2. $-3[2 + (-5)]$
3. $7[4 - (-3)] + 5[3^2 - (-4)]$
4. $(-7)^6$
5. $(-17)^5$
6. $(-104)^3$
7. -7^6
8. -17^5
9. -104^3
10. $\dfrac{38 - 178}{5 + 30}$
11. $\dfrac{311 - 17^2}{2 - 13}$
12. $785 - \dfrac{285 - 5^4}{17 + 3 \cdot 51}$

Simplify.

24. $52 \cdot 5 + 5^3 - (4^2 - 48 \div 4)$

25. $\dfrac{5 - 10 - 5 \cdot 23}{2^3 + 3^2 - 7}$

Answers on page A-6

A fraction bar can play the role of a grouping symbol, although such a symbol is not as evident as the others.

EXAMPLE 22 Simplify: $\dfrac{-64 \div (-16) \div (-2)}{2^3 - 3^2}$.

An equivalent expression with brackets as grouping symbols is

$$[-64 \div (-16) \div (-2)] \div [2^3 - 3^2].$$

This shows, in effect, that we do the calculations in the numerator and then in the denominator, and divide the results:

$$\frac{-64 \div (-16) \div (-2)}{2^3 - 3^2} = \frac{4 \div (-2)}{8 - 9} = \frac{-2}{-1} = 2.$$

Do Exercises 24 and 25.

Study Tips

TEST PREPARATION

You are probably ready to begin preparing for your first test. Here are some test-taking study tips.

- **Make up your own test questions as you study.** After you have done your homework over a particular objective, write one or two questions on your own that you think might be on a test. You will be amazed at the insight this will provide.
- **Do an overall review of the chapter, focusing on the objectives and the examples.** This should be accompanied by a study of any class notes you may have taken.
- **Do the review exercises at the end of the chapter.** Check your answers at the back of the book. If you have trouble with an exercise, use the objective symbol as a guide to go back and do further study of that objective.
- **Call the AW Math Tutor Center if you need extra help at 1-888-777-0463.**
- **Do the chapter test at the end of the chapter.** Check the answers and use the objective symbols at the back of the book as a reference for where to review.
- **Ask former students for old exams.** Working such exams can be very helpful and allows you to see what various professors think is important.
- **When taking a test, read each question carefully and try to do all the questions the first time through, but pace yourself.** Answer all the questions, and mark those to recheck if you have time at the end. Very often, your first hunch will be correct.
- **Try to write your test in a neat and orderly manner.** Very often, your instructor tries to give you partial credit when grading an exam. If your test paper is sloppy and disorderly, it is difficult to verify the partial credit. Doing your work neatly can ease such a task for the instructor.

1.8 EXERCISE SET

For Extra Help

Digital Video Tutor CD 2 Videotape 4 | InterAct Math | Math Tutor Center | MathXL | MyMathLab

a Find an equivalent expression without parentheses.

1. $-(2x + 7)$

2. $-(8x + 4)$

3. $-(8 - x)$

4. $-(a - b)$

5. $-(4a - 3b + 7c)$

6. $-(x - 4y - 3z)$

7. $-(6x - 8y + 5)$

8. $-(4x + 9y + 7)$

9. $-(3x - 5y - 6)$

10. $-(6a - 4b - 7)$

11. $-(-8x - 6y - 43)$

12. $-(-2a + 9b - 5c)$

b Remove parentheses and simplify.

13. $9x - (4x + 3)$

14. $4y - (2y + 5)$

15. $2a - (5a - 9)$

16. $12m - (4m - 6)$

17. $2x + 7x - (4x + 6)$

18. $3a + 2a - (4a + 7)$

19. $2x - 4y - 3(7x - 2y)$

20. $3a - 9b - 1(4a - 8b)$

21. $15x - y - 5(3x - 2y + 5z)$

22. $4a - b - 4(5a - 7b + 8c)$

23. $(3x + 2y) - 2(5x - 4y)$

24. $(-6a - b) - 5(2b + a)$

25. $(12a - 3b + 5c) - 5(-5a + 4b - 6c)$

26. $(-8x + 5y - 12) - 6(2x - 4y - 10)$

c Simplify.

27. $[9 - 2(5 - 4)]$

28. $[6 - 5(8 - 4)]$

29. $8[7 - 6(4 - 2)]$

30. $10[7 - 4(7 - 5)]$

31. $[4(9 - 6) + 11] - [14 - (6 + 4)]$

32. $[7(8 - 4) + 16] - [15 - (7 + 8)]$

33. $[10(x + 3) - 4] + [2(x - 1) + 6]$

34. $[9(x + 5) - 7] + [4(x - 12) + 9]$

35. $[7(x + 5) - 19] - [4(x - 6) + 10]$

36. $[6(x + 4) - 12] - [5(x - 8) + 14]$

37. $3\{[7(x - 2) + 4] - [2(2x - 5) + 6]\}$

38. $4\{[8(x - 3) + 9] - [4(3x - 2) + 6]\}$

39. $4\{[5(x - 3) + 2] - 3[2(x + 5) - 9]\}$

40. $3\{[6(x - 4) + 5] - 2[5(x + 8) - 3]\}$

d Simplify.

41. $8 - 2 \cdot 3 - 9$

42. $8 - (2 \cdot 3 - 9)$

43. $(8 - 2 \cdot 3) - 9$

44. $(8 - 2)(3 - 9)$

45. $[(-24) \div (-3)] \div \left(-\frac{1}{2}\right)$

46. $[32 \div (-2)] \div (-2)$

47. $16 \cdot (-24) + 50$

48. $10 \cdot 20 - 15 \cdot 24$

49. $2^4 + 2^3 - 10$

50. $40 - 3^2 - 2^3$

51. $5^3 + 26 \cdot 71 - (16 + 25 \cdot 3)$

52. $4^3 + 10 \cdot 20 + 8^2 - 23$

53. $4 \cdot 5 - 2 \cdot 6 + 4$

54. $4 \cdot (6 + 8)/(4 + 3)$

55. $4^3/8$

56. $5^3 - 7^2$

57. $8(-7) + 6(-5)$

58. $10(-5) + 1(-1)$

59. $19 - 5(-3) + 3$

60. $14 - 2(-6) + 7$

61. $9 \div (-3) + 16 \div 8$

62. $-32 - 8 \div 4 - (-2)$

63. $-4^2 + 6$

64. $-5^2 + 7$

65. $-8^2 - 3$

66. $-9^2 - 11$

67. $12 - 20^3$

68. $20 + 4^3 \div (-8)$

69. $2 \cdot 10^3 - 5000$

70. $-7(3^4) + 18$

71. $6[9 - (3 - 4)]$

72. $8[(6 - 13) - 11]$

73. $-1000 \div (-100) \div 10$

74. $256 \div (-32) \div (-4)$

75. $8 - (7 - 9)$

76. $(8 - 7) - 9$

77. $\dfrac{10 - 6^2}{9^2 + 3^2}$

78. $\dfrac{5^2 - 4^3 - 3}{9^2 - 2^2 - 1^5}$

79. $\dfrac{3(6-7)-5\cdot 4}{6\cdot 7-8(4-1)}$

80. $\dfrac{20(8-3)-4(10-3)}{10(2-6)-2(5+2)}$

81. $\dfrac{2^3-3^2+12\cdot 5}{-32\div(-16)\div(-4)}$

82. $\dfrac{|3-5|^2-|7-13|}{|12-9|+|11-14|}$

83. $^{\mathrm{D}}\mathrm{W}$ Jake keys in $18/2\cdot 3$ on his calculator and expects the result to be 3. What mistake is he making?

84. $^{\mathrm{D}}\mathrm{W}$ Determine whether $|-x|$ and $|x|$ are equivalent. Explain.

SKILL MAINTENANCE

85. Find the prime factorization of 236. [R.1a]

86. Find the LCM of 28 and 36. [R.1b]

87. Divide and simplify: $\dfrac{2}{3}\div\dfrac{5}{12}$. [R.2c]

88. Multiply and simplify: $\dfrac{2}{3}\cdot\dfrac{5}{12}$. [R.2c]

89. Add and simplify: $\dfrac{2}{3}+\dfrac{5}{12}$. [R.2c]

90. Subtract and simplify: $\dfrac{2}{3}-\dfrac{5}{12}$. [R.2c]

Evaluate. [R.5b]

91. 3^4

92. 10^3

93. 10^2

94. 15^2

SYNTHESIS

Find an equivalent expression by enclosing the last three terms in parentheses preceded by a minus sign.

95. $6y+2x-3a+c$

96. $x-y-a-b$

97. $6m+3n-5m+4b$

Simplify.

98. $z-\{2z-[3z-(4z-5z)-6z]-7z\}-8z$

99. $\{x-[f-(f-x)]+[x-f]\}-3x$

100. $x-\{x-1-[x-2-(x-3-\{x-4-[x-5-(x-6)]\})]\}$

101. Use your calculator to do the following.

a) Evaluate x^2+3 when $x=7$, when $x=-7$, and when $x=-5.013$.

b) Evaluate $1-x^2$ when $x=5$, when $x=-5$, and when $x=-10.455$.

102. Express $3^3+3^3+3^3$ as a power of 3.

Find the average.

103. $-15,\ 20,\ 50,\ -82,\ -7,\ -2$

104. $-1,\ 1,\ 2,\ -2,\ 3,\ -8,\ -10$

CHAPTER 1

Summary and Review

The review that follows is meant to prepare you for a chapter exam. It consists of two parts. The first part is a checklist of some of the Study Tips referred to in this and preceding chapters, as well as a list of important properties and formulas. The second part is the Review Exercises. These provide practice exercises for the exam, together with references to section objectives so you can go back and review. Before beginning, stop and look back over the skills you have obtained. What skills in mathematics do you have now that you did not have before studying this chapter?

STUDY TIPS CHECKLIST

The foundation of all your study skills is TIME!

- ☐ Are you approaching your study of mathematics with an assertive, positive attitude?
- ☐ Are you making use of the textbook supplements, such as the Math Tutor Center, the *Student's Solutions Manual,* and the videotapes?
- ☐ Have you determined the location of the learning resource centers on your campus, such as a math lab, tutor center, and your instructor's office?
- ☐ Are you stopping to work the margin exercises when directed to do so?
- ☐ Are you keeping one section ahead in your syllabus?

IMPORTANT PROPERTIES AND FORMULAS

PROPERTIES OF THE REAL-NUMBER SYSTEM

The Commutative Laws: $a + b = b + a, \quad ab = ba$

The Associative Laws: $a + (b + c) = (a + b) + c, \quad a(bc) = (ab)c$

The Identity Properties: $a + 0 = 0 + a = a, \quad a \cdot 1 = 1 \cdot a = a$

The Inverse Properties: For any real number a, there is an opposite $-a$ such that $a + (-a) = (-a) + a = 0$.

For any nonzero real number a, there is a reciprocal $\frac{1}{a}$ such that $a \cdot \frac{1}{a} = \frac{1}{a} \cdot a = 1$.

The Distributive Laws: $a(b + c) = ab + ac, \quad a(b - c) = ab - ac$

REVIEW EXERCISES

The review exercises that follow are for practice. Answers are at the back of the book. If you miss an exercise, restudy the objective indicated in blue after the exercise or the direction line that precedes it.

1. Evaluate $\frac{x - y}{3}$ when $x = 17$ and $y = 5$. [1.1a]

2. Translate to an algebraic expression: [1.1b]
Nineteen percent of some number.

3. Tell which integers correspond to this situation: [1.2a]
David has a debt of $45 and Joe has $72 in his savings account.

4. Find: $|-38|$. [1.2e]

Graph the number on a number line. [1.2b]

5. -2.5

6. $\frac{8}{9}$

Use either < or > for $\square$ to write a true sentence. [1.2d]

7. $-3 \square 10$

8. $-1 \square -6$

9. $0.126 \square -12.6$

10. $-\frac{2}{3} \square -\frac{1}{10}$

Find the opposite. [1.3b]

11. 3.8

12. $-\frac{3}{4}$

Find the reciprocal. [1.6b]

13. $\frac{3}{8}$

14. -7

15. Evaluate $-x$ when $x = -34$. [1.3b]

16. Evaluate $-(-x)$ when $x = 5$. [1.3b]

Compute and simplify.

17. $4 + (-7)$ [1.3a]

18. $6 + (-9) + (-8) + 7$ [1.3a]

19. $-3.8 + 5.1 + (-12) + (-4.3) + 10$ [1.3a]

20. $-3 - (-7)$ [1.4a]

21. $-\frac{9}{10} - \frac{1}{2}$ [1.4a]

22. $-3.8 - 4.1$ [1.4a]

23. $-9 \cdot (-6)$ [1.5a]

24. $-2.7(3.4)$ [1.5a]

25. $\frac{2}{3} \cdot \left(-\frac{3}{7}\right)$ [1.5a]

26. $3 \cdot (-7) \cdot (-2) \cdot (-5)$ [1.5a]

27. $35 \div (-5)$ [1.6a]

28. $-5.1 \div 1.7$ [1.6c]

29. $-\frac{3}{11} \div \left(-\frac{4}{11}\right)$ [1.6c]

30. $(-3.4 - 12.2) - 8(-7)$ [1.8d]

31. $\dfrac{-12(-3) - 2^3 - (-9)(-10)}{3 \cdot 10 + 1}$ [1.8d]

32. $-16 \div 4 - 30 \div (-5)$ [1.8d]

33. $9[(7 - 14) - 13]$ [1.8d]

Solve.

34. On the first, second, and third downs, a football team had these gains and losses: 5-yd gain, 12-yd loss, and 15-yd gain, respectively. Find the total gain (or loss). [1.3c]

35. Kaleb's total assets are \$170. He borrows \$300. What are his total assets now? [1.4b]

36. *Stock Price.* The value of EFX Corp. began the day at \$17.68 per share and dropped \$1.63 per hour for 8 hr. What was the price of the stock after 8 hr? [1.5b]

37. *Checking Account Balance.* Yuri had \$68 in his checking account. After writing checks to make seven purchases of DVDs at the same price for each, the balance in his account was −\$64.65. What was the price of each DVD? [1.6d]

Multiply. [1.7c]

38. $5(3x - 7)$

39. $-2(4x - 5)$

40. $10(0.4x + 1.5)$

41. $-8(3 - 6x)$

Factor. [1.7d]

42. $2x - 14$

43. $6x - 6$

44. $5x + 10$

45. $12 - 3x$

Collect like terms. [1.7e]

46. $11a + 2b - 4a - 5b$

47. $7x - 3y - 9x + 8y$

48. $6x + 3y - x - 4y$

49. $-3a + 9b + 2a - b$

Remove parentheses and simplify.

50. $2a - (5a - 9)$ [1.8b]

51. $3(b + 7) - 5b$ [1.8b]

52. $3[11 - 3(4 - 1)]$ [1.8c]

53. $2[6(y - 4) + 7]$ [1.8c]

54. $[8(x + 4) - 10] - [3(x - 2) + 4]$ [1.8c]

55. $5\{[6(x - 1) + 7] - [3(3x - 4) + 8]\}$ [1.8c]

Write true or false. [1.2d]

56. $-9 \leq 11$

57. $-11 \geq -3$

58. Write another inequality with the same meaning as $-3 < x$. [1.2d]

59. $^{D}_{W}$ Explain the notion of the opposite of a number in as many ways as possible. [1.3b]

60. $^{D}_{W}$ Is the absolute value of a number always positive? Why or why not? [1.2e]

SKILL MAINTENANCE

Certain objectives from four particular sections will be retested on the chapter test. The objectives are listed with the practice problems that follow.

61. Divide and simplify: $\frac{11}{12} \div \frac{7}{10}$. [R.2c]

62. Compute and simplify: $\frac{5^3 - 2^4}{5 \cdot 2 + 2^3}$. [R.5c]

63. Find the prime factorization of 648. [R.1a]

64. Convert to percent notation: $\frac{5}{8}$. [R.4d]

65. Convert to decimal notation: 5.67%. [R.4a]

66. Find the LCM of 15, 27, and 30. [R.1b]

SYNTHESIS

Simplify. [1.2e], [1.4a], [1.6a], [1.8d]

67. $-\left|\frac{7}{8} - \left(-\frac{1}{2}\right) - \frac{3}{4}\right|$

68. $(|2.7 - 3| + 3^2 - |-3|) \div (-3)$

69. $2000 - 1990 + 1980 - 1970 + \cdots - 20 + 10$

70. Find a formula for the perimeter of the following figure. [R.6a], [1.7e]

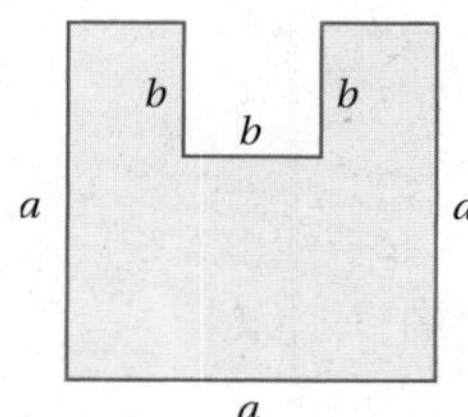

CHAPTER 1

Chapter Test

1. Evaluate $\frac{3x}{y}$ when $x = 10$ and $y = 5$.

2. Write an algebraic expression: Nine less than some number.

3. Find the area of a triangle when the height h is 30 ft and the base b is 16 ft.

Use either $<$ or $>$ for $\square$ to write a true sentence.

4. $-4 \square 0$

5. $-3 \square -8$

6. $-0.78 \square -0.87$

7. $-\frac{1}{8} \square \frac{1}{2}$

Find the absolute value.

8. $|-7|$

9. $\left|\frac{9}{4}\right|$

10. $|-2.7|$

Find the opposite.

11. $\frac{2}{3}$

12. -1.4

13. Evaluate $-x$ when $x = -8$.

Find the reciprocal.

14. -2

15. $\frac{4}{7}$

Compute and simplify.

16. $3.1 - (-4.7)$

17. $-8 + 4 + (-7) + 3$

18. $-\frac{1}{5} + \frac{3}{8}$

19. $2 - (-8)$

20. $3.2 - 5.7$

21. $\frac{1}{8} - \left(-\frac{3}{4}\right)$

22. $4 \cdot (-12)$

23. $-\frac{1}{2} \cdot \left(-\frac{3}{8}\right)$

24. $-45 \div 5$

25. $-\frac{3}{5} \div \left(-\frac{4}{5}\right)$

26. $4.864 \div (-0.5)$

27. $-2(16) - |2(-8) - 5^3|$

28. $-20 \div (-5) + 36 \div (-4)$

29. *Antarctica Highs and Lows.* The continent of Antarctica, which lies in the southern hemisphere, experiences winter in July. The average high temperature is $-67°F$ and the average low temperature is $-81°F$. How much higher is the average high than the average low?
Source: National Climatic Data Center

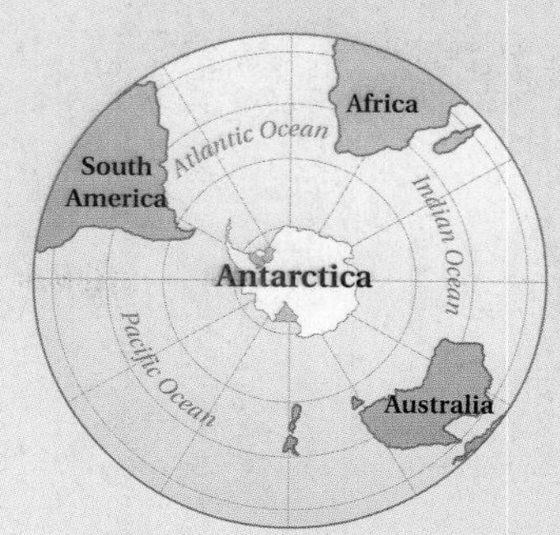

30. Maureen is a stockbroker. She kept track of the changes in the stock market over a period of 5 weeks. By how many points had the market risen or fallen over this time?

WEEK 1	WEEK 2	WEEK 3	WEEK 4	WEEK 5
Down 13 pts	Down 16 pts	Up 36 pts	Down 11 pts	Up 19 pts

31. *Population Decrease.* The population of a city was 18,600. It dropped 420 each year for 6 yr. What was the population of the city after 6 yr?

32. *Chemical Experiment.* During a chemical reaction, the temperature in the beaker decreased every minute by the same number of degrees. The temperature was 16°C at 11:08 A.M. By 11:43 A.M., the temperature had dropped to $-17°C$. By how many degrees did it drop each minute?

Multiply.

33. $3(6 - x)$

34. $-5(y - 1)$

Factor.

35. $12 - 22x$

36. $7x + 21 + 14y$

Simplify.

37. $6 + 7 - 4 - (-3)$

38. $5x - (3x - 7)$

39. $4(2a - 3b) + a - 7$

40. $4\{3[5(y - 3) + 9] + 2(y + 8)\}$

41. $256 \div (-16) \div 4$

42. $2^3 - 10[4 - (-2 + 18)3]$

43. Write an inequality with the same meaning as $x \leq -2$.

SKILL MAINTENANCE

44. Evaluate: $(1.2)^3$.

45. Convert to percent notation: $\frac{1}{8}$.

46. Find the prime factorization of 280.

47. Find the LCM of 16, 20, and 30.

SYNTHESIS

Simplify.

48. $|-27 - 3(4)| - |-36| + |-12|$

49. $a - \{3a - [4a - (2a - 4a)]\}$

50. Find a formula for the perimeter of the figure shown here.

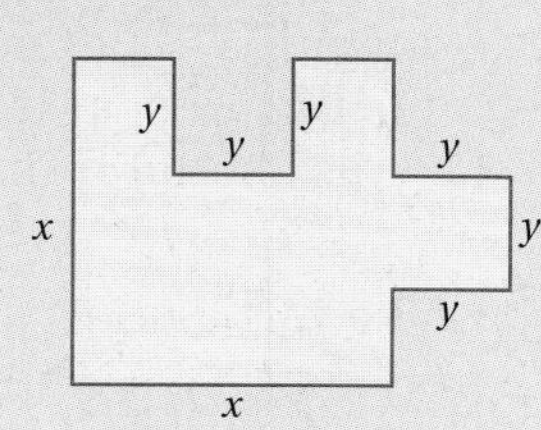

Solving Equations and Inequalities

Gateway to Chapter 2

In this chapter, we use the manipulations considered in Chapter 1 to solve equations and inequalities. We then use equations and inequalities to solve applied problems.

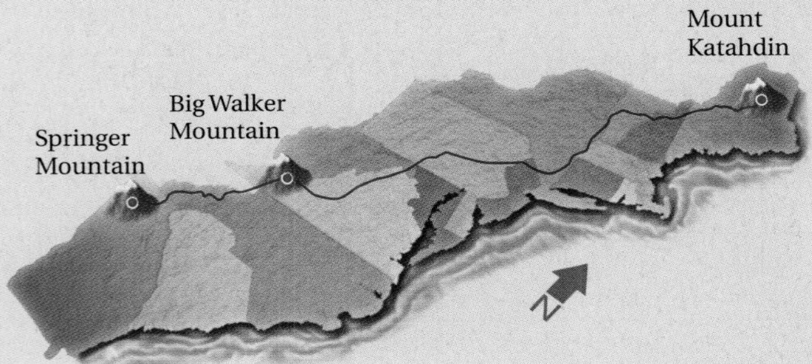

Real-World Application

In 1998, at age 79, Earl Shaffer became the oldest person to hike all 2100 miles of the Appalachian Trail—from Springer Mountain, Georgia, to Mount Katahdin, Maine. At one point, Shaffer stood atop Big Walker Mountain, Virginia, which is three times as far from the northern end as from the southern end. How far was Shaffer from each end of the trail?

Source: Appalachian Trail Conference

This problem appears as Example 1 in Section 2.6.

CHAPTER 2

CHAPTER

2

Pretest

Solve.

1. $4 + x = 12$ [2.1b]

2. $-7x = 49$ [2.2a]

3. $6a - 2 = 10$ [2.3a]

4. $4y + 9 = 2y + 7$ [2.3b]

5. $7 - 3(2x - 1) = 40$ [2.3d]

6. $\frac{4}{9}x - 1 = \frac{7}{8}$ [2.3a]

7. $1 + 2(a + 3) = 3(2a - 1) + 6$ [2.3d]

8. $y + 5 > 1$ [2.7c]

9. $-3x \leq 18$ [2.7d]

10. $5 - 2a < 7$ [2.7e]

11. $3x + 4 \geq 2x + 7$ [2.7c]

12. $8y < -18$ [2.7d]

13. $4 + x = x + 5$ [2.3c]

14. Solve for x: $y = Ax$. [2.4b]

15. Solve for a: $A = 3a - b$. [2.4b]

Solve.

16. **Oil-Painting Frame.** The perimeter of the ornate frame of an oil painting is 146 in. The width is 5 in. less than the length. Find the dimensions. [2.6a]

17. **Savings Investment.** Money is invested in a savings account at 4.25% simple interest. After 1 year, there is $479.55 in the account. How much was originally invested? [2.6a]

18. **Consecutive Page Numbers.** The sum of three consecutive page numbers is 246. Find the page numbers. [2.6a]

19. **Lengths of a Ballfield.** The width of a rectangular ballfield is 96 yd. Find all possible lengths so that the perimeter of the ballfield will be at least 540 yd. [2.8b]

Graph on a number line. [2.7b]

20. $x > -3$

21. $x \leq 4$

Solve. [2.5a]

22. What is 24% of 85?

23. 2.16 is 4% of what number?

24. What percent of 85 is 17?

25. **Job Opportunities.** There were 49 thousand manicurists in 1998. Job opportunities are expected to grow to 62 thousand by 2008. What is the predicted percent of increase?
Source: *Handbook of U.S. Labor Statistics*

2.1 SOLVING EQUATIONS: THE ADDITION PRINCIPLE

Objectives

 Determine whether a given number is a solution of a given equation.

 Solve equations using the addition principle.

a Equations and Solutions

In order to solve problems, we must learn to solve equations.

EQUATION

An **equation** is a number sentence that says that the expressions on either side of the equals sign, =, represent the same number.

Here are some examples:

$$3 + 2 = 5, \quad 14 - 10 = 1 + 3, \quad x + 6 = 13, \quad 3x - 2 = 7 - x.$$

Equations have expressions on each side of the equals sign. The sentence "$14 - 10 = 1 + 3$" asserts that the expressions $14 - 10$ and $1 + 3$ name the same number.

Some equations are true. Some are false. Some are neither true nor false.

EXAMPLES Determine whether the equation is true, false, or neither.

1. $3 + 2 = 5$ — The equation is *true.*
2. $7 - 2 = 4$ — The equation is *false.*
3. $x + 6 = 13$ — The equation is *neither* true nor false, because we do not know what number x represents.

Do Exercises 1–3.

Determine whether the equation is true, false, or neither.

1. $5 - 8 = -4$

2. $12 + 6 = 18$

3. $x + 6 = 7 - x$

SOLUTION OF AN EQUATION

Any replacement for the variable that makes an equation true is called a **solution** of the equation. To solve an equation means to find *all* of its solutions.

One way to determine whether a number is a solution of an equation is to evaluate the expression on each side of the equals sign by substitution. If the values are the same, then the number is a solution.

EXAMPLE 4 Determine whether 7 is a solution of $x + 6 = 13$.

We have

$x + 6 = 13$	Writing the equation
$7 + 6 \ ? \ 13$	Substituting 7 for x
$13 \ \vert$	TRUE

Since the left-hand and the right-hand sides are the same, we have a solution. No other number makes the equation true, so the only solution is the number 7.

Answers on page A-7

Determine whether the given number is a solution of the given equation.

4. 8; $x + 4 = 12$

5. 0; $x + 4 = 12$

6. -3; $7 + x = -4$

7. Solve using the addition principle:

$$x + 2 = 11.$$

Answers on page A-7

EXAMPLE 5 Determine whether 19 is a solution of $7x = 141$.

We have

$7x = 141$	Writing the equation
$7(19) \ ? \ 141$	Substituting 19 for x
$133 \ \vert$	FALSE

Since the left-hand and the right-hand sides are not the same, we do not have a solution.

Do Exercises 4–6.

b Using the Addition Principle

Consider the equation

$$x = 7.$$

We can easily see that the solution of this equation is 7. If we replace x with 7, we get

$$7 = 7, \quad \text{which is true.}$$

Now consider the equation of Example 4:

$$x + 6 = 13.$$

In Example 4, we discovered that the solution of this equation is also 7, but the fact that 7 is the solution is not as obvious. We now begin to consider principles that allow us to start with an equation like $x + 6 = 13$ and end up with an *equivalent equation,* like $x = 7$, in which the variable is alone on one side and for which the solution is easier to find.

EQUIVALENT EQUATIONS

Equations with the same solutions are called **equivalent equations.**

One of the principles that we use in solving equations involves adding. An equation $a = b$ says that a and b stand for the same number. Suppose this is true, and we add a number c to the number a. We get the same answer if we add c to b, because a and b are the same number.

THE ADDITION PRINCIPLE

For any real numbers a, b, and c,

$$a = b \quad \text{is equivalent to} \quad a + c = b + c.$$

Let's again solve the equation $x + 6 = 13$ using the addition principle. We want to get x alone on one side. To do so, we use the addition principle, choosing to add -6 because $6 + (-6) = 0$:

$x + 6 = 13$	
$x + 6 + (-6) = 13 + (-6)$	Using the addition principle: adding -6 on both sides
$x + 0 = 7$	Simplifying
$x = 7.$	Identity property of 0: $x + 0 = x$

The solution of $x + 6 = 13$ is 7.

Do Exercise 7 on the preceding page.

When we use the addition principle, we sometimes say that we "add the same number on both sides of the equation." This is also true for subtraction, since we can express every subtraction as an addition. That is, since

$$a - c = b - c \quad \text{is equivalent to} \quad a + (-c) = b + (-c),$$

the addition principle tells us that we can "subtract the same number on both sides of the equation."

EXAMPLE 6 Solve: $x + 5 = -7$.

We have

$x + 5 = -7$

$x + 5 - 5 = -7 - 5$ Using the addition principle: adding -5 on both sides or subtracting 5 on both sides

$x + 0 = -12$ Simplifying

$x = -12.$ Identity property of 0

We can see that the solution of $x = -12$ is the number -12. To check the answer, we substitute -12 in the original equation.

CHECK: $x + 5 = -7$

$-12 + 5 \ ? \ -7$

-7 | TRUE

The solution of the original equation is -12.

In Example 6, to get x alone, we used the addition principle and subtracted 5 on both sides. This eliminated the 5 on the left. We started with $x + 5 = -7$, and, using the addition principle, we found a simpler equation $x = -12$ for which it was easy to *"see"* the solution. The equations $x + 5 = -7$ and $x = -12$ are *equivalent.*

Do Exercise 8.

Now we use the addition principle to solve an equation that involves a subtraction.

EXAMPLE 7 Solve: $a - 4 = 10$.

We have

$a - 4 = 10$

$a - 4 + 4 = 10 + 4$ Using the addition principle: adding 4 on both sides

$a + 0 = 14$ Simplifying

$a = 14.$ Identity property of 0

CHECK: $a - 4 = 10$

$14 - 4 \ ? \ 10$

10 | TRUE

The solution is 14.

Do Exercise 9.

8. Solve using the addition principle, subtracting 5 on both sides:

$x + 5 = -8.$

9. Solve: $t - 3 = 19$.

Answers on page A-7

Solve.

10. $8.7 = n - 4.5$

11. $y + 17.4 = 10.9$

Solve.

12. $x + \frac{1}{2} = -\frac{3}{2}$

13. $t - \frac{13}{4} = \frac{5}{8}$

Answers on page A-7

EXAMPLE 8 Solve: $-6.5 = y - 8.4$.

We have

$$-6.5 = y - 8.4$$

$$-6.5 + 8.4 = y - 8.4 + 8.4 \quad \text{Using the addition principle: adding 8.4 on both sides to eliminate } -8.4 \text{ on the right}$$

$$1.9 = y.$$

CHECK:

$$\begin{array}{c|c} -6.5 & y - 8.4 \\ \hline -6.5 \ ? & 1.9 - 8.4 \\ & -6.5 \end{array} \qquad \text{TRUE}$$

The solution is 1.9.

Note that equations are reversible. That is, if $a = b$ is true, then $b = a$ is true. Thus when we solve $-6.5 = y - 8.4$, we can reverse it and solve $y - 8.4 = -6.5$ if we wish.

Do Exercises 10 and 11.

EXAMPLE 9 Solve: $-\frac{2}{3} + x = \frac{5}{2}$.

We have

$$-\frac{2}{3} + x = \frac{5}{2}$$

$$\frac{2}{3} - \frac{2}{3} + x = \frac{2}{3} + \frac{5}{2} \quad \text{Adding } \tfrac{2}{3} \text{ on both sides}$$

$$x = \frac{2}{3} \cdot \frac{2}{2} + \frac{5}{2} \cdot \frac{3}{3} \quad \text{Multiplying by 1 to obtain equivalent fractional expressions with the least common denominator 6}$$

$$x = \frac{4}{6} + \frac{15}{6}$$

$$x = \frac{19}{6}.$$

CHECK:

$$\begin{array}{c|c} -\frac{2}{3} + x & \frac{5}{2} \\ \hline -\frac{2}{3} + \frac{19}{6} \ ? & \frac{5}{2} \\ -\frac{4}{6} + \frac{19}{6} & \\ \frac{15}{6} & \\ \frac{5}{2} & \end{array} \qquad \text{TRUE}$$

The solution is $\frac{19}{6}$.

Do Exercises 12 and 13.

2.1 EXERCISE SET

For Extra Help

Digital Video Tutor CD 3 Videotape 5 | InterAct Math | Math Tutor Center | MathXL | MyMathLab

a Determine whether the given number is a solution of the given equation.

1. 15; $x + 17 = 32$

2. 35; $t + 17 = 53$

3. 21; $x - 7 = 12$

4. 36; $a - 19 = 17$

5. -7; $6x = 54$

6. -9; $8y = -72$

7. 30; $\frac{x}{6} = 5$

8. 49; $\frac{y}{8} = 6$

9. 19; $5x + 7 = 107$

10. 9; $9x + 5 = 86$

11. -11; $7(y - 1) = 63$

12. -18; $x + 3 = 3 + x$

b Solve using the addition principle. Don't forget to check!

13. $x + 2 = 6$

CHECK: $x + 2 = 6$?

14. $y + 4 = 11$

CHECK: $y + 4 = 11$?

15. $x + 15 = -5$

CHECK: $x + 15 = -5$?

16. $t + 10 = 44$

CHECK: $t + 10 = 44$?

17. $x + 6 = -8$

CHECK: $x + 6 = -8$?

18. $z + 9 = -14$

19. $x + 16 = -2$

20. $m + 18 = -13$

21. $x - 9 = 6$

22. $x - 11 = 12$

23. $x - 7 = -21$

24. $x - 3 = -14$

25. $5 + t = 7$

26. $8 + y = 12$

27. $-7 + y = 13$

28. $-8 + y = 17$

29. $-3 + t = -9$

30. $-8 + t = -24$

31. $x + \frac{1}{2} = 7$

32. $24 = -\frac{7}{10} + r$

33. $12 = a - 7.9$

34. $2.8 + y = 11$

35. $r + \frac{1}{3} = \frac{8}{3}$

36. $t + \frac{3}{8} = \frac{5}{8}$

37. $m + \frac{5}{6} = -\frac{11}{12}$

38. $x + \frac{2}{3} = -\frac{5}{6}$

39. $x - \frac{5}{6} = \frac{7}{8}$

40. $y - \frac{3}{4} = \frac{5}{6}$

41. $-\frac{1}{5} + z = -\frac{1}{4}$

42. $-\frac{1}{8} + y = -\frac{3}{4}$

43. $7.4 = x + 2.3$

44. $8.4 = 5.7 + y$

45. $7.6 = x - 4.8$

46. $8.6 = x - 7.4$

47. $-9.7 = -4.7 + y$

48. $-7.8 = 2.8 + x$

49. $5\frac{1}{6} + x = 7$

50. $5\frac{1}{4} = 4\frac{2}{3} + x$

51. $q + \frac{1}{3} = -\frac{1}{7}$

52. $52\frac{3}{8} = -84 + x$

53. D$_W$ Explain the difference between equivalent expressions and equivalent equations.

54. D$_W$ When solving an equation using the addition principle, how do you determine which number to add or subtract on both sides of the equation?

SKILL MAINTENANCE

55. Add: $-3 + (-8)$. [1.3a]

56. Subtract: $-3 - (-8)$. [1.4a]

57. Multiply: $-\frac{2}{3} \cdot \frac{5}{8}$. [1.5a]

58. Divide: $-\frac{3}{7} \div \left(-\frac{9}{7}\right)$. [1.6c]

59. Divide: $\frac{2}{3} \div \left(-\frac{4}{9}\right)$. [1.6c]

60. Add: $-8.6 + 3.4$. [1.3a]

61. Subtract: $-\frac{2}{3} - \left(-\frac{5}{8}\right)$. [1.4a]

62. Multiply: $(-25.4)(-6.8)$. [1.5a]

Translate to an algebraic expression. [1.1b]

63. Jane had \$83 before paying x dollars for a pair of tennis shoes. How much does she have left?

64. Justin drove his S-10 pickup truck 65 mph for t hours. How far did he drive?

SYNTHESIS

Solve.

65. 🖩 $-356.788 = -699.034 + t$

66. $-\frac{4}{5} + \frac{7}{10} = x - \frac{3}{4}$

67. $x + \frac{4}{5} = -\frac{2}{3} - \frac{4}{15}$

68. $8 - 25 = 8 + x - 21$

69. $16 + x - 22 = -16$

70. $x + x = x$

71. $x + 3 = 3 + x$

72. $x + 4 = 5 + x$

73. $-\frac{3}{2} + x = -\frac{5}{17} - \frac{3}{2}$

74. $|x| = 5$

75. $|x| + 6 = 19$

2.2 SOLVING EQUATIONS: THE MULTIPLICATION PRINCIPLE

a Using the Multiplication Principle

Suppose that $a = b$ is true, and we multiply a by some number c. We get the same number if we multiply b by c, because a and b are the same number.

THE MULTIPLICATION PRINCIPLE

For any real numbers a, b, and c, $c \neq 0$,

$a = b$ is equivalent to $a \cdot c = b \cdot c$.

When using the multiplication principle, we sometimes say that we "multiply on both sides of the equation by the same number."

EXAMPLE 1 Solve: $5x = 70$.

To get x alone, we multiply by the *multiplicative inverse*, or *reciprocal*, of 5. Then we get the *multiplicative identity* 1 times x, or $1 \cdot x$, which simplifies to x. This allows us to eliminate 5 on the left.

$5x = 70$	The reciprocal of 5 is $\frac{1}{5}$.
$\frac{1}{5} \cdot 5x = \frac{1}{5} \cdot 70$	Multiplying by $\frac{1}{5}$ to get $1 \cdot x$ and eliminate 5 on the left
$1 \cdot x = 14$	Simplifying
$x = 14$	Identity property of 1: $1 \cdot x = x$

CHECK:

$$\begin{array}{c|c} 5x & 70 \\ \hline 5 \cdot 14 \ ? & 70 \\ 70 & \end{array}$$

TRUE

The solution is 14.

The multiplication principle also tells us that we can "divide on both sides of the equation by a nonzero number." This is because division is the same as multiplying by a reciprocal. That is,

$$\frac{a}{c} = \frac{b}{c} \text{ is equivalent to } a \cdot \frac{1}{c} = b \cdot \frac{1}{c}, \text{ when } c \neq 0.$$

In an expression like $5x$ in Example 1, the number 5 is called the **coefficient.** Example 1 could be done as follows, dividing by 5, the coefficient of x, on both sides.

EXAMPLE 2 Solve: $5x = 70$.

We have

$5x = 70$	
$\frac{5x}{5} = \frac{70}{5}$	Dividing by 5 on both sides
$1 \cdot x = 14$	Simplifying
$x = 14.$	Identity property of 1

Objective

a Solve equations using the multiplication principle.

1. Solve. Multiply on both sides.

$6x = 90$

2. Solve. Divide on both sides.

$4x = -7$

Answers on page A-7

3. Solve: $-6x = 108$.

4. Solve: $-x = -10$.

5. Solve: $-x = -10$.

Answers on page A-7

Do Exercises 1 and 2 on the preceding page.

EXAMPLE 3 Solve: $-4x = 92$.

We have

$$-4x = 92$$

$$\frac{-4x}{-4} = \frac{92}{-4}$$ Using the multiplication principle. Dividing by -4 on both sides is the same as multiplying by $-\frac{1}{4}$.

$$1 \cdot x = -23$$ Simplifying

$$x = -23.$$ Identity property of 1

CHECK:

$$\begin{array}{r|l} -4x & = 92 \\ \hline -4(-23) & ?\ 92 \\ 92 & \end{array}$$ TRUE

The solution is -23.

Do Exercise 3.

EXAMPLE 4 Solve: $-x = 9$.

We have

$$-x = 9$$

$$-1 \cdot x = 9$$ Using the property of -1: $-x = -1 \cdot x$

$$\frac{-1 \cdot x}{-1} = \frac{9}{-1}$$ Dividing by -1 on both sides

$$1 \cdot x = -9$$

$$x = -9.$$

CHECK:

$$\begin{array}{r|l} -x & = 9 \\ \hline -(-9) & ?\ 9 \\ 9 & \end{array}$$ TRUE

The solution is -9.

Do Exercise 4.

We can also solve the equation $-x = 9$ by multiplying as follows.

EXAMPLE 5 Solve: $-x = 9$.

We have

$$-x = 9$$

$$-1(-x) = -1 \cdot 9$$ Multiplying by -1 on both sides

$$-1 \cdot (-1) \cdot x = -9$$

$$1 \cdot x = -9$$

$$x = -9.$$

The solution is -9.

Do Exercise 5.

In practice, it is generally more convenient to divide on both sides of the equation if the coefficient of the variable is in decimal notation or is an integer. If the coefficient is in fraction notation, it is more convenient to multiply by a reciprocal.

EXAMPLE 6 Solve: $\frac{3}{8} = -\frac{5}{4}x$.

$$\frac{3}{8} = -\frac{5}{4}x$$

The reciprocal of $-\frac{5}{4}$ is $-\frac{4}{5}$. There is no sign change.

$$-\frac{4}{5} \cdot \frac{3}{8} = -\frac{4}{5} \cdot \left(-\frac{5}{4}x\right)$$ Multiplying by $-\frac{4}{5}$ to get $1 \cdot x$ and eliminate $-\frac{5}{4}$ on the right

$$-\frac{12}{40} = 1 \cdot x$$

$$-\frac{3}{10} = 1 \cdot x$$ Simplifying

$$-\frac{3}{10} = x$$ Identity property of 1

CHECK: $\frac{3}{8} = -\frac{5}{4}x$

$\frac{3}{8} \; ? \; -\frac{5}{4}\left(-\frac{3}{10}\right)$

$\frac{3}{8}$ TRUE

The solution is $-\frac{3}{10}$.

Note that equations are reversible. That is, if $a = b$ is true, then $b = a$ is true. Thus when we solve $\frac{3}{8} = -\frac{5}{4}x$, we can reverse it and solve $-\frac{5}{4}x = \frac{3}{8}$ if we wish.

Do Exercise 6.

EXAMPLE 7 Solve: $1.16y = 9744$.

$$1.16y = 9744$$

$$\frac{1.16y}{1.16} = \frac{9744}{1.16}$$ Dividing by 1.16 on both sides

$$y = \frac{9744}{1.16}$$

$$y = 8400$$ Using a calculator to divide

CHECK: $1.16y = 9744$

$1.16(8400) \; ? \; 9744$

9744 TRUE

The solution is 8400.

Do Exercises 7 and 8.

6. Solve: $\frac{2}{3} = -\frac{5}{6}y$.

Solve.

7. $1.12x = 8736$

8. $6.3 = -2.1y$

Answers on page A-7

9. Solve: $-14 = \frac{-y}{2}$.

Answer on page A-7

Now we use the multiplication principle to solve an equation that involves division.

EXAMPLE 8 Solve: $\frac{-y}{9} = 14$.

$$\frac{-y}{9} = 14$$

$$9 \cdot \frac{-y}{9} = 9 \cdot 14 \quad \text{Multiplying by 9 on both sides}$$

$$-y = 126$$

$$-1 \cdot (-y) = -1 \cdot 126 \quad \text{Multiplying by } -1 \text{ on both sides}$$

$$y = -126$$

CHECK:

$$\frac{-y}{9} = 14$$

$\frac{-(-126)}{9}$	? 14
$\frac{126}{9}$	
14	TRUE

The solution is -126.

Do Exercise 9.

Study Tips

TIME MANAGEMENT (PART 1)

Time is the most critical factor in your success in learning mathematics. Have reasonable expectations about the time you need to study math.

- **Juggling time.** Working 40 hours per week and taking 12 credit hours is equivalent to working two full-time jobs. Can you handle such a load? Your ratio of number of work hours to number of credit hours should be about 40/3, 30/6, 20/9, 10/12, or 5/14.
- **A rule of thumb on study time.** Budget about 2–3 hours for homework and study per week for every hour of class time.
- **Scheduling your time.** Make an hour-by-hour schedule of your typical week. Include work, school, home, sleep, study, and leisure times. Try to schedule time for study when you are most alert. Choose a setting that will enable you to maximize your concentration. Plan for success and it will happen!

"You cannot increase the quality or quantity of your achievement or performance except to the degree in which you increase your ability to use time effectively."

Brian Tracy, motivational speaker

2.2 EXERCISE SET

For Extra Help

Digital Video Tutor CD 3 Videotape 5 | InterAct Math | Math Tutor Center | MathXL | MyMathLab

a Solve using the multiplication principle. Don't forget to check!

1. $6x = 36$

CHECK: $6x = 36$?

2. $3x = 51$

CHECK: $3x = 51$?

3. $5x = 45$

CHECK: $5x = 45$?

4. $8x = 72$

CHECK: $8x = 72$?

5. $84 = 7x$

6. $63 = 9x$

7. $-x = 40$

8. $53 = -x$

9. $-x = -1$

10. $-47 = -t$

11. $7x = -49$

12. $8x = -56$

13. $-12x = 72$

14. $-15x = 105$

15. $-21x = -126$

16. $-13x = -104$

17. $\frac{t}{7} = -9$

18. $\frac{y}{-8} = 11$

19. $\frac{3}{4}x = 27$

20. $\frac{4}{5}x = 16$

21. $\frac{-t}{3} = 7$

22. $\frac{-x}{6} = 9$

23. $-\frac{m}{3} = \frac{1}{5}$

24. $\frac{1}{8} = -\frac{y}{5}$

25. $-\frac{3}{5}r = \frac{9}{10}$

26. $\frac{2}{5}y = -\frac{4}{15}$

27. $-\frac{3}{2}r = -\frac{27}{4}$

28. $-\frac{3}{8}x = -\frac{15}{16}$

29. $6.3x = 44.1$

30. $2.7y = 54$

31. $-3.1y = 21.7$

32. $-3.3y = 6.6$

33. $38.7m = 309.6$

34. $29.4m = 235.2$

35. $-\frac{2}{3}y = -10.6$

36. $-\frac{9}{7}y = 12.06$

37. $\frac{-x}{5} = 10$

38. $\frac{-x}{8} = -16$

39. $-\frac{t}{2} = 7$

40. $\frac{m}{-3} = 10$

41. ^{D}W When solving an equation using the multiplication principle, how do you determine by what number to multiply or divide on both sides of the equation?

42. ^{D}W Are the equations $x = 5$ and $x^2 = 25$ equivalent? Why or why not?

SKILL MAINTENANCE

Collect like terms. [1.7e]

43. $3x + 4x$

44. $6x + 5 - 7x$

45. $-4x + 11 - 6x + 18x$

46. $8y - 16y - 24y$

Remove parentheses and simplify. [1.8b]

47. $3x - (4 + 2x)$

48. $2 - 5(x + 5)$

49. $8y - 6(3y + 7)$

50. $-2a - 4(5a - 1)$

Translate to an algebraic expression. [1.1b]

51. Patty drives her van for 8 hr at a speed of *r* mph. How far does she drive?

52. A triangle has a height of 10 meters and a base of *b* meters. What is the area of the triangle?

SYNTHESIS

Solve.

53. ▦ $-0.2344m = 2028.732$

54. $0 \cdot x = 0$

55. $0 \cdot x = 9$

56. $4|x| = 48$

57. $2|x| = -12$

Solve for x.

58. $ax = 5a$

59. $3x = \frac{b}{a}$

60. $cx = a^2 + 1$

61. $\frac{a}{b}x = 4$

62. A student makes a calculation and gets an answer of 22.5. On the last step, she multiplies by 0.3 when a division by 0.3 should have been done. What is the correct answer?

2.3 USING THE PRINCIPLES TOGETHER

a Applying Both Principles

Consider the equation $3x + 4 = 13$. It is more complicated than those we discussed in the preceding two sections. In order to solve such an equation, we first isolate the x-term, $3x$, using the addition principle. Then we apply the multiplication principle to get x by itself.

EXAMPLE 1 Solve: $3x + 4 = 13$.

$$3x + 4 = 13$$

$$3x + 4 - 4 = 13 - 4 \quad \text{Using the addition principle: subtracting 4 on both sides}$$

First isolate the x-term. → $3x = 9$ Simplifying

$$\frac{3x}{3} = \frac{9}{3} \quad \text{Using the multiplication principle: dividing by 3 on both sides}$$

Then isolate x. → $x = 3$ Simplifying

CHECK:

$3x + 4 = 13$	
$3 \cdot 3 + 4$?	13
$9 + 4$	
13	TRUE

We use the rules for order of operations to carry out the check. We find the product $3 \cdot 3$. Then we add 4.

The solution is 3.

Do Exercise 1.

EXAMPLE 2 Solve: $-5x - 6 = 16$.

$$-5x - 6 = 16$$

$$-5x - 6 + 6 = 16 + 6 \quad \text{Adding 6 on both sides}$$

$$-5x = 22$$

$$\frac{-5x}{-5} = \frac{22}{-5} \quad \text{Dividing by } -5 \text{ on both sides}$$

$$x = -\frac{22}{5}, \text{ or } -4\frac{2}{5} \quad \text{Simplifying}$$

CHECK:

$-5x - 6 = 16$	
$-5\left(-\frac{22}{5}\right) - 6$?	16
$22 - 6$	
16	TRUE

The solution is $-\frac{22}{5}$.

Do Exercises 2 and 3.

Objectives

a Solve equations using both the addition and the multiplication principles.

b Solve equations in which like terms may need to be collected.

c Solve equations by first removing parentheses and collecting like terms; solve equations with no solutions and equations with an infinite number of solutions.

1. Solve: $9x + 6 = 51$.

Solve.

2. $8x - 4 = 28$

3. $-\frac{1}{2}x + 3 = 1$

Answers on page A-8

4. Solve: $-18 - m = -57$.

Solve.

5. $-4 - 8x = 8$

6. $41.68 = 4.7 - 8.6y$

Solve.

7. $4x + 3x = -21$

8. $x - 0.09x = 728$

Answers on page A-8

EXAMPLE 3 Solve: $45 - t = 13$.

$$45 - t = 13$$
$$-45 + 45 - t = -45 + 13 \quad \text{Adding } -45 \text{ on both sides}$$
$$-t = -32$$
$$-1(-t) = -1(-32) \quad \text{Multiplying by } -1 \text{ on both sides}$$
$$t = 32$$

The number 32 checks and is the solution.

Do Exercise 4.

EXAMPLE 4 Solve: $16.3 - 7.2y = -8.18$.

$$16.3 - 7.2y = -8.18$$
$$-16.3 + 16.3 - 7.2y = -16.3 + (-8.18) \quad \text{Adding } -16.3 \text{ on both sides}$$
$$-7.2y = -24.48$$
$$\frac{-7.2y}{-7.2} = \frac{-24.48}{-7.2} \quad \text{Dividing by } -7.2 \text{ on both sides}$$
$$y = 3.4$$

CHECK:

$$16.3 - 7.2y = -8.18$$
$$16.3 - 7.2(3.4) \ ? \ -8.18$$
$$16.3 - 24.48$$
$$-8.18 \qquad \text{TRUE}$$

The solution is 3.4.

Do Exercises 5 and 6.

b Collecting Like Terms

If there are like terms on one side of the equation, we collect them before using the addition or the multiplication principle.

EXAMPLE 5 Solve: $3x + 4x = -14$.

$$3x + 4x = -14$$
$$7x = -14 \quad \text{Collecting like terms}$$
$$\frac{7x}{7} = \frac{-14}{7} \quad \text{Dividing by 7 on both sides}$$
$$x = -2$$

The number -2 checks, so the solution is -2.

Do Exercises 7 and 8.

If there are like terms on opposite sides of the equation, we get them on the same side by using the addition principle. Then we collect them. In other words, we get all terms with a variable on one side and all numbers on the other.

EXAMPLE 6 Solve: $2x - 2 = -3x + 3$.

$$
\begin{aligned}
2x - 2 &= -3x + 3 \\
2x - 2 + 2 &= -3x + 3 + 2 && \text{Adding 2} \\
2x &= -3x + 5 && \text{Collecting like terms} \\
2x + 3x &= -3x + 3x + 5 && \text{Adding } 3x \\
5x &= 5 && \text{Simplifying} \\
\frac{5x}{5} &= \frac{5}{5} && \text{Dividing by 5} \\
x &= 1 && \text{Simplifying}
\end{aligned}
$$

CHECK:

$$
\begin{array}{r|l}
\multicolumn{2}{c}{2x - 2 = -3x + 3} \\
\hline
2 \cdot 1 - 2 \; ? & -3 \cdot 1 + 3 \\
2 - 2 & -3 + 3 \\
0 & 0
\end{array}
\qquad \text{TRUE}
$$

The solution is 1.

Do Exercises 9 and 10.

In Example 6, we used the addition principle to get all terms with a variable on one side and all numbers on the other side. Then we collected like terms and proceeded as before. If there are like terms on one side at the outset, they should be collected before proceeding.

EXAMPLE 7 Solve: $6x + 5 - 7x = 10 - 4x + 3$.

$$
\begin{aligned}
6x + 5 - 7x &= 10 - 4x + 3 \\
-x + 5 &= 13 - 4x && \text{Collecting like terms} \\
4x - x + 5 &= 13 - 4x + 4x && \text{Adding } 4x \text{ to get all terms with a variable on one side} \\
3x + 5 &= 13 && \text{Simplifying; that is, collecting like terms} \\
3x + 5 - 5 &= 13 - 5 && \text{Subtracting 5} \\
3x &= 8 && \text{Simplifying} \\
\frac{3x}{3} &= \frac{8}{3} && \text{Dividing by 3} \\
x &= \frac{8}{3} && \text{Simplifying}
\end{aligned}
$$

The number $\frac{8}{3}$ checks, so it is the solution.

Do Exercises 11 and 12.

Clearing Fractions and Decimals

In general, equations are easier to solve if they do not contain fractions or decimals. Consider, for example,

$$\frac{1}{2}x + 5 = \frac{3}{4} \quad \text{and} \quad 2.3x + 7 = 5.4.$$

Solve.

9. $7y + 5 = 2y + 10$

10. $5 - 2y = 3y - 5$

Solve.

11. $7x - 17 + 2x = 2 - 8x + 15$

12. $3x - 15 = 5x + 2 - 4x$

Answers on page A-8

If we multiply by 4 on both sides of the first equation and by 10 on both sides of the second equation, we have

$$4\left(\frac{1}{2}x + 5\right) = 4 \cdot \frac{3}{4} \quad \text{and} \quad 10(2.3x + 7) = 10 \cdot 5.4$$

or

$$4 \cdot \frac{1}{2}x + 4 \cdot 5 = 4 \cdot \frac{3}{4} \quad \text{and} \quad 10 \cdot 2.3x + 10 \cdot 7 = 10 \cdot 5.4$$

or

$$2x + 20 = 3 \quad \text{and} \quad 23x + 70 = 54.$$

The first equation has been "cleared of fractions" and the second equation has been "cleared of decimals." Both resulting equations are equivalent to the original equations and are easier to solve. *It is your choice* whether to clear fractions or decimals, but doing so often eases computations.

The easiest way to clear an equation of fractions is to multiply *every term on both sides* by the **least common multiple of all the denominators.**

EXAMPLE 8 Solve: $\frac{2}{3}x - \frac{1}{6} + \frac{1}{2}x = \frac{7}{6} + 2x.$

The number 6 is the least common multiple of all the denominators. We multiply by 6 on both sides.

$6\left(\frac{2}{3}x - \frac{1}{6} + \frac{1}{2}x\right) = 6\left(\frac{7}{6} + 2x\right)$	Multiplying by 6 on both sides
$6 \cdot \frac{2}{3}x - 6 \cdot \frac{1}{6} + 6 \cdot \frac{1}{2}x = 6 \cdot \frac{7}{6} + 6 \cdot 2x$	Using the distributive law (*Caution*! Be sure to multiply *all* the terms by 6.)
$4x - 1 + 3x = 7 + 12x$	Simplifying. Note that the fractions are cleared.
$7x - 1 = 7 + 12x$	Collecting like terms
$7x - 1 - 12x = 7 + 12x - 12x$	Subtracting $12x$
$-5x - 1 = 7$	Collecting like terms
$-5x - 1 + 1 = 7 + 1$	Adding 1
$-5x = 8$	Collecting like terms
$\frac{-5x}{-5} = \frac{8}{-5}$	Dividing by -5
$x = -\frac{8}{5}$	

CHECK:

$$\frac{2}{3}x - \frac{1}{6} + \frac{1}{2}x = \frac{7}{6} + 2x$$

	?
$\frac{2}{3}\left(-\frac{8}{5}\right) - \frac{1}{6} + \frac{1}{2}\left(-\frac{8}{5}\right)$	$\frac{7}{6} + 2\left(-\frac{8}{5}\right)$
$-\frac{16}{15} - \frac{1}{6} - \frac{8}{10}$	$\frac{7}{6} - \frac{16}{5}$
$-\frac{32}{30} - \frac{5}{30} - \frac{24}{30}$	$\frac{35}{30} - \frac{96}{30}$
$\frac{-32 - 5 - 24}{30}$	$-\frac{61}{30}$
$-\frac{61}{30}$	

TRUE

The solution is $-\frac{8}{5}$.

Do Exercise 13.

To illustrate clearing decimals, we repeat Example 4, but this time we clear the equation of decimals first. Compare both methods.

To clear an equation of decimals, we count the greatest number of decimal places in any one number. If the greatest number of decimal places is 1, we multiply by 10; if it is 2, we multiply by 100; and so on.

EXAMPLE 9 Solve: $16.3 - 7.2y = -8.18$.

The greatest number of decimal places in any one number is *two.* Multiplying by 100, which has *two* 0's, will clear all decimals.

$100(16.3 - 7.2y) = 100(-8.18)$	Multiplying by 100 on both sides
$100(16.3) - 100(7.2y) = 100(-8.18)$	Using the distributive law
$1630 - 720y = -818$	Simplifying
$1630 - 720y - 1630 = -818 - 1630$	Subtracting 1630
$-720y = -2448$	Collecting like terms
$\frac{-720y}{-720} = \frac{-2448}{-720}$	Dividing by -720
$y = \frac{17}{5}$, or 3.4	

The number $\frac{17}{5}$, or 3.4, checks, so it is the solution.

Do Exercise 14.

C Equations Containing Parentheses

To solve certain kinds of equations that contain parentheses, we first use the distributive laws to remove the parentheses. Then we proceed as before.

EXAMPLE 10 Solve: $4x = 2(12 - 2x)$.

$4x = 2(12 - 2x)$	
$4x = 24 - 4x$	Using the distributive laws to multiply and remove parentheses
$4x + 4x = 24 - 4x + 4x$	Adding $4x$ to get all the x-terms on one side
$8x = 24$	Collecting like terms
$\frac{8x}{8} = \frac{24}{8}$	Dividing by 8
$x = 3$	

The number 3 checks, so the solution is 3.

Do Exercises 15 and 16.

13. Solve: $\frac{7}{8}x - \frac{1}{4} + \frac{1}{2}x = \frac{3}{4} + x$.

14. Solve: $41.68 = 4.7 - 8.6y$.

Solve.

15. $2(2y + 3) = 14$

16. $5(3x - 2) = 35$

Answers on page A-8

Solve.

17. $3(7 + 2x) = 30 + 7(x - 1)$

18. $4(3 + 5x) - 4 = 3 + 2(x - 2)$

Determine whether the given number is a solution of the given equation.

19. 10; $3 + x = x + 3$

20. -7; $3 + x = x + 3$

21. $\frac{1}{2}$; $3 + x = x + 3$

22. 0; $3 + x = x + 3$

Answers on page A-8

Here is a procedure for solving the types of equation discussed in this section.

AN EQUATION-SOLVING PROCEDURE

1. Multiply on both sides to clear the equation of fractions or decimals. (This is optional, but it can ease computations.)
2. If parentheses occur, multiply to remove them using the *distributive laws.*
3. Collect like terms on each side, if necessary.
4. Get all terms with variables on one side and all numbers (constant terms) on the other side, using the *addition principle.*
5. Collect like terms again, if necessary.
6. Multiply or divide to solve for the variable, using the *multiplication principle.*
7. Check all possible solutions in the original equation.

EXAMPLE 11 Solve: $2 - 5(x + 5) = 3(x - 2) - 1$.

$$2 - 5(x + 5) = 3(x - 2) - 1$$

$$2 - 5x - 25 = 3x - 6 - 1 \qquad \text{Using the distributive laws to multiply and remove parentheses}$$

$$-5x - 23 = 3x - 7 \qquad \text{Collecting like terms}$$

$$-5x - 23 + 5x = 3x - 7 + 5x \qquad \text{Adding } 5x$$

$$-23 = 8x - 7 \qquad \text{Collecting like terms}$$

$$-23 + 7 = 8x - 7 + 7 \qquad \text{Adding 7}$$

$$-16 = 8x \qquad \text{Collecting like terms}$$

$$\frac{-16}{8} = \frac{8x}{8} \qquad \text{Dividing by 8}$$

$$-2 = x$$

CHECK:

$$\begin{array}{r|l} 2 - 5(x + 5) & 3(x - 2) - 1 \\ \hline 2 - 5(-2 + 5) \ ? & 3(-2 - 2) - 1 \\ 2 - 5(3) & 3(-4) - 1 \\ 2 - 15 & -12 - 1 \\ -13 & -13 \end{array} \qquad \text{TRUE}$$

The solution is -2.

Do Exercises 17 and 18.

EQUATIONS WITH INFINITELY MANY SOLUTIONS

The types of equations we have considered thus far in Sections 2.1–2.3 have all had exactly one solution. We now look at two other possibilities.

Consider

$$3 + x = x + 3.$$

Let's explore the solutions in Margin Exercises 19–22.

Do Exercises 19–22.

We know by the commutative law that this equation holds for any replacement of x with a real number. (See Section 1.7.) We have confirmed some of these solutions in Margin Exercises 19–22. Suppose we try to solve this equation using the addition principle:

$$3 + x = x + 3$$
$$-x + 3 + x = -x + x + 3 \quad \text{Adding } -x$$
$$3 = 3. \quad \text{TRUE}$$

We end with a true equation. The original equation holds for all real-number replacements. Thus the number of solutions is **infinite.**

EXAMPLE 12 Solve: $7x - 17 = 4 + 7(x - 3)$.

$$7x - 17 = 4 + 7(x - 3)$$
$$7x - 17 = 4 + 7x - 21 \quad \text{Using the distributive law to multiply and remove parentheses}$$
$$7x - 17 = 7x - 17 \quad \text{Collecting like terms}$$
$$-7x + 7x - 17 = -7x + 7x - 17 \quad \text{Adding } -7x$$
$$-17 = -17 \quad \text{TRUE}$$

Every real number is a solution. There are infinitely many solutions.

EQUATIONS WITH NO SOLUTION

Now consider

$$3 + x = x + 8.$$

Let's explore the solutions in Margin Exercises 23–26.

Do Exercises 23–26.

None of the replacements in Margin Exercises 23–26 are solutions of the given equation. In fact, there are no solutions. Let's try to solve this equation using the addition principle:

$$3 + x = x + 8$$
$$-x + 3 + x = -x + x + 8 \quad \text{Adding } -x$$
$$3 = 8. \quad \text{FALSE}$$

We end with a false equation. The original equation is false for all real-number replacements. Thus it has **no** solutions.

EXAMPLE 13 Solve: $3x + 4(x + 2) = 11 + 7x$.

$$3x + 4(x + 2) = 11 + 7x$$
$$3x + 4x + 8 = 11 + 7x \quad \text{Using the distributive law to multiply and remove parentheses}$$
$$7x + 8 = 11 + 7x \quad \text{Collecting like terms}$$
$$7x + 8 - 7x = 11 + 7x - 7x \quad \text{Subtracting } 7x$$
$$8 = 11 \quad \text{FALSE}$$

There are no solutions.

Do Exercises 27 and 28.

Determine whether the given number is a solution of the given equation.

23. 10; $3 + x = x + 8$

24. -7; $3 + x = x + 8$

25. $\frac{1}{2}$; $3 + x = x + 8$

26. 0; $3 + x = x + 8$

Solve.

27. $30 + 5(x + 3) = -3 + 5x + 48$

28. $2x + 7(x - 4) = 13 + 9x$

Answers on page A-8

The following is a guideline for solving linear equations of the types that we have considered in Sections 2.1–2.3.

RESULTING EQUATION	NUMBER OF SOLUTIONS	SOLUTION(S)
$x = a$, where a is a real number	One	The number a
A true equation such as $3 = 3$, $-11 = -11$, or $0 = 0$	Infinitely many	Every real number is a solution.
A false equation such as $3 = 8$, $-4 = 5$, or $0 = -5$	Zero	There are no solutions.

CALCULATOR CORNER

Checking Possible Solutions To check the possible solutions of an equation on a calculator, we can substitute and carry out the calculations on each side of the equation just as we do when we check by hand. To check the possible solution -2 in Example 13, for instance, we first substitute -2 for x in the expression on the left side of the equation. We press [2] [−] [5] [(] [(−)] [2] [+] [5] [)] [ENTER]. We get -13. Next, we substitute -2 for x in the expression on the right side of the equation. We then press [3] [(] [(−)] [2] [−] [2] [)] [−] [1] [ENTER]. Again we get -13. Since the two sides of the equation have the same value when x is -2, we know that -2 is the solution of the equation.

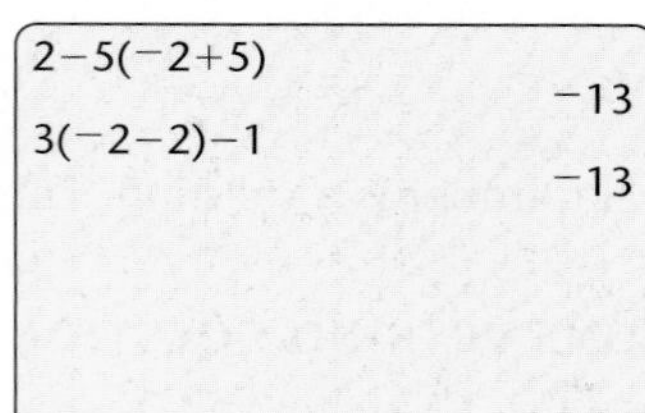
2−5(−2+5)
−13
3(−2−2)−1
−13

A table can also be used to check the possible solutions of an equation. First, we enter the left side and the right side of the equation on the Y = or equation editor screen. To do this, we first press [Y =]. If an expression for Y1 is currently entered, we place the cursor on it and press [CLEAR] to delete it. We do the same for any other entries that are present.

Next, we position the cursor to the right of Y1 = and enter the left side of the equation by pressing [2] [−] [5] [(] [X,T,θ,n] [+] [5] [)]. Then we position the cursor beside Y2 = and enter the right side of the equation by pressing [3] [(] [X,T,θ,n] [−] [2] [)] [−] [1]. Now we press [2nd] [TBLSET] to display the Table Setup screen. (TBLSET is the second operation associated with the [WINDOW] key.) On the Indpnt line, we position the cursor on "Ask" and press [ENTER] to set up a table in Ask mode. (The settings for TblStart and ΔTbl are irrelevant in Ask mode.)

Now we press [2nd] [TABLE] to display the table. (TABLE is the second operation associated with the [GRAPH] key.) We then enter the possible solution, -2, by pressing [(−)] [2] [ENTER]. We see that $Y1 = -13 = Y2$ for this value of x. This confirms that the left and right sides of the equation have the same value for $x = -2$, so -2 is the solution of the equation.

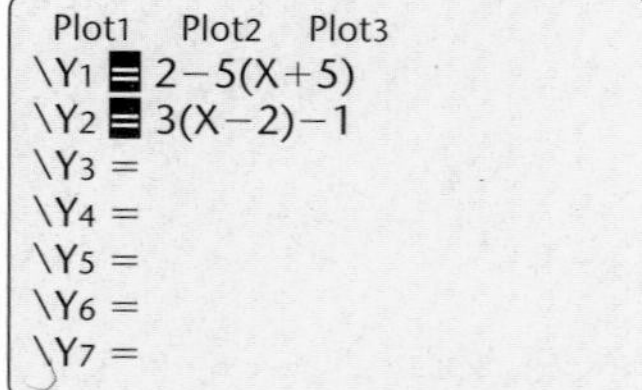
Plot1 Plot2 Plot3
\Y1 = 2−5(X+5)
\Y2 = 3(X−2)−1
\Y3 =
\Y4 =
\Y5 =
\Y6 =
\Y7 =

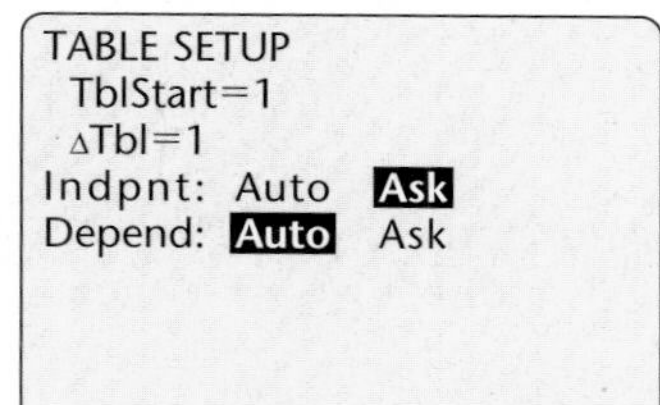
TABLE SETUP
TblStart=1
ΔTbl=1
Indpnt: Auto Ask
Depend: Auto Ask

X	Y1	Y2
−2	−13	−13

X =

Exercises:

1. Use substitution to check the solutions found in Examples 6, 7, 8, and 12.
2. Use a table set in Ask mode to check the solutions found in Examples 6, 7, 8, and 12.

2.3 EXERCISE SET

For Extra Help: Digital Video Tutor CD 3 Videotape 5 | InterAct Math | Math Tutor Center | MathXL | MyMathLab

a Solve. Don't forget to check!

1. $5x + 6 = 31$

CHECK: $5x + 6 = 31$ | ?

2. $7x + 6 = 13$

CHECK: $7x + 6 = 13$ | ?

3. $8x + 4 = 68$

CHECK: $8x + 4 = 68$ | ?

4. $4y + 10 = 46$

CHECK: $4y + 10 = 46$ | ?

5. $4x - 6 = 34$

6. $5y - 2 = 53$

7. $3x - 9 = 33$

8. $4x - 19 = 5$

9. $7x + 2 = -54$

10. $5x + 4 = -41$

11. $-45 = 3 + 6y$

12. $-91 = 9t + 8$

13. $-4x + 7 = 35$

14. $-5x - 7 = 108$

15. $-7x - 24 = -129$

16. $-6z - 18 = -132$

b Solve.

17. $5x + 7x = 72$

CHECK: $5x + 7x = 72$ | ?

18. $8x + 3x = 55$

CHECK: $8x + 3x = 55$ | ?

19. $8x + 7x = 60$

CHECK: $8x + 7x = 60$ | ?

20. $8x + 5x = 104$

CHECK: $8x + 5x = 104$ | ?

21. $4x + 3x = 42$

22. $7x + 18x = 125$

23. $-6y - 3y = 27$

24. $-5y - 7y = 144$

25. $-7y - 8y = -15$

26. $-10y - 3y = -39$

27. $x + \frac{1}{3}x = 8$

28. $x + \frac{1}{4}x = 10$

29. $10.2y - 7.3y = -58$

30. $6.8y - 2.4y = -88$

31. $8y - 35 = 3y$

32. $4x - 6 = 6x$

33. $8x - 1 = 23 - 4x$

34. $5y - 2 = 28 - y$

35. $2x - 1 = 4 + x$

36. $4 - 3x = 6 - 7x$

37. $6x + 3 = 2x + 11$

38. $14 - 6a = -2a + 3$

39. $5 - 2x = 3x - 7x + 25$

40. $-7z + 2z - 3z - 7 = 17$

41. $4 + 3x - 6 = 3x + 2 - x$

42. $5 + 4x - 7 = 4x - 2 - x$

43. $4y - 4 + y + 24 = 6y + 20 - 4y$

44. $5y - 7 + y = 7y + 21 - 5y$

Solve. Clear fractions or decimals first.

45. $\frac{7}{2}x + \frac{1}{2}x = 3x + \frac{3}{2} + \frac{5}{2}x$

46. $\frac{7}{8}x - \frac{1}{4} + \frac{3}{4}x = \frac{1}{16} + x$

47. $\frac{2}{3} + \frac{1}{4}t = \frac{1}{3}$

48. $-\frac{3}{2} + x = -\frac{5}{6} - \frac{4}{3}$

49. $\frac{2}{3} + 3y = 5y - \frac{2}{15}$

50. $\frac{1}{2} + 4m = 3m - \frac{5}{2}$

51. $\frac{5}{3} + \frac{2}{3}x = \frac{25}{12} + \frac{5}{4}x + \frac{3}{4}$

52. $1 - \frac{2}{3}y = \frac{9}{5} - \frac{y}{5} + \frac{3}{5}$

53. $2.1x + 45.2 = 3.2 - 8.4x$

54. $0.96y - 0.79 = 0.21y + 0.46$

55. $1.03 - 0.62x = 0.71 - 0.22x$

56. $1.7t + 8 - 1.62t = 0.4t - 0.32 + 8$

57. $\frac{2}{7}x - \frac{1}{2}x = \frac{3}{4}x + 1$

58. $\frac{5}{16}y + \frac{3}{8}y = 2 + \frac{1}{4}y$

C Solve.

59. $3(2y - 3) = 27$

60. $8(3x + 2) = 30$

61. $40 = 5(3x + 2)$

62. $9 = 3(5x - 2)$

63. $-23 + y = y + 25$

64. $17 - t = -t + 68$

65. $-23 + x = x - 23$

66. $y - \frac{2}{3} = -\frac{2}{3} + y$

67. $2(3 + 4m) - 9 = 45$

68. $5x + 5(4x - 1) = 20$

69. $5r - (2r + 8) = 16$

70. $6b - (3b + 8) = 16$

71. $6 - 2(3x - 1) = 2$

72. $10 - 3(2x - 1) = 1$

73. $5x + 5 - 7x = 15 - 12x + 10x - 10$

74. $3 - 7x + 10x - 14 = 9 - 6x + 9x - 20$

75. $22x - 5 - 15x + 3 = 10x - 4 - 3x + 11$

76. $11x - 6 - 4x + 1 = 9x - 8 - 2x + 12$

77. $5(d + 4) = 7(d - 2)$

78. $3(t - 2) = 9(t + 2)$

79. $8(2t + 1) = 4(7t + 7)$

80. $7(5x - 2) = 6(6x - 1)$

81. $3(r - 6) + 2 = 4(r + 2) - 21$

82. $5(t + 3) + 9 = 3(t - 2) + 6$

83. $19 - (2x + 3) = 2(x + 3) + x$

84. $13 - (2c + 2) = 2(c + 2) + 3c$

85. $2[4 - 2(3 - x)] - 1 = 4[2(4x - 3) + 7] - 25$

86. $5[3(7 - t) - 4(8 + 2t)] - 20 = -6[2(6 + 3t) - 4]$

87. $11 - 4(x + 1) - 3 = 11 + 2(4 - 2x) - 16$

88. $6(2x - 1) - 12 = 7 + 12(x - 1)$

89. $22x - 1 - 12x = 5(2x - 1) + 4$

90. $2 + 14x - 9 = 7(2x + 1) - 14$

91. $0.7(3x + 6) = 1.1 - (x + 2)$

92. $0.9(2x + 8) = 20 - (x + 5)$

93. $\mathrm{D_W}$ What procedure would you follow to solve an equation like $0.23x + \frac{17}{3} = -0.8 + \frac{3}{4}x$? Could your procedure be streamlined? If so, how?

94. $\mathrm{D_W}$ You are trying to explain to a classmate how equations can arise with infinitely many solutions and with no solutions. Give such an explanation. Does having no solution mean that 0 is a solution? Explain.

SKILL MAINTENANCE

95. Divide: $-22.1 \div 3.4$. [1.6c]

96. Multiply: $-22.1(3.4)$. [1.5a]

97. Factor: $7x - 21 - 14y$. [1.7d]

98. Factor: $8y - 88x + 8$. [1.7d]

SYNTHESIS

Solve.

99. 🖩 $0.008 + 9.62x - 42.8 = 0.944x + 0.0083 - x$

100. $\frac{1}{4}(8y + 4) - 17 = -\frac{1}{2}(4y - 8)$

101. $\frac{2}{3}\left(\frac{7}{8} - 4x\right) - \frac{5}{8} = \frac{3}{8}$

102. $\frac{4 - 3x}{7} = \frac{2 + 5x}{49} - \frac{x}{14}$

2.4 FORMULAS

a Evaluating Formulas

A **formula** is a "recipe" for doing a certain type of calculation. Formulas are often given as equations. When we replace the variables in an equation with numbers and calculate the result, we are **evaluating** the formula. We did some evaluating in Section 1.1.

Let's consider another example. A formula that has to do with weather is $M = \frac{1}{5}t$. You see a flash of lightning. After a few seconds you hear the thunder associated with that flash. How far away was the lightning?

Your distance from the storm is M miles. You can find that distance by counting the number of seconds t that it takes the sound of the thunder to reach you and then multiplying by $\frac{1}{5}$.

EXAMPLE 1 *Storm Distance.* Consider the formula $M = \frac{1}{5}t$. It takes 10 sec for the sound of thunder to reach you after you have seen a flash of lightning. How far away is the storm?

We substitute 10 for t and calculate M: $M = \frac{1}{5}t = \frac{1}{5}(10) = 2$. The storm is 2 mi away.

EXAMPLE 2 *Distance, Rate, and Time.* The distance d that a car will travel at a rate, or speed, r in time t is given by

$$d = rt.$$

A car travels at 75 miles per hour (mph) for 4.5 hr. How far will it travel?

We substitute 75 for r, 4.5 for t, and calculate d:

$$d = rt = (75)(4.5) = 337.5 \text{ mi}.$$

The car will travel 337.5 mi.

Do Exercises 1 and 2.

Objectives

a Evaluate a formula.

b Solve a formula for a specified letter.

1. **Storm Distance.** Suppose that it takes the sound of thunder 14 sec to reach you. How far away is the storm?

2. **Distance, Rate, and Time.** A car travels at 55 mph for 6.2 hr. How far will it travel?

Answers on page A-8

3. Solve for q: $B = \frac{1}{3}q$.

4. Distance, Rate, and Time. Solve for r: $d = rt$.

5. Electricity. Solve for I: $E = IR$. (This formula relates voltage E, current I, and resistance R.)

Solve for x.

6. $y = x + 5$

7. $y = x - 7$

8. $y = x - b$

Answers on page A-8

b Solving Formulas

Refer to Example 1. Suppose that we think we know how far we are from the storm and want to check by calculating the number of seconds it should take the sound of the thunder to reach us. We could substitute a number for M—say, 2—and solve for t:

$2 = \frac{1}{5}t$

$10 = t.$ Multiplying by 5

However, if we wanted to do this repeatedly, it might be easier to solve for t by getting it alone on one side. We "solve" the formula for t.

EXAMPLE 3 Solve for t: $M = \frac{1}{5}t$.

$M = \frac{1}{5}t$ We want this letter alone.

$5 \cdot M = 5 \cdot \frac{1}{5}t$ Multiplying by 5 on both sides

$5M = t$

In the above situation for $M = 2$, $t = 5M = 5(2)$, or 10.

EXAMPLE 4 *Distance, Rate, and Time.* Solve for t: $d = rt$.

$d = rt$ We want this letter alone.

$\frac{d}{r} = \frac{rt}{r}$ Dividing by r

$\frac{d}{r} = \frac{r}{r} \cdot t$

$\frac{d}{r} = t$ Simplifying

Do Exercises 3–5.

EXAMPLE 5 Solve for x: $y = x + 3$.

$y = x + 3$ We want this letter alone.

$y - 3 = x + 3 - 3$ Subtracting 3

$y - 3 = x$ Simplifying

EXAMPLE 6 Solve for x: $y = x - a$.

$y = x - a$ We want this letter alone.

$y + a = x - a + a$ Adding a

$y + a = x$ Simplifying

Do Exercises 6–8.

EXAMPLE 7 Solve for y: $6y = 3x$.

$6y = 3x$ We want this letter alone.

$\frac{6y}{6} = \frac{3x}{6}$ Dividing by 6

$y = \frac{1}{2}x$ Simplifying

EXAMPLE 8 Solve for y: $by = ax$.

$by = ax$ We want this letter alone.

$\frac{by}{b} = \frac{ax}{b}$ Dividing by b

$y = \frac{ax}{b}$ Simplifying

Do Exercises 9 and 10.

To see how the addition and multiplication principles apply to formulas, compare the following.

A. *Solve.* We carry this out as we did in Sections 2.1–2.3.

$5x + 2 = 12$ We want this letter alone.

$5x + 2 - 2 = 12 - 2$ Subtracting 2

$5x = 10$ Simplifying

$\frac{5x}{5} = \frac{10}{5}$ Dividing by 5

$x = 2$ Simplifying

B. *Solve.* We carry this out as we did in Sections 2.1–2.3, but we do not do as much simplifying or collecting like terms.

$$5x + 2 = 12$$
$$5x + 2 - 2 = 12 - 2$$
$$5x = 12 - 2$$
$$\frac{5x}{5} = \frac{12 - 2}{5}$$
$$x = \frac{12 - 2}{5}$$

C. *Solve for* x: $ax + b = c$. In this case, we cannot carry out any calculations because we have unknown letters.

$ax + b = c$ We want this letter alone.

$ax + b - b = c - b$ Subtracting b

$ax = c - b$ Simplifying

$\frac{ax}{a} = \frac{c - b}{a}$ Dividing by a

$x = \frac{c - b}{a}$ Simplifying

9. Solve for y: $9y = 5x$.

10. Solve for p: $ap = bq$.

11. Solve for x: $y = mx + b$.

12. Solve for Q: $tQ - p = a$.

Answers on page A-8

13. Circumference. Solve for D:

$$C = \pi D.$$

(This is a formula for the circumference C of a circle of diameter D.)

14. Averages. Solve for c:

$$A = \frac{a + b + c + d}{4}.$$

Answers on page A-8

Do Exercises 11 and 12 on the preceding page.

Solving Formulas

To solve a formula for a given letter, identify the letter and:

1. Multiply on both sides to clear fractions or decimals, if that is needed.
2. Collect like terms on each side, if necessary.
3. Get all terms with the letter to be solved for on one side of the equation and all other terms on the other side.
4. Collect like terms again, if necessary.
5. Solve for the letter in question.

EXAMPLE 9 *Circumference.* Solve for r: $C = 2\pi r$. This is a formula for the circumference C of a circle of radius r.

$C = 2\pi r$ We want this letter alone.

$\frac{C}{2\pi} = \frac{2\pi r}{2\pi}$ Dividing by 2π

$\frac{C}{2\pi} = r$

EXAMPLE 10 *Averages.* Solve for a: $A = \frac{a + b + c}{3}$. This is a formula for the average A of three numbers a, b, and c.

$A = \frac{a + b + c}{3}$ We want the letter a alone.

$3A = a + b + c$ Multiplying by 3 to clear the fraction

$3A - b - c = a$ Subtracting b and c

Do Exercises 13 and 14.

2.4 EXERCISE SET

For Extra Help

Digital Video Tutor CD 3 Videotape 5 | InterAct Math | Math Tutor Center | MathXL | MyMathLab

a, b Solve.

1. *Furnace Output.* The formula

$$B = 30a$$

is used in New England to estimate the minimum furnace output B, in Btu's, for a modern house with a square feet of flooring.

a) Determine the minimum furnace output for a 1900-ft^2 modern house.

b) Solve for a.

Source: U.S. Department of Energy

2. *Furnace Output.* The formula

$$B = 50a$$

is used in New England to estimate the minimum furnace output B, in Btu's, for an old, poorly insulated house with a square feet of flooring.

a) Determine the minimum furnace output for a 3200-ft^2 old, poorly insulated house.

b) Solve for a.

Source: U.S. Department of Energy

3. *Distance from a Storm.* The formula

$$M = \tfrac{1}{5}t$$

can be used to determine how far M, in miles, you are from lightning when its thunder takes t seconds to reach your ears.

a) It takes 8 sec for the sound of thunder to reach you after you have seen the lightning. How far away is the storm?

b) Solve for t.

4. *Electrical Power.* The power rating P, in watts, of an electrical appliance is determined by

$$P = I \cdot V,$$

where I is the current, in amperes, and V is measured in volts.

a) A kitchen requires 30 amps of current and the voltage in the house is 115 volts. What is the wattage of the kitchen?

b) Solve for I; for V.

5. *College Enrollment.* At many colleges, the number of "full-time-equivalent" students f is given by

$$f = \frac{n}{15},$$

where n is the total number of credits for which students have enrolled in a given semester.

a) Determine the number of full-time-equivalent students on a campus in which students registered for a total of 21,345 credits.

b) Solve for n.

6. *Surface Area of a Cube.* The surface area A of a cube with side s is given by

$$A = 6s^2.$$

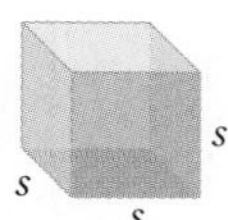

a) Find the surface area of a cube with sides of 3 in.

b) Solve for s^2.

7. *Calorie Density.* The calorie density D, in calories per ounce, of a food that contains c calories and weighs w ounces is given by

$$D = \frac{c}{w}.$$

Eight ounces of fat-free milk contains 84 calories. Find the calorie density of fat-free milk.

Source: *Nutrition Action Healthletter,* March 2000, p. 9. Center for Science in the Public Interest, Suite 300; 1875 Connecticut Ave NW, Washington, D.C. 20008.

8. *Wavelength of a Musical Note.* The wavelength w, in meters per cycle, of a musical note is given by

$$w = \frac{r}{f},$$

where r is the speed of the sound, in meters per second, and f is the frequency, in cycles per second. The speed of sound in air is 344 m/sec. What is the wavelength of a note whose frequency in air is 24 cycles per second?

9. *Size of a League Schedule.* When all n teams in a league play every other team twice, a total of N games are played, where

$$N = n^2 - n.$$

A soccer league has 7 teams and all teams play each other twice. How many games are played?

10. *Size of a League Schedule.* When all n teams in a league play every other team twice, a total of N games are played, where

$$N = n^2 - n.$$

A basketball league has 11 teams and all teams play each other twice. How many games are played?

b Solve for the indicated letter.

11. $y = 5x$, for x

12. $d = 55t$, for t

13. $a = bc$, for c

14. $y = mx$, for x

15. $y = 13 + x$, for x

16. $y = x - \frac{2}{3}$, for x

17. $y = x + b$, for x

18. $y = x - A$, for x

19. $y = 5 - x$, for x

20. $y = 10 - x$, for x

21. $y = a - x$, for x

22. $y = q - x$, for x

23. $8y = 5x$, for y

24. $10y = -5x$, for y

25. $By = Ax$, for x

26. $By = Ax$, for y

27. $W = mt + b$, for t

28. $W = mt - b$, for t

29. $y = bx + c$, for x

30. $y = bx - c$, for x

31. $A = \dfrac{a + b + c}{3}$, for b

32. $A = \dfrac{a + b + c}{3}$, for c

33. $A = at + b$, for t

34. $S = rx + s$, for x

35. *Area of a Parallelogram*:

$A = bh$, for h

(Area A, base b, height h)

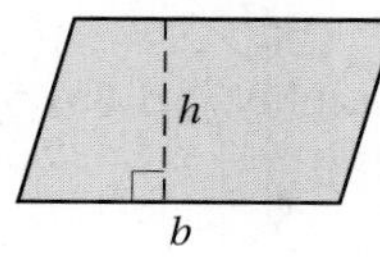

36. *Distance, Rate, Time*:

$d = rt$, for r

(Distance d, speed r, time t)

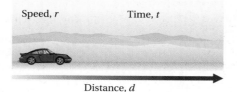

Speed, r Time, t

Distance, d

37. *Perimeter of a Rectangle*:

$P = 2l + 2w$, for w

(Perimeter P, length l, width w)

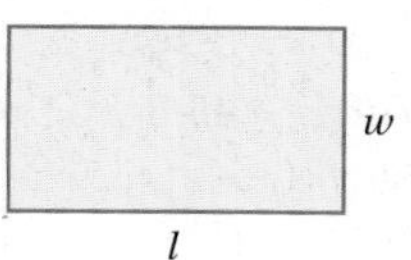

38. *Area of a Circle*:

$A = \pi r^2$, for r^2

(Area A, radius r)

39. *Average of Two Numbers*:

$A = \dfrac{a + b}{2}$, for a

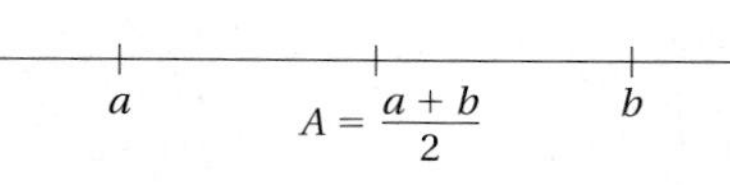

$A = \dfrac{a + b}{2}$

40. *Area of a Triangle*:

$A = \dfrac{1}{2}bh$, for b

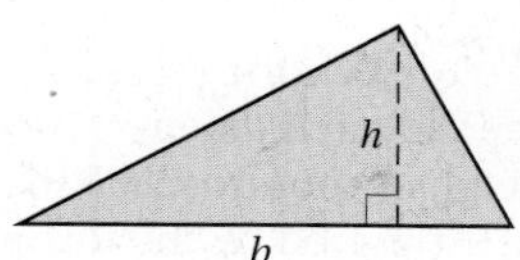

b

41. *Force*:

$F = ma$, for a

(Force F, mass m, acceleration a)

42. *Simple Interest*:

$I = Prt$, for P

(Interest I, principal P, interest rate r, time t)

43. *Relativity*:

$E = mc^2$, for c^2

(Energy E, mass m, speed of light c)

44. $Q = \dfrac{p - q}{2}$, for p

45. $Ax + By = c$, for x

46. $Ax + By = c$, for y

47. $v = \dfrac{3k}{t}$, for t

48. $P = \dfrac{ab}{c}$, for c

49. $\mathbf{D}_\mathbf{W}$ Devise an application in which it would be useful to solve the equation $d = rt$ for r. (See Exercise 36.)

50. $\mathbf{D}_\mathbf{W}$ The equations

$$P = 2l + 2w \quad \text{and} \quad w = \frac{P}{2} - l$$

are equivalent formulas involving the perimeter P, the length l, and the width w of a rectangle. (See Exercise 37.) Devise a problem for which the second of the two formulas would be more useful.

SKILL MAINTENANCE

51. Convert to decimal notation: $\frac{23}{25}$. [R.3a]

52. Add: $-23 + (-67)$. [1.3a]

53. Add: $0.082 + (-9.407)$. [1.3a]

54. Subtract: $-23 - (-67)$. [1.4a]

55. Subtract: $-45.8 - (-32.6)$. [1.4a]

56. Remove parentheses and simplify: [1.8b]

$$4a - 8b - 5(5a - 4b).$$

Convert to decimal notation. [R.4a]

57. 3.1%

58. 67.1%

59. Add: $-\frac{2}{3} + \frac{5}{6}$. [1.3a]

60. Subtract: $-\frac{2}{3} - \frac{5}{6}$. [1.4a]

SYNTHESIS

61. *Female Caloric Needs.* The number of calories K needed each day by a moderately active woman who weighs w pounds, is h inches tall, and is a years old can be estimated by the formula

$$K = 917 + 6(w + h - a).$$

a) Elaine is moderately active, weighs 120 lb, is 67 in. tall, and is 23 yr old. What are her caloric needs?

b) Solve the formula for a; for h; for w.

Source: Parker, M., *She Does Math.* Mathematical Association of America, p. 96

62. *Male Caloric Needs.* The number of calories K needed each day by a moderately active man who weighs w kilograms, is h centimeters tall, and is a years old can be estimated by the formula

$$K = 19.18w + 7h - 9.52a + 92.4.$$

a) Marv is moderately active, weighs 97 kg, is 185 cm tall, and is 55 yr old. What are his caloric needs?

b) Solve the formula for a; for h; for w.

Source: Parker, M., *She Does Math.* Mathematical Association of America, p. 96

Solve.

63. $A = \frac{1}{2}ah + \frac{1}{2}bh$, for b; for h

64. $P = 4m + 7mn$, for m

65. In $A = lw$, l and w both double. What is the effect on A?

66. In $P = 2a + 2b$, P doubles. Do a and b necessarily both double?

67. In $A = \frac{1}{2}bh$, b increases by 4 units and h does not change. What happens to A?

68. Solve for F: $D = \frac{1}{E + F}$.

2.5 APPLICATIONS OF PERCENT

a Translating and Solving

Many applied problems involve percent. Here we begin to see how equation solving can enhance our problem-solving skills. For background on the manipulative skills of percent notation, see Section R.4.

In solving percent problems, we first *translate* the problem to an equation. Then we *solve* the equation using the techniques discussed in Sections 2.1–2.3. The key words in the translation are as follows.

KEY WORDS IN PERCENT TRANSLATIONS

"**Of**" translates to "$\cdot$" or "$\times$". "**Is**" translates to "$=$".

"**What**" translates to any letter. **%** translates to "$\times \frac{1}{100}$" or "$\times$ 0.01".

EXAMPLE 1 Translate:

28%	of	5	is	what?
↓	↓	↓	↓	↓
28%	$\cdot$	5	$=$	a

This is a percent equation.

EXAMPLE 2 Translate:

45%	of	what	is	28?
↓	↓	↓	↓	↓
45%	$\times$	b	$=$	28

EXAMPLE 3 Translate:

What percent	of	90	is	7?
↓	↓	↓	↓	↓
n	$\cdot$	90	$=$	7

Do Exercises 1–6.

Percent problems are actually of three different types. Although the method we present does *not* require that you be able to identify which type we are studying, it is helpful to know them.

We know that

15 is 25% of 60, or

$15 = 25\% \times 60.$

We can think of this as:

Amount = Percent number $\times$ Base.

Objective

a Solve applied problems involving percent.

Translate to an equation. Do not solve.

1. 13% of 80 is what?

2. What is 60% of 70?

3. 43 is 20% of what?

4. 110% of what is 30?

5. 16 is what percent of 80?

6. What percent of 94 is 10.5?

Answers on page A-8

7. What is 2.4% of 80?

8. 25.3 is 22% of what number?

Answers on page A-8

Study Tips

USING THE SUPPLEMENTS

The new mathematical skills and concepts presented in the lectures will be of increased value to you if you begin the homework assignment as soon as possible after the lecture. Then if you still have difficulty with any of the exercises, you have time to access supplementary resources such as:

- *Student's Solutions Manual*
- Videotapes
- InterAct Math Tutorial CD-ROM
- AW Math Tutor Center
- MyMathLab
- MathXL

Each of the three types of percent problems depends on which of the three pieces of information is missing.

1. Finding the ***amount*** (the result of taking the percent)

   Example: What is 25% of 60?

   Translation: $y = 25\% \cdot 60$

2. Finding the ***base*** (the number you are taking the percent of)

   Example: 15 is 25% of what number?

   Translation: $15 = 25\% \cdot y$

3. Finding the ***percent number*** (the percent itself)

   Example: 15 is what percent of 60?

   Translation: $15 = y \cdot 60$

FINDING THE AMOUNT

EXAMPLE 4 What is 11% of 49?

What is 11% of 49?

Translate: $a = 11\% \times 49$

Solve: The letter is by itself. To solve the equation, we need only convert 11% to decimal notation and multiply:

$$a = 11\% \times 49 = 0.11 \times 49 = 5.39.$$

Thus, 5.39 is 11% of 49. The answer is 5.39.

Do Exercise 7.

FINDING THE BASE

EXAMPLE 5 3 is 16% of what?

3 is 16% of what number?

Translate: $3 = 16\% \times b$

$3 = 0.16 \times b$ Converting 16% to decimal notation

Solve: In this case, the letter is not by itself. To solve the equation, we divide by 0.16 on both sides:

$$3 = 0.16 \times b$$

$$\frac{3}{0.16} = \frac{0.16 \times b}{0.16} \quad \text{Dividing by 0.16}$$

$$18.75 = b. \quad \text{Simplifying}$$

The answer is 18.75.

Do Exercise 8.

FINDING THE PERCENT NUMBER

In solving these problems, you *must* remember to convert to percent notation after you have solved the equation.

EXAMPLE 6 $32 is what percent of $50?

$$\underbrace{\$32}\ \text{is}\ \underbrace{\text{what percent}}\ \text{of}\ \underbrace{\$50?}$$

Translate: $32 = p \times 50$

Solve: To solve the equation, we divide by 50 on both sides and convert the answer to percent notation:

$$32 = p \times 50$$

$$\frac{32}{50} = \frac{p \times 50}{50} \qquad \text{Dividing by 50}$$

$$0.64 = p$$

$$64\% = p. \qquad \text{Converting to percent notation}$$

Thus, 64% of $50 is $32. The answer is 64%.

Do Exercise 9.

EXAMPLE 7 *Coronary Heart Disease.* In 2001, there were 281 million people in the United States. About 2.5% of them had heart disease. How many had heart disease?

Source: American Heart Association

To solve the problem, we first reword and then translate. We let $a =$ the number of people in the United States with heart disease.

Rewording: What is 2.5% of 281?

Translate: $a = 2.5\% \times 281$

Solve: The letter is by itself. To solve the equation, we need only convert 2.5% to decimal notation and multiply:

$$a = 2.5\% \times 281 = 0.025 \times 281 = 7.025.$$

Thus, 7.025 million is 2.5% of 281 million, so in 2001 about 7.025 million people in the United States had heart disease.

Do Exercise 10.

EXAMPLE 8 *DVD Players.* At one time, Amazon.com had a Sharp DVD video player on sale for $899.98. This was 60% of the list price. What was the list price?

To solve the problem, we first reword and then translate. We let $L =$ the list price.

Rewording: $\underbrace{\$899.98}$ is 60% of $\underbrace{\text{what number?}}$

Translate: $899.98 = 60\% \times L$

9. What percent of $50 is $18?

10. Areas of Alaska and Arizona. The area of Arizona is 19% of the area of Alaska. The area of Alaska is 586,400 mi^2. What is the area of Arizona?

Answers on page A-8

11. **Population of Arizona.** The population of Arizona was 5.1 million in 2000. This was 130.6% of its population in 1990. What was the population in 1990?
Source: U.S. Bureau of the Census

12. **Job Opportunities.** There were 252 thousand medical assistants in 1998. Job opportunities are expected to grow to 398 thousand by 2008. What is the percent of increase?
Source: *Handbook of U.S. Labor Statistics*

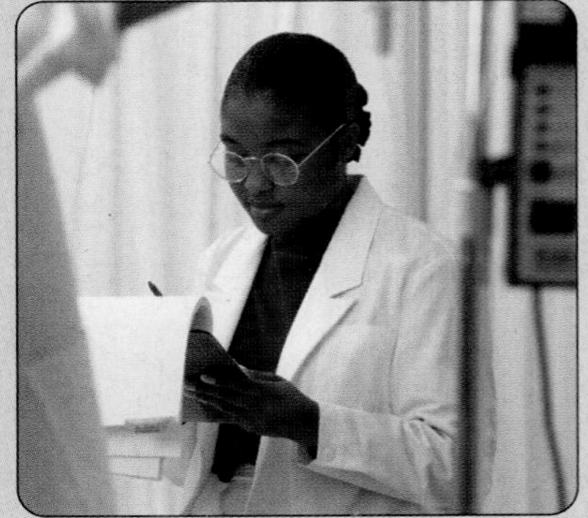

Answers on page A-8

Solve: To solve the equation, we convert 60% to decimal notation and divide by 0.60 on both sides:

$$899.98 = 60\% \times L$$

$$899.98 = 0.60 \times L \quad \text{Converting to decimal notation}$$

$$\frac{899.98}{0.60} = \frac{0.60 \times L}{0.60} \quad \text{Dividing by 0.60}$$

$$1499.97 \approx L. \quad \text{Simplifying using a calculator and rounding to the nearest cent}$$

The list price was about $1499.97.

Do Exercise 11.

EXAMPLE 9 *Compaq PAP.* A Compaq iPAQ PA-1 64 MB Personal Audio Player (PAP) was on sale on the Internet for $199.99, decreased from a normal list price of $249.99. What was the percent of decrease?

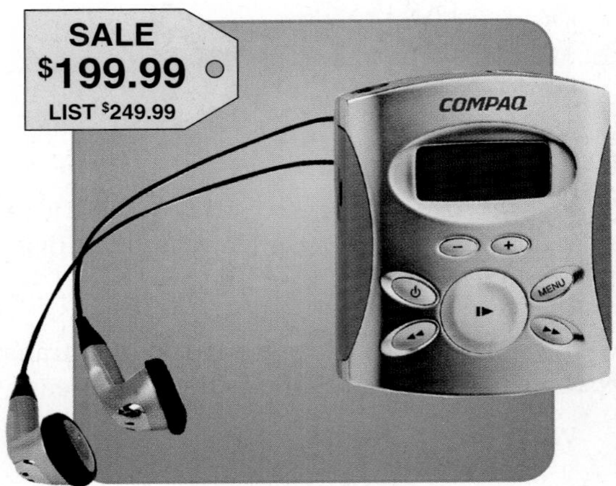

To solve the problem, we must first determine the amount of decrease from the original price:

Original price	minus	Sale price	=	Decrease
↓	↓	↓	↓	↓
$249.99	−	$199.99	=	$50.00.

Using the $50 decrease, we reword and translate. We let p = the percent of decrease. We want to know, "What percent of the *original* price is $50?"

Rewording:	$50	is	what percent	of	$249.99?
	↓	↓	↓	↓	↓
Translate:	50	=	p	×	249.99

Solve: To solve the equation, we divide by 249.99 on both sides and convert the answer to percent notation:

$$50 = p \times 249.99$$

$$\frac{50}{249.99} = \frac{p \times 249.99}{249.99} \quad \text{Dividing by 249.99}$$

$$0.20 \approx p \quad \text{Simplifying and converting to percent notation}$$

$$20\% = p.$$

Thus the percent of decrease was about 20%.

Do Exercise 12.

2.5 EXERCISE SET

For Extra Help

Digital Video Tutor CD 3 Videotape 6

InterAct Math

Math Tutor Center

MathXL

MyMathLab

a Solve.

1. What percent of 180 is 36?

2. What percent of 76 is 19?

3. 45 is 30% of what number?

4. 20.4 is 24% of what number?

5. What number is 65% of 840?

6. What is 50% of 50? (This was a \$500.00 question on the "Who Wants To Be a Millionaire?" television quiz show.)

7. 30 is what percent of 125?

8. 57 is what percent of 300?

9. 12% of what number is 0.3?

10. 7 is 175% of what number?

11. 2 is what percent of 40?

12. 40 is 2% of what number?

13. What percent of 68 is 17?

14. What percent of 150 is 39?

15. What number is 35% of 240?

16. What number is 1% of one million?

17. What percent of 125 is 30?

18. What percent of 60 is 75?

19. What percent of 300 is 48?

20. What percent of 70 is 70?

21. 14 is 30% of what number?

22. 54 is 24% of what number?

23. What is 2% of 40?

24. What is 40% of 2?

25. 0.8 is 16% of what number?

26. 25 is what percent of 50?

27. 54 is 135% of what number?

28. 8 is 2% of what number?

Costs of Owning a Dog. The American Pet Products Manufacturers Association estimates that the total cost of owning a dog for its lifetime is $6600. The following circle graph shows the relative costs of raising a dog from birth to death.

Costs of Owning a Dog

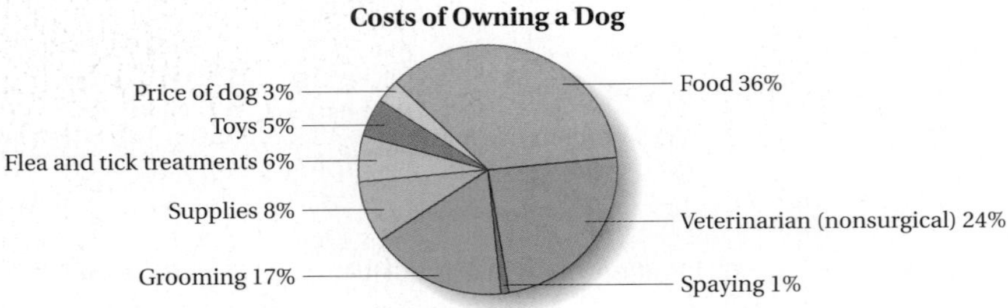

Source: The American Pet Products Manufacturers Association

Complete the following table of costs of owning a dog for its lifetime.

	EXPENSE ITEM	COST		EXPENSE ITEM	COST
29.	Price of dog		**30.**	Food	
31.	Veterinarian		**32.**	Grooming	
33.	Supplies		**34.**	Flea and tick treatments	

35. *Auto Sales.* In 2000, 17.4 million cars were sold in the United States. Of these, 11.9 million were manufactured in the United States, 4.5 million in Asia, and 1 million in Europe. What percent were manufactured in each region?
Source: *Autodata*

36. *Auto Sales.* In 1997, 15.2 million cars were sold in the United States. Of these, 10.8 million were manufactured in the United States, 3.7 million in Asia, and 0.7 million in Europe. What percent were manufactured in each region?
Source: *Autodata*

37. *Batting Average.* At one point in a recent season, Sammy Sosa of the Chicago Cubs had 193 hits. His batting average was 0.320, or 32%. That is, of the total number of at-bats, 32% were hits. How many at-bats did he have?
Source: Major League Baseball

38. *Pass Completions.* At one point in a recent season, Peyton Manning of the Indianapolis Colts had completed 357 passes. This was 62.5% of his attempts. How many attempts did he make?
Source: National Football League

39. *Student Loans.* To finance her community college education, Sarah takes out a Stafford loan for $3500. After a year, Sarah decides to pay off the interest, which is 8% of $3500. How much will she pay?

40. *Student Loans.* Paul takes out a subsidized federal Stafford loan for $2400. After a year, Paul decides to pay off the interest, which is 7% of $2400. How much will he pay?

41. *Tipping.* Leon left a $4 tip for a meal that cost $25.

a) What percent of the cost of the meal was the tip?
b) What was the total cost of the meal including the tip?

42. *Tipping.* Selena left a $12.76 tip for a meal that cost $58.

a) What percent of the cost of the meal was the tip?
b) What was the total cost of the meal including the tip?

43. *Tipping.* Leon left a 15% tip for a meal that cost \$25.

a) How much was the tip?
b) What was the total cost of the meal including the tip?

44. *Tipping.* Sam, Selena, Rachel, and Clement left a 15% tip for a meal that cost \$58.

a) How much was the tip?
b) What was the total cost of the meal including the tip?

45. *Tipping.* Leon left a 15% tip of \$4.32 for a meal.

a) What was the cost of the meal before the tip?
b) What was the total cost of the meal including the tip?

46. *Tipping.* Selena left a 15% tip of \$8.40 for a meal.

a) What was the cost of the meal before the tip?
b) What was the total cost of the meal including the tip?

47. In a medical study of a group of pregnant women with "poor" diets, 16 of the women, or 8%, had babies who were in good or excellent health. How many women were in the original study?

48. In a medical study of a group of pregnant women with "good-to-excellent" diets, 285 of the women, or 95%, had babies who were in good or excellent health. How many women were in the original study?

49. *Body Fat.* The author of this text exercises regularly at a local YMCA that recently offered a body-fat percentage test to its members. The device used measures the passage of a very low voltage of electricity through the body. The author's body-fat percentage was found to be 16.5% and he weighs 191 lb. What part, in pounds, of his body weight is fat?

50. *Junk Mail.* The U.S. Postal Service reports that we open and read 78% of the junk mail that we receive. A sports instructional videotape company sends out 10,500 advertising brochures.

a) How many of the brochures can it expect to be opened and read?
b) The company sells videos to 189 of the people who receive the brochure. What percent of the 10,500 people who receive the brochure buy the video?

Source: U.S. Postal Service

Life Insurance Rates for Smokers and Nonsmokers. The data in the following table illustrate how yearly rates (premiums) for a $500,000 term life insurance policy are increased for smokers. Complete the table by finding the missing numbers. Round to the nearest percent and dollar.

TYPICAL INSURANCE PREMIUMS (DOLLARS)

	AGE	RATE FOR NONSMOKER	RATE FOR SMOKER	RATE INCREASE	PERCENT OF INCREASE FOR SMOKER
	35	$345	$630	$285	83%
51.	40	$430	$735		
52.	45	$565			84%
53.	50	$780			100%
54.	55	$985	$2137		
55.	60	$1645	$2955		
56.	65	$2943			85%

Source: Pacific Life PL Protector Term Life Portfolio, OYT Rates

57. $^{\mathrm{D}}\mathrm{W}$ The 80/20 rule is commonly quoted in the field of business. It asserts that 80% of your results will come from 20% of your activities. Discuss how this might affect you as a student and as an employee.

58. $^{\mathrm{D}}\mathrm{W}$ Comment on the following quote by Yogi Berra, a famous Major League Hall of Fame baseball player: "Ninety percent of hitting is mental. The other half is physical."

SKILL MAINTENANCE

Compute. [R.3b]

59. $9.076 \div 0.05$

60. 9.076×0.05

61. $1.089 + 10.89 + 0.1089$

62. $1000.23 - 156.0893$

Remove parentheses and simplify. [1.8b]

63. $-5a + 3c - 2(c - 3a)$

64. $4(x - 2y) - (y - 3x)$

Add. [1.3a]

65. $-6.5 + 2.6$

66. $-\frac{3}{8} + (-5) + \frac{1}{4} + (-1)$

SYNTHESIS

67. It has been determined that at the age of 15, a boy has reached 96.1% of his final adult height. Jaraan is 6 ft 4 in. at the age of 15. What will his final adult height be?

68. It has been determined that at the age of 10, a girl has reached 84.4% of her final adult height. Dana is 4 ft 8 in. at the age of 10. What will her final adult height be?

2.6 APPLICATIONS AND PROBLEM SOLVING

Objective

a Solve applied problems by translating to equations.

a Five Steps for Solving Problems

We have discussed many new equation-solving tools in this chapter and used them for applications and problem solving. Here we consider a five-step strategy that can be very helpful in solving problems.

FIVE STEPS FOR PROBLEM SOLVING IN ALGEBRA

1. *Familiarize* yourself with the problem situation.
2. *Translate* the problem to an equation.
3. *Solve* the equation.
4. *Check* the answer in the original problem.
5. *State* the answer to the problem clearly.

Of the five steps, the most important is probably the first one: becoming familiar with the problem situation. The table below lists some hints for familiarization.

TO FAMILIARIZE YOURSELF WITH A PROBLEM:

- If a problem is given in words, read it carefully. Reread the problem, perhaps aloud. Try to verbalize the problem as if you were explaining it to someone else.
- Choose a variable (or variables) to represent the unknown and clearly state what the variable represents. Be descriptive! For example, let $L =$ the length, $d =$ the distance, and so on.
- Make a drawing and label it with known information, using specific units if given. Also, indicate unknown information.
- Find further information. Look up formulas or definitions with which you are not familiar. (Geometric formulas appear on the inside front cover of this text.) Consult a reference librarian or the Internet.
- Create a table that lists all the information you have available. Look for patterns that may help in the translation to an equation.
- Think of a possible answer and check the guess. Note the manner in which the guess is checked.

EXAMPLE 1 *Hiking.* In 1998, at age 79, Earl Shaffer became the oldest person to hike all 2100 miles of the Appalachian Trail—from Springer Mountain, Georgia, to Mount Katahdin, Maine. At one point, Shaffer stood atop Big Walker Mountain, Virginia, which is three times as far from the northern end as from the southern end. How far was Shaffer from each end of the trail?
Source: Appalachian Trail Conference

1. **Running.** In 1997, Yiannis Kouros of Australia set the record for the greatest distance run in 24 hr by running 188 mi. After 8 hr, he was approximately twice as far from the finish line as he was from the start. How far had he run?
Source: *Guinness World Records 2000 Millennium Edition*

Answer on page A-8

1. **Familiarize.** Let's consider a drawing.

To become familiar with the problem, let's guess a possible distance that Shaffer stood from Springer Mountain—say, 600 mi. Three times 600 mi is 1800 mi. Since 600 mi + 1800 mi = 2400 mi and 2400 mi is greater than 2100 mi, we see that our guess is too large. Rather than guess again, let's use the skills we have obtained in the ability to solve equations. We let

d = distance, in miles, to the southern end,

and

$3d$ = the distance, in miles, to the northern end.

(We could also let x = the distance to the northern end and $\frac{1}{3}x$ = the distance to the southern end.)

2. **Translate.** From the drawing, we see that the lengths of the two parts of the trail must add up to 2100 mi. This leads to our translation.

Distance to southern end	plus	Distance to northern end	is	2100 mi
↓	↓	↓	↓	↓
d	$+$	$3d$	$=$	2100

3. **Solve.** We solve the equation:

$$d + 3d = 2100$$
$$4d = 2100 \quad \text{Collecting like terms}$$
$$\frac{4d}{4} = \frac{2100}{4} \quad \text{Dividing by 4}$$
$$d = 525.$$

4. **Check.** As expected, d is less than 600 mi. If $d = 525$ mi, then $3d = 1575$ mi. Since 525 mi + 1575 mi = 2100 mi, we have a check.

5. **State.** Atop Big Walker Mountain, Shaffer stood 525 mi from Springer Mountain and 1575 mi from Mount Katahdin.

Do Exercise 1 on the preceding page.

EXAMPLE 2 *Gourmet Sandwiches.* A gourmet sandwich shop located near a college campus specializes in sandwiches prepared in buns of length 18 in. Suppose Jenny, Demi, and Sarah buy one of these sandwiches and take it back to their apartment. Since they have different appetites, Jenny cuts the sandwich in such a way that Demi gets half of what Jenny gets and Sarah gets three-fourths of what Jenny gets. Find the length of each person's sandwich.

1. **Familiarize.** We first make a drawing.

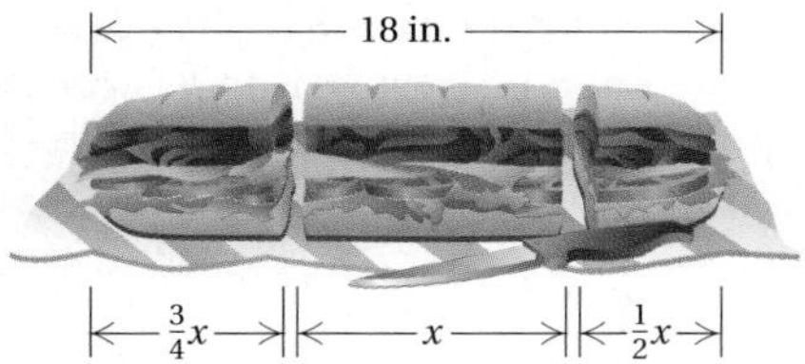

Because the sandwich lengths are expressed in terms of Jenny's sandwich, we let

$$x = \text{the length of Jenny's sandwich.}$$

Then $\frac{1}{2}x = $ the length of Demi's sandwich

and $\frac{3}{4}x = $ the length of Sarah's sandwich.

2. **Translate.** From the statement of the problem and the drawing, we see that the lengths add up to 18 in. That gives us our translation:

Length of Jenny's sandwich	plus	Length of Demi's sandwich	plus	Length of Sarah's sandwich	is	Total length
↓	↓	↓	↓	↓	↓	↓
x	$+$	$\frac{1}{2}x$	$+$	$\frac{3}{4}x$	$=$	$18.$

3. **Solve.** We begin by clearing fractions as follows:

$$x + \frac{1}{2}x + \frac{3}{4}x = 18 \qquad \text{The LCM of all the denominators is 4.}$$

$$4\left(x + \frac{1}{2}x + \frac{3}{4}x\right) = 4 \cdot 18 \qquad \text{Multiplying by the LCM, 4}$$

$$4 \cdot x + 4 \cdot \frac{1}{2}x + 4 \cdot \frac{3}{4}x = 4 \cdot 18 \qquad \text{Using the distributive law}$$

$$4x + 2x + 3x = 72 \qquad \text{Simplifying}$$

$$9x = 72 \qquad \text{Collecting like terms}$$

$$\frac{9x}{9} = \frac{72}{9} \qquad \text{Dividing by 9}$$

$$x = 8.$$

4. **Check.** Do we have an answer to the *problem*? If the length of Jenny's sandwich is 8 in., then the length of Demi's sandwich is $\frac{1}{2} \cdot 8$ in., or 4 in., and the length of Sarah's sandwich is $\frac{3}{4} \cdot 8$ in., or 6 in. These lengths add up to 18 in. Our answer checks.

2. **Rocket Sections.** A rocket is divided into three sections: the payload and navigation section in the top, the fuel section in the middle, and the rocket engine section in the bottom. The top section is one-sixth the length of the bottom section. The middle section is one-half the length of the bottom section. The total length is 240 ft. Find the length of each section.

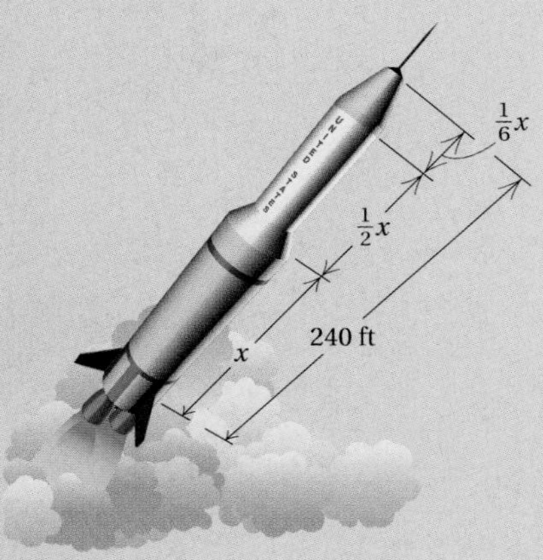

Answer on page A-8

5. **State.** The length of Jenny's sandwich is 8 in., the length of Demi's sandwich is 4 in., and the length of Sarah's sandwich is 6 in.

Do Exercise 2.

Recall that the

Set of integers $= \{\ldots, -5, -4, -3, -2, -1, 0, 1, 2, 3, 4, 5, \ldots\}$.

Before we solve the next problem, we need to learn some additional terminology regarding integers.

The following are examples of **consecutive integers:** 16, 17, 18, 19, 20; and -31, -30, -29, -28. Note that consecutive integers can be represented in the form x, $x + 1$, $x + 2$, and so on.

The following are examples of **consecutive even integers:** 16, 18, 20, 22, 24; and -52, -50, -48, -46. Note that consecutive even integers can be represented in the form x, $x + 2$, $x + 4$, and so on.

The following are examples of **consecutive odd integers:** 21, 23, 25, 27, 29; and -71, -69, -67, -65. Note that consecutive odd integers can be represented in the form x, $x + 2$, $x + 4$, and so on.

EXAMPLE 3 *Interstate Mile Markers.* If you are traveling on a U.S. interstate highway, you will notice numbered markers every mile to tell your location in case of an accident or other emergency. In many states, the numbers on the markers increase from west to east. The sum of two consecutive mile markers on I-70 in Kansas is 559. Find the numbers on the markers.

Source: Federal Highway Administration, Ed Rotalewski

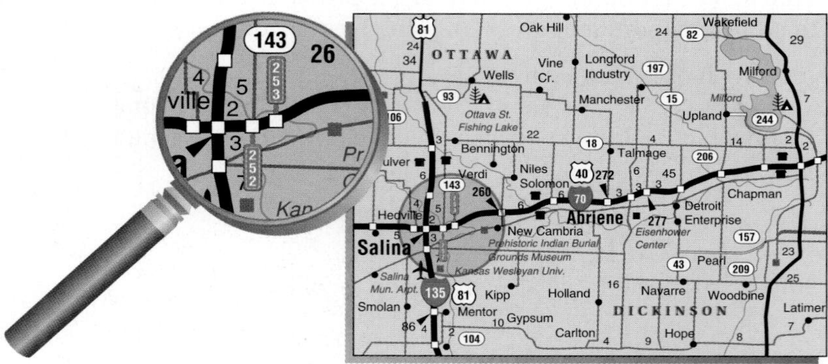

1. **Familiarize.** The numbers on the mile markers are consecutive positive integers. Thus if we let $x =$ the smaller number, then $x + 1 =$ the larger number.

To become familiar with the problem, we can make a table. First, we guess a value for x; then we find $x + 1$. Finally, we add the two numbers and check the sum.

x	$x + 1$	SUM OF x AND $x + 1$
114	115	229
252	253	505
302	303	605

From the table, we see that the first marker should be between 252 and 302. You might actually solve the problem this way, but let's work on developing our algebra skills.

2. **Translate.** We reword the problem and translate as follows.

First integer	plus	Second integer	is	559	Rewording
↓	↓	↓	↓	↓	
x	$+$	$(x + 1)$	$=$	559	Translating

3. **Solve.** We solve the equation:

$$\begin{aligned} x + (x + 1) &= 559 \\ 2x + 1 &= 559 && \text{Collecting like terms} \\ 2x + 1 - 1 &= 559 - 1 && \text{Subtracting 1} \\ 2x &= 558 \\ \frac{2x}{2} &= \frac{558}{2} && \text{Dividing by 2} \\ x &= 279. \end{aligned}$$

If x is 279, then $x + 1$ is 280.

4. **Check.** Our possible answers are 279 and 280. These are consecutive positive integers and $279 + 280 = 559$, so the answers check.

5. **State.** The mile markers are 279 and 280.

Do Exercise 3.

EXAMPLE 4 *IKON Copiers.* IKON Office Solutions rents a Canon IR330 copier for $225 per month plus 1.2¢ per copy. A law firm needs to lease a copy machine for use during a special case that they anticipate will take 3 months. If they allot a budget of $1100, how many copies can they make?

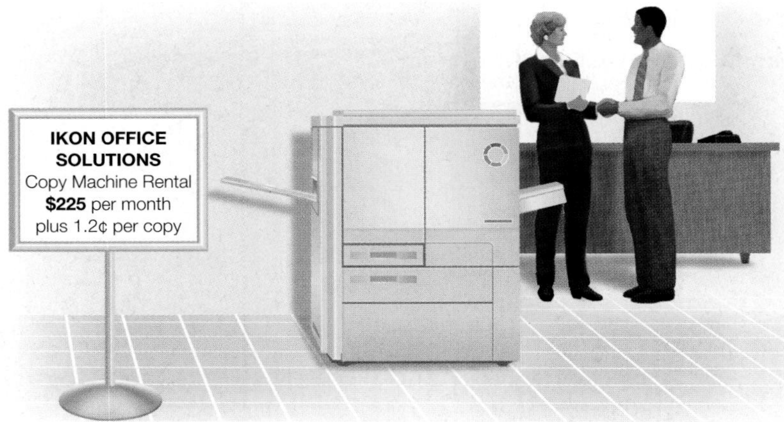

Source: IKON Office Solutions, Nathan DuMond, Sales Manager

1. **Familiarize.** Suppose that the law firm makes 20,000 copies. Then the cost is monthly charges plus copy charges, or

3($225)	plus	Cost per copy	times	Number of copies
↓	↓	↓	↓	↓
$675	$+$	$0.012	$\cdot$	20,000,

3. **Interstate Mile Markers.** The sum of two consecutive mile markers on I-90 in upstate New York is 627. (On I-90 in New York, the marker numbers increase from east to west.) Find the numbers on the markers.
Source: New York State Department of Transportation

Answer on page A-8

4. IKON Copiers. The law firm in Example 4 decides to raise its budget to $1400 for the 3-month period. How many copies can they make for $1400?

which is $915. This process familiarizes us with the way in which a calculation is made. Note that we convert 1.2¢ to $0.012 so that all information is in the same unit, dollars. Otherwise, we will not get the correct answer.

We let $c =$ the number of copies that can be made for $1100.

2. **Translate.** We reword the problem and translate as follows.

$$\underbrace{\text{Monthly cost}}_{3(\$225)} \ \underbrace{\text{plus}}_{+} \ \underbrace{\text{Cost per copy}}_{\$0.012} \ \underbrace{\text{times}}_{\cdot} \ \underbrace{\text{Number of copies}}_{c} \ \underbrace{\text{is}}_{=} \ \underbrace{\text{Cost}}_{\$1100}$$

3. **Solve.** We solve the equation:

$$3(225) + 0.012c = 1100$$
$$675 + 0.012c = 1100$$
$$0.012c = 425 \qquad \text{Subtracting 675}$$
$$\frac{0.012c}{0.012} = \frac{425}{0.012} \qquad \text{Dividing by 0.012}$$
$$c \approx 35{,}417. \qquad \text{Rounding to the nearest one}$$

4. **Check.** We check in the original problem. The cost for 35,417 pages is 35,417($0.012) = $425.004. The rental for 3 months is 3($225) = $675. The total cost is then $425.004 + $675 ≈ $1100, which is the $1100 that was allotted.

5. **State.** The law firm can make 35,417 copies on the copy rental allotment of $1100.

Do Exercise 4.

EXAMPLE 5 *Perimeter of NBA Court.* The perimeter of an NBA basketball court is 288 ft. The length is 44 ft longer than the width. Find the dimensions of the court.

Source: National Basketball Association

1. **Familiarize.** We first make a drawing.

We let $w =$ the width of the rectangle. Then $w + 44 =$ the length. The perimeter P of a rectangle is the distance around the rectangle and is given by the formula $2l + 2w = P$, where

$l =$ the length and $w =$ the width.

Answer on page A-8

2. **Translate.** To translate the problem, we substitute $w + 44$ for l and 288 for P:

$$2l + 2w = P$$
$$2(w + 44) + 2w = 288.$$

CAUTION! Parentheses are important here.

3. **Solve.** We solve the equation:

$$2(w + 44) + 2w = 288$$
$$2 \cdot w + 2 \cdot 44 + 2w = 288 \quad \text{Using the distributive law}$$
$$4w + 88 = 288 \quad \text{Collecting like terms}$$
$$4w + 88 - 88 = 288 - 88 \quad \text{Subtracting 88}$$
$$4w = 200$$
$$\frac{4w}{4} = \frac{200}{4} \quad \text{Dividing by 4}$$
$$w = 50.$$

Thus possible dimensions are

$w = 50$ ft and $l = w + 44 = 50 + 44$, or 94 ft.

4. **Check.** If the width is 50 ft and the length is 94 ft, then the perimeter is $2(50 \text{ ft}) + 2(94 \text{ ft})$, or 288 ft. This checks.
5. **State.** The width is 50 ft and the length is 94 ft.

Do Exercise 5.

5. **Perimeter of High School Basketball Court.** The perimeter of a standard high school basketball court is 268 ft. The length is 34 ft longer than the width. Find the dimensions of the court.

Source: Indiana High School Athletic Association

EXAMPLE 6 *Cross Section of a Roof.* In a triangular cross section of a roof, the second angle is twice as large as the first angle. The measure of the third angle is 20° greater than that of the first angle. How large are the angles?

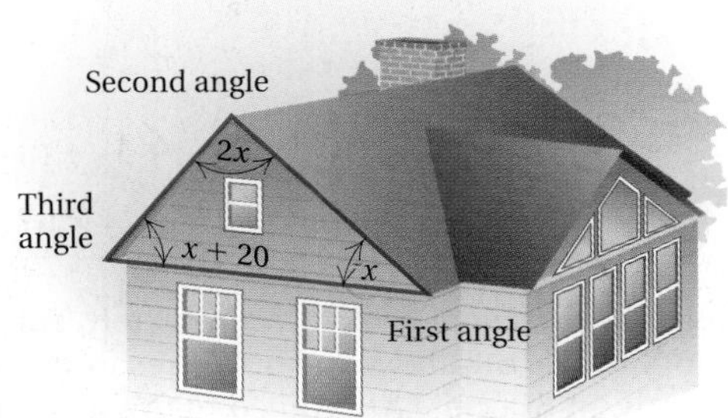

1. **Familiarize.** We first make a drawing as shown above. We let

measure of first angle $= x$.

Then measure of second angle $= 2x$

and measure of third angle $= x + 20$.

Answer on page A-8

6. The second angle of a triangle is three times as large as the first. The third angle measures 30° more than the first angle. Find the measures of the angles.

2. **Translate.** To translate, we need to recall a geometric fact. (You might, as part of step 1, look it up in a geometry book or in the list of formulas on the inside front cover.) Remember, the measures of the angles of a triangle total 180°.

Measure of first angle	plus	Measure of second angle	plus	Measure of third angle	is	180°
↓	↓	↓	↓	↓	↓	↓
x	$+$	$2x$	$+$	$(x + 20)$	$=$	180°

3. **Solve.** We solve the equation:

$$x + 2x + (x + 20) = 180$$
$$4x + 20 = 180$$
$$4x + 20 - 20 = 180 - 20$$
$$4x = 160$$
$$\frac{4x}{4} = \frac{160}{4}$$
$$x = 40.$$

Possible measures for the angles are as follows:

First angle: $x = 40°$;

Second angle: $2x = 2(40) = 80°$;

Third angle: $x + 20 = 40 + 20 = 60°$.

4. **Check.** Consider our answers: 40°, 80°, and 60°. The second is twice the first and the third is 20° greater than the first. The sum is 180°. The angles check.

5. **State.** The measures of the angles are 40°, 80°, and 60°.

CAUTION!

Units are important in answers. Remember to include them, where appropriate.

Do Exercise 6.

CAUTION!

Always be sure to answer the original problem completely. For instance, in Example 1, we need to find *two* numbers: the distances from *each* end of the trail to the hiker. Similarly, in Example 3, we need to find two mile markers, and in Example 5, we need to find two dimensions, not just the width.

EXAMPLE 7 *Simple Interest.* An investment is made at 6% simple interest for 1 year. It grows to $768.50. How much was originally invested (the principal)?

1. **Familiarize.** Suppose that $100 was invested. Recalling the formula for simple interest, $I = Prt$, we know that the interest for 1 year on $100 at 6% simple interest is given by $I = \$100 \cdot 0.06 \cdot 1 = \$6$. Then, at the end of the year, the amount in the account is found by adding the principal and the interest:

Principal	+	Interest	=	Amount
↓		↓		↓
$100	+	$6	=	$106.

Answer on page A-8

In this problem, we are working backward. We are trying to find the principal, which is the original investment. We let $x =$ the principal.

2. **Translate.** We reword the problem and then translate.

$$\begin{array}{ccccc} \text{Principal} & + & \text{Interest} & = & \text{Amount} \\ \downarrow & & \downarrow & & \downarrow \\ x & + & 6\%x & = & 768.50 \end{array}$$

Interest is 6% of the principal.

3. **Solve.** We solve the equation:

$$x + 6\%x = 768.50$$

$$x + 0.06x = 768.50 \quad \text{Converting to decimal notation}$$

$$1x + 0.06x = 768.50 \quad \text{Identity property of 1}$$

$$1.06x = 768.50 \quad \text{Collecting like terms}$$

$$\frac{1.06x}{1.06} = \frac{768.50}{1.06} \quad \text{Dividing by 1.06}$$

$$x = 725.$$

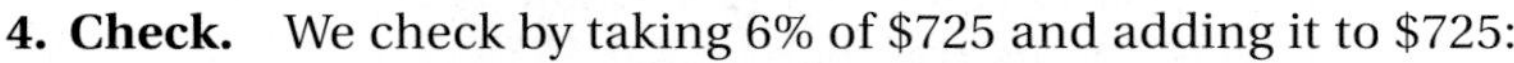

4. **Check.** We check by taking 6% of \$725 and adding it to \$725:

$$6\% \times \$725 = 0.06 \times 725 = \$43.50.$$

Then $\$725 + \$43.50 = \$768.50$, so \$725 checks.

5. **State.** The original investment was \$725.

Do Exercise 7.

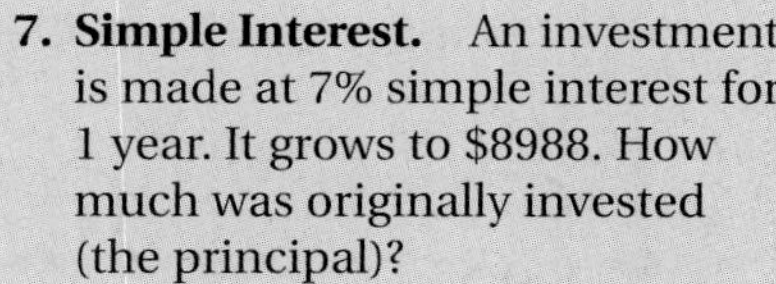

7. **Simple Interest.** An investment is made at 7% simple interest for 1 year. It grows to \$8988. How much was originally invested (the principal)?

EXAMPLE 8 *Selling a home.* The Landers are planning to sell their home. If they want to be left with \$117,500 after paying 6% of the selling price to a realtor as a commission, for how much must they sell the house?

1. **Familiarize.** Suppose the Landers sell the house for \$120,000. A 6% commission can be determined by finding 6% of \$120,000:

$$6\% \text{ of } \$120{,}000 = 0.06(\$120{,}000) = \$7200.$$

Subtracting this commission from \$120,000 would leave the Landers with

$$\$120{,}000 - \$7200 = \$112{,}800.$$

This shows that in order for the Landers to clear \$117,500, the house must sell for more than \$120,000. To determine what the sale price must be, we could check more guesses. Instead, we let $x =$ the selling price, in dollars. With a 6% commission, the realtor would receive $0.06x$.

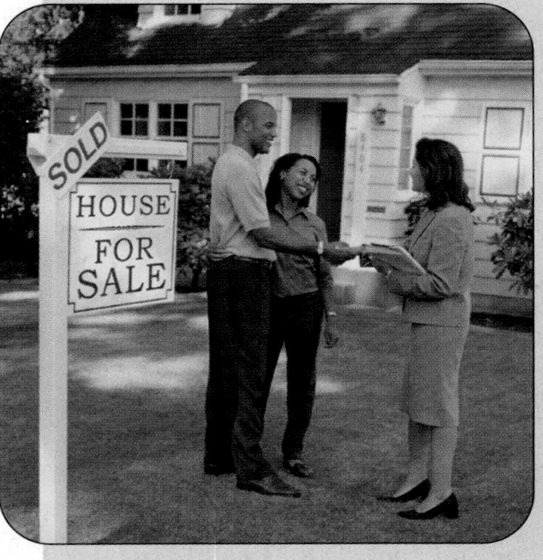

2. **Translate.** We reword the problem and translate as follows.

$$\begin{array}{ccccc} \underbrace{\text{Selling price}} & \text{less} & \underbrace{\text{Commission}} & \text{is} & \underbrace{\text{Amount remaining.}} \\ \downarrow & \downarrow & \downarrow & \downarrow & \downarrow \\ x & - & 0.06x & = & 117{,}500 \end{array}$$

Answer on page A-8

8. Price Before Sale. The price of a suit was decreased to a sale price of \$526.40. This was a 20% reduction. What was the former price?

Answer on page A-8

3. Solve. We solve the equation:

$$x - 0.06x = 117{,}500$$
$$1x - 0.06x = 117{,}500$$
$$0.94x = 117{,}500 \qquad \text{Collecting like terms. Had we noted that after the commission has been paid, 94\% remains, we could have begun with this equation.}$$
$$\frac{094x}{0.94} = \frac{117{,}500}{0.94} \qquad \text{Dividing by 0.94}$$
$$x = 125{,}000.$$

4. Check. To check, we first find 6% of \$125,000:

6% of \$125,000 = 0.06(\$125,000) = \$7500. This is the commission.

Next, we subtract the commission to find the remaining amount:

\$125,000 − \$7500 = \$117,500.

Since, after the commission, the Landers are left with \$117,500, our answer checks. Note that the \$125,000 selling price is greater than \$120,000, as predicted in the *Familiarize* step.

5. State. To be left with \$117,500, the Landers must sell the house for \$125,000.

Do Exercise 8.

CAUTION!

The problem in Example 8 is easy to solve with algebra. Without algebra, it is not. A common error in such a problem is to take 6% of the price after commission and then subtract or add. Note that 6% of the selling price (6% · \$125,000 = \$7500) is not equal to 6% of the amount that the Landers wanted to be left with (6% · \$117,500 = \$7050).

Study Tips

PROBLEM-SOLVING TIPS

The more problems you solve, the more your skills will improve.

1. Look for patterns when solving problems. Each time you study an example in a text, you may observe a pattern for problems that you will encounter later in the exercise sets or in other practical situations.
2. When translating in mathematics, consider the dimensions of the variables and constants in the equation. The variables that represent length should all be in the same unit, those that represent money should all be in dollars or all in cents, and so on.
3. Make sure that units appear in the answer whenever appropriate and that you have completely answered the original problem.

2.6 EXERCISE SET

For Extra Help

Digital Video Tutor CD 3 Videotape 6 | InterAct Math | Math Tutor Center | MathXL | MyMathLab

a Solve. *Even though you might find the answer quickly in some other way, practice using the five-step problem-solving strategy.*

1. *Pipe Cutting.* A 240-in. pipe is cut into two pieces. One piece is three times the length of the other. Find the lengths of the pieces.

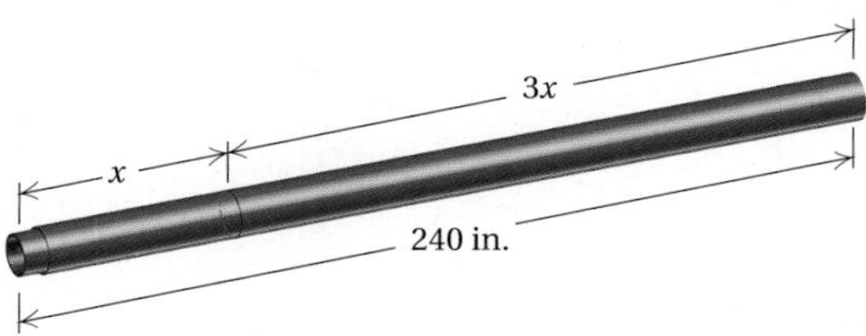

2. *Board Cutting.* A 72-in. board is cut into two pieces. One piece is 2 in. longer than the other. Find the lengths of the pieces.

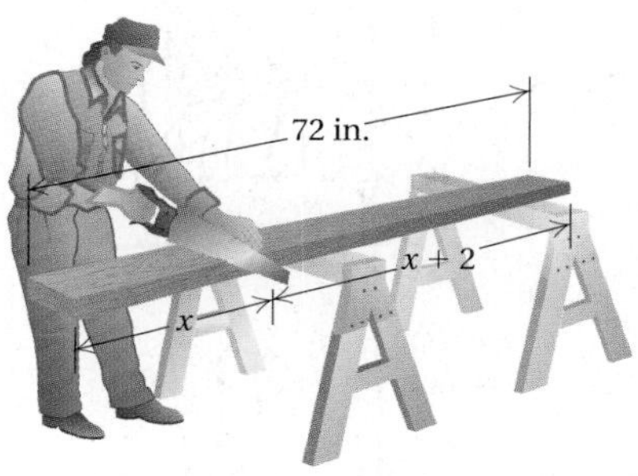

3. *Wheaties.* Recently, the cost of four 18-oz boxes of Wheaties cereal was $14.68. What was the cost of one box?

4. *Area of Lake Ontario.* The area of Lake Superior is about four times the area of Lake Ontario. The area of Lake Superior is 30,172 mi². What is the area of Lake Ontario?

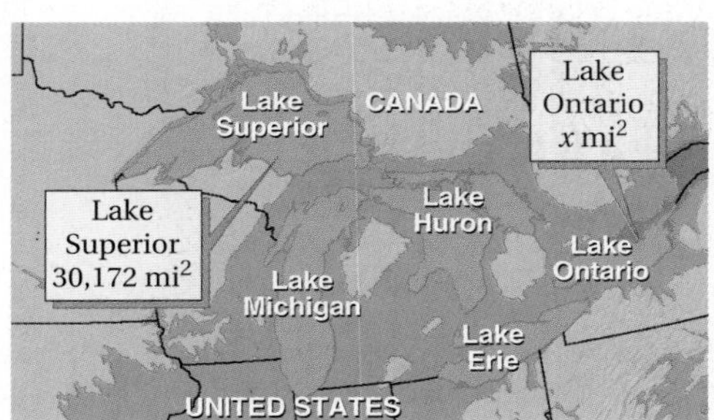

5. *Women's Dresses.* In a recent year, the total amount spent on women's blouses was $6.5 billion. This was $0.2 billion more than what was spent on women's dresses. How much was spent on women's dresses?

6. *Statue of Liberty.* The height of the Eiffel Tower is 974 ft, which is about 669 ft higher than the Statue of Liberty. What is the height of the Statue of Liberty?

7. *Iditarod Race.* The Iditarod sled dog race extends for 1049 mi from Anchorage to Nome. If a musher is twice as far from Anchorage as from Nome, how much of the race has the musher completed?
Source: Iditarod Trail Commission

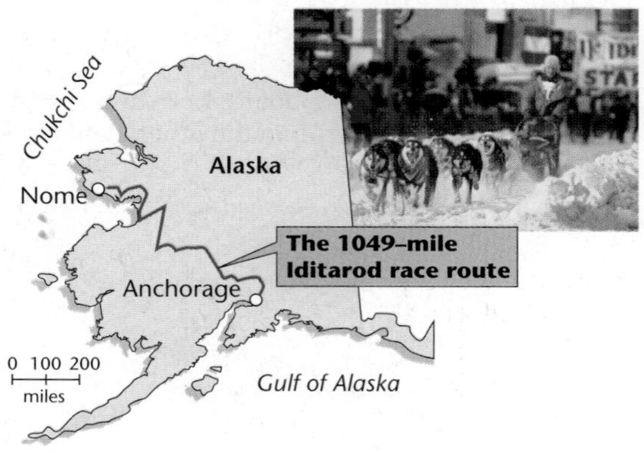

8. *Home Remodeling.* In a recent year, Americans spent a total of $35 billion to remodel bathrooms and kitchens. Twice as much was spent on kitchens as bathrooms. How much was spent on each?

9. *Consecutive Post Office Box Numbers.* The sum of the numbers on two consecutive post office boxes is 547. What are the numbers?

10. *Consecutive Page Numbers.* The sum of the page numbers on the facing pages of a book is 573. What are the page numbers?

11. *Consecutive Ticket Numbers.* The numbers on Sam's three raffle tickets are consecutive integers. The sum of the numbers is 126. What are the numbers?

12. *Consecutive Ages.* The ages of Whitney, Wesley, and Wanda are consecutive integers. The sum of their ages is 108. What are their ages?

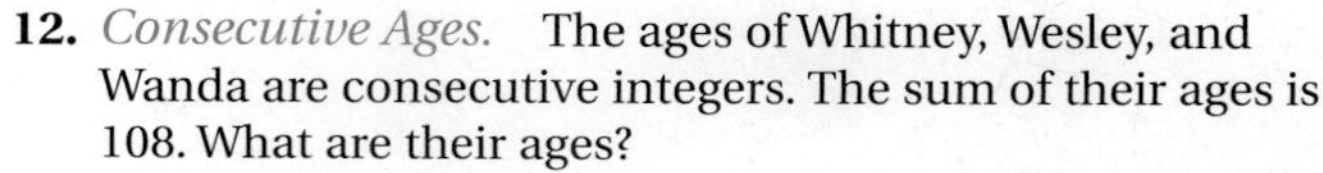

13. *Consecutive Odd Integers.* The sum of three consecutive odd integers is 189. What are the integers?

14. *Consecutive Integers.* Three consecutive integers are such that the first plus one-half the second plus seven less than twice the third is 2101. What are the integers?

15. *Standard Billboard Sign.* A standard rectangular highway billboard sign has a perimeter of 124 ft. The length is 6 ft more than three times the width. Find the dimensions of the sign.

16. *Two-by-Four.* The perimeter of a cross section of a "two-by-four" piece of lumber is $10\frac{1}{2}$ in. The length is twice the width. Find the actual dimensions of the cross section of a two-by-four.

17. *Price of Sneakers.* Amy paid \$63.75 for a pair of New Balance 903 running shoes during a 15%-off sale. What was the regular price?

18. *Price of a CD Player.* Doug paid \$72 for a shockproof portable CD player during a 20%-off sale. What was the regular price?

19. *Price of a Textbook.* Evelyn paid \$89.25, including 5% tax, for her biology textbook. How much did the book itself cost?

20. *Price of a Printer.* Jake paid \$100.70, including 6% tax, for a color printer. How much did the printer itself cost?

21. *Parking Costs.* A hospital parking lot charges \$1.50 for the first hour or part thereof, and \$1.00 for each additional hour or part thereof. A weekly pass costs \$27.00 and allows unlimited parking for 7 days. Suppose that each visit Ed makes to the hospital lasts $1\frac{1}{2}$ hr. What is the minimum number of times that Ed would have to visit per week to make it worthwhile for him to buy the pass?

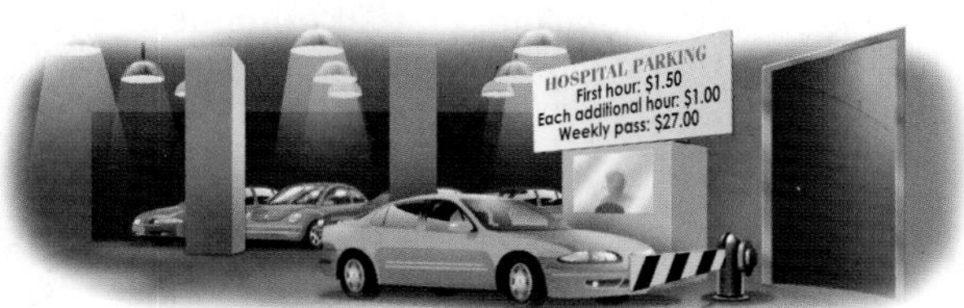

22. *Van Rental.* Value Rent-A-Car rents vans at a daily rate of \$84.95 plus 60 cents per mile. Molly rents a van to deliver electrical parts to her customers. She is allotted a daily budget of \$320. How many miles can she drive for \$320?

23. *Triangular Field.* The second angle of a triangular field is three times as large as the first angle. The third angle is 40° greater than the first angle. How large are the angles?

24. *Triangular Parking Lot.* The second angle of a triangular parking lot is four times as large as the first angle. The third angle is 45° less than the sum of the other two angles. How large are the angles?

25. *Triangular Backyard.* A home has a triangular backyard. The second angle of the triangle is 5° more than the first angle. The third angle is 10° more than three times the first angle. Find the angles of the triangular yard.

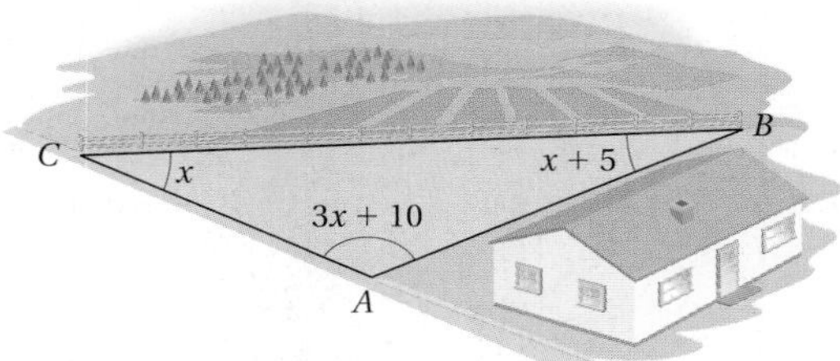

26. *Boarding Stable.* A rancher needs to form a triangular horse pen using ropes next to a stable. The second angle is three times the first angle. The third angle is 15° less than the first angle. Find the angles of the triangular pen.

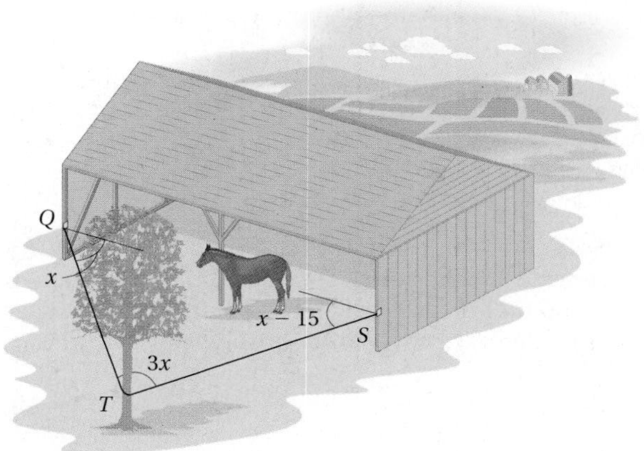

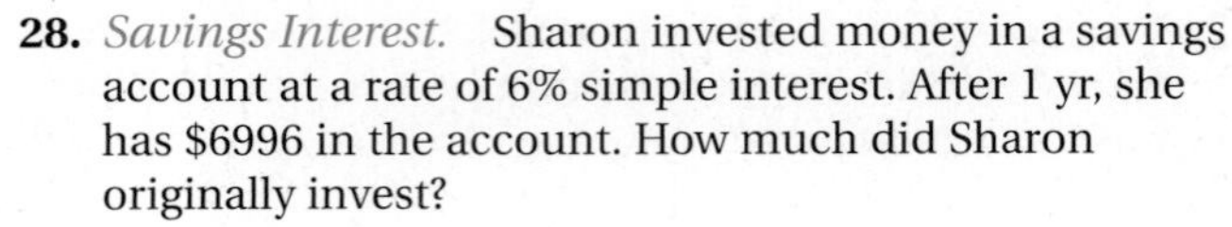

27. *Stock Prices.* Sarah's investment in AOL/Time Warner stock grew 28% to $448. How much did she invest?

28. *Savings Interest.* Sharon invested money in a savings account at a rate of 6% simple interest. After 1 yr, she has $6996 in the account. How much did Sharon originally invest?

29. *Credit Cards.* The balance in Will's Mastercard® account grew 2%, to $870, in one month. What was his balance at the beginning of the month?

30. *Loan Interest.* Alvin borrowed money from a cousin at a rate of 10% simple interest. After 1 yr, $7194 paid off the loan. How much did Alvin borrow?

31. *Taxi Fares.* In Beniford, taxis charge $3 plus 75¢ per mile for an airport pickup. How far from the airport can Courtney travel for $12?

32. *Taxi Fares.* In Cranston, taxis charge $4 plus 90¢ per mile for an airport pickup. How far from the airport can Ralph travel for $17.50?

33. *Tipping.* Leon left a 15% tip for a meal. The total cost of the meal, including the tip, was $41.40. What was the cost of the meal before the tip was added?

34. *Tipping.* Selena left an 18% tip for a meal. The total cost of the meal, including the tip, was $40.71. What was the cost of the meal before the tip was added?

35. D_W Erin returns a tent that she bought during a storewide 35% off sale that has ended. She is offered store credit for 125% of what she paid (not to be used on sale items). Is this fair to Erin? Why or why not?

36. D_W Write a problem for a classmate to solve so that it can be translated to the equation

$$\frac{2}{3}x + (x + 5) + x = 375.$$

SKILL MAINTENANCE

Calculate.

37. $-\frac{4}{5} - \frac{3}{8}$ [1.4a]

38. $-\frac{4}{5} + \frac{3}{8}$ [1.3a]

39. $-\frac{4}{5} \cdot \frac{3}{8}$ [1.5a]

40. $-\frac{4}{5} \div \frac{3}{8}$ [1.6c]

41. $\frac{1}{10} \div \left(-\frac{1}{100}\right)$ [1.6c]

42. $-25.6 \div (-16)$ [1.6c]

43. $-25.6(-16)$ [1.5a]

44. $-25.6 - (-16)$ [1.4a]

45. $-25.6 + (-16)$ [1.3a]

46. $(-0.02) \div (-0.2)$ [1.6c]

SYNTHESIS

47. Apples are collected in a basket for six people. One-third, one-fourth, one-eighth, and one-fifth are given to four people, respectively. The fifth person gets ten apples with one apple remaining for the sixth person. Find the original number of apples in the basket.

48. A student scored 78 on a test that had 4 seven-point fill-ins and 24 three-point multiple-choice questions. The student had one fill-in wrong. How many multiple-choice questions did the student answer correctly?

49. The area of this triangle is 2.9047 in^2. Find x.

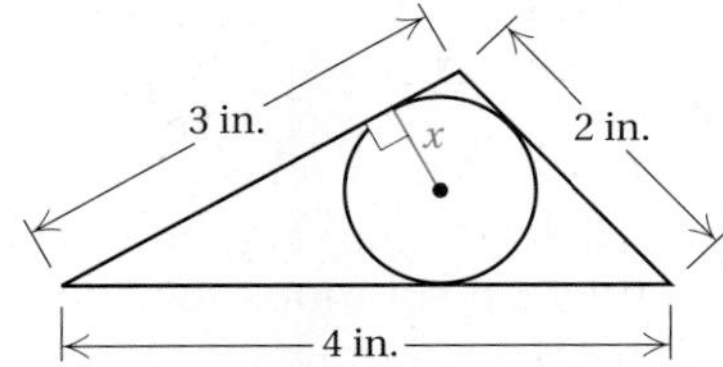

50. A storekeeper goes to the bank to get $10 worth of change. She requests twice as many quarters as half dollars, twice as many dimes as quarters, three times as many nickels as dimes, and no pennies or dollars. How many of each coin did the storekeeper get?

51. In one city, a sales tax of 9% was added to the price of gasoline as registered on the pump. Suppose a driver asked for $10 worth of gas. The attendant filled the tank until the pump read $9.10 and charged the driver $10. Something was wrong. Use algebra to correct the error.

2.7 SOLVING INEQUALITIES

Objectives

a. Determine whether a given number is a solution of an inequality.

b. Graph an inequality on a number line.

c. Solve inequalities using the addition principle.

d. Solve inequalities using the multiplication principle.

e. Solve inequalities using the addition and multiplication principles together.

Determine whether each number is a solution of the inequality.

1. $x > 3$

a) 2 **b)** 0

c) -5 **d)** 15.4

e) 3 **f)** $-\frac{2}{5}$

2. $x \leq 6$

a) 6 **b)** 0

c) -4.3 **d)** 25

e) -6 **f)** $\frac{5}{8}$

Answers on page A-9

We now extend our equation-solving principles to the solving of inequalities.

a Solutions of Inequalities

In Section 1.2, we defined the symbols $>$ (is greater than), $<$ (is less than), $\geq$ (is greater than or equal to), and $\leq$ (is less than or equal to). For example, $3 \leq 4$ and $3 \leq 3$ are both true, but $-3 \leq -4$ and $0 \geq 2$ are both false.

An **inequality** is a number sentence with $>$, $<$, $\geq$, or $\leq$ as its verb—for example,

$$-4 > t, \quad x < 3, \quad 2x + 5 \geq 0, \quad \text{and} \quad -3y + 7 \leq -8.$$

Some replacements for a variable in an inequality make it true and some make it false.

SOLUTION

A replacement that makes an inequality true is called a **solution.** The set of all solutions is called the **solution set.** When we have found the set of all solutions of an inequality, we say that we have **solved** the inequality.

EXAMPLES Determine whether the number is a solution of $x < 2$.

1. -2.7 Since $-2.7 < 2$ is true, -2.7 is a solution.
2. 2 Since $2 < 2$ is false, 2 is not a solution.

EXAMPLES Determine whether the number is a solution of $y \geq 6$.

3. 6 Since $6 \geq 6$ is true, 6 is a solution.
4. $-\frac{4}{3}$ Since $-\frac{4}{3} \geq 6$ is false, $-\frac{4}{3}$ is not a solution.

Do Exercises 1 and 2.

b Graphs of Inequalities

Some solutions of $x < 2$ are -3, 0, 1, 0.45, -8.9, $-\pi$, $\frac{5}{8}$, and so on. In fact, there are infinitely many real numbers that are solutions. Because we cannot list them all individually, it is helpful to make a drawing that represents all the solutions.

A **graph** of an inequality is a drawing that represents its solutions. An inequality in one variable can be graphed on a number line. An inequality in two variables can be graphed on a coordinate plane; we will study such graphs in Chapter 7.

EXAMPLE 5 Graph: $x < 2$.

The solutions of $x < 2$ are all those numbers less than 2. They are shown on the graph by shading all points to the left of 2. The open circle at 2 indicates that 2 is *not* part of the graph.

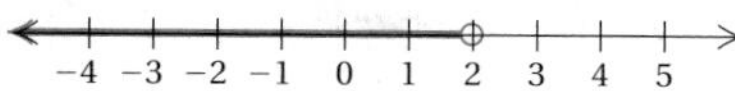

EXAMPLE 6 Graph: $x \geq -3$.

The solutions of $x \geq -3$ are shown on the number line by shading the point for -3 and all points to the right of -3. The closed circle at -3 indicates that -3 *is* part of the graph.

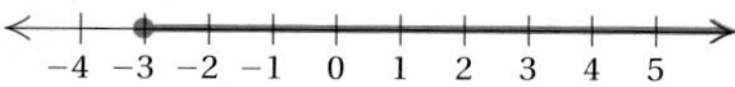

EXAMPLE 7 Graph: $-3 \leq x < 2$.

The inequality $-3 \leq x < 2$ is read "-3 is less than or equal to x *and* x is less than 2," or "x is greater than or equal to -3 *and* x is less than 2." In order to be a solution of this inequality, a number must be a solution of both $-3 \leq x$ and $x < 2$. The number 1 is a solution, as are -1.7, 0, 1.5, and $\frac{3}{8}$. We can see from the graphs below that the solution set consists of the numbers that overlap in the two solution sets in Examples 5 and 6:

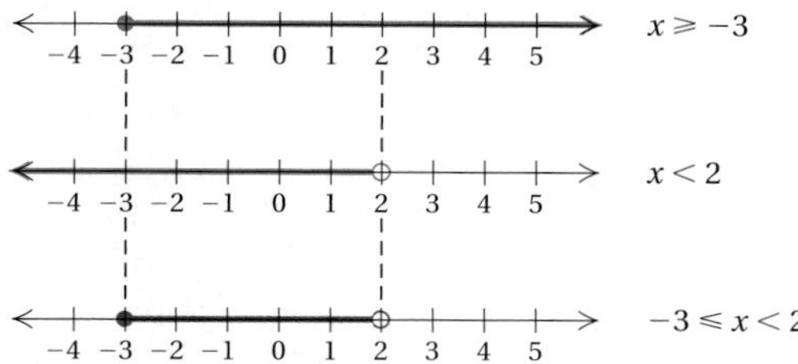

The open circle at 2 means that 2 is *not* part of the graph. The closed circle at -3 means that -3 *is* part of the graph. The other solutions are shaded.

Do Exercises 3–5.

C Solving Inequalities Using the Addition Principle

Consider the true inequality $3 < 7$. If we add 2 on both sides, we get another true inequality:

$$3 + 2 < 7 + 2, \quad \text{or} \quad 5 < 9.$$

Similarly, if we add -4 on both sides of $x + 4 < 10$, we get an *equivalent* inequality:

$$x + 4 + (-4) < 10 + (-4),$$

or

$$x < 6.$$

To say that $x + 4 < 10$ and $x < 6$ are **equivalent** is to say that they have the same solution set. For example, the number 3 is a solution of $x + 4 < 10$. It is also a solution of $x < 6$. The number -2 is a solution of $x < 6$. It is also a solution of $x + 4 < 10$. Any solution of one is a solution of the other—they are equivalent.

Graph.

3. $x \leq 4$

4. $x > -2$

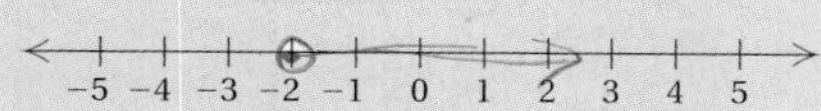

5. $-2 < x \leq 4$

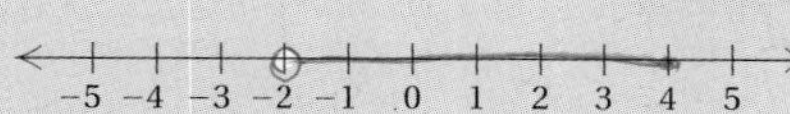

Answers on page A-9

Solve. Then graph.

6. $x + 3 > 5$

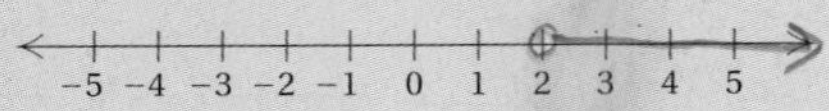

7. $x - 1 \leq 2$

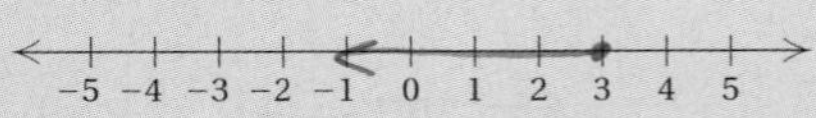

8. $5x + 1 < 4x - 2$

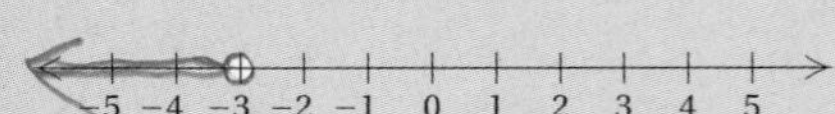

Answers on page A-9

THE ADDITION PRINCIPLE FOR INEQUALITIES

For any real numbers a, b, and c:

$a < b$ is equivalent to $a + c < b + c$;

$a > b$ is equivalent to $a + c > b + c$;

$a \leq b$ is equivalent to $a + c \leq b + c$;

$a \geq b$ is equivalent to $a + c \geq b + c$.

In other words, when we add or subtract the same number on both sides of an inequality, the direction of the inequality symbol is not changed.

As with equation solving, when solving inequalities, our goal is to isolate the variable on one side. Then it is easier to determine the solution set.

EXAMPLE 8 Solve: $x + 2 > 8$. Then graph.

We use the addition principle, subtracting 2 on both sides:

$$x + 2 - 2 > 8 - 2$$
$$x > 6.$$

From the inequality $x > 6$, we can determine the solutions directly. Any number greater than 6 makes the last sentence true and is a solution of that sentence. Any such number is also a solution of the original sentence. Thus the inequality is solved. The graph is as follows:

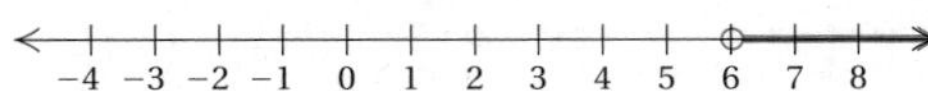

We cannot check all the solutions of an inequality by substitution, as we can check solutions of equations, because there are too many of them. A partial check can be done by substituting a number greater than 6—say, 7—into the original inequality:

$x + 2 > 8$	
$7 + 2$	8
9	

TRUE

Since $9 > 8$ is true, 7 is a solution. Any number greater than 6 is a solution.

EXAMPLE 9 Solve: $3x + 1 \leq 2x - 3$. Then graph.

We have

$$3x + 1 \leq 2x - 3$$
$$3x + 1 - 1 \leq 2x - 3 - 1 \quad \text{Subtracting 1}$$
$$3x \leq 2x - 4 \quad \text{Simplifying}$$
$$3x - 2x \leq 2x - 4 - 2x \quad \text{Subtracting } 2x$$
$$x \leq -4. \quad \text{Simplifying}$$

The graph is as follows:

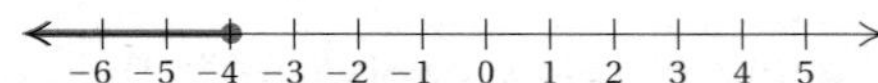

Remember that the graph is a drawing that represents the solutions of the original inequality.

In Example 9, any number less than or equal to -4 is a solution. The following are some solutions:

$$-4, \quad -5, \quad -6, \quad -\frac{13}{3}, \quad -204.5, \quad \text{and} \quad -18\pi.$$

Besides drawing a graph, we can also describe all the solutions of an inequality using **set notation.** We could just begin to list them in a set using roster notation (see p. 71), as follows:

$$\left\{-4, -5, -6, -\frac{13}{3}, -204.5, -18\pi, \ldots\right\}.$$

We can never list them all this way, however. Seeing this set without knowing the inequality makes it difficult for us to know what real numbers we are considering. There is, however, another kind of notation that we can use. It is

$$\{x \mid x \leq -4\},$$

which is read

"The set of all x such that x is less than or equal to -4."

This shorter notation for sets is called **set-builder notation.**

From now on, we will use this notation when solving inequalities.

Do Exercises 6–8 on the preceding page.

EXAMPLE 10 Solve: $x + \frac{1}{3} > \frac{5}{4}$.

We have

$$x + \tfrac{1}{3} > \tfrac{5}{4}$$

$$x + \tfrac{1}{3} - \tfrac{1}{3} > \tfrac{5}{4} - \tfrac{1}{3} \qquad \text{Subtracting } \tfrac{1}{3}$$

$$x > \tfrac{5}{4} \cdot \tfrac{3}{3} - \tfrac{1}{3} \cdot \tfrac{4}{4} \qquad \text{Multiplying by 1 to obtain a common denominator}$$

$$x > \tfrac{15}{12} - \tfrac{4}{12}$$

$$x > \tfrac{11}{12}.$$

Any number greater than $\frac{11}{12}$ is a solution. The solution set is

$$\{x \mid x > \tfrac{11}{12}\},$$

which is read

"The set of all x such that x is greater than $\frac{11}{12}$."

When solving inequalities, you may obtain an answer like $\frac{11}{12} < x$. Recall from Chapter 1 that this has the same meaning as $x > \frac{11}{12}$. Thus the solution set in Example 10 can be described as $\{x \mid \frac{11}{12} < x\}$ or as $\{x \mid x > \frac{11}{12}\}$. The latter is used most often.

Do Exercises 9 and 10.

d Solving Inequalities Using the Multiplication Principle

There is a multiplication principle for inequalities that is similar to that for equations, but it must be modified. When we are multiplying on both sides by a negative number, the direction of the inequality symbol must be changed.

Solve.

9. $x + \dfrac{2}{3} \geq \dfrac{4}{5}$

10. $5y + 2 \leq -1 + 4y$

Answers on page A-9

Solve. Then graph.

11. $8x < 64$

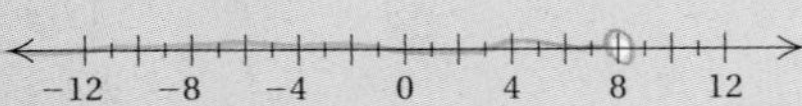

12. $5y \geq 160$

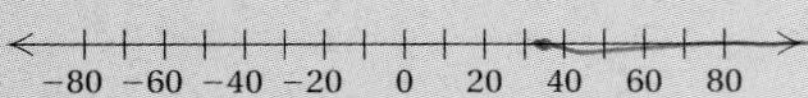

Answers on page A-9

Let's see what happens. Consider the true inequality $3 < 7$. If we multiply on both sides by a *positive* number, like 2, we get another true inequality:

$$3 \cdot 2 < 7 \cdot 2, \quad \text{or} \quad 6 < 14. \qquad \text{True}$$

If we multiply on both sides by a *negative* number, like -2, and we do not change the direction of the inequality symbol, we get a *false* inequality:

$$3 \cdot (-2) < 7 \cdot (-2), \quad \text{or} \quad -6 < -14. \qquad \text{False}$$

The fact that $6 < 14$ is true but $-6 < -14$ is false stems from the fact that the negative numbers, in a sense, mirror the positive numbers. That is, whereas 14 is to the *right* of 6 on a number line, the number -14 is to the *left* of -6. Thus, if we reverse (change the direction of) the inequality symbol, we get a *true* inequality: $-6 > -14$.

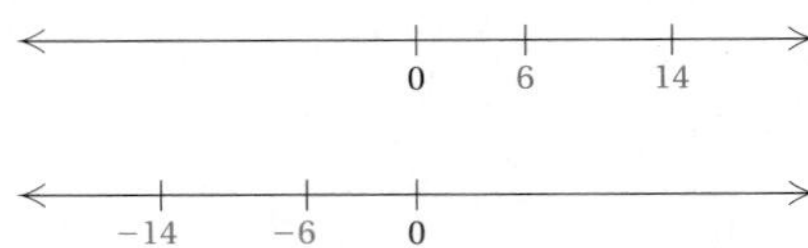

THE MULTIPLICATION PRINCIPLE FOR INEQUALITIES

For any real numbers a and b, and any *positive* number c:

$a < b$ is equivalent to $ac < bc$;

$a > b$ is equivalent to $ac > bc$.

For any real numbers a and b, and any *negative* number c:

$a < b$ is equivalent to $ac > bc$;

$a > b$ is equivalent to $ac < bc$.

Similar statements hold for $\leq$ and $\geq$.

In other words, when we multiply or divide by a positive number on both sides of an inequality, the direction of the inequality symbol stays the same. When we multiply or divide by a negative number on both sides of an inequality, the direction of the inequality symbol is reversed.

EXAMPLE 11 Solve: $4x < 28$. Then graph.

We have

$$4x < 28$$

$$\frac{4x}{4} < \frac{28}{4} \qquad \text{Dividing by 4}$$

The symbol stays the same.

$$x < 7. \qquad \text{Simplifying}$$

The solution set is $\{x | x < 7\}$. The graph is as follows:

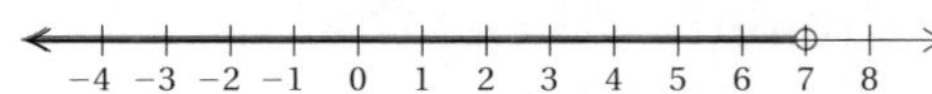

Do Exercises 11 and 12.

EXAMPLE 12 Solve: $-2y < 18$. Then graph.

We have

$$-2y < 18$$

$$\frac{-2y}{-2} > \frac{18}{-2} \quad \text{Dividing by } -2$$

The symbol must be reversed!

$$y > -9. \quad \text{Simplifying}$$

The solution set is $\{y \mid y > -9\}$. The graph is as follows:

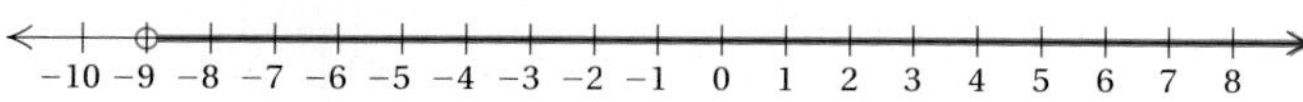

Do Exercises 13 and 14.

Solve.

13. $-4x \leq 24$

14. $-5y > 13$

e Using the Principles Together

All of the equation-solving techniques used in Sections 2.1–2.3 can be used with inequalities provided we remember to reverse the inequality symbol when multiplying or dividing on both sides by a negative number.

EXAMPLE 13 Solve: $6 - 5y > 7$.

We have

$$6 - 5y > 7$$

$$-6 + 6 - 5y > -6 + 7 \quad \text{Adding } -6\text{. The symbol stays the same.}$$

$$-5y > 1 \quad \text{Simplifying}$$

$$\frac{-5y}{-5} < \frac{1}{-5} \quad \text{Dividing by } -5$$

The symbol must be reversed because we are dividing by a *negative* number, −5.

$$y < -\frac{1}{5}. \quad \text{Simplifying}$$

The solution set is $\{y \mid y < -\frac{1}{5}\}$.

Do Exercise 15.

15. Solve: $7 - 4x < 8$.

EXAMPLE 14 Solve: $8y - 5 > 17 - 5y$.

$$-17 + 8y - 5 > -17 + 17 - 5y \quad \text{Adding } -17\text{. The symbol stays the same.}$$

$$8y - 22 > -5y \quad \text{Simplifying}$$

$$-8y + 8y - 22 > -8y - 5y \quad \text{Adding } -8y$$

$$-22 > -13y \quad \text{Simplifying}$$

$$\frac{-22}{-13} < \frac{-13y}{-13} \quad \text{Dividing by } -13$$

The symbol must be reversed because we are dividing by a *negative* number, −13.

$$\frac{22}{13} < y$$

The solution set is $\{y \mid \frac{22}{13} < y\}$, or $\{y \mid y > \frac{22}{13}\}$.

Answers on page A-9

16. Solve: $24 - 7y \le 11y - 14$.

17. Solve. Use a method like the one used in Example 15.

$$24 - 7y \le 11y - 14$$

18. Solve:

$$3(7 + 2x) \le 30 + 7(x - 1).$$

Answers on page A-9

We can often solve inequalities in such a way as to avoid having to reverse the inequality symbol. We add so that after like terms have been collected, the coefficient of the variable term is positive. We show this by solving the inequality in Example 14 a different way.

EXAMPLE 15 Solve: $8y - 5 > 17 - 5y$.

Note that if we add $5y$ on both sides, the coefficient of the y-term will be positive after like terms have been collected.

$$8y - 5 + 5y > 17 - 5y + 5y \quad \text{Adding } 5y$$
$$13y - 5 > 17 \quad \text{Simplifying}$$
$$13y - 5 + 5 > 17 + 5 \quad \text{Adding 5}$$
$$13y > 22 \quad \text{Simplifying}$$
$$\frac{13y}{13} > \frac{22}{13} \quad \text{Dividing by 13. We leave the inequality symbol the same because we are dividing by a positive number.}$$
$$y > \frac{22}{13}$$

The solution set is $\{y \mid y > \frac{22}{13}\}$.

Do Exercises 16 and 17.

EXAMPLE 16 Solve: $3(x - 2) - 1 < 2 - 5(x + 6)$.

$$3(x - 2) - 1 < 2 - 5(x + 6)$$
$$3x - 6 - 1 < 2 - 5x - 30 \quad \text{Using the distributive law to multiply and remove parentheses}$$
$$3x - 7 < -5x - 28 \quad \text{Simplifying}$$
$$3x + 5x < -28 + 7 \quad \text{Adding } 5x \text{ and 7 to get all } x\text{-terms on one side and all other terms on the other side}$$
$$8x < -21 \quad \text{Simplifying}$$
$$x < \frac{-21}{8}, \text{ or } -\frac{21}{8} \quad \text{Dividing by 8}$$

The solution set is $\{x \mid x < -\frac{21}{8}\}$.

Do Exercise 18.

EXAMPLE 17 Solve: $16.3 - 7.2p \leq -8.18$.

The greatest number of decimal places in any one number is *two.* Multiplying by 100, which has two 0's, will clear decimals. Then we proceed as before.

$$16.3 - 7.2p \leq -8.18$$

$$100(16.3 - 7.2p) \leq 100(-8.18) \quad \text{Multiplying by 100}$$

$$100(16.3) - 100(7.2p) \leq 100(-8.18) \quad \text{Using the distributive law}$$

$$1630 - 720p \leq -818 \quad \text{Simplifying}$$

$$1630 - 720p - 1630 \leq -818 - 1630 \quad \text{Subtracting 1630}$$

$$-720p \leq -2448 \quad \text{Simplifying}$$

$$\frac{-720p}{-720} \geq \frac{-2448}{-720} \quad \text{Dividing by } -720$$

The symbol must be reversed.

$$p \geq 3.4$$

The solution set is $\{p \mid p \geq 3.4\}$.

Do Exercise 19.

EXAMPLE 18 Solve: $\frac{2}{3}x - \frac{1}{6} + \frac{1}{2}x > \frac{7}{6} + 2x$.

The number 6 is the least common multiple of all the denominators. Thus we multiply by 6 on both sides.

$$\frac{2}{3}x - \frac{1}{6} + \frac{1}{2}x > \frac{7}{6} + 2x$$

$$6\left(\frac{2}{3}x - \frac{1}{6} + \frac{1}{2}x\right) > 6\left(\frac{7}{6} + 2x\right) \quad \text{Multiplying by 6 on both sides}$$

$$6 \cdot \frac{2}{3}x - 6 \cdot \frac{1}{6} + 6 \cdot \frac{1}{2}x > 6 \cdot \frac{7}{6} + 6 \cdot 2x \quad \text{Using the distributive law}$$

$$4x - 1 + 3x > 7 + 12x \quad \text{Simplifying}$$

$$7x - 1 > 7 + 12x \quad \text{Collecting like terms}$$

$$7x - 1 - 12x > 7 + 12x - 12x \quad \text{Subtracting } 12x$$

$$-5x - 1 > 7 \quad \text{Collecting like terms}$$

$$-5x - 1 + 1 > 7 + 1 \quad \text{Adding 1}$$

$$-5x > 8 \quad \text{Simplifying}$$

$$\frac{-5x}{-5} < \frac{8}{-5} \quad \text{Dividing by } -5$$

The symbol must be reversed.

$$x < -\frac{8}{5}$$

The solution set is $\{x \mid x < -\frac{8}{5}\}$.

Do Exercise 20.

19. Solve:

$2.1x + 43.2 \geq 1.2 - 8.4x.$

20. Solve.

$\frac{3}{4} + x < \frac{7}{8}x - \frac{1}{4} + \frac{1}{2}x.$

Answers on page A-9

2.7 EXERCISE SET

For Extra Help

Digital Video Tutor CD 3 Videotape 6 | InterAct Math | Math Tutor Center | MathXL | MyMathLab

a Determine whether each number is a solution of the given inequality.

1. $x > -4$
a) 4
b) 0
c) -4
d) 6
e) 5.6

2. $x \leq 5$
a) 0
b) 5
c) -1
d) -5
e) $7\frac{1}{4}$

3. $x \geq 6.8$
a) -6
b) 0
c) 6
d) 8
e) $-3\frac{1}{2}$

4. $x < 8$
a) 8
b) -10
c) 0
d) 11
e) -4.7

b Graph on a number line.

5. $x > 4$

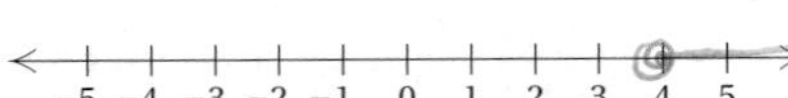

6. $x < 0$

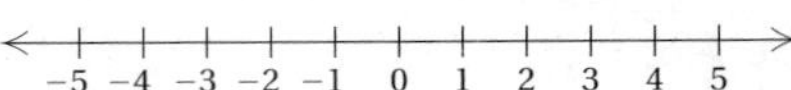

7. $t < -3$

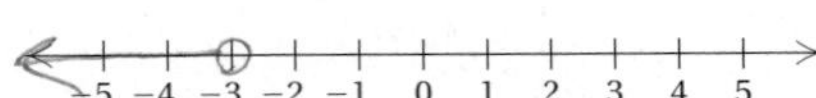

8. $y > 5$

9. $m \geq -1$

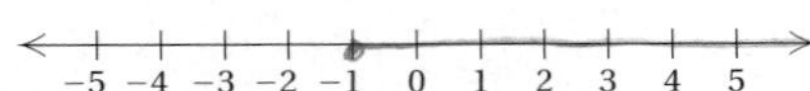

10. $x \leq -2$

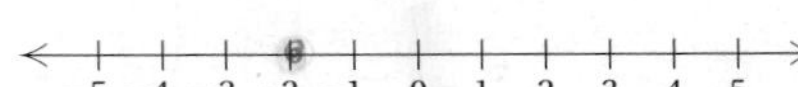

11. $-3 < x \leq 4$

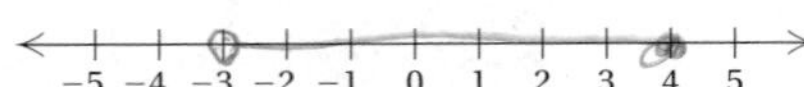

12. $-5 \leq x < 2$

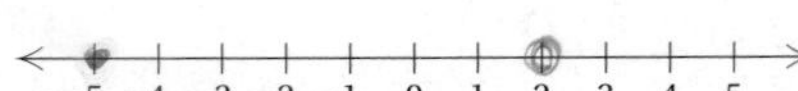

13. $0 < x < 3$

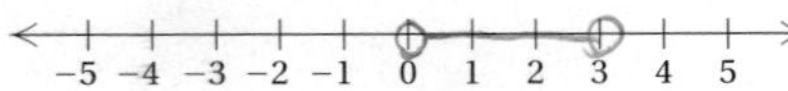

14. $-5 \leq x \leq 0$

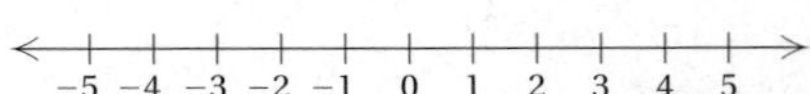

c Solve using the addition principle. Then graph.

15. $x + 7 > 2$

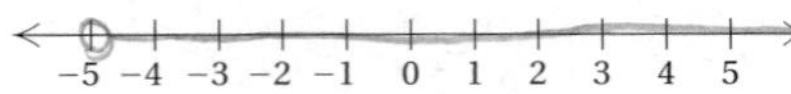

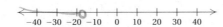

16. $x + 5 > 2$

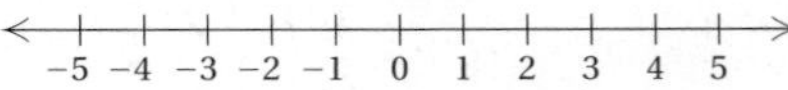

17. $x + 8 \leq -10$

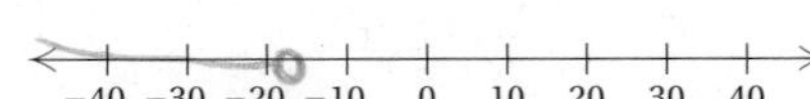

18. $x + 8 \leq -11$

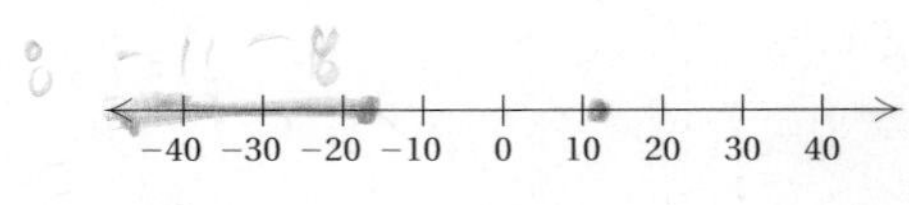

Solve using the addition principle.

19. $y - 7 > -12$

20. $y - 9 > -15$

21. $2x + 3 > x + 5$

22. $2x + 4 > x + 7$

23. $3x + 9 \le 2x + 6$

24. $3x + 18 \le 2x + 16$

25. $5x - 6 < 4x - 2$

26. $9x - 8 < 8x - 9$

27. $-9 + t > 5$

28. $-8 + p > 10$

29. $y + \frac{1}{4} \le \frac{1}{2}$

30. $x - \frac{1}{3} \le \frac{5}{6}$

31. $x - \frac{1}{3} > \frac{1}{4}$

32. $x + \frac{1}{8} > \frac{1}{2}$

d Solve using the multiplication principle. Then graph.

33. $5x < 35$

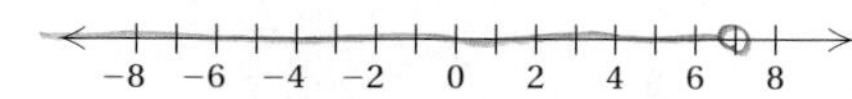

34. $8x \ge 32$

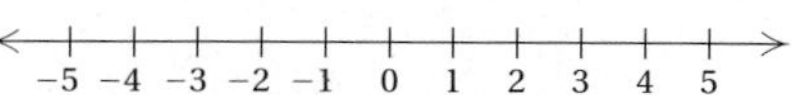

35. $-12x > -36$

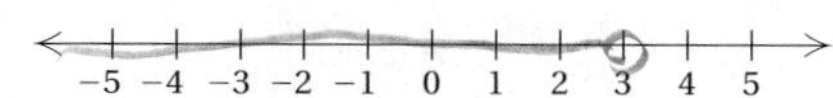

36. $-16x > -64$

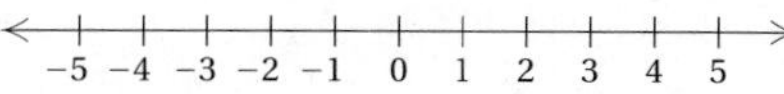

Solve using the multiplication principle.

37. $5y \ge -2$

38. $3x < -4$

39. $-2x \le 12$

40. $-3x \le 15$

41. $-4y \ge -16$

42. $-7x < -21$

43. $-3x < -17$

44. $-5y > -23$

45. $-2y > \frac{1}{7}$

46. $-4x \le \frac{1}{9}$

47. $-\frac{6}{5} \le -4x$

48. $-\frac{7}{9} > 63x$

e Solve using the addition and multiplication principles.

49. $4 + 3x < 28$

50. $3 + 4y < 35$

51. $3x - 5 \leq 13$

52. $5y - 9 \leq 21$

53. $13x - 7 < -46$

54. $8y - 6 < -54$

55. $30 > 3 - 9x$

56. $48 > 13 - 7y$

57. $4x + 2 - 3x \leq 9$

58. $15x + 5 - 14x \leq 9$

59. $-3 < 8x + 7 - 7x$

60. $-8 < 9x + 8 - 8x - 3$

61. $6 - 4y > 4 - 3y$

62. $9 - 8y > 5 - 7y + 2$

63. $5 - 9y \leq 2 - 8y$

64. $6 - 18x \leq 4 - 12x - 5x$

65. $19 - 7y - 3y < 39$

66. $18 - 6y - 4y < 63 + 5y$

67. $2.1x + 45.2 > 3.2 - 8.4x$

68. $0.96y - 0.79 \leq 0.21y + 0.46$

69. $\frac{x}{3} - 2 \leq 1$

70. $\frac{2}{3} + \frac{x}{5} < \frac{4}{15}$

71. $\frac{y}{5} + 1 \leq \frac{2}{5}$

72. $\frac{3x}{4} - \frac{7}{8} \geq -15$

73. $3(2y - 3) < 27$

74. $4(2y - 3) > 28$

75. $2(3 + 4m) - 9 \geq 45$

76. $3(5 + 3m) - 8 \leq 88$

77. $8(2t + 1) > 4(7t + 7)$

78. $7(5y - 2) > 6(6y - 1)$

79. $3(r - 6) + 2 < 4(r + 2) - 21$

80. $5(x + 3) + 9 \leq 3(x - 2) + 6$

81. $0.8(3x + 6) \geq 1.1 - (x + 2)$

82. $0.4(2x + 8) \geq 20 - (x + 5)$

83. $\frac{5}{3} + \frac{2}{3}x < \frac{25}{12} + \frac{5}{4}x + \frac{3}{4}$

84. $1 - \frac{2}{3}y \geq \frac{9}{5} - \frac{y}{5} + \frac{3}{5}$

85. DW Are the inequalities $3x - 4 < 10 - 4x$ and $2(x - 5) > 3(2x - 6)$ equivalent? Why or why not?

86. DW Explain in your own words why it is necessary to reverse the inequality symbol when multiplying on both sides of an inequality by a negative number.

SKILL MAINTENANCE

Add or subtract. [1.3a], [1.4a]

87. $-56 + (-18)$

88. $-2.3 + 7.1$

89. $-\frac{3}{4} + \frac{1}{8}$

90. $8.12 - 9.23$

91. $-56 - (-18)$

92. $-\frac{3}{4} - \frac{1}{8}$

93. $-2.3 - 7.1$

94. $-8.12 + 9.23$

Simplify.

95. $5 - 3^2 + (8 - 2)^2 \cdot 4$ [1.8d]

96. $10 \div 2 \cdot 5 - 3^2 + (-5)^2$ [1.8d]

97. $5(2x - 4) - 3(4x + 1)$ [1.8b]

98. $9(3 + 5x) - 4(7 + 2x)$ [1.8b]

SYNTHESIS

99. Determine whether each number is a solution of the inequality $|x| < 3$.

a) 0 b) -2
c) -3 d) 4
e) 3 f) 1.7
g) -2.8

100. Graph $|x| < 3$ on a number line.

−5 −4 −3 −2 −1 0 1 2 3 4 5

Solve.

101. $x + 3 \leq 3 + x$

102. $x + 4 < 3 + x$

2.8 APPLICATIONS AND PROBLEM SOLVING WITH INEQUALITIES

Objectives

a Translate number sentences to inequalities.

b Solve applied problems using inequalities.

Translate.

1. Maggie scored no less than 92 on her English exam.

2. The average credit card holder is at least $4000 in debt.

3. The price of that PT Cruiser is at most $21,900.

4. The time of the test was between 45 and 55 min.

5. Normandale Community College is more than 15 mi away.

6. Tania's weight is less than 110 lb.

7. That number is greater than −2.

8. The costs of production of that CD-ROM cannot exceed $12,500.

9. At most, 11.4% of all deaths in Arizona are from cancer.

10. Yesterday, at least 23 people got tickets for speeding.

Answers on page A-9

The five steps for problem solving can be used for problems involving inequalities.

a Translating to Inequalities

Before solving problems that involve inequalities, we list some important phrases to look for. Sample translations are listed as well.

IMPORTANT WORDS	SAMPLE SENTENCE	TRANSLATION
is at least	Bill is at least 21 years old.	$b \geq 21$
is at most	At most 5 students dropped the course.	$n \leq 5$
cannot exceed	To qualify, earnings cannot exceed $12,000.	$r \leq 12{,}000$
must exceed	The speed must exceed 15 mph.	$s > 15$
is less than	Tucker's weight is less than 50 lb.	$w < 50$
is more than	Boston is more than 200 miles away.	$d > 200$
is between	The film was between 90 and 100 minutes long.	$90 < t < 100$
no more than	Bing weighs no more than 90 lb.	$w \leq 90$
no less than	Valerie scored no less than 8.3.	$s \geq 8.3$

The following phrases deserve special attention.

TRANSLATING "AT LEAST" AND "AT MOST"

A quantity x is at least some amount q: $x \geq q$.
(If x is at least q, it cannot be less than q.)

A quantity x is at most some amount q: $x \leq q$.
(If x is at most q, it cannot be more than q.)

Do Exercises 1–10.

b Solving Problems

EXAMPLE 1 *Catering costs.* To cater a party, Curtis' Barbeque charges a $50 setup fee plus $15 per person. The cost of Hotel Pharmacy's end-of-season softball party cannot exceed $450. How many people can attend the party?
Source: Curtis' All American Barbeque, Putney, Vermont

1. **Familiarize.** Suppose that 20 people were to attend the party. The cost would then be $\$50 + \$15 \cdot 20$, or $350. This shows that more than 20 people could attend without exceeding $450. Instead of making another guess, we let $n =$ the number of people in attendance.

2. **Translate.** The cost of the party will be \$50 for the setup fee plus \$15 times the number of people attending. We can reword and translate to an inequality as follows:

			the cost of		
Rewording:	The setup fee	plus	the meals	cannot exceed	\$450.
	↓	↓	↓	↓	↓
Translating:	50	+	$15 \cdot n$	$\leq$	450

3. **Solve.** We solve the inequality for n:

$$50 + 15n \leq 450$$

$$50 + 15n - 50 \leq 450 - 50 \quad \text{Subtracting 50}$$

$$15n \leq 400 \quad \text{Simplifying}$$

$$\frac{15n}{15} \leq \frac{400}{15} \quad \text{Dividing by 15}$$

$$n \leq \frac{400}{15}$$

$$n \leq 26\tfrac{2}{3}. \quad \text{Simplifying}$$

4. **Check.** Although the solution set of the inequality is all numbers less than or equal to $26\frac{2}{3}$, since $n =$ the number of people in attendance, we round *down* to 26. If 26 people attend, the cost will be $\$50 + \$15 \cdot 26$, or \$440, and if 27 attend, the cost will exceed \$450.

5. **State.** At most 26 people can attend the party.

Do Exercise 11.

CAUTION!

Solutions of problems should always be checked using the original wording of the problem. In some cases, answers might need to be whole numbers or integers or rounded off in a particular direction.

EXAMPLE 2 *Nutrition.* The U.S. Department of Health and Human Services and the Department of Agriculture recommend that for a typical 2000-calorie daily diet, no more than 65 g of fat be consumed. In the first three days of a four-day vacation, Phil consumed 70 g, 62 g, and 80 g of fat. Determine (in terms of an inequality) how many grams of fat Phil can consume on the fourth day if he is to average no more than 65 g of fat per day.

Sources: U.S. Department of Health and Human Services and Department of Agriculture

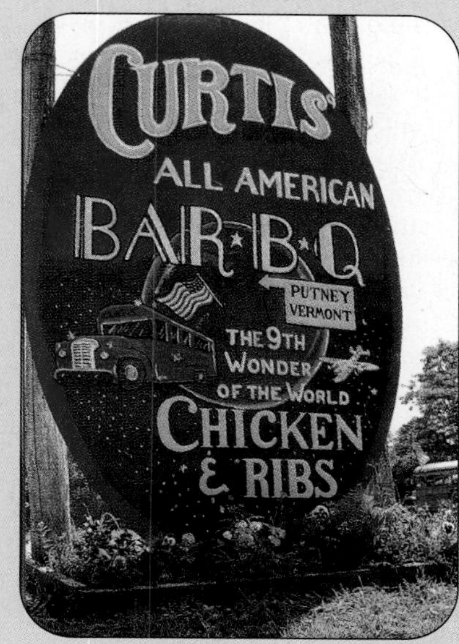

11. **Butter Temperatures.** Butter stays solid at Fahrenheit temperatures below 88°. The formula

$$F = \tfrac{9}{5}C + 32$$

can be used to convert Celsius temperatures C to Fahrenheit temperatures F. Determine (in terms of an inequality) those Celsius temperatures for which butter stays solid.

Answer on page A-9

Translate to an inequality and solve.

12. **Test Scores.** A pre-med student is taking a chemistry course in which four tests are to be given. To get an A, she must average at least 90 on the four tests. The student got scores of 91, 86, and 89 on the first three tests. Determine (in terms of an inequality) what scores on the last test will allow her to get an A.

1. **Familiarize.** Suppose Phil consumed 64 g of fat on the fourth day. His daily average for the vacation would then be

$$\frac{70\text{ g} + 62\text{ g} + 80\text{ g} + 64\text{ g}}{4} = 69\text{ g}.$$

This shows that Phil cannot consume 64 g of fat on the fourth day, if he is to average no more than 65 g of fat per day. We let $x =$ the number of grams of fat that Phil consumes on the fourth day.

2. **Translate.** We reword the problem and translate to an inequality as follows:

Rewording:	The average consumption of fat	should be no more than	65 g.
Translating:	$\frac{70 + 62 + 80 + x}{4}$	$\leq$	65

3. **Solve.** Because of the fraction expression, it is convenient to use the multiplication principle first to solve the inequality:

$$\frac{70 + 62 + 80 + x}{4} \leq 65$$

$$4\left(\frac{70 + 62 + 80 + x}{4}\right) \leq 4 \cdot 65 \qquad \text{Multiplying by 4}$$

$$70 + 62 + 80 + x \leq 260$$

$$212 + x \leq 260 \qquad \text{Simplifying}$$

$$x \leq 48. \qquad \text{Subtracting 212}$$

4. **Check.** As a partial check, we show that Phil can consume 48 g of fat on the fourth day and not exceed a 65-g average for the four days:

$$\frac{70 + 62 + 80 + 48}{4} = \frac{260}{4} = 65.$$

5. **State.** Phil's average fat intake for the vacation will not exceed 65 g per day if he consumes no more than 48 g of fat on the fourth day.

Do Exercise 12.

Answer on page A-9

2.8 EXERCISE SET

For Extra Help

Digital Video Tutor CD 3 Videotape 6

InterAct Math

Math Tutor Center

MathXL

MyMathLab

a Translate to an inequality.

1. A number is at least 7.

2. A number is greater than or equal to 5.

3. The baby weighs more than 2 kilograms (kg).

4. Between 75 and 100 people attended the concert.

5. The speed of the train was between 90 and 110 mph.

6. At least 400,000 people attended the Million Man March.

7. At most 1,200,000 people attended the Million Man March.

8. The amount of acid is not to exceed 40 liters (L).

9. The cost of gasoline is no less than \$1.50 per gallon.

10. The temperature is at most $-2°$.

11. A number is greater than 8.

12. A number is less than 5.

13. A number is less than or equal to -4.

14. A number is greater than or equal to 18.

15. The number of people is at least 1300.

16. The cost is at most \$4857.95.

17. The amount of acid is not to exceed 500 liters.

18. The cost of gasoline is no less than 94 cents per gallon.

19. Two more than three times a number is less than 13.

20. Five less than one-half a number is greater than 17.

b Solve.

21. *Test Scores.* A student is taking a literature course in which four tests are to be given. To get a B, he must average at least 80 on the four tests. The student got scores of 82, 76, and 78 on the first three tests. Determine (in terms of an inequality) what scores on the last test will allow him to get at least a B.

22. *Test Scores.* Your quiz grades are 73, 75, 89, and 91. Determine (in terms of an inequality) what scores on the last quiz will allow you to get an average quiz grade of at least 85.

23. *Gold Temperatures.* Gold stays solid at Fahrenheit temperatures below 1945.4°. Determine (in terms of an inequality) those Celsius temperatures for which gold stays solid. Use the formula given in Margin Exercise 11.

24. *Body Temperatures.* The human body is considered to be fevered when its temperature is higher than 98.6°F. Using the formula given in Margin Exercise 11, determine (in terms of an inequality) those Celsius temperatures for which the body is fevered.

25. *World Records in the 1500-m Run.* The formula

$$R = -0.075t + 3.85$$

can be used to predict the world record in the 1500-m run t years after 1930. Determine (in terms of an inequality) those years for which the world record will be less than 3.5 min.

26. *World Records in the 200-m Dash.* The formula

$$R = -0.028t + 20.8$$

can be used to predict the world record in the 200-m dash t years after 1920. Determine (in terms of an inequality) those years for which the world record will be less than 19.0 sec.

27. *Sizes of Envelopes.* Rhetoric Advertising is a direct-mail company. It determines that for a particular campaign, it can use any envelope with a fixed width of $3\frac{1}{2}$ in. and an area of at least $17\frac{1}{2}$ in^2. Determine (in terms of an inequality) those lengths that will satisfy the company constraints.

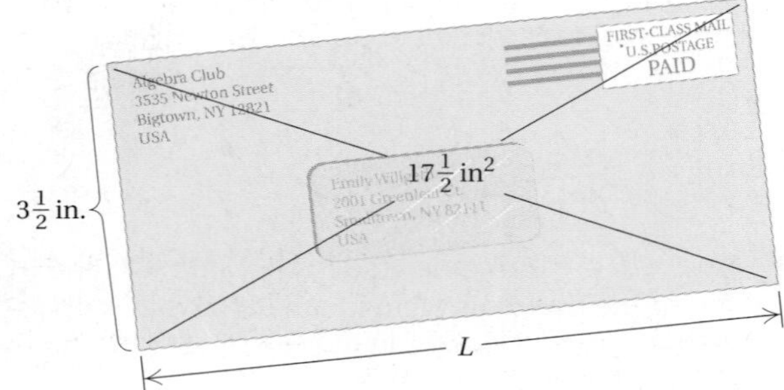

28. *Sizes of Packages.* An overnight delivery service accepts packages of up to 165 in. in length and girth combined. (Girth is the distance around the package.) A package has a fixed girth of 53 in. Determine (in terms of an inequality) those lengths for which a package is acceptable.

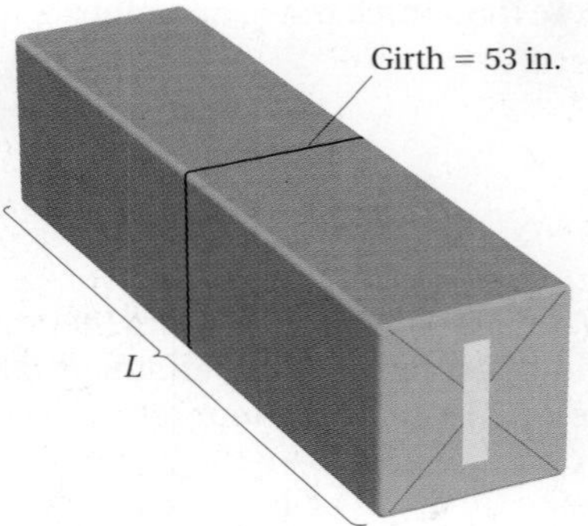

29. *Blueprints.* To make copies of blueprints, Vantage Reprographics charges a $5 setup fee plus $4 per copy. Myra can spend no more than $65 for the copying. What numbers of copies will allow her to stay within budget?

30. *Banquet Costs.* The women's volleyball team can spend at most $450 for its awards banquet at a local restaurant. If the restaurant charges a $40 setup fee plus $16 per person, at most how many can attend?

31. *Phone Costs.* Simon claims that it costs him at least $3.00 every time he calls an overseas customer. If his typical call costs 75¢ plus 45¢ for each minute, how long do his calls typically last?

32. *Parking Costs.* Laura is certain that every time she parks in the municipal garage it costs her at least $2.20. If the garage charges 45¢ plus 25¢ for each half hour, for how long is Laura's car generally parked?

33. *College Tuition.* Angelica's financial aid stipulates that her tuition not exceed $1000. If her local community college charges a $35 registration fee plus $375 per course, what is the greatest number of courses for which Angelica can register?

34. *Furnace Repairs.* RJ's Plumbing and Heating charges $25 plus $30 per hour for emergency service. Gary remembers being billed over $100 for an emergency call. How long was RJ's there?

35. *Nutrition.* Following the guidelines of the Food and Drug Administration, Dale tries to eat at least 5 servings of fruits or vegetables each day. For the first six days of one week, he had 4, 6, 7, 4, 6, and 4 servings. How many servings of fruits or vegetables should Dale eat on Saturday, in order to average at least 5 servings per day for the week?

36. *College Course Load.* To remain on financial aid, Millie needs to complete an average of at least 7 credits per quarter each year. In the first three quarters of 2001, Millie completed 5, 7, and 8 credits. How many credits of course work must Millie complete in the fourth quarter if she is to remain on financial aid?

37. *Perimeter of a Rectangle.* The width of a rectangle is fixed at 8 ft. What lengths will make the perimeter at least 200 ft? at most 200 ft?

38. *Perimeter of a Triangle.* One side of a triangle is 2 cm shorter than the base. The other side is 3 cm longer than the base. What lengths of the base will allow the perimeter to be greater than 19 cm?

39. *Area of a Rectangle.* The width of a rectangle is fixed at 4 cm. For what lengths will the area be less than 86 cm^2?

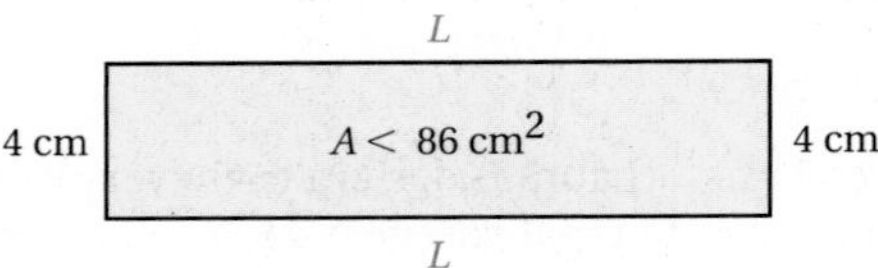

40. *Area of a Rectangle.* The width of a rectangle is fixed at 16 yd. For what lengths will the area be at least 264 yd^2?

41. *Insurance-covered Repairs.* Most insurance companies will replace a vehicle if an estimated repair exceeds 80% of the "blue-book" value of the vehicle. Michelle's insurance company paid \$8500 for repairs to her Subaru after an accident. What can be concluded about the blue-book value of the car?

42. *Insurance-covered Repairs.* Following an accident, Jeff's Ford pickup was replaced by his insurance company because the damage was so extensive. Before the damage, the blue-book value of the truck was \$21,000. How much would it have cost to repair the truck? (See Exercise 41.)

43. *Fat Content in Foods.* Reduced Fat Skippy® peanut butter contains 12 g of fat per serving. In order for a food to be labeled "reduced fat," it must have at least 25% less fat than the regular item. What can you conclude about the number of grams of fat in a serving of the regular Skippy peanut butter?
Source: Best Foods

44. *Fat Content in Foods.* Reduced Fat Chips Ahoy!® cookies contain 5 g of fat per serving. What can you conclude about the number of grams of fat in regular Chips Ahoy! cookies (see Exercise 43)?
Source: Nabisco Brands, Inc.

45. *Pond Depth.* On July 1, Garrett's Pond was 25 ft deep. Since that date, the water level has dropped $\frac{2}{3}$ ft per week. For what dates will the water level not exceed 21 ft?

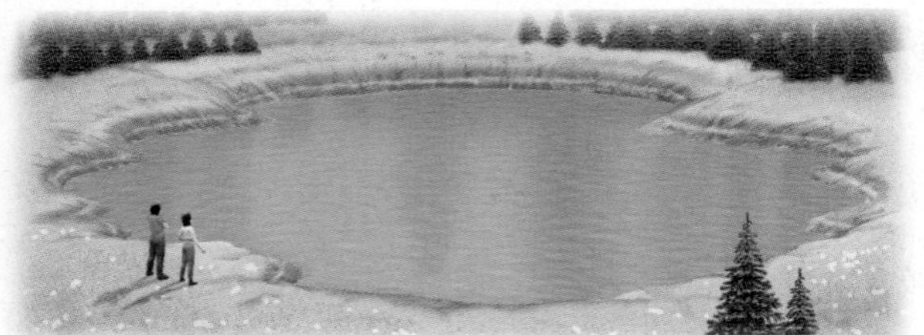

46. *Weight Gain.* A 3-lb puppy is gaining weight at a rate of $\frac{3}{4}$ lb per week. When will the puppy's weight exceed $22\frac{1}{2}$ lb?

47. *Area of a Triangular Flag.* As part of an outdoor education course, Wanda needs to make a bright-colored triangular flag with an area of at least 3 ft^2. What heights can the triangle be if the base is $1\frac{1}{2}$ ft?

48. *Area of a Triangular Sign.* Zoning laws in Harrington prohibit displaying signs with areas exceeding 12 ft^2. If Flo's Marina is ordering a triangular sign with an 8-ft base, how tall can the sign be?

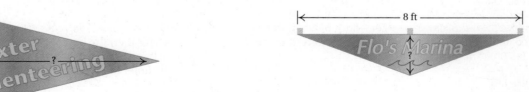

49. *Electrician Visits.* Dot's Electric made 17 customer calls last week and 22 calls this week. How many calls must be made next week in order to maintain an average of at least 20 for the three-week period?

50. *Volunteer Work.* George and Joan do volunteer work at a hospital. Joan worked 3 more hr than George, and together they worked more than 27 hr. What possible numbers of hours did each work?

51. ^{D}W If f represents Fran's age and t represents Todd's age, write a sentence that would translate to $t + 3 < f$.

52. ^{D}W Explain how the meanings of "Five more than a number" and "Five is more than a number" differ.

SKILL MAINTENANCE

Simplify.

53. $-3 + 2(-5)^2(-3) - 7$ [1.8d]

54. $3x + 2[4 - 5(2x - 1)]$ [1.8c]

55. $23(2x - 4) - 15(10 - 3x)$ [1.8b]

56. $256 \div 64 \div 4^2$ [1.8d]

SYNTHESIS

57. *Ski Wax.* Green ski wax works best between 5° and 15° Fahrenheit. Determine those Celsius temperatures for which green ski wax works best.

58. *Parking Fees.* Mack's Parking Garage charges $4.00 for the first hour and $2.50 for each additional hour. For how long has a car been parked when the charge exceeds $16.50?

59. *Nutritional Standards.* In order for a food to be labeled "lowfat," it must have fewer than 3 g of fat per serving. Reduced fat Tortilla Pops® contain 60% less fat than regular nacho cheese tortilla chips, but still cannot be labeled lowfat. What can you conclude about the fat content of a serving of nacho cheese tortilla chips?

60. *Parking Fees.* When asked how much the parking charge is for a certain car (see Exercise 58), Mack replies "between 14 and 24 dollars." For how long has the car been parked?

CHAPTER 2 Summary and Review

The review that follows is meant to prepare you for a chapter exam. It consists of two parts. The first part is a checklist of some of the Study Tips referred to in this and preceding chapters, as well as a list of important properties and formulas. The second part is the Review Exercises. These provide practice exercises for the exam, together with references to section objectives so you can go back and review. Before beginning, stop and look back over the skills you have obtained. What skills in mathematics do you have now that you did not have before studying this chapter?

STUDY TIPS CHECKLIST

The foundation of all your study skills is TIME!

- ☐ Are you making progress in learning to manage your time?
- ☐ Are you practicing the five-step problem-solving strategy?
- ☐ Are you doing the homework as soon as possible after class?
- ☐ Are you stopping to work the margin exercises when directed to do so?

IMPORTANT PROPERTIES AND FORMULAS

The Addition Principle for Equations: For any real numbers a, b, and c: $a = b$ is equivalent to $a + c = b + c$.

The Multiplication Principle for Equations: For any real numbers a, b, and c, $c \neq 0$: $a = b$ is equivalent to $a \cdot c = b \cdot c$.

The Addition Principle for Inequalities: For any real numbers a, b, and c:
$a < b$ is equivalent to $a + c < b + c$;
$a > b$ is equivalent to $a + c > b + c$;
$a \leq b$ is equivalent to $a + c \leq b + c$;
$a \geq b$ is equivalent to $a + c \geq b + c$.

The Multiplication Principle for Inequalities: For any real numbers a and b, and any *positive* number c:
$a < b$ is equivalent to $ac < bc$; $a > b$ is equivalent to $ac > bc$.
For any real numbers a and b, and any *negative* number c:
$a < b$ is equivalent to $ac > bc$; $a > b$ is equivalent to $ac < bc$.

REVIEW EXERCISES

Solve. [2.1b]

1. $x + 5 = -17$

2. $n - 7 = -6$

3. $x - 11 = 14$

4. $y - 0.9 = 9.09$

Solve. [2.2a]

5. $-\frac{2}{3}x = -\frac{1}{6}$

6. $-8x = -56$

7. $-\frac{x}{4} = 48$

8. $15x = -35$

9. $\frac{4}{5}y = -\frac{3}{16}$

Solve. [2.3a]

10. $5 - x = 13$

11. $\frac{1}{4}x - \frac{5}{8} = \frac{3}{8}$

Solve. [2.3b, c]

12. $5t + 9 = 3t - 1$

13. $7x - 6 = 25x$

14. $14y = 23y - 17 - 10$

15. $0.22y - 0.6 = 0.12y + 3 - 0.8y$

16. $\frac{1}{4}x - \frac{1}{8}x = 3 - \frac{1}{16}x$

17. $14y + 17 + 7y = 9 + 21y + 8$

Solve. [2.3c, d]

18. $4(x + 3) = 36$

19. $3(5x - 7) = -66$

20. $8(x - 2) - 5(x + 4) = 20 + x$

21. $-5x + 3(x + 8) = 16$

22. $6(x - 2) - 16 = 3(2x - 5) + 11$

Determine whether the given number is a solution of the inequality $x \leq 4$. [2.7a]

23. -3

24. 7

25. 4

Solve. Write set notation for the answers. [2.7c, d, e]

26. $y + \frac{2}{3} \geq \frac{1}{6}$

27. $9x \geq 63$

28. $2 + 6y > 14$

29. $7 - 3y \geq 27 + 2y$

30. $3x + 5 < 2x - 6$

31. $-4y < 28$

32. $4 - 8x < 13 + 3x$

33. $-4x \leq \frac{1}{3}$

Graph on a number line. [2.7b, e]

34. $4x - 6 < x + 3$

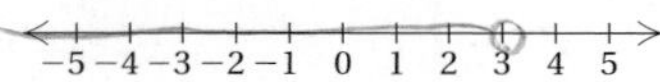

35. $-2 < x \leq 5$

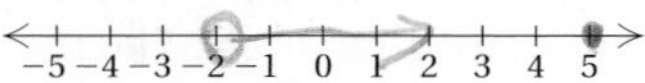

36. $y > 0$

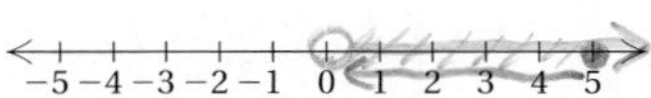

Solve. [2.4b]

37. $C = \pi d$, for d

38. $V = \frac{1}{3}Bh$, for B

39. $A = \frac{a + b}{2}$, for a

40. $y = mx + b$, for x

Solve. [2.6a]

41. *Dimensions of Wyoming.* The state of Wyoming is roughly in the shape of a rectangle whose perimeter is 1280 mi. The length is 90 mi more than the width. Find the dimensions.

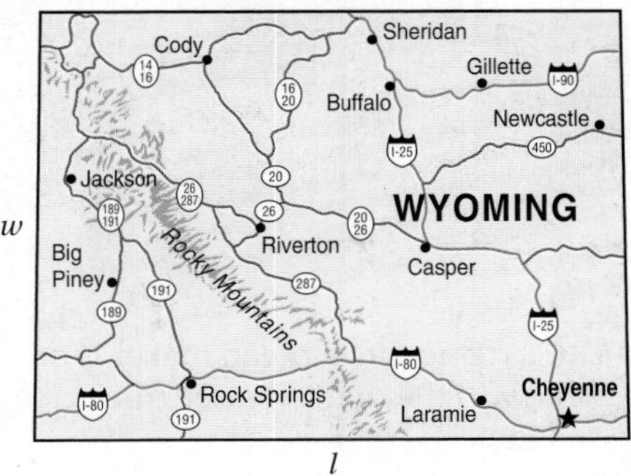

42. *Interstate Mile Markers.* The sum of two consecutive mile markers on I-5 in California is 691. Find the numbers on the markers.

43. An entertainment center sold for $2449 in June. This was $332 more than the cost in February. Find the cost in February.

44. Ty is paid a commission of $4 for each appliance he sells. One week, he received $108 in commissions. How many appliances did he sell?

45. The measure of the second angle of a triangle is 50° more than that of the first angle. The measure of the third angle is 10° less than twice the first angle. Find the measures of the angles.

Solve. [2.5a]

46. What is 20% of 75?

47. Fifteen is what percent of 80?

48. 18 is 3% of what number?

49. *Job Opportunities.* There were 905 thousand child-care workers in 1998. Job opportunities are expected to grow to 1141 thousand by 2008. What is the percent of increase?
Source: *Handbook of U.S. Labor Statistics*

Solve. [2.6a]

50. After a 30% reduction, a bread maker is on sale for $154. What was the marked price (the price before the reduction)?

51. A hotel manager's salary is $61,410, which is a 15% increase over the previous year's salary. What was the previous salary?

52. A tax-exempt charity received a bill of $145.90 for a sump pump. The bill incorrectly included sales tax of 5%. How much does the charity actually owe?

Solve. [2.8b]

53. *Test Scores.* Your test grades are 71, 75, 82, and 86. What is the lowest grade that you can get on the next test and still have an average test score of at least 80?

54. The length of a rectangle is 43 cm. What widths will make the perimeter greater than 120 cm?

55. $^{D}_{W}$ Would it be better to receive a 5% raise and then an 8% raise or the other way around? Why? [2.5a]

56. $^{D}_{W}$ Are the inequalities $x > -5$ and $-x < 5$ equivalent? Why or why not? [2.7c]

SKILL MAINTENANCE

Certain objectives from four particular sections will be retested on the chapter test. The objectives are listed with the practice problems that follow.

57. Evaluate $\dfrac{a+b}{4}$ when $a = 16$ and $b = 25$. [1.1a]

58. Translate to an algebraic expression: [1.1b]
Tricia drives her car at 58 mph for t hours. How far has she driven?

59. Add: $-12 + 10 + (-19) + (-24)$. [1.3a]

60. Remove parentheses and simplify: $5x - 8(6x - y)$. [1.8b]

SYNTHESIS

Solve.

61. $2|x| + 4 = 50$ [1.2e], [2.3a]

62. $|3x| = 60$ [1.2e], [2.2a]

63. $y = 2a - ab + 3$, for a [2.4b]

CHAPTER 2 Chapter Test

Solve.

1. $x + 7 = 15$

2. $t - 9 = 17$

3. $3x = -18$

4. $-\frac{4}{7}x = -28$

5. $3t + 7 = 2t - 5$

6. $\frac{1}{2}x - \frac{3}{5} = \frac{2}{5}$

7. $8 - y = 16$

8. $-\frac{2}{5} + x = -\frac{3}{4}$

9. $3(x + 2) = 27$

10. $-3x - 6(x - 4) = 9$

11. $0.4p + 0.2 = 4.2p - 7.8 - 0.6p$

12. $4(3x - 1) + 11 = 2(6x + 5) - 8$

13. $-2 + 7x + 6 = 5x + 4 + 2x$

Solve. Write set notation for the answers.

14. $x + 6 \leq 2$

15. $14x + 9 > 13x - 4$

16. $12x \leq 60$

17. $-2y \geq 26$

18. $-4y \leq -32$

19. $-5x \geq \frac{1}{4}$

20. $4 - 6x > 40$

21. $5 - 9x \geq 19 + 5x$

Graph on a number line.

22. $y \leq 9$

23. $6x - 3 < x + 2$

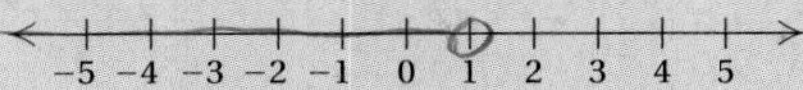

24. $-2 \leq x \leq 2$

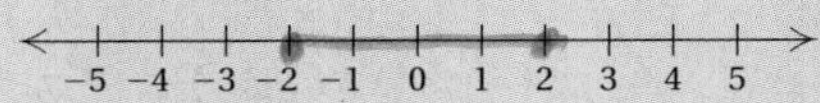

Solve.

25. What is 24% of 75?

26. 15.84 is what percent of 96?

27. 800 is 2% of what number?

28. *Job Opportunities.* Job opportunities for flight attendants are expected to increase from 99 thousand in 1998 to 129 thousand by 2008. What is the percent of increase?
Source: *Handbook of U.S. Labor Statistics*

29. *Perimeter of a Photograph.* The perimeter of a rectangular photograph is 36 cm. The length is 4 cm greater than the width. Find the width and the length.

30. *Charitable Contributions.* About 53% of all charitable contributions are made to religious organizations. In 1997, \$75 billion was given to religious organizations. How much was given to charities in general?
Source: *Statistical Abstract of the United States, 1999*

31. *Raffle Tickets.* The numbers on three raffle tickets are consecutive integers whose sum is 7530. Find the integers.

32. *Savings Account.* Money is invested in a savings account at 5% simple interest. After 1 year, there is \$924 in the account. How much was originally invested?

33. *Board Cutting.* An 8-m board is cut into two pieces. One piece is 2 m longer than the other. How long are the pieces?

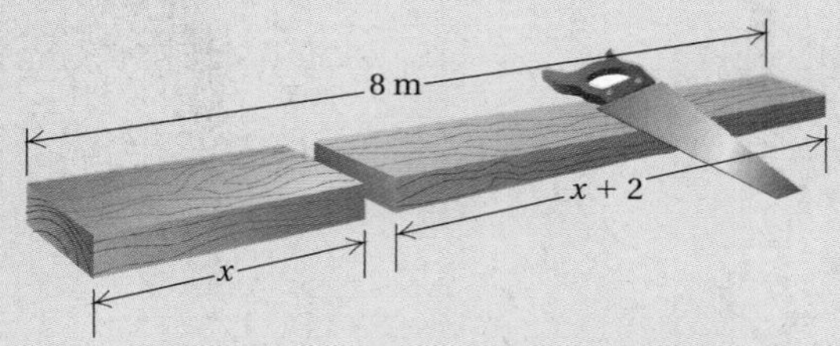

34. *Lengths of a Rectangle.* The width of a rectangle is 96 yd. Find all possible lengths such that the perimeter of the rectangle will be at least 540 yd.

35. *Budgeting.* Jason has budgeted an average of \$95 a month for entertainment. For the first five months of the year, he has spent \$98, \$89, \$110, \$85, and \$83. How much can Jason spend in the sixth month without exceeding his average budget?

36. *IKON Copiers.* IKON Office Solutions rents a Canon IR330 copier for \$225 per month plus 1.2¢ per copy. A catalog publisher needs to lease a copy machine for use during a special project that they anticipate will take 3 months. They decide to rent the copier, but must stay within a budget of \$2400 for copies. Determine (in terms of an inequality) the number of copies they can make and still remain within budget.
Source: Ikon Office Solutions, Nathan DuMond, Sales Manager

37. Solve $A = 2\pi rh$ for r.

38. Solve $y = 8x + b$ for x.

SKILL MAINTENANCE

39. Add: $\frac{2}{3} + \left(-\frac{8}{9}\right)$.

40. Evaluate $\frac{4x}{y}$ when $x = 2$ and $y = 3$.

41. Translate to an algebraic expression:
Seventy-three percent of p.

42. Simplify: $2x - 3y - 5(4x - 8y)$.

SYNTHESIS

43. Solve $c = \frac{1}{a - d}$ for d.

44. Solve: $3|w| - 8 = 37$.

45. A movie theater had a certain number of tickets to give away. Five people got the tickets. The first got one-third of the tickets, the second got one-fourth of the tickets, and the third got one-fifth of the tickets. The fourth person got eight tickets, and there were five tickets left for the fifth person. Find the total number of tickets given away.

CHAPTERS

1–2 Cumulative Review

Evaluate.

1. $\frac{y - x}{4}$, when $y = 12$ and $x = 6$

2. $\frac{3x}{y}$, when $x = 5$ and $y = 4$

3. $x - 3$, when $x = 3$

4. Translate to an algebraic expression: Four less than twice w.

Use $<$ or $>$ for $\square$ to write a true sentence.

5. $-4 \square -6$

6. $0 \square -5$

7. $-8 \square 7$

8. Find the opposite and the reciprocal of $\frac{2}{5}$.

Find the absolute value.

9. $|3|$

10. $\left|-\frac{3}{4}\right|$

11. $|0|$

Compute and simplify.

12. $-6.7 + 2.3$

13. $-\frac{1}{6} - \frac{7}{3}$

14. $-\frac{5}{8}\left(-\frac{4}{3}\right)$

15. $(-7)(5)(-6)(-0.5)$

16. $81 \div (-9)$

17. $-10.8 \div 3.6$

18. $-\frac{4}{5} \div -\frac{25}{8}$

Multiply.

19. $5(3x + 5y + 2z)$

20. $4(-3x - 2)$

21. $-6(2y - 4x)$

Factor.

22. $64 + 18x + 24y$

23. $16y - 56$

24. $5a - 15b + 25$

4 16 4,5 6

Collect like terms.

25. $9b + 18y + 6b + 4y$

26. $3y + 4 + 6z + 6y$

27. $-4d - 6a + 3a - 5d + 1$

28. $3.2x + 2.9y - 5.8x - 8.1y$

Simplify.

29. $7 - 2x - (-5x) - 8$

30. $-3x - (-x + y)$

31. $-3(x - 2) - 4x$

32. $10 - 2(5 - 4x)$

33. $[3(x + 6) - 10] - [5 - 2(x - 8)]$

Solve.

34. $x + 1.75 = 6.25$

35. $\frac{5}{2}y = \frac{2}{5}$

36. $-2.6 + x = 8.3$

37. $4\frac{1}{2} + y = 8\frac{1}{3}$

38. $-\frac{3}{4}x = 36$

39. $-2.2y = -26.4$

40. $5.8x = -35.96$

41. $-4x + 3 = 15$

42. $-3x + 5 = -8x - 7$

43. $4y - 4 + y = 6y + 20 - 4y$

44. $-3(x - 2) = -15$

45. $\frac{1}{3}x - \frac{5}{6} = \frac{1}{2} + 2x$

46. $-3.7x + 6.2 = -7.3x - 5.8$

47. $4(x + 2) = 4(x - 2) + 16$

48. $0(x + 3) + 4 = 0$

49. $5(7 + x) = (x + 7)5$

50. $3x - 1 < 2x + 1$

51. $5 - y \leq 2y - 7$

52. $3y + 7 > 5y + 13$

53. $H = 65 - m$, for m
(To determine the number of heating degree days H for a day with m degrees Fahrenheit as the average temperature)

54. $I = Prt$, for P
(Simple-interest formula, where I is interest, P is principal, r is interest rate, and t is time)

Solve.

55. What is 24% of 105?

56. 39.6 is what percent of 88?

57. $163.60 is 45% of what number?

58. *Overweight Americans.* In 2001, there were 281 million people in the United States. About 60% of them were considered overweight. How many people were overweight?
Source: U.S. Centers for Disease Control

59. *Grade Average.* Nadia is taking a literature course in which four tests are given. To get a B, a student must average at least 80 on the four tests. Nadia scored 82, 76, and 78 on the first three tests. What scores on the last test will earn her at least a B?

60. *Rollerblade Costs.* Susan and Melinda purchased rollerblades for a total of $107. Susan paid $17 more for her rollerblades than Melinda did. What did Melinda pay?

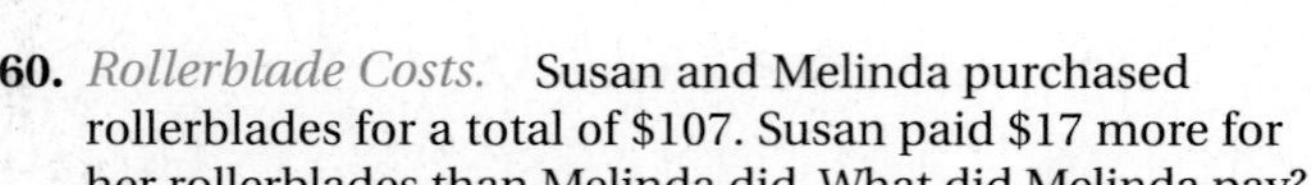

61. *Savings Investment.* Money is invested in a savings account at 8% simple interest. After 1 year, there is $1134 in the account. How much was originally invested?

62. *Wire Cutting.* A 143-m wire is cut into three pieces. The second piece is 3 m longer than the first. The third is four-fifths as long as the first. How long is each piece?

63. *Truck Rentals.* Truck-Rite Rentals rents trucks at a daily rate of \$49.95 plus 39¢ per mile. Concert Productions has budgeted \$100 for renting a truck to haul equipment to an upcoming concert. How far can they travel in one day and stay within their budget?

64. *Price Reduction.* After a 25% reduction, a tie is on sale for \$18.45. What was the price before reduction?

For each of Exercises 65–67, choose the correct answer from the selections given.

65. Simplify: $-125 \div 25 \cdot 625 \div 5$.

a) $-390{,}625$ b) -125 c) -625
d) 25 e) None

66. Remove parentheses and simplify:

$$[5(2x + 6) - 7] - [2(x + 4) + 5].$$

a) $8x + 36$ b) $8x - 10$ c) $8x + 8$
d) $8x + 10$ e) None

67. Solve $V = IR$ for I.

a) $I = V - R$ b) $I = \dfrac{V}{R}$ c) $I = \dfrac{R}{V}$

d) $I = VR$ e) None

SYNTHESIS

68. An engineer's salary at the end of a year is \$48,418.24. This reflects a 4% salary increase and a later 3% cost-of-living adjustment during the year. What was the salary at the beginning of the year?

69. Nadia needs to use a copier to reduce a drawing to fit on a page. The original drawing is 9 in. long and it must fit into a space that is 6.3 in. long. By what percent should she reduce the drawing on the copier?

Solve.

70. $4|x| - 13 = 3$

71. $\dfrac{2 + 5x}{4} = \dfrac{11}{28} + \dfrac{8x + 3}{7}$

72. $p = \dfrac{2}{m + Q}$, for Q

Graphs of Linear Equations

Gateway to Chapter 3

We now begin a study of graphs. First, we examine graphs as they commonly appear in newspapers or magazines and develop some terminology. Following that, we study graphs of linear equations. Finally, we consider the notion of the slope of a line and connect it to the concept of rate of change.

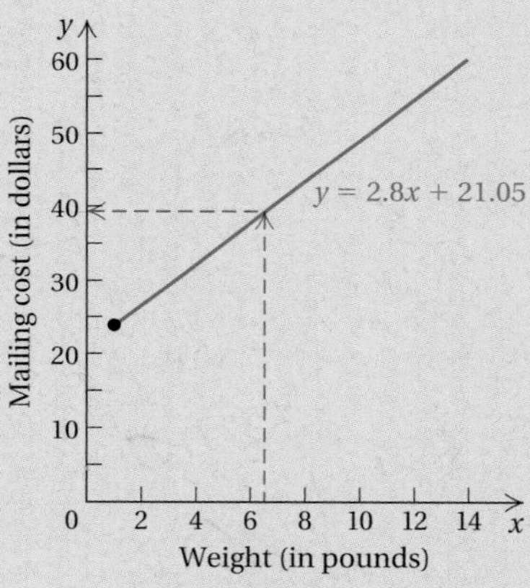

Real-World Application

The cost y, in dollars, of shipping a FedEx Priority Overnight package weighing 1 lb or more a distance of 1001 to 1400 mi is given by $y = 2.8x + 21.05$, where x is the number of pounds. Graph the equation and then use the graph to find the cost of shipping a $6\frac{1}{2}$-lb package.

Source: Federal Express Corporation

This problem appears as Example 8 in Section 3.2.

CHAPTER 3

CHAPTER

3 Pretest

Graph on a plane.

1. $y = -x$ [3.2b]

2. $x = -4$ [3.3b]

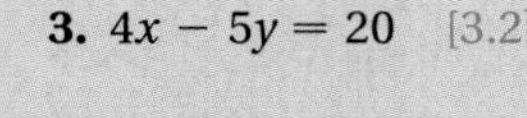

3. $4x - 5y = 20$ [3.2b]

4. $y = \frac{2}{3}x - 1$ [3.2b]

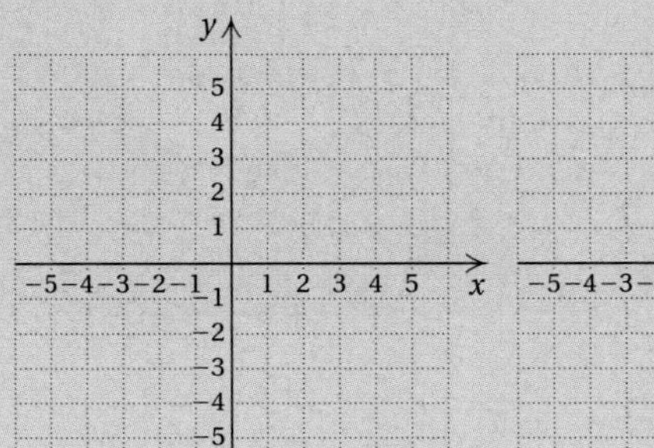

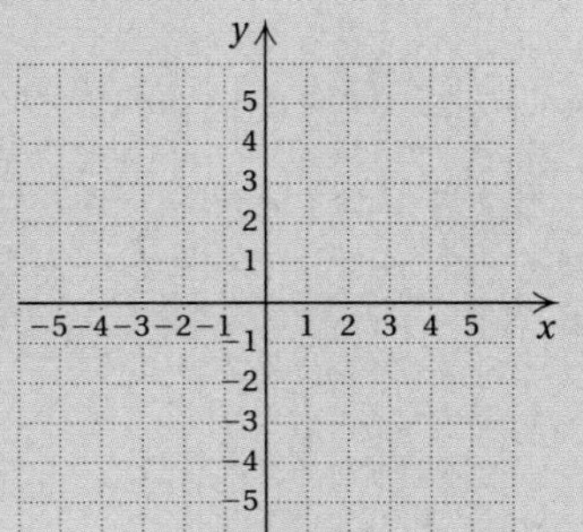

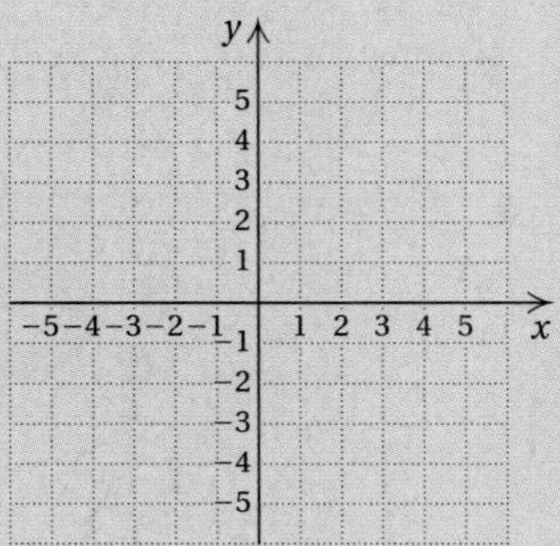

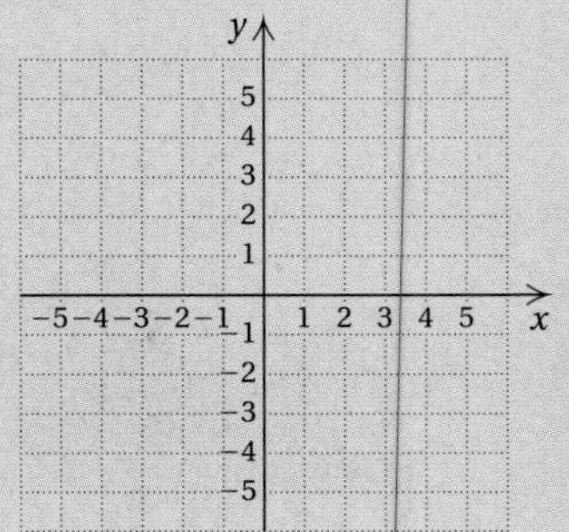

5. In which quadrant is the point $(-4, -1)$ located? [3.1c]

6. Determine whether the ordered pair $(-4, -1)$ is a solution of $4x - 5y = 20$. [3.2a]

7. Find the intercepts of the graph of $4x - 5y = 20$. [3.3a]

8. Find the y-intercept of $y = 3x - 8$. [3.2b]

9. Price of Printing. The price P, in cents, of a photocopied and bound lab manual is given by

$$P = \frac{7}{2}n + 20,$$

where n is the number of pages in the manual. Graph the equation and then use the graph to estimate the price of an 85-page manual. [3.2c]

10. Find the slope of the line. [3.4a]

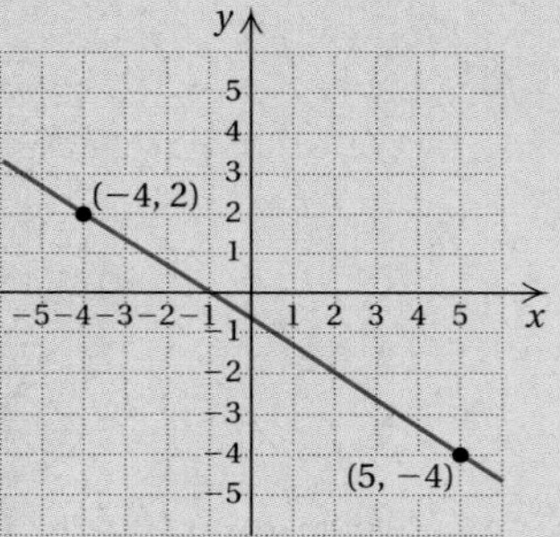

11. Find the rate of change. [3.4b]

TIME	NUMBER OF CARS PRODUCED
9:00 A.M.	11
11:00 A.M.	16

12. Find the rate at which a runner burns calories. [3.4a]

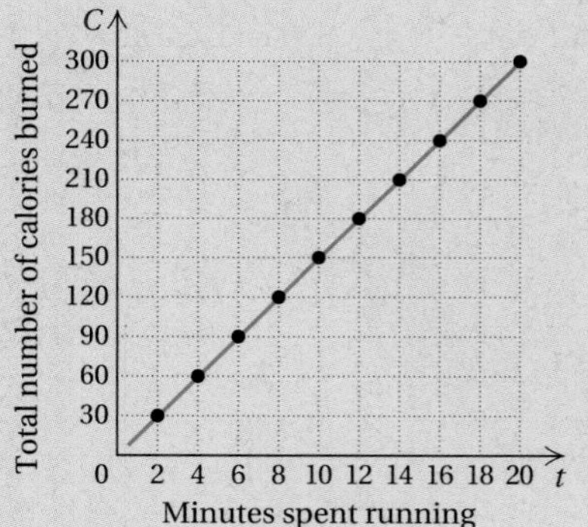

13. Find the slope, if it exists. [3.4c]

$$3x - 5y = 15$$

3.1 GRAPHS AND APPLICATIONS

Objectives

- **a** Solve applied problems involving circle, bar, and line graphs.
- **b** Plot points associated with ordered pairs of numbers.
- **c** Determine the quadrant in which a point lies.
- **d** Find the coordinates of a point on a graph.

Often data are available regarding an application in mathematics. We can use graphs to show the data and extract information about the data that can lead to making analyses and predictions.

Today's print and electronic media make extensive use of graphs. This is due in part to the ease with which some graphs can be prepared by computer and in part to the large quantity of information that a graph can display. We first consider applications with circle, bar, and line graphs.

a Applications with Graphs

CIRCLE GRAPHS

Circle graphs and *pie graphs,* or *charts,* are often used to show what percent of a whole each particular item in a group represents.

EXAMPLE 1 *Careers of Women Who Travel.* Consider all the businesswomen who travel in their careers. The following circle graph shows the percentage of these women in specific careers.

Jobs of Women Who Travel

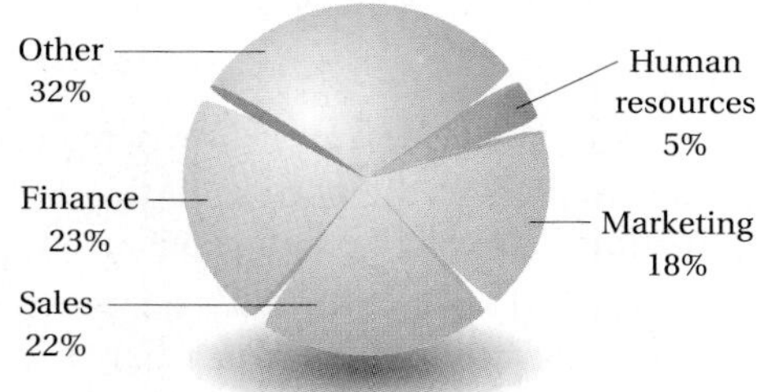

Source: Runzheimer Reports of Travel Management

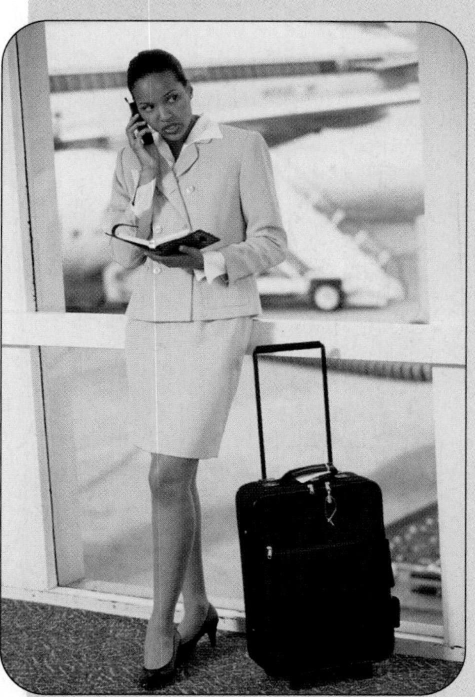

In a sample of 1500 women travelers, how many of them can be expected to be in sales?

1. **Familiarize.** The circle graph shows that 22% of these women are in sales. We let $y =$ the number of women who are in sales.
2. **Translate.** We reword and translate the problem as follows.

$$\begin{array}{lll} \text{What is } 22\% \text{ of } 1500? & & \text{Rewording} \\ \downarrow \quad \downarrow \downarrow \quad \downarrow \quad \downarrow & & \\ y \quad = 22\% \cdot 1500 & & \text{Translating} \end{array}$$

3. **Solve.** We solve the equation by carrying out the computation on the right:

$$y = 22\% \cdot 1500 = 0.22 \cdot 1500 = 330.$$

4. **Check.** We note that 330 is less than 1500, which we would expect. We can also repeat the calculation.
5. **State.** We expect that 330 of the women travelers are in sales.

Do Exercise 1.

1. Careers of Women Who Travel. Referring to the graph in Example 1, determine how many of the 1500 travelers are in marketing.

Answer on page A-10

2. **Tornado Touchdowns.** Referring to the graph in Example 2, determine the following.

   a) During which interval did the least number of touchdowns occur?

   b) During which intervals was the number of touchdowns less than 60?

Answers on page A-10

BAR GRAPHS

Bar graphs are convenient for showing comparisons. In every bar graph, certain categories are paired with certain numbers. Example 2 pairs intervals of time with the total number of reported cases of tornado touchdowns.

EXAMPLE 2 *Tornado Touchdowns.* The following bar graph shows the total number of tornado touchdowns by time of day in Indiana from 1950–1994.

Tornado Touchdowns in Indiana by Time of Day (1950–1994)

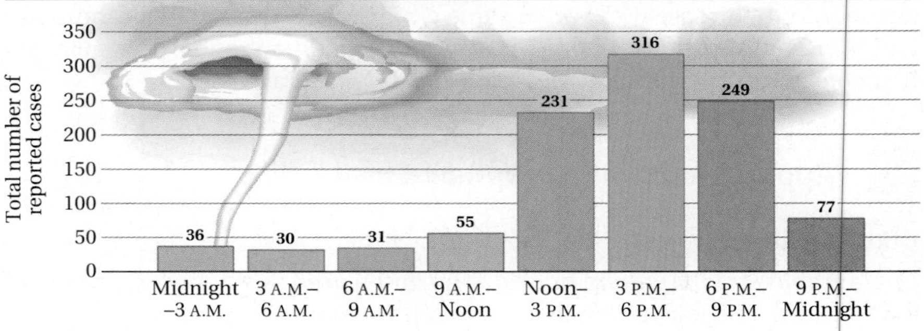

Time of day	Total number of reported cases
Midnight–3 A.M.	36
3 A.M.–6 A.M.	30
6 A.M.–9 A.M.	31
9 A.M.–Noon	55
Noon–3 P.M.	231
3 P.M.–6 P.M.	316
6 P.M.–9 P.M.	249
9 P.M.–Midnight	77

Source: National Weather Service

a) During which interval of time did the greatest number of tornado touchdowns occur?

b) During which intervals was the number of tornado touchdowns greater than 200?

We solve as follows.

a) In this bar graph, the values are written at the top of the bars. We see that 316 is the greatest number. We look at the bottom of that bar on the horizontal scale and see that the time interval of greatest occurrence is 3 P.M.–6 P.M.

b) We locate 200 on the vertical scale and move across the graph or draw a horizontal line. We note that the value on three bars exceeds 200. Then we look down at the horizontal scale and see that the corresponding time intervals are noon–3 P.M., 3 P.M.–6 P.M., and 6 P.M.–9 P.M.

Do Exercise 2.

LINE GRAPHS

Line graphs are often used to show change over time. Certain points are plotted to represent given information. When segments are drawn to connect the points, a line graph is formed.

Sometimes it is important to begin the labeling of horizontal or vertical values on the x- and y-axes with zero. When the values are large, as in Example 3, the symbol ≷ can be used to indicate a break in the list of values.

EXAMPLE 3 *Exercise and Pulse Rate.* The following line graph shows the relationship between a person's resting pulse rate and months of regular exercise.

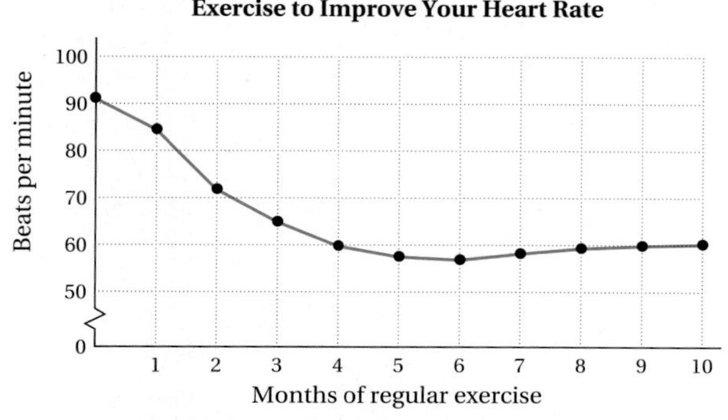

Source: Hughes, Martin, *Body Clock*. New York: Facts on File, Inc., p. 60

a) How many months of regular exercise are required to lower the pulse rate to its lowest point?

b) How many months of regular exercise are needed to achieve a pulse rate of 65 beats per minute?

We solve as follows.

a) The lowest point on the graph occurs above the number 6. Thus after 6 months of regular exercise, the pulse rate has been lowered as much as possible.

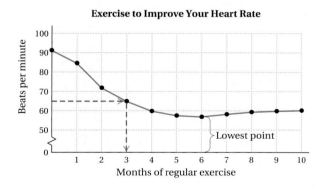

b) We locate 65 on the vertical scale and then move right until we reach the line. At that point, we move down to the horizontal scale and read the information we are seeking. The pulse rate is 65 beats per minute after 3 months of regular exercise.

Do Exercise 3.

3. Exercise and Pulse Rate. Referring to the graph in Example 3, determine the following.

a) About how many months of regular exercise are needed to achieve a pulse rate of about 72 beats per minute?

b) What pulse rate has been achieved after 10 months of exercise?

Answers on page A-10

Plot these points on the graph below.

4. $(4, 5)$ **5.** $(5, 4)$

6. $(-2, 5)$ **7.** $(-3, -4)$

8. $(5, -3)$ **9.** $(-2, -1)$

10. $(0, -3)$ **11.** $(2, 0)$

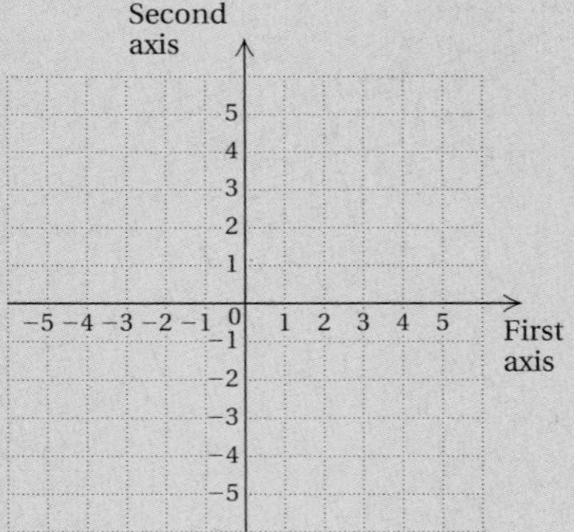

Answers on page A-11

b Plotting Ordered Pairs

The line graph in Example 3 is formed from a collection of points. Each point pairs a number of months of exercise with a pulse rate.

In Chapter 2, we graphed numbers and inequalities in one variable on a line. To enable us to graph an equation that contains two variables, we now learn to graph number pairs on a plane.

On a number line, each point is the graph of a number. On a plane, each point is the graph of a number pair. We use two perpendicular number lines called **axes.** They cross at a point called the **origin.** The arrows show the positive directions.

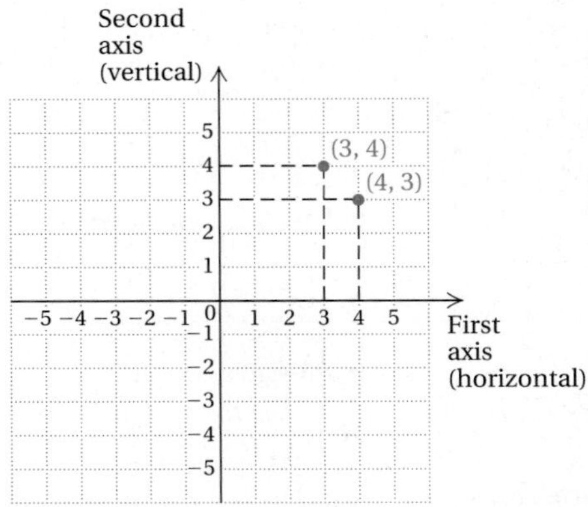

Consider the ordered pair $(3, 4)$. The numbers in an ordered pair are called **coordinates.** In $(3, 4)$, the **first coordinate (abscissa)** is 3 and the **second coordinate (ordinate)** is 4. To plot $(3, 4)$, we start at the origin and move horizontally to the 3. Then we move up vertically 4 units and make a "dot."

The point $(4, 3)$ is also plotted. Note that $(3, 4)$ and $(4, 3)$ give different points. The order of the numbers in the pair is indeed important. They are called **ordered pairs** because it makes a difference which number comes first. The coordinates of the origin are $(0, 0)$.

EXAMPLE 4 Plot the point $(-5, 2)$.

The first number, -5, is negative. Starting at the origin, we move -5 units in the horizontal direction (5 units to the left). The second number, 2, is positive. We move 2 units in the vertical direction (up).

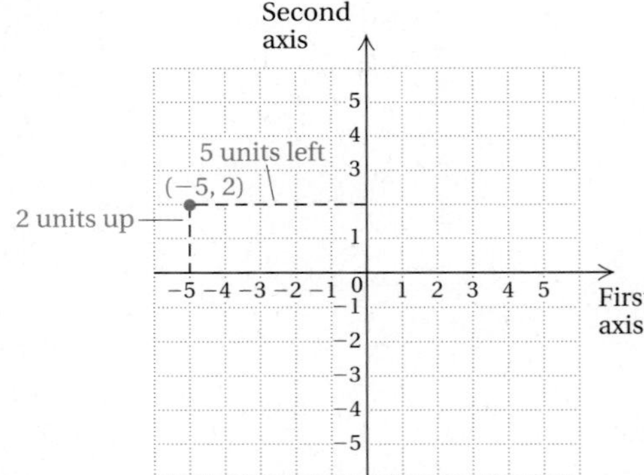

CAUTION!

The *first* coordinate of an ordered pair is always graphed in a *horizontal* direction and the *second* coordinate is always graphed in a *vertical* direction.

Do Exercises 4–11.

c Quadrants

The figure below shows some points and their coordinates. In region I (the *first quadrant*), both coordinates of any point are positive. In region II (the *second quadrant*), the first coordinate is negative and the second positive. In region III (the *third quadrant*), both coordinates are negative. In region IV (the *fourth quadrant*), the first coordinate is positive and the second is negative.

EXAMPLE 5 In which quadrant, if any, are the points $(-4, 5)$, $(5, -5)$, $(2, 4)$, $(-2, -5)$, and $(-5, 0)$ located?

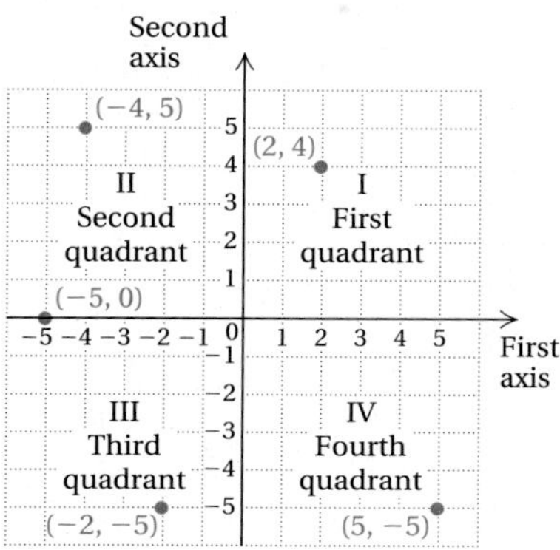

The point $(-4, 5)$ is in the second quadrant. The point $(5, -5)$ is in the fourth quadrant. The point $(2, 4)$ is in the first quadrant. The point $(-2, -5)$ is in the third quadrant. The point $(-5, 0)$ is on an axis and is *not* in any quadrant.

Do Exercises 12–18.

d Finding Coordinates

To find the coordinates of a point, we see how far to the right or left of zero it is located and how far up or down.

EXAMPLE 6 Find the coordinates of points *A, B, C, D, E, F,* and *G*.

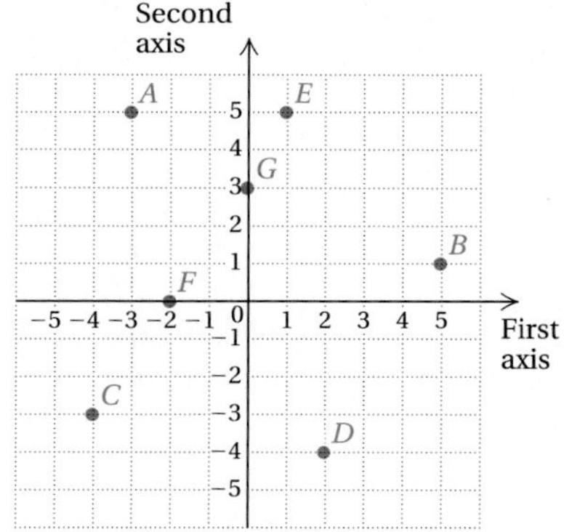

Point *A* is 3 units to the left (horizontal direction) and 5 units up (vertical direction). Its coordinates are $(-3, 5)$. Point *D* is 2 units to the right and 4 units down. Its coordinates are $(2, -4)$. The coordinates of the other points are as follows:

B: $(5, 1)$; *C*: $(-4, -3)$;
E: $(1, 5)$; *F*: $(-2, 0)$; *G*: $(0, 3)$.

Do Exercise 19.

12. What can you say about the coordinates of a point in the third quadrant?

13. What can you say about the coordinates of a point in the fourth quadrant?

In which quadrant, if any, is the point located?

14. $(5, 3)$

15. $(-6, -4)$

16. $(10, -14)$

17. $(-13, 9)$

18. $(0, -3)$

19. Find the coordinates of points *A, B, C, D, E, F,* and *G* on the graph below.

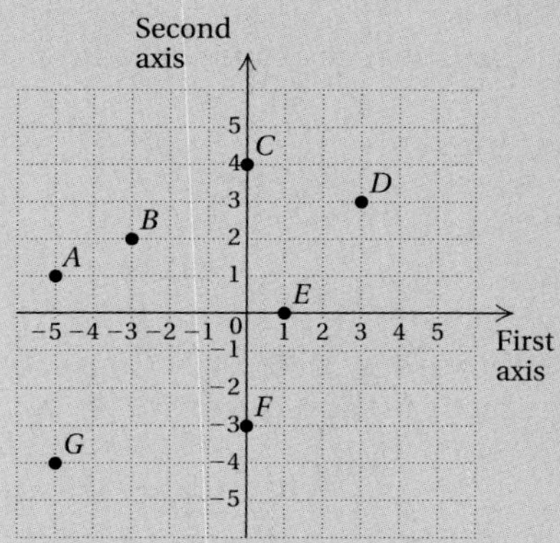

Answers on page A-11

3.1 EXERCISE SET

For Extra Help

a Solve.

Causes of Death. The following circle graphs show the leading causes of death for women and men ages 65 and over. Use the circle graphs for Exercises 1–8.

Leading Causes of Death Among Persons 65 Years of Age and Over

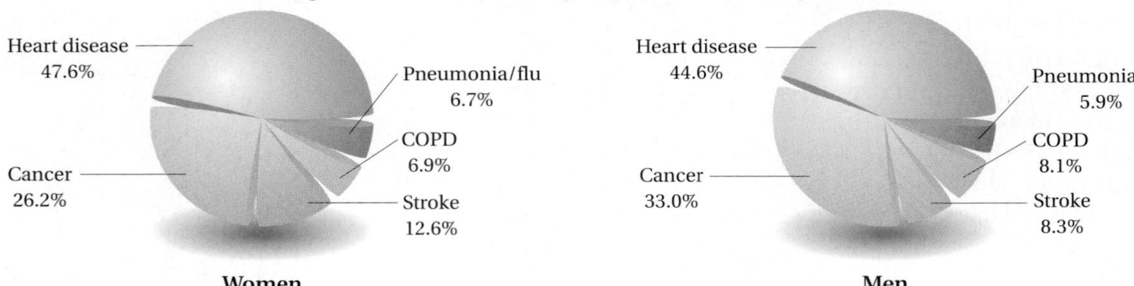

Sources: Centers for Disease Control and Prevention, National Center for Health Statistics, National Vital Statistics System, 1998.
Note: COPD is chronic obstructive pulmonary diseases such as emphysema, chronic bronchitis, and so on, often the result of cigarette smoking.

1. What percent of women age 65 and over die of heart disease?

2. What percent of men age 65 and over die of a stroke?

3. What percent of men age 65 and over die of heart disease?

4. What percent of women age 65 and over die of cancer?

5. In a group of 150,000 men age 65 and over, how many of them can be expected to die of cancer?

6. In a group of 150,000 women age 65 and over, how many of them can be expected to die of heart disease?

7. In a group of 150,000 women age 65 and over, how many of them can be expected to die of a stroke?

8. In a group of 150,000 men age 65 and over, how many of them can be expected to die of chronic obstructive pulmonary disease?

Driving While Intoxicated (DWI). State laws have determined that a blood alcohol level of at least 0.10% or higher indicates that an individual has consumed too much alcohol to drive safely. The following bar graph shows the number of drinks that a person of a certain weight would need to consume in order to reach a blood alcohol level of 0.10%. A 12-oz beer, a 5-oz glass of wine, or a cocktail containing $1\frac{1}{2}$ oz of distilled liquor all count as one drink. Use the bar graph for Exercises 9–14.

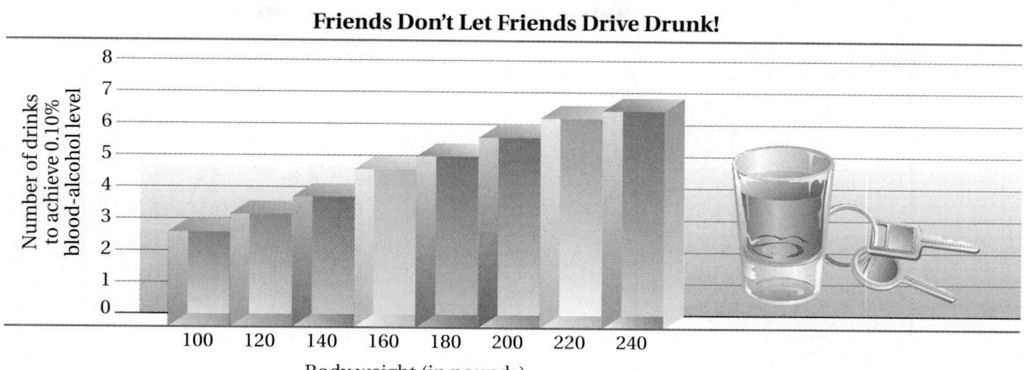

Source: Neighborhood Digest, 7, no. 12

9. Approximately how many drinks would a 200-lb person have consumed if he or she had a blood alcohol level of 0.10%?

10. What can be concluded about the weight of someone who can consume 4 drinks without reaching a blood alcohol level of 0.10%?

11. What can be concluded about the weight of someone who can consume 6 drinks without reaching a blood alcohol level of 0.10%?

12. Approximately how many drinks would a 160-lb person have consumed if he or she had a blood alcohol level of 0.10%?

13. What can be concluded about the weight of someone who has consumed $3\frac{1}{2}$ drinks without reaching a blood alcohol level of 0.10%?

14. What can be concluded about the weight of someone who has consumed $4\frac{1}{2}$ drinks without reaching a blood alcohol level of 0.10%?

Alcohol-Related Deaths. The data in the following line graph show the number of deaths from 1990 to 1998. Use the line graph for Exercises 15–20.

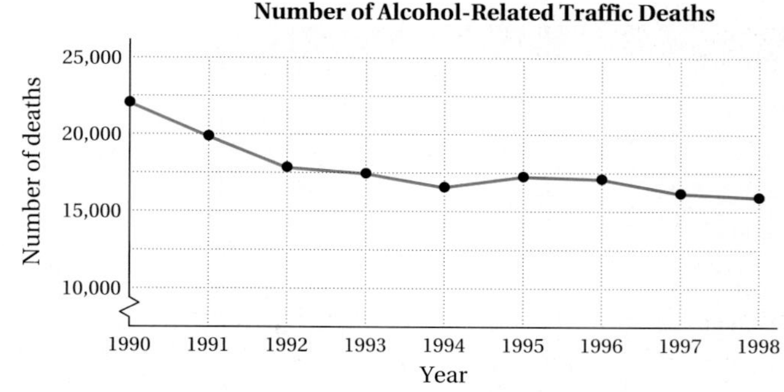

Source: National Highway Traffic Safety Administration

15. About how many alcohol-related deaths occurred in 1995?

16. About how many alcohol-related deaths occurred in 1998?

17. In what year did the lowest number of deaths occur?

18. In what years did fewer than 18,000 deaths occur?

19. By how much did the number of alcohol-related deaths decrease from 1995 to 1998?

20. By how much did the number of alcohol-related deaths increase from 1994 to 1995?

21. Plot these points.

$(2, 5)$ $(-1, 3)$ $(3, -2)$ $(-2, -4)$
$(0, 4)$ $(0, -5)$ $(5, 0)$ $(-5, 0)$

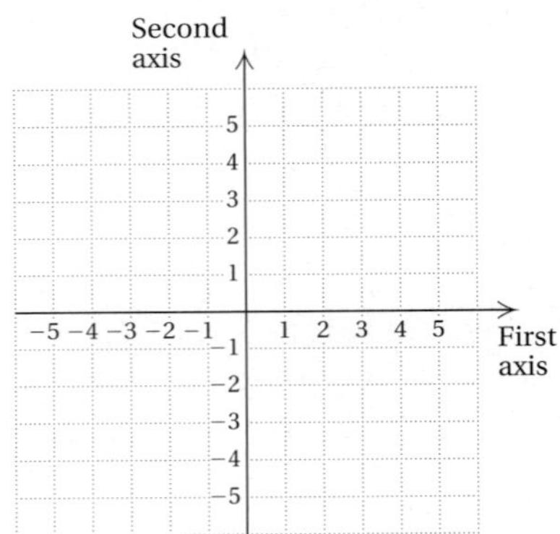

22. Plot these points.

$(4, 4)$ $(-2, 4)$ $(5, -3)$ $(-5, -5)$
$(0, 4)$ $(0, -4)$ $(3, 0)$ $(-4, 0)$

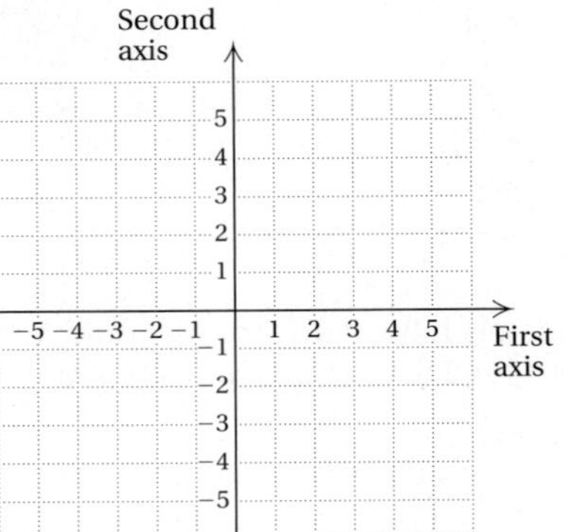

c In which quadrant is the point located?

23. $(-5, 3)$

24. $(1, -12)$

25. $(100, -1)$

26. $(-2.5, 35.6)$

27. $(-6, -29)$

28. $(3.6, 105.9)$

29. $(3.8, 9.2)$

30. $(-895, -492)$

31. $\left(-\frac{1}{3}, \frac{15}{7}\right)$

32. $\left(-\frac{2}{3}, -\frac{9}{8}\right)$

33. $\left(12\frac{7}{8}, -1\frac{1}{2}\right)$

34. $\left(23\frac{5}{8}, 81.74\right)$

For each of Exercises 35–38, complete the table regarding the signs of coordinates in certain quadrants.

	QUADRANT	FIRST COORDINATES	SECOND COORDINATES
35.		Positive	Positive
36.	III		Negative
37.	II	Negative	
38.		Positive	Negative

In which quadrant(s) can the point described be located?

39. The first coordinate is positive.

40. The second coordinate is negative.

41. The first and second coordinates are equal.

42. The first coordinate is the additive inverse of the second coordinate.

d

43. Find the coordinates of points A, B, C, D, and E.

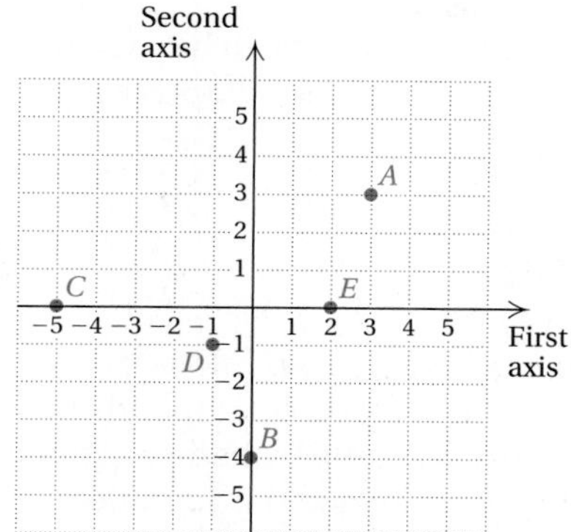

44. Find the coordinates of points A, B, C, D, and E.

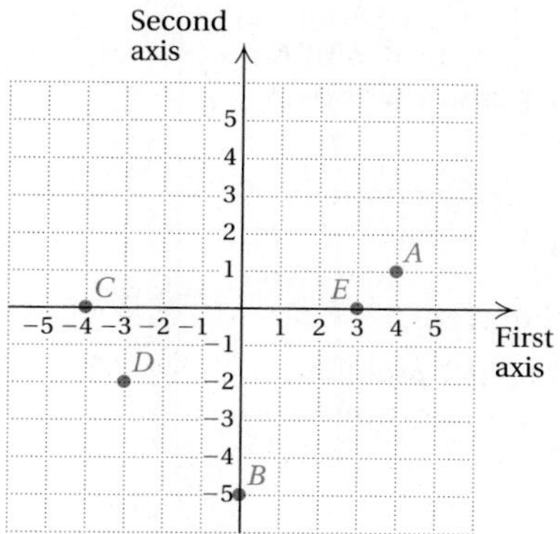

45. $\mathbf{D}_\mathbf{W}$ The sales of snow skis are highest in the winter months and lowest in the summer months. Sketch a line graph that might show the sales of a ski store and explain how an owner might use such a graph in decision making.

46. $\mathbf{D}_\mathbf{W}$ The graph in Example 3 tends to flatten out. Explain why the graph does not continue to decrease downward.

SKILL MAINTENANCE

Find the absolute value. [1.2e]

47. $|-12|$

48. $|4.89|$

49. $|0|$

50. $\left|-\frac{4}{5}\right|$

51. $|-3.4|$

52. $|\sqrt{2}|$

53. $\left|\frac{2}{3}\right|$

54. $\left|-\frac{7}{8}\right|$

Solve. [2.5a]

55. *Baseball Ticket Prices.* In 2001, the average price of a ticket to a Boston Red Sox baseball game was \$36.08, the highest in the major leagues. This price was an increase of 27.4% over the price in 2000. What was the price in 2000?
Source: Major League Baseball

56. *Tipping.* Erin left a 15% tip for a meal. The total cost of the meal, including the tip, was \$21.16. What was the cost of the meal before the tip was added?

SYNTHESIS

57. The points $(-1, 1)$, $(4, 1)$, and $(4, -5)$ are three vertices of a rectangle. Find the coordinates of the fourth vertex.

58. Three parallelograms share the vertices $(-2, -3)$, $(-1, 2)$, and $(4, -3)$. Find the fourth vertex of each parallelogram.

59. Graph eight points such that the sum of the coordinates in each pair is 6.

60. Graph eight points such that the first coordinate minus the second coordinate is 1.

61. Find the perimeter of a rectangle whose vertices have coordinates $(5, 3)$, $(5, -2)$, $(-3, -2)$, and $(-3, 3)$.

62. Find the area of a triangle whose vertices have coordinates $(0, 9)$, $(0, -4)$, and $(5, -4)$.

3.2 GRAPHING LINEAR EQUATIONS

Objectives

a. Determine whether an ordered pair is a solution of an equation with two variables.

b. Graph linear equations of the type $y = mx + b$ and $Ax + By = C$, identifying the y-intercept.

c. Solve applied problems involving graphs of linear equations.

We have seen how circle, bar, and line graphs can be used to represent the data in an application. Now we begin to learn how graphs can be used to represent solutions of equations.

a Solutions of Equations

When an equation contains two variables, the solutions of the equation are *ordered pairs* in which each number in the pair corresponds to a letter in the equation. Unless stated otherwise, to determine whether a pair is a solution, we use the first number in each pair to replace the variable that occurs first *alphabetically.*

EXAMPLE 1 Determine whether each of the following pairs is a solution of $4q - 3p = 22$: $(2, 7)$ and $(-1, 6)$.

For $(2, 7)$, we substitute 2 for p and 7 for q (using alphabetical order of variables):

$$\begin{array}{r|l} 4q - 3p & = 22 \\ \hline 4 \cdot 7 - 3 \cdot 2 & ?\ 22 \\ 28 - 6 & \\ 22 & \end{array} \quad \text{TRUE}$$

Thus, $(2, 7)$ is a solution of the equation.

For $(-1, 6)$, we substitute -1 for p and 6 for q:

$$\begin{array}{r|l} 4q - 3p & = 22 \\ \hline 4 \cdot 6 - 3 \cdot (-1) & ?\ 22 \\ 24 + 3 & \\ 27 & \end{array} \quad \text{FALSE}$$

Thus, $(-1, 6)$ is *not* a solution of the equation.

Do Exercises 1 and 2.

1. Determine whether $(2, -4)$ is a solution of $4q - 3p = 22$.

2. Determine whether $(2, -4)$ is a solution of $7a + 5b = -6$.

Answers on page A-11

EXAMPLE 2 Show that the pairs $(3, 7)$, $(0, 1)$, and $(-3, -5)$ are solutions of $y = 2x + 1$. Then graph the three points and use the graph to determine another pair that is a solution.

To show that a pair is a solution, we substitute, replacing x with the first coordinate and y with the second coordinate of each pair:

$$\begin{array}{r|l} y & = 2x + 1 \\ \hline 7 & ?\ 2 \cdot 3 + 1 \\ & 6 + 1 \\ & 7 \end{array} \quad \text{TRUE} \qquad \begin{array}{r|l} y & = 2x + 1 \\ \hline 1 & ?\ 2 \cdot 0 + 1 \\ & 0 + 1 \\ & 1 \end{array} \quad \text{TRUE}$$

$$\begin{array}{r|l} y & = 2x + 1 \\ \hline -5 & ?\ 2(-3) + 1 \\ & -6 + 1 \\ & -5 \end{array} \quad \text{TRUE}$$

In each of the three cases, the substitution results in a true equation. Thus the pairs are all solutions.

We plot the points as shown at right. The order of the points follows the alphabetical order of the variables. That is, x comes before y, so x-values are first coordinates and y-values are second coordinates. Similarly, we also label the horizontal axis as the x-axis and the vertical axis as the y-axis.

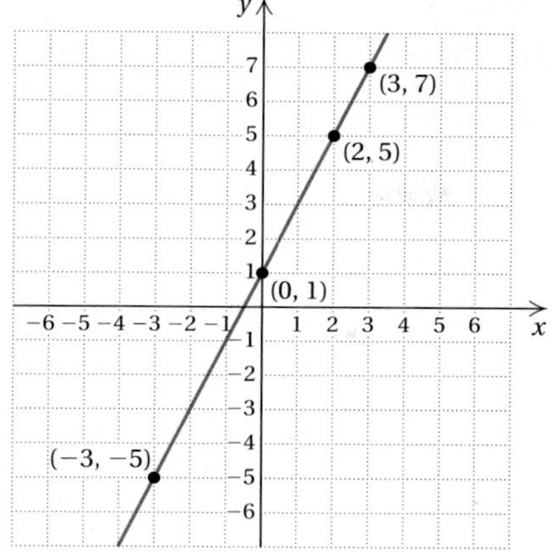

Note that the three points appear to "line up." That is, they appear to be on a straight line. Will other points that line up with these points also represent solutions of $y = 2x + 1$? To find out, we use a straightedge and lightly sketch a line passing through $(3, 7)$, $(0, 1)$, and $(-3, -5)$.

The line appears to also pass through $(2, 5)$. Let's see if this pair is a solution of $y = 2x + 1$:

$$\begin{array}{l|l} y & = 2x + 1 \\ \hline 5 & ?\ 2 \cdot 2 + 1 \\ & 4 + 1 \\ & 5 \end{array} \qquad \text{TRUE}$$

Thus, $(2, 5)$ is a solution.

Do Exercise 3.

Example 2 leads us to suspect that any point on the line that passes through $(3, 7)$, $(0, 1)$, and $(-3, -5)$ represents a solution of $y = 2x + 1$. In fact, every solution of $y = 2x + 1$ is represented by a point on that line and every point on that line represents a solution. The line is the *graph* of the equation.

GRAPH OF AN EQUATION

The **graph** of an equation is a drawing that represents all its solutions.

b Graphs of Linear Equations

Equations like $y = 2x + 1$ and $4q - 3p = 22$ are said to be **linear** because the graph of each equation is a straight line. In general, any equation equivalent to one of the form $y = mx + b$ or $Ax + By = C$, where m, b, A, B, and C are constants (not variables) and A and B are not both 0, is linear.

To graph a linear equation:

1. Select a value for one variable and calculate the corresponding value of the other variable. Form an ordered pair using alphabetical order as indicated by the variables.
2. Repeat step (1) to obtain at least two other ordered pairs. Two points are essential to determine a straight line. A third point serves as a check.
3. Plot the ordered pairs and draw a straight line passing through the points.

3. Use the graph in Example 2 to find at least two more points that are solutions of $y = 2x + 1$.

Answer on page A-11

Complete the table and graph.

4. $y = -2x$

x	y	(x, y)
-3		
-1		
0		
1		
3		

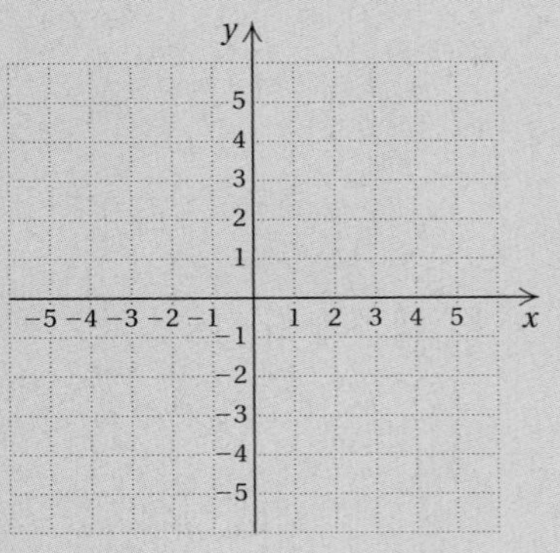

5. $y = \frac{1}{2}x$

x	y	(x, y)
4		
2		
0		
-2		
-4		
-1		

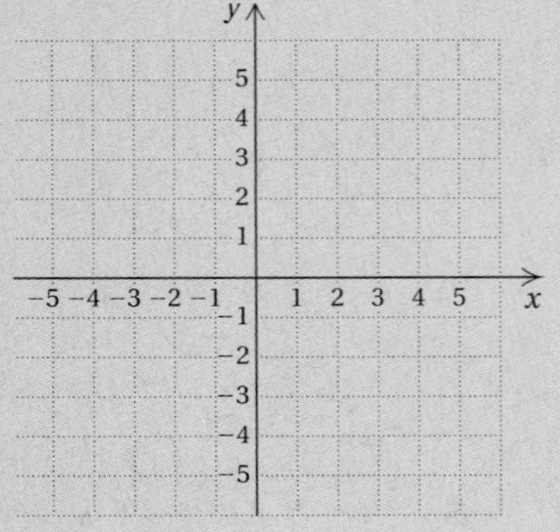

Answers on page A-11

In general, calculating three (or more) ordered pairs is not difficult for equations of the form $y = mx + b$. We simply substitute values for x and calculate the corresponding values for y.

EXAMPLE 3 Graph: $y = 2x$.

First, we find some ordered pairs that are solutions. We choose *any* number for x and then determine y by substitution. Since $y = 2x$, we find y by doubling x. Suppose that we choose 3 for x. Then

$$y = 2x = 2 \cdot 3 = 6.$$

We get a solution: the ordered pair $(3, 6)$.

Suppose that we choose 0 for x. Then

$$y = 2x = 2 \cdot 0 = 0.$$

We get another solution: the ordered pair $(0, 0)$.

For a third point, we make a negative choice for x. We now have enough points to plot the line, but if we wish, we can compute more. If a number takes us off the graph paper, we either do not use it or we use larger paper or rescale the axes. Continuing in this manner, we create a table like the one shown below.

Now we plot these points. We draw the line, or graph, with a straightedge and label it $y = 2x$.

	y	
x	$y = 2x$	(x, y)
3	6	$(3, 6)$
1	2	$(1, 2)$
0	0	$(0, 0)$
-2	-4	$(-2, -4)$
-3	-6	$(-3, -6)$

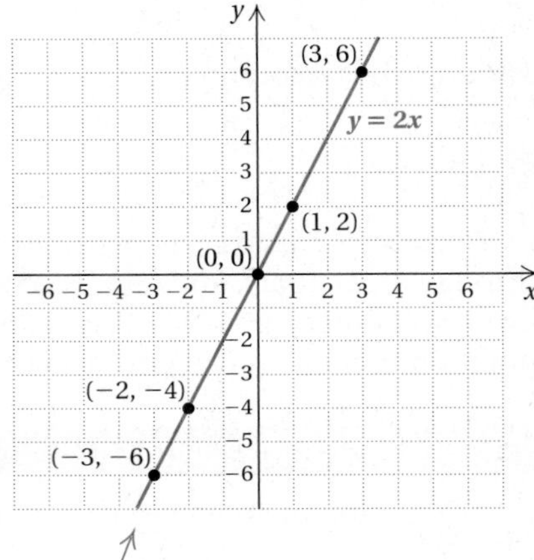

(1) Choose x.
(2) Compute y.
(3) Form the pair (x, y).
(4) Plot the points.

CAUTION!

Keep in mind that you can choose *any* number for x and then compute y. Our choice of certain numbers in the examples does not dictate the ones you can choose.

Do Exercises 4 and 5.

EXAMPLE 4 Graph: $y = -3x + 1$.

We select a value for x, compute y, and form an ordered pair. Then we repeat the process for other choices of x.

If $x = 2$, then $y = -3 \cdot 2 + 1 = -5$, and $(2, -5)$ is a solution.
If $x = 0$, then $y = -3 \cdot 0 + 1 = 1$, and $(0, 1)$ is a solution.
If $x = -1$, then $y = -3 \cdot (-1) + 1 = 4$, and $(-1, 4)$ is a solution.

Results are often listed in a table, as shown below. The points corresponding to each pair are then plotted.

x	y $y = -3x + 1$	(x, y)
2	-5	$(2, -5)$
0	1	$(0, 1)$
-1	4	$(-1, 4)$

(1) Choose x.
(2) Compute y.
(3) Form the pair (x, y).
(4) Plot the points.

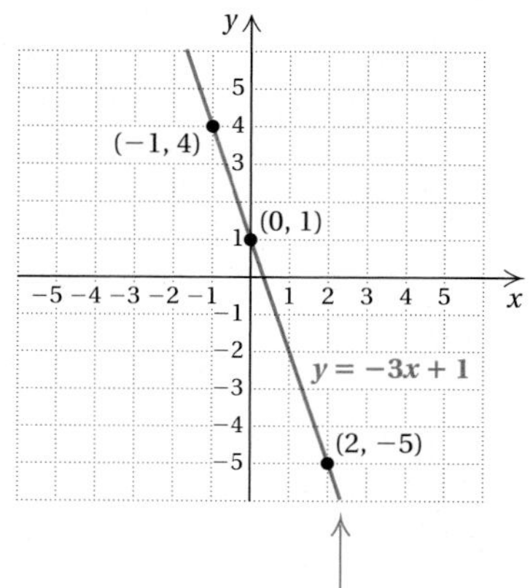

Note that all three points line up. If they did not, we would know that we had made a mistake. When only two points are plotted, a mistake is harder to detect. We use a ruler or other straightedge to draw a line through the points. Every point on the line represents a solution of $y = -3x + 1$.

Do Exercises 6 and 7.

In Example 3, we saw that $(0, 0)$ is a solution of $y = 2x$. It is also the point at which the graph crosses the y-axis. Similarly, in Example 4, we saw that $(0, 1)$ is a solution of $y = -3x + 1$. It is also the point at which the graph crosses the y-axis. A generalization can be made: If x is replaced with 0 in the equation $y = mx + b$, then the corresponding y-value is $m \cdot 0 + b$, or b. Thus any equation of the form $y = mx + b$ has a graph that passes through the point $(0, b)$. Since $(0, b)$ is the point at which the graph crosses the y-axis, it is called the **y-intercept.** Sometimes, for convenience, we simply refer to b as the y-intercept.

y-INTERCEPT

The graph of the equation $y = mx + b$ passes through the **y-intercept** $(0, b)$.

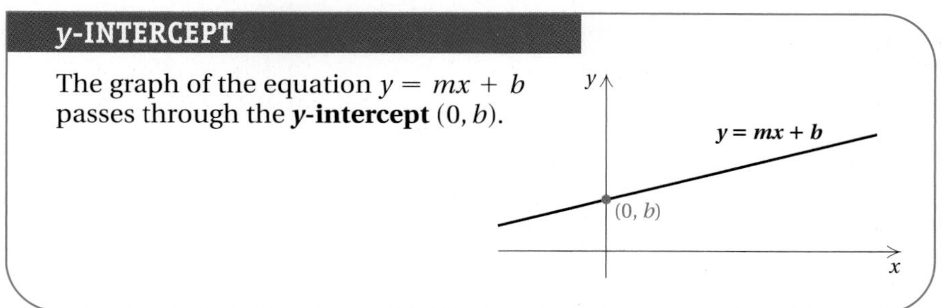

Graph.

6. $y = 2x + 3$

x	y	(x, y)

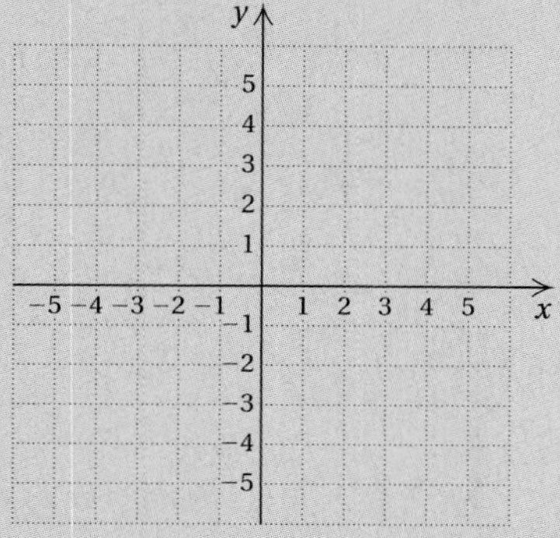

7. $y = -\frac{1}{2}x - 3$

x	y	(x, y)

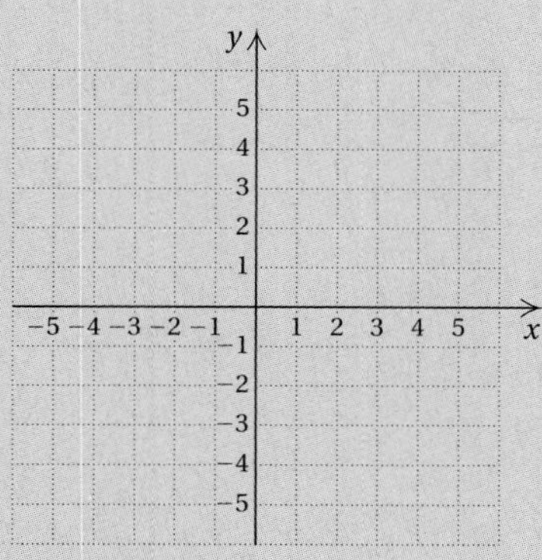

Answers on page A-11

CALCULATOR CORNER

Finding Solutions of Equations A table of values representing ordered pairs that are solutions of an equation can be displayed on a graphing calculator. To do this for the equation in Example 4, $y = -3x + 1$, we first press [Y =] to access the equation-editor screen. Then we clear any equations that are present. (See the Calculator Corner in Section 2.3 for the procedure for doing this.) Next, we enter the equation by positioning the cursor beside "Y1 =" and pressing [(−)] [3] [X, T, θ, n] [+] [1]. Now we press [2nd] [TblSet] to display the table set-up screen. (TblSet is the second function associated with the [WINDOW] key.) You can choose to supply the x-values yourself or you can set the calculator to supply them. To supply them yourself, follow the procedure for selecting ASK mode on p. 164. To have the calculator supply the x-values, set "Indpnt" to "Auto" by positioning the cursor over "Auto" and pressing [ENTER]. "Depend" should also be set to "Auto."

When "Indpnt" is set to "Auto," the graphing calculator will supply values of x, beginning with the value specified as TblStart and continuing by adding the value of △Tbl to the preceding value for x. Below, we show a table of values that starts with $x = -2$ and adds 1 to the preceding x-value. We press [(−)] [2] [▽] [1] or [(−)] [2] [ENTER] [1] to select a minimum x-value of -2 and an increment of 1. To display the table, we press [2nd] [TABLE]. (TABLE is the second operation associated with the [GRAPH] key.)

```
TABLE SETUP
 TblStart=−2
 ΔTbl=1
Indpnt:  Auto  Ask
Depend:  Auto  Ask
```

X	Y1
−2	7
−1	4
0	1
1	−2
2	−5
3	−8
4	−11

X = −2

We can use the [△] and [▽] keys to scroll up and down through the table to see other solutions of the equation.

Exercise

1. Create a table of ordered pairs that are solutions of the equations in Examples 3 and 5.

EXAMPLE 5 Graph $y = \frac{2}{5}x + 4$ and identify the y-intercept.

We select a value for x, compute y, and form an ordered pair. Then we repeat the process for other choices of x. In this case, using multiples of 5 avoids fractions. We try to avoid graphing ordered pairs with fractions because they are difficult to graph accurately.

If $x = 0$, then $y = \frac{2}{5} \cdot 0 + 4 = 4$, and $(0, 4)$ is a solution.

If $x = 5$, then $y = \frac{2}{5} \cdot 5 + 4 = 6$, and $(5, 6)$ is a solution.

If $x = -5$, then $y = \frac{2}{5} \cdot (-5) + 4 = 2$, and $(-5, 2)$ is a solution.

The following table lists these solutions. Next, we plot the points and see that they form a line. Finally, we draw and label the line.

x	y $y = \frac{2}{5}x + 4$	(x, y)
0	4	$(0, 4)$
5	6	$(5, 6)$
-5	2	$(-5, 2)$

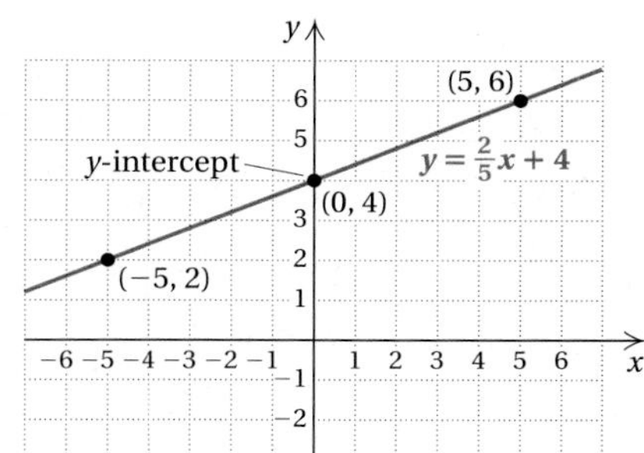

We see that $(0, 4)$ is a solution of $y = \frac{2}{5}x + 4$. It is the y-intercept. Because the equation is in the form $y = mx + b$, we can read the y-intercept directly from the equation as follows:

$$y = \frac{2}{5}x + 4 \qquad (0, 4) \text{ is the } y\text{-intercept.}$$

Do Exercises 8 and 9.

Calculating ordered pairs is generally easiest when y is isolated on one side of the equation, as in $y = mx + b$. To graph an equation in which y is not isolated, we can use the addition and multiplication principles to solve for y (see Section 2.3).

EXAMPLE 6 Graph $3y + 5x = 0$ and identify the y-intercept.

To find an equivalent equation in the form $y = mx + b$, we solve for y:

$$3y + 5x = 0$$

$$3y + 5x - 5x = 0 - 5x \qquad \text{Subtracting } 5x$$

$$3y = -5x \qquad \text{Collecting like terms}$$

$$\frac{3y}{3} = \frac{-5x}{3} \qquad \text{Dividing by 3}$$

$$y = -\frac{5}{3}x.$$

Graph the equation and identify the y-intercept.

8. $y = \frac{3}{5}x + 2$

x	y	(x, y)

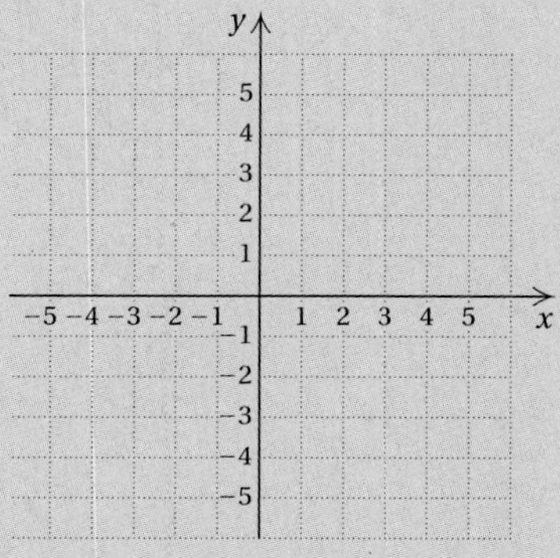

9. $y = -\frac{3}{5}x - 1$

x	y	(x, y)

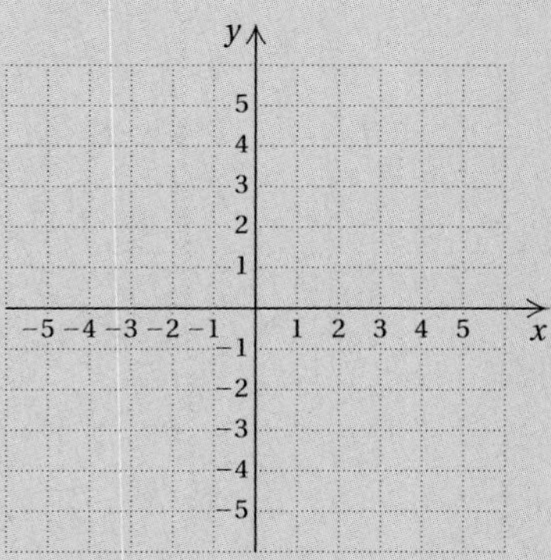

Answers on page A-11

Graph the equation and identify the y-intercept.

10. $5y + 4x = 0$

x	y	
0		← y-intercept

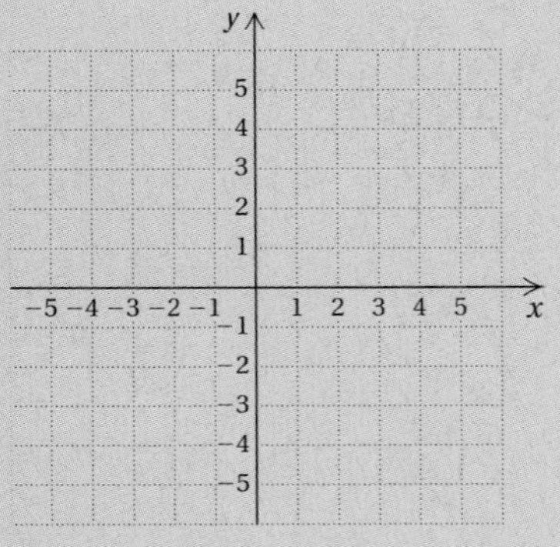

11. $4y = 3x$

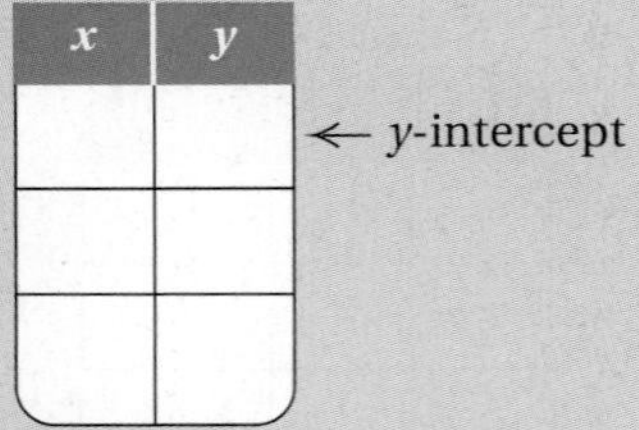

x	y	
		← y-intercept

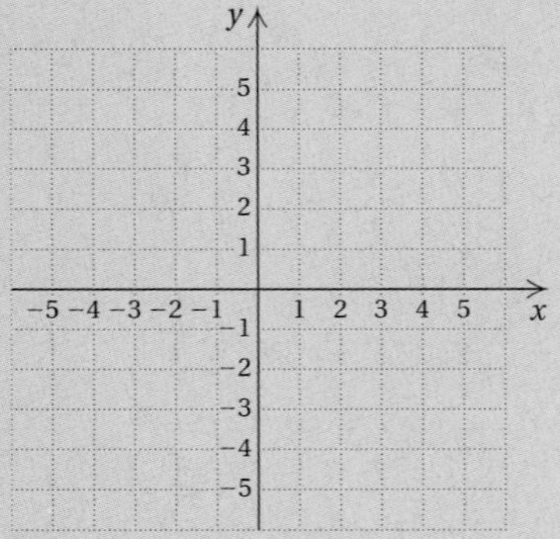

Answers on page A-11

Because all the equations above are equivalent, we can use $y = -\frac{5}{3}x$ to draw the graph of $3y + 5x = 0$. To graph $y = -\frac{5}{3}x$, we select x-values and compute y-values. In this case, if we select multiples of 3, we can avoid fractions.

$$\text{If } x = 0, \quad \text{then } y = -\frac{5}{3} \cdot 0 = 0.$$

$$\text{If } x = 3, \quad \text{then } y = -\frac{5}{3} \cdot 3 = -5.$$

$$\text{If } x = -3, \quad \text{then } y = -\frac{5}{3} \cdot (-3) = 5.$$

We list these solutions in a table. Next, we plot the points and see that they form a line. Finally, we draw and label the line. The y-intercept is $(0, 0)$.

x	y
0	0
3	−5
−3	5

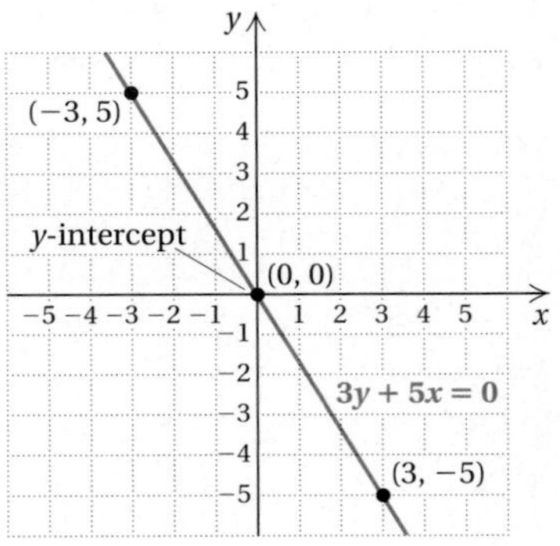

Do Exercises 10 and 11.

EXAMPLE 7 Graph $4y + 3x = -8$ and identify the y-intercept.

To find an equivalent equation in the form $y = mx + b$, we solve for y:

$$4y + 3x = -8$$

$$4y + 3x - 3x = -8 - 3x \qquad \text{Subtracting } 3x$$

$$4y = -3x - 8 \qquad \text{Simplifying}$$

$$\frac{1}{4} \cdot 4y = \frac{1}{4} \cdot (-3x - 8) \qquad \text{Multiplying by } \tfrac{1}{4} \text{ or dividing by 4}$$

$$y = \frac{1}{4} \cdot (-3x) - \frac{1}{4} \cdot 8 \qquad \text{Using the distributive law}$$

$$y = -\frac{3}{4}x - 2. \qquad \text{Simplifying}$$

Thus, $4y + 3x = -8$ is equivalent to $y = -\frac{3}{4}x - 2$. The y-intercept is $(0, -2)$. We find two other pairs using multiples of 4 for x to avoid fractions. We then complete and label the graph as shown.

x	y
0	−2
4	−5
−4	1

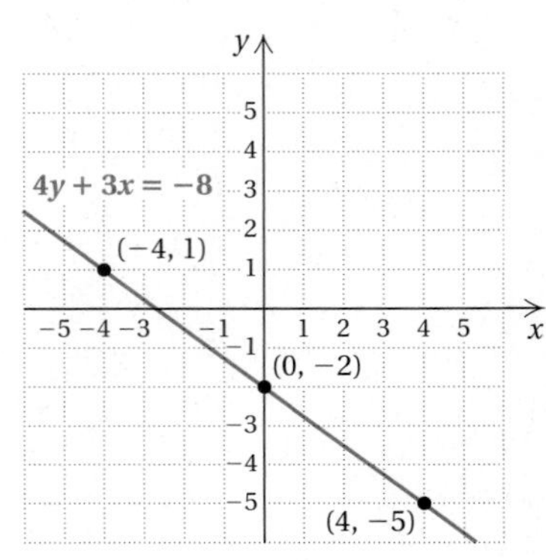

Do Exercises 12 and 13.

C Applications of Linear Equations

Mathematical concepts become more understandable through visualization. Throughout this text, you will occasionally see the heading AG Algebraic–Graphical Connection, as in Example 8, which follows. In this feature, the algebraic approach is enhanced and expanded with a graphical connection. Relating a solution of an equation to a graph can often give added meaning to the algebraic solution.

EXAMPLE 8 *FedEx Mailing Costs.* The cost y, in dollars, of shipping a FedEx Priority Overnight package weighing 1 lb or more a distance of 1001 to 1400 mi is given by

$$y = 2.8x + 21.05,$$

where x is the number of pounds.
Source: Federal Express Corporation

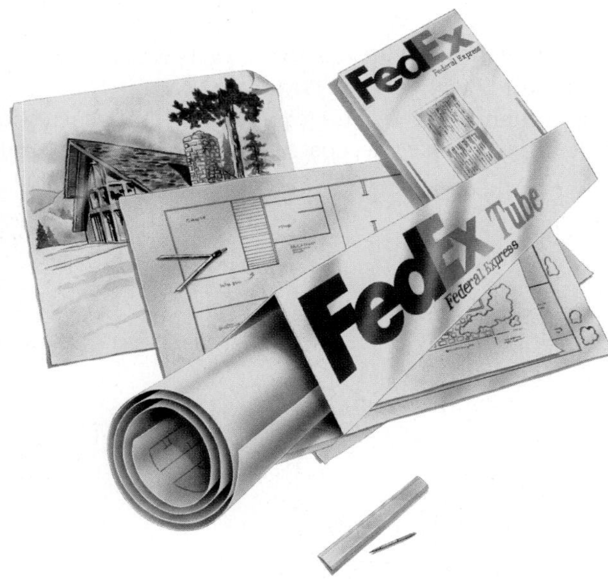

a) Find the cost of shipping packages weighing 2 lb, 5 lb, and 7 lb.
b) Graph the equation and then use the graph to estimate the cost of shipping a $6\frac{1}{2}$-lb package.
c) If a package costs \$239.45 to ship, how much does it weigh?

We solve as follows.

a) We substitute 2, 5, and 7 for x and then calculate y:

$$y = 2.8(2) + 21.05 = \$26.65;$$
$$y = 2.8(5) + 21.05 = \$35.05;$$
$$y = 2.8(7) + 21.05 = \$40.65.$$

It costs \$26.65, \$35.05, and \$40.65 to ship packages that weigh 2 lb, 5 lb, and 7 lb, respectively.

Graph the equation and identify the y-intercept.

12. $5y - 3x = -10$

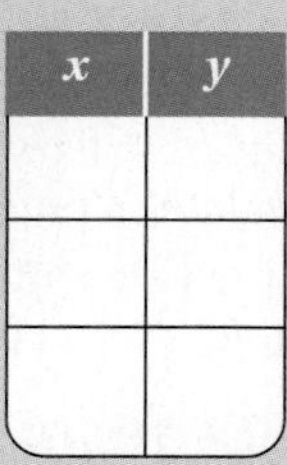

x	y

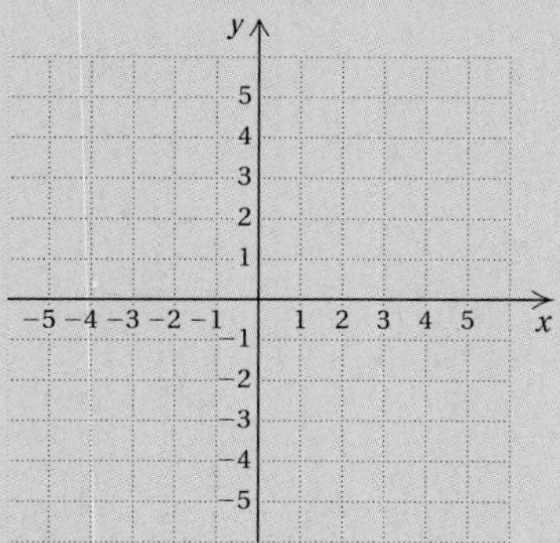

13. $5y + 3x = 20$

x	y

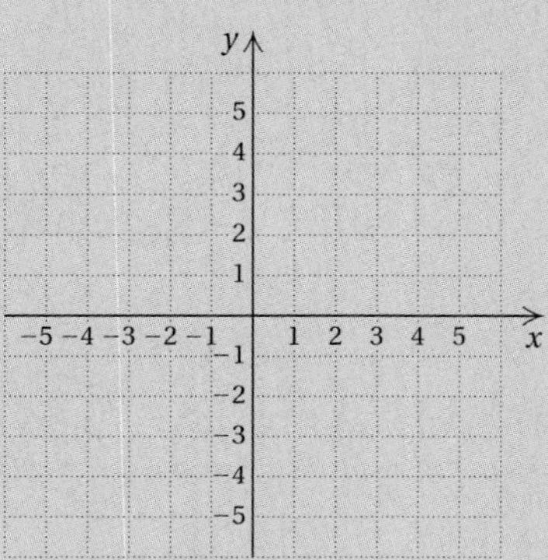

Answers on page A-11

14. Value of a Color Copier. The value of Dupliographic's color copier is given by

$$v = -0.68t + 3.4,$$

where v is the value, in thousands of dollars, t years from the date of purchase.

a) Find the value after 1 yr, 2 yr, 4 yr, and 5 yr.

t	v
1	
2	
4	
5	

b) Graph the equation and use the graph to estimate the value of the copier after $2\frac{1}{2}$ yr.

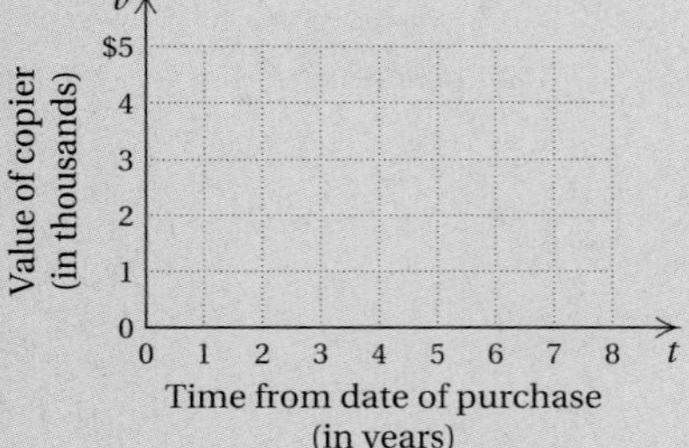

c) After what amount of time is the value of the copier $1500?

Answers on page A-12

ALGEBRAIC–GRAPHICAL CONNECTION

b) We have three ordered pairs from part (a). We plot these points and see that they line up. Thus our calculations are probably correct. Since this formula, $y = 2.8x + 21.05$, gives the cost of mailing packages that weigh 1 lb or more, we begin at $(1, 23.85)$ when drawing the graph.

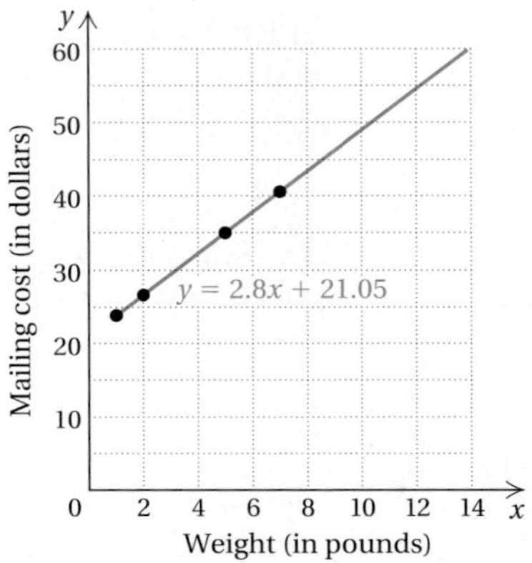

To estimate the cost of shipping a $6\frac{1}{2}$-lb package, we need to determine what y-value is paired with $x = 6\frac{1}{2}$. We locate the point on the line that is above $6\frac{1}{2}$ and then find the value on the y-axis that corresponds to that point. It appears that the cost of shipping a $6\frac{1}{2}$-lb package is about $39.

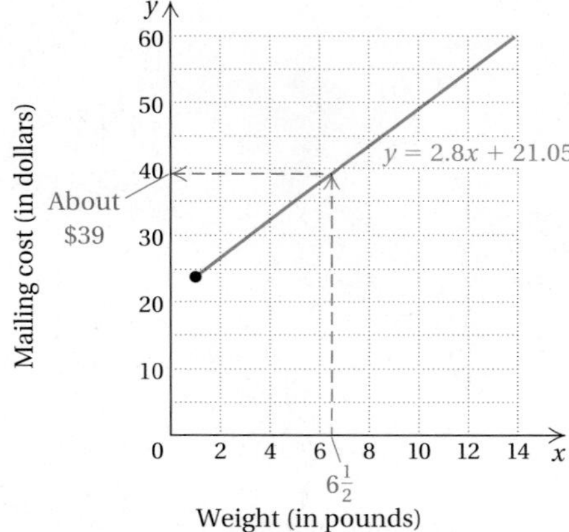

To obtain a more accurate cost, we can simply substitute into the equation:

$$y = 2.8(6.5) + 21.05 = \$39.25.$$

c) We substitute $239.45 for y and then solve for x:

$$y = 2.8x + 21.05$$

$239.45 = 2.8x + 21.05$	Substituting
$218.40 = 2.8x$	Subtracting 21.05
$78 = x.$	Dividing by 2.8

A package that costs $239.45 to ship weighs 78 lb.

Do Exercise 14.

Many equations in two variables have graphs that are not straight lines. Three such graphs are shown below. As before, each graph represents the solutions of the given equation. We are not going to develop methods of doing such graphing at this time, although such *nonlinear graphs* can be created very easily using a graphing calculator. We will cover such graphs in the optional Calculator Corners throughout the text and in Chapter 10.

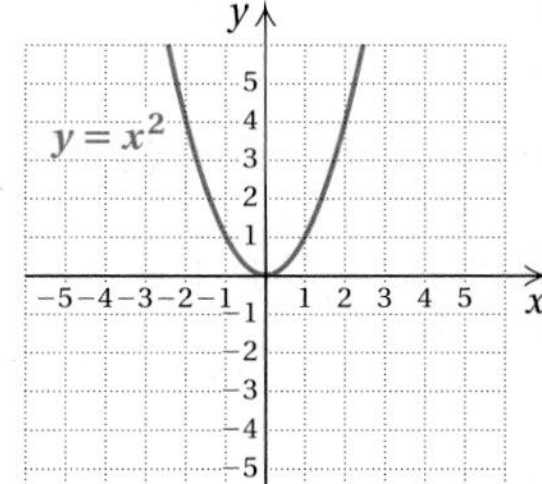

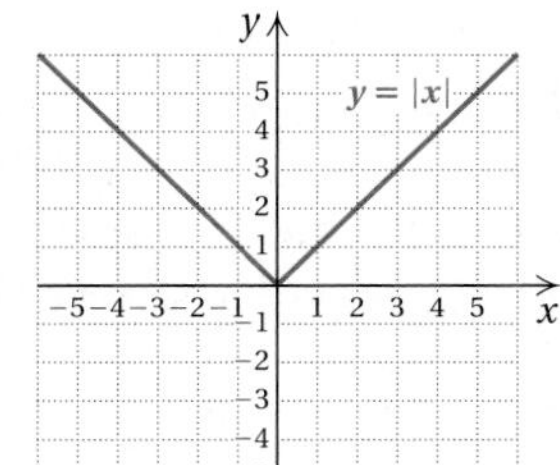

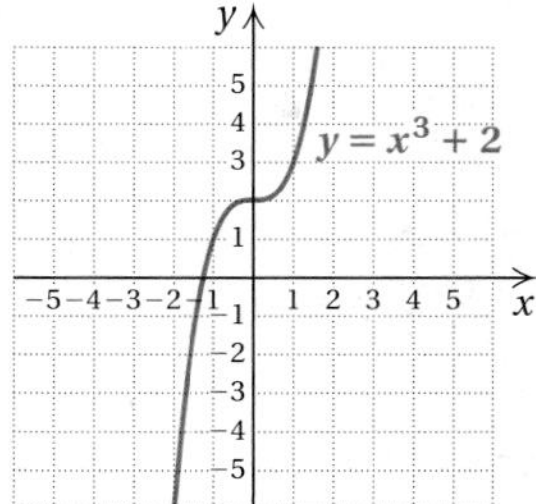

Study Tips

BETTER TEST TAKING

How often do you make the following statement after taking a test: "I was able to do the homework, but I froze during the test"? This can be an excuse for poor study habits. Here are two tips to help you with this difficulty. Both are intended to make test taking less stressful by getting you to practice good test-taking habits on a daily basis.

- **Treat every homework exercise as if it were a test question.** If you had to work a problem at your job with no backup answer provided, what would you do? You would probably work it very deliberately, checking and rechecking every step. You might work it more than one time, or you might try to work it another way to check the result. Try to use this approach when doing your homework. Treat every exercise as though it were a test question with no answer at the back of the book.
- **Be sure that you do questions without answers as part of every homework assignment whether or not the instructor has assigned them!** One reason a test may seem such a different task is that questions on a test lack answers. That is the reason for taking a test: to see if you can do the questions without assistance. As part of your test preparation, be sure you do some exercises for which you do not have the answers. Thus when you take a test, you are doing a more familiar task.

The purpose of doing your homework using these approaches is to give you more test-taking practice beforehand. Let's use a sports analogy: At a basketball game, the players take lots of practice shots before the game. They play the first half, go to the locker room, and come out for the second half. What do they do before the second half, even though they have just played 20 minutes of basketball? They shoot baskets again! We suggest the same approach here. Create more and more situations in which you practice taking test questions by treating each homework exercise like a test question and by doing exercises for which you have no answers. Good luck!

"He who does not venture has no luck."

Mexican proverb

CALCULATOR CORNER

Graphing Equations Graphs of equations are displayed in the **viewing window** of a graphing calculator. The viewing window is the portion of the coordinate plane that appears on the calculator's screen. It is defined by the minimum and maximum values of x and y: Xmin, Xmax, Ymin, and Ymax. The notation [Xmin, Xmax, Ymin, Ymax] is used to represent these window settings or dimensions. For example, $[-12, 12, -8, 8]$ denotes a window that displays the portion of the x-axis from -12 to 12 and the portion of the y-axis from -8 to 8. In addition, the distance between tick marks on the axes is defined by the settings Xscl and Yscl. The Xres setting indicates the pixel resolution. We usually select Xres = 1. The window corresponding to the settings $[-20, 30, -12, 20]$, Xscl = 5, Yscl = 2, Xres = 1, is shown on the left below. Press WINDOW on the top row of the keypad of your calculator to display the current window settings. The settings for the **standard viewing window** are shown on the right below.

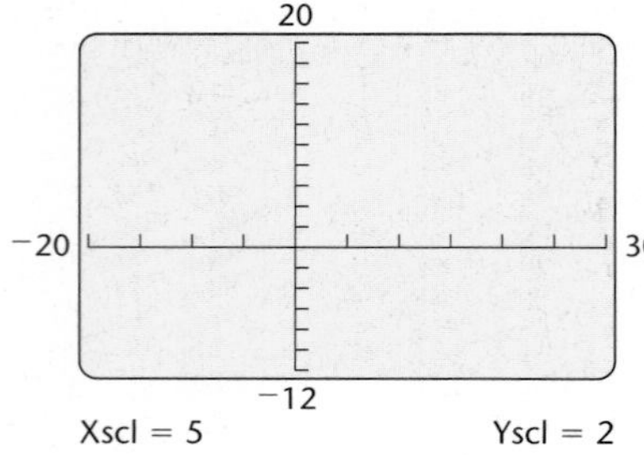

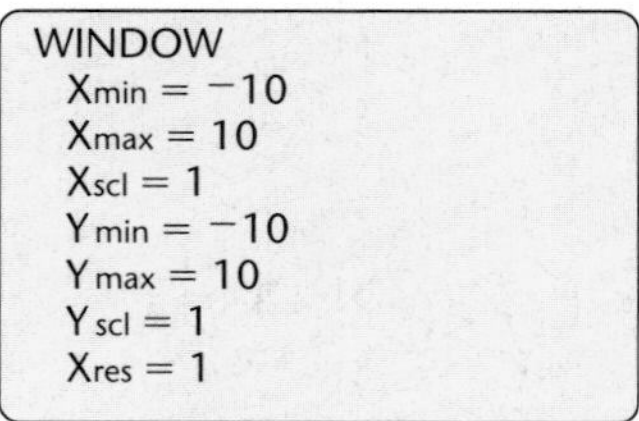

To change a setting, we position the cursor beside the setting we wish to change and enter the new value. For example, to change from the standard settings to $[-20, 30, -12, 20]$, Xscl = 5, Yscl = 2, on the WINDOW screen, we press (−) 2 0 ENTER 3 0 ENTER 5 ENTER (−) 1 2 ENTER 2 0 ENTER 2 ENTER. The ▽ key can be used instead of ENTER after typing each window setting. To see the window, we press GRAPH on the top row of the keypad. To return quickly to the standard window setting $[-10, 10, -10, 10]$, Xscl = 1, Yscl = 1, we press ZOOM 6.

Equations must be solved for y before they can be graphed on the TI-83 Plus. Consider the equation $3x + 2y = 6$. Solving for y, we get $y = \frac{6 - 3x}{2}$. We enter this equation as $y_1 = (6 - 3x)/2$ on the equation-editor screen as described in the Calculator Corner in Section 2.3 (see p. 164). Then we press ZOOM 6 to select the standard viewing window and display the graph.

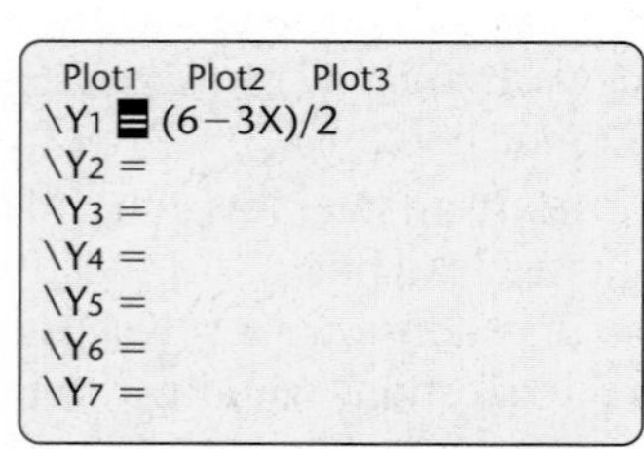

$y = (6 - 3x)/2$

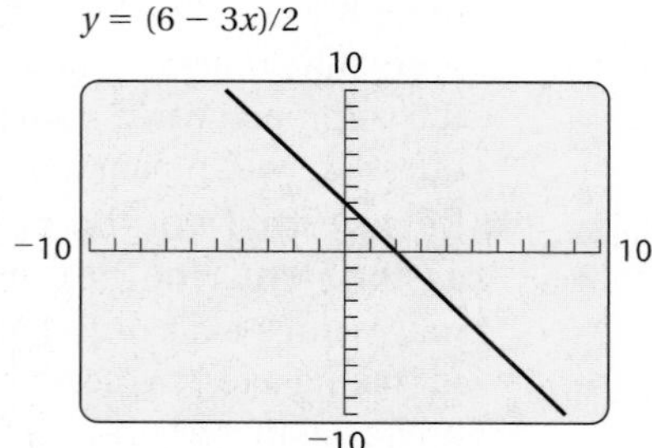

Exercises: Graph each equation in the standard viewing window $[-10, 10, -10, 10]$, Xscl = 1, Yscl = 1.

1. $y = 2x + 1$
2. $y = -3x + 1$
3. $y = -5x + 3$
4. $y = 4x - 5$
5. $y = \frac{4}{5}x + 2$
6. $y = -\frac{3}{5}x - 1$
7. $y = 2.085x + 5.08$
8. $y = -3.45x - 1.68$

For Extra Help

Digital Video Tutor CD 3 Videotape 7 | InterAct Math | Math Tutor Center | MathXL | MyMathLab

a Determine whether the given ordered pair is a solution of the equation.

1. $(2, 9)$; $y = 3x - 1$

2. $(1, 7)$; $y = 2x + 5$

3. $(4, 2)$; $2x + 3y = 12$

4. $(0, 5)$; $5x - 3y = 15$

5. $(3, -1)$; $3a - 4b = 13$

6. $(-5, 1)$; $2p - 3q = -13$

In Exercises 7–12, an equation and two ordered pairs are given. Show that each pair is a solution. Then use the graph of the two points to determine another solution. Answers may vary.

7. $y = x - 5$; $(4, -1)$ and $(1, -4)$

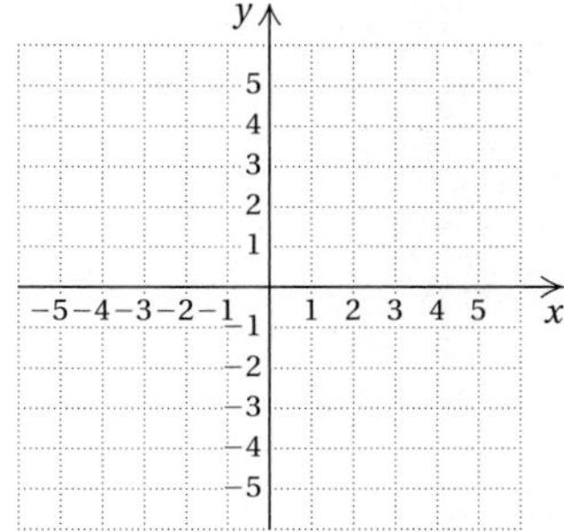

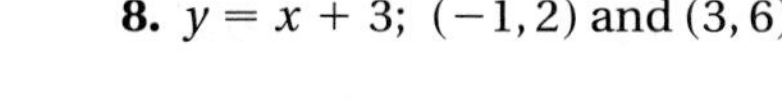

8. $y = x + 3$; $(-1, 2)$ and $(3, 6)$

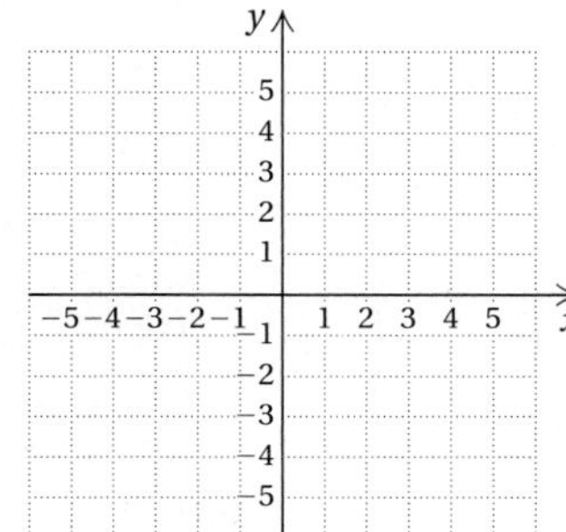

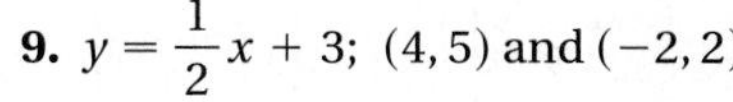

9. $y = \frac{1}{2}x + 3$; $(4, 5)$ and $(-2, 2)$

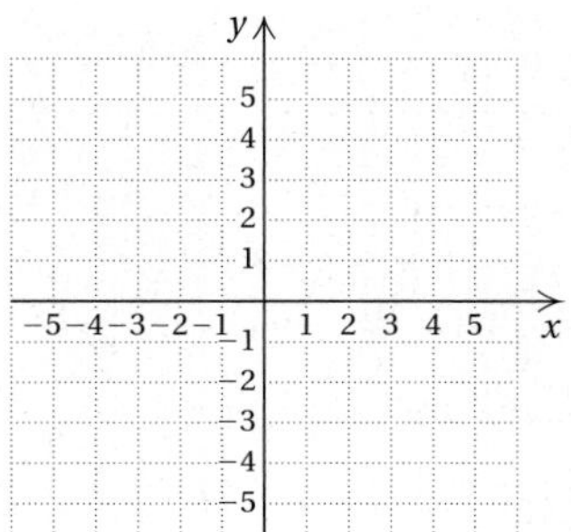

10. $3x + y = 7$; $(2, 1)$ and $(4, -5)$

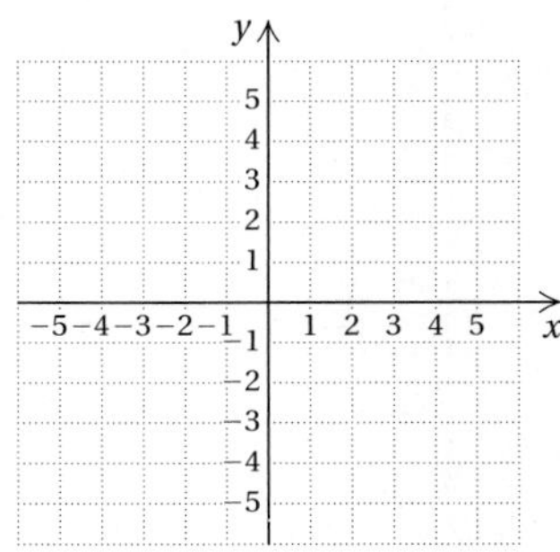

11. $4x - 2y = 10$; $(0, -5)$ and $(4, 3)$

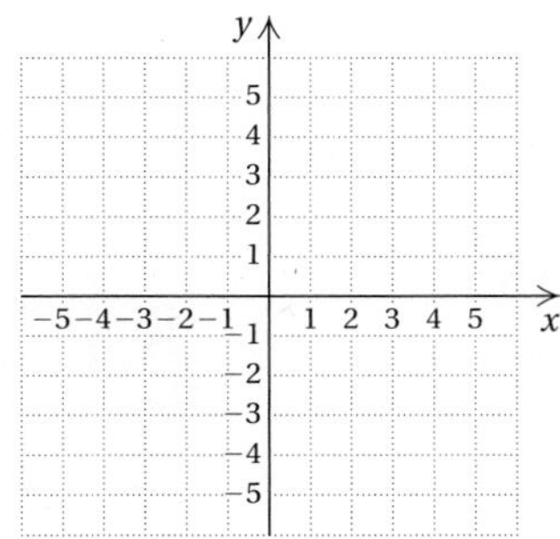

12. $6x - 3y = 3$; $(1, 1)$ and $(-1, -3)$

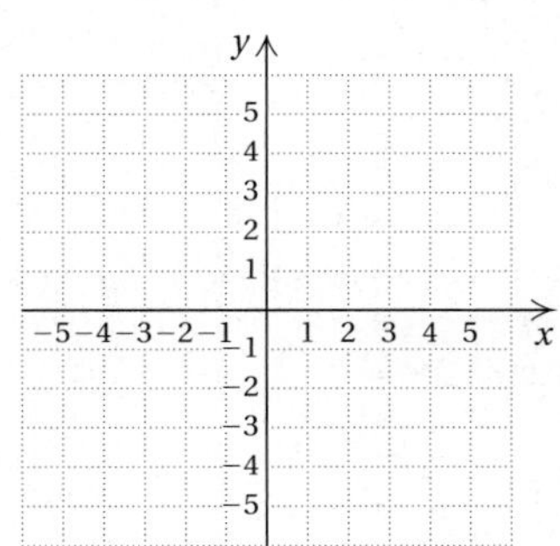

b Graph the equation and identify the y-intercept.

13. $y = x + 1$

x	y
-2	
-1	
0	
1	
2	
3	

14. $y = x - 1$

x	y
-2	
-1	
0	
1	
2	
3	

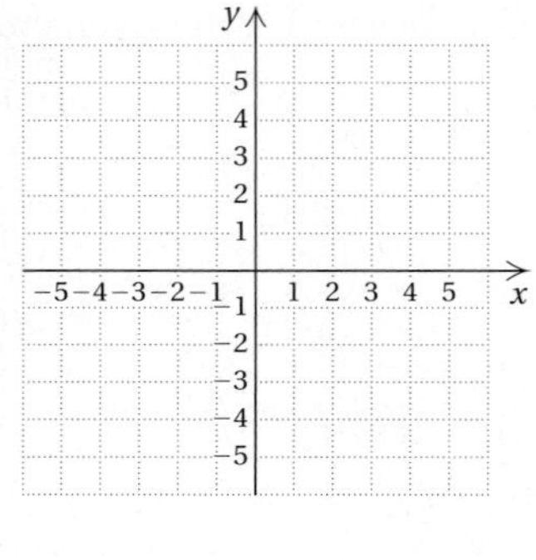

15. $y = x$

x	y
-2	
-1	
0	
1	
2	
3	

16. $y = -x$

x	y
-2	
-1	
0	
1	
2	
3	

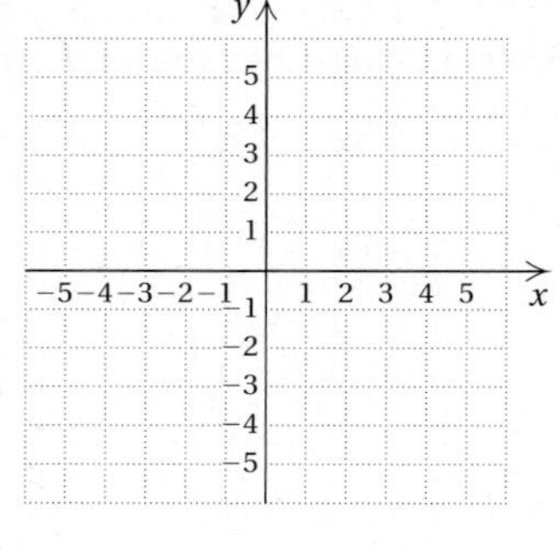

17. $y = \frac{1}{2}x$

x	y
-2	
0	
4	

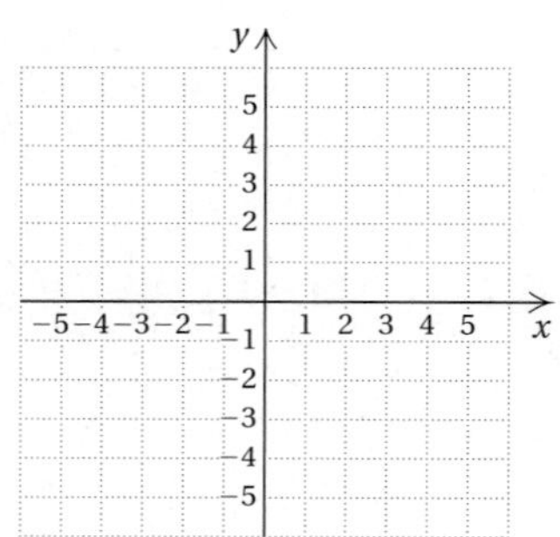

18. $y = \frac{1}{3}x$

x	y
-6	
0	
3	

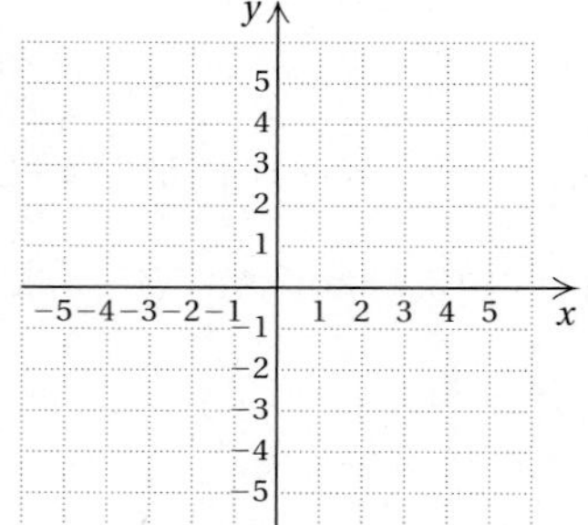

19. $y = x - 3$

x	y

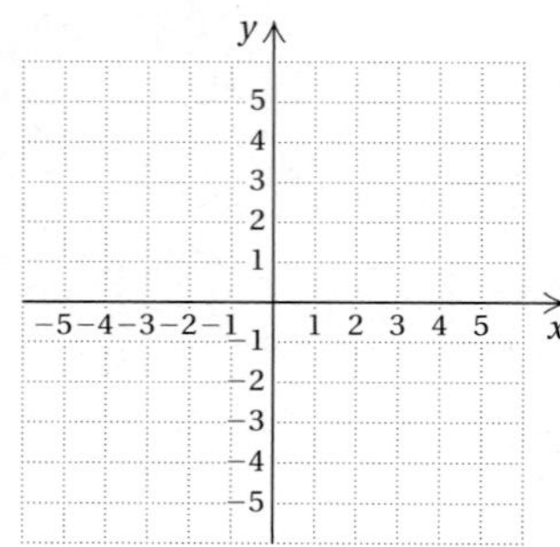

20. $y = x + 3$

x	y

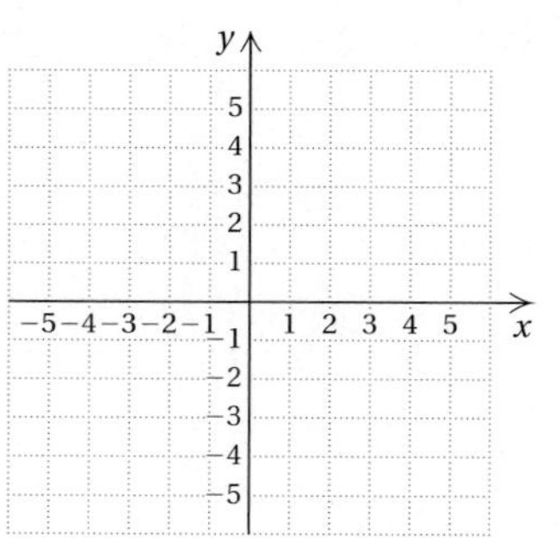

21. $y = 3x - 2$

x	y

22. $y = 2x + 2$

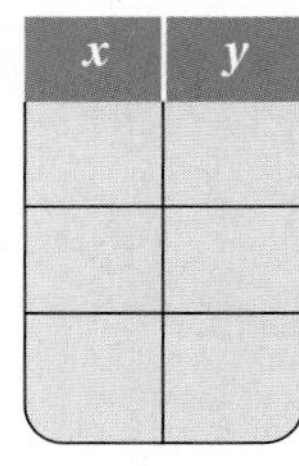

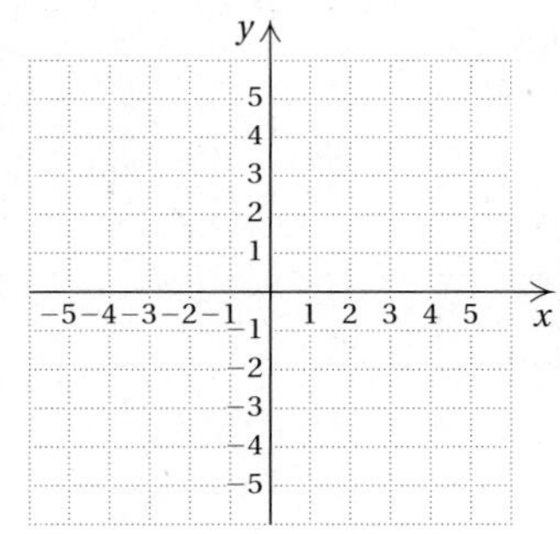

23. $y = \dfrac{1}{2}x + 1$

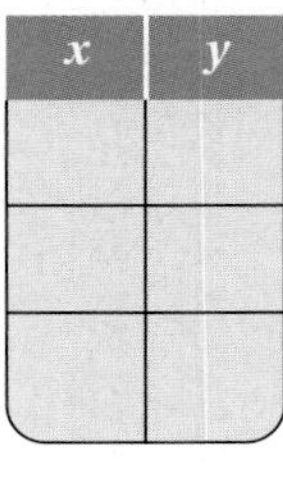

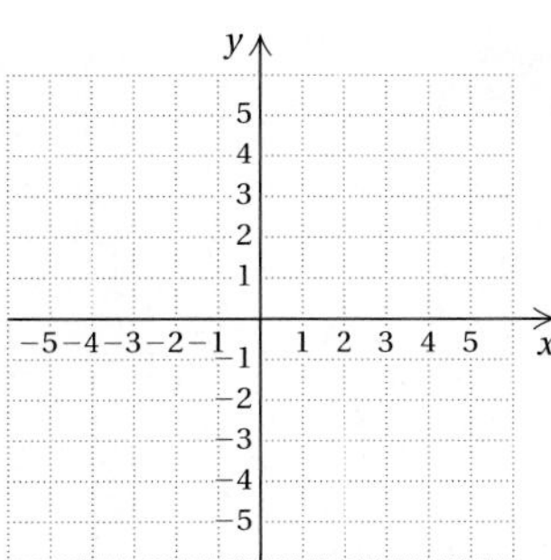

24. $y = \dfrac{1}{3}x - 4$

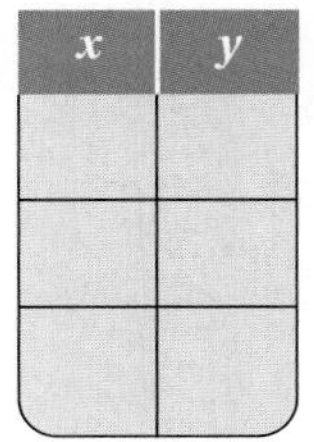

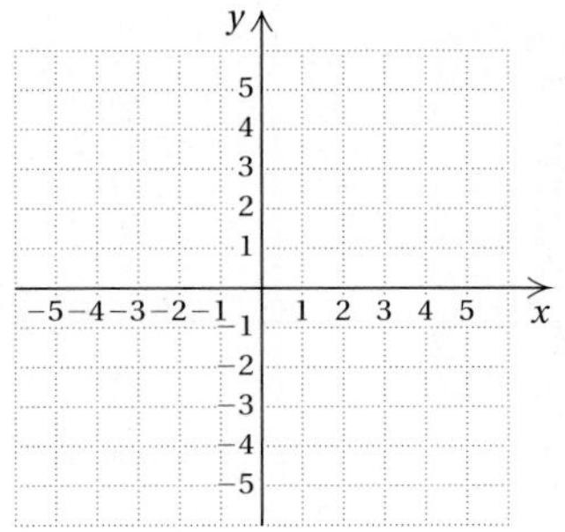

25. $x + y = -5$

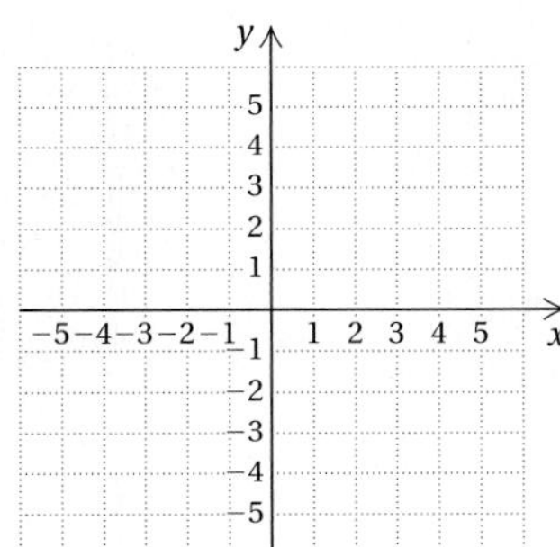

26. $x + y = 4$

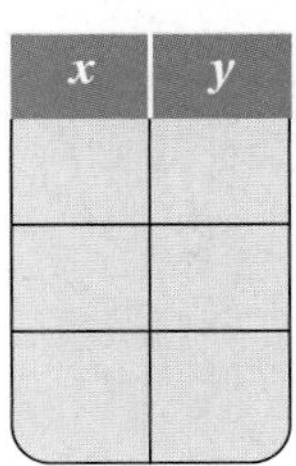

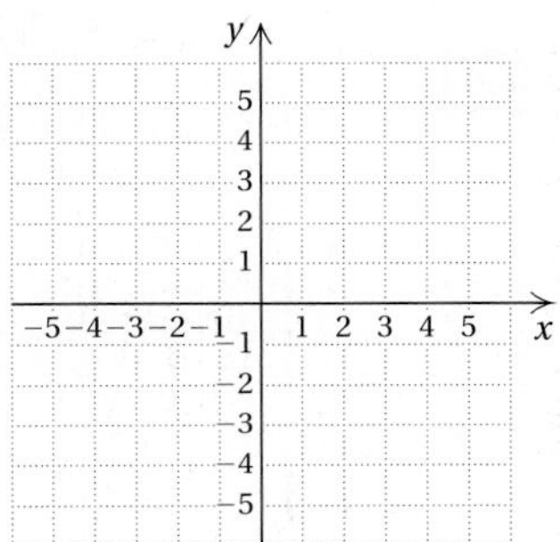

27. $y = \dfrac{5}{3}x - 2$

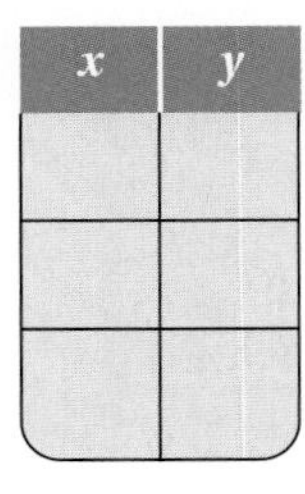

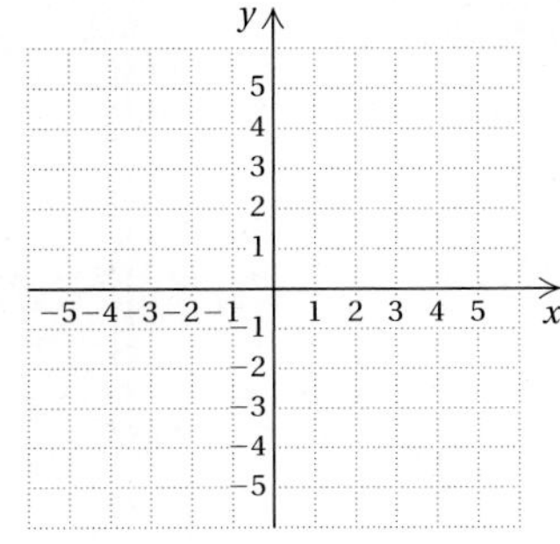

28. $y = \dfrac{5}{2}x + 3$

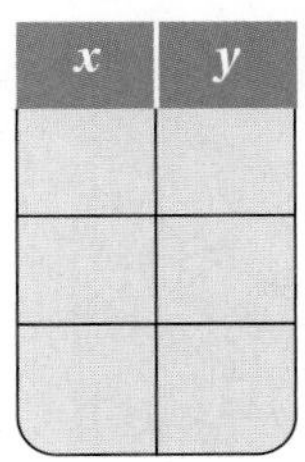

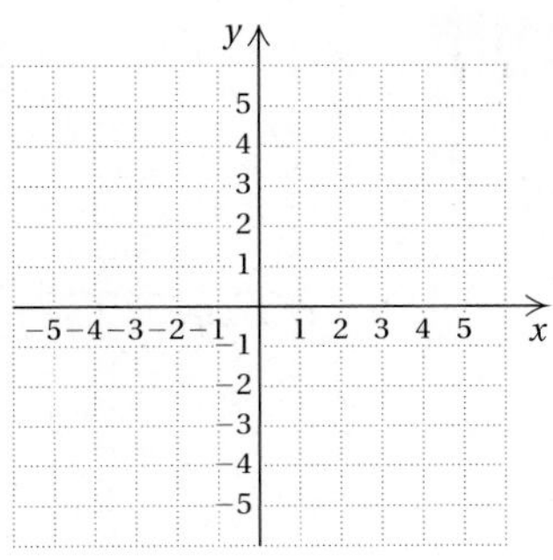

29. $x + 2y = 8$

x	y

30. $x + 2y = -6$

x	y

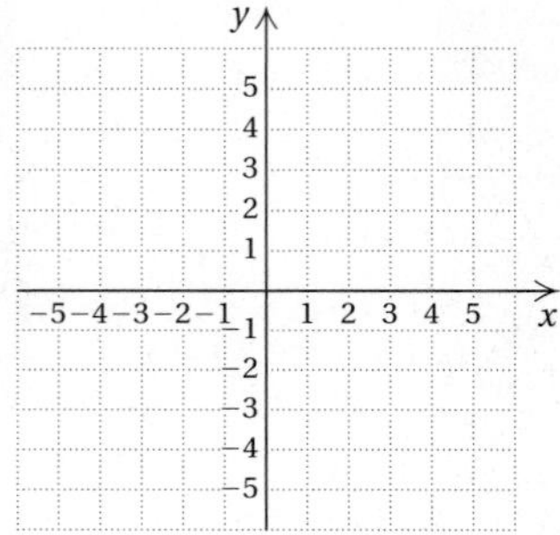

31. $y = \frac{3}{2}x + 1$

x	y

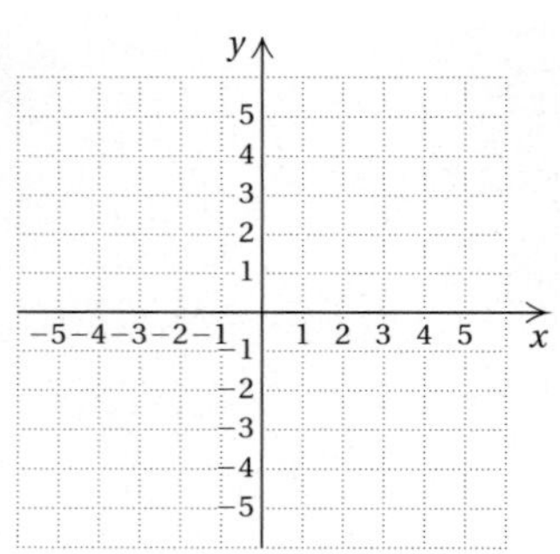

32. $y = -\frac{1}{2}x - 3$

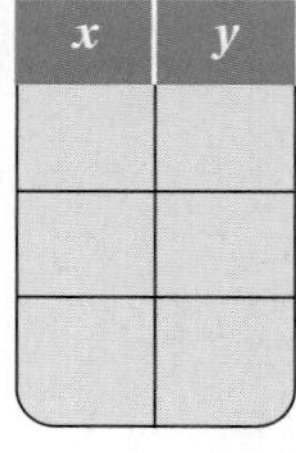

x	y

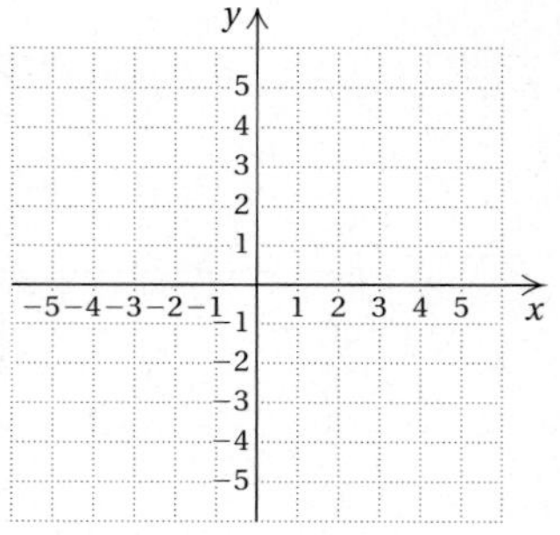

33. $8x - 2y = -10$

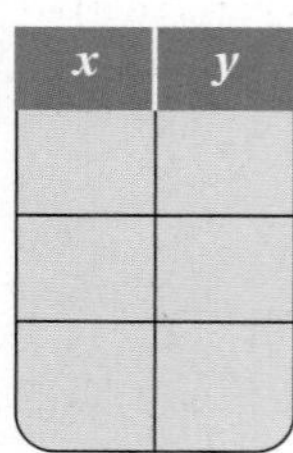

x	y

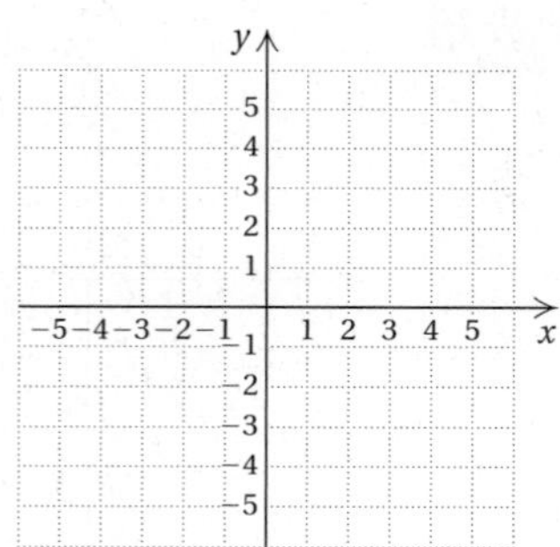

34. $6x - 3y = 9$

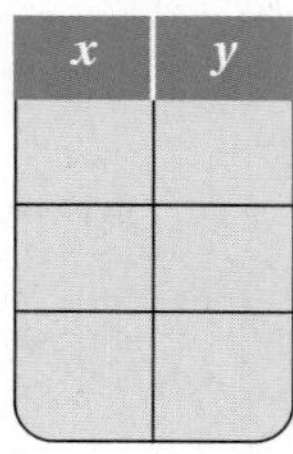

x	y

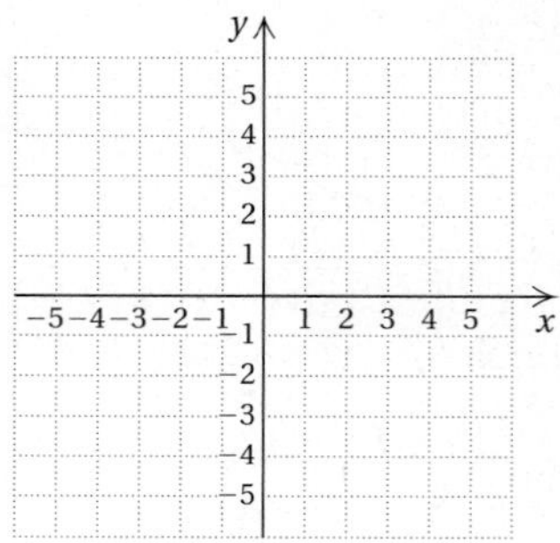

35. $8y + 2x = -4$

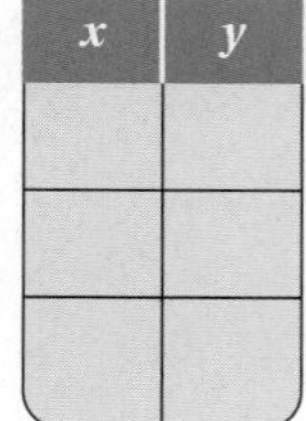

x	y

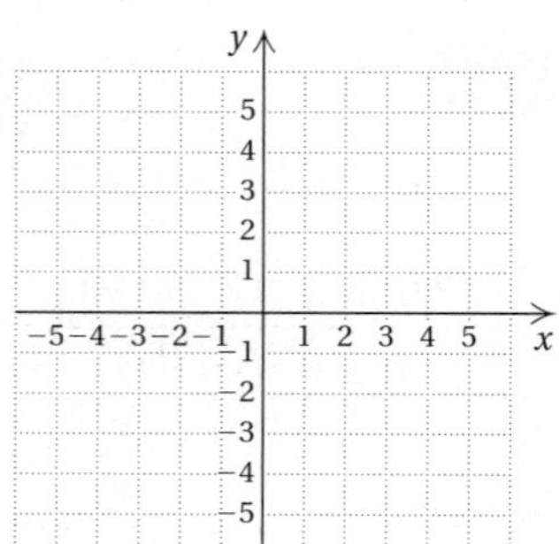

36. $6y + 2x = 8$

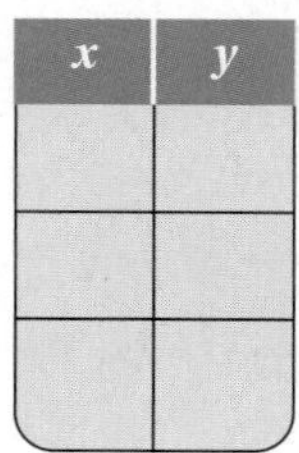

x	y

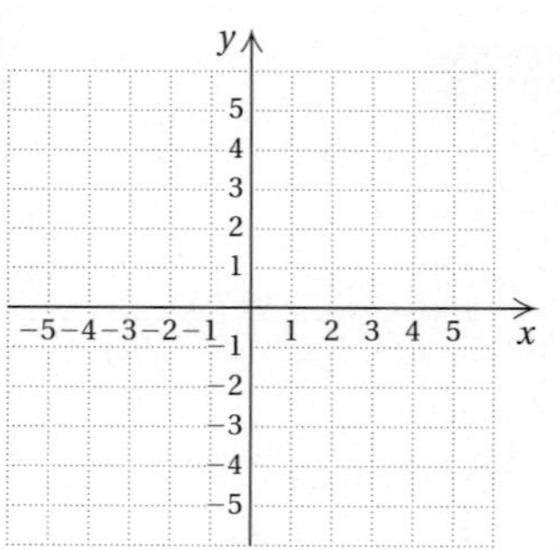

C Solve.

37. *Value of Computer Software.* The value V, in dollars, of a shopkeeper's inventory software program is given by $V = -50t + 300$, where t is the number of years since the shopkeeper first bought the program.

a) Find the value of the software after 0 yr, 4 yr, and 6 yr.

b) Graph the equation and then use the graph to estimate the value of the software after 5 yr.

t	V

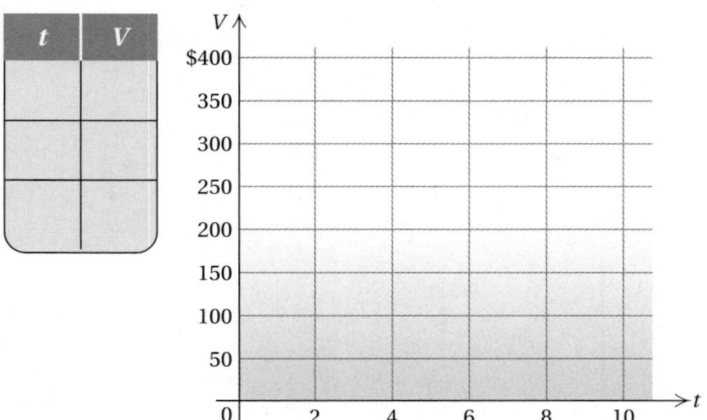

c) After how many years is the value of the software $150?

38. *College Costs.* The cost T, in dollars, of tuition and fees at many community colleges can be approximated by $T = 120c + 100$, where c is the number of credits for which a student registers.
Source: Community College of Vermont

a) Find the cost of tuition for a student who takes 8 hr, 12 hr, and 15 hr.

b) Graph the equation and then use the graph to estimate the cost of tuition and fees for 9 hr.

c	T

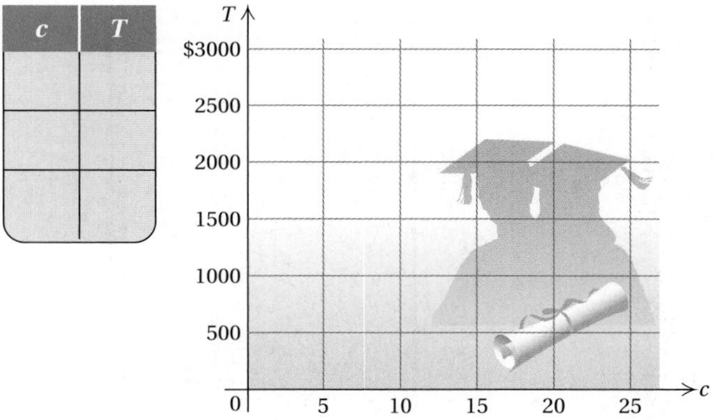

c) How many hours can a student take for $1420?

39. *Tea Consumption.* The number of gallons N of tea consumed each year by the average U.S. consumer can be approximated by $N = 0.1d + 7$, where d is the number of years since 1991.
Source: *Statistical Abstract of the United States*

a) Find the number of gallons of tea consumed in 1992 ($d = 1$), 1996, 2001, and 2011.

b) Graph the equation and use the graph to estimate what the tea consumption was in 1997.

d	N

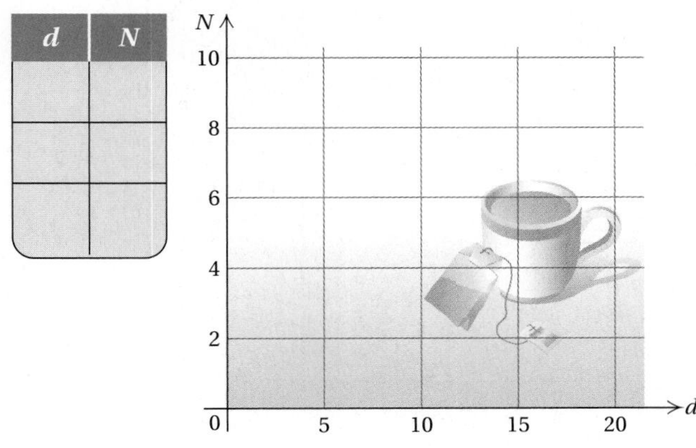

c) In what year will tea consumption be about 8.5 gal?

40. *Record Temperature Drop.* On 22 January 1943, the temperature T, in degrees Fahrenheit, in Spearfish, South Dakota, could be approximated by $T = -2.15m + 54$, where m is the number of minutes since 9:00 that morning.
Source: *Information Please Almanac*

a) Find the temperature at 9:01 A.M., 9:08 A.M., and 9:20 A.M.

b) Graph the equation and use the graph to estimate the temperature at 9:15 A.M.

m	T

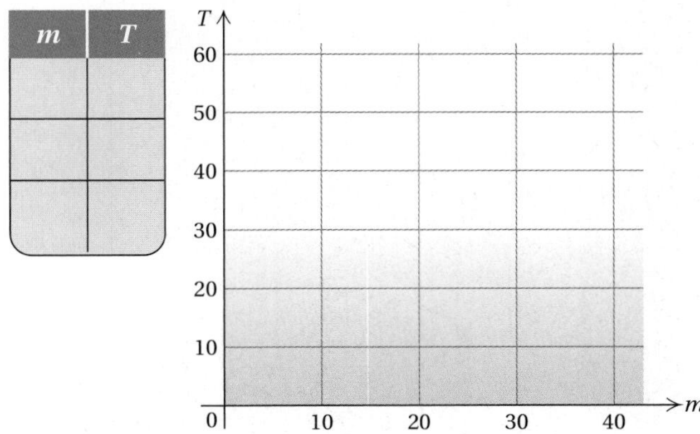

c) The temperature stopped dropping when it reached −4°F. At what time did this occur?

41. ^{D}W The equations $3x + 4y = 8$ and $y = -\frac{3}{4}x + 2$ are equivalent. Which equation is easier to graph and why?

42. ^{D}W Referring to Exercise 40, discuss why the linear equation no longer described the temperature after the temperature reached $-4°$.

SKILL MAINTENANCE

Round to the nearest thousand. [R.3c]

43. 2567.03

44. 124,748

45. 293.4572

46. 6,078,124

47. 3028

Convert to decimal notation. [1.2c]

48. $\frac{23}{32}$

49. $-\frac{7}{8}$

50. $-\frac{27}{12}$

51. $\frac{117}{64}$

52. $-\frac{17}{16}$

SYNTHESIS

In Exercises 53–56, find an equation for the graph shown.

53.

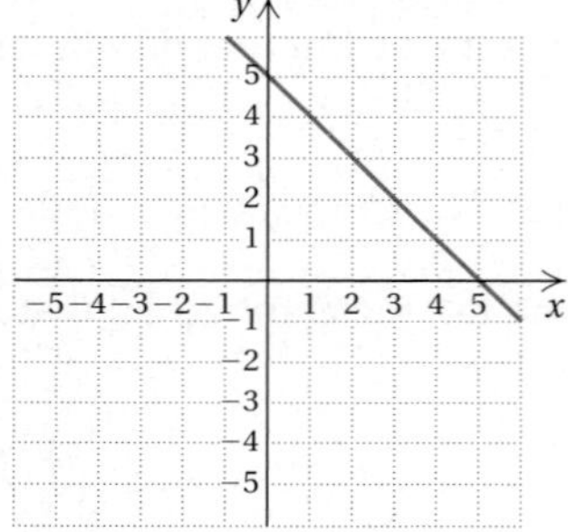

54.

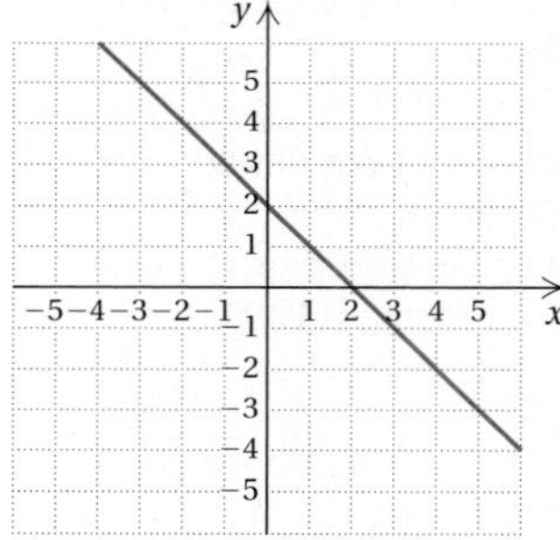

55.

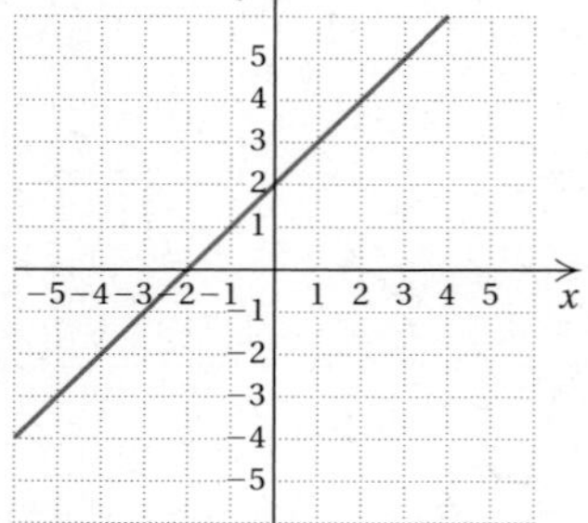

56.

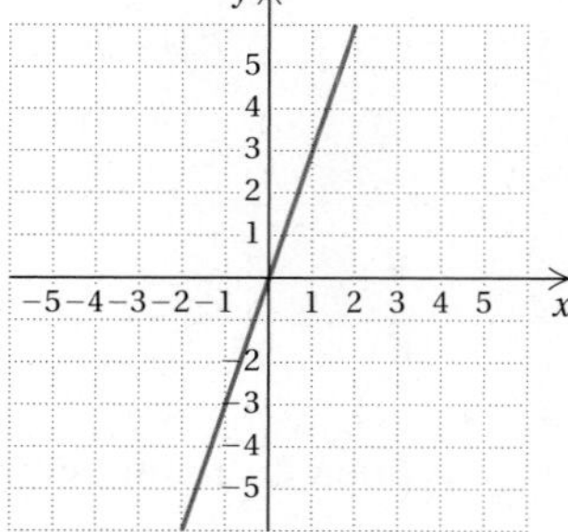

3.3 MORE WITH GRAPHING AND INTERCEPTS

Objectives

a Find the intercepts of a linear equation, and graph using intercepts.

b Graph equations equivalent to those of the type $x = a$ and $y = b$.

a Graphing Using Intercepts

In Section 3.2, we graphed linear equations of the form $Ax + By = C$ by first solving for y to find an equivalent equation in the form $y = mx + b$. We did so because it is then easier to calculate the y-value that corresponds to a given x-value. Another convenient way to graph $Ax + By = C$ is to use **intercepts.** Look at the graph of $-2x + y = 4$ shown below.

The y-intercept is $(0, 4)$. It occurs where the line crosses the y-axis and thus will always have 0 as the first coordinate. The x-intercept is $(-2, 0)$. It occurs where the line crosses the x-axis and thus will always have 0 as the second coordinate.

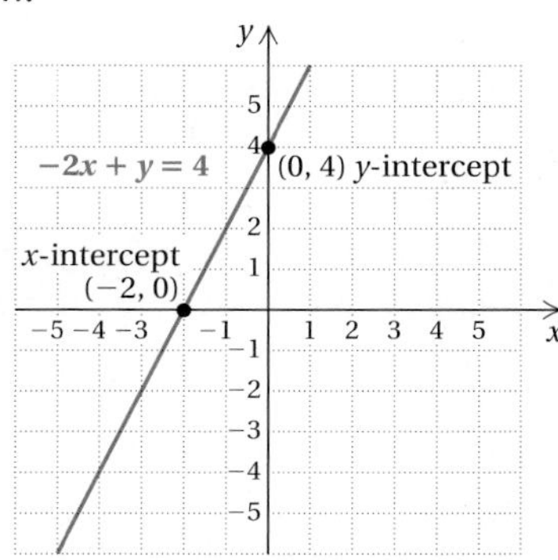

Do Exercise 1.

1. Look at the graph shown below.

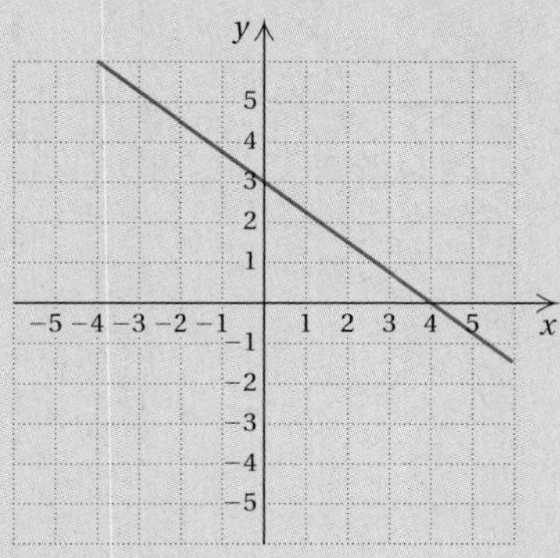

a) Find the coordinates of the y-intercept.

b) Find the coordinates of the x-intercept.

We find intercepts as follows.

INTERCEPTS

The **y-intercept** is $(0, b)$. To find b, let $x = 0$ and solve the original equation for y.

The **x-intercept** is $(a, 0)$. To find a, let $y = 0$ and solve the original equation for x.

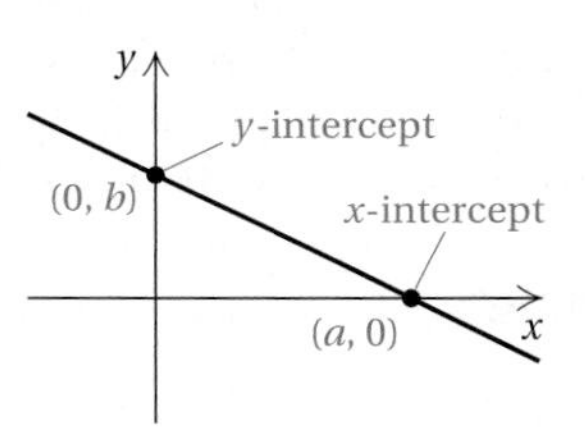

Now let's draw a graph using intercepts.

EXAMPLE 1 Consider $4x + 3y = 12$. Find the intercepts. Then graph the equation using the intercepts.

To find the y-intercept, we let $x = 0$. Then we solve for y:

$$4 \cdot 0 + 3y = 12$$
$$3y = 12$$
$$y = 4.$$

Thus, $(0, 4)$ is the y-intercept. Note that finding this intercept amounts to covering up the x-term and solving the rest of the equation.

To find the x-intercept, we let $y = 0$. Then we solve for x:

$$4x + 3 \cdot 0 = 12$$
$$4x = 12$$
$$x = 3.$$

Answers on page A-13

For each equation, find the intercepts. Then graph the equation using the intercepts.

2. $2x + 3y = 6$

x	y	
	0	← x-intercept
0		← y-intercept
		← Check point

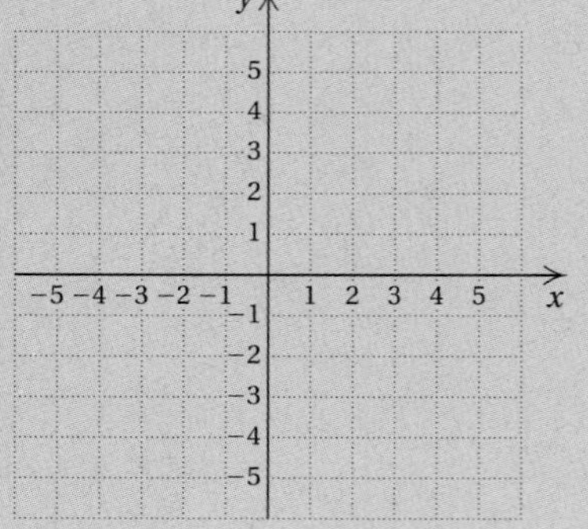

3. $3y - 4x = 12$

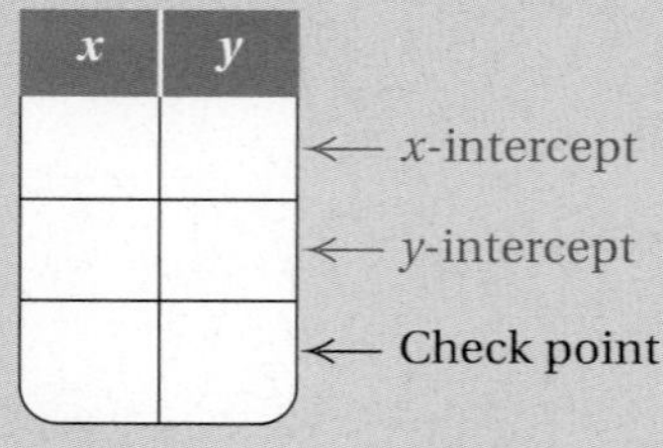

x	y	
		← x-intercept
		← y-intercept
		← Check point

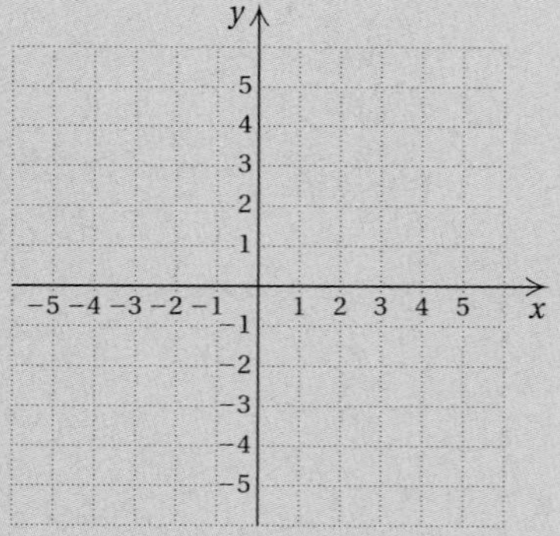

Thus, $(3, 0)$ is the x-intercept. Note that finding this intercept amounts to covering up the y-term and solving the rest of the equation.

We plot these points and draw the line, or graph.

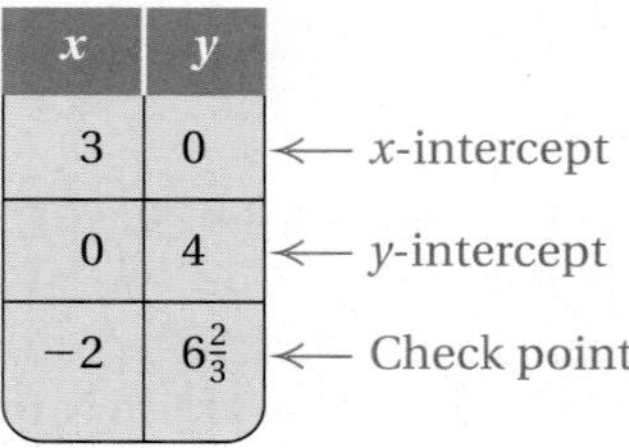

x	y	
3	0	← x-intercept
0	4	← y-intercept
-2	$6\frac{2}{3}$	← Check point

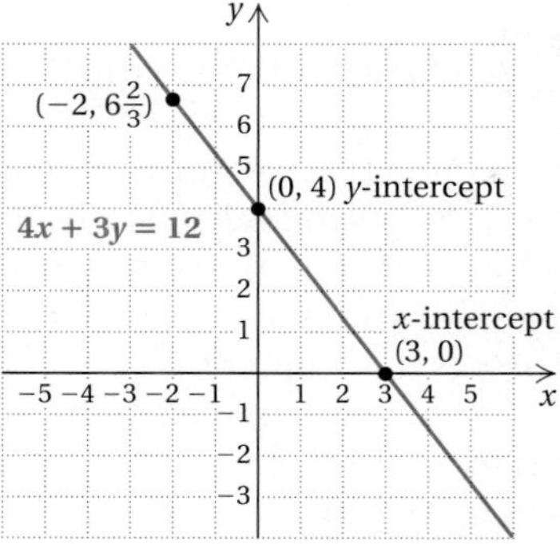

A third point should be used as a check. We substitute any convenient value for x and solve for y. In this case, we choose $x = -2$. Then

$$4(-2) + 3y = 12 \quad \text{Substituting } -2 \text{ for } x$$
$$-8 + 3y = 12$$
$$3y = 12 + 8 = 20$$
$$y = \tfrac{20}{3}, \text{ or } 6\tfrac{2}{3}. \quad \text{Solving for } y$$

It appears that the point $\left(-2, 6\frac{2}{3}\right)$ is on the graph, though graphing fraction values can be inexact. The graph is probably correct.

Do Exercises 2 and 3.

Graphs of equations of the type $y = mx$ pass through the origin. Thus the x-intercept and the y-intercept are the same, $(0, 0)$. In such cases, we must calculate another point in order to complete the graph. Another point would also have to be calculated if a check is desired.

EXAMPLE 2 Graph: $y = 3x$.

We know that $(0, 0)$ is both the x-intercept and the y-intercept. We calculate values at two other points and complete the graph, knowing that it passes through the origin $(0, 0)$.

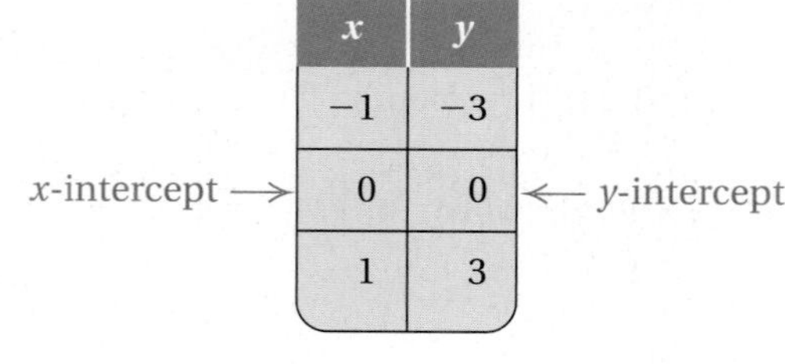

	x	y	
	-1	-3	
x-intercept →	0	0	← y-intercept
	1	3	

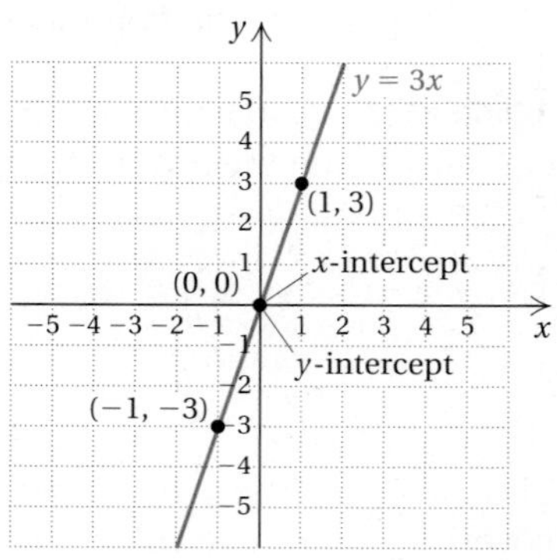

Do Exercises 4 and 5 on the following page.

Answers on page A-13

CALCULATOR CORNER

Viewing the Intercepts Knowing the intercepts of a linear equation helps us to determine a good viewing window for the graph of the equation. For example, when we graph the equation $y = -x + 15$ in the standard window, we see only a small portion of the graph in the upper righthand corner of the screen, as shown on the left below.

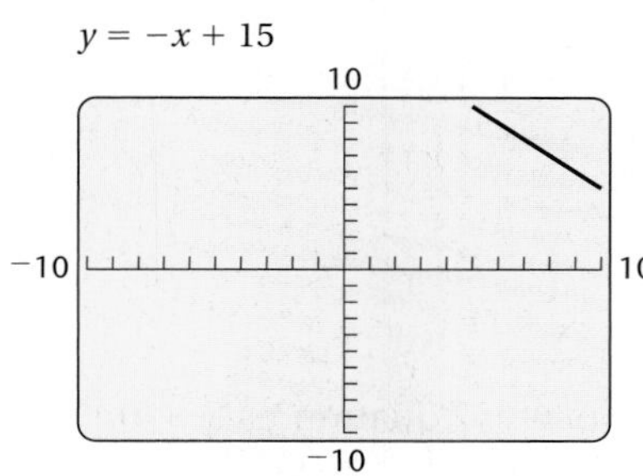

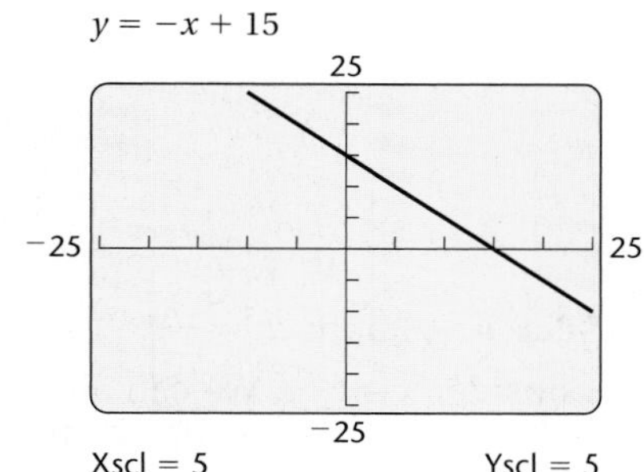

Using algebra, as we did in Example 1, we find that the intercepts of the graph of this equation are $(0, 15)$ and $(15, 0)$. This tells us that, if we are to see more of the graph than is shown on the left above, both Xmax and Ymax should be greater than 15. We can try different window settings until we find one that suits us. One good choice is $[-25, 25, -25, 25]$, Xscl $= 5$, Yscl $= 5$, shown on the right above.

Exercises: Find the intercepts of each equation algebraically. Then graph the equation on a graphing calculator, choosing window settings that allow the intercepts to be seen clearly. (Settings may vary.)

1. $y = -7.5x - 15$
2. $y - 2.15x = 43$
3. $6x - 5y = 150$
4. $y = 0.2x - 4$
5. $y = 1.5x - 15$
6. $5x - 4y = 2$

b Equations Whose Graphs Are Horizontal or Vertical Lines

EXAMPLE 3 Graph: $y = 3$.

Consider $y = 3$. We can also think of this equation as $0 \cdot x + y = 3$. No matter what number we choose for x, we find that y is 3. We make up a table with all 3's in the y-column.

x	y
	3
	3
	3

y must be 3.

Choose any number for x. ⟶

x	y
−2	3
0	3
4	3

← y-intercept (row 0, 3)

Graph.

4. $y = 2x$

x	y
−1	
0	
1	

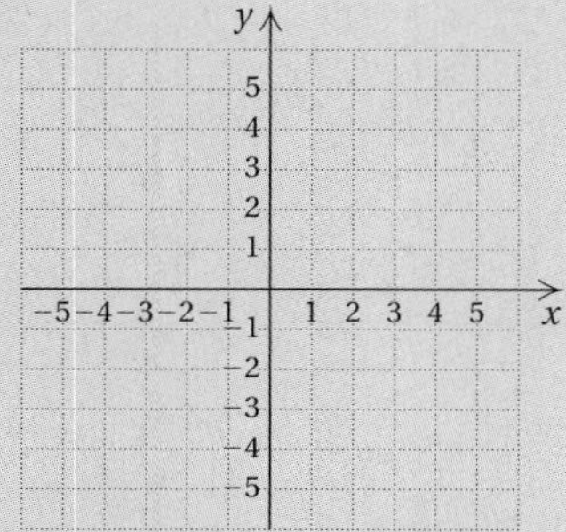

5. $y = -\frac{2}{3}x$

x	y

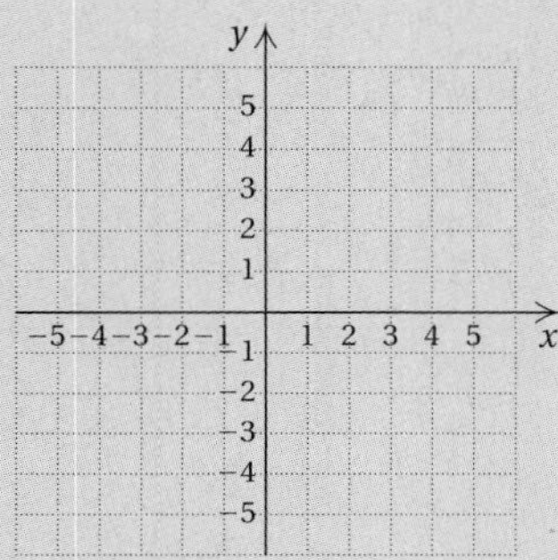

Answers on page A-13

Graph.

6. $x = 5$

x	y
5	
5	
5	

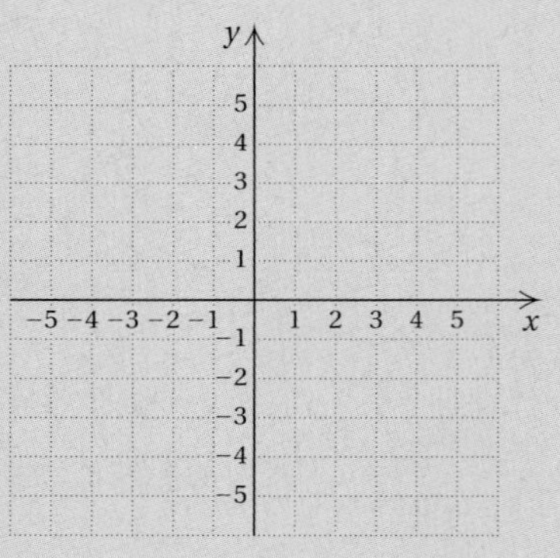

7. $y = -2$

x	y
	-2
	-2
	-2

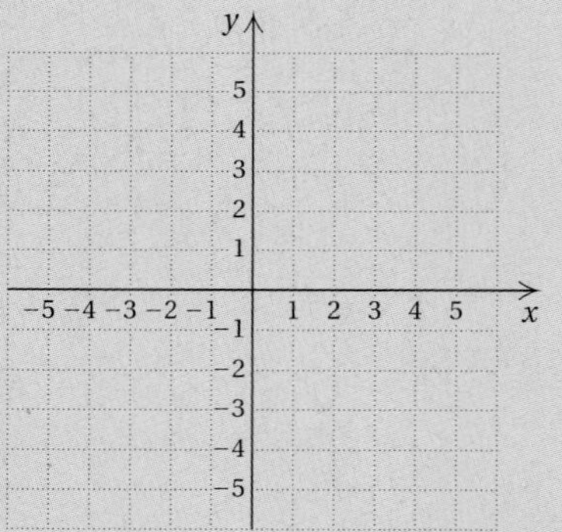

Answers on page A-13

When we plot the ordered pairs $(-2, 3)$, $(0, 3)$, and $(4, 3)$ and connect the points, we will obtain a horizontal line. Any ordered pair $(x, 3)$ is a solution. So the line is parallel to the x-axis with y-intercept $(0, 3)$.

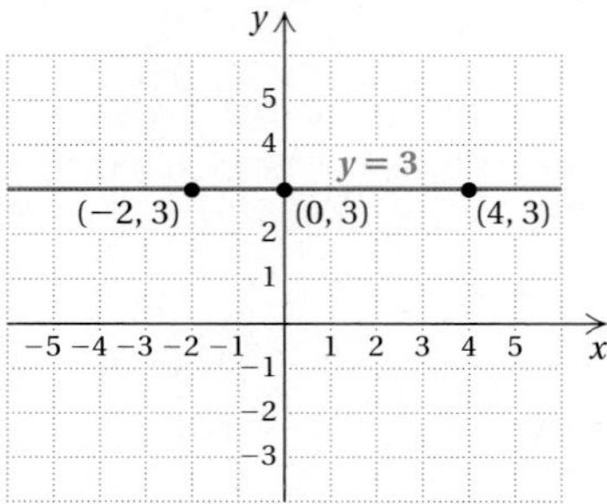

EXAMPLE 4 Graph: $x = -4$.

Consider $x = -4$. We can also think of this equation as $x + 0 \cdot y = -4$. We make up a table with all -4's in the x-column.

x	y
-4	
-4	
-4	
-4	

x must be -4.

Choose any number for y. →

x	y
-4	-5
-4	1
-4	3
-4	0

x-intercept → (last row)

When we plot the ordered pairs $(-4, -5)$, $(-4, 1)$, $(-4, 3)$, and $(-4, 0)$ and connect the points, we will obtain a vertical line. Any ordered pair $(-4, y)$ is a solution. So the line is parallel to the y-axis with x-intercept $(-4, 0)$.

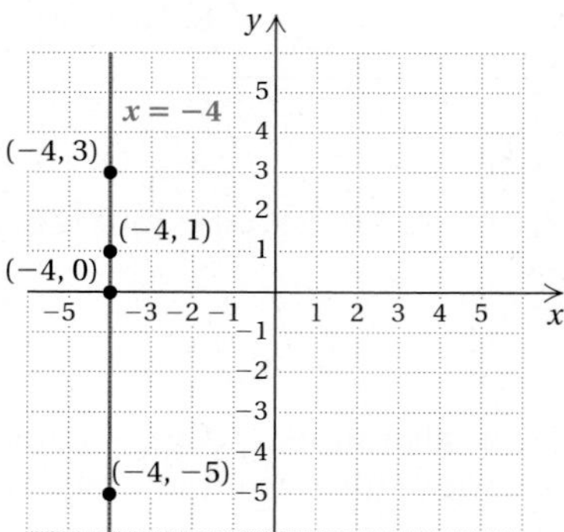

HORIZONTAL AND VERTICAL LINES

The graph of $y = b$ is a **horizontal line.** The y-intercept is $(0, b)$.
The graph of $x = a$ is a **vertical line.** The x-intercept is $(a, 0)$.

Do Exercises 6–9. (Exercises 8 and 9 are on the following page.)

The following is a general procedure for graphing linear equations.

GRAPHING LINEAR EQUATIONS

1. If the equation is of the type $x = a$ or $y = b$, the graph will be a line parallel to an axis; $x = a$ is vertical and $y = b$ is horizontal.

Examples.

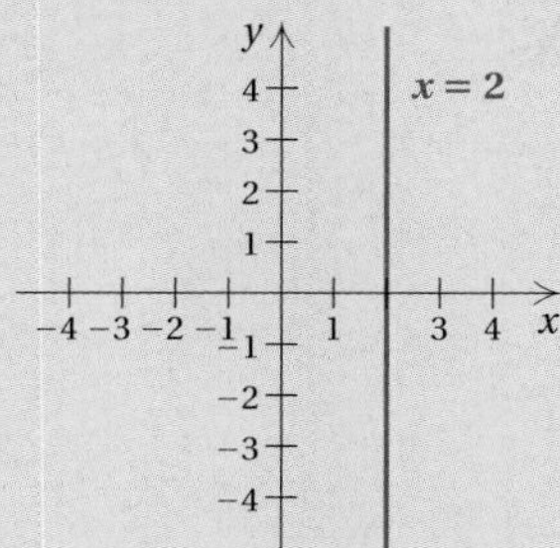

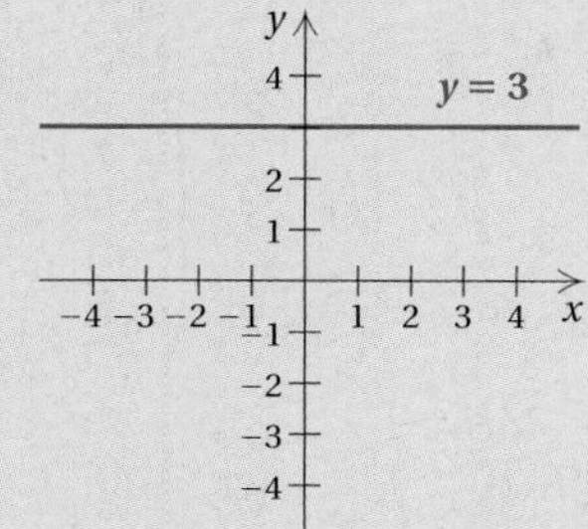

2. If the equation is of the type $y = mx$, both intercepts are the origin, $(0, 0)$. Plot $(0, 0)$ and two other points.

Example.

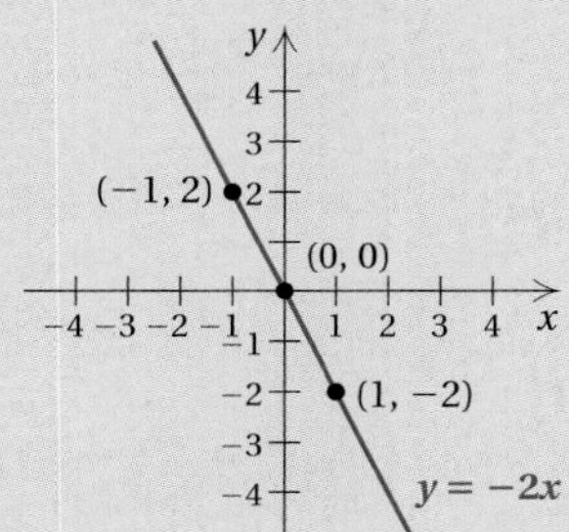

3. If the equation is of the type $y = mx + b$, plot the y-intercept $(0, b)$ and two other points.

Example.

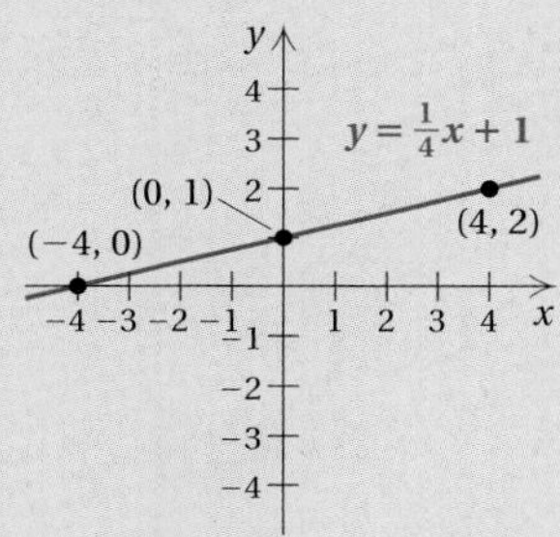

4. If the equation is of the type $Ax + By = C$, but not of the type $x = a$, $y = b$, $y = mx$, or $y = mx + b$, then either solve for y and proceed as with the equation $y = mx + b$, or graph using intercepts. If the intercepts are too close together, choose another point or points farther from the origin.

Examples.

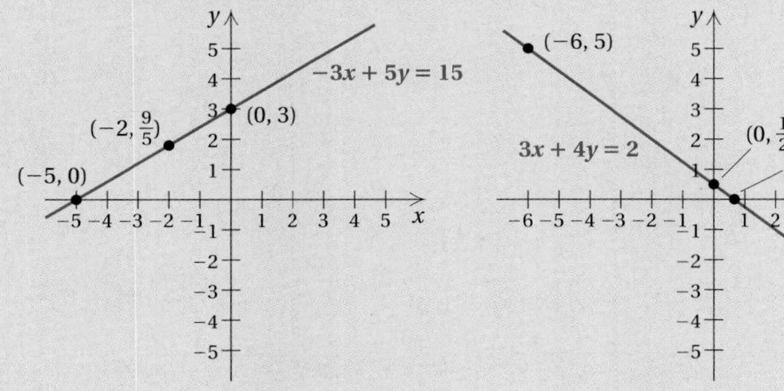

8. $x = 0$

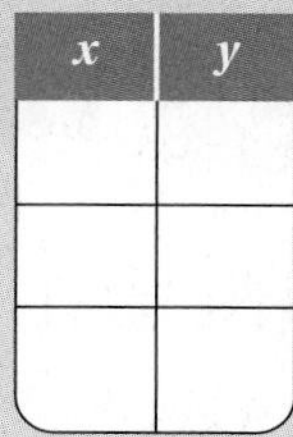

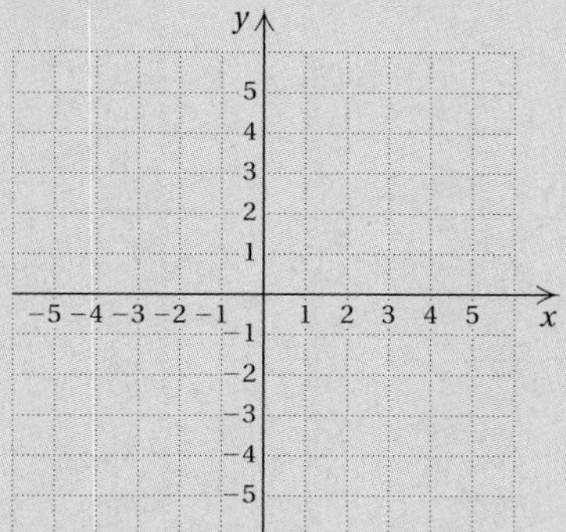

9. $x = -3$

x	y

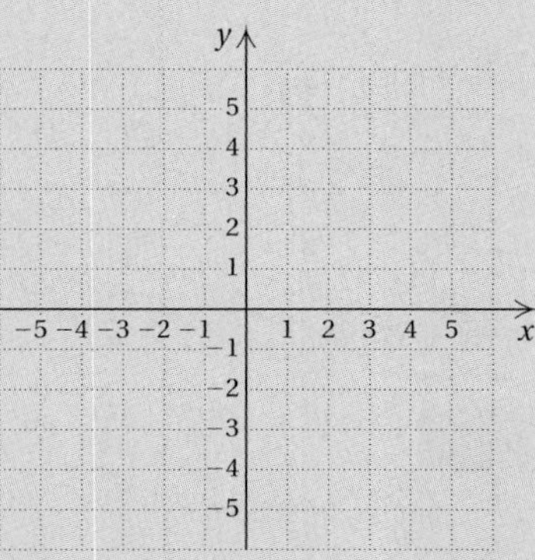

Answers on page A-14

3.3 EXERCISE SET

For Extra Help

Digital Video Tutor CD 4 Videotape 7 | InterAct Math | Math Tutor Center | MathXL | MyMathLab

a For Exercises 1–4, find (a) the coordinates of the y-intercept and (b) the coordinates of the x-intercept.

1.

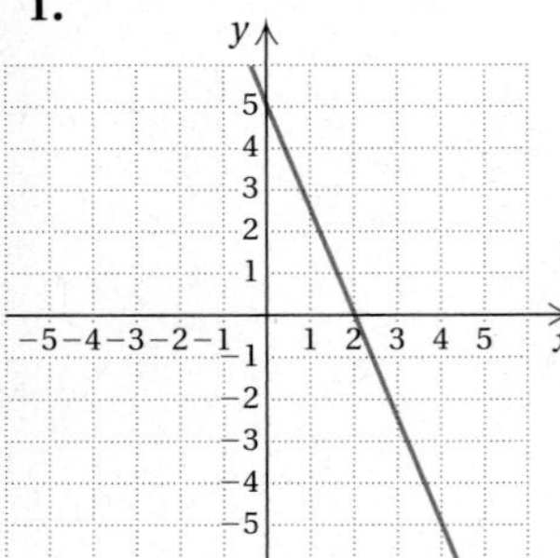

2.

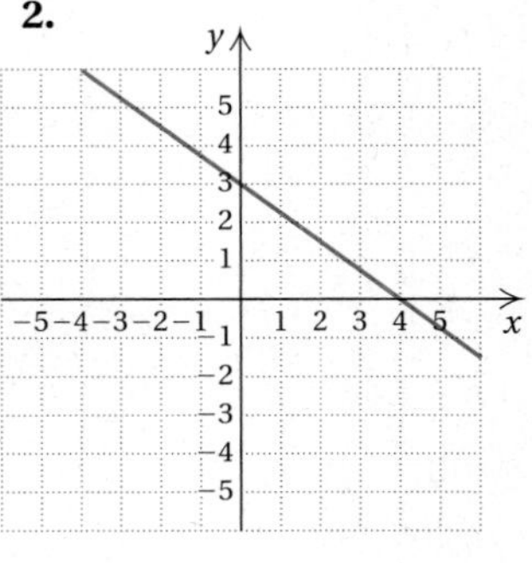

3.

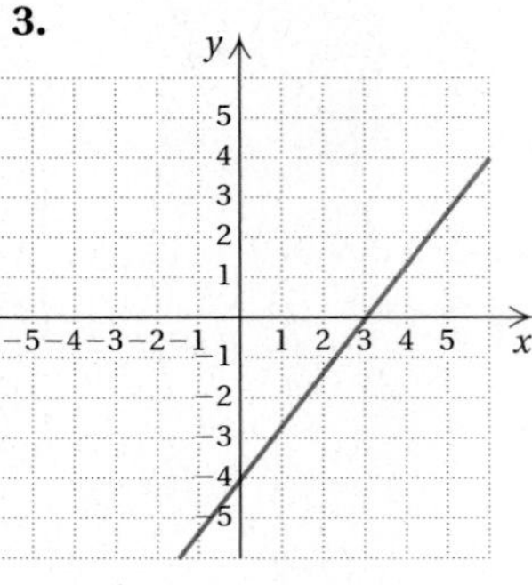

4.

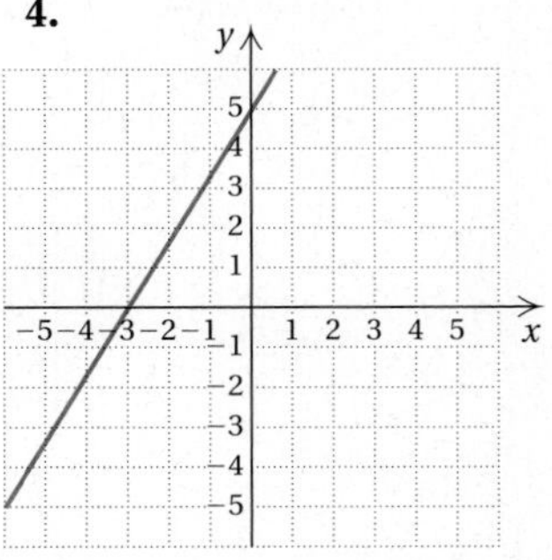

For Exercises 5–12, find (a) the coordinates of the y-intercept and (b) the coordinates of the x-intercept. Do not graph.

5. $3x + 5y = 15$

6. $5x + 2y = 20$

7. $7x - 2y = 28$

8. $3x - 4y = 24$

9. $-4x + 3y = 10$

10. $-2x + 3y = 7$

11. $6x - 3 = 9y$

12. $4y - 2 = 6x$

For each equation, find the intercepts. Then use the intercepts to graph the equation.

13. $x + 3y = 6$

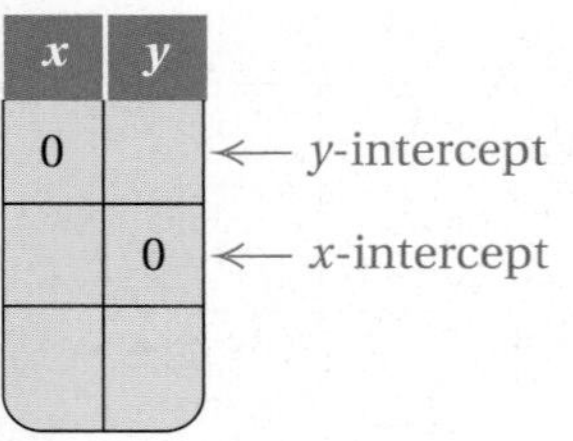

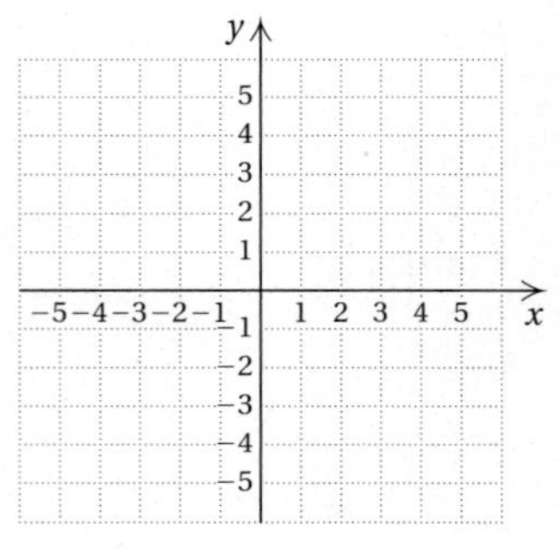

14. $x + 2y = 2$

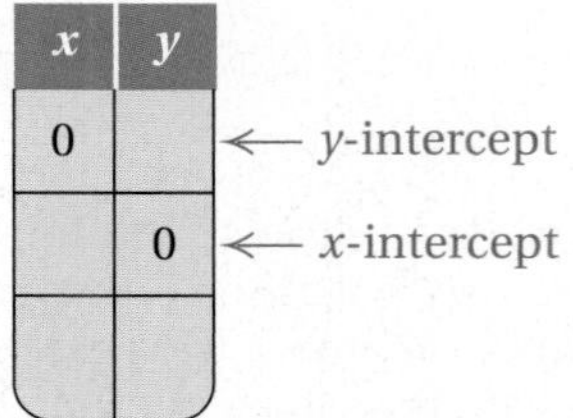

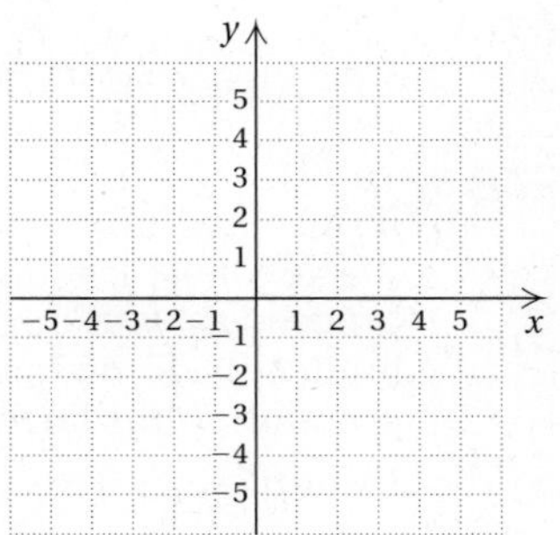

15. $-x + 2y = 4$

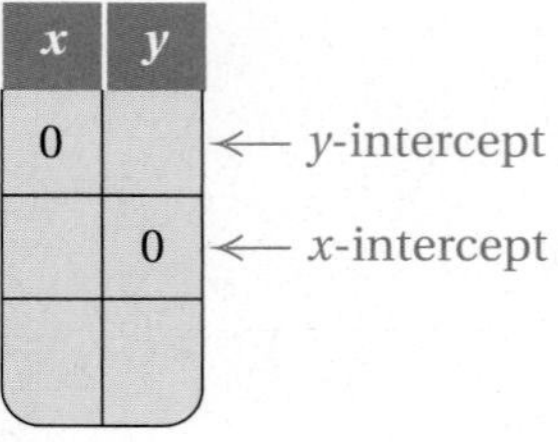

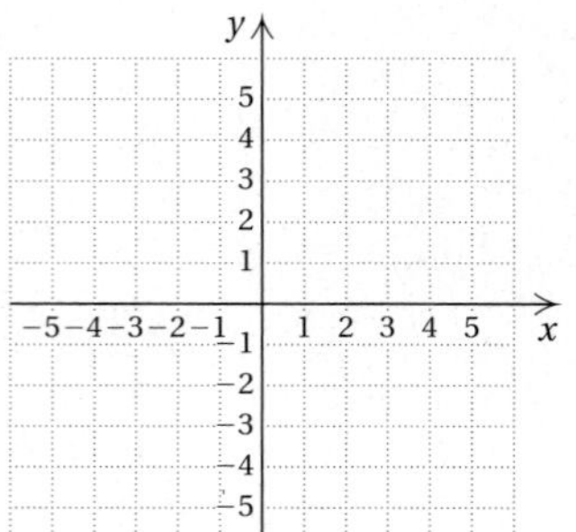

16. $-x + y = 5$

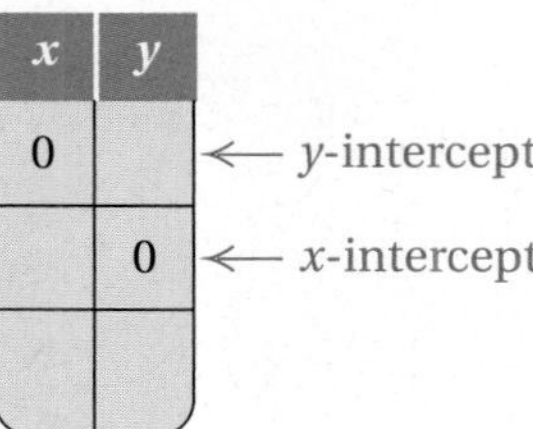

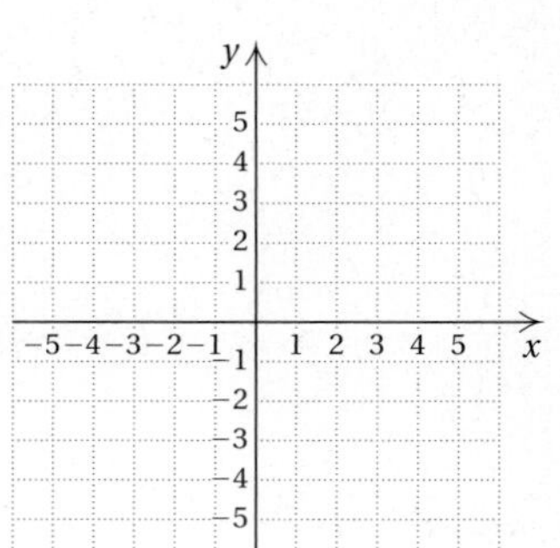

17. $3x + y = 6$

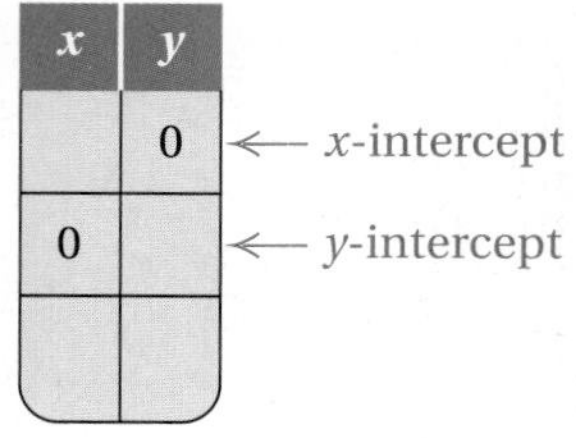

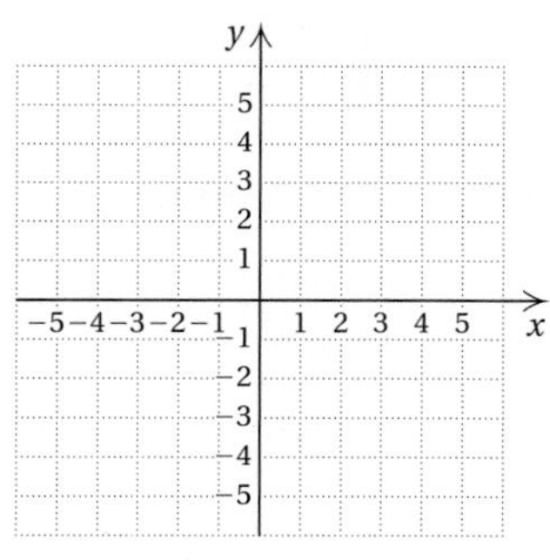

18. $2x + y = 6$

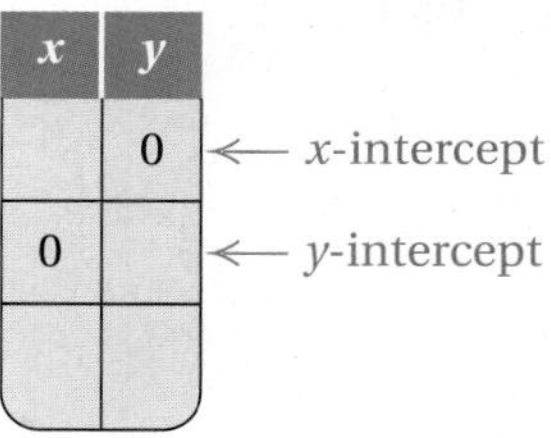

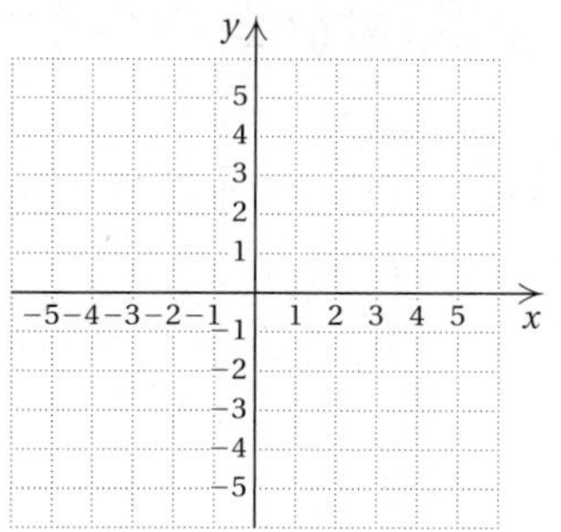

19. $2y - 2 = 6x$

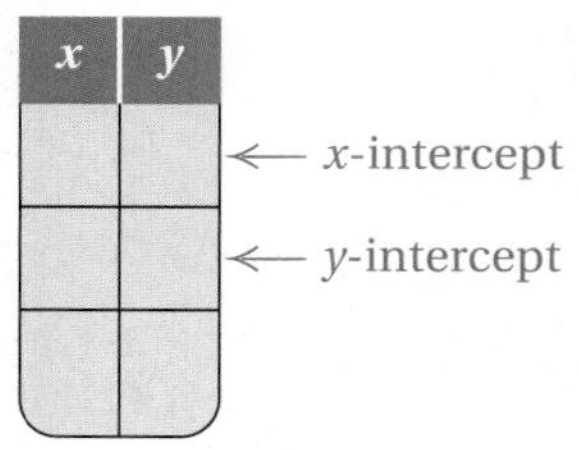

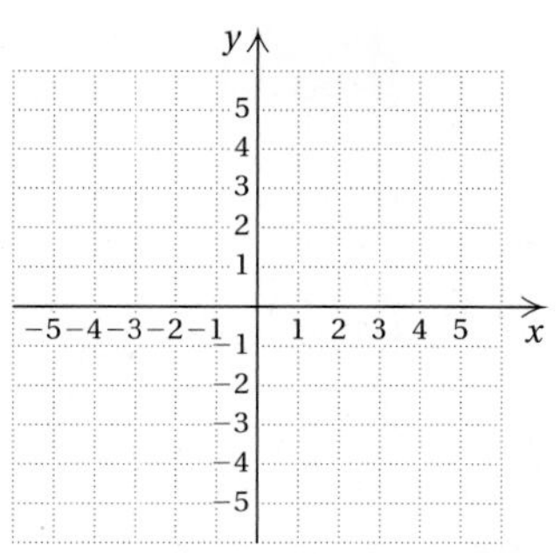

20. $3y - 6 = 9x$

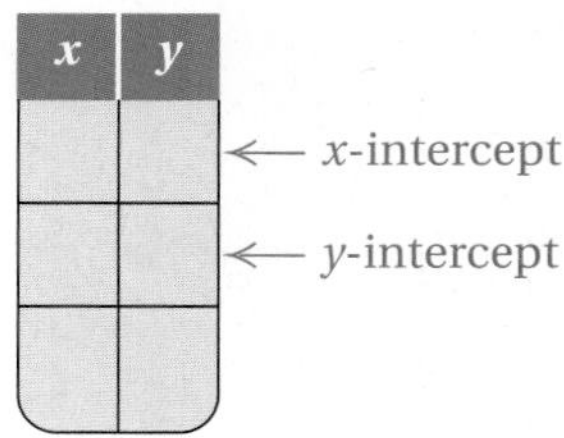

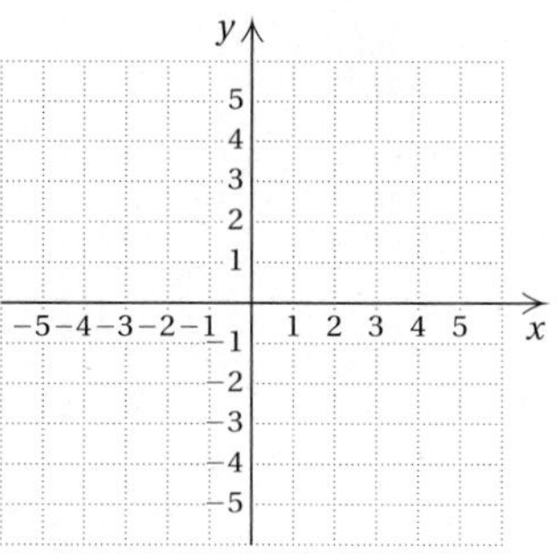

21. $3x - 9 = 3y$

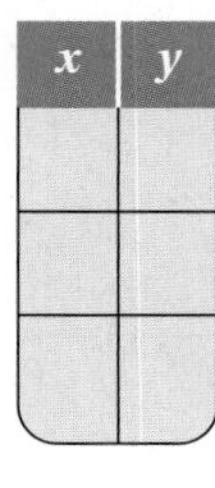

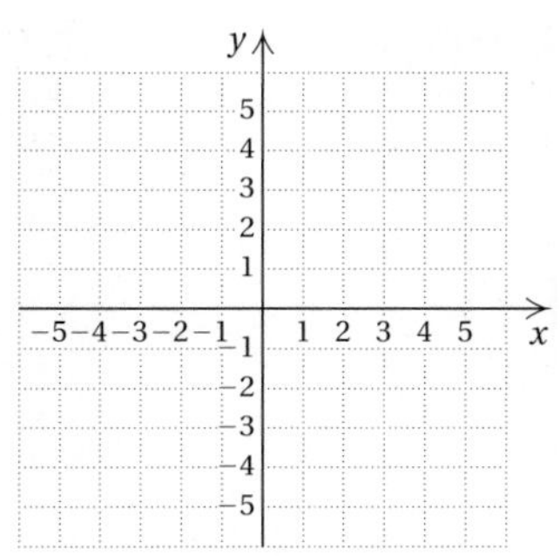

22. $5x - 10 = 5y$

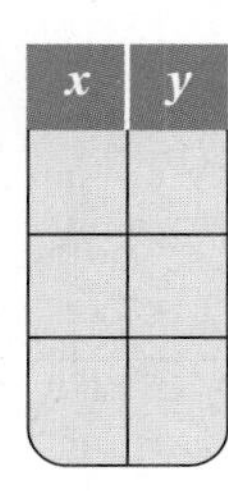

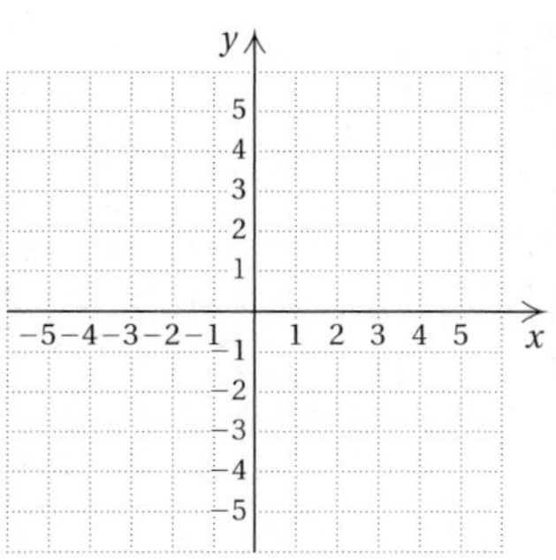

23. $2x - 3y = 6$

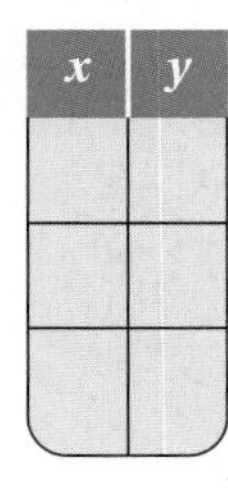

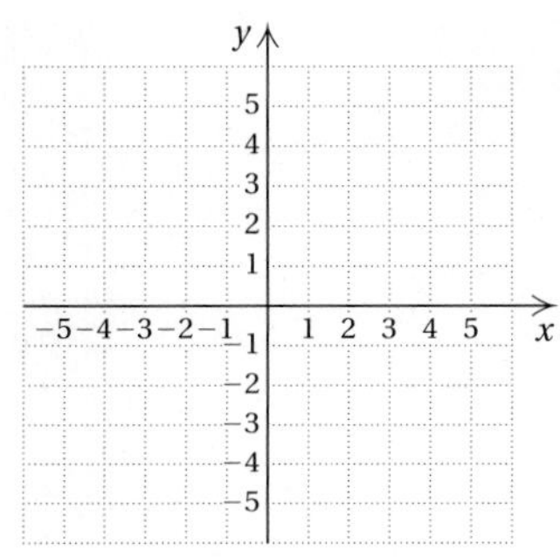

24. $2x - 5y = 10$

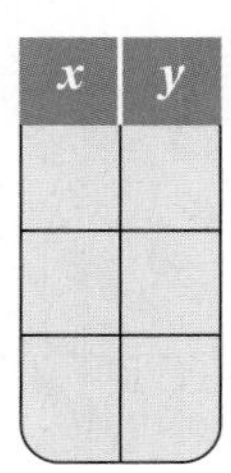

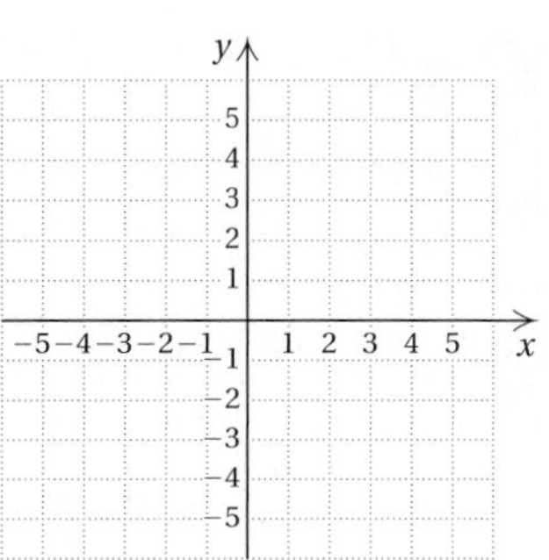

25. $4x + 5y = 20$

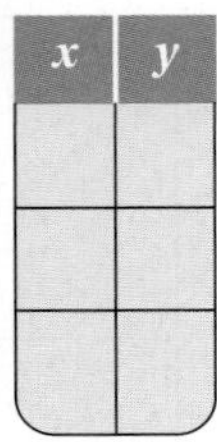

26. $2x + 6y = 12$

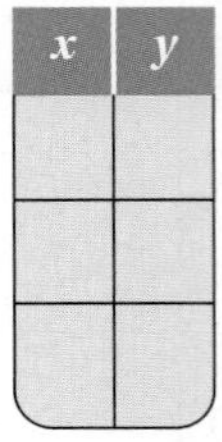

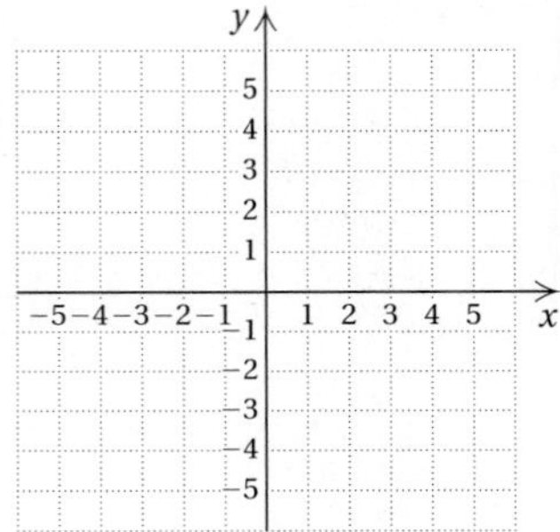

27. $2x + 3y = 8$

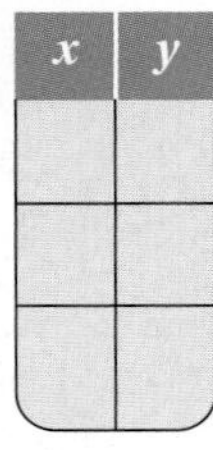

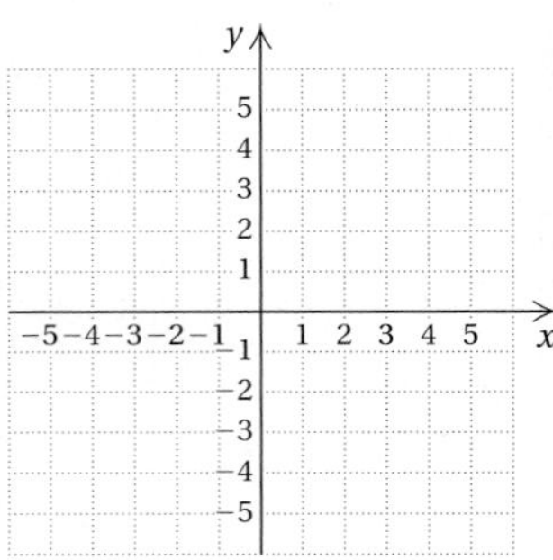

28. $x - 1 = y$

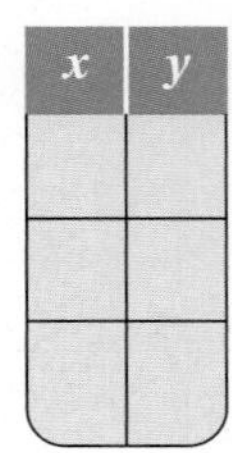

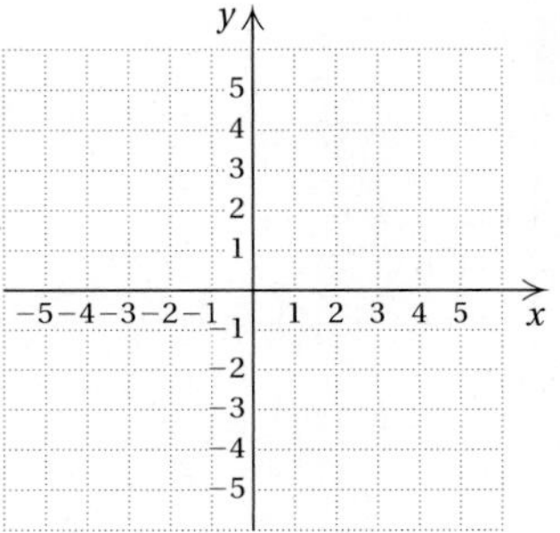

29. $x - 3 = y$

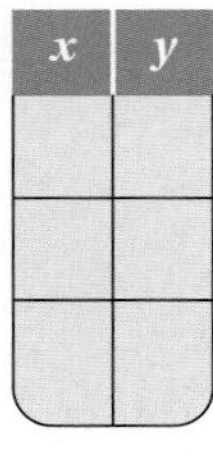

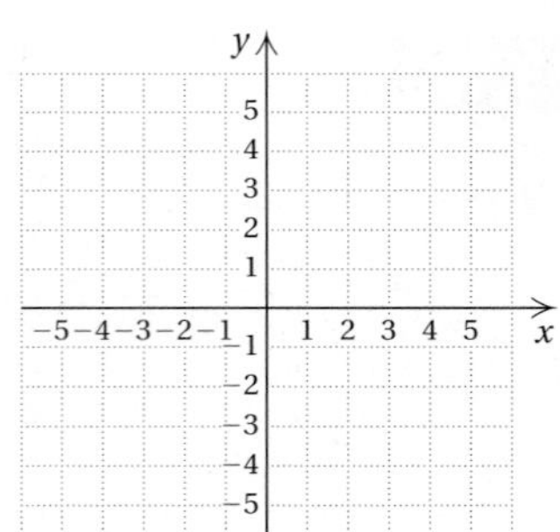

30. $2x - 1 = y$

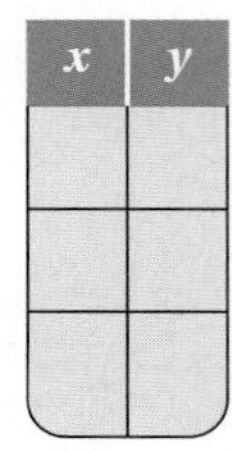

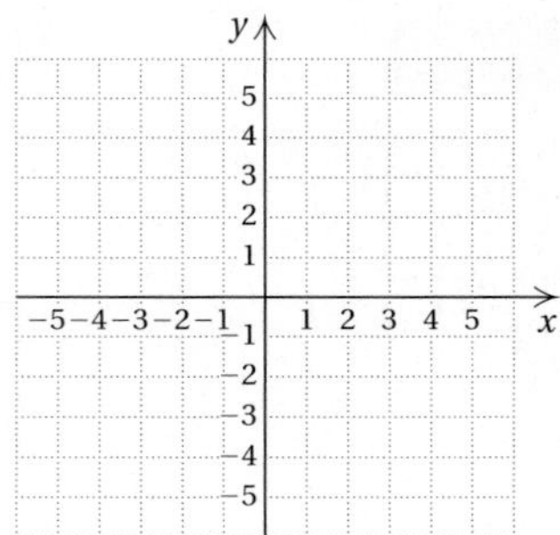

31. $3x - 2 = y$

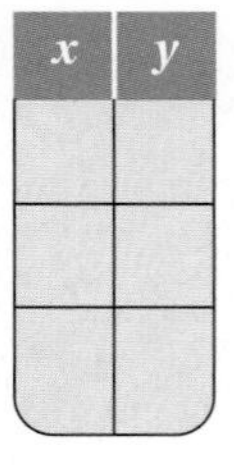

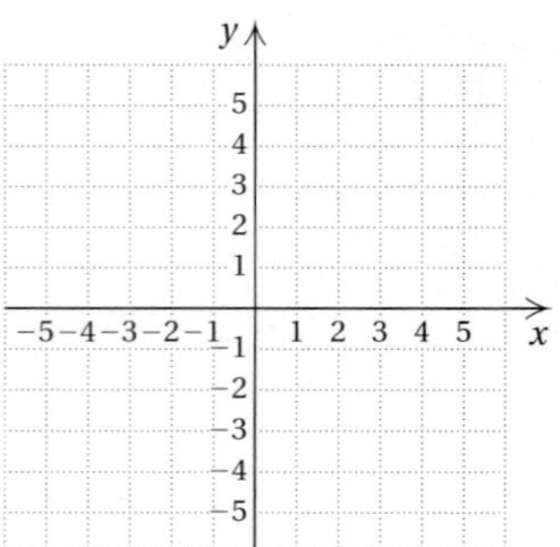

32. $4x - 3y = 12$

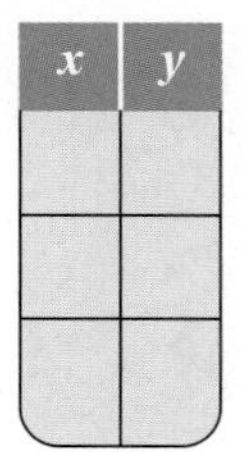

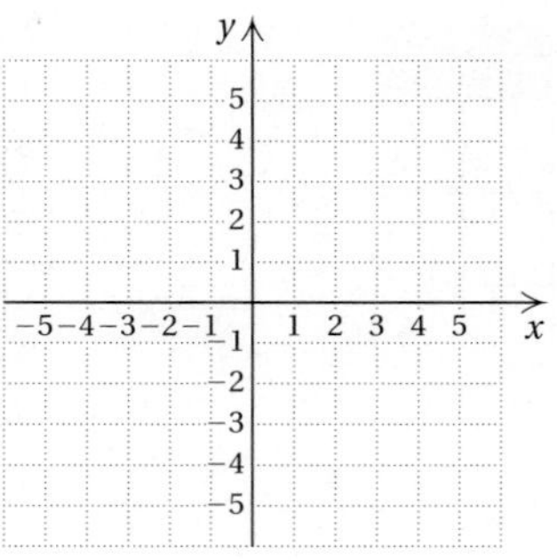

33. $6x - 2y = 12$

x	y

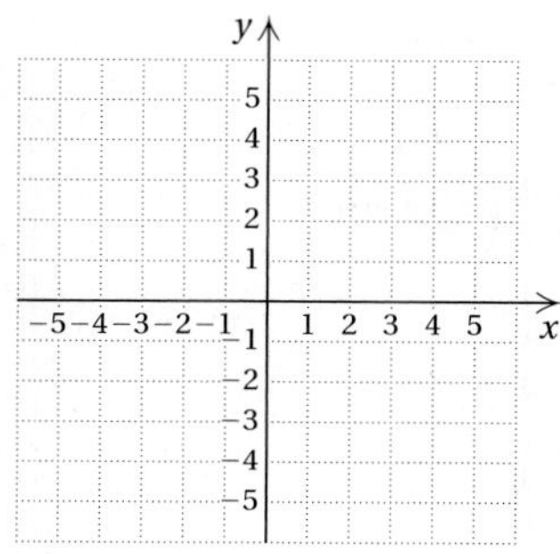

34. $7x + 2y = 6$

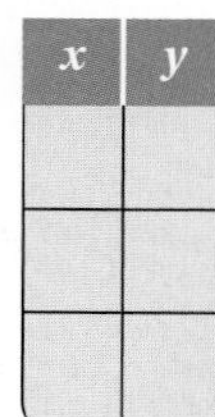

x	y

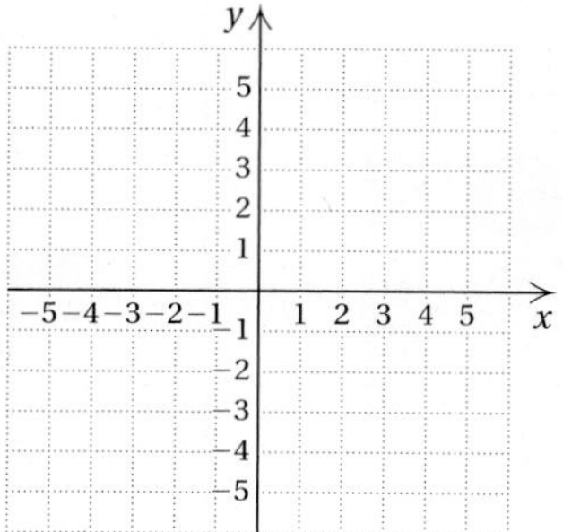

35. $3x + 4y = 5$

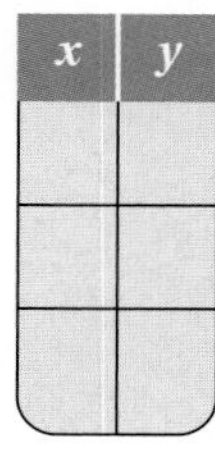

x	y

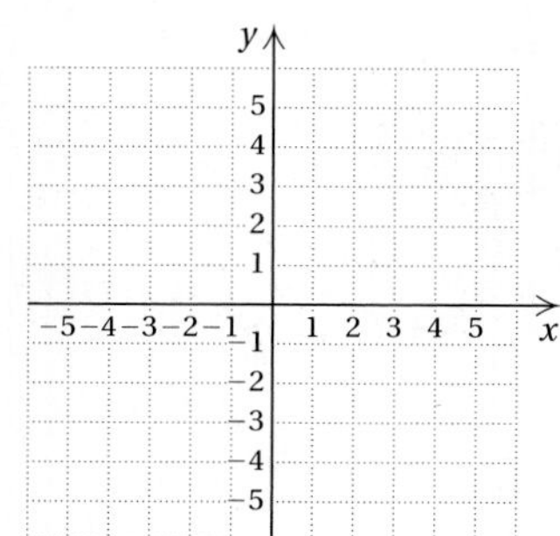

36. $y = -4 - 4x$

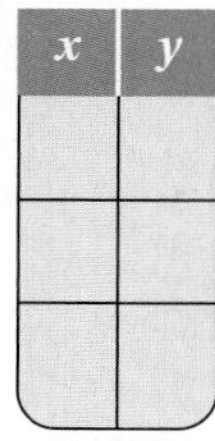

x	y

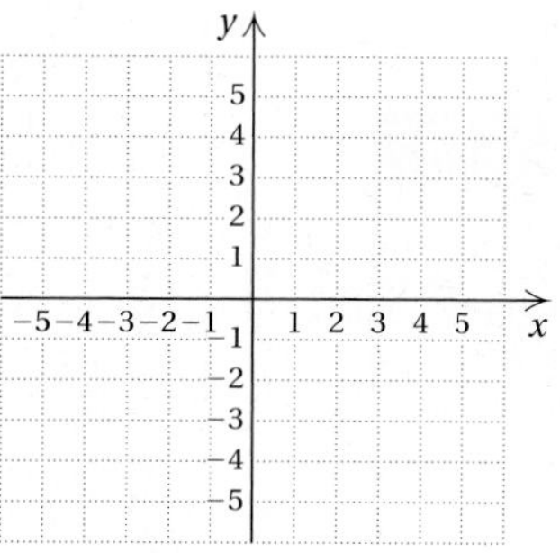

37. $y = -3 - 3x$

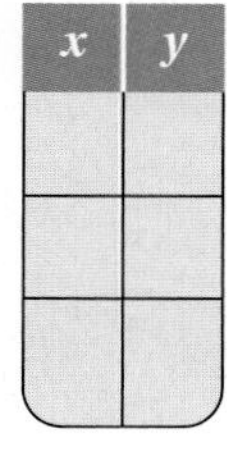

x	y

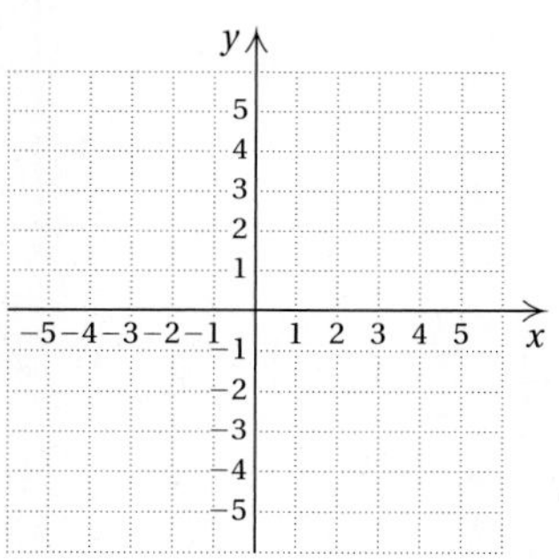

38. $-3x = 6y - 2$

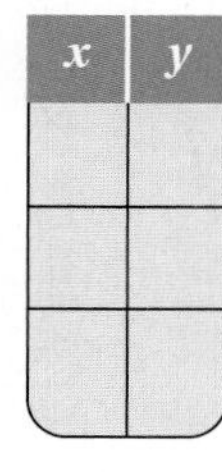

x	y

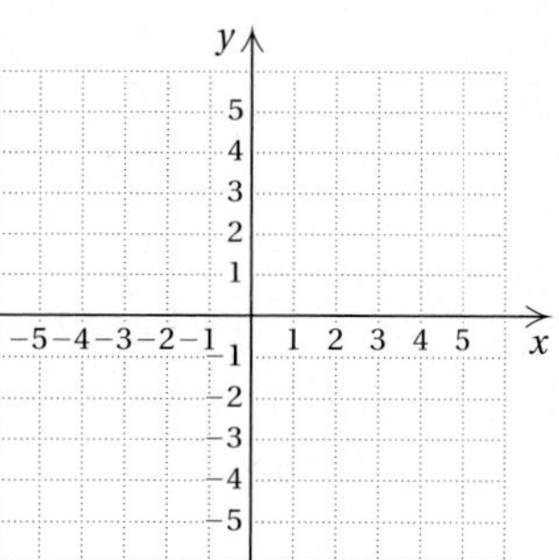

39. $y - 3x = 0$

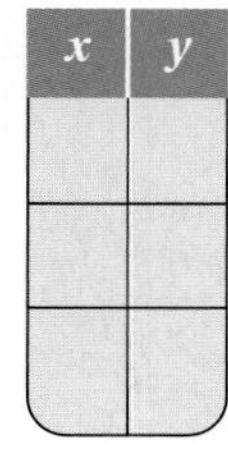

x	y

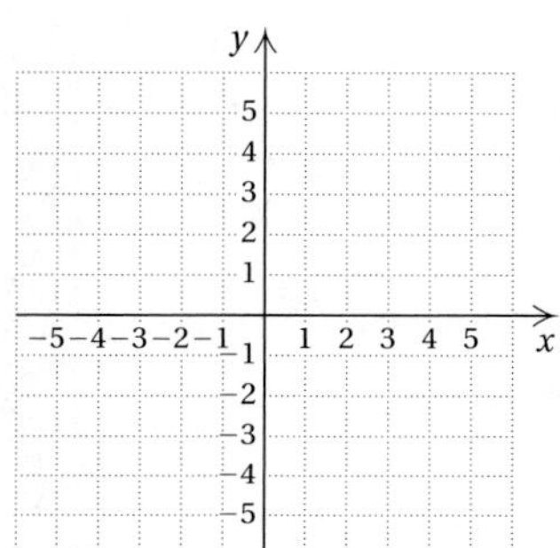

40. $x + 2y = 0$

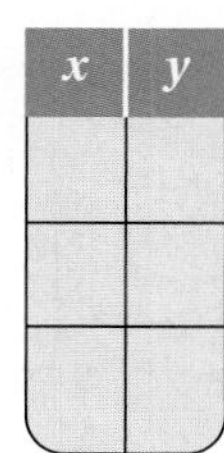

x	y

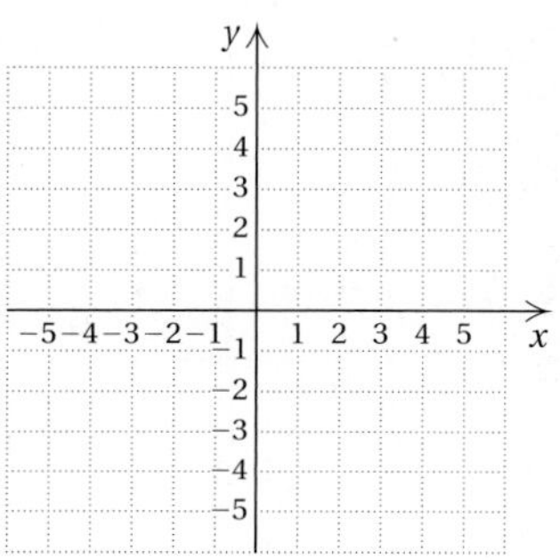

b Graph.

41. $x = -2$

x	y
-2	
-2	
-2	

42. $x = 1$

x	y
1	
1	
1	

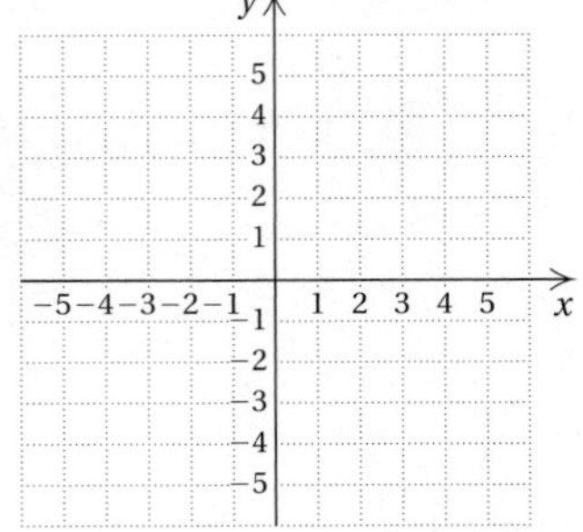

43. $y = 2$

x	y
	2
	2
	2

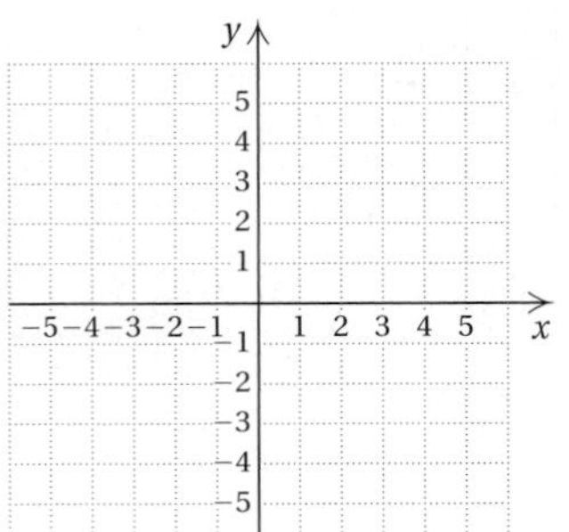

44. $y = -4$

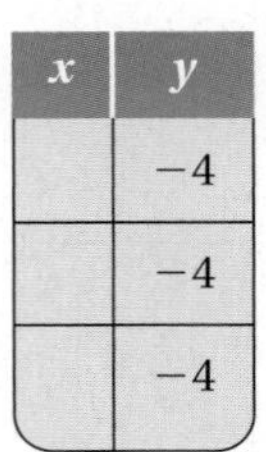

x	y
	-4
	-4
	-4

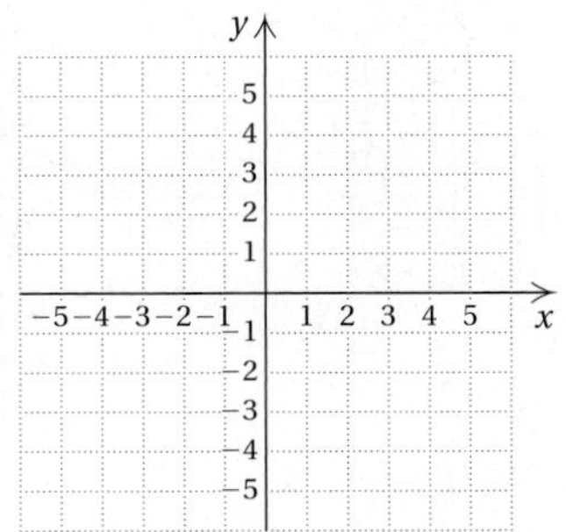

45. $x = 2$

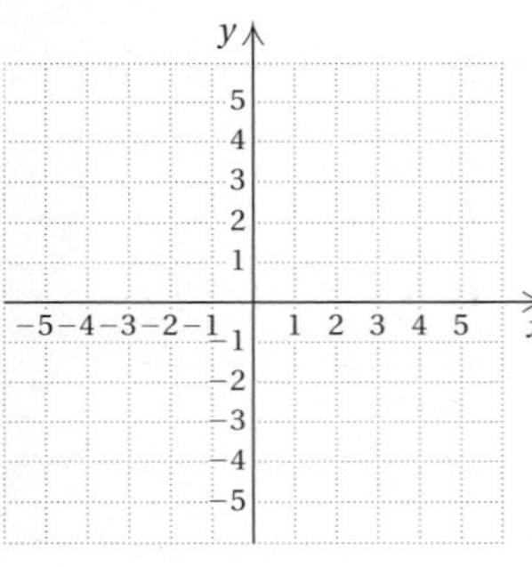

46. $x = 3$

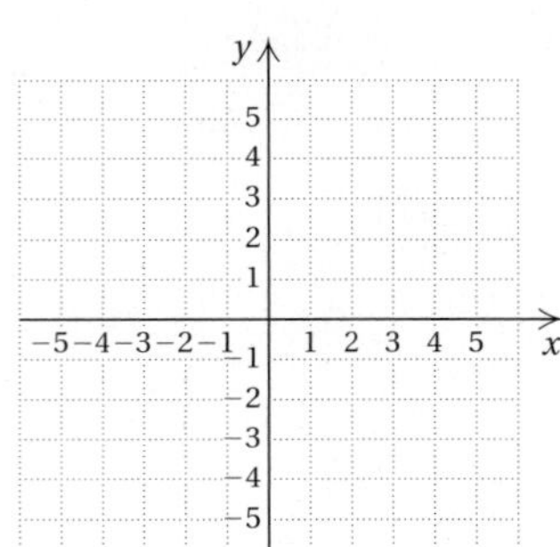

47. $y = 0$

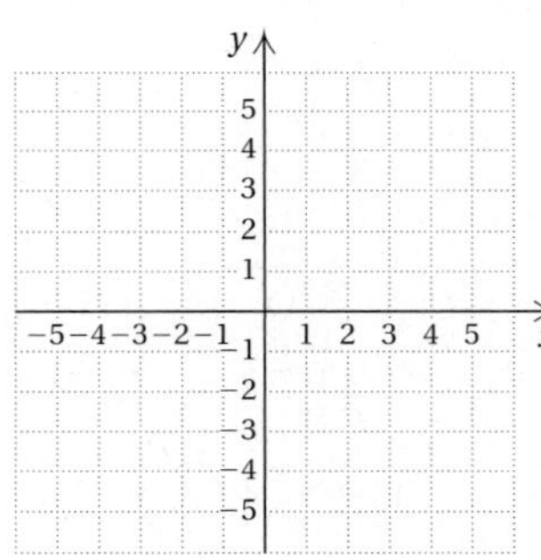

48. $y = -1$

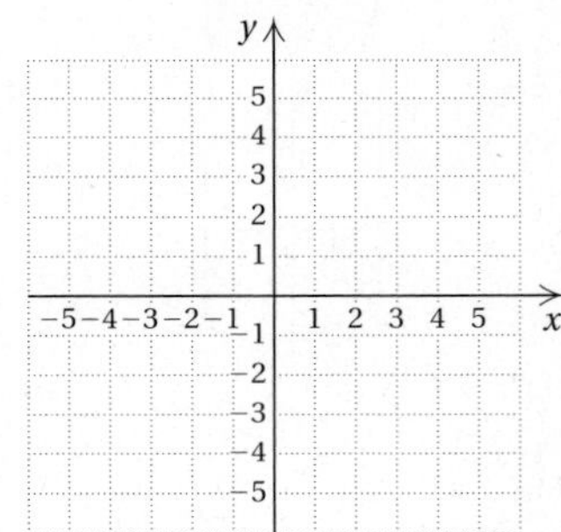

49. $x = \frac{3}{2}$

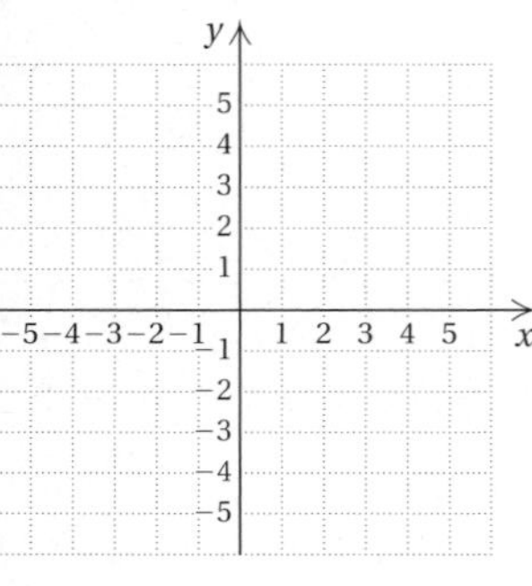

50. $x = -\frac{5}{2}$

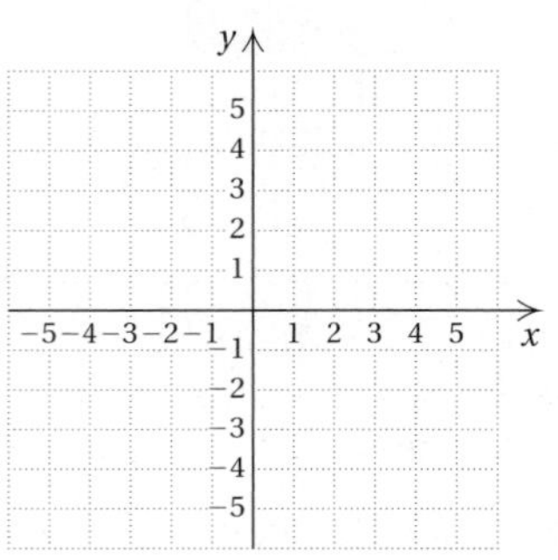

51. $3y = -5$

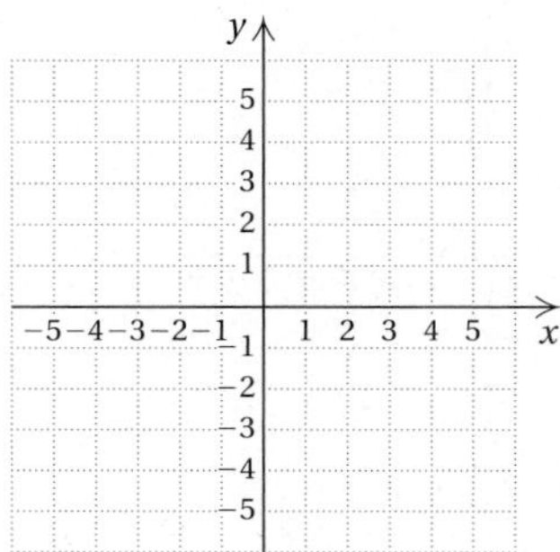

52. $12y = 45$

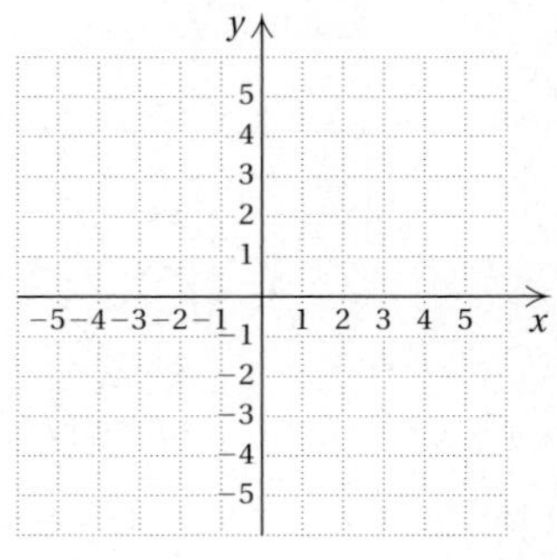

53. $4x + 3 = 0$

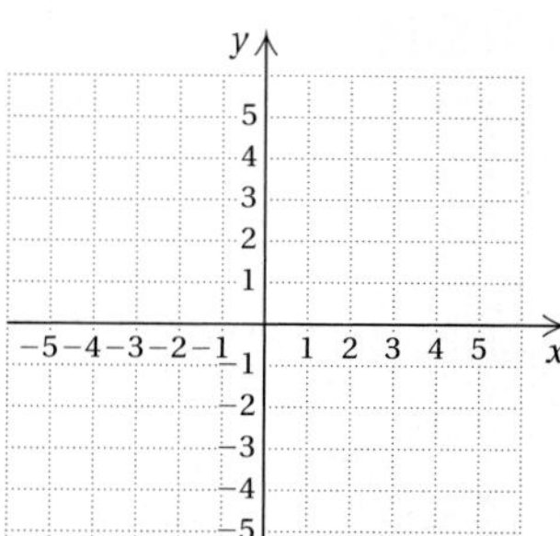

54. $-3x + 12 = 0$

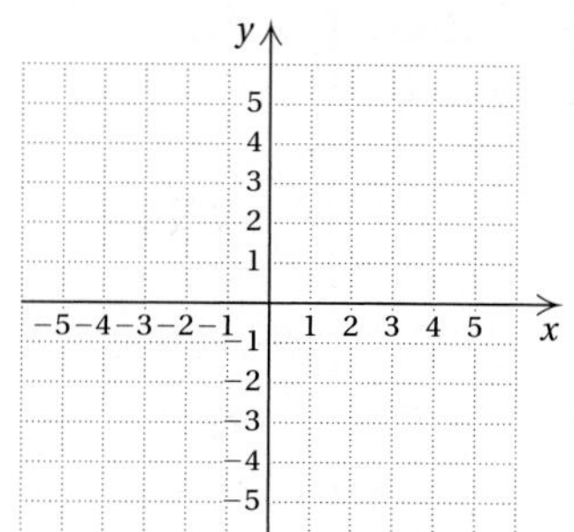

55. $48 - 3y = 0$

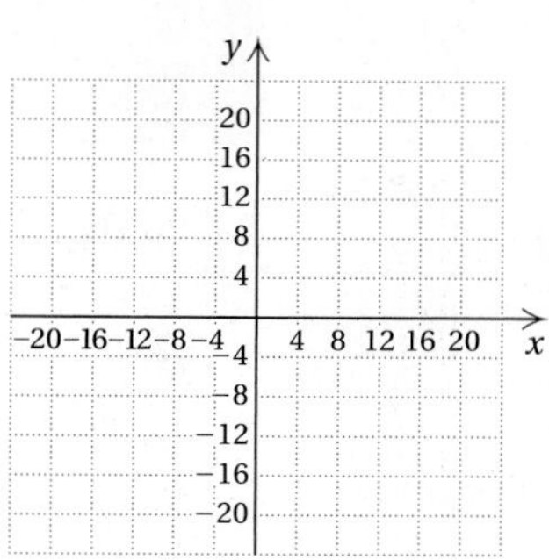

56. $63 + 7y = 0$

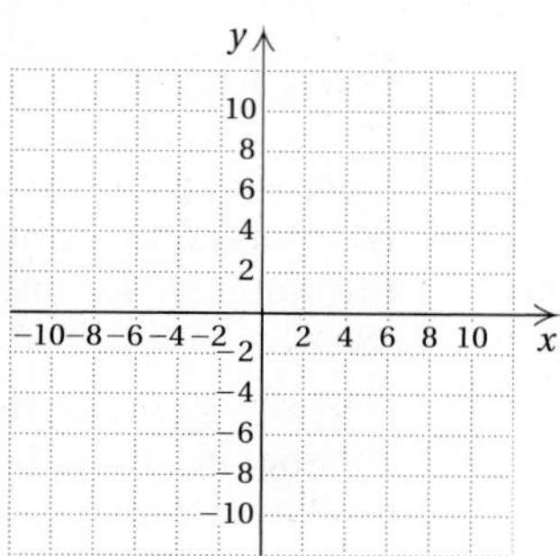

Write an equation for the graph shown.

57.

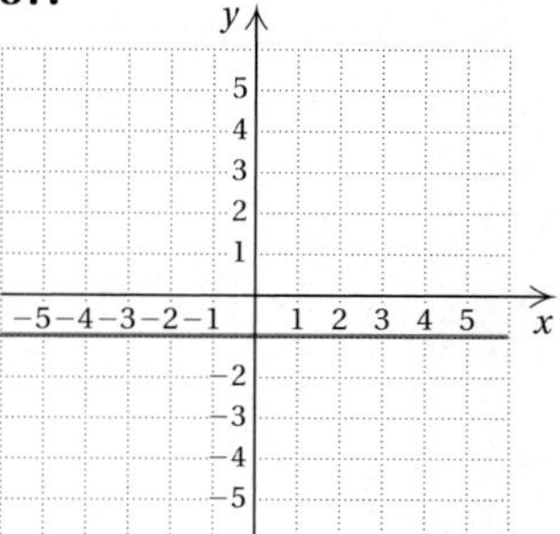

58.

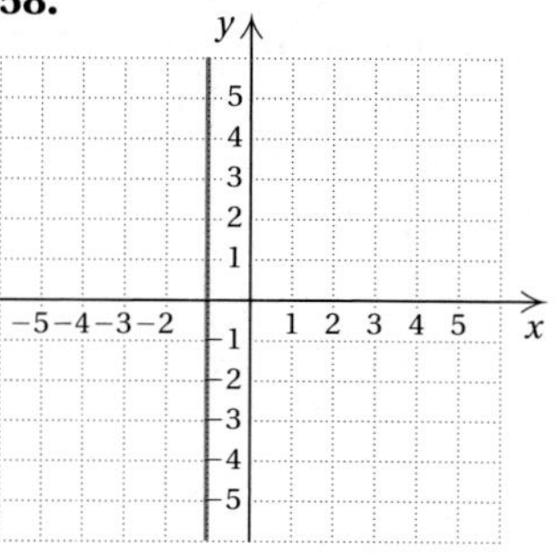

59.

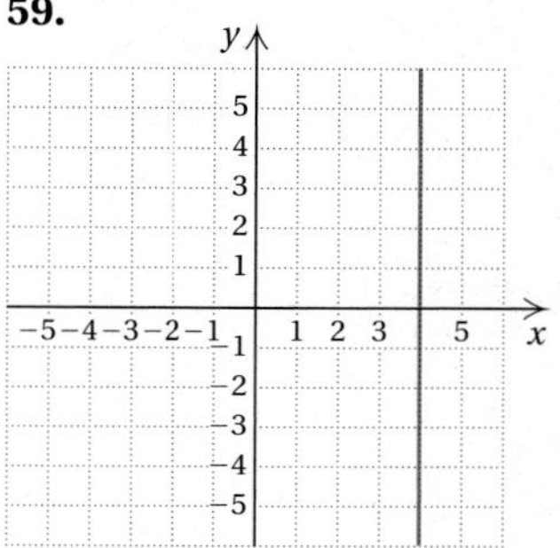

60.

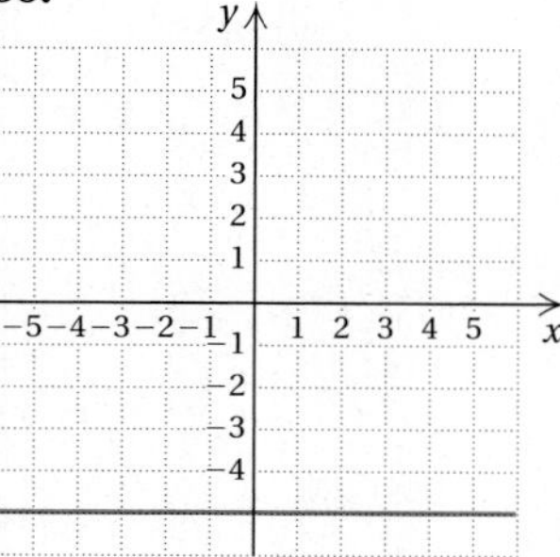

61. $\mathbf{D_W}$ If the graph of the equation $Ax + By = C$ is a horizontal line, what can you conclude about A? Why?

62. $\mathbf{D_W}$ Explain in your own words why the graph of $x = 7$ is a vertical line.

SKILL MAINTENANCE

Solve. [2.5a]

63. *Desserts.* If a restaurant sells 250 desserts in an evening, it is typical that 40 of them will be pie. What percent of the desserts sold will be pie?

64. *Tipping.* Harry left a 20% tip of \$6.50 for a meal. What was the cost of the meal before the tip?

Solve. [2.7e]

65. $-1.6x < 64$

66. $-12x - 71 \geq 13$

67. $x + (x - 1) < (x + 2) - (x + 1)$

68. $6 - 18x \leq 4 - 12x - 5x$

SYNTHESIS

69. Write an equation of a line parallel to the x-axis and passing through $(-3, -4)$.

70. Find the value of m such that the graph of $y = mx + 6$ has an x-intercept of $(2, 0)$.

71. Find the value of k such that the graph of $3x + k = 5y$ has an x-intercept of $(-4, 0)$.

72. Find the value of k such that the graph of $4x = k - 3y$ has a y-intercept of $(0, -8)$.

Objectives

a Given the coordinates of two points on a line, find the slope of the line, if it exists.

b Find the slope, or rate of change, in an applied problem involving slope.

c Find the slope of a line from an equation.

3.4 SLOPE AND APPLICATIONS

a Slope

We have considered two forms of a linear equation,

$$Ax + By = C \quad \text{and} \quad y = mx + b.$$

We found that from the form of the equation $y = mx + b$, we know certain information—namely, that the y-intercept of the line is $(0, b)$.

$y = mx + b$

? ← (from m) The y-intercept is $(0, b)$. (from b)

What about the constant m? Does it give us certain information about the line? Look at the following graphs and see if you can make any connection between the constant m and the "slant" of the line.

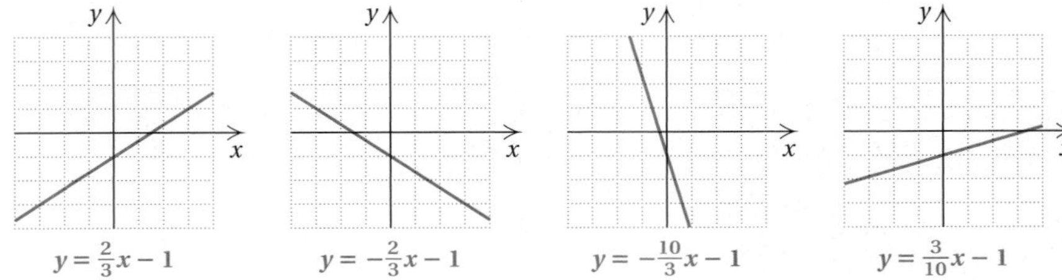

The graphs of some linear equations slant upward from left to right. Others slant downward. Some are vertical and some are horizontal. Some slant more steeply than others. We now look for a way to describe such possibilities with numbers.

Consider a line with two points marked P and Q. As we move from P to Q, the y-coordinate changes from 1 to 3 and the x-coordinate changes from 2 to 6. The change in y is $3 - 1$, or 2. The change in x is $6 - 2$, or 4.

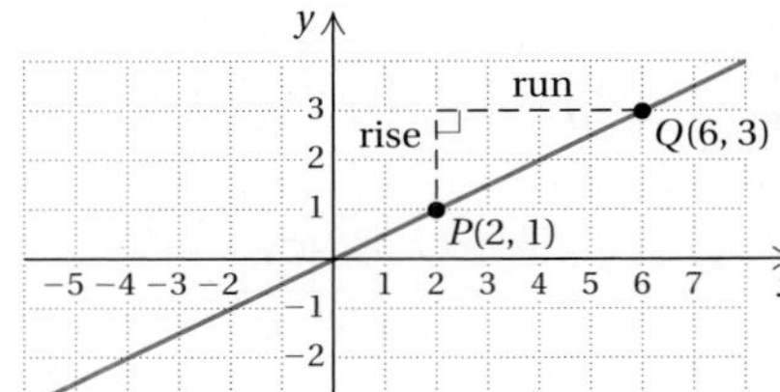

We call the change in y the **rise** and the change in x the **run.** The ratio rise/run is the same for any two points on a line. We call this ratio the **slope.** Slope describes the slant of a line. The slope of the line in the graph above is given by

$$\frac{\text{rise}}{\text{run}} = \frac{\text{the change in } y}{\text{the change in } x}, \text{ or } \frac{2}{4}, \text{ or } \frac{1}{2}.$$

SLOPE

The **slope** of a line containing points (x_1, y_1) and (x_2, y_2) is given by

$$m = \frac{\text{rise}}{\text{run}} = \frac{\text{the change in } y}{\text{the change in } x} = \frac{y_2 - y_1}{x_2 - x_1}.$$

In the preceding definition, (x_1, y_1) and (x_2, y_2)—read "x sub-one, y sub-one and x sub-two, y sub-two"— represent two different points on a line. It does not matter which point is considered (x_1, y_1) and which is considered (x_2, y_2) so long as coordinates are subtracted in the same order in both the numerator and the denominator — for example,

$$\frac{y_2 - y_1}{x_2 - x_1} = \frac{y_1 - y_2}{x_1 - x_2}.$$

EXAMPLE 1 Graph the line containing the points $(-4, 3)$ and $(2, -6)$ and find the slope.

The graph is shown below. We consider (x_1, y_1) to be $(-4, 3)$ and (x_2, y_2) to be $(2, -6)$. From $(-4, 3)$ and $(2, -6)$, we see that the change in y, or the rise, is $-6 - 3$, or -9. The change in x, or the run, is $2 - (-4)$, or 6.

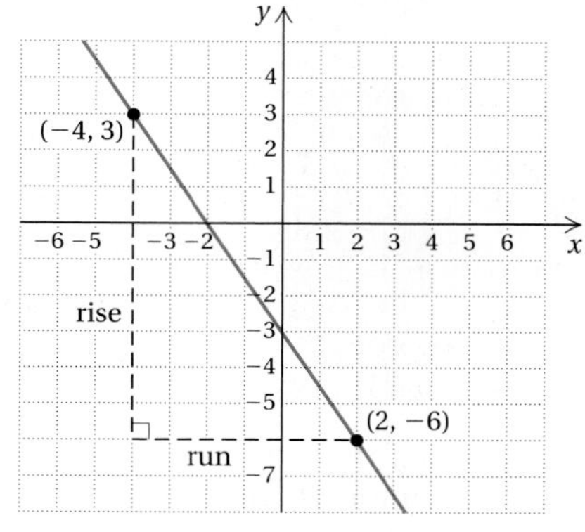

$$\text{Slope} = \frac{\text{rise}}{\text{run}} = \frac{\text{change in } y}{\text{change in } x}$$
$$= \frac{y_2 - y_1}{x_2 - x_1}$$
$$= \frac{-6 - 3}{2 - (-4)}$$
$$= \frac{-9}{6} = -\frac{9}{6}, \text{ or } -\frac{3}{2}.$$

When we use the formula

$$m = \frac{y_2 - y_1}{x_2 - x_1},$$

we can subtract in two ways. We must remember, however, to subtract the y-coordinates in the same order that we subtract the x-coordinates. Let's redo Example 1, where we consider (x_1, y_1) to be $(2, -6)$ and (x_2, y_2) to be $(-4, 3)$:

$$\text{Slope} = \frac{\text{change in } y}{\text{change in } x} = \frac{3 - (-6)}{-4 - 2} = \frac{9}{-6} = -\frac{3}{2}.$$

The slope of a line tells how it slants. A line with positive slope slants up from left to right. The larger the slope, the steeper the slant. A line with negative slope slants downward from left to right.

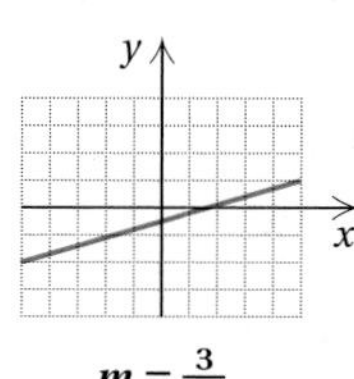

$m = \frac{3}{10}$

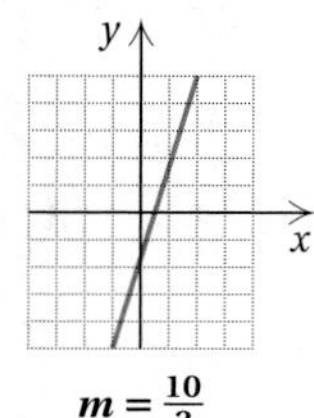

$m = \frac{10}{3}$

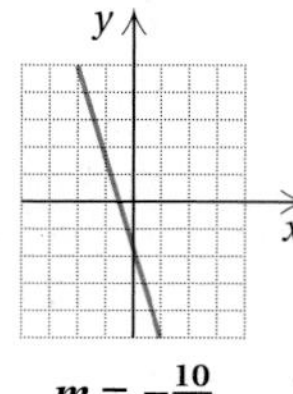

$m = -\frac{10}{3}$

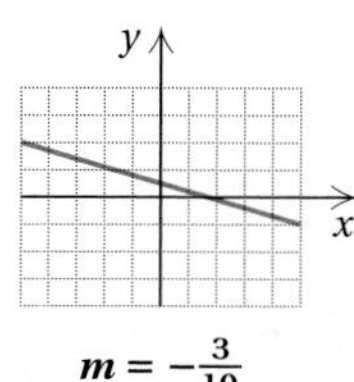

$m = -\frac{3}{10}$

Do Exercises 1 and 2.

Graph the line containing the points and find the slope in two different ways.

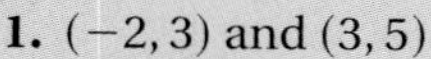

1. $(-2, 3)$ and $(3, 5)$

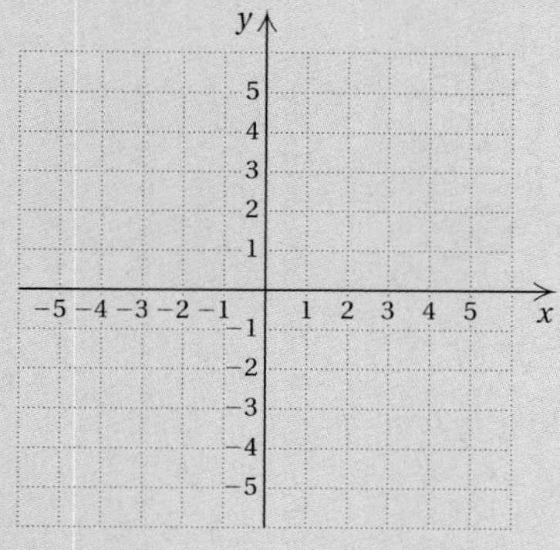

2. $(0, -3)$ and $(-3, 2)$

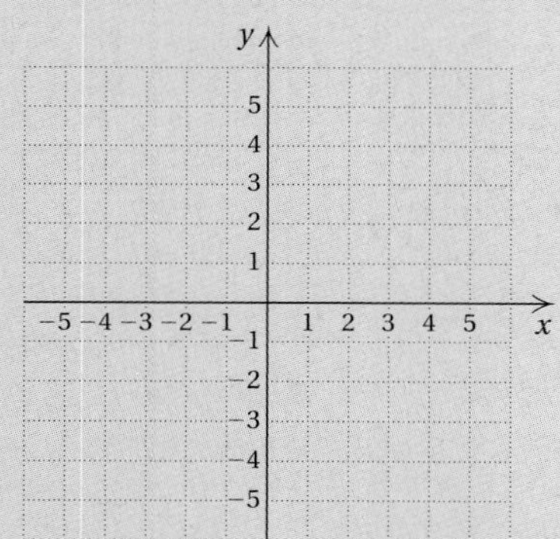

Answers on page A-15

3. **Construction.** Public buildings regularly include steps with 7-in. risers and 11-in. treads. Find the grade of such a stairway.

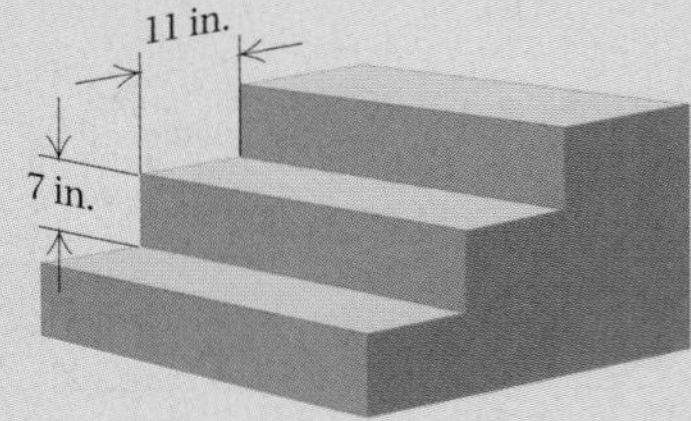

Answer on page A-15

b Applications of Slope; Rates of Change

Slope has many real-world applications. For example, numbers like 2%, 3%, and 6% are often used to represent the *grade* of a road, a measure of how steep a road on a hill or mountain is. For example, a 3% grade $\left(3\% = \frac{3}{100}\right)$ means that for every horizontal distance of 100 ft, the road rises 3 ft, and a −3% grade means that for every horizontal distance of 100 ft, the road drops 3 ft. (Road signs do not include negative signs. It's usually obvious whether you are climbing or descending.) The concept of grade also occurs in skiing or snowboarding, where a 4% grade is considered very tame, but a 40% grade is considered extremely steep. And in cardiology, a physician may change the grade of a treadmill to measure its effect on heartbeat.

Architects and carpenters use slope when designing and building stairs, ramps, or roof pitches. Another application occurs in hydrology. When a river flows, the strength or force of the river depends on how far the river falls vertically compared to how far it flows horizontally.

EXAMPLE 2 *Skiing.* Among the steepest skiable terrain in North America, the Headwall on Mount Washington, in New Hampshire, drops 720 ft over a horizontal distance of 900 ft. Find the grade of the Headwall.

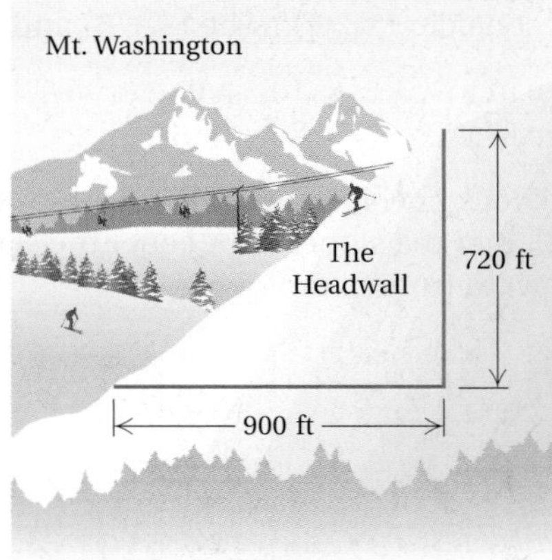

The grade of the Headwall is its slope, expressed as a percent:

$$m = \frac{720}{900} \begin{matrix}\longleftarrow \text{Vertical change} \\ \longleftarrow \text{Horizontal change}\end{matrix}$$

$$= \frac{8}{10} = 80\%.$$

Do Exercise 3.

Slope can also be considered as a **rate of change.**

EXAMPLE 3 *Haircutting.* Gary's Barber Shop has a graph displaying data from a recent day's work. Use the graph to determine the slope, or the rate of change, of the number of haircuts with respect to time.

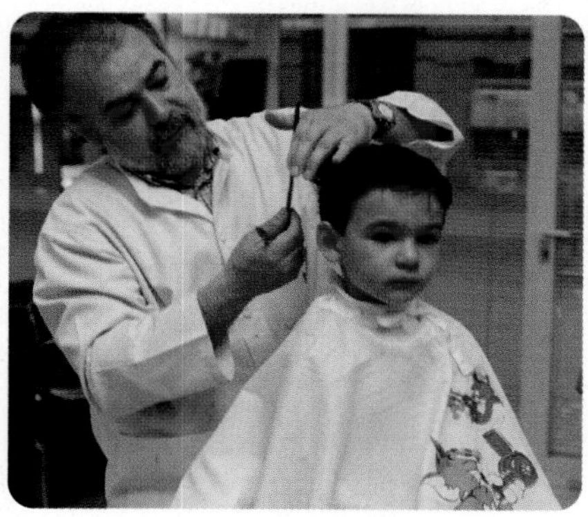

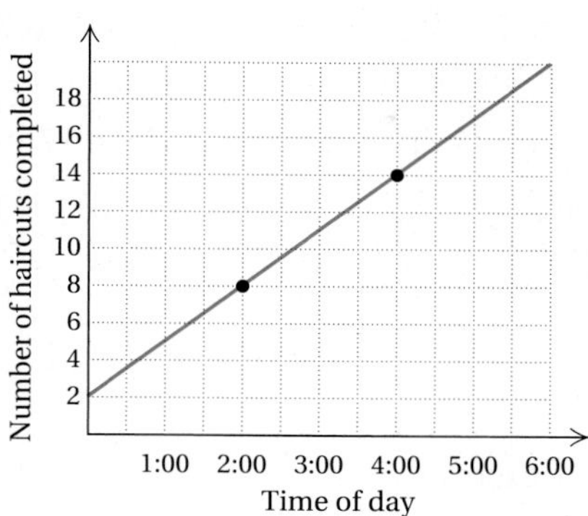

The vertical axis of the graph shows the number of haircuts and the horizontal axis the time, in units of one hour. We can describe the rate of change in the number of haircuts with respect to time as

$$\frac{\text{Haircuts}}{\text{Hour}}, \quad \text{or} \quad \textit{number of haircuts per hour.}$$

This value is the slope of the line. We determine two ordered pairs on the graph—in this case,

(2:00, 8 haircuts) and (4:00, 14 haircuts).

This tells us that in the 2 hr between 2:00 and 4:00, 6 haircuts were completed. Thus,

$$\text{Rate of change} = \frac{14 \text{ haircuts} - 8 \text{ haircuts}}{4{:}00 - 2{:}00} = \frac{6 \text{ haircuts}}{2 \text{ hours}} = 3 \text{ haircuts per hour.}$$

Do Exercise 4.

EXAMPLE 4 *Defense Spending.* Each year the United States spends a smaller percent of its annual budget on defense. Use the following graph to determine the slope, or rate of change, of the percent of the budget spent on defense with respect to time.

Source: U.S. Office of Management and Budget

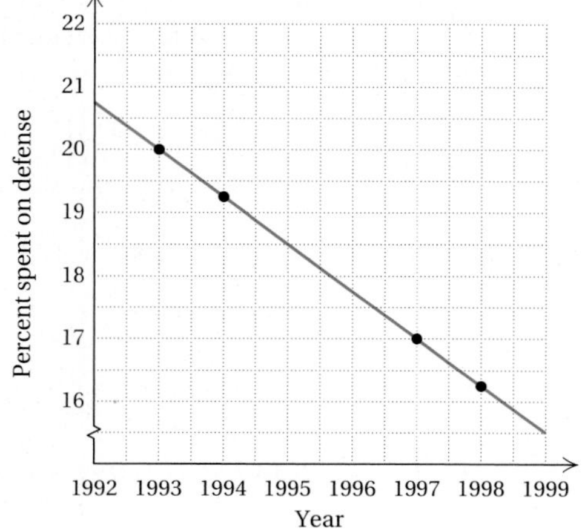

4. Cost of a Telephone Call. The following graph shows data concerning recent MCI phone calls from San Francisco to Pittsburgh. At what rate was the customer being billed?

Source: MCI

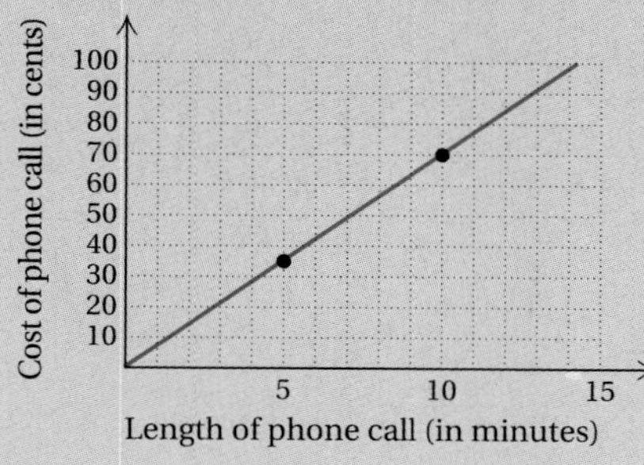

Answer on page A-15

5. Unemployment. Find the rate of change in the percent of U.S. workers that are unemployed.
Source: U.S. Bureau of Labor Statistics

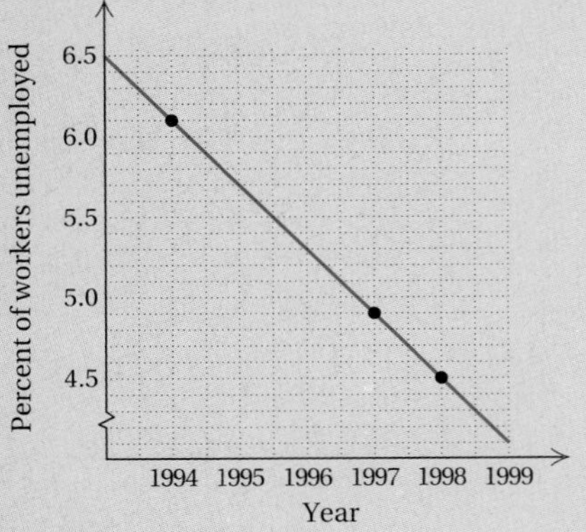

Answer on page A-15

The vertical axis of the graph shows the percent of the budget spent on defense and the horizontal axis the years. We can describe the rate of change in the percent spent with respect to time as

$$\frac{\text{Percent spent}}{\text{Years}}, \quad \text{or} \quad \textit{percent spent per year.}$$

This value is the slope of the line. We determine two ordered pairs on the graph—in this case,

$$(1993, 20) \quad \text{and} \quad (1997, 17).$$

This tells us that in the 4 yr from 1993 to 1997, the percent spent dropped from 20% to 17%. Thus,

$$\text{Rate of change} = \frac{17\% - 20\%}{1997 - 1993} = \frac{-3\%}{4 \text{ yr}} = -\tfrac{3}{4}\% \text{ per year.}$$

Do Exercise 5.

Study Tips

SMALL STEPS LEAD TO GREAT SUCCESS (PART 2)

Chris Widener is a popular motivational speaker and writer. In his article "A Little Equation That Creates Big Results," he proposes the following equation: "Your Short-Term Actions Multiplied By Time = Your Long-Term Accomplishments."

Think of the major or career toward which you are working as a long-term accomplishment. We (your authors and instructors) are at a point in life where we realize the long-term benefits of learning mathematics. For you as students, it may be more difficult to see those long-term results. But make an effort to do so.

Widener goes on to say, "We need to take action on our dreams and beliefs every day." Think of the long-term goal as you do the short-term tasks of homework in math, studying for tests, and completing this course so you can move on to what it prepares you for.

Who writes best-selling novels? The person who only dreams of becoming a best-selling author or the one who also spends 4 hours a day doing research and working at a computer?

Who loses weight? The person who thinks about being thin or the one who also plans a healthy diet and runs 3 miles a day?

Who is successful at math? The person who only knows all the benefits of math or the one who also spends 2 hours studying outside of class for every hour spent inside class?

"*The purpose of man is in action, not thought.*"

Thomas Carlyle, British historian/essayist

"*Prepare for your success in little ways and you will eventually see results in big ways. It's almost magical.*"

Tom Morris, public philosopher/speaker

C Finding the Slope from an Equation

It is possible to find the slope of a line from its equation. Let's consider the equation $y = 2x + 3$, which is in the form $y = mx + b$. We can find two points by choosing convenient values for x—say, 0 and 1—and substituting to find the corresponding y-values. We find the two points on the line to be $(0, 3)$ and $(1, 5)$. The slope of the line is found using the definition of slope:

$$m = \frac{\text{change in } y}{\text{change in } x} = \frac{5-3}{1-0} = \frac{2}{1} = 2.$$

The slope is 2. Note that this is also the coefficient of the x-term in the equation $y = 2x + 3$.

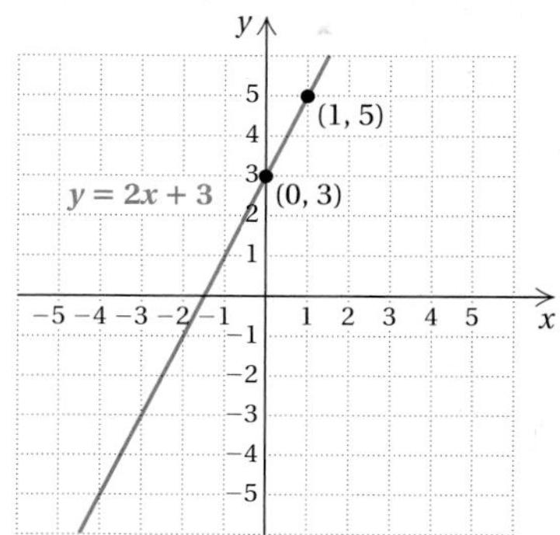

DETERMINING SLOPE FROM THE EQUATION $y = mx + b$

The slope of the line $y = mx + b$ is m. To find the slope of a nonvertical line, solve the linear equation in x and y for y and get the resulting equation in the form $y = mx + b$. The coefficient of the x-term, m, is the slope of the line.

EXAMPLES Find the slope of the line.

5. $y = -3x + \frac{2}{9}$

   $m = -3 =$ Slope

6. $y = \frac{4}{5}x$

   $m = \frac{4}{5} =$ Slope

7. $y = x + 6$

   $m = 1 =$ Slope

8. $y = -0.6x - 3.5$

   $m = -0.6 =$ Slope

Do Exercises 6–9.

To find slope from an equation, we may have to first find an equivalent form of the equation.

EXAMPLE 9 Find the slope of the line $2x + 3y = 7$.

We solve for y to get the equation in the form $y = mx + b$:

$$2x + 3y = 7$$
$$3y = -2x + 7$$
$$y = \frac{-2x + 7}{3}$$
$$y = -\frac{2}{3}x + \frac{7}{3}. \quad \text{This is } y = mx + b.$$

The slope is $-\frac{2}{3}$.

CALCULATOR CORNER

Visualizing Slope

Exercises: Graph each of the following sets of equations using the window settings $[-6, 6, -4, 4]$, Xscl = 1, Yscl = 1.

1. $y = x, y = 2x, y = 5x, y = 10x$

   What do you think the graph of $y = 123x$ will look like?

2. $y = x, y = \frac{3}{4}x, y = 0.38x, y = \frac{5}{32}x$

   What do you think the graph of $y = 0.000043x$ will look like?

Find the slope of the line.

6. $y = 4x + 11$

7. $y = -17x + 8$

8. $y = -x + \frac{1}{2}$

9. $y = \frac{2}{3}x - 1$

Find the slope of the line.

10. $4x + 4y = 7$

11. $5x - 4y = 8$

Answers on page A-15

CALCULATOR CORNER

Visualizing Slope

Exercises: Graph each of the following sets of equations using the window settings $[-6, 6, -4, 4]$, Xscl $= 1$, Yscl $= 1$.

1. $y = -x$, $y = -2x$, $y = -5x$, $y = -10x$

   What do you think the graph of $y = -123x$ will look like?

2. $y = -x$, $y = -\frac{3}{4}x$, $y = -0.38x$, $y = -\frac{5}{32}x$

   What do you think the graph of $y = -0.000043x$ will look like?

Find the slope, if it exists, of the line.

12. $x = 7$

13. $y = -5$

Answers on page A-15

Do Exercises 10 and 11 on the preceding page.

What about the slope of a horizontal or a vertical line?

EXAMPLE 10 Find the slope of the line $y = 5$.

We can think of $y = 5$ as $y = 0x + 5$. Then from this equation, we see that $m = 0$. Consider the points $(-3, 5)$ and $(4, 5)$, which are on the line. The change in $y = 5 - 5$, or 0. The change in $x = -3 - 4$, or -7. We have

$$m = \frac{5 - 5}{-3 - 4} = \frac{0}{-7} = 0.$$

Any two points on a horizontal line have the same y-coordinate. Thus the change in y is 0.

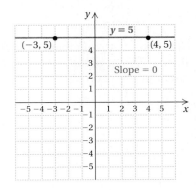

EXAMPLE 11 Find the slope of the line $x = -4$.

Consider the points $(-4, 3)$ and $(-4, -2)$, which are on the line. The change in $y = 3 - (-2)$, or 5. The change in $x = -4 - (-4)$, or 0. We have

$$m = \frac{3 - (-2)}{-4 - (-4)} = \frac{5}{0}. \quad \text{Not defined}$$

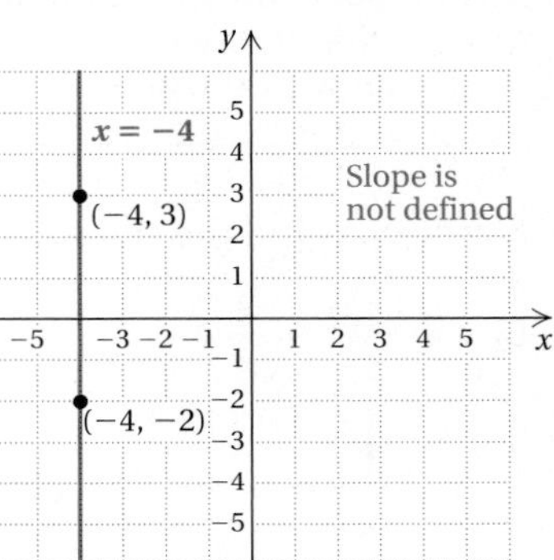

Since division by 0 is not defined, the slope of this line is not defined. The answer in this example is "The slope of this line is not defined."

SLOPE 0; SLOPE NOT DEFINED

The slope of a horizontal line is 0. The slope of a vertical line is not defined.

Do Exercises 12 and 13.

We will consider slope again and use it in graphing in Chapter 7.

3.4 EXERCISE SET

For Extra Help

Digital Video Tutor CD 4 Videotape 7 | InterAct Math | Math Tutor Center | MathXL | MyMathLab

a Find the slope, if it exists, of the line.

1.

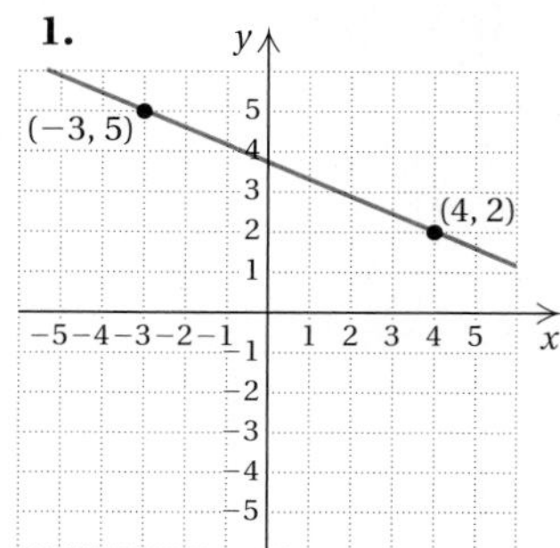

2.

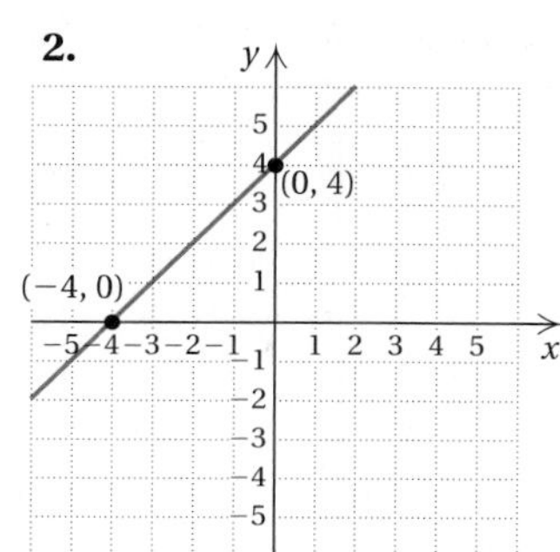

3.

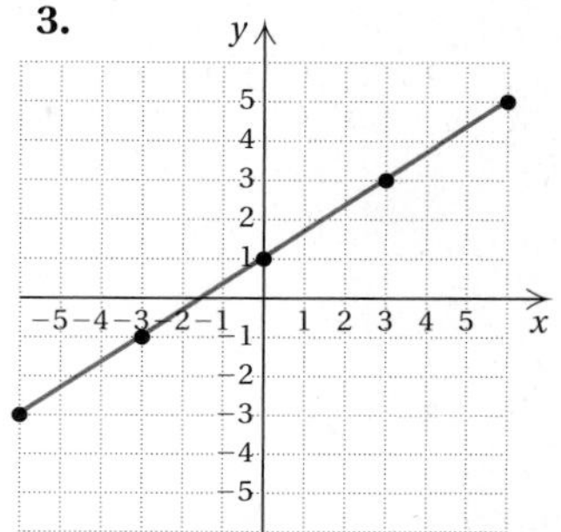

4.

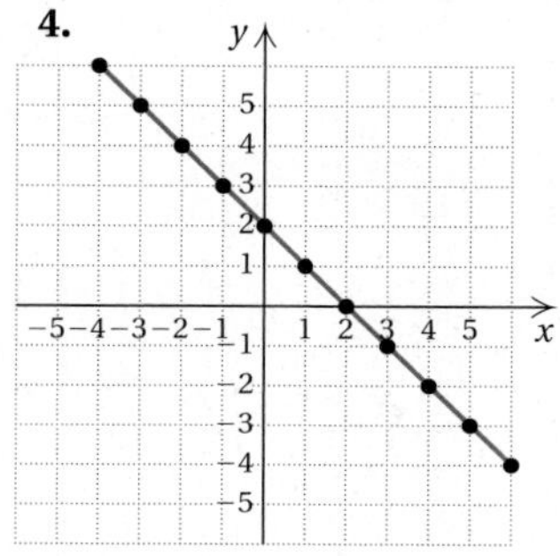

5.

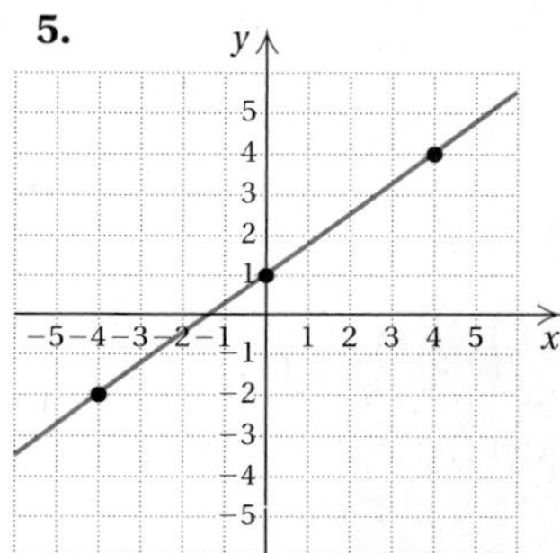

6.

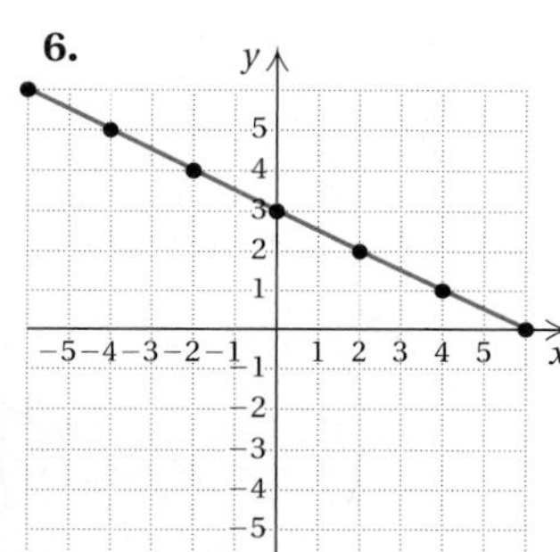

7.

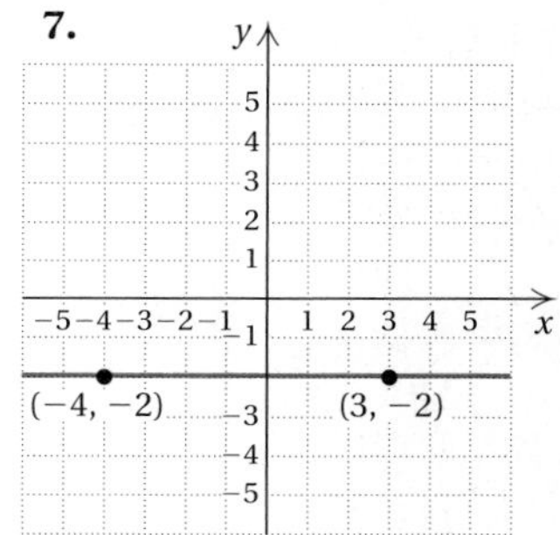

8.

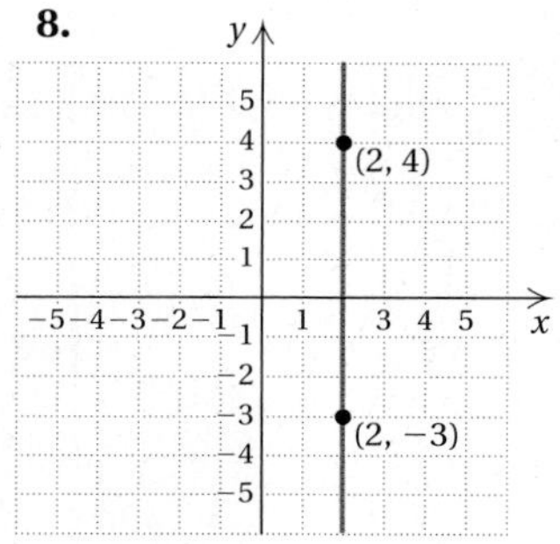

Graph the line containing the given pair of points and find the slope.

9. $(-2, 4), (3, 0)$

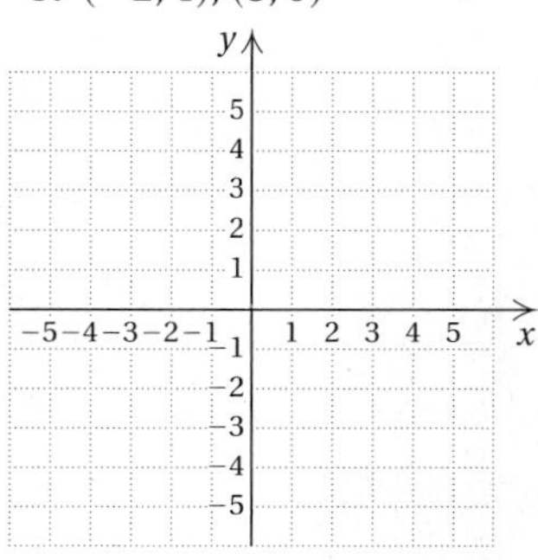

10. $(2, -4), (-3, 2)$

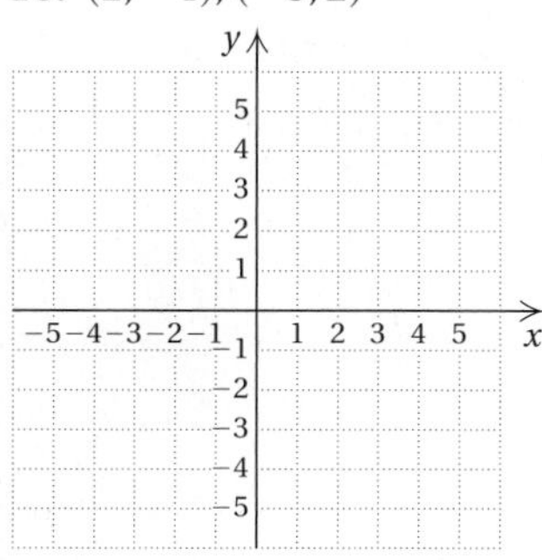

11. $(-4, 0), (-5, -3)$

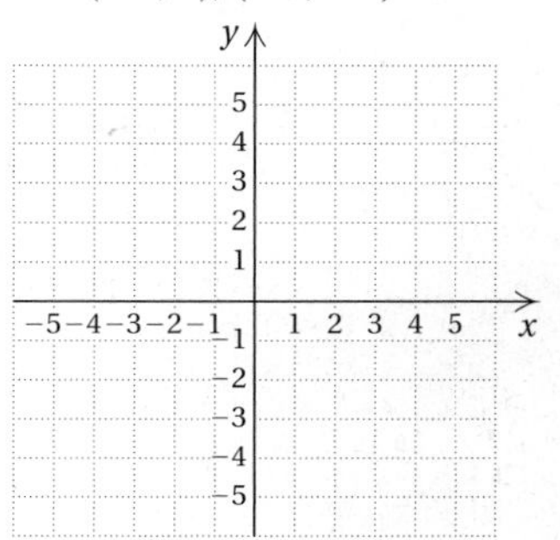

12. $(-3, 0), (-5, -2)$

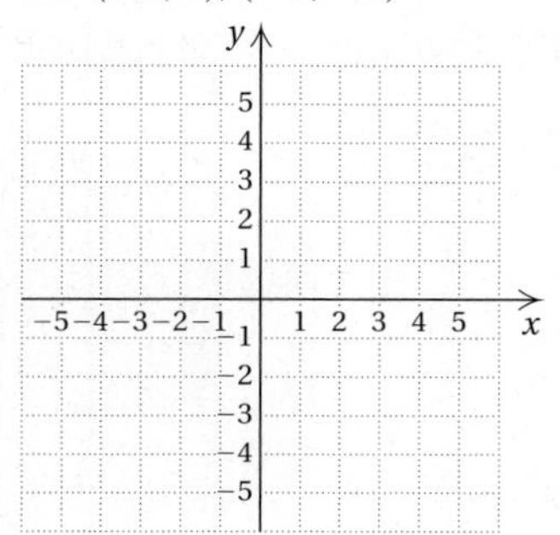

13. $(-4, 2), (2, -3)$

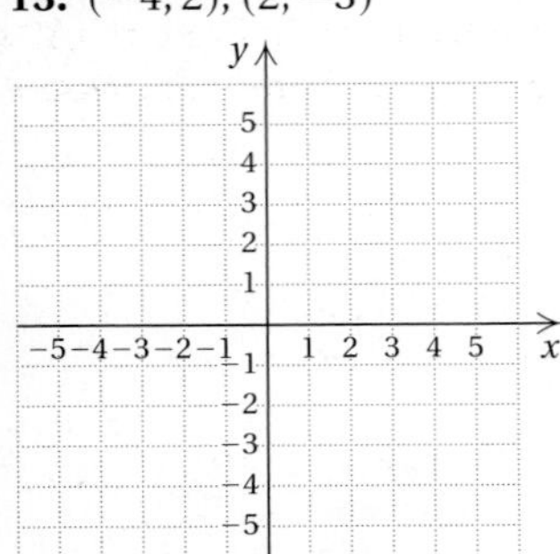

14. $(-3, 5), (4, -3)$

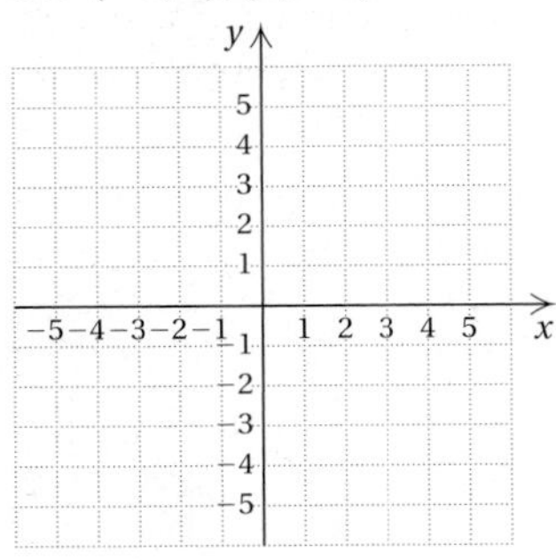

15. $(5, 3), (-3, -4)$

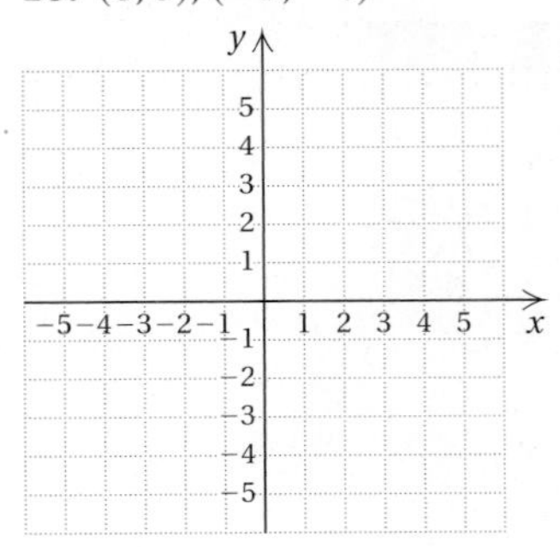

16. $(-4, -3), (2, 5)$

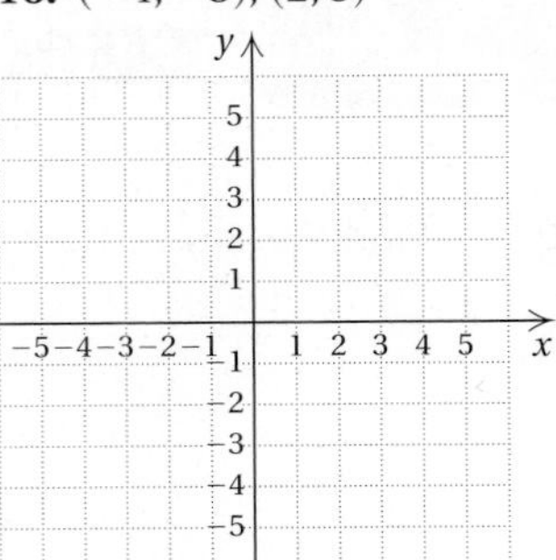

Find the slope, if it exists, of the line containing the given pair of points.

17. $\left(2, -\frac{1}{2}\right), \left(5, \frac{3}{2}\right)$

18. $\left(\frac{2}{3}, -1\right), \left(\frac{5}{3}, 2\right)$

19. $(4, -2), (4, 3)$

20. $(4, -3), (-2, -3)$

b In Exercises 21–24, find the slope (or rate of change).

21. Find the slope (or pitch) of the roof.

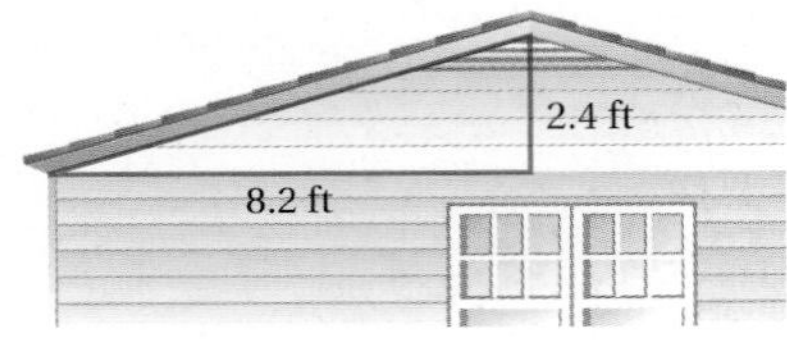

22. Find the slope (or grade) of the road.

23. Find the slope of the river.

24. Find the slope of the treadmill.

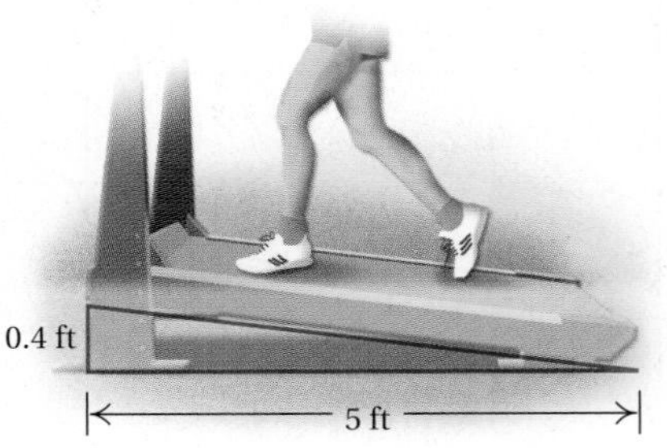

25. *Slope of Long's Peak.* From a base elevation of 9600 ft, Long's Peak in Colorado rises to a summit elevation of 14,255 ft over a horizontal distance of 15,840 ft. Find the grade of Long's Peak.

26. *Ramps for the Disabled.* In order to meet federal standards, a wheelchair ramp must not rise more than 1 ft over a horizontal distance of 12 ft. Express this slope as a grade.

In Exercises 27–32, use the graph to calculate a rate of change in which the units of the horizontal axis are used in the denominator.

27. *Gas Mileage.* The following graph shows data for a Honda Odyssey driven on interstate highways. Find the rate of change in miles per gallon, that is, the gas mileage.
Source: Honda Motor Company

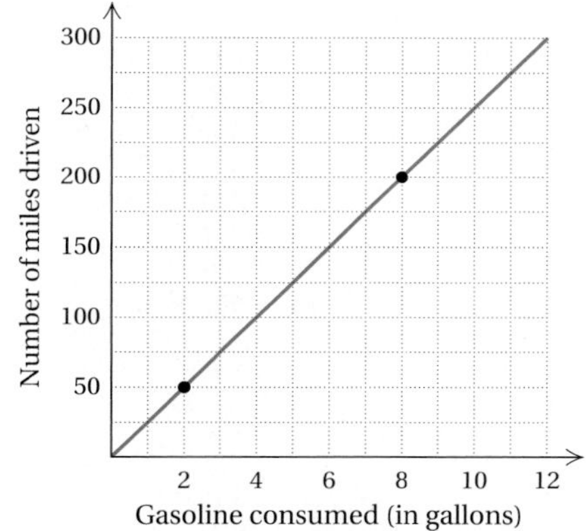

28. *Hairdresser.* Eve's Custom Cuts has a graph displaying data from a recent day of work. Find the rate of change of the number of haircuts with respect to time.

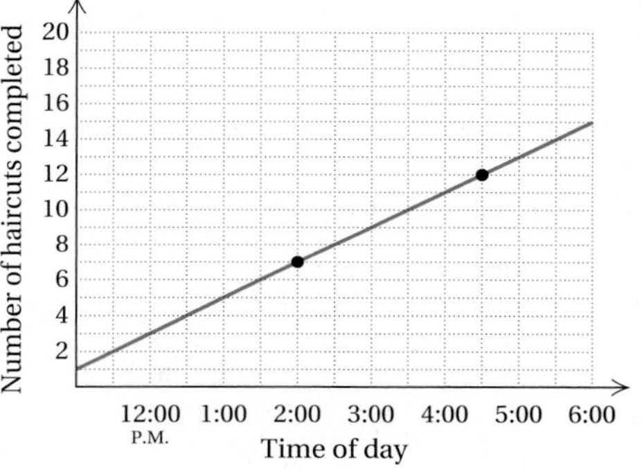

29. *Depreciation of an Office Machine.* The value of a particular color copier is represented in the following graph. Find the rate of change of the value with respect to time, in dollars per year.

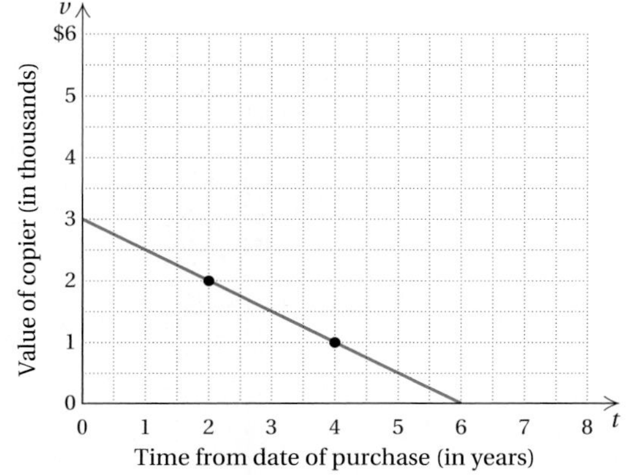

30. *NASA Spending.* The amount spent by the National Aeronautics and Space Administration (NASA) is represented in the following graph. Find the rate of change of spending with respect to time, in dollars per year.
Source: NASA

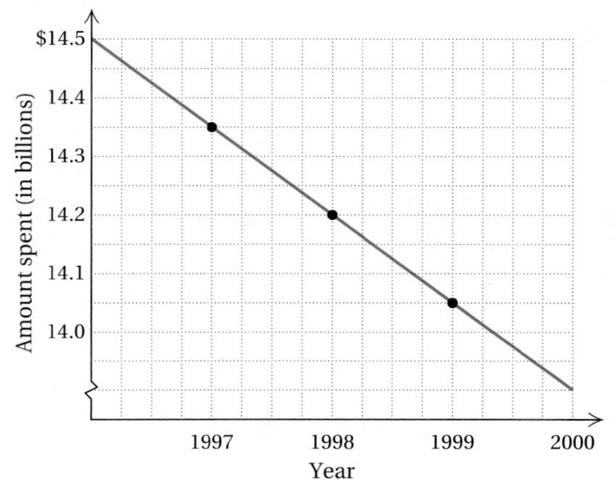

31. *Population Growth of Alaska.* The population of Alaska is shown in the following graph. Find the rate of change of the population with respect to time, in number of people per year.

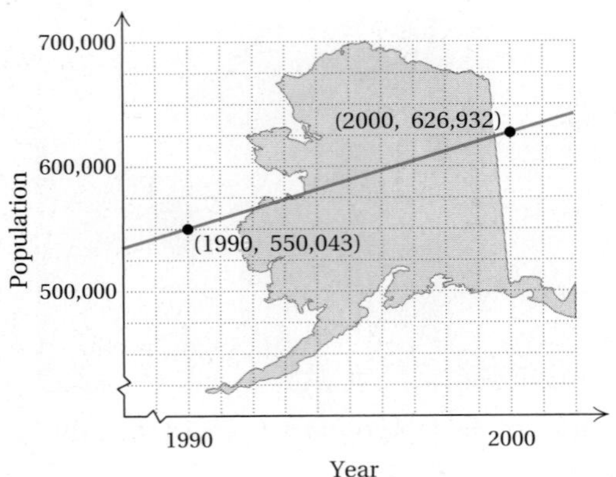

32. *Population Growth of Utah.* The population of Utah is shown in the following graph. Find the rate of change of the population, in number of people per year.

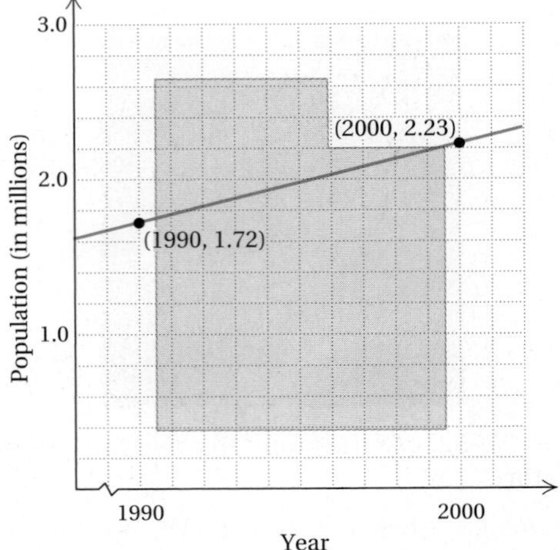

C Find the slope, if it exists.

33. $y = -10x + 7$

34. $y = \frac{10}{3}x - \frac{5}{7}$

35. $y = 3.78x - 4$

36. $y = -\frac{3}{5}x + 28$

37. $3x - y = 4$

38. $-2x + y = 8$

39. $x + 5y = 10$

40. $x - 4y = 8$

41. $3x + 2y = 6$

42. $2x - 4y = 8$

43. $5x - 7y = 14$

44. $3x - 6y = 10$

45. $y = -2.74x$

46. $y = \frac{219}{298}x - 6.7$

47. $9x = 3y + 5$

48. $4y = 9x - 7$

49. $5x - 4y + 12 = 0$

50. $16 + 2x - 8y = 0$

51. $y = 4$

52. $x = -3$

Assuming that the scales on each axis of each graph are the same, explain how you can estimate the slope of the line that contains segment *PQ* without knowing the coordinates of the points *P* and *Q*.

53. $\mathbf{D_W}$

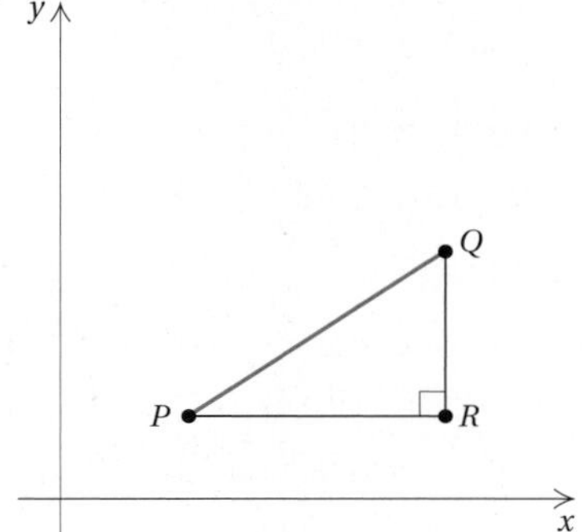

54. $\mathbf{D_W}$

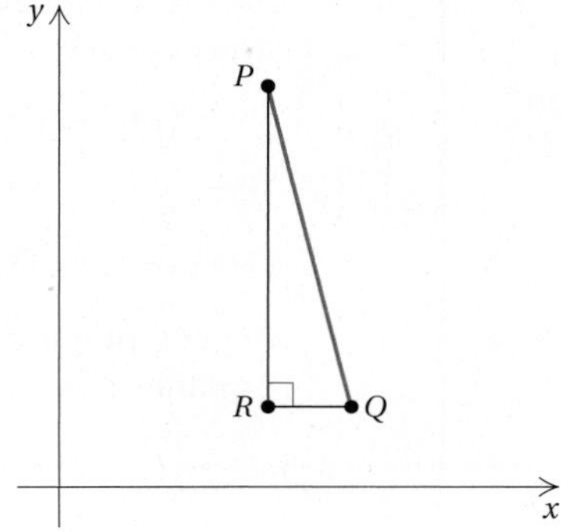

SKILL MAINTENANCE

Convert to fraction notation. [R.4b]

55. 16%

56. $33\frac{1}{3}\%$

57. 37.5%

58. 75%

Solve. [2.5a]

59. What is 15% of $23.80?

60. $7.29 is 15% of what number?

61. Jennifer left an $8.50 tip for a meal that cost $42.50. What percent of the cost of the meal was the tip?

62. Kristen left an 18% tip of $3.24 for a meal. What was the cost of the meal before the tip?

63. Juan left a 15% tip for a meal. The total cost of the meal, including the tip, was $51.92. What was the cost of the meal before the tip was added?

64. After a 25% reduction, a sweater is on sale for $41.25. What was the original price?

SYNTHESIS

Graph the equation using the standard viewing window. Then construct a table of *y*-values for *x*-values starting at $x = -10$ with $\Delta\text{Tbl} = 0.1$.

65. $y = 0.35x - 7$

66. $y = 5.6 - x^2$

67. $y = x^3 - 5$

68. $y = 4 + 3x - x^2$

CHAPTER 3 Summary and Review

The review that follows is meant to prepare you for a chapter exam. It consists of two parts. The first part is a checklist of some of the Study Tips referred to in this and preceding chapters. The second part is the Review Exercises. These provide practice exercises for the exam, together with references to section objectives so you can go back and review. Before beginning, stop and look back over the skills you have obtained. What skills in mathematics do you have now that you did not have before studying this chapter?

STUDY TIPS CHECKLIST

The foundation of all your study skills is TIME!

- [ ] Are you doing exercises without answers at the back of the book as part of every homework assignment to better prepare you to take tests?
- [ ] Have you been taking the primary responsibility for your learning?
- [ ] Have you established a learning relationship with your instructor?
- [ ] Have you used the videotapes to supplement your learning?
- [ ] Are you preparing for each homework assignment by reading the explanations and following the step-by-step examples in the text?

REVIEW EXERCISES

1. *Federal Spending.* The following pie chart shows how our federal income tax dollars are used. As a freelance graphic artist, Jennifer pays $3525 in taxes. How much of Jennifer's tax payment goes toward defense? toward social programs? [3.1a]

Where Your Tax Dollars Are Spent

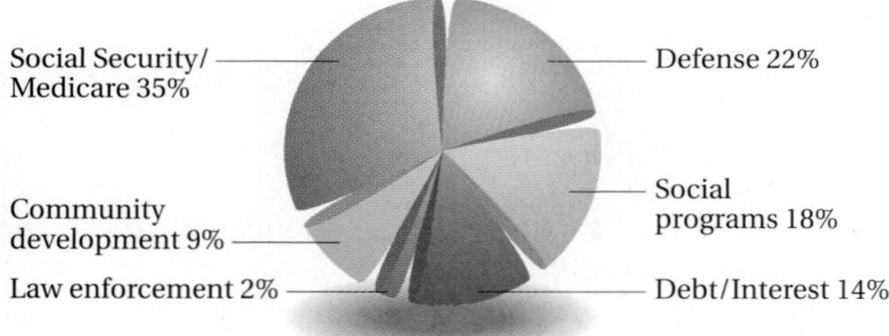

Source: U.S. Department of the Treasury

Chicken Consumption. The following line graph shows average chicken consumption from 1980 to 2000. Use the line graph for Exercises 2–6. [3.1a]

Chicken Consumption

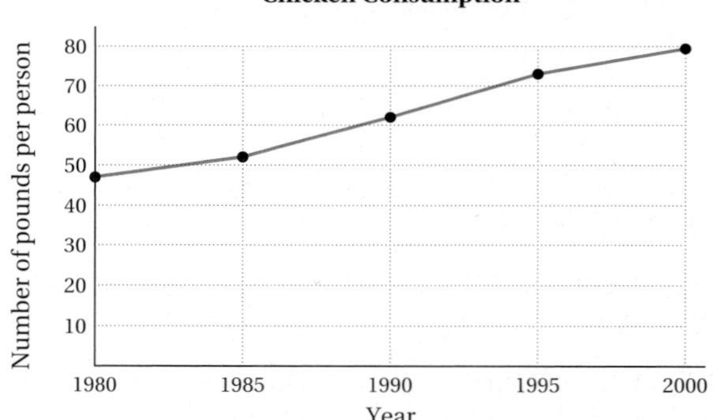

2. About how many pounds of chicken were consumed per person in 1980?
3. About how many pounds of chicken were consumed per person in 2000?
4. By what amount did chicken consumption increase from 1980 to 2000?
5. In what year did the consumption of chicken exceed 70 lb per person?
6. In what 5-yr period was the difference in consumption the greatest?

Water Usage. The following bar graph shows water usage, in gallons, for various tasks. Use the bar graph for Exercises 7–10. [3.1a]

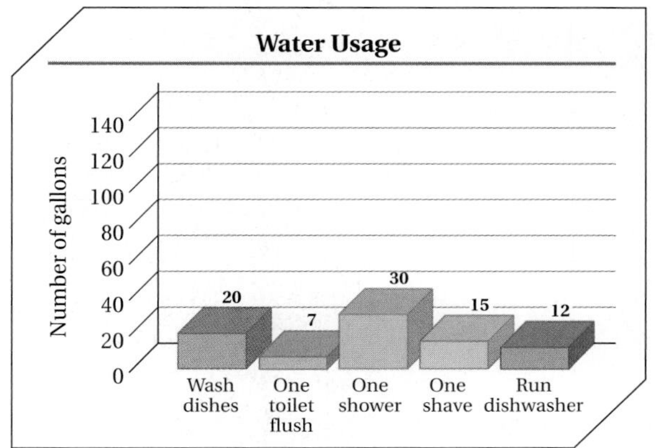

Source: American Water Works Association

7. Which task requires the most water?

8. Which task requires the least water?

9. Which tasks require 15 or more gallons?

10. Which task requires 7 gallons?

Find the coordinates of the point. [3.1d]

11. A **12.** B **13.** C

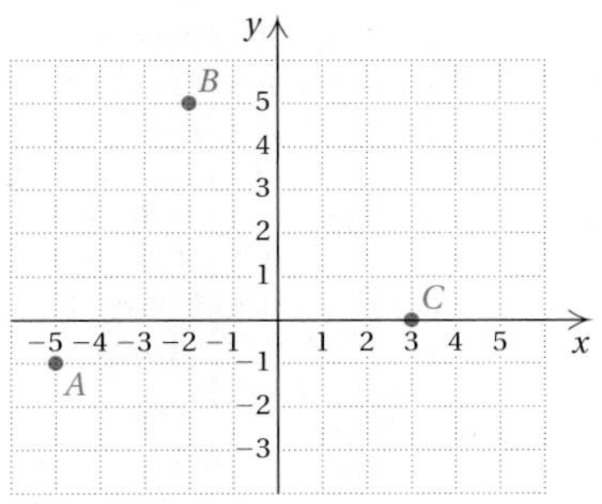

Plot the point. [3.1b]

14. $(2, 5)$ **15.** $(0, -3)$ **16.** $(-4, -2)$

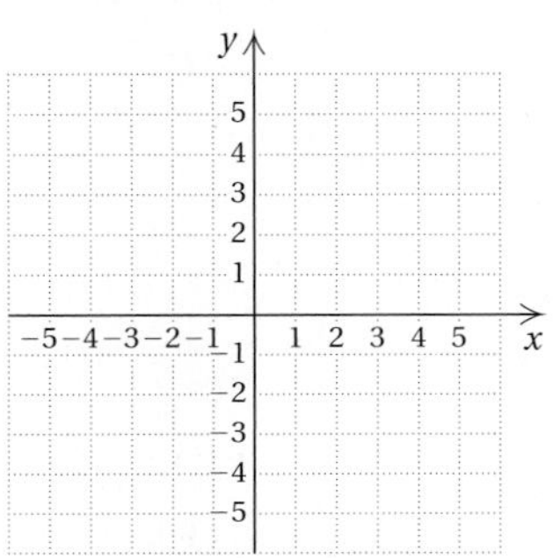

In which quadrant is the point located? [3.1c]

17. $(3, -8)$ **18.** $(-20, -14)$ **19.** $(4.9, 1.3)$

Determine whether the ordered pair is a solution of $2y - x = 10$. [3.2a]

20. $(2, -6)$ **21.** $(0, 5)$

22. Show that the ordered pairs $(0, -3)$ and $(2, 1)$ are solutions of the equation $2x - y = 3$. Then use the graph of the two points to determine another solution. Answers may vary. [3.2a]

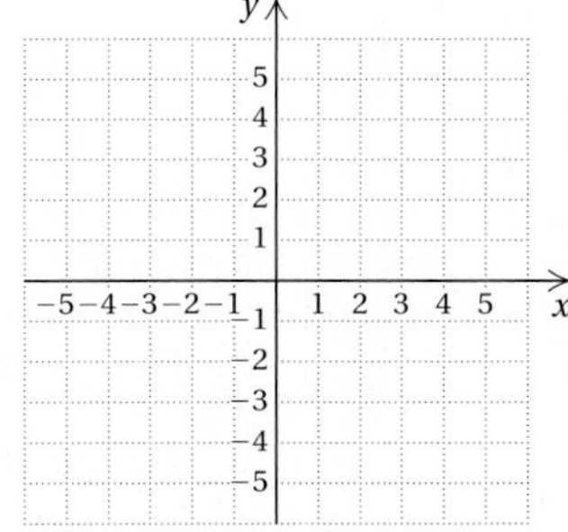

Graph the equation, identifying the y-intercept. [3.2b]

23. $y = 2x - 5$

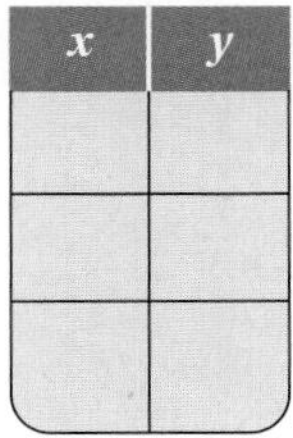

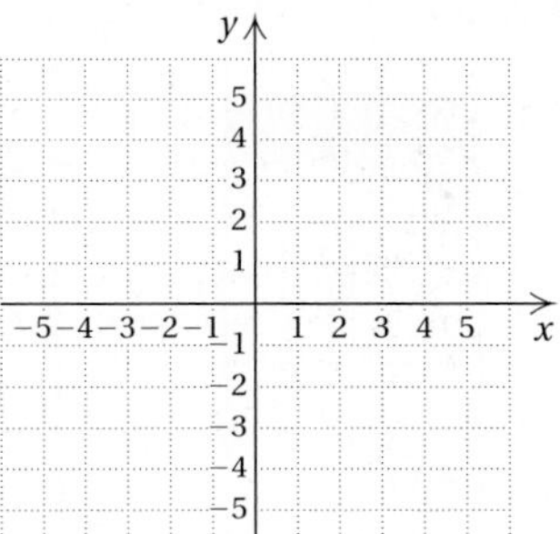

24. $y = -\frac{3}{4}x$

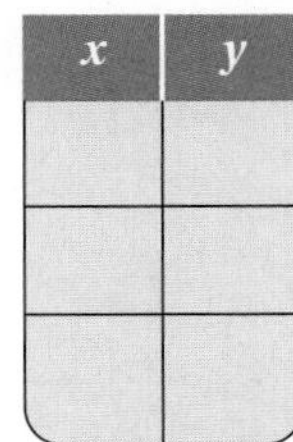

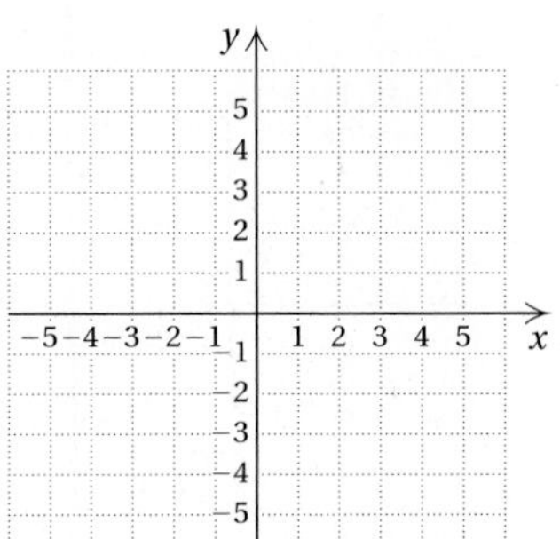

25. $y = -x + 4$

x	y

26. $y = 3 - 4x$

x	y

Graph the equation. [3.3b]

27. $y = 3$

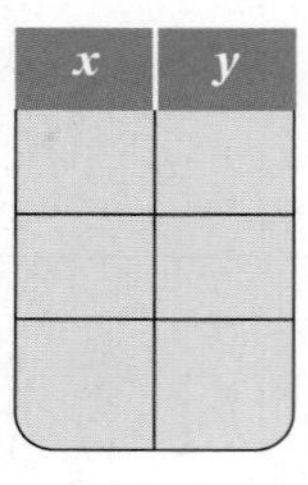

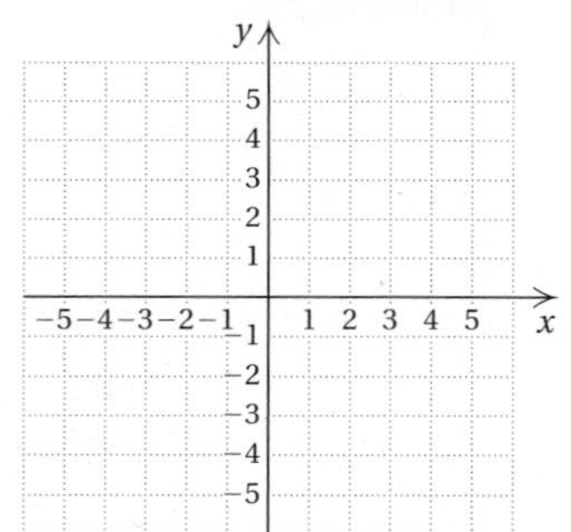

28. $5x - 4 = 0$

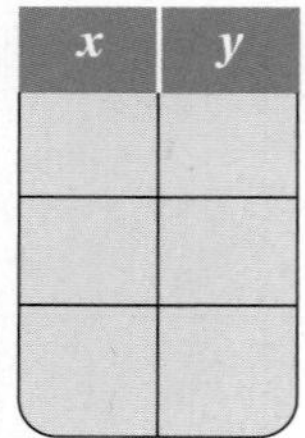

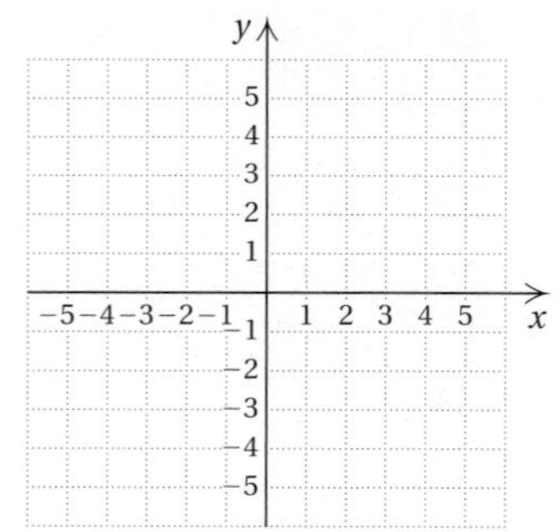

Find the intercepts of the equation. Then graph the equation. [3.3a]

29. $x - 2y = 6$

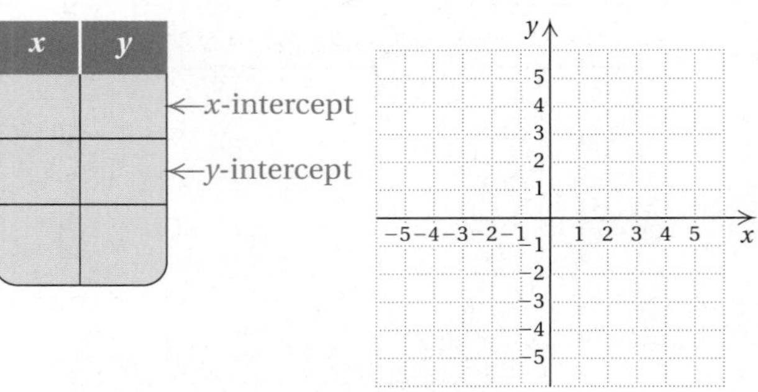

30. $5x - 2y = 10$

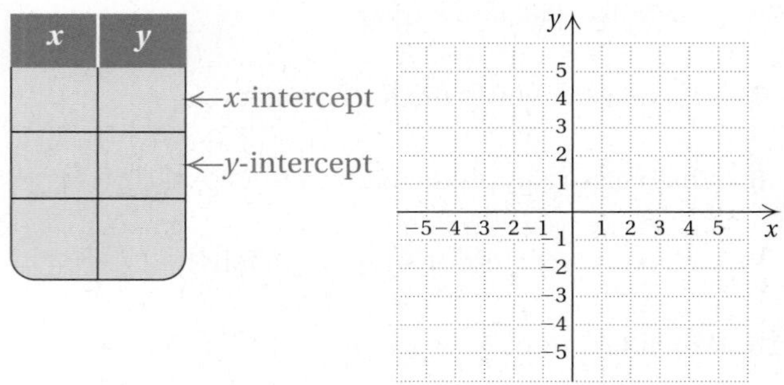

Solve. [3.2c]

31. *Kitchen Design.* Kitchen designers recommend that a refrigerator be selected on the basis of the number of people n in the household. The appropriate size S, in cubic feet, is given by

$$S = \frac{3}{2}n + 13.$$

a) Determine the recommended size of a refrigerator if the number of people is 1, 2, 5, and 10.

b) Graph the equation and use the graph to estimate the recommended size of a refrigerator for 3 people sharing an apartment.

c) A refrigerator is 22 ft^3. For how many residents is it the recommended size?

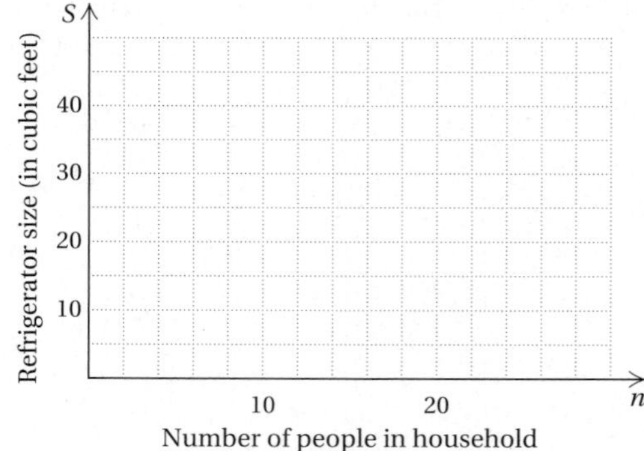

32. *Snow Removal.* By 3:00 P.M., Erin had plowed 7 driveways and by 5:30 P.M. had completed 13.

a) Find Erin's plowing rate in number of driveways per hour.

b) Find Erin's plowing rate in minutes per driveway. [3.4b]

33. *Manicures.* The following graph shows data from a recent day's work at the O'Hara School of Cosmetology. What is the rate of change, in number of manicures per hour? [3.4b]

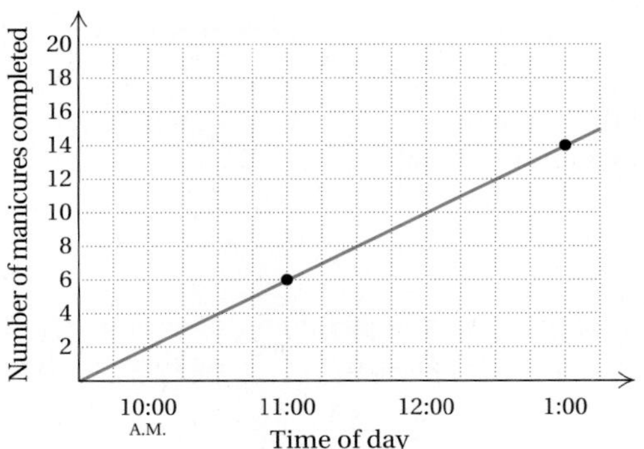

Find the slope. [3.4a]

34.

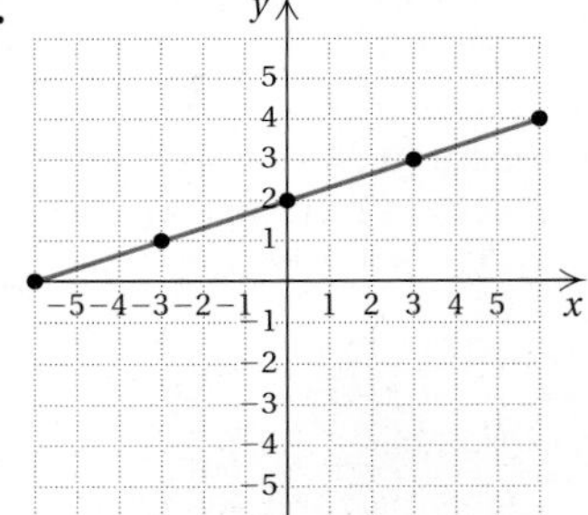

35.

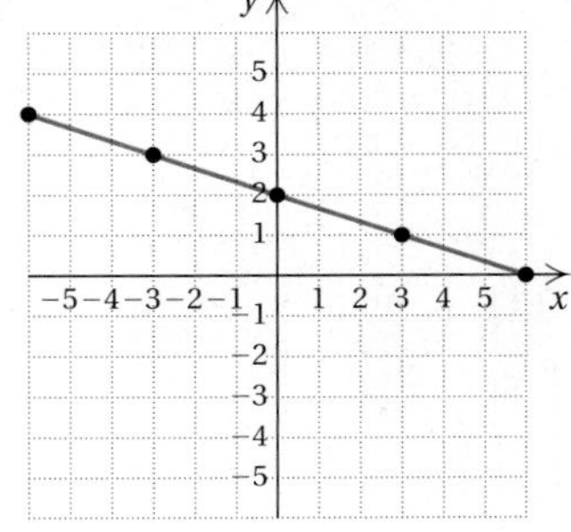

Graph the line containing the given pair of points and find the slope. [3.4a]

36. $(-5, -2), (5, 4)$

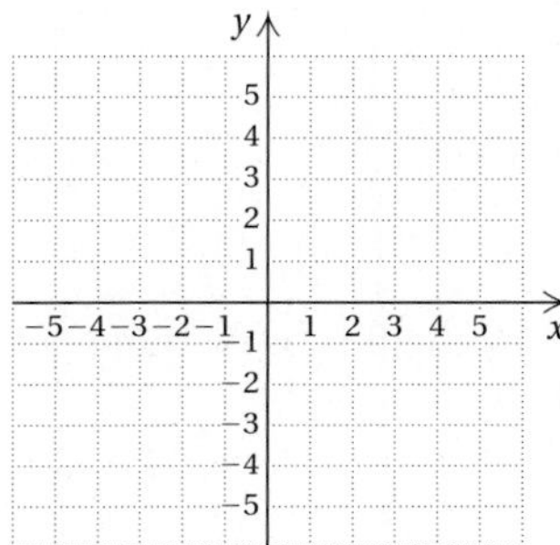

37. $(-5, 5), (4, -4)$

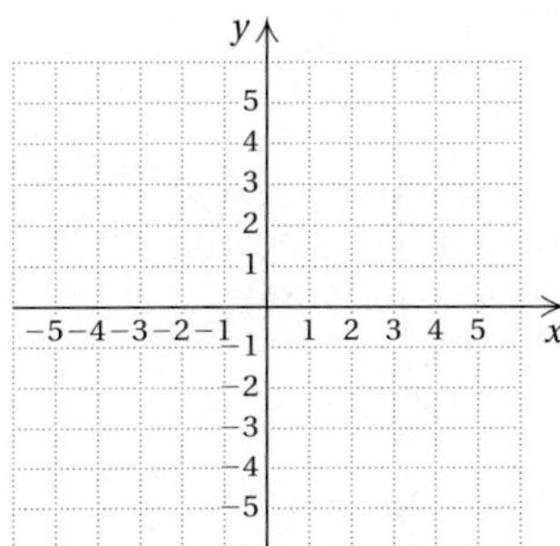

38. *Road Grade.* At one point, Beartooth Highway in Yellowstone National Park rises 315 ft over a horizontal distance of 4500 ft. Find the slope, or grade, of the road. [3.4b]

Find the slope, if it exists. [3.4c]

39. $y = -\frac{5}{8}x - 3$

40. $2x - 4y = 8$

41. $x = -2$

42. $y = 9$

43. $\mathbf{D_W}$ Describe two ways in which a small business might make use of graphs. [3.1a], [3.2c]

44. $\mathbf{D_W}$ Explain why the first coordinate of the *y*-intercept is always 0. [3.2b]

SKILL MAINTENANCE

Certain objectives from four particular sections will be retested on the chapter test. The objectives are listed with the practice problems that follow.

Convert to decimal notation. [1.2c]

45. $-\frac{11}{32}$

46. $\frac{8}{9}$

Find the absolute value. [1.2e]

47. $|-3.2|$

48. $\left|\frac{17}{19}\right|$

Round to the nearest hundredth. [R.3c]

49. 42.705

50. 112.5278

Solve. [2.5a]

51. An investment was made at 6% simple interest for 1 year. It grows to $10,340.40. How much was originally invested?

52. After a 20% reduction, a pair of slacks is on sale for $63.96. What was the original price (that is, the price before reduction)?

SYNTHESIS

53. Find the value of m in $y = mx + 3$ such that $(-2, 5)$ is on the graph. [3.2a]

54. Find the area and the perimeter of a rectangle for which $(-2, 2)$, $(7, 2)$, and $(7, -3)$ are three of the vertices. [3.1b]

55. *Mountaineering.* As part of an ill-fated expedition to climb Mount Everest in 1996, author Jon Krakauer departed "The Balcony," elevation 27,600 ft, at 7:00 A.M. and reached the summit, elevation 29,028 ft, at 1:25 P.M. [3.4b]

a) Find Krakauer's rate of ascent in feet per minute.
b) Find Krakauer's rate of ascent in minutes per foot.

Source: Jon Krakauer, *Into Thin Air.* New York: Villard, 1998.

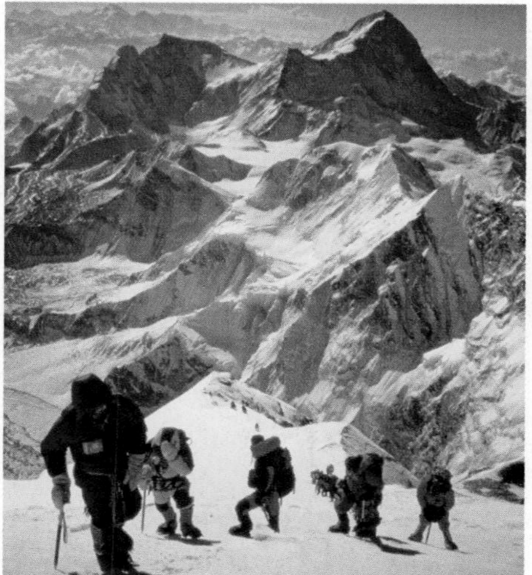

CHAPTER 3

Chapter Test

Toothpaste Sales. The following pie chart shows the percentages of sales of various toothpaste brands in the United States. In a recent year, total sales of toothpaste were $1,500,000,000. Use the pie chart for Questions 1–4.

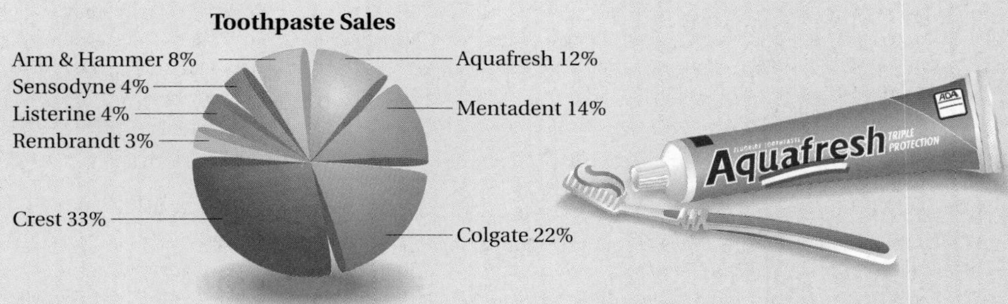

1. What were the total sales of Crest?

2. Which two brands together accounted for over half the sales?

3. Which brand had the greatest sales?

4. Which brand had sales of $120,000,000?

Tornado Touchdowns. The following bar graph shows the total number of tornado touchdowns by month in Indiana from 1950–1994. Use the bar graph for Questions 5–8.

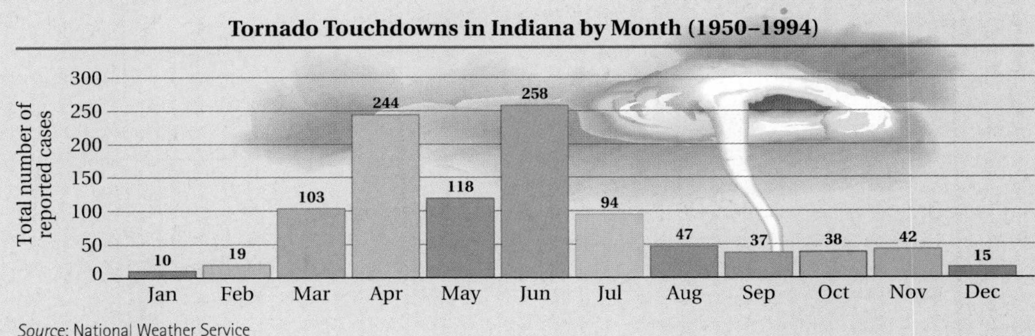

Source: National Weather Service

5. In which month did the greatest number of touchdowns occur?

6. In which month did the least number of touchdowns occur?

7. In which months was the number of touchdowns greater than 90?

8. In which month were there 47 touchdowns?

Radio Stations on the Internet. The line graph at right shows the number of radio stations in various years that were transmitting over the Internet. Use the line graph for Questions 9–14.

Number of Radio Stations on the Internet

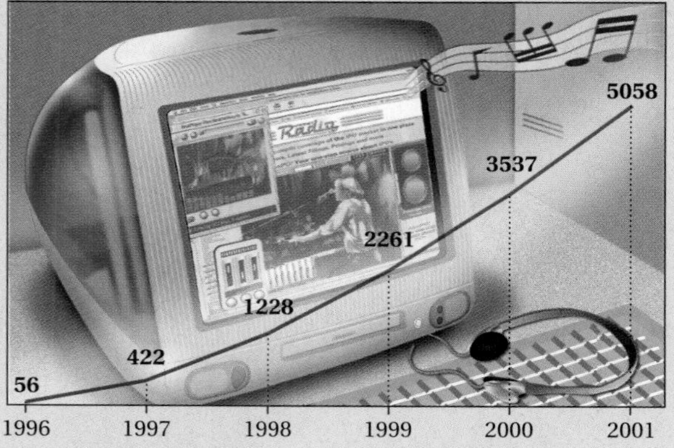

Source: BRS Media, Inc.

9. In which year were the greatest number of stations transmitting over the Internet?

10. In which year were the least number of stations transmitting over the Internet?

11. What is the difference between the greatest and least number of stations?

12. Between which two years was the increase in the number of stations the greatest?

13. By how much did the number of stations increase from 1996 to 2000?

14. By how much did the number of stations increase from 1998 to 2000?

In which quadrant is the point located?

15. $(-\frac{1}{2}, 7)$

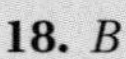

16. $(-5, -6)$

Find the coordinates of the point.

17. A

18. B

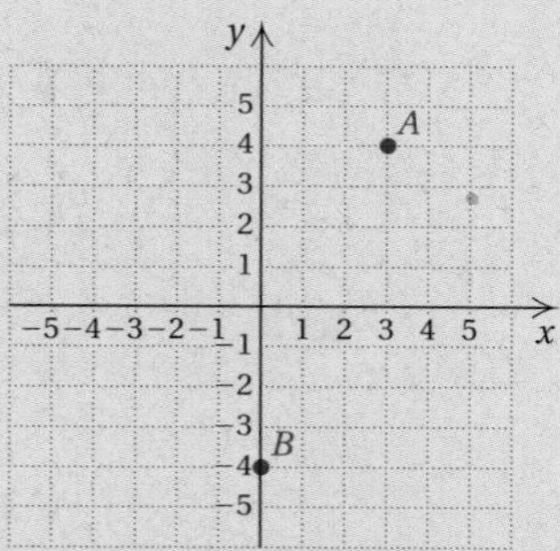

19. Show that the ordered pairs $(-4, -3)$ and $(-1, 3)$ are solutions of the equation $y - 2x = 5$. Then use the graph of the straight line containing the two points to determine another solution. Answers may vary.

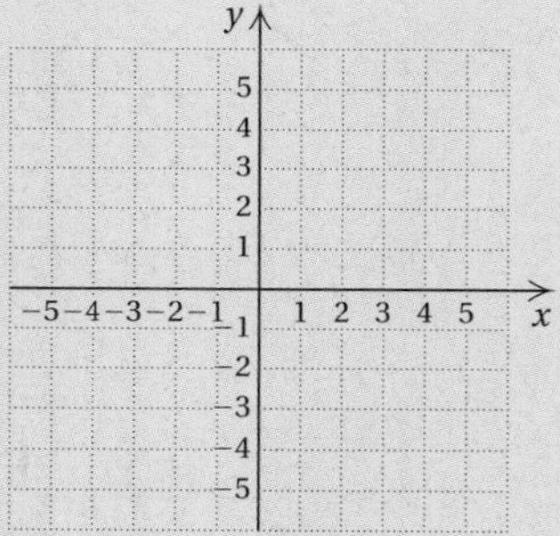

Graph the equation. Identify the y-intercept.

20. $y = 2x - 1$

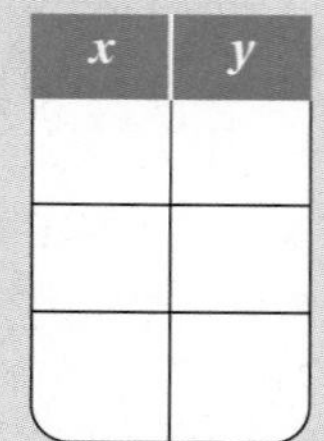

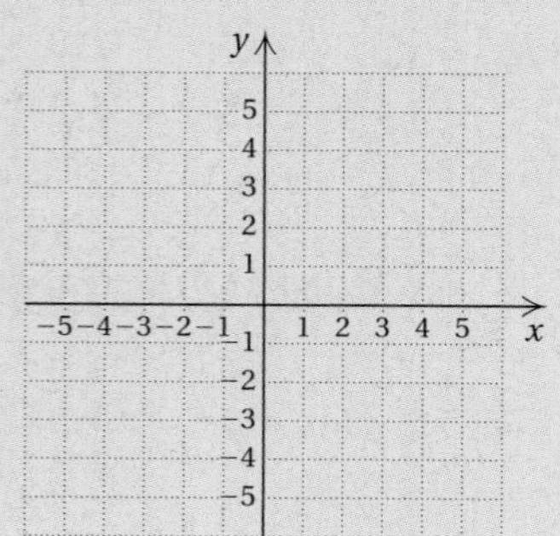

21. $y = -\frac{3}{2}x$

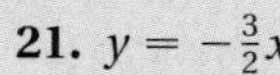

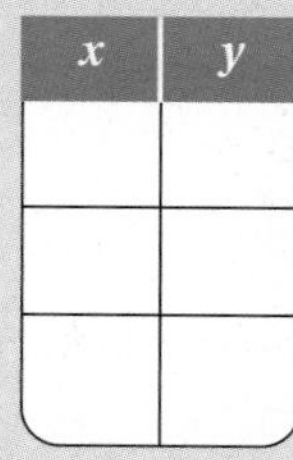

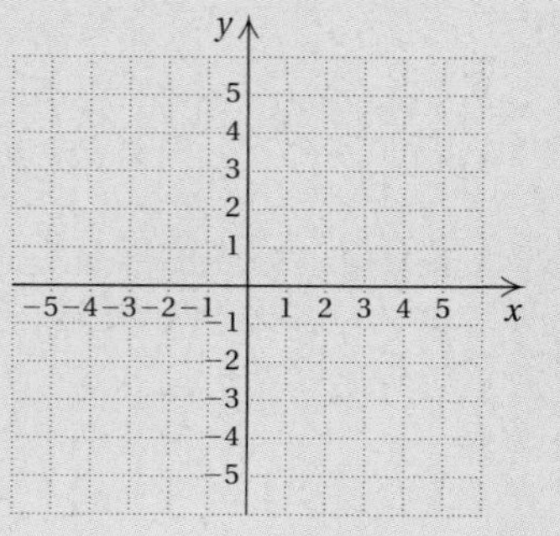

Graph the equation.

22. $2x + 8 = 0$

x	y

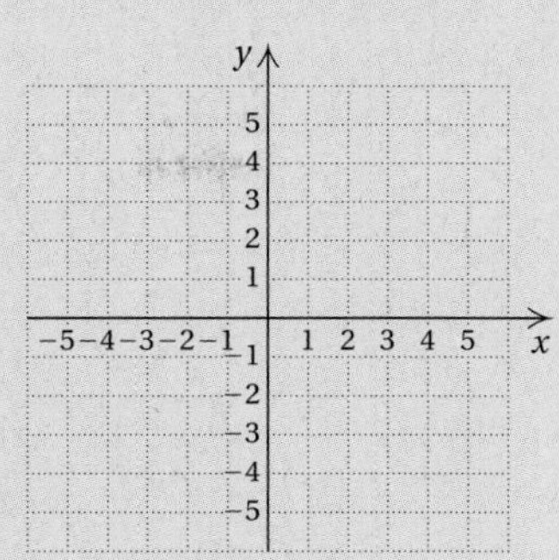

23. $y = 5$

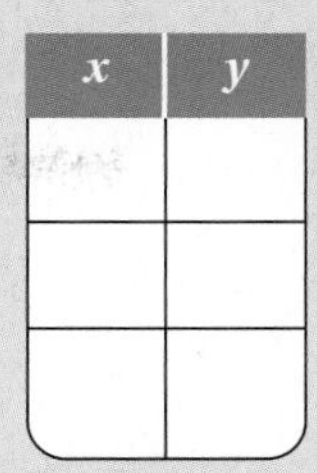

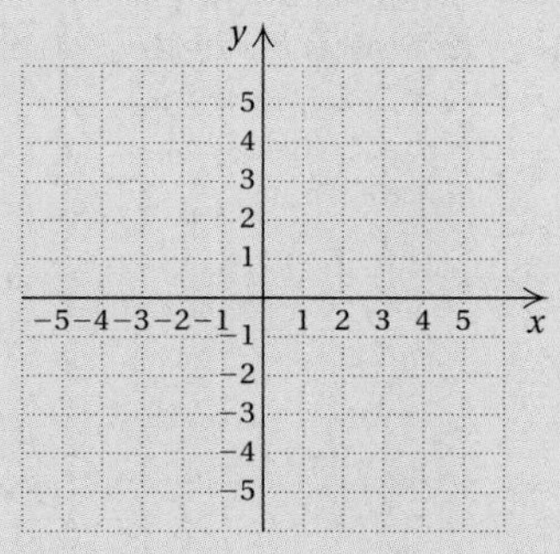

Find the intercepts of the equation. Then graph the equation.

24. $2x - 4y = -8$

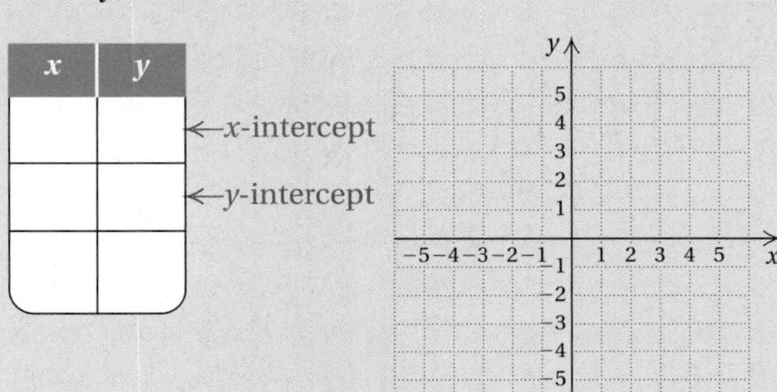

25. $2x - y = 3$

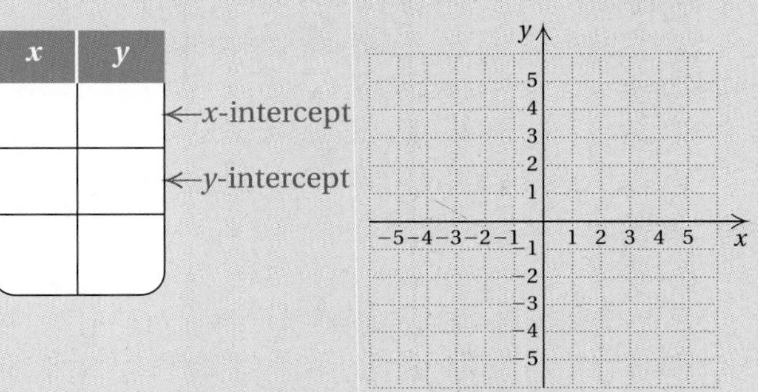

26. *Private-College Costs.* The cost T, in thousands of dollars, of tuition and fees at a private college (all expenses) can be approximated by

$$T = \frac{4}{5}n + 17,$$

where n is the number of years since 1992. That is, $n = 0$ corresponds to 1992, $n = 7$ corresponds to 1999, and so on.

a) Find the cost of tuition in 1992, 1995, 1999, and 2001.

b) Graph the equation and then use the graph to estimate the cost of tuition in 2005.

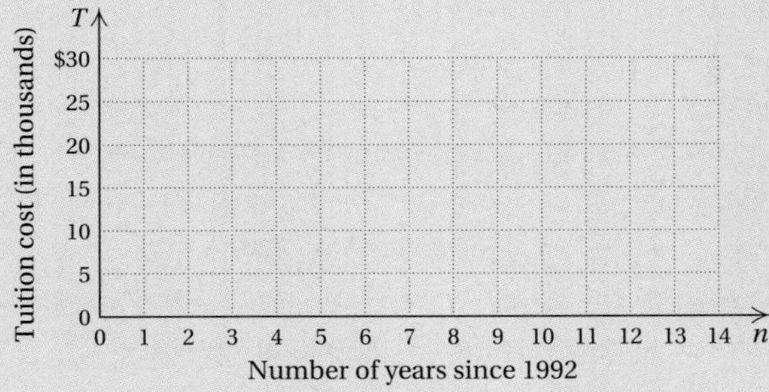

c) Estimate the year in which the cost of tuition will be $25,000.

Source: *Statistical Abstract of the United States*

27. *Elevators.* At 2:38, Serge entered an elevator on the 34th floor of the Regency Hotel. At 2:40, he stepped off at the 5th floor.

a) Find the elevator's average rate of travel in number of floors per minute.
b) Find the elevator's average rate of travel in seconds per floor.

28. *Train Travel.* The following graph shows data concerning a recent train ride from Denver to Kansas City. At what rate did the train travel?

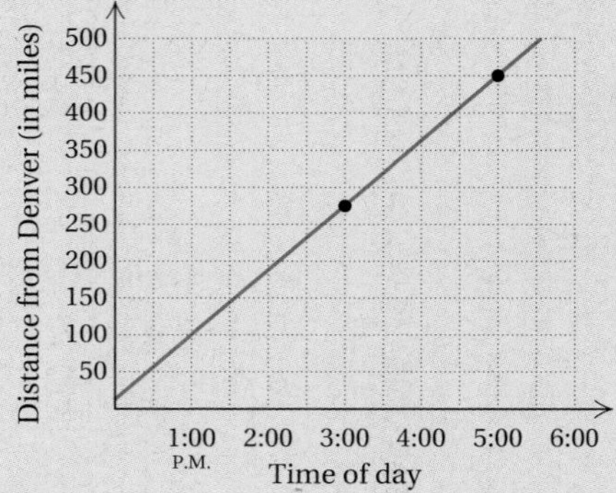

29. Find the slope.

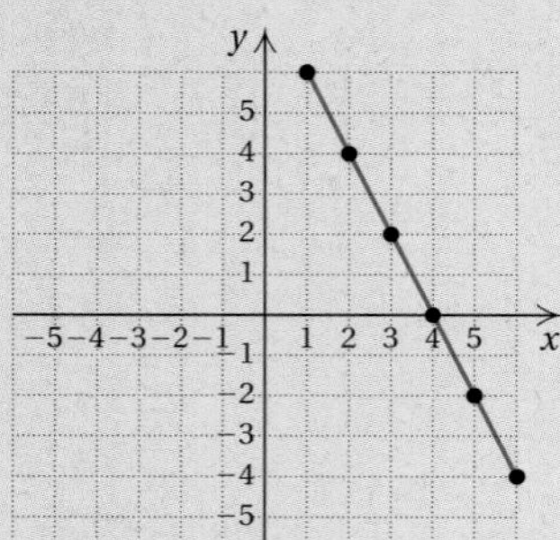

30. Graph the line containing $(-3, 1)$ and $(5, 4)$ and find the slope.

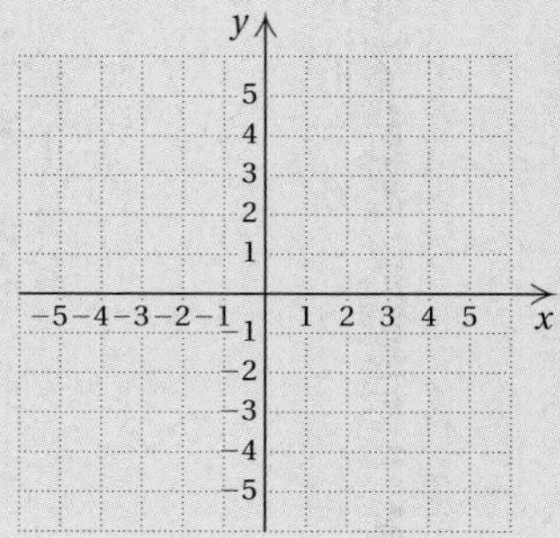

31. *Navigation.* Capital Rapids drops 54 ft vertically over a horizontal distance of 1080 ft. What is the slope of the rapids?

32. Find the slope, if it exists.

a) $2x - 5y = 10$
b) $x = -2$

SKILL MAINTENANCE

Convert to decimal notation.

33. $\frac{39}{40}$

34. $-\frac{13}{12}$

Find the absolute value.

35. $|71.2|$

36. $\left|-\frac{13}{47}\right|$

Round to the nearest thousandth.

37. 42.7047

38. 112.52702

Solve.

39. After a 24% reduction, a software game is on sale for \$64.22. What was the original price (that is, the price before reduction)?

40. An investment was made at 7% simple interest for 1 year. It grows to \$38,948. How much was originally invested?

SYNTHESIS

41. A diagonal of a square connects the points $(-3, -1)$ and $(2, 4)$. Find the area and the perimeter of the square.

42. Write an equation of a line parallel to the x-axis and 3 units above it.

CHAPTERS 1–3 Cumulative Review

1. Evaluate $\frac{x}{2y}$ when $x = 10$ and $y = 2$.

2. Multiply: $3(4x - 5y + 7)$.

3. Factor: $15x - 9y + 3$.

4. Find the prime factorization of 42.

5. Find decimal notation: $\frac{9}{20}$.

6. Find the absolute value: $|-4|$.

7. Find the opposite of -3.08.

8. Find the reciprocal of $-\frac{8}{7}$.

9. Collect like terms: $2x - 5y + (-3x) + 4y$.

10. Find decimal notation: 78.5%.

Simplify.

11. $\frac{3}{4} - \frac{5}{12}$

12. $3.4 + (-0.8)$

13. $(-2)(-1.4)(2.6)$

14. $\frac{3}{8} \div \left(-\frac{9}{10}\right)$

15. $2 - [32 \div (4 + 2^2)]$

16. $-5 + 16 \div 2 \cdot 4$

17. $y - (3y + 7)$

18. $3(x - 1) - 2[x - (2x + 7)]$

Solve.

19. $1.5 = 2.7 + x$

20. $\frac{2}{7}x = -6$

21. $5x - 9 = 36$

22. $\frac{5}{2}y = \frac{2}{5}$

23. $5.4 - 1.9x = 0.8x$

24. $x - \frac{7}{8} = \frac{3}{4}$

25. $2(2 - 3x) = 3(5x + 7)$

26. $\frac{1}{4}x - \frac{2}{3} = \frac{3}{4} + \frac{1}{3}x$

27. $y + 5 - 3y = 5y - 9$

28. $x - 28 < 20 - 2x$

29. $2(x + 2) \geq 5(2x + 3)$

30. $4(x + 2) = 4(x - 2) + 16$

31. $0(x + 3) + 4 = 0$

32. $5(7 + x) = (x + 7)5$

33. Solve $A = \frac{1}{2}h(b + c)$ for h.

34. In which quadrant is the point $(3, -1)$ located?

35. Graph on a number line: $-1 < x \leq 2$.

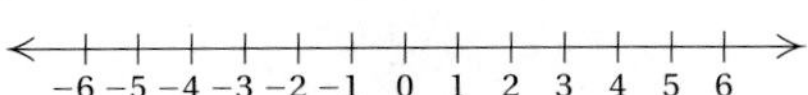

Graph.

36. $2x + 5y = 10$

x	y

37. $y = -2$

x	y

38. $y = -2x + 1$

x	y

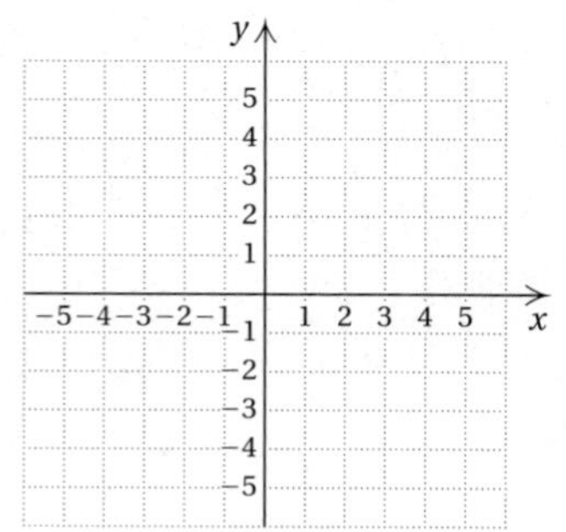

39. $y = \frac{2}{3}x - 2$

x	y

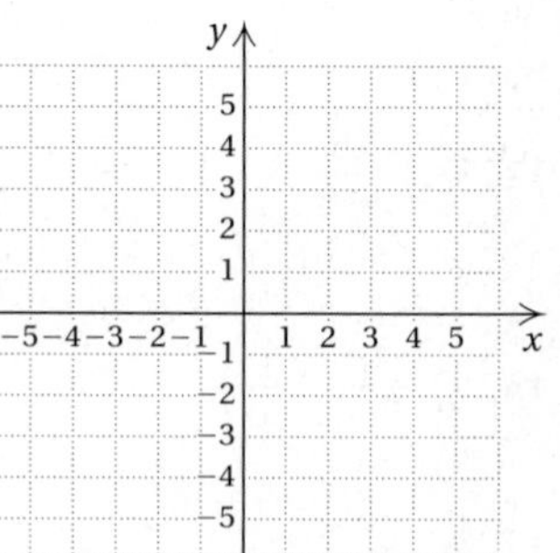

40. $y = \frac{2}{3}x$

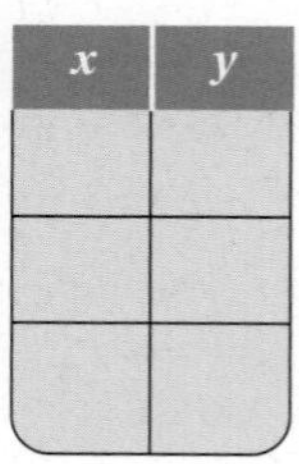

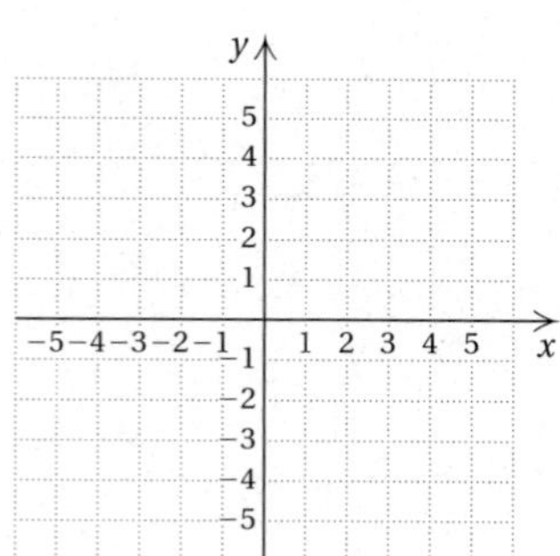

41. $y = -\frac{3}{2}x$

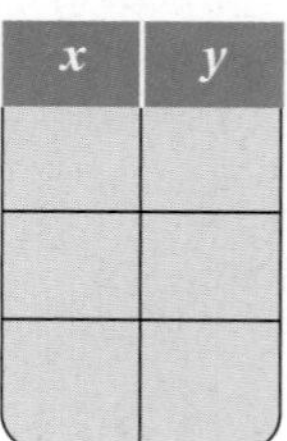

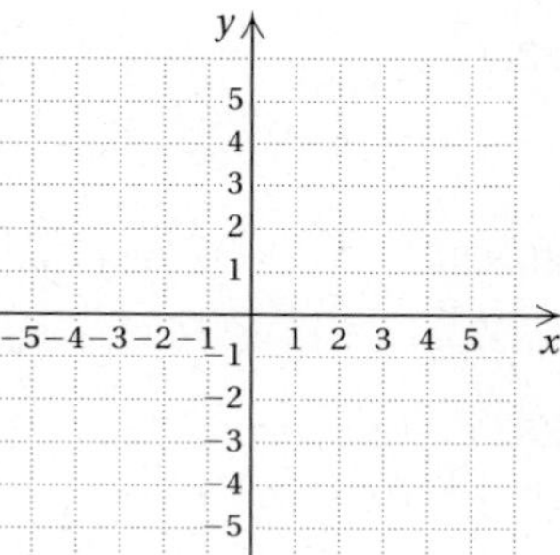

Find the intercepts. Do not graph.

42. $2x - 7y = 21$

43. $y = 4x + 5$

Solve.

44. *Blood Donors.* Each year, 8 million Americans donate blood. This is 5% of those healthy enough to do so. How many Americans are eligible to donate blood?

45. *Blood Types.* There are 117 million Americans with either O-positive or O-negative blood. Those with O-positive blood outnumber those with O-negative blood by 85.8 million. How many Americans have O-negative blood?

46. *Sales Tax.* Tina paid $126 for a cordless drill. This included 5% for sales tax. How much did the drill cost before tax?

47. *Wire Cutting.* A 143-m wire is cut into three pieces. The second piece is 3 m longer than the first. The third is four-fifths as long as the first. How long is each piece?

48. *Work Time.* Cory's contract stipulates that he cannot work more than 40 hr per week. For the first four days of one week, he worked 7, 10, 9, and 6 hr. Determine as an inequality the number of hours he can work on the fifth day without violating his contract.

49. *Telephone Line.* The cost P, in hundreds of dollars, of a telephone line for a business is given by $P = \frac{3}{4}n + 3$, where n is the number of months that the line has been in service.

a) Find the cost of the phone line for 1 month, 2 months, 3 months, and 7 months.

b) Graph the equation and use the graph to estimate the cost of the phone line for 10 months.

c) Find the number of months it will take for the cost to be $15,000.

50. *Gas Mileage.* The following graph shows data for a Ford Explorer driven on city streets. At what rate was the vehicle consuming gas?

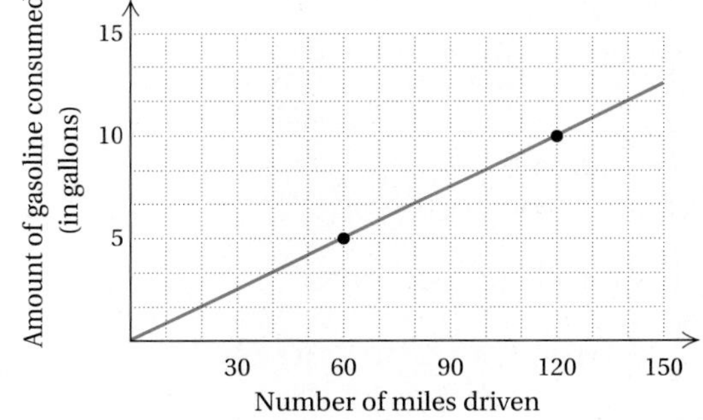

51. Find the slope of the line graphed below.

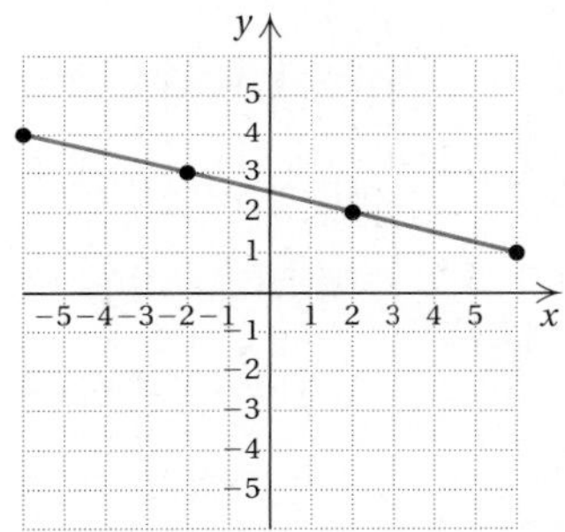

For each of Exercises 52–55, choose the correct answer from the selections given.

52. The x-intercept of the graph of $4x - 3y = 12$ is:

a) $(0, -4)$. **b)** $(3, 0)$. **c)** $(0, 0)$.
d) $(-4, 0)$. **e)** None of these.

53. Compute and simplify: $1000 \div 100 \cdot 10 - 10$.

a) 9 **b)** 0 **c)** −9
d) −90 **e)** None of these

54. Remove parentheses and simplify:

$$6a - 5b - (8a - 7b).$$

a) $-2a - 12b$ **b)** $a + 6b$ **c)** $-2a + 12b$
d) $-2a + 2b$ **e)** None of these

55. The slope of the line containing the points $(2, -7)$ and $(-4, 3)$ is:

a) $-\frac{5}{2}$. **b)** $-\frac{3}{5}$. **c)** $-\frac{2}{5}$.
d) $-\frac{5}{3}$. **e)** None of these.

SYNTHESIS

Solve.

56. $4|x| - 13 = 3$

57. $\dfrac{2 + 5x}{4} = \dfrac{11}{28} + \dfrac{8x + 3}{7}$

58. $p = \dfrac{2}{m + Q}$, for Q

Polynomials: Operations

Gateway to Chapter 4

An algebraic expression like $x^3 - 5x^2 + 4x - 11$ is called a polynomial. A polynomial equation like $y = x^3 - 5x^2 + 4x - 11$ contains a polynomial. The focus of this chapter is on manipulations such as adding, subtracting, multiplying, and dividing polynomials. Of particular importance is the study of quick ways to multiply polynomials, called special products. These manipulations will be helpful throughout most mathematics courses you will take.

Basic to our study of polynomials are the manipulations with exponential expressions considered in Sections 4.1 and 4.2. We will solve problems both using scientific notation and expressing situations using polynomials. Though we will not discuss how to graph polynomial equations, we will show how their graphs can help to solve applied problems.

Real-World Application

A strand of DNA (deoxyribonucleic acid) is about 150 cm long and 1.3×10^{-10} cm wide. How many times longer is DNA than it is wide?

Source: Human Genome Project Information

This problem appears as Example 24 in Section 4.2.

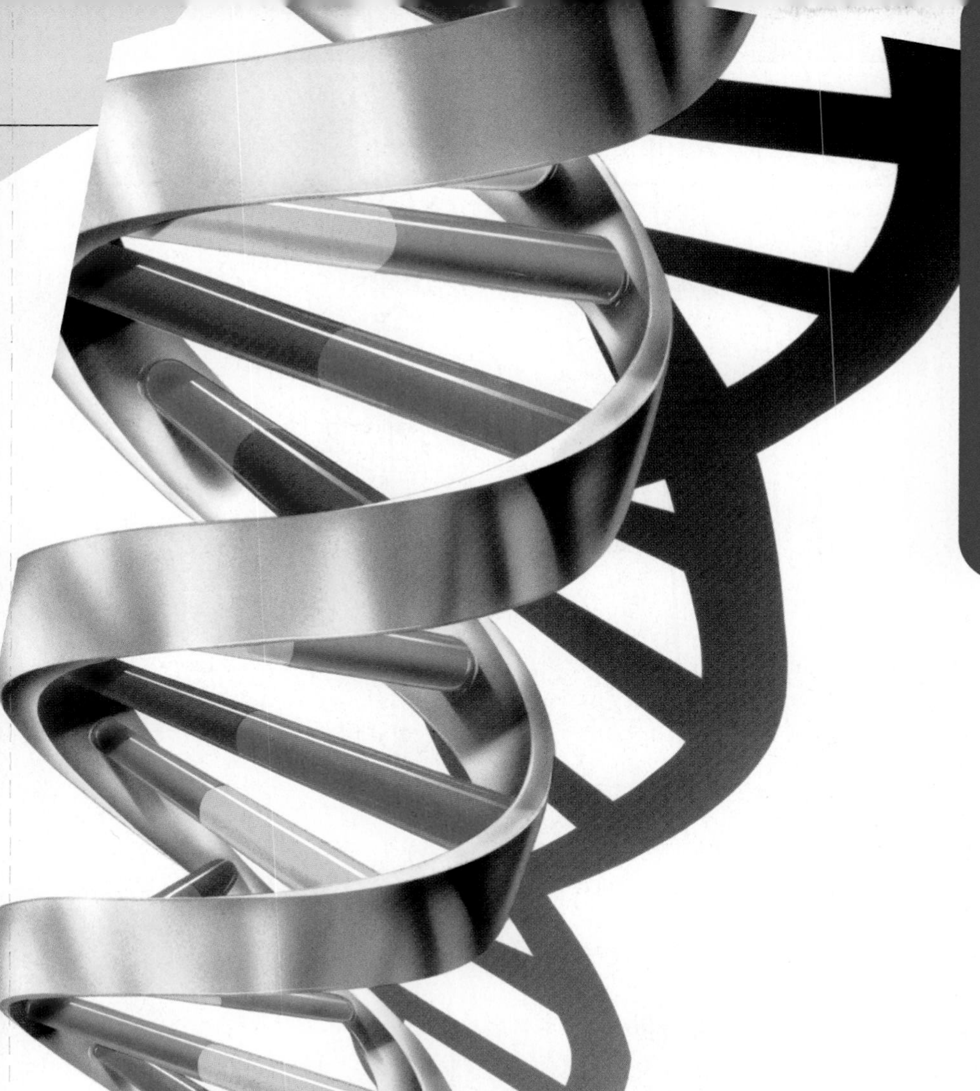

CHAPTER 4

CHAPTER

4 Pretest

1. Multiply: $x^{-3} \cdot x^5$. [4.1d]

2. Divide: $\frac{x^{-2}}{x^5}$. [4.1e]

3. Simplify: $(-4x^2y^{-3})^2$. [4.2b]

4. Express using a positive exponent: p^{-3}. [4.1f]

5. Convert to scientific notation: 0.000347. [4.2c]

6. Convert to decimal notation: 3.4×10^6. [4.2c]

Multiply and divide and express your results using scientific notation. [4.2d]

7. $(3.1 \times 10^5)(4.5 \times 10^{-3})$

8. $\frac{6.4 \times 10^{-7}}{8.0 \times 10^{-6}}$

9. Identify the degree of each term and the degree of the polynomial: [4.3g]

$2x^3 - 4x^2 + 3x - 5$.

10. Collect like terms: [4.3e]

$2a^3b - a^2b^2 + ab^3 + 9 - 5a^3b - a^2b^2 + 12b^3$.

11. Add: [4.4a]

$(5x^2 - 7x + 8) + (6x^2 + 11x - 19)$.

12. Subtract: [4.4c]

$(5x^2 - 7x + 8) - (6x^2 + 11x - 19)$.

Multiply.

13. $5x^2(3x^2 - 4x + 1)$ [4.5b]

14. $(x + 5)^2$ [4.6c]

15. $(x - 5)(x + 5)$ [4.6b]

16. $(x^3 + 6)(4x^3 - 5)$ [4.6a]

17. $(2x - 3y)(2x - 3y)$ [4.6c]

18. Divide: $(x^3 - x^2 + x + 2) \div (x - 2)$. [4.8b]

19. The length of a rectangle is 8 ft longer than the width. [4.4d]

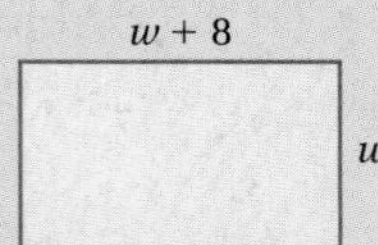

a) Find a polynomial for the perimeter.
b) Find a polynomial for the area.

4.1 INTEGERS AS EXPONENTS

We introduced integer exponents of 2 or higher in Section R.5. Here we consider 0 and 1, as well as negative integers, as exponents.

a Exponential Notation

An exponent of 2 or greater tells how many times the base is used as a factor. For example,

$$a \cdot a \cdot a \cdot a = a^4.$$

In this case, the **exponent** is 4 and the **base** is a. An expression for a power is called **exponential notation.**

$a^n \leftarrow$ This is the exponent.
$\uparrow$
This is the base.

EXAMPLE 1 What is the meaning of 3^5? of n^4? of $(2n)^3$? of $50x^2$? of $(-n)^3$?

3^5 means $3 \cdot 3 \cdot 3 \cdot 3 \cdot 3$; n^4 means $n \cdot n \cdot n \cdot n$;
$(2n)^3$ means $2n \cdot 2n \cdot 2n$; $50x^2$ means $50 \cdot x \cdot x$;
$(-n)^3$ means $(-n) \cdot (-n) \cdot (-n)$

Do Exercises 1–5.

We read exponential notation as follows: a^n is read the ***n*th power of *a*,** or simply ***a* to the *n*th,** or ***a* to the *n*.** We often read x^2 as "***x*-squared.**" The reason for this is that the area of a square of side x is $x \cdot x$, or x^2. We often read x^3 as "***x*-cubed.**" The reason for this is that the volume of a cube with length, width, and height x is $x \cdot x \cdot x$, or x^3.

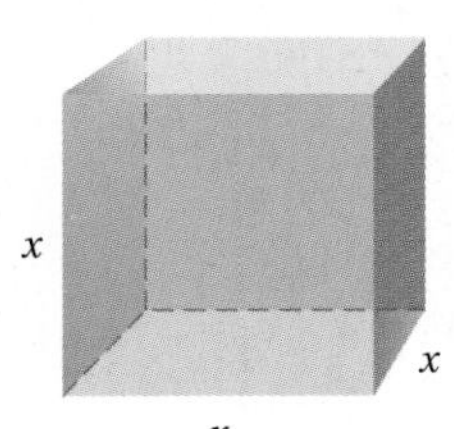

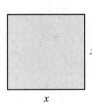

b One and Zero as Exponents

Look for a pattern in the following:

On each side, we divide by 8 at each step.

$$8 \cdot 8 \cdot 8 \cdot 8 = 8^4$$
$$8 \cdot 8 \cdot 8 = 8^3$$
$$8 \cdot 8 = 8^2$$
$$8 = 8^?$$
$$1 = 8^?.$$

On this side, the exponents decrease by 1.

To continue the pattern, we would say that

$$8 = 8^1$$

and $$1 = 8^0.$$

Objectives

- **a** Tell the meaning of exponential notation.
- **b** Evaluate exponential expressions with exponents of 0 and 1.
- **c** Evaluate algebraic expressions containing exponents.
- **d** Use the product rule to multiply exponential expressions with like bases.
- **e** Use the quotient rule to divide exponential expressions with like bases.
- **f** Express an exponential expression involving negative exponents with positive exponents.

What is the meaning of each of the following?

1. 5^4

2. x^5

3. $(3t)^2$

4. $3t^2$

5. $(-x)^4$

Answers on page A-18

Evaluate.

6. 6^1 6

7. 7^0 1

8. $(8.4)^1$ 8.4

9. 8654^0 1

Answers on page A-18

Study Tips

AUDIOTAPES

Your instructor can request a complete set of audiotapes designed to help lead you through each section of this textbook. If you have difficulty reading or if you want extra review, these tapes explain solution steps for examples, caution you about errors, give instructions to work margin exercises, and then review the solutions to the margin exercises. Ideal for use outside of class, these tapes can be played anywhere you have a tape player. To obtain these audiotapes, consult with your instructor and refer to the Preface of this text.

We make the following definition.

EXPONENTS OF 0 AND 1

$a^1 = a$, for any number a;
$a^0 = 1$, for any nonzero number a

We consider 0^0 to be not defined. We will explain why later in this section.

EXAMPLE 2 Evaluate 5^1, 8^1, 3^0, $(-7.3)^0$, and $(186{,}892{,}046)^0$.

$5^1 = 5$; $\quad 8^1 = 8$; $\quad 3^0 = 1$;
$(-7.3)^0 = 1$; $\quad (186{,}892{,}046)^0 = 1$

Do Exercises 6–9.

C Evaluating Algebraic Expressions

Algebraic expressions can involve exponential notation. For example, the following are algebraic expressions:

$$x^4, \qquad (3x)^3 - 2, \qquad a^2 + 2ab + b^2.$$

We evaluate algebraic expressions by replacing variables with numbers and following the rules for order of operations.

EXAMPLE 3 Evaluate x^4 when $x = 2$.

$$\begin{aligned} x^4 &= 2^4 && \text{Substituting} \\ &= 2 \cdot 2 \cdot 2 \cdot 2 = 16 \end{aligned}$$

EXAMPLE 4 *Area of a Compact Disc.* The standard compact disc used for software and music has a radius of 6 cm. Find the area of such a CD (ignoring the hole in the middle).

$$\begin{aligned} A &= \pi r^2 \\ &= \pi \cdot (6\text{ cm})^2 \\ &\approx 3.14 \times 36\text{ cm}^2 \\ &= 113.04\text{ cm}^2 \end{aligned}$$

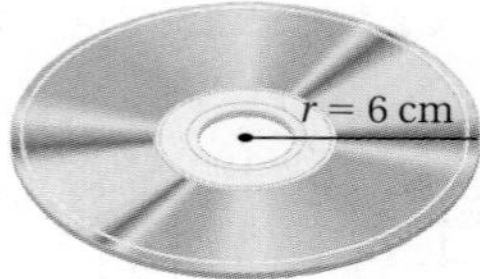

In Example 4, "cm^2" means "square centimeters" and "≈" means "is approximately equal to."

EXAMPLE 5 Evaluate $(5x)^3$ when $x = -2$.

When we evaluate with a negative number, we often use extra parentheses to show the substitution.

$$\begin{aligned} (5x)^3 &= [5 \cdot (-2)]^3 && \text{Substituting} \\ &= [-10]^3 && \text{Multiplying within brackets first} \\ &= [-10] \cdot [-10] \cdot [-10] \\ &= -1000 && \text{Evaluating the power} \end{aligned}$$

EXAMPLE 6 Evaluate $5x^3$ when $x = -2$.

$$\begin{aligned} 5x^3 &= 5 \cdot (-2)^3 && \text{Substituting} \\ &= 5(-8) && \text{Evaluating the power first} \\ &= -40 \end{aligned}$$

Recall that two expressions are equivalent if they have the same value for all meaningful replacements. Note that Examples 5 and 6 show that $(5x)^3$ and $5x^3$ are *not* equivalent—that is, $(5x)^3 \neq 5x^3$.

Do Exercises 10–14.

d Multiplying Powers with Like Bases

There are several rules for manipulating exponential notation to obtain equivalent expressions. We first consider multiplying powers with like bases:

$$a^3 \cdot a^2 = \underbrace{(a \cdot a \cdot a)}_{\text{3 factors}}\underbrace{(a \cdot a)}_{\text{2 factors}} = \underbrace{a \cdot a \cdot a \cdot a \cdot a}_{\text{5 factors}} = a^5.$$

Since an integer exponent greater than 1 tells how many times we use a base as a factor, then $(a \cdot a \cdot a)(a \cdot a) = a \cdot a \cdot a \cdot a \cdot a = a^5$ by the associative law. Note that the exponent in a^5 is the sum of those in $a^3 \cdot a^2$. That is, $3 + 2 = 5$. Likewise,

$$b^4 \cdot b^3 = (b \cdot b \cdot b \cdot b)(b \cdot b \cdot b) = b^7, \quad \text{where} \quad 4 + 3 = 7.$$

Adding the exponents gives the correct result.

THE PRODUCT RULE

For any number a and any positive integers m and n,

$$a^m \cdot a^n = a^{m+n}.$$

(When multiplying with exponential notation, if the bases are the same, keep the base and add the exponents.)

EXAMPLES Multiply and simplify. By simplify, we mean write the expression as one number to a nonnegative power.

7. $8^4 \cdot 8^3 = 8^{4+3}$ Adding exponents: $a^m \cdot a^n = a^{m+n}$
$= 8^7$

8. $x^2 \cdot x^9 = x^{2+9} = x^{11}$

9. $m^5m^{10}m^3 = m^{5+10+3} = m^{18}$

10. $x \cdot x^8 = x^1 \cdot x^8$ Writing x as x^1
$= x^{1+8}$
$= x^9$

11. $(a^3b^2)(a^3b^5) = (a^3a^3)(b^2b^5)$
$= a^6b^7$

Do Exercises 15–19.

10. Evaluate t^3 when $t = 5$.

11. Find the area of a circle when $r = 32$ cm. Use 3.14 for π.

12. Evaluate $200 - a^4$ when $a = 3$.

13. Evaluate $t^1 - 4$ and $t^0 - 4$ when $t = 7$.

14. a) Evaluate $(4t)^2$ when $t = -3$.

b) Evaluate $4t^2$ when $t = -3$.

c) Determine whether $(4t)^2$ and $4t^2$ are equivalent.

Multiply and simplify.

15. $3^5 \cdot 3^5$

16. $x^4 \cdot x^6$

17. $p^4p^{12}p^8$

18. $x \cdot x^4$

19. $(a^2b^3)(a^7b^5)$

Answers on page A-18

Divide and simplify.

20. $\frac{4^5}{4^2}$

21. $\frac{y^6}{y^2}$

22. $\frac{p^{10}}{p}$

23. $\frac{a^7b^6}{a^3b^4}$

Answers on page A-18

e Dividing Powers with Like Bases

The following suggests a rule for dividing powers with like bases, such as a^5/a^2:

$$\frac{a^5}{a^2} = \frac{a \cdot a \cdot a \cdot a \cdot a}{a \cdot a} = \frac{a \cdot a \cdot a \cdot a \cdot a}{1 \cdot a \cdot a} = \frac{a \cdot a \cdot a}{1} \cdot \frac{a \cdot a}{a \cdot a} = \frac{a \cdot a \cdot a}{1} \cdot 1$$

$$= a \cdot a \cdot a = a^3.$$

Note that the exponent in a^3 is the difference of those in $a^5 \div a^2$, that is, $5 - 2 = 3$. In a similar way, we have

$$\frac{t^9}{t^4} = \frac{t \cdot t \cdot t \cdot t \cdot t \cdot t \cdot t \cdot t \cdot t}{t \cdot t \cdot t \cdot t} = t^5, \quad \text{where } 9 - 4 = 5.$$

Subtracting exponents gives the correct answer.

THE QUOTIENT RULE

For any nonzero number a and any positive integers m and n,

$$\frac{a^m}{a^n} = a^{m-n}.$$

(When dividing with exponential notation, if the bases are the same, keep the base and subtract the exponent of the denominator from the exponent of the numerator.)

EXAMPLES Divide and simplify. By simplify, we mean write the expression as one number to a nonnegative power.

12. $\frac{6^5}{6^3} = 6^{5-3}$ Subtracting exponents
$= 6^2$

13. $\frac{x^8}{x^2} = x^{8-2}$
$= x^6$

14. $\frac{t^{12}}{t} = \frac{t^{12}}{t^1} = t^{12-1}$
$= t^{11}$

15. $\frac{p^5q^7}{p^2q^5} = \frac{p^5}{p^2} \cdot \frac{q^7}{q^5} = p^{5-2}q^{7-5}$
$= p^3q^2$

The quotient rule can also be used to explain the definition of 0 as an exponent. Consider the expression a^4/a^4, where a is nonzero:

$$\frac{a^4}{a^4} = \frac{a \cdot a \cdot a \cdot a}{a \cdot a \cdot a \cdot a} = 1.$$

This is true because the numerator and the denominator are the same. Now suppose we apply the rule for dividing powers with the same base:

$$\frac{a^4}{a^4} = a^{4-4} = a^0 = 1.$$

Since both expressions a^4/a^4 and a^{4-4} are equivalent to 1, it follows that $a^0 = 1$, when $a \neq 0$.

We can explain why we do not define 0^0 using the quotient rule. We know that 0^0 is 0^{1-1}. But 0^{1-1} is also equal to $0/0$. We have already seen that division by 0 is not defined, so 0^0 is also not defined.

Do Exercises 20–23.

f Negative Integers as Exponents

We can use the rule for dividing powers with like bases to lead us to a definition of exponential notation when the exponent is a negative integer. Consider $5^3/5^7$ and first simplify it using procedures we have learned for working with fractions:

$$\frac{5^3}{5^7} = \frac{5 \cdot 5 \cdot 5}{5 \cdot 5 \cdot 5 \cdot 5 \cdot 5 \cdot 5 \cdot 5} = \frac{5 \cdot 5 \cdot 5 \cdot 1}{5 \cdot 5 \cdot 5 \cdot 5 \cdot 5 \cdot 5 \cdot 5}$$

$$= \frac{5 \cdot 5 \cdot 5}{5 \cdot 5 \cdot 5} \cdot \frac{1}{5 \cdot 5 \cdot 5 \cdot 5} = \frac{1}{5^4}.$$

Now we apply the rule for dividing exponential expressions with the same bases. Then

$$\frac{5^3}{5^7} = 5^{3-7} = 5^{-4}.$$

From these two expressions for $5^3/5^7$, it follows that

$$5^{-4} = \frac{1}{5^4}.$$

This leads to our definition of negative exponents.

NEGATIVE EXPONENT

For any real number a that is nonzero and any integer n,

$$a^{-n} = \frac{1}{a^n}.$$

In fact, the numbers a^n and a^{-n} are reciprocals of each other because

$$a^n \cdot a^{-n} = a^n \cdot \frac{1}{a^n} = \frac{a^n}{a^n} = 1.$$

EXAMPLES Express using positive exponents. Then simplify.

16. $4^{-2} = \frac{1}{4^2} = \frac{1}{16}$

17. $(-3)^{-2} = \frac{1}{(-3)^2} = \frac{1}{(-3)(-3)} = \frac{1}{9}$

18. $m^{-3} = \frac{1}{m^3}$

19. $ab^{-1} = a\left(\frac{1}{b^1}\right) = a\left(\frac{1}{b}\right) = \frac{a}{b}$

20. $\frac{1}{x^{-3}} = x^{-(-3)} = x^3$

21. $3c^{-5} = 3\left(\frac{1}{c^5}\right) = \frac{3}{c^5}$

Example 20 might also be done as follows:

$$\frac{1}{x^{-3}} = \frac{1}{\frac{1}{x^3}} = 1 \cdot \frac{x^3}{1} = x^3.$$

CAUTION!

As shown in Examples 16 and 17, a negative exponent does not necessarily mean that an expression is negative.

Do Exercises 24–29.

Express with positive exponents. Then simplify.

24. 4^{-3}

25. 5^{-2}

26. 2^{-4}

27. $(-2)^{-3}$

28. $4p^{-3}$

29. $\frac{1}{x^{-2}}$

Answers on page A-18

Simplify.

30. $5^{-2} \cdot 5^4$

31. $x^{-3} \cdot x^{-4}$

32. $\frac{7^{-2}}{7^3}$

33. $\frac{b^{-2}}{b^{-3}}$

34. $\frac{t}{t^{-5}}$

Answers on page A-18

The rules for multiplying and dividing powers with like bases still hold when exponents are 0 or negative. We will state them in a summary at the end of this section.

EXAMPLES Simplify. By simplify, we generally mean write the expression as one number or variable to a nonnegative power.

22. $7^{-3} \cdot 7^6 = 7^{-3+6}$ Adding exponents
$= 7^3$

23. $x^4 \cdot x^{-3} = x^{4+(-3)} = x^1 = x$

24. $\frac{5^4}{5^{-2}} = 5^{4-(-2)}$ Subtracting exponents
$= 5^{4+2} = 5^6$

25. $\frac{x}{x^7} = x^{1-7} = x^{-6} = \frac{1}{x^6}$

26. $\frac{b^{-4}}{b^{-5}} = b^{-4-(-5)}$
$= b^{-4+5} = b^1 = b$

27. $y^{-4} \cdot y^{-8} = y^{-4+(-8)}$
$= y^{-12} = \frac{1}{y^{12}}$

Do Exercises 30–34.

The following is another way to arrive at the definition of negative exponents.

On each side, we divide by 5 at each step.

$$5 \cdot 5 \cdot 5 \cdot 5 = 5^4$$
$$5 \cdot 5 \cdot 5 = 5^3$$
$$5 \cdot 5 = 5^2$$
$$5 = 5^1$$
$$1 = 5^0$$
$$\frac{1}{5} = 5^?$$
$$\frac{1}{25} = 5^?$$

On this side, the exponents decrease by 1.

To continue the pattern, it should follow that

$$\frac{1}{5} = \frac{1}{5^1} = 5^{-1} \quad \text{and} \quad \frac{1}{25} = \frac{1}{5^2} = 5^{-2}.$$

The following is a summary of the definitions and rules for exponents that we have considered in this section.

DEFINITIONS AND RULES FOR EXPONENTS

1 as an exponent:	$a^1 = a$
0 as an exponent:	$a^0 = 1, a \neq 0$
Negative integers as exponents:	$a^{-n} = \frac{1}{a^n}, \frac{1}{a^{-n}} = a^n; a \neq 0$
Product Rule:	$a^m \cdot a^n = a^{m+n}$
Quotient Rule:	$\frac{a^m}{a^n} = a^{m-n}, a \neq 0$

4.1 EXERCISE SET

For Extra Help

Digital Video Tutor CD 4 Videotape 8 | InterAct Math | Math Tutor Center | MathXL | MyMathLab

a What is the meaning of each of the following?

1. 3^4

2. 4^3

3. $(-1.1)^5$

4. $(87.2)^6$

5. $\left(\frac{2}{3}\right)^4$

6. $\left(-\frac{5}{8}\right)^3$

7. $(7p)^2$

8. $(11c)^3$

9. $8k^3$

10. $17x^2$

b Evaluate.

11. $a^0, a \neq 0$

12. $t^0, t \neq 0$

13. b^1

14. c^1

15. $\left(\frac{2}{3}\right)^0$

16. $\left(-\frac{5}{8}\right)^0$

17. 8.38^0

18. 8.38^1

19. $(ab)^1$

20. $(ab)^0, a, b \neq 0$

21. ab^1

22. ab^0

c Evaluate.

23. m^3, when $m = 3$

24. x^6, when $x = 2$

25. p^1, when $p = 19$

26. x^{19}, when $x = 0$

27. x^4, when $x = 4$

28. y^{15}, when $y = 1$

29. $y^2 - 7$, when $y = -10$

30. $z^5 + 5$, when $z = -2$

31. $x^1 + 3$ and $x^0 + 3$, when $x = 7$

32. $y^0 - 8$ and $y^1 - 8$, when $y = -3$

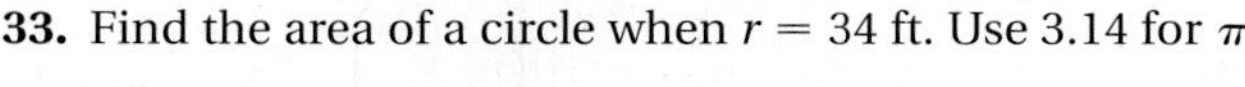

33. Find the area of a circle when $r = 34$ ft. Use 3.14 for π.

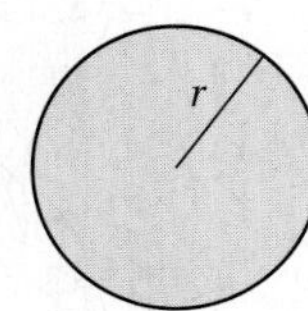

34. The area A of a square with sides of length s is given by $A = s^2$. Find the area of a square with sides of length 24 m.

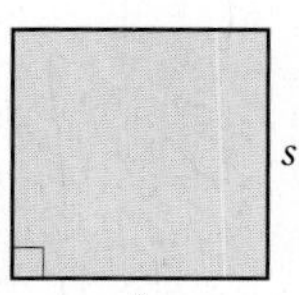

f Express using positive exponents. Then simplify.

35. 3^{-2}

36. 2^{-3}

37. 10^{-3}

38. 5^{-4}

39. 7^{-3}

40. 5^{-2}

41. a^{-3}

42. x^{-2}

43. $\frac{1}{8^{-2}}$

44. $\frac{1}{2^{-5}}$

45. $\frac{1}{y^{-4}}$

46. $\frac{1}{t^{-7}}$

47. $\frac{1}{z^{-n}}$

48. $\frac{1}{h^{-n}}$

Express using negative exponents.

49. $\frac{1}{4^3}$

50. $\frac{1}{5^2}$

51. $\frac{1}{x^3}$

52. $\frac{1}{y^2}$

53. $\frac{1}{a^5}$

54. $\frac{1}{b^7}$

d, **f** Multiply and simplify.

55. $2^4 \cdot 2^3$

56. $3^5 \cdot 3^2$

57. $8^5 \cdot 8^9$

58. $n^3 \cdot n^{20}$

59. $x^4 \cdot x^3$

60. $y^7 \cdot y^9$

61. $9^{17} \cdot 9^{21}$

62. $t^0 \cdot t^{16}$

63. $(3y)^4(3y)^8$

64. $(2t)^8(2t)^{17}$

65. $(7y)^1(7y)^{16}$

66. $(8x)^0(8x)^1$

67. $3^{-5} \cdot 3^8$

68. $5^{-8} \cdot 5^9$

69. $x^{-2} \cdot x$

70. $x \cdot x^{-1}$

71. $x^{14} \cdot x^3$

72. $x^9 \cdot x^4$

73. $x^{-7} \cdot x^{-6}$

74. $y^{-5} \cdot y^{-8}$

75. $a^{11} \cdot a^{-3} \cdot a^{-18}$

76. $a^{-11} \cdot a^{-3} \cdot a^{-7}$

77. $t^8 \cdot t^{-8}$

78. $m^{10} \cdot m^{-10}$

e, f Divide and simplify.

79. $\dfrac{7^5}{7^2}$

80. $\dfrac{5^8}{5^6}$

81. $\dfrac{8^{12}}{8^6}$

82. $\dfrac{8^{13}}{8^2}$

83. $\dfrac{y^9}{y^5}$

84. $\dfrac{x^{11}}{x^9}$

85. $\dfrac{16^2}{16^8}$

86. $\dfrac{7^2}{7^9}$

87. $\dfrac{m^6}{m^{12}}$

88. $\dfrac{a^3}{a^4}$

89. $\dfrac{(8x)^6}{(8x)^{10}}$

90. $\dfrac{(8t)^4}{(8t)^{11}}$

91. $\dfrac{(2y)^9}{(2y)^9}$

92. $\dfrac{(6y)^7}{(6y)^7}$

93. $\dfrac{x}{x^{-1}}$

94. $\dfrac{y^8}{y}$

95. $\dfrac{x^7}{x^{-2}}$

96. $\dfrac{t^8}{t^{-3}}$

97. $\dfrac{z^{-6}}{z^{-2}}$

98. $\dfrac{x^{-9}}{x^{-3}}$

99. $\dfrac{x^{-5}}{x^{-8}}$

100. $\dfrac{y^{-2}}{y^{-9}}$

101. $\dfrac{m^{-9}}{m^{-9}}$

102. $\dfrac{x^{-7}}{x^{-7}}$

Matching. In Exercises 103 and 104, match each item in the first column with the appropriate item in the second column by drawing connecting lines.

103.

5^2	$-\dfrac{1}{10}$
5^{-2}	$\dfrac{1}{10}$
$\left(\dfrac{1}{5}\right)^2$	$-\dfrac{1}{25}$
$\left(\dfrac{1}{5}\right)^{-2}$	10
-5^2	25
$(-5)^2$	-25
$-\left(-\dfrac{1}{5}\right)^2$	$\dfrac{1}{25}$
$\left(-\dfrac{1}{5}\right)^{-2}$	-10

104.

$-\left(\dfrac{1}{8}\right)^2$	16
$\left(\dfrac{1}{8}\right)^{-2}$	-16
8^{-2}	64
8^2	-64
-8^2	$\dfrac{1}{64}$
$(-8)^2$	$-\dfrac{1}{64}$
$\left(-\dfrac{1}{8}\right)^{-2}$	$-\dfrac{1}{16}$
$\left(-\dfrac{1}{8}\right)^2$	$\dfrac{1}{16}$

105. D_W Suppose that the width of a square is three times the width of a second square. How do the areas of the squares compare? Why?

106. D_W Suppose that the width of a cube is twice the width of a second cube. How do the volumes of the cubes compare? Why?

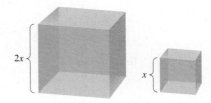

SKILL MAINTENANCE

107. Translate to an algebraic expression: Sixty-four percent of t. [1.1b]

108. Evaluate $\frac{3x}{y}$ when $x = 4$ and $y = 12$. [1.1a]

109. Divide: $1555.2 \div 24.3$. [1.6c]

110. Add: $1555.2 + 24.3$. [1.3a]

111. Solve: $3x - 4 + 5x - 10x = x - 8$. [2.3b]

112. Factor: $8x - 56$. [1.7d]

Solve. [2.6a]

113. *Cutting a Submarine Sandwich.* A 12-in. submarine sandwich is cut into two pieces. One piece is twice as long as the other. How long are the pieces?

114. *Book Pages.* A book is opened. The sum of the page numbers on the facing pages is 457. Find the page numbers.

SYNTHESIS

Determine whether each of the following is correct.

115. $(x + 1)^2 = x^2 + 1$

116. $(x - 1)^2 = x^2 - 2x + 1$

117. $(5x)^0 = 5x^0$

118. $\frac{x^3}{x^5} = x^2$

Simplify.

119. $(y^{2x})(y^{3x})$

120. $a^{5k} \div a^{3k}$

121. $\frac{a^{6t}(a^{7t})}{a^{9t}}$

122. $\frac{\left(\frac{1}{2}\right)^4}{\left(\frac{1}{2}\right)^5}$

123. $\frac{(0.8)^5}{(0.8)^3(0.8)^2}$

124. Determine whether $(a + b)^2$ and $a^2 + b^2$ are equivalent. (*Hint*: Choose values for a and b and evaluate.)

Use $>$, $<$, or $=$ for $\square$ to write a true sentence.

125. $3^5 \square 3^4$

126. $4^2 \square 4^3$

127. $4^3 \square 5^3$

128. $4^3 \square 3^4$

Find a value of the variable that shows that the two expressions are *not* equivalent.

129. $3x^2$; $(3x)^2$

130. $\frac{x + 2}{2}$; x

4.2 EXPONENTS AND SCIENTIFIC NOTATION

Objectives

a. Use the power rule to raise powers to powers.
b. Raise a product to a power and a quotient to a power.
c. Convert between scientific notation and decimal notation.
d. Multiply and divide using scientific notation.
e. Solve applied problems using scientific notation.

We now enhance our ability to manipulate exponential expressions by considering three more rules. The rules are also applied to a new way to name numbers called *scientific notation.*

a Raising Powers to Powers

Consider an expression like $(3^2)^4$. We are raising 3^2 to the fourth power:

$$\begin{aligned}(3^2)^4 &= (3^2)(3^2)(3^2)(3^2)\\ &= (3\cdot 3)(3\cdot 3)(3\cdot 3)(3\cdot 3)\\ &= 3\cdot 3\cdot 3\cdot 3\cdot 3\cdot 3\cdot 3\cdot 3\\ &= 3^8.\end{aligned}$$

Note that in this case we could have multiplied the exponents:

$$(3^2)^4 = 3^{2\cdot 4} = 3^8.$$

Likewise, $(y^8)^3 = (y^8)(y^8)(y^8) = y^{24}$. Once again, we get the same result if we multiply the exponents:

$$(y^8)^3 = y^{8\cdot 3} = y^{24}.$$

THE POWER RULE

For any real number a and any integers m and n,

$$(a^m)^n = a^{mn}.$$

(To raise a power to a power, multiply the exponents.)

EXAMPLES Simplify. Express the answers using positive exponents.

1. $(3^5)^4 = 3^{5\cdot 4}$ Multiplying exponents
   $= 3^{20}$
2. $(2^2)^5 = 2^{2\cdot 5} = 2^{10}$
3. $(y^{-5})^7 = y^{-5\cdot 7} = y^{-35} = \dfrac{1}{y^{35}}$
4. $(x^4)^{-2} = x^{4(-2)} = x^{-8} = \dfrac{1}{x^8}$
5. $(a^{-4})^{-6} = a^{(-4)(-6)} = a^{24}$

Do Exercises 1–4.

Simplify. Express the answers using positive exponents.

1. $(3^4)^5$
2. $(x^{-3})^4$
3. $(y^{-5})^{-3}$
4. $(x^4)^{-8}$

Answers on page A-18

b Raising a Product or a Quotient to a Power

When an expression inside parentheses is raised to a power, the inside expression is the base. Let's compare $2a^3$ and $(2a)^3$:

$2a^3 = 2\cdot a\cdot a\cdot a;$ The base is a.

$(2a)^3 = (2a)(2a)(2a)$ The base is $2a$.

$= (2\cdot 2\cdot 2)(a\cdot a\cdot a)$ Using the associative and commutative laws of multiplication to regroup the factors

$= 2^3a^3$

$= 8a^3.$

Simplify.

5. $(2x^5y^{-3})^4$

6. $(5x^5y^{-6}z^{-3})^2$

7. $[(-x)^{37}]^2$

8. $(3y^{-2}x^{-5}z^8)^3$

Simplify.

9. $\left(\frac{x^6}{5}\right)^2$

10. $\left(\frac{2t^5}{w^4}\right)^3$

11. $\left(\frac{x^4}{3}\right)^{-2}$

Do this two ways.

Answers on page A-18

We see that $2a^3$ and $(2a)^3$ are *not* equivalent. We also see that we can evaluate the power $(2a)^3$ by raising each factor to the power 3. This leads us to the following rule for raising a product to a power.

RAISING A PRODUCT TO A POWER

For any real numbers a and b and any integer n,

$$(ab)^n = a^nb^n.$$

(To raise a product to the nth power, raise each factor to the nth power.)

EXAMPLES Simplify.

6. $(4x^2)^3 = (4^1x^2)^3$ Since $4 = 4^1$

$= (4^1)^3 \cdot (x^2)^3$ Raising each factor to the third power

$= 4^3 \cdot x^6 = 64x^6$

7. $(5x^3y^5z^2)^4 = 5^4(x^3)^4(y^5)^4(z^2)^4$ Raising each factor to the fourth power

$= 625x^{12}y^{20}z^8$

8. $(-5x^4y^3)^3 = (-5)^3(x^4)^3(y^3)^3$

$= -125x^{12}y^9$

9. $[(-x)^{25}]^2 = (-x)^{50}$ Using the power rule

$= (-1 \cdot x)^{50}$ Using the property of -1 (Section 1.8)

$= (-1)^{50}x^{50}$

$= 1 \cdot x^{50}$ The product of an even number of negative factors is positive.

$= x^{50}$

10. $(5x^2y^{-2})^3 = 5^3(x^2)^3(y^{-2})^3 = 125x^6y^{-6}$ Be sure to raise *each* factor to the third power.

$= \frac{125x^6}{y^6}$

11. $(3x^3y^{-5}z^2)^4 = 3^4(x^3)^4(y^{-5})^4(z^2)^4 = 81x^{12}y^{-20}z^8 = \frac{81x^{12}z^8}{y^{20}}$

Do Exercises 5–8.

There is a similar rule for raising a quotient to a power.

RAISING A QUOTIENT TO A POWER

For any real numbers a and b, $b \neq 0$, and any integer n,

$$\left(\frac{a}{b}\right)^n = \frac{a^n}{b^n}.$$

(To raise a quotient to the nth power, raise both the numerator and the denominator to the nth power.) Also,

$$\left(\frac{a}{b}\right)^{-n} = \left(\frac{b}{a}\right)^n = \frac{b^n}{a^n},\ a \neq 0.$$

EXAMPLES Simplify.

12. $\left(\frac{x^2}{4}\right)^3 = \frac{(x^2)^3}{4^3} = \frac{x^6}{64}$

13. $\left(\frac{3a^4}{b^3}\right)^2 = \frac{(3a^4)^2}{(b^3)^2} = \frac{3^2(a^4)^2}{b^{3\cdot 2}} = \frac{9a^8}{b^6}$

14. $\left(\frac{y^3}{5}\right)^{-2} = \frac{(y^3)^{-2}}{5^{-2}} = \frac{y^{-6}}{5^{-2}} = \frac{\frac{1}{y^6}}{\frac{1}{5^2}} = \frac{1}{y^6} \div \frac{1}{5^2} = \frac{1}{y^6} \cdot \frac{5^2}{1} = \frac{25}{y^6}$

Example 14 might also be done as follows:

$$\left(\frac{y^3}{5}\right)^{-2} = \left(\frac{5}{y^3}\right)^2 = \frac{5^2}{(y^3)^2} = \frac{25}{y^6}.$$

Do Exercises 9–11 on the preceding page.

C Scientific Notation

There are many kinds of symbols, or notation, for numbers. You are already familiar with fraction notation, decimal notation, and percent notation. Now we study another, **scientific notation,** which makes use of exponential notation. Scientific notation is especially useful when calculations involve very large or very small numbers. The following are examples of scientific notation:

①

②

③

① *Niagara Falls*: On the Canadian side, during the summer the amount of water that spills over the falls in 1 day is about

4.9793×10^{10} gal = 49,793,000,000 gal.

② *The mass of the earth*:

6.615×10^{21} = 6,615,000,000,000,000,000,000 tons.

③ *The mass of a hydrogen atom*:

1.7×10^{-24} g = 0.0000000000000000000000017 g.

SCIENTIFIC NOTATION

Scientific notation for a number is an expression of the type

$$M \times 10^n,$$

where n is an integer, M is greater than or equal to 1 and less than 10 ($1 \le M < 10$), and M is expressed in decimal notation. 10^n is also considered to be scientific notation when $M = 1$.

Convert to scientific notation.

12. 0.000517

13. 523,000,000

Answers on page A-18

You should try to make conversions to scientific notation mentally as much as possible. Here is a handy mental device.

A positive exponent in scientific notation indicates a large number (greater than or equal to 10) and a negative exponent indicates a small number (between 0 and 1).

EXAMPLES Convert to scientific notation.

15. $78{,}000 = 7.8 \times 10^4$

7.8,000.

4 places

Large number, so the exponent is positive.

16. $0.0000057 = 5.7 \times 10^{-6}$

0.000005.7

6 places

Small number, so the exponent is negative.

Each of the following is *not* scientific notation.

$\underbrace{12.46}_{\uparrow} \times 10^7$ — This number is greater than 10.

$\underbrace{0.347}_{\uparrow} \times 10^{-5}$ — This number is less than 1.

Do Exercises 12 and 13.

Study Tips

TIME MANAGEMENT (PART 2)

Here are some additional tips to help you with time management. (See also the Study Tips on time management in Sections 2.2 and 5.2.)

- **Avoid "time killers."** We live in a media age, and the Internet, e-mail, television, and movies are all time killers. Allow yourself a break to enjoy some college and outside activities. But keep track of the time you spend on such activities and compare it to the time you spend studying.
- **Prioritize your tasks.** Be careful about taking on too many college activities that fall outside of academics. Examples of such activities are decorating a homecoming float, joining a fraternity or sorority, and participating on a student council committee. Any of these is important but keep them to a minimum to be sure that you have enough time for your studies.
- **Be aggressive about your study tasks.** Instead of worrying over your math homework or test preparation, do something to get yourself started. Work a problem here and a problem there, and before long you will accomplish the task at hand. If the task is large, break it down into smaller parts, and do one at a time. You will be surprised at how quickly the large task can then be completed.

"Time is more valuable than money. You can get more money, but you can't get more time."

Jim Rohn, motivational speaker

EXAMPLES Convert mentally to decimal notation.

17. $7.893 \times 10^5 = 789{,}300$

7.89300.
5 places

Positive exponent, so the answer is a large number.

18. $4.7 \times 10^{-8} = 0.000000047$

.00000004.7
8 places

Negative exponent, so the answer is a small number.

Do Exercises 14 and 15.

Convert to decimal notation.

14. 6.893×10^{11}

15. 5.67×10^{-5}

d Multiplying and Dividing Using Scientific Notation

MULTIPLYING

Consider the product

$$400 \cdot 2000 = 800{,}000.$$

In scientific notation, this is

$$(4 \times 10^2) \cdot (2 \times 10^3) = (4 \cdot 2)(10^2 \cdot 10^3) = 8 \times 10^5.$$

By applying the commutative and associative laws, we can find this product by multiplying $4 \cdot 2$, to get 8, and $10^2 \cdot 10^3$, to get 10^5 (we do this by adding the exponents).

EXAMPLE 19 Multiply: $(1.8 \times 10^6) \cdot (2.3 \times 10^{-4})$.

We apply the commutative and associative laws to get

$$\begin{aligned}(1.8 \times 10^6) \cdot (2.3 \times 10^{-4}) &= (1.8 \cdot 2.3) \times (10^6 \cdot 10^{-4}) \\ &= 4.14 \times 10^{6+(-4)} \\ &= 4.14 \times 10^2.\end{aligned}$$

We get 4.14 by multiplying 1.8 and 2.3. We get 10^2 by adding the exponents 6 and -4.

EXAMPLE 20 Multiply: $(3.1 \times 10^5) \cdot (4.5 \times 10^{-3})$.

We have

$$\begin{aligned}(3.1 \times 10^5) \cdot (4.5 \times 10^{-3}) &= (3.1 \times 4.5)(10^5 \cdot 10^{-3}) \\ &= 13.95 \times 10^2 && \text{Not scientific notation. 13.95 is greater than 10.} \\ &= (1.395 \times 10^1) \times 10^2 && \text{Substituting } 1.395 \times 10^1 \text{ for } 13.95 \\ &= 1.395 \times (10^1 \times 10^2) && \text{Associative law} \\ &= 1.395 \times 10^3. && \text{Adding exponents. The answer is now in scientific notation.}\end{aligned}$$

Do Exercises 16 and 17.

Multiply and write scientific notation for the result.

16. $(1.12 \times 10^{-8})(5 \times 10^{-7})$

17. $(9.1 \times 10^{-17})(8.2 \times 10^3)$

Answers on page A-18

Divide and write scientific notation for the result.

18. $\dfrac{4.2 \times 10^5}{2.1 \times 10^2}$

19. $\dfrac{1.1 \times 10^{-4}}{2.0 \times 10^{-7}}$

Answers on page A-18

DIVIDING

Consider the quotient

$$800{,}000 \div 400 = 2000.$$

In scientific notation, this is

$$(8 \times 10^5) \div (4 \times 10^2) = \frac{8 \times 10^5}{4 \times 10^2} = \frac{8}{4} \times \frac{10^5}{10^2} = 2 \times 10^3.$$

We can find this product by dividing 8 by 4, to get 2, and 10^5 by 10^2, to get 10^3 (we do this by subtracting the exponents.)

EXAMPLE 21 Divide: $(3.41 \times 10^5) \div (1.1 \times 10^{-3})$.

We have

$$\begin{aligned}(3.41 \times 10^5) \div (1.1 \times 10^{-3}) &= \frac{3.41 \times 10^5}{1.1 \times 10^{-3}} = \frac{3.41}{1.1} \times \frac{10^5}{10^{-3}} \\ &= 3.1 \times 10^{5-(-3)} \\ &= 3.1 \times 10^8.\end{aligned}$$

CALCULATOR CORNER

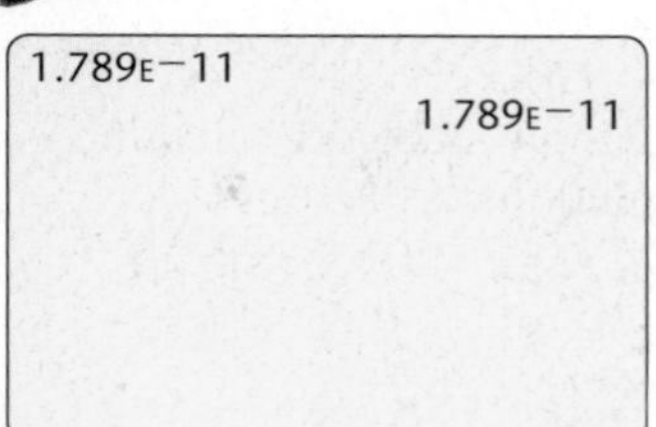

Normal **Sci** Eng
Float 0123456789
Radian Degree
Func Par Pol Seq
Connected Dot
Sequential Simul
Real a+bi re^θi
Full Horiz G−T

1.8E6*2.3E−4
4.14E2

Scientific Notation To enter a number in scientific notation on a graphing calculator, we first type the decimal portion of the number and then press [2nd] [EE]. (EE is the second operation associated with the [,] key.) Finally, we type the exponent, which can be at most two digits. For example, to enter 1.789×10^{-11} in scientific notation, we press [1] [.] [7] [8] [9] [2nd] [EE] [(−)] [1] [1] [ENTER]. The decimal portion of the number appears before a small E and the exponent follows the E.

The graphing calculator can be used to perform computations using scientific notation. To find the product in Example 19 and express the result in scientific notation, we first set the calculator in Scientific mode by pressing [MODE], positioning the cursor over Sci on the first line, and pressing [ENTER]. Then we press [2nd] [QUIT] to go to the home screen and enter the computation by pressing [1] [.] [8] [2nd] [EE] [6] [×] [2] [.] [3] [2nd] [EE] [(−)] [4] [ENTER].

Exercises: Multiply or divide and express the answer in scientific notation.

1. $(3.15 \times 10^7)(4.3 \times 10^{-12})$
2. $(4.76 \times 10^{-5})(1.9 \times 10^{10})$
3. $(8 \times 10^9)(4 \times 10^{-5})$
4. $(4 \times 10^4)(9 \times 10^7)$
5. $\dfrac{4.5 \times 10^6}{1.5 \times 10^{12}}$
6. $\dfrac{6.4 \times 10^{-5}}{1.6 \times 10^{-10}}$
7. $\dfrac{4 \times 10^{-9}}{5 \times 10^{16}}$
8. $\dfrac{9 \times 10^{11}}{3 \times 10^{-2}}$

EXAMPLE 22 Divide: $(6.4 \times 10^{-7}) \div (8.0 \times 10^{6})$.

We have

$$\begin{aligned}
(6.4 \times 10^{-7}) \div (8.0 \times 10^{6}) &= \frac{6.4 \times 10^{-7}}{8.0 \times 10^{6}} \\
&= \frac{6.4}{8.0} \times \frac{10^{-7}}{10^{6}} \\
&= 0.8 \times 10^{-7-6} \\
&= 0.8 \times 10^{-13} && \text{Not scientific notation. 0.8 is less than 1.} \\
&= (8.0 \times 10^{-1}) \times 10^{-13} && \text{Substituting } 8.0 \times 10^{-1} \text{ for } 0.8 \\
&= 8.0 \times (10^{-1} \times 10^{-13}) && \text{Associative law} \\
&= 8.0 \times 10^{-14}. && \text{Adding exponents}
\end{aligned}$$

Do Exercises 18 and 19 on the preceding page.

e Applications with Scientific Notation

EXAMPLE 23 *Distance from the Sun to Earth.* Light from the sun traveling at a rate of 300,000 kilometers per second (km/s) reaches Earth in 499 sec. Find the distance, expressed in scientific notation, from the sun to Earth.

The time t that it takes for light to reach Earth from the sun is 4.99×10^{2} sec (s). The speed is 3.0×10^{5} km/s. Recall that distance can be expressed in terms of speed and time as

Distance = Speed · Time

$d = rt.$

We substitute 3.0×10^{5} for r and 4.99×10^{2} for t:

$$\begin{aligned}
d &= rt \\
&= (3.0 \times 10^{5})(4.99 \times 10^{2}) && \text{Substituting} \\
&= 14.97 \times 10^{7} \\
&= (1.497 \times 10^{1}) \times 10^{7} \\
&= 1.497 \times (10^{1} \times 10^{7}) \\
&= 1.497 \times 10^{8} \text{ km.} && \text{Converting to scientific notation}
\end{aligned}$$

Thus the distance from the sun to Earth is 1.497×10^{8} km.

Do Exercise 20.

20. Niagara Falls Water Flow. On the Canadian side, during the summer the amount of water that spills over the falls in 1 min is about

$$1.3088 \times 10^{8} \text{ L}.$$

How much water spills over the falls in one day? Express the answer in scientific notation.

Source: *Collier's Encyclopedia*, 1997, Vol. 17

Answer on page A-18

Study Tips

WRITING ALL THE STEPS

Take the time to include all the steps when working your homework problems. Doing so will help you organize your thinking and avoid computational errors. If you find a wrong answer, having all the steps allows easier checking of your work. It will also give you complete, step-by-step solutions of the exercises that can be used to study for an exam.

Writing down all the steps and keeping your work organized may also give you a better chance of getting partial credit.

"Success comes before work only in the dictionary."

Anonymous

21. Earth vs. Saturn. The mass of Earth is about 6×10^{21} metric tons. The mass of Saturn is about 5.7×10^{23} metric tons. About how many times the mass of Earth is the mass of Saturn? Express the answer in scientific notation.

Answer on page A-18

EXAMPLE 24 *DNA.* A strand of DNA (deoxyribonucleic acid) is about 150 cm long and 1.3×10^{-10} cm wide. How many times longer is DNA than it is wide?
Source: Human Genome Project Information

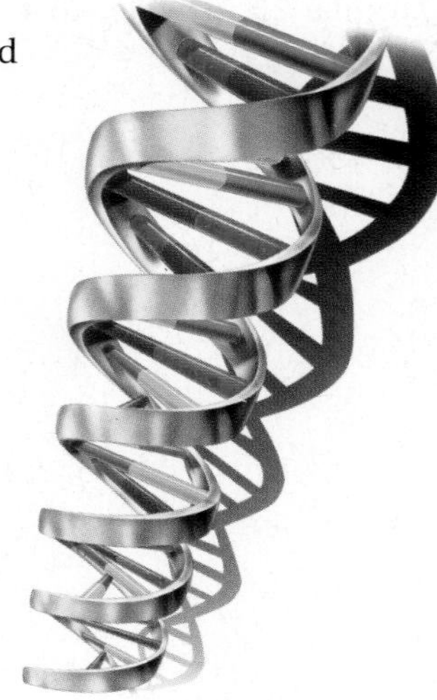

To determine how many times longer (N) DNA is than it is wide, we divide the length by the width:

$$\begin{aligned} N = \frac{150}{1.3 \times 10^{-10}} &= \frac{150}{1.3} \times \frac{1}{10^{-10}} \\ &\approx 115.385 \times 10^{10} \\ &= (1.15385 \times 10^{2}) \times 10^{10} \\ &= 1.15385 \times 10^{12}. \end{aligned}$$

Thus the length of DNA is about 1.15385×10^{12} times its width.

Do Exercise 21.

The following is a summary of the definitions and rules for exponents that we have considered in this section and the preceding one.

DEFINITIONS AND RULES FOR EXPONENTS

Exponent of 1:	$a^1 = a$
Exponent of 0:	$a^0 = 1, a \neq 0$
Negative exponents:	$a^{-n} = \frac{1}{a^n}, \frac{1}{a^{-n}} = a^n, a \neq 0$
Product Rule:	$a^m \cdot a^n = a^{m+n}$
Quotient Rule:	$\frac{a^m}{a^n} = a^{m-n}, a \neq 0$
Power Rule:	$(a^m)^n = a^{mn}$
Raising a product to a power:	$(ab)^n = a^n b^n$
Raising a quotient to a power:	$\left(\frac{a}{b}\right)^n = \frac{a^n}{b^n}, b \neq 0;$
	$\left(\frac{a}{b}\right)^{-n} = \frac{b^n}{a^n}, b \neq 0, a \neq 0$
Scientific notation:	$M \times 10^n$, or 10^n, where $1 \leq M < 10$

4.2 EXERCISE SET

For Extra Help

Digital Video Tutor CD 4 Videotape 8 | InterAct Math | Math Tutor Center | MathXL | MyMathLab

a, b Simplify.

1. $(2^3)^2$
2. $(5^2)^4$
3. $(5^2)^{-3}$
4. $(7^{-3})^5$
5. $(x^{-3})^{-4}$
6. $(a^{-5})^{-6}$
7. $(a^{-2})^9$
8. $(x^{-5})^6$
9. $(t^{-3})^{-6}$
10. $(a^{-4})^{-7}$
11. $(t^4)^{-3}$
12. $(t^5)^{-2}$
13. $(x^{-2})^{-4}$
14. $(t^{-6})^{-5}$
15. $(ab)^3$
16. $(xy)^2$
17. $(ab)^{-3}$
18. $(xy)^{-6}$
19. $(mn^2)^{-3}$
20. $(x^3y)^{-2}$
21. $(4x^3)^2$
22. $4(x^3)^2$
23. $(3x^{-4})^2$
24. $(2a^{-5})^3$
25. $(x^4y^5)^{-3}$
26. $(t^5x^3)^{-4}$
27. $(x^{-6}y^{-2})^{-4}$
28. $(x^{-2}y^{-7})^{-5}$
29. $(a^{-2}b^7)^{-5}$
30. $(q^5r^{-1})^{-3}$
31. $(5r^{-4}t^3)^2$
32. $(4x^5y^{-6})^3$
33. $(a^{-5}b^7c^{-2})^3$
34. $(x^{-4}y^{-2}z^9)^2$
35. $(3x^3y^{-8}z^{-3})^2$
36. $(2a^2y^{-4}z^{-5})^3$
37. $\left(\frac{y^3}{2}\right)^2$
38. $\left(\frac{a^5}{3}\right)^3$
39. $\left(\frac{a^2}{b^3}\right)^4$
40. $\left(\frac{x^3}{y^4}\right)^5$
41. $\left(\frac{y^2}{2}\right)^{-3}$
42. $\left(\frac{a^4}{3}\right)^{-2}$
43. $\left(\frac{7}{x^{-3}}\right)^2$
44. $\left(\frac{3}{a^{-2}}\right)^3$
45. $\left(\frac{x^2y}{z}\right)^3$
46. $\left(\frac{m}{n^4p}\right)^3$
47. $\left(\frac{a^2b}{cd^3}\right)^{-2}$
48. $\left(\frac{2a^2}{3b^4}\right)^{-3}$

c Convert to scientific notation.

49. 28,000,000,000

50. 4,900,000,000,000

51. 907,000,000,000,000,000

52. 168,000,000,000,000

53. 0.00000304

54. 0.000000000865

55. 0.000000018

56. 0.00000000002

57. 100,000,000,000

58. 0.0000001

Convert the number in the sentence to scientific notation.

59. *Population of the United States.* After the 2000 census, the population of the United States was 281 million (1 million $= 10^6$).
Source: U.S. Bureau of the Census

60. *NASCAR.* Total revenue of NASCAR (National Association of Stock Car Automobile Racing) is expected to be \$3423 million by 2006.
Source: NASCAR

61. *State Lottery.* Typically, the probability of winning a state lottery is about 1/10,000,000.

62. *Cancer Death Rate.* In Michigan, the death rate due to cancer is about 127.1/1000.
Source: AARP

Convert to decimal notation.

63. 8.74×10^7

64. 1.85×10^8

65. 5.704×10^{-8}

66. 8.043×10^{-4}

67. 10^7

68. 10^6

69. 10^{-5}

70. 10^{-8}

d Multiply or divide and write scientific notation for the result.

71. $(3 \times 10^4)(2 \times 10^5)$

72. $(3.9 \times 10^8)(8.4 \times 10^{-3})$

73. $(5.2 \times 10^5)(6.5 \times 10^{-2})$

74. $(7.1 \times 10^{-7})(8.6 \times 10^{-5})$

75. $(9.9 \times 10^{-6})(8.23 \times 10^{-8})$

76. $(1.123 \times 10^4) \times 10^{-9}$

77. $\dfrac{8.5 \times 10^{8}}{3.4 \times 10^{-5}}$

78. $\dfrac{5.6 \times 10^{-2}}{2.5 \times 10^{5}}$

79. $(3.0 \times 10^{6}) \div (6.0 \times 10^{9})$

80. $(1.5 \times 10^{-3}) \div (1.6 \times 10^{-6})$

81. $\dfrac{7.5 \times 10^{-9}}{2.5 \times 10^{12}}$

82. $\dfrac{4.0 \times 10^{-3}}{8.0 \times 10^{20}}$

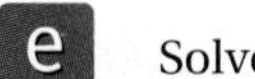

Solve.

83. *River Discharge.* The average discharge at the mouths of the Amazon River is 4,200,000 cubic feet per second. How much water is discharged from the Amazon River in 1 yr? Express the answer in scientific notation.

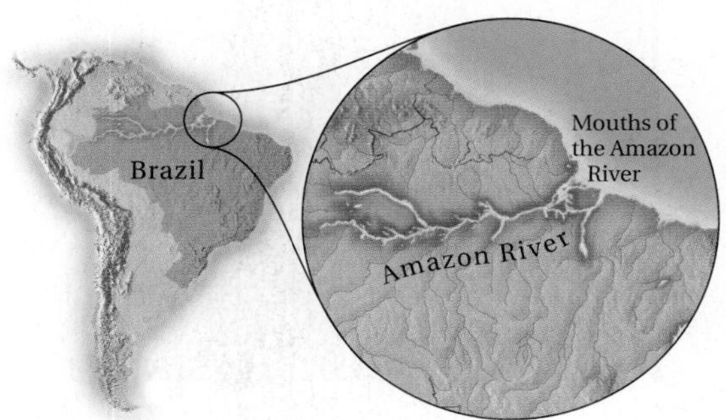

84. *Computers.* A gigabyte is a measure of a computer's storage capacity. One gigabyte holds about one billion bytes of information. If a firm's computer network contains 2500 gigabytes of memory, how many bytes are in the network? Express the answer in scientific notation.

85. *Earth vs. Jupiter.* The mass of Earth is about 6×10^{21} metric tons. The mass of Jupiter is about 1.908×10^{24} metric tons. About how many times the mass of Earth is the mass of Jupiter? Express the answer in scientific notation.

86. *Water Contamination.* In the United States, 200 million gal of used motor oil is improperly disposed of each year. One gallon of used oil can contaminate one million gallons of drinking water. How many gallons of drinking water can 200 million gallons of oil contaminate? Express the answer in scientific notation.
Source: *The Macmillan Visual Almanac*

87. *Stars.* It is estimated that there are 10 billion trillion stars in the known universe. Express the number of stars in scientific notation (1 billion $= 10^9$; 1 trillion $= 10^{12}$).

88. *Closest Star.* Excluding the sun, the closest star to Earth is Proxima Centauri, which is 4.3 light-years away (one light-year $= 5.88 \times 10^{12}$ mi). How far, in miles, is Proxima Centauri from Earth? Express the answer in scientific notation.

89. *Earth vs. Sun.* The mass of Earth is about 6×10^{21} metric tons. The mass of the sun is about 1.998×10^{27} metric tons. About how many times the mass of Earth is the mass of the sun? Express the answer in scientific notation.

90. *Red Light.* The wavelength of light is given by the velocity divided by the frequency. The velocity of red light is 300,000,000 m/sec, and its frequency is 400,000,000,000,000 cycles per second. What is the wavelength of red light? Express the answer in scientific notation.

Space Travel. Use the following information for Exercises 91 and 92.

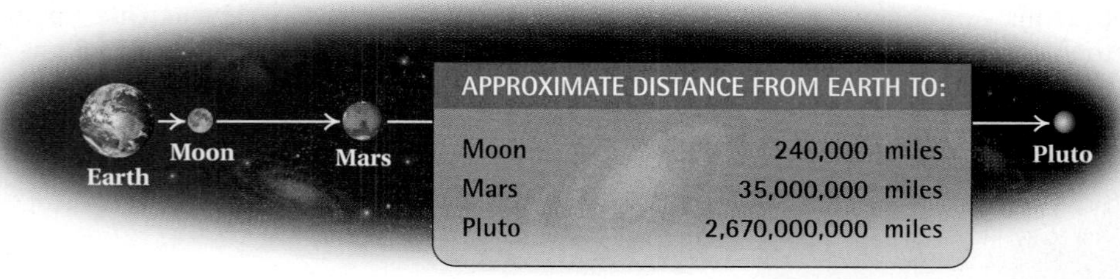

Earth
Moon
Mars
Pluto
APPROXIMATE DISTANCE FROM EARTH TO:
Moon 240,000 miles
Mars 35,000,000 miles
Pluto 2,670,000,000 miles

91. *Time to Reach Mars.* Suppose that it takes about 3 days for a space vehicle to travel from Earth to the moon. About how long would it take the same space vehicle traveling at the same speed to reach Mars? Express the answer in scientific notation.

92. *Time to Reach Pluto.* Suppose that it takes about 3 days for a space vehicle to travel from Earth to the moon. About how long would it take the same space vehicle traveling at the same speed to reach Pluto? Express the answer in scientific notation.

93. $^{D}_{W}$ Explain in your own words when exponents should be added and when they should be multiplied.

94. $^{D}_{W}$ Without performing actual computations, explain why 3^{-29} is smaller than 2^{-29}.

SKILL MAINTENANCE

Factor. [1.7d]

95. $9x - 36$

96. $4x - 2y + 16$

97. $3s + 3t + 24$

98. $-7x - 14$

Solve. [2.3b]

99. $2x - 4 - 5x + 8 = x - 3$

100. $8x + 7 - 9x = 12 - 6x + 5$

Solve. [2.3c]

101. $8(2x + 3) - 2(x - 5) = 10$

102. $4(x - 3) + 5 = 6(x + 2) - 8$

Graph. [3.2b], [3.3a]

103. $y = x - 5$

104. $2x + y = 8$

SYNTHESIS

105. Carry out the indicated operations. Express the result in scientific notation.

$$\frac{(5.2 \times 10^{6})(6.1 \times 10^{-11})}{1.28 \times 10^{-3}}$$

106. Find the reciprocal and express it in scientific notation.

$$6.25 \times 10^{-3}$$

Simplify.

107. $\dfrac{(5^{12})^2}{5^{25}}$

108. $\dfrac{a^{22}}{(a^2)^{11}}$

109. $\dfrac{(3^5)^4}{3^5 \cdot 3^4}$

110. $\left(\dfrac{5x^{-2}}{3y^{-2}z}\right)^0$

111. $\dfrac{49^{18}}{7^{35}}$

112. $\left(\dfrac{1}{a}\right)^{-n}$

113. $\dfrac{(0.4)^5}{[(0.4^3)]^2}$

114. $\left(\dfrac{4a^3b^{-2}}{5c^{-3}}\right)^1$

Determine whether each of the following is true for any pairs of integers m and n and any positive numbers x and y.

115. $x^m \cdot y^n = (xy)^{mn}$

116. $x^m \cdot y^m = (xy)^{2m}$

117. $(x - y)^m = x^m - y^m$

4.3 INTRODUCTION TO POLYNOMIALS

Objectives

- **a** Evaluate a polynomial for a given value of the variable.
- **b** Identify the terms of a polynomial.
- **c** Identify the like terms of a polynomial.
- **d** Identify the coefficients of a polynomial.
- **e** Collect the like terms of a polynomial.
- **f** Arrange a polynomial in descending order, or collect the like terms and then arrange in descending order.
- **g** Identify the degree of each term of a polynomial and the degree of the polynomial.
- **h** Identify the missing terms of a polynomial.
- **i** Classify a polynomial as a monomial, binomial, trinomial, or none of these.

We have already learned to evaluate and to manipulate certain kinds of algebraic expressions. We will now consider algebraic expressions called *polynomials*.

The following are examples of *monomials in one variable*:

$$3x^2, \quad 2x, \quad -5, \quad 37p^4, \quad 0.$$

Each expression is a constant or a constant times some variable to a nonnegative integer power.

MONOMIAL

A **monomial** is an expression of the type ax^n, where a is a real-number constant and n is a nonnegative integer.

Algebraic expressions like the following are **polynomials:**

$$\tfrac{3}{4}y^5, \quad -2, \quad 5y+3, \quad 3x^2+2x-5, \quad -7a^3+\tfrac{1}{2}a, \quad 6x, \quad 37p^4, \quad x, \quad 0.$$

POLYNOMIAL

A **polynomial** is a monomial or a combination of sums and/or differences of monomials.

The following algebraic expressions are *not* polynomials:

$$\text{(1)}\ \frac{x+3}{x-4}, \qquad \text{(2)}\ 5x^3-2x^2+\frac{1}{x}, \qquad \text{(3)}\ \frac{1}{x^3-2}.$$

Expressions (1) and (3) are not polynomials because they represent quotients, not sums. Expression (2) is not a polynomial because

$$\frac{1}{x} = x^{-1},$$

and this is not a monomial because the exponent is negative.

Do Exercise 1.

1. Write three polynomials.

Answer on page A-19

a Evaluating Polynomials and Applications

When we replace the variable in a polynomial with a number, the polynomial then represents a number called a **value** of the polynomial. Finding that number, or value, is called **evaluating the polynomial.** We evaluate a polynomial using the rules for order of operations (Section 1.8).

EXAMPLE 1 Evaluate the polynomial when $x = 2$.

a) $3x + 5 = 3 \cdot 2 + 5$
$= 6 + 5$
$= 11$

b) $2x^2 - 7x + 3 = 2 \cdot 2^2 - 7 \cdot 2 + 3$
$= 2 \cdot 4 - 7 \cdot 2 + 3$
$= 8 - 14 + 3$
$= -3$

EXAMPLE 2 Evaluate the polynomial when $x = -4$.

a) $2 - x^3 = 2 - (-4)^3 = 2 - (-64)$
$= 2 + 64$
$= 66$

b) $-x^2 - 3x + 1 = -(-4)^2 - 3(-4) + 1$
$= -16 + 12 + 1$
$= -3$

Do Exercises 2–5.

ALGEBRAIC–GRAPHICAL CONNECTION

An equation like $y = 2x - 2$, which has a polynomial on one side and y on the other, is called a **polynomial equation.** Here and in many places throughout the book, we will connect graphs to related concepts.

Recall from Chapter 3 that in order to plot points before graphing an equation, we choose values for x and compute the corresponding y-values. If the equation has y on one side and a polynomial involving x on the other, then determining y is the same as evaluating the polynomial. Once the graph of such an equation has been drawn, we can evaluate the polynomial for a given x-value by finding the y-value that is paired with it on the graph.

EXAMPLE 3 Use *only* the given graph of $y = 2x - 2$ to evaluate the polynomial $2x - 2$ when $x = 3$.

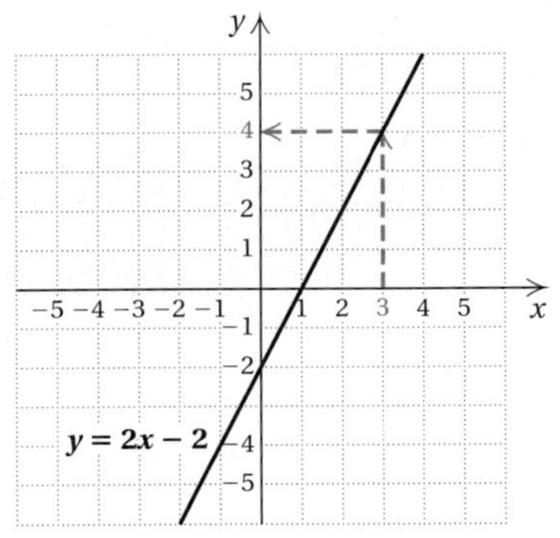

First, we locate 3 on the x-axis. From there we move vertically to the graph of the equation and then horizontally to the y-axis. There we locate the y-value that is paired with 3. Although our drawing may not be precise, it appears that the y-value 4 is paired with 3. Thus the value of $2x - 2$ is 4 when $x = 3$.

Do Exercise 6.

Polynomial equations can be used to model many real-world situations.

EXAMPLE 4 *Games in a Sports League.* In a sports league of x teams in which each team plays every other team twice, the total number of games N to be played is given by the polynomial equation

$$N = x^2 - x.$$

A women's slow-pitch softball league has 10 teams. What is the total number of games to be played?

We evaluate the polynomial when $x = 10$:

$$N = x^2 - x = 10^2 - 10 = 100 - 10 = 90.$$

The league plays 90 games.

Do Exercises 7 and 8.

Evaluate the polynomial when $x = 3$.

2. $-4x - 7$

3. $-5x^3 + 7x + 10$

Evaluate the polynomial when $x = -5$.

4. $5x + 7$

5. $2x^2 + 5x - 4$

6. Use *only* the graph shown in Example 3 to evaluate the polynomial $2x - 2$ when $x = 4$ and when $x = -1$.

7. Referring to Example 4, determine the total number of games to be played in a league of 12 teams.

8. Perimeter of a Baseball Diamond. The perimeter P of a square of side x is given by the polynomial equation $P = 4x$.

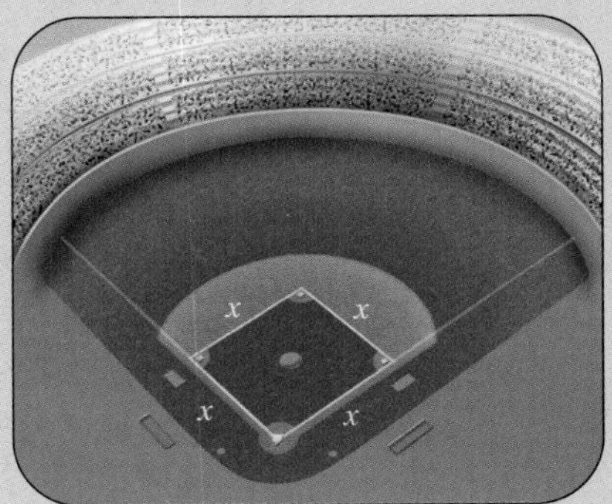

A baseball diamond is a square 90 ft on a side. Find the perimeter of a baseball diamond.

Answers on page A-19

EXAMPLE 5 *Medical Dosage.* The concentration C, in parts per million, of a certain antibiotic in the bloodstream after t hours is given by the polynomial equation

$$C = -0.05t^2 + 2t + 2.$$

Find the concentration after 2 hr.

To find the concentration after 2 hr, we evaluate the polynomial when $t = 2$:

$$\begin{aligned} C &= -0.05t^2 + 2t + 2 \\ &= -0.05(2)^2 + 2(2) + 2 \\ &= -0.05(4) + 2(2) + 2 \\ &= -0.2 + 4 + 2 \\ &= -0.2 + 6 \\ &= 5.8. \end{aligned}$$

Carrying out the calculation using the rules for order of operations

The concentration after 2 hr is 5.8 parts per million.

ALGEBRAIC–GRAPHICAL CONNECTION

The polynomial equation in Example 5 can be graphed if we evaluate the polynomial for several values of t. We list the values in a table and show the graph below. Note that the concentration peaks at the 20-hr mark and after slightly more than 40 hr, the concentration is 0. Since neither time nor concentration can be negative, our graph uses only the first quadrant.

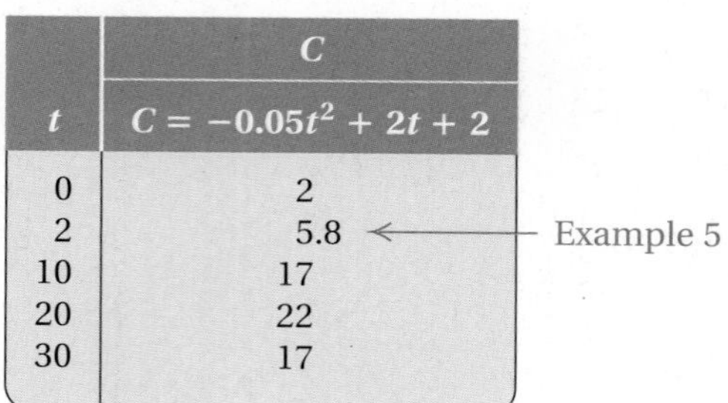

t	C $C = -0.05t^2 + 2t + 2$
0	2
2	5.8 ← Example 5
10	17
20	22
30	17

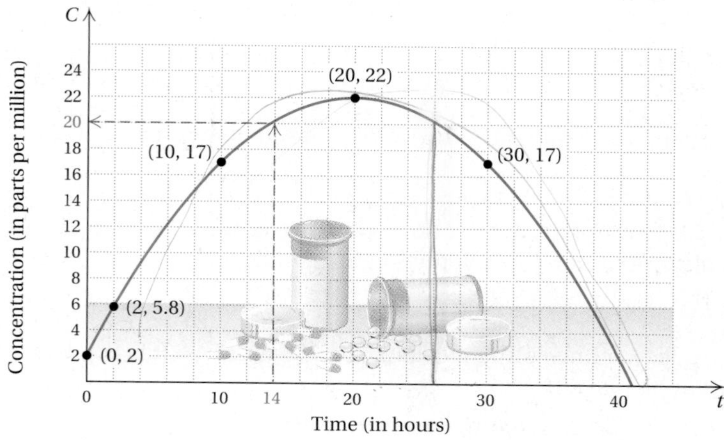

Do Exercises 9 and 10.

9. Medical Dosage.

a) Referring to Example 5, determine the concentration after 3 hr by evaluating the polynomial when $t = 3$.

b) Use *only* the graph showing medical dosage to check the value found in part (a).

10. Medical Dosage. Referring to Example 5, use *only* the graph showing medical dosage to estimate the value of the polynomial when $t = 26$.

Answers on page A-19

CALCULATOR CORNER

Evaluating Polynomials (*Note*: If you set your graphing calculator in Sci (scientific) mode to do the exercises in Section 4.2, return it to Normal mode now.)

There are several ways to evaluate polynomials on a graphing calculator. One method uses a table. To evaluate the polynomial in Example 2(b), $-x^2 - 3x + 1$, when $x = -4$, we first enter $y_1 = -x^2 - 3x + 1$ on the equation-editor screen. Then we set up a table in ASK mode (see p. 164) and enter the value -4 for x. We see that when $x = -4$, the value of Y1 is -3. This is the value of the polynomial when $x = -4$.

X	Y1	
−4	−3	

X =

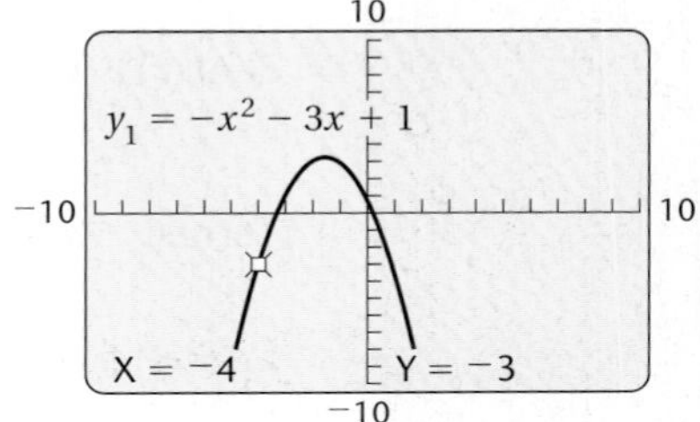

We can also use the Value feature from the CALC menu to evaluate this polynomial. First, we graph $y_1 = -x^2 - 3x + 1$ in a window that includes the x-value -4. We will use the standard window (see p. 250). Then we press [2nd] [CALC] [1] or [2nd] [CALC] [ENTER] to access the CALC menu and select item 1, Value. Now we supply the desired x-value by pressing [(−)] [4]. We then press [ENTER] to see X = −4, Y = −3 at the bottom of the screen. Thus, when $x = -4$, the value of $-x^2 - 3x + 1$ is -3.

Exercises: Use the Value feature to evaluate the polynomial for the given values of x.

1. $-x^2 - 3x + 1$, when $x = -2$, $x = -0.5$, and $x = 4$
2. $3x^2 - 5x + 2$, when $x = -3$, $x = 1$, and $x = 2.6$
3. $2x^2 - x - 8$, when $x = -3$, $x = 1.8$, and $x = 3$
4. $-5x^2 + 3x + 7$, when $x = -1$, $x = 2$, and $x = 3.4$

b Identifying Terms

As we saw in Section 1.4, subtractions can be rewritten as additions. For any polynomial that has some subtractions, we can find an equivalent polynomial using only additions.

EXAMPLES Find an equivalent polynomial using only additions.

6. $-5x^2 - x = -5x^2 + (-x)$
7. $4x^5 - 2x^6 - 4x + 7 = 4x^5 + (-2x^6) + (-4x) + 7$

Do Exercises 11 and 12.

Find an equivalent polynomial using only additions.

11. $-9x^3 - 4x^5$

12. $-2y^3 + 3y^7 - 7y$

Answers on page A-19

Identify the terms of the polynomial.

13. $3x^2 + 6x + \frac{1}{2}$

14. $-4y^5 + 7y^2 - 3y - 2$

Identify the like terms in the polynomial.

15. $4x^3 - x^3 + 2$

16. $4t^4 - 9t^3 - 7t^4 + 10t^3$

17. $5x^2 + 3x - 10 + 7x^2 - 8x + 11$

18. Identify the coefficient of each term in the polynomial

$2x^4 - 7x^3 - 8.5x^2 + 10x - 4.$

Answers on page A-19

When a polynomial has only additions, the monomials being added are called **terms.** In Example 6, the terms are $-5x^2$ and $-x$. In Example 7, the terms are $4x^5$, $-2x^6$, $-4x$, and 7.

EXAMPLE 8 Identify the terms of the polynomial

$$4x^7 + 3x + 12 + 8x^3 + 5x.$$

Terms: $4x^7$, $3x$, 12, $8x^3$, and $5x$.

If there are subtractions, you can *think* of them as additions without rewriting.

EXAMPLE 9 Identify the terms of the polynomial

$$3t^4 - 5t^6 - 4t + 2.$$

Terms: $3t^4$, $-5t^6$, $-4t$, and 2.

Do Exercises 13 and 14.

c Like Terms

When terms have the same variable and the variable is raised to the same power, we say that they are **like terms.**

EXAMPLES Identify the like terms in the polynomials.

10. $4x^3 + 5x - 4x^2 + 2x^3 + x^2$

Like terms: $4x^3$ and $2x^3$ — Same variable and exponent
Like terms: $-4x^2$ and x^2 — Same variable and exponent

11. $6 - 3a^2 + 8 - a - 5a$

Like terms: 6 and 8 — Constant terms are like terms because $6 = 6x^0$ and $8 = 8x^0$.
Like terms: $-a$ and $-5a$

Do Exercises 15–17.

d Coefficients

The coefficient of the term $5x^3$ is 5. In the following polynomial, the color numbers are the **coefficients,** 3, −2, 5, and 4:

$$3x^5 - 2x^3 + 5x + 4.$$

EXAMPLE 12 Identify the coefficient of each term in the polynomial

$$3x^4 - 4x^3 + 7x^2 + x - 8.$$

The coefficient of the first term is 3.
The coefficient of the second term is −4.
The coefficient of the third term is 7.
The coefficient of the fourth term is 1.
The coefficient of the fifth term is −8.

Do Exercise 18.

e Collecting Like Terms

We can often simplify polynomials by **collecting like terms,** or **combining like terms.** To do this, we use the distributive laws. We factor out the variable expression and add or subtract the coefficients. We try to do this mentally as much as possible.

EXAMPLES Collect like terms.

13. $2x^3 - 6x^3 = (2 - 6)x^3 = -4x^3$ Using a distributive law

14. $5x^2 + 7 + 4x^4 + 2x^2 - 11 - 2x^4 = (5 + 2)x^2 + (4 - 2)x^4 + (7 - 11)$
$= 7x^2 + 2x^4 - 4$

Note that using the distributive laws in this manner allows us to collect like terms by adding or subtracting the coefficients. Often the middle step is omitted and we add or subtract mentally, writing just the answer. In collecting like terms, we may get 0.

EXAMPLES Collect like terms.

15. $5x^3 - 5x^3 = (5 - 5)x^3 = 0x^3 = 0$

16. $3x^4 + 2x^2 - 3x^4 + 8 = (3 - 3)x^4 + 2x^2 + 8$
$= 0x^4 + 2x^2 + 8 = 2x^2 + 8$

Do Exercises 19–24.

Expressing a term like x^2 by showing 1 as a factor may make it easier to understand how to factor or collect like terms.

EXAMPLES Collect like terms.

17. $5x^2 + x^2 = 5x^2 + 1x^2$ Replacing x^2 with $1x^2$
$= (5 + 1)x^2$ Using a distributive law
$= 6x^2$

18. $5x^4 - 6x^3 - x^4 = 5x^4 - 6x^3 - 1x^4$ $x^4 = 1x^4$
$= (5 - 1)x^4 - 6x^3$
$= 4x^4 - 6x^3$

19. $\frac{2}{3}x^4 - x^3 - \frac{1}{6}x^4 + \frac{2}{5}x^3 - \frac{3}{10}x^3 = \left(\frac{2}{3} - \frac{1}{6}\right)x^4 + \left(-1 + \frac{2}{5} - \frac{3}{10}\right)x^3$
$= \left(\frac{4}{6} - \frac{1}{6}\right)x^4 + \left(-\frac{10}{10} + \frac{4}{10} - \frac{3}{10}\right)x^3$
$= \frac{3}{6}x^4 - \frac{9}{10}x^3 = \frac{1}{2}x^4 - \frac{9}{10}x^3$

Do Exercises 25–28.

f Descending and Ascending Order

Note in the following polynomial that the exponents decrease from left to right. We say that the polynomial is arranged in **descending order:**

$$2x^4 - 8x^3 + 5x^2 - x + 3.$$

The term with the largest exponent is first. The term with the next largest exponent is second, and so on. The associative and commutative laws allow us to arrange the terms of a polynomial in descending order.

Collect like terms.

19. $3x^2 + 5x^2$

20. $4x^3 - 2x^3 + 2 + 5$

21. $\frac{1}{2}x^5 - \frac{3}{4}x^5 + 4x^2 - 2x^2$

22. $24 - 4x^3 - 24$

23. $5x^3 - 8x^5 + 8x^5$

24. $-2x^4 + 16 + 2x^4 + 9 - 3x^5$

Collect like terms.

25. $7x - x$

26. $5x^3 - x^3 + 4$

27. $\frac{3}{4}x^3 + 4x^2 - x^3 + 7$

28. $8x^2 - x^2 + x^3 - 1 - 4x^2 + 10$

Answers on page A-19

EXAMPLES Arrange the polynomial in descending order.

20. $6x^5 + 4x^7 + x^2 + 2x^3 = 4x^7 + 6x^5 + 2x^3 + x^2$

21. $\frac{2}{3} + 4x^5 - 8x^2 + 5x - 3x^3 = 4x^5 - 3x^3 - 8x^2 + 5x + \frac{2}{3}$

Do Exercises 29–31.

EXAMPLE 22 Collect like terms and then arrange in descending order:

$2x^2 - 4x^3 + 3 - x^2 - 2x^3.$

$2x^2 - 4x^3 + 3 - x^2 - 2x^3 = x^2 - 6x^3 + 3$ Collecting like terms

$= -6x^3 + x^2 + 3$ Arranging in descending order

Do Exercises 32 and 33.

We usually arrange polynomials in descending order, but not always. The opposite order is called **ascending order.** Generally, if an exercise is written in a certain order, we give the answer in that same order.

g Degrees

The **degree** of a term is the exponent of the variable. The degree of the term $5x^3$ is 3.

EXAMPLE 23 Identify the degree of each term of $8x^4 + 3x + 7$.

The degree of $8x^4$ is 4.

The degree of $3x$ is 1. Recall that $x = x^1$.

The degree of 7 is 0. Think of 7 as $7x^0$. Recall that $x^0 = 1$.

The **degree of a polynomial** is the largest of the degrees of the terms, unless it is the polynomial 0. The polynomial 0 is a special case. We agree that it has *no* degree either as a term or as a polynomial. This is because we can express 0 as $0 = 0x^5 = 0x^7$, and so on, using any exponent we wish.

EXAMPLE 24 Identify the degree of the polynomial $5x^3 - 6x^4 + 7$.

$5x^3 - 6x^4 + 7.$ The largest exponent is 4.

The degree of the polynomial is 4.

Do Exercise 34.

Let's summarize the terminology that we have learned, using the polynomial $3x^4 - 8x^3 + 5x^2 + 7x - 6$.

TERM	COEFFICIENT	DEGREE OF THE TERM	DEGREE OF THE POLYNOMIAL
$3x^4$	3	4	
$-8x^3$	−8	3	
$5x^2$	5	2	4
$7x$	7	1	
-6	−6	0	

Arrange the polynomial in descending order.

29. $x + 3x^5 + 4x^3 + 5x^2 + 6x^7 - 2x^4$

30. $4x^2 - 3 + 7x^5 + 2x^3 - 5x^4$

31. $-14 + 7t^2 - 10t^5 + 14t^7$

Collect like terms and then arrange in descending order.

32. $3x^2 - 2x + 3 - 5x^2 - 1 - x$

33. $-x + \frac{1}{2} + 14x^4 - 7x - 1 - 4x^4$

34. Identify the degree of each term and the degree of the polynomial

$-6x^4 + 8x^2 - 2x + 9.$

Answers on page A-19

h Missing Terms

If a coefficient is 0, we generally do not write the term. We say that we have a **missing term.**

EXAMPLE 25 Identify the missing terms in the polynomial

$$8x^5 - 2x^3 + 5x^2 + 7x + 8.$$

There is no term with x^4. We say that the x^4-term is missing.

Do Exercises 35–38.

For certain skills or manipulations, we can write missing terms with zero coefficients or leave space.

EXAMPLE 26 Write the polynomial $x^4 - 6x^3 + 2x - 1$ in two ways: with its missing terms and by leaving space for them.

a) $x^4 - 6x^3 + 2x - 1 = x^4 - 6x^3 + 0x^2 + 2x - 1$

b) $x^4 - 6x^3 + 2x - 1 = x^4 - 6x^3 \qquad\quad + 2x - 1$

EXAMPLE 27 Write the polynomial $y^5 - 1$ in two ways: with its missing terms and by leaving space for them.

a) $y^5 - 1 = y^5 + 0y^4 + 0y^3 + 0y^2 + 0y - 1$

b) $y^5 - 1 = y^5 \qquad\qquad\qquad\qquad\quad - 1$

Do Exercises 39 and 40.

i Classifying Polynomials

Polynomials with just one term are called **monomials.** Polynomials with just two terms are called **binomials.** Those with just three terms are called **trinomials.** Those with more than three terms are generally not specified with a name.

EXAMPLE 28

MONOMIALS	BINOMIALS	TRINOMIALS	NONE OF THESE
$4x^2$	$2x + 4$	$3x^3 + 4x + 7$	$4x^3 - 5x^2 + x - 8$
9	$3x^5 + 6x$	$6x^7 - 7x^2 + 4$	$z^5 + 2z^4 - z^3 + 7z + 3$
$-23x^{19}$	$-9x^7 - 6$	$4x^2 - 6x - \frac{1}{2}$	$4x^6 - 3x^5 + x^4 - x^3 + 2x - 1$

Do Exercises 41–44.

Identify the missing terms in the polynomial.

35. $2x^3 + 4x^2 - 2$

36. $-3x^4$

37. $x^3 + 1$

38. $x^4 - x^2 + 3x + 0.25$

Write the polynomial in two ways: with its missing terms and by leaving space for them.

39. $2x^3 + 4x^2 - 2$

40. $a^4 + 10$

Classify the polynomial as a monomial, binomial, trinomial, or none of these.

41. $5x^4$

42. $4x^3 - 3x^2 + 4x + 2$

43. $3x^2 + x$

44. $3x^2 + 2x - 4$

Answers on page A-19

4.3 EXERCISE SET

a Evaluate the polynomial when $x = 4$ and when $x = -1$.

1. $-5x + 2$

2. $-8x + 1$

3. $2x^2 - 5x + 7$

4. $3x^2 + x - 7$

5. $x^3 - 5x^2 + x$

6. $7 - x + 3x^2$

Evaluate the polynomial when $x = -2$ and when $x = 0$.

7. $3x + 5$

8. $8 - 4x$

9. $x^2 - 2x + 1$

10. $5x + 6 - x^2$

11. $-3x^3 + 7x^2 - 3x - 2$

12. $-2x^3 + 5x^2 - 4x + 3$

13. *Skydiving.* During the first 13 sec of a jump, the number of feet S that a skydiver falls in t seconds can be approximated by the polynomial equation

$$S = 11.12t^2.$$

Approximately how far has a skydiver fallen 10 sec after having jumped from a plane?

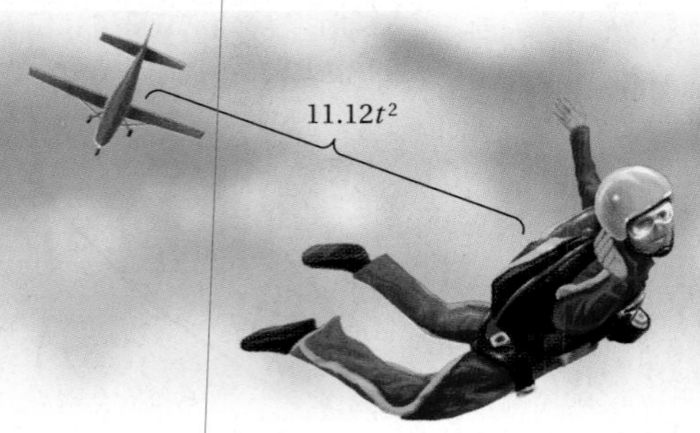

14. *Skydiving.* For jumps that exceed 13 sec, the polynomial equation

$$S = 173t - 369$$

can be used to approximate the distance S, in feet, that a skydiver has fallen in t seconds. Approximately how far has a skydiver fallen 20 sec after having jumped from a plane?

15. *Electricity Consumption.* The consumption of electricity in the United States can be approximated by the polynomial equation

$$E = 0.19t + 3.93,$$

where E is the electricity consumption, in millions of gigawatt hours, and t is the number of years since 2000—that is, $t = 0$ corresponds to 2000, $t = 5$ corresponds to 2005, and so on.

Source: Cambridge Energy Research Associates

a) Approximate the electricity consumption in 2000, 2001, 2003, 2005, 2008, and 2010.

b) Check the results of part (a) using the graph below.

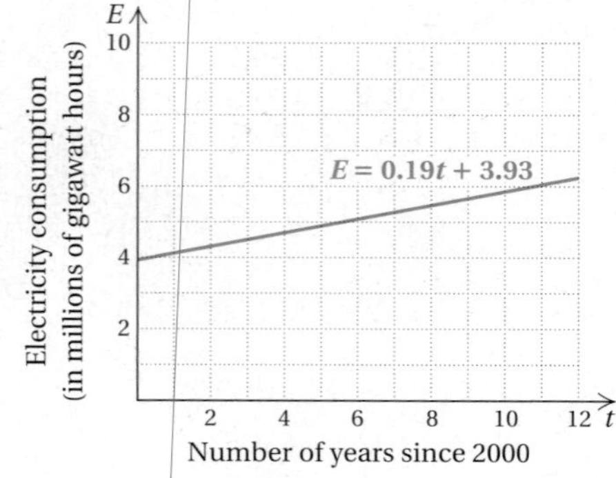

16. *Consumer Debt.* The amount of debt D, in trillions of dollars, held by consumers in the United States is given by the polynomial equation

$$D = 0.15t + 1.42,$$

where t is the number of years since 2000—that is, $t = 0$ corresponds to 2000, $t = 5$ corresponds to 2005, and so on.

Source: Federal Reserve

a) Find consumer debt in 2000, 2001, 2003, 2005, 2008, and 2010.

b) Check the results of part (a) using the graph below.

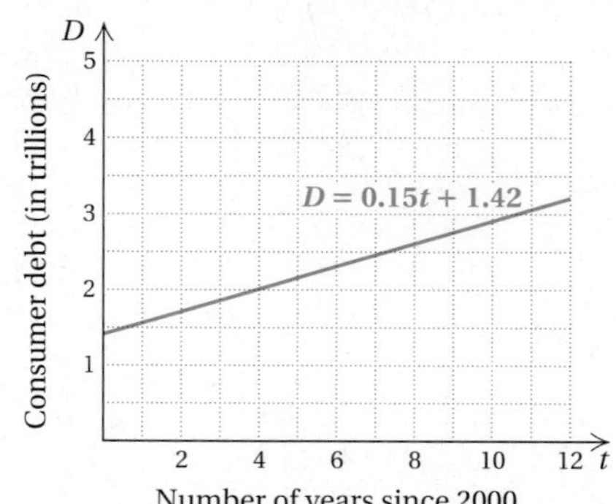

17. *Total Revenue.* Hadley Electronics is marketing a new kind of high-density TV. The firm determines that when it sells x TVs, its total revenue R (the total amount of money taken in) will be

$$R = 280x - 0.4x^2 \text{ dollars.}$$

What is the total revenue from the sale of 75 TVs? 100 TVs?

18. *Total Cost.* Hadley Electronics determines that the total cost C of producing x high-density TVs is given by

$$C = 5000 + 0.6x^2 \text{ dollars.}$$

What is the total cost of producing 500 TVs? 650 TVs?

19. The graph of the polynomial equation $y = 5 - x^2$ is shown below. Use *only* the graph to estimate the value of the polynomial when $x = -3$, $x = -1$, $x = 0$, $x = 1.5$, and $x = 2$.

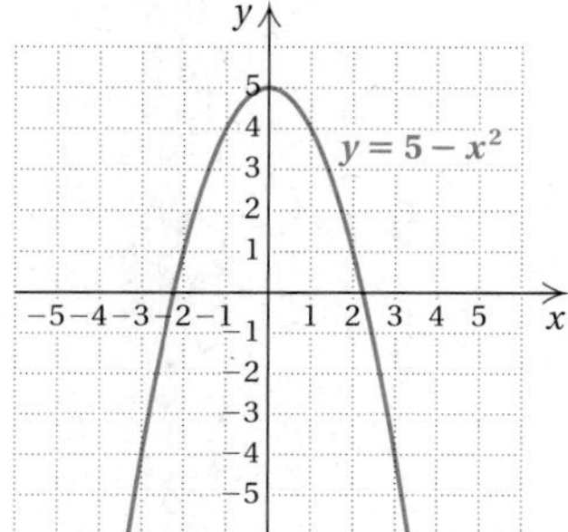

20. The graph of the polynomial equation $y = 6x^3 - 6x$ is shown below. Use *only* the graph to estimate the value of the polynomial when $x = -1$, $x = -0.5$, $x = 0.5$, $x = 1$, and $x = 1.1$.

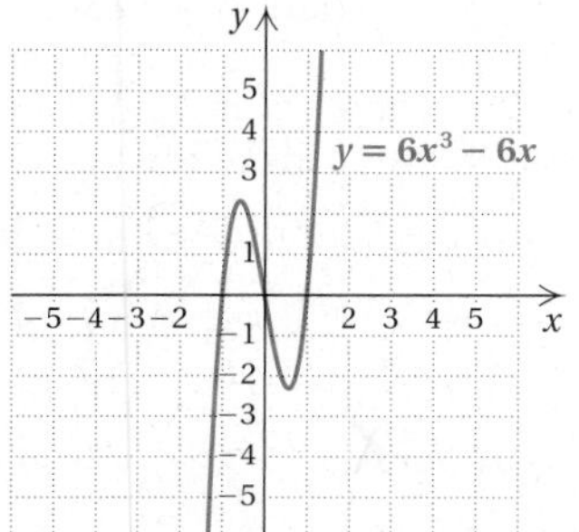

Hearing-Impaired Americans. The number N, in millions, of hearing-impaired Americans of age x can be approximated by the polynomial equation

$$N = -0.00006x^3 + 0.006x^2 - 0.1x + 1.9.$$

The graph of this equation is shown at right. Use either the graph or the polynomial equation for Exercises 21 and 22.

Source: American Speech-Language Hearing Association

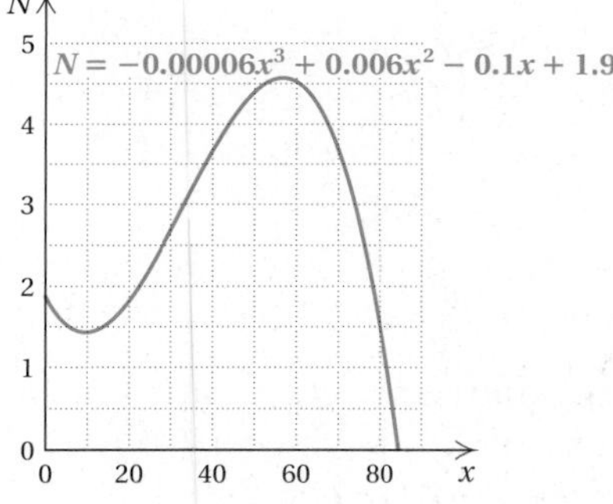

21. Approximate the number of hearing-impaired Americans of ages 20 and 40.

22. Approximate the number of hearing-impaired Americans of ages 50 and 60.

Memorizing words. Participants in a psychology experiment were able to memorize an average of M words in t minutes, where $M = -0.001t^3 + 0.1t^2$. Use the graph below for Exercises 23–28.

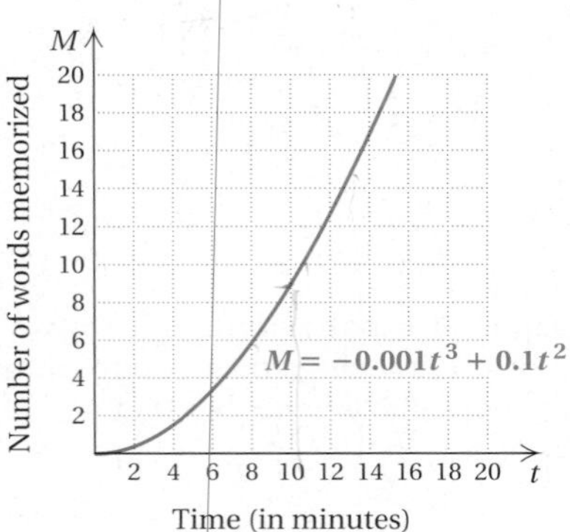

23. Estimate the number of words memorized after 10 min.

24. Estimate the number of words memorized after 14 min.

25. Find the approximate value of M for $t = 8$.

26. Find the approximate value of M for $t = 12$.

27. Estimate the value of M when t is 13.

28. Estimate the value of M when t is 7.

b Identify the terms of the polynomial.

29. $2 - 3x + x^2$

30. $2x^2 + 3x - 4$

c Identify the like terms in the polynomial.

31. $5x^3 + 6x^2 - 3x^2$

32. $3x^2 + 4x^3 - 2x^2$

33. $2x^4 + 5x - 7x - 3x^4$

34. $-3t + t^3 - 2t - 5t^3$

35. $3x^5 - 7x + 8 + 14x^5 - 2x - 9$

36. $8x^3 + 7x^2 - 11 - 4x^3 - 8x^2 - 29$

d Identify the coefficient of each term of the polynomial.

37. $-3x + 6$

38. $2x - 4$

39. $5x^2 + 3x + 3$

40. $3x^2 - 5x + 2$

41. $-5x^4 + 6x^3 - 3x^2 + 8x - 2$

42. $7x^3 - 4x^2 - 4x + 5$

e Collect like terms.

43. $2x - 5x$

44. $2x^2 + 8x^2$

45. $x - 9x$

46. $x - 5x$

47. $5x^3 + 6x^3 + 4$

48. $6x^4 - 2x^4 + 5$

49. $5x^3 + 6x - 4x^3 - 7x$

50. $3a^4 - 2a + 2a + a^4$

51. $6b^5 + 3b^2 - 2b^5 - 3b^2$

52. $2x^2 - 6x + 3x + 4x^2$

53. $\frac{1}{4}x^5 - 5 + \frac{1}{2}x^5 - 2x - 37$

54. $\frac{1}{3}x^3 + 2x - \frac{1}{6}x^3 + 4 - 16$

55. $6x^2 + 2x^4 - 2x^2 - x^4 - 4x^2$

56. $8x^2 + 2x^3 - 3x^3 - 4x^2 - 4x^2$

57. $\frac{1}{4}x^3 - x^2 - \frac{1}{6}x^2 + \frac{3}{8}x^3 + \frac{5}{16}x^3$

58. $\frac{1}{5}x^4 + \frac{1}{5} - 2x^2 + \frac{1}{10} - \frac{3}{15}x^4 + 2x^2 - \frac{3}{10}$

f Arrange the polynomial in descending order.

59. $x^5 + x + 6x^3 + 1 + 2x^2$

60. $3 + 2x^2 - 5x^6 - 2x^3 + 3x$

61. $5y^3 + 15y^9 + y - y^2 + 7y^8$

62. $9p - 5 + 6p^3 - 5p^4 + p^5$

Collect like terms and then arrange in descending order.

63. $3x^4 - 5x^6 - 2x^4 + 6x^6$

64. $-1 + 5x^3 - 3 - 7x^3 + x^4 + 5$

65. $-2x + 4x^3 - 7x + 9x^3 + 8$

66. $-6x^2 + x - 5x + 7x^2 + 1$

67. $3x + 3x + 3x - x^2 - 4x^2$

68. $-2x - 2x - 2x + x^3 - 5x^3$

69. $-x + \frac{3}{4} + 15x^4 - x - \frac{1}{2} - 3x^4$

70. $2x - \frac{5}{6} + 4x^3 + x + \frac{1}{3} - 2x$

g Identify the degree of each term of the polynomial and the degree of the polynomial.

71. $2x - 4$

72. $6 - 3x$

73. $3x^2 - 5x + 2$

74. $5x^3 - 2x^2 + 3$

75. $-7x^3 + 6x^2 + 3x + 7$

76. $5x^4 + x^2 - x + 2$

77. $x^2 - 3x + x^6 - 9x^4$

78. $8x - 3x^2 + 9 - 8x^3$

79. Complete the following table for the polynomial $-7x^4 + 6x^3 - 3x^2 + 8x - 2$.

TERM	COEFFICIENT	DEGREE OF THE TERM	DEGREE OF THE POLYNOMIAL
$-7x^4$			
$6x^3$	6		
		2	
$8x$		1	
	−2		

80. Complete the following table for the polynomial $3x^2 + 8x^5 - 46x^3 + 6x - 2.4 - \frac{1}{2}x^4$.

TERM	COEFFICIENT	DEGREE OF THE TERM	DEGREE OF THE POLYNOMIAL
		5	
$-\frac{1}{2}x^4$		4	
	−46		
$3x^2$		2	
	6		
−2.4			

h Identify the missing terms in the polynomial.

81. $x^3 - 27$

82. $x^5 + x$

83. $x^4 - x$

84. $5x^4 - 7x + 2$

85. $2x^3 - 5x^2 + x - 3$

86. $-6x^3$

Write the polynomial in two ways: with its missing terms and by leaving space for them.

87. $x^3 - 27$

88. $x^5 + x$

89. $x^4 - x$

90. $5x^4 - 7x + 2$

91. $2x^3 - 5x^2 + x - 3$

92. $-6x^3$

i Classify the polynomial as a monomial, binomial, trinomial, or none of these.

93. $x^2 - 10x + 25$

94. $-6x^4$

95. $x^3 - 7x^2 + 2x - 4$

96. $x^2 - 9$

97. $4x^2 - 25$

98. $2x^4 - 7x^3 + x^2 + x - 6$

99. $40x$

100. $4x^2 + 12x + 9$

101. ^{D}W Is it better to evaluate a polynomial before or after like terms have been collected? Why?

102. ^{D}W Explain why an understanding of the rules for order of operations is essential when evaluating polynomials.

SKILL MAINTENANCE

103. Three tired campers stopped for the night. All they had to eat was a bag of apples. During the night, one awoke and ate one-third of the apples. Later, a second camper awoke and ate one-third of the apples that remained. Much later, the third camper awoke and ate one-third of those apples yet remaining after the other two had eaten. When they got up the next morning, 8 apples were left. How many apples did they begin with? [2.6a]

Subtract. [1.4a]

104. $1 - 20$

105. $\frac{1}{8} - \frac{5}{6}$

106. $\frac{3}{8} - \left(-\frac{1}{4}\right)$

107. $5.6 - 8.2$

108. Solve: $3(x + 2) = 5x - 9.$ [2.3c]

109. Solve $C = ab - r$ for b. [2.4b]

110. A nut dealer has 1800 lb of peanuts, 1500 lb of cashews, and 700 lb of almonds. What percent of the total is peanuts? cashews? almonds? [2.5a]

111. Factor: $3x - 15y + 63.$ [1.7d]

SYNTHESIS

Collect like terms.

112. $\frac{9}{2}x^8 + \frac{1}{9}x^2 + \frac{1}{2}x^9 + \frac{9}{2}x + \frac{9}{2}x^9 + \frac{8}{9}x^2 + \frac{1}{2}x - \frac{1}{2}x^8$

113. $(3x^2)^3 + 4x^2 \cdot 4x^4 - x^4(2x)^2 + ((2x)^2)^3 - 100x^2(x^2)^2$

114. Construct a polynomial in x (meaning that x is the variable) of degree 5 with four terms and coefficients that are integers.

115. What is the degree of $(5m^5)^2$?

116. A polynomial in x has degree 3. The coefficient of x^2 is 3 less than the coefficient of x^3. The coefficient of x is three times the coefficient of x^2. The remaining coefficient is 2 more than the coefficient of x^3. The sum of the coefficients is -4. Find the polynomial.

Use the CALC feature and choose VALUE on your graphing calculator to find the values in each of the following.

117. Exercise 19

118. Exercise 20

119. Exercise 27

120. Exercise 28

4.4 ADDITION AND SUBTRACTION OF POLYNOMIALS

Objectives

a Add polynomials.

b Simplify the opposite of a polynomial.

c Subtract polynomials.

d Use polynomials to represent perimeter and area.

a Addition of Polynomials

To add two polynomials, we can write a plus sign between them and then collect like terms. Depending on the situation, you may see polynomials written in descending order, ascending order, or neither. Generally, if an exercise is written in a particular order, we write the answer in that same order.

EXAMPLE 1 Add: $(-3x^3 + 2x - 4) + (4x^3 + 3x^2 + 2)$.

$$(-3x^3 + 2x - 4) + (4x^3 + 3x^2 + 2)$$
$$= (-3 + 4)x^3 + 3x^2 + 2x + (-4 + 2) \quad \text{Collecting like terms (No signs are changed.)}$$
$$= x^3 + 3x^2 + 2x - 2$$

EXAMPLE 2 Add:

$$\left(\tfrac{2}{3}x^4 + 3x^2 - 2x + \tfrac{1}{2}\right) + \left(-\tfrac{1}{3}x^4 + 5x^3 - 3x^2 + 3x - \tfrac{1}{2}\right).$$

We have

$$\left(\tfrac{2}{3}x^4 + 3x^2 - 2x + \tfrac{1}{2}\right) + \left(-\tfrac{1}{3}x^4 + 5x^3 - 3x^2 + 3x - \tfrac{1}{2}\right)$$
$$= \left(\tfrac{2}{3} - \tfrac{1}{3}\right)x^4 + 5x^3 + (3 - 3)x^2 + (-2 + 3)x + \left(\tfrac{1}{2} - \tfrac{1}{2}\right) \quad \text{Collecting like terms}$$
$$= \tfrac{1}{3}x^4 + 5x^3 + x.$$

We can add polynomials as we do because they represent numbers. After some practice, you will be able to add mentally.

Do Exercises 1–4.

EXAMPLE 3 Add: $(3x^2 - 2x + 2) + (5x^3 - 2x^2 + 3x - 4)$.

$$(3x^2 - 2x + 2) + (5x^3 - 2x^2 + 3x - 4)$$
$$= 5x^3 + (3 - 2)x^2 + (-2 + 3)x + (2 - 4) \quad \text{You might do this step mentally.}$$
$$= 5x^3 + x^2 + x - 2 \quad \text{Then you would write only this.}$$

Do Exercises 5 and 6.

We can also add polynomials by writing like terms in columns.

EXAMPLE 4 Add: $9x^5 - 2x^3 + 6x^2 + 3$ and $5x^4 - 7x^2 + 6$ and $3x^6 - 5x^5 + x^2 + 5$.

We arrange the polynomials with the like terms in columns.

$$\begin{array}{rrrrrrr} & 9x^5 & & -\,2x^3 & +\,6x^2 & +\ 3 & \\ & & 5x^4 & & -\,7x^2 & +\ 6 & \text{We leave spaces for missing terms.} \\ 3x^6 & -\,5x^5 & & & +\ \ x^2 & +\ 5 & \\ \hline 3x^6 & +\,4x^5 & +\,5x^4 & -\,2x^3 & & +\,14 & \text{Adding} \end{array}$$

We write the answer as $3x^6 + 4x^5 + 5x^4 - 2x^3 + 14$ without the space.

Add.

1. $(3x^2 + 2x - 2) + (-2x^2 + 5x + 5)$

2. $(-4x^5 + x^3 + 4) + (7x^4 + 2x^2)$

3. $(31x^4 + x^2 + 2x - 1) + (-7x^4 + 5x^3 - 2x + 2)$

4. $(17x^3 - x^2 + 3x + 4) + \left(-15x^3 + x^2 - 3x - \frac{2}{3}\right)$

Add mentally. Try to write just the answer.

5. $(4x^2 - 5x + 3) + (-2x^2 + 2x - 4)$

6. $(3x^3 - 4x^2 - 5x + 3) + \left(5x^3 + 2x^2 - 3x - \frac{1}{2}\right)$

Answers on page A-20

Do Exercises 7 and 8.

b Opposites of Polynomials

In Section 1.8, we used the property of -1 to show that we can find the opposite of an expression like

$$-(x - 2y + 5)$$

by changing the sign of every term:

$$-(x - 2y + 5) = -x + 2y - 5.$$

This applies to polynomials as well.

OPPOSITES OF POLYNOMIALS

To find an equivalent polynomial for the **opposite,** or **additive inverse,** of a polynomial, change the sign of every term. This is the same as multiplying by -1.

EXAMPLE 5 Simplify: $-(x^2 - 3x + 4)$.

$$-(x^2 - 3x + 4) = -x^2 + 3x - 4$$

EXAMPLE 6 Simplify: $-(-t^3 - 6t^2 - t + 4)$.

$$-(-t^3 - 6t^2 - t + 4) = t^3 + 6t^2 + t - 4$$

EXAMPLE 7 Simplify: $-\left(-7x^4 - \frac{5}{9}x^3 + 8x^2 - x + 67\right)$.

$$-\left(-7x^4 - \tfrac{5}{9}x^3 + 8x^2 - x + 67\right) = 7x^4 + \tfrac{5}{9}x^3 - 8x^2 + x - 67$$

Do Exercises 9–11.

c Subtraction of Polynomials

Recall that we can subtract a real number by adding its opposite, or additive inverse: $a - b = a + (-b)$. This allows us to subtract polynomials.

EXAMPLE 8 Subtract:

$$(9x^5 + x^3 - 2x^2 + 4) - (2x^5 + x^4 - 4x^3 - 3x^2).$$

We have

$(9x^5 + x^3 - 2x^2 + 4) - (2x^5 + x^4 - 4x^3 - 3x^2)$

$= 9x^5 + x^3 - 2x^2 + 4 + [-(2x^5 + x^4 - 4x^3 - 3x^2)]$ Adding the opposite

$= 9x^5 + x^3 - 2x^2 + 4 - 2x^5 - x^4 + 4x^3 + 3x^2$ Finding the opposite by changing the sign of *each* term

$= 7x^5 - x^4 + 5x^3 + x^2 + 4.$ Adding (collecting like terms)

Do Exercises 12 and 13.

Add.

7.
$$\begin{array}{r} -2x^3 + 5x^2 - 2x + 4 \\ x^4 \qquad\quad + 6x^2 + 7x - 10 \\ -9x^4 + 6x^3 + x^2 \qquad\quad - 2 \\ \hline \end{array}$$

8. $-3x^3 + 5x + 2$ and
$x^3 + x^2 + 5$ and
$x^3 - 2x - 4$

Simplify.

9. $-(4x^3 - 6x + 3)$

10. $-(5x^4 + 3x^2 + 7x - 5)$

11. $-\left(14x^{10} - \frac{1}{2}x^5 + 5x^3 - x^2 + 3x\right)$

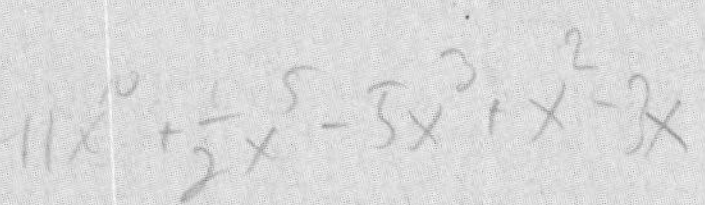

Subtract.

12. $(7x^3 + 2x + 4) - (5x^3 - 4)$

13. $(-3x^2 + 5x - 4) - (-4x^2 + 11x - 2)$

Answers on page A-20

Subtract.

14. $(-6x^4 + 3x^2 + 6) - (2x^4 + 5x^3 - 5x^2 + 7)$

15. $\left(\frac{3}{2}x^3 - \frac{1}{2}x^2 + 0.3\right) - \left(\frac{1}{2}x^3 + \frac{1}{2}x^2 + \frac{4}{3}x + 1.2\right)$

Write in columns and subtract.

16. $(4x^3 + 2x^2 - 2x - 3) - (2x^3 - 3x^2 + 2)$

17. $(2x^3 + x^2 - 6x + 2) - (x^5 + 4x^3 - 2x^2 - 4x)$

Answers on page A-20

As with similar work in Section 1.8, we combine steps by changing the sign of each term of the polynomial being subtracted and collecting like terms. Try to do this mentally as much as possible.

EXAMPLE 9 Subtract: $(9x^5 + x^3 - 2x) - (-2x^5 + 5x^3 + 6)$.

$(9x^5 + x^3 - 2x) - (-2x^5 + 5x^3 + 6)$

$= 9x^5 + x^3 - 2x + 2x^5 - 5x^3 - 6$ Finding the opposite by changing the sign of each term

$= 11x^5 - 4x^3 - 2x - 6$ Adding (collecting like terms)

Do Exercises 14 and 15.

We can use columns to subtract. We replace coefficients with their opposites, as shown in Example 8.

EXAMPLE 10 Write in columns and subtract:

$(5x^2 - 3x + 6) - (9x^2 - 5x - 3)$.

a) $\begin{array}{r} 5x^2 - 3x + 6 \\ \underline{-(9x^2 - 5x - 3)} \end{array}$ Writing like terms in columns

b) $\begin{array}{r} 5x^2 - 3x + 6 \\ \underline{-9x^2 + 5x + 3} \end{array}$ Changing signs

c) $\begin{array}{r} 5x^2 - 3x + 6 \\ \underline{-9x^2 + 5x + 3} \\ -4x^2 + 2x + 9 \end{array}$ Adding

If you can do so without error, you can arrange the polynomials in columns and write just the answer, remembering to change the signs and add.

EXAMPLE 11 Write in columns and subtract:

$(x^3 + x^2 + 2x - 12) - (-2x^3 + x^2 - 3x)$.

$\begin{array}{r} x^3 + x^2 + 2x - 12 \\ \underline{-(-2x^3 + x^2 - 3x \qquad)} \\ 3x^3 \qquad + 5x - 12. \end{array}$ Leaving space for the missing term; Adding

Do Exercises 16 and 17.

d Polynomials and Geometry

EXAMPLE 12 Find a polynomial for the sum of the areas of these rectangles.

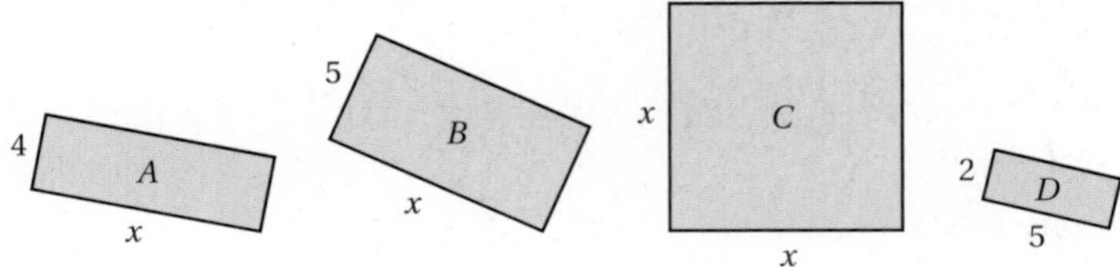

Recall that the area of a rectangle is the product of the length and the width. The sum of the areas is a sum of products. We find these products and then collect like terms.

Area of A	plus	Area of B	plus	Area of C	plus	Area of D
$4x$	$+$	$5x$	$+$	$x \cdot x$	$+$	$2 \cdot 5$

We collect like terms:

$$4x + 5x + x^2 + 10 = x^2 + 9x + 10.$$

Do Exercise 18.

EXAMPLE 13 *Lawn Area.* A water fountain with a 4-ft by 4-ft square base is placed on a square grassy park area that is x ft on a side. To determine the amount of grass seed needed for the lawn, find a polynomial for the grassy area.

We make a drawing of the situation as shown here. We then reword the problem and write the polynomial as follows.

Area of park	$-$	Area of base of fountain	$=$	Area left over
$x \cdot x$	$-$	$4 \cdot 4$	$=$	Area left over

Then $x^2 - 16\ \text{ft}^2 =$ Area left over.

Do Exercise 19.

18. Find a polynomial for the sum of the perimeters and the areas of the rectangles.

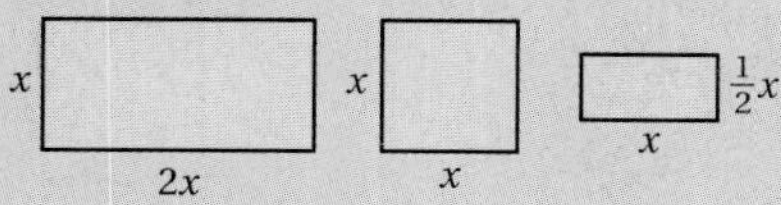

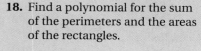

19. Lawn Area. An 8-ft by 8-ft shed is placed on a lawn x ft on a side. Find a polynomial for the remaining area.

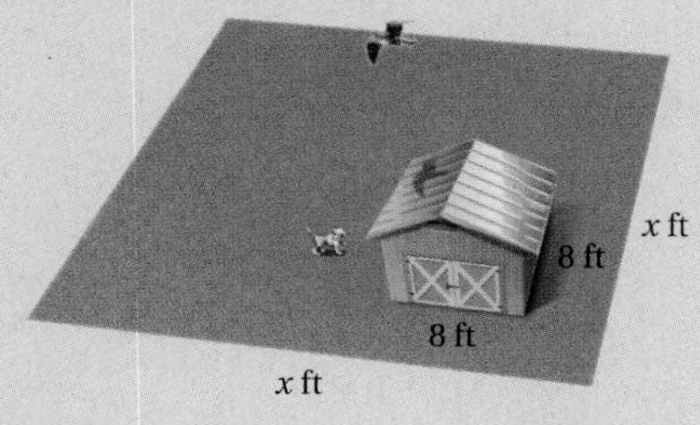

Answers on page A-20

CALCULATOR CORNER

Checking Addition and Subtraction of Polynomials A table set in AUTO mode can be used to perform a partial check that polynomials have been added or subtracted correctly. To check Example 3, we enter $y_1 = (3x^2 - 2x + 2) + (5x^3 - 2x^2 + 3x - 4)$ and $y_2 = 5x^3 + x^2 + x - 2$. If the addition has been done correctly, the values of y_1 and y_2 will be the same regardless of the table settings used.

A graph can also be used to check addition and subtraction. See the Calculator Corner on p. 343 for the procedure.

X	Y1	Y2
−2	−40	−40
−1	−7	−7
0	−2	−2
1	5	5
2	44	44
3	145	145
4	338	338
X = −2		

Exercises: Use a table to determine whether the sum or difference is correct.

1. $(-3x^3 + 2x - 4) + (4x^3 + 3x^2 + 2) = x^3 + 3x^2 + 2x - 2$
2. $(x^3 - 2x^2 + 3x - 7) + (3x^2 - 4x + 5) = x^3 + x^2 - x - 2$
3. $(5x^2 - 7x + 4) + (2x^2 + 3x - 6) = 7x^2 + 4x - 2$
4. $(9x^5 + x^3 - 2x) - (-2x^5 + 5x^3 + 6) = 11x^5 - 4x^3 - 2x - 6$
5. $(3x^4 - 2x^2 - 1) - (2x^4 - 3x^2 - 4) = x^4 + x^2 - 5$
6. $(-2x^3 + 3x^2 - 4x + 5) - (3x^2 + 2x + 8) = -2x^3 - 6x - 3$

4.4 EXERCISE SET

For Extra Help

Digital Video Tutor CD 4 Videotape 8 | InterAct Math | Math Tutor Center | MathXL | MyMathLab

a Add.

1. $(3x + 2) + (-4x + 3)$

2. $(6x + 1) + (-7x + 2)$

3. $(-6x + 2) + (x^2 + x - 3)$

4. $(x^2 - 5x + 4) + (8x - 9)$

5. $(x^2 - 9) + (x^2 + 9)$

6. $(x^3 + x^2) + (2x^3 - 5x^2)$

7. $(3x^2 - 5x + 10) + (2x^2 + 8x - 40)$

8. $(6x^4 + 3x^3 - 1) + (4x^2 - 3x + 3)$

9. $(1.2x^3 + 4.5x^2 - 3.8x) + (-3.4x^3 - 4.7x^2 + 23)$

10. $(0.5x^4 - 0.6x^2 + 0.7) + (2.3x^4 + 1.8x - 3.9)$

11. $(1 + 4x + 6x^2 + 7x^3) + (5 - 4x + 6x^2 - 7x^3)$

12. $(3x^4 - 6x - 5x^2 + 5) + (6x^2 - 4x^3 - 1 + 7x)$

13. $\left(\frac{1}{4}x^4 + \frac{2}{3}x^3 + \frac{5}{8}x^2 + 7\right) + \left(-\frac{3}{4}x^4 + \frac{3}{8}x^2 - 7\right)$

14. $\left(\frac{1}{3}x^9 + \frac{1}{5}x^5 - \frac{1}{2}x^2 + 7\right) + \left(-\frac{1}{5}x^9 + \frac{1}{4}x^4 - \frac{3}{5}x^5 + \frac{3}{4}x^2 + \frac{1}{2}\right)$

15. $(0.02x^5 - 0.2x^3 + x + 0.08) + (-0.01x^5 + x^4 - 0.8x - 0.02)$

16. $(0.03x^6 + 0.05x^3 + 0.22x + 0.05) + \left(\frac{7}{100}x^6 - \frac{3}{100}x^3 + 0.5\right)$

17. $(9x^8 - 7x^4 + 2x^2 + 5) + (8x^7 + 4x^4 - 2x) + (-3x^4 + 6x^2 + 2x - 1)$

18. $(4x^5 - 6x^3 - 9x + 1) + (6x^3 + 9x^2 + 9x) + (-4x^3 + 8x^2 + 3x - 2)$

19.
$$\begin{array}{rrrrr}
0.15x^4 & + 0.10x^3 & - 0.9x^2 & & \\
 & - 0.01x^3 & + 0.01x^2 & + x & \\
1.25x^4 & & + 0.11x^2 & & + 0.01 \\
 & 0.27x^3 & & & + 0.99 \\
-0.35x^4 & & + 15x^2 & & - 0.03 \\
\hline
\end{array}$$

20.
$$\begin{array}{rrrrr}
0.05x^4 & + 0.12x^3 & - 0.5x^2 & & \\
 & - 0.02x^3 & + 0.02x^2 & + 2x & \\
1.5x^4 & & + 0.01x^2 & & + 0.15 \\
 & 0.25x^3 & & & + 0.85 \\
-0.25x^4 & & + 10x^2 & & - 0.04 \\
\hline
\end{array}$$

b Simplify.

21. $-(-5x)$

22. $-(x^2 - 3x)$

23. $-(-x^2 + 10x - 2)$

24. $-(-4x^3 - x^2 - x)$

25. $-(12x^4 - 3x^3 + 3)$

26. $-(4x^3 - 6x^2 - 8x + 1)$

27. $-(3x - 7)$

28. $-(-2x + 4)$

29. $-(4x^2 - 3x + 2)$

30. $-(-6a^3 + 2a^2 - 9a + 1)$

31. $-\left(-4x^4 + 6x^2 + \frac{3}{4}x - 8\right)$

32. $-(-5x^4 + 4x^3 - x^2 + 0.9)$

c Subtract.

33. $(3x + 2) - (-4x + 3)$

34. $(6x + 1) - (-7x + 2)$

35. $(-6x + 2) - (x^2 + x - 3)$

36. $(x^2 - 5x + 4) - (8x - 9)$

37. $(x^2 - 9) - (x^2 + 9)$

38. $(x^3 + x^2) - (2x^3 - 5x^2)$

39. $(6x^4 + 3x^3 - 1) - (4x^2 - 3x + 3)$

40. $(-4x^2 + 2x) - (3x^3 - 5x^2 + 3)$

41. $(1.2x^3 + 4.5x^2 - 3.8x) - (-3.4x^3 - 4.7x^2 + 23)$

42. $(0.5x^4 - 0.6x^2 + 0.7) - (2.3x^4 + 1.8x - 3.9)$

43. $\left(\frac{5}{8}x^3 - \frac{1}{4}x - \frac{1}{3}\right) - \left(-\frac{1}{8}x^3 + \frac{1}{4}x - \frac{1}{3}\right)$

44. $\left(\frac{1}{5}x^3 + 2x^2 - 0.1\right) - \left(-\frac{2}{5}x^3 + 2x^2 + 0.01\right)$

45. $(0.08x^3 - 0.02x^2 + 0.01x) - (0.02x^3 + 0.03x^2 - 1)$

46. $(0.8x^4 + 0.2x - 1) - \left(\frac{7}{10}x^4 + \frac{1}{5}x - 0.1\right)$

Subtract.

47. $$\begin{array}{r} x^2 + 5x + 6 \\ \underline{-(x^2 + 2x)} \end{array}$$

48. $$\begin{array}{r} x^3 \qquad + 1 \\ \underline{-(x^3 + x^2 \qquad)} \end{array}$$

49. $$\begin{array}{r} 5x^4 + 6x^3 - 9x^2 \qquad\qquad \\ \underline{-(-6x^4 - 6x^3 \qquad + 8x + 9)} \end{array}$$

50. $$\begin{array}{r} 5x^4 \qquad + 6x^2 - 3x + 6 \\ \underline{-(\quad 6x^3 + 7x^2 - 8x - 9)} \end{array}$$

51. $$\begin{array}{r} x^5 \qquad\qquad\qquad\qquad - 1 \\ \underline{-(x^5 - x^4 + x^3 - x^2 + x - 1)} \end{array}$$

52. $$\begin{array}{r} x^5 + x^4 - x^3 + x^2 - x + 2 \\ \underline{-(x^5 - x^4 + x^3 - x^2 - x + 2)} \end{array}$$

Solve.

53. Find a polynomial for the sum of the areas of these rectangles.

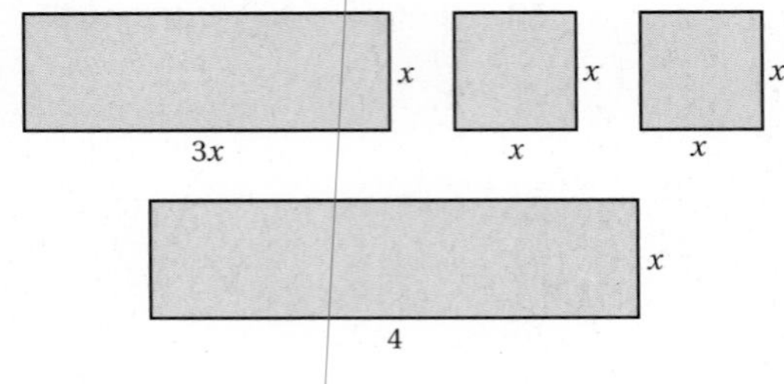

54. Find a polynomial for the sum of the areas of these circles.

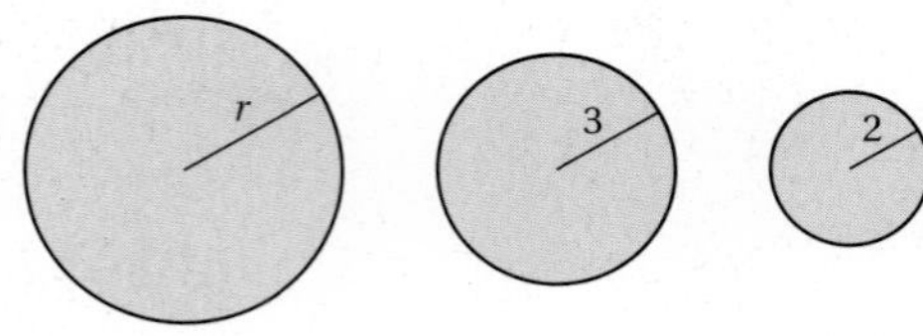

Find a polynomial for the perimeter of the figure.

55.

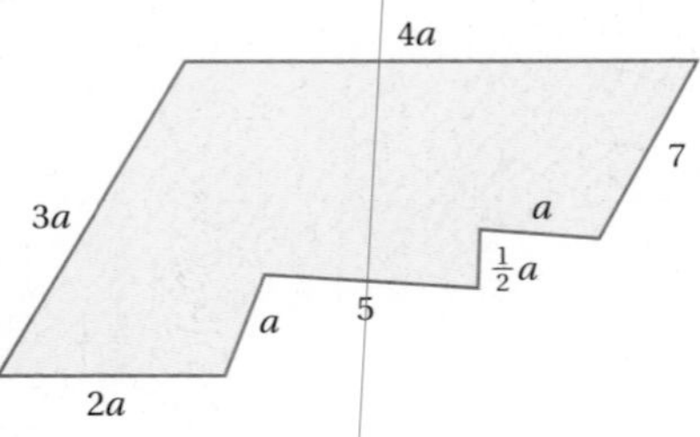

56.

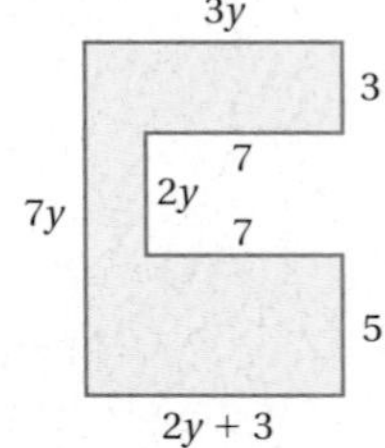

Find two algebraic expressions for the area of each figure. First, regard the figure as one large rectangle, and then regard the figure as a sum of four smaller rectangles.

57.

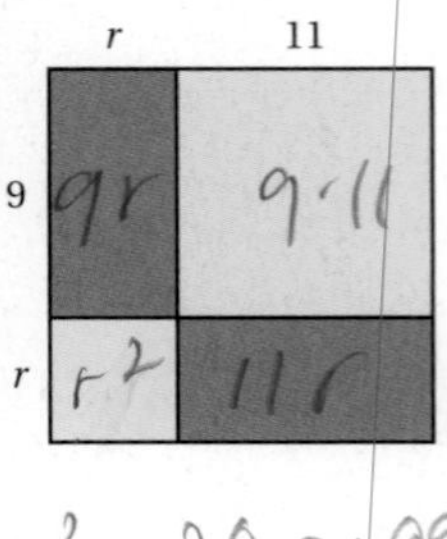

58.

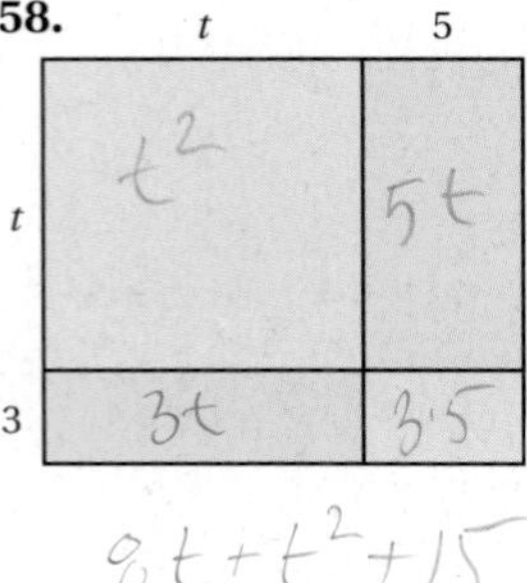

59.

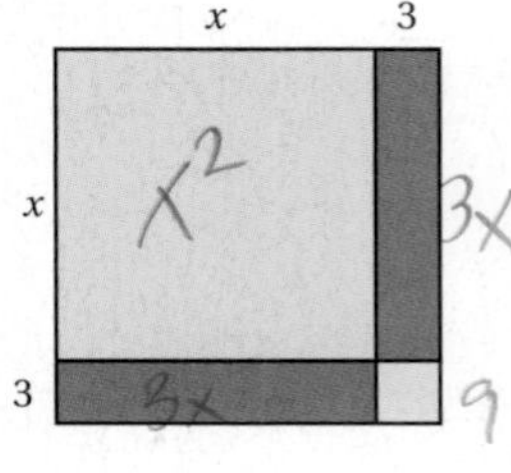

60.

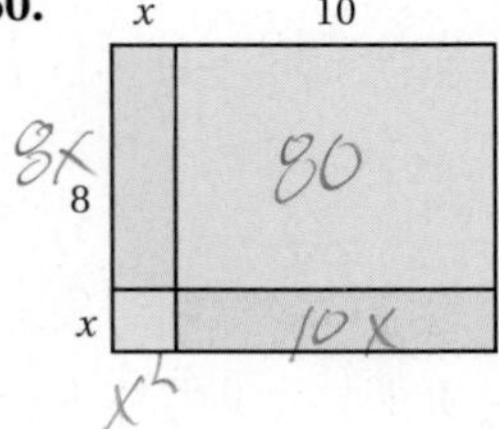

Find a polynomial for the shaded area of each figure.

61.

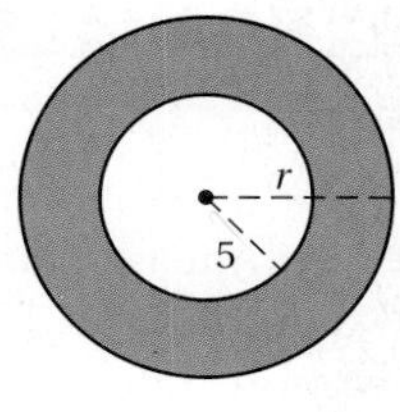

62.

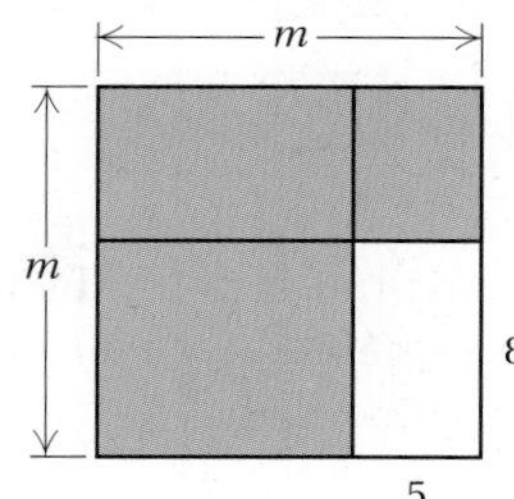

63.

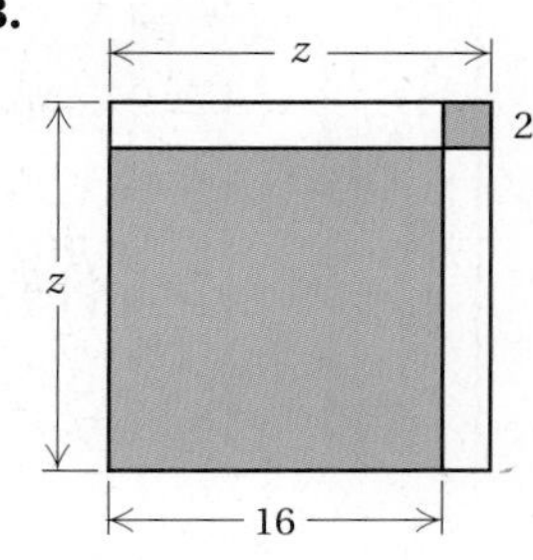

64.

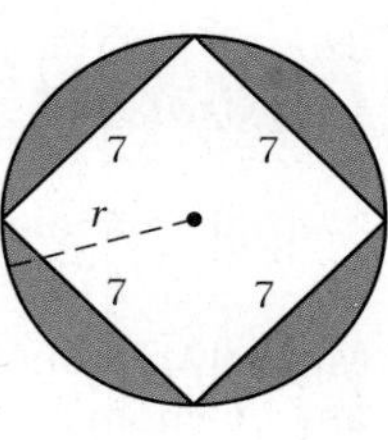

65. D_W Is the sum of two binomials ever a trinomial? Why or why not?

66. D_W Which, if any, of the commutative, associative, and distributive laws are needed for adding polynomials? Why?

SKILL MAINTENANCE

Solve. [2.3b]

67. $8x + 3x = 66$

68. $5x - 7x = 38$

69. $\frac{3}{8}x + \frac{1}{4} - \frac{3}{4}x = \frac{11}{16} + x$

70. $5x - 4 = 26 - x$

71. $1.5x - 2.7x = 22 - 5.6x$

72. $3x - 3 = -4x + 4$

Solve. [2.3c]

73. $6(y - 3) - 8 = 4(y + 2) + 5$

74. $8(5x + 2) = 7(6x - 3)$

Solve. [2.7e]

75. $3x - 7 \leq 5x + 13$

76. $2(x - 4) > 5(x - 3) + 7$

SYNTHESIS

Find a polynomial for the surface area of the right rectangular solid.

77.

78.

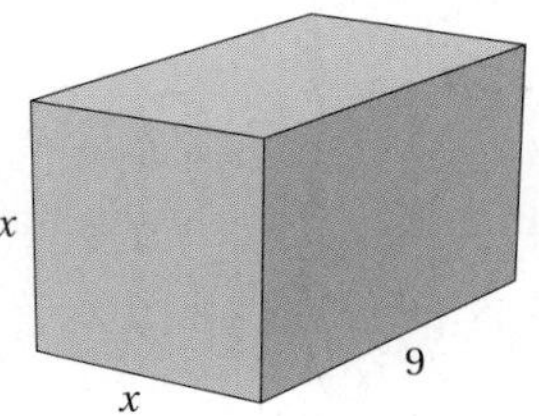

79.

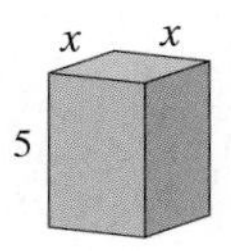

80.

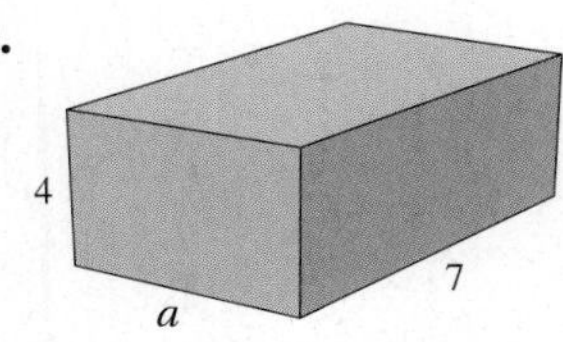

81. Find $(y - 2)^2$ using the four parts of this square.

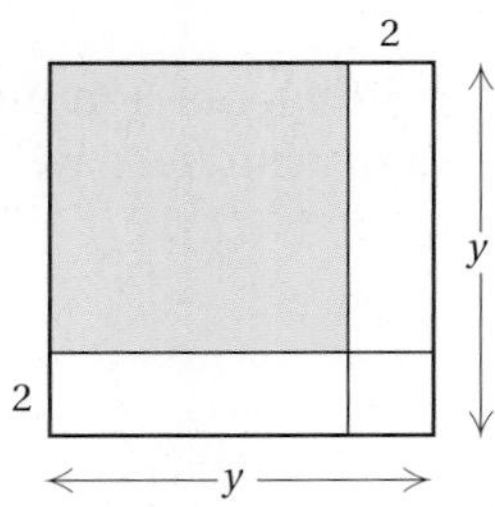

Simplify.

82. $(3x^2 - 4x + 6) - (-2x^2 + 4) + (-5x - 3)$

83. $(7y^2 - 5y + 6) - (3y^2 + 8y - 12) + (8y^2 - 10y + 3)$

84. $(-4 + x^2 + 2x^3) - (-6 - x + 3x^3) - (-x^2 - 5x^3)$

85. $(-y^4 - 7y^3 + y^2) + (-2y^4 + 5y - 2) - (-6y^3 + y^2)$

4.5 MULTIPLICATION OF POLYNOMIALS

Objectives

a Multiply monomials.

b Multiply a monomial and any polynomial.

c Multiply two binomials.

d Multiply any two polynomials.

Multiply.

1. $(3x)(-5)$

2. $(-x) \cdot x$

3. $(-x)(-x)$

4. $(-x^2)(x^3)$

5. $3x^5 \cdot 4x^2$

6. $(4y^5)(-2y^6)$

7. $(-7y^4)(-y)$

8. $7x^5 \cdot 0$

Answers on page A-20

We now multiply polynomials using techniques based, for the most part, on the distributive laws, but also on the associative and commutative laws. As we proceed in this chapter, we will develop special ways to find certain products.

a Multiplying Monomials

Consider $(3x)(4x)$. We multiply as follows:

$(3x)(4x) = 3 \cdot x \cdot 4 \cdot x$ — By the associative law of multiplication

$= 3 \cdot 4 \cdot x \cdot x$ — By the commutative law of multiplication

$= (3 \cdot 4)(x \cdot x)$ — By the associative law

$= 12x^2.$ — Using the product rule for exponents

MULTIPLYING MONOMIALS

To find an equivalent expression for the product of two monomials, multiply the coefficients and then multiply the variables using the product rule for exponents.

EXAMPLES Multiply.

1. $5x \cdot 6x = (5 \cdot 6)(x \cdot x)$ — By the associative and commutative laws

$= 30x^2$ — Multiplying the coefficients and multiplying the variables

2. $(3x)(-x) = (3x)(-1x)$

$= (3)(-1)(x \cdot x) = -3x^2$

3. $(-7x^5)(4x^3) = (-7 \cdot 4)(x^5 \cdot x^3)$

$= -28x^{5+3}$ — Adding the exponents

$= -28x^8$ — Simplifying

After some practice, you can do this mentally. Multiply the coefficients and then the variables by keeping the base and adding the exponents. Write only the answer.

Do Exercises 1–8.

b Multiplying a Monomial and Any Polynomial

To find an equivalent expression for the product of a monomial, such as $2x$, and a binomial, such as $5x + 3$, we use a distributive law and multiply each term of $5x + 3$ by $2x$.

EXAMPLE 4 Multiply: $2x(5x + 3)$.

$2x(5x + 3) = (2x)(5x) + (2x)(3)$ — Using a distributive law

$= 10x^2 + 6x$ — Multiplying the monomials

EXAMPLE 5 Multiply: $5x(2x^2 - 3x + 4)$.

$$5x(2x^2 - 3x + 4) = (5x)(2x^2) - (5x)(3x) + (5x)(4)$$
$$= 10x^3 - 15x^2 + 20x$$

MULTIPLYING A MONOMIAL AND A POLYNOMIAL

To multiply a monomial and a polynomial, multiply each term of the polynomial by the monomial.

EXAMPLE 6 Multiply: $-2x^2(x^3 - 7x^2 + 10x - 4)$.

$$-2x^2(x^3 - 7x^2 + 10x - 4) = -2x^5 + 14x^4 - 20x^3 + 8x^2$$

Do Exercises 9–11.

C Multiplying Two Binomials

To find an equivalent expression for the product of two binomials, we use the distributive laws more than once. In Example 7, we use a distributive law three times.

EXAMPLE 7 Multiply: $(x + 5)(x + 4)$.

$(x + 5)(x + 4) = x(x + 4) + 5(x + 4)$ Using a distributive law

$= x \cdot x + x \cdot 4 + 5 \cdot x + 5 \cdot 4$ Using a distributive law on each part

$= x^2 + 4x + 5x + 20$ Multiplying the monomials

$= x^2 + 9x + 20$ Collecting like terms

To visualize the product in Example 7, consider a rectangle of length $x + 5$ and width $x + 4$.

	x	5
4	$4x$	20
x	x^2	$5x$

($x + 4$ by $x + 5$)

The total area can be expressed as $(x + 5)(x + 4)$ or, by adding the four smaller areas, $x^2 + 5x + 4x + 20$.

Do Exercises 12–14.

Multiply.

9. $4x(2x + 4)$

10. $3t^2(-5t + 2)$

11. $-5x^3(x^3 + 5x^2 - 6x + 8)$

12. Multiply: $(y + 2)(y + 7)$.

a) Fill in the blanks in the steps of the solution below.

$(y + 2)(y + 7)$

$= y \cdot$ _____ $+ 2 \cdot$ _____

$= y \cdot$ _____ $+ y \cdot$ _____

$+ 2 \cdot$ _____ $+ 2 \cdot$ _____

$=$ _____ $+$ _____

$+$ _____ $+$ _____

$= y^2 +$ _____ $+ 14$

b) Write an algebraic expression that represents the area of the four smaller rectangles in the figure shown here.

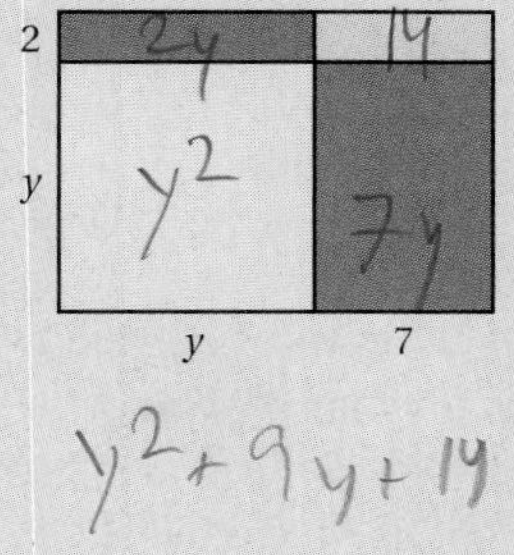

Multiply.

13. $(x + 8)(x + 5)$

14. $(x + 5)(x - 4)$

Answers on page A-20

Multiply.

15. $(5x + 3)(x - 4)$

16. $(2x - 3)(3x - 5)$

Multiply.

17. $(x^2 + 3x - 4)(x^2 + 5)$

18. $(3y^2 - 7)(2y^3 - 2y + 5)$

Answers on page A-20

EXAMPLE 8 Multiply: $(4x + 3)(x - 2)$.

$(4x + 3)(x - 2) = 4x(x - 2) + 3(x - 2)$	Using a distributive law
$= 4x \cdot x - 4x \cdot 2 + 3 \cdot x - 3 \cdot 2$	Using a distributive law on each part
$= 4x^2 - 8x + 3x - 6$	Multiplying the monomials
$= 4x^2 - 5x - 6$	Collecting like terms

Do Exercises 15 and 16.

d Multiplying Any Two Polynomials

Let's consider the product of a binomial and a trinomial. We use a distributive law four times. You may see ways to skip some steps and do the work mentally.

EXAMPLE 9 Multiply: $(x^2 + 2x - 3)(x^2 + 4)$.

$$\begin{aligned}(x^2 + 2x - 3)(x^2 + 4) &= x^2(x^2 + 4) + 2x(x^2 + 4) - 3(x^2 + 4)\\ &= x^2 \cdot x^2 + x^2 \cdot 4 + 2x \cdot x^2 + 2x \cdot 4 - 3 \cdot x^2 - 3 \cdot 4\\ &= x^4 + 4x^2 + 2x^3 + 8x - 3x^2 - 12\\ &= x^4 + 2x^3 + x^2 + 8x - 12\end{aligned}$$

Do Exercises 17 and 18.

PRODUCT OF TWO POLYNOMIALS

To multiply two polynomials P and Q, select one of the polynomials—say, P. Then multiply each term of P by every term of Q and collect like terms.

To use columns for long multiplication, multiply each term in the top row by every term in the bottom row. We write like terms in columns, and then add the results. Such multiplication is like multiplying with whole numbers:

3 2 1	300 + 20 + 1	
× 1 2	× 10 + 2	
6 4 2	600 + 40 + 2	Multiplying the top row by 2
3 2 1	3000 + 200 + 10	Multiplying the top row by 10
3 8 5 2	3000 + 800 + 50 + 2	Adding

EXAMPLE 10 Multiply: $(4x^3 - 2x^2 + 3x)(x^2 + 2x)$.

$4x^3 - 2x^2 + 3x$	
$x^2 + 2x$	
$8x^4 - 4x^3 + 6x^2$	Multiplying the top row by $2x$
$4x^5 - 2x^4 + 3x^3$	Multiplying the top row by x^2
$4x^5 + 6x^4 - x^3 + 6x^2$	Collecting like terms
	Line up like terms in columns.

EXAMPLE 11 Multiply: $(5x^3 - 3x + 4)(-2x^2 - 3)$.

When missing terms occur, it helps to leave spaces for them and align like terms as we multiply.

$$\begin{array}{rrrrrl} & 5x^3 & & -3x & +4 & \\ & & -2x^2 & & -3 & \\ \hline & -15x^3 & & +9x & -12 & \text{Multiplying by } -3 \\ -10x^5 & +6x^3 & -8x^2 & & & \text{Multiplying by } -2x^2 \\ \hline -10x^5 & -9x^3 & -8x^2 & +9x & -12 & \text{Collecting like terms} \end{array}$$

Do Exercises 19 and 20.

EXAMPLE 12 Multiply: $(2x^2 + 3x - 4)(2x^2 - x + 3)$.

$$\begin{array}{rrrrrl} & & 2x^2 & +3x & -4 & \\ & & 2x^2 & -x & +3 & \\ \hline & & 6x^2 & +9x & -12 & \text{Multiplying by } 3 \\ & -2x^3 & -3x^2 & +4x & & \text{Multiplying by } -x \\ 4x^4 & +6x^3 & -8x^2 & & & \text{Multiplying by } 2x^2 \\ \hline 4x^4 & +4x^3 & -5x^2 & +13x & -12 & \text{Collecting like terms} \end{array}$$

Do Exercise 21.

Multiply.

19. $\begin{array}{r} 3x^2 - 2x + 4 \\ x + 5 \\ \hline \end{array}$

20. $\begin{array}{r} -5x^2 + 4x + 2 \\ -4x^2 - 8 \\ \hline \end{array}$

21. Multiply.

$\begin{array}{r} 3x^2 - 2x - 5 \\ 2x^2 + x - 2 \\ \hline \end{array}$

Answers on page A-20

CALCULATOR CORNER

Checking Multiplication of Polynomials A partial check of multiplication of polynomials can be performed graphically on the TI-83 Plus graphing calculator. Consider the product $(x + 3)(x - 2) = x^2 + x - 6$. We will use two graph styles to determine whether this product is correct. First, we press MODE to determine whether Sequential mode is selected. If it is not, we position the blinking cursor over Sequential and then press ENTER. Next, on the Y = screen, we enter $y_1 = (x + 3)(x - 2)$ and $y_2 = x^2 + x - 6$. We will select the line-graph style for y_1 and the path style for y_2. To select these graph styles, we use ◁ to position the cursor over the icon to the left of the equation and press ENTER repeatedly until the desired style of icon appears, as shown below.

Plot1 Plot2 Plot3
\Y1 ■ (X+3)(X−2)
-0Y2 ■ X2+X−6
\Y3 =
\Y4 =
\Y5 =
\Y6 =
\Y7 =

$y_1 = (x + 3)(x - 2)$, $y_2 = x^2 + x - 6$

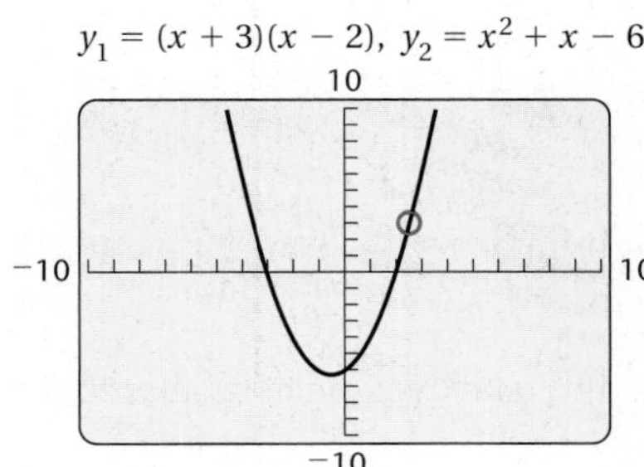

The graphing calculator will graph y_1 first as a solid line. Then it will graph y_2 as the circular cursor traces the leading edge of the graph, allowing us to determine visually whether the graphs coincide. In this case, the graphs appear to coincide, so the factorization is probably correct.

A table can also be used to perform a partial check of a product. See the Calculator Corner on p. 335 for the procedure.

Exercises: Determine graphically whether the following products are correct.

1. $(x + 5)(x + 4) = x^2 + 9x + 20$
2. $(4x + 3)(x - 2) = 4x^2 - 5x - 6$
3. $(5x + 3)(x - 4) = 5x^2 + 17x - 12$
4. $(2x - 3)(3x - 5) = 6x^2 - 19x - 15$

4.5 EXERCISE SET

For Extra Help

Digital Video Tutor CD 4 Videotape 9 | InterAct Math | Math Tutor Center | MathXL | MyMathLab

a Multiply.

1. $(8x^2)(5)$

2. $(4x^2)(-2)$

3. $(-x^2)(-x)$

4. $(-x^3)(x^2)$

5. $(8x^5)(4x^3)$

6. $(10a^2)(2a^2)$

7. $(0.1x^6)(0.3x^5)$

8. $(0.3x^4)(-0.8x^6)$

9. $\left(-\frac{1}{5}x^3\right)\left(-\frac{1}{3}x\right)$

10. $\left(-\frac{1}{4}x^4\right)\left(\frac{1}{5}x^8\right)$

11. $(-4x^2)(0)$

12. $(-4m^5)(-1)$

13. $(3x^2)(-4x^3)(2x^6)$

14. $(-2y^5)(10y^4)(-3y^3)$

b Multiply.

15. $2x(-x + 5)$

16. $3x(4x - 6)$

17. $-5x(x - 1)$

18. $-3x(-x - 1)$

19. $x^2(x^3 + 1)$

20. $-2x^3(x^2 - 1)$

21. $3x(2x^2 - 6x + 1)$

22. $-4x(2x^3 - 6x^2 - 5x + 1)$

23. $(-6x^2)(x^2 + x)$

24. $(-4x^2)(x^2 - x)$

25. $(3y^2)(6y^4 + 8y^3)$

26. $(4y^4)(y^3 - 6y^2)$

c Multiply.

27. $(x + 6)(x + 3)$

28. $(x + 5)(x + 2)$

29. $(x + 5)(x - 2)$

30. $(x + 6)(x - 2)$

31. $(x - 4)(x - 3)$

32. $(x - 7)(x - 3)$

33. $(x + 3)(x - 3)$

34. $(x + 6)(x - 6)$

35. $(5 - x)(5 - 2x)$

36. $(3 + x)(6 + 2x)$

37. $(2x + 5)(2x + 5)$

38. $(3x - 4)(3x - 4)$

39. $\left(x - \frac{5}{2}\right)\left(x + \frac{2}{5}\right)$

40. $\left(x + \frac{4}{3}\right)\left(x + \frac{3}{2}\right)$

41. $(x - 2.3)(x + 4.7)$

42. $(2x + 0.13)(2x - 0.13)$

Draw and label rectangles similar to the one following Example 7 to illustrate each product.

43. $x(x + 5)$

44. $x(x + 2)$

45. $(x + 1)(x + 2)$

46. $(x + 3)(x + 1)$

47. $(x + 5)(x + 3)$

48. $(x + 4)(x + 6)$

49. $(3x + 2)(3x + 2)$

50. $(5x + 3)(5x + 3)$

d Multiply.

51. $(x^2 + x + 1)(x - 1)$

52. $(x^2 + x - 2)(x + 2)$

53. $(2x + 1)(2x^2 + 6x + 1)$

54. $(3x - 1)(4x^2 - 2x - 1)$

55. $(y^2 - 3)(3y^2 - 6y + 2)$

56. $(3y^2 - 3)(y^2 + 6y + 1)$

57. $(x^3 + x^2)(x^3 + x^2 - x)$

58. $(x^3 - x^2)(x^3 - x^2 + x)$

59. $(-5x^3 - 7x^2 + 1)(2x^2 - x)$

60. $(-4x^3 + 5x^2 - 2)(5x^2 + 1)$

61. $(1 + x + x^2)(-1 - x + x^2)$

62. $(1 - x + x^2)(1 - x + x^2)$

63. $(2t^2 - t - 4)(3t^2 + 2t - 1)$

64. $(3a^2 - 5a + 2)(2a^2 - 3a + 4)$

65. $(x - x^3 + x^5)(x^2 - 1 + x^4)$

66. $(x - x^3 + x^5)(3x^2 + 3x^6 + 3x^4)$

67. $(x^3 + x^2 + x + 1)(x - 1)$

68. $(x + 2)(x^3 - x^2 + x - 2)$

69. $(x + 1)(x^3 + 7x^2 + 5x + 4)$

70. $(x + 2)(x^3 + 5x^2 + 9x + 3)$

71. $(x - \frac{1}{2})(2x^3 - 4x^2 + 3x - \frac{2}{5})$

72. $(x + \frac{1}{3})(6x^3 - 12x^2 - 5x + \frac{1}{2})$

73. ^{D}W Under what conditions will the product of two binomials be a trinomial?

74. ^{D}W How can the following figure be used to show that $(x + 3)^2 \neq x^2 + 9$?

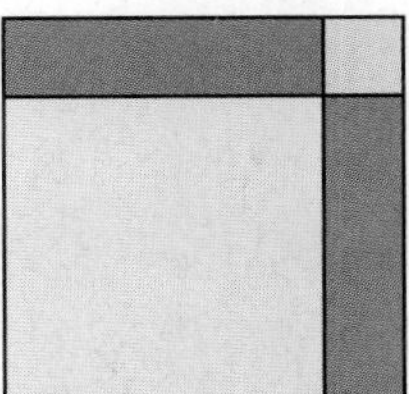

SKILL MAINTENANCE

Simplify.

75. $-\frac{1}{4} - \frac{1}{2}$ [1.4a]

76. $-3.8 - (-10.2)$ [1.4a]

77. $(10 - 2)(10 + 2)$ [1.8d]

78. $10 - 2 + (-6)^2 \div 3 \cdot 2$ [1.8d]

Factor. [1.7d]

79. $15x - 18y + 12$

80. $16x - 24y + 36$

81. $-9x - 45y + 15$

82. $100x - 100y + 1000a$

83. Graph: $y = \frac{1}{2}x - 3$. [3.2b]

84. Solve: $4(x - 3) = 5(2 - 3x) + 1$. [2.3c]

SYNTHESIS

Find a polynomial for the shaded area of each figure.

85.

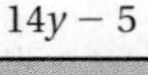

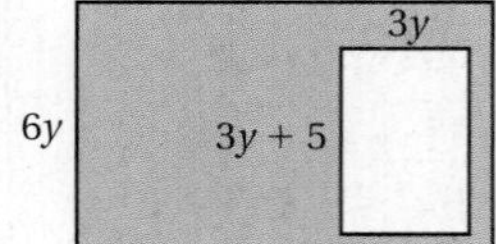

86.

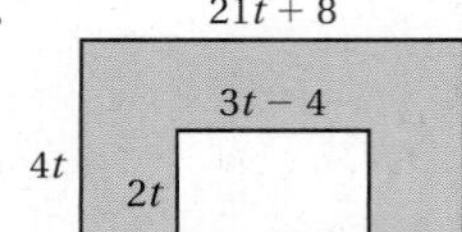

87. A box with a square bottom is to be made from a 12-in.-square piece of cardboard. Squares with side x are cut out of the corners and the sides are folded up. Find the polynomials for the volume and the outside surface area of the box.

For each figure, determine what the missing number must be in order for the figure to have the given area.

88. Area $= x^2 + 7x + 10$

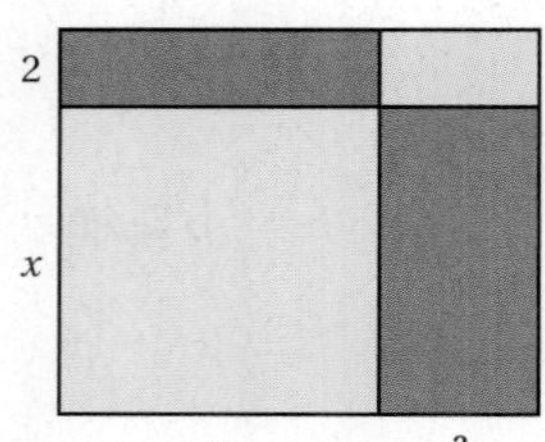

89. Area $= x^2 + 8x + 15$

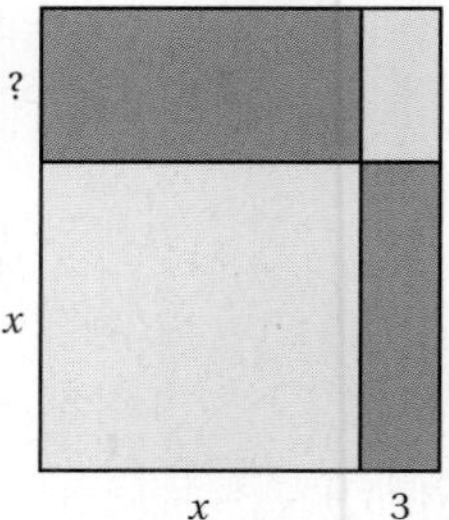

90. An open wooden box is a cube with side x cm. The box, including its bottom, is made of wood that is 1 cm thick. Find a polynomial for the interior volume of the cube.

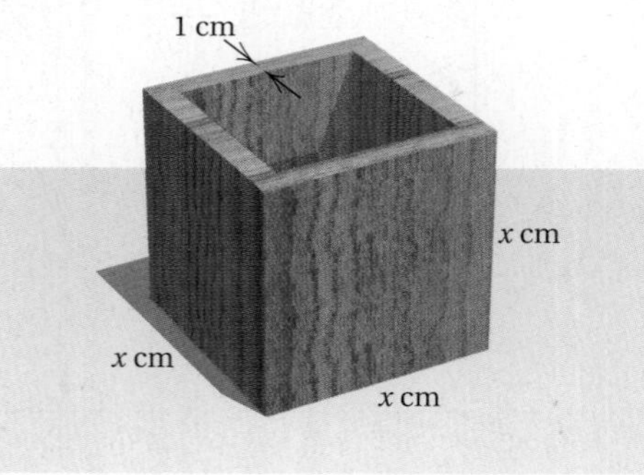

91. Find a polynomial for the volume of the solid shown below.

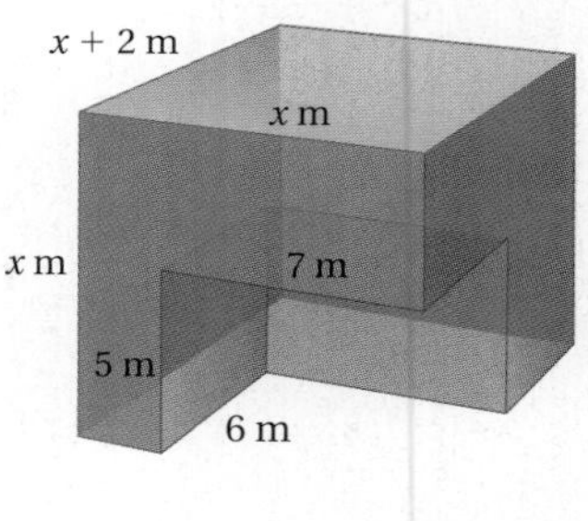

Compute and simplify.

92. $(x + 3)(x + 6) + (x + 3)(x + 6)$

93. $(x - 2)(x - 7) - (x - 7)(x - 2)$

94. $(x + 5)^2 - (x - 3)^2$

95. Extend the pattern and simplify

$$(x - a)(x - b)(x - c)(x - d) \cdots (x - z).$$

96. Use a graphing calculator to check your answers to Exercises 15, 29, and 51. Use graphs, tables, or both, as directed by your instructor.

4.6 SPECIAL PRODUCTS

Objectives

a. Multiply two binomials mentally using the FOIL method.

b. Multiply the sum and the difference of two terms mentally.

c. Square a binomial mentally.

d. Find special products when polynomial products are mixed together.

We encounter certain products so often that it is helpful to have faster methods of computing. We now consider special ways of multiplying any two binomials. Such techniques are called *special products*.

a Products of Two Binomials Using FOIL

To multiply two binomials, we can select one binomial and multiply each term of that binomial by every term of the other. Then we collect like terms. Consider the product $(x + 5)(x + 4)$:

$$\begin{aligned}(x + 5)(x + 4) &= x \cdot x + 5 \cdot x + x \cdot 4 + 5 \cdot 4\\ &= x^2 + 5x + 4x + 20\\ &= x^2 + 9x + 20.\end{aligned}$$

We can rewrite the first line of this product to show a special technique for finding the product of two binomials:

First terms, Outside terms, Inside terms, Last terms

$$(x + 5)(x + 4) = x \cdot x + 4 \cdot x + 5 \cdot x + 5 \cdot 4.$$

To remember this method of multiplying, we use the initials **FOIL.**

THE FOIL METHOD

To multiply two binomials, $A + B$ and $C + D$, multiply the First terms AC, the Outside terms AD, the Inside terms BC, and then the Last terms BD. Then collect like terms, if possible.

$$(A + B)(C + D) = AC + AD + BC + BD$$

1. Multiply First terms: AC.
2. Multiply Outside terms: AD.
3. Multiply Inside terms: BC.
4. Multiply Last terms: BD.

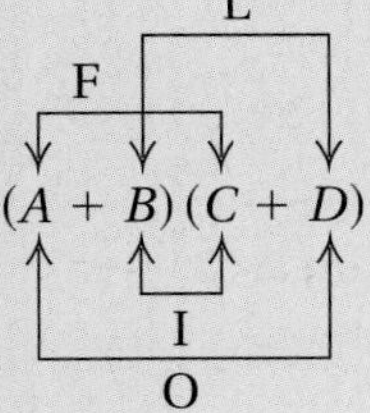

EXAMPLE 1 Multiply: $(x + 8)(x^2 - 5)$.

We have

$$\begin{aligned}(x + 8)(x^2 - 5) &= \overset{F}{x \cdot x^2} + \overset{O}{x \cdot (-5)} + \overset{I}{8 \cdot x^2} + \overset{L}{8(-5)}\\ &= x^3 - 5x + 8x^2 - 40\\ &= x^3 + 8x^2 - 5x - 40.\end{aligned}$$

Since each of the original binomials is in descending order, we write the product in descending order, as is customary, but this is not a "must."

Multiply mentally, if possible. If you need extra steps, be sure to use them.

1. $(x + 3)(x + 4)$

2. $(x + 3)(x - 5)$

3. $(2x - 1)(x - 4)$

4. $(2x^2 - 3)(x - 2)$

5. $(6x^2 + 5)(2x^3 + 1)$

6. $(y^3 + 7)(y^3 - 7)$

7. $(t + 5)(t + 3)$

8. $(2x^4 + x^2)(-x^3 + x)$

Multiply.

9. $\left(x + \frac{4}{5}\right)\left(x - \frac{4}{5}\right)$

10. $(x^3 - 0.5)(x^2 + 0.5)$

11. $(2 + 3x^2)(4 - 5x^2)$

12. $(6x^3 - 3x^2)(5x^2 - 2x)$

Answers on page A-20

Often we can collect like terms after we have multiplied.

EXAMPLES Multiply.

2. $(x + 6)(x - 6) = x^2 - 6x + 6x - 36$ Using FOIL
$= x^2 - 36$ Collecting like terms

3. $(x + 7)(x + 4) = x^2 + 4x + 7x + 28$
$= x^2 + 11x + 28$

4. $(y - 3)(y - 2) = y^2 - 2y - 3y + 6$
$= y^2 - 5y + 6$

5. $(x^3 - 5)(x^3 + 5) = x^6 + 5x^3 - 5x^3 - 25$
$= x^6 - 25$

Do Exercises 1–8.

EXAMPLES Multiply.

6. $(4t^3 + 5)(3t^2 - 2) = 12t^5 - 8t^3 + 15t^2 - 10$

7. $\left(x - \frac{2}{3}\right)\left(x + \frac{2}{3}\right) = x^2 + \frac{2}{3}x - \frac{2}{3}x - \frac{4}{9}$
$= x^2 - \frac{4}{9}$

8. $(x^2 - 0.3)(x^2 - 0.3) = x^4 - 0.3x^2 - 0.3x^2 + 0.09$
$= x^4 - 0.6x^2 + 0.09$

9. $(3 - 4x)(7 - 5x^3) = 21 - 15x^3 - 28x + 20x^4$
$= 21 - 28x - 15x^3 + 20x^4$

(*Note*: If the original polynomials are in ascending order, it is natural to write the product in ascending order, but this is not a "must.")

10. $(5x^4 + 2x^3)(3x^2 - 7x) = 15x^6 - 35x^5 + 6x^5 - 14x^4$
$= 15x^6 - 29x^5 - 14x^4$

Do Exercises 9–12.

We can show the FOIL method geometrically as follows.

The area of the large rectangle is $(A + B)(C + D)$.

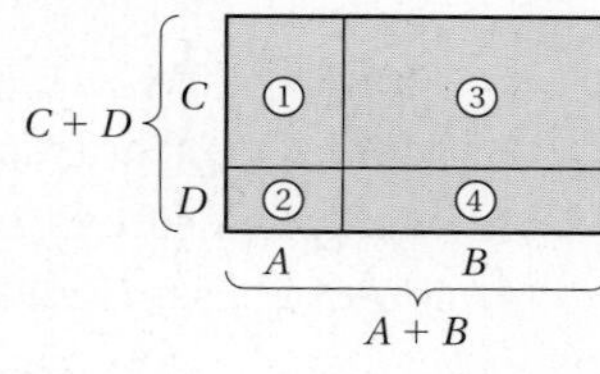

The area of rectangle ① is AC.
The area of rectangle ② is AD.
The area of rectangle ③ is BC.
The area of rectangle ④ is BD.

The area of the large rectangle is the sum of the areas of the smaller rectangles. Thus,

$$(A + B)(C + D) = AC + AD + BC + BD.$$

b Multiplying Sums and Differences of Two Terms

Consider the product of the sum and the difference of the same two terms, such as

$$(x + 2)(x - 2).$$

Since this is the product of two binomials, we can use FOIL. This type of product occurs so often, however, that it would be valuable if we could use an even faster method. To find a faster way to compute such a product, look for a pattern in the following:

a) $(x + 2)(x - 2) = x^2 - 2x + 2x - 4$
$= x^2 - 4;$

b) $(3x - 5)(3x + 5) = 9x^2 + 15x - 15x - 25$
$= 9x^2 - 25.$

Do Exercises 13 and 14.

Perhaps you discovered in each case that when you multiply the two binomials, two terms are opposites, or additive inverses, which add to 0 and "drop out."

PRODUCT OF THE SUM AND THE DIFFERENCE

The product of the sum and the difference of the same two terms is the square of the first term minus the square of the second term:

$$(A + B)(A - B) = A^2 - B^2.$$

It is helpful to memorize this rule in both words and symbols. (If you do forget it, you can, of course, use FOIL.)

EXAMPLES Multiply. (Carry out the rule and say the words as you go.)

$(A + B)(A - B) = A^2 - B^2$

11. $(x + 4)(x - 4) = x^2 - 4^2$ "The square of the first term, x^2, minus the square of the second, 4^2"
$= x^2 - 16$ Simplifying

12. $(5 + 2w)(5 - 2w) = 5^2 - (2w)^2$
$= 25 - 4w^2$

13. $(3x^2 - 7)(3x^2 + 7) = (3x^2)^2 - 7^2$
$= 9x^4 - 49$

14. $(-4x - 10)(-4x + 10) = (-4x)^2 - 10^2$
$= 16x^2 - 100$

15. $\left(x + \frac{3}{8}\right)\left(x - \frac{3}{8}\right) = x^2 - \left(\frac{3}{8}\right)^2 = x^2 - \frac{9}{64}$

Do Exercises 15–19.

C Squaring Binomials

Consider the square of a binomial, such as $(x + 3)^2$. This can be expressed as $(x + 3)(x + 3)$. Since this is the product of two binomials, we can again use FOIL. But again, this type of product occurs so often that we would like to use an even faster method. Look for a pattern in the following:

Multiply.

13. $(x + 5)(x - 5)$

14. $(2x - 3)(2x + 3)$

Multiply.

15. $(x + 2)(x - 2)$

16. $(x - 7)(x + 7)$

17. $(6 - 4y)(6 + 4y)$

18. $(2x^3 - 1)(2x^3 + 1)$

19. $\left(x - \frac{2}{5}\right)\left(x + \frac{2}{5}\right)$

Answers on page A-21

Multiply.

20. $(x + 8)(x + 8)$

21. $(x - 5)(x - 5)$

Multiply.

22. $(x + 2)^2$

23. $(a - 4)^2$

24. $(2x + 5)^2$

25. $(4x^2 - 3x)^2$

26. $(7.8 + 1.2y)(7.8 + 1.2y)$

27. $(3x^2 - 5)(3x^2 - 5)$

Answers on page A-21

a) $(x + 3)^2 = (x + 3)(x + 3)$
$= x^2 + 3x + 3x + 9$
$= x^2 + 6x + 9;$

b) $(x - 3)^2 = (x - 3)(x - 3)$
$= x^2 - 3x - 3x + 9$
$= x^2 - 6x + 9;$

c) $(5 + 3p)^2 = (5 + 3p)(5 + 3p)$
$= 25 + 15p + 15p + 9p^2$
$= 25 + 30p + 9p^2;$

d) $(3x - 5)^2 = (3x - 5)(3x - 5)$
$= 9x^2 - 15x - 15x + 25$
$= 9x^2 - 30x + 25.$

Do Exercises 20 and 21.

When squaring a binomial, we multiply a binomial by itself. Perhaps you noticed that two terms are the same and when added give twice the product of the terms in the binomial. The other two terms are squares.

SQUARE OF A BINOMIAL

The square of a sum or a difference of two terms is the square of the first term, plus or minus twice the product of the two terms, plus the square of the last term:

$$(A + B)^2 = A^2 + 2AB + B^2; \qquad (A - B)^2 = A^2 - 2AB + B^2.$$

It is helpful to memorize this rule in both words and symbols.

EXAMPLES Multiply. (Carry out the rule and say the words as you go.)

$$(A + B)^2 = A^2 + 2 \cdot A \cdot B + B^2$$

16. $(x + 3)^2 = x^2 + 2 \cdot x \cdot 3 + 3^2$ "x^2 plus 2 times x times 3 plus 3^2"
$= x^2 + 6x + 9$

$$(A - B)^2 = A^2 - 2 \cdot A \cdot B + B^2$$

17. $(t - 5)^2 = t^2 - 2 \cdot t \cdot 5 + 5^2$ "t^2 minus 2 times t times 5 plus 5^2"
$= t^2 - 10t + 25$

18. $(2x + 7)^2 = (2x)^2 + 2 \cdot 2x \cdot 7 + 7^2 = 4x^2 + 28x + 49$

19. $(5x - 3x^2)^2 = (5x)^2 - 2 \cdot 5x \cdot 3x^2 + (3x^2)^2 = 25x^2 - 30x^3 + 9x^4$

20. $(2.3 - 5.4m)^2 = 2.3^2 - 2(2.3)(5.4m) + (5.4m)^2$
$= 5.29 - 24.84m + 29.16m^2$

Do Exercises 22–27.

CAUTION!

Although the square of a product is the product of the squares, the square of a sum is *not* the sum of the squares. That is, $(AB)^2 = A^2B^2$, but

The term $2AB$ is missing.

$$(A + B)^2 \neq A^2 + B^2.$$

To confirm this inequality, note, using the rules for order of operations, that

$$(7 + 5)^2 = 12^2 = 144,$$

whereas

$$7^2 + 5^2 = 49 + 25 = 74, \quad \text{and} \quad 74 \neq 144.$$

We can look at the rule for finding $(A + B)^2$ geometrically as follows. The area of the large square is

$$(A + B)(A + B) = (A + B)^2.$$

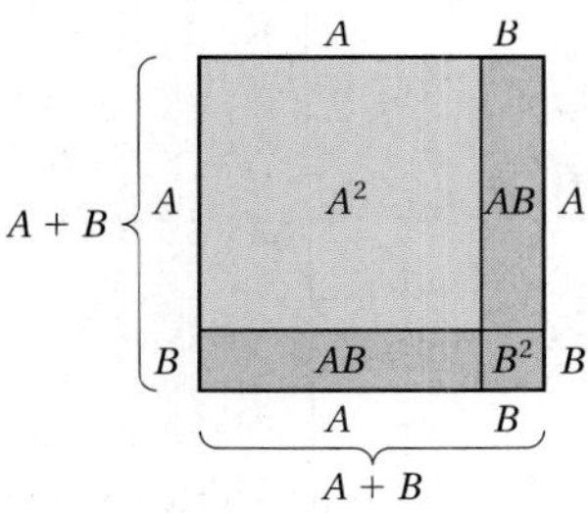

This is equal to the sum of the areas of the smaller rectangles:

$$A^2 + AB + AB + B^2 = A^2 + 2AB + B^2.$$

Thus, $(A + B)^2 = A^2 + 2AB + B^2$.

d Multiplication of Various Types

We have considered how to quickly multiply certain kinds of polynomials. Let's now try several types of multiplications mixed together so that we can learn to sort them out. When you multiply, first see what kind of multiplication you have. Then use the best method. The formulas you should know and the questions you should ask yourself are as follows:

MULTIPLYING TWO POLYNOMIALS

1. Is it the product of a monomial and a polynomial? If so, multiply each term of the polynomial by the monomial.
   Example: $5x(x + 7) = 5x \cdot x + 5x \cdot 7 = 5x^2 + 35x$
2. Is it the product of the sum and the difference of the *same* two terms? If so, use the following:
   $$(A + B)(A - B) = A^2 - B^2.$$
   The product of the sum and the difference of the same two terms is the difference of the squares. [The answer has 2 terms.]
   Example: $(x + 7)(x - 7) = x^2 - 7^2 = x^2 - 49$
3. Is the product the square of a binomial? If so, use the following:
   $$(A + B)(A + B) = (A + B)^2 = A^2 + 2AB + B^2,$$
   or $(A - B)(A - B) = (A - B)^2 = A^2 - 2AB + B^2.$
   The square of a binomial is the square of the first term, plus or minus *twice* the product of the two terms, plus the square of the last term. [The answer has 3 terms.]
   Example: $(x + 7)(x + 7) = (x + 7)^2$
   $= x^2 + 2 \cdot x \cdot 7 + 7^2 = x^2 + 14x + 49$
4. Is it the product of two binomials other than those above? If so, use FOIL. [The answer will have 3 or 4 terms.]
   Example: $(x + 7)(x - 4) = x^2 - 4x + 7x - 28 = x^2 + 3x - 28$
5. Is it the product of two polynomials other than those above? If so, multiply each term of one by every term of the other. Use columns if you wish. [The answer will have 2 or more terms, usually more than 2 terms.]
   Example:
   $$\begin{aligned}(x^2 - 3x + 2)(x + 7) &= x^2(x + 7) - 3x(x + 7) + 2(x + 7)\\ &= x^2 \cdot x + x^2 \cdot 7 - 3x \cdot x - 3x \cdot 7\\ &\quad + 2 \cdot x + 2 \cdot 7\\ &= x^3 + 7x^2 - 3x^2 - 21x + 2x + 14\\ &= x^3 + 4x^2 - 19x + 14\end{aligned}$$

Remember that FOIL will *always* work for two binomials. You can use it instead of either of rules 2 and 3, but those rules will make your work go faster.

Study Tips

MEMORIZING FORMULAS

Memorizing can be a very helpful tool in the study of mathematics. Don't underestimate its power as you consider the special products. Consider putting the rules, in words and in math symbols, on index cards and go over them many times.

Multiply.

28. $(x + 5)(x + 6)$

29. $(t - 4)(t + 4)$

30. $4x^2(-2x^3 + 5x^2 + 10)$

31. $(9x^2 + 1)^2$

32. $(2a - 5)(2a + 8)$

33. $\left(5x + \frac{1}{2}\right)^2$

34. $\left(2x - \frac{1}{2}\right)^2$

35. $(x^2 - x + 4)(x - 2)$

Answers on page A-21

EXAMPLE 21 Multiply: $(x + 3)(x - 3)$.

$(x + 3)(x - 3) = x^2 - 9$ Using method 2 (the product of the sum and the difference of two terms)

EXAMPLE 22 Multiply: $(t + 7)(t - 5)$.

$(t + 7)(t - 5) = t^2 + 2t - 35$ Using method 4 (the product of two binomials, but neither the square of a binomial nor the product of the sum and the difference of two terms)

EXAMPLE 23 Multiply: $(x + 6)(x + 6)$.

$(x + 6)(x + 6) = x^2 + 2(6)x + 36$ Using method 3 (the square of a binomial sum)

$= x^2 + 12x + 36$

EXAMPLE 24 Multiply: $2x^3(9x^2 + x - 7)$.

$2x^3(9x^2 + x - 7) = 18x^5 + 2x^4 - 14x^3$ Using method 1 (the product of a monomial and a trinomial; multiplying each term of the trinomial by the monomial)

EXAMPLE 25 Multiply: $(5x^3 - 7x)^2$.

$(5x^3 - 7x)^2 = 25x^6 - 2(5x^3)(7x) + 49x^2$ Using method 3 (the square of a binomial difference)

$= 25x^6 - 70x^4 + 49x^2$

EXAMPLE 26 Multiply: $(3x + \frac{1}{4})^2$.

$(3x + \frac{1}{4})^2 = 9x^2 + 2(3x)(\frac{1}{4}) + \frac{1}{16}$ Using method 3 (the square of a binomial sum. To get the middle term, we multiply $3x$ by $\frac{1}{4}$ and double.)

$= 9x^2 + \frac{3}{2}x + \frac{1}{16}$

EXAMPLE 27 Multiply: $(4x - \frac{3}{4})^2$.

$(4x - \frac{3}{4})^2 = 16x^2 - 2(4x)(\frac{3}{4}) + \frac{9}{16}$ Using method 3 (the square of a binomial difference)

$= 16x^2 - 6x + \frac{9}{16}$

EXAMPLE 28 Multiply: $(p + 3)(p^2 + 2p - 1)$.

$$\begin{array}{rl}
p^2 + 2p - 1 & \text{Using method 5 (the product of two polynomials)} \\
p + 3 & \\
\hline
3p^2 + 6p - 3 & \text{Multiplying by 3} \\
p^3 + 2p^2 - p \phantom{{}- 3} & \text{Multiplying by } p \\
\hline
p^3 + 5p^2 + 5p - 3 &
\end{array}$$

Do Exercises 28–35.

4.6 EXERCISE SET

For Extra Help

Digital Video Tutor CD 4 Videotape 9 | InterAct Math | Math Tutor Center | MathXL | MyMathLab

a Multiply. Try to write only the answer. If you need more steps, be sure to use them.

1. $(x + 1)(x^2 + 3)$

2. $(x^2 - 3)(x - 1)$

3. $(x^3 + 2)(x + 1)$

4. $(x^4 + 2)(x + 10)$

5. $(y + 2)(y - 3)$

6. $(a + 2)(a + 3)$

7. $(3x + 2)(3x + 2)$

8. $(4x + 1)(4x + 1)$

9. $(5x - 6)(x + 2)$

10. $(x - 8)(x + 8)$

11. $(3t - 1)(3t + 1)$

12. $(2m + 3)(2m + 3)$

13. $(4x - 2)(x - 1)$

14. $(2x - 1)(3x + 1)$

15. $\left(p - \frac{1}{4}\right)\left(p + \frac{1}{4}\right)$

16. $\left(q + \frac{3}{4}\right)\left(q + \frac{3}{4}\right)$

17. $(x - 0.1)(x + 0.1)$

18. $(x + 0.3)(x - 0.4)$

19. $(2x^2 + 6)(x + 1)$

20. $(2x^2 + 3)(2x - 1)$

21. $(-2x + 1)(x + 6)$

22. $(3x + 4)(2x - 4)$

23. $(a + 7)(a + 7)$

24. $(2y + 5)(2y + 5)$

25. $(1 + 2x)(1 - 3x)$

26. $(-3x - 2)(x + 1)$

27. $(x^2 + 3)(x^3 - 1)$

28. $(x^4 - 3)(2x + 1)$

29. $(3x^2 - 2)(x^4 - 2)$

30. $(x^{10} + 3)(x^{10} - 3)$

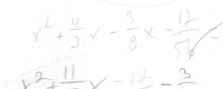

31. $(2.8x - 1.5)(4.7x + 9.3)$

32. $\left(x - \frac{3}{8}\right)\left(x + \frac{4}{7}\right)$

33. $(3x^5 + 2)(2x^2 + 6)$

34. $(1 - 2x)(1 + 3x^2)$

35. $(8x^3 + 1)(x^3 + 8)$

36. $(4 - 2x)(5 - 2x^2)$

37. $(4x^2 + 3)(x - 3)$

38. $(7x - 2)(2x - 7)$

39. $(4y^4 + y^2)(y^2 + y)$

40. $(5y^6 + 3y^3)(2y^6 + 2y^3)$

b Multiply mentally, if possible. If you need extra steps, be sure to use them.

41. $(x + 4)(x - 4)$

42. $(x + 1)(x - 1)$

43. $(2x + 1)(2x - 1)$

44. $(x^2 + 1)(x^2 - 1)$

45. $(5m - 2)(5m + 2)$

46. $(3x^4 + 2)(3x^4 - 2)$

47. $(2x^2 + 3)(2x^2 - 3)$

48. $(6x^5 - 5)(6x^5 + 5)$

49. $(3x^4 - 4)(3x^4 + 4)$

50. $(t^2 - 0.2)(t^2 + 0.2)$

51. $(x^6 - x^2)(x^6 + x^2)$

52. $(2x^3 - 0.3)(2x^3 + 0.3)$

53. $(x^4 + 3x)(x^4 - 3x)$

54. $\left(\frac{3}{4} + 2x^3\right)\left(\frac{3}{4} - 2x^3\right)$

55. $(x^{12} - 3)(x^{12} + 3)$

56. $(12 - 3x^2)(12 + 3x^2)$

57. $(2y^8 + 3)(2y^8 - 3)$

58. $\left(m - \frac{2}{3}\right)\left(m + \frac{2}{3}\right)$

59. $\left(\frac{5}{8}x - 4.3\right)\left(\frac{5}{8}x + 4.3\right)$

60. $(10.7 - x^3)(10.7 + x^3)$

c Multiply mentally, if possible. If you need extra steps, be sure to use them.

61. $(x + 2)^2$

62. $(2x - 1)^2$

63. $(3x^2 + 1)^2$

64. $\left(3x + \frac{3}{4}\right)^2$

65. $\left(a - \frac{1}{2}\right)^2$

66. $\left(2a - \frac{1}{5}\right)^2$

67. $(3 + x)^2$

68. $(x^3 - 1)^2$

69. $(x^2 + 1)^2$

70. $(8x - x^2)^2$

71. $(2 - 3x^4)^2$

72. $(6x^3 - 2)^2$

73. $(5 + 6t^2)^2$

74. $(3p^2 - p)^2$

75. $\left(x - \frac{5}{8}\right)^2$

76. $(0.3y + 2.4)^2$

d Multiply mentally, if possible.

77. $(3 - 2x^3)^2$

78. $(x - 4x^3)^2$

79. $4x(x^2 + 6x - 3)$

80. $8x(-x^5 + 6x^2 + 9)$

81. $\left(2x^2 - \frac{1}{2}\right)\left(2x^2 - \frac{1}{2}\right)$

82. $(-x^2 + 1)^2$

83. $(-1 + 3p)(1 + 3p)$

84. $(-3q + 2)(3q + 2)$

85. $3t^2(5t^3 - t^2 + t)$

86. $-6x^2(x^3 + 8x - 9)$

87. $(6x^4 + 4)^2$

88. $(8a + 5)^2$

89. $(3x + 2)(4x^2 + 5)$

90. $(2x^2 - 7)(3x^2 + 9)$

91. $(8 - 6x^4)^2$

92. $\left(\frac{1}{5}x^2 + 9\right)\left(\frac{3}{5}x^2 - 7\right)$

93. $(t - 1)(t^2 + t + 1)$

94. $(y + 5)(y^2 - 5y + 25)$

Compute each of the following and compare.

95. $3^2 + 4^2$; $(3 + 4)^2$

96. $6^2 + 7^2$; $(6 + 7)^2$

97. $9^2 - 5^2$; $(9 - 5)^2$

98. $11^2 - 4^2$; $(11 - 4)^2$

Find the total area of all the shaded rectangles.

99.
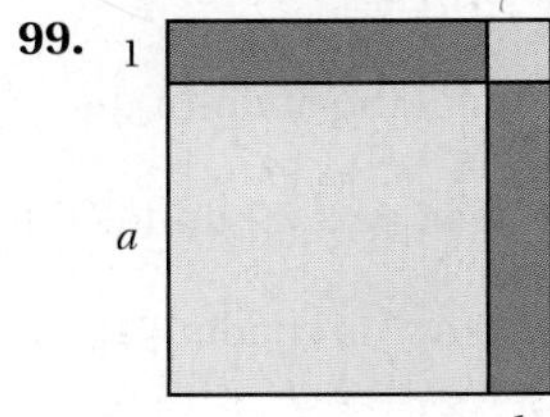

100.
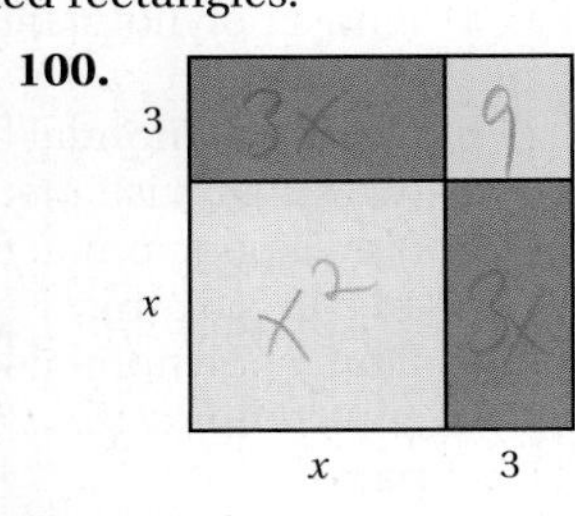

101.
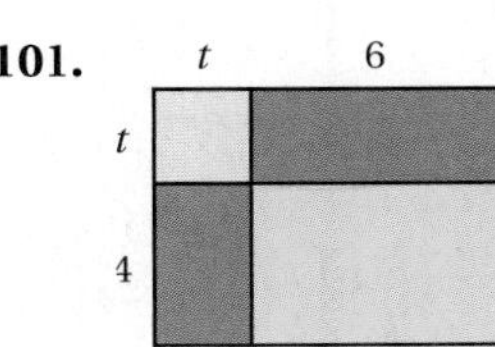

102.
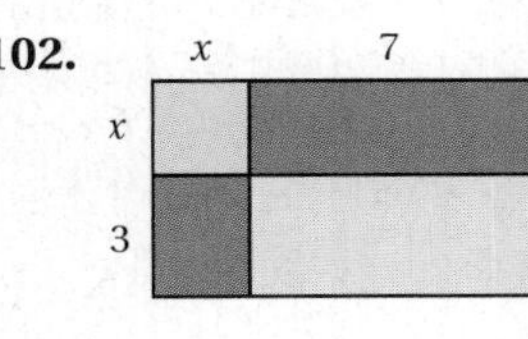

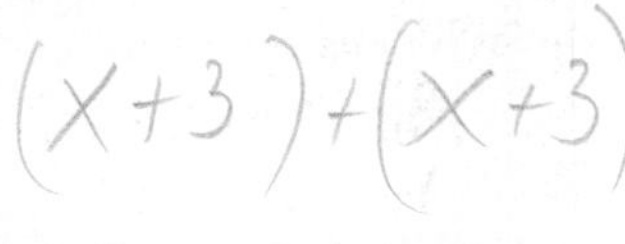

103. $^{\text{D}}\text{W}$ Under what conditions is the product of two binomials a binomial?

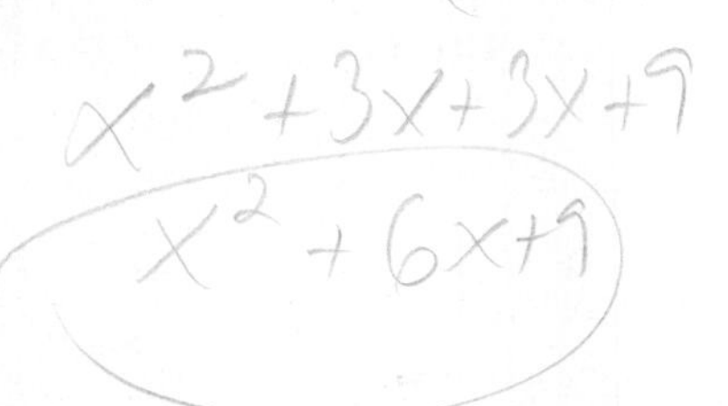

104. $^{\text{D}}\text{W}$ Brittney feels that since the FOIL method can be used to find the product of any two binomials, she needn't study the other special products. What advice would you give her?

SKILL MAINTENANCE

105. *Electricity Usage.* In apartment 3B, lamps, an air conditioner, and a television set are all operating at the same time. The lamps use 10 times as many watts of electricity as the television set, and the air conditioner uses 40 times as many watts as the television set. The total wattage used in the apartment is 2550. How many watts are used by each appliance? [2.6a]

Solve. [2.3c]

106. $3x - 8x = 4(7 - 8x)$

107. $3(x - 2) = 5(2x + 7)$

108. $5(2x - 3) - 2(3x - 4) = 20$

Solve. [2.4b]

109. $3x - 2y = 12$, for y

110. $3a - 5d = 4$, for a

SYNTHESIS

Multiply.

111. $5x(3x - 1)(2x + 3)$

112. $[(2x - 3)(2x + 3)](4x^2 + 9)$

113. $[(a - 5)(a + 5)]^2$

114. $(a - 3)^2(a + 3)^2$
(*Hint*: Examine Exercise 113.)

115. $(3t^4 - 2)^2(3t^4 + 2)^2$
(*Hint*: Examine Exercise 113.)

116. $[3a - (2a - 3)][3a + (2a - 3)]$

Solve.

117. $(x + 2)(x - 5) = (x + 1)(x - 3)$

118. $(2x + 5)(x - 4) = (x + 5)(2x - 4)$

119. *Factors and Sums.* To *factor* a number is to express it as a product. Since $12 = 4 \cdot 3$, we say that 12 is *factored* and that 4 and 3 are *factors* of 12. In the following table, the top number has been factored in such a way that the sum of the factors is the bottom number. For example, in the first column, 40 has been factored as $5 \cdot 8$, and $5 + 8 = 13$, the bottom number. Such thinking is important in algebra when we factor trinomials of the type $x^2 + bx + c$. Find the missing numbers in the table.

Product	40	63	36	72	−140	−96	48	168	110			
Factor	5									−9	−24	−3
Factor	8									−10	18	
Sum	13	16	−20	−38	−4	4	−14	−29	−21			18

120. A factored polynomial for the shaded area in this rectangle is $(A + B)(A - B)$.

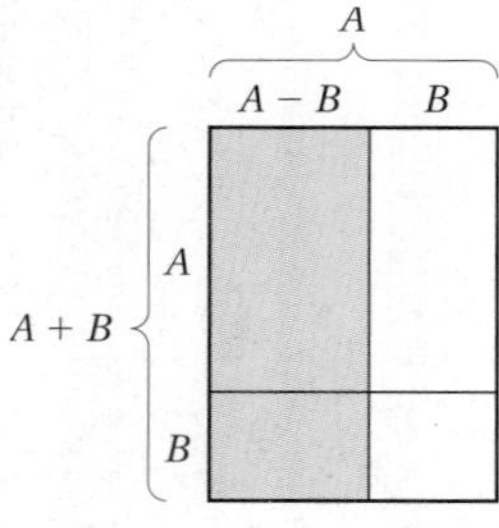

a) Find a polynomial for the area of the entire rectangle.
b) Find a polynomial for the sum of the areas of the two small unshaded rectangles.
c) Find a polynomial for the area in part (a) minus the area in part (b).
d) Find a polynomial for the area of the shaded region and compare this with the polynomial found in part (c).

Use the TABLE or GRAPH feature to check whether each of the following is correct.

121. $(x - 1)^2 = x^2 - 2x + 1$

122. $(x - 2)^2 = x^2 - 4x - 4$

123. $(x - 3)(x + 3) = x^2 - 6$

124. $(x - 3)(x + 2) = x^2 - x - 6$

4.7 OPERATIONS WITH POLYNOMIALS IN SEVERAL VARIABLES

The polynomials that we have been studying have only one variable. A **polynomial in several variables** is an expression like those you have already seen, but with more than one variable. Here are two examples:

$$3x + xy^2 + 5y + 4, \qquad 8xy^2z - 2x^3z - 13x^4y^2 + 15.$$

a Evaluating Polynomials

EXAMPLE 1 Evaluate the polynomial $4 + 3x + xy^2 + 8x^3y^3$ when $x = -2$ and $y = 5$.

We replace x with -2 and y with 5:

$$\begin{aligned} 4 + 3x + xy^2 + 8x^3y^3 &= 4 + 3(-2) + (-2) \cdot 5^2 + 8(-2)^3 \cdot 5^3 \\ &= 4 - 6 - 50 - 8000 \\ &= -8052. \end{aligned}$$

EXAMPLE 2 *Male Caloric Needs.* The number of calories needed each day by a moderately active man who weighs w kilograms, is h centimeters tall, and is a years old can be estimated by the polynomial

$$19.18w + 7h - 9.52a + 92.4.$$

The author of this text is moderately active, weighs 87 kg, is 185 cm tall, and is 59 yr old. What are his daily caloric needs?

Source: Parker, M., *She Does Math.* Mathematical Association of America

We evaluate the polynomial for $w = 87$, $h = 185$, and $a = 59$:

$$\begin{aligned} &19.18w + 7h - 9.52a + 92.4 \\ &\quad = 19.18(87) + 7(185) - 9.52(59) + 92.4 \qquad \text{Substituting} \\ &\quad = 2494.38. \end{aligned}$$

His daily caloric need is about 2494 calories.

Do Exercises 1–3.

Objectives

a. Evaluate a polynomial in several variables for given values of the variables.

b. Identify the coefficients and the degrees of the terms of a polynomial and the degree of a polynomial.

c. Collect like terms of a polynomial.

d. Add polynomials.

e. Subtract polynomials.

f. Multiply polynomials.

1. Evaluate the polynomial

$$4 + 3x + xy^2 + 8x^3y^3$$

when $x = 2$ and $y = -5$.

2. Evaluate the polynomial

$$8xy^2 - 2x^3z - 13x^4y^2 + 5$$

when $x = -1$, $y = 3$, and $z = 4$.

3. Female Caloric Needs. The number of calories needed each day by a moderately active woman who weighs w pounds, is h inches tall, and is a years old can be estimated by the polynomial

$$917 + 6w + 6h - 6a.$$

Christine is moderately active, weighs 125 lb, is 64 in. tall, and is 27 yr old. What are her daily caloric needs?

Source: Parker, M., *She Does Math.* Mathematical Association of America

Answers on page A-21

4. Identify the coefficient of each term:

$-3xy^2 + 3x^2y - 2y^3 + xy + 2.$

5. Identify the degree of each term and the degree of the polynomial

$4xy^2 + 7x^2y^3z^2 - 5x + 2y + 4.$

Collect like terms.

6. $4x^2y + 3xy - 2x^2y$

7. $-3pq - 5pqr^3 - 12 + 8pq + 5pqr^3 + 4$

Answers on page A-21

b Coefficients and Degrees

The **degree** of a term is the sum of the exponents of the variables. The **degree of a polynomial** is the degree of the term of highest degree.

EXAMPLE 3 Identify the coefficient and the degree of each term and the degree of the polynomial

$$9x^2y^3 - 14xy^2z^3 + xy + 4y + 5x^2 + 7.$$

TERM	COEFFICIENT	DEGREE	DEGREE OF THE POLYNOMIAL
$9x^2y^3$	9	5	
$-14xy^2z^3$	-14	6	6
xy	1	2	
$4y$	4	1	Think: $4y = 4y^1$.
$5x^2$	5	2	
7	7	0	Think: $7 = 7x^0$, or $7x^0y^0z^0$.

Do Exercises 4 and 5.

c Collecting Like Terms

Like terms have exactly the same variables with exactly the same exponents. For example,

$3x^2y^3$ and $-7x^2y^3$ are like terms;
$9x^4z^7$ and $12x^4z^7$ are like terms.

But

$13xy^5$ and $-2x^2y^5$ are *not* like terms, because the x-factors have different exponents;

and

$3xyz^2$ and $4xy$ are *not* like terms, because there is no factor of z^2 in the second expression.

Collecting like terms is based on the distributive laws.

EXAMPLES Collect like terms.

4. $5x^2y + 3xy^2 - 5x^2y - xy^2 = (5 - 5)x^2y + (3 - 1)xy^2 = 2xy^2$
5. $8a^2 - 2ab + 7b^2 + 4a^2 - 9ab - 17b^2 = 12a^2 - 11ab - 10b^2$
6. $7xy - 5xy^2 + 3xy^2 - 7 + 6x^3 + 9xy - 11x^3 + y - 1$
   $= -2xy^2 + 16xy - 5x^3 + y - 8$

Do Exercises 6 and 7.

d Addition

We can find the sum of two polynomials in several variables by writing a plus sign between them and then collecting like terms.

EXAMPLE 7 Add: $(-5x^3 + 3y - 5y^2) + (8x^3 + 4x^2 + 7y^2)$.

$$(-5x^3 + 3y - 5y^2) + (8x^3 + 4x^2 + 7y^2)$$
$$= (-5 + 8)x^3 + 4x^2 + 3y + (-5 + 7)y^2$$
$$= 3x^3 + 4x^2 + 3y + 2y^2$$

EXAMPLE 8 Add:

$$(5xy^2 - 4x^2y + 5x^3 + 2) + (3xy^2 - 2x^2y + 3x^3y - 5).$$

We first look for like terms. They are $5xy^2$ and $3xy^2$, $-4x^2y$ and $-2x^2y$, and 2 and -5. We collect these. Since there are no more like terms, the answer is

$$8xy^2 - 6x^2y + 5x^3 + 3x^3y - 3.$$

Do Exercises 8–10.

Add.

8. $(4x^3 + 4x^2 - 8y - 3) + (-8x^3 - 2x^2 + 4y + 5)$

9. $(13x^3y + 3x^2y - 5y) + (x^3y + 4x^2y - 3xy + 3y)$

10. $(-5p^2q^4 + 2p^2q^2 + 3q) + (6pq^2 + 3p^2q + 5)$

e Subtraction

We subtract a polynomial by adding its opposite, or additive inverse. The opposite of the polynomial $4x^2y - 6x^3y^2 + x^2y^2 - 5y$ is

$$-(4x^2y - 6x^3y^2 + x^2y^2 - 5y) = -4x^2y + 6x^3y^2 - x^2y^2 + 5y.$$

EXAMPLE 9 Subtract:

$$(4x^2y + x^3y^2 + 3x^2y^3 + 6y + 10) - (4x^2y - 6x^3y^2 + x^2y^2 - 5y - 8).$$

We have

$$(4x^2y + x^3y^2 + 3x^2y^3 + 6y + 10) - (4x^2y - 6x^3y^2 + x^2y^2 - 5y - 8)$$
$$= 4x^2y + x^3y^2 + 3x^2y^3 + 6y + 10 - 4x^2y + 6x^3y^2 - x^2y^2 + 5y + 8$$

Finding the opposite by changing the sign of each term

$$= 7x^3y^2 + 3x^2y^3 - x^2y^2 + 11y + 18.$$

Collecting like terms. (Try to write just the answer!)

Do Exercises 11 and 12.

Subtract.

11. $(-4s^4t + s^3t^2 + 2s^2t^3) - (4s^4t - 5s^3t^2 + s^2t^2)$

12. $(-5p^4q + 5p^3q^2 - 3p^2q^3 - 7q^4 - 2) - (4p^4q - 4p^3q^2 + p^2q^3 + 2q^4 - 7)$

Answers on page A-21

Multiply.

13. $(x^2y^3 + 2x)(x^3y^2 + 3x)$

14. $(p^4q - 2p^3q^2 + 3q^3)(p + 2q)$

Multiply.

15. $(3xy + 2x)(x^2 + 2xy^2)$

16. $(x - 3y)(2x - 5y)$

17. $(4x + 5y)^2$

18. $(3x^2 - 2xy^2)^2$

19. $(2xy^2 + 3x)(2xy^2 - 3x)$

20. $(3xy^2 + 4y)(-3xy^2 + 4y)$

21. $(3y + 4 - 3x)(3y + 4 + 3x)$

22. $(2a + 5b + c)(2a - 5b - c)$

Answers on page A-21

f Multiplication

To multiply polynomials in several variables, we can multiply each term of one by every term of the other. We can use columns for long multiplications as with polynomials in one variable. We multiply each term at the top by every term at the bottom. We write like terms in columns, and then we add the results.

EXAMPLE 10 Multiply: $(3x^2y - 2xy + 3y)(xy + 2y)$.

$$\begin{array}{rrrrl}
 & 3x^2y & -\ 2xy & +\ 3y & \\
 & & xy & +\ 2y & \\
\hline
 & 6x^2y^2 & -\ 4xy^2 & +\ 6y^2 & \text{Multiplying by } 2y \\
3x^3y^2 & -\ 2x^2y^2 & +\ 3xy^2 & & \text{Multiplying by } xy \\
\hline
3x^3y^2 & +\ 4x^2y^2 & -\ xy^2 & +\ 6y^2 & \text{Adding}
\end{array}$$

Do Exercises 13 and 14.

Where appropriate, we use the special products that we have learned.

EXAMPLES Multiply.

11. $(x^2y + 2x)(xy^2 + y^2) = x^3y^3 + x^2y^3 + 2x^2y^2 + 2xy^2$ (F O I L)

12. $(p + 5q)(2p - 3q) = 2p^2 - 3pq + 10pq - 15q^2$
$= 2p^2 + 7pq - 15q^2$

$(A + B)^2 = A^2 + 2 \cdot A \cdot B + B^2$

13. $(3x + 2y)^2 = (3x)^2 + 2(3x)(2y) + (2y)^2$
$= 9x^2 + 12xy + 4y^2$

$(A - B)^2 = A^2 - 2 \cdot A \cdot B + B^2$

14. $(2y^2 - 5x^2y)^2 = (2y^2)^2 - 2(2y^2)(5x^2y) + (5x^2y)^2$
$= 4y^4 - 20x^2y^3 + 25x^4y^2$

$(A + B)(A - B) = A^2 - B^2$

15. $(3x^2y + 2y)(3x^2y - 2y) = (3x^2y)^2 - (2y)^2$
$= 9x^4y^2 - 4y^2$

16. $(-2x^3y^2 + 5t)(2x^3y^2 + 5t) = (5t - 2x^3y^2)(5t + 2x^3y^2)$
$= (5t)^2 - (2x^3y^2)^2$
$= 25t^2 - 4x^6y^4$

$(A - B)(A + B) = A^2 - B^2$

17. $(2x + 3 - 2y)(2x + 3 + 2y) = (2x + 3)^2 - (2y)^2$
$= 4x^2 + 12x + 9 - 4y^2$

Do Exercises 15–22.

4.7 EXERCISE SET

For Extra Help

Digital Video Tutor CD 4 Videotape 9 | InterAct Math | Math Tutor Center | MathXL | MyMathLab

a Evaluate the polynomial when $x = 3$, $y = -2$, and $z = -5$.

1. $x^2 - y^2 + xy$

2. $x^2 + y^2 - xy$

3. $x^2 - 3y^2 + 2xy$

4. $x^2 - 4xy + 5y^2$

5. $8xyz$

6. $-3xyz^2$

7. $xyz^2 - z$

8. $xy - xz + yz$

Lung Capacity. The polynomial equation

$$C = 0.041h - 0.018A - 2.69$$

can be used to estimate the lung capacity C, in liters, of a female of height h, in centimeters, and age A, in years.

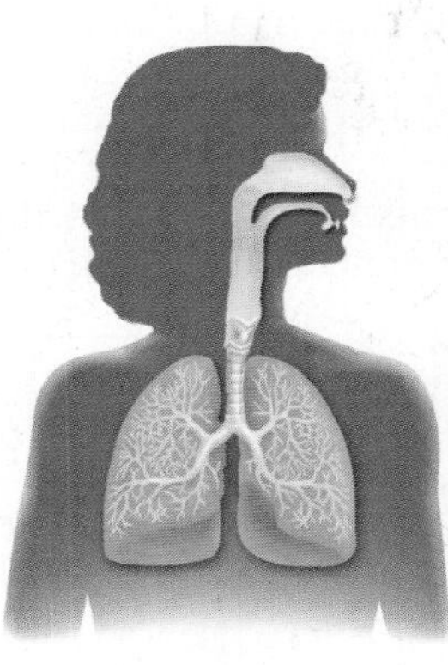

9. Find the lung capacity of a 20-yr-old woman who is 165 cm tall.

10. Find the lung capacity of a 50-yr-old woman who is 160 cm tall.

Altitude of a Launched Object. The altitude h, in meters, of a launched object is given by the polynomial equation

$$h = h_0 + vt - 4.9t^2,$$

where h_0 is the height, in meters, from which the launch occurs, v is the initial upward speed (or velocity), in meters per second (m/s), and t is the number of seconds for which the object is airborne.

11. A model rocket is launched from the top of the Leaning Tower of Pisa, 50 m above the ground. The upward speed is 40 m/s. How high will the rocket be 2 sec after the blastoff?

12. A golf ball is thrown upward with an initial speed of 30 m/s by a golfer atop the Washington Monument, which is 160 m above the ground. How high above the ground will the ball be after 3 sec?

Surface Area of a Right Circular Cylinder. The surface area S of a right circular cylinder is given by the polynomial equation

$$S = 2\pi rh + 2\pi r^2,$$

where h is the height and r is the radius of the base.

13. A 12-oz beverage can has a height of 4.7 in. and a radius of 1.2 in. Evaluate the polynomial when $h = 4.7$ and $r = 1.2$ to find the area of the can. Use 3.14 for π.

14. A 26-oz coffee can has a height of 6.5 in. and a radius of 2.5 in. Evaluate the polynomial when $h = 6.5$ and $r = 2.5$ to find the area of the can. Use 3.14 for π.

Surface Area of a Silo. A silo is a structure that is shaped like a right circular cylinder with a half sphere on top. The surface area S of a silo of height h and radius r (including the area of the base) is given by the polynomial equation $S = 2\pi rh + \pi r^2$.

15. A container of tennis balls is silo-shaped, with a height of $7\frac{1}{2}$ in. and a radius of $1\frac{1}{4}$ in. Find the surface area of the container. Use 3.14 for π.

16. A $1\frac{1}{2}$-oz bottle of roll-on deodorant has a height of 4 in. and a radius of $\frac{3}{4}$ in. Find the surface area of the bottle if the bottle is shaped like a silo. Use 3.14 for π.

b Identify the coefficient and the degree of each term of the polynomial. Then find the degree of the polynomial.

17. $x^3y - 2xy + 3x^2 - 5$

18. $5y^3 - y^2 + 15y + 1$

19. $17x^2y^3 - 3x^3yz - 7$

20. $6 - xy + 8x^2y^2 - y^5$

c Collect like terms.

21. $a + b - 2a - 3b$

22. $y^2 - 1 + y - 6 - y^2$

23. $3x^2y - 2xy^2 + x^2$

24. $m^3 + 2m^2n - 3m^2 + 3mn^2$

25. $6au + 3av + 14au + 7av$

26. $3x^2y - 2z^2y + 3xy^2 + 5z^2y$

27. $2u^2v - 3uv^2 + 6u^2v - 2uv^2$

28. $3x^2 + 6xy + 3y^2 - 5x^2 - 10xy - 5y^2$

d Add.

29. $(2x^2 - xy + y^2) + (-x^2 - 3xy + 2y^2)$

30. $(2z - z^2 + 5) + (z^2 - 3z + 1)$

31. $(r - 2s + 3) + (2r + s) + (s + 4)$

32. $(ab - 2a + 3b) + (5a - 4b) + (3a + 7ab - 8b)$

33. $(b^3a^2 - 2b^2a^3 + 3ba + 4) + (b^2a^3 - 4b^3a^2 + 2ba - 1)$

34. $(2x^2 - 3xy + y^2) + (-4x^2 - 6xy - y^2) + (x^2 + xy - y^2)$

e Subtract.

35. $(a^3 + b^3) - (a^2b - ab^2 + b^3 + a^3)$

36. $(x^3 - y^3) - (-2x^3 + x^2y - xy^2 + 2y^3)$

37. $(xy - ab - 8) - (xy - 3ab - 6)$

38. $(3y^4x^2 + 2y^3x - 3y - 7) - (2y^4x^2 + 2y^3x - 4y - 2x + 5)$

39. $(-2a + 7b - c) - (-3b + 4c - 8d)$

40. Find the sum of $2a + b$ and $3a - b$. Then subtract $5a + 2b$.

f Multiply.

41. $(3z - u)(2z + 3u)$

42. $(a - b)(a^2 + b^2 + 2ab)$

43. $(a^2b - 2)(a^2b - 5)$

44. $(xy + 7)(xy - 4)$

45. $(a^3 + bc)(a^3 - bc)$

46. $(m^2 + n^2 - mn)(m^2 + mn + n^2)$

47. $(y^4x + y^2 + 1)(y^2 + 1)$

48. $(a - b)(a^2 + ab + b^2)$

49. $(3xy - 1)(4xy + 2)$

50. $(m^3n + 8)(m^3n - 6)$

51. $(3 - c^2d^2)(4 + c^2d^2)$

52. $(6x - 2y)(5x - 3y)$

53. $(m^2 - n^2)(m + n)$

54. $(pq + 0.2)(0.4pq - 0.1)$

55. $(xy + x^5y^5)(x^4y^4 - xy)$

56. $(x - y^3)(2y^3 + x)$

57. $(x + h)^2$

58. $(3a + 2b)^2$

59. $(r^3t^2 - 4)^2$

60. $(3a^2b - b^2)^2$

61. $(p^4 + m^2n^2)^2$

62. $(2ab - cd)^2$

63. $\left(2a^3 - \frac{1}{2}b^3\right)^2$

64. $-3x(x + 8y)^2$

65. $3a(a - 2b)^2$

66. $(a^2 + b + 2)^2$

67. $(2a - b)(2a + b)$

68. $(x - y)(x + y)$

69. $(c^2 - d)(c^2 + d)$

70. $(p^3 - 5q)(p^3 + 5q)$

71. $(ab + cd^2)(ab - cd^2)$

72. $(xy + pq)(xy - pq)$

73. $(x + y - 3)(x + y + 3)$

74. $(p + q + 4)(p + q - 4)$

75. $[x + y + z][x - (y + z)]$

76. $[a + b + c][a - (b + c)]$

77. $(a + b + c)(a - b - c)$

78. $(3x + 2 - 5y)(3x + 2 + 5y)$

79. D_W Is it possible for a polynomial in four variables to have a degree less than 4? Why or why not?

80. D_W Can the sum of two trinomials in several variables be a trinomial in one variable? Why or why not?

SKILL MAINTENANCE

In which quadrant is the point located? [3.1c]

81. $(2, -5)$

82. $(-8, -9)$

83. $(16, 23)$

84. $(-3, 2)$

Graph. [3.3b]

85. $2x = -10$

86. $y = -4$

87. $8y - 16 = 0$

88. $x = 4$

SYNTHESIS

Find a polynomial for the shaded area. (Leave results in terms of π where appropriate.)

89.

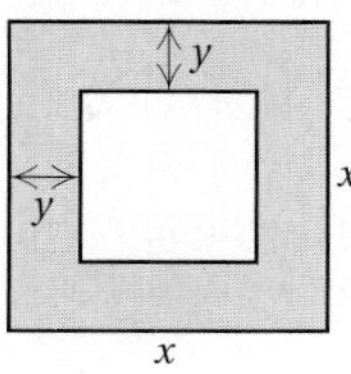

90.

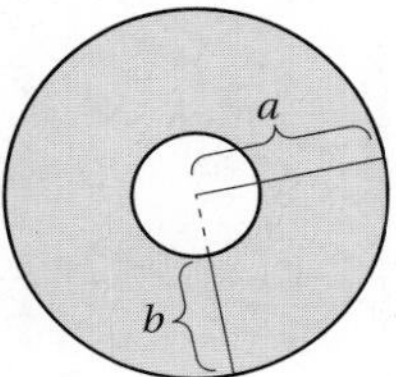

91.

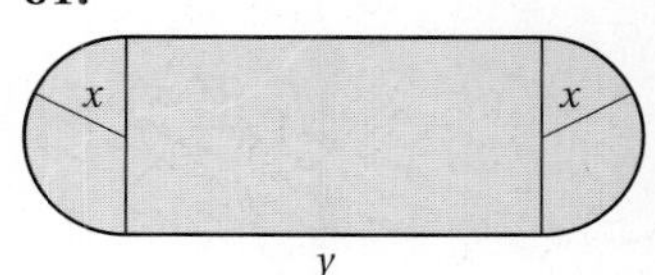

92.

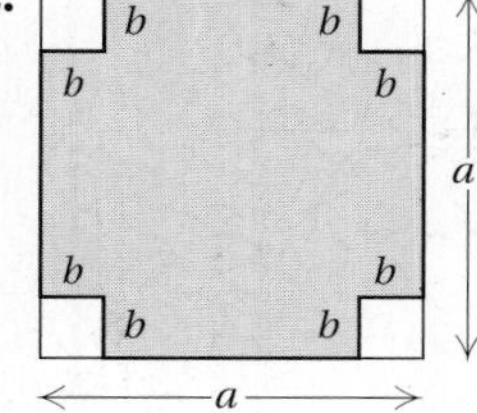

Find a formula for the surface area of the solid object. Leave results in terms of π.

93.

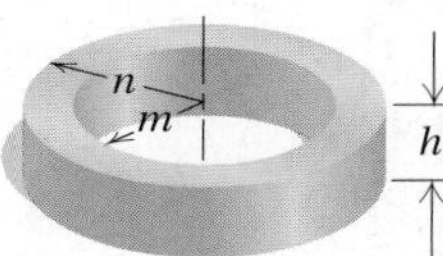

94.

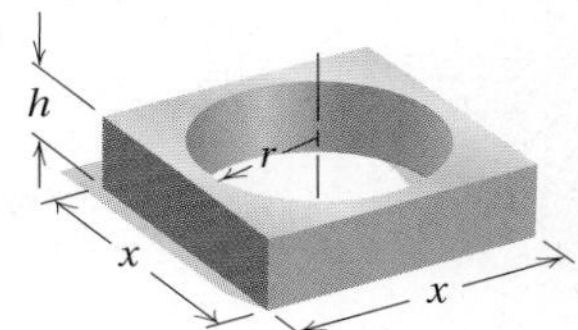

95. *Observatory Paint Costs.* The observatory at Danville University is shaped like a silo that is 40 ft high and 30 ft wide (see Exercise 15). The Heavenly Bodies Astronomy Club is to paint the exterior of the observatory using paint that covers 250 ft^2 per gallon. How many gallons should they purchase?

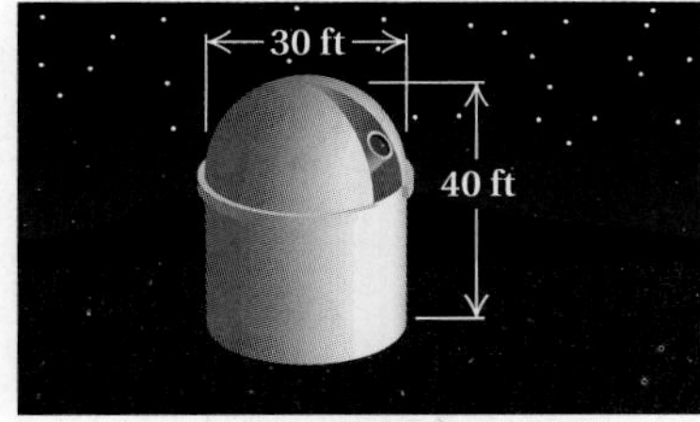

96. *Interest Compounded Annually.* An amount of money P that is invested at the yearly interest rate r grows to the amount

$$P(1 + r)^t$$

after t years. Find a polynomial that can be used to determine the amount to which P will grow after 2 yr.

97. Suppose that \$10,400 is invested at 8.5% compounded annually. How much is in the account at the end of 5 yr? (See Exercise 96.)

98. Multiply: $(x + a)(x - b)(x - a)(x + b)$.

4.8 DIVISION OF POLYNOMIALS

Objectives

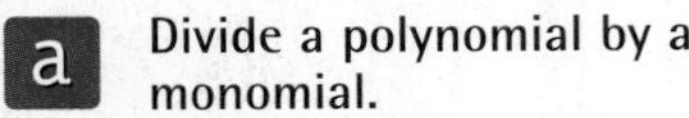
Divide a polynomial by a monomial.

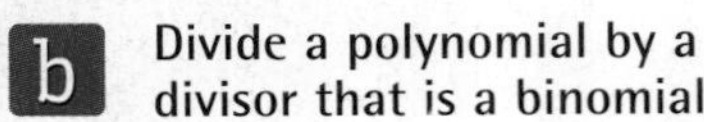
Divide a polynomial by a divisor that is a binomial.

In this section, we consider division of polynomials. You will see that such division is similar to what is done in arithmetic.

a Divisor a Monomial

We first consider division by a monomial. When dividing a monomial by a monomial, we use the quotient rule of Section 4.1 to subtract exponents when the bases are the same. We also divide the coefficients.

EXAMPLES Divide.

1. $\dfrac{10x^2}{2x} = \dfrac{10}{2} \cdot \dfrac{x^2}{x} = 5x^{2-1} = 5x$

2. $\dfrac{x^9}{3x^2} = \dfrac{1x^9}{3x^2} = \dfrac{1}{3} \cdot \dfrac{x^9}{x^2} = \dfrac{1}{3}x^{9-2} = \dfrac{1}{3}x^7$

3. $\dfrac{-18x^{10}}{3x^3} = \dfrac{-18}{3} \cdot \dfrac{x^{10}}{x^3} = -6x^{10-3} = -6x^7$

4. $\dfrac{42a^2b^5}{-3ab^2} = \dfrac{42}{-3} \cdot \dfrac{a^2}{a} \cdot \dfrac{b^5}{b^2} = -14a^{2-1}b^{5-2} = -14ab^3$

CAUTION!

The coefficients are divided but the exponents are subtracted.

Do Exercises 1–4.

Divide.

1. $\dfrac{20x^3}{5x}$

2. $\dfrac{-28x^{14}}{4x^3}$

3. $\dfrac{-56p^5q^7}{2p^2q^6}$

4. $\dfrac{x^5}{4x}$

Answers on page A-22

To divide a polynomial by a monomial, we note that since

$$\frac{A}{C} + \frac{B}{C} = \frac{A + B}{C},$$

it follows that

$$\frac{A + B}{C} = \frac{A}{C} + \frac{B}{C}.$$ Switching the left and right sides of the equation

This is actually the procedure we use when performing divisions like $86 \div 2$. Although we might write

$$\frac{86}{2} = 43,$$

we could also calculate as follows:

$$\frac{86}{2} = \frac{80 + 6}{2} = \frac{80}{2} + \frac{6}{2} = 40 + 3 = 43.$$

Similarly, to divide a polynomial by a monomial, we divide each term by the monomial.

EXAMPLE 5 Divide: $(9x^8 + 12x^6) \div 3x^2$.

We have

$$(9x^8 + 12x^6) \div 3x^2 = \frac{9x^8 + 12x^6}{3x^2}$$

$$= \frac{9x^8}{3x^2} + \frac{12x^6}{3x^2}.$$ To see this, add and get the original expression.

We now perform the separate divisions:

$$\frac{9x^8}{3x^2} + \frac{12x^6}{3x^2} = \frac{9}{3} \cdot \frac{x^8}{x^2} + \frac{12}{3} \cdot \frac{x^6}{x^2}$$
$$= 3x^{8-2} + 4x^{6-2}$$
$$= 3x^6 + 4x^4.$$

CAUTION!

The coefficients are *divided*, but the exponents are *subtracted*.

To check, we multiply the quotient $3x^6 + 4x^4$ by the divisor $3x^2$:

$$3x^2(3x^6 + 4x^4) = (3x^2)(3x^6) + (3x^2)(4x^4) = 9x^8 + 12x^6.$$

This is the polynomial that was being divided, so our answer is $3x^6 + 4x^4$.

Do Exercises 5–7.

EXAMPLE 6 Divide and check: $(10a^5b^4 - 2a^3b^2 + 6a^2b) \div (2a^2b)$.

$$\frac{10a^5b^4 - 2a^3b^2 + 6a^2b}{2a^2b} = \frac{10a^5b^4}{2a^2b} - \frac{2a^3b^2}{2a^2b} + \frac{6a^2b}{2a^2b}$$
$$= \frac{10}{2}a^{5-2}b^{4-1} - \frac{2}{2}a^{3-2}b^{2-1} + \frac{6}{2}$$
$$= 5a^3b^3 - ab + 3$$

CHECK:

$$2a^2b(5a^3b^3 - ab + 3) = 2a^2b \cdot 5a^3b^3 - 2a^2b \cdot ab + 2a^2b \cdot 3$$
$$= 10a^5b^4 - 2a^3b^2 + 6a^2b$$

Our answer, $5a^3b^3 - ab + 3$, checks.

To divide a polynomial by a monomial, divide each term by the monomial.

Do Exercises 8 and 9.

b Divisor a Binomial

Let's first consider long division as it is performed in arithmetic. When we divide, we repeat the following procedure.

To carry out long division:

1. Divide,
2. Multiply,
3. Subtract, and
4. Bring down the next term.

We review this by considering the division $3711 \div 8$.

```
       4    <—— ① Divide: 37 ÷ 8 ≈ 4.
8 ) 3 7 1 1
    3 2     <—— ② Multiply: 4 × 8 = 32.
      5 1       ③ Subtract: 37 − 32 = 5.
        ↑—————— ④ Bring down the 1.
```

```
       4 6 3
8 ) 3 7 1 1
    3 2
      5 1
      4 8
        3 1
        2 4
          7
```

5. Divide: $(28x^7 + 32x^5) \div 4x^3$. Check the result.

6. Divide: $(2x^3 + 6x^2 + 4x) \div 2x$. Check the result.

7. Divide: $(6x^2 + 3x - 2) \div 3$. Check the result.

Divide and check.

8. $(8x^2 - 3x + 1) \div 2$

9. $\dfrac{2x^4y^6 - 3x^3y^4 + 5x^2y^3}{x^2y^2}$

Answers on page A-22

10. Divide and check:

$(x^2 + x - 6) \div (x + 3)$.

Answer on page A-22

Study Tips

FORMING A STUDY GROUP

Consider forming a study group with some of your fellow students. Exchange e-mail addresses, telephone numbers, and schedules so that you can coordinate study time for homework and tests.

Next, we repeat the process two more times. We obtain the complete division as shown on the right above. The quotient is 463. The remainder is 7, expressed as R = 7. We write the answer as

$$463 \text{ R } 7 \qquad \text{or} \qquad 463 + \frac{7}{8} = 463\frac{7}{8}.$$

We check by multiplying the quotient, 463, by the divisor, 8, and adding the remainder, 7:

$$8 \cdot 463 + 7 = 3704 + 7 = 3711.$$

Now let's look at long division with polynomials. We use this procedure when the divisor is not a monomial. We write polynomials in descending order and then write in missing terms.

EXAMPLE 7 Divide $x^2 + 5x + 6$ by $x + 2$.

$$\begin{array}{r|l} x & \leftarrow \text{Divide the first term by the first term: } x^2/x = x. \\ x + 2\overline{)x^2 + 5x + 6} & \text{Ignore the term 2.} \\ \underline{x^2 + 2x} & \leftarrow \text{Multiply } x \text{ above by the divisor, } x + 2. \\ 3x & \leftarrow \text{Subtract: } (x^2 + 5x) - (x^2 + 2x) = x^2 + 5x - x^2 - 2x \\ & \qquad = 3x. \end{array}$$

We now "bring down" the next term of the dividend—in this case, 6.

$$\begin{array}{r|l} x + 3 & \leftarrow \text{Divide the first term by the first term: } 3x/x = 3. \\ x + 2\overline{)x^2 + 5x + 6} & \\ \underline{x^2 + 2x} & \\ 3x + 6 & \leftarrow \text{The 6 has been "brought down."} \\ \underline{3x + 6} & \leftarrow \text{Multiply 3 by the divisor, } x + 2. \\ 0 & \leftarrow \text{Subtract: } (3x + 6) - (3x + 6) = 3x + 6 - 3x - 6 = 0. \end{array}$$

The quotient is $x + 3$. The remainder is 0, expressed as R = 0. A remainder of 0 is generally not listed in an answer.

To check, we multiply the quotient by the divisor and add the remainder, if any, to see if we get the dividend:

$$\underbrace{(x + 2)}_{\text{Divisor}} \cdot \underbrace{(x + 3)}_{\text{Quotient}} + \underbrace{0}_{\text{Remainder}} = \underbrace{x^2 + 5x + 6}_{\text{Dividend}}. \qquad \text{The division checks.}$$

Do Exercise 10.

EXAMPLE 8 Divide and check: $(x^2 + 2x - 12) \div (x - 3)$.

$$\begin{array}{r|l} x & \leftarrow \text{Divide the first term by the first term: } x^2/x = x. \\ x - 3\overline{)x^2 + 2x - 12} & \\ \underline{x^2 - 3x} & \leftarrow \text{Multiply } x \text{ above by the divisor, } x - 3. \\ 5x & \leftarrow \text{Subtract: } (x^2 + 2x) - (x^2 - 3x) = x^2 + 2x - x^2 + 3x \\ & \qquad = 5x. \end{array}$$

We now "bring down" the next term of the dividend—in this case, -12.

$$\begin{array}{r|l} x + 5 & \leftarrow \text{Divide the first term by the first term: } 5x/x = 5. \\ x - 3\overline{)x^2 + 2x - 12} & \\ \underline{x^2 - 3x} & \\ 5x - 12 & \leftarrow \text{Bring down the } -12. \\ \underline{5x - 15} & \leftarrow \text{Multiply 5 above by the divisor, } x - 3. \\ 3 & \leftarrow \text{Subtract:} \\ & (5x - 12) - (5x - 15) = 5x - 12 - 5x + 15 \\ & \qquad = 3. \end{array}$$

The answer is $x + 5$ with $R = 3$, or

$$\underbrace{x + 5}_{\text{Quotient}} + \frac{3 \rightarrow \text{Remainder}}{\underbrace{x - 3}_{\text{Divisor}}}$$

(This is the way answers will be given at the back of the book.)

CHECK: We can check by multiplying the divisor by the quotient and adding the remainder, as follows:

$$\begin{aligned}(x - 3)(x + 5) + 3 &= x^2 + 2x - 15 + 3\\ &= x^2 + 2x - 12.\end{aligned}$$

When dividing, an answer may "come out even" (that is, have a remainder of 0, as in Example 7), or it may not (as in Example 8). If a remainder is not 0, we continue dividing until the degree of the remainder is less than the degree of the divisor. Check this in each of Examples 7 and 8.

Do Exercises 11 and 12.

EXAMPLE 9 Divide and check: $(x^3 + 1) \div (x + 1)$.

$$\begin{array}{rll}
 & x^2 - x + 1 & \\
x + 1 \,\big)\!\! & \overline{x^3 + 0x^2 + 0x + 1} & \leftarrow \text{Fill in the missing terms (see Section 4.3).}\\
 & \underline{x^3 + x^2} & \leftarrow \text{Subtract: } x^3 - (x^3 + x^2) = -x^2.\\
 & -x^2 + 0x & \\
 & \underline{-x^2 - x} & \leftarrow \text{Subtract: } -x^2 - (-x^2 - x) = x.\\
 & x + 1 & \\
 & \underline{x + 1} & \leftarrow \text{Subtract: } (x + 1) - (x + 1) = 0.\\
 & 0 &
\end{array}$$

The answer is $x^2 - x + 1$. The check is left to the student.

EXAMPLE 10 Divide and check: $(x^4 - 3x^2 + 1) \div (x - 4)$.

$$\begin{array}{rll}
 & x^3 + 4x^2 + 13x + 52 & \\
x - 4 \,\big)\!\! & \overline{x^4 + 0x^3 - 3x^2 + 0x + 1} & \leftarrow \text{Fill in the missing terms.}\\
 & \underline{x^4 - 4x^3} & \leftarrow \text{Subtract: } x^4 - (x^4 - 4x^3) = 4x^3.\\
 & 4x^3 - 3x^2 & \\
 & \underline{4x^3 - 16x^2} & \leftarrow \text{Subtract: } (4x^3 - 3x^2) - (4x^3 - 16x^2) = 13x^2.\\
 & 13x^2 + 0x & \\
 & \underline{13x^2 - 52x} & \leftarrow \text{Subtract: } 13x^2 - (13x^2 - 52x) = 52x.\\
 & 52x + 1 & \\
 & \underline{52x - 208} & \leftarrow \text{Subtract: } (52x + 1) - (52x - 208) = 209.\\
 & 209 &
\end{array}$$

The answer is $x^3 + 4x^2 + 13x + 52$, with $R = 209$, or

$$x^3 + 4x^2 + 13x + 52 + \frac{209}{x - 4}.$$

CHECK:
$$\begin{aligned}&(x - 4)(x^3 + 4x^2 + 13x + 52) + 209\\ &\quad = -4x^3 - 16x^2 - 52x - 208 + x^4 + 4x^3 + 13x^2 + 52x + 209\\ &\quad = x^4 - 3x^2 + 1\end{aligned}$$

Do Exercise 13.

Divide and check.

11. $x - 2 \overline{\smash{)}x^2 + 2x - 8}$

12. $x + 3 \overline{\smash{)}x^2 + 7x + 10}$

13. Divide and check:

$(x^3 - 1) \div (x - 1)$.

Answers on page A-22

For Extra Help

Digital Video Tutor CD 4 Videotape 9 | InterAct Math | Math Tutor Center | MathXL | MyMathLab

4.8 EXERCISE SET

a Divide and check.

1. $\frac{24x^4}{8}$

2. $\frac{-2u^2}{u}$

3. $\frac{25x^3}{5x^2}$

4. $\frac{16x^7}{-2x^2}$

5. $\frac{-54x^{11}}{-3x^8}$

6. $\frac{-75a^{10}}{3a^2}$

7. $\frac{64a^5b^4}{16a^2b^3}$

8. $\frac{-34p^{10}q^{11}}{-17pq^9}$

9. $\frac{24x^4 - 4x^3 + x^2 - 16}{8}$

10. $\frac{12a^4 - 3a^2 + a - 6}{6}$

11. $\frac{u - 2u^2 - u^5}{u}$

12. $\frac{50x^5 - 7x^4 + x^2}{x}$

13. $(15t^3 + 24t^2 - 6t) \div (3t)$

14. $(25t^3 + 15t^2 - 30t) \div (5t)$

15. $(20x^6 - 20x^4 - 5x^2) \div (-5x^2)$

16. $(24x^6 + 32x^5 - 8x^2) \div (-8x^2)$

17. $(24x^5 - 40x^4 + 6x^3) \div (4x^3)$

18. $(18x^6 - 27x^5 - 3x^3) \div (9x^3)$

19. $\frac{18x^2 - 5x + 2}{2}$

20. $\frac{15x^2 - 30x + 6}{3}$

21. $\frac{12x^3 + 26x^2 + 8x}{2x}$

22. $\frac{2x^4 - 3x^3 + 5x^2}{x^2}$

23. $\frac{9r^2s^2 + 3r^2s - 6rs^2}{3rs}$

24. $\frac{4x^4y - 8x^6y^2 + 12x^8y^6}{4x^4y}$

b Divide.

25. $(x^2 + 4x + 4) \div (x + 2)$

26. $(x^2 - 6x + 9) \div (x - 3)$

27. $(x^2 - 10x - 25) \div (x - 5)$

28. $(x^2 + 8x - 16) \div (x + 4)$

29. $(x^2 + 4x - 14) \div (x + 6)$

30. $(x^2 + 5x - 9) \div (x - 2)$

31. $\dfrac{x^2 - 9}{x + 3}$

32. $\dfrac{x^2 - 25}{x - 5}$

33. $\dfrac{x^5 + 1}{x + 1}$

34. $\dfrac{x^5 - 1}{x - 1}$

35. $\dfrac{8x^3 - 22x^2 - 5x + 12}{4x + 3}$

36. $\dfrac{2x^3 - 9x^2 + 11x - 3}{2x - 3}$

37. $(x^6 - 13x^3 + 42) \div (x^3 - 7)$

38. $(x^6 + 5x^3 - 24) \div (x^3 - 3)$

39. $(x^4 - 16) \div (x - 2)$

40. $(x^4 - 81) \div (x - 3)$

41. $(t^3 - t^2 + t - 1) \div (t - 1)$

42. $(t^3 - t^2 + t - 1) \div (t + 1)$

43. ^{D}W How is the distributive law used when dividing a polynomial by a binomial?

44. ^{D}W On an assignment, Emma *incorrectly* writes

$$\frac{12x^3 - 6x}{3x} = 4x^2 - 6x.$$

What mistake do you think she is making and how might you convince her that a mistake has been made?

SKILL MAINTENANCE

Subtract. [1.4a]

45. $17 - 45$

46. $-14 - 45$

47. $-2.3 - (-9.1)$

48. $-\dfrac{5}{8} - \dfrac{3}{4}$

Solve. [2.6a]

49. The perimeter of a rectangle is 640 ft. The length is 15 ft more than the width. Find the area of the rectangle.

50. The first angle of a triangle is 24° more than the second. The third angle is twice the first. Find the measures of the angles of the triangle.

Solve. [2.3c]

51. $-6(2 - x) + 10(5x - 7) = 10$

52. $-10(x - 4) = 5(2x + 5) - 7$

Factor. [1.7d]

53. $4x - 12 + 24y$

54. $256 - 2a - 4b$

SYNTHESIS

Divide.

55. $(x^4 + 9x^2 + 20) \div (x^2 + 4)$

56. $(y^4 + a^2) \div (y + a)$

57. $(5a^3 + 8a^2 - 23a - 1) \div (5a^2 - 7a - 2)$

58. $(15y^3 - 30y + 7 - 19y^2) \div (3y^2 - 2 - 5y)$

59. $(6x^5 - 13x^3 + 5x + 3 - 4x^2 + 3x^4) \div (3x^3 - 2x - 1)$

60. $(5x^7 - 3x^4 + 2x^2 - 10x + 2) \div (x^2 - x + 1)$

61. $(a^6 - b^6) \div (a - b)$

62. $(x^5 + y^5) \div (x + y)$

If the remainder is 0 when one polynomial is divided by another, the divisor is a *factor* of the dividend. Find the value(s) of c for which $x - 1$ is a factor of the polynomial.

63. $x^2 + 4x + c$

64. $2x^2 + 3cx - 8$

65. $c^2x^2 - 2cx + 1$

CHAPTER

4 Summary and Review

The review that follows is meant to prepare you for a chapter exam. It consists of two parts. The first part is a checklist of some of the Study Tips referred to in this and preceding chapters, as well as a list of important properties and formulas. The second part is the Review Exercises. These provide practice exercises for the exam, together with references to section objectives so you can go back and review. Before beginning, stop and look back over the skills you have obtained. What skills in mathematics do you have now that you did not have before studying this chapter?

STUDY TIPS CHECKLIST

The foundation of all your study skills is TIME!

- ☐ Have you tried using the audiotapes?
- ☐ Are you taking the time to include all the steps when working your homework and the tests?
- ☐ Are you using the time-management suggestions we have given so you have the proper amount of time to study mathematics?
- ☐ Have you been using the supplements for the text such as the *Student's Solutions Manual* and the Math Tutor Center?
- ☐ Have you memorized the rules for special products of polynomials and for manipulating expressions with exponents?

IMPORTANT PROPERTIES AND FORMULAS

FOIL: $(A + B)(C + D) = AC + AD + BC + BD$

Square of a Sum: $(A + B)(A + B) = (A + B)^2 = A^2 + 2AB + B^2$

Square of a Difference: $(A - B)(A - B) = (A - B)^2 = A^2 - 2AB + B^2$

Product of a Sum and a Difference: $(A + B)(A - B) = A^2 - B^2$

Definitions and Rules for Exponents: See p. 312.

REVIEW EXERCISES

Multiply and simplify. [4.1d, f]

1. $7^2 \cdot 7^{-4}$

2. $y^7 \cdot y^3 \cdot y$

3. $(3x)^5 \cdot (3x)^9$

4. $t^8 \cdot t^0$

Divide and simplify. [4.1e, f]

5. $\frac{4^5}{4^2}$

6. $\frac{a^5}{a^8}$

7. $\frac{(7x)^4}{(7x)^4}$

Simplify.

8. $(3t^4)^2$ [4.2a, b]

9. $(2x^3)^2(-3x)^2$ [4.1d], [4.2a, b]

10. $\left(\frac{2x}{y}\right)^{-3}$ [4.2b]

11. Express using a negative exponent: $\frac{1}{t^5}$. [4.1f]

12. Express using a positive exponent: y^{-4}. [4.1f]

13. Convert to scientific notation: 0.0000328. [4.2c]

14. Convert to decimal notation: 8.3×10^6. [4.2c]

Multiply or divide and write scientific notation for the result. [4.2d]

15. $(3.8 \times 10^4)(5.5 \times 10^{-1})$

16. $\dfrac{1.28 \times 10^{-8}}{2.5 \times 10^{-4}}$

17. *Diet-Drink Consumption.* It has been estimated that there will be 292 million people in the United States by 2005 and that on average, each of them will drink 15.3 gal of diet drinks that year. How many gallons of diet drinks will be consumed by the entire population in 2005? Express the answer in scientific notation. [4.2e]
Source: U.S. Department of Agriculture

18. Evaluate the polynomial $x^2 - 3x + 6$ when $x = -1$. [4.3a]

19. Identify the terms of the polynomial $-4y^5 + 7y^2 - 3y - 2$. [4.3b]

20. Identify the missing terms in $x^3 + x$. [4.3h]

21. Identify the degree of each term and the degree of the polynomial $4x^3 + 6x^2 - 5x + \frac{5}{3}$. [4.3g]

Classify the polynomial as a monomial, binomial, trinomial, or none of these. [4.3i]

22. $4x^3 - 1$

23. $4 - 9t^3 - 7t^4 + 10t^2$

24. $7y^2$

Collect like terms and then arrange in descending order. [4.3f]

25. $3x^2 - 2x + 3 - 5x^2 - 1 - x$

26. $-x + \frac{1}{2} + 14x^4 - 7x^2 - 1 - 4x^4$

Add. [4.4a]

27. $(3x^4 - x^3 + x - 4) + (x^5 + 7x^3 - 3x^2 - 5) + (-5x^4 + 6x^2 - x)$

28. $(3x^5 - 4x^4 + x^3 - 3) + (3x^4 - 5x^3 + 3x^2) + (-5x^5 - 5x^2) + (-5x^4 + 2x^3 + 5)$

Subtract. [4.4c]

29. $(5x^2 - 4x + 1) - (3x^2 + 1)$

30. $(3x^5 - 4x^4 + 3x^2 + 3) - (2x^5 - 4x^4 + 3x^3 + 4x^2 - 5)$

31. Find a polynomial for the perimeter and for the area. [4.4d], [4.5b]

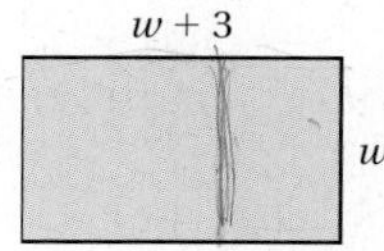

32. Find two algebraic expressions for the area of this figure. First, regard the figure as one large rectangle, and then regard the figure as a sum of four smaller rectangles. [4.4d]

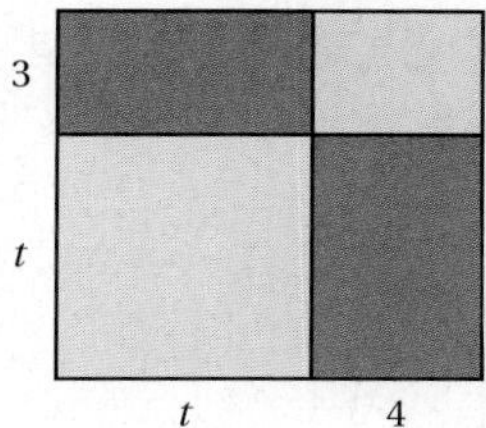

Multiply.

33. $\left(x + \frac{2}{3}\right)\left(x + \frac{1}{2}\right)$ [4.6a]

34. $(7x + 1)^2$ [4.6c]

35. $(4x^2 - 5x + 1)(3x - 2)$ [4.5d]

36. $(3x^2 + 4)(3x^2 - 4)$ [4.6b]

37. $5x^4(3x^3 - 8x^2 + 10x + 2)$ [4.5b]

38. $(x + 4)(x - 7)$ [4.6a]

39. $(3y^2 - 2y)^2$ [4.6c]

40. $(2t^2 + 3)(t^2 - 7)$ [4.6a]

41. Evaluate the polynomial

$$2 - 5xy + y^2 - 4xy^3 + x^6$$

when $x = -1$ and $y = 2$. [4.7a]

42. Identify the coefficient and the degree of each term of the polynomial

$$x^5y - 7xy + 9x^2 - 8.$$

Then find the degree of the polynomial. [4.7b]

Collect like terms. [4.7c]

43. $y + w - 2y + 8w - 5$

44. $m^6 - 2m^2n + m^2n^2 + n^2m - 6m^3 + m^2n^2 + 7n^2m$

45. Add: [4.7d]

$(5x^2 - 7xy + y^2) + (-6x^2 - 3xy - y^2) + (x^2 + xy - 2y^2).$

46. Subtract: [4.7e]

$(6x^3y^2 - 4x^2y - 6x) - (-5x^3y^2 + 4x^2y + 6x^2 - 6).$

Multiply. [4.7f]

47. $(p - q)(p^2 + pq + q^2)$

48. $\left(3a^4 - \frac{1}{3}b^3\right)^2$

Divide.

49. $(10x^3 - x^2 + 6x) \div (2x)$ [4.8a]

50. $(6x^3 - 5x^2 - 13x + 13) \div (2x + 3)$ [4.8b]

51. The graph of the polynomial equation $y = 10x^3 - 10x$ is shown below. Use *only* the graph to estimate the value of the polynomial when $x = -1$, $x = -0.5$, $x = 0.5$, $x = 1$, and $x = 1.1$. [4.3a]

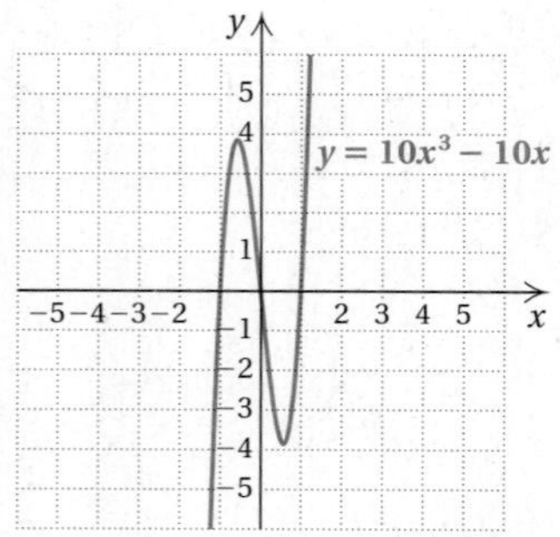

52. DW Explain why the expression 578.6×10^{-7} is not in scientific notation. [4.2c]

53. DW Write a short explanation of the difference between a monomial, a binomial, a trinomial, and a general polynomial. [4.3i]

SKILL MAINTENANCE

Certain objectives from four particular sections will be retested on the chapter test. The objectives are listed with the practice problems that follow.

54. Factor: $25t - 50 + 100m$. [1.7d]

55. Solve: $7x + 6 - 8x = 11 - 5x + 4$. [2.3b]

56. Solve: $3(x - 2) + 6 = 5(x + 3) + 9$. [2.3c]

57. Subtract: $-3.4 - 7.8$. [1.4a]

58. The perimeter of a rectangle is 540 m. The width is 19 m less than the length. Find the width and the length. [2.6a]

SYNTHESIS

Find a polynomial for the shaded area. [4.4d], [4.6b]

59.

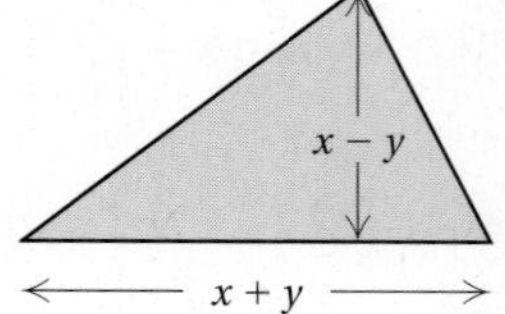

60.

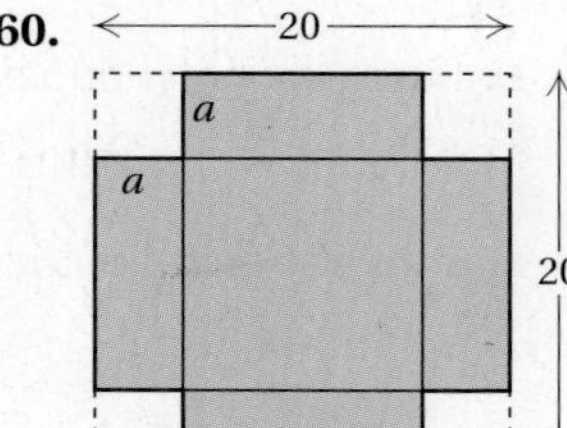

61. Collect like terms: [4.1d], [4.2a], [4.3e]

$$-3x^5 \cdot 3x^3 - x^6(2x)^2 + (3x^4)^2 + (2x^2)^4 - 40x^2(x^3)^2.$$

62. Solve: [2.3b], [4.6a]

$$(x - 7)(x + 10) = (x - 4)(x - 6).$$

63. The product of two polynomials is $x^5 - 1$. One of the polynomials is $x - 1$. Find the other. [4.8b]

64. A rectangular garden is twice as long as it is wide and is surrounded by a sidewalk that is 4 ft wide (see the figure below). The area of the sidewalk is 256 ft^2. Find the dimensions of the garden. [2.3b], [4.4d], [4.5a], [4.6a]

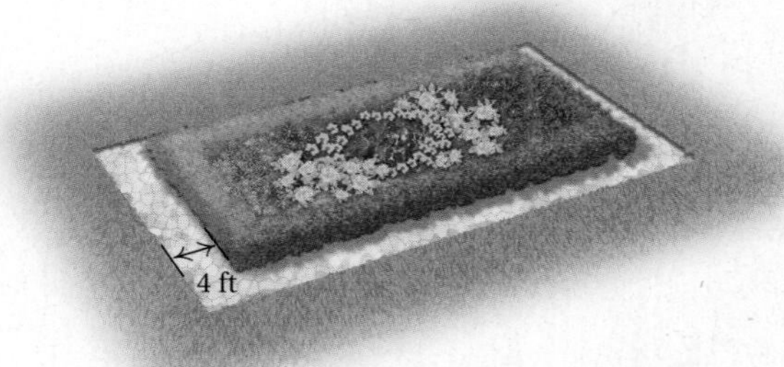

CHAPTER 4

Chapter Test

Multiply and simplify.

1. $6^{-2} \cdot 6^{-3}$

2. $x^6 \cdot x^2 \cdot x$

3. $(4a)^3 \cdot (4a)^8$

Divide and simplify.

4. $\dfrac{3^5}{3^2}$

5. $\dfrac{x^3}{x^8}$

6. $\dfrac{(2x)^5}{(2x)^5}$

Simplify.

7. $(x^3)^2$

8. $(-3y^2)^3$

9. $(2a^3b)^4$

10. $\left(\dfrac{ab}{c}\right)^3$

11. $(3x^2)^3(-2x^5)^3$

12. $3(x^2)^3(-2x^5)^3$

13. $2x^2(-3x^2)^4$

14. $(2x)^2(-3x^2)^4$

15. Express using a positive exponent:
5^{-3}.

16. Express using a negative exponent:
$\dfrac{1}{y^8}$.

17. Convert to scientific notation:
3,900,000,000.

18. Convert to decimal notation:
5×10^{-8}.

Multiply or divide and write scientific notation for the answer.

19. $\dfrac{5.6 \times 10^6}{3.2 \times 10^{-11}}$

20. $(2.4 \times 10^5)(5.4 \times 10^{16})$

21. *CD-ROM Memory.* A CD-ROM can contain about 600 million pieces of information (bytes). How many sound files, each containing 40,000 bytes, can a CD-ROM hold? Express the answer in scientific notation.

22. Evaluate the polynomial $x^5 + 5x - 1$ when $x = -2$.

23. Identify the coefficient of each term of the polynomial $\frac{1}{3}x^5 - x + 7$.

24. Identify the degree of each term and the degree of the polynomial $2x^3 - 4 + 5x + 3x^6$.

25. Classify the polynomial $7 - x$ as a monomial, a binomial, a trinomial, or none of these.

Collect like terms.

26. $4a^2 - 6 + a^2$

27. $y^2 - 3y - y + \dfrac{3}{4}y^2$

28. Collect like terms and then arrange in descending order:
$3 - x^2 + 2x^3 + 5x^2 - 6x - 2x + x^5$.

Add.

29. $(3x^5 + 5x^3 - 5x^2 - 3) +$
$(x^5 + x^4 - 3x^3 - 3x^2 + 2x - 4)$

30. $\left(x^4 + \dfrac{2}{3}x + 5\right) + \left(4x^4 + 5x^2 + \dfrac{1}{3}x\right)$

Subtract.

31. $(2x^4 + x^3 - 8x^2 - 6x - 3) - (6x^4 - 8x^2 + 2x)$

32. $(x^3 - 0.4x^2 - 12) - (x^5 + 0.3x^3 + 0.4x^2 + 9)$

Multiply.

33. $-3x^2(4x^2 - 3x - 5)$

34. $\left(x - \frac{1}{3}\right)^2$

35. $(3x + 10)(3x - 10)$

36. $(3b + 5)(b - 3)$

37. $(x^6 - 4)(x^8 + 4)$

38. $(8 - y)(6 + 5y)$

39. $(2x + 1)(3x^2 - 5x - 3)$

40. $(5t + 2)^2$

41. Collect like terms: $x^3y - y^3 + xy^3 + 8 - 6x^3y - x^2y^2 + 11$.

42. Subtract: $(8a^2b^2 - ab + b^3) - (-6ab^2 - 7ab - ab^3 + 5b^3)$.

43. Multiply: $(3x^5 - 4y^5)(3x^5 + 4y^5)$.

Divide.

44. $(12x^4 + 9x^3 - 15x^2) \div (3x^2)$

45. $(6x^3 - 8x^2 - 14x + 13) \div (3x + 2)$

46. The graph of the polynomial equation $y = x^3 - 5x - 1$ is shown at right. Use *only* the graph to estimate the value of the polynomial when $x = -1$, $x = -0.5$, $x = 0.5$, $x = 1$, and $x = 1.1$.

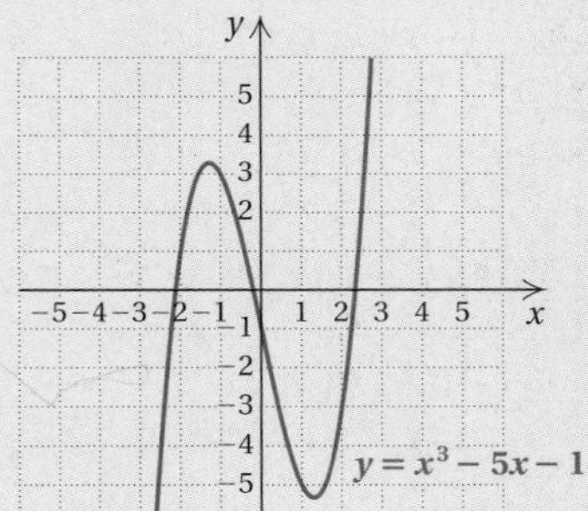

47. Find a polynomial for the surface area of this right rectangular solid.

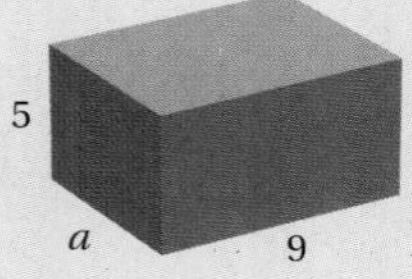

48. Find two algebraic expressions for the area of this figure. First, regard the figure as one large rectangle, and then regard the figure as a sum of four smaller rectangles.

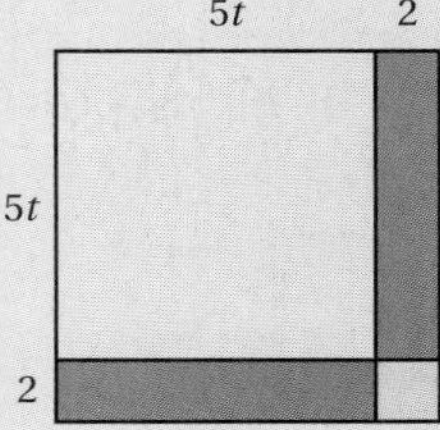

SKILL MAINTENANCE

49. Solve: $7x - 4x - 2 = 37$.

50. Solve: $4(x + 2) - 21 = 3(x - 6) + 2$.

51. Factor: $64t - 32m + 16$.

52. Subtract: $\frac{2}{5} - \left(-\frac{3}{4}\right)$.

53. The first angle of a triangle is four times as large as the second. The measure of the third angle is 30° greater than that of the second. How large are the angles?

SYNTHESIS

54. The height of a box is 1 less than its length, and the length is 2 more than its width. Find the volume in terms of the length.

55. Solve: $(x - 5)(x + 5) = (x + 6)^2$.

CHAPTERS

1–4 Cumulative Review

1. *Path of the Olympic Arrow.* The Olympic flame at the 1992 Summer Olympics was lit by a flaming arrow. As the arrow moved d feet horizontally from the archer, its height h, in feet, could be approximated by the polynomial equation

$$h = -0.002d^2 + 0.8d + 6.6.$$

The graph of this equation is shown at right. Use either the graph or the polynomial to approximate the height of the arrow after it has traveled horizontally for 100 ft, 200 ft, 300 ft, and 350 ft.

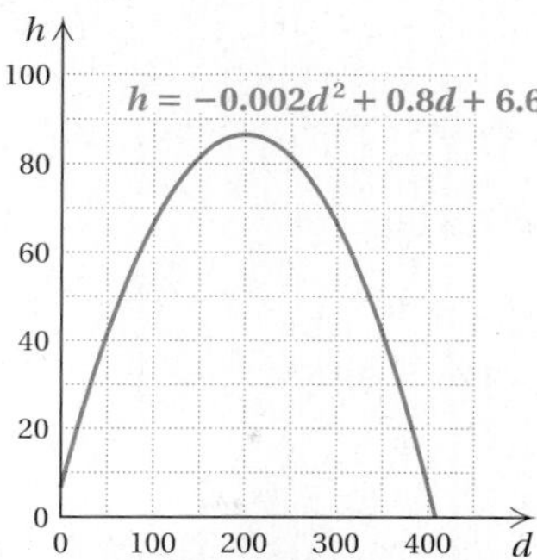

2. Evaluate $\frac{x}{2y}$ when $x = 10$ and $y = 2$.

3. Evaluate $2x^3 + x^2 - 3$ when $x = -1$.

4. Evaluate $x^3y^2 + xy + 2xy^2$ when $x = -1$ and $y = 2$.

5. Find the absolute value: $|-4|$.

6. Find the reciprocal of 5.

Compute and simplify.

7. $-\frac{3}{5} + \frac{5}{12}$

8. $3.4 - (-0.8)$

9. $(-2)(-1.4)(2.6)$

10. $\frac{3}{8} \div \left(-\frac{9}{10}\right)$

11. $(1.1 \times 10^{10})(2 \times 10^{12})$

12. $(3.2 \times 10^{-10}) \div (8 \times 10^{-6})$

Simplify.

13. $\frac{-9x}{3x}$

14. $y - (3y + 7)$

15. $3(x - 1) - 2[x - (2x + 7)]$

16. $2 - [32 \div (4 + 2^2)]$

Add.

17. $(x^4 + 3x^3 - x + 7) + (2x^5 - 3x^4 + x - 5)$

18. $(x^2 + 2xy) + (y^2 - xy) + (2x^2 - 3y^2)$

Subtract.

19. $(x^3 + 3x^2 - 4) - (-2x^2 + x + 3)$

20. $\left(\frac{1}{3}x^2 - \frac{1}{4}x - \frac{1}{5}\right) - \left(\frac{2}{3}x^2 + \frac{1}{2}x - \frac{1}{5}\right)$

Multiply.

21. $3(4x - 5y + 7)$

22. $(-2x^3)(-3x^5)$

23. $2x^2(x^3 - 2x^2 + 4x - 5)$

24. $(y^2 - 2)(3y^2 + 5y + 6)$

25. $(2p^3 + p^2q + pq^2)(p - pq + q)$

26. $(2x + 3)(3x + 2)$

27. $(3x^2 + 1)^2$

28. $\left(t + \frac{1}{2}\right)\left(t - \frac{1}{2}\right)$

29. $(2y^2 + 5)(2y^2 - 5)$

30. $(2x^4 - 3)(2x^2 + 3)$

31. $(t - 2t^2)^2$

32. $(3p + q)(5p - 2q)$

Divide.

33. $(18x^3 + 6x^2 - 9x) \div 3x$

34. $(3x^3 + 7x^2 - 13x - 21) \div (x + 3)$

Solve.

35. $1.5 = 2.7 + x$

36. $\frac{2}{7}x = -6$

37. $5x - 9 = 36$

38. $\frac{2}{3} = \frac{-m}{10}$

39. $5.4 - 1.9x = 0.8x$

40. $x - \frac{7}{8} = \frac{3}{4}$

41. $2(2 - 3x) = 3(5x + 7)$

42. $\frac{1}{4}x - \frac{2}{3} = \frac{3}{4} + \frac{1}{3}x$

43. $y + 5 - 3y = 5y - 9$

44. $\frac{1}{4}x - 7 < 5 - \frac{1}{2}x$

45. $2(x + 2) \geq 5(2x + 3)$

46. $A = Qx + P$, for x

47. A 6-ft by 3-ft raft is floating in a swimming pool of radius r. Find a polynomial for the area of the surface of the pool not covered by the raft.

Solve.

48. *Consecutive Page Numbers.* The sum of the page numbers on the facing pages of a book is 37. What are the page numbers?

49. *Room Perimeter.* The perimeter of a room is 88 ft. The width is 4 ft less than the length. Find the width and the length.

50. The second angle of a triangle is five times as large as the first. The third angle is twice the sum of the other two angles. Find the measure of the first angle.

51. *Discount.* A bookstore sells books at a price that is 80% higher than the price the store pays for the books. A book is priced for sale at \$6.30. How much did the store pay for the book?

52. *Coffee Consumption.* It has been estimated that there will be 292 million people in the United States by 2005 and that on average, each of them will drink 21.1 gal of coffee that year. How many gallons of coffee will be consumed by the entire population in 2005? Express the answer in scientific notation.
Source: U.S. Department of Agriculture

Simplify.

53. $y^2 \cdot y^{-6} \cdot y^8$

54. $\frac{x^6}{x^7}$

55. $(-3x^3y^{-2})^3$

56. $\frac{x^3x^{-4}}{x^{-5}x}$

57. Identify the coefficient of each term of the polynomial $\frac{2}{3}x^2 + 4x - 6$.

58. Identify the degree of each term and the degree of the polynomial $2x^4 + 3x^2 + 2x + 1$.

Classify the polynomial as a monomial, a binomial, a trinomial, or none of these.

59. $2x^2 + 1$

60. $2x^2 + x + 1$

61. Find the intercepts of $4x - 5y = 20$.

62. Graph: $4x - 5y = 20$.

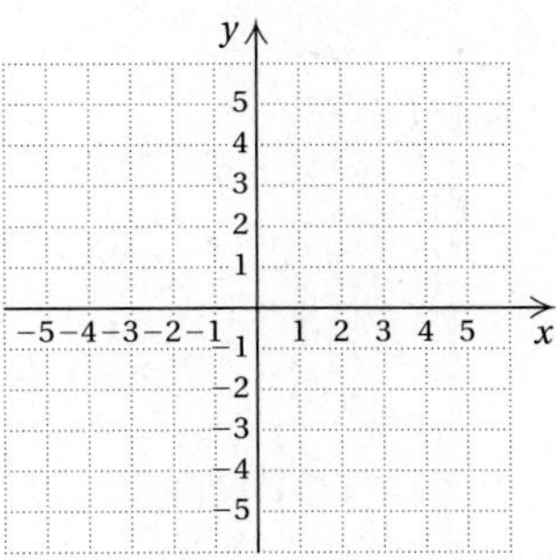

63. Find two algebraic expressions for the area of this figure. First, regard the figure as one large rectangle, and then regard the figure as a sum of four smaller rectangles.

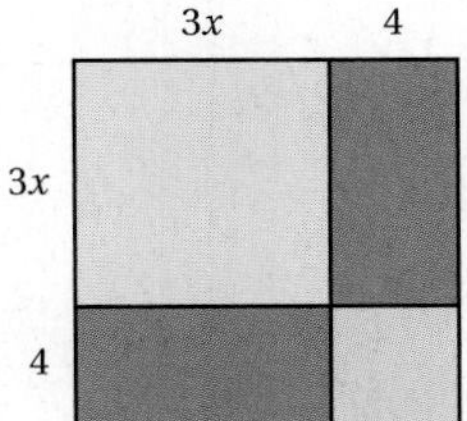

64. *Matching.* Match each item in the first column with the appropriate item in the second column by drawing connecting lines.

3^2	$\frac{1}{6}$
3^{-2}	$-\frac{1}{9}$
$\left(\frac{1}{3}\right)^2$	6
$\left(\frac{1}{3}\right)^{-2}$	9
-3^2	-9
$(-3)^2$	$\frac{1}{9}$
$\left(-\frac{1}{3}\right)^2$	-6
$\left(-\frac{1}{3}\right)^{-2}$	12

SYNTHESIS

65. A picture frame is x inches square. The picture that it frames is 2 in. shorter than the frame in both length and width. Find a polynomial for the area of the frame.

Add.

66. $[(2x)^2 - (3x)^3 + 2x^2x^3 + (x^2)^2] + [5x^2(2x^3) - ((2x)^2)^2]$

67. $(x - 3)^2 + (2x + 1)^2$

68. $[(3x^3 + 11x^2 + 11x + 15) \div (x + 3)] + [(2x^3 - 7x^2 + 2) \div (2x + 1)]$

Solve.

69. $(x + 3)(2x - 5) + (x - 1)^2 = (3x + 1)(x - 3)$

70. $(2x^2 + x - 6) \div (2x - 3) = (2x^2 - 9x - 5) \div (x - 5)$

71. $20 - 3|x| = 5$

72. $(x - 3)(x + 4) = (x^3 - 4x^2 - 17x + 60) \div (x - 5)$

73. A side of a cube is $(x + 2)$ cm long. Find a polynomial for the volume of the cube.

Polynomials: Factoring

Gateway to Chapter 5

Factoring is the reverse of multiplying. To factor a polynomial or other algebraic expression is to find an equivalent expression that is a product. In this chapter, we study the very important skill of factoring polynomials. To learn to factor quickly, we use the quick methods for multiplication that we learned in Chapter 4.

In this chapter, we introduce a new equation-solving technique to be used to solve equations involving quadratic, or second-degree, polynomials. This leads us to new ways of problem solving in Section 5.8. Problems we consider there could not have been solved easily without these new skills.

Real-World Application

The height of a triangular sail on a racing sailboat is 9 ft more than the base. The area of the triangle is 110 ft^2. Find the height and the base of the sail.

Source: Whitney Gladstone, North Graphics, San Diego, California

This problem appears as Example 2 in Section 5.8.

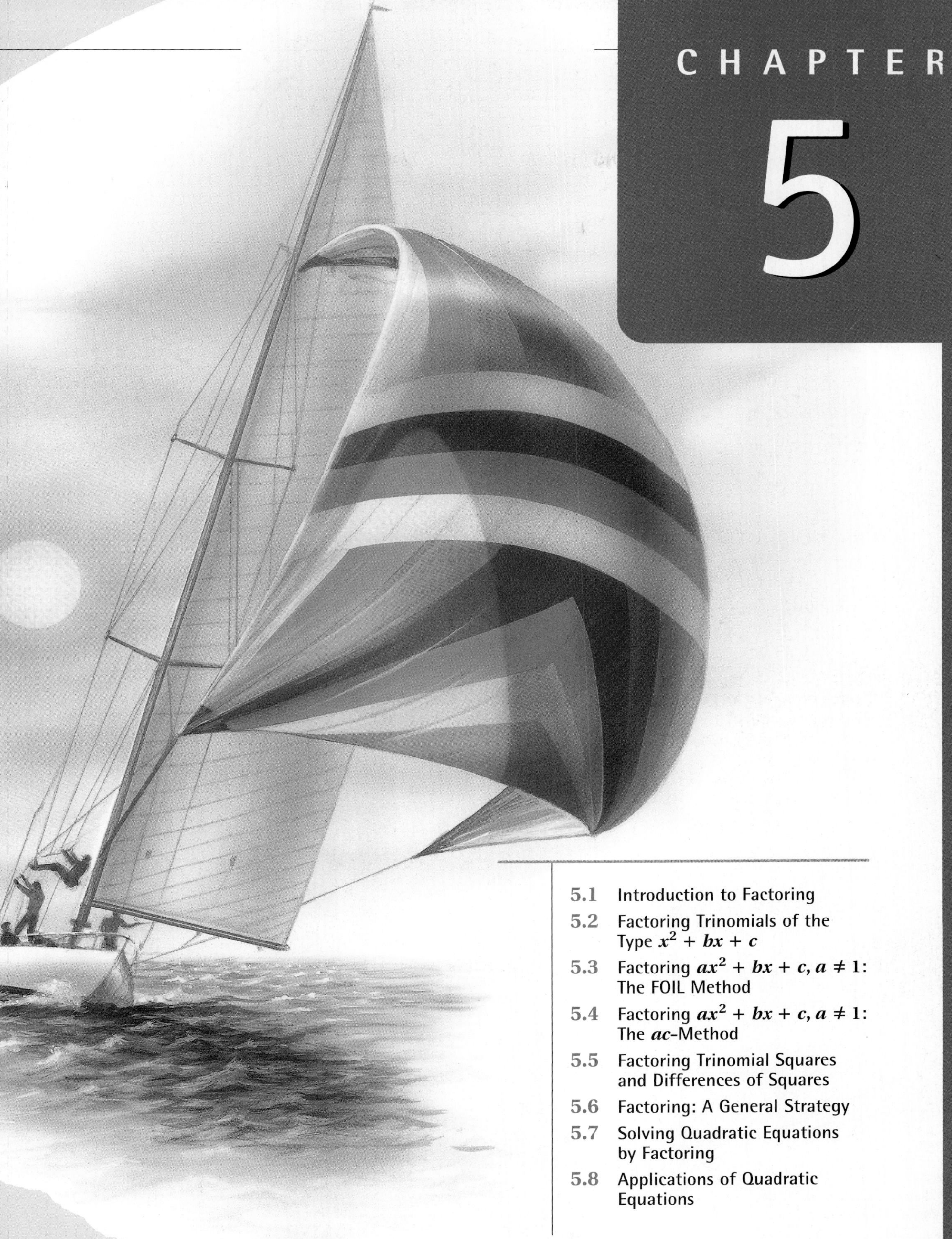

CHAPTER 5

CHAPTER 5

Pretest

1. Find three factorizations of $-20x^6$. [5.1a]

Factor.

2. $2x^2 + 4x + 2$ [5.5b]

3. $x^2 + 6x + 8$ [5.2a]

4. $8a^5 + 4a^3 - 20a$ [5.1b]

5. $-6 + 5x^2 - 13x$ [5.3a], [5.4a]

6. $81 - z^4$ [5.5d]

7. $y^6 - 4y^3 + 4$ [5.5b]

8. $3x^3 + 2x^2 + 12x + 8$ [5.1c]

9. $p^2 - p - 30$ [5.2a]

10. $x^4y^2 - 64$ [5.5d]

11. $2p^2 + 7pq - 4q^2$ [5.3a], [5.4a]

Solve.

12. $x^2 - 5x = 0$ [5.7b]

13. $(x - 4)(5x - 3) = 0$ [5.7a]

14. $3x^2 + 10x - 8 = 0$ [5.7b]

15. $(x + 2)(x - 2) = 5$ [5.7b]

Solve. [5.8a]

16. Dimensions of a Triangle. The height of a triangle is 3 cm longer than the base. The area of the triangle is 44 cm^2. Find the base and the height.

17. Framing. A rectangular picture frame is twice as long as it is wide. The area of the frame is 162 in^2. Find the dimensions of the frame.

18. Right-Triangle Geometry. Find the length of the missing side of this right triangle.

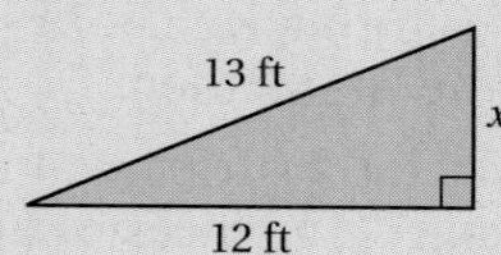

5.1 INTRODUCTION TO FACTORING

Objectives

a Factor monomials.

b Factor polynomials when the terms have a common factor, factoring out the largest common factor.

c Factor certain expressions with four terms using factoring by grouping.

To solve certain types of algebraic equations involving polynomials of second degree, we must learn to factor polynomials.

Consider the product $15 = 3 \cdot 5$. We say that 3 and 5 are **factors** of 15 and that $3 \cdot 5$ is a **factorization** of 15. Since $15 = 15 \cdot 1$, we also know that 15 and 1 are factors of 15 and that $15 \cdot 1$ is a factorization of 15. We use the word "factor" as both a verb and a noun.

FACTOR; FACTORIZATION

To **factor** a polynomial is to express it as a product.

A **factor** of a polynomial P is a polynomial that can be used to express P as a product.

A **factorization** of a polynomial is an expression that names that polynomial as a product.

a Factoring Monomials

To factor a monomial, we find two monomials whose product is equivalent to the original monomial. Compare.

Multiplying	*Factoring*
a) $(4x)(5x) = 20x^2$	$20x^2 = (4x)(5x)$
b) $(2x)(10x) = 20x^2$	$20x^2 = (2x)(10x)$
c) $(-4x)(-5x) = 20x^2$	$20x^2 = (-4x)(-5x)$
d) $(x)(20x) = 20x^2$	$20x^2 = (x)(20x)$

In each case, we have expressed $20x^2$ as a *product.*

You can see that the monomial $20x^2$ has many factorizations. There are still other ways to factor $20x^2$.

Do Exercises 1 and 2.

EXAMPLE 1 Find three factorizations of $15x^3$.

a) $15x^3 = (3 \cdot 5)(x \cdot x^2)$
$= (3x)(5x^2)$ ← This is a *product.*

b) $15x^3 = (3 \cdot 5)(x^2 \cdot x)$
$= (3x^2)(5x)$

c) $15x^3 = (-15)(-1)x^3$
$= (-15)(-x^3)$

Do Exercises 3–5.

b Factoring When Terms Have a Common Factor

To multiply a monomial and a polynomial with more than one term, we multiply each term of the polynomial by the monomial using the distributive laws,

$$a(b + c) = ab + ac \quad \text{and} \quad a(b - c) = ab - ac.$$

1. a) Multiply: $(3x)(4x)$.

b) Factor: $12x^2$.

2. a) Multiply: $(2x)(-8x^2)$.

b) Factor: $-16x^3$.

Find three factorizations of the monomial.

3. $8x^4$

4. $-21x^2$

5. $6x^5$

Answers on page A-23

6. a) Multiply: $3(x + 2)$.

b) Factor: $3x + 6$.

7. a) Multiply: $2x(x^2 + 5x + 4)$.

b) Factor: $2x^3 + 10x^2 + 8x$.

Answers on page A-23

To factor, we do the reverse. We express a polynomial as a product using the distributive laws in reverse:

$$ab + ac = a(b + c) \quad \text{and} \quad ab - ac = a(b - c).$$

Compare.

Multiply	*Factor*
$3x(x^2 + 2x - 4)$	$3x^3 + 6x^2 - 12x$
$= 3x \cdot x^2 + 3x \cdot 2x - 3x \cdot 4$	$= 3x \cdot x^2 + 3x \cdot 2x - 3x \cdot 4$
$= 3x^3 + 6x^2 - 12x$	$= 3x(x^2 + 2x - 4)$

Do Exercises 6 and 7.

CAUTION!

Consider the following:

$$3x^3 + 6x^2 - 12x = 3 \cdot x \cdot x \cdot x + 2 \cdot 3 \cdot x \cdot x - 2 \cdot 2 \cdot 3 \cdot x.$$

The terms of the polynomial, $3x^3$, $6x^2$, and $-12x$, have been factored but the polynomial itself has not been factored. This is not what we mean by a factorization of the polynomial. The *factorization* is

$$3x(x^2 + 2x - 4).$$

The expressions $3x$ and $x^2 + 2x - 4$ are *factors* of $3x^3 + 6x^2 - 12x$.

To factor, we first try to find a factor common to all terms. There may not be one other than 1. When there is, we generally use the factor with the largest possible coefficient and the largest possible exponent.

EXAMPLE 2 Factor: $7x^2 + 14$.

We have

$$7x^2 + 14 = 7 \cdot x^2 + 7 \cdot 2 \qquad \text{Factoring each term}$$
$$= 7(x^2 + 2). \qquad \text{Factoring out the common factor 7}$$

CHECK: We multiply to check:

$$7(x^2 + 2) = 7 \cdot x^2 + 7 \cdot 2 = 7x^2 + 14.$$

EXAMPLE 3 Factor: $16x^3 + 20x^2$.

$$16x^3 + 20x^2 = (4x^2)(4x) + (4x^2)(5) \qquad \text{Factoring each term}$$
$$= 4x^2(4x + 5) \qquad \text{Factoring out the common factor } 4x^2$$

Suppose in Example 3 that you had not recognized the largest common factor and removed only part of it, as follows:

$$16x^3 + 20x^2 = (2x^2)(8x) + (2x^2)(10)$$
$$= 2x^2(8x + 10).$$

Note that $8x + 10$ still has a common factor of 2. You need not begin again. Just continue factoring out common factors, as follows, until finished:

$$= 2x^2[2(4x + 5)]$$
$$= 4x^2(4x + 5).$$

EXAMPLE 4 Factor: $15x^5 - 12x^4 + 27x^3 - 3x^2$.

$$15x^5 - 12x^4 + 27x^3 - 3x^2 = (3x^2)(5x^3) - (3x^2)(4x^2) + (3x^2)(9x) - (3x^2)(1)$$
$$= 3x^2(5x^3 - 4x^2 + 9x - 1) \quad \text{Factoring out } 3x^2$$

CAUTION!
Don't forget the term -1.

CHECK: We multiply to check:

$$3x^2(5x^3 - 4x^2 + 9x - 1)$$
$$= (3x^2)(5x^3) - (3x^2)(4x^2) + (3x^2)(9x) - (3x^2)(1)$$
$$= 15x^5 - 12x^4 + 27x^3 - 3x^2.$$

As you become more familiar with factoring, you will be able to spot the largest common factor without factoring each term. Then you can write just the answer.

EXAMPLES Factor.

5. $8m^3 - 16m = 8m(m^2 - 2)$
6. $14p^2y^3 - 8py^2 + 2py = 2py(7py^2 - 4y + 1)$
7. $\frac{4}{5}x^2 + \frac{1}{5}x + \frac{2}{5} = \frac{1}{5}(4x^2 + x + 2)$
8. $2.4x^2 + 1.2x - 3.6 = 1.2(2x^2 + x - 3)$

Do Exercises 8–13.

There are two important points to keep in mind as we study this chapter.

TIPS FOR FACTORING

- Before doing any other kind of factoring, first try to factor out the largest common factor.
- Always check the result of factoring by multiplying.

C Factoring by Grouping: Four Terms

Certain polynomials with four terms can be factored using a method called *factoring by grouping.*

EXAMPLE 9 Factor: $x^2(x + 1) + 2(x + 1)$.

The binomial $x + 1$ is common to both terms:

$$x^2(x + 1) + 2(x + 1) = (x^2 + 2)(x + 1).$$

The factorization is $(x^2 + 2)(x + 1)$.

Do Exercises 14 and 15.

Factor. Check by multiplying.

8. $x^2 + 3x$

9. $3y^6 - 5y^3 + 2y^2$

10. $9x^4 - 15x^3 + 3x^2$

11. $\frac{3}{4}t^3 + \frac{5}{4}t^2 + \frac{7}{4}t + \frac{1}{4}$

12. $35x^7 - 49x^6 + 14x^5 - 63x^3$

13. $8.4x^2 - 5.6x + 2.8$

Factor.

14. $x^2(x + 7) + 3(x + 7)$

15. $x^2(a + b) + 2(a + b)$

Answers on page A-23

Factor by grouping.

16. $x^3 + 7x^2 + 3x + 21$

17. $8t^3 + 2t^2 + 12t + 3$

18. $3m^5 - 15m^3 + 2m^2 - 10$

19. $3x^3 - 6x^2 - x + 2$

20. $4x^3 - 6x^2 - 6x + 9$

21. $y^4 - 2y^3 - 2y - 10$

Answers on page A-23

Consider the four-term polynomial

$$x^3 + x^2 + 2x + 2.$$

There is no factor other than 1 that is common to all the terms. We can, however, factor $x^3 + x^2$ and $2x + 2$ separately:

$x^3 + x^2 = x^2(x + 1)$;   Factoring $x^3 + x^2$

$2x + 2 = 2(x + 1)$.   Factoring $2x + 2$

We have grouped certain terms and factored each polynomial separately:

$$\begin{aligned} x^3 + x^2 + 2x + 2 &= (x^3 + x^2) + (2x + 2) \\ &= x^2(x + 1) + 2(x + 1) \\ &= (x^2 + 2)(x + 1), \end{aligned}$$

as in Example 9. This method is called **factoring by grouping.** We began with a polynomial with four terms. After grouping and removing common factors, we obtained a polynomial with two parts, each having a common factor $x + 1$. Not all polynomials with four terms can be factored by this procedure, but it does give us a method to try.

EXAMPLES Factor by grouping.

10. $6x^3 - 9x^2 + 4x - 6$

$= (6x^3 - 9x^2) + (4x - 6)$

$= 3x^2(2x - 3) + 2(2x - 3)$ Factoring each binomial

$= (3x^2 + 2)(2x - 3)$ Factoring out the common factor $2x - 3$

We think through this process as follows:

$$6x^3 - 9x^2 + 4x - 6 = 3x^2(2x - 3) \square (2x - 3)$$

(1) Factor the first two terms.

(2) This factor, $2x - 3$, gives us a hint to the factorization of the last two terms.

(3) Now we ask ourselves, "What needs to be here to enable us to get $4x - 6$ when we multiply?"

CAUTION! Don't forget the 1.

11. $x^3 + x^2 + x + 1 = (x^3 + x^2) + (x + 1)$

$= x^2(x + 1) + 1(x + 1)$ Factoring each binomial

$= (x^2 + 1)(x + 1)$ Factoring out the common factor $x + 1$

12. $2x^3 - 6x^2 - x + 3$

$= (2x^3 - 6x^2) + (-x + 3)$

$= 2x^2(x - 3) - 1(x - 3)$ *Check*: $-1(x - 3) = -x + 3$.

$= (2x^2 - 1)(x - 3)$ Factoring out the common factor $x - 3$

13. $12x^5 + 20x^2 - 21x^3 - 35 = 4x^2(3x^3 + 5) - 7(3x^3 + 5)$

$= (4x^2 - 7)(3x^3 + 5)$

14. $x^3 + x^2 + 2x - 2 = x^2(x + 1) + 2(x - 1)$

This polynomial is not factorable using factoring by grouping. It may be factorable, but not by methods that we will consider in this text.

Do Exercises 16–21.

5.1 EXERCISE SET

For Extra Help

Digital Video Tutor CD 5 Videotape 10 | InterAct Math | Math Tutor Center | MathXL | MyMathLab

a Find three factorizations for the monomial.

1. $8x^3$

2. $6x^4$

3. $-10a^6$

4. $-8y^5$

5. $24x^4$

6. $15x^5$

b Factor. Check by multiplying.

7. $x^2 - 6x$

8. $x^2 + 5x$

9. $2x^2 + 6x$

10. $8y^2 - 8y$

11. $x^3 + 6x^2$

12. $3x^4 - x^2$

13. $8x^4 - 24x^2$

14. $5x^5 + 10x^3$

15. $2x^2 + 2x - 8$

16. $8x^2 - 4x - 20$

17. $17x^5y^3 + 34x^3y^2 + 51xy$

18. $16p^6q^4 + 32p^5q^3 - 48pq^2$

19. $6x^4 - 10x^3 + 3x^2$

20. $5x^5 + 10x^2 - 8x$

21. $x^5y^5 + x^4y^3 + x^3y^3 - x^2y^2$

22. $x^9y^6 - x^7y^5 + x^4y^4 + x^3y^3$

23. $2x^7 - 2x^6 - 64x^5 + 4x^3$

24. $8y^3 - 20y^2 + 12y - 16$

25. $1.6x^4 - 2.4x^3 + 3.2x^2 + 6.4x$

26. $2.5x^6 - 0.5x^4 + 5x^3 + 10x^2$

27. $\frac{5}{3}x^6 + \frac{4}{3}x^5 + \frac{1}{3}x^4 + \frac{1}{3}x^3$

28. $\frac{5}{9}x^7 + \frac{2}{9}x^5 - \frac{4}{9}x^3 - \frac{1}{9}x$

c Factor.

29. $x^2(x + 3) + 2(x + 3)$

30. $3z^2(2z + 1) + (2z + 1)$

31. $5a^3(2a - 7) - (2a - 7)$

32. $m^4(8 - 3m) - 7(8 - 3m)$

Factor by grouping.

33. $x^3 + 3x^2 + 2x + 6$

34. $6z^3 + 3z^2 + 2z + 1$

35. $2x^3 + 6x^2 + x + 3$

36. $3x^3 + 2x^2 + 3x + 2$

37. $8x^3 - 12x^2 + 6x - 9$

38. $10x^3 - 25x^2 + 4x - 10$

39. $12x^3 - 16x^2 + 3x - 4$

40. $18x^3 - 21x^2 + 30x - 35$

41. $5x^3 - 5x^2 - x + 1$

42. $7x^3 - 14x^2 - x + 2$

43. $x^3 + 8x^2 - 3x - 24$

44. $2x^3 + 12x^2 - 5x - 30$

45. $2x^3 - 8x^2 - 9x + 36$

46. $20g^3 - 4g^2 - 25g + 5$

47. $\mathbf{D_W}$ Josh says that there is no need to print answers for Exercises 1–46 at the back of the book. Is he correct in saying this? Why or why not?

48. $\mathbf{D_W}$ Explain how one could construct a polynomial with four terms that can be factored by grouping.

SKILL MAINTENANCE

Solve.

49. $-2x < 48$ [2.7d]

50. $4x - 8x + 16 \geq 6(x - 2)$ [2.7e]

51. Divide: $\frac{-108}{-4}$. [1.6a]

52. Solve $A = \frac{p + q}{2}$ for p. [2.4b]

Multiply. [4.6d]

53. $(y + 5)(y + 7)$

54. $(y + 7)^2$

55. $(y + 7)(y - 7)$

56. $(y - 7)^2$

Find the intercepts of the equation. Then graph the equation. [3.3a]

57. $x + y = 4$

58. $x - y = 3$

59. $5x - 3y = 15$

60. $y - 3x = 6$

SYNTHESIS

Factor.

61. $4x^5 + 6x^3 + 6x^2 + 9$

62. $x^6 + x^4 + x^2 + 1$

63. $x^{12} + x^7 + x^5 + 1$

64. $x^3 - x^2 - 2x + 5$

65. $p^3 + p^2 - 3p + 10$

5.2 FACTORING TRINOMIALS OF THE TYPE $x^2 + bx + c$

Objective

a Factor trinomials of the type $x^2 + bx + c$ by examining the constant term c.

a Factoring $x^2 + bx + c$

We now begin a study of the factoring of trinomials. We first factor trinomials like

$$x^2 + 5x + 6 \quad \text{and} \quad x^2 + 3x - 10$$

by a refined *trial-and-error process*. In this section, we restrict our attention to trinomials of the type $ax^2 + bx + c$, where $a = 1$. The coefficient a is often called the **leading coefficient.**

To understand the factoring that follows, compare the following multiplications:

$$\begin{aligned}
& \quad\ \ \text{F} \quad\ \ \text{O} \quad\ \ \text{I} \quad\ \ \text{L} \\
(x+2)(x+5) &= x^2 + 5x + 2x + 2 \cdot 5 \\
&= x^2 + 7x + 10; \\
(x-2)(x-5) &= x^2 - 5x - 2x + 2 \cdot 5 \\
&= x^2 - 7x + 10; \\
(x+3)(x-7) &= x^2 - 7x + 3x + 3(-7) \\
&= x^2 - 4x - 21; \\
(x-3)(x+7) &= x^2 + 7x - 3x + (-3)7 \\
&= x^2 + 4x - 21.
\end{aligned}$$

Note that for all four products:

- The product of the two binomials is a trinomial.
- The coefficient of x in the trinomial is the sum of the constant terms in the binomials.
- The constant term in the trinomial is the product of the constant terms in the binomials.

These observations lead to a method for factoring certain trinomials. The first type we consider has a positive constant term, just as in the first two multiplications above.

CONSTANT TERM POSITIVE

To factor $x^2 + 7x + 10$, we think of FOIL in reverse. We multiplied x times x to get the first term of the trinomial, so we know that the first term of each binomial factor is x. Next, we look for numbers p and q such that

$$x^2 + 7x + 10 = (x + p)(x + q).$$

To get the middle term and the last term of the trinomial, we look for two numbers p and q whose product is 10 and whose sum is 7. Those numbers are 2 and 5. Thus the factorization is

$$(x + 2)(x + 5).$$

CHECK: $(x + 2)(x + 5) = x^2 + 5x + 2x + 10$
$= x^2 + 7x + 10.$

1. Consider the trinomial $x^2 + 7x + 12$.

a) Complete the following table.

PAIRS OF FACTORS	SUMS OF FACTORS
1, 12	13
−1, −12	
2, 6	
−2, −6	
3, 4	
−3, −4	

b) Explain why you need to consider only positive factors, as in the following table.

PAIRS OF FACTORS	SUMS OF FACTORS
1, 12	
2, 6	
3, 4	

c) Factor: $x^2 + 7x + 12$.

2. Factor: $x^2 + 13x + 36$.

Answers on page A-24

3. Explain why you would *not* consider the pairs of factors listed below in factoring $y^2 - 8y + 12$.

PAIRS OF FACTORS	SUMS OF FACTORS
1, 12	
2, 6	
3, 4	

Factor.

4. $x^2 - 8x + 15$

5. $t^2 - 9t + 20$

Answers on page A-24

EXAMPLE 1 Factor: $x^2 + 5x + 6$.

Think of FOIL in reverse. The first term of each factor is x: $(x + \square)(x + \square)$. Next, we look for two numbers whose product is 6 and whose sum is 5. All the pairs of factors of 6 are shown in the table on the left below. Since both the product, 6, and the sum, 5, of the pair of numbers must be positive, we need consider only the positive factors, listed in the table on the right.

PAIRS OF FACTORS	SUMS OF FACTORS
1, 6	7
−1, −6	−7
2, 3	5
−2, −3	−5

PAIRS OF FACTORS	SUMS OF FACTORS
1, 6	7
2, 3	5

The numbers we need are 2 and 3.

The factorization is $(x + 2)(x + 3)$. We can check by multiplying to see whether we get the original trinomial.

CHECK: $(x + 2)(x + 3) = x^2 + 3x + 2x + 6 = x^2 + 5x + 6$.

Do Exercises 1 and 2 on the preceding page.

Compare these multiplications:

$$(x - 2)(x - 5) = x^2 - 5x - 2x + 10 = x^2 - 7x + 10;$$
$$(x + 2)(x + 5) = x^2 + 5x + 2x + 10 = x^2 + 7x + 10.$$

TO FACTOR $x^2 + bx + c$ WHEN c IS POSITIVE

When the constant term of a trinomial is positive, look for two numbers with the same sign. The sign is that of the middle term:

$$x^2 - 7x + 10 = (x - 2)(x - 5);$$

$$x^2 + 7x + 10 = (x + 2)(x + 5).$$

EXAMPLE 2 Factor: $y^2 - 8y + 12$.

Since the constant term, 12, is positive and the coefficient of the middle term, −8, is negative, we look for a factorization of 12 in which both factors are negative. Their sum must be −8.

PAIRS OF FACTORS	SUMS OF FACTORS
−1, −12	−13
−2, −6	−8
−3, −4	−7

The numbers we need are −2 and −6.

The factorization is $(y - 2)(y - 6)$.

Do Exercises 3–5.

CONSTANT TERM NEGATIVE

As we saw in two of the multiplications earlier in this section, the product of two binomials can have a negative constant term:

$$(x + 3)(x - 7) = x^2 - 4x - 21$$

and

$$(x - 3)(x + 7) = x^2 + 4x - 21.$$

Note that when the signs of the constants in the binomials are reversed, only the sign of the middle term in the product changes.

EXAMPLE 3 Factor: $x^2 - 8x - 20$.

The constant term, -20, must be expressed as the product of a negative number and a positive number. Since the sum of these two numbers must be negative (specifically, -8), the negative number must have the greater absolute value.

PAIRS OF FACTORS	SUMS OF FACTORS
1, −20	−19
2, −10	−8
4, −5	−1
5, −4	1
10, −2	8
20, −1	19

The numbers we need are 2 and −10.

Because these sums are all positive, for this problem all of the corresponding pairs can be disregarded. Note that in all three pairs, the positive number has the greater absolute value.

The numbers that we are looking for are 2 and -10. The factorization is $(x + 2)(x - 10)$.

CHECK: $(x + 2)(x - 10) = x^2 - 10x + 2x - 20$
$= x^2 - 8x - 20.$

TO FACTOR $x^2 + bx + c$ WHEN c IS NEGATIVE

When the constant term of a trinomial is negative, look for two numbers whose product is negative. One must be positive and the other negative:

$$x^2 - 4x - 21 = (x + 3)(x - 7);$$

$$x^2 + 4x - 21 = (x - 3)(x + 7).$$

Consider pairs of numbers for which the number with the larger absolute value has the same sign as b, the coefficient of the middle term.

Do Exercises 6 and 7. (Exercise 7 is on the following page.)

6. Consider $x^2 - 5x - 24$.

a) Explain why you would *not* consider the pairs of factors listed below in factoring $x^2 - 5x - 24$.

PAIRS OF FACTORS	SUMS OF FACTORS
−1, 24	
−2, 12	
−3, 8	
−4, 6	

b) Explain why you *would* consider the pairs of factors listed below in factoring $x^2 - 5x - 24$.

PAIRS OF FACTORS	SUMS OF FACTORS
1, −24	
2, −12	
3, −8	
4, −6	

c) Factor: $x^2 - 5x - 24$.

Answers on page A-24

7. Consider $x^2 + 10x - 24$.

a) Explain why you would *not* consider the pairs of factors listed below in factoring $x^2 + 10x - 24$.

PAIRS OF FACTORS	SUMS OF FACTORS
1, −24	
2, −12	
3, −8	
4, −6	

b) Explain why you *would* consider the pairs of factors listed below in factoring $x^2 + 10x - 24$.

PAIRS OF FACTORS	SUMS OF FACTORS
−1, 24	
−2, 12	
−3, 8	
−4, 6	

c) Factor: $x^2 + 10x - 24$.

Factor.

8. $a^2 - 24 + 10a$

9. $-24 - 10t + t^2$

Answers on page A-24

EXAMPLE 4 Factor: $t^2 - 24 + 5t$.

It helps to first write the trinomial in descending order: $t^2 + 5t - 24$. Since the constant term, -24, is negative, we look for a factorization of -24 in which one factor is positive and one factor is negative. Their sum must be 5, so the positive factor must have the larger absolute value. Thus we consider only pairs of factors in which the positive term has the larger absolute value.

PAIRS OF FACTORS	SUMS OF FACTORS
−1, 24	23
−2, 12	10
−3, 8	5 ← The numbers we need are −3 and 8.
−4, 6	2

The factorization is $(t - 3)(t + 8)$. The check is left to the student.

Do Exercises 8 and 9.

EXAMPLE 5 Factor: $x^4 - x^2 - 110$.

Consider this trinomial as $(x^2)^2 - x^2 - 110$. We look for numbers p and q such that

$$x^4 - x^2 - 110 = (x^2 + p)(x^2 + q).$$

Since the constant term, -110, is negative, we look for a factorization of -110 in which one factor is positive and one factor is negative. Their sum must be -1. The middle-term coefficient, -1, is small compared to -110. This tells us that the desired factors are close to each other in absolute value. The numbers we want are 10 and -11. The factorization is

$$(x^2 + 10)(x^2 - 11).$$

EXAMPLE 6 Factor: $a^2 + 4ab - 21b^2$.

We consider the trinomial in the equivalent form

$$a^2 + 4ba - 21b^2.$$

This way we think of $-21b^2$ as the "constant" term and $4b$ as the "coefficient" of the middle term. Then we try to express $-21b^2$ as a product of two factors whose sum is $4b$. Those factors are $-3b$ and $7b$. The factorization is $(a - 3b)(a + 7b)$.

CHECK: $(a - 3b)(a + 7b) = a^2 + 7ab - 3ba - 21b^2$
$= a^2 + 4ab - 21b^2$.

There are polynomials that are not factorable.

EXAMPLE 7 Factor: $x^2 - x + 5$.

Since 5 has very few factors, we can easily check all possibilities.

PAIRS OF FACTORS	SUMS OF FACTORS
5, 1	6
−5, −1	−6

There are no factors whose sum is -1. Thus the polynomial is *not* factorable into factors that are polynomials.

In this text, a polynomial like $x^2 - x + 5$ that cannot be factored further is said to be **prime.** In more advanced courses, polynomials like $x^2 - x + 5$ can be factored and are not considered prime.

Do Exercises 10–12.

Often factoring requires two or more steps. In general, when told to factor, we should *factor completely.* This means that the final factorization should not contain any factors that can be factored further.

EXAMPLE 8 Factor: $2x^3 - 20x^2 + 50x$.

Always look first for a common factor. This time there is one, $2x$, which we factor out first:

$$2x^3 - 20x^2 + 50x = 2x(x^2 - 10x + 25).$$

Now consider $x^2 - 10x + 25$. Since the constant term is positive and the coefficient of the middle term is negative, we look for a factorization of 25 in which both factors are negative. Their sum must be -10.

PAIRS OF FACTORS	SUMS OF FACTORS
$-25, -1$	-26
$-5, -5$	-10 ← The numbers we need are -5 and -5.

The factorization of $x^2 - 10x + 25$ is $(x - 5)(x - 5)$, or $(x - 5)^2$. The final factorization is $2x(x - 5)^2$.

Do Exercises 13–15.

Once any common factors have been factored out, the following summary can be used to factor $x^2 + bx + c$.

TO FACTOR $x^2 + bx + c$

1. First arrange in descending order.
2. Use a trial-and-error process that looks for factors of c whose sum is b.
3. If c is positive, the signs of the factors are the same as the sign of b.
4. If c is negative, one factor is positive and the other is negative. If the sum of two factors is the opposite of b, changing the sign of each factor will give the desired factors whose sum is b.
5. Check by multiplying.

Factor.

10. $y^2 - 12 - 4y$

11. $t^4 + 5t^2 - 14$

12. $x^2 + 2x + 7$

Factor.

13. $x^3 + 4x^2 - 12x$

14. $p^2 - pq - 3pq^2$

15. $3x^3 + 24x^2 + 48x$

Answers on page A-24

Factor.

16. $14 + 5x - x^2$

17. $-x^2 + 3x + 18$

Answers on page A-24

LEADING COEFFICIENT −1

EXAMPLE 9 Factor: $10 - 3x - x^2$.

Note that the polynomial is written in ascending order. When we write it in descending order, we get

$$-x^2 - 3x + 10,$$

which has a leading coefficient of -1. Before factoring in such a case, we can factor out a -1, as follows:

$$-x^2 - 3x + 10 = -1(x^2 + 3x - 10).$$

Then we proceed to factor $x^2 + 3x - 10$. We get

$$-x^2 - 3x + 10 = -1(x^2 + 3x - 10) = -1(x + 5)(x - 2).$$

We can also express this answer in two other ways by multiplying either binomial by -1. Thus each of the following is a correct answer:

$$\begin{aligned} -x^2 - 3x + 10 &= -1(x + 5)(x - 2) \\ &= (-x - 5)(x - 2) && \text{Multiplying } x + 5 \text{ by } -1 \\ &= (x + 5)(-x + 2). && \text{Multiplying } x - 2 \text{ by } -1 \end{aligned}$$

Do Exercises 16 and 17.

Study Tips

TIME MANAGEMENT (PART 3)

Here are some additional tips to help you with time management. (See also the Study Tips on time management in Sections 2.2 and 4.2.)

- **Are you a morning or an evening person?** If you are an evening person, it might be best to avoid scheduling early-morning classes. If you are a morning person, do the opposite, but go to bed earlier to compensate. Nothing can drain your study time and effectiveness like fatigue.
- **Keep on schedule.** Your course syllabus provides a plan for the semester's schedule. Use a write-on calendar, daily planner, laptop computer, or personal digital assistant to outline your time for the semester. Be sure to note deadlines involving term papers and exams so you can begin a task early, breaking it down into smaller segments that can be accomplished more easily.
- **Balance your class schedule.** You may be someone who prefers large blocks of time for study on the off days. In that case, it might be advantageous for you to take courses that meet only three days a week. Keep in mind, however, that this might be a problem when tests in more than one course are scheduled for the same day.

"Time is our most important asset, yet we tend to waste it, kill it, and spend it rather than invest it."

Jim Rohn, motivational speaker

5.2 EXERCISE SET

For Extra Help

Digital Video Tutor CD 5 Videotape 10

InterAct Math

Math Tutor Center

MathXL

MyMathLab

a Factor. Remember that you can check by multiplying.

1. $x^2 + 8x + 15$

PAIRS OF FACTORS	SUMS OF FACTORS

2. $x^2 + 5x + 6$

PAIRS OF FACTORS	SUMS OF FACTORS

3. $x^2 + 7x + 12$

PAIRS OF FACTORS	SUMS OF FACTORS

4. $x^2 + 9x + 8$

PAIRS OF FACTORS	SUMS OF FACTORS

5. $x^2 - 6x + 9$

PAIRS OF FACTORS	SUMS OF FACTORS

6. $y^2 - 11y + 28$

PAIRS OF FACTORS	SUMS OF FACTORS

7. $x^2 - 5x - 14$

PAIRS OF FACTORS	SUMS OF FACTORS

8. $a^2 + 7a - 30$

PAIRS OF FACTORS	SUMS OF FACTORS

9. $b^2 + 5b + 4$

PAIRS OF FACTORS	SUMS OF FACTORS

10. $z^2 - 8z + 7$

PAIRS OF FACTORS	SUMS OF FACTORS

11. $x^2 + \frac{2}{3}x + \frac{1}{9}$

PAIRS OF FACTORS	SUMS OF FACTORS

12. $x^2 - \frac{2}{5}x + \frac{1}{25}$

PAIRS OF FACTORS	SUMS OF FACTORS

13. $d^2 - 7d + 10$

14. $t^2 - 12t + 35$

15. $y^2 - 11y + 10$

16. $x^2 - 4x - 21$

17. $x^2 + x + 1$

18. $x^2 + 5x + 3$

19. $x^2 - 7x - 18$

20. $y^2 - 3y - 28$

21. $x^3 - 6x^2 - 16x$

22. $x^3 - x^2 - 42x$

23. $y^3 - 4y^2 - 45y$

24. $x^3 - 7x^2 - 60x$

25. $-2x - 99 + x^2$

26. $x^2 - 72 + 6x$

27. $c^4 + c^2 - 56$

28. $b^4 + 5b^2 - 24$

29. $a^4 + 2a^2 - 35$

30. $x^4 - x^2 - 6$

31. $x^2 + x - 42$

32. $x^2 + 2x - 15$

33. $7 - 2p + p^2$

34. $11 - 3w + w^2$

35. $x^2 + 20x + 100$

36. $a^2 + 19a + 88$

37. $30 + 7x - x^2$

38. $45 + 4x - x^2$

39. $24 - a^2 - 10a$

40. $-z^2 + 36 - 9z$

41. $x^4 - 21x^3 - 100x^2$

42. $x^4 - 20x^3 + 96x^2$

43. $x^2 - 21x - 72$

44. $4x^2 + 40x + 100$

45. $x^2 - 25x + 144$

46. $y^2 - 21y + 108$

47. $a^2 + a - 132$

48. $a^2 + 9a - 90$

49. $120 - 23x + x^2$

50. $96 + 22d + d^2$

51. $108 - 3x - x^2$

52. $112 + 9y - y^2$

53. $y^2 - 0.2y - 0.08$

54. $t^2 - 0.3t - 0.10$

55. $p^2 + 3pq - 10q^2$

56. $a^2 + 2ab - 3b^2$

57. $84 - 8t - t^2$

58. $72 - 6m - m^2$

59. $m^2 + 5mn + 4n^2$

60. $x^2 + 11xy + 24y^2$

61. $s^2 - 2st - 15t^2$

62. $p^2 + 5pq - 24q^2$

63. $6a^{10} - 30a^9 - 84a^8$

64. $7x^9 - 28x^8 - 35x^7$

65. $\mathbf{D}_{\mathbf{W}}$ Gwyneth factors $x^3 - 8x^2 + 15x$ as $(x^2 - 5x)(x - 3)$. Is she wrong? Why or why not? What advice would you offer?

66. $\mathbf{D}_{\mathbf{W}}$ When searching for a factorization, why do we list pairs of numbers with the correct *product* instead of pairs of numbers with the correct *sum*?

67. $\mathbf{D}_{\mathbf{W}}$ Without multiplying $(x - 17)(x - 18)$, explain why it cannot possibly be a factorization of $x^2 + 35x + 306$.

68. $\mathbf{D}_{\mathbf{W}}$ What is the advantage of writing out the prime factorization of c when factoring $x^2 + bx + c$ with a large value of c?

SKILL MAINTENANCE

Multiply. [4.6d]

69. $8x(2x^2 - 6x + 1)$

70. $(7w + 6)(4w - 11)$

71. $(7w + 6)^2$

72. $(4w - 11)^2$

73. $(4w - 11)(4w + 11)$

74. Simplify: $(3x^4)^3$. [4.2a, b]

Solve. [2.3a]

75. $3x - 8 = 0$

76. $2x + 7 = 0$

Solve.

77. *Arrests for Counterfeiting.* In a recent year, 29,200 people were arrested for counterfeiting. This number was down 1.2% from the preceding year. How many people were arrested the preceding year? [2.5a]

78. The first angle of a triangle is four times as large as the second. The measure of the third angle is 30° greater than that of the second. Find the angle measures. [2.6a]

SYNTHESIS

79. Find all integers m for which $y^2 + my + 50$ can be factored.

80. Find all integers b for which $a^2 + ba - 50$ can be factored.

Factor completely.

81. $x^2 - \frac{1}{2}x - \frac{3}{16}$

82. $x^2 - \frac{1}{4}x - \frac{1}{8}$

83. $x^2 + \frac{30}{7}x - \frac{25}{7}$

84. $\frac{1}{3}x^3 + \frac{1}{3}x^2 - 2x$

85. $b^{2n} + 7b^n + 10$

86. $a^{2m} - 11a^m + 28$

Find a polynomial in factored form for the shaded area. (Leave answers in terms of π.)

87.

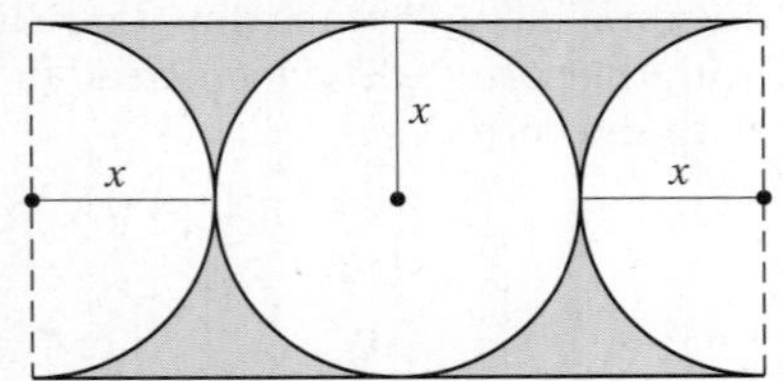

88.

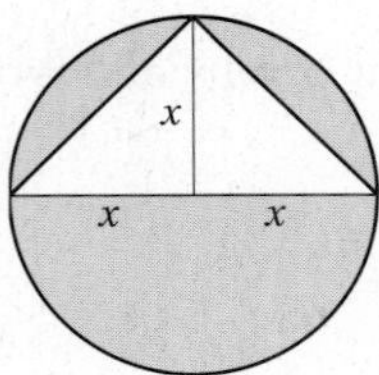

89. A census taker asks a woman, "How many children do you have?" "Three," she answers. "What are their ages?" She responds, "The product of their ages is 36. The sum of their ages is the house number next door." The math-savvy census taker walks next door, reads the house number, appears puzzled, and returns to the woman, asking, "Is there something you forgot to tell me?" "Oh yes," says the woman. "I'm sorry. The oldest child is at the park." The census taker records the three ages, thanks the woman for her time, and leaves. How old is each child? Explain how you reached this conclusion. (*Hint*: Consider factorizations.)

Source: Adapted from Anita Harnadek, *Classroom Quickies.* Pacific Grove, CA: Critical Thinking Press and Software

5.3 FACTORING $ax^2 + bx + c, a \neq 1$: THE FOIL METHOD

Objective

a Factor trinomials of the type $ax^2 + bx + c$, $a \neq 1$, using the FOIL method.

In Section 5.2, we learned a trial-and-error method to factor trinomials of the type $x^2 + bx + c$. In this section, we factor trinomials in which the coefficient of the leading term x^2 is not 1. The procedure we learn is a refined trial-and-error method.

a The FOIL Method

We want to factor trinomials of the type $ax^2 + bx + c$. Consider the following multiplication:

$$\begin{aligned} & \quad\ \ \text{F} \qquad \text{O} \qquad \text{I} \qquad \text{L} \\ (2x + 5)(3x + 4) &= 6x^2 + 8x + 15x + 20 \\ &= 6x^2 + 23x + 20 \end{aligned}$$

F	O + I	L
$2 \cdot 3$	$2 \cdot 4 \quad 5 \cdot 3$	$5 \cdot 4$

To factor $6x^2 + 23x + 20$, we reverse the above multiplication, using what we might call an "unFOIL" process. We look for two binomials $rx + p$ and $sx + q$ whose product is $(rx + p)(sx + q) = 6x^2 + 23x + 20$. The product of the First terms must be $6x^2$. The product of the Outside terms plus the product of the Inside terms must be $23x$. The product of the Last terms must be 20. We know from the preceding discussion that the answer is $(2x + 5)(3x + 4)$. Generally, however, finding such an answer is a refined trial-and-error process. It turns out that $(-2x - 5)(-3x - 4)$ is also a correct answer, but we generally choose an answer in which the first coefficients are positive.

We will use the following trial-and-error method.

THE FOIL METHOD

To factor $ax^2 + bx + c$, $a \neq 1$, using the FOIL method:

1. Factor out the largest common factor, if one exists.
2. Find two First terms whose product is ax^2.

   $(\square x + \quad)(\square x + \quad) = ax^2 + bx + c.$ (FOIL)

3. Find two Last terms whose product is c:

   $(\quad x + \square)(\quad x + \square) = ax^2 + bx + c.$ (FOIL)

4. Look for Outer and Inner products resulting from steps (2) and (3) for which the sum is bx:

   $(\square x + \square)(\square x + \square) = ax^2 + bx + c.$ (I, O; FOIL)

5. Always check by multiplying.

To the student: *In Section 5.4, we will consider an alternative method for the same kind of factoring. It involves factoring by grouping and is called the ac-method.*

To the instructor: *We present two ways to factor general trinomials in Sections 5.3 and 5.4: the FOIL method in Section 5.3 and the ac-method in Section 5.4. You can teach both methods and let the student use the one that he or she prefers or you can select just one. In the latter case, the exercise set that is not studied can be used for extra practice.*

Factor.

1. $2x^2 - x - 15$

2. $12x^2 - 17x - 5$

Answers on page A-25

Study Tips

READING EXAMPLES

A careful study of the examples in these sections on factoring is critical. *Read them carefully* to ensure success!

EXAMPLE 1 Factor: $3x^2 - 10x - 8$.

1) First, we check for a common factor. Here there is none (other than 1 or -1).
2) Find two **F**irst terms whose product is $3x^2$.

The only possibilities for the **F**irst terms are $3x$ and x, so any factorization must be of the form

$$(3x + \square)(x + \square).$$

3) Find two **L**ast terms whose product is -8.

Possible factorizations of -8 are

$$(-8) \cdot 1, \quad 8 \cdot (-1), \quad (-2) \cdot 4, \quad \text{and} \quad 2 \cdot (-4).$$

Since the First terms are not identical, we must also consider

$$1 \cdot (-8), \quad (-1) \cdot 8, \quad 4 \cdot (-2), \quad \text{and} \quad (-4) \cdot 2.$$

4) Inspect the **O**uter and **I**nner products resulting from steps (2) and (3). Look for a combination in which the sum of the products is the middle term, $-10x$:

Trial	*Product*	
$(3x - 8)(x + 1)$	$3x^2 + 3x - 8x - 8$ $= 3x^2 - 5x - 8$	← Wrong middle term
$(3x + 8)(x - 1)$	$3x^2 - 3x + 8x - 8$ $= 3x^2 + 5x - 8$	← Wrong middle term
$(3x - 2)(x + 4)$	$3x^2 + 12x - 2x - 8$ $= 3x^2 + 10x - 8$	← Wrong middle term
$(3x + 2)(x - 4)$	$3x^2 - 12x + 2x - 8$ $= 3x^2 - 10x - 8$	← **Correct middle term!**
$(3x + 1)(x - 8)$	$3x^2 - 24x + x - 8$ $= 3x^2 - 23x - 8$	← Wrong middle term
$(3x - 1)(x + 8)$	$3x^2 + 24x - x - 8$ $= 3x^2 + 23x - 8$	← Wrong middle term
$(3x + 4)(x - 2)$	$3x^2 - 6x + 4x - 8$ $= 3x^2 - 2x - 8$	← Wrong middle term
$(3x - 4)(x + 2)$	$3x^2 + 6x - 4x - 8$ $= 3x^2 + 2x - 8$	← Wrong middle term

The correct factorization is $(3x + 2)(x - 4)$.

5) CHECK: $(3x + 2)(x - 4) = 3x^2 - 10x - 8$.

Two observations can be made from Example 1. First, we listed all possible trials even though we could have stopped after having found the correct factorization. We did this to show that each trial differs only in the middle term of the product. **Second, note that as in Section 5.2, only the sign of the middle term changes when the signs in the binomials are reversed:**

Plus Minus

$$(3x + 4)(x - 2) = 3x^2 - 2x - 8$$

Minus Plus ← Middle term changes sign

$$(3x - 4)(x + 2) = 3x^2 + 2x - 8.$$

Do Exercises 1 and 2.

EXAMPLE 2 Factor: $24x^2 - 76x + 40$.

1) First, we factor out the largest common factor, 4:

$$4(6x^2 - 19x + 10).$$

Now we factor the trinomial $6x^2 - 19x + 10$.

2) Because $6x^2$ can be factored as $3x \cdot 2x$ or $6x \cdot x$, we have these possibilities for factorizations:

$$(3x + \quad)(2x + \quad) \quad \text{or} \quad (6x + \quad)(x + \quad).$$

3) There are four pairs of factors of 10 and they each can be listed in two ways:

$10, 1 \qquad -10, -1 \qquad 5, 2 \qquad -5, -2$

and

$1, 10 \qquad -1, -10 \qquad 2, 5 \qquad -2, -5.$

4) The two possibilities from step (2) and the eight possibilities from step (3) give $2 \cdot 8$, or 16 possibilities for factorizations. We look for **O**uter and **I**nner products resulting from steps (2) and (3) for which the sum is the middle term, $-19x$. Since the sign of the middle term is negative, but the sign of the last term, 10, is positive, the two factors of 10 must both be negative. This means only four pairings from step (3) need be considered. We first try these factors with $(3x + \quad)(2x + \quad)$. If none gives the correct factorization, we will consider $(6x + \quad)(x + \quad)$.

Trial	*Product*	
$(3x - 10)(2x - 1)$	$6x^2 - 3x - 20x + 10$	
	$= 6x^2 - 23x + 10$	← Wrong middle term
$(3x - 1)(2x - 10)$	$6x^2 - 30x - 2x + 10$	
	$= 6x^2 - 32x + 10$	← Wrong middle term
$(3x - 5)(2x - 2)$	$6x^2 - 6x - 10x + 10$	
	$= 6x^2 - 16x + 10$	← Wrong middle term
$(3x - 2)(2x - 5)$	$6x^2 - 15x - 4x + 10$	
	$= 6x^2 - 19x + 10$	← **Correct middle term!**

Since we have a correct factorization, we need not consider

$$(6x + \quad)(x + \quad).$$

The factorization of $6x^2 - 19x + 10$ is $(3x - 2)(2x - 5)$, but *do not forget the common factor*! We must include it in order to factor the original trinomial:

$$24x^2 - 76x + 40 = 4(6x^2 - 19x + 10)$$
$$= 4(3x - 2)(2x - 5).$$

5) CHECK: $4(3x - 2)(2x - 5) = 4(6x^2 - 19x + 10) = 24x^2 - 76x + 40.$

CAUTION!

When factoring any polynomial, always look for a common factor. Failure to do so is such a common error that this caution bears repeating.

In Example 2, look again at the possibility $(3x - 5)(2x - 2)$. Without multiplying, we can reject such a possibility. To see why, consider the following:

$$(3x - 5)(2x - 2) = 2(3x - 5)(x - 1).$$

Factor.

3. $3x^2 - 19x + 20$

4. $20x^2 - 46x + 24$

5. Factor: $6x^2 + 7x + 2$.

Answers on page A-25

The expression $2x - 2$ has a common factor, 2. But we removed the *largest* common factor in the first step. If $2x - 2$ were one of the factors, then 2 would have to be a common factor in addition to the original 4. Thus, $(2x - 2)$ cannot be part of the factorization of the original trinomial.

Given that the largest common factor is factored out at the outset, we need not consider factorizations that have a common factor.

Do Exercises 3 and 4.

EXAMPLE 3 Factor: $10x^2 + 37x + 7$.

1) There is no common factor (other than 1 or -1).

2) Because $10x^2$ factors as $10x \cdot x$ or $5x \cdot 2x$, we have these possibilities for factorizations:

$$(10x + \quad)(x + \quad) \quad \text{or} \quad (5x + \quad)(2x + \quad).$$

3) There are two pairs of factors of 7 and they each can be listed in two ways:

$$1, 7 \quad -1, -7 \qquad \text{and} \qquad 7, 1 \quad -7, -1.$$

4) From steps (2) and (3), we see that there are 8 possibilities for factorizations. Look for **O**uter and **I**nner products for which the sum is the middle term. Because all coefficients in $10x^2 + 37x + 7$ are positive, we need consider only positive factors of 7. The possibilities are

$(10x + 1)(x + 7) = 10x^2 + 71x + 7,$

$(10x + 7)(x + 1) = 10x^2 + 17x + 7,$

$(5x + 7)(2x + 1) = 10x^2 + 19x + 7,$

$(5x + 1)(2x + 7) = 10x^2 + 37x + 7.$ ← **Correct middle term**

The factorization is $(5x + 1)(2x + 7)$.

5) CHECK: $(5x + 1)(2x + 7) = 10x^2 + 37x + 7$.

Keep in mind that this method of factoring trinomials of the type $ax^2 + bx + c$ involves *trial and error.* As you practice, you will find that you can make better and better guesses.

Do Exercise 5.

TIPS FOR FACTORING $ax^2 + bx + c$, $a \neq 1$

- Always factor out the largest common factor, if one exists.
- Once the common factor has been factored out of the original trinomial, no binomial factor can contain a common factor (other than 1 or -1).
- If c is positive, then the signs in both binomial factors must match the sign of b. (This assumes that $a > 0$.)
- Reversing the signs in the binomials reverses the sign of the middle term of their product.
- Organize your work so that you can keep track of which possibilities have or have not been checked.
- Always check by multiplying.

EXAMPLE 4 Factor: $10x + 8 - 3x^2$.

An important problem-solving strategy is to find a way to make new problems look like problems we already know how to solve. (See Example 9 in Section 5.2.) The factoring tips above apply only to trinomials of the form $ax^2 + bx + c$, with $a > 0$. This leads us to rewrite $10x + 8 - 3x^2$ in descending order:

$$10x + 8 - 3x^2 = -3x^2 + 10x + 8. \qquad \text{Writing in descending order}$$

Although $-3x^2 + 10x + 8$ looks similar to the trinomials we have factored, the tips above require a positive leading coefficient. This can be attained by factoring out -1:

$$-3x^2 + 10x + 8 = -1(3x^2 - 10x - 8) \qquad \text{Factoring out } -1 \text{ changes the signs of the coefficients.}$$

$$= -1(3x + 2)(x - 4). \qquad \text{Using the result from Example 1}$$

The factorization of $10x + 8 - 3x^2$ is $-1(3x + 2)(x - 4)$. Other correct answers are

$$10x + 8 - 3x^2 = (3x + 2)(-x + 4) \qquad \text{Multiplying } x - 4 \text{ by } -1$$

$$= (-3x - 2)(x - 4). \qquad \text{Multiplying } 3x + 2 \text{ by } -1$$

Do Exercises 6 and 7.

Factor.

6. $2 - x - 6x^2$

7. $2x + 8 - 6x^2$

EXAMPLE 5 Factor: $6p^2 - 13pq - 28q^2$.

1) Factor out a common factor, if any.
   There is none (other than 1 or -1).
2) Factor the first term, $6p^2$.
   Possibilities are $2p$, $3p$ and $6p$, p. We have these as possibilities for factorizations:
   $$(2p + \quad)(3p + \quad) \quad \text{or} \quad (6p + \quad)(p + \quad).$$
3) Factor the last term, $-28q^2$, which has a negative coefficient.
   The possibilities are $-14q$, $2q$ and $14q$, $-2q$; $-28q$, q and $28q$, $-q$; and $-7q$, $4q$ and $7q$, $-4q$.
4) The coefficient of the middle term is negative, so we look for combinations of factors from steps (2) and (3) such that the sum of their products has a negative coefficient. We try some possibilities:
   $$(2p + q)(3p - 28q) = 6p^2 - 53pq - 28q^2,$$
   $$(2p - 7q)(3p + 4q) = 6p^2 - 13pq - 28q^2. \qquad \leftarrow \textbf{Correct middle term}$$
   The factorization of $6p^2 - 13pq - 28q^2$ is $(2p - 7q)(3p + 4q)$.
5) The check is left to the student.

Do Exercises 8 and 9.

Factor.

8. $6a^2 - 5ab + b^2$

9. $6x^2 + 15xy + 9y^2$

Answers on page A-25

CALCULATOR CORNER

Checking Factorizations A partial check of a factorization can be performed using a table or a graph. To check the factorization $6x^3 - 9x^2 + 4x - 6 = (3x^2 + 2)(2x - 3)$, for example, we enter $y_1 = 6x^3 - 9x^2 + 4x - 6$ and $y_2 = (3x^2 + 2)(2x - 3)$ on the equation-editor screen (see page 244). Then we set up a table in AUTO mode (see page 250). If the factorization is correct, the values of y_1 and y_2 will be the same regardless of the table settings used.

X	Y1	Y2
−3	−261	−261
−2	−98	−98
−1	−25	−25
0	−6	−6
1	−5	−5
2	14	14
3	87	87
X = −3		

We can also graph $y_1 = 6x^3 - 9x^2 + 4x - 6$ and $y_2 = (3x^2 + 2)(2x - 3)$. If the graphs appear to coincide, the factorization is probably correct.

$y_1 = 6x^3 - 9x^2 + 4x - 6,$
$y_2 = (3x^2 + 2)(2x - 3)$

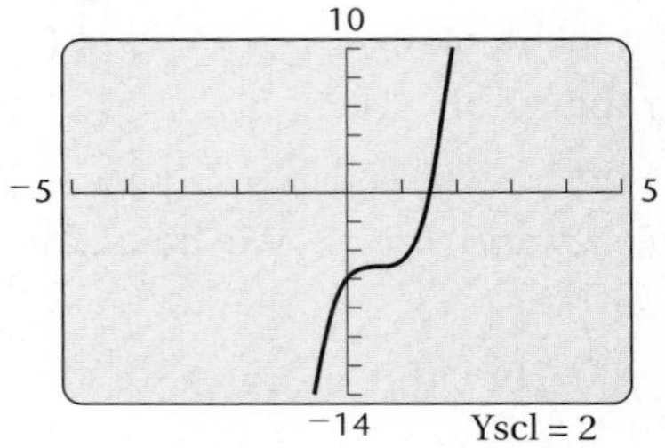

Keep in mind that these procedures provide only a partial check since we cannot view all possible values of x in a table nor can we see the entire graph.

Exercises: Use a table or a graph to determine whether the factorization is correct.

1. $24x^2 - 76x + 40 = 4(3x - 2)(2x - 5)$
2. $4x^2 - 5x - 6 = (4x + 3)(x - 2)$
3. $5x^2 + 17x - 12 = (5x + 3)(x - 4)$
4. $10x^2 + 37x + 7 = (5x - 1)(2x + 7)$
5. $12x^2 - 17x - 5 = (6x + 1)(2x - 5)$
6. $12x^2 - 17x - 5 = (4x + 1)(3x - 5)$
7. $x^2 - 4 = (x - 2)(x - 2)$
8. $x^2 - 4 = (x + 2)(x - 2)$

Study Tips

SKILL MAINTENANCE EXERCISES

It is never too soon to begin reviewing for the final examination. The Skill Maintenance exercises found in each exercise set review and reinforce skills taught in earlier sections. Include all of these exercises in your weekly preparation. Answers to both odd-numbered and even-numbered exercises along with section references appear at the back of the book.

5.3 EXERCISE SET

For Extra Help

Digital Video Tutor CD 5 Videotape 10 | InterAct Math | Math Tutor Center | MathXL | MyMathLab

a Factor.

1. $2x^2 - 7x - 4$

2. $3x^2 - x - 4$

3. $5x^2 - x - 18$

4. $4x^2 - 17x + 15$

5. $6x^2 + 23x + 7$

6. $6x^2 - 23x + 7$

7. $3x^2 + 4x + 1$

8. $7x^2 + 15x + 2$

9. $4x^2 + 4x - 15$

10. $9x^2 + 6x - 8$

11. $2x^2 - x - 1$

12. $15x^2 - 19x - 10$

13. $9x^2 + 18x - 16$

14. $2x^2 + 5x + 2$

15. $3x^2 - 5x - 2$

16. $18x^2 - 3x - 10$

17. $12x^2 + 31x + 20$

18. $15x^2 + 19x - 10$

19. $14x^2 + 19x - 3$

20. $35x^2 + 34x + 8$

21. $9x^2 + 18x + 8$

22. $6 - 13x + 6x^2$

23. $49 - 42x + 9x^2$

24. $16 + 36x^2 + 48x$

25. $24x^2 + 47x - 2$

26. $16p^2 - 78p + 27$

27. $35x^2 - 57x - 44$

28. $9a^2 + 12a - 5$

29. $20 + 6x - 2x^2$

30. $15 + x - 2x^2$

31. $12x^2 + 28x - 24$

32. $6x^2 + 33x + 15$

33. $30x^2 - 24x - 54$

34. $18t^2 - 24t + 6$

35. $4y + 6y^2 - 10$

36. $-9 + 18x^2 - 21x$

37. $3x^2 - 4x + 1$

38. $6t^2 + 13t + 6$

39. $12x^2 - 28x - 24$

40. $6x^2 - 33x + 15$

41. $-1 + 2x^2 - x$

42. $-19x + 15x^2 + 6$

43. $9x^2 - 18x - 16$

44. $14y^2 + 35y + 14$

45. $15x^2 - 25x - 10$

46. $18x^2 + 3x - 10$

47. $12p^3 + 31p^2 + 20p$

48. $15x^3 + 19x^2 - 10x$

49. $16 + 18x - 9x^2$

50. $33t - 15 - 6t^2$

51. $-15x^2 + 19x - 6$

52. $1 + p - 2p^2$

53. $14x^4 + 19x^3 - 3x^2$

54. $70x^4 + 68x^3 + 16x^2$

55. $168x^3 - 45x^2 + 3x$

56. $144x^5 + 168x^4 + 48x^3$

57. $15x^4 - 19x^2 + 6$

58. $9x^4 + 18x^2 + 8$

59. $25t^2 + 80t + 64$

60. $9x^2 - 42x + 49$

61. $6x^3 + 4x^2 - 10x$

62. $18x^3 - 21x^2 - 9x$

63. $25x^2 + 79x + 64$

64. $9y^2 + 42y + 47$

65. $6x^2 - 19x - 5$

66. $2x^2 + 11x - 9$

67. $12m^2 - mn - 20n^2$

68. $12a^2 - 17ab + 6b^2$

69. $6a^2 - ab - 15b^2$

70. $3p^2 - 16pq - 12q^2$

71. $9a^2 + 18ab + 8b^2$

72. $10s^2 + 4st - 6t^2$

73. $35p^2 + 34pq + 8q^2$

74. $30a^2 + 87ab + 30b^2$

75. $18x^2 - 6xy - 24y^2$

76. $15a^2 - 5ab - 20b^2$

77. D_W Explain how the factoring in Exercise 21 can be used to aid the factoring in Exercise 71.

78. D_W A student presents the following work:

$$4x^2 + 28x + 48 = (2x + 6)(2x + 8) = 2(x + 3)(x + 4).$$

Is it correct? Explain.

SKILL MAINTENANCE

Solve. [2.4b]

79. $A = pq - 7$, for q

80. $y = mx + b$, for x

81. $3x + 2y = 6$, for y

82. $p - q + r = 2$, for q

Solve. [2.7e]

83. $5 - 4x < -11$

84. $2x - 4(x + 3x) \geq 6x - 8 - 9x$

85. Graph: $y = \frac{2}{5}x - 1$. [3.2b]

86. Divide: $\frac{y^{12}}{y^4}$. [4.1e]

Multiply. [4.6d]

87. $(3x - 5)(3x + 5)$

88. $(4a - 3)^2$

SYNTHESIS

Factor.

89. $20x^{2n} + 16x^n + 3$

90. $-15x^{2m} + 26x^m - 8$

91. $3x^{6a} - 2x^{3a} - 1$

92. $x^{2n+1} - 2x^{n+1} + x$

93.–102. Use the TABLE feature to check the factoring in Exercises 15–24.

Objective

a Factor trinomials of the type $ax^2 + bx + c$, $a \neq 1$, using the ac-method.

5.4 FACTORING $ax^2 + bx + c$, $a \neq 1$: THE ac-METHOD

a The ac-Method

Another method for factoring trinomials of the type $ax^2 + bx + c$, $a \neq 1$, involves the product, ac, of the leading coefficient a and the last term c. It is called the ***ac*-method**. Because it uses factoring by grouping, it is also referred to as the **grouping method**.

We know how to factor the trinomial $x^2 + 5x + 6$. We look for factors of the constant term, 6, whose sum is the coefficient of the middle term, 5:

$x^2 + 5x + 6.$

(1) Factor: $6 = 2 \cdot 3$

(2) Sum: $2 + 3 = 5$

What happens when the leading coefficient is not 1? To factor a trinomial like $3x^2 - 10x - 8$, we can use a method similar to what we used for $x^2 + 5x + 6$, but we need two more steps. That method is outlined as follows.

THE ac-METHOD

To factor $ax^2 + bx + c$, $a \neq 1$, using the ac-method:

1. Factor out a common factor, if any.
2. Multiply the leading coefficient a and the constant c.
3. Try to factor the product ac so that the sum of the factors is b. That is, find integers p and q such that $pq = ac$ and $p + q = b$.
4. Split the middle term. That is, write it as a sum using the factors found in step (3).
5. Factor by grouping.
6. Check by multiplying.

EXAMPLE 1 Factor: $3x^2 - 10x - 8$.

1) First, we factor out a common factor, if any. There is none (other than 1 or -1).

2) We multiply the leading coefficient, 3, and the constant, -8:

$3(-8) = -24.$

3) Then we look for a factorization of -24 in which the sum of the factors is the coefficient of the middle term, -10.

PAIRS OF FACTORS	SUMS OF FACTORS
-1, 24	23
1, -24	-23
-2, 12	10
2, -12	-10 ← $2 + (-12) = -10$
-3, 8	5
3, -8	-5
-4, 6	2
4, -6	-2

4) Next, we split the middle term as a sum or a difference using the factors found in step (3): $-10x = 2x - 12x$.

5) Finally, we factor by grouping, as follows:

$3x^2 - 10x - 8 = 3x^2 + 2x - 12x - 8$ Substituting $2x - 12x$ for $-10x$

$= x(3x + 2) - 4(3x + 2)$ Factoring by grouping

$= (x - 4)(3x + 2)$.

We can also split the middle term as $-12x + 2x$. We still get the same factorization, although the factors may be in a different order. Note the following:

$3x^2 - 10x - 8 = 3x^2 - 12x + 2x - 8$ Substituting $-12x + 2x$ for $-10x$

$= 3x(x - 4) + 2(x - 4)$ Factoring by grouping

$= (3x + 2)(x - 4)$.

6) CHECK: $(3x + 2)(x - 4) = 3x^2 - 10x - 8$.

Do Exercises 1 and 2.

EXAMPLE 2 Factor: $8x^2 + 8x - 6$.

1) First, we factor out a common factor, if any. The number 2 is common to all three terms, so we factor it out: $2(4x^2 + 4x - 3)$.

2) Next, we factor the trinomial $4x^2 + 4x - 3$. We multiply the leading coefficient and the constant, 4 and -3: $4(-3) = -12$.

3) We try to factor -12 so that the sum of the factors is 4.

PAIRS OF FACTORS	SUMS OF FACTORS
$-1,\ 12$	11
$1,\ -12$	-11
$-2,\ 6$	4 ← $-2 + 6 = 4$
$2,\ -6$	-4
$-3,\ 4$	1
$3,\ -4$	-1

4) Then we split the middle term, $4x$, as follows: $4x = -2x + 6x$.

5) Finally, we factor by grouping:

$4x^2 + 4x - 3 = 4x^2 - 2x + 6x - 3$ Substituting $-2x + 6x$ for $4x$

$= 2x(2x - 1) + 3(2x - 1)$ Factoring by grouping

$= (2x + 3)(2x - 1)$.

The factorization of $4x^2 + 4x - 3$ is $(2x + 3)(2x - 1)$. But don't forget the common factor! We must include it to get a factorization of the original trinomial: $8x^2 + 8x - 6 = 2(2x + 3)(2x - 1)$.

6) CHECK: $2(2x + 3)(2x - 1) = 2(4x^2 + 4x - 3) = 8x^2 + 8x - 6$.

Do Exercises 3 and 4.

Factor.

1. $6x^2 + 7x + 2$

2. $12x^2 - 17x - 5$

Factor.

3. $6x^2 + 15x + 9$

4. $20x^2 - 46x + 24$

Answers on page A-25

5.4 EXERCISE SET

For Extra Help

Digital Video Tutor CD 5 Videotape 10 | InterAct Math | Math Tutor Center | MathXL | MyMathLab

a Factor. Note that the middle term has already been split.

1. $x^2 + 2x + 7x + 14$

2. $x^2 + 3x + x + 3$

3. $x^2 - 4x - x + 4$

4. $a^2 + 5a - 2a - 10$

5. $6x^2 + 4x + 9x + 6$

6. $3x^2 - 2x + 3x - 2$

7. $3x^2 - 4x - 12x + 16$

8. $24 - 18y - 20y + 15y^2$

9. $35x^2 - 40x + 21x - 24$

10. $8x^2 - 6x - 28x + 21$

11. $4x^2 + 6x - 6x - 9$

12. $2x^4 - 6x^2 - 5x^2 + 15$

13. $2x^4 + 6x^2 + 5x^2 + 15$

14. $9x^4 - 6x^2 - 6x^2 + 4$

Factor by grouping.

15. $2x^2 + 7x - 4$

16. $5x^2 + x - 18$

17. $3x^2 - 4x - 15$

18. $3x^2 + x - 4$

19. $6x^2 + 23x + 7$

20. $6x^2 + 13x + 6$

21. $3x^2 - 4x + 1$

22. $7x^2 - 15x + 2$

23. $4x^2 - 4x - 15$

24. $9x^2 - 6x - 8$

25. $2x^2 + x - 1$

26. $15x^2 + 19x - 10$

27. $9x^2 - 18x - 16$

28. $2x^2 - 5x + 2$

29. $3x^2 + 5x - 2$

30. $18x^2 + 3x - 10$

31. $12x^2 - 31x + 20$

32. $15x^2 - 19x - 10$

33. $14x^2 - 19x - 3$

34. $35x^2 - 34x + 8$

35. $9x^2 + 18x + 8$

36. $6 - 13x + 6x^2$

37. $49 - 42x + 9x^2$

38. $25x^2 + 40x + 16$

39. $24x^2 - 47x - 2$

40. $16a^2 + 78a + 27$

41. $5 - 9a^2 - 12a$

42. $17x - 4x^2 + 15$

43. $20 + 6x - 2x^2$

44. $15 + x - 2x^2$

45. $12x^2 + 28x - 24$

46. $6x^2 + 33x + 15$

47. $30x^2 - 24x - 54$

48. $18t^2 - 24t + 6$

49. $4y + 6y^2 - 10$

50. $-9 + 18x^2 - 21x$

51. $3x^2 - 4x + 1$

52. $6t^2 + t - 15$

53. $12x^2 - 28x - 24$

54. $6x^2 - 33x + 15$

55. $-1 + 2x^2 - x$

56. $-19x + 15x^2 + 6$

57. $9x^2 + 18x - 16$

58. $14y^2 + 35y + 14$

59. $15x^2 - 25x - 10$

60. $18x^2 + 3x - 10$

61. $12p^3 + 31p^2 + 20p$

62. $15x^3 + 19x^2 - 10x$

63. $4 - x - 5x^2$

64. $1 - p - 2p^2$

65. $33t - 15 - 6t^2$

66. $-15x^2 - 19x - 6$

67. $14x^4 + 19x^3 - 3x^2$

68. $70x^4 + 68x^3 + 16x^2$

69. $168x^3 - 45x^2 + 3x$

70. $144x^5 + 168x^4 + 48x^3$

71. $15x^4 - 19x^2 + 6$

72. $9x^4 + 18x^2 + 8$

73. $25t^2 + 80t + 64$

74. $9x^2 - 42x + 49$

75. $6x^3 + 4x^2 - 10x$

76. $18x^3 - 21x^2 - 9x$

77. $25x^2 + 79x + 64$

78. $9y^2 + 42y + 47$

79. $6x^2 - 19x - 5$

80. $2x^2 + 11x - 9$

81. $12m^2 - mn - 20n^2$

82. $12a^2 - 17ab + 6b^2$

83. $6a^2 - ab - 15b^2$

84. $3p^2 - 16pq - 12q^2$

85. $9a^2 - 18ab + 8b^2$

86. $10s^2 + 4st - 6t^2$

87. $35p^2 + 34pq + 8q^2$

88. $30a^2 + 87ab + 30b^2$

89. $18x^2 - 6xy - 24y^2$

90. $15a^2 - 5ab - 20b^2$

91. $60x + 18x^2 - 6x^3$

92. $60x + 4x^2 - 8x^3$

93. $35x^5 - 57x^4 - 44x^3$

94. $15x^3 + 33x^4 + 6x^5$

95. ^{D}W If you have studied both the FOIL and the *ac*-methods of factoring $ax^2 + bx + c$, $a \neq 1$, decide which method you think is better and explain why.

96. ^{D}W Explain factoring $ax^2 + bx + c$, $a \neq 1$, using the *ac*-method as though you were teaching a fellow student.

SKILL MAINTENANCE

Solve. [2.7d, e]

97. $-10x > 1000$

98. $-3.8x \leq -824.6$

99. $6 - 3x \geq -18$

100. $3 - 2x - 4x > -9$

101. $\frac{1}{2}x - 6x + 10 \leq x - 5x$

102. $-2(x + 7) > -4(x - 5)$

103. $3x - 6x + 2(x - 4) > 2(9 - 4x)$

104. $-6(x - 4) + 8(4 - x) \leq 3(x - 7)$

Solve. [2.6a]

105. The earth is a sphere (or ball) that is about 40,000 km in circumference. Find the radius of the earth, in kilometers and in miles. Use 3.14 for π. (*Hint*: 1 km $\approx$ 0.62 mi.)

106. The second angle of a triangle is 10° less than twice the first. The third angle is 15° more than four times the first. Find the measure of the second angle.

SYNTHESIS

Factor.

107. $9x^{10} - 12x^5 + 4$

108. $24x^{2n} + 22x^n + 3$

109. $16x^{10} + 8x^5 + 1$

110. $(a + 4)^2 - 2(a + 4) + 1$

111.–120. Use the TABLE feature to check the factoring in Exercises 15–24.

Objectives

- **a** Recognize trinomial squares.
- **b** Factor trinomial squares.
- **c** Recognize differences of squares.
- **d** Factor differences of squares, being careful to factor completely.

5.5 FACTORING TRINOMIAL SQUARES AND DIFFERENCES OF SQUARES

In this section, we first learn to factor trinomials that are squares of binomials. Then we factor binomials that are differences of squares.

a Recognizing Trinomial Squares

Some trinomials are squares of binomials. For example, the trinomial $x^2 + 10x + 25$ is the square of the binomial $x + 5$. To see this, we can calculate $(x + 5)^2$. It is $x^2 + 2 \cdot x \cdot 5 + 5^2$, or $x^2 + 10x + 25$. A trinomial that is the square of a binomial is called a **trinomial square,** or a **perfect-square trinomial.**

In Chapter 4, we considered squaring binomials as special-product rules:

$$(A + B)^2 = A^2 + 2AB + B^2;$$
$$(A - B)^2 = A^2 - 2AB + B^2.$$

We can use these equations in reverse to factor trinomial squares.

TRINOMIAL SQUARES

$$A^2 + 2AB + B^2 = (A + B)^2;$$
$$A^2 - 2AB + B^2 = (A - B)^2$$

How can we recognize when an expression to be factored is a trinomial square? Look at $A^2 + 2AB + B^2$ and $A^2 - 2AB + B^2$. In order for an expression to be a trinomial square:

a) The two expressions A^2 and B^2 must be squares, such as

$$4, \quad x^2, \quad 25x^4, \quad 16t^2.$$

When the coefficient is a perfect square and the power(s) of the variable(s) is (are) even, then the expression is a perfect square.

b) There must be no minus sign before A^2 or B^2.

c) If we multiply A and B and double the result, we get either the remaining term $2 \cdot A \cdot B$, or its opposite, $-2 \cdot A \cdot B$.

EXAMPLE 1 Determine whether $x^2 + 6x + 9$ is a trinomial square.

a) We know that x^2 and 9 are squares.

b) There is no minus sign before x^2 or 9.

c) If we multiply the square roots, x and 3, and double the product, we get the remaining term: $2 \cdot x \cdot 3 = 6x$.

Thus, $x^2 + 6x + 9$ is the square of a binomial. In fact, $x^2 + 6x + 9 = (x + 3)^2$.

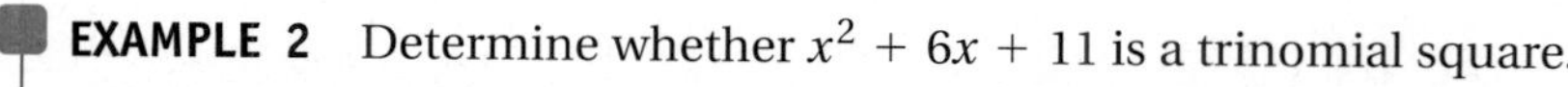

EXAMPLE 2 Determine whether $x^2 + 6x + 11$ is a trinomial square.

The answer is no, because only one term is a square.

EXAMPLE 3 Determine whether $16x^2 + 49 - 56x$ is a trinomial square.

It helps to first write the trinomial in descending order:

$$16x^2 - 56x + 49.$$

a) We know that $16x^2$ and 49 are squares.

b) There is no minus sign before $16x^2$ or 49.

c) If we multiply the square roots, $4x$ and 7, and double the product, we get the opposite of the remaining term: $2 \cdot 4x \cdot 7 = 56x$; $56x$ is the opposite of $-56x$.

Thus, $16x^2 + 49 - 56x$ is a trinomial square. In fact, $16x^2 - 56x + 49 = (4x - 7)^2$.

Do Exercises 1–8.

b Factoring Trinomial Squares

We can use the trial-and-error or grouping methods from Sections 5.2–5.4 to factor trinomial squares, but there is a faster method using the following equations.

FACTORING TRINOMIAL SQUARES

$A^2 + 2AB + B^2 = (A + B)^2$;
$A^2 - 2AB + B^2 = (A - B)^2$

We consider 3 to be a square root of 9 because $3^2 = 9$. Similarly, A is a square root of A^2. We use square roots of the squared terms and the sign of the remaining term to factor a trinomial square.

EXAMPLE 4 Factor: $x^2 + 6x + 9$.

$$x^2 + 6x + 9 = x^2 + 2 \cdot x \cdot 3 + 3^2 = (x + 3)^2$$
$$A^2 + 2\ A\ B + B^2 = (A + B)^2$$

The sign of the middle term is positive.

EXAMPLE 5 Factor: $x^2 + 49 - 14x$.

$$x^2 + 49 - 14x = x^2 - 14x + 49 \quad \text{Changing to descending order}$$
$$= x^2 - 2 \cdot x \cdot 7 + 7^2 \quad \text{The sign of the middle term is negative.}$$
$$= (x - 7)^2$$

EXAMPLE 6 Factor: $16x^2 - 40x + 25$.

$$16x^2 - 40x + 25 = (4x)^2 - 2 \cdot 4x \cdot 5 + 5^2 = (4x - 5)^2$$
$$A^2\ -2\ A\ B + B^2 = (A - B)^2$$

Do Exercises 9–13.

Determine whether each is a trinomial square. Write "yes" or "no."

1. $x^2 + 8x + 16$

2. $25 - x^2 + 10x$

3. $t^2 - 12t + 4$

4. $25 + 20y + 4y^2$

5. $5x^2 + 16 - 14x$

6. $16x^2 + 40x + 25$

7. $p^2 + 6p - 9$

8. $25a^2 + 9 - 30a$

Factor.

9. $x^2 + 2x + 1$

10. $1 - 2x + x^2$

11. $4 + t^2 + 4t$

12. $25x^2 - 70x + 49$

13. $49 - 56y + 16y^2$

Answers on page A-25

Factor.

14. $48m^2 + 75 + 120m$

15. $p^4 + 18p^2 + 81$

16. $4z^5 - 20z^4 + 25z^3$

17. $9a^2 + 30ab + 25b^2$

Answers on page A-25

EXAMPLE 7 Factor: $t^4 + 20t^2 + 100$.

$$t^4 + 20t^2 + 100 = (t^2)^2 + 2(t^2)(10) + 10^2$$
$$= (t^2 + 10)^2$$

EXAMPLE 8 Factor: $75m^3 + 210m^2 + 147m$.

Always look first for a common factor. This time there is one, $3m$:

$$75m^3 + 210m^2 + 147m = 3m[25m^2 + 70m + 49]$$
$$= 3m[(5m)^2 + 2(5m)(7) + 7^2]$$
$$= 3m(5m + 7)^2.$$

EXAMPLE 9 Factor: $4p^2 - 12pq + 9q^2$.

$$4p^2 - 12pq + 9q^2 = (2p)^2 - 2(2p)(3q) + (3q)^2$$
$$= (2p - 3q)^2$$

Do Exercises 14–17.

C Recognizing Differences of Squares

The following polynomials are *differences of squares*:

$$x^2 - 9, \quad 4t^2 - 49, \quad a^2 - 25b^2.$$

To factor a difference of squares such as $x^2 - 9$, think about the formula we used in Chapter 4:

$$\mathbf{(A + B)(A - B) = A^2 - B^2.}$$

Equations are reversible, so we also know the following.

DIFFERENCE OF SQUARES

$$A^2 - B^2 = (A + B)(A - B)$$

Thus,

$$x^2 - 9 = (x + 3)(x - 3).$$

To use this formula, we must be able to recognize when it applies. A **difference of squares** is an expression like the following:

$$A^2 - B^2.$$

How can we recognize such expressions? Look at $A^2 - B^2$. In order for a binomial to be a difference of squares:

a) There must be two expressions, both squares, such as

$$4x^2, \quad 9, \quad 25t^4, \quad 1, \quad x^6, \quad 49y^8.$$

b) The terms must have different signs.

EXAMPLE 10 Is $9x^2 - 64$ a difference of squares?

a) The first expression is a square: $9x^2 = (3x)^2$.
The second expression is a square: $64 = 8^2$.

b) The terms have different signs.

Thus we have a difference of squares, $(3x)^2 - 8^2$.

EXAMPLE 11 Is $25 - t^3$ a difference of squares?

a) The expression t^3 is not a square.

The expression is not a difference of squares.

EXAMPLE 12 Is $-4x^2 + 16$ a difference of squares?

a) The expressions $4x^2$ and 16 are squares: $4x^2 = (2x)^2$ and $16 = 4^2$.

b) The terms have different signs.

Thus we have a difference of squares. We can also see this by rewriting in the equivalent form: $16 - 4x^2$.

Do Exercises 18–24.

d Factoring Differences of Squares

To factor a difference of squares, we use the following equation.

FACTORING A DIFFERENCE OF SQUARES

$A^2 - B^2 = (A + B)(A - B)$

To factor a difference of squares $A^2 - B^2$, we find A and B, which are square roots of the expressions A^2 and B^2. We then use A and B to form two factors. One is the sum $A + B$, and the other is the difference $A - B$.

EXAMPLE 13 Factor: $x^2 - 4$.

$$x^2 - 4 = x^2 - 2^2 = (x + 2)(x - 2)$$
$$A^2 - B^2 = (A + B)(A - B)$$

EXAMPLE 14 Factor: $9 - 16t^4$.

$$9 - 16t^4 = 3^2 - (4t^2)^2 = (3 + 4t^2)(3 - 4t^2)$$
$$A^2 - B^2 = (A + B)(A - B)$$

Determine whether each is a difference of squares. Write "yes" or "no."

18. $x^2 - 25$

19. $t^2 - 24$

20. $y^2 + 36$

21. $4x^2 - 15$

22. $16x^4 - 49$

23. $9w^6 - 1$

24. $-49 + 25t^2$

Answers on page A-25

Factor.

25. $x^2 - 9$

26. $64 - 4t^2$

27. $a^2 - 25b^2$

28. $64x^4 - 25x^6$

29. $5 - 20t^6$
[*Hint*: $1 = 1^2$, $t^6 = (t^3)^2$.]

Answers on page A-25

EXAMPLE 15 Factor: $m^2 - 4p^2$.

$$m^2 - 4p^2 = m^2 - (2p)^2 = (m + 2p)(m - 2p)$$

EXAMPLE 16 Factor: $x^2 - \frac{1}{9}$.

$$x^2 - \frac{1}{9} = x^2 - \left(\frac{1}{3}\right)^2 = \left(x + \frac{1}{3}\right)\left(x - \frac{1}{3}\right)$$

EXAMPLE 17 Factor: $18x^2 - 50x^6$.

Always look first for a factor common to all terms. This time there is one, $2x^2$.

$$\begin{aligned} 18x^2 - 50x^6 &= 2x^2(9 - 25x^4) \\ &= 2x^2[3^2 - (5x^2)^2] \\ &= 2x^2(3 + 5x^2)(3 - 5x^2) \end{aligned}$$

EXAMPLE 18 Factor: $49x^4 - 9x^6$.

$$49x^4 - 9x^6 = x^4(49 - 9x^2) = x^4(7 + 3x)(7 - 3x)$$

Do Exercises 25–29.

CAUTION!

Note carefully in these examples that a difference of squares is *not* the square of the difference; that is,

$$A^2 - B^2 \neq (A - B)^2.$$

For example,

$$(45 - 5)^2 = 40^2 = 1600,$$

but

$$45^2 - 5^2 = 2025 - 25 = 2000.$$

Similarly,

$$A^2 - 2AB + B^2 \neq (A - B)(A + B).$$

For example,

$$(10 - 3)(10 + 3) = 7 \cdot 13 = 91,$$

but

$$\begin{aligned} 10^2 - 2 \cdot 10 \cdot 3 + 3^2 &= 100 - 2 \cdot 10 \cdot 3 + 9 \\ &= 100 - 60 + 9 \\ &= 49. \end{aligned}$$

FACTORING COMPLETELY

If a factor with more than one term can still be factored, you should do so. When no factor can be factored further, you have **factored completely.** Always factor completely whenever told to factor.

EXAMPLE 19 Factor: $p^4 - 16$.

$$p^4 - 16 = (p^2)^2 - 4^2$$
$$= (p^2 + 4)(p^2 - 4)$$ Factoring a difference of squares
$$= (p^2 + 4)(p + 2)(p - 2)$$ Factoring further. The factor $p^2 - 4$ is a difference of squares.

The polynomial $p^2 + 4$ cannot be factored further into polynomials with real coefficients.

CAUTION!

Apart from possibly removing a common factor, you cannot factor a sum of squares. In particular,

$$A^2 + B^2 \neq (A + B)^2.$$

Consider $25x^2 + 100$. Here a sum of squares has a common factor, 25. Factoring, we get $25(x^2 + 4)$, where $x^2 + 4$ is prime. For example,

$$x^2 + 4 \neq (x + 2)^2.$$

EXAMPLE 20 Factor: $y^4 - 16x^{12}$.

$$y^4 - 16x^{12} = (y^2 + 4x^6)(y^2 - 4x^6)$$ Factoring a difference of squares
$$= (y^2 + 4x^6)(y + 2x^3)(y - 2x^3)$$ Factoring further. The factor $y^2 - 4x^6$ is a difference of squares.

As you proceed through the exercises, these suggestions may prove helpful.

TIPS FOR FACTORING

- Always look first for a common factor! If there is one, factor it out.
- Be alert for trinomial squares and differences of squares. Once recognized, they can be factored without trial and error.
- Always factor completely.
- Check by multiplying.

Do Exercises 30 and 31.

Factor completely.

30. $81x^4 - 1$

31. $49p^4 - 25q^6$

Answers on page A-26

5.5 EXERCISE SET

For Extra Help

Digital Video Tutor CD 5 Videotape 11 | InterAct Math | Math Tutor Center | MathXL | MyMathLab

a Determine whether each of the following is a trinomial square.

1. $x^2 - 14x + 49$

2. $x^2 - 16x + 64$

3. $x^2 + 16x - 64$

4. $x^2 - 14x - 49$

5. $x^2 - 2x + 4$

6. $x^2 + 3x + 9$

7. $9x^2 - 36x + 24$

8. $36x^2 - 24x + 16$

b Factor completely. Remember to look first for a common factor and to check by multiplying.

9. $x^2 - 14x + 49$

10. $x^2 - 20x + 100$

11. $x^2 + 16x + 64$

12. $x^2 + 20x + 100$

13. $x^2 - 2x + 1$

14. $x^2 + 2x + 1$

15. $4 + 4x + x^2$

16. $4 + x^2 - 4x$

17. $q^4 - 6q^2 + 9$

18. $64 + 16a^2 + a^4$

19. $49 + 56y + 16y^2$

20. $75 + 48a^2 - 120a$

21. $2x^2 - 4x + 2$

22. $2x^2 - 40x + 200$

23. $x^3 - 18x^2 + 81x$

24. $x^3 + 24x^2 + 144x$

25. $12q^2 - 36q + 27$

26. $20p^2 + 100p + 125$

27. $49 - 42x + 9x^2$

28. $64 - 112x + 49x^2$

29. $5y^4 + 10y^2 + 5$

30. $a^4 + 14a^2 + 49$

31. $1 + 4x^4 + 4x^2$

32. $1 - 2a^5 + a^{10}$

33. $4p^2 + 12pq + 9q^2$

34. $25m^2 + 20mn + 4n^2$

35. $a^2 - 6ab + 9b^2$

36. $x^2 - 14xy + 49y^2$

37. $81a^2 - 18ab + b^2$

38. $64p^2 + 16pq + q^2$

39. $36a^2 + 96ab + 64b^2$

40. $16m^2 - 40mn + 25n^2$

c Determine whether each of the following is a difference of squares.

41. $x^2 - 4$

42. $x^2 - 36$

43. $x^2 + 25$

44. $x^2 + 9$

45. $x^2 - 45$

46. $x^2 - 80y^2$

47. $16x^2 - 25y^2$

48. $-1 + 36x^2$

d Factor completely. Remember to look first for a common factor.

49. $y^2 - 4$

50. $q^2 - 1$

51. $p^2 - 9$

52. $x^2 - 36$

53. $-49 + t^2$

54. $-64 + m^2$

55. $a^2 - b^2$

56. $p^2 - q^2$

57. $25t^2 - m^2$

58. $w^2 - 49z^2$

59. $100 - k^2$

60. $81 - w^2$

61. $16a^2 - 9$

62. $25x^2 - 4$

63. $4x^2 - 25y^2$

64. $9a^2 - 16b^2$

65. $8x^2 - 98$

66. $24x^2 - 54$

67. $36x - 49x^3$

68. $16x - 81x^3$

69. $49a^4 - 81$

70. $25a^4 - 9$

71. $a^4 - 16$

72. $y^4 - 1$

73. $5x^4 - 405$

74. $4x^4 - 64$

75. $1 - y^8$

76. $x^8 - 1$

77. $x^{12} - 16$

78. $x^8 - 81$

79. $y^2 - \frac{1}{16}$

80. $x^2 - \frac{1}{25}$

81. $25 - \frac{1}{49}x^2$

82. $\frac{1}{4} - 9q^2$

83. $16m^4 - t^4$

84. $p^4q^4 - 1$

85. D$_W$ Explain in your own words how to determine whether a polynomial is a trinomial square.

86. D$_W$ Spiro concludes that since $x^2 - 9 = (x - 3)(x + 3)$, it must follow that $x^2 + 9 = (x + 3)(x + 3)$. What mistake is the student making? How would you go about correcting the misunderstanding?

SKILL MAINTENANCE

Divide. [1.6a, c]

87. $(-110) \div 10$

88. $-1000 \div (-2.5)$

89. $\left(-\frac{2}{3}\right) \div \frac{4}{5}$

90. $8.1 \div (-9)$

91. $-64 \div (-32)$

92. $-256 \div 1.6$

Find a polynomial for the shaded area. (Leave results in terms of π where appropriate.) [4.4d]

93.

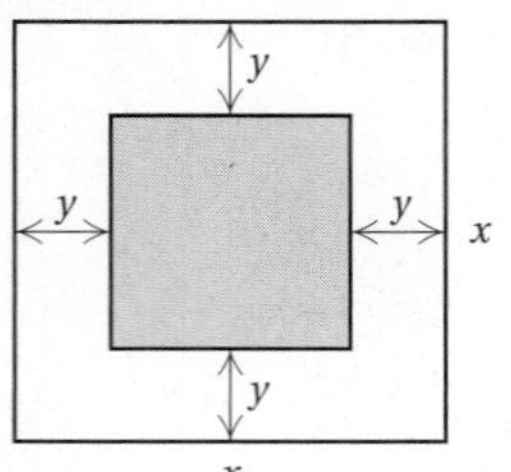

94.

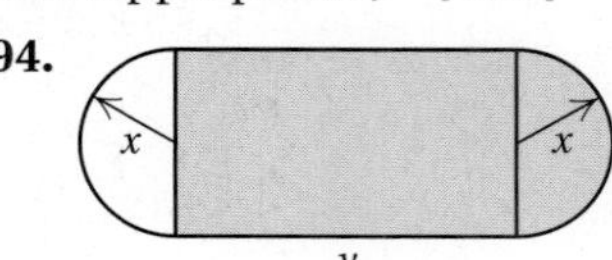

Simplify.

95. $y^5 \cdot y^7$ [4.1d]

96. $(5a^2b^3)^2$ [4.2a, b]

Find the intercepts. Then graph the equation. [3.3a]

97. $y - 6x = 6$

98. $3x - 5y = 15$

SYNTHESIS

Factor completely, if possible.

99. $49x^2 - 216$

100. $27x^3 - 13x$

101. $x^2 + 22x + 121$

102. $x^2 - 5x + 25$

103. $18x^3 + 12x^2 + 2x$

104. $162x^2 - 82$

105. $x^8 - 2^8$

106. $4x^4 - 4x^2$

107. $3x^5 - 12x^3$

108. $3x^2 - \frac{1}{3}$

109. $18x^3 - \frac{8}{25}x$

110. $x^2 - 2.25$

111. $0.49p - p^3$

112. $3.24x^2 - 0.81$

113. $0.64x^2 - 1.21$

114. $1.28x^2 - 2$

115. $(x + 3)^2 - 9$

116. $(y - 5)^2 - 36q^2$

117. $x^2 - \left(\frac{1}{x}\right)^2$

118. $a^{2n} - 49b^{2n}$

119. $81 - b^{4k}$

120. $9x^{18} + 48x^9 + 64$

121. $9b^{2n} + 12b^n + 4$

122. $(x + 7)^2 - 4x - 24$

123. $(y + 3)^2 + 2(y + 3) + 1$

124. $49(x + 1)^2 - 42(x + 1) + 9$

Find c such that the polynomial is the square of a binomial.

125. $cy^2 + 6y + 1$

126. $cy^2 - 24y + 9$

Use the TABLE feature to determine whether the factorization is correct.

127. $x^2 + 9 = (x + 3)(x + 3)$

128. $x^2 - 49 = (x - 7)(x + 7)$

129. $x^2 + 9 = (x + 3)^2$

130. $x^2 - 49 = (x - 7)^2$

Objective

a Factor polynomials completely using any of the methods considered in this chapter.

5.6 FACTORING: A GENERAL STRATEGY

a We now combine all of our factoring techniques and consider a general strategy for factoring polynomials. Here we will encounter polynomials of all the types we have considered, in random order, so you will have the opportunity to determine which method to use.

FACTORING STRATEGY

To factor a polynomial:

a) Always look first for a common factor. If there is one, factor out the largest common factor.

b) Then look at the number of terms.

Two terms: Determine whether you have a difference of squares. Do not try to factor a sum of squares: $A^2 + B^2$.

Three terms: Determine whether the trinomial is a square. If it is, you know how to factor. If not, try trial and error, using FOIL or the *ac*-method.

Four terms: Try factoring by grouping.

c) *Always factor completely.* If a factor with more than one term can still be factored, you should factor it. When no factor can be factored further, you have finished.

d) Check by multiplying.

EXAMPLE 1 Factor: $5t^4 - 80$.

a) We look for a common factor:

$$5t^4 - 80 = 5(t^4 - 16).$$

b) The factor $t^4 - 16$ has only two terms. It is a difference of squares: $(t^2)^2 - 4^2$. We factor $t^4 - 16$ and then include the common factor:

$$5(t^2 + 4)(t^2 - 4).$$

c) We see that one of the factors is again a difference of squares. We factor it:

$$5(t^2 + 4)(t + 2)(t - 2).$$

↑ This is a sum of squares. It cannot be factored!

We have factored completely because no factor with more than one term can be factored further.

d) CHECK:
$$\begin{aligned} 5(t^2 + 4)(t + 2)(t - 2) &= 5(t^2 + 4)(t^2 - 4) \\ &= 5(t^4 - 16) \\ &= 5t^4 - 80. \end{aligned}$$

EXAMPLE 2 Factor: $2x^3 + 10x^2 + x + 5$.

a) We look for a common factor. There isn't one.

b) There are four terms. We try factoring by grouping:

$$\begin{aligned} &2x^3 + 10x^2 + x + 5 \\ &\quad = (2x^3 + 10x^2) + (x + 5) && \text{Separating into two binomials} \\ &\quad = 2x^2(x + 5) + 1(x + 5) && \text{Factoring each binomial} \\ &\quad = (2x^2 + 1)(x + 5). && \text{Factoring out the common factor } x + 5 \end{aligned}$$

c) None of these factors can be factored further, so we have factored completely.

d) CHECK: $(2x^2 + 1)(x + 5) = 2x^2 \cdot x + 2x^2 \cdot 5 + 1 \cdot x + 1 \cdot 5 = 2x^3 + 10x^2 + x + 5.$

EXAMPLE 3 Factor: $x^5 - 2x^4 - 35x^3$.

a) We look first for a common factor. This time there is one, x^3:

$$x^5 - 2x^4 - 35x^3 = x^3(x^2 - 2x - 35).$$

b) The factor $x^2 - 2x - 35$ has three terms, but it is not a trinomial square. We factor it using trial and error (FOIL):

$$x^5 - 2x^4 - 35x^3 = x^3(x^2 - 2x - 35) = x^3(x - 7)(x + 5).$$

Don't forget to include the common factor in the final answer!

c) No factor with more than one term can be factored further, so we have factored completely.

d) CHECK: $x^3(x - 7)(x + 5) = x^3(x^2 - 2x - 35) = x^5 - 2x^4 - 35x^3.$

EXAMPLE 4 Factor: $x^4 - 10x^2 + 25$.

a) We look first for a common factor. There isn't one.

b) There are three terms. We see that this polynomial is a trinomial square. We factor it:

$$x^4 - 10x^2 + 25 = (x^2)^2 - 2 \cdot x^2 \cdot 5 + 5^2 = (x^2 - 5)^2.$$

We could use FOIL if we have not recognized that we have a trinomial square.

c) Since $x^2 - 5$ cannot be factored further, we have factored completely.

d) CHECK: $(x^2 - 5)^2 = (x^2)^2 - 2(x^2)(5) + 5^2 = x^4 - 10x^2 + 25.$

Do Exercises 1–5.

EXAMPLE 5 Factor: $6x^2y^4 - 21x^3y^5 + 3x^2y^6$.

a) We look first for a common factor:

$$6x^2y^4 - 21x^3y^5 + 3x^2y^6 = 3x^2y^4(2 - 7xy + y^2).$$

Factor.

1. $3m^4 - 3$

2. $x^6 + 8x^3 + 16$

3. $2x^4 + 8x^3 + 6x^2$

4. $3x^3 + 12x^2 - 2x - 8$

5. $8x^3 - 200x$

Answers on page A-26

b) There are three terms in $2 - 7xy + y^2$. We determine whether the trinomial is a square. Since only y^2 is a square, we do not have a trinomial square. Can the trinomial be factored by trial and error? A key to the answer is that x is only in the term $-7xy$. The polynomial might be in a form like $(1 - y)(2 + y)$, but there would be no x in the middle term. Thus, $2 - 7xy + y^2$ cannot be factored.

c) Have we factored completely? Yes because no factor with more than one term can be factored further.

d) The check is left to the student.

EXAMPLE 6 Factor: $(p + q)(x + 2) + (p + q)(x + y)$.

a) We look for a common factor:

$$\begin{aligned}(p + q)(x + 2) + (p + q)(x + y) &= (p + q)[(x + 2) + (x + y)] \\ &= (p + q)(2x + y + 2).\end{aligned}$$

b) There are three terms in $2x + y + 2$, but this trinomial cannot be factored further.

c) Neither factor can be factored further, so we have factored completely.

d) The check is left to the student.

EXAMPLE 7 Factor: $px + py + qx + qy$.

a) We look first for a common factor. There isn't one.

b) There are four terms. We try factoring by grouping:

$$\begin{aligned}px + py + qx + qy &= p(x + y) + q(x + y) \\ &= (p + q)(x + y).\end{aligned}$$

c) Have we factored completely? Since neither factor can be factored further, we have factored completely.

d) CHECK: $(p + q)(x + y) = px + py + qx + qy$.

EXAMPLE 8 Factor: $25x^2 + 20xy + 4y^2$.

a) We look first for a common factor. There isn't one.

b) There are three terms. We determine whether the trinomial is a square. The first term and the last term are squares:

$$25x^2 = (5x)^2 \quad \text{and} \quad 4y^2 = (2y)^2.$$

Since twice the product of $5x$ and $2y$ is the other term,

$$2 \cdot 5x \cdot 2y = 20xy,$$

the trinomial is a perfect square.

We factor by writing the square roots of the square terms and the sign of the middle term:

$$25x^2 + 20xy + 4y^2 = (5x + 2y)^2.$$

c) Since $5x + 2y$ cannot be factored further, we have factored completely.

d) CHECK: $(5x + 2y)^2 = (5x)^2 + 2(5x)(2y) + (2y)^2$
$= 25x^2 + 20xy + 4y^2.$

EXAMPLE 9 Factor: $p^2q^2 + 7pq + 12$.

a) We look first for a common factor. There isn't one.

b) There are three terms. We determine whether the trinomial is a square. The first term is a square, but neither of the other terms is a square, so we do not have a trinomial square. We factor, thinking of the product pq as a single variable. We consider this possibility for factorization:

$$(pq + \quad)(pq + \quad).$$

We factor the last term, 12. All the signs are positive, so we consider only positive factors. Possibilities are 1, 12 and 2, 6 and 3, 4. The pair 3, 4 gives a sum of 7 for the coefficient of the middle term. Thus,

$$p^2q^2 + 7pq + 12 = (pq + 3)(pq + 4).$$

c) No factor with more than one term can be factored further, so we have factored completely.

d) CHECK: $(pq + 3)(pq + 4) = (pq)(pq) + 4 \cdot pq + 3 \cdot pq + 3 \cdot 4$
$= p^2q^2 + 7pq + 12.$

EXAMPLE 10 Factor: $8x^4 - 20x^2y - 12y^2$.

a) We look first for a common factor:

$$8x^4 - 20x^2y - 12y^2 = 4(2x^4 - 5x^2y - 3y^2).$$

b) There are three terms in $2x^4 - 5x^2y - 3y^2$. We determine whether the trinomial is a square. Since none of the terms is a square, we do not have a trinomial square. We factor $2x^4$. Possibilities are $2x^2$, x^2 and $2x$, x^3 and others. We also factor the last term, $-3y^2$. Possibilities are $3y$, $-y$ and $-3y$, y and others. We look for factors such that the sum of their products is the middle term. The x^2 in the middle term, $-5x^2y$, should lead us to try $(2x^2)(x^2)$. We try some possibilities:

$$(2x^2 - y)(x^2 + 3y) = 2x^4 + 5x^2y - 3y^2,$$
$$(2x^2 + y)(x^2 - 3y) = 2x^4 - 5x^2y - 3y^2.$$

c) No factor with more than one term can be factored further, so we have factored completely. The factorization, including the common factor, is

$$4(2x^2 + y)(x^2 - 3y).$$

d) CHECK: $4(2x^2 + y)(x^2 - 3y) = 4[(2x^2)(x^2) + 2x^2(-3y) + yx^2 + y(-3y)]$
$= 4[2x^4 - 6x^2y + x^2y - 3y^2]$
$= 4(2x^4 - 5x^2y - 3y^2)$
$= 8x^4 - 20x^2y - 12y^2.$

EXAMPLE 11 Factor: $a^4 - 16b^4$.

a) We look first for a common factor. There isn't one.

b) There are two terms. Since $a^4 = (a^2)^2$ and $16b^4 = (4b^2)^2$, we see that we do have a difference of squares. Thus,

$$a^4 - 16b^4 = (a^2 + 4b^2)(a^2 - 4b^2).$$

c) The last factor can be factored further. It is also a difference of squares. Thus,

$$a^4 - 16b^4 = (a^2 + 4b^2)(a + 2b)(a - 2b).$$

d) CHECK: $(a^2 + 4b^2)(a + 2b)(a - 2b) = (a^2 + 4b^2)(a^2 - 4b^2)$
$= a^4 - 16b^4.$

Do Exercises 6–12.

Factor.

6. $x^4y^2 + 2x^3y + 3x^2y$

7. $10p^6q^2 + 4p^5q^3 + 2p^4q^4$

8. $(a - b)(x + 5) + (a - b)(x + y^2)$

9. $ax^2 + ay + bx^2 + by$

10. $x^4 + 2x^2y^2 + y^4$

11. $x^2y^2 + 5xy + 4$

12. $p^4 - 81q^4$

Answers on page A-26

5.6 EXERCISE SET

For Extra Help

Digital Video Tutor CD 5 Videotape 11 | InterAct Math | Math Tutor Center | MathXL | MyMathLab

a Factor completely.

1. $3x^2 - 192$

2. $2t^2 - 18$

3. $a^2 + 25 - 10a$

4. $y^2 + 49 + 14y$

5. $2x^2 - 11x + 12$

6. $8y^2 - 18y - 5$

7. $x^3 + 24x^2 + 144x$

8. $x^3 - 18x^2 + 81x$

9. $x^3 + 3x^2 - 4x - 12$

10. $x^3 - 5x^2 - 25x + 125$

11. $48x^2 - 3$

12. $50x^2 - 32$

13. $9x^3 + 12x^2 - 45x$

14. $20x^3 - 4x^2 - 72x$

15. $x^2 + 4$

16. $t^2 + 25$

17. $x^4 + 7x^2 - 3x^3 - 21x$

18. $m^4 + 8m^3 + 8m^2 + 64m$

19. $x^5 - 14x^4 + 49x^3$

20. $2x^6 + 8x^5 + 8x^4$

21. $20 - 6x - 2x^2$

22. $45 - 3x - 6x^2$

23. $x^2 - 6x + 1$

24. $x^2 + 8x + 5$

25. $4x^4 - 64$

26. $5x^5 - 80x$

27. $1 - y^8$

28. $t^8 - 1$

29. $x^5 - 4x^4 + 3x^3$

30. $x^6 - 2x^5 + 7x^4$

31. $\frac{1}{81}x^6 - \frac{8}{27}x^3 + \frac{16}{9}$

32. $36a^2 - 15a + \frac{25}{16}$

33. $mx^2 + my^2$

34. $12p^2 + 24q^3$

35. $9x^2y^2 - 36xy$

36. $x^2y - xy^2$

37. $2\pi rh + 2\pi r^2$

38. $10p^4q^4 + 35p^3q^3 + 10p^2q^2$

39. $(a + b)(x - 3) + (a + b)(x + 4)$

40. $5c(a^3 + b) - (a^3 + b)$

41. $(x - 1)(x + 1) - y(x + 1)$

42. $3(p - q) - q^2(p - q)$

43. $n^2 + 2n + np + 2p$

44. $a^2 - 3a + ay - 3y$

45. $6q^2 - 3q + 2pq - p$

46. $2x^2 - 4x + xy - 2y$

47. $4b^2 + a^2 - 4ab$

48. $x^2 + y^2 - 2xy$

49. $16x^2 + 24xy + 9y^2$

50. $9c^2 + 6cd + d^2$

51. $49m^4 - 112m^2n + 64n^2$

52. $4x^2y^2 + 12xyz + 9z^2$

53. $y^4 + 10y^2z^2 + 25z^4$

54. $0.01x^4 - 0.1x^2y^2 + 0.25y^4$

55. $\frac{1}{4}a^2 + \frac{1}{3}ab + \frac{1}{9}b^2$

56. $4p^2q + pq^2 + 4p^3$

57. $a^2 - ab - 2b^2$

58. $3b^2 - 17ab - 6a^2$

59. $2mn - 360n^2 + m^2$

60. $15 + x^2y^2 + 8xy$

61. $m^2n^2 - 4mn - 32$

62. $p^2q^2 + 7pq + 6$

63. $r^5s^2 - 10r^4s + 16r^3$

64. $p^5q^2 + 3p^4q - 10p^3$

65. $a^5 + 4a^4b - 5a^3b^2$

66. $2s^6t^2 + 10s^3t^3 + 12t^4$

67. $a^2 - \frac{1}{25}b^2$

68. $p^2 - \frac{1}{49}b^2$

69. $x^2 - y^2$

70. $p^2q^2 - r^2$

71. $16 - p^4q^4$

72. $15a^4 - 15b^4$

73. $1 - 16x^{12}y^{12}$

74. $81a^4 - b^4$

75. $q^3 + 8q^2 - q - 8$

76. $m^3 - 7m^2 - 4m + 28$

77. $112xy + 49x^2 + 64y^2$

78. $4ab^5 - 32b^4 + a^2b^6$

79. $\mathbf{D_W}$ Kelly factored $16 - 8x + x^2$ as $(x - 4)^2$, while Tony factored it as $(4 - x)^2$. Evaluate each expression for several values of x. Then explain why both answers are correct.

80. $\mathbf{D_W}$ Describe in your own words a strategy that can be used to factor polynomials.

SKILL MAINTENANCE

CD Sales. The line graph below charts data concerning the number of music CDs sold in recent years. Use it for Exercises 81–86. [3.1a]

Growing CD Sales

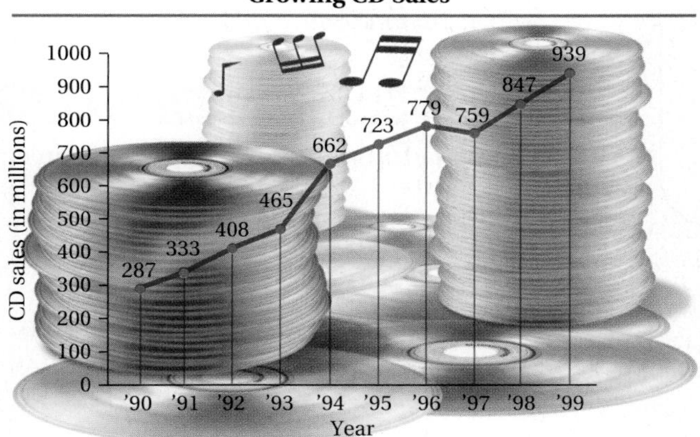

Source: Recording Industry Association of America

81. In which year were CD sales highest?

82. In which year were CD sales lowest?

83. In which year were CD sales 779 million?

84. What were CD sales in 1998?

85. What was the percent of increase in sales from 1998 to 1999? [2.5a]

86. What was the percent of decrease in sales from 1996 to 1997? [2.5a]

87. Divide: $\frac{7}{5} \div \left(-\frac{11}{10}\right)$. [1.6c]

88. Multiply: $(5x - t)^2$. [4.6d]

89. Solve $A = aX + bX - 7$ for X. [2.4b]

90. Solve: $4(x - 9) - 2(x + 7) < 14$. [2.7e]

SYNTHESIS

Factor completely.

91. $a^4 - 2a^2 + 1$

92. $x^4 + 9$

93. $12.25x^2 - 7x + 1$

94. $\frac{1}{5}x^2 - x + \frac{4}{5}$

95. $5x^2 + 13x + 7.2$

96. $x^3 - (x - 3x^2) - 3$

97. $18 + y^3 - 9y - 2y^2$

98. $-(x^4 - 7x^2 - 18)$

99. $a^3 + 4a^2 + a + 4$

100. $x^3 + x^2 - (4x + 4)$

101. $x^3 - x^2 - 4x + 4$

102. $3x^4 - 15x^2 + 12$

103. $y^2(y - 1) - 2y(y - 1) + (y - 1)$

104. $y^2(y + 1) - 4y(y + 1) - 21(y + 1)$

105. $(y + 4)^2 + 2x(y + 4) + x^2$

106. $6(x - 1)^2 + 7y(x - 1) - 3y^2$

5.7 SOLVING QUADRATIC EQUATIONS BY FACTORING

Objectives

a Solve equations (already factored) using the principle of zero products.

b Solve quadratic equations by factoring and then using the principle of zero products.

Second-degree equations like $x^2 + x - 156 = 0$ and $9 - x^2 = 0$ are examples of *quadratic equations*.

QUADRATIC EQUATION

A **quadratic equation** is an equation equivalent to an equation of the type

$$ax^2 + bx + c = 0, \quad a \neq 0.$$

In order to solve quadratic equations, we need a new equation-solving principle.

a The Principle of Zero Products

The product of two numbers is 0 if one or both of the numbers is 0. Furthermore, *if any product is* 0, *then a factor must be* 0. For example:

If $7x = 0$, then we know that $x = 0$.

If $x(2x - 9) = 0$, then we know that $x = 0$ or $2x - 9 = 0$.

If $(x + 3)(x - 2) = 0$, then we know that $x + 3 = 0$ or $x - 2 = 0$.

In a product such as $ab = 24$, we cannot conclude with certainty that a is 24 or that b is 24, but if $ab = 0$, we can conclude that $a = 0$ or $b = 0$.

EXAMPLE 1 Solve: $(x + 3)(x - 2) = 0$.

We have a product of 0. This equation will be true when either factor is 0. Thus it is true when

$$x + 3 = 0 \quad \text{or} \quad x - 2 = 0.$$

Here we have two simple equations that we know how to solve:

$$x = -3 \quad \text{or} \quad x = 2.$$

Each of the numbers -3 and 2 is a solution of the original equation, as we can see in the following checks.

CHECK: For -3:

$$\begin{array}{r|l} (x + 3)(x - 2) & = 0 \\ \hline (-3 + 3)(-3 - 2) & ?\ 0 \\ 0(-5) & \\ 0 & \quad \text{TRUE} \end{array}$$

For 2:

$$\begin{array}{r|l} (x + 3)(x - 2) & = 0 \\ \hline (2 + 3)(2 - 2) & ?\ 0 \\ 5(0) & \\ 0 & \quad \text{TRUE} \end{array}$$

We now have a principle to help in solving quadratic equations.

THE PRINCIPLE OF ZERO PRODUCTS

An equation $ab = 0$ is true if and only if $a = 0$ is true or $b = 0$ is true, or both are true. (A product is 0 if and only if one or both of the factors is 0.)

Study Tips

WORKING WITH A CLASSMATE

If you are finding it difficult to master a particular topic or concept, try talking about it with a classmate. Verbalizing your questions about the material might help clarify it. If your classmate is also finding the material difficult, it is possible that the majority of the people in your class are confused and you can ask your instructor to explain the concept again.

EXAMPLE 2 Solve: $(5x + 1)(x - 7) = 0$.

$$(5x + 1)(x - 7) = 0$$

$5x + 1 = 0$ *or* $x - 7 = 0$ — Using the principle of zero products

$5x = -1$ *or* $x = 7$ — Solving the two equations separately

$x = -\frac{1}{5}$ *or* $x = 7$

CHECK: For $-\frac{1}{5}$:

$$\begin{array}{c|c} (5x + 1)(x - 7) & 0 \\ \hline (5(-\frac{1}{5}) + 1)(-\frac{1}{5} - 7) & ?\ 0 \\ (-1 + 1)(-7\frac{1}{5}) & \\ 0(-7\frac{1}{5}) & \\ 0 & \end{array}$$

TRUE

For 7:

$$\begin{array}{c|c} (5x + 1)(x - 7) & 0 \\ \hline (5(7) + 1)(7 - 7) & ?\ 0 \\ (35 + 1) \cdot 0 & \\ 36 \cdot 0 & \\ 0 & \end{array}$$

TRUE

The solutions are $-\frac{1}{5}$ and 7.

When you solve an equation using the principle of zero products, a check by substitution, as in Examples 1 and 2, will detect errors in solving.

Do Exercises 1–3.

When some factors have only one term, you can still use the principle of zero products.

EXAMPLE 3 Solve: $x(2x - 9) = 0$.

We have

$$x(2x - 9) = 0$$

$x = 0$ *or* $2x - 9 = 0$ — Using the principle of zero products

$x = 0$ *or* $2x = 9$

$x = 0$ *or* $x = \frac{9}{2}$.

The solutions are 0 and $\frac{9}{2}$. The check is left to the student.

Do Exercise 4.

Solve using the principle of zero products.

1. $(x - 3)(x + 4) = 0$

2. $(x - 7)(x - 3) = 0$

3. $(4t + 1)(3t - 2) = 0$

4. Solve: $y(3y - 17) = 0$.

Answers on page A-26

5. Solve: $x^2 - x - 6 = 0$.

Solve.

6. $x^2 - 3x = 28$

7. $x^2 = 6x - 9$

Solve.

8. $x^2 - 4x = 0$

9. $9x^2 = 16$

Answers on page A-26

b Using Factoring to Solve Equations

Using factoring and the principle of zero products, we can solve some new kinds of equations. Thus we have extended our equation-solving abilities.

EXAMPLE 4 Solve: $x^2 + 5x + 6 = 0$.

Compare this equation to those that we know how to solve from Chapter 2. There are no like terms to collect, and we have a squared term. We first factor the polynomial. Then we use the principle of zero products.

$$x^2 + 5x + 6 = 0$$

$$(x + 2)(x + 3) = 0 \qquad \text{Factoring}$$

$$x + 2 = 0 \quad \text{or} \quad x + 3 = 0 \qquad \text{Using the principle of zero products}$$

$$x = -2 \quad \text{or} \quad x = -3$$

CHECK: For -2:

$$\begin{array}{c|c} x^2 + 5x + 6 & 0 \\ \hline (-2)^2 + 5(-2) + 6 \;?\; & 0 \\ 4 - 10 + 6 & \\ -6 + 6 & \\ 0 & \end{array} \quad \text{TRUE}$$

For -3:

$$\begin{array}{c|c} x^2 + 5x + 6 & 0 \\ \hline (-3)^2 + 5(-3) + 6 \;?\; & 0 \\ 9 - 15 + 6 & \\ -6 + 6 & \\ 0 & \end{array} \quad \text{TRUE}$$

The solutions are -2 and -3.

CAUTION!

Keep in mind that you *must* have 0 on one side of the equation before you can use the principle of zero products. Get all nonzero terms on one side and 0 on the other.

Do Exercise 5.

EXAMPLE 5 Solve: $x^2 - 8x = -16$.

We first add 16 to get a 0 on one side:

$$x^2 - 8x = -16$$

$$x^2 - 8x + 16 = 0 \qquad \text{Adding 16}$$

$$(x - 4)(x - 4) = 0 \qquad \text{Factoring}$$

$$x - 4 = 0 \quad \text{or} \quad x - 4 = 0 \qquad \text{Using the principle of zero products}$$

$$x = 4 \quad \text{or} \quad x = 4. \qquad \text{Solving each equation}$$

There is only one solution, 4. The check is left to the student.

Do Exercises 6 and 7.

EXAMPLE 6 Solve: $x^2 + 5x = 0$.

$$x^2 + 5x = 0$$

$$x(x + 5) = 0 \qquad \text{Factoring out a common factor}$$

$$x = 0 \quad \text{or} \quad x + 5 = 0 \qquad \text{Using the principle of zero products}$$

$$x = 0 \quad \text{or} \quad x = -5$$

The solutions are 0 and -5. The check is left to the student.

EXAMPLE 7 Solve: $4x^2 = 25$.

$$4x^2 = 25$$
$$4x^2 - 25 = 0 \quad \text{Subtracting 25 on both sides to get 0 on one side}$$
$$(2x - 5)(2x + 5) = 0 \quad \text{Factoring a difference of squares}$$
$$2x - 5 = 0 \quad or \quad 2x + 5 = 0$$
$$2x = 5 \quad or \quad 2x = -5 \quad \text{Solving each equation}$$
$$x = \frac{5}{2} \quad or \quad x = -\frac{5}{2}$$

The solutions are $\frac{5}{2}$ and $-\frac{5}{2}$. The check is left to the student.

Do Exercises 8 and 9 on the preceding page.

EXAMPLE 8 Solve: $-5x^2 + 2x + 3 = 0$.

In this case, the leading coefficient of the trinomial is negative. Thus we first multiply by -1 and then proceed as we have in Examples 1–7.

$$-5x^2 + 2x + 3 = 0$$
$$-1(-5x^2 + 2x + 3) = -1 \cdot 0 \quad \text{Multiplying by } -1$$
$$5x^2 - 2x - 3 = 0 \quad \text{Simplifying}$$
$$(5x + 3)(x - 1) = 0 \quad \text{Factoring}$$
$$5x + 3 = 0 \quad or \quad x - 1 = 0 \quad \text{Using the principle of zero products}$$
$$5x = -3 \quad or \quad x = 1$$
$$x = -\frac{3}{5} \quad or \quad x = 1$$

The solutions are $-\frac{3}{5}$ and 1. The check is left to the student.

Do Exercises 10 and 11.

EXAMPLE 9 Solve: $(x + 2)(x - 2) = 5$.

Be careful with an equation like this one! It might be tempting to set each factor equal to 5. Remember: We must have a 0 on one side. We first carry out the product on the left. Then we subtract 5 on both sides to get 0 on one side. Then we proceed with the principle of zero products.

$$(x + 2)(x - 2) = 5$$
$$x^2 - 4 = 5 \quad \text{Multiplying on the left}$$
$$x^2 - 4 - 5 = 5 - 5 \quad \text{Subtracting 5}$$
$$x^2 - 9 = 0 \quad \text{Simplifying}$$
$$(x + 3)(x - 3) = 0 \quad \text{Factoring}$$
$$x + 3 = 0 \quad or \quad x - 3 = 0 \quad \text{Using the principle of zero products}$$
$$x = -3 \quad or \quad x = 3$$

The solutions are -3 and 3. The check is left to the student.

Do Exercise 12.

Solve.

10. $-2x^2 + 13x - 21 = 0$

11. $10 - 3x - x^2 = 0$

12. Solve: $(x + 1)(x - 1) = 8$.

Answers on page A-26

13. Find the x-intercepts of the graph shown below.

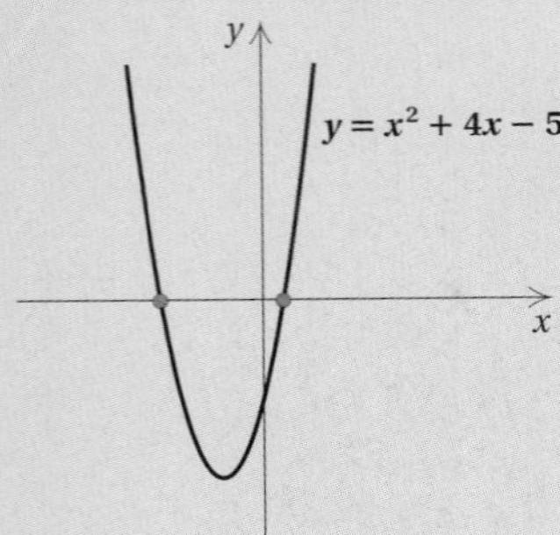

14. Use *only* the graph shown below to solve $3x - x^2 = 0$.

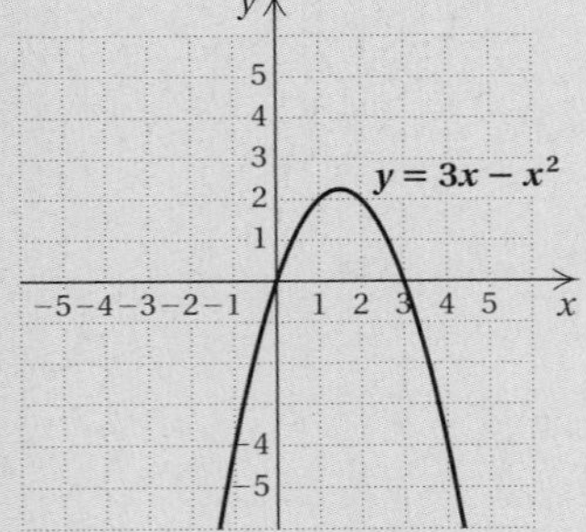

Answers on page A-26

AG ALGEBRAIC–GRAPHICAL CONNECTION

In Chapter 3, we graphed linear equations of the type $y = mx + b$ and $Ax + By = C$. Recall that to find the x-intercept, we replaced y with 0 and solved for x. This procedure can also be used to find the x-intercepts when an equation of the form $y = ax^2 + bx + c$, $a \neq 0$, is to be graphed. Although the details of creating such graphs will be left to Chapter 10, we consider them briefly here from the standpoint of finding the x-intercepts. The graphs are shaped like the following curves. Note that each x-intercept represents a solution of $ax^2 + bx + c = 0$.

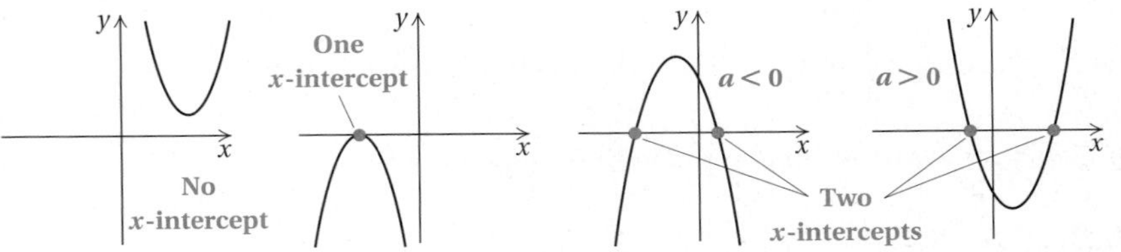

EXAMPLE 10 Find the x-intercepts of the graph of $y = x^2 - 4x - 5$ shown at right. (The grid is intentionally not included.)

To find the x-intercepts, we let $y = 0$ and solve for x:

$$0 = x^2 - 4x - 5 \qquad \text{Substituting 0 for } y$$

$$0 = (x - 5)(x + 1) \qquad \text{Factoring}$$

$$x - 5 = 0 \quad or \quad x + 1 = 0 \qquad \text{Using the principle of zero products}$$

$$x = 5 \quad or \quad x = -1.$$

The x-intercepts are $(5, 0)$ and $(-1, 0)$. We can now label them on the graph.

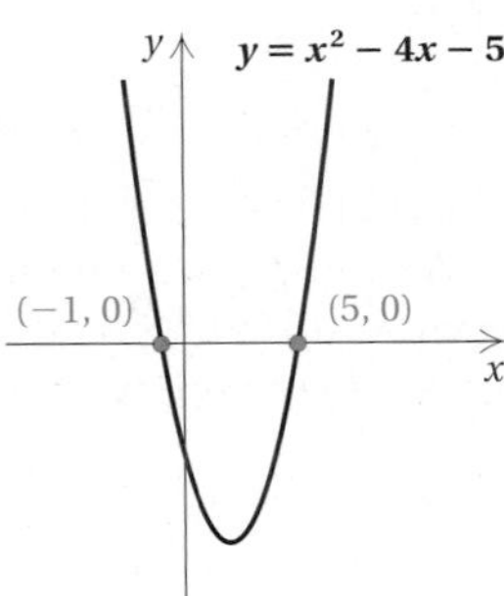

Do Exercises 13 and 14.

CALCULATOR CORNER

Solving Quadratic Equations We can solve quadratic equations graphically. Consider the equation $x^2 + 2x = 8$. First, we must write the equation with 0 on one side. To do this, we subtract 8 on both sides of the equation; we get $x^2 + 2x - 8 = 0$. Next, we graph $y = x^2 + 2x - 8$ in a window that shows the x-intercepts. The standard window works well in this case.

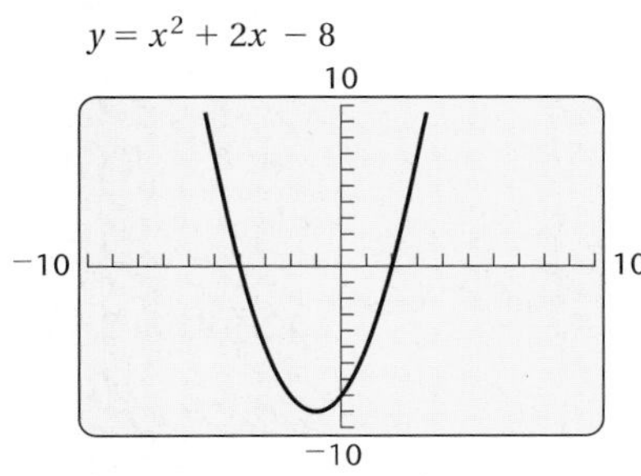

The solutions of the equation are the values of x for which $x^2 + 2x - 8 = 0$. These are also the first coordinates of the x-intercepts of the graph. We use the ZERO feature from the CALC menu to find these numbers. To find the solution corresponding to the leftmost x-intercept, we first press [2nd] [CALC] [2] to select the ZERO feature. The prompt "Left Bound?" appears. Next, we use the [◁] or the [▷] key to move the cursor to the left of the intercept and press [ENTER]. Now the prompt "Right Bound?" appears. Then we move the cursor to the right of the intercept and press [ENTER]. The prompt "Guess?" appears. We move the cursor close to the intercept and press [ENTER] again. We now see the cursor positioned at the leftmost x-intercept and the coordinates of that point, $x = -4$, $y = 0$, are displayed. Thus, $x^2 + 2x - 8 = 0$ when $x = -4$. This is one solution of the equation.

We can repeat this procedure to find the first coordinate of the other x-intercept. We see that $x = 2$ at that point. Thus the solutions of the equation are -4 and 2.

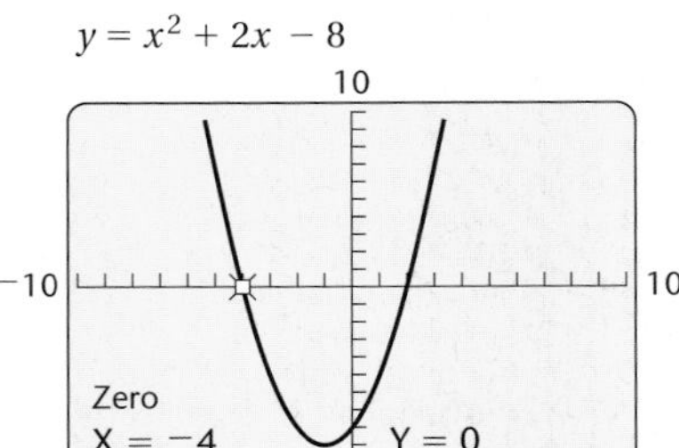

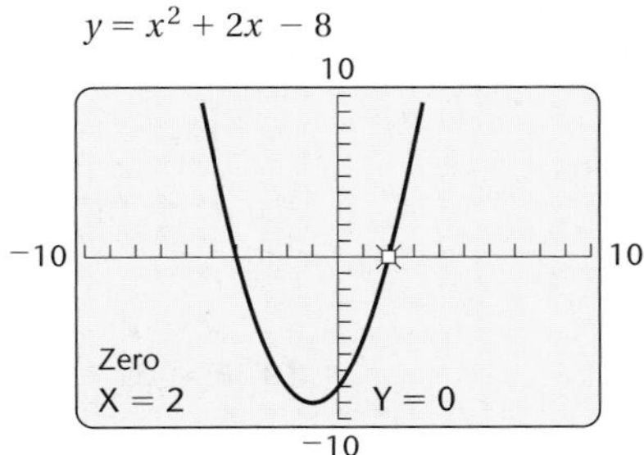

Exercises:

1. Solve each of the equations in Examples 4–8 graphically.

5.7 EXERCISE SET

For Extra Help

Digital Video Tutor CD 5 Videotape 11 | InterAct Math | Math Tutor Center | MathXL | MyMathLab

a Solve using the principle of zero products.

1. $(x + 4)(x + 9) = 0$

2. $(x + 2)(x - 7) = 0$

3. $(x + 3)(x - 8) = 0$

4. $(x + 6)(x - 8) = 0$

5. $(x + 12)(x - 11) = 0$

6. $(x - 13)(x + 53) = 0$

7. $x(x + 3) = 0$

8. $y(y + 5) = 0$

9. $0 = y(y + 18)$

10. $0 = x(x - 19)$

11. $(2x + 5)(x + 4) = 0$

12. $(2x + 9)(x + 8) = 0$

13. $(5x + 1)(4x - 12) = 0$

14. $(4x + 9)(14x - 7) = 0$

15. $(7x - 28)(28x - 7) = 0$

16. $(13x + 14)(6x - 5) = 0$

17. $2x(3x - 2) = 0$

18. $55x(8x - 9) = 0$

19. $\left(\frac{1}{5} + 2x\right)\left(\frac{1}{9} - 3x\right) = 0$

20. $\left(\frac{7}{4}x - \frac{1}{16}\right)\left(\frac{2}{3}x - \frac{16}{15}\right) = 0$

21. $(0.3x - 0.1)(0.05x + 1) = 0$

22. $(0.1x + 0.3)(0.4x - 20) = 0$

23. $9x(3x - 2)(2x - 1) = 0$

24. $(x + 5)(x - 75)(5x - 1) = 0$

b Solve by factoring and using the principle of zero products. Remember to check.

25. $x^2 + 6x + 5 = 0$

26. $x^2 + 7x + 6 = 0$

27. $x^2 + 7x - 18 = 0$

28. $x^2 + 4x - 21 = 0$

29. $x^2 - 8x + 15 = 0$

30. $x^2 - 9x + 14 = 0$

31. $x^2 - 8x = 0$

32. $x^2 - 3x = 0$

33. $x^2 + 18x = 0$

34. $x^2 + 16x = 0$

35. $x^2 = 16$

36. $100 = x^2$

37. $9x^2 - 4 = 0$

38. $4x^2 - 9 = 0$

39. $0 = 6x + x^2 + 9$

40. $0 = 25 + x^2 + 10x$

41. $x^2 + 16 = 8x$

42. $1 + x^2 = 2x$

43. $5x^2 = 6x$

44. $7x^2 = 8x$

45. $6x^2 - 4x = 10$

46. $3x^2 - 7x = 20$

47. $12y^2 - 5y = 2$

48. $2y^2 + 12y = -10$

49. $t(3t + 1) = 2$

50. $x(x - 5) = 14$

51. $100y^2 = 49$

52. $64a^2 = 81$

53. $x^2 - 5x = 18 + 2x$

54. $3x^2 + 8x = 9 + 2x$

55. $10x^2 - 23x + 12 = 0$

56. $12x^2 + 17x - 5 = 0$

Find the x-intercepts for the graph of the equation. (The grids are intentionally not included.)

57.

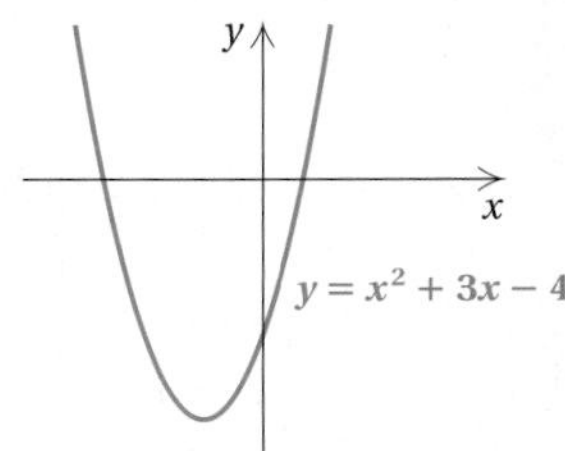

58.

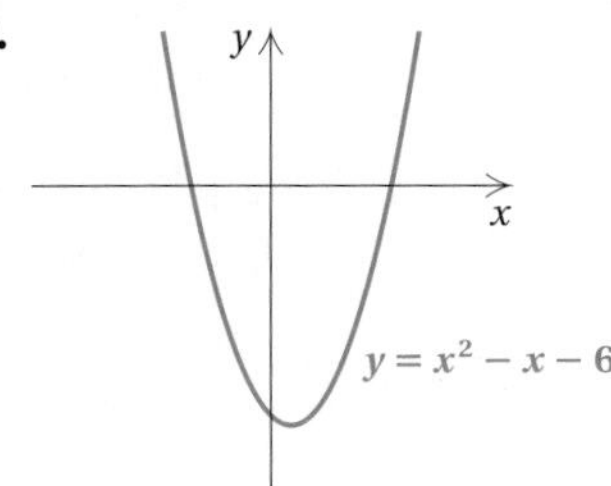

59.

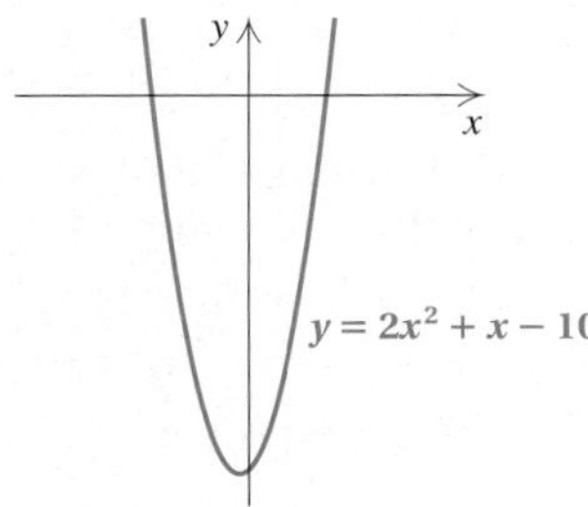

60.

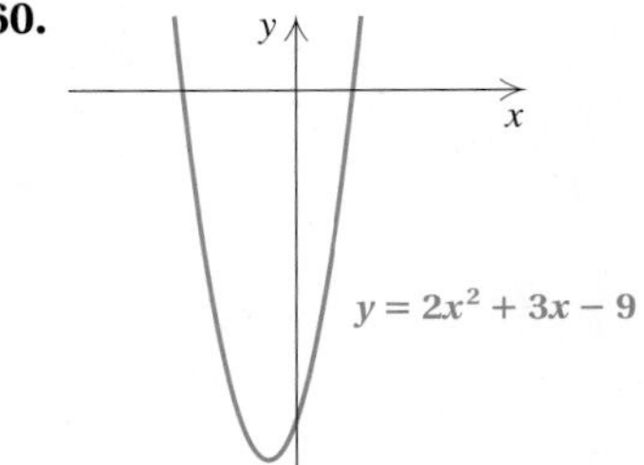

61.

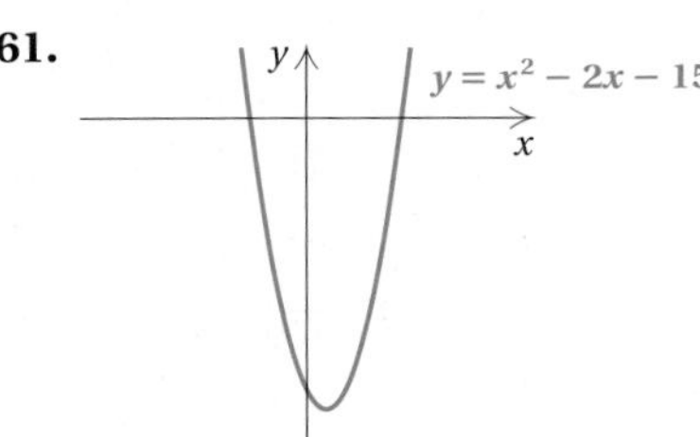

62.

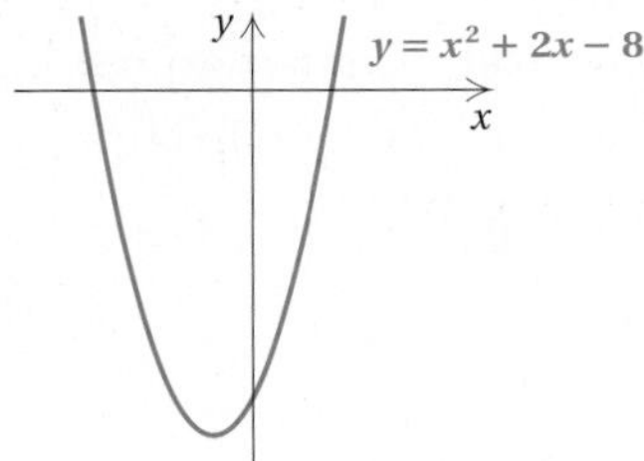

63. Use the following graph to solve $x^2 - 3x - 4 = 0$.

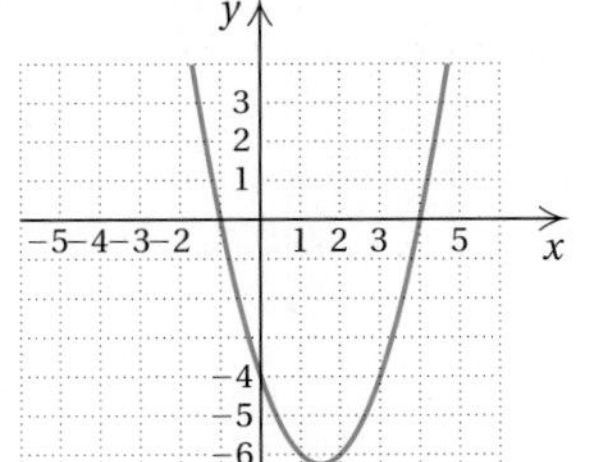

64. Use the following graph to solve $x^2 + x - 6 = 0$.

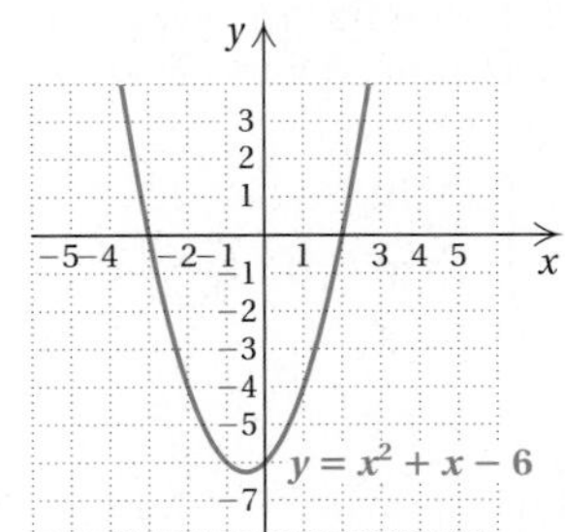

65. Use the following graph to solve $-x^2 + 2x + 3 = 0$.

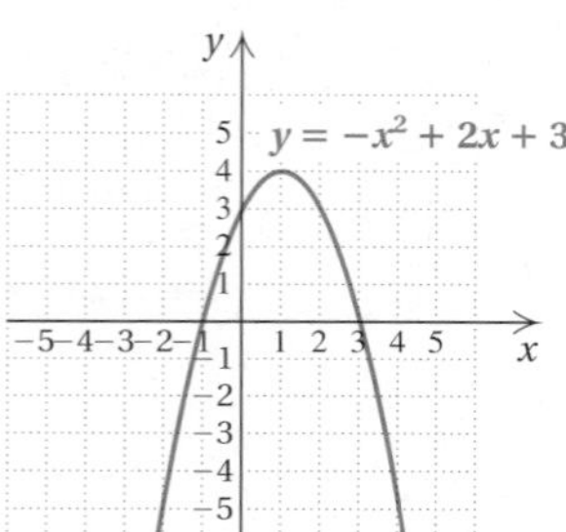

66. Use the following graph to solve $-x^2 - x + 6 = 0$.

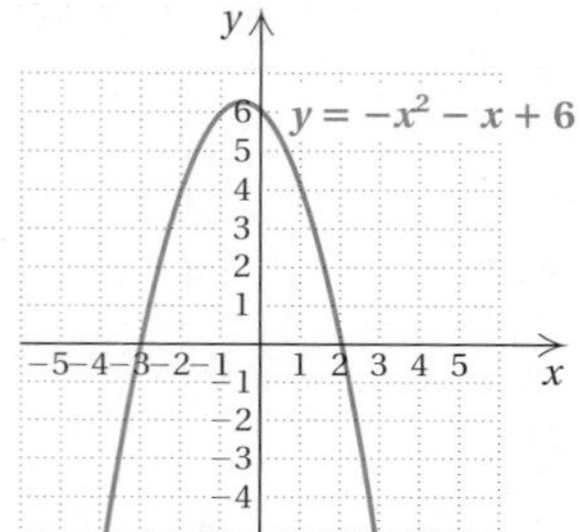

67. D$_W$ What is wrong with the following? Explain the correct method of solution.

$$(x - 3)(x + 4) = 8$$
$$x - 3 = 8 \quad or \quad x + 4 = 8$$
$$x = 11 \quad or \quad x = 4$$

68. D$_W$ What is incorrect about solving $x^2 = 3x$ by dividing both sides by x?

SKILL MAINTENANCE

Translate to an algebraic expression. [1.1b]

69. The square of the sum of a and b

70. The sum of the squares of a and b

Divide. [1.6c]

71. $144 \div (-9)$

72. $-24.3 \div 5.4$

73. $-\frac{5}{8} \div \frac{3}{16}$

74. $-\frac{3}{16} \div \left(-\frac{5}{8}\right)$

SYNTHESIS

Solve.

75. $b(b + 9) = 4(5 + 2b)$

76. $y(y + 8) = 16(y - 1)$

77. $(t - 3)^2 = 36$

78. $(t - 5)^2 = 2(5 - t)$

79. $x^2 - \frac{1}{64} = 0$

80. $x^2 - \frac{25}{36} = 0$

81. $\frac{5}{16}x^2 = 5$

82. $\frac{27}{25}x^2 = \frac{1}{3}$

83. Find an equation that has the given numbers as solutions. For example, 3 and -2 are solutions to $x^2 - x - 6 = 0$.

a) $-3, 4$ **b)** $-3, -4$ **c)** $\frac{1}{2}, \frac{1}{2}$

d) $5, -5$ **e)** $0, 0.1, \frac{1}{4}$

84. *Matching.* Match each equation in the first column with the equivalent equation in the second column.

$x^2 + 10x - 2 = 0$	$4x^2 + 8x + 36 = 0$
$(x - 6)(x + 3) = 0$	$(2x + 8)(2x - 5) = 0$
$5x^2 - 5 = 0$	$9x^2 - 12x + 24 = 0$
$(2x - 5)(x + 4) = 0$	$(x + 1)(5x - 5) = 0$
$x^2 + 2x + 9 = 0$	$x^2 - 3x - 18 = 0$
$3x^2 - 4x + 8 = 0$	$2x^2 + 20x - 4 = 0$

Use a graphing calculator to find the solutions of each equation. Round solutions to the nearest hundredth.

85. $x^2 - 9.10x + 15.77 = 0$

86. $x^2 + 1.80x - 5.69 = 0$

87. $x^2 + 13.74x + 42.00 = 0$

88. $-x^2 + 0.63x + 0.22 = 0$

89. $0.84x^2 - 2.30x = 0$

90. $6.4x^2 - 8.45x - 94.06 = 0$

5.8 APPLICATIONS OF QUADRATIC EQUATIONS

a Applied Problems, Quadratic Equations, and Factoring

We can now use our new method for solving quadratic equations and the five steps for solving problems.

EXAMPLE 1 *Manufacturing.* Wooden Work, Ltd., builds cutting boards that are twice as long as they are wide. The most popular board that Wooden Work makes has an area of 800 cm^2. What are the dimensions of the board?

1. **Familiarize.** We first make a drawing. Recall that the area of any rectangle is Length · Width. We let $x =$ the width of the board, in centimeters. The length is then $2x$.

2. **Translate.** We reword and translate as follows:

Rewording: The area of the rectangle is 800 cm^2.

Translating: $2x \cdot x = 800$

3. **Solve.** We solve the equation as follows:

$$2x \cdot x = 800$$
$$2x^2 = 800$$
$$2x^2 - 800 = 0 \quad \text{Subtracting 800 to get 0 on one side}$$
$$2(x^2 - 400) = 0 \quad \text{Removing a common factor of 2}$$
$$2(x - 20)(x + 20) = 0 \quad \text{Factoring a difference of squares}$$
$$(x - 20)(x + 20) = 0 \quad \text{Dividing by 2}$$
$$x - 20 = 0 \quad or \quad x + 20 = 0 \quad \text{Using the principle of zero products}$$
$$x = 20 \quad or \quad x = -20. \quad \text{Solving each equation}$$

4. **Check.** The solutions of the equation are 20 and -20. Since the width must be positive, -20 cannot be a solution. To check 20 cm, we note that if the width is 20 cm, then the length is $2 \cdot 20$ cm $= 40$ cm and the area is 20 cm · 40 cm $= 800$ cm^2. Thus the solution 20 checks.

5. **State.** The cutting board is 20 cm wide and 40 cm long.

Do Exercise 1.

Objective

a Solve applied problems involving quadratic equations that can be solved by factoring.

1. **Framing.** A rectangular picture frame is twice as long as it is wide. If the area of the frame is 288 in^2, find its dimensions.

Answer on page A-27

2. **Dimensions of a Sail.** The mainsail of Stacey's lightning-styled sailboat has an area of 125 ft^2. The sail is 15 ft taller than it is wide. Find the height and the width of the sail.

Answer on page A-27

EXAMPLE 2 *Racing Sailboat.* The height of a triangular sail on a racing sailboat is 9 ft more than the base. The area of the triangle is 110 ft^2. Find the height and the base of the sail.

Source: Whitney Gladstone, North Graphics, San Diego, CA

1. **Familiarize.** We first make a drawing. If you don't remember the formula for the area of a triangle, look it up on the inside front cover of this book or in a geometry book. The area is $\frac{1}{2}$(base)(height).

   We let b = the base of the triangle. Then $b + 9$ = the height.

2. **Translate.** It helps to reword this problem before translating:

$\frac{1}{2}$	times	Base	times	Height	is	110.	Rewording
↓	↓	↓	↓	↓	↓	↓	
$\frac{1}{2}$	$\cdot$	b	$\cdot$	$(b + 9)$	$=$	110	Translating

3. **Solve.** We solve the equation as follows:

$$\frac{1}{2} \cdot b \cdot (b + 9) = 110$$

$$\frac{1}{2}(b^2 + 9b) = 110 \qquad \text{Multiplying}$$

$$2 \cdot \frac{1}{2}(b^2 + 9b) = 2 \cdot 110 \qquad \text{Multiplying by 2}$$

$$b^2 + 9b = 220 \qquad \text{Simplifying}$$

$$b^2 + 9b - 220 = 220 - 220 \qquad \text{Subtracting 220 to get 0 on one side}$$

$$b^2 + 9b - 220 = 0$$

$$(b - 11)(b + 20) = 0 \qquad \text{Factoring}$$

$$b - 11 = 0 \quad or \quad b + 20 = 0 \qquad \text{Using the principle of zero products}$$

$$b = 11 \quad or \quad b = -20.$$

4. **Check.** The base of a triangle cannot have a negative length, so -20 cannot be a solution. Suppose the base is 11 ft. The height is 9 ft more than the base, so the height is 20 ft and the area is $\frac{1}{2}(11)(20)$, or 110 ft^2. These numbers check in the original problem.

5. **State.** The height is 20 ft and the base is 11 ft.

Do Exercise 2.

EXAMPLE 3 *Games in a Sports League.* In a sports league of x teams in which each team plays every other team twice, the total number N of games to be played is given by

$$x^2 - x = N.$$

Maggie's basketball league plays a total of 240 games. How many teams are in the league?

1., 2. Familiarize and **Translate.** We are given that x is the number of teams in a league and N is the number of games. To familiarize yourself with this problem, reread Example 4 in Section 4.3 where we first considered it. To find the number of teams x in a league in which 240 games are played, we substitute 240 for N in the equation:

$$x^2 - x = 240. \qquad \text{Substituting 240 for } N$$

3. Solve. We solve the equation as follows:

$$x^2 - x = 240$$

$$x^2 - x - 240 = 240 - 240 \qquad \text{Subtracting 240 to get 0 on one side}$$

$$x^2 - x - 240 = 0$$

$$(x - 16)(x + 15) = 0 \qquad \text{Factoring}$$

$$x - 16 = 0 \quad or \quad x + 15 = 0 \qquad \text{Using the principle of zero products}$$

$$x = 16 \quad or \quad x = -15.$$

4. Check. The solutions of the equation are 16 and -15. Since the number of teams cannot be negative, -15 cannot be a solution. But 16 checks, since $16^2 - 16 = 256 - 16 = 240$.

5. State. There are 16 teams in the league.

Do Exercise 3.

3. Use $N = x^2 - x$ for the following.

a) **Volleyball League.** Amy's volleyball league has 19 teams. What is the total number of games to be played?

b) **Softball League.** Barry's slow-pitch softball league plays a total of 72 games. How many teams are in the league?

Answers on page A-27

Study Tips

FIVE STEPS FOR PROBLEM SOLVING

Are you remembering to use the five steps for problem solving that were developed in Section 2.6?

1. **Familiarize** yourself with the situation.
   a) Carefully read and reread until you understand *what* you are being asked to find.
   b) Draw a diagram or see if there is a formula that applies.
   c) Assign a letter, or *variable,* to the unknown.
2. **Translate** the problem to an equation using the letter or variable.
3. **Solve** the equation.
4. **Check** the answer in the original wording of the problem.
5. **State** the answer to the problem clearly with appropriate units.

"Most worthwhile achievements are the result of many little things done in a simple direction."

Nido Quebin, speaker/entrepreneur

4. **Page Numbers.** The product of the page numbers on two facing pages of a book is 506. Find the page numbers.

Answer on page A-27

Study Tips

ASKING QUESTIONS

Don't be afraid to ask questions in class. Most instructors welcome this and encourage students to ask them. Other students probably have the same questions you do.

"Better to ask twice than lose your way once."

Danish proverb

EXAMPLE 4 *Athletic Numbers.* The product of the numbers of two consecutive entrants in a marathon race is 156. Find the numbers.

1. **Familiarize.** The numbers are consecutive integers. Recall that consecutive integers are next to each other, such as 49 and 50, or -6 and -5. Let $x =$ the smaller integer; then $x + 1 =$ the larger integer.
2. **Translate.** It helps to reword the problem before translating:

First integer	times	Second integer	is	156.	Rewording
↓	↓	↓	↓	↓	
x	$\cdot$	$(x + 1)$	$=$	156	Translating

3. **Solve.** We solve the equation as follows:

$$x(x + 1) = 156$$
$$x^2 + x = 156 \qquad \text{Multiplying}$$
$$x^2 + x - 156 = 156 - 156 \qquad \text{Subtracting 156 to get 0 on one side}$$
$$x^2 + x - 156 = 0 \qquad \text{Simplifying}$$
$$(x - 12)(x + 13) = 0 \qquad \text{Factoring}$$
$$x - 12 = 0 \quad or \quad x + 13 = 0 \qquad \text{Using the principle of zero products}$$
$$x = 12 \quad or \quad x = -13.$$

4. **Check.** The solutions of the equation are 12 and -13. When x is 12, then $x + 1$ is 13, and $12 \cdot 13 = 156$. The numbers 12 and 13 are consecutive integers that are solutions to the problem. When x is -13, then $x + 1$ is -12, and $(-13)(-12) = 156$. The numbers -13 and -12 are consecutive integers, but they are not solutions of the problem because negative numbers are not used as entry numbers.
5. **State.** The entry numbers are 12 and 13.

Do Exercise 4.

b The Pythagorean Theorem

The following problems involve the Pythagorean theorem, which relates the lengths of the sides of a *right* triangle. A triangle is a **right triangle** if it has a 90°, or *right,* angle. The side opposite the 90° angle is called the **hypotenuse**. The other sides are called **legs**.

THE PYTHAGOREAN THEOREM

In any right triangle, if a and b are the lengths of the legs and c is the length of the hypotenuse, then

$$a^2 + b^2 = c^2.$$

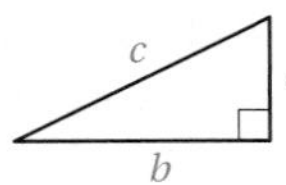

The symbol ⌐ denotes a 90° angle.

EXAMPLE 5 *Lookout Tower.* The diagonal braces in a lookout tower are 15 ft long and span a distance of 12 ft. How high does each brace reach vertically?

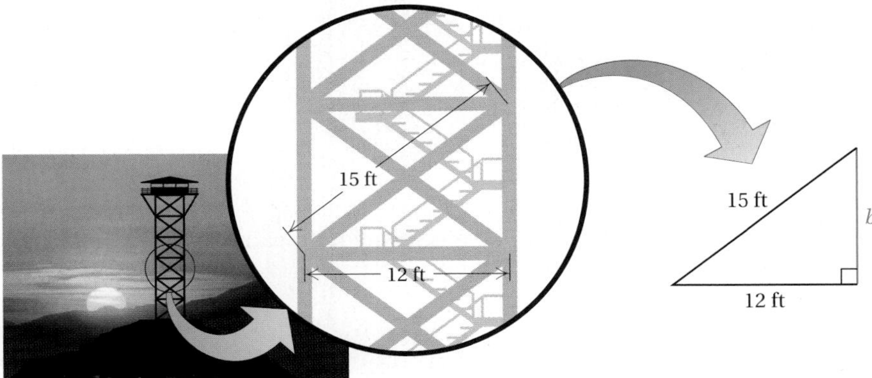

1. **Familiarize.** We make a drawing as shown above. We let b = the height of the vertical part of the brace.
2. **Translate.** Since a right triangle is formed, we can use the Pythagorean theorem:

$$a^2 + b^2 = c^2$$

$$12^2 + b^2 = 15^2. \quad \text{Substituting}$$

3. **Solve.** We solve the equation as follows:

$$12^2 + b^2 = 15^2$$

$$144 + b^2 = 225 \quad \text{Squaring 12 and 15}$$

$$b^2 - 81 = 0 \quad \text{Subtracting 225}$$

$$(b - 9)(b + 9) = 0 \quad \text{Factoring}$$

$$b - 9 = 0 \quad or \quad b + 9 = 0 \quad \text{Using the principle of zero products}$$

$$b = 9 \quad or \quad b = -9.$$

4. **Check.** Since the height cannot be negative, -9 cannot be a solution. If the height is 9 ft, we have $12^2 + 9^2 = 144 + 81 = 225$, which is 15^2. Thus, 9 checks and is a solution.
5. **Solve.** The vertical height of the brace is 9 ft.

Do Exercise 5.

5. **Reach of a Ladder.** Twila has a 26-ft ladder leaning against her house. If the bottom of the ladder is 10 ft from the base of the house, how high does the ladder reach?

Answer on page A-27

EXAMPLE 6 *Ladder Settings.* A ladder of length 13 ft is placed against a building in such a way that the distance from the top of the ladder to the ground is 7 ft more than the distance from the bottom of the ladder to the building. Find both distances.

1. **Familiarize.** We first make a drawing. The ladder and the missing distances form the hypotenuse and legs of a right triangle. We let $x =$ the length of the side (leg) across the bottom. Then $x + 7 =$ the length of the other side (leg). The hypotenuse has length 13 ft.

2. **Translate.** Since a right triangle is formed, we can use the Pythagorean theorem:

$$a^2 + b^2 = c^2$$
$$x^2 + (x + 7)^2 = 13^2. \quad \text{Substituting}$$

3. **Solve.** We solve the equation as follows:

$$x^2 + (x^2 + 14x + 49) = 169 \quad \text{Squaring the binomial and 13}$$
$$2x^2 + 14x + 49 = 169 \quad \text{Collecting like terms}$$
$$2x^2 + 14x + 49 - 169 = 169 - 169 \quad \text{Subtracting 169 to get 0 on one side}$$
$$2x^2 + 14x - 120 = 0 \quad \text{Simplifying}$$
$$2(x^2 + 7x - 60) = 0 \quad \text{Factoring out a common factor}$$
$$x^2 + 7x - 60 = 0 \quad \text{Dividing by 2}$$
$$(x + 12)(x - 5) = 0 \quad \text{Factoring}$$
$$x + 12 = 0 \quad or \quad x - 5 = 0 \quad \text{Using the principle of zero products}$$
$$x = -12 \quad or \quad x = 5.$$

4. **Check.** The negative integer -12 cannot be the length of a side. When $x = 5$, $x + 7 = 12$, and $5^2 + 12^2 = 13^2$. So 5 and 12 check.

5. **State.** The distance from the top of the ladder to the ground is 12 ft. The distance from the bottom of the ladder to the building is 5 ft.

Do Exercise 6.

6. **Right-Triangle Geometry.** The length of one leg of a right triangle is 1 m longer than the other. The length of the hypotenuse is 5 m. Find the lengths of the legs.

Answer on page A-27

For Extra Help

Digital Video Tutor CD 5 Videotape 11 | InterAct Math | Math Tutor Center | MathXL | MyMathLab

a Solve.

1. *Furnishings.* A rectangular table in Arlo's House of Tunes is six times as long as it is wide. The area of the table is 24 ft^2. Find the length and the width of the table.

2. *Framing.* A rectangular picture frame is three times as long as it is wide. The area of the frame is 588 in^2. Find the dimensions of the frame.

3. *Design.* The keypad and viewing window of the TI83 graphing calculator is rectangular. The length of the rectangle is 2 cm more than twice the width, and the area of the rectangle is 144 cm^2. Find the length and the width.

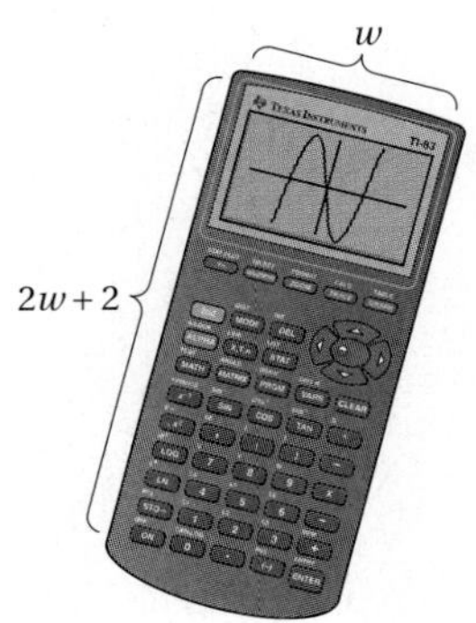

4. *Area of a Garden.* The length of a rectangular garden is 4 m greater than the width. The area of the garden is 96 m^2. Find the length and the width.

5. *Dimensions of a Triangle.* A triangle is 10 cm wider than it is tall. The area is 28 cm^2. Find the height and the base.

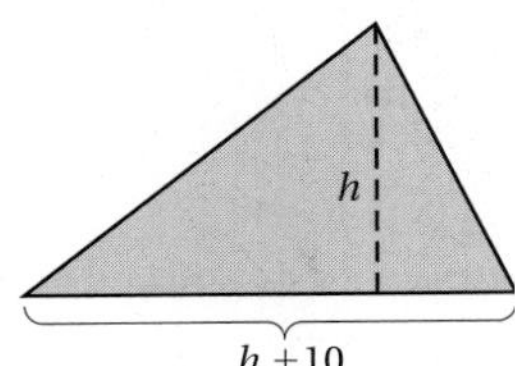

6. *Dimensions of a Triangle.* The height of a triangle is 3 cm less than the length of the base. The area of the triangle is 35 cm^2. Find the height and the length of the base.

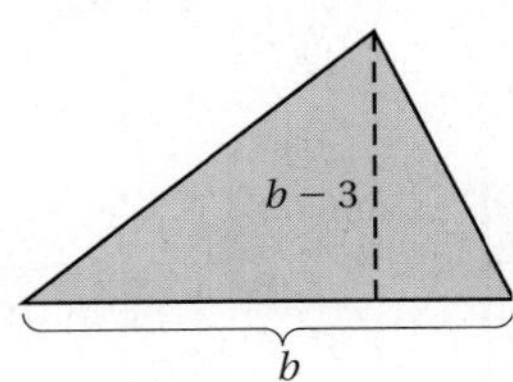

7. *Road Design.* A triangular traffic island has a base half as long as its height. The island has an area of 64 m^2. Find the base and the height.

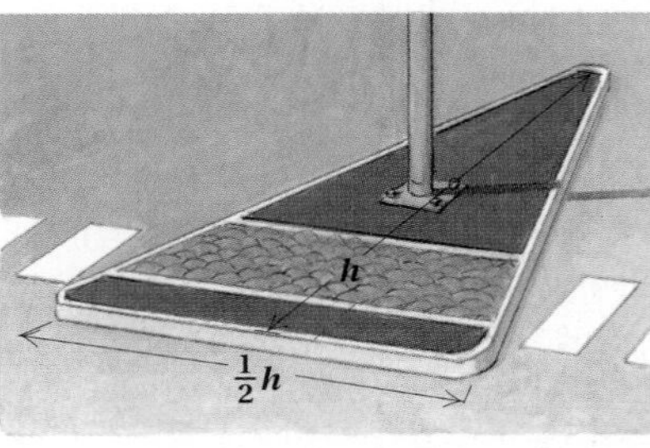

8. *Dimensions of a Sail.* The height of the jib sail on a Lightning sailboat is 5 ft greater than the length of its "foot." The area of the sail is 42 ft^2. Find the length of the foot and the height of the sail.

Games in a League. Use $x^2 - x = N$ for Exercises 9–12.

9. A chess league has 14 teams. What is the total number of games to be played?

10. A women's volleyball league has 23 teams. What is the total number of games to be played?

11. A slow-pitch softball league plays a total of 132 games. How many teams are in the league?

12. A basketball league plays a total of 90 games. How many teams are in the league?

Handshakes. A researcher wants to investigate the potential spread of germs by contact. She knows that the number of possible handshakes within a group of x people is given by

$$N = \frac{1}{2}(x^2 - x).$$

13. There are 100 people at a party. How many handshakes are possible?

14. There are 40 people at a meeting. How many handshakes are possible?

15. Everyone at a meeting shook hands. There were 300 handshakes in all. How many people were at the meeting?

16. Everyone at a party shook hands. There were 153 handshakes in all. How many people were at the party?

17. *Toasting.* During a toast at a party, there were 190 "clicks" of glasses. How many people took part in the toast?

18. *High-fives.* After winning the championship, all Los Angeles Laker teammates exchanged "high-fives." Altogether there were 66 high-fives. How many players were there?

19. *Consecutive Page Numbers.* The product of the page numbers on two facing pages of a book is 210. Find the page numbers.

20. *Consecutive Page Numbers.* The product of the page numbers on two facing pages of a book is 420. Find the page numbers.

21. The product of two consecutive even integers is 168. Find the integers. (See Section 2.6.)

22. The product of two consecutive even integers is 224. Find the integers. (See Section 2.6.)

23. The product of two consecutive odd integers is 255. Find the integers.

24. The product of two consecutive odd integers is 143. Find the integers.

25. *Right-Triangle Geometry.* The length of one leg of a right triangle is 8 ft. The length of the hypotenuse is 2 ft longer than the other leg. Find the length of the hypotenuse and the other leg.

26. *Right-Triangle Geometry.* The length of one leg of a right triangle is 24 ft. The length of the other leg is 16 ft shorter than the hypotenuse. Find the length of the hypotenuse and the other leg.

27. *Roadway Design.* Elliott Street is 24 ft wide when it ends at Main Street in Brattleboro, Vermont. A 40-ft long diagonal crosswalk allows pedestrians to cross Main Street to or from either corner of Elliott Street (see the figure). Determine the width of Main Street.

28. *Sailing.* The mainsail of a Lightning sailboat is a right triangle in which the hypotenuse is called the leech. If a 24-ft tall mainsail has a leech length of 26 ft and if Dacron® sailcloth costs $10 per square foot, find the cost of a new mainsail.

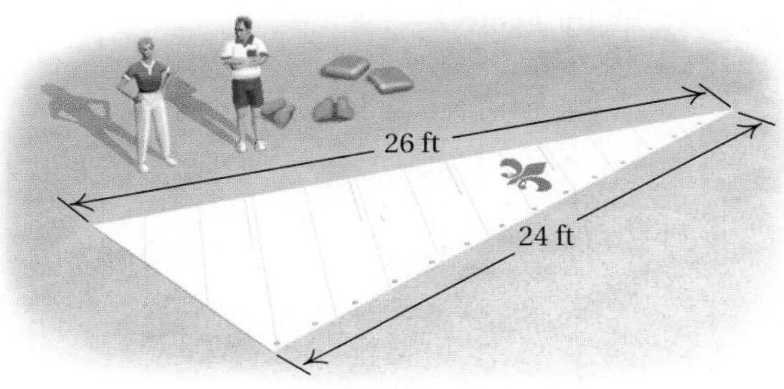

29. *Physical Education.* An outdoor-education ropes course includes a cable that slopes downward from a height of 37 ft to a height of 30 ft. The trees that the cable connects are 24 ft apart. How long is the cable?

30. *Aviation.* Engine failure forced Geraldine to pilot her Cessna 150 to an emergency landing. To land, Geraldine's plane glided 17,000 ft over a 15,000-ft stretch of deserted highway. From what altitude did the descent begin?

31. *Architecture.* An architect has allocated a rectangular space of 264 ft^2 for a square dining room and a 10-ft wide kitchen, as shown in the figure. Find the dimensions of each room.

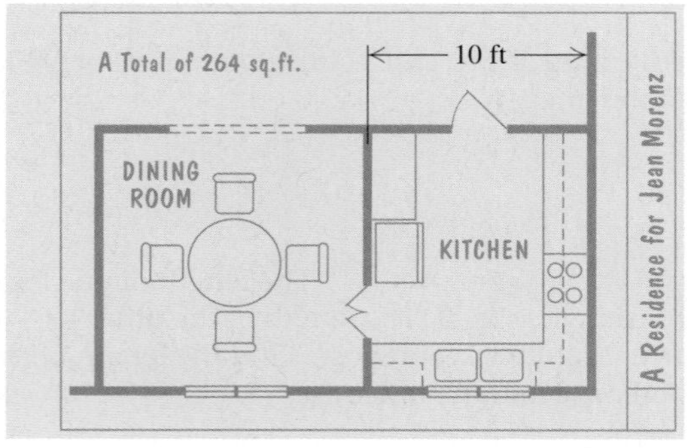

32. *Guy Wire.* The guy wire on a TV antenna is 1 m longer than the height of the antenna. If the guy wire is anchored 3 m from the foot of the antenna, how tall is the antenna?

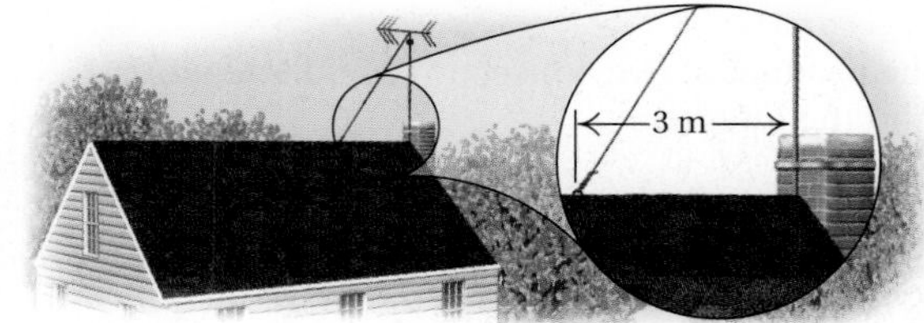

Rocket Launch. A model water rocket is launched with an initial velocity of 180 ft/sec. Its height h, in feet, after t seconds is given by the formula

$$h = 180t - 16t^2.$$

Use this formula for Exercises 33 and 34.

33. After how many seconds will the rocket first reach a height of 464 ft?

34. After how many seconds from launching will the rocket again be at that same height of 464 ft? (See Exercise 33.)

35. The sum of the squares of two consecutive odd positive integers is 74. Find the integers.

36. The sum of the squares of two consecutive odd positive integers is 130. Find the integers.

37. $^{\mathrm{D}}_{\mathrm{W}}$ An archaeologist has measuring sticks of 3 ft, 4 ft, and 5 ft. Explain how she could draw a 7-ft by 9-ft rectangle on a piece of land being excavated.

38. $^{\mathrm{D}}_{\mathrm{W}}$ Write a problem for a classmate to solve such that only one of the two solutions of a quadratic equation can be used as an answer.

SKILL MAINTENANCE

Multiply. [4.6d], [4.7f]

39. $(3x - 5y)(3x + 5y)$

40. $(3x - 5y)^2$

41. $(3x + 5y)^2$

42. $(3x - 5y)(2x + 7y)$

Find the intercepts of the equation. [3.3a]

43. $4x - 16y = 64$

44. $4x + 16y = 64$

45. $x - 1.3y = 6.5$

46. $\frac{2}{3}x + \frac{5}{8}y = \frac{5}{12}$

47. $y = 4 - 5x$

48. $y = 2x - 5$

SYNTHESIS

49. *Telephone Service.* Use the information in the figure below to determine the height of the telephone pole.

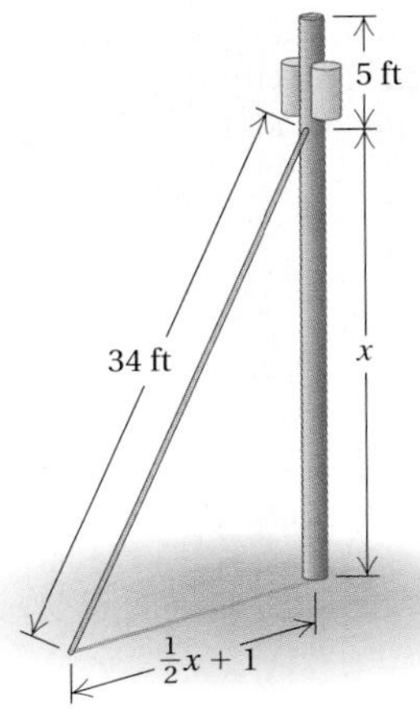

50. *Roofing.* A *square* of shingles covers 100 ft^2 of surface area. How many squares will be needed to reshingle the house shown?

51. *Pool Sidewalk.* A cement walk of constant width is built around a 20-ft by 40-ft rectangular pool. The total area of the pool and the walk is 1500 ft^2. Find the width of the walk.

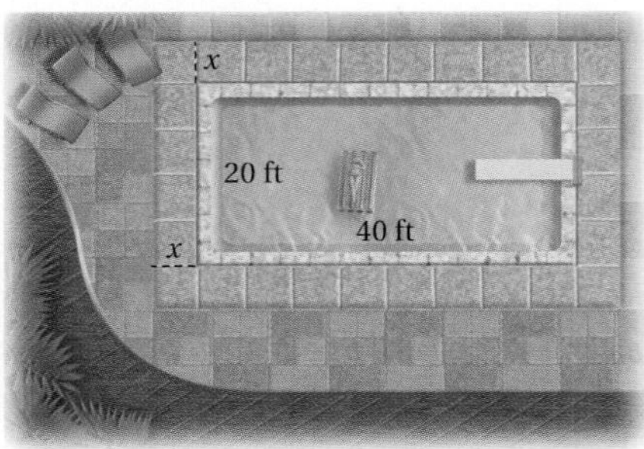

52. *Rain-gutter Design.* An open rectangular gutter is made by turning up the sides of a piece of metal 20 in. wide. The area of the cross-section of the gutter is 50 in^2. Find the depth of the gutter.

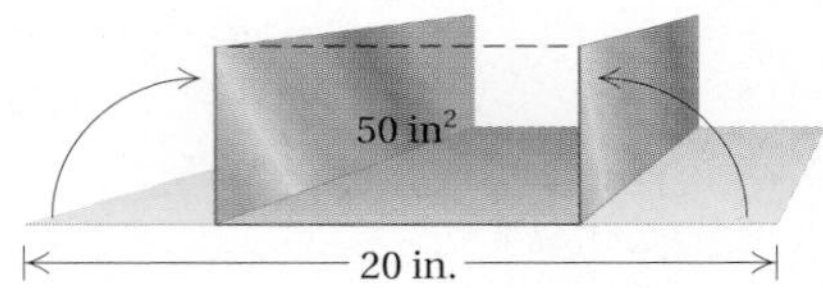

53. *Dimensions of an Open Box.* A rectangular piece of cardboard is twice as long as it is wide. A 4-cm square is cut out of each corner, and the sides are turned up to make a box with an open top. The volume of the box is 616 cm^3. Find the original dimensions of the cardboard.

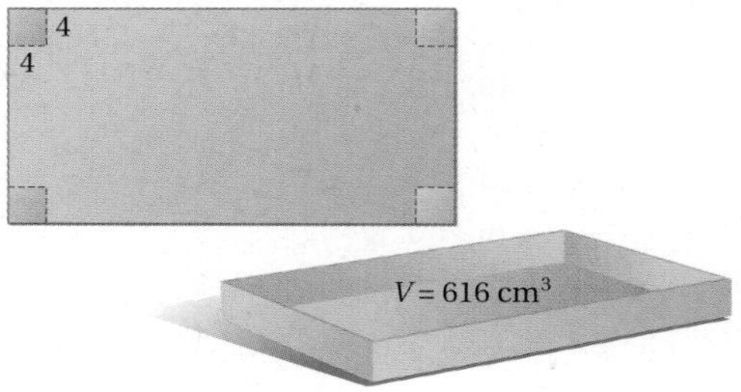

54. *Dimensions of a Closed Box.* The total surface area of a closed box is 350 m^2. The box is 9 m high and has a square base and lid. Find the length of a side of the base.

55. Solve for x.

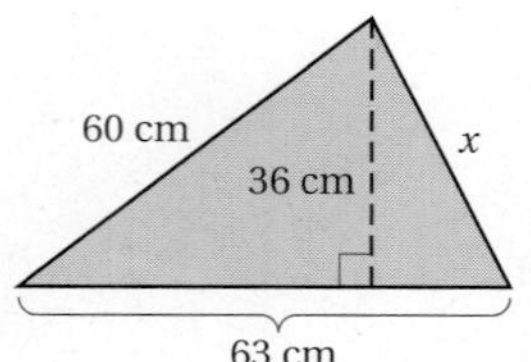

56. The ones digit of a number less than 100 is 4 greater than the tens digit. The sum of the number and the product of the digits is 58. Find the number.

CHAPTER 5

Summary and Review

The review that follows is meant to prepare you for a chapter exam. It consists of two parts. The first part is a checklist of some of the Study Tips referred to in this and preceding chapters, as well as a list of important properties and formulas. The second part is the Review Exercises. These provide practice exercises for the exam, together with references to section objectives so you can go back and review. Before beginning, stop and look back over the skills you have obtained. What skills in mathematics do you have now that you did not have before studying this chapter?

STUDY TIPS CHECKLIST

The foundation of all your study skills is TIME!

- ☐ Are you staying on schedule and on time for class and adapting your study time and class schedule to your personality?
- ☐ Did you study the examples in this chapter carefully?
- ☐ Did you use the five steps for problem solving as you did the applications in Section 5.8?
- ☐ Are you asking questions at appropriate times in class and with your tutors?
- ☐ Are you doing exercises without answers as part of every homework assignment to prepare you for tests?

IMPORTANT PROPERTIES AND FORMULAS

Factoring Formulas: $A^2 - B^2 = (A + B)(A - B)$,
$A^2 + 2AB + B^2 = (A + B)^2$,
$A^2 - 2AB + B^2 = (A - B)^2$

The Principle of Zero Products: An equation $ab = 0$ is true if and only if $a = 0$ is true or $b = 0$ is true, or both are true.

Pythagorean Theorem: $a^2 + b^2 = c^2$

c
a
b

REVIEW EXERCISES

Find three factorizations of the monomial. [5.1a]

1. $-10x^2$

2. $36x^5$

Factor completely. [5.6a]

3. $5 - 20x^6$

4. $x^2 - 3x$

5. $9x^2 - 4$

6. $x^2 + 4x - 12$

7. $x^2 + 14x + 49$

8. $6x^3 + 12x^2 + 3x$

9. $x^3 + x^2 + 3x + 3$

10. $6x^2 - 5x + 1$

11. $x^4 - 81$

12. $9x^3 + 12x^2 - 45x$

13. $2x^2 - 50$

14. $x^4 + 4x^3 - 2x - 8$

15. $16x^4 - 1$

16. $8x^6 - 32x^5 + 4x^4$

17. $75 + 12x^2 + 60x$

18. $x^2 + 9$

19. $x^3 - x^2 - 30x$

20. $4x^2 - 25$

21. $9x^2 + 25 - 30x$

22. $6x^2 - 28x - 48$

23. $x^2 - 6x + 9$

24. $2x^2 - 7x - 4$

25. $18x^2 - 12x + 2$

26. $3x^2 - 27$

27. $15 - 8x + x^2$

28. $25x^2 - 20x + 4$

29. $49b^{10} + 4a^8 - 28a^4b^5$

30. $x^2y^2 + xy - 12$

31. $12a^2 + 84ab + 147b^2$

32. $m^2 + 5m + mt + 5t$

33. $32x^4 - 128y^4z^4$

Solve. [5.7a], [5.7b]

34. $(x - 1)(x + 3) = 0$

35. $x^2 + 2x - 35 = 0$

36. $x^2 + x - 12 = 0$

37. $3x^2 + 2 = 5x$

38. $2x^2 + 5x = 12$

39. $16 = x(x - 6)$

Solve. [5.8a]

40. *Sharks' Teeth.* Sharks' teeth are shaped like triangles. The height of a tooth of a great white shark is 1 cm longer than the base. The area is 15 cm^2. Find the height and the base.

41. The product of two consecutive even integers is 288. Find the integers.

42. The product of two consecutive odd integers is 323. Find the integers.

43. *Antenna Guy Wire.* The guy wires for a television antenna are 2 m longer than the height of the antenna. The guy wires are anchored 4 m from the foot of the antenna. How tall is the antenna?

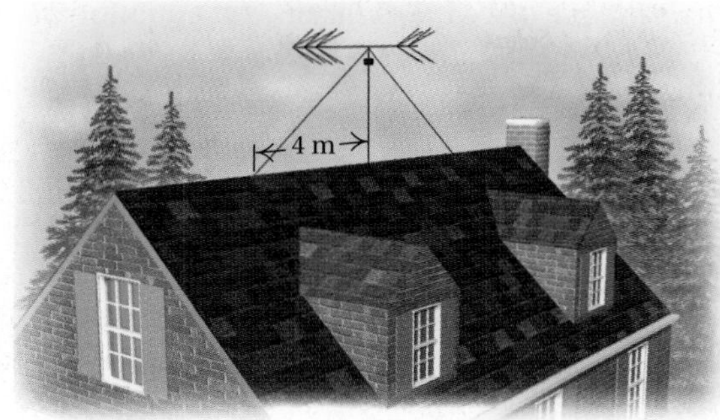

44. If the sides of a square are lengthened by 3 km, the area becomes 81 km^2. Find the length of a side of the original square.

Find the x-intercepts for the graph of the equation. [5.7b]

45. $y = x^2 + 9x + 20$

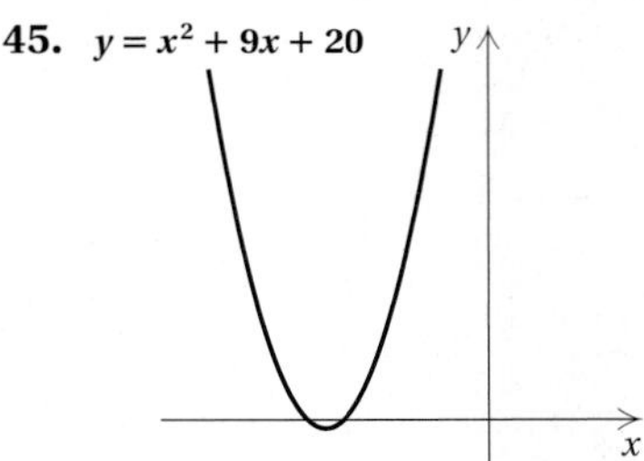

46. $y = 2x^2 - 7x - 15$

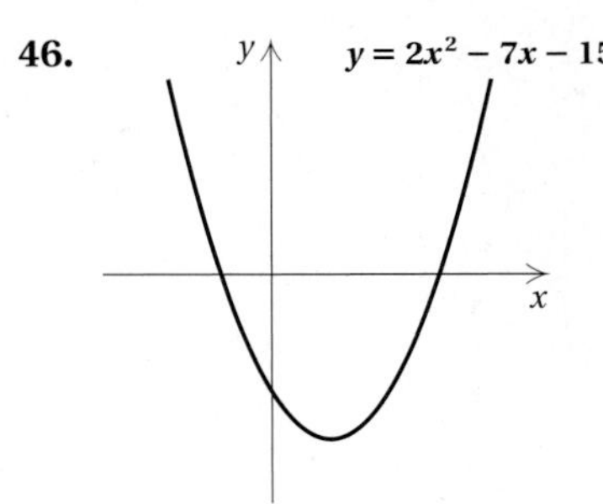

47. $\mathbf{D_W}$ On a quiz, Sheri writes the factorization of $4x^2 - 100$ as $(2x - 10)(2x + 10)$. If this were a 10-point question, how many points would you give Sheri? Why? [5.5d]

48. $\mathbf{D_W}$ How do the equations solved in this chapter differ from those solved in previous chapters? [5.7b]

SKILL MAINTENANCE

Certain objectives from four particular sections will be retested on the chapter test. The objectives are listed with the practice problems that follow.

49. Divide: $-\frac{12}{25} \div \left(-\frac{21}{10}\right)$. [1.6c]

50. Solve: $20 - (3x + 2) \geq 2(x + 5) + x$. [2.7e]

51. Multiply: $(2a - 3)(2a + 3)$. [4.6d]

52. Find the intercepts. Then graph the equation. [3.3a]

$$3y - 4x = -12$$

SYNTHESIS

Solve. [5.8a]

53. The pages of a book measure 15 cm by 20 cm. Margins of equal width surround the printing on each page and constitute one-half of the area of the page. Find the width of the margins.

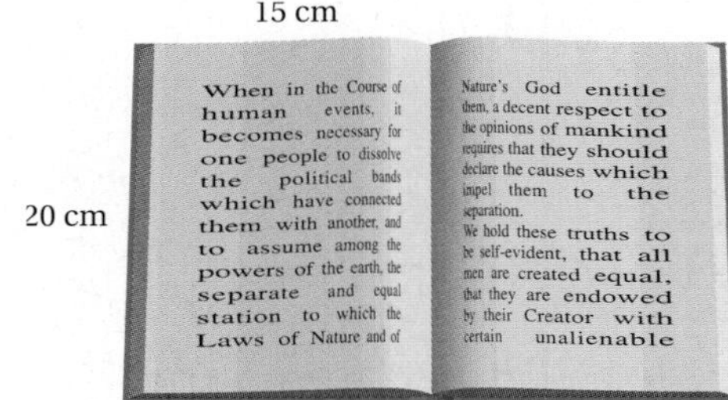

54. The cube of a number is the same as twice the square of the number. Find all such numbers.

55. The length of a rectangle is two times its width. When the length is increased by 20 and the width decreased by 1, the area is 160. Find the original length and width.

Solve. [5.7b]

56. $x^2 + 25 = 0$

57. $(x - 2)(x + 3)(2x - 5) = 0$

58. $(x - 3)4x^2 + 3x(x - 3) - (x - 3)10 = 0$

59. Find a polynomial for the shaded area in the figure below.

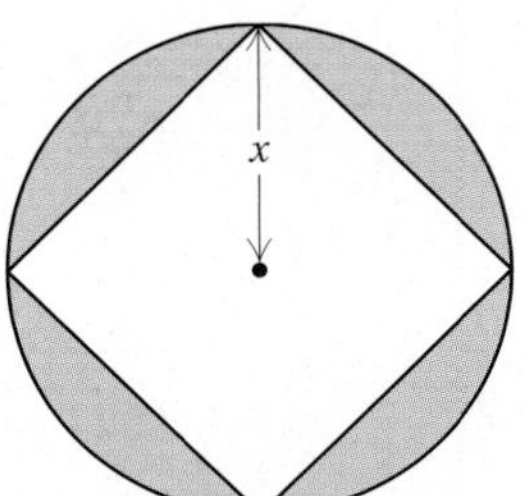

CHAPTER 5

Chapter Test

1. Find three factorizations of $4x^3$.

Factor completely.

2. $x^2 - 7x + 10$

3. $x^2 + 25 - 10x$

4. $6y^2 - 8y^3 + 4y^4$

5. $x^3 + x^2 + 2x + 2$

6. $x^2 - 5x$

7. $x^3 + 2x^2 - 3x$

8. $28x - 48 + 10x^2$

9. $4x^2 - 9$

10. $x^2 - x - 12$

11. $6m^3 + 9m^2 + 3m$

12. $3w^2 - 75$

13. $60x + 45x^2 + 20$

14. $3x^4 - 48$

15. $49x^2 - 84x + 36$

16. $5x^2 - 26x + 5$

17. $x^4 + 2x^3 - 3x - 6$

18. $80 - 5x^4$

19. $4x^2 - 4x - 15$

20. $6t^3 + 9t^2 - 15t$

21. $3m^2 - 9mn - 30n^2$

Solve.

22. $x^2 - x - 20 = 0$

23. $2x^2 + 7x = 15$

24. $x(x - 3) = 28$

Solve.

25. The length of a rectangle is 2 m more than the width. The area of the rectangle is 48 m^2. Find the length and the width.

26. The base of a triangle is 6 cm greater than twice the height. The area is 28 cm^2. Find the height and the base.

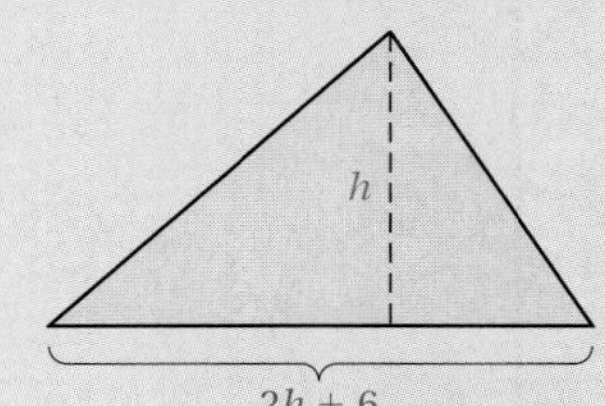

27. *Masonry Corner.* A mason wants to be sure she has a right corner in a building's foundation. She marks a point 3 ft from the corner along one wall and another point 4 ft from the corner along the other wall. If the corner is a right angle, what should the distance be between the two marked points?

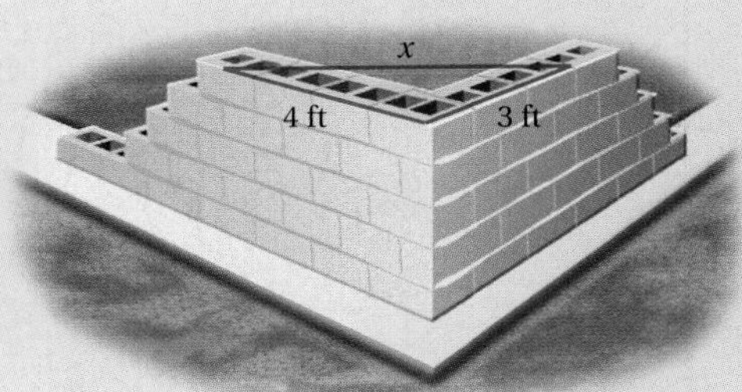

Find the x-intercepts for the graph of the equation.

28.

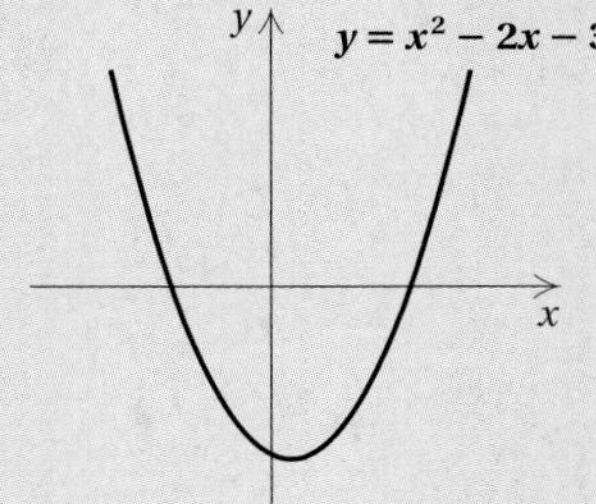

29.

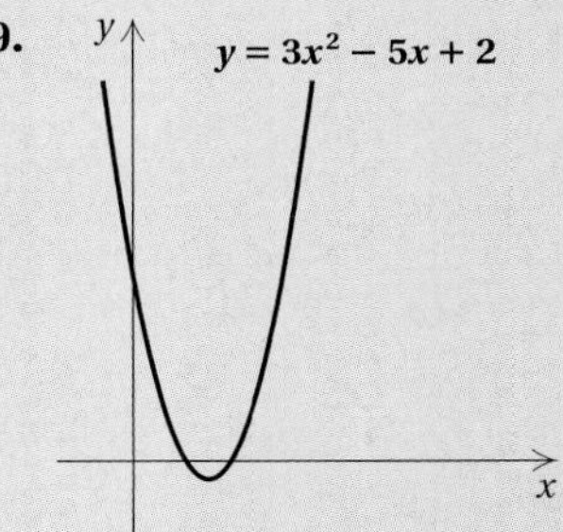

SKILL MAINTENANCE

30. Divide: $\frac{5}{8} \div \left(-\frac{11}{16}\right)$.

31. Solve: $10(x - 3) < 4(x + 2)$.

32. Find the intercepts. Then graph the equation.

$$2y - 5x = 10$$

33. Multiply: $(5x^2 - 7)^2$.

SYNTHESIS

34. The length of a rectangle is five times its width. When the length is decreased by 3 and the width is increased by 2, the area of the new rectangle is 60. Find the original length and width.

35. Factor: $(a + 3)^2 - 2(a + 3) - 35$.

36. Solve: $20x(x + 2)(x - 1) = 5x^3 - 24x - 14x^2$.

37. If $x + y = 4$ and $x - y = 6$, then $x^2 - y^2 = ?$

a) 2 **b)** 10
c) 34 **d)** 24

CHAPTERS 1–5

Cumulative Review

Find two algebraic expressions for the total area of each rectangle.

1. 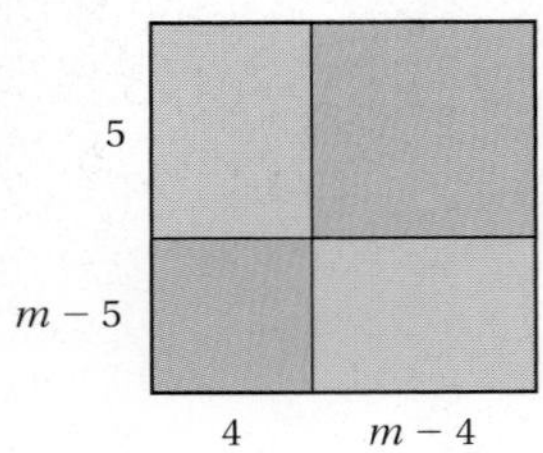

2. 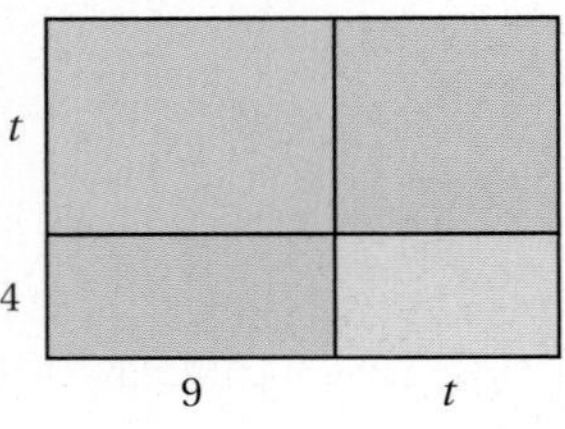

3. 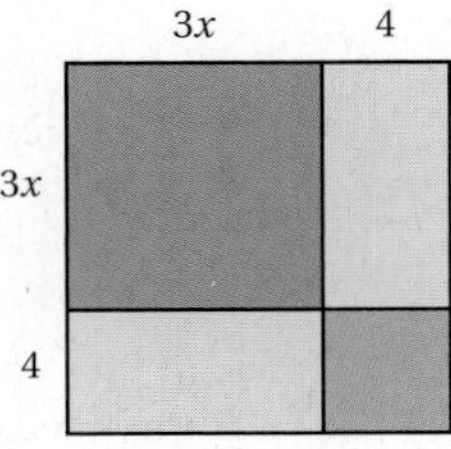

Use either < or > for □ to write a true sentence.

4. $\frac{2}{3} \square \frac{5}{7}$

5. $-\frac{4}{7} \square -\frac{8}{11}$

Compute and simplify.

6. $2.06 + (-4.79) - (-3.08)$

7. $5.652 \div (-3.6)$

8. $\left(\frac{2}{9}\right)\left(-\frac{3}{8}\right)\left(\frac{6}{7}\right)$

9. $\frac{21}{5} \div \left(-\frac{7}{2}\right)$

Simplify.

10. $[3x + 2(x - 1)] - [2x - (x + 3)]$

11. $1 - [14 + 28 \div 7 - (6 + 9 \div 3)]$

12. $(2x^2y^{-1})^3$

13. $\frac{3x^5}{4x^3} \cdot \frac{-2x^{-3}}{9x^2}$

14. Add:
$$(2x^2 - 3x^3 + x - 4) + (x^4 - x - 5x^2).$$

15. Subtract:
$$(2x^2y^2 + xy - 2xy^2) - (2xy - 2xy^2 + x^2y).$$

16. Divide: $(x^3 + 2x^2 - x + 1) \div (x - 1)$.

Multiply.

17. $(2t - 3)^2$

18. $(x^2 - 3)(x^2 + 3)$

19. $(2x + 4)(3x - 4)$

20. $2x(x^3 + 3x^2 + 4x)$

21. $(2y - 1)(2y^2 + 3y + 4)$

22. $\left(x + \frac{2}{3}\right)\left(x - \frac{2}{3}\right)$

Factor.

23. $x^2 + 2x - 8$

24. $4x^2 - 25$

25. $3x^3 - 4x^2 + 3x - 4$

26. $x^2 - 26x + 169$

27. $75x^2 - 108y^2$

28. $6x^2 - 13x - 63$

29. $x^4 - 2x^2 - 3$

30. $4y^3 - 6y^2 - 4y + 6$

31. $6p^2 + pq - q^2$

32. $10x^3 + 52x^2 + 10x$

33. $49x^3 - 42x^2 + 9x$

34. $3x^2 + 5x - 4$

35. $75x^3 + 27x$

36. $3x^8 - 48y^8$

37. $14x^2 + 28 + 42x$

38. $x^5 - x^3 + x^2 - 1$

Solve.

39. $3x - 5 = 2x + 10$

40. $3y + 4 > 5y - 8$

41. $(x - 15)\left(x + \frac{1}{4}\right) = 0$

42. $-98x(x + 37) = 0$

43. $x^3 + x^2 = 25x + 25$

44. $2x^2 = 72$

45. $9x^2 + 1 = 6x$

46. $x^2 + 17x + 70 = 0$

47. $14y^2 = 21y$

48. $1.6 - 3.5x = 0.9$

49. $(x + 3)(x - 4) = 8$

50. $1.5x - 3.6 \leq 1.3x + 0.4$

51. $2x - [3x - (2x + 3)] = 3x + [4 - (2x + 1)]$

52. $y = mx + b$, for m

Solve.

53. The sum of two consecutive even integers is 102. Find the integers.

54. The product of two consecutive even integers is 360. Find the integers.

55. *Window Dimensions.* The length of a rectangular window is 3 ft longer than the height. The area of the window is 18 ft^2. Find the length and the height.

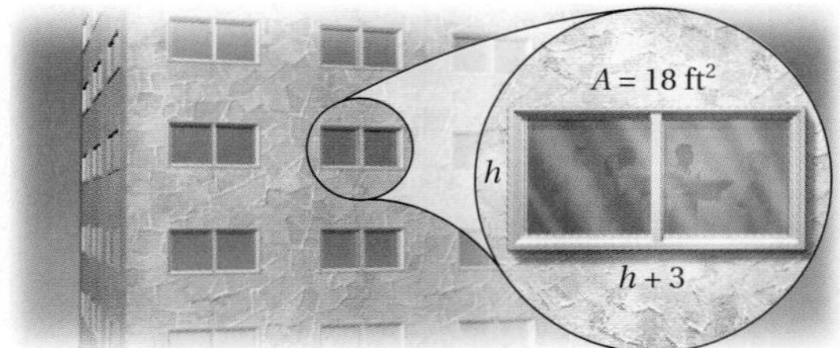

56. *Lot Dimensions.* The length of a rectangular lot is 200 m longer than the width. The perimeter of the lot is 1000 m. Find the dimensions of the lot.

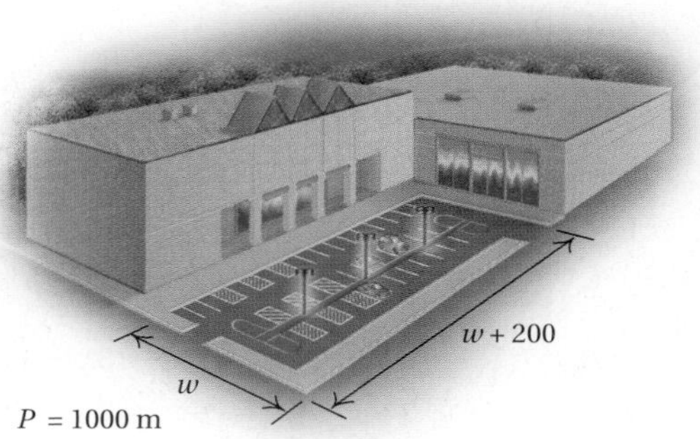

57. *PVC Pipe.* A 100-m PVC pipe is cut into three pieces. The second piece is twice as long as the first piece. The third piece is one-third as long as the first piece. How long is each piece?

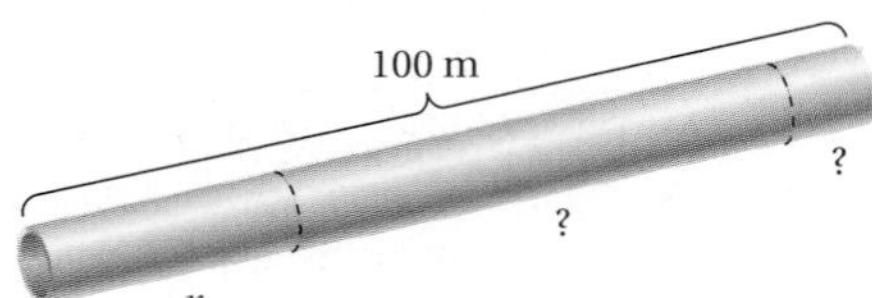

58. *Right-Triangle Geometry.* The length of one leg of a right triangle is 15 m. The length of the other leg is 9 m shorter than the length of the hypotenuse. Find the length of the hypotenuse.

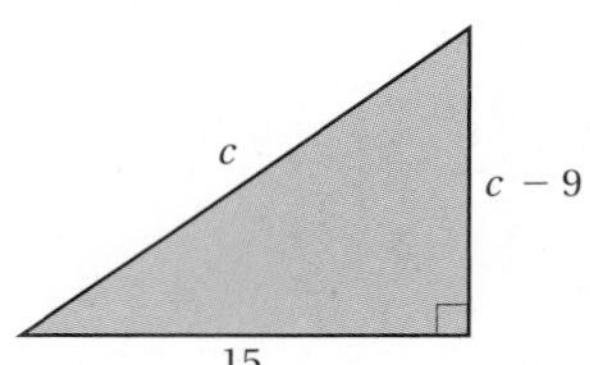

59. Money is borrowed at 12% simple interest. After 1 year, $7280 pays off the loan. How much was originally borrowed?

60. After a 25% price reduction, a pair of shoes is on sale for $21.75. What was the price before reduction?

61. The height of a triangle is 2 cm more than the base. The area of the triangle is 144 cm^2. Find the height and the base.

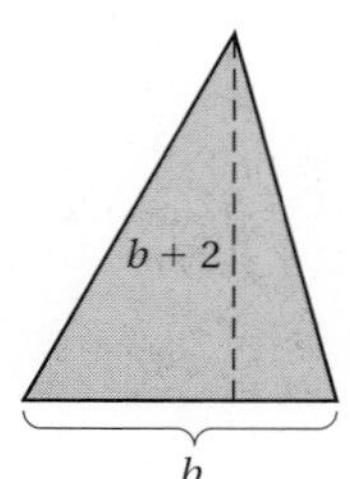

62. Find the intercepts. Then graph the equation.

$$3x + 4y = -12$$

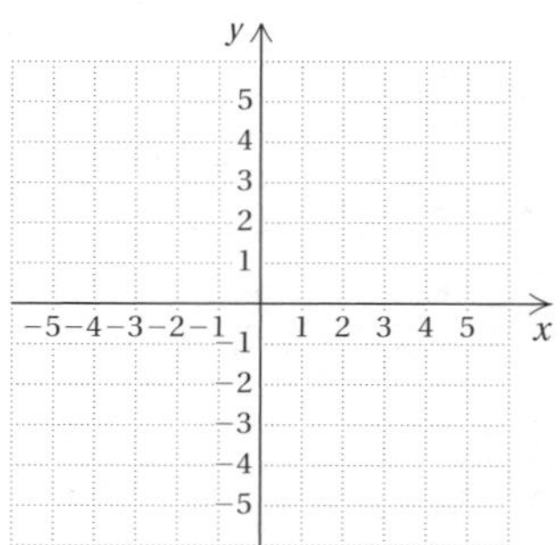

63. Find the x-intercepts for the graph of the equation.

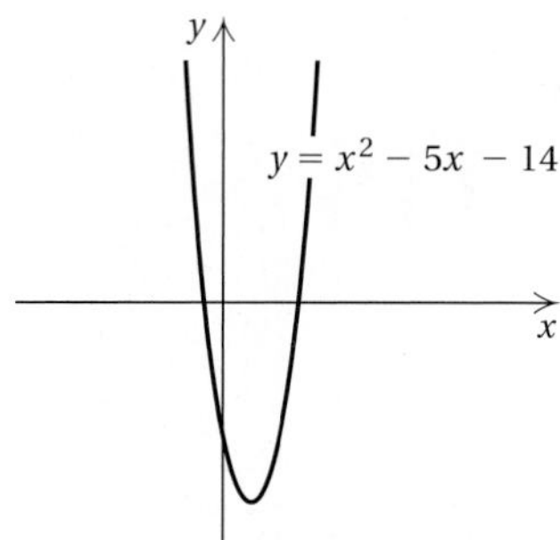

64. *Matching.* Match each item in the first column with the appropriate item in the second column by drawing connecting lines.

$\left(\frac{1}{8}\right)^3$	24
$\left(\frac{1}{8}\right)^{-3}$	-24
8^{-3}	64
8^3	-64
-8^3	$\frac{1}{64}$
$(-8)^3$	$-\frac{1}{64}$
$\left(-\frac{1}{8}\right)^{-3}$	$\frac{1}{24}$
$\left(-\frac{1}{8}\right)^3$	$-\frac{1}{24}$
	512
	$-\frac{1}{512}$
	-512
	$\frac{1}{512}$

For each of Exercises 65–68, choose the correct answer from the selections given.

65. The x-intercept of the graph of $4x - 3y = -12$ is:

a) $(0, -3)$. **b)** $(3, 0)$. **c)** $(0, 0)$.
d) $(-4, 0)$. **e)** None of these.

66. Compute and simplify: $1000 \cdot 100 \div 10 \cdot 10$.

a) 1000 **b)** 100,000 **c)** 10
d) 100 **e)** None of these.

67. If $x^2 - 16 = 10 \cdot 18$, then one possibility for x is:

a) 12. **b)** 14. **c)** 16.
d) 18. **e)** None of these.

68. The slope of the line containing the points $(2, -6)$ and $(-4, 3)$ is:

a) $\frac{2}{3}$. **b)** $-\frac{2}{3}$. **c)** $\frac{3}{2}$.
d) $-\frac{3}{2}$. **e)** None of these.

SYNTHESIS

Solve.

69. $(x + 3)(x - 5) \le (x + 2)(x - 1)$

70. $\frac{x - 3}{2} - \frac{2x + 5}{26} = \frac{4x + 11}{13}$

71. $(x + 1)^2 = 25$

Factor.

72. $x^2(x - 3) - x(x - 3) - 2(x - 3)$

73. $4a^2 - 4a + 1 - 9b^2 - 24b - 16$

Solve.

74. Find c such that the polynomial will be the square of a binomial: $cx^2 - 40x + 16$.

75. The length of the radius of a circle is increased by 2 cm to form a new circle. The area of the new circle is four times the area of the original circle. Find the length of the radius of the original circle.

Rational Expressions and Equations

Gateway to Chapter 6

A rational expression is the ratio or quotient of two polynomials. In this chapter, we learn to add, subtract, multiply, and divide rational expressions. Then we apply these manipulative skills to the solving of rational equations. These equations require careful thinking because certain numbers that seem to be solutions must be checked carefully.

Finally, we apply our new equation-solving skills to new kinds of applications, such as work and motion problems, and to applications involving ratio and proportion. You may have studied ratio and proportion problems in earlier mathematics courses.

Real-World Application

A cheetah can run 20 mph faster than a lion. A cheetah can run 7 mi in the same time that a lion can run 5 mi. Find the speed of each animal.

Source: Barbara Ann Kipfer, *The Order of Things*. New York: Random House, 1998.

This problem appears as Example 2 in Section 6.7.

CHAPTER 6

CHAPTER 6

Pretest

1. Find the LCM of $x^2 + 5x + 6$ and $x^2 + 6x + 9$. [6.3c]

Perform the indicated operations and simplify.

2. $\dfrac{b-1}{2-b} + \dfrac{b^2-3}{b^2-4}$ [6.4a]

3. $\dfrac{4y-4}{y^2-y-2} - \dfrac{3y-5}{y^2-y-2}$ [6.5a]

4. $\dfrac{4}{a+2} + \dfrac{3}{a}$ [6.4a]

5. $\dfrac{x}{x+1} - \dfrac{x}{x-1} + \dfrac{2x^2}{x^2-1}$ [6.5b]

6. $\dfrac{4x+8}{x+1} \cdot \dfrac{x^2-2x-3}{2x^2-8}$ [6.1d]

7. $\dfrac{x+3}{x^2-9} \div \dfrac{x+3}{x^2-6x+9}$ [6.2b]

8. Simplify: $\dfrac{\frac{1}{x}+\frac{1}{y}}{\frac{1}{x}-\frac{1}{y}}$. [6.8a]

Solve. [6.6a]

9. $\dfrac{1}{x+4} = \dfrac{5}{x}$

10. $\dfrac{3}{x-2} + \dfrac{x}{2} = \dfrac{6}{2x-4}$

11. Mercedes-Benz Cabriolet Gas Mileage. A Mercedes-Benz Cabriolet travels 297 miles on 16.5 gal of gas. How much gas would it take to drive 1000 mi? [6.7b]
Source: Mercedes-Benz

12. Paper Delivery. It takes 6 hr for a paper carrier to deliver 200 papers. At this rate, how long would it take to deliver 350 papers? [6.7b]

13. Data Entry. One data-entry clerk can key in a report in 6 hr. Another can key in the same report in 5 hr. How long would it take them, working together, to key in the same report? [6.7a]

14. Car Speeds. One car travels 20 mph faster than another. While one car travels 300 mi, the other travels 400 mi. Find the speed of each car. [6.7a]

6.1 MULTIPLYING AND SIMPLIFYING RATIONAL EXPRESSIONS

a Rational Expressions and Replacements

Rational numbers are quotients of integers. Some examples are

$$\frac{2}{3}, \quad \frac{4}{-5}, \quad \frac{-8}{17}, \quad \frac{563}{1}.$$

The following are called **rational expressions** or **fractional expressions.** They are quotients, or ratios, of polynomials:

$$\frac{3}{4}, \quad \frac{z}{6}, \quad \frac{5}{x+2}, \quad \frac{t^2+3t-10}{7t^2-4}.$$

A rational expression is also a division. For example,

$$\frac{3}{4} \text{ means } 3 \div 4 \quad \text{and} \quad \frac{x-8}{x+2} \text{ means } (x-8) \div (x+2).$$

Because rational expressions indicate division, we must be careful to avoid denominators of zero. When a variable is replaced with a number that produces a denominator equal to zero, the rational expression is not defined. For example, in the expression

$$\frac{x-8}{x+2},$$

when x is replaced with -2, the denominator is 0, and the expression is *not* defined:

$$\frac{x-8}{x+2} = \frac{-2-8}{-2+2} = \frac{-10}{0} \leftarrow \text{Division by 0 is not defined.}$$

When x is replaced with a number other than -2, such as 3, the expression *is* defined because the denominator is nonzero:

$$\frac{x-8}{x+2} = \frac{3-8}{3+2} = \frac{-5}{5} = -1.$$

EXAMPLE 1 Find all numbers for which the rational expression

$$\frac{x+4}{x^2-3x-10}$$

is not defined.

The value of the numerator has no bearing on whether or not a rational expression is defined. To determine which numbers make the rational expression not defined, we set the *denominator* equal to 0 and solve:

$$x^2 - 3x - 10 = 0$$

$$(x-5)(x+2) = 0 \qquad \text{Factoring}$$

$$x-5=0 \quad or \quad x+2=0 \qquad \text{Using the principle of zero products (see Section 5.7)}$$

$$x = 5 \quad or \quad x = -2.$$

The expression is not defined for the replacement numbers 5 and -2.

Do Exercises 1–3.

Objectives

a Find all numbers for which a rational expression is not defined.

b Multiply a rational expression by 1, using an expression such as A/A.

c Simplify rational expressions by factoring the numerator and the denominator and removing factors of 1.

d Multiply rational expressions and simplify.

Find all numbers for which the rational expression is not defined.

1. $\dfrac{16}{x-3}$

2. $\dfrac{2x-7}{x^2+5x-24}$

3. $\dfrac{x+5}{8}$

Answers on page A-28

Multiply.

4. $\dfrac{2x+1}{3x-2} \cdot \dfrac{x}{x}$

5. $\dfrac{x+1}{x-2} \cdot \dfrac{x+2}{x+2}$

6. $\dfrac{x-8}{x-y} \cdot \dfrac{-1}{-1}$

Answers on page A-28

b Multiplying by 1

We multiply rational expressions in the same way that we multiply fraction notation in arithmetic. For a review, see Section R.2. We saw that

$$\frac{3}{7} \cdot \frac{2}{5} = \frac{3 \cdot 2}{7 \cdot 5} = \frac{6}{35}.$$

MULTIPLYING RATIONAL EXPRESSIONS

To multiply rational expressions, multiply numerators and multiply denominators:

$$\frac{A}{B} \cdot \frac{C}{D} = \frac{AC}{BD}.$$

For example,

$$\frac{x-2}{3} \cdot \frac{x+2}{x+7} = \frac{(x-2)(x+2)}{3(x+7)}.$$ Multiplying the numerators and the denominators

Note that we leave the numerator, $(x-2)(x+2)$, and the denominator, $3(x+7)$, in factored form because it is easier to simplify if we do not multiply. In order to learn to simplify, we first need to consider multiplying the rational expression by 1.

Any rational expression with the same numerator and denominator is a symbol for 1:

$$\frac{19}{19} = 1, \quad \frac{x+8}{x+8} = 1, \quad \frac{3x^2-4}{3x^2-4} = 1, \quad \frac{-1}{-1} = 1.$$

EQUIVALENT EXPRESSIONS

Expressions that have the same value for all allowable (or meaningful) replacements are called **equivalent expressions.**

We can multiply by 1 to obtain an *equivalent expression*. At this point, we select expressions for 1 arbitrarily. Later, we will have a system for our choices when we add and subtract.

EXAMPLES Multiply.

2. $\dfrac{3x+2}{x+1} \cdot 1 = \dfrac{3x+2}{x+1} \cdot \dfrac{2x}{2x} = \dfrac{(3x+2)2x}{(x+1)2x}$ Using the identity property of 1. We arbitrarily choose $2x/2x$ as a symbol for 1.

3. $\dfrac{x+2}{x-7} \cdot \dfrac{x+3}{x+3} = \dfrac{(x+2)(x+3)}{(x-7)(x+3)}$ We arbitrarily choose $(x+3)/(x+3)$ as a symbol for 1.

4. $\dfrac{2+x}{2-x} \cdot \dfrac{-1}{-1} = \dfrac{(2+x)(-1)}{(2-x)(-1)}$

Do Exercises 4–6.

C Simplifying Rational Expressions

Simplifying rational expressions is similar to simplifying fractional expressions in arithmetic. For a review, see Section R.2. We saw there, for example, that an expression like $\frac{15}{40}$ can be simplified as follows:

$$\frac{15}{40} = \frac{3 \cdot 5}{8 \cdot 5}$$ Factoring the numerator and the denominator. Note the common factor of 5.

$$= \frac{3}{8} \cdot \frac{5}{5}$$ Factoring the fractional expression

$$= \frac{3}{8} \cdot 1$$ $\frac{5}{5} = 1$

$$= \frac{3}{8}.$$ Using the identity property of 1, or "removing a factor of 1"

Similar steps are followed when simplifying rational expressions: We factor and remove a factor of 1, using the fact that

$$\frac{ab}{cb} = \frac{a}{c} \cdot \frac{b}{b} = \frac{a}{c} \cdot 1 = \frac{a}{c}.$$

In algebra, instead of simplifying

$$\frac{15}{40},$$

we may need to simplify an expression like

$$\frac{x^2 - 16}{x + 4}.$$

Just as factoring is important in simplifying in arithmetic, so too is it important in simplifying rational expressions. The factoring we use most is the factoring of polynomials, which we studied in Chapter 5.

To simplify, we can do the reverse of multiplying. We factor the numerator and the denominator and "remove" a factor of 1.

EXAMPLE 5 Simplify: $\frac{8x^2}{24x}$.

$$\frac{8x^2}{24x} = \frac{8 \cdot x \cdot x}{3 \cdot 8 \cdot x}$$ Factoring the numerator and the denominator

$$= \frac{8x}{8x} \cdot \frac{x}{3}$$ Factoring the rational expression

$$= 1 \cdot \frac{x}{3}$$ $\frac{8x}{8x} = 1$

$$= \frac{x}{3}$$ We removed a factor of 1.

Do Exercises 7 and 8.

Simplify.

7. $\frac{5y}{y}$

8. $\frac{9x^2}{36x}$

Answers on page A-28

Study Tips

TUNE OUT DISTRACTIONS

Do you generally study in noisy places? If there is constant noise in your home, dorm, or other study area, consider finding a quiet place in the library—maybe an uncrowded place so you are not distracted with people-watching!

"The ability to concentrate and use your time well is everything."

Lee Iacocca, former CEO of Chrysler Corporation

Simplify.

9. $\dfrac{2x^2 + x}{3x^2 + 2x}$

10. $\dfrac{x^2 - 1}{2x^2 - x - 1}$

11. $\dfrac{7x + 14}{7}$

12. $\dfrac{12y + 24}{48}$

Answers on page A-28

EXAMPLES Simplify.

6. $\dfrac{5a + 15}{10} = \dfrac{5(a + 3)}{5 \cdot 2}$ Factoring the numerator and the denominator

$= \dfrac{5}{5} \cdot \dfrac{a + 3}{2}$ Factoring the rational expression

$= 1 \cdot \dfrac{a + 3}{2}$ $\dfrac{5}{5} = 1$

$= \dfrac{a + 3}{2}$ Removing a factor of 1

7. $\dfrac{6a + 12}{7a + 14} = \dfrac{6(a + 2)}{7(a + 2)}$ Factoring the numerator and the denominator

$= \dfrac{6}{7} \cdot \dfrac{a + 2}{a + 2}$ Factoring the rational expression

$= \dfrac{6}{7} \cdot 1$ $\dfrac{a + 2}{a + 2} = 1$

$= \dfrac{6}{7}$ Removing a factor of 1

8. $\dfrac{6x^2 + 4x}{2x^2 + 2x} = \dfrac{2x(3x + 2)}{2x(x + 1)}$ Factoring the numerator and the denominator

$= \dfrac{2x}{2x} \cdot \dfrac{3x + 2}{x + 1}$ Factoring the rational expression

$= 1 \cdot \dfrac{3x + 2}{x + 1}$ $\dfrac{2x}{2x} = 1$

$= \dfrac{3x + 2}{x + 1}$ Removing a factor of 1

9. $\dfrac{x^2 + 3x + 2}{x^2 - 1} = \dfrac{(x + 2)(x + 1)}{(x + 1)(x - 1)}$

$= \dfrac{x + 1}{x + 1} \cdot \dfrac{x + 2}{x - 1}$

$= 1 \cdot \dfrac{x + 2}{x - 1}$

$= \dfrac{x + 2}{x - 1}$

CAUTION! Note in this step that you *cannot* remove the x's because x is not a factor of the entire numerator and the entire denominator.

CANCELING

You may have encountered canceling when working with rational expressions. With great concern, we mention it as a possible way to speed up your work. Our concern is that canceling be done with care and understanding. Example 9 might have been done faster as follows:

$$\frac{x^2 + 3x + 2}{x^2 - 1} = \frac{(x + 2)(x + 1)}{(x + 1)(x - 1)}$$ Factoring the numerator and the denominator

$$= \frac{(x + 2)\cancel{(x + 1)}}{\cancel{(x + 1)}(x - 1)}$$ When a factor of 1 is noted, it is canceled, as shown: $\frac{x + 1}{x + 1} = 1$.

$$= \frac{x + 2}{x - 1}.$$ Simplifying

CAUTION!

The difficulty with canceling is that it is often applied incorrectly, as in the following situations:

$$\frac{\cancel{x} + 3}{\cancel{x}} = 3; \qquad \frac{\cancel{4} + 1}{\cancel{4} + 2} = \frac{1}{2}; \qquad \frac{1\cancel{5}}{\cancel{5}4} = \frac{1}{4}.$$

Wrong! Wrong! Wrong!

In each of these situations, the expressions canceled were *not* factors of 1. Factors are parts of products. For example, in $2 \cdot 3$, 2 and 3 are factors, but in $2 + 3$, 2 and 3 are *not* factors. If you can't factor, you can't cancel. If in doubt, don't cancel!

Do Exercises 9–12 on the preceding page.

OPPOSITES IN RATIONAL EXPRESSIONS

Expressions of the form $a - b$ and $b - a$ are opposites of each other. When either of these binomials is multiplied by -1, the result is the other binomial:

$$-1(a - b) = -a + b = b + (-a) = b - a;$$
$$-1(b - a) = -b + a = a + (-b) = a - b.$$

Multiplication by -1 reverses the order in which subtraction occurs.

Consider, for example,

$$\frac{x - 4}{4 - x}.$$

At first glance, it appears as though the numerator and the denominator do not have any common factors other than 1. But $x - 4$ and $4 - x$ are opposites, or additive inverses, of each other. Thus we can rewrite one as the opposite of the other by factoring out a -1.

EXAMPLE 10 Simplify: $\frac{x - 4}{4 - x}$.

$$\frac{x - 4}{4 - x} = \frac{x - 4}{-(x - 4)} = \frac{x - 4}{-1(x - 4)}$$ $4 - x = -(x - 4)$; $4 - x$ and $x - 4$ are opposites.

$$= -1 \cdot \frac{x - 4}{x - 4}$$
$$= -1 \cdot 1$$
$$= -1$$

Do Exercises 13–15.

Simplify.

13. $\frac{x - 8}{8 - x}$

14. $\frac{c - d}{d - c}$

15. $\frac{-x - 7}{x + 7}$

Answers on page A-28

16. Multiply and simplify:

$$\frac{a^2 - 4a + 4}{a^2 - 9} \cdot \frac{a + 3}{a - 2}.$$

17. Multiply and simplify:

$$\frac{x^2 - 25}{6} \cdot \frac{3}{x + 5}.$$

Answers on page A-28

d Multiplying and Simplifying

We try to simplify after we multiply. That is why we leave the numerator and the denominator in factored form.

EXAMPLE 11 Multiply and simplify: $\dfrac{5a^3}{4} \cdot \dfrac{2}{5a}$.

$$\frac{5a^3}{4} \cdot \frac{2}{5a} = \frac{5a^3(2)}{4(5a)} \quad \text{Multiplying the numerators and the denominators}$$

$$= \frac{2 \cdot 5 \cdot a \cdot a \cdot a}{2 \cdot 2 \cdot 5 \cdot a} \quad \text{Factoring the numerator and the denominator}$$

$$= \frac{\cancel{2} \cdot \cancel{5} \cdot \cancel{a} \cdot a \cdot a}{\cancel{2} \cdot 2 \cdot \cancel{5} \cdot \cancel{a}} \quad \text{Removing a factor of 1: } \frac{2 \cdot 5 \cdot a}{2 \cdot 5 \cdot a} = 1$$

$$= \frac{a^2}{2} \quad \text{Simplifying}$$

EXAMPLE 12 Multiply and simplify: $\dfrac{x^2 + 6x + 9}{x^2 - 4} \cdot \dfrac{x - 2}{x + 3}$.

$$\frac{x^2 + 6x + 9}{x^2 - 4} \cdot \frac{x - 2}{x + 3} = \frac{(x^2 + 6x + 9)(x - 2)}{(x^2 - 4)(x + 3)} \quad \text{Multiplying the numerators and the denominators}$$

$$= \frac{(x + 3)(x + 3)(x - 2)}{(x + 2)(x - 2)(x + 3)} \quad \text{Factoring the numerator and the denominator}$$

$$= \frac{\cancel{(x + 3)}(x + 3)\cancel{(x - 2)}}{(x + 2)\cancel{(x - 2)}\cancel{(x + 3)}} \quad \text{Removing a factor of 1: } \frac{(x + 3)(x - 2)}{(x + 3)(x - 2)} = 1$$

$$= \frac{x + 3}{x + 2} \quad \text{Simplifying}$$

Do Exercise 16.

EXAMPLE 13 Multiply and simplify: $\dfrac{x^2 + x - 2}{15} \cdot \dfrac{5}{2x^2 - 3x + 1}$.

$$\frac{x^2 + x - 2}{15} \cdot \frac{5}{2x^2 - 3x + 1} = \frac{(x^2 + x - 2)5}{15(2x^2 - 3x + 1)} \quad \text{Multiplying the numerators and the denominators}$$

$$= \frac{(x + 2)(x - 1)5}{5(3)(x - 1)(2x - 1)} \quad \text{Factoring the numerator and the denominator}$$

$$= \frac{(x + 2)\cancel{(x - 1)}\cancel{5}}{\cancel{5}(3)\cancel{(x - 1)}(2x - 1)} \quad \text{Removing a factor of 1: } \frac{(x - 1)5}{(x - 1)5} = 1$$

$$= \frac{x + 2}{3(2x - 1)} \quad \text{Simplifying}$$

You need not carry out this multiplication.

Do Exercise 17.

CALCULATOR CORNER

Checking Multiplication and Simplification We can use the TABLE feature as a partial check that rational expressions have been multiplied and/or simplified correctly. To check the simplification in Example 9,

$$\frac{x^2 + 3x + 2}{x^2 - 1} = \frac{x + 2}{x - 1},$$

we first enter $y_1 = (x^2 + 3x + 2)/(x^2 - 1)$ and $y_2 = (x + 2)/(x - 1)$. Then, using AUTO mode, we look at a table of values of y_1 and y_2. If the simplification is correct, the values should be the same for all allowable replacements.

X	Y1	Y2
−4	.4	.4
−3	.25	.25
−2	0	0
−1	ERROR	−.5
0	−2	−2
1	ERROR	ERROR
2	4	4
X = −4		

The ERROR messages indicate that −1 and 1 are not allowable replacements in the first expression, and 1 is not an allowable replacement in the second. For all other numbers, we see that y_1 and y_2 are the same, so the simplification appears to be correct. Remember, this is only a partial check since we cannot check all possible values.

Exercises: Use the TABLE feature to determine whether each of the following appears to be correct.

1. $\dfrac{8x^2}{24x} = \dfrac{x}{3}$
2. $\dfrac{5x + 15}{10} = \dfrac{x + 3}{2}$
3. $\dfrac{x + 3}{x} = 3$
4. $\dfrac{x^2 + 3x - 4}{x^2 - 16} = \dfrac{x - 1}{x + 4}$
5. $\dfrac{x^2 + 2x - 3}{x^2 - 4} \cdot \dfrac{4x - 8}{x + 3} = \dfrac{4x - 1}{x + 2}$
6. $\dfrac{x^2 - 25}{6} \cdot \dfrac{3}{x + 5} = \dfrac{x - 5}{3}$
7. $\dfrac{x^2 + 6x + 9}{x^2 - 4} \cdot \dfrac{x - 2}{x + 3} = \dfrac{x + 3}{x + 2}$
8. $\dfrac{x^2}{x^2 - 3x} \cdot \dfrac{x^2 - 9}{3} = \dfrac{x(x + 3)}{3}$

6.1 EXERCISE SET

For Extra Help

Digital Video Tutor CD 6 Videotape 12 | InterAct Math | Math Tutor Center | MathXL | MyMathLab

a Find all numbers for which the rational expression is not defined.

1. $\dfrac{-3}{2x}$

2. $\dfrac{24}{-8y}$

3. $\dfrac{5}{x-8}$

4. $\dfrac{y-4}{y+6}$

5. $\dfrac{3}{2y+5}$

6. $\dfrac{x^2-9}{4x-12}$

7. $\dfrac{x^2+11}{x^2-3x-28}$

8. $\dfrac{p^2-9}{p^2-7p+10}$

9. $\dfrac{m^3-2m}{m^2-25}$

10. $\dfrac{7-3x+x^2}{49-x^2}$

11. $\dfrac{x-4}{3}$

12. $\dfrac{x^2-25}{14}$

b Multiply. Do not simplify. Note that in each case you are multiplying by 1.

13. $\dfrac{4x}{4x} \cdot \dfrac{3x^2}{5y}$

14. $\dfrac{5x^2}{5x^2} \cdot \dfrac{6y^3}{3z^4}$

15. $\dfrac{2x}{2x} \cdot \dfrac{x-1}{x+4}$

16. $\dfrac{2a-3}{5a+2} \cdot \dfrac{a}{a}$

17. $\dfrac{3-x}{4-x} \cdot \dfrac{-1}{-1}$

18. $\dfrac{x-5}{5-x} \cdot \dfrac{-1}{-1}$

19. $\dfrac{y+6}{y+6} \cdot \dfrac{y-7}{y+2}$

20. $\dfrac{x^2+1}{x^3-2} \cdot \dfrac{x-4}{x-4}$

c Simplify.

21. $\dfrac{8x^3}{32x}$

22. $\dfrac{4x^2}{20x}$

23. $\dfrac{48p^7q^5}{18p^5q^4}$

24. $\dfrac{-76x^8y^3}{-24x^4y^3}$

25. $\dfrac{4x-12}{4x}$

26. $\dfrac{5a-40}{5}$

27. $\dfrac{3m^2 + 3m}{6m^2 + 9m}$

28. $\dfrac{4y^2 - 2y}{5y^2 - 5y}$

29. $\dfrac{a^2 - 9}{a^2 + 5a + 6}$

30. $\dfrac{t^2 - 25}{t^2 + t - 20}$

31. $\dfrac{a^2 - 10a + 21}{a^2 - 11a + 28}$

32. $\dfrac{x^2 - 2x - 8}{x^2 - x - 6}$

33. $\dfrac{x^2 - 25}{x^2 - 10x + 25}$

34. $\dfrac{x^2 + 8x + 16}{x^2 - 16}$

35. $\dfrac{a^2 - 1}{a - 1}$

36. $\dfrac{t^2 - 1}{t + 1}$

37. $\dfrac{x^2 + 1}{x + 1}$

38. $\dfrac{m^2 + 9}{m + 3}$

39. $\dfrac{6x^2 - 54}{4x^2 - 36}$

40. $\dfrac{8x^2 - 32}{4x^2 - 16}$

41. $\dfrac{6t + 12}{t^2 - t - 6}$

42. $\dfrac{4x + 32}{x^2 + 9x + 8}$

43. $\dfrac{2t^2 + 6t + 4}{4t^2 - 12t - 16}$

44. $\dfrac{3a^2 - 9a - 12}{6a^2 + 30a + 24}$

45. $\dfrac{t^2 - 4}{(t + 2)^2}$

46. $\dfrac{m^2 - 10m + 25}{m^2 - 25}$

47. $\dfrac{6 - x}{x - 6}$

48. $\dfrac{t - 3}{3 - t}$

49. $\dfrac{a - b}{b - a}$

50. $\dfrac{y - x}{-x + y}$

51. $\dfrac{6t - 12}{2 - t}$

52. $\dfrac{5a - 15}{3 - a}$

53. $\dfrac{x^2 - 1}{1 - x}$

54. $\dfrac{a^2 - b^2}{b^2 - a^2}$

d Multiply and simplify.

55. $\dfrac{4x^3}{3x} \cdot \dfrac{14}{x}$

56. $\dfrac{18}{x^3} \cdot \dfrac{5x^2}{6}$

57. $\dfrac{3c}{d^2} \cdot \dfrac{4d}{6c^3}$

58. $\dfrac{3x^2y}{2} \cdot \dfrac{4}{xy^3}$

59. $\dfrac{x^2 - 3x - 10}{x^2 - 4x + 4} \cdot \dfrac{x - 2}{x - 5}$

60. $\dfrac{t^2}{t^2 - 4} \cdot \dfrac{t^2 - 5t + 6}{t^2 - 3t}$

61. $\dfrac{a^2 - 9}{a^2} \cdot \dfrac{a^2 - 3a}{a^2 + a - 12}$

62. $\dfrac{x^2 + 10x - 11}{x^2 - 1} \cdot \dfrac{x + 1}{x + 11}$

63. $\dfrac{4a^2}{3a^2 - 12a + 12} \cdot \dfrac{3a - 6}{2a}$

64. $\dfrac{5v + 5}{v - 2} \cdot \dfrac{v^2 - 4v + 4}{v^2 - 1}$

65. $\dfrac{t^4 - 16}{t^4 - 1} \cdot \dfrac{t^2 + 1}{t^2 + 4}$

66. $\dfrac{x^4 - 1}{x^4 - 81} \cdot \dfrac{x^2 + 9}{x^2 + 1}$

67. $\dfrac{(x+4)^3}{(x+2)^3} \cdot \dfrac{x^2+4x+4}{x^2+8x+16}$

68. $\dfrac{(t-2)^3}{(t-1)^3} \cdot \dfrac{t^2-2t+1}{t^2-4t+4}$

69. $\dfrac{5a^2-180}{10a^2-10} \cdot \dfrac{20a+20}{2a-12}$

70. $\dfrac{2t^2-98}{4t^2-4} \cdot \dfrac{8t+8}{16t-112}$

71. $^{\text{D}}_{\text{W}}$ How is the process of canceling related to the identity property of 1?

72. $^{\text{D}}_{\text{W}}$ Explain how a rational expression can be formed for which -3 and 4 are not allowable replacements.

SKILL MAINTENANCE

Solve.

73. *Consecutive Even Integers.* The product of two consecutive even integers is 360. Find the integers. [5.8a]

74. *Chemistry.* About 5 L of oxygen can be dissolved in 100 L of water at 0°C. This is 1.6 times the amount that can be dissolved in the same volume of water at 20°C. How much oxygen can be dissolved in 100 L at 20°C? [2.6a]

Factor. [5.6a]

75. $x^2 - x - 56$

76. $a^2 - 16a + 64$

77. $x^5 - 2x^4 - 35x^3$

78. $2y^3 - 10y^2 + y - 5$

79. $16 - t^4$

80. $10x^2 + 80x + 70$

81. $x^2 - 9x + 14$

82. $x^2 + x + 7$

83. $16x^2 - 40xy + 25y^2$

84. $a^2 - 9ab + 14b^2$

SYNTHESIS

Simplify.

85. $\dfrac{x^4-16y^4}{(x^2+4y^2)(x-2y)}$

86. $\dfrac{(a-b)^2}{b^2-a^2}$

87. $\dfrac{t^4-1}{t^4-81} \cdot \dfrac{t^2-9}{t^2+1} \cdot \dfrac{(t-9)^2}{(t+1)^2}$

88. $\dfrac{(t+2)^3}{(t+1)^3} \cdot \dfrac{t^2+2t+1}{t^2+4t+4} \cdot \dfrac{t+1}{t+2}$

89. $\dfrac{x^2-y^2}{(x-y)^2} \cdot \dfrac{x^2-2xy+y^2}{x^2-4xy-5y^2}$

90. $\dfrac{x-1}{x^2+1} \cdot \dfrac{x^4-1}{(x-1)^2} \cdot \dfrac{x^2-1}{x^4-2x^2+1}$

91. Select any number x, multiply by 2, add 5, multiply by 5, subtract 25, and divide by 10. What do you get? Explain how this procedure can be used for a number trick.

Objectives

a Find the reciprocal of a rational expression.

b Divide rational expressions and simplify.

Find the reciprocal.

1. $\frac{7}{2}$

2. $\frac{x^2 + 5}{2x^3 - 1}$

3. $x - 5$

4. $\frac{1}{x^2 - 3}$

5. Divide: $\frac{3}{5} \div \frac{7}{10}$.

Answers on page A-29

6.2 DIVISION AND RECIPROCALS

There is a similarity between what we do with rational expressions and what we do with rational numbers. In fact, after variables have been replaced with rational numbers, a rational expression represents a rational number.

a Finding Reciprocals

Two expressions are reciprocals of each other if their product is 1. The reciprocal of a rational expression is found by interchanging the numerator and the denominator.

EXAMPLES

1. The reciprocal of $\frac{2}{5}$ is $\frac{5}{2}$. $\left(\text{This is because } \frac{2}{5} \cdot \frac{5}{2} = \frac{10}{10} = 1.\right)$
2. The reciprocal of $\frac{2x^2 - 3}{x + 4}$ is $\frac{x + 4}{2x^2 - 3}$.
3. The reciprocal of $x + 2$ is $\frac{1}{x + 2}$. $\left(\text{Think of } x + 2 \text{ as } \frac{x + 2}{1}.\right)$

Do Exercises 1–4.

b Division

We divide rational expressions in the same way that we divide fraction notation in arithmetic. For a review, see Section R.2.

DIVIDING RATIONAL EXPRESSIONS

To divide by a rational expression, multiply by its reciprocal:

$$\frac{A}{B} \div \frac{C}{D} = \frac{A}{B} \cdot \frac{D}{C} = \frac{AD}{BC}.$$

Then factor and, if possible, simplify.

EXAMPLE 4 Divide: $\frac{3}{4} \div \frac{9}{5}$.

$\frac{3}{4} \div \frac{9}{5} = \frac{3}{4} \cdot \frac{5}{9}$ — Multiplying by the reciprocal of the divisor

$= \frac{3 \cdot 5}{4 \cdot 9} = \frac{3 \cdot 5}{2 \cdot 2 \cdot 3 \cdot 3}$ — Factoring

$= \frac{\cancel{3} \cdot 5}{2 \cdot 2 \cdot \cancel{3} \cdot 3}$ — Removing a factor of 1: $\frac{3}{3} = 1$

$= \frac{5}{12}$ — Simplifying

Do Exercise 5.

EXAMPLE 5 Divide: $\frac{2}{x} \div \frac{3}{x}$.

$$\frac{2}{x} \div \frac{3}{x} = \frac{2}{x} \cdot \frac{x}{3}$$ Multiplying by the reciprocal of the divisor

$$= \frac{2 \cdot x}{x \cdot 3}$$

$$= \frac{2 \cdot \cancel{x}}{\cancel{x} \cdot 3}$$ Removing a factor of 1: $\frac{x}{x} = 1$

$$= \frac{2}{3}$$

Do Exercise 6.

EXAMPLE 6 Divide: $\frac{x+1}{x+2} \div \frac{x-1}{x+3}$.

$$\frac{x+1}{x+2} \div \frac{x-1}{x+3} = \frac{x+1}{x+2} \cdot \frac{x+3}{x-1}$$ Multiplying by the reciprocal of the divisor

$$= \frac{(x+1)(x+3)}{(x+2)(x-1)}$$

We usually do not carry out the multiplication in the numerator or the denominator. It is not wrong to do so, but the factored form is often more useful.

Do Exercise 7.

EXAMPLE 7 Divide and simplify: $\frac{x+1}{x^2-1} \div \frac{x+1}{x^2-2x+1}$.

$$\frac{x+1}{x^2-1} \div \frac{x+1}{x^2-2x+1}$$

$$= \frac{x+1}{x^2-1} \cdot \frac{x^2-2x+1}{x+1}$$ Multiplying by the reciprocal

$$= \frac{(x+1)(x^2-2x+1)}{(x^2-1)(x+1)}$$

$$= \frac{(x+1)(x-1)(x-1)}{(x-1)(x+1)(x+1)}$$ Factoring the numerator and the denominator

$$= \frac{\cancel{(x+1)}\cancel{(x-1)}(x-1)}{\cancel{(x-1)}\cancel{(x+1)}(x+1)}$$ Removing a factor of 1: $\frac{(x+1)(x-1)}{(x+1)(x-1)} = 1$

$$= \frac{x-1}{x+1}$$

EXAMPLE 8 Divide and simplify: $\frac{x^2-2x-3}{x^2-4} \div \frac{x+1}{x+5}$.

$$\frac{x^2-2x-3}{x^2-4} \div \frac{x+1}{x+5}$$

$$= \frac{x^2-2x-3}{x^2-4} \cdot \frac{x+5}{x+1}$$ Multiplying by the reciprocal

6. Divide: $\frac{x}{8} \div \frac{x}{5}$.

7. Divide:

$$\frac{x-3}{x+5} \div \frac{x+5}{x-2}.$$

Answers on page A-29

CALCULATOR CORNER

Checking Division Use the TABLE feature, as described on p. 471, to check the divisions in Examples 5–8. Then check your answers to Margin Exercises 7–9.

Divide and simplify.

8. $\dfrac{x-3}{x+5} \div \dfrac{x+2}{x+5}$

9. $\dfrac{x^2-5x+6}{x+5} \div \dfrac{x+2}{x+5}$

10. $\dfrac{y^2-1}{y+1} \div \dfrac{y^2-2y+1}{y+1}$

Answers on page A-29

Then

$$= \frac{(x^2-2x-3)(x+5)}{(x^2-4)(x+1)}$$

$$= \frac{(x-3)(x+1)(x+5)}{(x-2)(x+2)(x+1)} \quad \text{Factoring the numerator and the denominator}$$

$$= \frac{(x-3)\cancel{(x+1)}(x+5)}{(x-2)(x+2)\cancel{(x+1)}} \quad \text{Removing a factor of 1: } \frac{x+1}{x+1}=1$$

$$= \frac{(x-3)(x+5)}{(x-2)(x+2)}.$$

You need not carry out the multiplications in the numerator and the denominator.

Do Exercises 8–10.

Study Tips

TURNING NEGATIVES INTO POSITIVES (PART 1)

Dusty Baker, presently manager of the San Francisco Giants, was cut from Little League by his father because he had a poor attitude. Dusty turned that negative into many positives, eventually earning his way back into Little League. He went on to an outstanding Major League career primarily with the Atlanta Braves and Los Angeles Dodgers. Dusty has been the manager of the San Francisco Giants since 1993, winning the National League Manager of the Year Award in 1993, 1997, and 2000.

Dusty has amazing people skills. He speaks fluent Spanish and easily relates to players from a variety of backgrounds. Rod Beck, former relief pitcher with the Giants, once said of Dusty, "If you can't play for Dusty Baker, you can't play for anybody."

Dusty is an avid reader, finding inspiration in books on positive thinking, because he wants all the creative aids he can find to counsel his ballplayers. He refuses to let his players consider the word "slump," referring to such experiences as "unfortunate time periods." Major-league players face many failures on a daily basis. Think about it: A player hitting .300 gets 3 hits in 10 at-bats, but also fails 7 out of 10 times at bat! A .300 hitter is considered highly successful.

What might Dusty say to you as a math student? You may have had some negative experiences, or "unfortunate time periods," in your past with mathematics, or maybe even as you take this course. What can you do? One answer is to keep picking yourself up with an intense resolve to keep on with the task, not letting negative thinking drag you down. Think about how Dusty and others overcame obstacles to achieve positive results.

In 1993, your author had the pleasure of coauthoring with Dusty a book, a three-part video, and a CD-ROM titled *You Can Teach Hitting*. For more on Dusty, visit www.dustybaker.com.

6.2 EXERCISE SET

For Extra Help

Digital Video Tutor CD 6 Videotape 12 | InterAct Math | Math Tutor Center | MathXL | MyMathLab

a Find the reciprocal.

1. $\frac{4}{x}$

2. $\frac{a + 3}{a - 1}$

3. $x^2 - y^2$

4. $x^2 - 5x + 7$

5. $\frac{1}{a + b}$

6. $\frac{x^2}{x^2 - 3}$

7. $\frac{x^2 + 2x - 5}{x^2 - 4x + 7}$

8. $\frac{(a - b)(a + b)}{(a + 4)(a - 5)}$

b Divide and simplify.

9. $\frac{2}{5} \div \frac{4}{3}$

10. $\frac{3}{10} \div \frac{3}{2}$

11. $\frac{2}{x} \div \frac{8}{x}$

12. $\frac{t}{3} \div \frac{t}{15}$

13. $\frac{a}{b^2} \div \frac{a^2}{b^3}$

14. $\frac{x^2}{y} \div \frac{x^3}{y^3}$

15. $\frac{a + 2}{a - 3} \div \frac{a - 1}{a + 3}$

16. $\frac{x - 8}{x + 9} \div \frac{x + 2}{x - 1}$

17. $\frac{x^2 - 1}{x} \div \frac{x + 1}{x - 1}$

18. $\frac{4y - 8}{y + 2} \div \frac{y - 2}{y^2 - 4}$

19. $\frac{x + 1}{6} \div \frac{x + 1}{3}$

20. $\frac{a}{a - b} \div \frac{b}{a - b}$

21. $\frac{5x - 5}{16} \div \frac{x - 1}{6}$

22. $\frac{4y - 12}{12} \div \frac{y - 3}{3}$

23. $\frac{-6 + 3x}{5} \div \frac{4x - 8}{25}$

24. $\frac{-12 + 4x}{4} \div \frac{-6 + 2x}{6}$

25. $\frac{a + 2}{a - 1} \div \frac{3a + 6}{a - 5}$

26. $\frac{t - 3}{t + 2} \div \frac{4t - 12}{t + 1}$

27. $\frac{x^2 - 4}{x} \div \frac{x - 2}{x + 2}$

28. $\frac{x + y}{x - y} \div \frac{x^2 + y}{x^2 - y^2}$

29. $\frac{x^2 - 9}{4x + 12} \div \frac{x - 3}{6}$

30. $\frac{a - b}{2a} \div \frac{a^2 - b^2}{8a^3}$

31. $\frac{c^2 + 3c}{c^2 + 2c - 3} \div \frac{c}{c + 1}$

32. $\frac{y + 5}{2y} \div \frac{y^2 - 25}{4y^2}$

33. $\dfrac{2y^2 - 7y + 3}{2y^2 + 3y - 2} \div \dfrac{6y^2 - 5y + 1}{3y^2 + 5y - 2}$

34. $\dfrac{x^2 + x - 20}{x^2 - 7x + 12} \div \dfrac{x^2 + 10x + 25}{x^2 - 6x + 9}$

35. $\dfrac{x^2 - 1}{4x + 4} \div \dfrac{2x^2 - 4x + 2}{8x + 8}$

36. $\dfrac{5t^2 + 5t - 30}{10t + 30} \div \dfrac{2t^2 - 8}{6t^2 + 36t + 54}$

37. $^{\mathrm{D}}\mathrm{W}$ Is the reciprocal of a product the product of the reciprocals? Why or why not?

38. $^{\mathrm{D}}\mathrm{W}$ Explain why 5, -1, and 7 are *not* allowable replacements in the division

$$\frac{x + 3}{x - 5} \div \frac{x - 7}{x + 1}.$$

SKILL MAINTENANCE

Solve.

39. Bonnie is taking an astronomy course. In order to receive an A, she must average at least 90 after four exams. Bonnie scored 96, 98, and 89 on the first three tests. Determine (in terms of an inequality) what scores on the last test will earn her an A. [2.8b]

40. *Triangle Dimensions.* The base of a triangle is 6 cm greater than twice the height. The area is 28 cm^2. Find the height and the base. [5.8a]

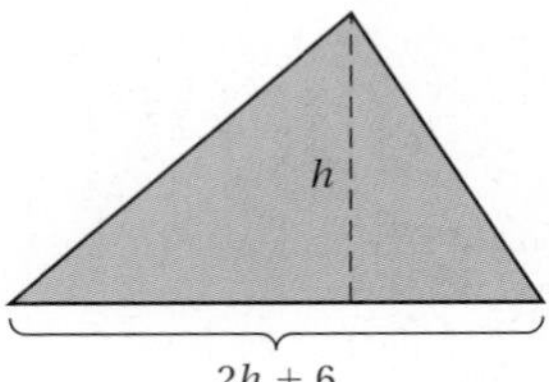

Subtract. [4.4c]

41. $(8x^3 - 3x^2 + 7) - (8x^2 + 3x - 5)$

42. $(3p^2 - 6pq + 7q^2) - (5p^2 - 10pq + 11q^2)$

Simplify. [4.2b]

43. $(2x^{-3}y^4)^2$

44. $(5x^6y^{-4})^3$

45. $\left(\dfrac{2x^3}{y^5}\right)^2$

46. $\left(\dfrac{a^{-3}}{b^4}\right)^5$

SYNTHESIS

Simplify.

47. $\dfrac{3a^2 - 5ab - 12b^2}{3ab + 4b^2} \div (3b^2 - ab)$

48. $\dfrac{3x + 3y + 3}{9x} \div \dfrac{x^2 + 2xy + y^2 - 1}{x^4 + x^2}$

49. The volume of this rectangular solid is $x - 3$. What is its height?

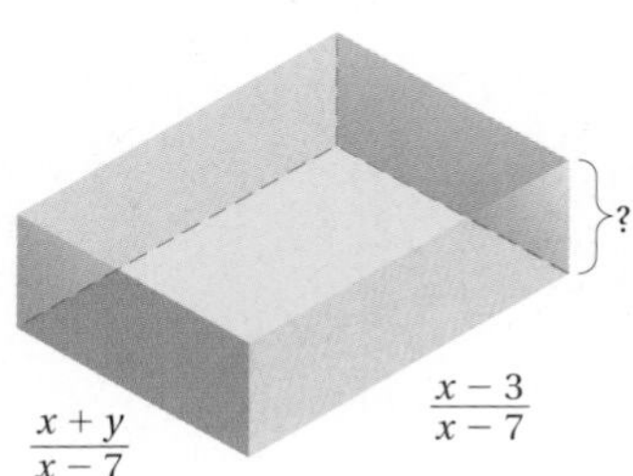

6.3 LEAST COMMON MULTIPLES AND DENOMINATORS

Objectives

a. Find the LCM of several numbers by factoring.
b. Add fractions, first finding the LCD.
c. Find the LCM of algebraic expressions by factoring.

a Least Common Multiples

To add when denominators are different, we first find a common denominator. For a review, see Sections R.1 and R.2. We saw there, for example, that to add $\frac{5}{12}$ and $\frac{7}{30}$, we first look for the **least common multiple, LCM,** of both 12 and 30. That number becomes the **least common denominator, LCD.** To find the LCM of 12 and 30, we factor:

$$12 = 2 \cdot 2 \cdot 3;$$
$$30 = 2 \cdot 3 \cdot 5.$$

The LCM is the number that has 2 as a factor twice, 3 as a factor once, and 5 as a factor once:

$$\text{LCM} = 2 \cdot 2 \cdot 3 \cdot 5 = 60.$$

12 is a factor of the LCM.

30 is a factor of the LCM.

FINDING LCMS

To find the LCM, use each factor the greatest number of times that it appears in any one factorization.

EXAMPLE 1 Find the LCM of 24 and 36.

$$\left.\begin{aligned} 24 &= 2 \cdot 2 \cdot 2 \cdot 3 \\ 36 &= 2 \cdot 2 \cdot 3 \cdot 3 \end{aligned}\right\} \quad \text{LCM} = 2 \cdot 2 \cdot 2 \cdot 3 \cdot 3, \text{ or } 72$$

Do Exercises 1–4.

Find the LCM by factoring.

1. 16, 18

2. 6, 12

3. 2, 5

4. 24, 30, 20

Answers on page A-29

b Adding Using the LCD

Let's finish adding $\frac{5}{12}$ and $\frac{7}{30}$:

$$\frac{5}{12} + \frac{7}{30} = \frac{5}{2 \cdot 2 \cdot 3} + \frac{7}{2 \cdot 3 \cdot 5}.$$

The least common denominator, LCD, is $2 \cdot 2 \cdot 3 \cdot 5$. To get the LCD in the first denominator, we need a 5. To get the LCD in the second denominator, we need another 2. We get these numbers by multiplying by 1:

$$\frac{5}{12} + \frac{7}{30} = \frac{5}{2 \cdot 2 \cdot 3} \cdot \frac{5}{5} + \frac{7}{2 \cdot 3 \cdot 5} \cdot \frac{2}{2}$$ Multiplying by 1

$$= \frac{25}{2 \cdot 2 \cdot 3 \cdot 5} + \frac{14}{2 \cdot 3 \cdot 5 \cdot 2}$$ The denominators are now the LCD.

$$= \frac{39}{2 \cdot 2 \cdot 3 \cdot 5}$$ Adding the numerators and keeping the LCD

$$= \frac{\not{3} \cdot 13}{2 \cdot 2 \cdot \not{3} \cdot 5}$$ Factoring the numerator and removing a factor of 1: $\frac{3}{3} = 1$

$$= \frac{13}{20}.$$ Simplifying

Add, first finding the LCD. Simplify if possible.

5. $\frac{3}{16} + \frac{1}{18}$

6. $\frac{1}{6} + \frac{1}{12}$

7. $\frac{1}{2} + \frac{3}{5}$

8. $\frac{1}{24} + \frac{1}{30} + \frac{3}{20}$

Find the LCM.

9. $12xy^2,\ 15x^3y$

10. $y^2 + 5y + 4,\ y^2 + 2y + 1$

11. $t^2 + 16,\ t - 2,\ 7$

12. $x^2 + 2x + 1,\ 3x^2 - 3x,\ x^2 - 1$

Answers on page A-29

EXAMPLE 2 Add: $\frac{5}{12} + \frac{11}{18}$.

$$\left.\begin{aligned} 12 &= 2 \cdot 2 \cdot 3 \\ 18 &= 2 \cdot 3 \cdot 3 \end{aligned}\right\} \quad \text{LCD} = 2 \cdot 2 \cdot 3 \cdot 3, \text{ or } 36$$

$$\frac{5}{12} + \frac{11}{18} = \frac{5}{2 \cdot 2 \cdot 3} \cdot \frac{3}{3} + \frac{11}{2 \cdot 3 \cdot 3} \cdot \frac{2}{2} = \frac{15 + 22}{2 \cdot 2 \cdot 3 \cdot 3} = \frac{37}{36}$$

Do Exercises 5–8.

C LCMs of Algebraic Expressions

To find the LCM of two or more algebraic expressions, we factor them. Then we use each factor the greatest number of times that it occurs in any one expression. In Section 6.4, each LCM will become an LCD used to add rational expressions.

EXAMPLE 3 Find the LCM of $12x$, $16y$, and $8xyz$.

$$\left.\begin{aligned} 12x &= 2 \cdot 2 \cdot 3 \cdot x \\ 16y &= 2 \cdot 2 \cdot 2 \cdot 2 \cdot y \\ 8xyz &= 2 \cdot 2 \cdot 2 \cdot x \cdot y \cdot z \end{aligned}\right\} \quad \begin{aligned} \text{LCM} &= 2 \cdot 2 \cdot 2 \cdot 2 \cdot 3 \cdot x \cdot y \cdot z \\ &= 48xyz \end{aligned}$$

EXAMPLE 4 Find the LCM of $x^2 + 5x - 6$ and $x^2 - 1$.

$$\left.\begin{aligned} x^2 + 5x - 6 &= (x + 6)(x - 1) \\ x^2 - 1 &= (x + 1)(x - 1) \end{aligned}\right\} \quad \text{LCM} = (x + 6)(x - 1)(x + 1)$$

EXAMPLE 5 Find the LCM of $x^2 + 4$, $x + 1$, and 5.

These expressions do not share a common factor other than 1, so the LCM is their product:

$$5(x^2 + 4)(x + 1).$$

EXAMPLE 6 Find the LCM of $x^2 - 25$ and $2x - 10$.

$$\left.\begin{aligned} x^2 - 25 &= (x + 5)(x - 5) \\ 2x - 10 &= 2(x - 5) \end{aligned}\right\} \quad \text{LCM} = 2(x + 5)(x - 5)$$

EXAMPLE 7 Find the LCM of $x^2 - 4y^2$, $x^2 - 4xy + 4y^2$, and $x - 2y$.

$$\left.\begin{aligned} x^2 - 4y^2 &= (x - 2y)(x + 2y) \\ x^2 - 4xy + 4y^2 &= (x - 2y)(x - 2y) \\ x - 2y &= x - 2y \end{aligned}\right\} \quad \begin{aligned} \text{LCM} &= (x + 2y)(x - 2y)(x - 2y) \\ &= (x + 2y)(x - 2y)^2 \end{aligned}$$

Do Exercises 9–12.

6.3 EXERCISE SET

For Extra Help

Digital Video Tutor CD 6 Videotape 12 | InterAct Math | Math Tutor Center | MathXL | MyMathLab

a Find the LCM.

1. 12, 27

2. 10, 15

3. 8, 9

4. 12, 18

5. 6, 9, 21

6. 8, 36, 40

7. 24, 36, 40

8. 4, 5, 20

9. 10, 100, 500

10. 28, 42, 60

b Add, first finding the LCD. Simplify if possible.

11. $\frac{7}{24} + \frac{11}{18}$

12. $\frac{7}{60} + \frac{2}{25}$

13. $\frac{1}{6} + \frac{3}{40}$

14. $\frac{5}{24} + \frac{3}{20}$

15. $\frac{1}{20} + \frac{1}{30} + \frac{2}{45}$

16. $\frac{2}{15} + \frac{5}{9} + \frac{3}{20}$

c Find the LCM.

17. $6x^2,\ 12x^3$

18. $2a^2b,\ 8ab^3$

19. $2x^2,\ 6xy,\ 18y^2$

20. $p^3q,\ p^2q,\ pq^2$

21. $2(y-3),\ 6(y-3)$

22. $5(m+2),\ 15(m+2)$

23. $t,\ t+2,\ t-2$

24. $y,\ y-5,\ y+5$

25. $x^2-4,\ x^2+5x+6$

26. $x^2-4,\ x^2-x-2$

27. $t^3+4t^2+4t,\ t^2-4t$

28. $m^4-m^2,\ m^3-m^2$

29. $a+1,\ (a-1)^2,\ a^2-1$

30. $a^2-2ab+b^2,\ a^2-b^2,\ 3a+3b$

31. $m^2-5m+6,\ m^2-4m+4$

32. $2x^2+5x+2,\ 2x^2-x-1$

33. $2+3x,\ 4-9x^2,\ 2-3x$

34. $9-4x^2,\ 3+2x,\ 3-2x$

35. $10v^2+30v,\ 5v^2+35v+60$

36. $12a^2+24a,\ 4a^2+20a+24$

37. $9x^3-9x^2-18x,\ 6x^5-24x^4+24x^3$

38. $x^5-4x^3,\ x^3+4x^2+4x$

39. $x^5+4x^4+4x^3,\ 3x^2-12,\ 2x+4$

40. $x^5+2x^4+x^3,\ 2x^3-2x,\ 5x-5$

41. D_W If the LCM of a binomial and a trinomial is the trinomial, what relationship exists between the two expressions?

42. D_W Explain why the product of two numbers is not always their least common multiple.

SKILL MAINTENANCE

Factor. [5.6a]

43. $x^2 - 6x + 9$

44. $6x^2 + 4x$

45. $x^2 - 9$

46. $x^2 + 4x - 21$

47. $x^2 + 6x + 9$

48. $x^2 - 4x - 21$

Divorce Rate. The graph at right is that of the equation

$$D = 0.00509x^2 - 19.17x + 18{,}065.305$$

for values of x ranging from 1900 to 2010. It shows the percentage of couples who are married in a given year, x, whose marriages, it is predicted, will end in divorce. Use *only* the graph to answer the questions in Exercises 49–54. [3.1a], [4.3a]

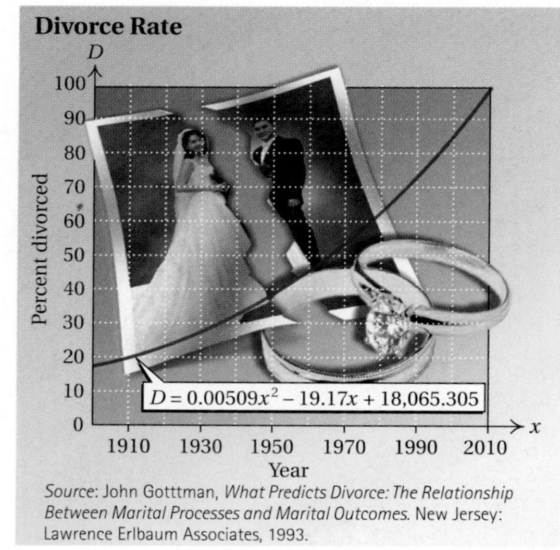

Source: John Gotttman, *What Predicts Divorce: The Relationship Between Marital Processes and Marital Outcomes.* New Jersey: Lawrence Erlbaum Associates, 1993.

49. Estimate the divorce percentage of those married in 1970.

50. Estimate the divorce percentage of those married in 1980.

51. Estimate the divorce percentage of those married in 1990.

52. Estimate the divorce percentage of those married in 2010. Does this seem reasonable?

53. In what year was the divorce percentage about 50%?

54. In what year was the divorce percentage about 84%?

SYNTHESIS

55. *Running.* Pedro and Maria leave the starting point of a fitness loop at the same time. Pedro jogs a lap in 6 min and Maria jogs one in 8 min. Assuming they continue to run at the same pace, when will they next meet at the starting place?

6.4 ADDING RATIONAL EXPRESSIONS

a Adding Rational Expressions

We add rational expressions as we do rational numbers.

ADDING RATIONAL EXPRESSIONS WITH LIKE DENOMINATORS

To add when the denominators are the same, add the numerators and keep the same denominator. Then simplify if possible.

EXAMPLES Add.

1. $\dfrac{x}{x+1} + \dfrac{2}{x+1} = \dfrac{x+2}{x+1}$

2. $\dfrac{2x^2+3x-7}{2x+1} + \dfrac{x^2+x-8}{2x+1} = \dfrac{(2x^2+3x-7)+(x^2+x-8)}{2x+1}$

$$= \frac{3x^2+4x-15}{2x+1}$$

3. $\dfrac{x-5}{x^2-9} + \dfrac{2}{x^2-9} = \dfrac{(x-5)+2}{x^2-9} = \dfrac{x-3}{x^2-9}$

$$= \frac{x-3}{(x-3)(x+3)} \quad \text{Factoring}$$

$$= \frac{\cancel{x-3}}{\cancel{(x-3)}(x+3)} \quad \text{Removing a factor of 1: } \frac{x-3}{x-3} = 1$$

$$= \frac{1}{x+3} \quad \text{Simplifying}$$

As in Example 3, simplifying should be done if possible after adding.

Do Exercises 1–3.

When denominators are different, we find the least common denominator, LCD. The procedure we use is as follows.

ADDING RATIONAL EXPRESSIONS WITH DIFFERENT DENOMINATORS

To add rational expressions with different denominators:

1. Find the LCM of the denominators. This is the least common denominator (LCD).
2. For each rational expression, find an equivalent expression with the LCD. To do so, multiply by 1 using an expression for 1 made up of factors of the LCD that are missing from the original denominator.
3. Add the numerators. Write the sum over the LCD.
4. Simplify if possible.

Objective

a Add rational expressions.

Add.

1. $\dfrac{5}{9} + \dfrac{2}{9}$

2. $\dfrac{3}{x-2} + \dfrac{x}{x-2}$

3. $\dfrac{4x+5}{x-1} + \dfrac{2x-1}{x-1}$

Answers on page A-29

Add.

4. $\dfrac{3x}{16} + \dfrac{5x^2}{24}$

5. $\dfrac{3}{16x} + \dfrac{5}{24x^2}$

6. Add:

$$\frac{3}{x^3 - x} + \frac{4}{x^2 + 2x + 1}.$$

Answers on page A-29

EXAMPLE 4 Add: $\dfrac{5x^2}{8} + \dfrac{7x}{12}$.

First, we find the LCD:

$$\left.\begin{aligned} 8 &= 2 \cdot 2 \cdot 2 \\ 12 &= 2 \cdot 2 \cdot 3 \end{aligned}\right\} \quad \text{LCD} = 2 \cdot 2 \cdot 2 \cdot 3, \text{ or } 24.$$

Compare the factorization $8 = 2 \cdot 2 \cdot 2$ with the factorization of the LCD, $24 = 2 \cdot 2 \cdot 2 \cdot 3$. The factor of 24 that is missing from 8 is 3. Compare $12 = 2 \cdot 2 \cdot 3$ and $24 = 2 \cdot 2 \cdot 2 \cdot 3$. The factor of 24 that is missing from 12 is 2.

We multiply by a symbol for 1 to get the LCD in each expression, and then add and, if possible, simplify:

$$\begin{aligned} \frac{5x^2}{8} + \frac{7x}{12} &= \frac{5x^2}{2 \cdot 2 \cdot 2} + \frac{7x}{2 \cdot 2 \cdot 3} \\ &= \frac{5x^2}{2 \cdot 2 \cdot 2} \cdot \frac{3}{3} + \frac{7x}{2 \cdot 2 \cdot 3} \cdot \frac{2}{2} \qquad \text{Multiplying by 1 to get the same denominators} \\ &= \frac{15x^2}{24} + \frac{14x}{24} = \frac{15x^2 + 14x}{24}. \end{aligned}$$

EXAMPLE 5 Add: $\dfrac{3}{8x} + \dfrac{5}{12x^2}$.

First, we find the LCD:

$$\left.\begin{aligned} 8x &= 2 \cdot 2 \cdot 2 \cdot x \\ 12x^2 &= 2 \cdot 2 \cdot 3 \cdot x \cdot x \end{aligned}\right\} \quad \text{LCD} = 2 \cdot 2 \cdot 2 \cdot 3 \cdot x \cdot x, \text{ or } 24x^2.$$

The factors of the LCD missing from $8x$ are 3 and x. The factor of the LCD missing from $12x^2$ is 2. We multiply by 1 to get the LCD in each expression, and then add and, if possible, simplify:

$$\begin{aligned} \frac{3}{8x} + \frac{5}{12x^2} &= \frac{3}{8x} \cdot \frac{3 \cdot x}{3 \cdot x} + \frac{5}{12x^2} \cdot \frac{2}{2} \\ &= \frac{9x}{24x^2} + \frac{10}{24x^2} = \frac{9x + 10}{24x^2}. \end{aligned}$$

Do Exercises 4 and 5.

EXAMPLE 6 Add: $\dfrac{2a}{a^2 - 1} + \dfrac{1}{a^2 + a}$.

First, we find the LCD:

$$\left.\begin{aligned} a^2 - 1 &= (a - 1)(a + 1) \\ a^2 + a &= a(a + 1) \end{aligned}\right\} \quad \text{LCD} = a(a - 1)(a + 1).$$

We multiply by 1 to get the LCD in each expression, and then add and simplify:

$$\begin{aligned} &\frac{2a}{(a - 1)(a + 1)} \cdot \frac{a}{a} + \frac{1}{a(a + 1)} \cdot \frac{a - 1}{a - 1} \\ &\quad = \frac{2a^2}{a(a - 1)(a + 1)} + \frac{a - 1}{a(a - 1)(a + 1)} \\ &\quad = \frac{2a^2 + a - 1}{a(a - 1)(a + 1)} \\ &\quad = \frac{(a + 1)(2a - 1)}{a(a - 1)(a + 1)}. \qquad \text{Factoring the numerator in order to simplify} \end{aligned}$$

Then

$$= \frac{(a+1)(2a-1)}{a(a-1)(a+1)} \quad \text{Removing a factor of 1: } \frac{a+1}{a+1} = 1$$
$$= \frac{2a-1}{a(a-1)}.$$

Do Exercise 6 on the preceding page.

EXAMPLE 7 Add: $\frac{x+4}{x-2} + \frac{x-7}{x+5}$.

First, we find the LCD. It is just the product of the denominators:

LCD $= (x-2)(x+5)$.

We multiply by 1 to get the LCD in each expression, and then add and simplify:

$$\frac{x+4}{x-2} \cdot \frac{x+5}{x+5} + \frac{x-7}{x+5} \cdot \frac{x-2}{x-2}$$
$$= \frac{(x+4)(x+5)}{(x-2)(x+5)} + \frac{(x-7)(x-2)}{(x-2)(x+5)}$$
$$= \frac{x^2+9x+20}{(x-2)(x+5)} + \frac{x^2-9x+14}{(x-2)(x+5)}$$
$$= \frac{x^2+9x+20+x^2-9x+14}{(x-2)(x+5)} = \frac{2x^2+34}{(x-2)(x+5)}.$$

Do Exercise 7.

EXAMPLE 8 Add: $\frac{x}{x^2+11x+30} + \frac{-5}{x^2+9x+20}$.

$$\frac{x}{x^2+11x+30} + \frac{-5}{x^2+9x+20}$$
$$= \frac{x}{(x+5)(x+6)} + \frac{-5}{(x+5)(x+4)}$$
Factoring the denominators in order to find the LCD. The LCD is $(x+4)(x+5)(x+6)$.
$$= \frac{x}{(x+5)(x+6)} \cdot \frac{x+4}{x+4} + \frac{-5}{(x+5)(x+4)} \cdot \frac{x+6}{x+6} \quad \text{Multiplying by 1}$$
$$= \frac{x(x+4)+(-5)(x+6)}{(x+4)(x+5)(x+6)} = \frac{x^2+4x-5x-30}{(x+4)(x+5)(x+6)}$$
$$= \frac{x^2-x-30}{(x+4)(x+5)(x+6)}$$
$$= \frac{(x-6)(x+5)}{(x+4)(x+5)(x+6)}$$
$$= \frac{(x-6)}{(x+4)(x+6)}$$
Always simplify at the end if possible: $\frac{x+5}{x+5} = 1$.

Do Exercise 8.

DENOMINATORS THAT ARE OPPOSITES

When one denominator is the opposite of the other, we can first multiply either expression by 1 using $-1/-1$.

7. Add:

$$\frac{x-2}{x+3} + \frac{x+7}{x+8}.$$

8. Add:

$$\frac{5}{x^2+17x+16} + \frac{3}{x^2+9x+8}.$$

CALCULATOR CORNER

Checking Addition Use the TABLE feature, as described on p. 471, to check the additions in Examples 7 and 9. Then check your answers to Margin Exercises 7–9.

Answers on page A-29

Add.

9. $\dfrac{x}{4} + \dfrac{5}{-4}$

10. $\dfrac{2x+1}{x-3} + \dfrac{x+2}{3-x}$

11. Add:

$$\frac{x+3}{x^2-16} + \frac{5}{12-3x}.$$

Answers on page A-29

EXAMPLES

9. $\dfrac{x}{2} + \dfrac{3}{-2} = \dfrac{x}{2} + \dfrac{3}{-2} \cdot \dfrac{-1}{-1}$ Multiplying by 1 using $\dfrac{-1}{-1}$

$= \dfrac{x}{2} + \dfrac{-3}{2}$ The denominators are now the same.

$= \dfrac{x+(-3)}{2} = \dfrac{x-3}{2}$

10. $\dfrac{3x+4}{x-2} + \dfrac{x-7}{2-x} = \dfrac{3x+4}{x-2} + \dfrac{x-7}{2-x} \cdot \dfrac{-1}{-1}$

We could have chosen to multiply this expression by $-1/-1$. We multiply only one expression, *not* both.

$= \dfrac{3x+4}{x-2} + \dfrac{-x+7}{x-2}$ *Note:* $(2-x)(-1) = -2 + x = x - 2.$

$= \dfrac{(3x+4)+(-x+7)}{x-2} = \dfrac{2x+11}{x-2}$

Do Exercises 9 and 10.

FACTORS THAT ARE OPPOSITES

Suppose that when we factor to find the LCD, we find factors that are opposites. The easiest way to handle this is to first go back and multiply by $-1/-1$ appropriately to change factors so that they are not opposites.

EXAMPLE 11 Add: $\dfrac{x}{x^2-25} + \dfrac{3}{10-2x}.$

First, we factor to find the LCD:

$x^2 - 25 = (x-5)(x+5);$

$10 - 2x = 2(5-x).$

We note that there is an $x - 5$ as one factor of $x^2 - 25$ and a $5 - x$ as one factor of $10 - 2x$. If the denominator of the second expression were $2x - 10$, this situation would not occur. To rewrite the second expression with a denominator of $2x - 10$, we multiply by 1 using $-1/-1$, and then continue as before:

$$\frac{x}{x^2-25} + \frac{3}{10-2x} = \frac{x}{(x-5)(x+5)} + \frac{3}{10-2x} \cdot \frac{-1}{-1}$$

$$= \frac{x}{(x-5)(x+5)} + \frac{-3}{2x-10}$$

$$= \frac{x}{(x-5)(x+5)} + \frac{-3}{2(x-5)} \qquad \text{LCD} = 2(x-5)(x+5)$$

$$= \frac{x}{(x-5)(x+5)} \cdot \frac{2}{2} + \frac{-3}{2(x-5)} \cdot \frac{x+5}{x+5}$$

$$= \frac{2x}{2(x-5)(x+5)} + \frac{-3(x+5)}{2(x-5)(x+5)}$$

$$= \frac{2x-3(x+5)}{2(x-5)(x+5)} = \frac{2x-3x-15}{2(x-5)(x+5)}$$

$$= \frac{-x-15}{2(x-5)(x+5)}. \qquad \text{Collecting like terms}$$

Do Exercise 11.

6.4 EXERCISE SET

For Extra Help

Digital Video Tutor CD 6 Videotape 12 | InterAct Math | Math Tutor Center | MathXL | MyMathLab

a Add. Simplify if possible.

1. $\dfrac{5}{8} + \dfrac{3}{8}$

2. $\dfrac{3}{16} + \dfrac{5}{16}$

3. $\dfrac{1}{3 + x} + \dfrac{5}{3 + x}$

4. $\dfrac{4x + 6}{2x - 1} + \dfrac{5 - 8x}{-1 + 2x}$

5. $\dfrac{x^2 + 7x}{x^2 - 5x} + \dfrac{x^2 - 4x}{x^2 - 5x}$

6. $\dfrac{4}{x + y} + \dfrac{9}{y + x}$

7. $\dfrac{2}{x} + \dfrac{5}{x^2}$

8. $\dfrac{3}{y^2} + \dfrac{6}{y}$

9. $\dfrac{5}{6r} + \dfrac{7}{8r}$

10. $\dfrac{13}{18x} + \dfrac{7}{24x}$

11. $\dfrac{4}{xy^2} + \dfrac{6}{x^2y}$

12. $\dfrac{8}{ab^3} + \dfrac{3}{a^2b}$

13. $\dfrac{2}{9t^3} + \dfrac{1}{6t^2}$

14. $\dfrac{5}{c^2d^3} + \dfrac{-4}{7cd^2}$

15. $\dfrac{x + y}{xy^2} + \dfrac{3x + y}{x^2y}$

16. $\dfrac{2c - d}{c^2d} + \dfrac{c + d}{cd^2}$

17. $\dfrac{3}{x - 2} + \dfrac{3}{x + 2}$

18. $\dfrac{2}{y + 1} + \dfrac{2}{y - 1}$

19. $\dfrac{3}{x + 1} + \dfrac{2}{3x}$

20. $\dfrac{4}{5y} + \dfrac{7}{y - 2}$

21. $\dfrac{2x}{x^2 - 16} + \dfrac{x}{x - 4}$

22. $\frac{4x}{x^2 - 25} + \frac{x}{x + 5}$

23. $\frac{5}{z + 4} + \frac{3}{3z + 12}$

24. $\frac{t}{t - 3} + \frac{5}{4t - 12}$

25. $\frac{3}{x - 1} + \frac{2}{(x - 1)^2}$

26. $\frac{8}{(y + 3)^2} + \frac{5}{y + 3}$

27. $\frac{4a}{5a - 10} + \frac{3a}{10a - 20}$

28. $\frac{9x}{6x - 30} + \frac{3x}{4x - 20}$

29. $\frac{x + 4}{x} + \frac{x}{x + 4}$

30. $\frac{a}{a - 3} + \frac{a - 3}{a}$

31. $\frac{4}{a^2 - a - 2} + \frac{3}{a^2 + 4a + 3}$

32. $\frac{a}{a^2 - 2a + 1} + \frac{1}{a^2 - 5a + 4}$

33. $\frac{x + 3}{x - 5} + \frac{x - 5}{x + 3}$

34. $\frac{3x}{2y - 3} + \frac{2x}{3y - 2}$

35. $\frac{a}{a^2 - 1} + \frac{2a}{a^2 - a}$

36. $\frac{3x + 2}{3x + 6} + \frac{x - 2}{x^2 - 4}$

37. $\frac{7}{8} + \frac{5}{-8}$

38. $\frac{5}{-3} + \frac{11}{3}$

39. $\frac{3}{t} + \frac{4}{-t}$

40. $\frac{5}{-a} + \frac{8}{a}$

41. $\frac{2x + 7}{x - 6} + \frac{3x}{6 - x}$

42. $\frac{2x - 7}{5x - 8} + \frac{6 + 10x}{8 - 5x}$

43. $\frac{y^2}{y-3} + \frac{9}{3-y}$

44. $\frac{t^2}{t-2} + \frac{4}{2-t}$

45. $\frac{b-7}{b^2-16} + \frac{7-b}{16-b^2}$

46. $\frac{a-3}{a^2-25} + \frac{a-3}{25-a^2}$

47. $\frac{a^2}{a-b} + \frac{b^2}{b-a}$

48. $\frac{x^2}{x-7} + \frac{49}{7-x}$

49. $\frac{x+3}{x-5} + \frac{2x-1}{5-x} + \frac{2(3x-1)}{x-5}$

50. $\frac{3(x-2)}{2x-3} + \frac{5(2x+1)}{2x-3} + \frac{3(x+1)}{3-2x}$

51. $\frac{2(4x+1)}{5x-7} + \frac{3(x-2)}{7-5x} + \frac{-10x-1}{5x-7}$

52. $\frac{5(x-2)}{3x-4} + \frac{2(x-3)}{4-3x} + \frac{3(5x+1)}{4-3x}$

53. $\frac{x+1}{(x+3)(x-3)} + \frac{4(x-3)}{(x-3)(x+3)} + \frac{(x-1)(x-3)}{(3-x)(x+3)}$

54. $\frac{2(x+5)}{(2x-3)(x-1)} + \frac{3x+4}{(2x-3)(1-x)} + \frac{x-5}{(3-2x)(x-1)}$

55. $\frac{6}{x-y} + \frac{4x}{y^2-x^2}$

56. $\frac{a-2}{3-a} + \frac{4-a^2}{a^2-9}$

57. $\frac{4-a}{25-a^2} + \frac{a+1}{a-5}$

58. $\frac{x+2}{x-7} + \frac{3-x}{49-x^2}$

59. $\frac{2}{t^2+t-6} + \frac{3}{t^2-9}$

60. $\frac{10}{a^2-a-6} + \frac{3a}{a^2+4a+4}$

61. ^{D}W Explain why the expressions

$$\frac{1}{3-x} \text{ and } \frac{1}{x-3}$$

are opposites.

62. ^{D}W A student insists on finding a common denominator by always multiplying the denominators of the expressions being added. How could this approach be improved?

SKILL MAINTENANCE

Subtract. [4.4c]

63. $(x^2 + x) - (x + 1)$

64. $(4y^3 - 5y^2 + 7y - 24) - (-9y^3 + 9y^2 - 5y + 49)$

Simplify. [4.2b]

65. $(2x^4y^3)^{-3}$

66. $\left(\dfrac{x^3}{5y}\right)^2$

67. $\left(\dfrac{x^{-4}}{y^7}\right)^3$

68. $(5x^{-2}y^{-3})^2$

Graph.

69. $y = \dfrac{1}{2}x - 5$ [3.2b], [3.3a]

70. $2y + x + 10 = 0$ [3.2b], [3.3a]

71. $y = 3$ [3.3b]

72. $x = -5$ [3.3b]

Solve.

73. $3x - 7 = 5x + 9$ [2.3b]

74. $2a + 8 = 13 - 4a$ [2.3b]

75. $x^2 - 8x + 15 = 0$ [5.7b]

76. $x^2 - 7x = 18$ [5.7b]

SYNTHESIS

Find the perimeter and the area of the figure.

77.

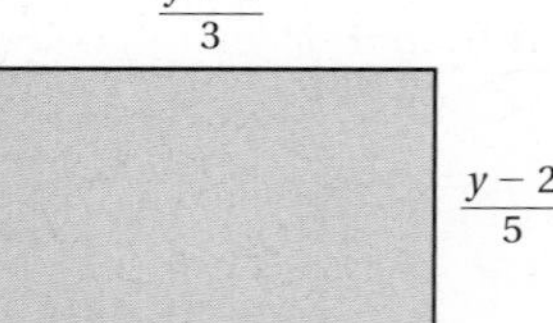

78.

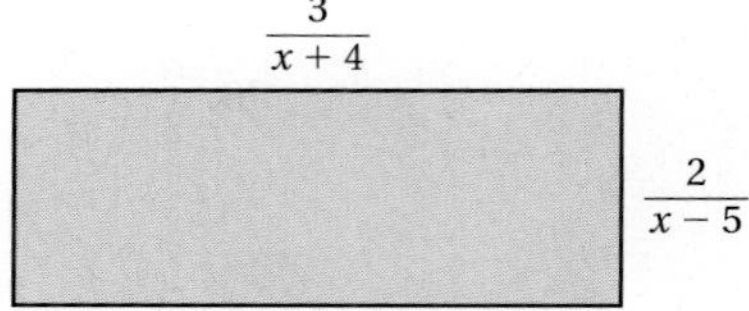

Add. Simplify if possible.

79. $\dfrac{5}{z+2} + \dfrac{4z}{z^2 - 4} + 2$

80. $\dfrac{-2}{y^2 - 9} + \dfrac{4y}{(y-3)^2} + \dfrac{6}{3-y}$

81. $\dfrac{3z^2}{z^4 - 4} + \dfrac{5z^2 - 3}{2z^4 + z^2 - 6}$

82. Find an expression equivalent to

$$\frac{a - 3b}{a - b}$$

that is a sum of two rational expressions. Answers may vary.

83.–86. Use the TABLE feature to check the additions in Exercises 29–32.

6.5 SUBTRACTING RATIONAL EXPRESSIONS

a Subtracting Rational Expressions

We subtract rational expressions as we do rational numbers.

SUBTRACTING RATIONAL EXPRESSIONS WITH LIKE DENOMINATORS

To subtract when the denominators are the same, subtract the numerators and keep the same denominator. Then simplify if possible.

EXAMPLE 1 Subtract: $\frac{8}{x} - \frac{3}{x}$.

$$\frac{8}{x} - \frac{3}{x} = \frac{8-3}{x} = \frac{5}{x}$$

EXAMPLE 2 Subtract: $\frac{3x}{x+2} - \frac{x-2}{x+2}$.

$$\frac{3x}{x+2} - \frac{x-2}{x+2} = \frac{3x-(x-2)}{x+2}$$

CAUTION! The parentheses are important to make sure that you subtract the entire numerator.

$$= \frac{3x-x+2}{x+2} \quad \text{Removing parentheses}$$

$$= \frac{2x+2}{x+2}$$

Do Exercises 1–3.

To subtract rational expressions with different denominators, we use a procedure similar to what we used for addition, except that we subtract numerators and write the difference over the LCD.

SUBTRACTING RATIONAL EXPRESSIONS WITH DIFFERENT DENOMINATORS

To subtract rational expressions with different denominators:

1. Find the LCM of the denominators. This is the least common denominator (LCD).
2. For each rational expression, find an equivalent expression with the LCD. To do so, multiply by 1 using a symbol for 1 made up of factors of the LCD that are missing from the original denominator.
3. Subtract the numerators. Write the difference over the LCD.
4. Simplify if possible.

Objectives

a Subtract rational expressions.

b Simplify combined additions and subtractions of rational expressions.

Subtract.

1. $\frac{7}{11} - \frac{3}{11}$

2. $\frac{7}{y} - \frac{2}{y}$

3. $\frac{2x^2+3x-7}{2x+1} - \frac{x^2+x-8}{2x+1}$

Answers on page A-30

4. Subtract:

$$\frac{x-2}{3x} - \frac{2x-1}{5x}.$$

5. Subtract:

$$\frac{x}{x^2+15x+56} - \frac{6}{x^2+13x+42}.$$

Answers on page A-30

EXAMPLE 3 Subtract: $\dfrac{x+2}{x-4} - \dfrac{x+1}{x+4}$.

The LCD $= (x-4)(x+4)$.

$$\frac{x+2}{x-4}\cdot\frac{x+4}{x+4} - \frac{x+1}{x+4}\cdot\frac{x-4}{x-4} \quad \text{Multiplying by 1}$$

$$= \frac{(x+2)(x+4)}{(x-4)(x+4)} - \frac{(x+1)(x-4)}{(x-4)(x+4)}$$

$$= \frac{x^2+6x+8}{(x-4)(x+4)} - \frac{x^2-3x-4}{(x-4)(x+4)}$$

Subtracting this numerator. Don't forget the parentheses.

$$= \frac{x^2+6x+8-(x^2-3x-4)}{(x-4)(x+4)}$$

$$= \frac{x^2+6x+8-x^2+3x+4}{(x-4)(x+4)} \quad \text{Removing parentheses}$$

$$= \frac{9x+12}{(x-4)(x+4)}$$

Do Exercise 4.

EXAMPLE 4 Subtract: $\dfrac{x}{x^2+5x+6} - \dfrac{2}{x^2+3x+2}$.

$$\frac{x}{x^2+5x+6} - \frac{2}{x^2+3x+2}$$

$$= \frac{x}{(x+2)(x+3)} - \frac{2}{(x+2)(x+1)} \quad \text{LCD} = (x+1)(x+2)(x+3)$$

$$= \frac{x}{(x+2)(x+3)}\cdot\frac{x+1}{x+1} - \frac{2}{(x+2)(x+1)}\cdot\frac{x+3}{x+3}$$

$$= \frac{x^2+x}{(x+1)(x+2)(x+3)} - \frac{2x+6}{(x+1)(x+2)(x+3)}$$

Subtracting this numerator. Don't forget the parentheses.

$$= \frac{x^2+x-(2x+6)}{(x+1)(x+2)(x+3)}$$

$$= \frac{x^2+x-2x-6}{(x+1)(x+2)(x+3)}$$

$$= \frac{x^2-x-6}{(x+1)(x+2)(x+3)}$$

$$= \frac{(x+2)(x-3)}{(x+1)(x+2)(x+3)}$$

$$= \frac{\cancel{(x+2)}(x-3)}{(x+1)\cancel{(x+2)}(x+3)} \quad \text{Simplifying by removing a factor of 1: } \frac{x+2}{x+2} = 1$$

$$= \frac{x-3}{(x+1)(x+3)}.$$

Do Exercise 5.

DENOMINATORS THAT ARE OPPOSITES

When one denominator is the opposite of the other, we can first multiply one expression by $-1/-1$ to obtain a common denominator.

EXAMPLE 5 Subtract: $\frac{x}{5} - \frac{3x-4}{-5}$.

$$\frac{x}{5} - \frac{3x-4}{-5} = \frac{x}{5} - \frac{3x-4}{-5} \cdot \frac{-1}{-1}$$ Multiplying by 1 using $\frac{-1}{-1}$ ← This is equal to 1 (not −1).

$$= \frac{x}{5} - \frac{(3x-4)(-1)}{(-5)(-1)}$$

$$= \frac{x}{5} - \frac{4-3x}{5}$$

$$= \frac{x-(4-3x)}{5}$$ Remember the parentheses!

$$= \frac{x-4+3x}{5} = \frac{4x-4}{5}$$

EXAMPLE 6 Subtract: $\frac{5y}{y-5} - \frac{2y-3}{5-y}$.

$$\frac{5y}{y-5} - \frac{2y-3}{5-y} = \frac{5y}{y-5} - \frac{2y-3}{5-y} \cdot \frac{-1}{-1}$$

$$= \frac{5y}{y-5} - \frac{(2y-3)(-1)}{(5-y)(-1)}$$

$$= \frac{5y}{y-5} - \frac{3-2y}{y-5}$$

$$= \frac{5y-(3-2y)}{y-5}$$ Remember the parentheses!

$$= \frac{5y-3+2y}{y-5} = \frac{7y-3}{y-5}$$

Do Exercises 6 and 7.

Subtract.

6. $\frac{x}{3} - \frac{2x-1}{-3}$

7. $\frac{3x}{x-2} - \frac{x-3}{2-x}$

Answers on page A-30

FACTORS THAT ARE OPPOSITES

Suppose that when we factor to find the LCD, we find factors that are opposites. Then we multiply by $-1/-1$ appropriately to change factors so that they are not opposites.

EXAMPLE 7 Subtract: $\frac{p}{64-p^2} - \frac{5}{p-8}$.

Factoring $64 - p^2$, we get $(8-p)(8+p)$. Note that the factors $8 - p$ in the first denominator and $p - 8$ in the second denominator are opposites. We multiply the first expression by $-1/-1$ to avoid this situation. Then we proceed as before.

$$\frac{p}{64-p^2} - \frac{5}{p-8} = \frac{p}{64-p^2} \cdot \frac{-1}{-1} - \frac{5}{p-8}$$

$$= \frac{-p}{p^2-64} - \frac{5}{p-8}$$

$$= \frac{-p}{(p-8)(p+8)} - \frac{5}{p-8}$$ LCD $= (p-8)(p+8)$

$$= \frac{-p}{(p-8)(p+8)} - \frac{5}{p-8} \cdot \frac{p+8}{p+8}$$

CALCULATOR CORNER

Checking Subtraction
Use the TABLE feature, as described on p. 471, to check the subtractions in Examples 5 and 6. Then check your answers to Margin Exercises 6 and 7.

8. Subtract:

$$\frac{y}{16-y^2}-\frac{7}{y-4}.$$

9. Perform the indicated operations and simplify:

$$\frac{x+2}{x^2-9}-\frac{x-7}{9-x^2}+\frac{-8-x}{x^2-9}.$$

10. Perform the indicated operations and simplify:

$$\frac{1}{x}-\frac{5}{3x}+\frac{2x}{x+1}.$$

Answers on page A-30

Then

$$=\frac{-p}{(p-8)(p+8)}-\frac{5p+40}{(p-8)(p+8)}$$

Subtracting this numerator. Don't forget the parentheses.

$$=\frac{-p-(5p+40)}{(p-8)(p+8)}$$

$$=\frac{-p-5p-40}{(p-8)(p+8)}=\frac{-6p-40}{(p-8)(p+8)}.$$

Do Exercise 8.

b Combined Additions and Subtractions

Now let's look at some combined additions and subtractions.

EXAMPLE 8 Perform the indicated operations and simplify:

$$\frac{x+9}{x^2-4}+\frac{5-x}{4-x^2}-\frac{2+x}{x^2-4}.$$

$$\frac{x+9}{x^2-4}+\frac{5-x}{4-x^2}-\frac{2+x}{x^2-4}$$

$$=\frac{x+9}{x^2-4}+\frac{5-x}{4-x^2}\cdot\frac{-1}{-1}-\frac{2+x}{x^2-4}$$

$$=\frac{x+9}{x^2-4}+\frac{x-5}{x^2-4}-\frac{2+x}{x^2-4}=\frac{(x+9)+(x-5)-(2+x)}{x^2-4}$$

$$=\frac{x+9+x-5-2-x}{x^2-4}=\frac{x+2}{x^2-4}$$

$$=\frac{\cancel{(x+2)}\cdot 1}{\cancel{(x+2)}(x-2)}=\frac{1}{x-2}.$$

Do Exercise 9.

EXAMPLE 9 Perform the indicated operations and simplify:

$$\frac{1}{x}-\frac{1}{x^2}+\frac{2}{x+1}.$$

The LCD $= x\cdot x(x+1)$, or $x^2(x+1)$.

$$\frac{1}{x}\cdot\frac{x(x+1)}{x(x+1)}-\frac{1}{x^2}\cdot\frac{(x+1)}{(x+1)}+\frac{2}{x+1}\cdot\frac{x^2}{x^2}$$

$$=\frac{x(x+1)}{x^2(x+1)}-\frac{x+1}{x^2(x+1)}+\frac{2x^2}{x^2(x+1)}$$

Subtracting this numerator. Don't forget the parentheses.

$$=\frac{x(x+1)-(x+1)+2x^2}{x^2(x+1)}$$

$$=\frac{x^2+x-x-1+2x^2}{x^2(x+1)}$$ Removing parentheses

$$=\frac{3x^2-1}{x^2(x+1)}$$

Do Exercise 10.

6.5 EXERCISE SET

For Extra Help

Digital Video Tutor CD 6 Videotape 13 | InterAct Math | Math Tutor Center | MathXL | MyMathLab

a Subtract. Simplify if possible.

1. $\dfrac{7}{x} - \dfrac{3}{x}$

2. $\dfrac{5}{a} - \dfrac{8}{a}$

3. $\dfrac{y}{y-4} - \dfrac{4}{y-4}$

4. $\dfrac{t^2}{t+5} - \dfrac{25}{t+5}$

5. $\dfrac{2x-3}{x^2+3x-4} - \dfrac{x-7}{x^2+3x-4}$

6. $\dfrac{x+1}{x^2-2x+1} - \dfrac{5-3x}{x^2-2x+1}$

7. $\dfrac{a-2}{10} - \dfrac{a+1}{5}$

8. $\dfrac{y+3}{2} - \dfrac{y-4}{4}$

9. $\dfrac{4z-9}{3z} - \dfrac{3z-8}{4z}$

10. $\dfrac{a-1}{4a} - \dfrac{2a+3}{a}$

11. $\dfrac{4x+2t}{3xt^2} - \dfrac{5x-3t}{x^2t}$

12. $\dfrac{5x+3y}{2x^2y} - \dfrac{3x+4y}{xy^2}$

13. $\dfrac{5}{x+5} - \dfrac{3}{x-5}$

14. $\dfrac{3t}{t-1} - \dfrac{8t}{t+1}$

15. $\dfrac{3}{2t^2-2t} - \dfrac{5}{2t-2}$

16. $\dfrac{11}{x^2-4} - \dfrac{8}{x+2}$

17. $\dfrac{2s}{t^2-s^2} - \dfrac{s}{t-s}$

18. $\dfrac{3}{12+x-x^2} - \dfrac{2}{x^2-9}$

19. $\dfrac{y-5}{y} - \dfrac{3y-1}{4y}$

20. $\dfrac{3x-2}{4x} - \dfrac{3x+1}{6x}$

21. $\dfrac{a}{x+a} - \dfrac{a}{x-a}$

22. $\dfrac{a}{a-b} - \dfrac{a}{a+b}$

23. $\dfrac{11}{6} - \dfrac{5}{-6}$

24. $\dfrac{5}{9} - \dfrac{7}{-9}$

25. $\dfrac{5}{a} - \dfrac{8}{-a}$

26. $\dfrac{8}{x} - \dfrac{3}{-x}$

27. $\dfrac{4}{y-1} - \dfrac{4}{1-y}$

28. $\dfrac{5}{a-2} - \dfrac{3}{2-a}$

29. $\dfrac{3-x}{x-7} - \dfrac{2x-5}{7-x}$

30. $\dfrac{t^2}{t-2} - \dfrac{4}{2-t}$

31. $\dfrac{a-2}{a^2-25} - \dfrac{6-a}{25-a^2}$

32. $\dfrac{x-8}{x^2-16} - \dfrac{x-8}{16-x^2}$

33. $\dfrac{4-x}{x-9} - \dfrac{3x-8}{9-x}$

34. $\dfrac{4x-6}{x-5} - \dfrac{7-2x}{5-x}$

35. $\dfrac{5x}{x^2-9} - \dfrac{4}{3-x}$

36. $\dfrac{8x}{16-x^2} - \dfrac{5}{x-4}$

37. $\dfrac{t^2}{2t^2 - 2t} - \dfrac{1}{2t - 2}$

38. $\dfrac{4}{5a^2 - 5a} - \dfrac{2}{5a - 5}$

39. $\dfrac{x}{x^2 + 5x + 6} - \dfrac{2}{x^2 + 3x + 2}$

40. $\dfrac{a}{a^2 + 11a + 30} - \dfrac{5}{a^2 + 9a + 20}$

b Perform the indicated operations and simplify.

41. $\dfrac{3(2x + 5)}{x - 1} - \dfrac{3(2x - 3)}{1 - x} + \dfrac{6x - 1}{x - 1}$

42. $\dfrac{a - 2b}{b - a} - \dfrac{3a - 3b}{a - b} + \dfrac{2a - b}{a - b}$

43. $\dfrac{x - y}{x^2 - y^2} + \dfrac{x + y}{x^2 - y^2} - \dfrac{2x}{x^2 - y^2}$

44. $\dfrac{x - 3y}{2(y - x)} + \dfrac{x + y}{2(x - y)} - \dfrac{2x - 2y}{2(x - y)}$

45. $\dfrac{2(x - 1)}{2x - 3} - \dfrac{3(x + 2)}{2x - 3} - \dfrac{x - 1}{3 - 2x}$

46. $\dfrac{5(2y + 1)}{2y - 3} - \dfrac{3(y - 1)}{3 - 2y} - \dfrac{3(y - 2)}{2y - 3}$

47. $\dfrac{10}{2y - 1} - \dfrac{6}{1 - 2y} + \dfrac{y}{2y - 1} + \dfrac{y - 4}{1 - 2y}$

48. $\dfrac{(x + 1)(2x - 1)}{(2x - 3)(x - 3)} - \dfrac{(x - 3)(x + 1)}{(3 - x)(3 - 2x)} + \dfrac{(2x + 1)(x + 3)}{(3 - 2x)(x - 3)}$

49. $\dfrac{a + 6}{4 - a^2} - \dfrac{a + 3}{a + 2} + \dfrac{a - 3}{2 - a}$

50. $\dfrac{4t}{t^2 - 1} - \dfrac{2}{t} - \dfrac{2}{t + 1}$

51. $\dfrac{2z}{1 - 2z} + \dfrac{3z}{2z + 1} - \dfrac{3}{4z^2 - 1}$

52. $\dfrac{1}{x - y} - \dfrac{2x}{x^2 - y^2} + \dfrac{1}{x + y}$

53. $\dfrac{1}{x + y} - \dfrac{1}{x - y} + \dfrac{2x}{x^2 - y^2}$

54. $\dfrac{2b}{a^2 - b^2} - \dfrac{1}{a + b} + \dfrac{1}{a - b}$

55. $\mathbf{D_W}$ Are parentheses as important when adding rational expressions as they are when subtracting? Why or why not?

56. $\mathbf{D_W}$ Is it possible to add or subtract rational expressions without knowing how to factor? Why or why not?

SKILL MAINTENANCE

Simplify.

57. $\dfrac{x^8}{x^3}$ [4.1e]

58. $3x^4 \cdot 10x^8$ [4.1d]

59. $(a^2b^{-5})^{-4}$ [4.2b]

60. $\dfrac{54x^{10}}{3x^7}$ [4.1e]

61. $\dfrac{66x^2}{11x^5}$ [4.1e]

62. $5x^{-7} \cdot 2x^4$ [4.1d]

Find a polynomial for the shaded area of the figure. [4.4d]

63.

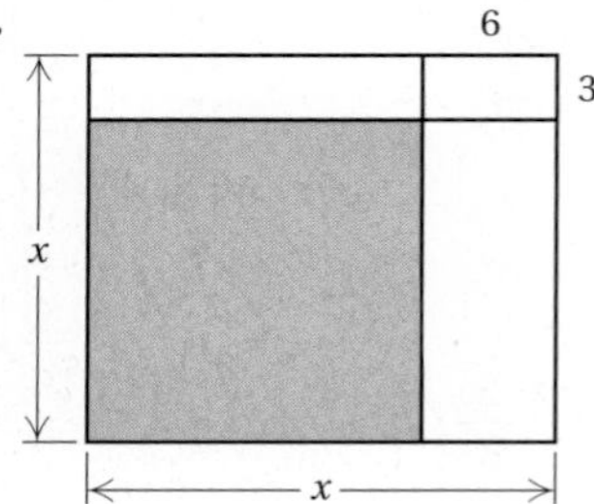

64.

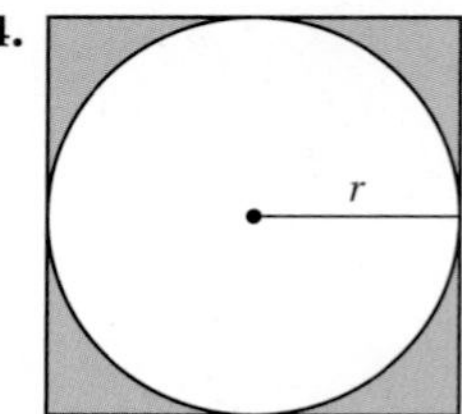

SYNTHESIS

Perform the indicated operations and simplify.

65. $\dfrac{2x+11}{x-3} \cdot \dfrac{3}{x+4} + \dfrac{2x+1}{4+x} \cdot \dfrac{3}{3-x}$

66. $\dfrac{x^2}{3x^2-5x-2} - \dfrac{2x}{3x+1} \cdot \dfrac{1}{x-2}$

67. $\dfrac{x}{x^4-y^4} - \left(\dfrac{1}{x+y}\right)^2$

68. $\left(\dfrac{a}{a-b} + \dfrac{b}{a+b}\right)\left(\dfrac{1}{3a+b} + \dfrac{2a+6b}{9a^2-b^2}\right)$

69. The perimeter of the following right triangle is $2a+5$. Find the length of the missing side and the area.

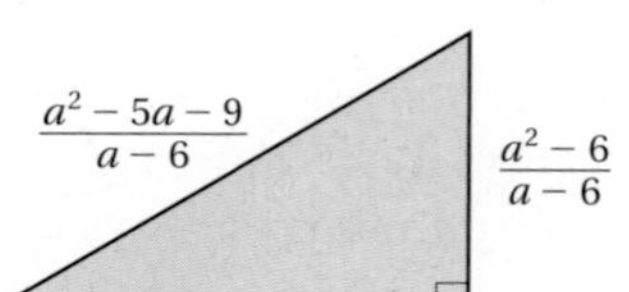

70.–73. Use the TABLE feature to check the subtractions in Exercises 15, 16, 19, and 20.

6.6 SOLVING RATIONAL EQUATIONS

a Rational Equations

In Sections 6.1–6.5, we studied operations with *rational expressions.* These expressions have no equals signs. We can add, subtract, multiply, or divide and simplify expressions, but we cannot solve if there are no equals signs—as, for example, in

$$\frac{x^2+6x+9}{x^2-4}\cdot\frac{x-2}{x+3},\qquad \frac{x+y}{x-y}\div\frac{x^2+y}{x^2-y^2},\quad\text{and}\quad \frac{a+3}{a^2-16}+\frac{5}{12-3a}.$$

Operation signs occur. There are no equals signs!

Most often, the result of our calculation is another rational expression that has not been cleared of fractions.

Equations *do have* equals signs, and we can clear them of fractions as we did in Section 2.3. A **rational,** or **fractional, equation** is an equation containing one or more rational expressions. Here are some examples:

$$\frac{2}{3}+\frac{5}{6}=\frac{x}{9},\qquad x+\frac{6}{x}=-5,\quad\text{and}\quad \frac{x^2}{x-1}=\frac{1}{x-1}.$$

There are equals signs as well as operation signs.

SOLVING RATIONAL EQUATIONS

To solve a rational equation, the first step is to clear the equation of fractions. To do this, multiply all terms on both sides of the equation by the LCM of all the denominators. Then carry out the equation-solving process as we learned it in Chapter 2.

When clearing an equation of fractions, we use the terminology LCM instead of LCD because we are *not* adding or subtracting rational expressions.

EXAMPLE 1 Solve: $\frac{2}{3}+\frac{5}{6}=\frac{x}{9}$.

The LCM of all denominators is $2\cdot 3\cdot 3$, or 18. We multiply all terms on both sides by 18:

$$18\left(\frac{2}{3}+\frac{5}{6}\right)=18\cdot\frac{x}{9}$$ Multiplying both sides by the LCM

$$18\cdot\frac{2}{3}+18\cdot\frac{5}{6}=18\cdot\frac{x}{9}$$ Multiplying each term by the LCM to remove parentheses

$$12+15=2x$$ Simplifying. Note that we have now cleared fractions.

$$27=2x$$

$$\frac{27}{2}=x.$$

The solution is $\frac{27}{2}$.

Do Exercise 1.

Objective

a Solve rational equations.

1. Solve: $\frac{3}{4}+\frac{5}{8}=\frac{x}{12}$.

Answer on page A-30

Study Tips

TAPING YOUR LECTURES

Consider recording your notes and playing them back when convenient, for example, while commuting to campus. It can even be advantageous to record math lectures. (Be sure to get permission from your instructor before doing so, however.) Important points can be emphasized verbally. We consider this idea so worthwhile that we provide a series of audiotapes that accompany the book. (See the Preface for more information.)

2. Solve: $\frac{x}{4} - \frac{x}{6} = \frac{1}{8}$.

3. Solve: $\frac{1}{x} = \frac{1}{6 - x}$.

Answers on page A-30

EXAMPLE 2 Solve: $\frac{x}{6} - \frac{x}{8} = \frac{1}{12}$.

The LCM is 24. We multiply all terms on both sides by 24:

$$\frac{x}{6} - \frac{x}{8} = \frac{1}{12}$$

$$24\left(\frac{x}{6} - \frac{x}{8}\right) = 24 \cdot \frac{1}{12} \quad \text{Multiplying both sides by the LCM}$$

$$24 \cdot \frac{x}{6} - 24 \cdot \frac{x}{8} = 24 \cdot \frac{1}{12} \quad \text{Multiplying to remove parentheses}$$

Be sure to multiply *each* term by the LCM.

$$4x - 3x = 2 \quad \text{Simplifying}$$

$$x = 2.$$

CHECK:

$\frac{x}{6} - \frac{x}{8}$	$= \frac{1}{12}$
$\frac{2}{6} - \frac{2}{8}$	$\frac{1}{12}$
$\frac{1}{3} - \frac{1}{4}$	
$\frac{4}{12} - \frac{3}{12}$	
$\frac{1}{12}$	TRUE

This checks, so the solution is 2.

Do Exercise 2.

EXAMPLE 3 Solve: $\frac{1}{x} = \frac{1}{4 - x}$.

The LCM is $x(4 - x)$. We multiply all terms on both sides by $x(4 - x)$:

$$\frac{1}{x} = \frac{1}{4 - x}$$

$$x(4 - x) \cdot \frac{1}{x} = x(4 - x) \cdot \frac{1}{4 - x} \quad \text{Multiplying both sides by the LCM}$$

$$4 - x = x \quad \text{Simplifying}$$

$$4 = 2x$$

$$x = 2.$$

CHECK:

$\frac{1}{x}$	$= \frac{1}{4 - x}$
$\frac{1}{2}$	$\frac{1}{4 - 2}$
	$\frac{1}{2}$ TRUE

This checks, so the solution is 2.

Do Exercise 3.

AG ALGEBRAIC–GRAPHICAL CONNECTION

We can obtain a visual check of the solutions of a rational equation by graphing. For example, consider the equation

$$\frac{x}{4} + \frac{x}{2} = 6.$$

We can examine the solution by graphing the equations

$$y = \frac{x}{4} + \frac{x}{2} \quad \text{and} \quad y = 6$$

using the same set of axes.

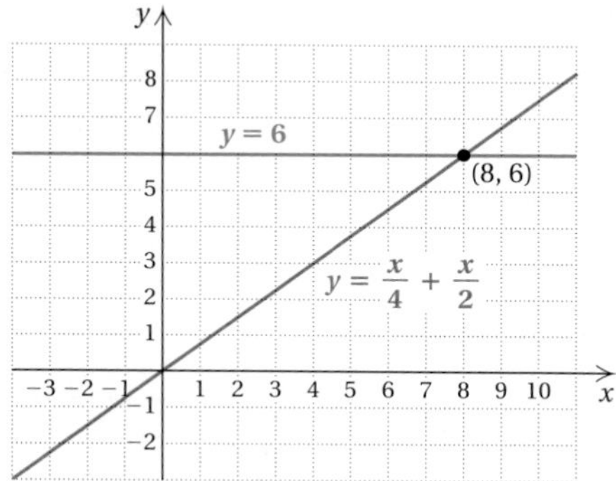

The y-values for each equation will be the same where the graphs intersect. The x-value of that point will yield that value, so it will be the solution of the equation. It appears from the graph that when $x = 8$, the value of $x/4 + x/2$ is 6. We can check by substitution:

$$\frac{x}{4} + \frac{x}{2} = \frac{8}{4} + \frac{8}{2} = 2 + 4 = 6.$$

Thus the solution is 8.

4. Solve: $\dfrac{1}{2x} + \dfrac{1}{x} = -12.$

EXAMPLE 4 Solve: $\dfrac{2}{3x} + \dfrac{1}{x} = 10.$

The LCM is $3x$. We multiply all terms on both sides by $3x$:

$$\frac{2}{3x} + \frac{1}{x} = 10$$

$$3x\left(\frac{2}{3x} + \frac{1}{x}\right) = 3x \cdot 10 \qquad \text{Multiplying both sides by the LCM}$$

$$3x \cdot \frac{2}{3x} + 3x \cdot \frac{1}{x} = 3x \cdot 10 \qquad \text{Multiplying to remove parentheses}$$

$$2 + 3 = 30x \qquad \text{Simplifying}$$

$$5 = 30x$$

$$\frac{5}{30} = x$$

$$\frac{1}{6} = x.$$

The check is left to the student. The solution is $\frac{1}{6}$.

Do Exercise 4.

Answer on page A-30

5. Solve: $x + \frac{1}{x} = 2.$

Answer on page A-30

EXAMPLE 5 Solve: $x + \frac{6}{x} = -5.$

The LCM is x. We multiply all terms on both sides by x:

$$x + \frac{6}{x} = -5$$

$$x\left(x + \frac{6}{x}\right) = -5x \qquad \text{Multiplying both sides by } x$$

$$x \cdot x + x \cdot \frac{6}{x} = -5x \qquad \text{Note that each rational expression on the left is now multiplied by } x.$$

$$x^2 + 6 = -5x \qquad \text{Simplifying}$$

$$x^2 + 5x + 6 = 0 \qquad \text{Adding } 5x \text{ to get a 0 on one side}$$

$$(x + 3)(x + 2) = 0 \qquad \text{Factoring}$$

$$x + 3 = 0 \quad \text{or} \quad x + 2 = 0 \qquad \text{Using the principle of zero products}$$

$$x = -3 \quad \text{or} \quad x = -2.$$

CHECK: For -3:

$$x + \frac{6}{x} = -5$$

$-3 + \frac{6}{-3}$	-5
$-3 - 2$	
-5	TRUE

For -2:

$$x + \frac{6}{x} = -5$$

$-2 + \frac{6}{-2}$	-5
$-2 - 3$	
-5	TRUE

Both of these check, so there are two solutions, -3 and -2.

Do Exercise 5.

CALCULATOR CORNER

Checking Solutions of Rational Equations We can use a table to check possible solutions of rational equations. Consider the equation in Example 6,

$$\frac{x^2}{x-1} = \frac{1}{x-1},$$

and the possible solutions that were found, 1 and -1. To check these solutions, we enter $y_1 = x^2/(x - 1)$ and $y_2 = 1/(x - 1)$ on the equation-editor screen. Then, with a table set in ASK mode, we enter $x = 1$. The ERROR messages indicate that 1 is not a solution because it is not an allowable replacement for x in the equation. Next, we enter $x = -1$. Since y_1 and y_2 have the same value, we know that the equation is true when $x = -1$, and thus -1 is a solution.

X	Y1	Y2
1	ERROR	ERROR
−1	−.5	−.5

X =

Exercises: Use a graphing calculator to check the possible solutions in each of the following.

1. Examples 1, 3, 5, and 7

2. Margin Exercises 1, 3, 6, and 7

CHECKING POSSIBLE SOLUTIONS

When we multiply both sides of an equation by the LCM, the resulting equation might have solutions that are *not* solutions of the original equation. Thus we must *always* check possible solutions in the original equation.

1. If you have carried out all algebraic procedures correctly, you need only check if a number makes a denominator 0 in the original equation. If it does make a denominator 0, it is *not* a solution.
2. To be sure that no computational errors have been made and that you indeed have a solution, a complete check is necessary, as we did in Chapter 2.

Example 6 illustrates the importance of checking all possible solutions.

EXAMPLE 6 Solve: $\dfrac{x^2}{x-1} = \dfrac{1}{x-1}$.

The LCM is $x - 1$. We multiply all terms on both sides by $x - 1$:

$$\frac{x^2}{x-1} = \frac{1}{x-1}$$

$$(x-1)\cdot\frac{x^2}{x-1} = (x-1)\cdot\frac{1}{x-1} \quad \text{Multiplying both sides by } x-1$$

$$x^2 = 1 \quad \text{Simplifying}$$

$$x^2 - 1 = 0 \quad \text{Subtracting 1 to get a 0 on one side}$$

$$(x-1)(x+1) = 0 \quad \text{Factoring}$$

$$x - 1 = 0 \quad or \quad x + 1 = 0 \quad \text{Using the principle of zero products}$$

$$x = 1 \quad or \quad x = -1.$$

The numbers 1 and -1 are possible solutions. We look at the original equation and see that 1 makes a denominator 0 and is therefore not a solution. The number -1 checks and is a solution.

Do Exercise 6.

EXAMPLE 7 Solve: $\dfrac{3}{x-5} + \dfrac{1}{x+5} = \dfrac{2}{x^2-25}$.

The LCM is $(x-5)(x+5)$. We multiply all terms on both sides by $(x-5)(x+5)$:

$$(x-5)(x+5)\left(\frac{3}{x-5} + \frac{1}{x+5}\right) = (x-5)(x+5)\left(\frac{2}{x^2-25}\right)$$

Multiplying both sides by the LCM

$$(x-5)(x+5)\cdot\frac{3}{x-5} + (x-5)(x+5)\cdot\frac{1}{x+5} = (x-5)(x+5)\cdot\frac{2}{x^2-25}$$

$$3(x+5) + (x-5) = 2 \quad \text{Simplifying}$$

$$3x + 15 + x - 5 = 2 \quad \text{Removing parentheses}$$

$$4x + 10 = 2$$

$$4x = -8$$

$$x = -2.$$

The check is left to the student. The number -2 checks and is the solution.

Do Exercise 7.

6. Solve: $\dfrac{x^2}{x+2} = \dfrac{4}{x+2}$.

7. Solve: $\dfrac{4}{x-2} + \dfrac{1}{x+2} = \dfrac{26}{x^2-4}$.

CAUTION!

We have introduced a new use of the LCM in this section. We previously used the LCM in adding or subtracting rational expressions. *Now* we have equations with equals signs. We clear fractions by multiplying both sides of the equation by the LCM. This eliminates the denominators. Do *not* make the mistake of trying to clear fractions when you do not have an equation.

Answers on page A-30

Study Tips

ARE YOU CALCULATING OR SOLVING?

One of the common difficulties with this chapter is knowing for sure the task at hand. Are you combining expressions using operations to get another *rational expression,* or are you solving equations for which the results are numbers that are *solutions* of an equation? To learn to make these decisions, complete the following list by writing in the blank the type of answer you should get: "Rational expression" or "Solutions." You need not complete the mathematical operations.

Task	*Answer* *(Just write "Rational expression" or "Solutions.")*
1. Add: $\frac{4}{x-2} + \frac{1}{x+2}$.	
2. Solve: $\frac{4}{x-2} = \frac{1}{x+2}$.	
3. Subtract: $\frac{4}{x-2} - \frac{1}{x+2}$.	
4. Multiply: $\frac{4}{x-2} \cdot \frac{1}{x+2}$.	
5. Divide: $\frac{4}{x-2} \div \frac{1}{x+2}$.	
6. Solve: $\frac{4}{x-2} + \frac{1}{x+2} = \frac{26}{x^2-4}$.	
7. Perform the indicated operations and simplify: $\frac{4}{x-2} + \frac{1}{x+2} - \frac{26}{x^2-4}$.	
8. Solve: $\frac{x^2}{x-1} = \frac{1}{x-1}$.	
9. Solve: $\frac{2}{y^2-25} = \frac{3}{y-5} + \frac{1}{y-5}$.	
10. Solve: $\frac{x}{x+4} - \frac{4}{x-4} = \frac{x^2+16}{x^2-16}$.	
11. Perform the indicated operations and simplify: $\frac{x}{x+4} - \frac{4}{x-4} - \frac{x^2+16}{x^2-16}$.	
12. Solve: $\frac{5}{y-3} - \frac{30}{y^2-9} = 1$.	
13. Add: $\frac{5}{y-3} + \frac{30}{y^2-9} + 1$.	

6.6 EXERCISE SET

For Extra Help

Digital Video Tutor CD 6 Videotape 13 | InterAct Math | Math Tutor Center | MathXL | MyMathLab

a Solve. Don't forget to check!

1. $\frac{4}{5} - \frac{2}{3} = \frac{x}{9}$

2. $\frac{x}{20} = \frac{3}{8} - \frac{4}{5}$

3. $\frac{3}{5} + \frac{1}{8} = \frac{1}{x}$

4. $\frac{2}{3} + \frac{5}{6} = \frac{1}{x}$

5. $\frac{3}{8} + \frac{4}{5} = \frac{x}{20}$

6. $\frac{3}{5} + \frac{2}{3} = \frac{x}{9}$

7. $\frac{1}{x} = \frac{2}{3} - \frac{5}{6}$

8. $\frac{1}{x} = \frac{1}{8} - \frac{3}{5}$

9. $\frac{1}{6} + \frac{1}{8} = \frac{1}{t}$

10. $\frac{1}{8} + \frac{1}{12} = \frac{1}{t}$

11. $x + \frac{4}{x} = -5$

12. $\frac{10}{x} - x = 3$

13. $\frac{x}{4} - \frac{4}{x} = 0$

14. $\frac{x}{5} - \frac{5}{x} = 0$

15. $\frac{5}{x} = \frac{6}{x} - \frac{1}{3}$

16. $\frac{4}{x} = \frac{5}{x} - \frac{1}{2}$

17. $\frac{5}{3x} + \frac{3}{x} = 1$

18. $\frac{5}{2y} + \frac{8}{y} = 1$

19. $\frac{t-2}{t+3} = \frac{3}{8}$

20. $\frac{x-7}{x+2} = \frac{1}{4}$

21. $\frac{2}{x+1} = \frac{1}{x-2}$

22. $\frac{8}{y-3} = \frac{6}{y+4}$

23. $\frac{x}{6} - \frac{x}{10} = \frac{1}{6}$

24. $\frac{x}{8} - \frac{x}{12} = \frac{1}{8}$

25. $\frac{t+2}{5} - \frac{t-2}{4} = 1$

26. $\frac{x+1}{3} - \frac{x-1}{2} = 1$

27. $\frac{5}{x-1} = \frac{3}{x+2}$

28. $\frac{x-7}{x-9} = \frac{2}{x-9}$

29. $\frac{a-3}{3a+2} = \frac{1}{5}$

30. $\frac{x+7}{8x-5} = \frac{2}{3}$

31. $\dfrac{x-1}{x-5} = \dfrac{4}{x-5}$

32. $\dfrac{y+11}{y+8} = \dfrac{3}{y+8}$

33. $\dfrac{2}{x+3} = \dfrac{5}{x}$

34. $\dfrac{6}{y} = \dfrac{5}{y-8}$

35. $\dfrac{x-2}{x-3} = \dfrac{x-1}{x+1}$

36. $\dfrac{t+5}{t-2} = \dfrac{t-2}{t+4}$

37. $\dfrac{1}{x+3} + \dfrac{1}{x-3} = \dfrac{1}{x^2-9}$

38. $\dfrac{4}{x-3} + \dfrac{2x}{x^2-9} = \dfrac{1}{x+3}$

39. $\dfrac{x}{x+4} - \dfrac{4}{x-4} = \dfrac{x^2+16}{x^2-16}$

40. $\dfrac{5}{y-3} - \dfrac{30}{y^2-9} = 1$

41. $\frac{4 - a}{8 - a} = \frac{4}{a - 8}$

42. $\frac{3}{x - 7} = \frac{x + 10}{x - 7}$

43. $2 - \frac{a - 2}{a + 3} = \frac{a^2 - 4}{a + 3}$

44. $\frac{5}{x - 1} + x + 1 = \frac{5x + 4}{x - 1}$

45. **DW** Why is it especially important to check the possible solutions to a rational equation?

46. **DW** How can a graph be used to determine how many solutions an equation has?

SKILL MAINTENANCE

Simplify.

47. $(a^2b^5)^{-3}$ [4.2b]

48. $(x^{-2}y^{-3})^{-4}$ [4.2b]

49. $\left(\frac{2x}{t^2}\right)^4$ [4.2b]

50. $\left(\frac{y^3}{w^2}\right)^{-2}$ [4.2b]

51. $4x^{-5} \cdot 8x^{11}$ [4.1d]

52. $(8x^5y^{-4})^2$ [4.2b]

Find the intercepts. Then graph the equation. [3.3a]

53. $5x + 10y = 20$

54. $2x - 4y = 8$

55. $10y - 4x = -20$

56. $y - 5x = 5$

SYNTHESIS

Solve.

57. $\frac{4}{y - 2} - \frac{2y - 3}{y^2 - 4} = \frac{5}{y + 2}$

58. $\frac{x}{x^2 + 3x - 4} + \frac{x + 1}{x^2 + 6x + 8} = \frac{2x}{x^2 + x - 2}$

59. $\frac{x + 1}{x + 2} = \frac{x + 3}{x + 4}$

60. $\frac{x^2}{x^2 - 4} = \frac{x}{x + 2} - \frac{2x}{2 - x}$

61. $4a - 3 = \frac{a + 13}{a + 1}$

62. $\frac{3x - 9}{x - 3} = \frac{5x - 4}{2}$

63. $\frac{y^2 - 4}{y + 3} = 2 - \frac{y - 2}{y + 3}$

64. $\frac{3a - 5}{a^2 + 4a + 3} + \frac{2a + 2}{a + 3} = \frac{a - 3}{a + 1}$

65. Use a graphing calculator to check the solutions to Exercises 1–4.

66. Use a graphing calculator to check the solutions to Exercises 13, 15, and 25.

6.7 APPLICATIONS USING RATIONAL EQUATIONS AND PROPORTIONS

Objectives

a Solve applied problems using rational equations.

b Solve proportion problems.

In many areas of study, applications involving rates, proportions, or reciprocals translate to rational equations. By using the five steps for problem solving and the skills of Sections 6.1–6.6, we can now solve such problems.

a Solving Applied Problems

PROBLEMS INVOLVING WORK

EXAMPLE 1 *Recyclable Work.* Erin and Nick work as volunteers at a community recycling depot. Erin can sort a morning's accumulation of recyclables in 4 hr, while Nick requires 6 hr to do the same job. How long would it take them, working together, to sort the recyclables?

1. **Familiarize.** We familiarize ourselves with the problem by considering two *incorrect* ways of translating the problem to mathematical language.

   a) A common *incorrect* way to translate the problem is to add the two times: 4 hr + 6 hr = 10 hr. Let's think about this. Erin can do the job alone in 4 hr. If Erin and Nick work together, whatever time it takes them should be *less* than 4 hr. Thus we reject 10 hr as a solution, but we do have a partial check on any answer we get. The answer should be less than 4 hr.

   b) Another *incorrect* way to translate the problem is as follows. Suppose the two people split up the sorting job in such a way that Erin does half the sorting and Nick does the other half. Then

   Erin sorts $\frac{1}{2}$ the recyclables in $\frac{1}{2}(4\text{ hr})$, or 2 hr,

   and Nick sorts $\frac{1}{2}$ the recyclables in $\frac{1}{2}(6\text{ hr})$, or 3 hr.

   But time is wasted since Erin would finish 1 hr earlier than Nick. In effect, they have not worked together to get the job done as fast as possible. If Erin helps Nick after completing her half, the entire job could be done in a time somewhere between 2 hr and 3 hr.

   We proceed to a translation by considering how much of the job is finished in 1 hr, 2 hr, 3 hr, and so on. It takes Erin 4 hr to do the sorting job alone. Then, in 1 hr, she can do $\frac{1}{4}$ of the job. It takes Nick 6 hr to do the job alone. Then, in 1 hr, he can do $\frac{1}{6}$ of the job. Working together, they can do

   $\frac{1}{4} + \frac{1}{6}$, or $\frac{5}{12}$ of the job in 1 hr.

   In 2 hr, Erin can do $2\left(\frac{1}{4}\right)$ of the job and Nick can do $2\left(\frac{1}{6}\right)$ of the job. Working together, they can do

   $2\left(\frac{1}{4}\right) + 2\left(\frac{1}{6}\right)$, or $\frac{5}{6}$ of the job in 2 hr.

Study Tips

BEING A TUTOR

Try being a tutor for a fellow student. You can maximize your understanding and retention of concepts if you explain the material to someone else.

Continuing this reasoning, we can create a table like the following one.

TIME	FRACTION OF THE JOB COMPLETED		
	Erin	Nick	Together
1 hr	$\frac{1}{4}$	$\frac{1}{6}$	$\frac{1}{4}+\frac{1}{6}$, or $\frac{5}{12}$
2 hr	$2\left(\frac{1}{4}\right)$	$2\left(\frac{1}{6}\right)$	$2\left(\frac{1}{4}\right)+2\left(\frac{1}{6}\right)$, or $\frac{5}{6}$
3 hr	$3\left(\frac{1}{4}\right)$	$3\left(\frac{1}{6}\right)$	$3\left(\frac{1}{4}\right)+3\left(\frac{1}{6}\right)$, or $1\frac{1}{4}$
t hr	$t\left(\frac{1}{4}\right)$	$t\left(\frac{1}{6}\right)$	$t\left(\frac{1}{4}\right)+t\left(\frac{1}{6}\right)$

From the table, we see that if they work 3 hr, the fraction of the job completed is $1\frac{1}{4}$, which is more of the job than needs to be done. We see again that the answer is somewhere between 2 hr and 3 hr. What we want is a number t such that the fraction of the job that gets completed is 1; that is, the job is just completed.

2. **Translate.** From the table, we see that the time we want is some number t for which

$$t\left(\frac{1}{4}\right)+t\left(\frac{1}{6}\right)=1, \quad \text{or} \quad \frac{t}{4}+\frac{t}{6}=1,$$

where 1 represents the idea that the entire job is completed in time t.

3. **Solve.** We solve the equation:

$$12\left(\frac{t}{4}+\frac{t}{6}\right)=12\cdot 1$$ Multiplying by the LCM, which is $2\cdot 2\cdot 3$, or 12

$$12\cdot\frac{t}{4}+12\cdot\frac{t}{6}=12$$

$$3t+2t=12$$

$$5t=12$$

$$t=\frac{12}{5}, \text{ or } 2\frac{2}{5}\text{ hr.}$$

4. **Check.** The check can be done by recalculating:

$$\frac{12}{5}\left(\frac{1}{4}\right)+\frac{12}{5}\left(\frac{1}{6}\right)=\frac{3}{5}+\frac{2}{5}=\frac{5}{5}=1.$$

We also have another check in what we learned from the *Familiarize* step. The answer, $2\frac{2}{5}$ hr, is between 2 hr and 3 hr (see the table), and it is less than 4 hr, the time it takes Erin working alone.

5. **State.** It takes $2\frac{2}{5}$ hr for them to do the sorting, working together.

THE WORK PRINCIPLE

Suppose $a =$ the time it takes A to do a job, $b =$ the time it takes B to do the same job, and $t =$ the time it takes them to do the same job working together. Then

$$\frac{t}{a} + \frac{t}{b} = 1, \quad \text{or} \quad \frac{1}{a} + \frac{1}{b} = \frac{1}{t}.$$

Do Exercise 1.

1. **Wall Construction.** By checking work records, a contractor finds that it takes Eduardo 6 hr to construct a wall of a certain size. It takes Yolanda 8 hr to construct the same wall. How long would it take if they worked together?

Answer on page A-30

PROBLEMS INVOLVING MOTION

Problems that deal with distance, speed (or rate), and time are called **motion problems.** Translation of these problems involves the distance formula, $d = r \cdot t$, and/or the equivalent formulas $r = d/t$ and $t = d/r$.

MOTION FORMULAS

The following are the formulas for motion problems:

$d = rt$; Distance = Rate · Time (basic formula)

$r = \frac{d}{t}$; Rate = Distance/Time

$t = \frac{d}{r}$. Time = Distance/Rate

EXAMPLE 2 *Animal Speeds.* A cheetah can run 20 mph faster than a lion. A cheetah can run 7 mi in the same time that a lion can run 5 mi. Find the speed of each animal.

Source: Barbara Ann Kipfer, *The Order of Things.* New York: Random House, 1998.

1. **Familiarize.** We first make a drawing. Let $r =$ the speed of the lion. Then $r + 20 =$ the speed of the cheetah.

Recall that sometimes we need to find a formula in order to solve an application. A formula that relates the notions of distance, speed, and time is $d = rt$, or

Distance = Speed · Time.

(Indeed, you may need to look up such a formula.)

Study Tips

TEST TAKING: MORE ON DOING EVEN-NUMBERED EXERCISES

In an earlier study tip (p. 249), as a way to improve your test-taking skills, we encouraged you to build some even-numbered exercises into your homework. Here we explore this issue further.

Working a test is different from working your homework, when the answers are provided. When taking the test, you are "on your own," so to speak. Keep the following tips in mind when taking your next test or quiz.

1. Work a bit slower and deliberately, taking a fresh piece of paper to redo the problem. Check your work against your previous work to see if there is a difference and why. This is especially helpful if you finish the test early and have extra time.
2. Use estimation techniques to solve the problem as a check.
3. Do the checks to applied problems that we so often discuss in the book.

2. Driving speed. Nancy drives 20 mph faster than her father, Greg. In the same time that Nancy travels 180 mi, her father travels 120 mi. Find their speeds.

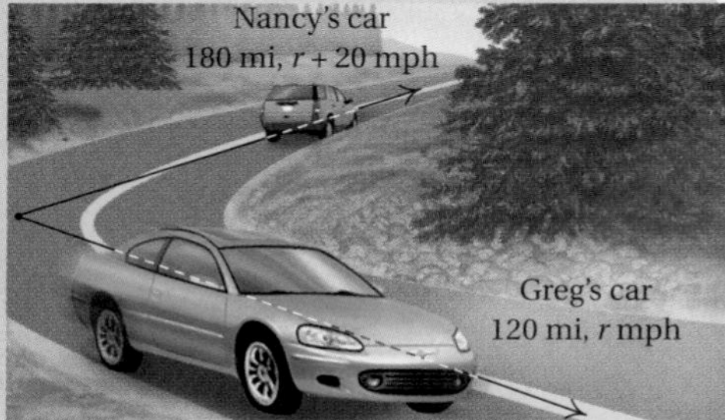

Answer on page A-30

Since each animal travels the same length of time, we can use just t for time. We organize the information in a chart, as follows.

$$d = r \cdot t$$

	DISTANCE	SPEED	TIME	
Lion	5	r	t	$\rightarrow 5 = rt$
Cheetah	7	$r + 20$	t	$\rightarrow 7 = (r + 20)t$

2. **Translate.** We can apply the formula $d = rt$ along the rows of the table to obtain two equations:

$$5 = rt, \quad \textbf{(1)}$$
$$7 = (r + 20)t. \quad \textbf{(2)}$$

We know that the animals travel for the same length of time. Thus if we solve each equation for t and set the results equal to each other, we get an equation in terms of r.

Solving $5 = rt$ for t: $\quad t = \dfrac{5}{r}$

Solving $7 = (r + 20)t$ for t: $\quad t = \dfrac{7}{r + 20}$

Since the times are the same, we have the following equation:

$$\frac{5}{r} = \frac{7}{r + 20}.$$

3. **Solve.** To solve the equation, we first multiply both sides by the LCM, which is $r(r + 20)$:

$$r(r + 20) \cdot \frac{5}{r} = r(r + 20) \cdot \frac{7}{r + 20} \quad \text{Multiplying both sides by the LCM, which is } r(r + 20)$$
$$5(r + 20) = 7r \quad \text{Simplifying}$$
$$5r + 100 = 7r \quad \text{Removing parentheses}$$
$$100 = 2r$$
$$50 = r.$$

We now have a possible solution. The speed of the lion is 50 mph, and the speed of the cheetah is $r + 20 = 50 + 20$, or 70 mph.

4. **Check.** We first reread the problem to see what we were to find. We check the speeds of 50 for the lion and 70 for the cheetah. The cheetah does travel 20 mph faster than the lion and will travel farther than the lion, which runs at a slower speed. If the cheetah runs 7 mi at 70 mph, the time it has traveled is $\frac{7}{70}$, or $\frac{1}{10}$ hr. If the lion runs 5 mi at 50 mph, the time it has traveled is $\frac{5}{50}$, or $\frac{1}{10}$ hr. Since the times are the same, the speeds check.
5. **State.** The speed of the lion is 50 mph and the speed of the cheetah is 70 mph.

Do Exercise 2.

b Applications Involving Proportions

We now consider applications with proportions. A **proportion** involves ratios. A **ratio** of two quantities is their quotient. For example, 73% is the ratio of 73 to 100, $\frac{73}{100}$. The ratio of two different kinds of measure is called a **rate.** Suppose an animal travels 720 ft in 2.5 hr. Its **rate,** or **speed,** is then

$$\frac{720 \text{ ft}}{2.5 \text{ hr}} = 288\frac{\text{ft}}{\text{hr}}.$$

Do Exercises 3–6.

PROPORTION

An equality of ratios, $A/B = C/D$, is called a **proportion.** The numbers within a proportion are said to be **proportional** to each other.

Proportions can be used to solve a variety of applied problems.

EXAMPLE 3 *Mileage.* A Honda Insight is a gasoline–electric car that travels 280 mi on 4 gal of gas. Find the amount of gas required for a 700-mi trip.
Source: American Honda Motor Company

1. **Familiarize.** We know that the Honda can travel 280 mi on 4 gal of gas. Thus we can set up a ratio, letting x = the amount of gas required to drive 700 mi.
2. **Translate.** We assume that the car uses gas at the same rate throughout the 700-mi trip. Thus the ratios are the same and we can write a proportion. Note that the units of *mileage* are in the numerators and the units of *gasoline* are in the denominators.

$$\begin{matrix}\text{Miles} \rightarrow \\ \text{Gas} \rightarrow\end{matrix} \frac{280}{4} = \frac{700}{x} \begin{matrix}\leftarrow \text{Miles} \\ \leftarrow \text{Gas}\end{matrix}$$

3. **Solve.** To solve for x, we multiply both sides by the LCM, which is $4x$:

$$4x \cdot \frac{280}{4} = 4x \cdot \frac{700}{x} \quad \text{Multiplying by } 4x$$
$$280x = 2800 \quad \text{Simplifying}$$
$$\frac{280x}{280} = \frac{2800}{280} \quad \text{Dividing by 280}$$
$$x = 10. \quad \text{Simplifying}$$

We can also use cross products to solve the proportion:

$$\frac{280}{4} = \frac{700}{x} \quad 280x \text{ and } 4 \cdot 700 \text{ are cross products.}$$

$$280x = 4 \cdot 700 \quad \text{Equating cross products}$$
$$\frac{280x}{280} = \frac{4 \cdot 700}{280} \quad \text{Dividing by 280}$$
$$x = 10.$$

4. **Check.** The check is left for the student.
5. **State.** The Honda Insight will require 10 gal of gas for 700 mi of driving.

3. Find the ratio of 145 km to 2.5 liters (L).

4. **Batting Average.** Recently, a baseball player got 7 hits in 25 times at bat. What was the rate, or batting average, in hits per times at bat?

5. Impulses in nerve fibers travel 310 km in 2.5 hr. What is the rate, or speed, in kilometers per hour?

6. A lake of area 550 yd^2 contains 1320 fish. What is the population density of the lake in number of fish per square yard?

7. **Automotive Mileage.** In highway driving, a Chrysler PT Cruiser will travel 377 mi on 14.5 gal of gasoline. How much gas will be required for a 900-mi trip?
Source: DaimlerChrysler Corporation

Answers on page A-30

8. **Environmental Science.** To determine the number of humpback whales in a pod, a marine biologist, using tail markings, identifies 27 members of the pod. Several weeks later, 40 whales from the pod are randomly sighted. Of the 40 sighted, 12 are from the 27 originally identified. Estimate the number of whales in the pod.

Do Exercise 7 on the preceding page.

EXAMPLE 4 *Environmental Science.* To determine the number of fish in a lake, a park ranger catches 225 fish, tags them, and throws them back into the lake. Later, 108 fish are caught, and 15 of them are found to be tagged. Estimate how many fish are in the lake.

1. **Familiarize.** The ratio of the number of fish tagged to the total number of fish in the lake, F, is $\frac{225}{F}$. Of the 108 fish caught later, 15 fish were tagged. The ratio of fish tagged to fish caught is $\frac{15}{108}$.
2. **Translate.** Assuming that the two ratios are the same, we can translate to a proportion.

$$\text{Fish tagged originally} \longrightarrow \frac{225}{F} = \frac{15}{108} \longleftarrow \text{Tagged fish caught later}$$
Fish in lake $\longrightarrow$ (denominator F); Fish caught later $\longleftarrow$ (denominator 108)

3. **Solve.** We solve the proportion. We multiply by the LCM, which is $108F$:

$$108F \cdot \frac{225}{F} = 108F \cdot \frac{15}{108} \quad \text{Multiplying by } 108F$$
$$108 \cdot 225 = F \cdot 15$$
$$\frac{108 \cdot 225}{15} = F \quad \text{Dividing by 15}$$
$$1620 = F.$$

4. **Check.** The check is left to the student.
5. **State.** We estimate that there are about 1620 fish in the lake.

Do Exercise 8.

In the following example, we predict whether an important home-run record can be broken.

EXAMPLE 5 *Pursuit of Baseball's Home-Run Record.* Mark McGwire hit 70 home runs in 1998 to claim the major-league season home-run record. In 2001, Barry Bonds of the San Francisco Giants hit 31 home runs in the first 58 games of the season, which consists of 162 games. At this rate, could it be predicted that Bonds would break McGwire's record?
Source: Major League Baseball

1. **Familiarize.** Let's assume that Bond's rate of hitting 31 home runs in 58 games will continue for the entire 162-game season. We let $H =$ the number of home runs that Bonds can hit in 162 games.

Answer on page A-30

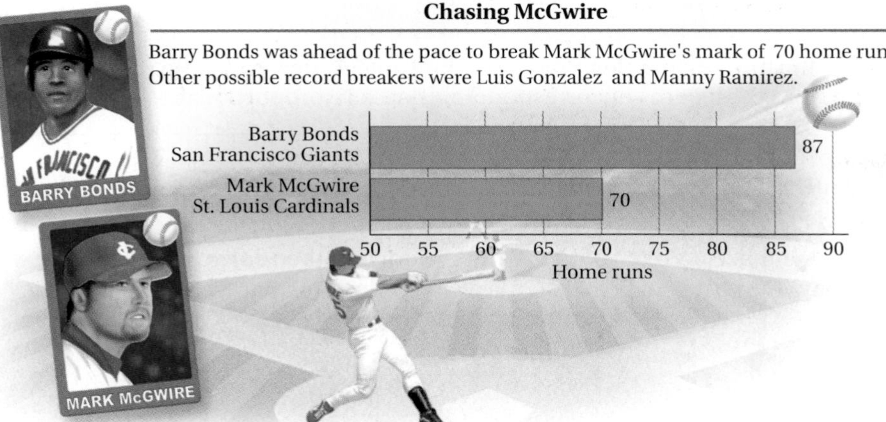

2. **Translate.** Assuming the rate continues, the ratios are the same, and we have the proportion

$$\text{Number of home runs} \rightarrow \frac{H}{162} = \frac{31}{58} \leftarrow \text{Number of home runs}$$
$$\text{Number of games} \rightarrow \qquad\qquad \leftarrow \text{Number of games}$$

3. **Solve.** We solve the equation:

$$\frac{H}{162} = \frac{31}{58}$$
$$58H = 162 \cdot 31 \quad \text{Equating cross products}$$
$$\frac{58H}{58} = \frac{162 \cdot 31}{58} \quad \text{Dividing by 58}$$
$$H \approx 87.$$

4. **Check.** The check is left to the student.
5. **State.** We can indeed predict that Bonds will hit 87 home runs and break McGwire's record. (Bonds actually completed the season with 73 home runs and broke McGwire's record.)

Do Exercise 9.

9. **Baseball's Home-Run Record.** In 2001, Luis Gonzalez of the Arizona Diamondbacks hit 22 home runs in the first 59 games. The season consists of 162 games. At this rate, could it be predicted that Gonzalez would break McGwire's record?
Source: Major League Baseball

SIMILAR TRIANGLES

Proportions arise in geometry when we are studying *similar triangles.* If two triangles are **similar,** then their corresponding angles have the same measure and their corresponding sides are proportional. To illustrate, if triangle ABC is similar to triangle RST, then angles A and R have the same measure, angles B and S have the same measure, angles C and T have the same measure, and

$$\frac{a}{r} = \frac{b}{s} = \frac{c}{t}.$$

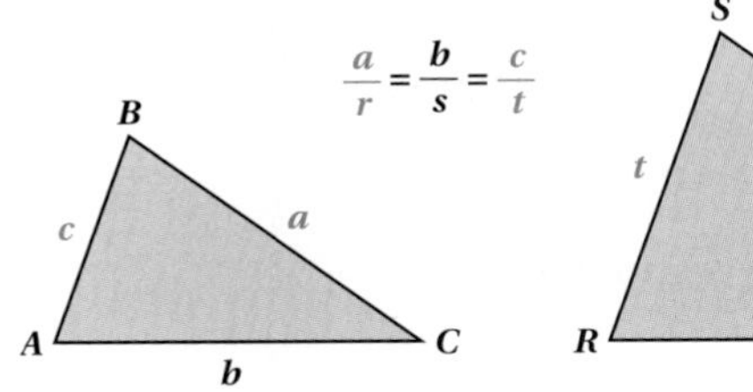

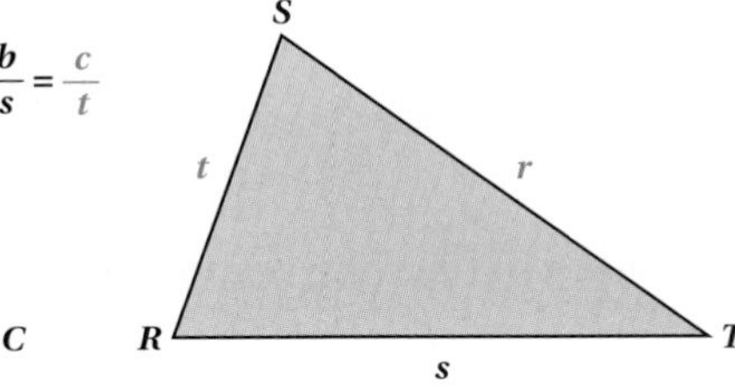

Answer on page A-30

10. Height of a Flagpole. How high is a flagpole that casts a 45-ft shadow at the same time that a 5.5-ft woman casts a 10-ft shadow?

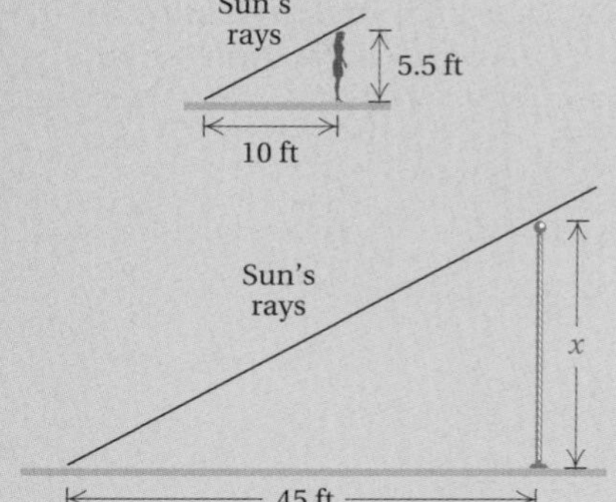

11. F-106 Blueprint. Referring to Example 7, find the length x on the plane.

Answers on page A-30

SIMILAR TRIANGLES

In **similar triangles,** corresponding angles have the same measure and the lengths of corresponding sides are proportional.

EXAMPLE 6 *Similar Triangles.* Triangles ABC and XYZ below are similar triangles. Solve for z if $x = 10$, $a = 8$, and $c = 5$.

We make a sketch, write a proportion, and then solve. Note that side a is always opposite angle A, side x is always opposite angle X, and so on.

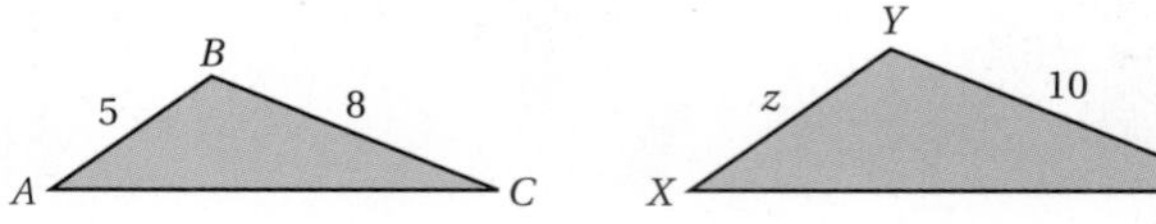

We have

$$\frac{z}{5} = \frac{10}{8} \qquad \text{The proportion } \frac{5}{z} = \frac{8}{10} \text{ could also be used.}$$

$$40 \cdot \frac{z}{5} = 40 \cdot \frac{10}{8} \qquad \text{Multiplying by 40}$$

$$8z = 50$$

$$z = \frac{50}{8}, \text{ or } 6.25. \qquad \text{Dividing by 8}$$

EXAMPLE 7 *F-106 Blueprint.* A blueprint for an F-106 Delta Dart fighter plane is a scale drawing, as shown below. Each wing has a triangular shape. The blueprint shows similar triangles. Find the length of side a of the wing.

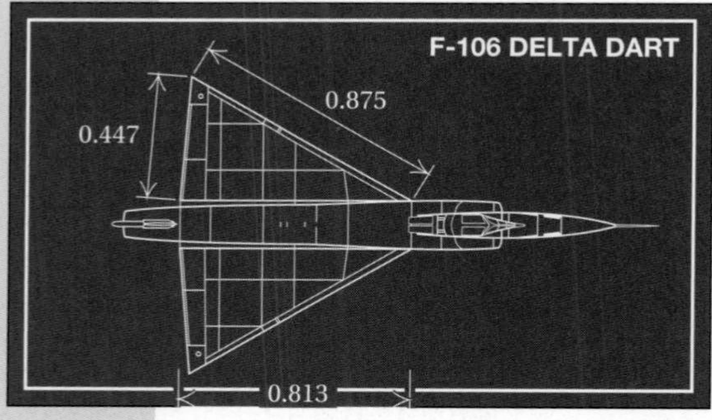

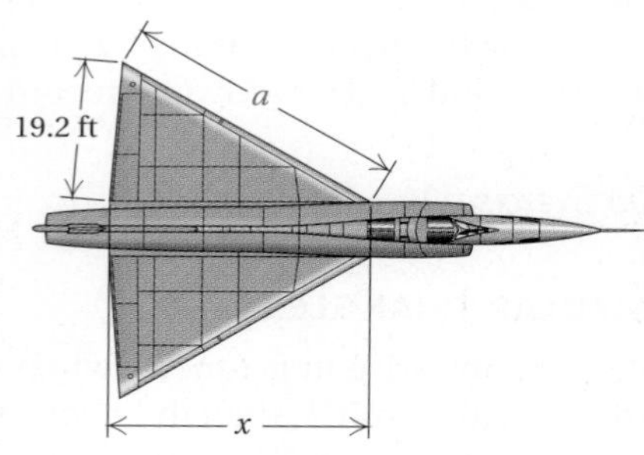

We let $a =$ the length of the wing. Thus we have the proportion

$$\begin{matrix}\text{Length on the blueprint} \rightarrow \\ \text{Length of the wing} \rightarrow\end{matrix} \frac{0.447}{19.2} = \frac{0.875}{a}. \begin{matrix}\leftarrow \text{Length on the blueprint} \\ \leftarrow \text{Length of the wing}\end{matrix}$$

Solve: $0.447 \cdot a = 19.2 \cdot 0.875$ Equating cross products

$$a = \frac{19.2 \cdot 0.875}{0.447} \qquad \text{Dividing by 0.447}$$

$$a \approx 37.6 \text{ ft}$$

The length of side a of the wing is about 37.6 ft.

Do Exercises 10 and 11.

6.7 EXERCISE SET

For Extra Help

Digital Video Tutor CD 6 Videotape 13 | InterAct Math | Math Tutor Center | MathXL | MyMathLab

a Solve.

1. *Construction.* It takes Mandy 4 hr to put up paneling in a room. Omar takes 5 hr to do the same job. How long would it take them, working together, to panel the room?

2. *Carpentry.* By checking work records, a carpenter finds that Juanita can build a small shed in 12 hr. Anton can do the same job in 16 hr. How long would it take if they worked together?

3. *Shoveling.* Vern can shovel the snow from his driveway in 45 min. Nina can do the same job in 60 min. How long would it take Nina and Vern to shovel the driveway if they worked together?

4. *Raking.* Zoë can rake her yard in 4 hr. Steffi does the same job in 3 hr. How long would it take the two of them, working together, to rake the yard?

5. *Wiring.* By checking work records, a contractor finds that Kenny Dewitt can wire a room addition in 9 hr. It takes Betty Wohnt 7 hr to wire the same room. How long would it take if they worked together?

6. *Plumbing.* By checking work records, a plumber finds that Raul can plumb a house in 48 hr. Mira can do the same job in 36 hr. How long would it take if they worked together?

7. *Gardening.* Nicole can weed her vegetable garden in 50 min. Glen can weed the same garden in 40 min. How long would it take if they worked together?

8. *Harvesting.* Bobbi can pick a quart of raspberries in 20 min. Blanche can pick a quart in 25 min. How long would it take if Bobbi and Blanche worked together?

9. *Computer Printers.* The HP OfficeJetG85 printer can copy Charlotte's dissertation in 12 min. The HP LaserJet 3200se can copy the same document in 20 min. If the two machines work together, how long would they take to copy the dissertation?

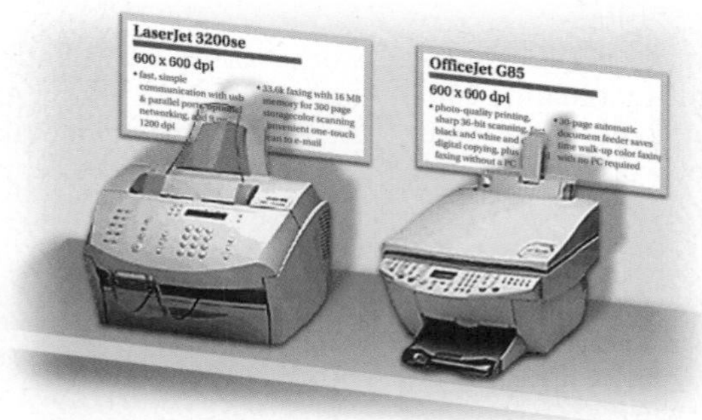

10. *Fax Machines.* The Brother MFC4500® can fax a year-end report in 10 min while the Xerox 850® can fax the same report in 8 min. How long would it take the two machines, working together, to fax the report? (Assume that the recipient has at least two machines for incoming faxes.)

11. *Car Speed.* Rick drives his four-wheel-drive truck 40 km/h faster than Sarah drives her Saturn. While Sarah travels 150 km, Rick travels 350 km. Find their speeds.

Complete this table and the equations as part of the *Familiarize* step.

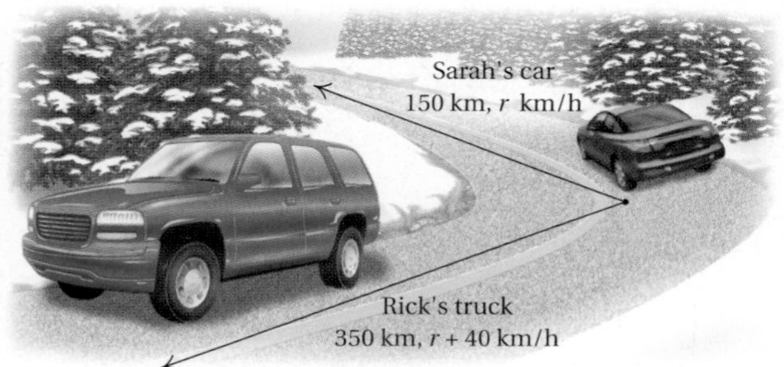

$d = r \cdot t$

	DISTANCE	SPEED	TIME	
Car	150	r		$\to 150 = r(\ \ )$
Truck	350		t	$\to 350 = (\ \ )t$

12. *Car Speed.* A passenger car travels 30 km/h faster than a delivery truck. While the car goes 400 km, the truck goes 250 km. Find their speeds.

13. *Train Speed.* The speed of a B & M freight train is 14 mph slower than the speed of an Amtrak passenger train. The freight train travels 330 mi in the same time that it takes the passenger train to travel 400 mi. Find the speed of each train.

Complete this table and the equations as part of the *Familiarize* step.

$d = r \cdot t$

	DISTANCE	SPEED	TIME	
B & M	330		t	$\to 330 = (\ \ )t$
Amtrak	400	r		$\to 400 = r(\ \ )$

14. *Train Speed.* The speed of a freight train is 15 mph slower than the speed of a passenger train. The freight train travels 390 mi in the same time that it takes the passenger train to travel 480 mi. Find the speed of each train.

15. *Trucking Speed.* A long-distance trucker traveled 120 mi in one direction during a snowstorm. The return trip in rainy weather was accomplished at double the speed and took 3 hr less time. Find the speed going.

16. *Car Speed.* After making a trip of 126 mi, a person found that the trip would have taken 1 hr less time by increasing the speed by 8 mph. What was the actual speed?

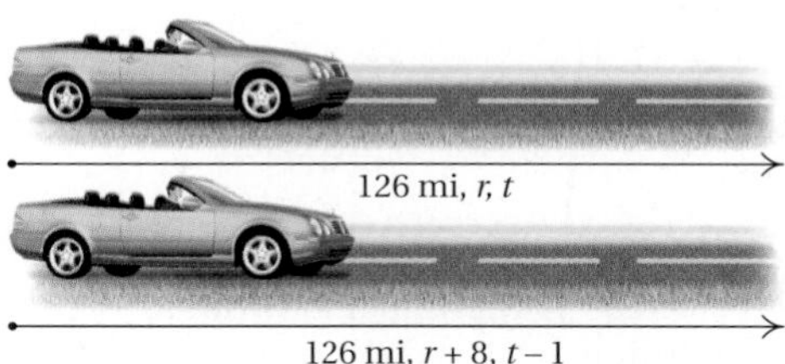

17. *Bicycle Speed.* Hank bicycles 5 km/h slower than Kelly. In the time that it takes Hank to bicycle 42 km, Kelly can bicycle 57 km. How fast does each bicyclist travel?

18. *Driving Speed.* Hillary's Lexus travels 30 mph faster than Bill's Harley. In the same time that Bill travels 75 mi, Hillary travels 120 mi. Find their speeds.

19. *Walking Speed.* Bonnie power walks 3 km/h faster than Ralph. In the time that it takes Ralph to walk 7.5 km, Bonnie walks 12 km. Find their speeds.

20. *Cross-Country Skiing.* Gerard cross-country skis 4 km/h faster than Sally. In the time that it takes Sally to ski 18 km, Gerard skis 24 km. Find their speeds.

21. *Tractor Speed.* Manley's tractor is just as fast as Caledonia's. It takes Manley 1 hr more than it takes Caledonia to drive to town. If Manley is 20 mi from town and Caledonia is 15 mi from town, how long does it take Caledonia to drive to town?

22. *Boat Speed.* Tory and Emilio's motorboats both travel at the same speed. Tory pilots her boat 40 km before docking. Emilio continues for another 2 hr, traveling a total of 100 km before docking. How long did it take Tory to navigate the 40 km?

b Find the ratio of the following. Simplify if possible.

23. 10 divorces, 18 marriages

24. 800 mi, 50 gal

25. *Speed of Black Racer.* A black racer snake travels 4.6 km in 2 hr. What is the speed in kilometers per hour?

26. *Speed of Light.* Light travels 558,000 mi in 3 sec. What is the speed in miles per second?

Solve.

27. *Protein Needs.* A 120-lb person should eat a minimum of 44 g of protein each day. How much protein should a 180-lb person eat each day?

28. *Coffee Beans.* The coffee beans from 14 trees are required to produce 7.7 kg of coffee (this is the average amount that each person in the United States drinks each year). How many trees are required to produce 320 kg of coffee?

29. *Hemoglobin.* A normal 10-cc specimen of human blood contains 1.2 g of hemoglobin. How much hemoglobin would 16 cc of the same blood contain?

30. *Walking Speed.* Wanda walked 234 km in 14 days. At this rate, how far would she walk in 42 days?

31. *Honey Bees.* Making 1 lb of honey requires 20,000 trips by bees to flowers to gather nectar. How many pounds of honey would 35,000 trips produce?
Source: Tom Turpin, Professor of Entomology, Purdue University

32. *Cockroaches and Horses.* A cockroach can run about 2 mi/hr (mph). The average body length of a cockroach is 1 in. The average body length of a horse is 8 ft (96 in.). If we assume that a horse's speed-to-length ratio is the same as that of a cockroach, how fast can a horse run?
Source: Tom Turpin, Professor of Entomology, Purdue University

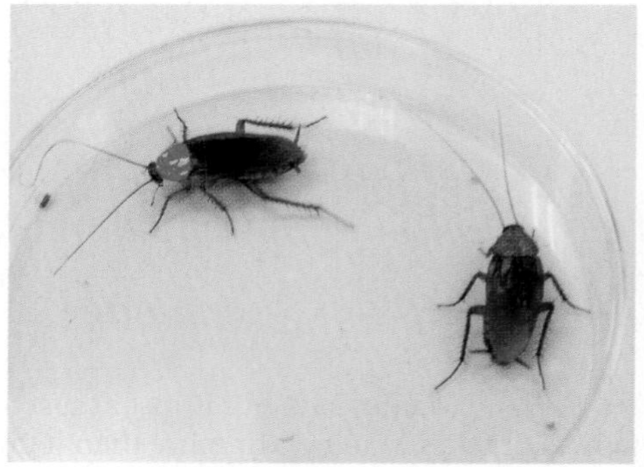

Professor Turpin founded the annual cockroach race at Purdue University.

33. *Money.* The ratio of the weight of copper to the weight of zinc in a U.S. penny is $\frac{1}{39}$. If 50 kg of zinc is being turned into pennies, how much copper is needed?

34. *Baking.* In a potato bread recipe, the ratio of milk to flour is $\frac{3}{13}$. If 5 cups of milk are used, how many cups of flour are used?

35. *Ichiro Suzuki.* In the 2001 major-league baseball season, Ichiro Suzuki, a rookie from Japan playing for the Seattle Mariners, led the American League in hitting by collecting 96 hits in 266 at-bats in the first 58 games.

a) The ratio of number of hits to number of at-bats, rounded to the nearest thousandth, is a player's *batting average.* What was Suzuki's batting average?

b) Based on the ratio of number of hits to number of games, how many hits would he get in the 162-game season?

c) Based on the ratio of number of hits to number of at-bats and assuming he bats 560 times in 2001, how many hits would he get?

36. *Rich Aurilia.* In the 2001 major-league baseball season, Rich Aurilia, playing for the San Francisco Giants, led the National League in hitting by collecting 79 hits in 213 at-bats in the first 55 games.

a) The ratio of number of hits to number of at-bats, rounded to the nearest thousandth, is a player's *batting average.* What was Aurilia's batting average?

b) Based on the ratio of number of hits to number of games, how many hits would he get in the 162-game season?

c) Based on the ratio of number of hits to number of at-bats and assuming he bats 550 times in 2001, how many hits would he get?

Hat Sizes. Hat sizes are determined by measuring the circumference of one's head in either inches or centimeters. Use ratio and proportion to complete the missing parts of the following table.

	HAT SIZE	HEAD CIRCUMFERENCE (in inches)	HEAD CIRCUMFERENCE (in centimeters)
	$6\frac{3}{4}$	$21\frac{1}{5}$ in.	53.8 cm
37.	7		
38.			56.8 cm
39.		$22\frac{4}{5}$ in.	
40.	$7\frac{3}{8}$		
41.			59.8 cm
42.		24 in.	

43. *Estimating Whale Population.* To determine the number of blue whales in the world's oceans, marine biologists tag 500 blue whales in various parts of the world. Later, 400 blue whales are checked, and it is found that 20 of them are tagged. Estimate the blue whale population.

44. *Estimating Trout Population.* To determine the number of trout in a lake, a conservationist catches 112 trout, tags them, and throws them back into the lake. Later, 82 trout are caught; 32 of them are tagged. Estimate the number of trout in the lake.

45. *Weight on Mars.* The ratio of the weight of an object on Mars to the weight of an object on Earth is 0.4 to 1.

a) How much would a 12-ton rocket weigh on Mars?
b) How much would a 120-lb astronaut weigh on Mars?

46. *Weight on Moon.* The ratio of the weight of an object on the moon to the weight of an object on Earth is 0.16 to 1.

a) How much would a 12-ton rocket weigh on the moon?
b) How much would a 180-lb astronaut weigh on the moon?

47. *Quality Control.* A sample of 144 firecrackers contained 9 "duds." How many duds would you expect in a sample of 3200 firecrackers?

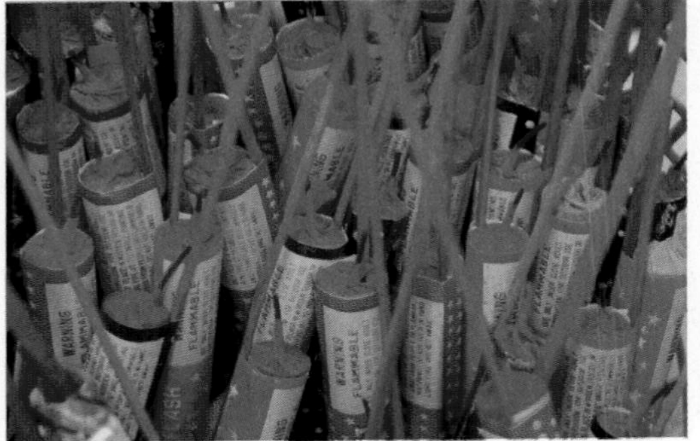

48. *Grass Seed.* It takes 60 oz of grass seed to seed 3000 ft^2 of lawn. At this rate, how much would be needed to seed 5000 ft^2 of lawn?

Geometry. For each pair of similar triangles, find the length of the indicated side.

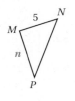

49. b:

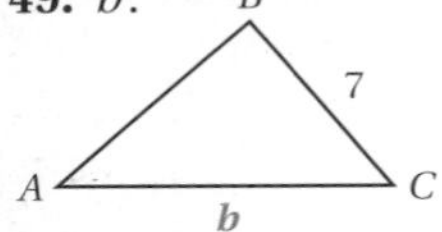

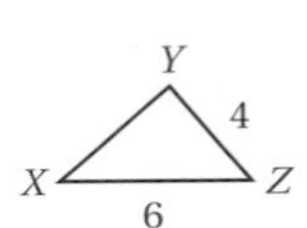

50. a:

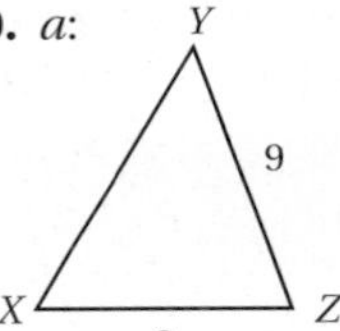

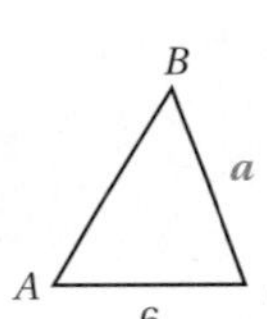

51. f:

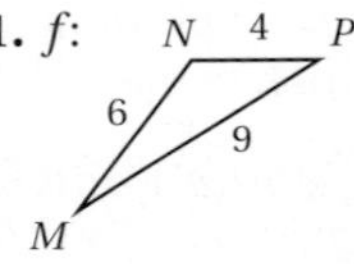

G f H
4
6
F

52. r:

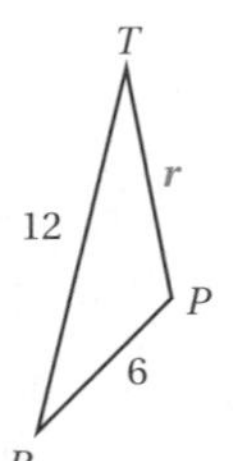

N
10
16
O
8
M

53. h:

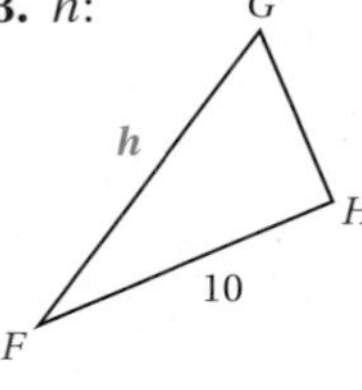

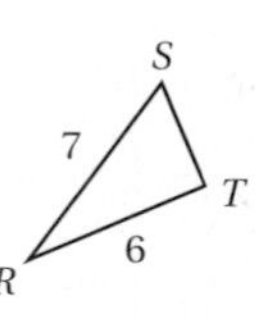

54. n:

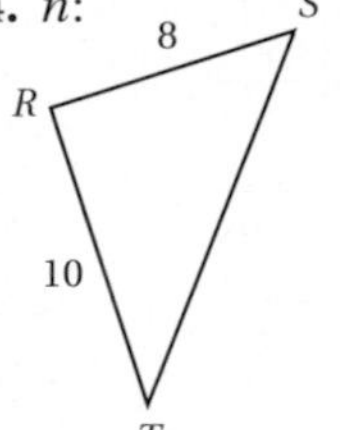

5
N
M
n
P

55. l:

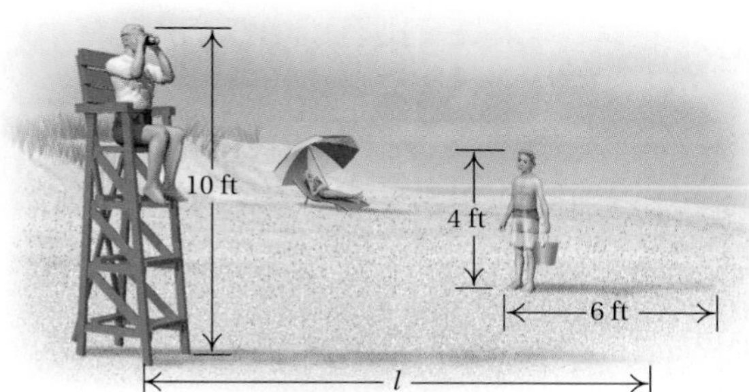

56. h:

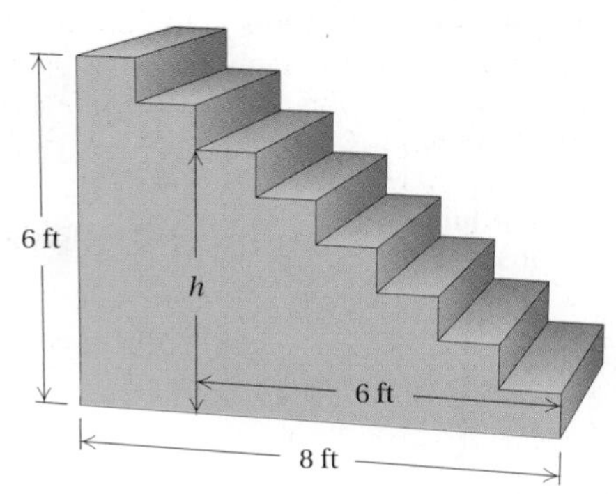

57. DW Explain why it is incorrect to assume that two workers can complete a task twice as quickly as one person working alone.

58. DW Write a problem similar to Example 1 or Margin Exercise 1 for a classmate to solve.

SKILL MAINTENANCE

Simplify. [4.1d]

59. $x^5 \cdot x^6$

60. $x^{-5} \cdot x^6$

61. $x^{-5} \cdot x^{-6}$

62. $x^5 \cdot x^{-6}$

Graph. [3.2b]

63. $y = 2x - 6$

64. $y = -2x + 6$

65. $3x + 2y = 12$

66. $x - 3y = 6$

67. $y = -\frac{3}{4}x + 2$

68. $y = \frac{2}{5}x - 4$

SYNTHESIS

69. Ann and Betty work together and complete a sales report in 4 hr. It would take Betty 6 hr longer, working alone, to do the job than it would Ann. How long would it take each of them to do the job working alone?

71. How soon after 5 o'clock will the hands on a clock first be together?

70. Express 100 as the sum of two numbers for which the ratio of one number, increased by 5, to the other number, decreased by 5, is 4.

72. Rachel allows herself 1 hr to reach a sales appointment 50 mi away. After she has driven 30 mi, she realizes that she must increase her speed by 15 mph in order to get there on time. What was her speed for the first 30 mi?

73. Solve $\frac{t}{a} + \frac{t}{b} = 1$ for t.

Objective

a Simplify complex rational expressions.

To the instructor and the student: Students can be instructed to either try both methods and then choose the one that works best for them, or use the method chosen by the instructor.

6.8 COMPLEX RATIONAL EXPRESSIONS

a Simplifying Complex Rational Expressions

A **complex rational expression,** or **complex fractional expression,** is a rational expression that has one or more rational expressions within its numerator or denominator. Here are some examples:

$$\frac{1+\frac{2}{x}}{3}, \quad \frac{\frac{x+y}{2}}{\frac{2x}{x+1}}, \quad \frac{\frac{1}{3}+\frac{1}{5}}{\frac{2}{x}-\frac{x}{y}}.$$

These are rational expressions within the complex rational expression.

There are two methods to simplify complex rational expressions. We will consider them both.

METHOD 1

MULTIPLYING BY THE LCM OF ALL THE DENOMINATORS

To simplify a complex rational expression:

1. First, find the LCM of all the denominators of all the rational expressions occurring *within* both the numerator and the denominator of the complex rational expression.
2. Then multiply by 1 using LCM/LCM.
3. If possible, simplify by removing a factor of 1.

EXAMPLE 1 Simplify: $\dfrac{\frac{1}{2}+\frac{3}{4}}{\frac{5}{6}-\frac{3}{8}}$.

We have

$$\frac{\frac{1}{2}+\frac{3}{4}}{\frac{5}{6}-\frac{3}{8}}$$

The denominators *within* the complex rational expression are 2, 4, 6, and 8. The LCM of these denominators is 24. We multiply by 1 using $\frac{24}{24}$. This amounts to multiplying both the numerator *and* the denominator by 24.

$$= \frac{\frac{1}{2}+\frac{3}{4}}{\frac{5}{6}-\frac{3}{8}} \cdot \frac{24}{24}$$ Multiplying by 1

$$= \frac{\left(\frac{1}{2}+\frac{3}{4}\right)24}{\left(\frac{5}{6}-\frac{3}{8}\right)24}$$

← Multiplying the numerator by 24

← Multiplying the denominator by 24

Using the distributive laws, we carry out the multiplications:

$$= \frac{\frac{1}{2}(24) + \frac{3}{4}(24)}{\frac{5}{6}(24) - \frac{3}{8}(24)}$$

$$= \frac{12 + 18}{20 - 9} \qquad \text{Simplifying}$$

$$= \frac{30}{11}.$$

Multiplying in this manner has the effect of clearing fractions in both the top and the bottom of the complex rational expression.

Do Exercise 1.

EXAMPLE 2 Simplify: $\dfrac{\frac{3}{x} + \frac{1}{2x}}{\frac{1}{3x} - \frac{3}{4x}}$.

The denominators within the complex expression are x, $2x$, $3x$, and $4x$. The LCM of these denominators is $12x$. We multiply by 1 using $12x/12x$.

$$\frac{\frac{3}{x} + \frac{1}{2x}}{\frac{1}{3x} - \frac{3}{4x}} \cdot \frac{12x}{12x} = \frac{\left(\frac{3}{x} + \frac{1}{2x}\right)12x}{\left(\frac{1}{3x} - \frac{3}{4x}\right)12x} = \frac{\frac{3}{x}(12x) + \frac{1}{2x}(12x)}{\frac{1}{3x}(12x) - \frac{3}{4x}(12x)}$$

$$= \frac{36 + 6}{4 - 9} = -\frac{42}{5}$$

Do Exercise 2.

EXAMPLE 3 Simplify: $\dfrac{1 - \frac{1}{x}}{1 - \frac{1}{x^2}}$.

The denominators within the complex expression are x and x^2. The LCM of these denominators is x^2. We multiply by 1 using x^2/x^2. Then, after obtaining a single rational expression, we simplify:

$$\frac{1 - \frac{1}{x}}{1 - \frac{1}{x^2}} \cdot \frac{x^2}{x^2} = \frac{\left(1 - \frac{1}{x}\right)x^2}{\left(1 - \frac{1}{x^2}\right)x^2} = \frac{1(x^2) - \frac{1}{x}(x^2)}{1(x^2) - \frac{1}{x^2}(x^2)} = \frac{x^2 - x}{x^2 - 1}$$

$$= \frac{x\cancel{(x - 1)}}{(x + 1)\cancel{(x - 1)}} = \frac{x}{x + 1}.$$

Do Exercise 3.

1. Simplify. Use method 1.

$$\frac{\frac{1}{3} + \frac{4}{5}}{\frac{7}{8} - \frac{5}{6}}$$

2. Simplify. Use method 1.

$$\frac{\frac{x}{2} + \frac{2x}{3}}{\frac{1}{x} - \frac{x}{2}}$$

3. Simplify. Use method 1.

$$\frac{1 + \frac{1}{x}}{1 - \frac{1}{x^2}}$$

Answers on page A-31

4. Simplify. Use method 2.

$$\frac{\frac{1}{3}+\frac{4}{5}}{\frac{7}{8}-\frac{5}{6}}$$

Answer on page A-31

METHOD 2

ADDING IN THE NUMERATOR AND THE DENOMINATOR

To simplify a complex rational expression:

1. Add or subtract, as necessary, to get a single rational expression in the numerator.
2. Add or subtract, as necessary, to get a single rational expression in the denominator.
3. Divide the numerator by the denominator.
4. If possible, simplify by removing a factor of 1.

We will redo Examples 1–3 using this method.

EXAMPLE 4 Simplify: $\dfrac{\frac{1}{2}+\frac{3}{4}}{\frac{5}{6}-\frac{3}{8}}$.

The LCM of 2 and 4 in the numerator is 4. The LCM of 6 and 8 in the denominator is 24. We have

$$\frac{\frac{1}{2}+\frac{3}{4}}{\frac{5}{6}-\frac{3}{8}}=\frac{\frac{1}{2}\cdot\frac{2}{2}+\frac{3}{4}}{\frac{5}{6}\cdot\frac{4}{4}-\frac{3}{8}\cdot\frac{3}{3}}$$

Multiplying the $\frac{1}{2}$ by 1 to get the common denominator, 4

Multiplying the $\frac{5}{6}$ and the $\frac{3}{8}$ by 1 to get the common denominator, 24

$$=\frac{\frac{2}{4}+\frac{3}{4}}{\frac{20}{24}-\frac{9}{24}}$$

$$=\frac{\frac{5}{4}}{\frac{11}{24}}$$ Adding in the numerator; subtracting in the denominator

$$=\frac{5}{4}\cdot\frac{24}{11}$$ Multiplying by the reciprocal of the divisor

$$=\frac{5\cdot 3\cdot 2\cdot 2\cdot 2}{2\cdot 2\cdot 11}$$ Factoring

$$=\frac{5\cdot 3\cdot 2\cdot \cancel{2}\cdot \cancel{2}}{\cancel{2}\cdot \cancel{2}\cdot 11}$$ Removing a factor of 1: $\frac{2\cdot 2}{2\cdot 2}=1$

$$=\frac{30}{11}.$$

Do Exercise 4.

EXAMPLE 5 Simplify: $\dfrac{\dfrac{3}{x}+\dfrac{1}{2x}}{\dfrac{1}{3x}-\dfrac{3}{4x}}$.

We have

$$\frac{\frac{3}{x}+\frac{1}{2x}}{\frac{1}{3x}-\frac{3}{4x}} = \frac{\frac{3}{x}\cdot\frac{2}{2}+\frac{1}{2x}}{\frac{1}{3x}\cdot\frac{4}{4}-\frac{3}{4x}\cdot\frac{3}{3}}$$

Finding the LCD, $2x$, and multiplying by 1 in the numerator

Finding the LCD, $12x$, and multiplying by 1 in the denominator

$$= \frac{\frac{6}{2x}+\frac{1}{2x}}{\frac{4}{12x}-\frac{9}{12x}} = \frac{\frac{7}{2x}}{\frac{-5}{12x}}$$

Adding in the numerator and subtracting in the denominator

$$= \frac{7}{2x}\cdot\frac{12x}{-5}$$

Multiplying by the reciprocal of the divisor

$$= \frac{7}{2x}\cdot\frac{6(2x)}{-5}$$

Factoring

$$= \frac{7}{\cancel{2x}}\cdot\frac{6(\cancel{2x})}{-5}$$

Removing a factor of 1: $\frac{2x}{2x}=1$

$$= \frac{42}{-5} = -\frac{42}{5}.$$

Do Exercise 5.

EXAMPLE 6 Simplify: $\dfrac{1-\dfrac{1}{x}}{1-\dfrac{1}{x^2}}$.

We have

$$\frac{1-\frac{1}{x}}{1-\frac{1}{x^2}} = \frac{1\cdot\frac{x}{x}-\frac{1}{x}}{1\cdot\frac{x^2}{x^2}-\frac{1}{x^2}}$$

Finding the LCD, x, and multiplying by 1 in the numerator

Finding the LCD, x^2, and multiplying by 1 in the denominator

$$= \frac{\frac{x-1}{x}}{\frac{x^2-1}{x^2}}$$

Subtracting in the numerator and subtracting in the denominator

$$= \frac{x-1}{x}\cdot\frac{x^2}{x^2-1}$$

Multiplying by the reciprocal of the divisor

$$= \frac{(x-1)x\cdot x}{x(x-1)(x+1)}$$

Factoring

$$= \frac{\cancel{(x-1)}\cancel{x}\cdot x}{\cancel{x}\cancel{(x-1)}(x+1)}$$

Removing a factor of 1: $\frac{x(x-1)}{x(x-1)}=1$

$$= \frac{x}{x+1}.$$

Do Exercise 6.

5. Simplify. Use method 2.

$$\frac{\frac{x}{2}+\frac{2x}{3}}{\frac{1}{x}-\frac{x}{2}}$$

6. Simplify. Use method 2.

$$\frac{1+\frac{1}{x}}{1-\frac{1}{x^2}}$$

Answers on page A-31

6.8 EXERCISE SET

For Extra Help

Digital Video Tutor CD 6 Videotape 13 | InterAct Math | Math Tutor Center | MathXL | MyMathLab

a Simplify.

1. $\dfrac{1+\dfrac{9}{16}}{1-\dfrac{3}{4}}$

2. $\dfrac{6-\dfrac{3}{8}}{4+\dfrac{5}{6}}$

3. $\dfrac{1-\dfrac{3}{5}}{1+\dfrac{1}{5}}$

4. $\dfrac{2+\dfrac{2}{3}}{2-\dfrac{2}{3}}$

5. $\dfrac{\dfrac{1}{2}+\dfrac{3}{4}}{\dfrac{5}{8}-\dfrac{5}{6}}$

6. $\dfrac{\dfrac{3}{4}+\dfrac{7}{8}}{\dfrac{2}{3}-\dfrac{5}{6}}$

7. $\dfrac{\dfrac{1}{x}+3}{\dfrac{1}{x}-5}$

8. $\dfrac{2-\dfrac{1}{a}}{4+\dfrac{1}{a}}$

9. $\dfrac{4-\dfrac{1}{x^2}}{2-\dfrac{1}{x}}$

10. $\dfrac{\dfrac{2}{y}+\dfrac{1}{2y}}{y+\dfrac{y}{2}}$

11. $\dfrac{8+\dfrac{8}{d}}{1+\dfrac{1}{d}}$

12. $\dfrac{3+\dfrac{2}{t}}{3-\dfrac{2}{t}}$

13. $\dfrac{\dfrac{x}{8}-\dfrac{8}{x}}{\dfrac{1}{8}+\dfrac{1}{x}}$

14. $\dfrac{\dfrac{2}{m}+\dfrac{m}{2}}{\dfrac{m}{3}-\dfrac{3}{m}}$

15. $\dfrac{1+\dfrac{1}{y}}{1-\dfrac{1}{y^2}}$

16. $\dfrac{\dfrac{1}{q^2}-1}{\dfrac{1}{q}+1}$

17. $\dfrac{\dfrac{1}{5}-\dfrac{1}{a}}{\dfrac{5-a}{5}}$

18. $\dfrac{\dfrac{4}{t}}{4+\dfrac{1}{t}}$

19. $\dfrac{\dfrac{1}{a}+\dfrac{1}{b}}{\dfrac{1}{a^2}-\dfrac{1}{b^2}}$

20. $\dfrac{\dfrac{1}{x^2}-\dfrac{1}{y^2}}{\dfrac{2}{x}-\dfrac{2}{y}}$

21. $\dfrac{\dfrac{p}{q}+\dfrac{q}{p}}{\dfrac{1}{p}+\dfrac{1}{q}}$

22. $\dfrac{x-3+\dfrac{2}{x}}{x-4+\dfrac{3}{x}}$

23. $\dfrac{\dfrac{2}{a}+\dfrac{4}{a^2}}{\dfrac{5}{a^3}-\dfrac{3}{a}}$

24. $\dfrac{\dfrac{5}{x^3}-\dfrac{1}{x^2}}{\dfrac{2}{x}+\dfrac{3}{x^2}}$

25. $\dfrac{\dfrac{2}{7a^4} - \dfrac{1}{14a}}{\dfrac{3}{5a^2} + \dfrac{2}{15a}}$

26. $\dfrac{\dfrac{5}{4x^3} - \dfrac{3}{8x}}{\dfrac{3}{2x} + \dfrac{3}{4x^3}}$

27. $\dfrac{\dfrac{a}{b} + \dfrac{c}{d}}{\dfrac{b}{a} + \dfrac{d}{c}}$

28. $\dfrac{\dfrac{a}{b} - \dfrac{c}{d}}{\dfrac{b}{a} - \dfrac{d}{c}}$

29. $\dfrac{\dfrac{x}{5y^3} + \dfrac{3}{10y}}{\dfrac{3}{10y} + \dfrac{x}{5y^3}}$

30. $\dfrac{\dfrac{a}{6b^3} + \dfrac{4}{9b^2}}{\dfrac{5}{6b} - \dfrac{1}{9b^3}}$

31. $\dfrac{\dfrac{3}{x+1} + \dfrac{1}{x}}{\dfrac{2}{x+1} + \dfrac{3}{x}}$

32. $\dfrac{x - 7 + \dfrac{5}{x-1}}{x - 3 + \dfrac{1}{x-1}}$

33. $^{\text{D}}\text{W}$ Why is factoring an important skill when simplifying complex rational expressions?

34. $^{\text{D}}\text{W}$ Why is the distributive law especially important when using method 1 of this section?

SKILL MAINTENANCE

Add. [4.4a]

35. $(2x^3 - 4x^2 + x - 7) + (4x^4 + x^3 + 4x^2 + x)$

36. $(2x^3 - 4x^2 + x - 7) + (-2x^3 + 4x^2 - x + 7)$

Factor. [5.6a]

37. $p^2 - 10p + 25$

38. $p^2 + 10p + 25$

39. $50p^2 - 100$

40. $5p^2 - 40p - 100$

Solve. [5.8a]

41. *Perimeter of a Rectangle.* The length of a rectangle is 3 yd greater than the width. The area of the rectangle is 10 yd^2. Find the perimeter.

42. *Ladder Distances.* A ladder of length 13 ft is placed against a building in such a way that the distance from the top of the ladder to the ground is 7 ft more than the distance from the bottom of the ladder to the building. Find these distances.

SYNTHESIS

43. Find the reciprocal of $\dfrac{2}{x-1} - \dfrac{1}{3x-2}$.

Simplify.

44. $\left[\dfrac{\dfrac{x+1}{x-1} + 1}{\dfrac{x+1}{x-1} - 1}\right]^5$

45. $1 + \dfrac{1}{1 + \dfrac{1}{1 + \dfrac{1}{1 + \dfrac{1}{x}}}}$

46. $\dfrac{\dfrac{z}{1 - \dfrac{z}{2 + 2z}} - 2z}{\dfrac{2z}{5z - 2} - 3}$

CHAPTER 6

Summary and Review

The review that follows is meant to prepare you for a chapter exam. It consists of two parts. The first part is a checklist of some of the Study Tips referred to in this and preceding chapters. The second part is the Review Exercises. These provide practice exercises for the exam, together with references to section objectives so you can go back and review. Before beginning, stop and look back over the skills you have obtained. What skills in mathematics do you have now that you did not have before studying this chapter?

STUDY TIPS CHECKLIST

The foundation of all your study skills is TIME!

- ☐ Are you finding quiet, nondistracting places to study?
- ☐ Have you tried taping your lectures, with your instructor's permission?
- ☐ Have you tried tutoring a fellow student?
- ☐ Did you use the five-step problem-solving strategy when doing the applications in Section 6.7?
- ☐ Are you doing more even-numbered exercises and using strategies to check your work, such as doing the problem a second time, checking, or estimating?

REVIEW EXERCISES

Find all numbers for which the rational expression is not defined. [6.1a]

1. $\dfrac{3}{x}$

2. $\dfrac{4}{x-6}$

3. $\dfrac{x+5}{x^2-36}$

4. $\dfrac{x^2-3x+2}{x^2+x-30}$

5. $\dfrac{-4}{(x+2)^2}$

6. $\dfrac{x-5}{x^3-8x^2+15x}$

Simplify. [6.1c]

7. $\dfrac{4x^2-8x}{4x^2+4x}$

8. $\dfrac{14x^2-x-3}{2x^2-7x+3}$

9. $\dfrac{(y-5)^2}{y^2-25}$

Multiply and simplify. [6.1d]

10. $\dfrac{a^2-36}{10a}\cdot\dfrac{2a}{a+6}$

11. $\dfrac{6t-6}{2t^2+t-1}\cdot\dfrac{t^2-1}{t^2-2t+1}$

Divide and simplify. [6.2b]

12. $\dfrac{10-5t}{3}\div\dfrac{t-2}{12t}$

13. $\dfrac{4x^4}{x^2-1}\div\dfrac{2x^3}{x^2-2x+1}$

Find the LCM. [6.3c]

14. $3x^2,\ 10xy,\ 15y^2$

15. $a - 2,\ 4a - 8$

16. $y^2 - y - 2,\ y^2 - 4$

Add and simplify. [6.4a]

17. $\frac{x+8}{x+7} + \frac{10-4x}{x+7}$

18. $\frac{3}{3x-9} + \frac{x-2}{3-x}$

19. $\frac{2a}{a+1} + \frac{4a}{a^2-1}$

20. $\frac{d^2}{d-c} + \frac{c^2}{c-d}$

Subtract and simplify. [6.5a]

21. $\frac{6x-3}{x^2-x-12} - \frac{2x-15}{x^2-x-12}$

22. $\frac{3x-1}{2x} - \frac{x-3}{x}$

23. $\frac{x+3}{x-2} - \frac{x}{2-x}$

24. $\frac{1}{x^2-25} - \frac{x-5}{x^2-4x-5}$

25. Perform the indicated operations and simplify: [6.5b]

$$\frac{3x}{x+2} - \frac{x}{x-2} + \frac{8}{x^2-4}.$$

Simplify. [6.8a]

26. $\dfrac{\frac{1}{z}+1}{\frac{1}{z^2}-1}$

27. $\dfrac{\frac{c}{d}-\frac{d}{c}}{\frac{1}{c}+\frac{1}{d}}$

Solve. [6.6a]

28. $\frac{3}{y} - \frac{1}{4} = \frac{1}{y}$

29. $\frac{15}{x} - \frac{15}{x+2} = 2$

Solve. [6.7a]

30. *Highway Work.* In checking records, a contractor finds that crew A can pave a certain length of highway in 9 hr, while crew B can do the same job in 12 hr. How long would it take if they worked together?

31. *Train Speed.* A manufacturer is testing two high-speed trains. One train travels 40 km/h faster than the other. While one train travels 70 km, the other travels 60 km. Find the speed of each train.

32. *Airplane Speed.* One plane travels 80 mph faster than another. While one travels 1750 mi, the other travels 950 mi. Find the speed of each plane.

Solve. [6.7b]

33. *Quality Control.* A sample of 250 calculators contained 8 defective calculators. How many defective calculators would you expect to find in a sample of 5000?

34. *Pizza Proportions.* A certain kind of pizza at Finnelli's Pizzeria uses the following ratio: 5 parts sausage to 7 parts cheese, 6 parts onion to 13 parts green pepper, and 9 parts pepperoni to 14 parts cheese.

a) Finnelli's makes several pizzas with green pepper and onion. They use 2 cups of green pepper. How much onion would they use?

b) Finnelli's makes several pizzas with sausage and cheese. They use 3 cups of sausage. How much cheese would they use?

c) Finnelli's makes several pizzas with pepperoni and cheese. They use 6 cups of pepperoni. How much cheese would they use?

35. *Frog Population.* To estimate how many frogs there are in a rain forest, a research team tags 600 frogs and then releases them. Later the team catches 300 frogs and notes that 25 of them have been tagged. Estimate the total frog population in the rain forest.

36. Triangles ABC and XYZ below are similar. Find the value of x.

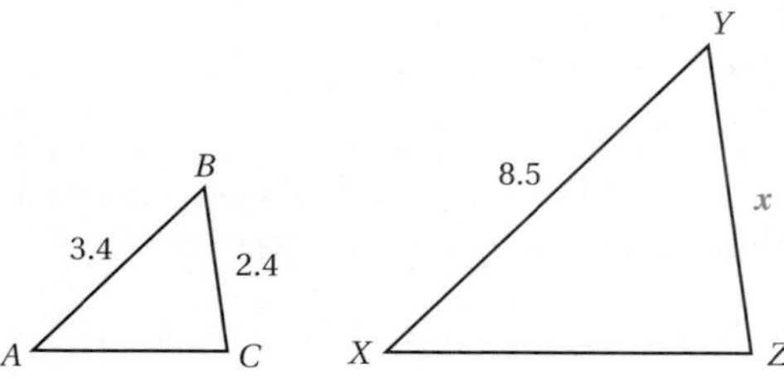

D$_W$ Carry out the direction for each of the following. Explain the use of the LCM in each case.

37. Add: $\dfrac{4}{x-2} + \dfrac{1}{x+2}$. [6.4a]

38. Subtract: $\dfrac{4}{x-2} - \dfrac{1}{x+2}$. [6.5a]

39. Solve: $\dfrac{4}{x-2} + \dfrac{1}{x+2} = \dfrac{26}{x^2-4}$. [6.6a]

40. Simplify: $\dfrac{1-\dfrac{2}{x}}{1+\dfrac{x}{4}}$. [6.8a]

SKILL MAINTENANCE

Certain objectives from four particular sections will be retested on the chapter test. The objectives are listed with the practice problems that follow.

41. Factor: $5x^3 + 20x^2 - 3x - 12$. [5.6a]

42. Simplify: $(5x^3y^2)^{-3}$. [4.2b]

43. Subtract: [4.4c]

$$(5x^3 - 4x^2 + 3x - 4) - (7x^3 - 7x^2 - 9x + 14).$$

44. *Rectangle Dimensions.* The width of a rectangle is 2 cm less than the length. The area is 15 cm^2. Find the dimensions and the perimeter of the rectangle. [5.8a]

SYNTHESIS

Simplify.

45. $\dfrac{2a^2+5a-3}{a^2} \cdot \dfrac{5a^3+30a^2}{2a^2+7a-4} \div \dfrac{a^2+6a}{a^2+7a+12}$

[6.1d], [6.2b]

46. $\dfrac{12a}{(a-b)(b-c)} - \dfrac{2a}{(b-a)(c-b)}$ [6.5a]

47. Compare

$$\frac{A+B}{B} = \frac{C+D}{D}$$

with the proportion

$$\frac{A}{B} = \frac{C}{D}.$$

[6.7b]

CHAPTER 6

Chapter Test

Find all numbers for which the rational expression is not defined.

1. $\frac{8}{2x}$

2. $\frac{5}{x + 8}$

3. $\frac{x - 7}{x^2 - 49}$

4. $\frac{x^2 + x - 30}{x^2 - 3x + 2}$

5. $\frac{11}{(x - 1)^2}$

6. $\frac{x + 2}{x^3 + 8x^2 + 15x}$

7. Simplify:

$$\frac{6x^2 + 17x + 7}{2x^2 + 7x + 3}.$$

8. Multiply and simplify:

$$\frac{a^2 - 25}{6a} \cdot \frac{3a}{a - 5}.$$

9. Divide and simplify:

$$\frac{25x^2 - 1}{9x^2 - 6x} \div \frac{5x^2 + 9x - 2}{3x^2 + x - 2}.$$

10. Find the LCM:

$$y^2 - 9,\ y^2 + 10y + 21,\ y^2 + 4y - 21.$$

Add or subtract. Simplify if possible.

11. $\frac{16 + x}{x^3} + \frac{7 - 4x}{x^3}$

12. $\frac{5 - t}{t^2 + 1} - \frac{t - 3}{t^2 + 1}$

13. $\frac{x - 4}{x - 3} + \frac{x - 1}{3 - x}$

14. $\frac{x - 4}{x - 3} - \frac{x - 1}{3 - x}$

15. $\frac{5}{t - 1} + \frac{3}{t}$

16. $\frac{1}{x^2 - 16} - \frac{x + 4}{x^2 - 3x - 4}$

17. $\frac{1}{x - 1} + \frac{4}{x^2 - 1} - \frac{2}{x^2 - 2x + 1}$

18. Simplify: $\dfrac{9 - \dfrac{1}{y^2}}{3 - \dfrac{1}{y}}.$

Solve.

19. $\frac{7}{y} - \frac{1}{3} = \frac{1}{4}$

20. $\frac{15}{x} - \frac{15}{x-2} = -2$

Solve.

21. *Quality Control.* A sample of 125 spark plugs contained 4 defective spark plugs. How many defective spark plugs would you expect to find in a sample of 500?

22. *Zebra Population.* A game warden catches, tags, and then releases 15 zebras. A month later, a sample of 20 zebras is collected and 6 of them have tags. Use this information to estimate the size of the zebra population in that area.

23. *Copying Time.* Kopy Kwik has 2 copiers. One can copy a year-end report in 20 min. The other can copy the same document in 30 min. How long would it take both machines, working together, to copy the report?

24. *Driving Speed.* Craig drives 20 km/h faster than Marilyn. In the same time that Marilyn drives 225 km, Craig drives 325 km. Find the speed of each car.

25. This pair of triangles is similar. Find the missing length x.

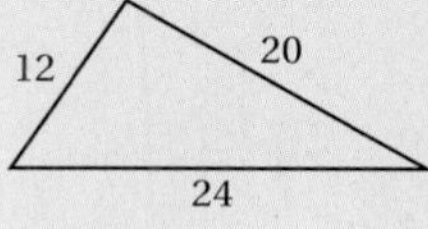

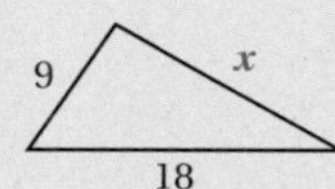

SKILL MAINTENANCE

26. Factor: $16a^2 - 49$.

27. Simplify: $\left(\frac{3x^2}{y^3}\right)^{-4}$.

28. Subtract:

$(5x^2 - 19x + 34) - (-8x^2 + 10x - 42)$.

29. The product of two consecutive integers is 462. Find the integers.

SYNTHESIS

30. Reggie and Rema work together to mulch the flower beds around an office complex in $2\frac{6}{7}$ hr. Working alone, it would take Reggie 6 hr more than it would take Rema. How long would it take each of them to complete the landscaping working alone?

31. Simplify: $1 + \cfrac{1}{1 + \cfrac{1}{1 + \cfrac{1}{a}}}$.

CHAPTERS 1–6

Cumulative Review

Solve.

1. *NFL Salaries.* In 2000, the average salary of an NFL player was $1.2 million. This was an increase of 279% over the average paid in 1990. What was the average salary in 1990?
Source: National Football League

2. *Cycling Distance.* A bicyclist traveled 197 mi in 7 days. At this rate, how many miles could the cyclist travel in 30 days?

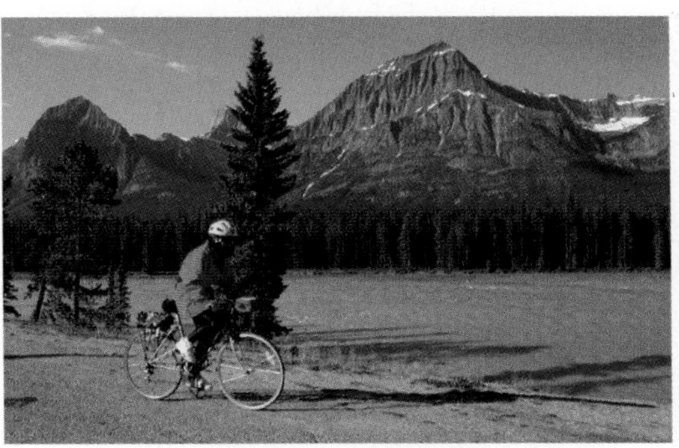

3. *Bicycling.* The speed of one bicyclist is 2 km/h faster than the speed of another bicyclist. The first bicyclist travels 60 km in the same time that it takes the second to travel 50 km. Find the speed of each bicyclist.

4. *Filling Time.* A swimming pool can be filled in 5 hr by hose A alone and in 6 hr by hose B alone. How long would it take to fill the tank if both hoses were working?

5. Linnae has $36 budgeted for stationery. Engraved stationery costs $20 for the first 25 sheets and $0.08 for each additional sheet. How many sheets of stationery can Linnae order and still stay within her budget?

6. If the sides of a square are increased by 2 ft, the area of the original square plus the area of the enlarged square is 452 ft^2. Find the length of a side of the original square.

7. The sum of two consecutive even integers is -554. Find the integers.

8. The sum of the squares of two consecutive odd positive integers is 202. Find the integers.

Evaluate.

9. $\frac{2x + 5}{y - 10}$, when $x = 2$ and $y = 5$

10. $4 - x^3$, when $x = -2$

Simplify.

11. $x - [x - 2(x + 3)]$

12. $(2x^{-2})^{-2}(3x)^3$

13. $\dfrac{24x^8}{18x^{-2}}$

14. $\dfrac{2t^2 + 8t - 42}{2t^2 + 13t - 7}$

15. $\dfrac{\frac{2}{x} + 1}{\frac{x}{x + 2}}$

16. $\dfrac{a^2 - 16}{a^2 - 8a + 16}$

Add. Simplify if possible.

17. $\dfrac{9}{14} + \left(-\dfrac{5}{21}\right)$

18. $\dfrac{2x + y}{x^2y} + \dfrac{x + 2y}{xy^2}$

19. $\dfrac{z}{z^2 - 1} + \dfrac{2}{z + 1}$

20. $(2x^4 + 5x^3 + 4) + (3x^3 - 2x + 5)$

Subtract. Simplify if possible.

21. $1.53 - (-0.8)$

22. $(x^2 - xy - y^2) - (x^2 - y^2)$

23. $\dfrac{3}{x^2 - 9} - \dfrac{x}{9 - x^2}$

24. $\dfrac{2x}{x^2 - x - 20} - \dfrac{4}{x^2 - 10x + 25}$

Multiply. Simplify if possible.

25. $(1.3)(-0.5)(2)$

26. $3x^2(2x^2 + 4x - 5)$

27. $\left(3t + \frac{1}{2}\right)\left(3t - \frac{1}{2}\right)$

28. $(2p - q)^2$

29. $(3x + 5)(x - 4)$

30. $(2x^2 + 1)(2x^2 - 1)$

31. $\dfrac{6t + 6}{t^3 - 2t^2} \cdot \dfrac{t^3 - 3t^2 + 2t}{3t + 3}$

32. $\dfrac{a^2 - 1}{a^2} \cdot \dfrac{2a}{1 - a}$

Divide. Simplify if possible.

33. $(3x^3 - 7x^2 + 9x - 5) \div (x - 1)$

34. $-\frac{21}{25} \div \frac{28}{15}$

35. $\frac{x^2 - x - 2}{4x^3 + 8x^2} \div \frac{x^2 - 2x - 3}{2x^2 + 4x}$

36. $\frac{3 - 3x}{x^2} \div \frac{x - 1}{4x}$

Factor completely.

37. $4x^3 + 12x^2 - 9x - 27$

38. $x^2 + 7x - 8$

39. $3x^2 - 14x - 5$

40. $16y^2 + 40xy + 25x^2$

41. $3x^3 + 24x^2 + 45x$

42. $2x^2 - 2$

43. $x^2 - 28x + 196$

44. $4y^3 + 10y^2 + 12y + 30$

Solve.

45. $2(x - 3) = 5(x + 3)$

46. $2x(3x + 4) = 0$

47. $x^2 = 8x$

48. $x^2 + 16 = 8x$

49. $x - 5 \leq 2x + 4$

50. $3x^2 = 27$

51. $\frac{1}{3}x - \frac{2}{5} = \frac{4}{5}x + \frac{1}{3}$

52. $\frac{x}{3} = \frac{3}{x}$

53. $\frac{x + 5}{2x + 1} = \frac{x - 7}{2x - 1}$

54. $\frac{1}{3}x\left(2x - \frac{1}{5}\right) = 0$

55. $\frac{3 - x}{x - 1} = \frac{2}{x - 1}$

56. $\frac{3}{2x + 5} = \frac{2}{5 - x}$

57. Find the intercepts. Then graph the equation.

$$5y - 2x = -10$$

SYNTHESIS

Solve.

58. $(2x - 1)^2 = (x + 3)^2$

59. $\frac{x + 2}{3x + 2} = \frac{1}{x}$

60. $\frac{2 + \frac{2}{x}}{x + 2 + \frac{1}{x}} = \frac{x + 2}{3}$

61. $\frac{x^6x^4}{x^9x^{-1}} = \frac{5^{14}}{25^6}$

62. Find the reciprocal of $\frac{1 - x}{x + 3} + \frac{x + 1}{2 - x}$.

63. Find the reciprocal of 2.0×10^{-8} and express in scientific notation.

Graphs, Slope, and Applications

Gateway to Chapter 7

We began our study of graphs in Chapter 3, where we focused on linear equations, intercepts, and an introduction to slope. In this chapter, we review and expand on those concepts, considering the slope–intercept equation and a way to graph using the slope and the y-intercept. We also consider graphing of inequalities in two variables and applications such as variation.

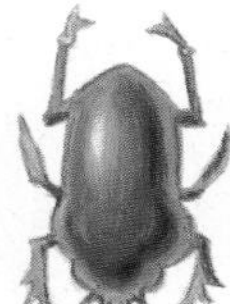

Real-World Application

The beauty of gold has inspired artisans throughout the centuries, as shown by this magnificent sarcophagus, which contained Tutankhamen's mummy. Today we know that the karat rating K of a gold object varies directly as the actual percentage P of gold in the object. A 14-karat gold ring is 58.25% gold. What is the percentage of gold in a 10-karat chain?

Source: Barbara Ann Kipfer, *The Order of Things*. New York: Random House, 1998.

This problem appears as Example 3 in Section 7.5.

CHAPTER 7

CHAPTER 7

Pretest

1. Find the slope and the y-intercept of the line $x - 3y = 7$. [7.1a]

2. Find an equation of the line with slope -4.7 and y-intercept $(0, 8)$. [7.1a]

3. Find an equation of the line containing the points $(3, -1)$ and $(1, -3)$. [7.1c]

4. Find an equation of the line containing the point $(-1, 3)$ and having slope 4. [7.1b]

5. Find an equation of variation in which y varies directly as x and $y = 10$ when $x = 4$. Then find the value of y when $x = 50$. [7.5a]

6. Find an equation of variation in which y varies inversely as x and $y = 10$ when $x = 4$. Then find the value of y when $x = 100$. [7.5c]

7. Draw a line that has slope $-\frac{2}{3}$ and y-intercept $(0, 4)$. [7.2a]

8. Graph $y = \frac{3}{4}x + 5$ using the slope and the y-intercept. [7.2a]

9. **Steak Servings.** The number of servings S of meat that can be obtained from round steak varies directly as the weight W. From 9 kg of round steak, one can get 70 servings of meat. [7.5b]
   a) Find an equation of variation.
   b) How many servings can one get from 12 kg of round steak?

10. **Computer Files.** The number of files N of the same size that a computer's hard drive will hold varies inversely as the size S of the files. Loretta's hard drive will hold 1600 files if each is 50,000 bytes long. [7.5d]
    a) Find an equation of variation.
    b) How many files will the drive hold if each is 125,000 bytes long?

Graph on a plane. [7.4b]

11. $y < x + 2$

12. $2y - 3x \geq 6$

Determine whether the graphs of the equations are parallel, perpendicular, or neither. [7.3a, b]

13. $y - 3x = 9,$
    $y - 3x = 7$

14. $-x + 2y = 7,$
    $2x + y = 4$

15. $y = \dfrac{2}{3}x - 5,$
    $y = -\dfrac{3}{2}x + 4$

16. Determine whether the ordered pair $(-3, 4)$ is a solution of $2x + 5y < 17$. [7.4a]

17. **Consumer Spending on Software.** The line graph at right describes the amount of consumer spending S, in billions, on software in recent years. [7.1c]
    a) Find an equation of the line.
    b) What is the rate of change in software spending with respect to time?
    c) Use the equation to predict consumer spending on software in 2005.

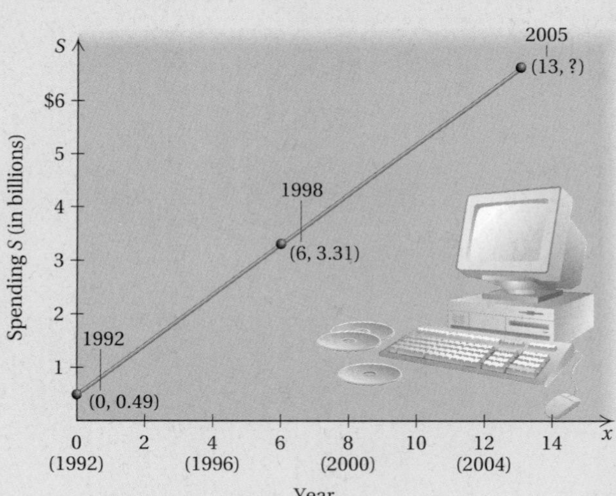

Source: Veronis, Suhler & Associates, PC Data

7.1 THE SLOPE–INTERCEPT EQUATION

We began our study of graphs of lines in Chapter 3, where we learned that the slope of a line and the y-intercept can be read directly from an equation in the form $y = mx + b$. The **slope** is m and the **y-intercept** is $(0, b)$. (It may be helpful to you to review that chapter now.)

Let's review the concepts of Chapter 3. Suppose the equation $y = \frac{3}{2}x + 4$ describes the automobile production y in a Michigan plant after time x, in hours.

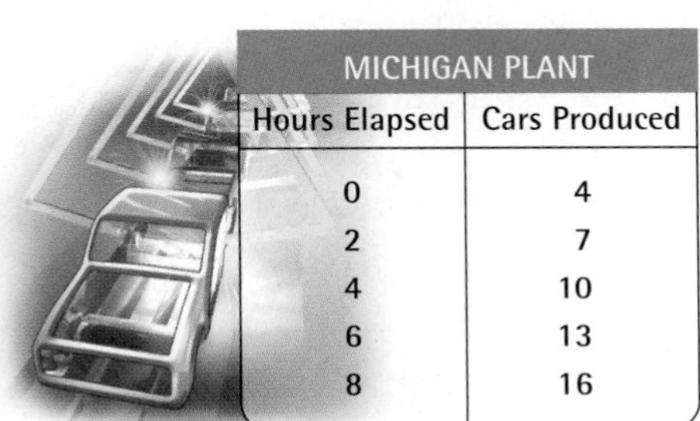

MICHIGAN PLANT	
Hours Elapsed	Cars Produced
0	4
2	7
4	10
6	13
8	16

Michigan Plant

y; Number of cars produced; 2, 4, 6, 8, 10, 12, 14, 16, 18

(0, 4) (2, 7) (4, 10) (6, 13) (8, 16) $y = \frac{3}{2}x + 4$

x; 2 4 6 8 10 12 14; Number of hours elapsed

We know that slope indicates how a line slants and can be thought of as a rate of change. Slope is the ratio of the change in y to the change in x, or rise to run.

SLOPE

$$m = \text{Slope} = \frac{\text{Rise}}{\text{Run}} = \frac{\text{Change in } y}{\text{Change in } x} = \text{Rate of change}$$

Using the points (6, 13) and (2, 7), we can compute the slope of the line $y = \frac{3}{2}x + 4$ as follows:

$$m = \frac{y_2 - y_1}{x_2 - x_1} = \frac{13 - 7}{6 - 2} = \frac{6}{4} = \frac{3}{2} \text{ cars per hour.}$$

The rate of change is $\frac{3}{2}$ cars per hour.

We also see that the graph crosses the y-axis at the point (0, 4). This point is the y-intercept. Both the slope and the y-intercept can be read directly from the equation as follows:

$$y = \frac{3}{2}x + 4.$$

Slope $= \frac{3}{2}$ y-intercept: (0, 4)

Do Exercise 1.

Objectives

a Given an equation in the form $y = mx + b$, find the slope and the y-intercept; and find an equation of a line when the slope and the y-intercept are given.

b Find an equation of a line when the slope and a point on the line are given.

c Find an equation of a line when two points on the line are given.

Margin Exercise 1 is review. If you have difficulty working the exercise, you might consider reviewing Chapter 3.

1. **Calories Burned.** The equation

$$y = 15x$$

describes the number of calories y burned by a runner after time x, in minutes.

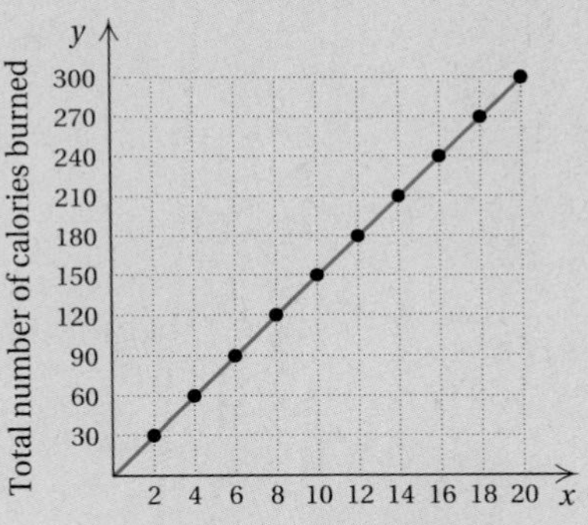

a) Use the points (14, 210) and (6, 90) to compute the slope using the formula:

$$m = \frac{y_2 - y_1}{x_2 - x_1}.$$

b) Find the rate of change of number of calories burned with respect to time.

c) Find the slope and the y-intercept directly from the equation $y = 15x$.

Answers on page A-33

Find the slope and the y-intercept.

2. $y = 5x$

3. $y = -\frac{3}{2}x - 6$

4. $3x + 4y = 15$

5. $2y = 4x - 17$

6. $-7x - 5y = 22$

Answers on page A-33

a Finding an Equation of a Line When the Slope and the y-Intercept Are Given

We know from Chapter 3 that in the equation $y = mx + b$ the slope is m and the y-intercept is $(0, b)$. Thus we call the equation $y = mx + b$ the **slope–intercept equation.**

THE SLOPE–INTERCEPT EQUATION: $y = mx + b$

The equation $y = mx + b$ is called the **slope–intercept equation.** The slope is m and the y-intercept is $(0, b)$.

EXAMPLE 1 Find the slope and the y-intercept of $2x - 3y = 8$.

We first solve for y:

$$2x - 3y = 8$$

$$-3y = -2x + 8 \quad \text{Subtracting } 2x$$

$$\frac{-3y}{-3} = \frac{-2x + 8}{-3} \quad \text{Dividing by } -3$$

$$y = \frac{-2x}{-3} + \frac{8}{-3}$$

$$y = \frac{2}{3}x - \frac{8}{3}$$

The slope is $\frac{2}{3}$. The y-intercept is $\left(0, -\frac{8}{3}\right)$.

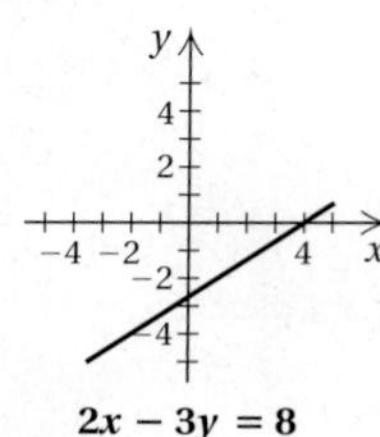

$2x - 3y = 8$

Do Exercises 2–6.

EXAMPLE 2 A line has slope -2.4 and y-intercept $(0, 11)$. Find an equation of the line.

We use the slope–intercept equation and substitute -2.4 for m and 11 for b:

$$y = mx + b$$

$$y = -2.4x + 11. \quad \text{Substituting}$$

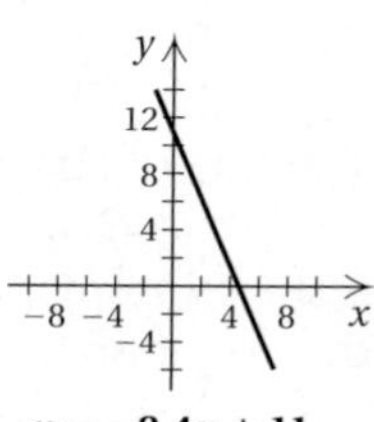

$y = -2.4x + 11$

EXAMPLE 3 A line has slope 0 and y-intercept $(0, -6)$. Find an equation of the line.

We use the slope–intercept equation and substitute 0 for m and -6 for b:

$$y = mx + b$$
$$y = 0x + (-6) \quad \text{Substituting}$$
$$y = -6.$$

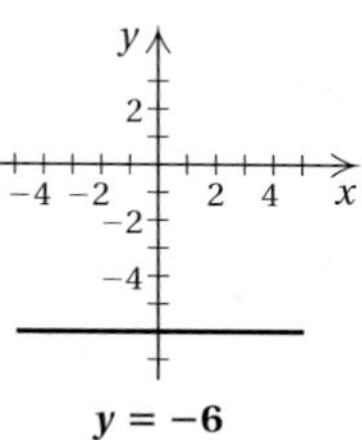

EXAMPLE 4 A line has slope $-\frac{5}{3}$ and y-intercept $(0, 0)$. Find an equation of the line.

We use the slope–intercept equation and substitute $-\frac{5}{3}$ for m and 0 for b:

$$y = mx + b$$
$$y = -\tfrac{5}{3}x + 0 \quad \text{Substituting}$$
$$y = -\tfrac{5}{3}x.$$

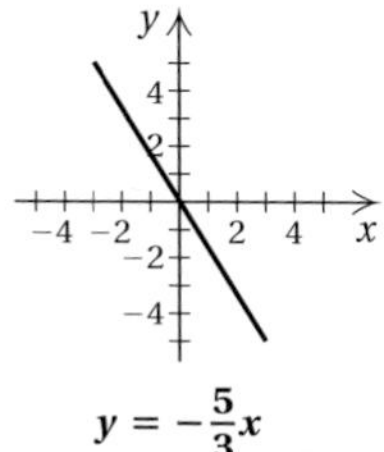

Do Exercises 7–9.

7. A line has slope 3.5 and y-intercept $(0, -23)$. Find an equation of the line.

8. A line has slope 0 and y-intercept $(0, 13)$. Find an equation of the line.

9. A line has slope -7.29 and y-intercept $(0, 0)$. Find an equation of the line.

Answers on page A-33

b Finding an Equation of a Line When the Slope and a Point Are Given

Suppose we know the slope of a line and a certain point on that line. We can use the slope–intercept equation $y = mx + b$ to find an equation of the line. To write an equation in this form, we need to know the slope (m) and the y-intercept (b).

EXAMPLE 5 Find an equation of the line with slope 3 that contains the point $(4, 1)$.

We know that the slope is 3, so the equation is $y = 3x + b$. Using the point $(4, 1)$, we substitute 4 for x and 1 for y in $y = 3x + b$. Then we solve for b:

$$y = 3x + b \quad \text{Substituting 3 for } m \text{ in } y = mx + b$$
$$1 = 3(4) + b \quad \text{Substituting 4 for } x \text{ and 1 for } y$$
$$1 = 12 + b$$
$$-11 = b. \quad \text{Solving for } b\text{, the } y\text{-intercept}$$

We use the equation $y = mx + b$ and substitute 3 for m and -11 for b:

$$y = 3x - 11.$$

Find an equation of the line that contains the given point and has the given slope.

10. $(4, 2)$, $m = 5$

11. $(-2, 1)$, $m = -3$

12. $(3, 5)$, $m = 6$

13. $(1, 4)$, $m = -\frac{2}{3}$

Answers on page A-33

This is the equation of the line with slope 3 and y-intercept $(0, -11)$.

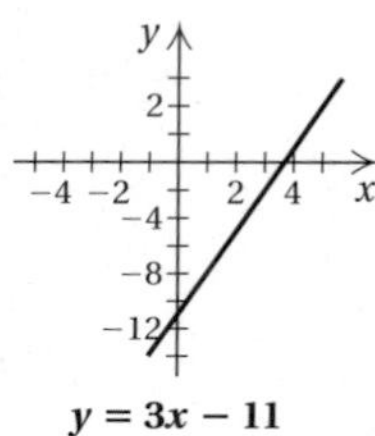

$y = 3x - 11$

EXAMPLE 6 Find an equation of the line with slope -5 that contains the point $(-2, 3)$.

We know that the slope is -5, so the equation is $y = -5x + b$. Using the point $(-2, 3)$, we substitute -2 for x and 3 for y in $y = -5x + b$. Then we solve for b:

$$y = -5x + b \quad \text{Substituting } -5 \text{ for } m \text{ in } y = mx + b$$
$$3 = -5(-2) + b \quad \text{Substituting } -2 \text{ for } x \text{ and } 3 \text{ for } y$$
$$3 = 10 + b$$
$$-7 = b. \quad \text{Solving for } b$$

We use the equation $y = mx + b$ and substitute -5 for m and -7 for b:

$$y = -5x - 7.$$

This is the equation of the line with slope -5 and y-intercept $(0, -7)$.

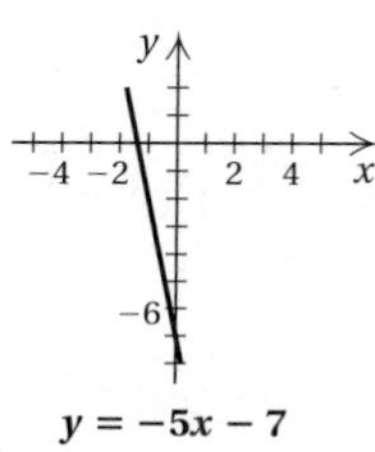

$y = -5x - 7$

Do Exercises 10–13.

C Finding an Equation of a Line When Two Points Are Given

We can also use the slope–intercept equation to find an equation of a line when two points are given.

EXAMPLE 7 Find an equation of the line containing the points $(2, 3)$ and $(-6, 1)$.

First, we find the slope:

$$m = \frac{3 - 1}{2 - (-6)} = \frac{2}{8}, \text{ or } \frac{1}{4}.$$

Thus, $y = \frac{1}{4}x + b$. We then proceed as we did in Example 6, using either point to find b.

We choose (2, 3) and substitute 2 for x and 3 for y:

$$y = \frac{1}{4}x + b \qquad \text{Substituting } \frac{1}{4} \text{ for } m \text{ in } y = mx + b$$

$$3 = \frac{1}{4} \cdot 2 + b \qquad \text{Substituting 2 for } x \text{ and 3 for } y$$

$$3 = \frac{1}{2} + b$$

$$\frac{5}{2} = b. \qquad \text{Solving for } b$$

We use the equation $y = mx + b$ and substitute $\frac{1}{4}$ for m and $\frac{5}{2}$ for b:

$$y = \frac{1}{4}x + \frac{5}{2}.$$

This is the equation of the line with slope $\frac{1}{4}$ and y-intercept $\left(0, \frac{5}{2}\right)$.

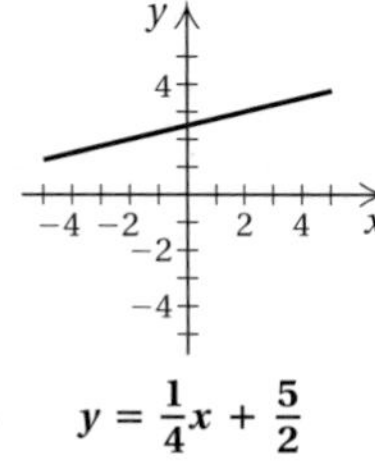

$y = \frac{1}{4}x + \frac{5}{2}$

Do Exercises 14 and 15.

Find an equation of the line containing the given points.

14. (2, 4) and (3, 5)

15. (−1, 2) and (−3, −2)

Answers on page A-33

Study Tips

SPECIAL VIDEOTAPES

In addition to the videotaped lectures for the books, there are two special tapes, *Math Study Skills for Students* and *Math Problem Solving in the Real World*. Check with your instructor to see whether these tapes are available on your campus.

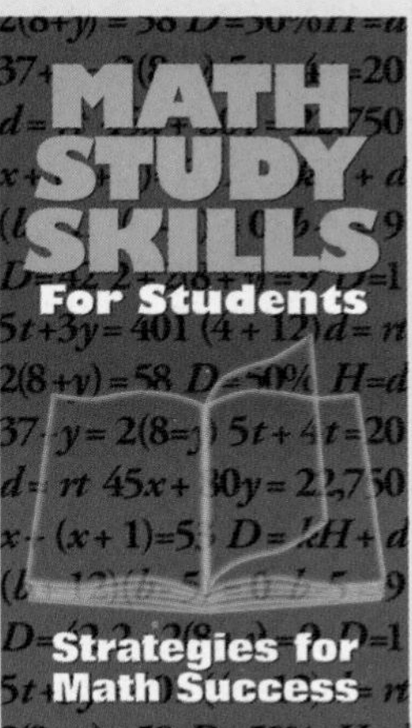

7.1 EXERCISE SET

For Extra Help

Digital Video Tutor CD 7 Videotape 14 | InterAct Math | Math Tutor Center | MathXL | MyMathLab

a Find the slope and the y-intercept.

1. $y = -4x - 9$

2. $y = -2x + 3$

3. $y = 1.8x$

4. $y = -27.4x$

5. $-8x - 7y = 21$

6. $-2x - 8y = 16$

7. $4x = 9y + 7$

8. $5x + 4y = 12$

9. $-6x = 4y + 2$

10. $4.8x - 1.2y = 36$

11. $y = -17$

12. $y = 28$

Find an equation of the line with the given slope and y-intercept.

13. Slope $= -7$, y-intercept $= (0, -13)$

14. Slope $= 73$, y-intercept $= (0, 54)$

15. Slope $= 1.01$, y-intercept $= (0, -2.6)$

16. Slope $= -\frac{3}{8}$, y-intercept $= \left(0, \frac{7}{11}\right)$

b Find an equation of the line containing the given point and having the given slope.

17. $(-3, 0)$, $m = -2$

18. $(2, 5)$, $m = 5$

19. $(2, 4)$, $m = \dfrac{3}{4}$

20. $\left(\dfrac{1}{2}, 2\right)$, $m = -1$

21. $(2, -6)$, $m = 1$

22. $(4, -2)$, $m = 6$

23. $(0, 3)$, $m = -3$

24. $(-2, -4)$, $m = 0$

c Find an equation of the line that contains the given pair of points.

25. $(12, 16)$ and $(1, 5)$

26. $(-6, 1)$ and $(2, 3)$

27. $(0, 4)$ and $(4, 2)$

28. $(0, 0)$ and $(4, 2)$

29. $(3, 2)$ and $(1, 5)$

30. $(-4, 1)$ and $(-1, 4)$

31. $(-4, 5)$ and $(-2, -3)$

32. $(-2, -4)$ and $(2, -1)$

33. *Aerobic Exercise.* The line graph below describes the *target heart rate T,* in beats per minute, of a person of age a, who is exercising. The goal is to get the number of beats per minute to this target level.

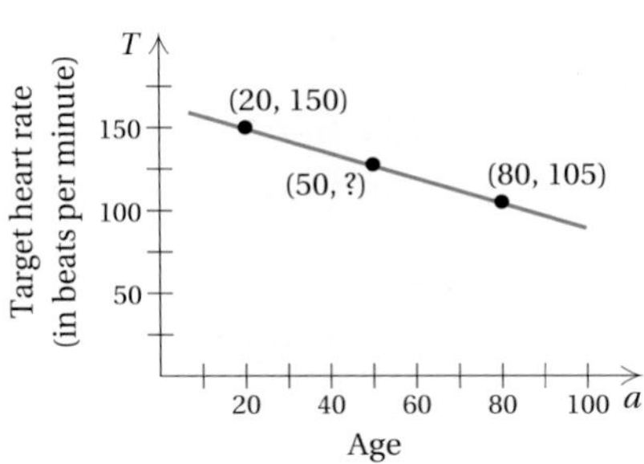

a) Find an equation of the line.
b) What is the rate of change of target heart rate with respect to time?
c) Use the equation to calculate the target heart rate of a person of age 50.

34. *Diabetes Cases.* The line graph below describes the number N, in millions, of cases of diabetes in this country in years x since 1983.

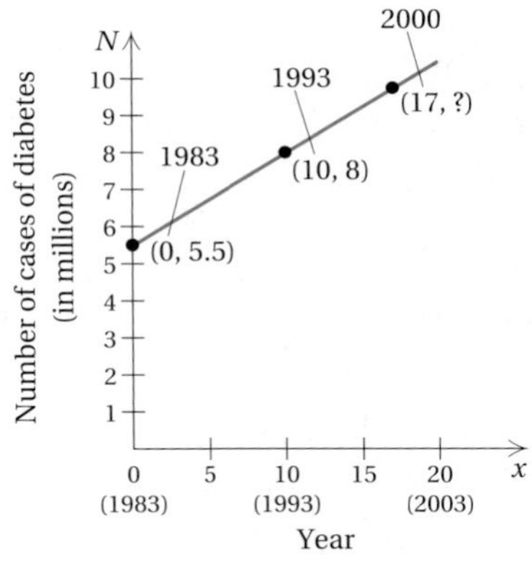

Source: U.S. National Center for Health Statistics

a) Find an equation of the line.
b) What is the rate of change of the number of cases of diabetes with respect to time?
c) Use the equation to find the number of cases of diabetes in 2000.

35. $^{\mathrm{D}}_{\mathrm{W}}$ Do all graphs of linear equations have y-intercepts? Why or why not?

36. $^{\mathrm{D}}_{\mathrm{W}}$ Do all graphs of linear equations have x-intercepts? Why or why not?

SKILL MAINTENANCE

Solve. [5.7b]

37. $2x^2 + 6x = 0$

38. $x^2 - 49 = 0$

39. $x^2 - x - 6 = 0$

40. $x^2 + 4x - 5 = 0$

41. $2x^2 + 11x = 21$

42. $5x^2 = 14x + 24$

43. $x^2 + 5x - 14 = 0$

44. $12x^2 + 16x - 16 = 0$

Solve. [2.3c]

45. $3x - 4(9 - x) = 17$

46. $2(5 + 2y) + 4y = 13$

47. $40(2x - 7) = 50(4 - 6x)$

48. $\frac{2}{3}(x - 5) = \frac{3}{8}(x + 5)$

SYNTHESIS

49. Find an equation of the line that contains the point $(2, -3)$ and has the same slope as the line $3x - y + 4 = 0$.

50. Find an equation of the line that has the same y-intercept as the line $x - 3y = 6$ and contains the point $(5, -1)$.

51. Find an equation of the line with the same slope as the line $3x - 2y = 8$ and the same y-intercept as the line $2y + 3x = -4$.

7.2 GRAPHING USING THE SLOPE AND THE y-INTERCEPT

Objective

a Use the slope and the y-intercept to graph a line.

1. Draw a line that has slope $\frac{2}{5}$ and y-intercept $(0, -3)$. What equation is graphed?

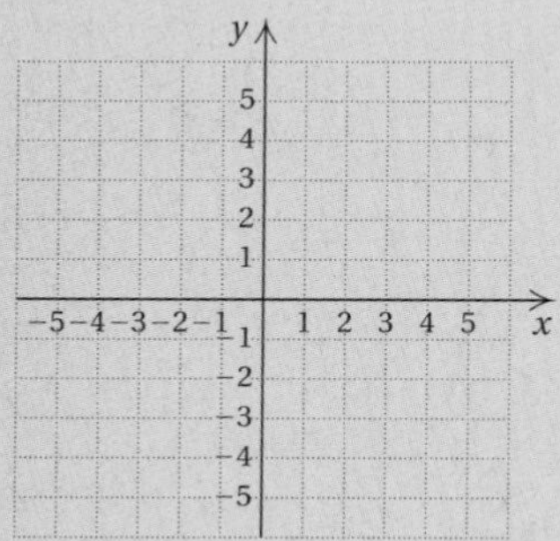

2. Draw a line that has slope $-\frac{2}{5}$ and y-intercept $(0, -3)$. What equation is graphed?

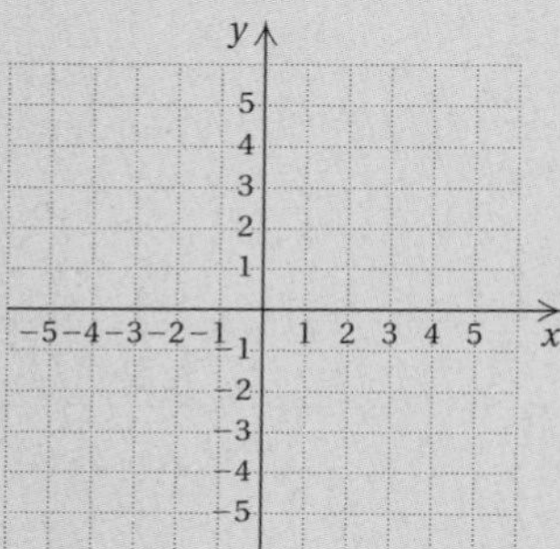

3. Draw a line that has slope 6 and y-intercept $(0, -3)$. Think of 6 as $\frac{6}{1}$. What equation is graphed?

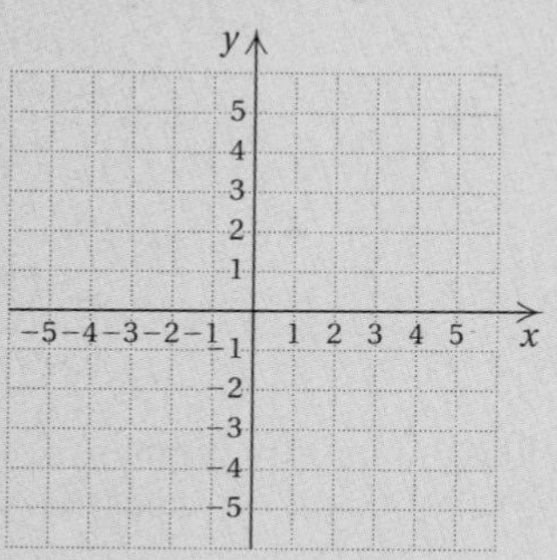

Answers on page A-33

a Graphs Using the Slope and the y-Intercept

We can graph a line if we know the coordinates of two points on that line. We can also graph a line if we know the slope and the y-intercept.

EXAMPLE 1 Draw a line that has slope $\frac{1}{4}$ and y-intercept $(0, 2)$.

We plot $(0, 2)$ and from there move *up* 1 unit (since the numerator is *positive* and corresponds to the change in y) and *to the right* 4 units (since the denominator is *positive* and corresponds to the change in x). This locates the point $(4, 3)$. We plot $(4, 3)$ and draw a line passing through $(0, 2)$ and $(4, 3)$, as shown on the right below.

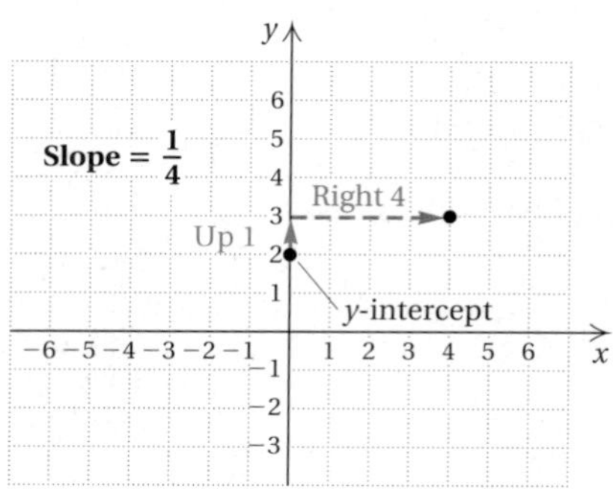

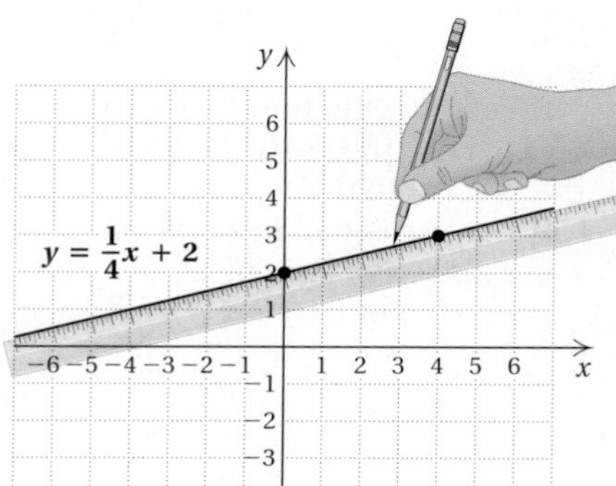

We are actually graphing the equation $y = \frac{1}{4}x + 2$.

EXAMPLE 2 Draw a line that has slope $-\frac{2}{3}$ and y-intercept $(0, 4)$.

We can think of $-\frac{2}{3}$ as $\frac{-2}{3}$. We plot $(0, 4)$ and from there move *down* 2 units (since the numerator is *negative*) and *to the right* 3 units (since the denominator is *positive*). We plot the point $(3, 2)$ and draw a line passing through $(0, 4)$ and $(3, 2)$.

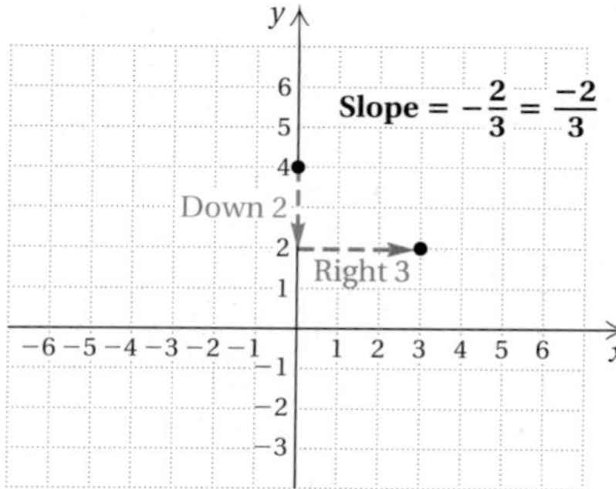

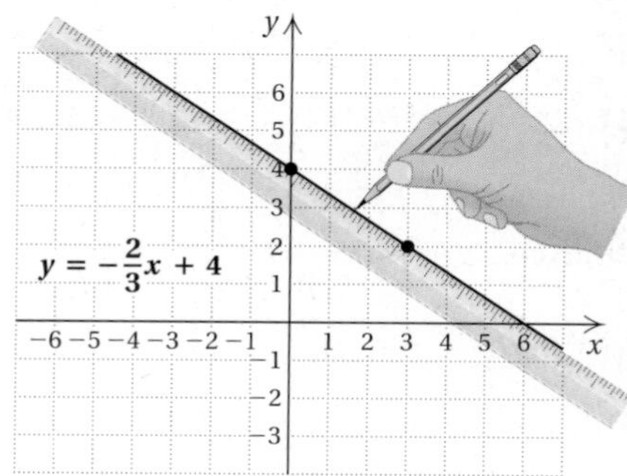

We are actually graphing the equation $y = -\frac{2}{3}x + 4$.

Do Exercises 1–3.

We now use our knowledge of the slope–intercept equation to graph linear equations.

EXAMPLE 3 Graph $y = \frac{3}{4}x + 5$ using the slope and the y-intercept.

From the equation $y = \frac{3}{4}x + 5$, we see that the slope of the graph is $\frac{3}{4}$ and the y-intercept is $(0, 5)$. We plot $(0, 5)$ and then consider the slope, $\frac{3}{4}$. Starting at $(0, 5)$, we plot a second point by moving *up* 3 units (since the numerator is *positive*) and *to the right* 4 units (since the denominator is *positive*). We reach a new point, $(4, 8)$.

We can also rewrite the slope as $\frac{-3}{-4}$. We again start at the y-intercept, $(0, 5)$, but move *down* 3 units (since the numerator is *negative* and corresponds to the change in y) and *to the left* 4 units (since the denominator is *negative* and corresponds to the change in x). We reach another point, $(-4, 2)$. Once two or three points have been plotted, the line representing all solutions of $y = \frac{3}{4}x + 5$ can be drawn.

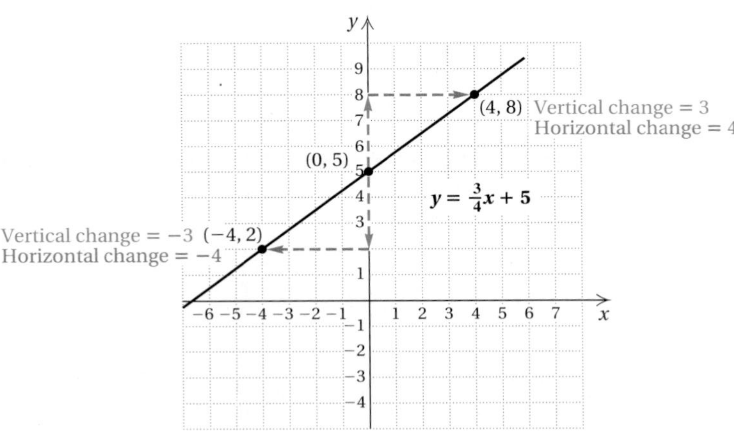

Do Exercise 4.

4. Graph $y = \frac{3}{5}x - 4$ using the slope and the y-intercept.

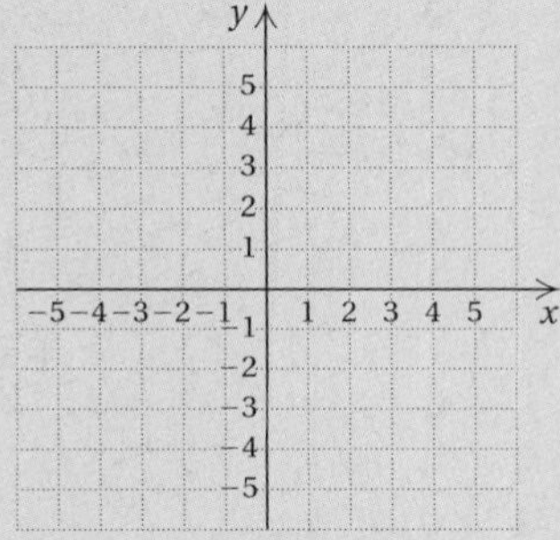

EXAMPLE 4 Graph $2x + 3y = 3$ using the slope and the y-intercept.

To graph $2x + 3y = 3$, we first rewrite the equation in slope–intercept form:

$$2x + 3y = 3$$

$$3y = -2x + 3 \qquad \text{Adding } -2x$$

$$\frac{1}{3} \cdot 3y = \frac{1}{3}(-2x + 3) \qquad \text{Multiplying by } \frac{1}{3}$$

$$y = -\frac{2}{3}x + 1. \qquad \text{Simplifying}$$

To graph $y = -\frac{2}{3}x + 1$, we first plot the y-intercept, $(0, 1)$. We can think of the slope as $\frac{-2}{3}$. Starting at $(0, 1)$ and using the slope, we find a second point by moving *down* 2 units (since the numerator is *negative*) and *to the right* 3 units (since the denominator is *positive*). We plot the new point, $(3, -1)$. In a similar manner, we can move from the point $(3, -1)$ to locate a third point, $(6, -3)$. The line can then be drawn.

Answer on page A-33

5. Graph: $3x + 4y = 12$.

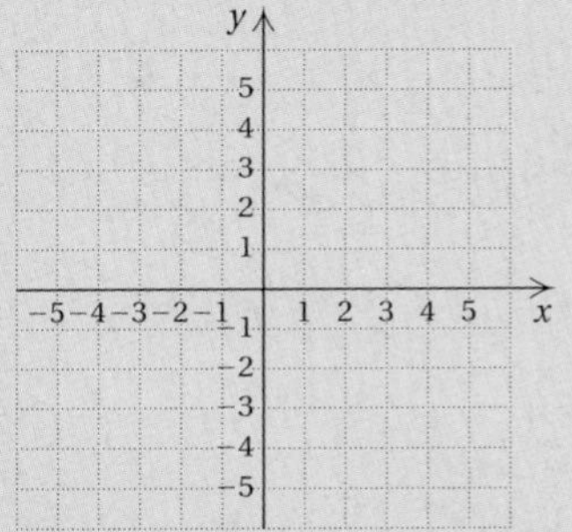

Since $-\frac{2}{3} = \frac{2}{-3}$, an alternative approach is to again plot $(0, 1)$, but this time move *up* 2 units (since the numerator is *positive*) and *to the left* 3 units (since the denominator is *negative*). This leads to another point on the graph, $(-3, 3)$.

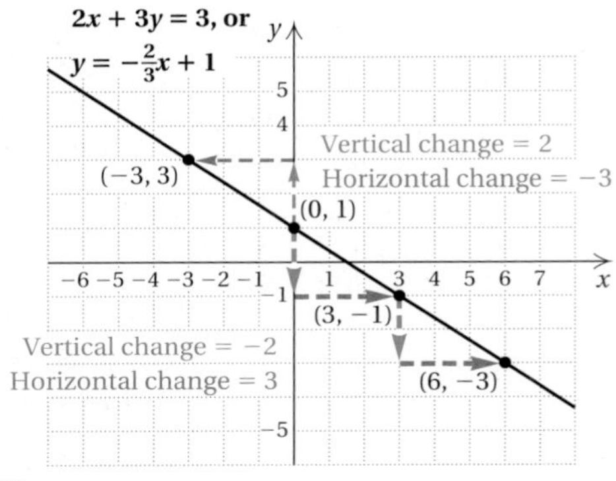

It helps to use both $\frac{2}{-3}$ and $\frac{-2}{3}$ to draw the graph.

Do Exercise 5.

Answer on page A-33

Study Tips

TURNING NEGATIVES INTO POSITIVES (PART 2)

B. C. Forbes said, "History has demonstrated that notable *winners* usually encountered heartbreaking obstacles before they triumphed. They won because they refused to become discouraged by their defeats."

Here are some anecdotes about well-known people who turned what could have been a negative experience into a positive outcome.

- *Richard Bach* sold more than 7 million copies of his story about a "soaring" seagull, Jonathan Livingston Seagull. His work was turned down by 18 publishers before Macmillan finally published it in 1970.
- *Walt Disney* was once fired by a newspaper for what they said was his "lack of ideas." He went bankrupt several times before he built an entertainment empire that now includes Disneyland and Disney World.
- *Erik Weihenmayer*, a blind man, has climbed the tallest mountains in Africa and North and South America and recently climbed the tallest mountain in the world, Mt. Everest.
- *Hank Aaron* holds the all-time Major League home run record with a total of 755, topping the former record holder, Babe Ruth, who had 714. But Aaron also held the all-time record for many years for striking out 1383 times, also topping Babe Ruth, who struck out 1330 times!
- At the age of 15, *Michael Jordan* was cut from his school basketball team. He was told he was too small to play. Yet in 2000, he was selected by ESPN as the top athlete of the 20th century.
- *Albert Einstein* didn't speak until he was 4 years old and was not able to read until he was 7. He is now recognized as one of the greatest physicists of all time, having developed the famous theory of relativity.

In an article entitled "*Mistakes—Important Teacher*," Josh Hinds writes, "Another approach (to negative experiences) is to remind ourselves that failures are not always failures, rather they are lessons. I would challenge you to find one occurrence in your own life where you have learned from a past mistake. While it can be true that we don't gain direct rewards from them, we still gain something of great importance. Therefore we need to explore our failures and take the time to use them as our teachers ..."

"*Whoever makes no mistakes is doing nothing.*"

Dutch proverb

7.2 EXERCISE SET

For Extra Help

Digital Video Tutor CD 7 Videotape 14 | InterAct Math | Math Tutor Center | MathXL | MyMathLab

a Draw a line that has the given slope and y-intercept.

1. Slope $\frac{2}{5}$; y-intercept $(0, 1)$

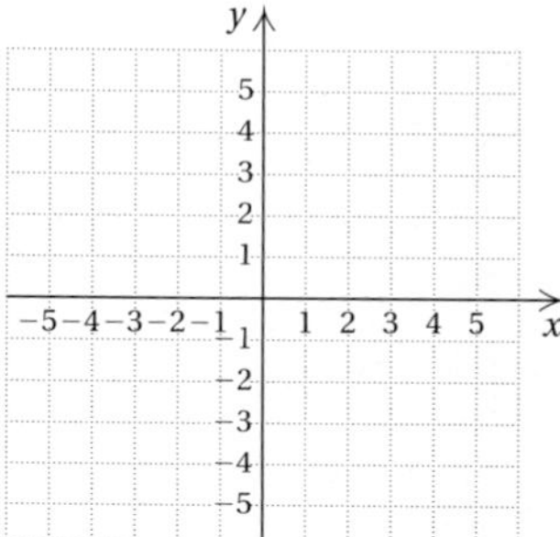

2. Slope $\frac{3}{5}$; y-intercept $(0, -1)$

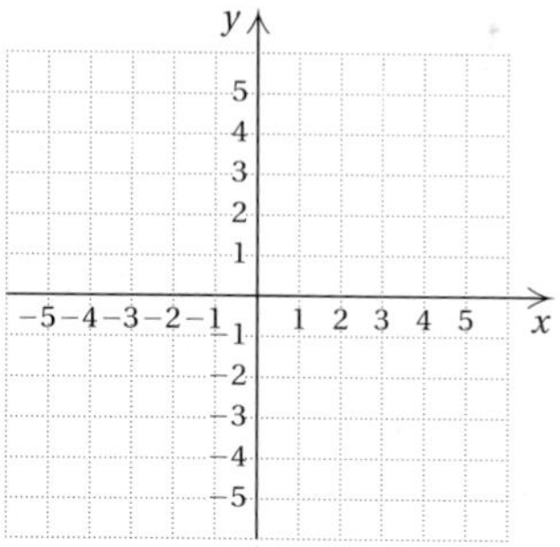

3. Slope $\frac{5}{3}$; y-intercept $(0, -2)$

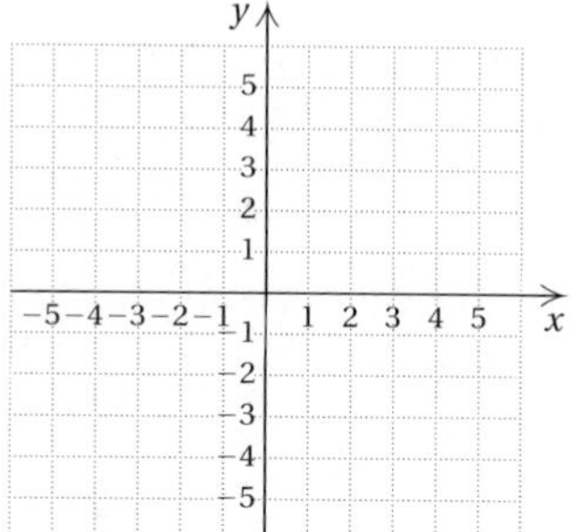

4. Slope $\frac{5}{2}$; y-intercept $(0, 1)$

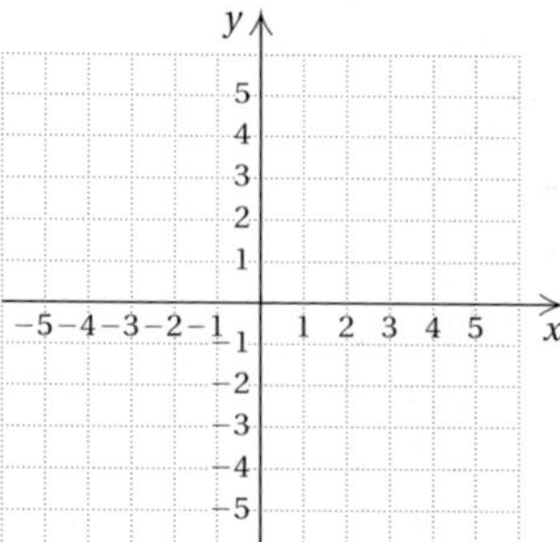

5. Slope $-\frac{3}{4}$; y-intercept $(0, 5)$

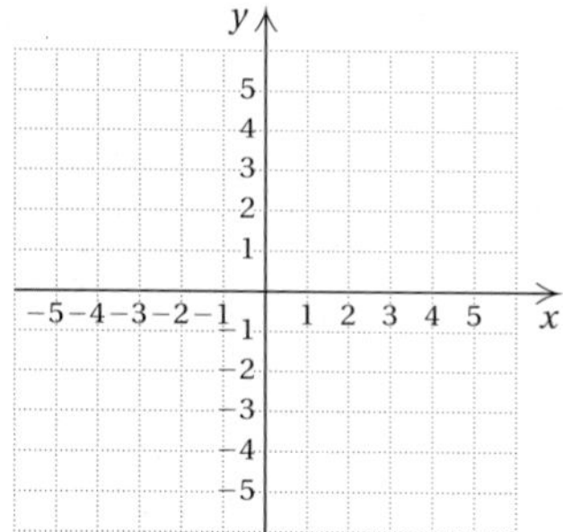

6. Slope $-\frac{4}{5}$; y-intercept $(0, 6)$

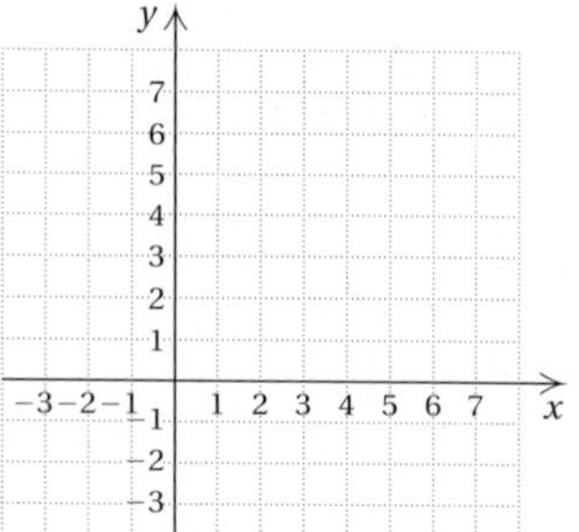

7. Slope $-\frac{1}{2}$; y-intercept $(0, 3)$

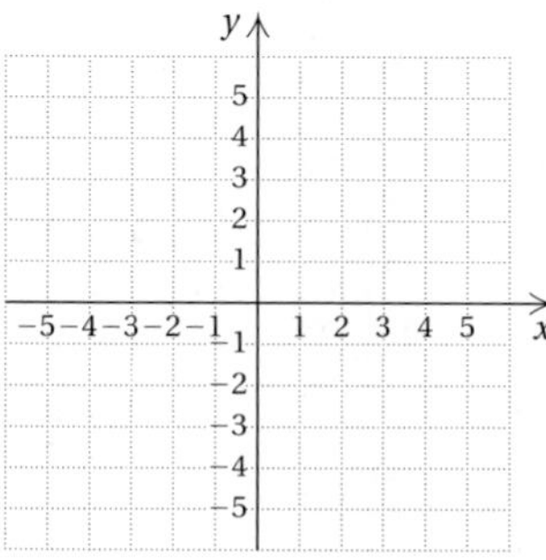

8. Slope $\frac{1}{3}$; y-intercept $(0, -4)$

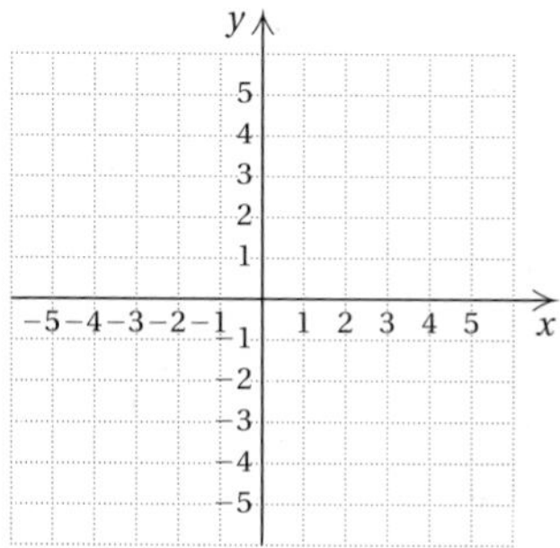

9. Slope 2; y-intercept $(0, -4)$

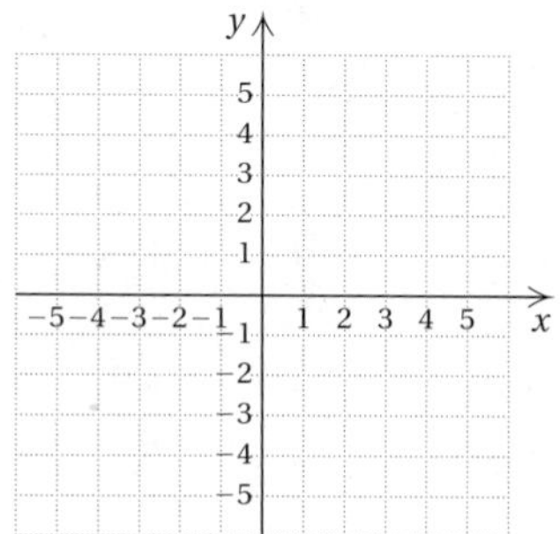

10. Slope -2; y-intercept $(0, -3)$

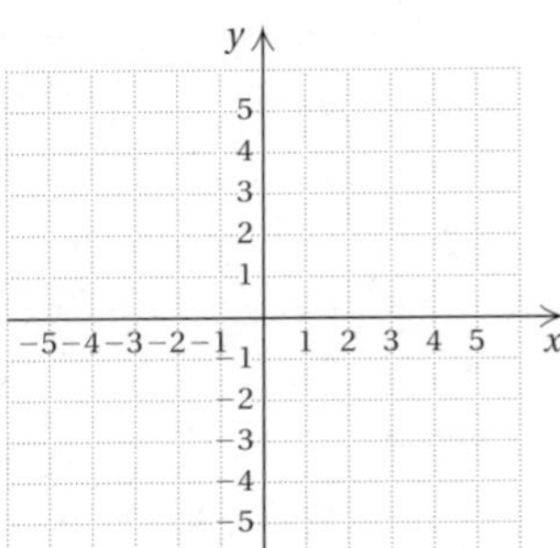

11. Slope -3; y-intercept $(0, 2)$

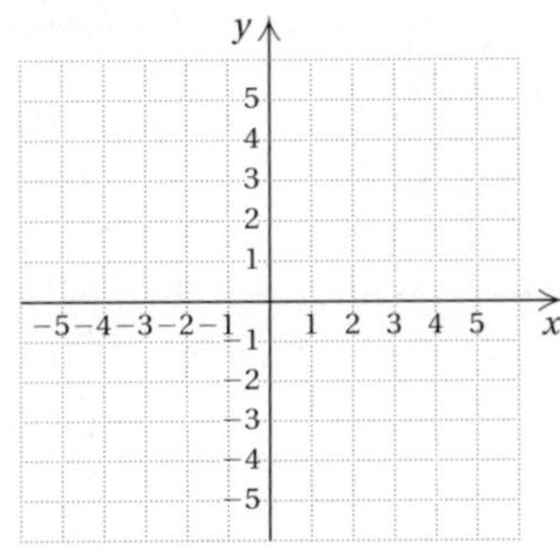

12. Slope 3; y-intercept $(0, 4)$

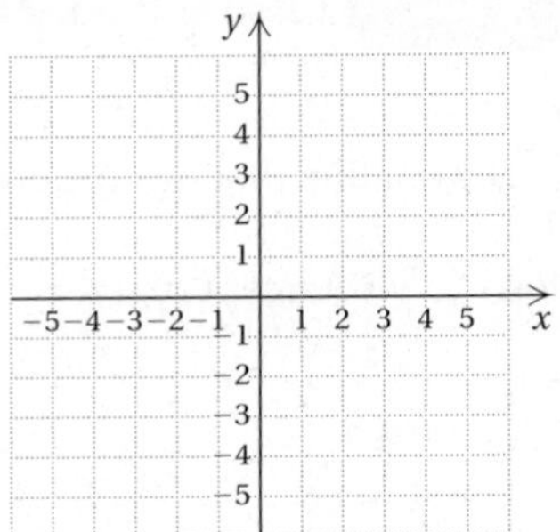

Graph using the slope and the y-intercept.

13. $y = \frac{3}{5}x + 2$

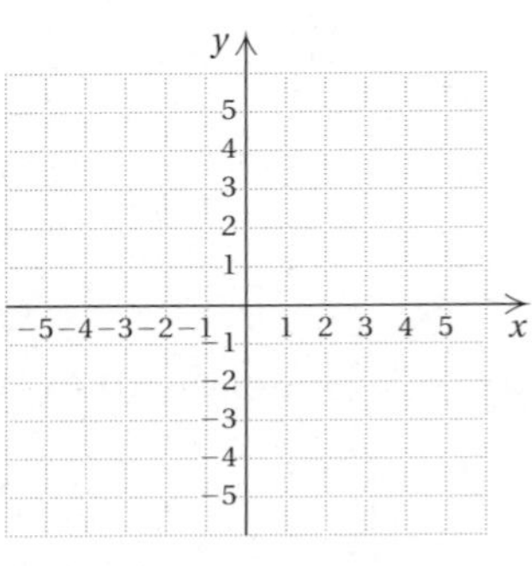

14. $y = -\frac{3}{5}x - 1$

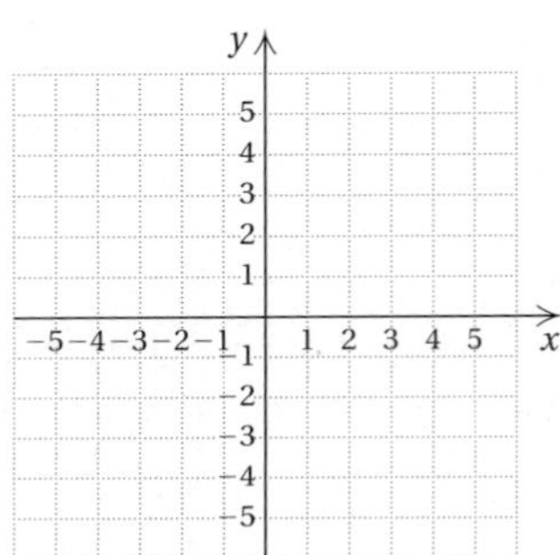

15. $y = -\frac{3}{5}x + 1$

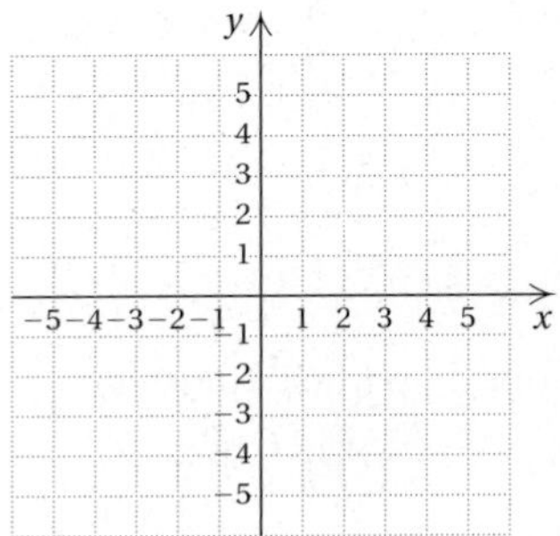

16. $y = \frac{3}{5}x - 2$

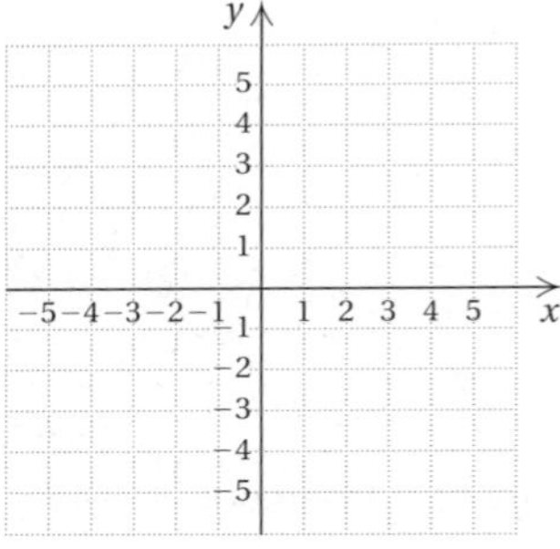

17. $y = \frac{5}{3}x + 3$

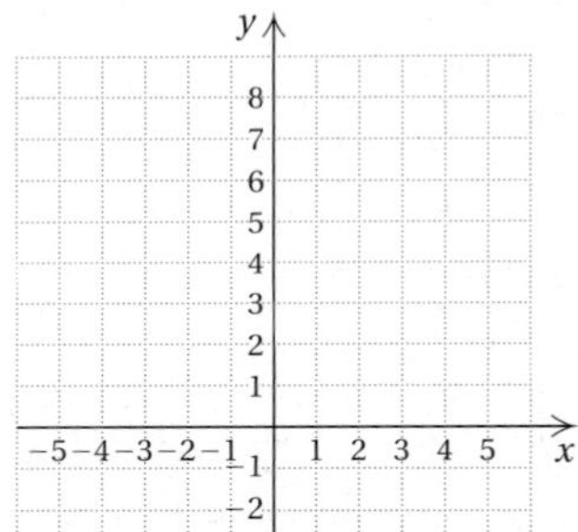

18. $y = \frac{5}{3}x - 2$

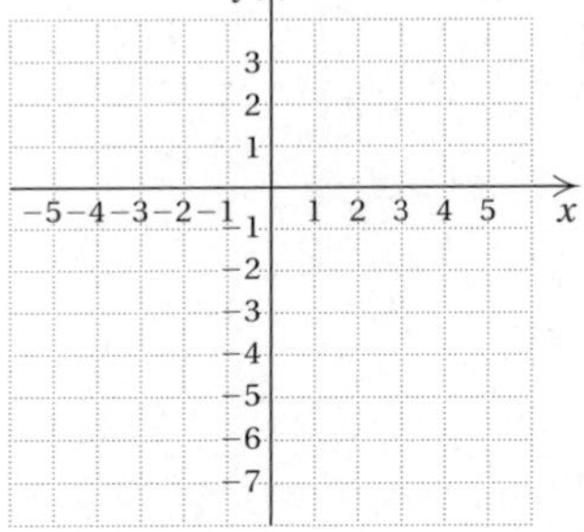

19. $y = -\frac{3}{2}x - 2$

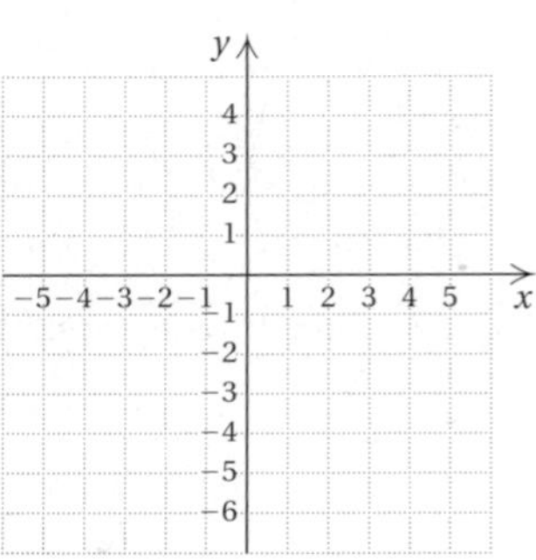

20. $y = -\frac{4}{3}x + 3$

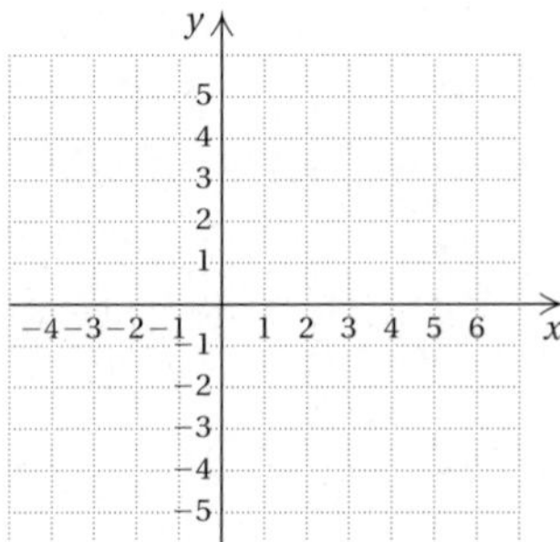

21. $2x + y = 1$

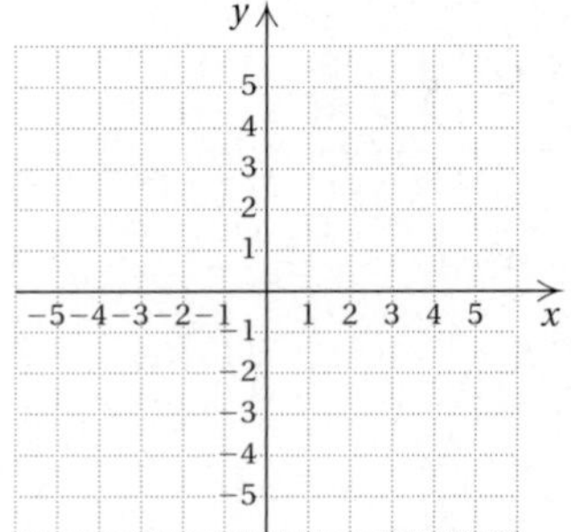

22. $3x + y = 2$

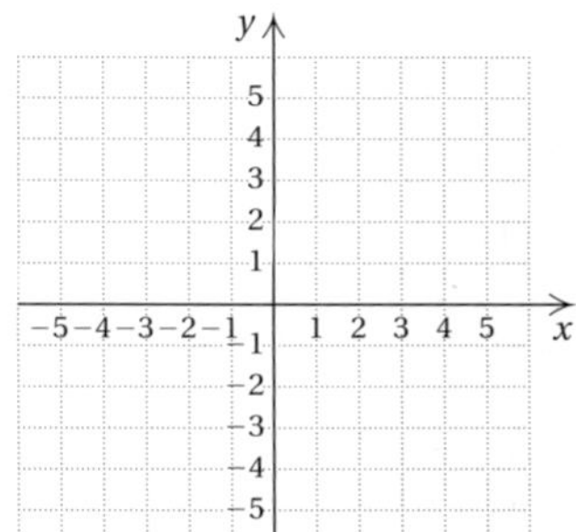

23. $3x - y = 4$

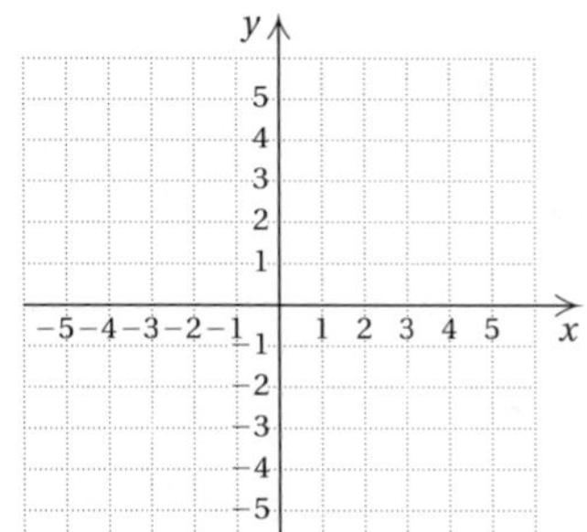

24. $2x - y = 5$

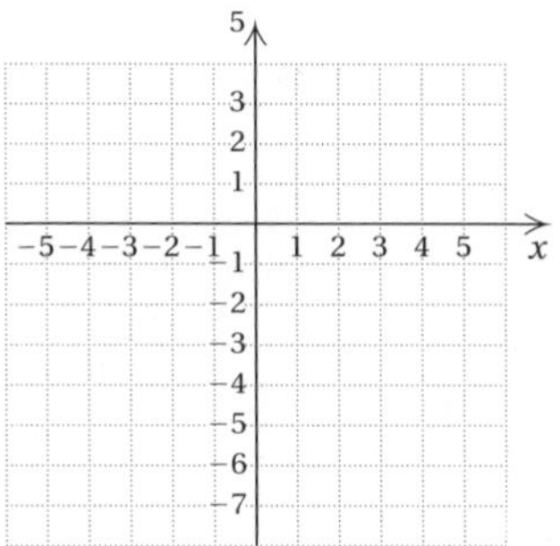

25. $2x + 3y = 9$

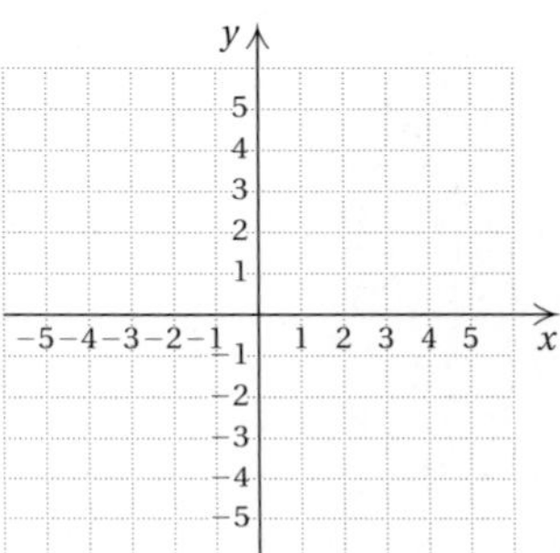

26. $4x + 5y = 15$

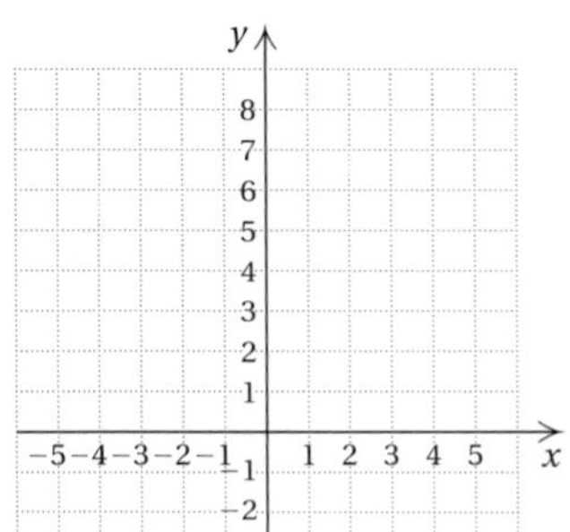

27. $x - 4y = 12$

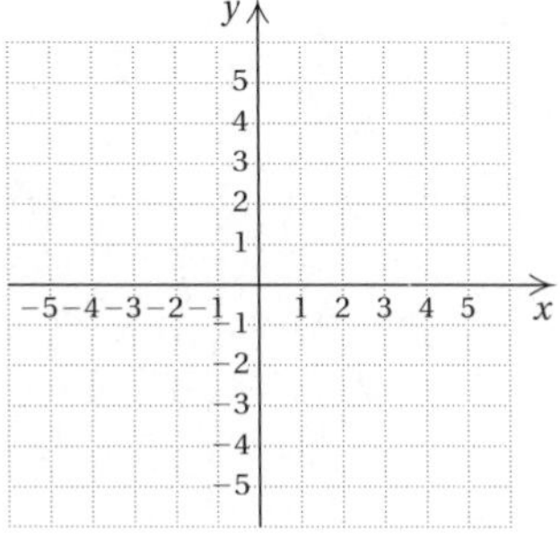

28. $x + 5y = 20$

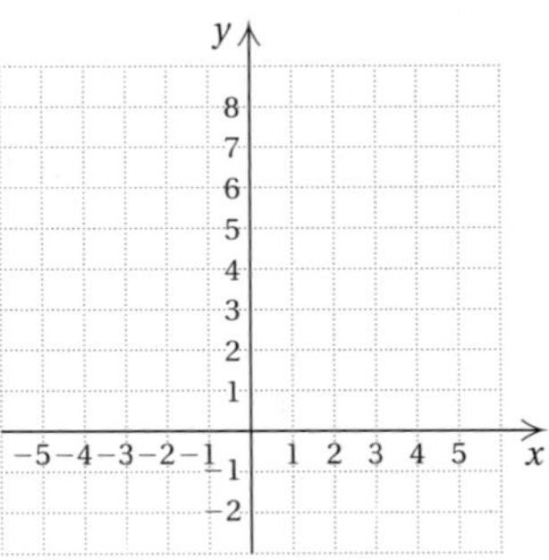

29. $x + 2y = 6$

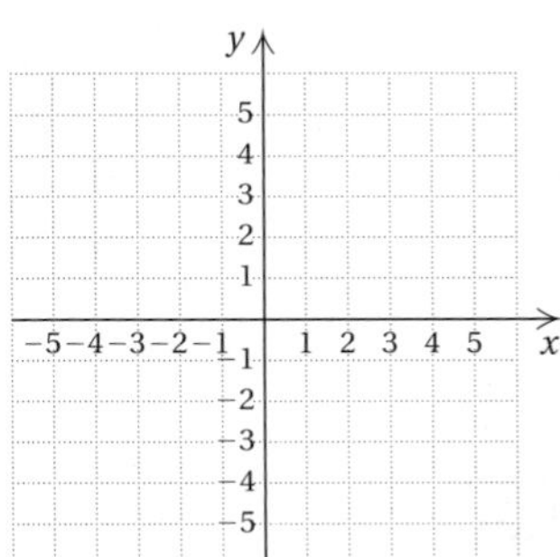

30. $x - 3y = 9$

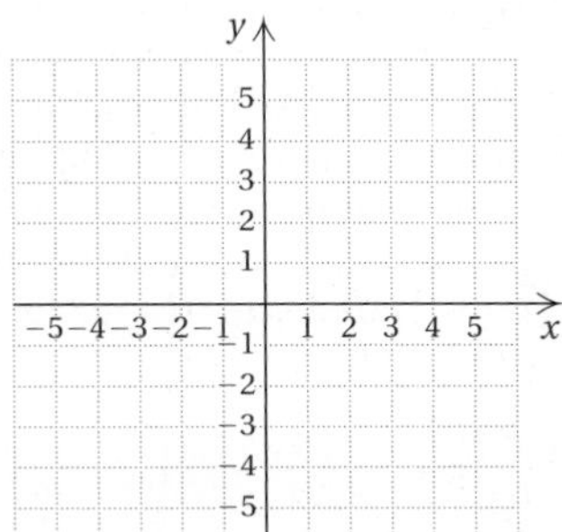

31. ^{D}W Can a horizontal line be graphed using the method of Examples 3 and 4? Why or why not?

32. ^{D}W Can a vertical line be graphed using the method of Examples 3 and 4? Why or why not?

SKILL MAINTENANCE

Find the slope of the line containing the given pair of points. [3.4a]

33. $(-2, -6), (8, 7)$

34. $(2, -6), (8, -7)$

35. $(4.5, -2.3), (14.5, 4.6)$

36. $(-0.8, -2.3), (-4.8, 0.1)$

37. $(-2, -6), (8, -6)$

38. $(-2, -6), (-2, 7)$

39. $(11, -1), (11, -4)$

40. $(-3, 5), (8, 5)$

41. *SUV, Pickup Truck, and Minivan Sales.* Sales of sport utility vehicles, pickup trucks, and minivans have increased in recent years, as shown in the following graph. Find the rate of change of sales with respect to time. Find the slope of the graph. [3.4b]

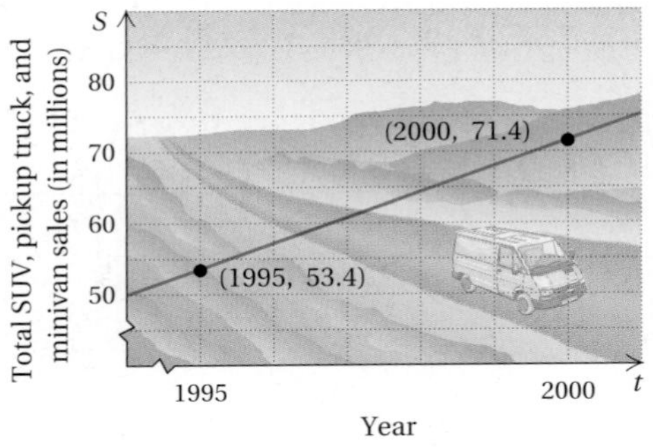

42. *Car Sales.* Sales of cars have increased in recent years, as shown in the following graph. Find the rate of change of sales with respect to time. Find the slope of the graph. [3.4b]

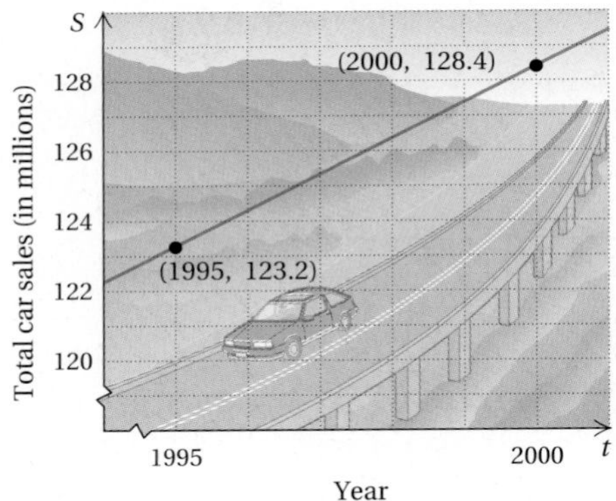

SYNTHESIS

43. *Refrigerator Size.* Kitchen designers recommend that a refrigerator be selected on the basis of the number of people in the household. For 1–2 people, a 16 ft^3 model is suggested. For each additional person, an additional 1.5 ft^3 is recommended. If x is the number of residents in excess of 2, find the slope–intercept equation for the recommended size of a refrigerator.

44. *Telephone Service.* In a recent promotion, AT&T charged a monthly fee of \$4.95 plus 7¢ for each minute of long-distance phone calls. If x is the number of minutes of long-distance calls, find the slope–intercept equation for the monthly bill.

45. Graph the line with slope 2 that passes through the point $(-3, 1)$.

7.3 PARALLEL AND PERPENDICULAR LINES

Objectives

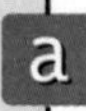
Determine whether the graphs of two linear equations are parallel.

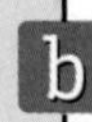
Determine whether the graphs of two linear equations are perpendicular.

When we graph a pair of linear equations, there are three possibilities:

1. The graphs are the same.
2. The graphs intersect at exactly one point.
3. The graphs are parallel (they do not intersect).

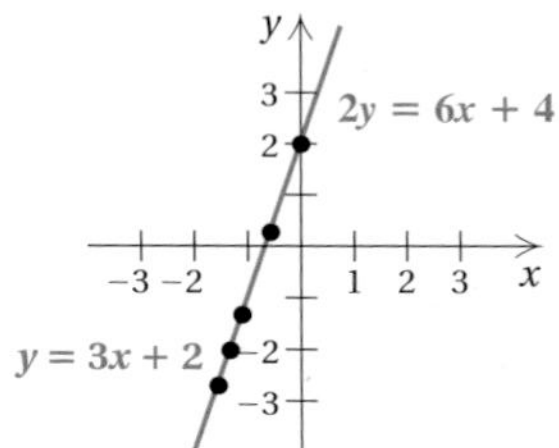

Equations have the same graph.

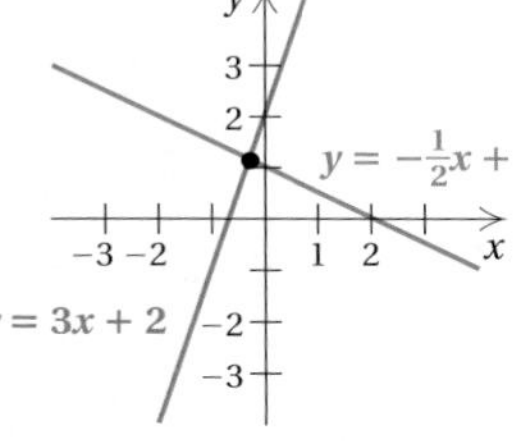

Graphs intersect at exactly one point.

$y = 3x - 1$

$y = 3x + 2$

Graphs are parallel.

a Parallel Lines

The graphs shown below are of the linear equations

$$y = 2x + 5 \quad \text{and} \quad y = 2x - 3.$$

The slope of each line is 2. The y-intercepts are $(0, 5)$ and $(0, -3)$ and are different. The lines do not intersect and are parallel.

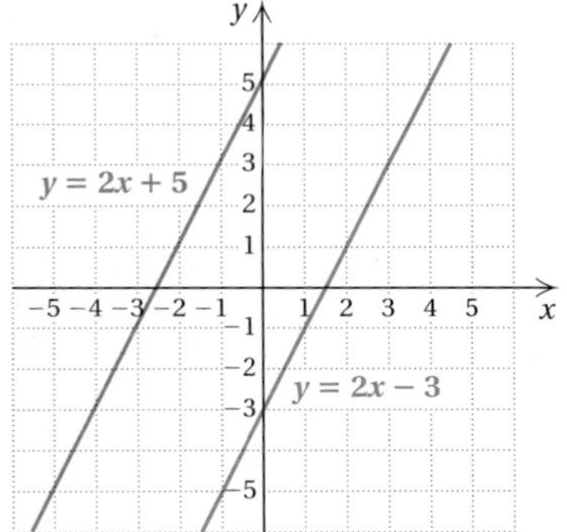

PARALLEL LINES

- Parallel nonvertical lines have the *same* slope, $m_1 = m_2$, and *different* y-intercepts, $b_1 \neq b_2$.
- Parallel horizontal lines have equations $y = p$ and $y = q$, where $p \neq q$.
- Parallel vertical lines have equations $x = p$ and $x = q$, where $p \neq q$.

By simply graphing, we may find it difficult to determine whether lines are parallel. Sometimes they may intersect only very far from the origin. We can use the preceding statements about slopes, y-intercepts, and parallel lines to determine for certain whether lines are parallel.

Determine whether the graphs of the pair of equations are parallel.

1. $y - 3x = 1,$
   $-2y = 3x + 2$

2. $3x - y = -5,$
   $y - 3x = -2$

Answers on page A-34

CALCULATOR CORNER

Parallel Lines Graph each pair of equations in Margin Exercises 1 and 2 in the standard viewing window, $[-10, 10, -10, 10]$. (Note that each equation must be solved for y so that it can be entered in "$y =$" form on the graphing calculator.) Determine whether the lines appear to be parallel.

EXAMPLE 1 Determine whether the graphs of the lines $y = -3x + 4$ and $6x + 2y = -10$ are parallel.

The graphs of these equations are shown below, but they are not necessary in order to determine whether the lines are parallel.

We first solve each equation for y. In this case, the first equation is already solved for y.

a) $y = -3x + 4$

b) $6x + 2y = -10$

$$2y = -6x - 10$$

$$y = \frac{1}{2}(-6x - 10)$$

$$y = -3x - 5$$

The slope of each line is -3. The y-intercepts are $(0, 4)$ and $(0, -5)$ and are different. The lines are parallel.

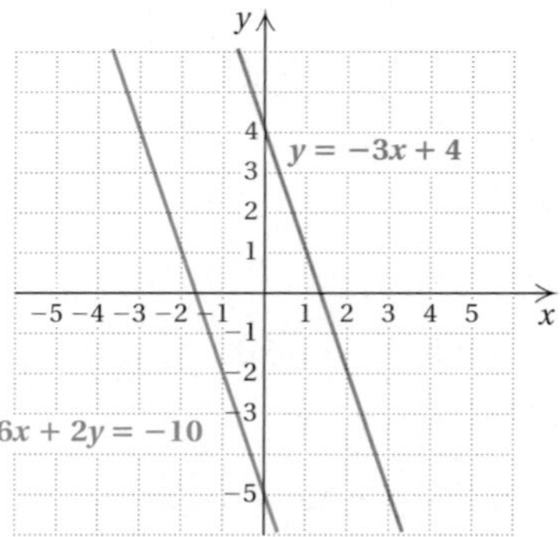

Do Exercises 1 and 2.

b Perpendicular Lines

Perpendicular lines in a plane are lines that intersect at a right angle. The measure of a right angle is 90°. The lines whose graphs are shown below are perpendicular. You can check this approximately by using a protractor or placing a rectangular piece of paper at the intersection.

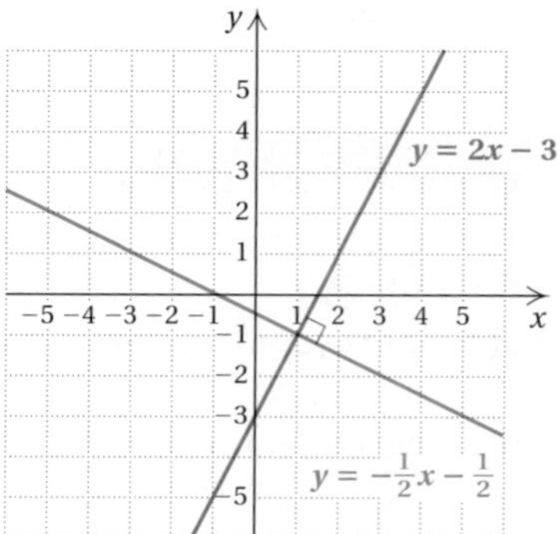

The slopes of the lines are 2 and $-\frac{1}{2}$. Note that $2\left(-\frac{1}{2}\right) = -1$. That is, the product of the slopes is -1.

PERPENDICULAR LINES

- Two nonvertical lines are perpendicular if the product of their slopes is -1, $m_1 \cdot m_2 = -1$. (If one line has slope m, the slope of the line perpendicular to it is $-1/m$.)
- If one equation in a pair of perpendicular lines is vertical, then the other is horizontal. These equations are of the form $x = a$ and $y = b$.

EXAMPLE 2 Determine whether the graphs of the lines $3y = 9x + 3$ and $6y + 2x = 6$ are perpendicular.

The graphs are shown below, but they are not necessary in order to determine whether the lines are perpendicular.

We first solve each equation for y in order to determine the slopes:

a) $3y = 9x + 3$

$y = \frac{1}{3}(9x + 3)$

$y = 3x + 1$;

b) $6y + 2x = 6$

$6y = -2x + 6$

$y = \frac{1}{6}(-2x + 6)$

$y = -\frac{1}{3}x + 1$.

The slopes are 3 and $-\frac{1}{3}$. The product of the slopes is $3\left(-\frac{1}{3}\right) = -1$. The lines are perpendicular.

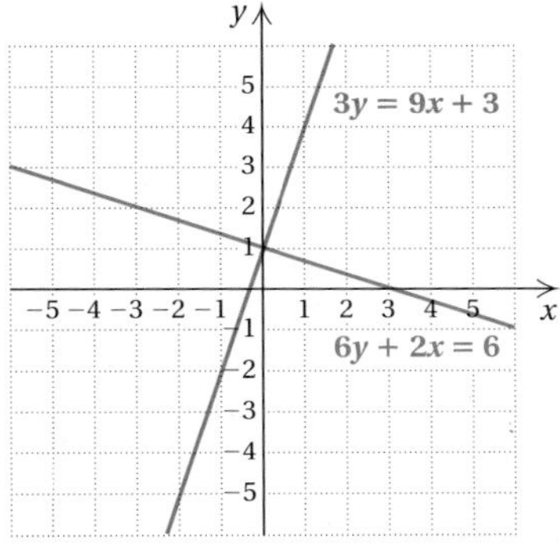

Do Exercises 3 and 4.

Determine whether the graphs of the pair of equations are perpendicular.

3. $y = -\frac{3}{4}x + 7,$
$y = \frac{4}{3}x - 9$

4. $4x - 5y = 8,$
$6x + 9y = -12$

Answers on page A-34

CALCULATOR CORNER

Perpendicular Lines
Graph each pair of equations in Margin Exercises 3 and 4 in the window $[-9, 9, -6, 6]$. (Note that the equations in Margin Exercise 4 must be solved for y so that they can be entered in "$y =$" form on the graphing calculator.) Determine whether the lines appear to be perpendicular. Note (in the viewing window) that more of the x-axis is shown than the y-axis. The dimensions were chosen to more accurately reflect the slopes of the lines.

7.3 EXERCISE SET

For Extra Help

Digital Video Tutor CD 7 Videotape 14 | InterAct Math | Math Tutor Center | MathXL | MyMathLab

a Determine whether the graphs of the equations are parallel lines.

1. $x + 4 = y,$
$y - x = -3$

2. $3x - 4 = y,$
$y - 3x = 8$

3. $y + 3 = 6x,$
$-6x - y = 2$

4. $y = -4x + 2,$
$-5 = -2y + 8x$

5. $10y + 32x = 16.4,$
$y + 3.5 = 0.3125x$

6. $y = 6.4x + 8.9,$
$5y - 32x = 5$

7. $y = 2x + 7,$
$5y + 10x = 20$

8. $y + 5x = -6,$
$3y + 5x = -15$

9. $3x - y = -9,$
$2y - 6x = -2$

10. $y - 6 = -6x,$
$-2x + y = 5$

11. $x = 3,$
$x = 4$

12. $y = 1,$
$y = -2$

b Determine whether the graphs of the equations are perpendicular lines.

13. $y = -4x + 3,$
$4y + x = -1$

14. $y = -\frac{2}{3}x + 4,$
$3x + 2y = 1$

15. $x + y = 6,$
$4y - 4x = 12$

16. $2x - 5y = -3,$
$5x + 2y = 6$

17. $y = -0.3125x + 11,$
$y - 3.2x = -14$

18. $y = -6.4x - 7,$
$64y - 5x = 32$

19. $y = -x + 8,$
$x - y = -1$

20. $2x + 6y = -3,$
$12y = 4x + 20$

21. $\frac{3}{8}x - \frac{y}{2} = 1,$
$\frac{4}{3}x - y + 1 = 0$

22. $\frac{1}{2}x + \frac{3}{4}y = 6,$
$-\frac{3}{2}x + y = 4$

23. $x = 0,$
$y = -2$

24. $x = -3,$
$y = 5$

25. ^{D}W Consider two equations of the type $Ax + By = C$. Explain how you would go about showing that their graphs are perpendicular.

26. ^{D}W Consider two equations of the type $Ax + By = C$. Explain how you would go about showing that their graphs are parallel.

SKILL MAINTENANCE

Solve. [6.7a]

27. *Train Travel.* A train leaves a station and travels west at 70 km/h. Two hours later, a second train leaves on a parallel track and travels west at 90 km/h. When will it overtake the first train?

28. *Car Travel.* One car travels 10 km/h faster than another. While one car travels 130 km, the other travels 140 km. What is the speed of each car?

Solve. [6.6a]

29. $\dfrac{x^2}{x+4} = \dfrac{16}{x+4}$

30. $\dfrac{2}{3} - \dfrac{5}{6} = \dfrac{1}{x}$

31. $\dfrac{t}{3} + \dfrac{t}{10} = 1$

32. $\dfrac{5}{x-4} = \dfrac{3}{x+2}$

33. $\dfrac{4}{x-2} + \dfrac{7}{x-3} = \dfrac{10}{x^2 - 5x + 6}$

34. $\dfrac{3}{x-5} + \dfrac{4}{x+5} = \dfrac{2}{x^2 - 25}$

SYNTHESIS

35.–40. Check the results of Exercises 1–6 by graphing each pair of equations using the window settings $[-6, 6, -4, 4]$, Xscl = 1, Yscl = 1.

41.–46. Check the results of Exercises 13–18 by graphing each pair of equations using the window settings $[-24, 24, -16, 16]$, Xscl = 1, Yscl = 1.

47. Find an equation of a line that contains the point $(0, 6)$ and is parallel to $y - 3x = 4$.

48. Find an equation of the line that contains the point $(-2, 4)$ and is parallel to $y = 2x - 3$.

49. Find an equation of the line that contains the point $(0, 2)$ and is perpendicular to $3y - x = 0$.

50. Find an equation of the line that contains the point $(1, 0)$ and is perpendicular to $2x + y = -4$.

51. Find an equation of the line that has x-intercept $(-2, 0)$ and is parallel to $4x - 8y = 12$.

52. Find the value of k such that $4y = kx - 6$ and $5x + 20y = 12$ are parallel.

53. Find the value of k such that $4y = kx - 6$ and $5x + 20y = 12$ are perpendicular.

The lines in the graphs in Exercises 54 and 55 are perpendicular and the lines in the graph in Exercise 56 are parallel. Find an equation of each line.

54.

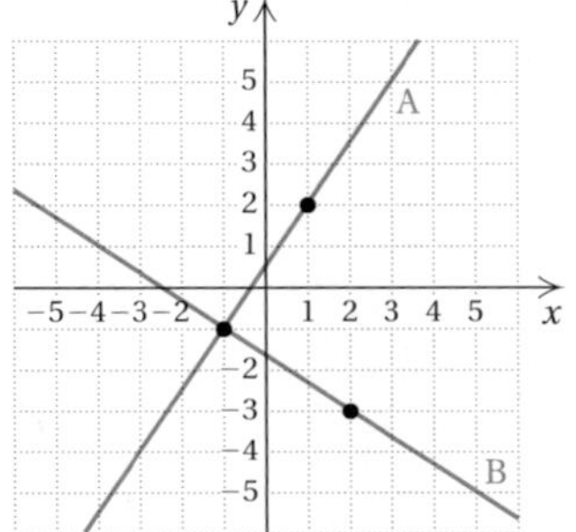

55.

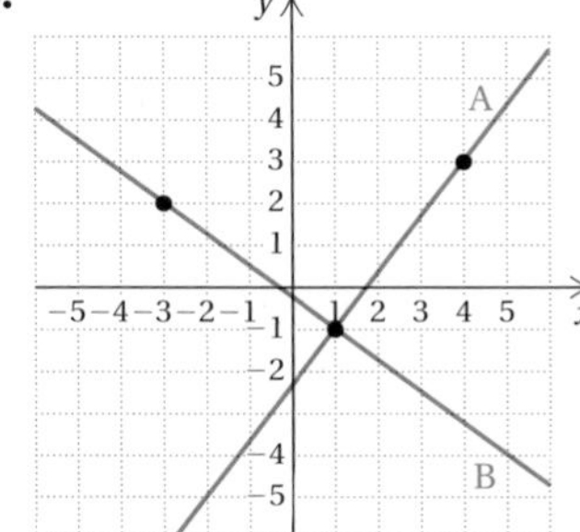

56.

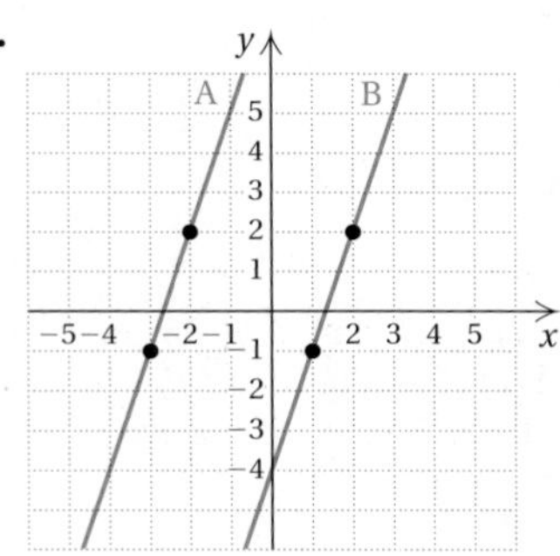

7.4 GRAPHING INEQUALITIES IN TWO VARIABLES

Objectives

a Determine whether an ordered pair of numbers is a solution of an inequality in two variables.

b Graph linear inequalities.

A graph of an inequality is a drawing that represents its solutions. An inequality in one variable can be graphed on a number line. An inequality in two variables can be graphed on a coordinate plane.

a Solutions of Inequalities in Two Variables

The solutions of inequalities in two variables are ordered pairs.

EXAMPLE 1 Determine whether $(-3, 2)$ is a solution of $5x + 4y < 13$.

We use alphabetical order to replace x with -3 and y with 2.

$$\begin{array}{r|l} 5x + 4y & < 13 \\ \hline 5(-3) + 4 \cdot 2 & ?\ 13 \\ -15 + 8 & \\ -7 & \end{array} \quad \text{TRUE}$$

Since $-7 < 13$ is true, $(-3, 2)$ is a solution.

EXAMPLE 2 Determine whether $(6, 8)$ is a solution of $5x + 4y < 13$.

We use alphabetical order to replace x with 6 and y with 8.

$$\begin{array}{r|l} 5x + 4y & < 13 \\ \hline 5(6) + 4(8) & ?\ 13 \\ 30 + 32 & \\ 62 & \end{array} \quad \text{FALSE}$$

Since $62 < 13$ is false, $(6, 8)$ is not a solution.

Do Exercises 1 and 2.

1. Determine whether $(4, 3)$ is a solution of $3x - 2y < 1$.

2. Determine whether $(2, -5)$ is a solution of $4x + 7y \geq 12$.

Answers on page A-34

b Graphing Inequalities in Two Variables

EXAMPLE 3 Graph: $y > x$.

We first graph the line $y = x$. Every solution of $y = x$ is an ordered pair like $(3, 3)$. The first and second coordinates are the same. We draw the line $y = x$ dashed because its points (as shown on the left below) are *not* solutions of $y > x$.

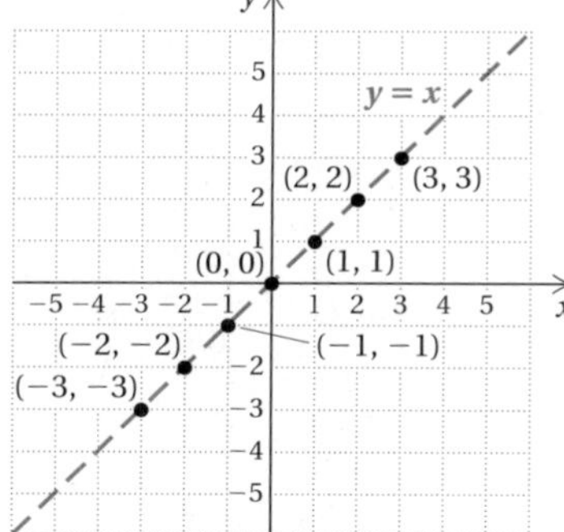

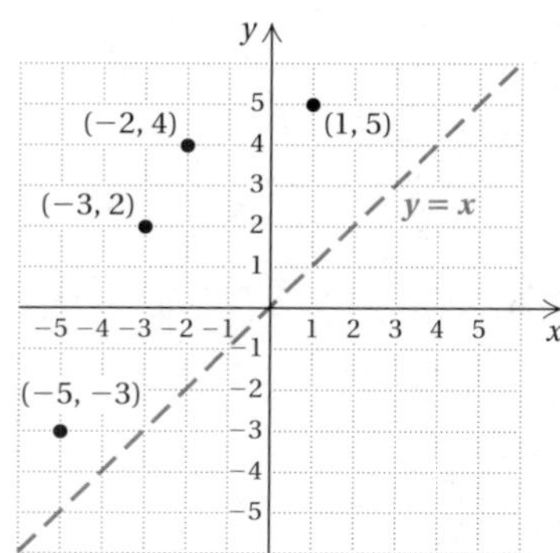

Now look at the graph on the right on the preceding page. Several ordered pairs are plotted in the half-plane above the line $y = x$. Each is a solution of $y > x$.

We can check a pair such as $(-2, 4)$ as follows:

$$\begin{array}{c|c} y > x & \\ \hline 4 \; ? \; -2 & \text{TRUE} \end{array}$$

It turns out that any point on the same side of $y = x$ as $(-2, 4)$ is also a solution. *If we know that one point in a half-plane is a solution, then all points in that half-plane are solutions.* We could have chosen other points to check. The graph of $y > x$ is shown below. (Solutions are indicated by color shading throughout.) We shade the half-plane above $y = x$.

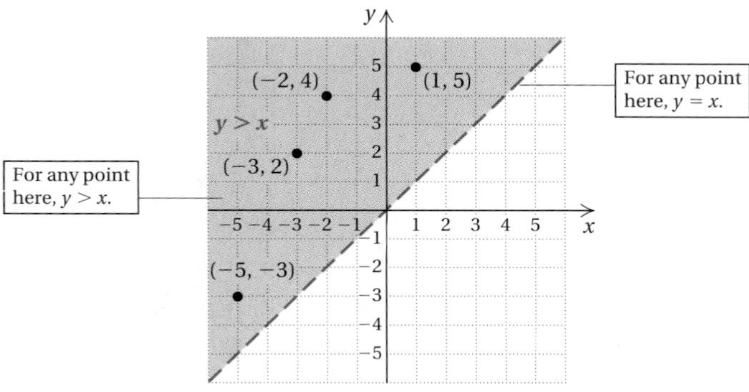

Do Exercise 3.

A **linear inequality** is one that we can get from a linear equation by changing the equals symbol to an inequality symbol. Every linear equation has a graph that is a straight line. The graph of a linear inequality is a half-plane, sometimes including the line along the edge.

To graph an inequality in two variables:

1. Replace the inequality symbol with an equals sign and graph this related equation.
2. If the inequality symbol is $<$ or $>$, draw the line dashed. If the inequality symbol is $\leq$ or $\geq$, draw the line solid.
3. The graph consists of a half-plane, either above or below or left or right of the line, and, if the line is solid, the line as well. To determine which half-plane to shade, choose a point not on the line as a test point. Substitute to find whether that point is a solution of the inequality. If it is, shade the half-plane containing that point. If it is not, shade the half-plane on the opposite side of the line.

3. Graph: $y < x$.

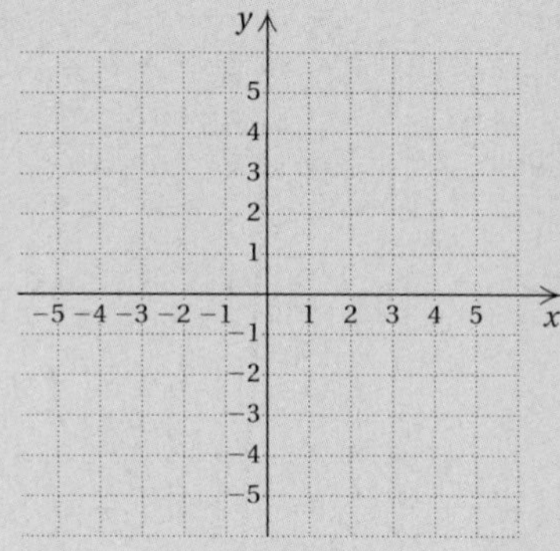

Answer on page A-34

4. Graph: $2x + 4y < 8$.

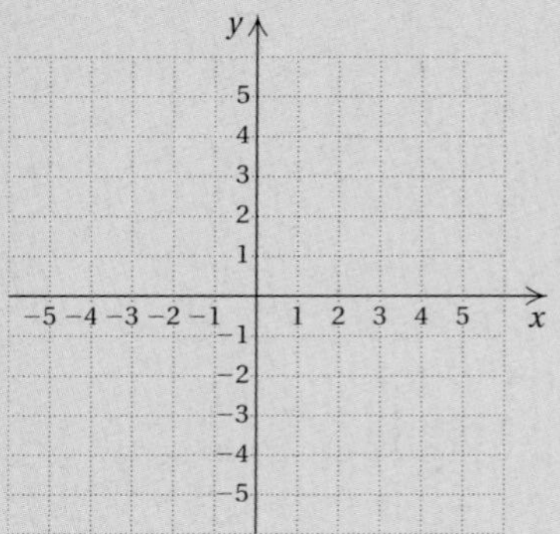

Graph.

5. $3x - 5y < 15$

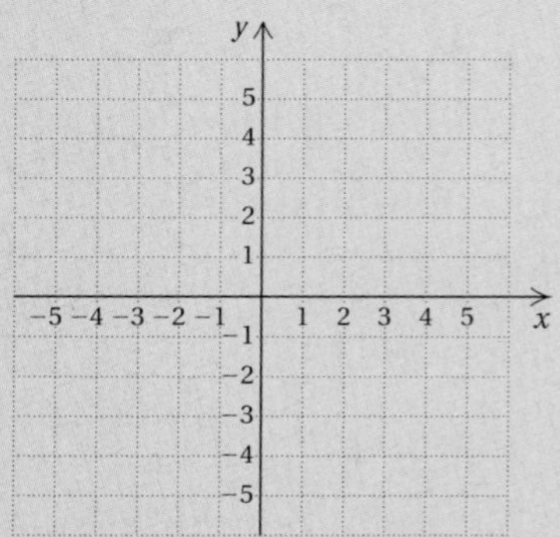

6. $2x + 3y \geq 12$

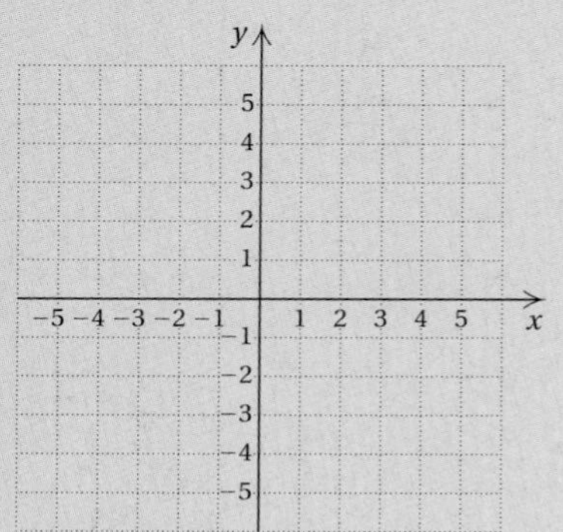

Answers on page A-34

EXAMPLE 4 Graph: $5x - 2y < 10$.

1. We first graph the line $5x - 2y = 10$. The intercepts are $(0, -5)$ and $(2, 0)$. This line forms the boundary of the solutions of the inequality.
2. Since the inequality contains the $<$ symbol, points on the line are not solutions of the inequality, so we draw a dashed line.
3. To determine which half-plane to shade, we consider a test point *not* on the line. We try $(3, -2)$ and substitute:

$$\begin{array}{r|l} 5x - 2y & < 10 \\ \hline 5(3) - 2(-2) & ?\ 10 \\ 15 + 4 & \\ 19 & \end{array} \quad \text{FALSE}$$

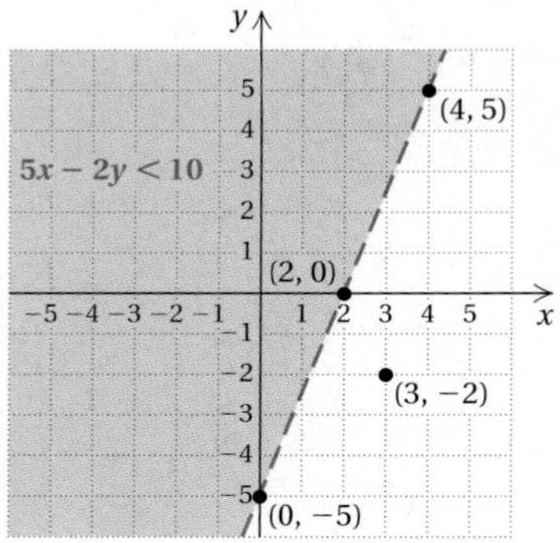

Since this inequality is false, the point $(3, -2)$ is *not* a solution; no point in the half-plane containing $(3, -2)$ is a solution. Thus the points in the opposite half-plane are solutions. The graph is shown above.

Do Exercise 4.

EXAMPLE 5 Graph: $2x + 3y \leq 6$.

1. First, we graph the line $2x + 3y = 6$. The intercepts are $(0, 2)$ and $(3, 0)$.
2. Since the inequality contains the $\leq$ symbol, we draw the line solid to indicate that any pair on the line is a solution.
3. Next, we choose a test point that does not belong to the line. We substitute to determine whether this point is a solution. The origin $(0, 0)$ is generally an easy one to use:

$$\begin{array}{r|l} 2x + 3y & \leq 6 \\ \hline 2 \cdot 0 + 3 \cdot 0 & ?\ 6 \\ 0 & \end{array} \quad \text{TRUE}$$

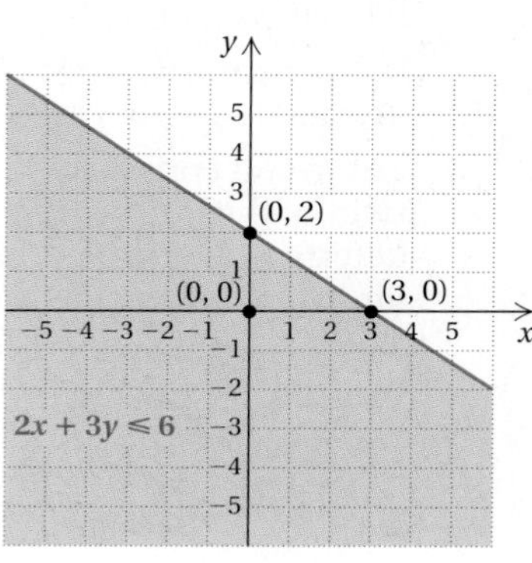

We see that $(0, 0)$ is a solution, so we shade the lower half-plane. Had the substitution given us a false inequality, we would have shaded the other half-plane.

Do Exercises 5 and 6.

EXAMPLE 6 Graph $x < 3$ on a plane.

There is no y-term in this inequality, but we can rewrite this inequality as $x + 0y < 3$. We use the same technique that we have used with the other examples.

1. We graph the related equation $x = 3$ on the plane.
2. Since the inequality symbol is $<$, we use a dashed line.
3. The graph is a half-plane either to the left or to the right of the line $x = 3$. To determine which, we consider a test point, $(-4, 5)$:

$$\begin{array}{c|c} x + 0y & 3 \\ \hline -4 + 0(5) & ?\ 3 \\ -4 & \end{array} \quad \text{TRUE}$$

We see that $(-4, 5)$ is a solution, so all the pairs in the half-plane containing $(-4, 5)$ are solutions. We shade that half-plane.

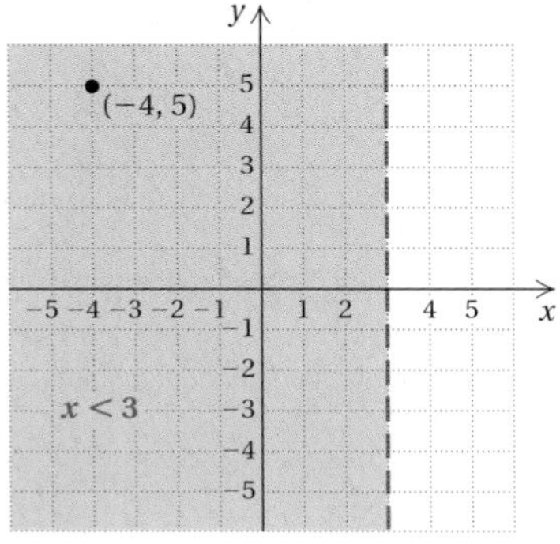

We see from the graph that the solutions of $x < 3$ are all those ordered pairs whose first coordinates are less than 3.

If we graph the inequality in Example 6 on a line rather than on a plane, its graph is as follows:

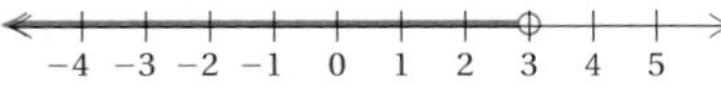

Graph.

7. $x > -3$

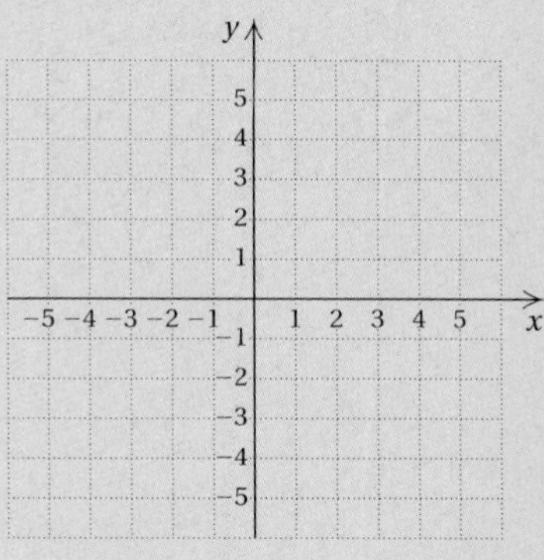

8. $y \leq 4$

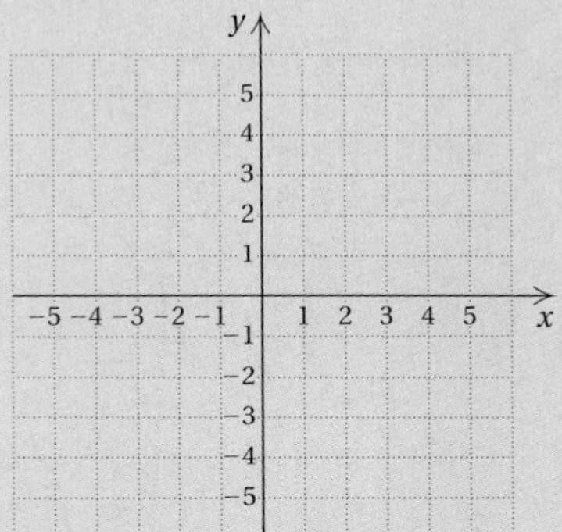

Answers on page A-34

EXAMPLE 7 Graph $y \geq -4$.

1. We first graph $y = -4$.
2. We use a solid line to indicate that all points on the line are solutions.
3. We then use $(2, 3)$ as a test point and substitute:

$$\begin{array}{c|c} 0x + y \geq -4 & \\ \hline 0(2) + 3 \ ? \ -4 & \\ 3 & \end{array} \quad \text{TRUE}$$

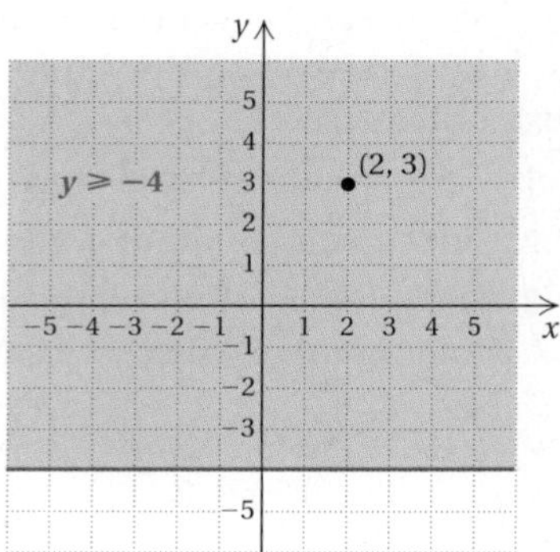

Since $(2, 3)$ is a solution, all points in the half-plane containing $(2, 3)$ are solutions. Note that this half-plane consists of all ordered pairs whose second coordinate is greater than or equal to -4.

Do Exercises 7 and 8 on the preceding page.

CALCULATOR CORNER

Graphs of Inequalities We can graph inequalities on a graphing calculator, shading the region of the solution set. To graph the inequality in Example 5,

$$2x + 3y \leq 6,$$

we first graph the line $2x + 3y = 6$. Solving for y, we get $y = \frac{6 - 2x}{3}$, or $y = -\frac{2}{3} + 2$. We enter this equation on the equation-editor screen.

After determining algebraically that the solution set consists of all points below the line, we use the graphing calculator's "shade below" graph style to shade this region. On the equation-editor screen, we position the cursor over the graphstyle icon to the left of the equation and press ENTER repeatedly until the "shade below" icon appears. Then we press GRAPH to display the graph of the inequality.

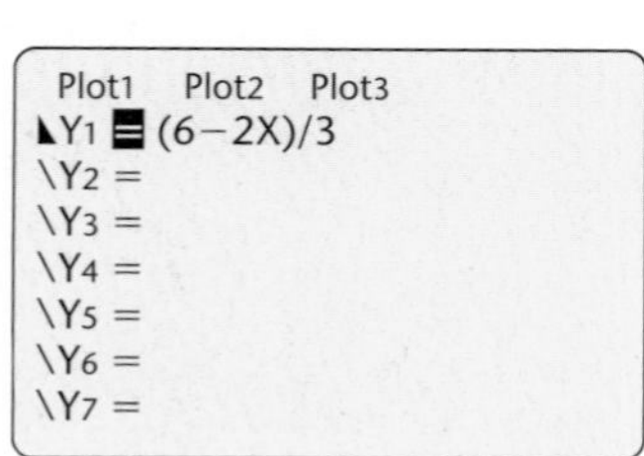

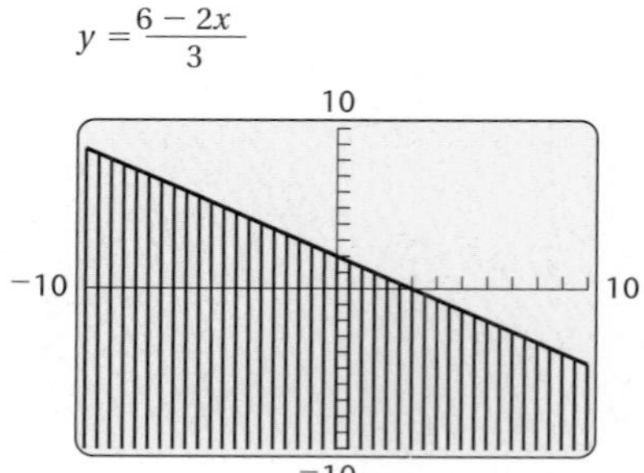

Note that we cannot graph an inequality like the one in Example 6, $x < 3$, on a graphing calculator because the related equation has no y-term and thus cannot be entered in "$y =$" form.

Exercises

1. Use a graphing calculator to graph the inequalities in Margin Exercises 6 and 8.
2. Use a graphing calculator to graph the inequality in Example 7.

7.4 EXERCISE SET

For Extra Help

Digital Video Tutor CD 7 Videotape 14 | InterAct Math | Math Tutor Center | MathXL | MyMathLab

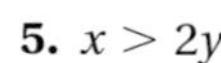

1. Determine whether $(-3, -5)$ is a solution of $-x - 3y < 18$.

2. Determine whether $(2, -3)$ is a solution of $5x - 4y \geq 1$.

3. Determine whether $\left(\frac{1}{2}, -\frac{1}{4}\right)$ is a solution of $7y - 9x \leq -3$.

4. Determine whether $(-8, 5)$ is a solution of $x + 0 \cdot y > 4$.

b Graph on a plane.

5. $x > 2y$

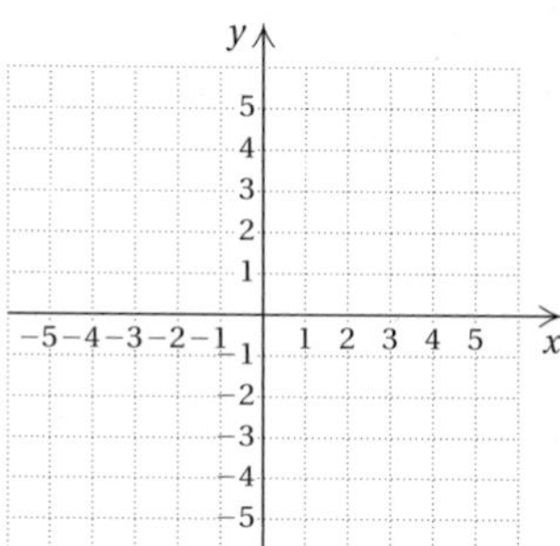

6. $x > 3y$

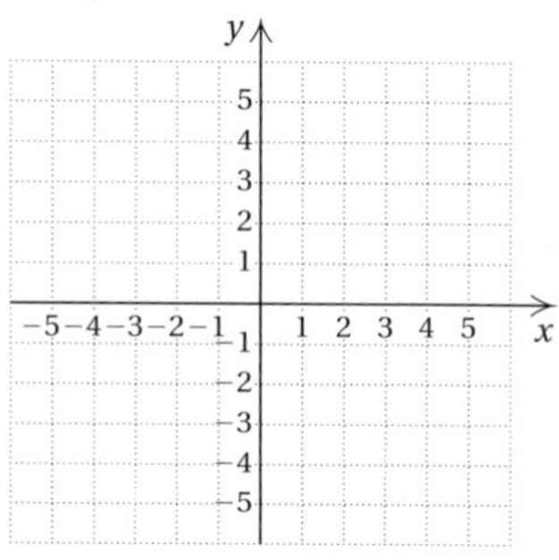

7. $y \leq x - 3$

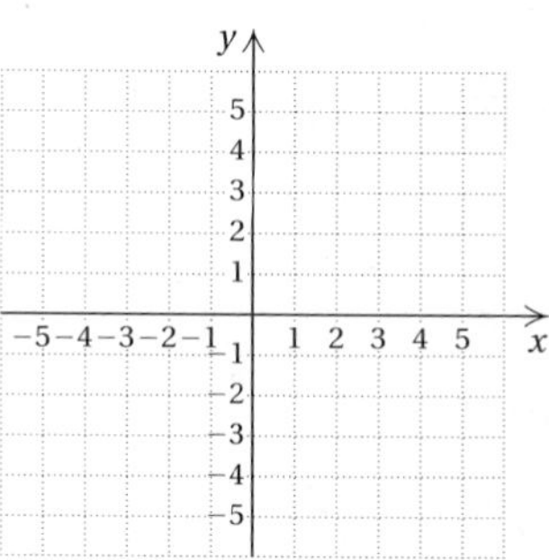

8. $y \leq x - 5$

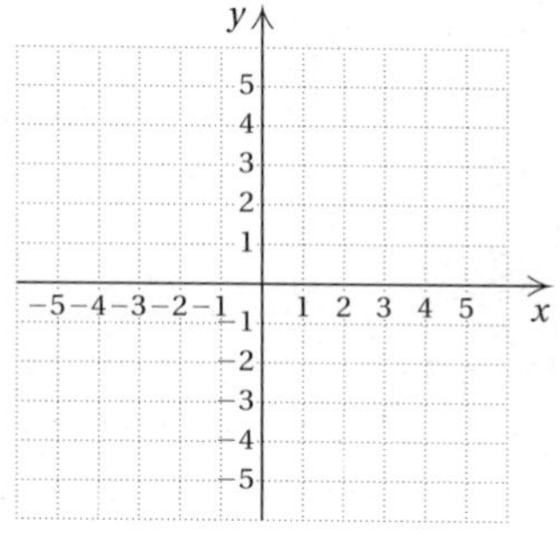

9. $y < x + 1$

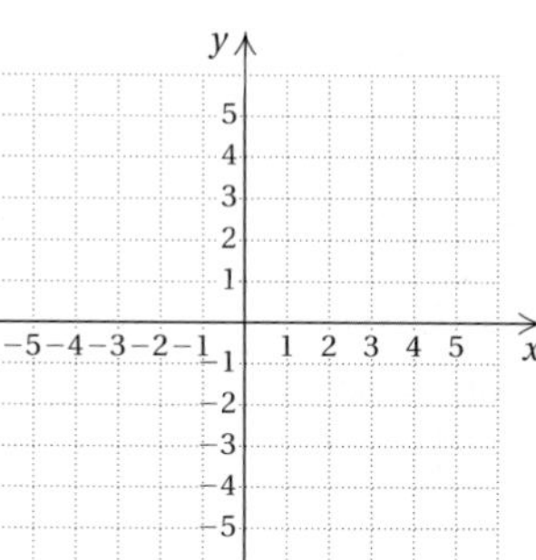

10. $y < x + 4$

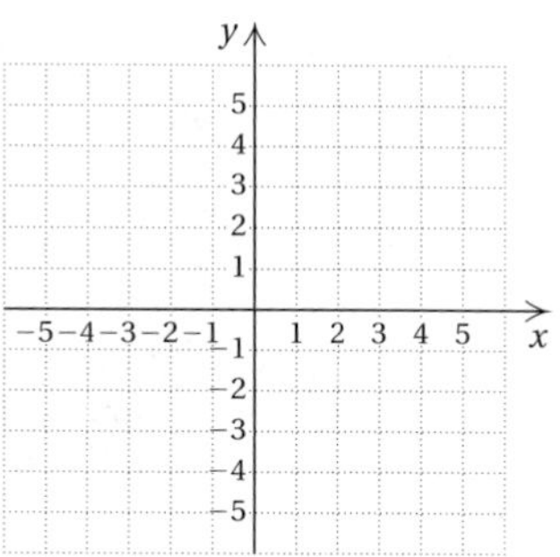

11. $y \geq x - 2$

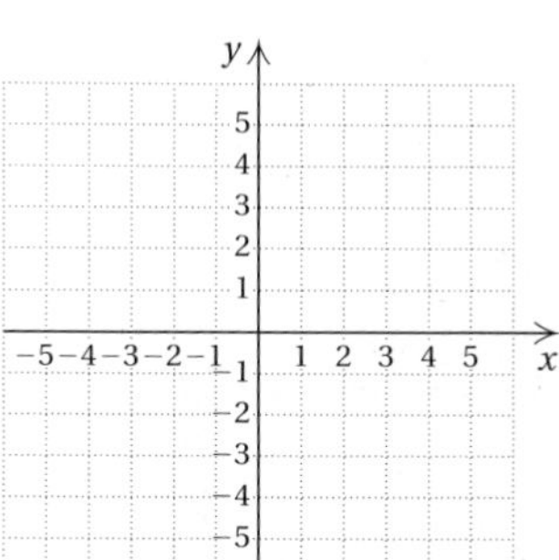

12. $y \geq x - 1$

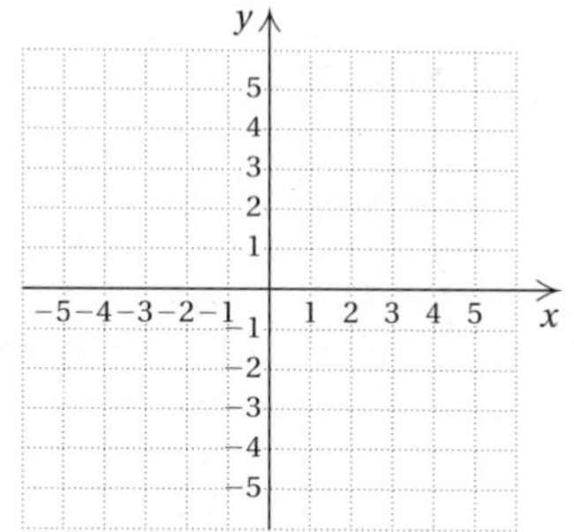

13. $y \leq 2x - 1$

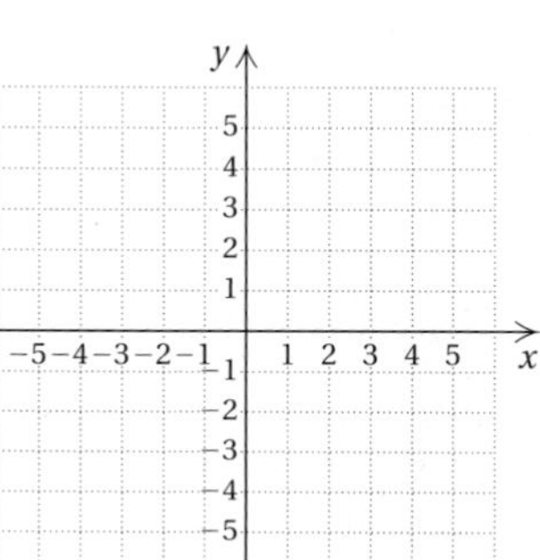

14. $y \leq 3x + 2$

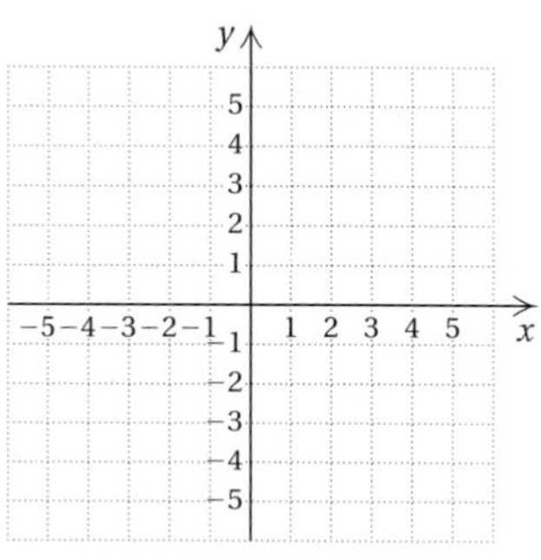

15. $x + y \leq 3$

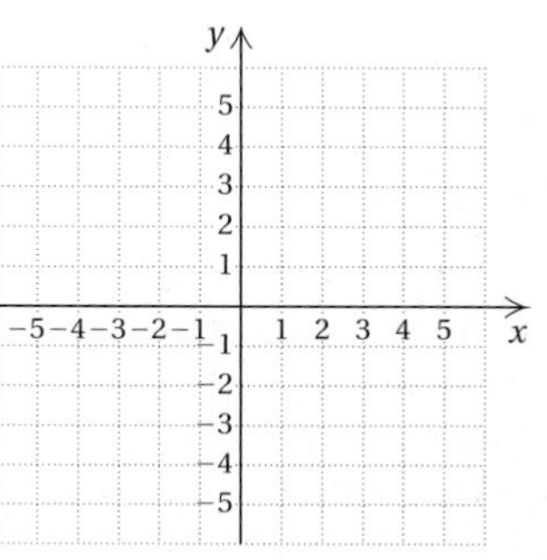

16. $x + y \leq 4$

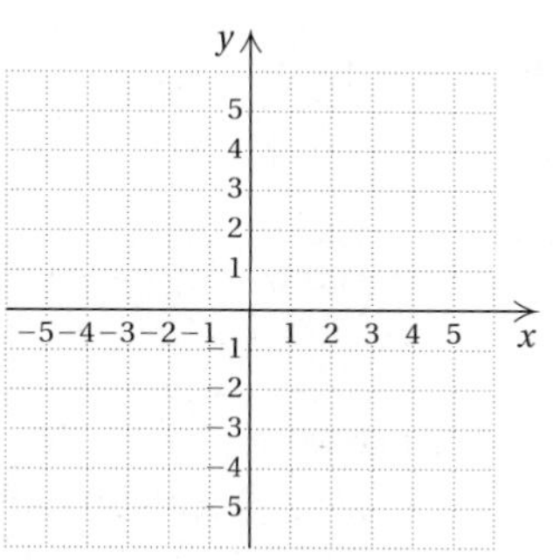

17. $x - y > 7$

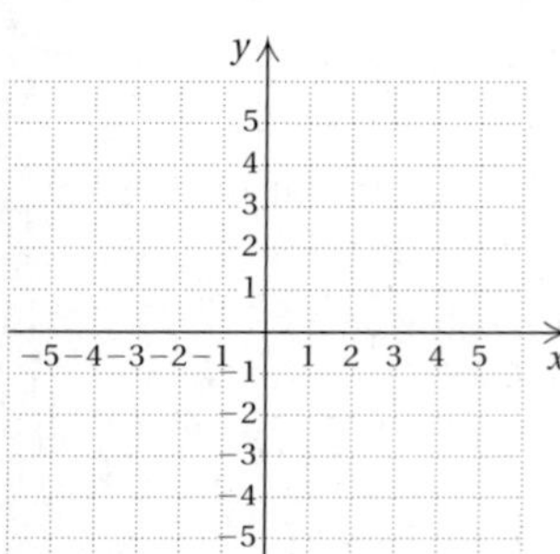

18. $x - y > -2$

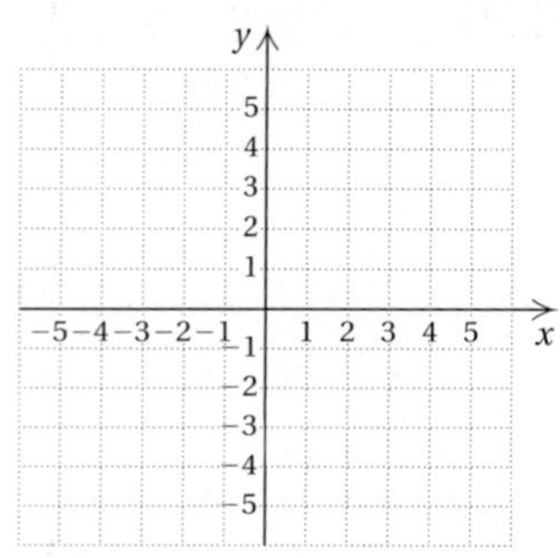

19. $2x + 3y \leq 12$

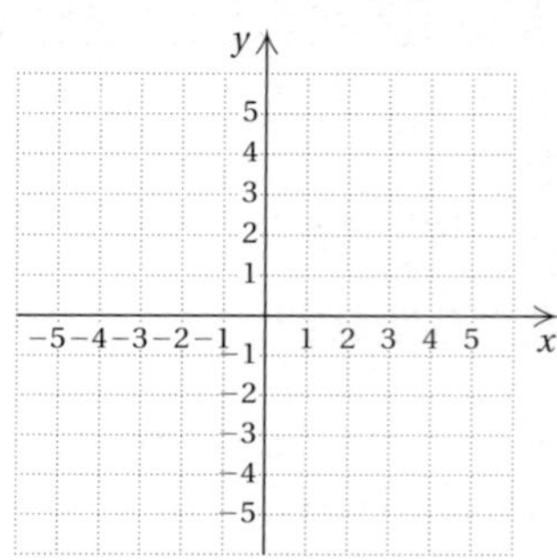

20. $5x + 4y \geq 20$

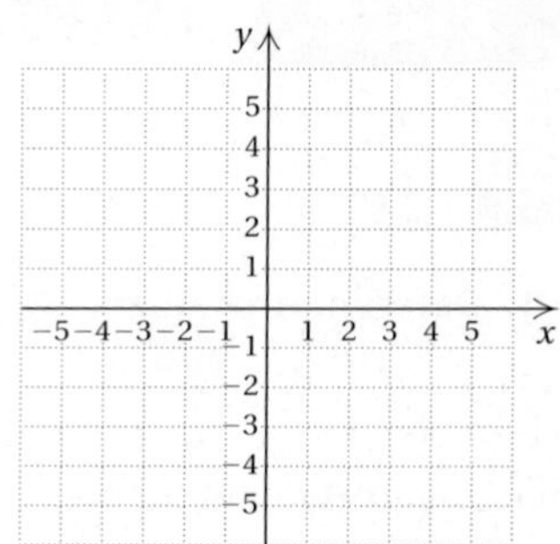

21. $y \geq 1 - 2x$

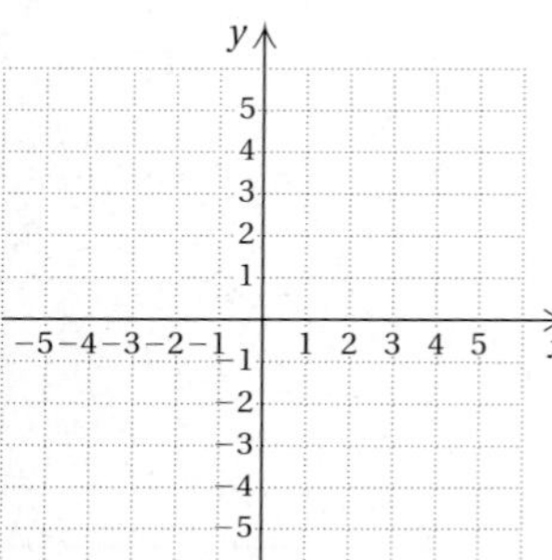

22. $y - 2x \leq -1$

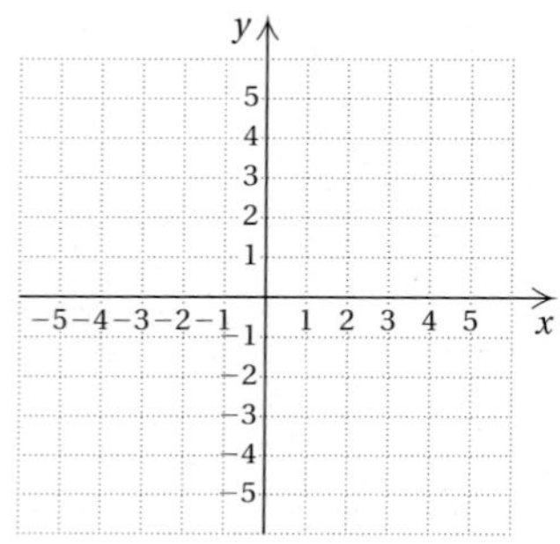

23. $2x - 3y > 6$

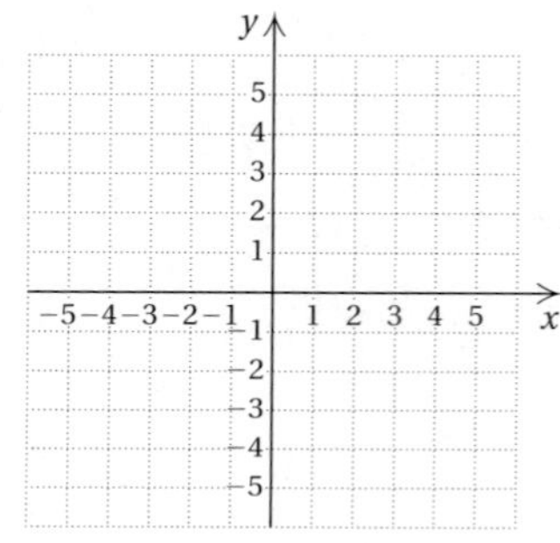

24. $5y - 2x \leq 10$

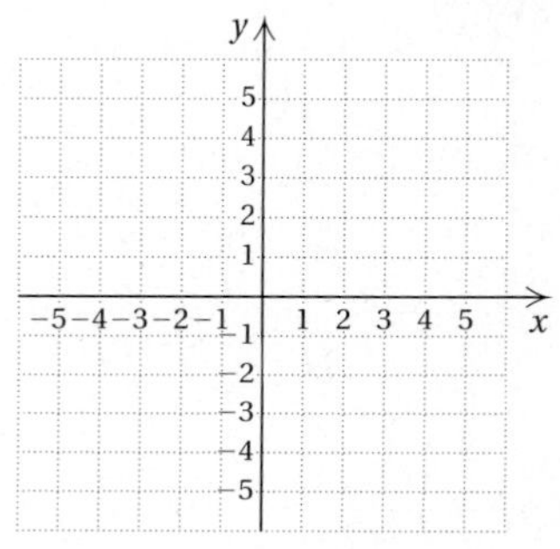

25. $y \leq 3$

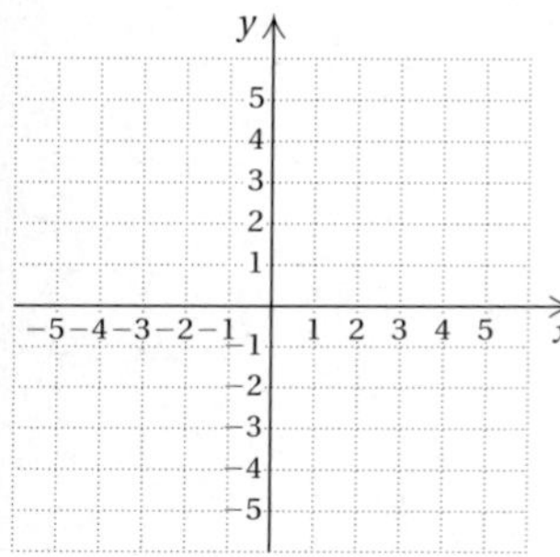

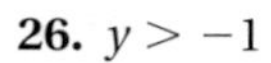

26. $y > -1$

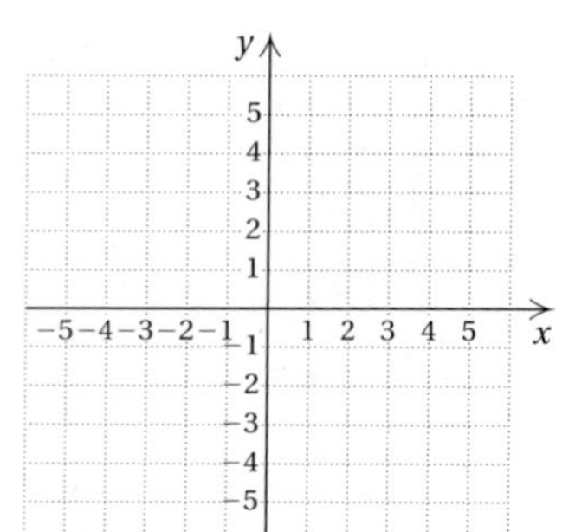

27. $x \geq -1$

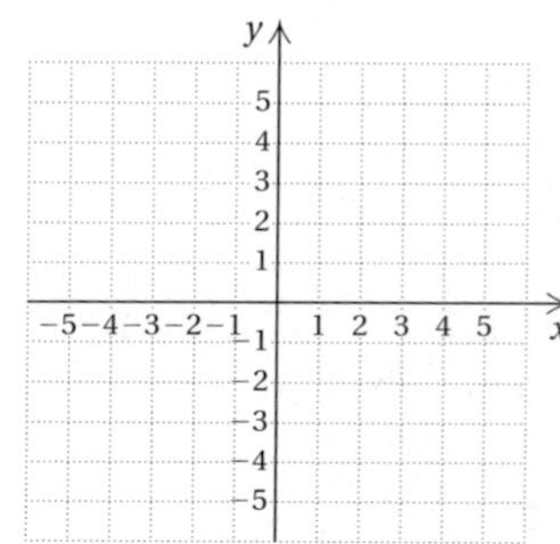

28. $x < 0$

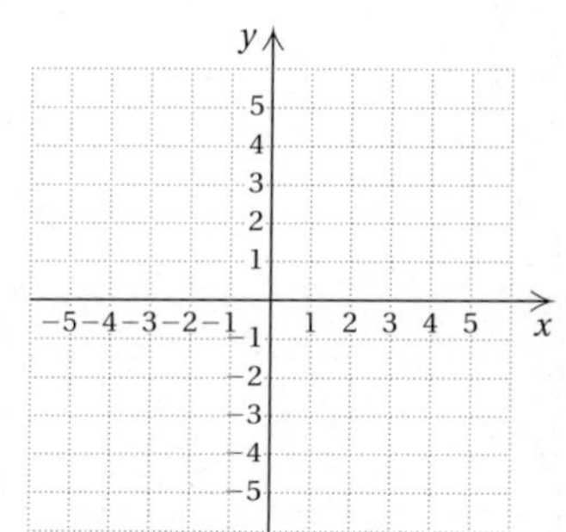

29. ^{D}W Why is (0, 0) such a "convenient" test point?

30. ^{D}W Is the graph of any inequality in the form $y > mx + b$ shaded *above* the line $y = mx + b$? Why or why not?

SKILL MAINTENANCE

Solve. [6.6a]

31. $\frac{12}{x} = \frac{48}{x + 9}$

32. $x + 5 = -\frac{6}{x}$

Solve. [5.7b]

33. $x^2 + 16 = 8x$

34. $12x^2 + 17x = 5$

SYNTHESIS

35. *Elevators.* Many elevators have a capacity of 1 metric ton (1000 kg). Suppose c children, each weighing 35 kg, and a adults, each weighing 75 kg, are on an elevator. Find and graph an inequality that asserts that the elevator is overloaded.

36. *Hockey Wins and Losses.* A hockey team determines that it needs at least 60 points for the season in order to make the playoffs. A win w is worth 2 points and a tie t is worth 1 point. Find and graph an inequality that describes the situation.

7.5 DIRECT AND INVERSE VARIATION

Objectives

a Find an equation of direct variation given a pair of values of the variables.

b Solve applied problems involving direct variation.

c Find an equation of inverse variation given a pair of values of the variables.

d Solve applied problems involving inverse variation.

a Equations of Direct Variation

A bicycle is traveling at a speed of 15 km/h. In 1 hr, it goes 15 km; in 2 hr, it goes 30 km; in 3 hr, it goes 45 km; and so on. We can form a set of ordered pairs using the number of hours as the first coordinate and the number of kilometers traveled as the second coordinate. These determine a set of ordered pairs:

$$(1, 15), \ (2, 30), \ (3, 45), \ (4, 60), \quad \text{and so on.}$$

Note that the second coordinate is always 15 times the first.

In this example, distance is a constant multiple of time, so we say that there is *direct variation* and that distance *varies directly* as time. The *equation of variation* is $d = 15t$.

DIRECT VARIATION

When a situation translates to an equation described by $y = kx$, with k a positive constant, we say that **y varies directly as x.** The equation $y = kx$ is called an **equation of direct variation.**

In direct variation, as one variable increases, the other variable increases as well. This is shown in the graph above.

The terminologies

"y varies as x,"

"y is directly proportional to x," and

"y is proportional to x"

also imply direct variation and are used in many situations. The constant k is called the **constant of proportionality** or the **variation constant.** It can be found if one pair of values of x and y is known. Once k is known, other pairs can be determined.

Study Tips

MAKING APPLICATIONS REAL

Newspapers and magazines are full of mathematical applications. Find such an application and share it with your class. As you develop more skills in mathematics, you will find yourself observing the world from a different perspective, seeing mathematics everywhere. Math courses become more interesting when we connect the concepts to the real world.

1. Find an equation of variation in which y varies directly as x and $y = 84$ when $x = 12$. Then find the value of y when $x = 41$.

EXAMPLE 1 Find an equation of variation in which y varies directly as x and $y = 7$ when $x = 25$.

We first substitute to find k:

$y = kx$

$7 = k \cdot 25$ Substituting 25 for x and 7 for y

$\frac{7}{25} = k$, or $k = 0.28$. Solving for k, the variation constant

Then the equation of variation is

$$y = 0.28x.$$

The answer is the equation $y = 0.28x$, *not* $k = 0.28$. We can visualize the example by looking at the graph.

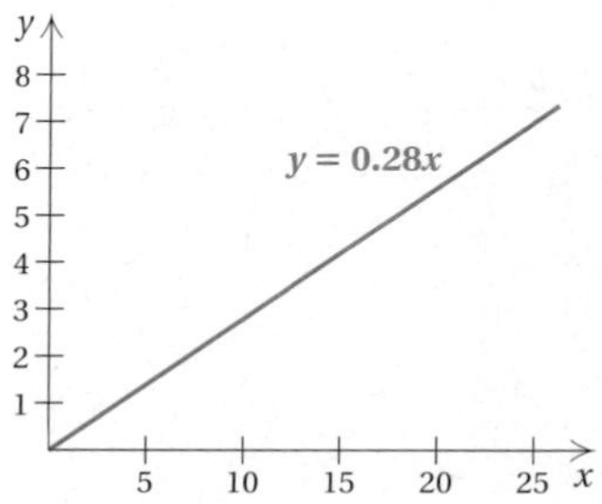

We see that when y varies directly as x, the constant of proportionality is also the slope of the associated graph—the rate at which y changes with respect to x.

2. Find an equation of variation in which y varies directly as x and $y = 50$ when $x = 80$. Then find the value of y when $x = 20$.

EXAMPLE 2 Find an equation in which s varies directly as t and $s = 10$ when $t = 15$. Then find the value of s when $t = 32$.

We have

$s = kt$ We know that s varies directly as t.

$10 = k \cdot 15$ Substituting 10 for s and 15 for t

$\frac{10}{15} = k$, or $k = \frac{2}{3}$. Solving for k

Thus the equation of variation is $s = \frac{2}{3}t$.

$s = \frac{2}{3}t$

$s = \frac{2}{3} \cdot 32$ Substituting 32 for t in the equation of variation

$s = \frac{64}{3}$, or $21\frac{1}{3}$.

The value of s is $21\frac{1}{3}$ when $t = 32$.

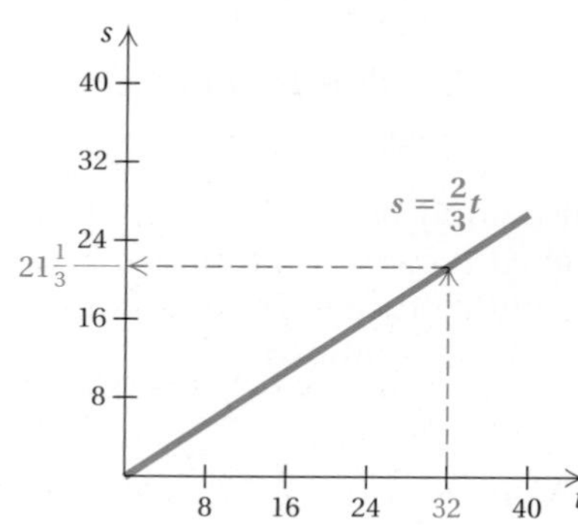

Do Exercises 1 and 2.

Answers on page A-35

b Applications of Direct Variation

EXAMPLE 3 *Karat Ratings of Gold Objects.* The beauty of gold has inspired artisans throughout the centuries, as shown by this magnificent sarcophagus, which contained Tutankhamen's mummy. Today we know that the karat rating K of a gold object varies directly as the actual percentage P of gold in the object. A 14-karat gold ring is 58.25% gold. What is the percentage of gold in a 10-karat chain?

Source: Barbara Ann Kipfer, *The Order of Things.* New York: Random House, 1998.

1., 2. Familiarize and **Translate.** The problem states that we have direct variation between the variables K and P. Thus an equation $K = kP$, $k > 0$, applies. As the percentage of gold increases, the karat rating increases. The letters K and k represent different quantities.

3. Solve. The mathematical manipulation has two parts. First, we determine the equation of variation by substituting known values for K and P to find the variation constant k. Second, we compute the percentage of gold in a 10-karat chain.

a) First, we find an equation of variation:

$$K = kP$$

$$14 = k(0.5825) \qquad \text{Substituting 14 for } K \text{ and 58.25\%, or 0.5825, for } P$$

$$\frac{14}{0.5825} = k$$

$$24.03 \approx k. \qquad \text{Dividing and rounding to the nearest hundredth}$$

The equation of variation is $K = 24.03P$.

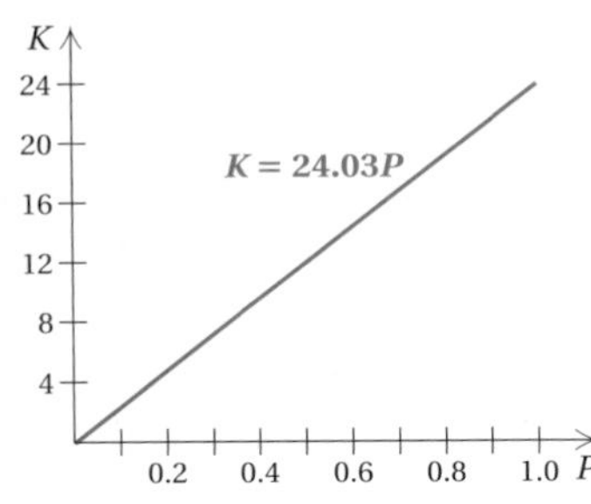

b) We then use the equation to find the percentage of gold in a 10-karat chain:

$$K = 24.03P$$

$$10 = 24.03P \qquad \text{Substituting 10 for } K$$

$$\frac{10}{24.03} = P$$

$$0.416 \approx P$$

$$41.6\% \approx P.$$

4. Check. The check might be done by repeating the computations. You might also do some reasoning about the answer. The karat rating decreased from 14 to 10. Similarly, the percentage decreased from 58.25% to 41.6%.

5. State. A 10-karat chain is 41.6% gold.

Do Exercises 3 and 4.

3. Electricity Costs. The cost C of operating a television varies directly as the number n of hours that it is in operation. It costs \$14.00 to operate a standard-size color TV continuously for 30 days.

a) Find an equation of variation.

b) At this rate, how much would it cost to operate the TV for 1 day? for 1 hour?

4. Weight on Venus. The weight V of an object on Venus varies directly as its weight E on Earth. A person weighing 165 lb on Earth would weigh 145.2 lb on Venus.

a) Find an equation of variation.

b) How much would a person weighing 198 lb on Earth weigh on Venus?

Answers on page A-35

Let's consider direct variation from the standpoint of a graph. The graph of $y = kx$, $k > 0$, always goes through the origin and rises from left to right. Note that as x increases, y increases; and as x decreases, y decreases. This is why the terminology "direct" is used. What one variable does, the other does as well.

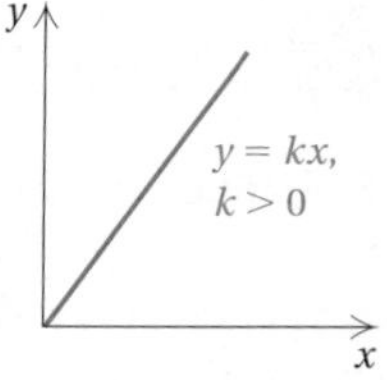

C Equations of Inverse Variation

A car is traveling a distance of 20 mi. At a speed of 5 mph, it will take 4 hr; at 20 mph, it will take 1 hr; at 40 mph, it will take $\frac{1}{2}$ hr; and so on. We use speed as the first coordinate and the time as the second coordinate. These determine a set of ordered pairs:

$$(5, 4),\ (20, 1),\ \left(40, \tfrac{1}{2}\right),\ \left(60, \tfrac{1}{3}\right),\quad \text{and so on.}$$

Note that the product of speed and time for each of these pairs is 20. Note too that as the speed *increases,* the time *decreases.*

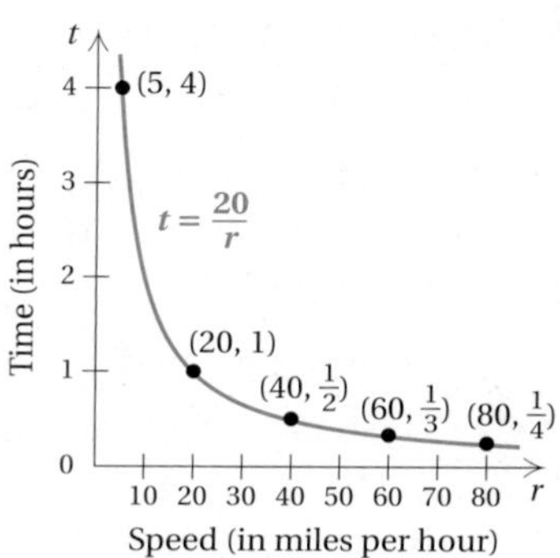

In this case, the product of speed and time is constant so we say that there is *inverse variation* and that time *varies inversely* as speed. The equation of variation is

$$rt = 20 \text{ (a constant)}, \quad \text{or} \quad t = \frac{20}{r}.$$

INVERSE VARIATION

When a situation translates to an equation described by $y = k/x$, with k a positive constant, we say that **y varies inversely as x.** The equation $y = k/x$ is called an **equation of inverse variation.**

In inverse variation, as one variable increases, the other variable decreases.

The terminology

"y is inversely proportional to x"

also implies inverse variation and is used in some situations. The constant k is again called the **constant of proportionality** or the **variation constant.**

EXAMPLE 4 Find an equation of variation in which y varies inversely as x and $y = 145$ when $x = 0.8$. Then find the value of y when $x = 25$.

We first substitute to find k:

$$y = \frac{k}{x}$$
$$145 = \frac{k}{0.8}$$
$$(0.8)145 = k$$
$$116 = k.$$

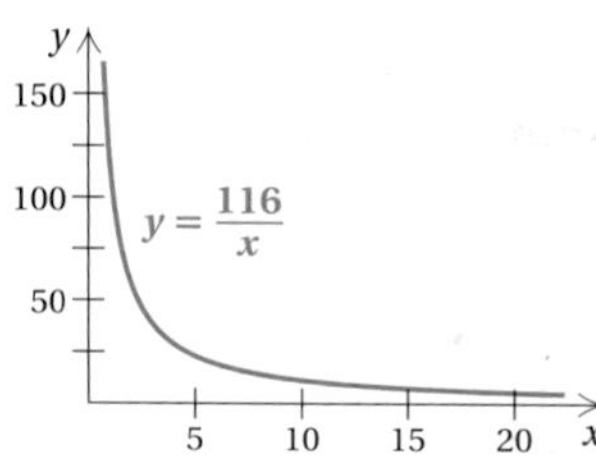

The equation of variation is $y = 116/x$. The answer is the equation $y = 116/x$, *not* $k = 116$.

When $x = 25$, we have

$$y = \frac{116}{x}$$
$$y = \frac{116}{25} \quad \text{Substituting 25 for } x$$
$$y = 4.64.$$

The value of y is 4.64 when $x = 25$.

Do Exercises 5 and 6.

The graph of $y = k/x$, $k > 0$, is shaped like the figure at right for positive values of x. (You need not know how to graph such equations at this time.) Note that as x increases, y decreases; and as x decreases, y increases. This is why the terminology "inverse" is used. One variable does the opposite of what the other does.

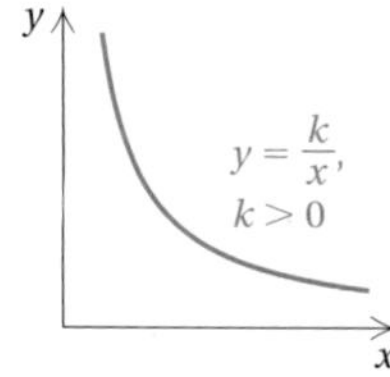

d Applications of Inverse Variation

Often in an applied situation we must decide which kind of variation, if any, might apply to the problem.

EXAMPLE 5 *Work Time.* Molly is a maintenance supervisor. She notes that it takes 4 hr for 20 people to wash and wax the floors in a building. How long would it then take 25 people to do the job?

1. **Familiarize.** Think about the problem situation. What kind of variation would be used? It seems reasonable that the more people there are working on the job, the less time it will take to finish. (One might argue that too many people in a crowded area would be counterproductive, but we will disregard that possibility.) Thus inverse variation might apply. We let $T =$ the time to do the job, in hours, and $N =$ the number of people. Assuming inverse variation, we know that an equation $T = k/N$, $k > 0$, applies. As the number of people increases, the time it takes to do the job decreases.

5. Find an equation of variation in which y varies inversely as x and $y = 105$ when $x = 0.6$. Then find the value of y when $x = 20$.

6. Find an equation of variation in which y varies inversely as x and $y = 45$ when $x = 20$. Then find the value of y when $x = 1.6$.

Answers on page A-35

7. Referring to Example 5, determine how long it would take 10 people to do the job.

8. **Time of Travel.** The time t required to drive a fixed distance varies inversely as the speed r. It takes 5 hr at 60 km/h to drive a fixed distance.

a) Find an equation of variation.

b) How long would it take at 40 km/h?

Answers on page A-35

2. **Translate.** We write an equation of variation:

$$T = \frac{k}{N}.$$

Time varies inversely as the number of people involved.

3. **Solve.** The mathematical manipulation has two parts. First, we find the equation of variation by substituting known values for T and N to find k. Second, we compute the amount of time it would take 25 people to do the job.

a) First, we find an equation of variation:

$$T = \frac{k}{N}$$

$$4 = \frac{k}{20} \quad \text{Substituting 4 for } T \text{ and 20 for } N$$

$$20 \cdot 4 = k.$$

$$80 = k.$$

T
150
100
50
$T = \frac{80}{N}$
5 10 15 20 N

The equation of variation is $T = \frac{80}{N}$.

b) We then use the equation to find the amount of time that it takes 25 people to do the job:

$$T = \frac{80}{N}$$

$$= \frac{80}{25} \quad \text{Substituting 25 for } N$$

$$= 3.2.$$

4. **Check.** The check might be done by repeating the computations. We might also analyze the results. The number of people increased from 20 to 25. Did the time decrease? It did, and this confirms what we expect with inverse variation.

5. **State.** It should take 3.2 hr for 25 people to complete the job.

Do Exercises 7 and 8.

7.5 EXERCISE SET

For Extra Help

Digital Video Tutor CD 7 Videotape 14 | InterAct Math | Math Tutor Center | MathXL | MyMathLab

a Find an equation of variation in which y varies directly as x and the following are true. Then find the value of y when $x = 20$.

1. $y = 36$ when $x = 9$

2. $y = 60$ when $x = 16$

3. $y = 0.8$ when $x = 0.5$

4. $y = 0.7$ when $x = 0.4$

5. $y = 630$ when $x = 175$

6. $y = 400$ when $x = 125$

7. $y = 500$ when $x = 60$

8. $y = 200$ when $x = 300$

b Solve.

9. *Wages and Work Time.* A person's paycheck P varies directly as the number H of hours worked. For working 15 hr, the pay is \$84.

a) Find an equation of variation.
b) Find the pay for 35 hr of work.

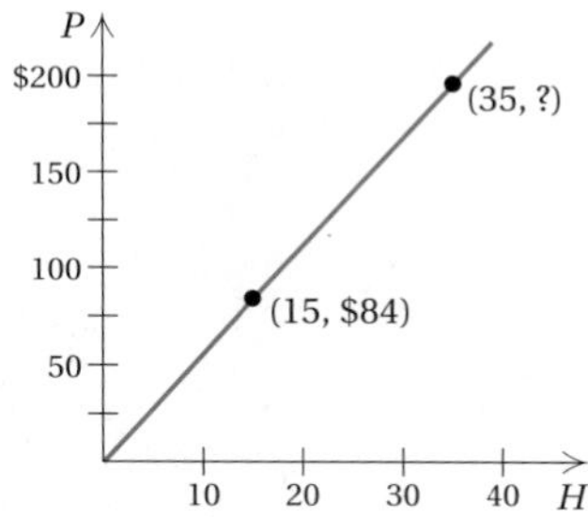

10. *Interest and Interest Rate.* The interest I earned in 1 yr on a fixed principal varies directly as the interest rate r. An investment earns \$53.55 at an interest rate of 4.25%.

a) Find an equation of variation.
b) How much will the investment earn at a rate of 5.75%?

11. *Cost of Sand.* The cost C to fill a sandbox varies directly as the depth S of the sand. Lucinda checks at her local hardware store and finds that it would cost \$75 to fill the box with 6 in. of sand. She decides to fill the sandbox to a depth of 8 in.

a) Find an equation of variation.
b) How much will the sand cost Lucinda?

12. *Cost of Cement.* The cost C of cement needed to pave a driveway varies directly as the depth D of the driveway. John checks at his local building materials store and finds that it costs \$500 to install his driveway with a depth of 8 in. He decides to build a stronger driveway at a depth of 12 in.

a) Find an equation of variation.
b) How much will it cost for the cement?

13. *Lunar Weight.* The weight M of an object on the moon varies directly as its weight E on Earth. Your author, Marv Bittinger, weighs 192 lb, but would weigh only 32 lb on the moon.

a) Find an equation of variation.
b) Marv's wife, Elaine, weighs 110 lb on Earth. How much would she weigh on the moon?
c) Marv's granddaughter, Maggie, would weigh only 5 lb on the moon. How much does Maggie weigh on Earth?

14. *Mars Weight.* The weight M of an object on Mars varies directly as its weight E on Earth. In 1999, Chen Yanqing, who weighs 128 lb, set a world record for her weight class with a lift (snatch) of 231 lb. On Mars, this lift would be only 88 lb.

a) Find an equation of variation.
b) How much would Yanqing weigh on Mars?

Source: *The Guinness Book of Records*, 2001

15. *Computer Megahertz.* The number of computer instructions N per second varies directly as the speed S of its internal processor. A processor with a speed of 25 megahertz can perform 2,000,000 instructions per second.

a) Find an equation of variation.
b) How many instructions will the same processor perform if it is running at a speed of 200 megahertz?

16. *Water in Human Body.* The number of kilograms W of water in a human body varies directly as the total body weight B. A person who weighs 75 kg contains 54 kg of water.

a) Find an equation of variation.
b) How many kilograms of water are in a person who weighs 95 kg?

17. *Steak Servings.* The number of servings S of meat that can be obtained from round steak varies directly as the weight W. From 9 kg of round steak, one can get 70 servings of meat. How many servings can one get from 12 kg of round steak?

18. *Turkey Servings.* A chef is planning meals in a refreshment tent at a golf tournament. The number of servings S of meat that can be obtained from a turkey varies directly as its weight W. From a turkey weighing 30.8 lb, one can get 40 servings of meat. How many servings can be obtained from a 19.8-lb turkey?

Find an equation of variation in which y varies inversely as x and the following are true. Then find the value of y when $x = 10$.

19. $y = 3$ when $x = 25$

20. $y = 2$ when $x = 45$

21. $y = 10$ when $x = 8$

22. $y = 10$ when $x = 7$

23. $y = 6.25$ when $x = 0.16$

24. $y = 0.125$ when $x = 8$

25. $y = 50$ when $x = 42$

26. $y = 25$ when $x = 42$

27. $y = 0.2$ when $x = 0.3$

28. $y = 0.4$ when $x = 0.6$

b, d Solve.

29. *Production and Time.* A production line produces 15 compact disc players every 8 hr. How many players can it produce in 37 hr?

a) What kind of variation might apply to this situation?
b) Solve the problem.

30. *Wages and Work Time.* A person works for 15 hr and makes $93.75. How much will the person make by working 35 hr?

a) What kind of variation might apply to this situation?
b) Solve the problem.

31. *Cooking Time.* It takes 4 hr for 9 cooks to prepare the food for a wedding rehearsal dinner. How long will it take 8 cooks to prepare the dinner?

a) What kind of variation might apply to this situation?
b) Solve the problem.

32. *Work Time.* It takes 16 hr for 2 people to resurface a tennis court. How long will it take 6 people to do the job?

a) What kind of variation might apply to this situation?
b) Solve the problem.

33. *Miles per Gallon.* To travel a fixed distance, the number of gallons N of gasoline needed is inversely proportional to the miles-per-gallon rating P of the car. A car that gets 20 miles per gallon (mpg) needs 14 gal to travel the distance.

a) Find an equation of variation.
b) How much gas will be needed for a car that gets 28 mpg?

34. *Miles per Gallon.* To travel a fixed distance, the number of gallons N of gasoline needed is inversely proportional to the miles-per-gallon rating P of the car. A car that gets 25 miles per gallon (mpg) needs 12 gal to travel the distance.

a) Find an equation of variation.
b) How much gas will be needed for a car that gets 20 mpg?

35. *Electrical Current.* The current I in an electrical conductor varies inversely as the resistance R of the conductor. The current is 96 amperes when the resistance is 20 ohms.

a) Find an equation of variation.
b) What is the current when the resistance is 60 ohms?

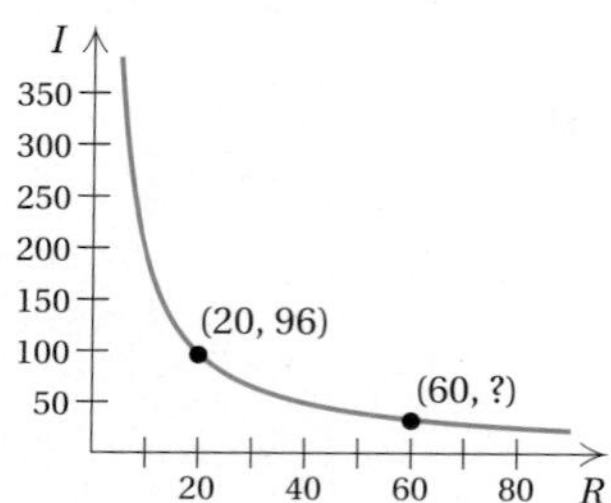

36. *Gas Volume.* The volume V of a gas varies inversely as the pressure P on it. The volume of a gas is 200 cm^3 under a pressure of 32 kg/cm^2.

a) Find an equation of variation.
b) What will be its volume under a pressure of 20 kg/cm^2?

37. *Answering Questions.* For a fixed time limit for a quiz, the number of minutes m that a student should allow for each question on a quiz is inversely proportional to the number of questions n on the quiz. A 16-question quiz means that students have 2.5 min per question.

a) Find an equation of variation.
b) How many questions would appear on a quiz in which students have 4 min per question?

38. *Pumping Time.* The time t required to empty a tank varies inversely as the rate r of pumping. A pump can empty a tank in 90 min at a rate of 1200 L/min.

a) Find an equation of variation.
b) How long will it take the pump to empty the tank at a rate of 2000 L/min?

39. *Apparent Size.* The apparent size A of an object varies inversely as the distance d of the object from the eye. A flagpole 30 ft from an observer appears to be 27.5 ft tall. How tall will the same flagpole appear to be if it is 100 ft from the eye?

40. *Driving Time.* The time t required to drive a fixed distance varies inversely as the speed r. It takes 5 hr at 55 mph to drive a fixed distance. How long would it take at 40 mph?

^D_W In Exercises 41–44, determine whether the situation represents direct variation, inverse variation, or neither. Give a reason for your answer.

41. The cost of mailing a package in the United States and the weight of the package

42. The number of hours that a student watches TV per week and the student's grade point average

43. The weight of a turkey and the cooking time

44. The number of plays that it takes to go 80 yd for a touchdown and the average gain per play

SKILL MAINTENANCE

Solve. [6.6a]

45. $\frac{x+2}{x+5} = \frac{x-4}{x-6}$

46. $\frac{x-3}{x-5} = \frac{x+5}{x+1}$

Solve. [5.7b]

47. $x^2 - 25x + 144 = 0$

48. $t^2 + 21t + 108 = 0$

49. $35x^2 + 8 = 34x$

50. $14x^2 - 19x - 3 = 0$

Calculate. [1.8d]

51. $3^7 \div 3^4 \div 3^3 \div 3$

52. $\frac{37 - 5(4-6)}{2 \cdot 6 + 8}$

53. $-5^2 + 4 \cdot 6$

54. $(-5)^2 + 4 \cdot 6$

SYNTHESIS

55. Graph the equation that corresponds to Exercise 12. Then use the TABLE feature to create a table with TblStart $= 1$ and ΔTbl $= 1$. What happens to the y-values as the x-values become larger?

56. Graph the equation that corresponds to Exercise 17. Then use the TABLE feature to create a table with TblStart $= 1$ and ΔTbl $= 1$. What happens to the y-values as the x-values become larger?

Write an equation of variation for the situation.

57. The square of the pitch P of a vibrating string varies directly as the tension t on the string.

58. In a stream, the amount S of salt carried varies directly as the sixth power of the speed V of the stream.

59. The power P in a windmill varies directly as the cube of the wind speed V.

60. The volume V of a sphere varies directly as the cube of the radius r.

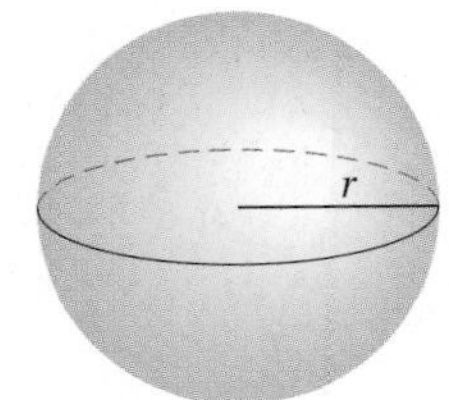

CHAPTER 7

Summary and Review

The review that follows is meant to prepare you for a chapter exam. It consists of two parts. The first part is a checklist of some of the Study Tips referred to in this and preceding chapters, as well as a list of important properties and formulas. The second part is the Review Exercises. These provide practice exercises for the exam, together with references to section objectives so you can go back and review. Before beginning, stop and look back over the skills you have obtained. What skills in mathematics do you have now that you did not have before studying this chapter?

STUDY TIPS CHECKLIST

The foundation of all your study skills is TIME!

- ☐ Have you been using the supplements for the text such as the *Student's Solutions Manual,* MyMathLab, and the Math Tutor Center?
- ☐ As you do your homework, go to class, and prepare for tests, do you keep in mind that completing many short tasks can add up to great success?
- ☐ Are you approaching your study of mathematics with a positive attitude?
- ☐ Have you looked for applications of linear equations in newspapers or magazines?
- ☐ Did you study the examples in this chapter carefully?

IMPORTANT PROPERTIES AND FORMULAS

Slope $= m = \dfrac{y_2 - y_1}{x_2 - x_1}$

Slope–Intercept Equation: $y = mx + b$

Parallel Lines: Slopes equal, y-intercepts different

Perpendicular Lines: Product of slopes $= -1$

Equation of Direct Variation: $y = kx, k > 0$

Equation of Inverse Variation: $y = \dfrac{k}{x}, k > 0$

REVIEW EXERCISES

Find the slope and the y-intercept. [7.1a]

1. $y = -9x + 46$

2. $x + y = 9$

3. $3x - 5y = 4$

Find an equation of the line with the given slope and y-intercept. [7.1a]

4. Slope $= -2.8$; y-intercept: $(0, 19)$

5. Slope $= \frac{5}{8}$; y-intercept: $\left(0, -\frac{7}{8}\right)$

Find an equation of the line containing the given point and with the given slope. [7.1b]

6. $(1, 2),\ m = 3$

7. $(-2, -5),\ m = \frac{2}{3}$

8. $(0, -4),\ m = -2$

Find an equation of the line containing the given pair of points. [7.1c]

9. $(5, 7)$ and $(-1, 1)$

10. $(2, 0)$ and $(-4, -3)$

Solve. [7.1c]

11. *Median Age of Cars.* People are driving cars for longer periods of time. The line graph below describes the *median age of cars A,* in years, for years since 1990.

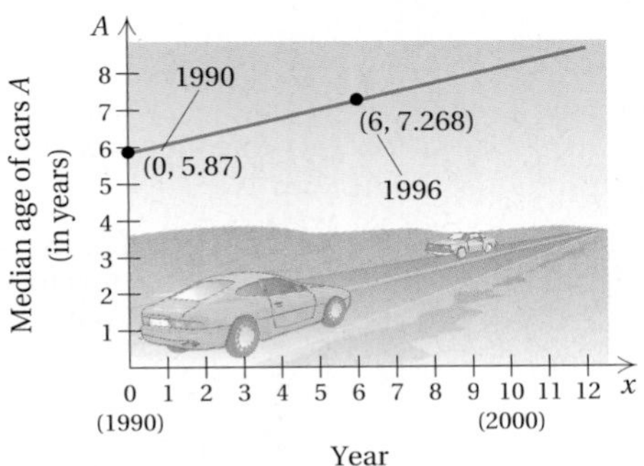

Source: The Polk Co.

a) Find an equation of the line.

b) What is the rate of change in the median age of cars with respect to time?

c) Use the equation to find the median age of cars in 2000.

12. Draw a line that has slope -1 and y-intercept $(0, 4)$. [7.2a]

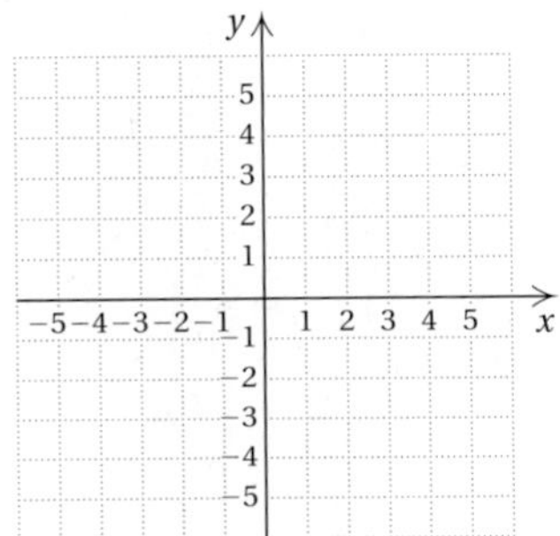

13. Draw a line that has slope $\frac{5}{3}$ and y-intercept $(0, -3)$. [7.2a]

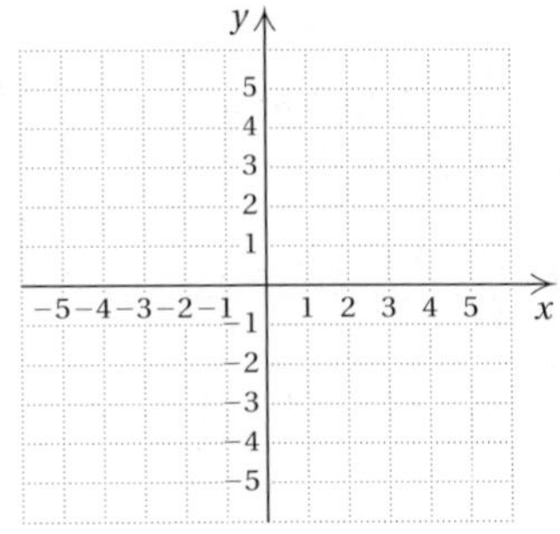

14. Graph $y = -\frac{3}{5}x + 2$ using the slope and the y-intercept. [7.2a]

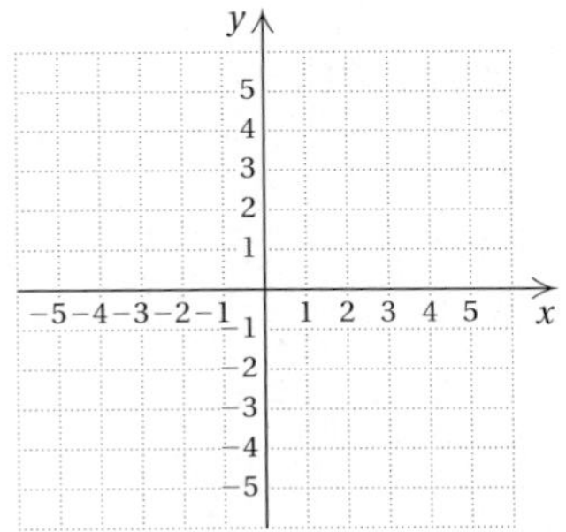

15. Graph $2y - 3x = 6$ using the slope and the y-intercept. [7.2a]

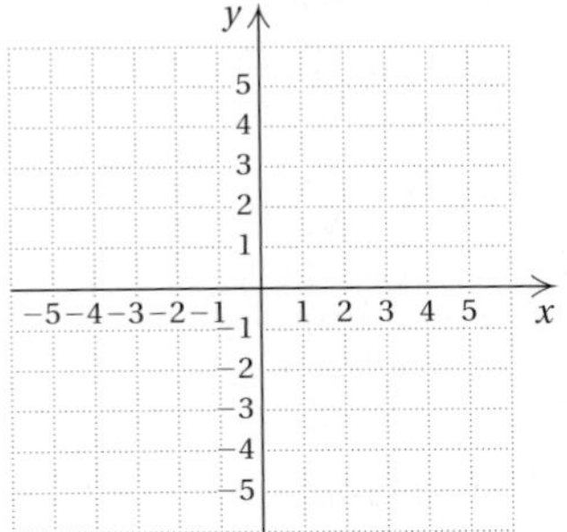

Determine whether the graphs of the equations are parallel, perpendicular, or neither. [7.3a, b]

16. $4x + y = 6,$
$4x + y = 8$

17. $2x + y = 10,$
$y = \frac{1}{2}x - 4$

18. $x + 4y = 8,$
$x = -4y - 10$

19. $3x - y = 6,$
$3x + y = 8$

Determine whether the given point is a solution of the inequality $x - 2y > 1$. [7.4a]

20. $(0, 0)$

21. $(1, 3)$

22. $(4, -1)$

Graph on a plane. [7.4b]

23. $x < y$

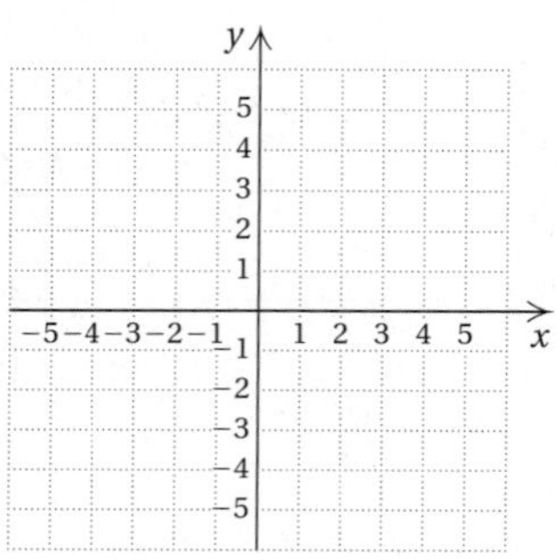

24. $x + 2y \geq 4$

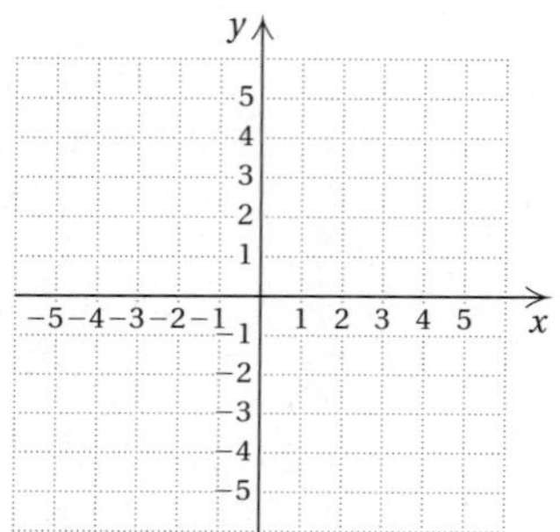

25. $x > -2$

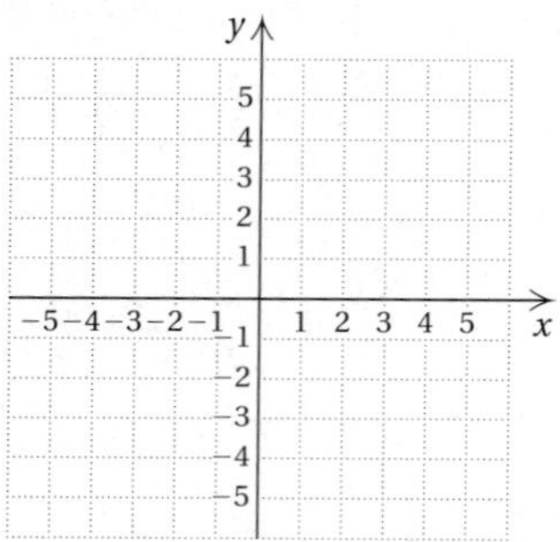

Find an equation of variation in which y varies directly as x and the following are true. Then find the value of y when $x = 20$. [7.5a]

26. $y = 12$ when $x = 4$

27. $y = 4$ when $x = 8$

28. $y = 0.4$ when $x = 0.5$

Find an equation of variation in which y varies inversely as x and the following are true. Then find the value of y when $x = 5$. [7.5c]

29. $y = 5$ when $x = 6$

30. $y = 0.5$ when $x = 2$

31. $y = 1.3$ when $x = 0.5$

Solve.

32. *Wages.* A person's paycheck P varies directly as the number H of hours worked. The pay is $165.00 for working 20 hr. Find the pay for 35 hr of work. [7.5b]

33. *Washing Time.* It takes 5 hr for 2 washing machines to wash a fixed amount of laundry. How long would it take 10 washing machines to do the same job? (The number of hours varies inversely as the number of washing machines.) [7.5d]

34. D_W Describe how you would graph $y = 0.37x + 2458$ using the slope and the y-intercept. You need not actually draw the graph. [7.2a]

35. D_W Graph $x < 1$ on both a number line and a plane, and explain the difference between the graphs. [7.4b]

SKILL MAINTENANCE

Certain objectives from four particular sections will be retested on the chapter test. The objectives are listed with the practice problems that follow.

36. *Painting Time.* Judd can paint a shed alone in 5 hr. Bud can paint the same shed in 10 hr. How long would it take both of them, working together, to paint the fence? [6.7a]

37. Compute: $13 \cdot 6 \div 3 \cdot 26 \div 13$. [1.8d]

Solve.

38. $\dfrac{x^2}{x - 4} = \dfrac{16}{x - 4}$ [6.6a]

39. $a^2 + 6a - 55 = 0$ [5.7b]

SYNTHESIS

40. In chess, the knight can move to any of the eight squares shown by a, b, c, d, e, f, g, and h below. If lines are drawn from the beginning to the end of the move, what slopes are possible for these lines? [7.1a]

CHAPTER 7

Chapter Test

1. Draw a graph of the line with slope $-\frac{3}{2}$ and y-intercept $(0, 1)$.

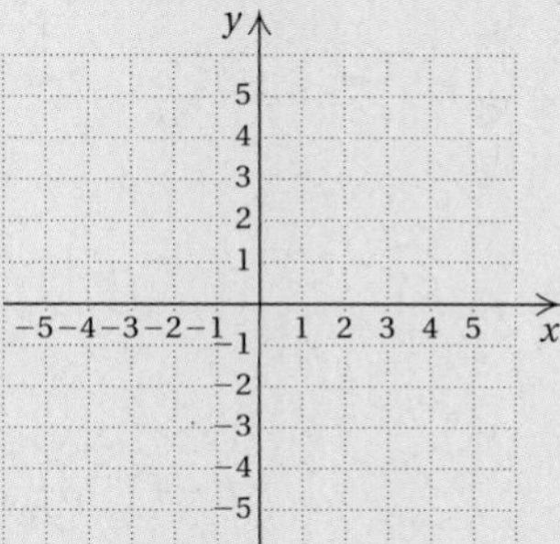

2. Graph $y = 2x - 3$ using the slope and the y-intercept.

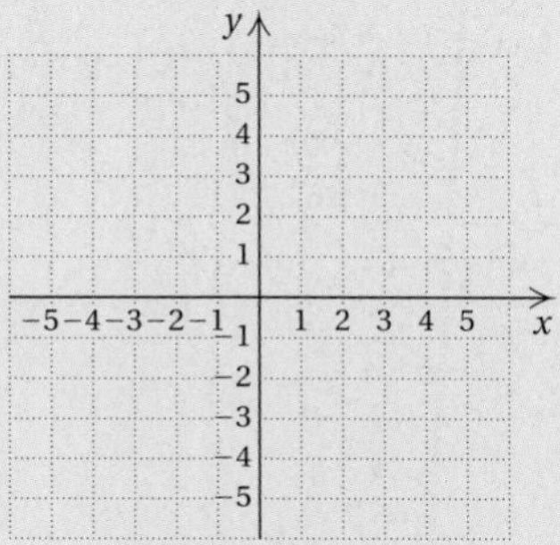

Find the slope and the y-intercept.

3. $y = 2x - \frac{1}{4}$

4. $-4x + 3y = -6$

Find an equation of the line with the given slope and y-intercept.

5. Slope $= 1.8$; y-intercept: $(0, -7)$

6. Slope $= -\frac{3}{8}$; y-intercept: $\left(0, -\frac{1}{8}\right)$

Find an equation of the line containing the given point and with the given slope.

7. $(3, 5)$, $m = 1$

8. $(-2, 0)$, $m = -3$

Find an equation of the line containing the given pair of points.

9. $(1, 1)$ and $(2, -2)$

10. $(4, -1)$ and $(-4, -3)$

11. *Cancer Research.* Increasing amounts of money are being spent each year on cancer research. The line graph at right describes the amount spent on cancer research M, in millions of dollars, for years since 1992.
    a) Find an equation of the line.
    b) What is the rate of change in the amount spent on cancer research with respect to time?
    c) Use the equation to find the amount spent on cancer research in 2000.

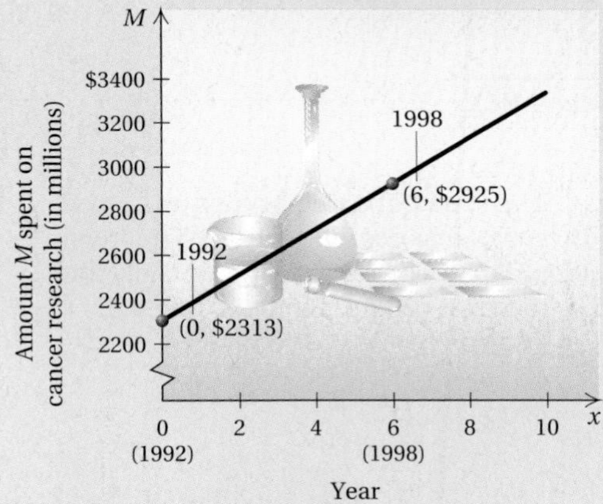

Source: The New England Journal of Medicine

Determine whether the graphs of the equations are parallel, perpendicular, or neither.

12. $2x + y = 8,$
    $2x + y = 4$

13. $2x + 5y = 2,$
    $y = 2x + 4$

14. $x + 2y = 8,$
    $-2x + y = 8$

Determine whether the given point is a solution of the inequality $3y - 2x < -2$.

15. $(0, 0)$

16. $(-4, -10)$

Graph on a plane.

17. $y > x - 1$

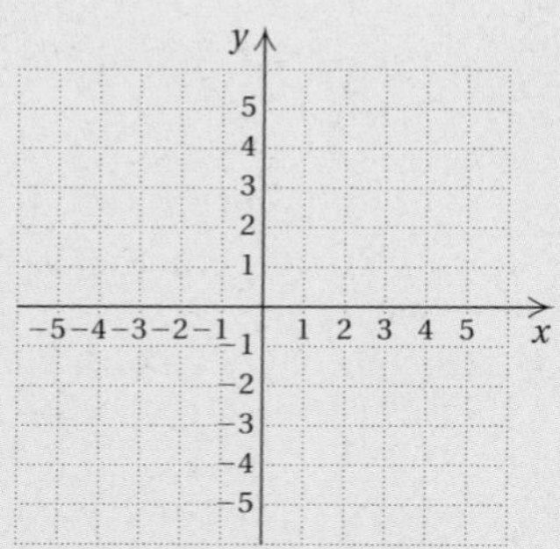

18. $2x - y \leq 4$

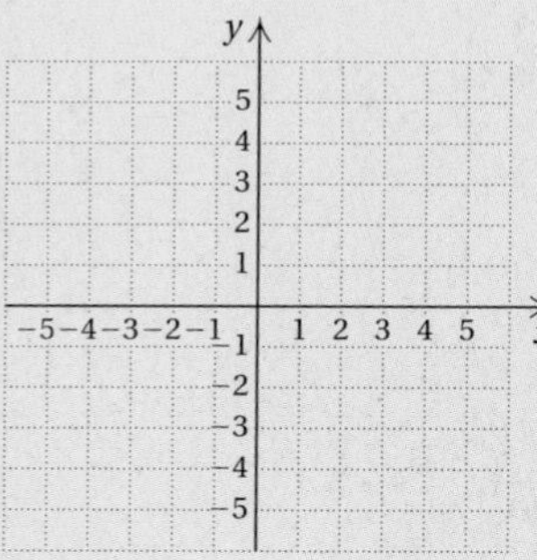

Find an equation of variation in which y varies directly as x and the following are true. Then find the value of y when $x = 25$.

19. $y = 6$ when $x = 3$

20. $y = 1.5$ when $x = 3$

Find an equation of variation in which y varies inversely as x and the following are true. Then find the value of y when $x = 100$.

21. $y = 6$ when $x = 3$

22. $y = 11$ when $x = 2$

Solve.

23. *Train Travel.* The distance d traveled by a train varies directly as the time t that it travels. The train travels 60 km in $\frac{1}{2}$ hr. How far will it travel in 2 hr?

24. *Concrete Work.* It takes 3 hr for 2 concrete mixers to mix a fixed amount of concrete. The number of hours varies inversely as the number of concrete mixers used. How long would it take 5 concrete mixers to do the same job?

SKILL MAINTENANCE

25. *Train Speeds.* The speed of a freight train is 15 mph slower than the speed of a passenger train. The freight train travels 360 mi in the same time that it takes the passenger train to travel 420 mi. Find the speed of each train.

26. Compute: $\dfrac{3^2 - 2^3}{2^2 + 3 - 12 \div 2}$.

Solve.

27. $\dfrac{x^2}{x + 10} = \dfrac{100}{x + 10}$

28. $a^2 + 3a - 28 = 0$

SYNTHESIS

29. Find the value of k such that $3x + 7y = 14$ and $ky - 7x = -3$ are perpendicular.

30. Find the slope–intercept equation of the line that contains the point $(-4, 1)$ and has the same slope as the line $2x - 3y = -6$.

CHAPTERS

1–7 Cumulative Review

1. Find the absolute value: $|3.5|$.

2. Identify the coefficient of each term of the polynomial
$$x^3 - 2x^2 + x - 1.$$

3. Identify the degree of each term and the degree of the polynomial
$$x^3 - 2x^2 + x - 1.$$

4. Classify this polynomial as a monomial, binomial, trinomial, or none of these:
$$x^3 - 2x^2 + x - 1.$$

5. *Ticket Prices.* The average price of a ticket to a major league baseball game in 2001 was \$18.90, an increase of 12.9% over the previous year. What was the average price in 2000?
Source: Major League Baseball

6. *Indy 500.* Forces of gravity affect race-car drivers dramatically at speeds over 200 mph. A driver whose head normally weighs 7.5 lb would experience a head weight of 22.5 lb at a speed over 200 mph. A driver has a head weight of 9 lb. How much would his head seem to weigh at a speed over 200 mph?
Source: Indianapolis Motor Speedway

7. *Shoveling Time.* It takes Dina 50 min to shovel 9 in. of snow from her driveway. It takes Nell 75 min to do the same job. How long would it take if they worked together?

8. *Muscle Weight.* The number of pounds of muscle in the human body varies directly as its body weight. A person who weighs 175 lb has a muscle weight of 70 lb.
   a) Define the variables and write an equation of variation that describes this situation.
   b) Mike weighs 192 lb. How much is his muscle weight?

9. *Principal Borrowed.* Money is borrowed at 6% simple interest. After 1 yr, \$2650 pays off the loan. How much was originally borrowed?

10. *Car Travel.* One car travels 105 mi in the same time that a car traveling 10 mph slower travels 75 mi. Find the speed of each car.

11. *Areas.* If the sides of a square are increased by 2 ft, the sum of the areas of the two squares is 452 ft^2. Find the length of a side of the original square.

12. *Page Numbers.* The sum of the page numbers on the facing pages of a book is 69. What are the page numbers?

13. Collect like terms: $x^2 - 3x^3 - 4x^2 + 5x^3 - 2$.

Simplify.

14. $\frac{1}{2}x - \left[\frac{3}{8}x - \left(\frac{2}{3} + \frac{1}{4}x\right) - \frac{1}{3}\right]$

15. $\left(\frac{2x^3}{3x^{-1}}\right)^{-2}$

16. $\dfrac{\frac{4}{x} - \frac{6}{x^2}}{\frac{5}{x} + \frac{7}{2x}}$

Perform the indicated operations. Simplify if possible.

17. $(5xy^2 - 6x^2y^2 - 3xy^3) - (-4xy^3 + 7xy^2 - 2x^2y^2)$

18. $(4x^4 + 6x^3 - 6x^2 - 4) + (2x^5 + 2x^4 - 4x^3 - 4x^2 + 3x - 5)$

19. $\dfrac{2y+4}{21} \cdot \dfrac{7}{y^2+4y+4}$

20. $\dfrac{x^2-9}{x^2+8x+15} \div \dfrac{x-3}{2x+10}$

21. $\dfrac{x^2}{x-4} + \dfrac{16}{4-x}$

22. $\dfrac{5x}{x^2-4} - \dfrac{-3}{2-x}$

Multiply.

23. $(2.5a+7.5)(0.4a-1.2)$

24. $(6x-5)^2$

25. $(2x^3+1)(2x^3-1)$

Factor.

26. $9a^2+52a-12$

27. $9x^2-30xy+25y^2$

28. $49x^2-1$

Solve.

29. $x-[x-(x-1)]=2$

30. $2x^2+7x=4$

31. $x^2+x-20=0$

32. $3(x-2) \le 4(x+5)$

33. $x(x-4)=0$

34. $x^2=10x$

35. $2x^2=800$

36. $t=ax+ay$, for a

37. $\dfrac{5x-2}{4} - \dfrac{4x-5}{3} = 1$

38. $\dfrac{2x}{x-3} - \dfrac{6}{x} = \dfrac{18}{x^2-3x}$

Graph on a plane.

39. $y=\dfrac{1}{2}x$

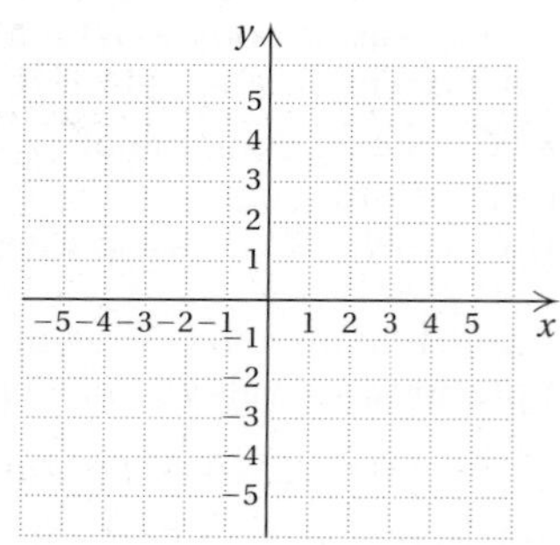

40. $3x-5y=15$

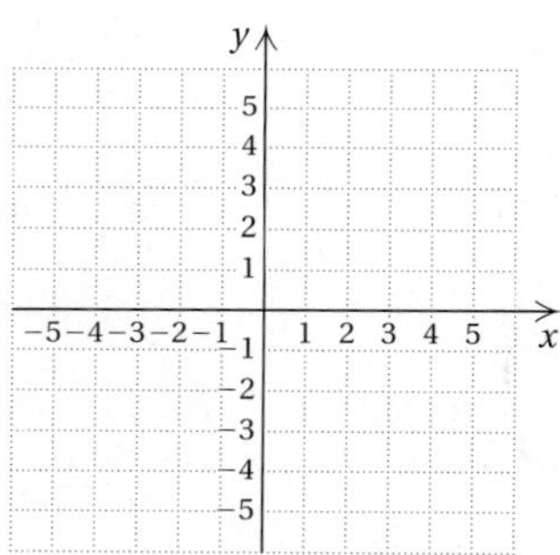

41. $y=1$

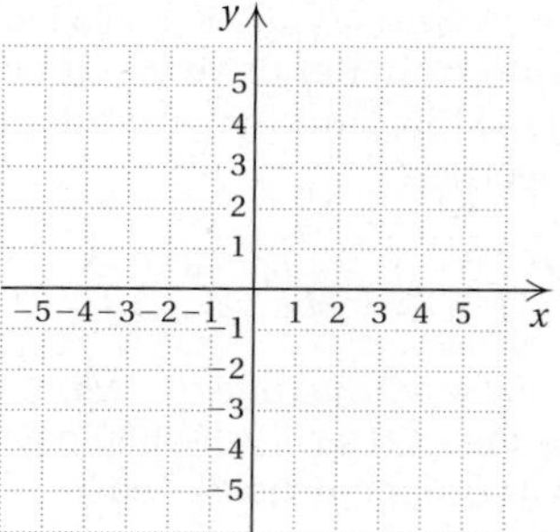

42. $x=-3$

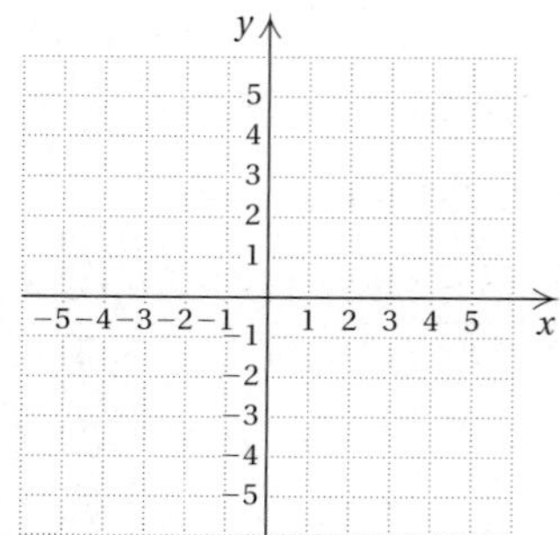

43. $y<-x-2$

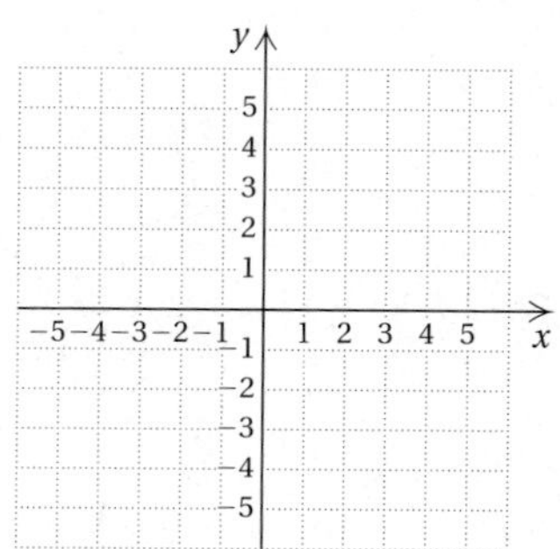

44. $x \le -3$

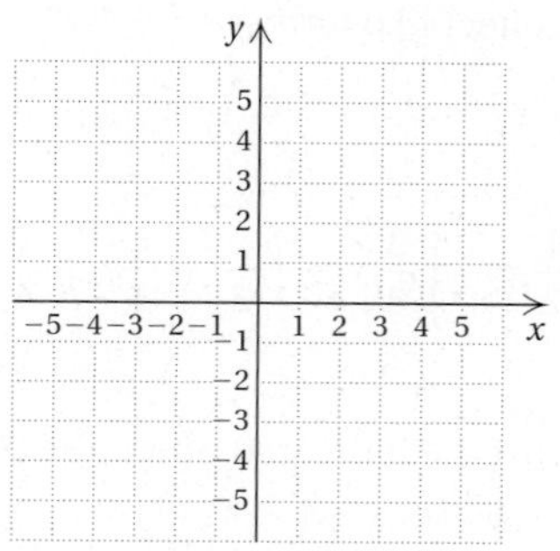

45. Find an equation of variation in which y varies directly as x and $y = 8$ when $x = 12$. Then find the value of y when $x = 125$.

46. Find an equation of variation in which y varies inversely as x and $y = 20$ when $x = 0.5$. Then find the value of y when $x = 125$.

Find the slope, if it exists, of the line containing the given pair of points.

47. $(-2, 6)$ and $(-2, -1)$

48. $(-4, 1)$ and $(3, -2)$

49. Find the slope and the y-intercept of $4x - 3y = 6$.

50. Find an equation for the line containing the point $(2, -3)$ and having slope $m = -4$.

51. Find an equation of the line containing the points $(-1, -3)$ and $(5, -2)$.

Determine whether the graphs of the equations are parallel, perpendicular, or neither.

52. $2x = 7 - 3y$,
$7 + 2x = 3y$

53. $x - y = 4$,
$y = x + 5$

Cost of a Formal Wedding. The average cost C of a formal wedding is increasing over the years, as shown in the graph at right, where x is the number of years since 1995.
Source: *Modern Bride Magazine*

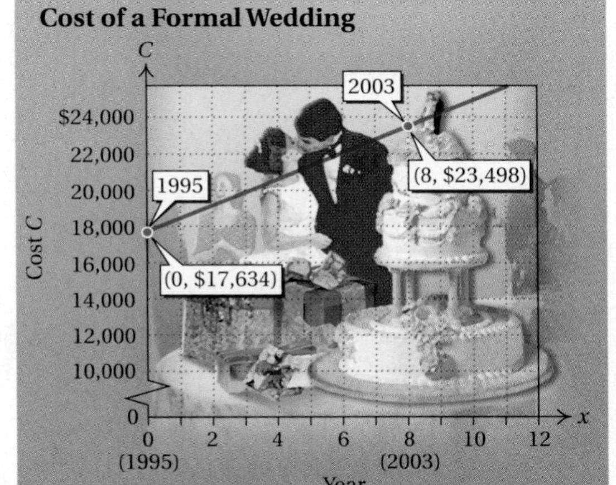

54. Find the slope of the line.

55. Find the y-intercept of the line.

56. Find the rate of change of the cost of a formal wedding with respect to time.

57. Find an equation of the line.

58. Use the equation to predict the cost of a formal wedding in 2010.

59. In what year will the average cost of a formal wedding be $50,000?

Choose the correct answer from the selections given.

60. A circle with a diameter of 1 ft will have a circumference of approximately how many feet? (This was a $32,000 question on the "Who Wants To Be a Millionaire?" television quiz show.)

a) 2 **b)** 3 **c)** 5 **d)** 6 **e)** None of these

SYNTHESIS

61. Simplify: $(x + 7)(x - 4) - (x + 8)(x - 5)$.

62. Multiply: $[4y^3 - (y^2 - 3)][4y^3 + (y^2 - 3)]$.

63. Factor: $2a^{32} - 13{,}122b^{40}$.

64. Solve: $(x - 4)(x + 7)(x - 12) = 0$.

65. Find an equation of the line that contains the point $(-3, -2)$ and is parallel to the line $2x - 3y = -12$.

66. Find all numbers for which the following complex rational expression is not defined:

$$\frac{\frac{1}{x} + x}{2 + \frac{1}{x - 3}}.$$

Systems of Equations

Gateway to Chapter 8

We now consider how the graphs of two linear equations might intersect. Such a point of intersection is a solution of what is called a system of equations. Many applications and problems involve two facts about two quantities and are easier to solve by translating to a system of two equations in two variables.

Systems of equations have applications in many fields such as psychology, sociology, business, education, engineering, and science.

Real-World Application

In most areas of the United States, gas stations offer three grades of gasoline, indicated by octane ratings on the pumps, such as 87, 89, and 93. When a tanker delivers gas, it brings only two grades of gasoline, the highest and lowest, filling two large underground tanks. If you purchase the middle grade, the pump's computer mixes the other two grades appropriately. How much 87-octane gas and 93-octane gas should be blended in order to make 18 gal of 89-octane gas?

Source: Exxon

This problem appears as Exercise 35 in Section 8.4.

CHAPTER 8

CHAPTER 8

Pretest

1. Determine whether the ordered pair $(-1, 1)$ is a solution of the system of equations [8.1a]

$$2x + y = -1,$$
$$3x - 2y = -5.$$

2. Solve this system by graphing. Show your work. [8.1b]

$$x + y = 6,$$
$$x = y + 2$$

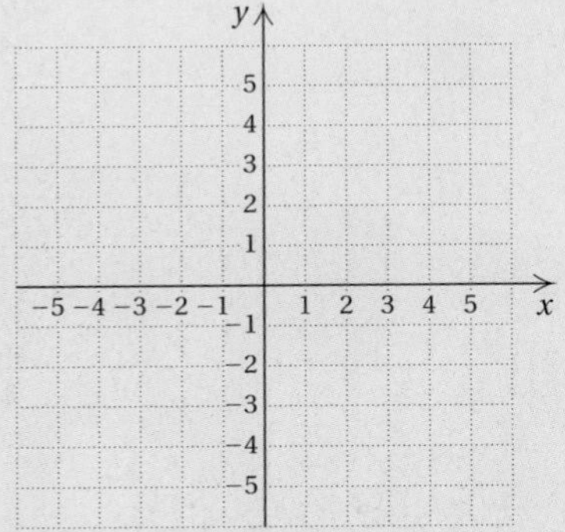

Solve by the substitution method.

3. $x + y = 7,$
$x = 2y + 1$ [8.2a]

4. $2x - 3y = 7,$
$x + y = 1$ [8.2b]

Solve by the elimination method.

5. $2x - y = 1,$
$2x + y = 2$ [8.3a]

6. $2x - 3y = -4,$
$3x - 4y = -7$ [8.3b]

7. $\frac{3}{5}x - \frac{1}{4}y = 4,$
$\frac{1}{5}x + \frac{3}{4}y = 8$ [8.3b]

8. **Ticket Sales.** An amusement park charged $11 each for adult tickets and $4 each for children's tickets. One weekend, 947 tickets were sold and the income from admissions was $6028. How many adults and how many children bought tickets? [8.4a]

9. **Basketball Scoring.** Former NBA star Wilt Chamberlain holds the all-time single-game record of 100 points. He scored a total of 64 baskets, of which some were foul shots worth 1 point each and some were field goals worth 2 points each. (There were no three-point shots at that time.) How many of each type of shot did he take? [8.4a]
Source: National Basketball Association

10. **Train Travel.** A train leaves a station and travels north at 96 mph. Two hours later, a second train leaves on a parallel track and travels north at 120 mph. When will it overtake the first train? [8.5a]

11. **Complementary Angles.** Complementary angles are angles whose sum is 90°. Two complementary angles are such that one angle is 15° more than twice the other. Find the angles. [8.4a]

8.1 SYSTEMS OF EQUATIONS IN TWO VARIABLES

Objectives

a. Determine whether an ordered pair is a solution of a system of equations.

b. Solve systems of two linear equations in two variables by graphing.

a Systems of Equations and Solutions

Many problems can be solved more easily by translating to two equations in two variables. The following is such a **system of equations:**

$$x + y = 8,$$
$$2x - y = 1.$$

SOLUTION OF A SYSTEM OF EQUATIONS

A **solution** of a system of two equations is an ordered pair that makes both equations true.

Look at the graphs shown at right. Recall that a graph of an equation is a drawing that represents its solution set. Each point on the graph corresponds to a solution of that equation. Which points (ordered pairs) are solutions of *both* equations?

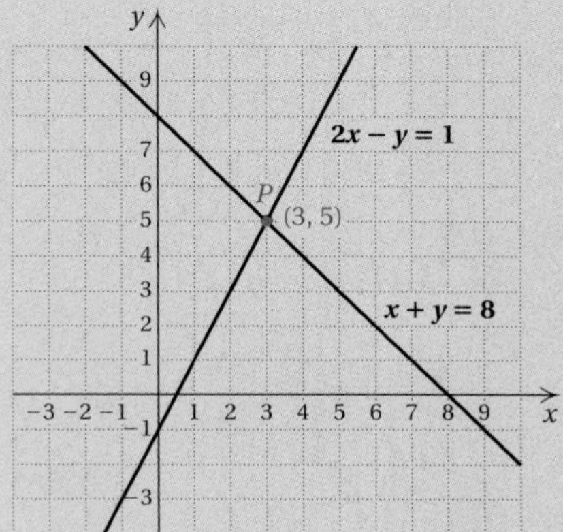

The graph shows that there is only one. It is the point P where the graphs cross. This point looks as if its coordinates are $(3, 5)$. We check to see if $(3, 5)$ is a solution of *both* equations, substituting 3 for x and 5 for y.

CHECK:

$$\begin{array}{c|c} x + y & 8 \\ \hline 3 + 5 \;?\; & 8 \\ 8 & \end{array} \quad \text{TRUE} \qquad \begin{array}{c|c} 2x - y & 1 \\ \hline 2 \cdot 3 - 5 \;?\; & 1 \\ 6 - 5 & \\ 1 & \end{array} \quad \text{TRUE}$$

There is just one solution of the system of equations. It is $(3, 5)$. In other words, $x = 3$ and $y = 5$.

EXAMPLE 1 Determine whether $(1, 2)$ is a solution of the system

$$y = x + 1,$$
$$2x + y = 4.$$

We check by substituting alphabetically 1 for x and 2 for y.

CHECK:

$$\begin{array}{c|c} y & x + 1 \\ \hline 2 \;?\; & 1 + 1 \\ & 2 \end{array} \quad \text{TRUE} \qquad \begin{array}{c|c} 2x + y & 4 \\ \hline 2 \cdot 1 + 2 \;?\; & 4 \\ 2 + 2 & \\ 4 & \end{array} \quad \text{TRUE}$$

This checks, so $(1, 2)$ is a solution of the system.

Determine whether the given ordered pair is a solution of the system of equations.

1. $(2, -3)$; $x = 2y + 8$,
$2x + y = 1$

CHECK:

$$\begin{array}{c|c} x = 2y + 8 & \\ \hline & ? \end{array}$$

$$\begin{array}{c|c} 2x + y = 1 & \\ \hline & ? \end{array}$$

2. $(20, 40)$; $a = \frac{1}{2}b$,
$b - a = 60$

CHECK:

$$\begin{array}{c|c} a = \frac{1}{2}b & \\ \hline & ? \end{array}$$

$$\begin{array}{c|c} b - a = 60 & \\ \hline & ? \end{array}$$

3. Solve this system by graphing:

$2x + y = 1$,

$x = 2y + 8$.

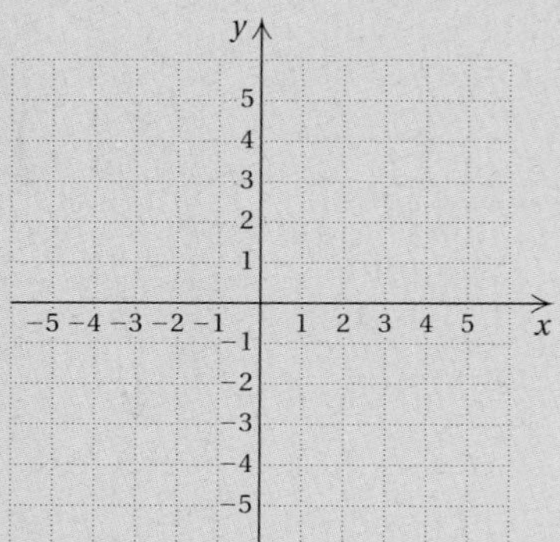

Answers on page A-38

EXAMPLE 2 Determine whether $(-3, 2)$ is a solution of the system

$p + q = -1$,

$q + 3p = 4$.

We check by substituting alphabetically -3 for p and 2 for q.

CHECK:

$$\begin{array}{c|c} p + q = -1 & \\ \hline -3 + 2 \ ? \ -1 & \\ -1 & \end{array} \quad \text{TRUE}$$

$$\begin{array}{c|c} q + 3p = 4 & \\ \hline 2 + 3(-3) \ ? \ 4 & \\ 2 - 9 & \\ -7 & \end{array} \quad \text{FALSE}$$

The point $(-3, 2)$ is not a solution of $q + 3p = 4$. Thus it is not a solution of the system.

Example 2 illustrates that an ordered pair may be a solution of one equation while *not* a solution of *both* equations. If that is the case, it is *not* a solution of the system.

Do Exercises 1 and 2.

b Graphing Systems of Equations

Recall that the **graph** of an equation is a drawing that represents its solution set. If the graph of an equation is a line, then every point on the line corresponds to an ordered pair that is a solution of the equation. If we graph a **system** of two linear equations, we graph both equations and find the coordinates of the points of intersection, if any exist.

EXAMPLE 3 Solve this system of equations by graphing:

$x + y = 6$,

$x = y + 2$.

We graph the equations using any of the methods studied in Chapters 3 and 7. Point P with coordinates $(4, 2)$ looks as if it is the solution.

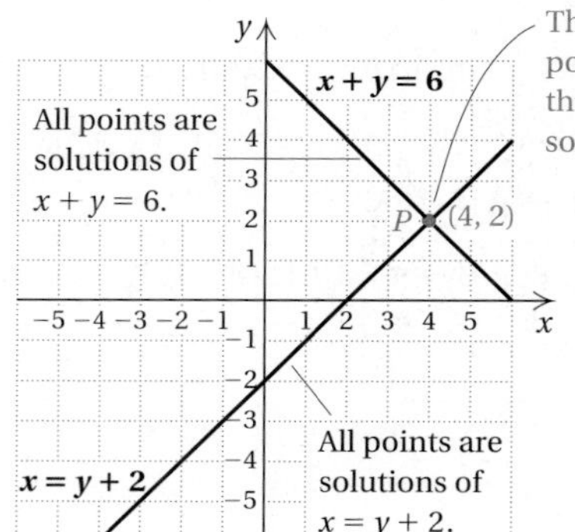

We check the pair as follows.

CHECK:

$$\begin{array}{c|c} x + y = 6 & \\ \hline 4 + 2 \ ? \ 6 & \\ 6 & \end{array} \quad \text{TRUE}$$

$$\begin{array}{c|c} x = y + 2 & \\ \hline 4 \ ? \ 2 + 2 & \\ & 4 \end{array} \quad \text{TRUE}$$

The solution is $(4, 2)$.

Do Exercise 3.

EXAMPLE 4 Solve this system of equations by graphing:

$x = 2,$

$y = -3.$

The graph of $x = 2$ is a vertical line, and the graph of $y = -3$ is a horizontal line. They intersect at the point $(2, -3)$.

The solution is $(2, -3)$.

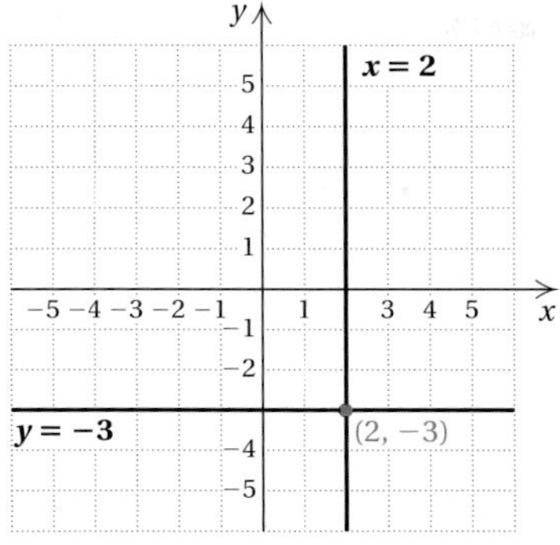

Do Exercise 4.

Sometimes the equations in a system have graphs that are parallel lines.

EXAMPLE 5 Solve this system of equations by graphing:

$y = 3x + 4,$

$y = 3x - 3.$

We graph the equations, again using any of the methods studied in Chapters 3 and 7. The lines have the same slope, 3, and different y-intercepts, $(0, 4)$ and $(0, -3)$, so they are parallel.

There is no point at which the lines cross, so the system has no solution. The solution set is the empty set, denoted $\varnothing$, or $\{\ \}$.

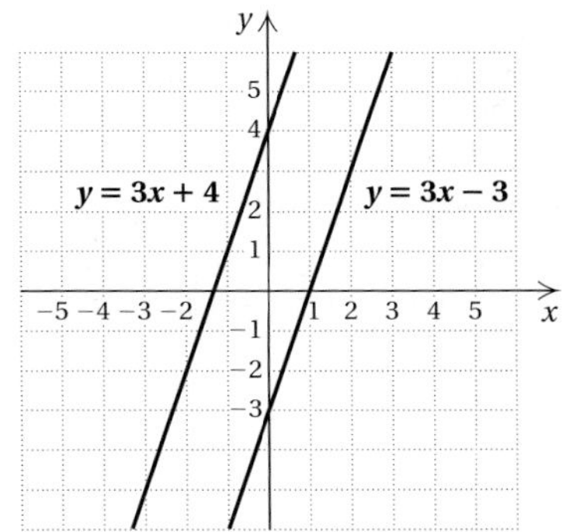

Do Exercise 5.

Sometimes the equations in a system have the same graph.

EXAMPLE 6 Solve this system of equations by graphing:

$2x + 3y = 6,$

$-8x - 12y = -24.$

We graph the equations and see that the graphs are the same. Thus any solution of one of the equations is a solution of the other. Each equation has an infinite number of solutions, some of which are indicated on the graph.

We check one such solution, $(0, 2)$: the y-intercept of each equation.

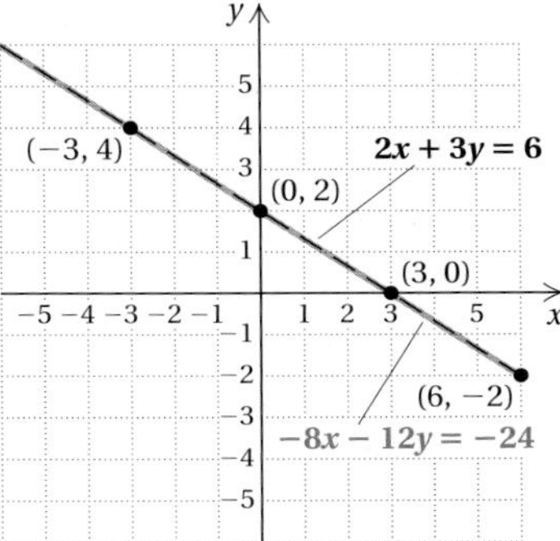

4. Solve this system by graphing:

$x = -4,$

$y = 3.$

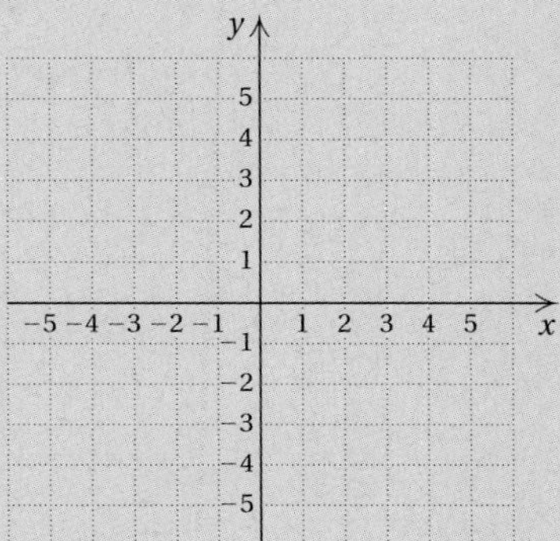

5. Solve this system by graphing:

$y + 4 = x,$

$x - y = -2.$

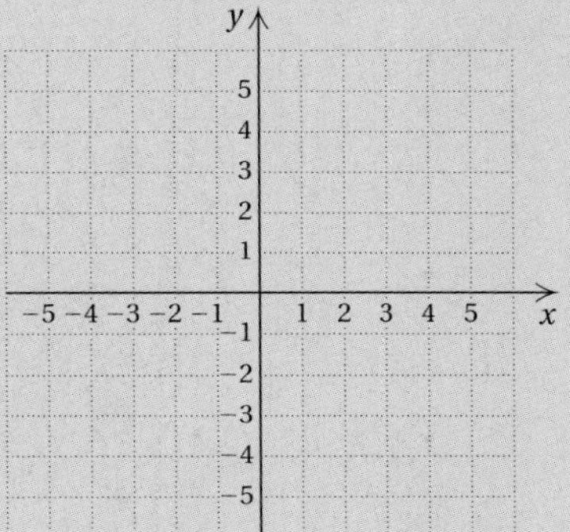

6. Solve this system by graphing:

$2x + y = 4,$

$-6x - 3y = -12.$

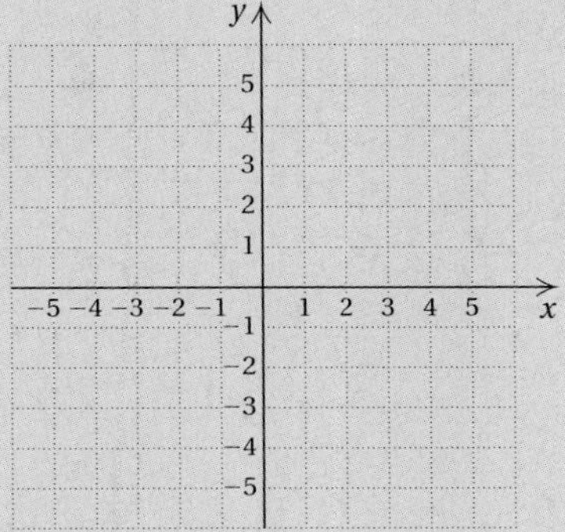

Answers on page A-38

7. a) Solve $2x - 1 = 8 - x$ algebraically.

b) Solve $2x - 1 = 8 - x$ graphically using method 1.

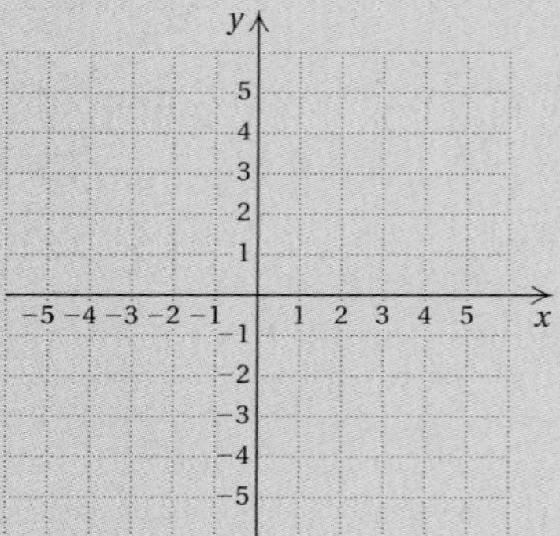

c) Compare your answers to parts (a) and (b).

Answers on page A-38

CHECK:

$$\begin{array}{c|c} 2x + 3y = 6 & \\ \hline 2(0) + 3(2) \ ? \ 6 & \\ 0 + 6 & \\ 6 & \end{array} \quad \text{TRUE} \qquad \begin{array}{c|c} -8x - 12y = -24 & \\ \hline -8(0) - 12(2) \ ? \ -24 & \\ 0 - 24 & \\ -24 & \end{array} \quad \text{TRUE}$$

We leave it to the student to check that $(-3, 4)$ is also a solution of the system. If $(0, 2)$ and $(-3, 4)$ are solutions, then all points on the line containing them are solutions. The system has an infinite number of solutions.

Do Exercise 6 on the preceding page.

When we graph a system of two equations in two variables, we obtain one of the following three results.

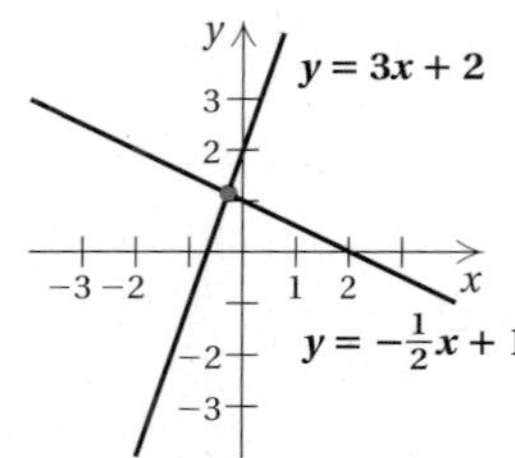

One solution.
Graphs intersect.

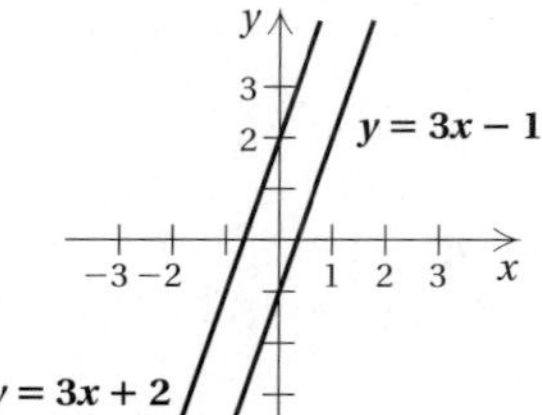

No solution.
Graphs are parallel.

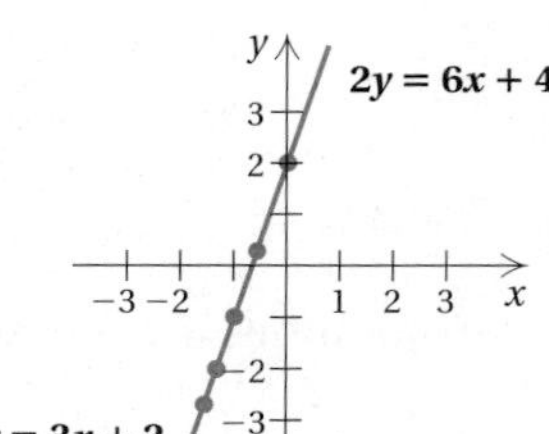

Infinitely many solutions.
Equations have the same graph.

ALGEBRAIC–GRAPHICAL CONNECTION

To bring together the concepts of Chapters 1–8, let's take an algebraic–graphical look at equation solving. Such interpretation is useful when using a graphing calculator or computer graphing software.

Consider the equation $6 - x = x - 2$. Let's solve it algebraically as we did in Chapter 2:

$$\begin{aligned} 6 - x &= x - 2 \\ 6 &= 2x - 2 && \text{Adding } x \\ 8 &= 2x && \text{Adding 2} \\ 4 &= x. && \text{Dividing by 2} \end{aligned}$$

Can we also solve the equation graphically? We can, as we see in the following two methods.

METHOD 1 Solve $6 - x = x - 2$ graphically.

We let $y = 6 - x$ and $y = x - 2$. Graphing the system of equations gives us the graph at right. The point of intersection is $(4, 2)$. Note that the x-coordinate of the intersection is 4. This value for x is also the *solution* of the equation $6 - x = x - 2$.

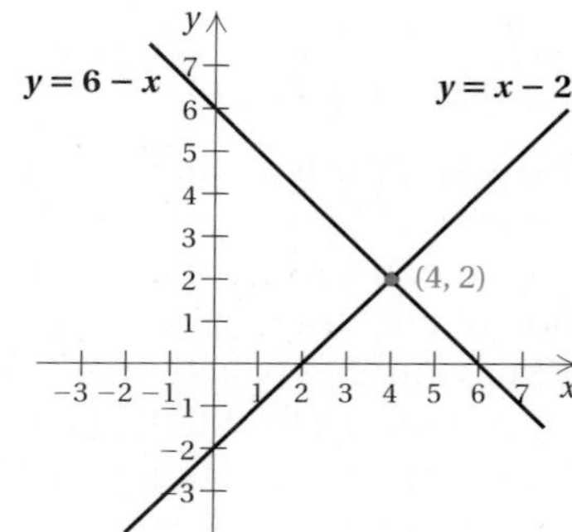

Do Exercise 7.

METHOD 2 Solve $6 - x = x - 2$ graphically.

Adding x and -6 on both sides, we obtain the form $0 = 2x - 8$. In this case, we let $y = 0$ and $y = 2x - 8$. Since $y = 0$ is the x-axis, we need only graph $y = 2x - 8$ and see where it crosses the x-axis. Note that the x-intercept of $y = 2x - 8$ is $(4, 0)$, or just 4. This x-value is also the *solution* of the equation $6 - x = x - 2$.

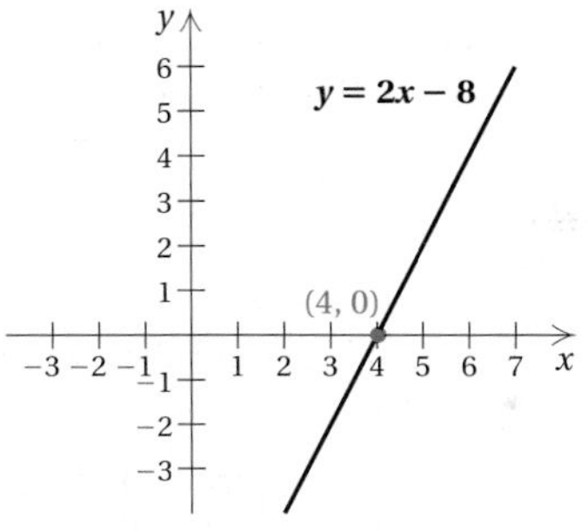

Do Exercise 8.

Let's compare the two methods. Using method 1, we graph two equations. The solution of the original equation is the x-coordinate of the point of intersection. Using method 2, we find that the solution of the original equation is the x-intercept of the graph.

8. a) Solve $2x - 1 = 8 - x$ graphically using method 2.

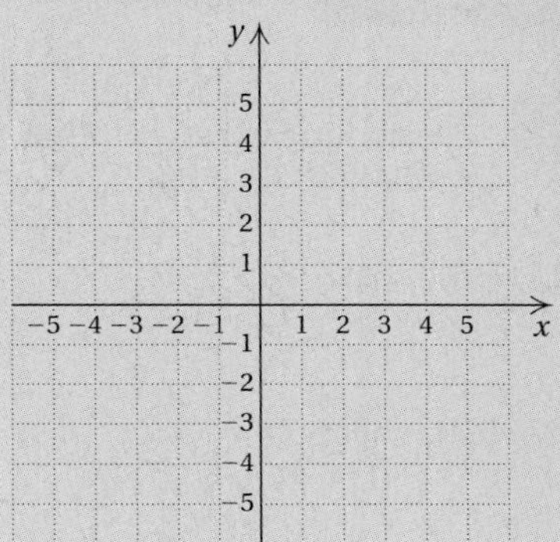

b) Compare your answers to Margin Exercises 7(a), 7(b), and 8(a).

Answers on page A-38

CALCULATOR CORNER

Solving Systems of Equations We can solve a system of two equations in two variables on a graphing calculator. Consider the system of equations in Example 3,

$$x + y = 6,$$
$$x = y + 2.$$

First, we solve the equations for y, obtaining $y = -x + 6$ and $y = x - 2$. Then we enter $y_1 = -x + 6$ and $y_2 = x - 2$ on the equation-editor screen and graph the equations. We can use the standard viewing window, $[-10, 10, -10, 10]$.

We will use the INTERSECT feature to find the coordinates of the point of intersection of the lines. To access this feature, we press [2nd] [CALC] [5]. (CALC is the second operation associated with the [TRACE] key.) The query "First curve?" appears on the graph screen. The blinking cursor is positioned on the graph of y_1. We press [ENTER] to indicate that this is the first curve involved in the intersection. Next, the query "Second curve?" appears and the blinking cursor is positioned on the graph of y_2. We press [ENTER] to indicate that this is the second curve. Now the query "Guess?" appears, so we use the [▷] and [◁] keys to move the cursor close to the point of intersection and press [ENTER]. The coordinates of the point of intersection of the graphs, $x = 4$, $y = 2$, appear at the bottom of the screen. Thus the solution of the system of equations is $(4, 2)$.

$y_1 = -x + 6,\ y_2 = x - 2$

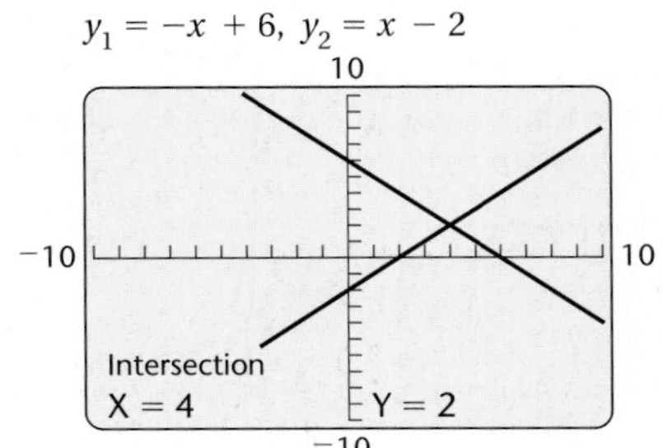

Exercises: Use a graphing calculator to solve the system of equations.

1. $x + y = 2,$
   $y = x + 4$
2. $y = x + 5,$
   $2x + y = 5$
3. $x - y = 5,$
   $y = 2x - 7$
4. $x + 3y = -1,$
   $x - y = -5$

8.1 EXERCISE SET

For Extra Help

Digital Video Tutor CD 7 Videotape 15 | InterAct Math | Math Tutor Center | MathXL | MyMathLab

a Determine whether the given ordered pair is a solution of the system of equations. Use alphabetical order of the variables.

1. $(1, 5)$; $5x - 2y = -5,$
$3x - 7y = -32$

2. $(3, 2)$; $2x + 3y = 12,$
$x - 4y = -5$

3. $(4, 2)$; $3b - 2a = -2,$
$b + 2a = 8$

4. $(6, -6)$; $t + 2s = 6,$
$t - s = -12$

5. $(15, 20)$; $3x - 2y = 5,$
$6x - 5y = -10$

6. $(-1, -5)$; $4r + s = -9,$
$3r = 2 + s$

7. $(-1, 1)$; $x = -1,$
$x - y = -2$

8. $(-3, 4)$; $2x = -y - 2,$
$y = -4$

9. $(18, 3)$; $y = \frac{1}{6}x,$
$2x - y = 33$

10. $(-3, 1)$; $y = -\frac{1}{3}x,$
$3y = -5x - 12$

b Solve the system of equations by graphing.

11. $x - y = 2,$
$x + y = 6$

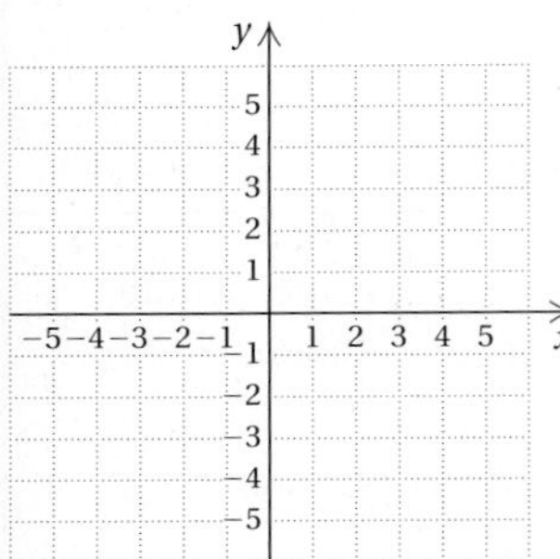

12. $x + y = 3,$
$x - y = 1$

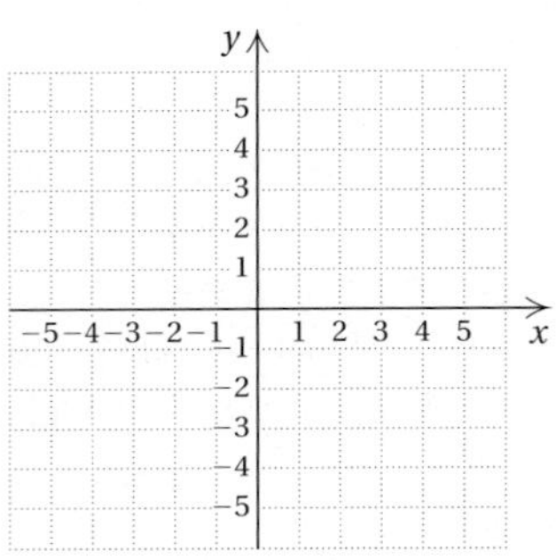

13. $8x - y = 29,$
$2x + y = 11$

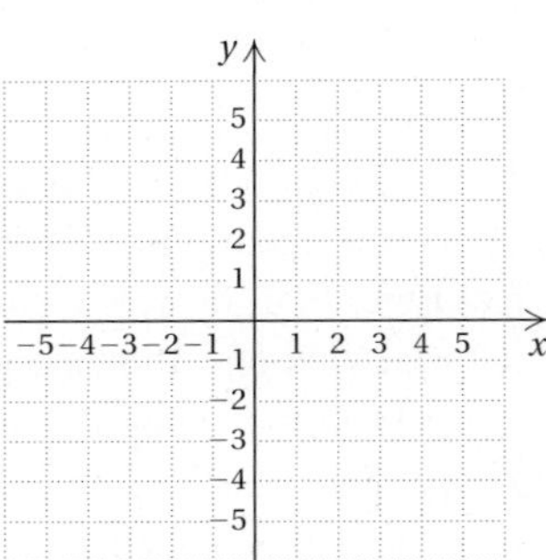

14. $4x - y = 10,$
$3x + 5y = 19$

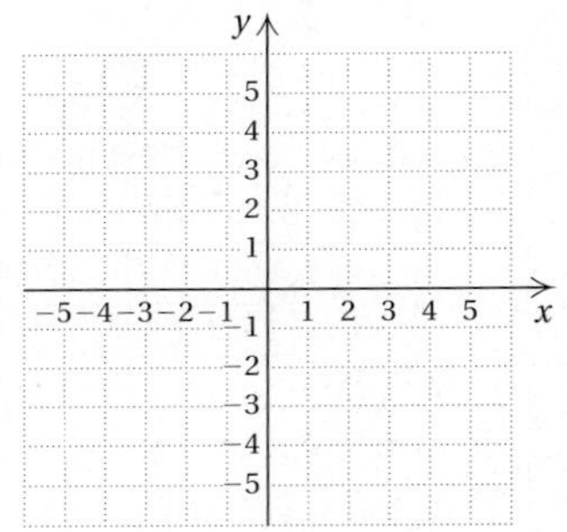

15. $u = v,$
$4u = 2v - 6$

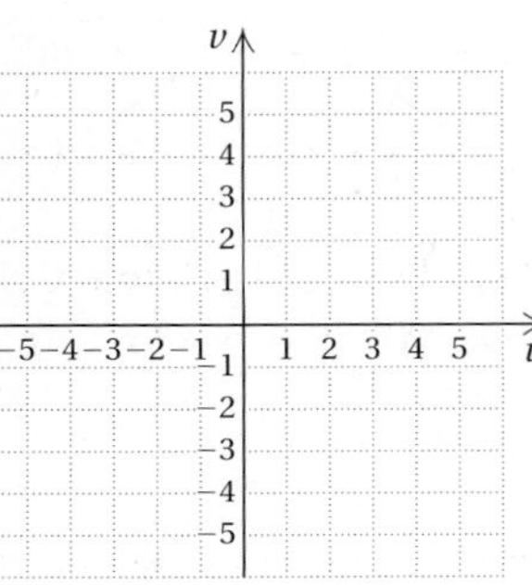

16. $x = 3y,$
$3y - 6 = 2x$

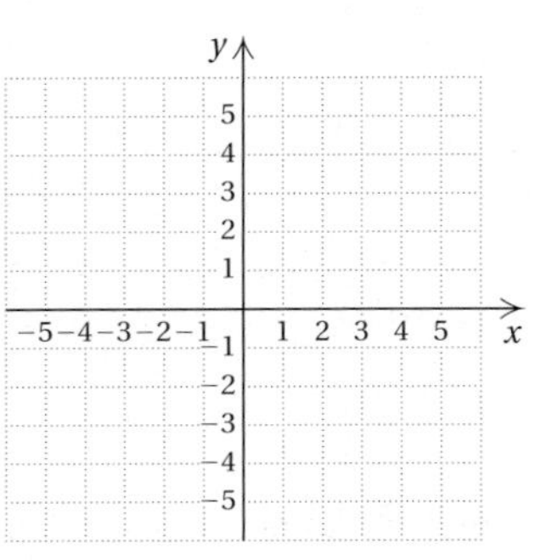

17. $x = -y,$
$x + y = 4$

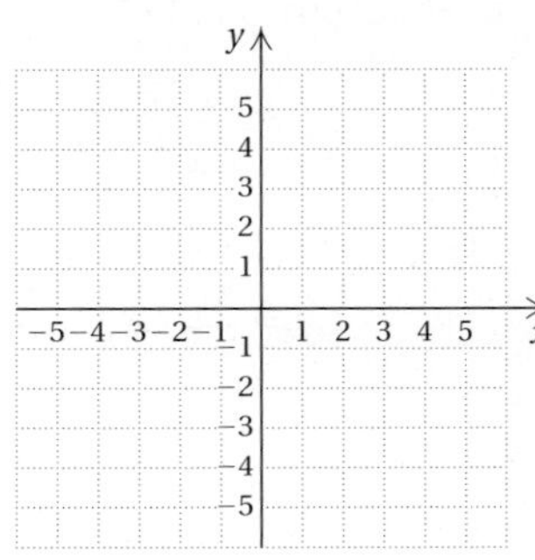

18. $-3x = 5 - y,$
$2y = 6x + 10$

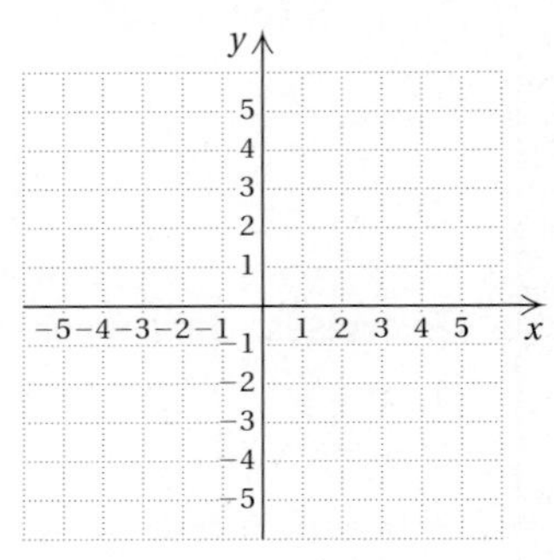

19. $a = \frac{1}{2}b + 1,$
$a - 2b = -2$

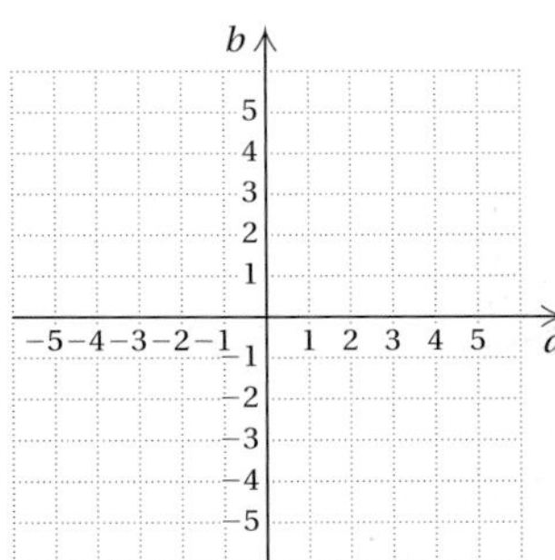

20. 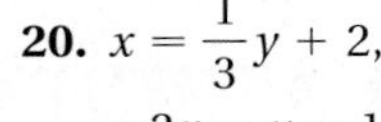
$x = \frac{1}{3}y + 2,$
$-2x - y = 1$

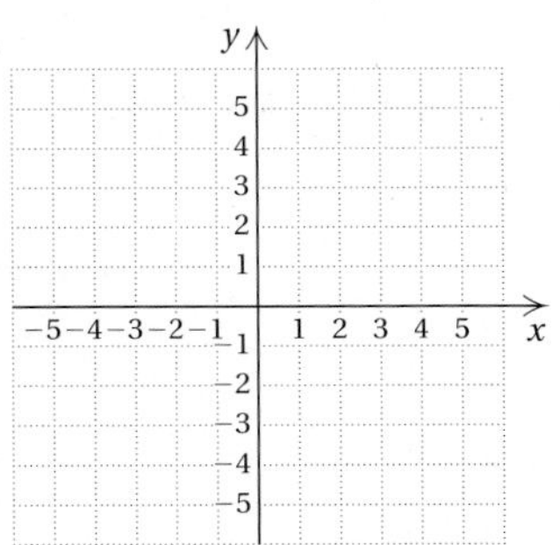

21. $y - 2x = 0,$
$y = 6x - 2$

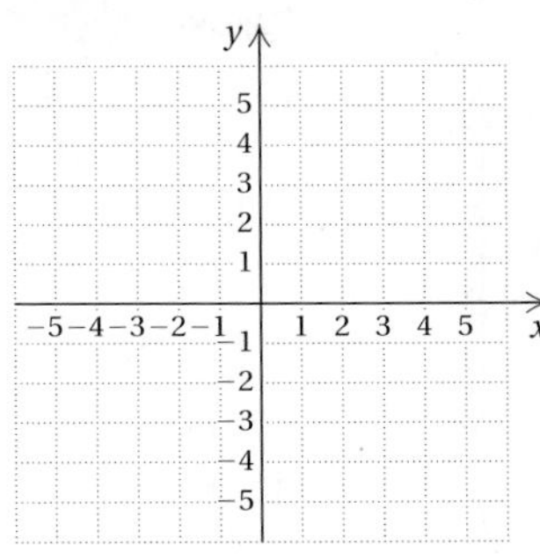

22. $y = 3x,$
$y = -3x + 2$

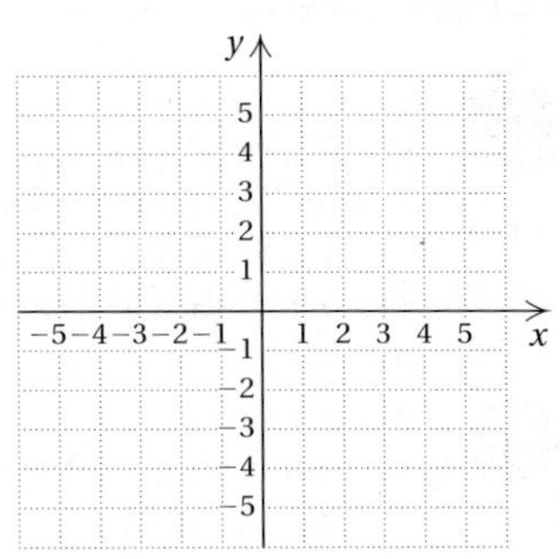

23. $x + y = 9,$
$3x + 3y = 27$

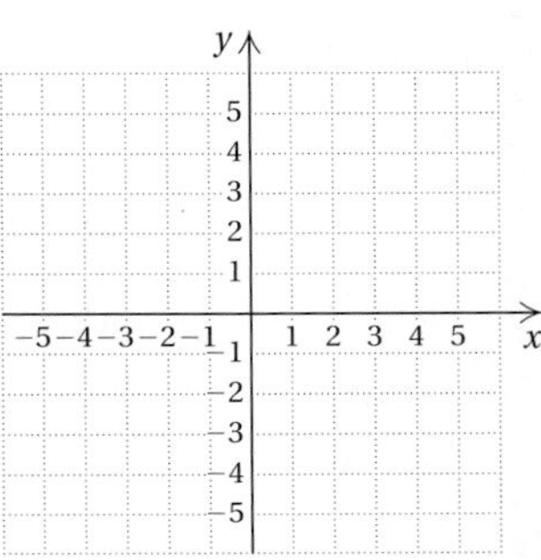

24. $x + y = 4,$
$x + y = -4$

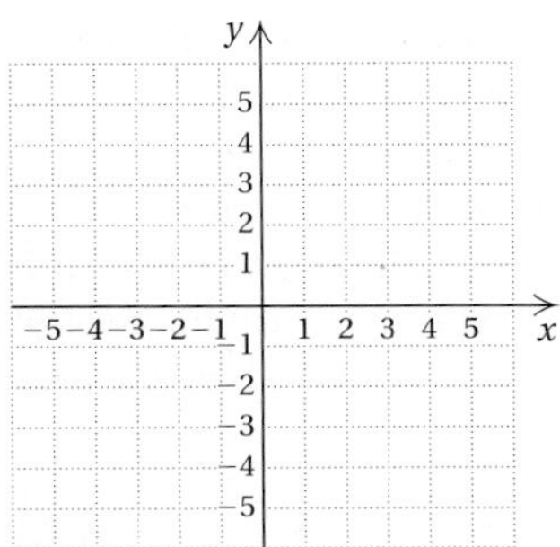

25. $x = 5,$
$y = -3$

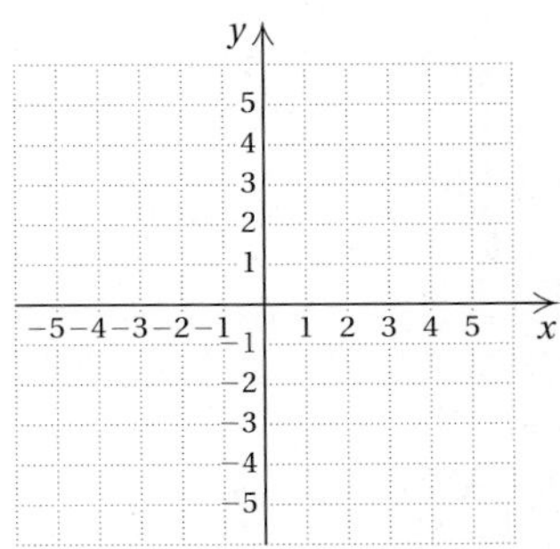

26. $y = 2,$
$y = -4$

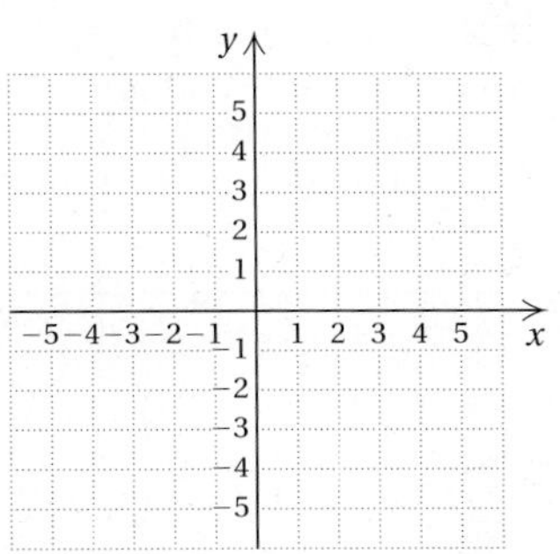

27. ^{D}W Suppose you have shown that the solution of the equation $3x - 1 = 9 - 2x$ is 2. How can this result be used to determine where the graphs of $y = 3x - 1$ and $y = 9 - 2x$ intersect?

28. ^{D}W Graph this system of equations. What happens when you try to determine a solution from the graph?

$x - 2y = 6,$
$3x + 2y = 4$

SKILL MAINTENANCE

Simplify.

29. $\frac{1}{x} - \frac{1}{x^2} + \frac{1}{x+1}$ [6.5b]

30. $\frac{3-x}{x-2} - \frac{x-7}{2-x}$ [6.5a]

31. $\frac{x+2}{x-4} - \frac{x+1}{x+4}$ [6.5a]

32. $\frac{2x^2 - x - 15}{x^2 - 9}$ [6.1c]

Classify the polynomial as a monomial, binomial, trinomial, or none of these. [4.3i]

33. $5x^2 - 3x + 7$

34. $4x^3 - 2x^2$

35. $1.8x^5$

36. $x^3 + 2x^2 - 3x + 1$

SYNTHESIS

37. The solution of the following system is $(2, -3)$. Find A and B.

$Ax - 3y = 13,$
$x - By = 8$

38. Find an equation to go with $5x + 2y = 11$ such that the solution of the system is $(3, -2)$. Answers may vary.

39. Find a system of equations with $(6, -2)$ as a solution. Answers may vary.

40.–47. Use the TABLE feature on a graphing calculator to check your answers to Exercises 11–18.

8.2 THE SUBSTITUTION METHOD

Objectives

a Solve a system of two equations in two variables by the substitution method when one of the equations has a variable alone on one side.

b Solve a system of two equations in two variables by the substitution method when neither equation has a variable alone on one side.

c Solve applied problems by translating to a system of two equations and then solving using the substitution method.

Consider the following system of equations:

$$3x + 7y = 5,$$
$$6x - 7y = 1.$$

Suppose we try to solve this system graphically. We obtain the graph shown at right.

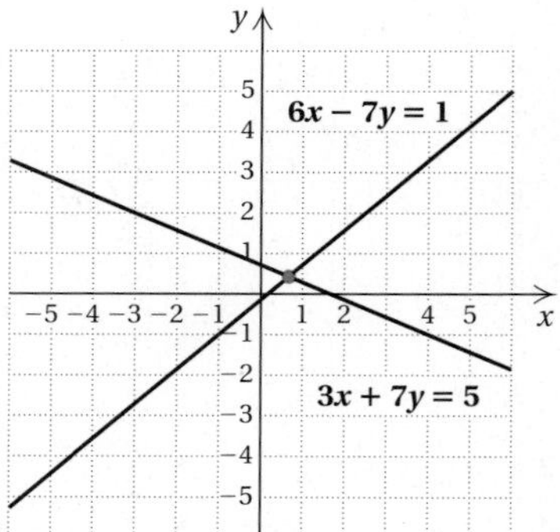

What is the solution? It is rather difficult to tell exactly. It would appear that the coordinates of the point are not integers. It turns out that the solution is $\left(\frac{2}{3}, \frac{3}{7}\right)$. We need techniques involving algebra to determine the solution exactly. Graphing helps us picture the solution of a system of equations, but solving by graphing, though practical in many applications, is not always fast or accurate in cases where solutions are not integers. We now learn other methods using algebra. Because they use algebra, they are called **algebraic.**

a Solving by the Substitution Method

One nongraphical method for solving systems is known as the **substitution method.** In Example 1, we use the substitution method to solve a system we graphed in Example 3 of Section 8.1.

EXAMPLE 1 Solve the system

$$x + y = 6, \quad (1)$$
$$x = y + 2. \quad (2)$$

Equation (2) says that x and $y + 2$ name the same thing. Thus in equation (1), we can substitute $y + 2$ for x:

$x + y = 6$ Equation (1)

$(y + 2) + y = 6.$ Substituting $y + 2$ for x

This last equation has only one variable. We solve it:

$y + 2 + y = 6$ Removing parentheses

$2y + 2 = 6$ Collecting like terms

$2y + 2 - 2 = 6 - 2$ Subtracting 2 on both sides

$2y = 4$ Simplifying

$\frac{2y}{2} = \frac{4}{2}$ Dividing by 2

$y = 2.$ Simplifying

We have found the y-value of the solution. To find the x-value, we return to the original pair of equations. Substituting into either equation will give us the x-value.

Study Tips

BEGINNING TO STUDY FOR THE FINAL EXAM (PART 1)

It is never too soon to begin to study for the final examination. Take a few minutes each week to review the highlighted information, such as formulas, properties, and procedures. Make special use of the Summary and Reviews, Chapter Tests, and Cumulative Reviews, as well as the supplements such as Interact Math Tutorial software and MathXL. The Cumulative Review/Final Examination for Chapters 1–11 is a sample final exam.

"Practice does not make perfect; practice makes permanent."

Dr. Richard Chase, former president, Wheaton College

We choose equation (2) because it has x alone on one side:

$$
\begin{aligned}
x &= y + 2 && \text{Equation (2)} \\
&= 2 + 2 && \text{Substituting 2 for } y \\
&= 4.
\end{aligned}
$$

The ordered pair $(4, 2)$ may be a solution. We check.

CHECK:

$x + y = 6$		$x = y + 2$	
$4 + 2 \stackrel{?}{=} 6$		$4 \stackrel{?}{=} 2 + 2$	
6	TRUE	4	TRUE

Since $(4, 2)$ checks, we have the solution. We could also express the answer as $x = 4$, $y = 2$.

Note in Example 1 that substituting 2 for y in equation (1) will also give us the x-value of the solution:

$$
\begin{aligned}
x + y &= 6 \\
x + 2 &= 6 \\
x &= 4.
\end{aligned}
$$

Note also that we are using alphabetical order in listing the coordinates in an ordered pair. That is, since x precedes y, we list 4 before 2 in the pair $(4, 2)$.

Do Exercise 1.

1. Solve by the substitution method. Do not graph.

$$
\begin{aligned}
x + y &= 5, \\
x &= y + 1
\end{aligned}
$$

EXAMPLE 2 Solve the system

$$
\begin{aligned}
t &= 1 - 3s, && \text{(1)} \\
s - t &= 11. && \text{(2)}
\end{aligned}
$$

We substitute $1 - 3s$ for t in equation (2):

$$
\begin{aligned}
s - t &= 11 && \text{Equation (2)} \\
s - (1 - 3s) &= 11. && \text{Substituting } 1 - 3s \text{ for } t
\end{aligned}
$$

Remember to use parentheses when you substitute.

Now we solve for s:

$$
\begin{aligned}
s - 1 + 3s &= 11 && \text{Removing parentheses} \\
4s - 1 &= 11 && \text{Collecting like terms} \\
4s &= 12 && \text{Adding 1} \\
s &= 3. && \text{Dividing by 4}
\end{aligned}
$$

Next, we substitute 3 for s in equation (1) of the original system:

$$
\begin{aligned}
t &= 1 - 3s && \text{Equation (1)} \\
&= 1 - 3 \cdot 3 && \text{Substituting 3 for } s \\
&= -8.
\end{aligned}
$$

The pair $(3, -8)$ checks and is the solution. Remember: We list the answer in alphabetical order, (s, t). That is, since s comes before t in the alphabet, 3 is listed first and -8 second.

Do Exercise 2.

2. Solve by the substitution method:

$$
\begin{aligned}
a - b &= 4, \\
b &= 2 - a.
\end{aligned}
$$

Answers on page A-38

3. Solve:

$$x - 2y = 8,$$
$$2x + y = 8.$$

Answer on page A-38

CAUTION! A solution of a system of equations in two variables is an ordered *pair* of numbers. Once you have solved for one variable, don't forget the other. A common mistake is to solve for only one variable.

b Solving for the Variable First

Sometimes neither equation of a pair has a variable alone on one side. Then we solve one equation for one of the variables and proceed as before, substituting into the *other* equation. If possible, we solve in either equation for a variable that has a coefficient of 1.

EXAMPLE 3 Solve the system

$$x - 2y = 6, \quad (1)$$
$$3x + 2y = 4. \quad (2)$$

We solve one equation for one variable. Since the coefficient of x is 1 in equation (1), it is easier to solve that equation for x:

$$x - 2y = 6 \qquad \text{Equation (1)}$$
$$x = 6 + 2y. \qquad \text{Adding } 2y \qquad (3)$$

We substitute $6 + 2y$ for x in equation (2) of the original pair and solve for y:

$$3x + 2y = 4 \qquad \text{Equation (2)}$$
$$3(6 + 2y) + 2y = 4 \qquad \text{Substituting } 6 + 2y \text{ for } x$$
$$18 + 6y + 2y = 4 \qquad \text{Removing parentheses}$$
$$18 + 8y = 4 \qquad \text{Collecting like terms}$$
$$8y = -14 \qquad \text{Subtracting 18}$$
$$y = \frac{-14}{8}, \text{ or } -\frac{7}{4}. \qquad \text{Dividing by 8}$$

To find x, we go back to either of the original equations (1) or (2) or to equation (3), which we solved for x. It is generally easier to use an equation like equation (3) where we have solved for a specific variable. We substitute $-\frac{7}{4}$ for y in equation (3) and compute x:

$$x = 6 + 2y \qquad \text{Equation (3)}$$
$$= 6 + 2\left(-\tfrac{7}{4}\right) \qquad \text{Substituting } -\tfrac{7}{4} \text{ for } y$$
$$= 6 - \tfrac{7}{2} = \tfrac{5}{2}.$$

We check the ordered pair $\left(\frac{5}{2}, -\frac{7}{4}\right)$.

CHECK:

$$\begin{array}{c|c} x - 2y & 6 \\ \hline \frac{5}{2} - 2\left(-\frac{7}{4}\right) & ?\ 6 \\ \frac{5}{2} + \frac{7}{2} & \\ \frac{12}{2} & \\ 6 & \end{array} \quad \text{TRUE}$$

$$\begin{array}{c|c} 3x + 2y & 4 \\ \hline 3 \cdot \frac{5}{2} + 2\left(-\frac{7}{4}\right) & ?\ 4 \\ \frac{15}{2} - \frac{7}{2} & \\ \frac{8}{2} & \\ 4 & \end{array} \quad \text{TRUE}$$

Since $\left(\frac{5}{2}, -\frac{7}{4}\right)$ checks, it is the solution. This solution would have been difficult to find graphically because it involves fractions.

Do Exercise 3.

c Solving Applied Problems

Now let's solve an applied problem using systems of equations and the substitution method.

EXAMPLE 4 *Standard Billboard.* A standard rectangular highway billboard has a perimeter of 124 ft. The length is 34 ft more than the width. Find the length and the width.
Source: Eller Sign Company

1. **Familiarize.** We make a drawing and label it. We let l = the length and w = the width.

2. **Translate.** The perimeter of the rectangle is given by the formula $2l + 2w$. We translate each statement, as follows.

The perimeter is 124 ft.

$$2l + 2w = 124$$

The length is 34 ft longer than the width.

$$l = 34 + w$$

We now have a system of equations:

$$2l + 2w = 124, \quad (1)$$
$$l = 34 + w. \quad (2)$$

3. **Solve.** We solve the system. To begin, we substitute $34 + w$ for l in the first equation and solve:

$2(34 + w) + 2w = 124$ Substituting $34 + w$ for l in equation (1)
$2 \cdot 34 + 2 \cdot w + 2w = 124$ Removing parentheses
$4w + 68 = 124$ Collecting like terms
$4w = 56$ Subtracting 68
$w = 14.$ Dividing by 4

We go back to one of the original equations and substitute 14 for w:

$l = 34 + w = 34 + 14 = 48.$ Substituting in equation (2)

4. **Check.** If the length is 48 ft and the width is 14 ft, then the length is 34 ft more than the width and the perimeter is 2(48 ft) + 2(14 ft), or 124 ft. Thus these dimensions check in the original problem.
5. **State.** The width is 14 ft and the length is 48 ft.

The problem in Example 4 illustrates that many problems that can be solved by translating to *one* equation in *one* variable are actually easier to solve by translating to *two* equations in *two* variables.

Do Exercise 4.

4. **Community Garden.** A rectangular community garden is to be enclosed with 92 m of fencing. In order to allow for compost storage, the garden must be 4 m longer than it is wide. Determine the dimensions of the garden.

Answer on page A-38

8.2 EXERCISE SET

For Extra Help

Digital Video Tutor CD 7 Videotape 15 | InterAct Math | Math Tutor Center | MathXL | MyMathLab

a Solve using the substitution method.

1. $x + y = 10,$
$y = x + 8$

2. $x + y = 4,$
$y = 2x + 1$

3. $y = x - 6,$
$x + y = -2$

4. $y = x + 1,$
$2x + y = 4$

5. $y = 2x - 5,$
$3y - x = 5$

6. $y = 2x + 1,$
$x + y = -2$

7. $x = -2y,$
$x + 4y = 2$

8. $r = -3s,$
$r + 4s = 10$

b Solve using the substitution method. First, solve one equation for one variable.

9. $x - y = 6,$
$x + y = -2$

10. $s + t = -4,$
$s - t = 2$

11. $y - 2x = -6,$
$2y - x = 5$

12. $x - y = 5,$
$x + 2y = 7$

13. $2x + 3y = -2,$
$2x - y = 9$

14. $x + 2y = 10,$
$3x + 4y = 8$

15. $x - y = -3,$
$2x + 3y = -6$

16. $3b + 2a = 2,$
$-2b + a = 8$

17. $r - 2s = 0,$
$4r - 3s = 15$

18. $y - 2x = 0,$
$3x + 7y = 17$

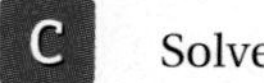

Solve.

19. *Perimeter of NBA Court.* The perimeter of an NBA-sized basketball court is 288 ft. The length is 44 ft longer than the width. Find the dimensions of the court.
Source: National Basketball Association

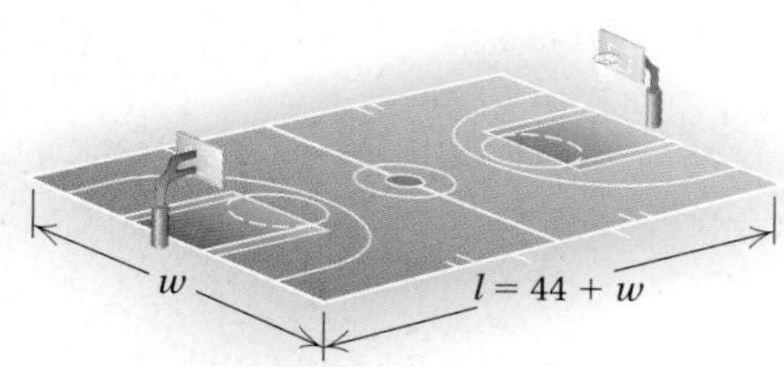

20. *Perimeter of High School Court.* The perimeter of a standard high school basketball court is 268 ft. The length is 34 ft longer than the width. Find the dimensions of the court.
Source: Indiana High School Athletic Association

21. *Two-by-Four.* The perimeter of a cross section of a "two-by-four" piece of lumber is $10\frac{1}{2}$ in. The length is twice the width. Find the actual dimensions of the cross section of a two-by-four.

22. *Rose Garden.* The perimeter of a rectangular rose garden is 400 m. The length is 3 m more than twice the width. Find the length and the width.

23. *Dimensions of Wyoming.* The state of Wyoming is a rectangle with a perimeter of 1280 mi. The width is 90 mi less than the length. Find the length and the width.

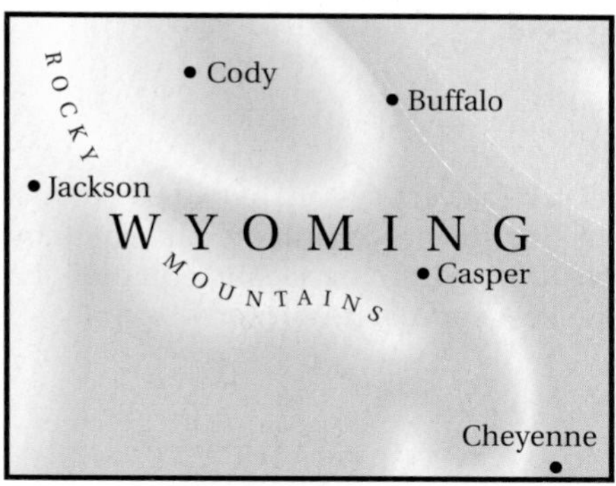

24. *Dimensions of Colorado.* The state of Colorado is roughly in the shape of a rectangle whose perimeter is 1300 mi. The width is 110 mi less than the length. Find the length and the width.

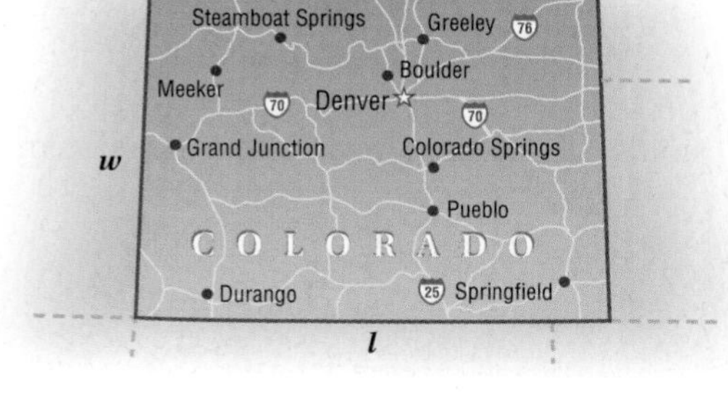

25. *Racquetball.* A regulation racquetball court should have a perimeter of 120 ft, with a length that is twice the width. Find the length and the width of a court.

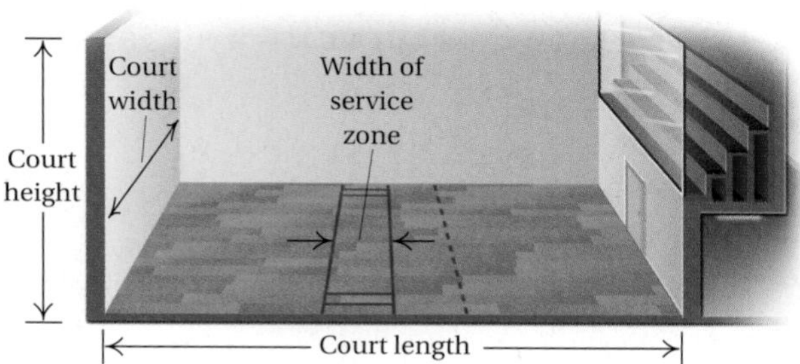

26. *Racquetball.* The height of the front wall of a standard racquetball court is four times the width of the service zone (see the figure). Together, these measurements total 25 ft. Find the height and the width.

27. *Lacrosse.* The perimeter of a lacrosse field is 340 yd. The length is 10 yd less than twice the width. Find the length and the width.

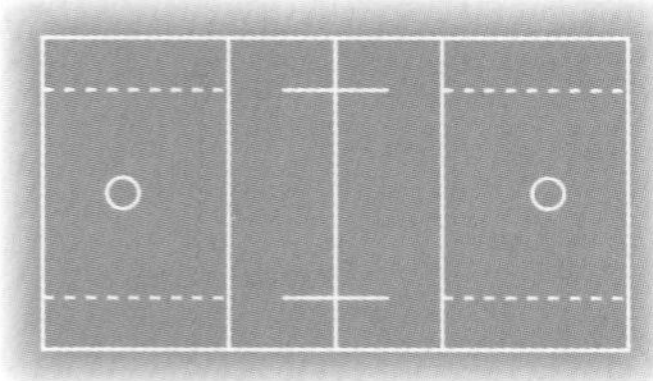

28. *Soccer.* The perimeter of a soccer field is 280 yd. The width is 5 more than half the length. Find the length and the width.

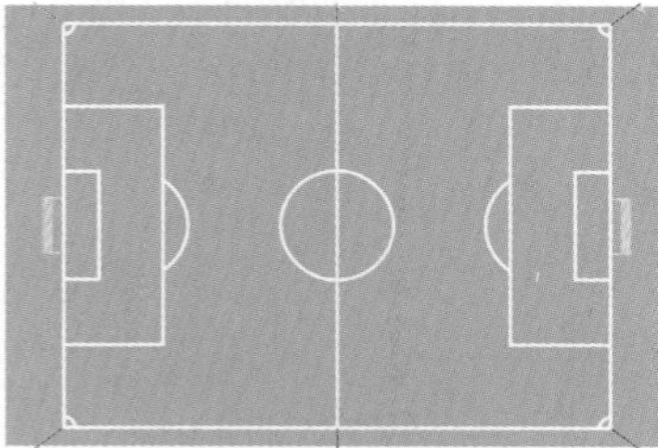

29. The sum of two numbers is 37. One number is 5 more than the other. Find the numbers.

30. The sum of two numbers is 26. One number is 12 more than the other. Find the numbers.

31. Find two numbers whose sum is 52 and whose difference is 28.

32. Find two numbers whose sum is 63 and whose difference is 5.

33. The difference between two numbers is 12. Two times the larger is five times the smaller. What are the numbers?

34. The difference between two numbers is 18. Twice the smaller number plus three times the larger is 74. What are the numbers?

35. $\mathbf{D}_{\mathbf{W}}$ Janine can tell by inspection that the system

$$x = 2y - 1,$$
$$x = 2y + 3$$

has no solution. How can she tell?

36. $\mathbf{D}_{\mathbf{W}}$ Joel solves every system of two equations (in x and y) by first solving for y in the first equation and then substituting into the second equation. Is he using the best approach? Why or why not?

SKILL MAINTENANCE

Graph. [3.3a, b]

37. $2x - 3y = 6$

38. $2x + 3y = 6$

39. $y = 2x - 5$

40. $y = 4$

Factor completely. [5.6a]

41. $6x^2 - 13x + 6$

42. $4p^2 - p - 3$

43. $4x^2 + 3x + 2$

44. $9a^2 - 25$

SYNTHESIS

Solve using a graphing calculator and its CALC-INTERSECT feature. Then solve algebraically and decide which method you prefer to use.

45. $x - y = 5,$
$x + 2y = 7$

46. $y - 2x = -6,$
$2y - x = 5$

47. $y - 2.35x = -5.97,$
$2.14y - x = 4.88$

48. $y = 1.2x - 32.7,$
$y = -0.7x + 46.15$

49. *Softball.* The perimeter of a softball diamond is two-thirds of the perimeter of a baseball diamond. Together, the two perimeters measure 200 yd. Find the distance between the bases in each sport.

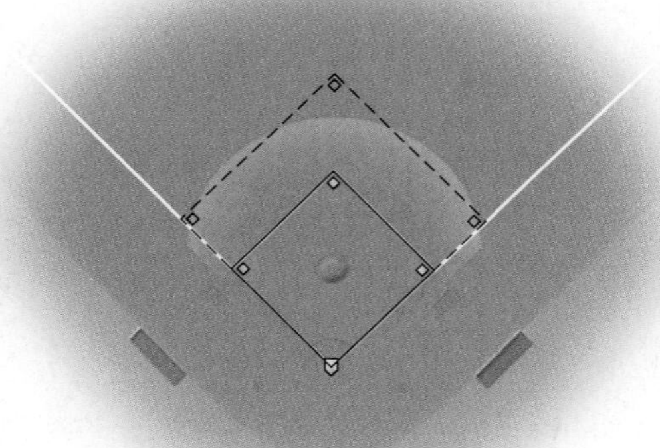

50. Write a system of two linear equations that can be solved more quickly—but still precisely—by a graphing calculator than by substitution. Time yourself using both methods to solve the system.

8.3 THE ELIMINATION METHOD

a Solving by the Elimination Method

The **elimination method** for solving systems of equations makes use of the *addition principle*. Some systems are much easier to solve using this method. For example, to solve the system

$$2x + 3y = 13, \quad (1)$$
$$4x - 3y = 17 \quad (2)$$

by substitution, we would need to first solve for a variable in one of the equations. Were we to solve equation (1) for y, we would find (after several steps) that $y = \frac{13}{3} - \frac{2}{3}x$. We could then use the expression $\frac{13}{3} - \frac{2}{3}x$ in equation (2) as a replacement for y:

$$4x - 3\left(\tfrac{13}{3} - \tfrac{2}{3}x\right) = 17.$$

As you can see, although substitution could be used to solve this system, doing so is not easy. Fortunately, another method, elimination, can be used to solve systems and, on problems like this, is simpler to use.

EXAMPLE 1 Solve the system

$$2x + 3y = 13, \quad (1)$$
$$4x - 3y = 17. \quad (2)$$

The key to the advantage of the elimination method for solving this system involves the $3y$ in one equation and the $-3y$ in the other. The terms are opposites. If we add the terms on the sides of the equations, the y-terms will add to 0, and in effect, the variable y will be eliminated.

We will use the addition principle for equations. According to equation (2), $4x - 3y$ and 17 are the same number. Thus we can use a vertical form and add $4x - 3y$ to the left side of equation (1) and 17 to the right side—in effect, adding the same number on both sides of equation (1):

$$\begin{aligned} 2x + 3y &= 13 && (1)\\ 4x - 3y &= 17 && (2)\\ \hline 6x + 0y &= 30. && \text{Adding} \end{aligned}$$

We have "eliminated" one variable. This is why we call this the **elimination method.** We now have an equation with just one variable that can be solved for x:

$$\begin{aligned} 6x &= 30\\ x &= 5. \end{aligned}$$

Next, we substitute 5 for x in either of the original equations:

$$\begin{aligned} 2x + 3y &= 13 && \text{Equation (1)}\\ 2(5) + 3y &= 13 && \text{Substituting 5 for } x\\ 10 + 3y &= 13\\ 3y &= 3\\ y &= 1. && \text{Solving for } y \end{aligned}$$

Objectives

- **a** Solve a system of two equations in two variables using the elimination method when no multiplication is necessary.
- **b** Solve a system of two equations in two variables using the elimination method when multiplication is necessary.

Solve using the elimination method.

1. $x + y = 5,$
   $2x - y = 4$

2. $-2x + y = -4,$
   $2x - 5y = 12$

Answers on page A-39

We check the ordered pair (5, 1).

CHECK:

$$\begin{array}{r|l} 2x + 3y & 13 \\ \hline 2(5) + 3(1) & 13 \\ 10 + 3 & \\ 13 & \end{array} \quad \text{TRUE}$$

$$\begin{array}{r|l} 4x - 3y & 17 \\ \hline 4(5) - 3(1) & 17 \\ 20 - 3 & \\ 17 & \end{array} \quad \text{TRUE}$$

Since (5, 1) checks, it is the solution. We can see the solution in the graph shown below.

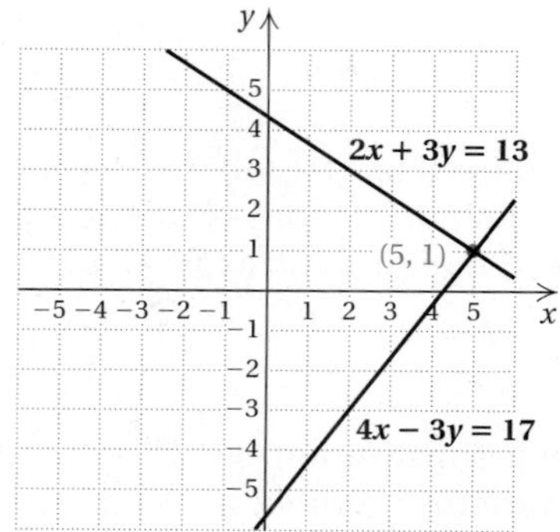

Do Exercises 1 and 2.

b Using the Multiplication Principle First

The elimination method allows us to eliminate a variable. We may need to multiply by certain numbers first, however, so that terms become opposites.

EXAMPLE 2 Solve the system

$2x + 3y = 8,$ (1)

$x + 3y = 7.$ (2)

If we add, we will not eliminate a variable. However, if the $3y$ were $-3y$ in one equation, we could eliminate y. Thus we multiply both sides of equation (2) by -1 and then add, using a vertical form:

$$\begin{aligned} 2x + 3y &= 8 && \text{Equation (1)} \\ -x - 3y &= -7 && \text{Multiplying equation (2) by } -1 \\ \hline x \quad &= 1. && \text{Adding} \end{aligned}$$

Next, we substitute 1 for x in one of the original equations:

$$\begin{aligned} x + 3y &= 7 && \text{Equation (2)} \\ 1 + 3y &= 7 && \text{Substituting 1 for } x \\ 3y &= 6 \\ y &= 2. && \text{Solving for } y \end{aligned}$$

We check the ordered pair $(1, 2)$.

CHECK:

$$\begin{array}{c|l} 2x + 3y = 8 & \\ \hline 2 \cdot 1 + 3 \cdot 2 \ ? \ 8 & \\ 2 + 6 & \\ 8 & \text{TRUE} \end{array} \qquad \begin{array}{c|l} x + 3y = 7 & \\ \hline 1 + 3 \cdot 2 \ ? \ 7 & \\ 1 + 6 & \\ 7 & \text{TRUE} \end{array}$$

Since $(1, 2)$ checks, it is the solution.

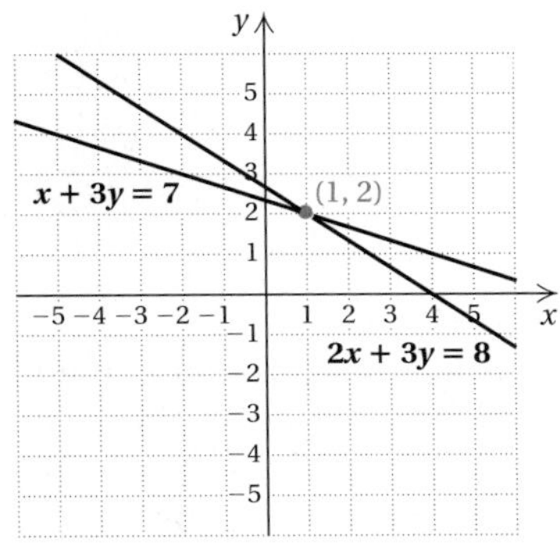

Do Exercises 3 and 4.

In Example 2, we used the multiplication principle, multiplying by -1. However, we often need to multiply by something other than -1.

EXAMPLE 3 Solve the system

$$3x + 6y = -6, \quad (1)$$
$$5x - 2y = 14. \quad (2)$$

Looking at the terms with variables, we see that if $-2y$ were $-6y$, we would have terms that are opposites. We can achieve this by multiplying both sides of equation (2) by 3. Then we add and solve for x:

$$\begin{aligned} 3x + 6y &= -6 && \text{Equation (1)} \\ 15x - 6y &= 42 && \text{Multiplying equation (2) by 3} \\ \hline 18x \quad &= 36 && \text{Adding} \\ x &= 2. && \text{Solving for } x \end{aligned}$$

Next, we substitute 2 for x in either of the original equations. We choose the first:

$$\begin{aligned} 3x + 6y &= -6 \\ 3 \cdot 2 + 6y &= -6 && \text{Substituting} \\ 6 + 6y &= -6 \\ 6y &= -12 \\ y &= -2. && \text{Solving for } y \end{aligned}$$

We check the ordered pair $(2, -2)$.

CHECK:

$$\begin{array}{c|l} 3x + 6y = -6 & \\ \hline 3 \cdot 2 + 6 \cdot (-2) \ ? \ -6 & \\ 6 + (-12) & \\ -6 & \text{TRUE} \end{array} \qquad \begin{array}{c|l} 5x - 2y = 14 & \\ \hline 5 \cdot 2 - 2 \cdot (-2) \ ? \ 14 & \\ 10 - (-4) & \\ 14 & \text{TRUE} \end{array}$$

3. Solve. Multiply one equation by -1 first.

$$5x + 3y = 17,$$
$$5x - 2y = -3$$

4. Solve the system

$$3x - 2y = -30,$$
$$5x - 2y = -46.$$

Answers on page A-39

Solve the system.

5. $4a + 7b = 11,$
$2a + 3b = 5$

6. $3x - 8y = 2,$
$5x + 2y = -12$

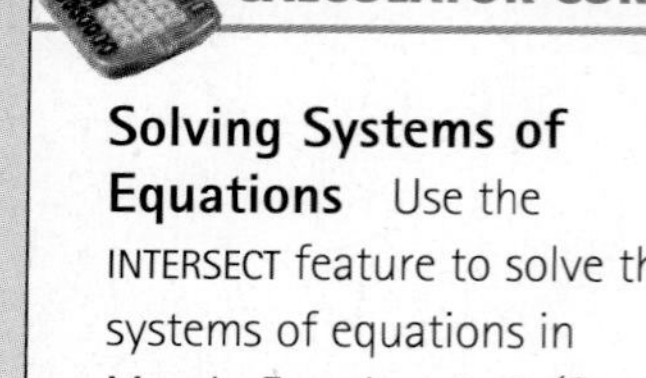

CALCULATOR CORNER

Solving Systems of Equations Use the INTERSECT feature to solve the systems of equations in Margin Exercises 1–6. (See the Calculator Corner on p. 595 for the procedure.)

Answers on page A-39

CAUTION!

Solving a *system* of equations in two variables requires finding an ordered *pair* of numbers. Once you have solved for one variable, don't forget the other, and remember to list the ordered-pair solution using alphabetical order.

Since $(2, -2)$ checks, it is the solution.

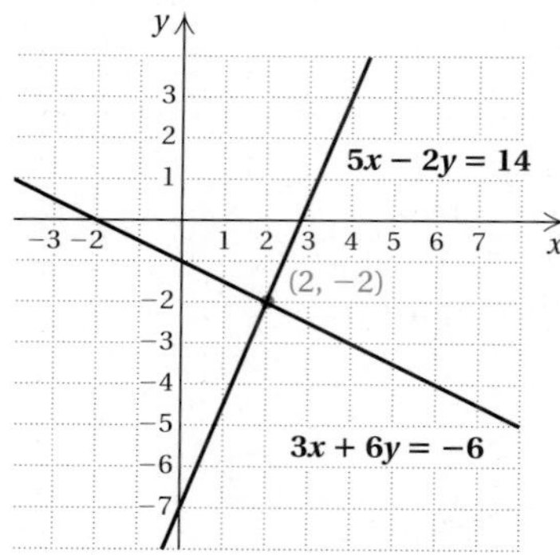

Do Exercises 5 and 6.

Part of the strategy in using the elimination method is making a decision about which variable to eliminate. So long as the algebra has been carried out correctly, the solution can be found by eliminating *either* variable. We multiply so that terms involving the variable to be eliminated are opposites. It is helpful to first get each equation in a form equivalent to $Ax + By = C$.

EXAMPLE 4 Solve the system

$$3y + 1 + 2x = 0, \quad (1)$$
$$5x = 7 - 4y. \quad (2)$$

We first rewrite each equation in a form equivalent to $Ax + By = C$:

$$2x + 3y = -1, \quad (1) \qquad \text{Subtracting 1 on both sides and rearranging terms}$$
$$5x + 4y = 7. \quad (2) \qquad \text{Adding } 4y \text{ on both sides}$$

We decide to eliminate the x-term. We do this by multiplying both sides of equation (1) by 5 and both sides of equation (2) by -2. Then we add and solve for y:

$$10x + 15y = -5 \qquad \text{Multiplying both sides of equation (1) by 5}$$
$$-10x - 8y = -14 \qquad \text{Multiplying both sides of equation (2) by } -2$$
$$7y = -19 \qquad \text{Adding}$$
$$y = \frac{-19}{7}, \text{ or } -\frac{19}{7}. \qquad \text{Solving for } y$$

Next, we substitute $-\frac{19}{7}$ for y in one of the original equations:

$$2x + 3y = -1 \qquad \text{Equation (1)}$$
$$2x + 3\left(-\tfrac{19}{7}\right) = -1 \qquad \text{Substituting } -\tfrac{19}{7} \text{ for } y$$
$$2x - \tfrac{57}{7} = -1$$
$$2x = -1 + \tfrac{57}{7}$$
$$2x = -\tfrac{7}{7} + \tfrac{57}{7}$$
$$2x = \tfrac{50}{7}$$
$$x = \tfrac{50}{7} \cdot \tfrac{1}{2}, \text{ or } \tfrac{25}{7}. \qquad \text{Solving for } x$$

We check the ordered pair $\left(\frac{25}{7}, -\frac{19}{7}\right)$.

CHECK:

$$\begin{array}{c|c} 3y + 1 + 2x = 0 & \\ \hline 3\left(-\frac{19}{7}\right) + 1 + 2\left(\frac{25}{7}\right) \; ? \; 0 & \\ -\frac{57}{7} + \frac{7}{7} + \frac{50}{7} & \\ 0 & \text{TRUE} \end{array}$$

$$\begin{array}{c|c} 5x = 7 - 4y & \\ \hline 5\left(\frac{25}{7}\right) \; ? \; 7 - 4\left(-\frac{19}{7}\right) & \\ \frac{125}{7} & \frac{49}{7} + \frac{76}{7} \\ & \frac{125}{7} \quad \text{TRUE} \end{array}$$

The solution is $\left(\frac{25}{7}, -\frac{19}{7}\right)$.

Do Exercise 7.

Let's consider a system with no solution and see what happens when we apply the elimination method.

EXAMPLE 5 Solve the system

$$y - 3x = 2, \quad (1)$$
$$y - 3x = 1. \quad (2)$$

We multiply both sides of equation (2) by -1 and then add:

$$\begin{aligned} y - 3x &= 2 \\ -y + 3x &= -1 \quad \text{Multiplying by } -1 \\ \hline 0 &= 1. \quad \text{Adding} \end{aligned}$$

We obtain a false equation, $0 = 1$, so there is *no solution*. (See Section 2.3c.) The slope–intercept forms of these equations are

$$y = 3x + 2,$$
$$y = 3x + 1.$$

The slopes are the same and the y-intercepts are different. Thus the lines are parallel. They do not intersect.

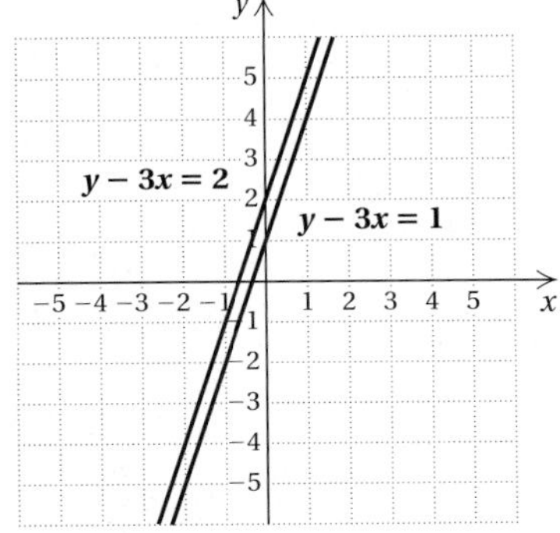

Do Exercise 8.

7. Solve the system

$$3x = 5 + 2y,$$
$$2x + 3y - 1 = 0.$$

8. Solve the system

$$2x + y = 15,$$
$$4x + 2y = 23.$$

Answers on page A-39

9. Solve the system

$$5x - 2y = 3,$$
$$-15x + 6y = -9.$$

CALCULATOR CORNER

Solving Systems of Equations

1. Consider the system of equations in Example 5. In order to enter these equations on a graphing calculator, we must first solve each for y. What happens when we do this? What does this indicate about the nature of the solutions of the system of equations?
2. Consider the system of equations in Example 6. In order to enter these equations on a graphing calculator, we must first solve each for y. What happens when we do this? What does this indicate about the nature of the solutions of the system of equations?

Answer on page A-39

Sometimes there is an infinite number of solutions. Let's look at a system that we graphed in Example 6 of Section 8.1.

EXAMPLE 6 Solve the system

$$2x + 3y = 6, \qquad (1)$$
$$-8x - 12y = -24. \qquad (2)$$

We multiply both sides of equation (1) by 4 and then add the two equations:

$$8x + 12y = 24 \qquad \text{Multiplying by 4}$$
$$-8x - 12y = -24$$
$$0 = 0. \qquad \text{Adding}$$

We have eliminated both variables, and what remains, $0 = 0$, is an equation easily seen to be true. If this happens when we use the elimination method, we have an infinite number of solutions. (See Section 2.3c.)

Do Exercise 9.

When decimals or fractions appear, we first multiply to clear them. Then we proceed as before.

EXAMPLE 7 Solve the system

$$\frac{1}{3}x + \frac{1}{2}y = -\frac{1}{6}, \qquad (1)$$
$$\frac{1}{2}x + \frac{2}{5}y = \frac{7}{10}. \qquad (2)$$

The number 6 is a multiple of all the denominators of equation (1). The number 10 is a multiple of all the denominators of equation (2). We multiply both sides of equation (1) by 6 and both sides of equation (2) by 10:

$$6\left(\frac{1}{3}x + \frac{1}{2}y\right) = 6\left(-\frac{1}{6}\right) \qquad 10\left(\frac{1}{2}x + \frac{2}{5}y\right) = 10\left(\frac{7}{10}\right)$$
$$6 \cdot \frac{1}{3}x + 6 \cdot \frac{1}{2}y = -1 \qquad 10 \cdot \frac{1}{2}x + 10 \cdot \frac{2}{5}y = 7$$
$$2x + 3y = -1; \qquad 5x + 4y = 7.$$

The resulting system is

$$2x + 3y = -1,$$
$$5x + 4y = 7.$$

As we saw in Example 4, the solution of this system is $\left(\frac{25}{7}, -\frac{19}{7}\right)$.

Do Exercises 10 and 11 on the following page.

The following is a summary that compares the graphical, substitution, and elimination methods for solving systems of equations.

METHOD	STRENGTHS	WEAKNESSES
Graphical	Can "see" solution.	Inexact when solution involves numbers that are not integers or are very large and off the graph.
Substitution	Works well when solutions are not integers. Easy to use when a variable is alone on one side.	Introduces extensive computations with fractions for more complicated systems where coefficients are not 1 or −1. Cannot "see" solution.
Elimination	Works well when solutions are not integers, when coefficients are not 1 or −1, and when coefficients involve decimals or fractions.	Cannot "see" solution.

When deciding which method to use, consider the preceding chart and directions from your instructor. The situation is like having a piece of wood to cut and three saws with which to cut it. The saw you use depends on the type of wood, the type of cut you are making, and how you want the wood to turn out.

Solve the system.

10. $\frac{1}{2}x + \frac{3}{10}y = \frac{1}{5},$
$\frac{3}{5}x + y = -\frac{2}{5}$

11. $3.3x + 6.6y = -6.6,$
$0.1x - 0.04y = 0.28$

Answers on page A-39

Study Tips

THE FOURTEEN BEST JOBS: HOW MATH STACKS UP

Although this does not qualify as a Study Tip, you can use the information to motivate your study of mathematics. The book *Jobs Related Almanac* by Les Krantz lists 250 jobs, ranked from best to worst, according to six criteria: income, stress, physical demands, potential growth, job security, and work environment.

We list the Top 14 best jobs here and note whether math is an important aspect or requirement of the job. Note that math is significant in 11 of the 14 jobs and has at least some use in all the jobs!

	JOB	MATH EMPHASIS	MID-LEVEL SALARY
1.	Financial planner	Yes	$107,000
2.	Web site manager	Yes	68,000
3.	Computer systems analyst	Yes	54,000
4.	Actuary	Yes	71,000
4.	Computer programmer (tie)	Yes	57,000
6.	Software engineer	Yes	53,000
7.	Meteorologist	Some	57,000
8.	Biologist	Some	53,000
9.	Astronomer	Yes	74,000
10.	Paralegal assistant	Some	37,000
11.	Statistician	Yes	55,000
12.	Hospital administrator	Yes	64,000
13.	Dietician	Yes	39,000
14.	Mathematician	Yes	45,000

Source: Les Krantz, *Jobs Related Almanac.* New York: St. Martin's Press, 2000

8.3 EXERCISE SET

For Extra Help

Digital Video Tutor CD 7 Videotape 15 | InterAct Math | Math Tutor Center | MathXL | MyMathLab

a Solve using the elimination method.

1. $x - y = 7,$
$x + y = 5$

2. $x + y = 11,$
$x - y = 7$

3. $x + y = 8,$
$-x + 2y = 7$

4. $x + y = 6,$
$-x + 3y = -2$

5. $5x - y = 5,$
$3x + y = 11$

6. $2x - y = 8,$
$3x + y = 12$

7. $4a + 3b = 7,$
$-4a + b = 5$

8. $7c + 5d = 18,$
$c - 5d = -2$

9. $8x - 5y = -9,$
$3x + 5y = -2$

10. $3a - 3b = -15,$
$-3a - 3b = -3$

11. $4x - 5y = 7,$
$-4x + 5y = 7$

12. $2x + 3y = 4,$
$-2x - 3y = -4$

b Solve using the multiplication principle first. Then add.

13. $x + y = -7,$
$3x + y = -9$

14. $-x - y = 8,$
$2x - y = -1$

15. $3x - y = 8,$
$x + 2y = 5$

16. $x + 3y = 19,$
$x - y = -1$

17. $x - y = 5,$
$4x - 5y = 17$

18. $x + y = 4,$
$5x - 3y = 12$

19. $2w - 3z = -1,$
$3w + 4z = 24$

20. $7p + 5q = 2,$
$8p - 9q = 17$

21. $2a + 3b = -1,$
$3a + 5b = -2$

22. $3x - 4y = 16,$
$5x + 6y = 14$

23. $x = 3y,$
$5x + 14 = y$

24. $5a = 2b,$
$2a + 11 = 3b$

25. $2x + 5y = 16,$
$3x - 2y = 5$

26. $3p - 2q = 8,$
$5p + 3q = 7$

27. $p = 32 + q,$
$3p = 8q + 6$

28. $3x = 8y + 11,$
$x + 6y - 8 = 0$

29. $3x - 2y = 10,$
$-6x + 4y = -20$

30. $2x + y = 13,$
$4x + 2y = 23$

31. $0.06x + 0.05y = 0.07,$
$0.4x - 0.3y = 1.1$

32. $1.8x - 2y = 0.9,$
$0.04x + 0.18y = 0.15$

33. $\frac{1}{3}x + \frac{3}{2}y = \frac{5}{4},$
$\frac{3}{4}x - \frac{5}{6}y = \frac{3}{8}$

34. $x - \frac{3}{2}y = 13,$
$\frac{3}{2}x - y = 17$

35. $-4.5x + 7.5y = 6,$
$-x + 1.5y = 5$

36. $0.75x + 0.6y = -0.3,$
$3.9x + 5.2y = 96.2$

37. D_W The following lists the steps a student uses to solve a system of equations, but an error has been made. Find and describe the error and correct the answer.

$$\begin{aligned} 3x - y &= 4 \\ 2x + y &= 16 \\ \hline 5x \quad &= 20 \\ x &= 4 \end{aligned}$$

$$\begin{aligned} 3x - y &= 4 \\ 3(4) - y &= 4 \\ y &= 4 - 12 \\ y &= -8 \end{aligned}$$

The solution is $(4, -8)$.

38. D_W Explain how the addition and multiplication principles are used in this section. Then count the number of times that these principles are used in Example 4.

SKILL MAINTENANCE

Simplify. [4.1d, e, f]

39. $x^{-2} \cdot x^{-5}$

40. $x^{-2} \cdot x^{5}$

41. $x^{2} \cdot x^{-5}$

42. $x^{2} \cdot x^{5}$

43. $\frac{x^{-2}}{x^{-5}}$

44. $\frac{x^{2}}{x^{-5}}$

45. $(a^2b^{-3})(a^5b^{-6})$

46. $\frac{a^2b^{-3}}{a^5b^{-6}}$

Simplify. [6.1c]

47. $\frac{x^2 - 5x + 6}{x^2 - 4}$

48. $\frac{x^2 - 25}{x^2 - 10x + 25}$

Subtract. [6.5a]

49. $\frac{x - 2}{x + 3} - \frac{2x - 5}{x - 4}$

50. $\frac{x + 7}{x^2 - 1} - \frac{3}{x + 1}$

SYNTHESIS

51.–60. Use the TABLE feature to check the possible solutions to Exercises 1–10.

61.–70. Use a graphing calculator and the CALC-INTERSECT feature to solve the systems in Exercises 21–30.

Solve using the substitution method, the elimination method, or the graphing method.

71. $3(x - y) = 9,$
$x + y = 7$

72. $2(x - y) = 3 + x,$
$x = 3y + 4$

73. $2(5a - 5b) = 10,$
$-5(6a + 2b) = 10$

74. $\frac{x}{3} + \frac{y}{2} = 1\frac{1}{3},$
$x + 0.05y = 4$

75. $y = -\frac{2}{7}x + 3,$
$y = \frac{4}{5}x + 3$

76. $y = \frac{2}{5}x - 7,$
$y = \frac{2}{5}x + 4$

Solve for x and y.

77. $y = ax + b,$
$y = x + c$

78. $ax + by + c = 0,$
$ax + cy + b = 0$

8.4 APPLICATIONS AND PROBLEM SOLVING

Objective

a Solve applied problems by translating to a system of two equations in two variables.

a We now use systems of equations to solve applied problems that involve two equations in two variables.

EXAMPLE 1 *Pizza and Soda Prices.* A campus vendor charges \$3.50 for one slice of pizza and one medium soda and \$9.15 for three slices of pizza and two medium sodas. Determine the price of one medium soda and the price of one slice of pizza.

1. **Familiarize.** We let p = the price of one slice of pizza and s = the price of one medium soda.
2. **Translate.** The price of one slice of pizza and one medium soda is \$3.50. This gives us one equation:

   $$p + s = 3.50.$$

   The price of three slices of pizza and two medium sodas is \$9.15. This gives us another equation:

   $$3p + 2s = 9.15.$$

3. **Solve.** We solve the system of equations

   $$p + s = 3.50, \quad (1)$$
   $$3p + 2s = 9.15. \quad (2)$$

   Which method should we use? As we discussed in Section 8.3, any method can be used. Each has its advantages and disadvantages. We decide to proceed with the elimination method, because we see that if we multiply each side of equation (1) by -2 and add, the s-terms can be eliminated. (We could also multiply equation (1) by -3 and eliminate p.)

   $$-2p - 2s = -7.00 \quad \text{Multiplying equation (1) by } -2$$
   $$3p + 2s = 9.15 \quad \text{Equation (2)}$$
   $$p = 2.15. \quad \text{Adding}$$

   Next, we substitute 2.15 for p in equation (1) and solve for s:

   $$p + s = 3.50$$
   $$2.15 + s = 3.50$$
   $$s = 1.35.$$

4. **Check.** The sum of the prices for one slice of pizza and one medium soda is

   $$\$2.15 + \$1.35, \quad \text{or} \quad \$3.50.$$

   Three times the price of one slice of pizza plus twice the price of a medium soda is

   $$3(\$2.15) + 2(\$1.35), \quad \text{or} \quad \$9.15.$$

   The prices check.

5. **State.** The price of one slice of pizza is \$2.15, and the price of one medium soda is \$1.35.

Do Exercise 1.

1. **Chicken and Hamburger Prices.** Fast Rick's Burger restaurant decides to include chicken on its menu. It offers a special two-and-one promotion. The price of one hamburger and two pieces of chicken is \$5.39, and the price of two hamburgers and one piece of chicken is \$5.68. Find the price of one hamburger and the price of one piece of chicken.

Answer on page A-39

EXAMPLE 2 *Imax Movie Prices.* There were 270 people at a recent showing of the IMAX 3D movie *Antarctica*. Admission was $8.00 each for adults and $4.75 each for children, and receipts totaled $2088.50. How many adults and how many children attended?

1. Familiarize. There are many ways in which to familiarize ourselves with a problem situation. This time, let's make a guess and do some calculations. The total number of people at the movie was 270, so we choose numbers that total 270. Let's try

220 adults and
50 children.

How much money was taken in? The problem says that adults paid $8.00 each, so the total amount of money collected from the adults was

220($8), or $1760.

Children paid $4.75 each, so the total amount of money collected from the children was

50($4.75), or $237.50.

This makes the total receipts $1760 + $237.50, or $1997.50.

Our guess is not the answer to the problem because the total taken in, according to the problem, was $2088.50. If we were to continue guessing, we would need to add more adults and fewer children, since our first guess gave us an amount of total receipts that was lower than $2088.50. The steps we have used to see if our guesses are correct help us to understand the actual steps involved in solving the problem.

Let's list the information in a table. That usually helps in the familiarization process. We let a = the number of adults and c = the number of children.

	ADULTS	CHILDREN	TOTAL	
Admission	$8.00	$4.75		
Number Attending	a	c	270	→ $a + c = 270$
Money Taken In	$8.00a$	$4.75c$	$2088.50	→ $8.00a + 4.75c = 2088.50$

2. Translate. The total number of people attending was 270, so

$$a + c = 270.$$

The amount taken in from the adults was $8.00a$, and the amount taken in from the children was $4.75c$. These amounts are in dollars. The total was $2088.50, so we have

$$8.00a + 4.75c = 2088.50.$$

We can multiply both sides by 100 to clear decimals. Thus we have a translation to a system of equations:

$$a + c = 270, \qquad (1)$$
$$800a + 475c = 208{,}850. \qquad (2)$$

3. **Solve.** We solve the system. We use the elimination method since the equations are both in the form $Ax + By = C$. (A case can certainly be made for using the substitution method since we can solve for one of the variables quite easily in the first equation. Very often a decision is just a matter of choice.) We multiply both sides of equation (1) by -475 and then add and solve for a:

$$\begin{aligned} -475a - 475c &= -128{,}250 \qquad \text{Multiplying by } -475 \\ 800a + 475c &= 208{,}850 \\ \hline 325a \qquad &= 80{,}600 \qquad \text{Adding} \\ a &= \frac{80{,}600}{325} \qquad \text{Dividing by 325} \\ a &= 248. \end{aligned}$$

Next, we go back to equation (1), substituting 248 for a, and solve for c:

$$\begin{aligned} a + c &= 270 \\ 248 + c &= 270 \\ c &= 22. \end{aligned}$$

4. **Check.** The check is left to the student. It is similar to what we did in the *Familiarize* step.
5. **State.** Attending the showing were 248 adults and 22 children.

Do Exercise 2.

EXAMPLE 3 *Mixture of Solutions.* A chemist has one solution that is 80% acid (that is, 8 parts are acid and 2 parts are water) and another solution that is 30% acid. What is needed is 200 L of a solution that is 62% acid. The chemist will prepare it by mixing the two solutions. How much of each should be used?

1. **Familiarize.** We can make a drawing of the situation. The chemist uses x liters of the first solution and y liters of the second solution.

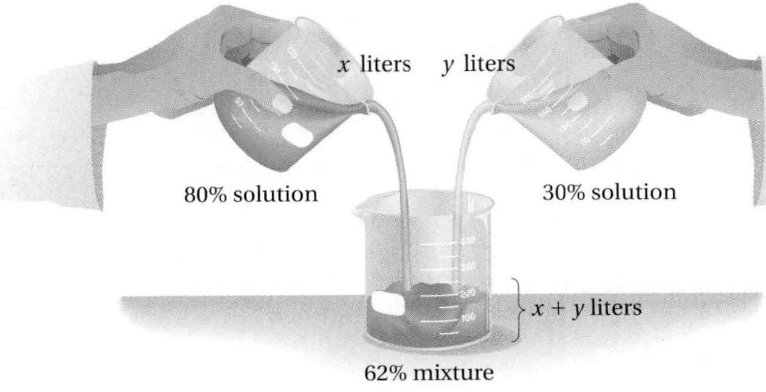

We can also arrange the information in a table.

	FIRST SOLUTION	SECOND SOLUTION	MIXTURE	
Amount of Solution	x	y	200 L	$\rightarrow x + y = 200$
Percent of Acid	80%	30%	62%	
Amount of Acid in Solution	$80\%x$	$30\%y$	$62\% \times 200$, or 124 L	$\rightarrow 80\%x + 30\%y = 124$

2. **Game Admissions.** There were 166 paid admissions to a game. The price was \$3.10 each for adults and \$1.75 each for children. The amount taken in was \$459.25. How many adults and how many children attended?

Complete the following table to aid with the familiarization.

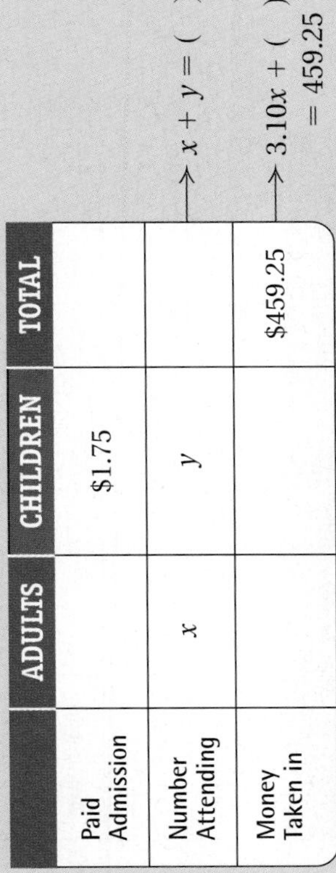

	ADULTS	CHILDREN	TOTAL	
Paid Admission		\$1.75		
Number Attending	x	y		$\rightarrow x + y = (\ \ )$
Money Taken in			\$459.25	$\rightarrow 3.10x + (\ \ ) = 459.25$

Answer on page A-39

3. Mixture of Solutions. One solution is 50% alcohol and a second is 70% alcohol. How much of each should be mixed in order to make 30 L of a solution that is 55% alcohol?

Complete the following table to aid in the familiarization.

	FIRST SOLUTION	SECOND SOLUTION	MIXTURE	
Amount of Solution	x	y		→ $x + y = (\ \ )$
Percent of Alcohol		70%	55%	
Amount of Alcohol in Solution				→ $(\ \ ) + 70\%y = (\ \ )$

Answer on page A-39

2. **Translate.** The chemist uses x liters of the first solution and y liters of the second. Since the total is to be 200 L, we have

$$\textit{Total amount of solution:}\quad x + y = 200.$$

The amount of acid in the new mixture is to be 62% of 200 L, or 124 L. The amounts of acid from the two solutions are $80\%x$ and $30\%y$. Thus,

$$\textit{Total amount of acid:}\quad 80\%x + 30\%y = 124$$

or

$$0.8x + 0.3y = 124.$$

We clear decimals by multiplying both sides by 10:

$$10(0.8x + 0.3y) = 10 \cdot 124$$
$$8x + 3y = 1240.$$

Thus we have a translation to a system of equations:

$$x + y = 200, \quad (1)$$
$$8x + 3y = 1240. \quad (2)$$

3. **Solve.** We solve the system. We use the elimination method, again because equations are in the form $Ax + By = C$ and a multiplication in one equation will allow us to eliminate a variable, but substitution would also work. We multiply both sides of equation (1) by -3 and then add and solve for x:

$$-3x - 3y = -600 \quad \text{Multiplying by } -3$$
$$8x + 3y = 1240$$
$$5x = 640 \quad \text{Adding}$$
$$x = \frac{640}{5} \quad \text{Dividing by 5}$$
$$x = 128.$$

Next, we go back to equation (1) and substitute 128 for x:

$$x + y = 200$$
$$128 + y = 200$$
$$y = 72.$$

The solution is $x = 128$ and $y = 72$.

4. **Check.** The sum of 128 and 72 is 200. Also, 80% of 128 is 102.4 and 30% of 72 is 21.6. These add up to 124.

5. **State.** The chemist should use 128 L of the 80%-acid solution and 72 L of the 30%-acid solution.

Do Exercise 3.

EXAMPLE 4 *Candy Mixtures.* A bulk wholesaler wishes to mix some candy worth 45 cents per pound and some worth 80 cents per pound to make 350 lb of a mixture worth 65 cents per pound. How much of each type of candy should be used?

1. **Familiarize.** Arranging the information in a table will help. We let $x =$ the amount of 45-cents candy and $y =$ the amount of 80-cents candy.

	INEXPENSIVE CANDY	EXPENSIVE CANDY	MIXTURE	
Cost of Candy	45 cents	80 cents	65 cents	
Amount (in pounds)	x	y	350	$\rightarrow x + y = 350$
Total Cost	$45x$	$80y$	65 cents · (350), or 22,750 cents	$\rightarrow 45x + 80y = 22{,}750$

Note the similarity of this problem to Example 2. Here we consider types of candy instead of groups of people.

2. **Translate.** We translate as follows. From the second row of the table, we find that

Total amount of candy: $x + y = 350$.

Our second equation will come from the costs. The value of the inexpensive candy, in cents, is $45x$ (x pounds at 45 cents per pound). The value of the expensive candy is $80y$, and the value of the mixture is 65×350, or 22,750 cents. Thus we have

Total cost of mixture: $45x + 80y = 22{,}750$.

Remember the problem-solving tip about dimension symbols. In this last equation, all expressions are given in cents. We could have expressed them all in dollars, but we do not want some in cents and some in dollars. Thus we have a translation to a system of equations:

$$x + y = 350, \quad (1)$$
$$45x + 80y = 22{,}750. \quad (2)$$

3. **Solve.** We solve the system using the elimination method again. We multiply both sides of equation (1) by -45 and then add and solve for y:

$$\begin{aligned} -45x - 45y &= -15{,}750 \quad \text{Multiplying by } -45 \\ 45x + 80y &= 22{,}750 \\ \hline 35y &= 7{,}000 \quad \text{Adding} \\ y &= \frac{7{,}000}{35} \\ y &= 200. \end{aligned}$$

Next, we go back to equation (1), substituting 200 for y, and solve for x:

$$\begin{aligned} x + y &= 350 \\ x + 200 &= 350 \\ x &= 150. \end{aligned}$$

4. **Check.** We consider $x = 150$ lb and $y = 200$ lb. The sum is 350 lb. The value of the candy is $45(150) + 80(200)$, or 22,750 cents and each pound of the mixture is worth $22{,}750 \div 350$, or 65 cents. These values check.

5. **State.** The grocer should mix 150 lb of the 45-cents candy with 200 lb of the 80-cents candy.

Do Exercise 4.

4. Mixture of Grass Seeds. Grass seed A is worth \$1.40 per pound and seed B is worth \$1.75 per pound. How much of each should be mixed in order to make 50 lb of a mixture worth \$1.54 per pound?

Complete the following table to aid in the familiarization.

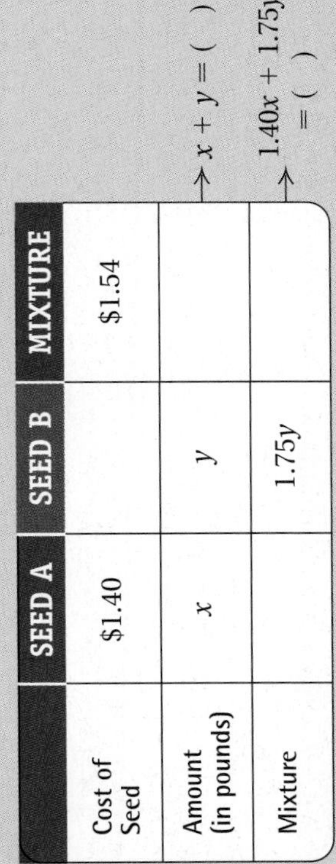

	SEED A	SEED B	MIXTURE	
Cost of Seed	\$1.40		\$1.54	
Amount (in pounds)	x	y		$\rightarrow x + y = (\ )$
Mixture		$1.75y$		$\rightarrow 1.40x + 1.75y = (\ )$

Answer on page A-39

5. **Coin Value.** On a table are 20 coins, quarters and dimes. Their value is \$3.05. How many of each kind of coin are there?

Answer on page A-39

EXAMPLE 5 *Coin Value.* A student assistant at the university copy center has some nickels and dimes to use for change when students make copies. The value of the coins is \$7.40. There are 26 more dimes than nickels. How many of each kind of coin are there?

1. **Familiarize.** We let $d =$ the number of dimes and $n =$ the number of nickels.
2. **Translate.** We have one equation at once:

   $d = n + 26.$

   The value of the nickels, in cents, is $5n$, since each coin is worth 5 cents. The value of the dimes, in cents, is $10d$, since each coin is worth 10 cents. The total value is given as \$7.40. Since we have the values of the nickels and dimes *in cents,* we must use cents for the total value. This is 740. This gives us another equation:

   $10d + 5n = 740.$

   We now have a system of equations:

   $d = n + 26,$ (1)

   $10d + 5n = 740.$ (2)

3. **Solve.** Since we have d alone on one side of one equation, we use the substitution method. We substitute $n + 26$ for d in equation (2):

   $$\begin{aligned} 10d + 5n &= 740 \\ 10(n + 26) + 5n &= 740 && \text{Substituting } n + 26 \text{ for } d \\ 10n + 260 + 5n &= 740 && \text{Removing parentheses} \\ 15n + 260 &= 740 && \text{Collecting like terms} \\ 15n &= 480 && \text{Subtracting 260} \\ n &= \frac{480}{15}, \text{ or } 32. && \text{Dividing by 15} \end{aligned}$$

   Next, we substitute 32 for n in either of the original equations to find d. We use equation (1):

   $d = n + 26 = 32 + 26 = 58.$

4. **Check.** We have 58 dimes and 32 nickels. There are 26 more dimes than nickels. The value of the coins is 58(\$0.10) + 32(\$0.05), which is \$7.40. This checks.
5. **State.** The student assistant has 58 dimes and 32 nickels.

Do Exercise 5 on the preceding page.

You should look back over Examples 2–5. The problems are quite similar in their structure. Compare them and try to see the similarities. The problems in Examples 2–5 are often called *mixture problems.* These problems provide a pattern, or model, for many related problems.

PROBLEM-SOLVING TIP

When solving problems, see if they are patterned or modeled after other problems that you have studied.

Study Tips

TROUBLE SPOTS

By now you have probably encountered certain topics that gave you more difficulty than others. It is important to know that this happens to every person who studies mathematics. Unfortunately, frustration is often part of the learning process and it is important not to give up when difficulty arises.

One source of frustration for many students is not being able to set aside sufficient time for studying. Family commitments, work schedules, and extracurricular activities are just a few of the time demands that many students face. Couple these demands with a math lesson that seems to require a greater than usual amount of study time, and it is no wonder that many students often feel frustrated. Below are some study tips that might be useful if and when troubles arise.

- **Realize that everyone—even your instructor—has been stumped at times when studying math.** You are not the first person, nor will you be the last, to encounter a "roadblock."
- **Whether working alone or with a classmate, try to allow enough study time so that you won't need to constantly glance at a clock.** Difficult material is best mastered when your mind is completely focused on the subject matter. Thus, if you are tired, it is usually best to study early the next morning or to take a ten-minute "power-nap" in order to make the most productive use of your time.
- **Talk about your trouble spot with a classmate.** It is possible that she or he is also having difficulty with the same material. If that is the case, perhaps the majority of your class is confused and your instructor's coverage of the topic is not yet finished. If your classmate *does* understand the topic that is troubling you, patiently allow him or her to explain it to you. By verbalizing the math in question, your classmate may help clarify the material for both of you. Perhaps you will be able to return the favor for your classmate when he or she is struggling with a topic that you understand.
- **Try to study in a "controlled" environment.** What we mean by this is that you can often put yourself in a setting that will enable you to maximize your powers of concentration. For example, some students may succeed in studying at home or in a dorm room, but for many these settings are filled with distractions. Consider a trip to a library, classroom building, or perhaps the attic or basement if such a setting is more conducive to studying. If you plan on working with a classmate, try to find a location in which conversation will not be bothersome to others.
- **When working on difficult material, it is often helpful to first "back up" and review the most recent material that did make sense.** This can build your confidence and create a momentum that can often carry you through the roadblock. Sometimes a small piece of information that appeared in a previous section is all that is needed for your problem spot to disappear. When the difficult material is finally mastered, try to make use of what is fresh in your mind by taking a "sneak preview" of what your next topic for study will be.

8.4 EXERCISE SET

For Extra Help

Digital Video Tutor CD 7 Videotape 15 | InterAct Math | Math Tutor Center | MathXL | MyMathLab

a Solve.

1. *Basketball Scoring.* In the final game of the 2000 basketball season, the Los Angeles Lakers scored 96 of their points on a combination of 43 two- and three-point baskets. How many of each type of shot were made?
Source: National Basketball Association

2. *Basketball Scoring.* Shaquille O'Neill of the Los Angeles Lakers once scored 36 points on 22 shots in an NBA game, shooting only two-pointers and foul shots (one point). How many of each type of shot did he make?
Source: National Basketball Association

3. *Film Processing.* Photoworks.com charges $7.00 for processing a 24-exposure roll and $10.00 for processing a 36-exposure roll. After Karen's photography field trip, she sent 19 rolls of film to Photoworks and paid $151 for processing. How many rolls of each type were processed?

PHOTOWorks.com
HOME | SHOP | HELP
24 Exposures, $7
Total # of rolls
36 Exposures, $10
Total # of rolls
Order Now
$ Total
Home | Terms & Conditions | Site Map | Privacy Policy

4. *Zoo Admissions.* During the summer months, the Bronx Zoo charges $9 each for adults and $5 each for children and seniors. One July day, a total of $6320 was collected from 960 admissions. How many adult admissions were there?
Source: Bronx Zoo

5. *Grain Mixtures for Horses.* Irene is a barn manager at a horse stable. She needs to calculate the correct mix of grain and hay to feed her horse. On the basis of her horse's age, weight, and workload, she determines that he needs to eat 15 lb of feed per day, with an average protein content of 8%. Hay contains 6% protein, whereas grain has a 12% protein content. How many pounds of hay and grain should she feed her horse each day?
Source: *Michael Plumb's Horse Journal*, February 1996: 26–29

6. *Paint Mixtures.* At a local "paint swap," Gayle found large supplies of Skylite Pink (12.5% red pigment) and MacIntosh Red (20% red pigment). How many gallons of each color should Gayle pick up in order to mix a gallon of Summer Rose (17% red pigment)?

7. *Food Prices.* Mr. Cholesterol's Pizza Parlor charges \$3.70 for a slice of pizza and a soda and \$9.65 for three slices of pizza and two sodas. Determine the cost of one soda and the cost of one slice of pizza.

8. *Investments.* Cassandra has a number of \$50 and \$100 savings bonds to use for part of her college expenses. The total value of the bonds is \$1250. There are 7 more \$50 bonds than \$100 bonds. How many of each type of bond does she have?

9. *Ticket Sales.* There were 203 tickets sold for a volleyball game. For activity-card holders, the price was \$2.25 each, and for non-cardholders, the price was \$3 each. The total amount of money collected was \$513. How many of each type of ticket were sold?

10. *Paid Admissions.* There were 429 people at a play. Admission was \$8 each for adults and \$4.50 each for children. The total receipts were \$2641. How many adults and how many children attended?

11. *Paid Admissions.* Following the baseball season, the players on a junior college team decided to go to a major-league baseball game. Ticket prices for the game are shown in the table below. They bought 29 tickets of two types, Upper Box and Lower Reserved. The cost of all the tickets was \$318. How many of each kind of ticket did they buy?

12. *Paid Admissions.* Referring to Exercise 11, suppose a faculty group bought tickets for the game, but they bought 54 tickets of two types, Lower Box and Upper Box. The cost of all their tickets was \$745.50. How many of each kind of ticket did they buy?

TICKET INFORMATION	
Lower Box	\$18.50
Upper Box	\$12.00
Lower Reserved	\$ 9.50
Upper Reserved	\$ 8.00
General Admission	\$ 6.50

13. *Mixture of Solutions.* Solution A is 50% acid and solution B is 80% acid. How many liters of each should be used in order to make 100 L of a solution that is 68% acid? Complete the following table to aid in the familiarization.

	SEED A	SEED B	MIXTURE	
Amount of Solution	x	y	L	$\rightarrow x + y = (\quad)$
Percent of Acid	50%		68%	
Amount of Acid in Solution		80%y	68% × 100, or L	$\rightarrow$ 50%x + ( ) = ( )

14. *Mixture of Solutions.* Solution A is 30% alcohol and solution B is 75% alcohol. How much of each should be used in order to make 100 L of a solution that is 50% alcohol?

15. *Coin Value.* A parking meter contains dimes and quarters worth \$15.25. There are 103 coins in all. How many of each type of coin are there?

16. *Coin Value.* A vending machine contains nickels and dimes worth \$14.50. There are 95 more nickels than dimes. How many of each type of coin are there?

17. *Coffee Blends.* Cafebucks coffee shop mixes Brazilian coffee worth \$19 per pound with Turkish coffee worth \$22 per pound. The mixture is to sell for \$20 per pound. How much of each type of coffee should be used in order to make a 300-lb mixture? Complete the following table to aid in the familiarization.

	BRAZILIAN COFFEE	TURKISH COFFEE	MIXTURE	
Cost of Coffee	\$19		\$20	
Amount (in pounds)	x	y	300	→ $x + y = (\quad)$
Mixture		$22y$	20(300), or \$6000	→ $19x + (\quad) = 6000$

18. *Coffee Blends.* The Java Joint wishes to mix Kenyan coffee beans that sell for \$7.25 per pound with Venezuelan beans that sell for \$8.50 per pound in order to form a 50-lb batch of Morning Blend that sells for \$8.00 per pound. How many pounds of Kenyan beans and how many pounds of Venezuelan beans should be used to make the blend?

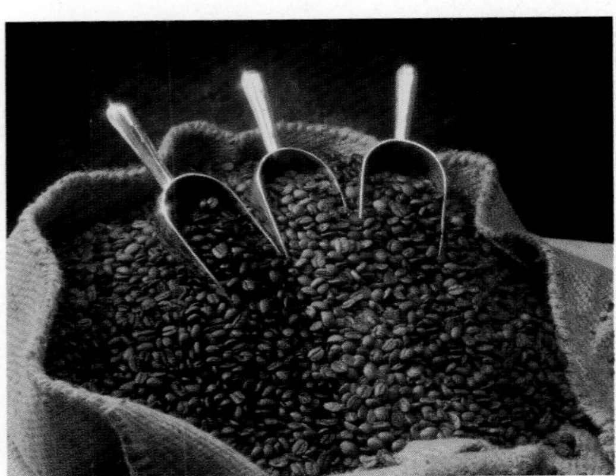

19. *Horticulture.* A solution containing 28% fungicide is to be mixed with a solution containing 40% fungicide to make 300 L of a solution containing 36% fungicide. How much of each solution should be used?

20. *Production.* Clear Shine window cleaner is 12% alcohol and Sunstream window cleaner is 30% alcohol. How much of each should be used to make 90 oz of a cleaner that is 20% alcohol?

21. *Printing.* A printer knows that a page of print contains 1300 words if large type is used and 1850 words if small type is used. A document containing 18,526 words fills exactly 12 pages. How many pages are in the large type? in the small type?

22. *Paint Mixture.* A merchant has two kinds of paint. If 9 gal of the inexpensive paint is mixed with 7 gal of the expensive paint, the mixture will be worth \$19.70 per gallon. If 3 gal of the inexpensive paint is mixed with 5 gal of the expensive paint, the mixture will be worth \$19.825 per gallon. What is the price per gallon of each type of paint?

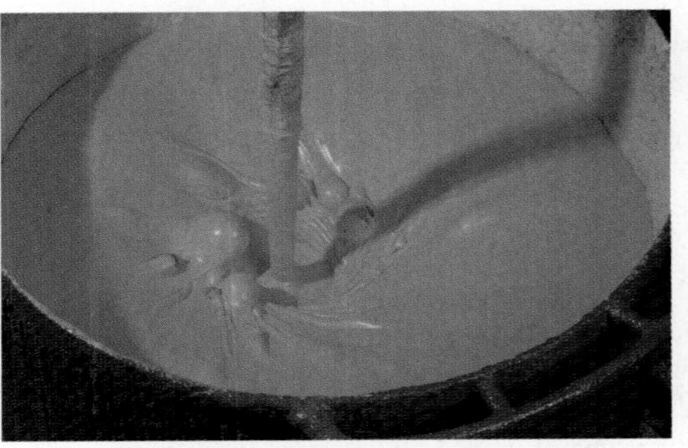

23. *Mixture of Grass Seeds.* Grass seed A is worth \$2.50 per pound and seed B is worth \$1.75 per pound. How much of each would you use in order to make 75 lb of a mixture worth \$2.14 per pound?

24. *Mixed Nuts.* A customer has asked a caterer to provide 60 lb of nuts, 60% of which are to be cashews. The caterer has available mixtures of 70% cashews and 45% cashews. How many pounds of each mixture should be used?

25. *Test Scores.* You are taking a test in which items of type A are worth 10 points and items of type B are worth 15 points. It takes 3 min to complete each item of type A and 6 min to complete each item of type B. The total time allowed is 60 min and you do exactly 16 questions. How many questions of each type did you complete? Assuming that all your answers were correct, what was your score?

26. *Gold Alloys.* A goldsmith has two alloys that are different purities of gold. The first is three-fourths pure gold and the second is five-twelfths pure gold. How many ounces of each should be melted and mixed in order to obtain a 6-oz mixture that is two-thirds pure gold?

27. *Ages.* The Kuyatts' house is twice as old as the Marconis' house. Eight years ago, the Kuyatts' house was three times as old as the Marconis' house. How old is each house?

28. *Ages.* David is twice as old as his daughter. In 4 yr, David's age will be three times what his daughter's age was 6 yr ago. How old are they now?

29. *Ages.* Randy is four times as old as Mandy. In 12 yr, Mandy's age will be half of Randy's. How old are they now?

30. *Ages.* Jennifer is twice as old as Ramon. The sum of their ages 7 yr ago was 13. How old are they now?

31. *Supplementary Angles.* **Supplementary angles** are angles whose sum is 180°. Two supplementary angles are such that one is 30° more than two times the other. Find the angles.

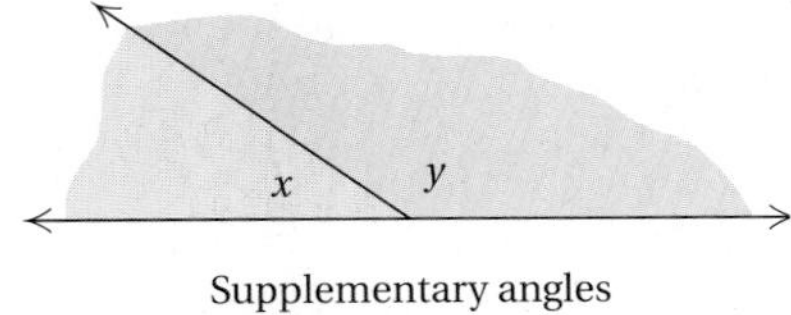

Supplementary angles
$x + y = 180°$

32. *Supplementary Angles.* Two supplementary angles are such that one is 8° less than three times the other. Find the angles.

33. *Complementary Angles.* **Complementary angles** are angles whose sum is 90°. Two complementary angles are such that their difference is 34°. Find the angles.

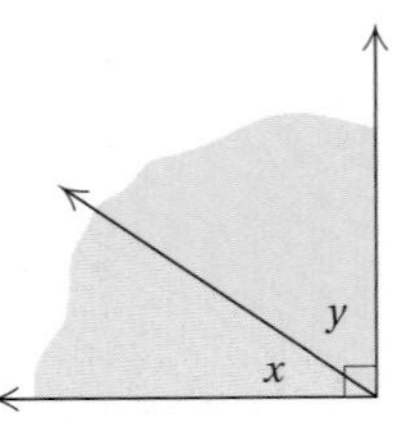

Complementary angles
$x + y = 90°$

34. *Complementary Angles.* Two angles are complementary. One angle is 42° more than one-half the other. Find the angles.

35. *Octane Ratings.* In most areas of the United States, gas stations offer three grades of gasoline, indicated by octane ratings on the pumps, such as 87, 89, and 93. When a tanker delivers gas, it brings only two grades of gasoline, the highest and the lowest, filling two large underground tanks. If you purchase the middle grade, the pump's computer mixes the other two grades appropriately. How much 87-octane gas and 93-octane gas should be blended in order to make 18 gal of 89-octane gas?
Source: Exxon

36. *Octane Ratings.* Referring to Exercise 35, suppose the pump grades offered are 85, 87, and 91. How much 85-octane gas and 91-octane gas should be blended in order to make 12 gal of 87-octane gas?
Source: Exxon

37. *Suntan Lotion.* Lisa has a tube of Kinney's suntan lotion that is rated 15 spf and a second tube of Coppertone that is 30 spf. How many fluid ounces of each type of lotion should be mixed in order to create 50 fluid ounces of sunblock that is rated 20 spf?

38. *Cough Syrup.* Dr. Zeke's cough syrup is 2% alcohol. Vitabrite cough syrup is 5% alcohol. How much of each type should be used in order to prepare an 80-oz batch of cough syrup that is 3% alcohol?

39. D_W What characteristics do Examples 1–3 share when they are translated to systems of equations?

40. D_W Which of the five problem-solving steps have you found the most challenging? Why?

SKILL MAINTENANCE

Factor. [5.6a]

41. $25x^2 - 81$

42. $36 - a^2$

43. $4x^2 + 100$

44. $4x^2 - 100$

Find the intercepts. Then graph the equation. [3.3a]

45. $y = -2x - 3$

46. $y = -0.1x + 0.4$

47. $5x - 2y = -10$

48. $2.5x + 4y = 10$

SYNTHESIS

49. *Milk Mixture.* A farmer has 100 L of milk that is 4.6% butterfat. How much skim milk (no butterfat) should be mixed with it in order to make milk that is 3.2% butterfat?

50. *Investments.* Eduardo invested $54,000, part of it at 6% and the rest at 6.5%. The total yield after 1 yr is $3385. How much was invested at each rate?

51. *Automobile Maintenance.* An automobile radiator contains 16 L of antifreeze and water. This mixture is 30% antifreeze. How much of this mixture should be drained and replaced with pure antifreeze so that the mixture will be 50% antifreeze?

52. *Employer Payroll.* An employer has a daily payroll of $1225 when employing some workers at $80 per day and others at $85 per day. When the number of $80 workers is increased by 50% and the number of $85 workers is decreased by $\frac{1}{5}$, the new daily payroll is $1540. How many were originally employed at each rate?

53. A flavored-drink manufacturer mixes flavoring worth $1.45 per ounce with sugar worth $0.05 per ounce. The mixture sells for $0.106 per ounce. How much of each should be mixed in order to fill a 20-oz can?

54. A two-digit number is six times the sum of its digits. The tens digit is 1 more than the units digit. Find the number.

55. One year, Shannon made $288 from two investments: $1100 was invested at one yearly rate and $1800 at a rate that was 1.5% higher. Find the two rates of interest.

8.5 APPLICATIONS WITH MOTION

Objective

a Solve motion problems using the formula $d = rt$.

a We first studied problems involving motion in Chapter 6. Here we extend our problem-solving skills by solving certain motion problems whose solutions can be found using systems of equations. Recall the motion formula.

THE MOTION FORMULA

Distance = Rate (or speed) · Time

$$d = rt$$

We have five steps for problem solving. The tips in the margin at right are also helpful when solving motion problems.

As we saw in Chapter 6, there are motion problems that can be solved with just one equation. Let's start with another such problem.

TIPS FOR SOLVING MOTION PROBLEMS

1. Draw a diagram using an arrow or arrows to represent distance and the direction of each object in motion.
2. Organize the information in a chart.
3. Look for as many things as you can that are the same so that you can write equations.

EXAMPLE 1 *Car Travel.* Two cars leave York at the same time traveling in opposite directions. One travels at 60 mph and the other at 30 mph. In how many hours will they be 150 mi apart?

1. **Familiarize.** We first make a drawing.

From the wording of the problem and the drawing, we see that the distances may *not* be the same. But the times that the cars travel are the same, so we can use just t for time. We can organize the information in a chart.

$$d \quad = \quad r \quad \cdot \quad t$$

	DISTANCE	SPEED	TIME
Fast Car	Distance of fast car	60	t
Slow Car	Distance of slow car	30	t
Total	150		

Study Tips

VARYING ORDER IN TEST PREPARATION

When studying for a test, it is common for students to study material in the order in which it appears in the book. Educational research has shown that when students do so, they tend to remember material at the beginning and at the end the best and the material in the middle the least. So when you're preparing for a test, especially when you allow yourself two or three days to study (as you should), vary what you study at the beginning and at the end of your study sessions.

2. Translate. From the drawing, we see that

(Distance of fast car) + (Distance of slow car) = 150.

Then using $d = rt$ in each row of the table, we get $60t + 30t = 150$.

3. Solve. We solve the equation:

$$60t + 30t = 150$$

$$90t = 150 \qquad \text{Collecting like terms}$$

$$t = \frac{150}{90}, \text{ or } \frac{5}{3}, \text{ or } 1\frac{2}{3} \text{ hr.} \qquad \text{Dividing by 90}$$

4. Check. When $t = \frac{5}{3}$ hr,

$$\text{(Distance of fast car)} + \text{(Distance of slow car)} = 60\left(\frac{5}{3}\right) + 30\left(\frac{5}{3}\right)$$

$$= 100 + 50, \text{ or } 150 \text{ mi.}$$

Thus the time of $\frac{5}{3}$ hr, or $1\frac{2}{3}$ hr, checks.

5. State. In $1\frac{2}{3}$ hr, the cars will be 150 mi apart.

Do Exercises 1 and 2.

Now let's solve some motion problems using systems of equations.

EXAMPLE 2 *Train Travel.* A train leaves Stanton traveling east at 35 miles per hour (mph). An hour later, another train leaves Stanton on a parallel track at 40 mph. How far from Stanton will the second (or faster) train catch up with the first (or slower) train?

1. Familiarize. We first make a drawing.

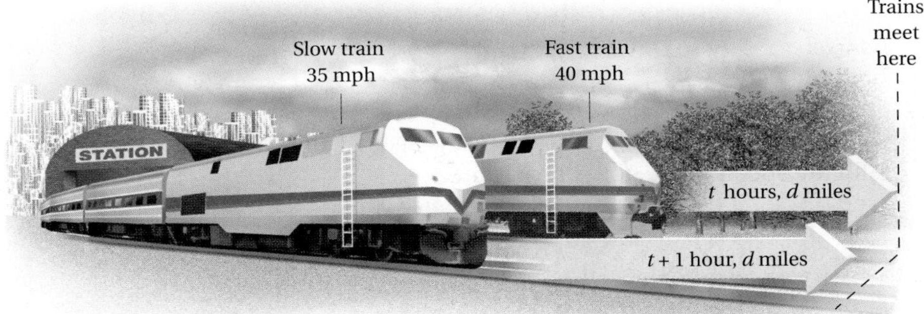

From the drawing, we see that the distances are the same. Let's call the distance d. We don't know the times. We let t = the time for the faster train. Then the time for the slower train = $t + 1$, since it left 1 hr earlier. We can organize the information in a chart.

$d = r \cdot t$

	DISTANCE	SPEED	TIME	
Slow Train	d	35	$t + 1$	→ $d = 35(t + 1)$
Fast Train	d	40	t	→ $d = 40t$

1. Car Travel. Two cars leave town at the same time traveling in opposite directions. One travels at 48 mph and the other at 60 mph. How far apart will they be 3 hr later? (*Hint*: The times are the same. Be *sure* to make a drawing.)

2. Car Travel. Two cars leave town at the same time traveling in the same direction. One travels at 35 mph and the other at 40 mph. In how many hours will they be 15 mi apart? (*Hint*: The times are the same. Be *sure* to make a drawing.)

Answers on page A-39

3. **Car Travel.** A car leaves Spokane traveling north at 56 km/h. Another car leaves Spokane 1 hr later traveling north at 84 km/h. How far from Spokane will the second car catch up with the first? (*Hint*: The cars travel the same distance.)

Answers on page A-39

2. **Translate.** In motion problems, we look for things that are the same so that we can write equations. From each row of the chart, we get an equation, $d = rt$. Thus we have two equations:

$$d = 35(t + 1), \quad (1)$$
$$d = 40t. \quad (2)$$

3. **Solve.** Since we have a variable alone on one side, we solve the system using the substitution method:

$$35(t + 1) = 40t \quad \text{Using the substitution method (substituting } 35(t+1) \text{ for } d \text{ in equation 2)}$$
$$35t + 35 = 40t \quad \text{Removing parentheses}$$
$$35 = 5t \quad \text{Subtracting } 35t$$
$$\frac{35}{5} = t \quad \text{Dividing by 5}$$
$$7 = t.$$

The problem asks us to find how far from Stanton the fast train catches up with the other. Thus we need to find d. We can do this by substituting 7 for t in the equation $d = 40t$:

$$d = 40(7)$$
$$= 280.$$

4. **Check.** If the time is 7 hr, then the distance that the slow train travels is $35(7 + 1)$, or 280 mi. The fast train travels $40(7)$, or 280 mi. Since the distances are the same, we know how far from Stanton the trains will be when the fast train catches up with the other.

5. **State.** The fast train will catch up with the slow train 280 mi from Stanton.

Do Exercise 3.

EXAMPLE 3 *Boat Travel.* A motorboat took 3 hr to make a downstream trip with a 6-km/h current. The return trip against the same current took 5 hr. Find the speed of the boat in still water.

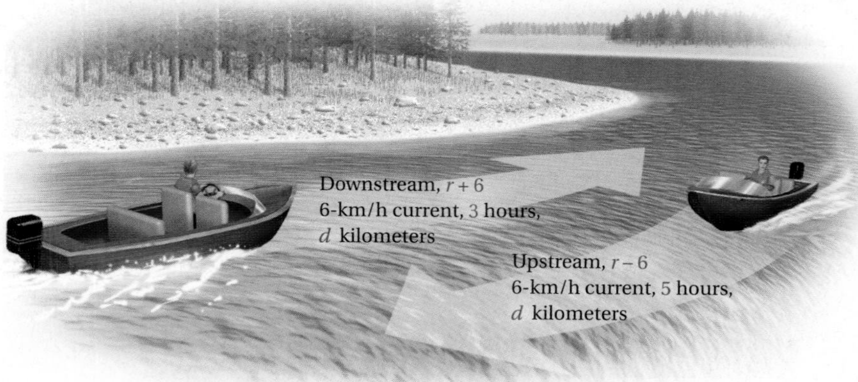

1. **Familiarize.** We first make a drawing. From the drawing, we see that the distances are the same. Let's call the distance d. We let $r =$ the speed of the boat in still water. Then, when the boat is traveling downstream, its speed is $r + 6$ (the current helps the boat along). When it is traveling upstream, its speed is $r - 6$ (the current holds the boat back).

We can organize the information in a chart. In this case, the distances are the same, so we use the formula $d = rt$.

$$d \quad = \quad r \quad \cdot \quad t$$

	DISTANCE	SPEED	TIME	
Downstream	d	$r + 6$	3	$\rightarrow d = (r + 6)3$
Upstream	d	$r - 6$	5	$\rightarrow d = (r - 6)5$

2. **Translate.** From each row of the chart, we get an equation, $d = rt$:

$d = (r + 6)3$, **(1)**

$d = (r - 6)5$. **(2)**

3. **Solve.** Since there is a variable alone on one side of an equation, we solve the system using substitution:

$(r + 6)3 = (r - 6)5$ Substituting $(r + 6)3$ for d in equation (2)

$3r + 18 = 5r - 30$ Removing parentheses

$-2r + 18 = -30$ Subtracting $5r$

$-2r = -48$ Subtracting 18

$r = \frac{-48}{-2}$, or 24. Dividing by -2

4. **Check.** When $r = 24$, $r + 6 = 30$, and $30 \cdot 3 = 90$, the distance downstream. When $r = 24$, $r - 6 = 18$, and $18 \cdot 5 = 90$, the distance upstream. In both cases, we get the same distance.

5. **State.** The speed in still water is 24 km/h.

MORE TIPS FOR SOLVING MOTION PROBLEMS

1. Translating to a system of equations eases the solution of many motion problems.
2. At the end of the problem, always ask yourself, "Have I found what the problem asked for?" You might have solved for a certain variable but still not have answered the question of the original problem. For example, in Example 2 we solve for t but the question of the original problem asks for d. Thus we need to continue the *Solve* step.

Do Exercise 4.

4. **Air Travel.** An airplane flew for 5 hr with a 25-km/h tail wind. The return flight against the same wind took 6 hr. Find the speed of the airplane in still air. (*Hint*: The distance is the same both ways. The speeds are $r + 25$ and $r - 25$, where r is the speed in still air.)

Answers on page A-39

8.5 EXERCISE SET

For Extra Help

Digital Video Tutor CD 7 Videotape 15 | InterAct Math | Math Tutor Center | MathXL | MyMathLab

a Solve. In Exercises 1–6, complete the table to aid the translation.

1. *Car Travel.* Two cars leave town at the same time going in the same direction. One travels at 30 mph and the other travels at 46 mph. In how many hours will they be 72 mi apart?

$d = r \cdot t$

	DISTANCE	SPEED	TIME
Slow Car	Distance of slow car		t
Fast Car	Distance of fast car	46	

2. *Car and Truck Travel.* A truck and a car leave a service station at the same time and travel in the same direction. The truck travels at 55 mph and the car at 40 mph. They can maintain CB radio contact within a range of 10 mi. When will they lose contact?

$d = r \cdot t$

	DISTANCE	SPEED	TIME
Truck	Distance of truck	55	
Car	Distance of car		t

3. *Train Travel.* A train leaves a station and travels east at 72 mph. Three hours later, a second train leaves on a parallel track and travels east at 120 mph. When will it overtake the first train?

$d = r \cdot t$

	DISTANCE	SPEED	TIME	
Slow Train	d		$t + 3$	→ $d = 72(\quad)$
Fast Train	d	120		→ $d = (\quad)t$

4. *Airplane Travel.* A private airplane leaves an airport and flies due south at 192 mph. Two hours later, a jet leaves the same airport and flies due south at 960 mph. When will the jet overtake the plane?

$d = r \cdot t$

	DISTANCE	SPEED	TIME	
Private Plane	d	192		→ $d = 192(\ )$
Jet	d		$t - 2$	→ $d = (\quad)(t - 2)$

5. *Canoeing.* A canoeist paddled for 4 hr with a 6-km/h current to reach a campsite. The return trip against the same current took 10 hr. Find the speed of the canoe in still water.

$d = r \cdot t$

	DISTANCE	SPEED	TIME	
Downstream	d	$r + 6$		→ $d = (\quad)4$
Upstream	d		10	→ $= (r - 6)10$

6. *Airplane Travel.* An airplane flew for 4 hr with a 20-km/h tail wind. The return flight against the same wind took 5 hr. Find the speed of the plane in still air.

$d = r \cdot t$

	DISTANCE	SPEED	TIME	
With Wind	d		4	→ $d = (\quad)4$
Against Wind	d	$r - 20$		→ $d = (\quad)5$

7. *Train Travel.* It takes a passenger train 2 hr less time than it takes a freight train to make the trip from Central City to Clear Creek. The passenger train averages 96 km/h, while the freight train averages 64 km/h. How far is it from Central City to Clear Creek?

8. *Airplane Travel.* It takes a small jet 4 hr less time than it takes a propeller-driven plane to travel from Glen Rock to Oakville. The jet averages 637 km/h, while the propeller plane averages 273 km/h. How far is it from Glen Rock to Oakville?

9. *Motorboat Travel.* On a weekend outing, Antoine rents a motorboat for 8 hr to travel down the river and back. The rental operator tells him to go for 3 hr downstream, leaving him 5 hr to return upstream.

   a) If the river current flows at a speed of 6 mph, how fast must Antoine travel in order to return in 8 hr?

   b) How far downstream did Antoine travel before he turned back?

10. *Airplane Travel.* For spring break some students flew to Cancun. From Mexico City, the airplane took 2 hr to fly 600 mi against a head wind. The return trip with the wind took $1\frac{2}{3}$ hr. Find the speed of the plane in still air.

11. *Running.* A toddler takes off running down the sidewalk at 230 ft/min. One minute later, a worried mother runs after the child at 660 ft/min. When will the mother overtake the toddler?

12. *Airplane Travel.* Two airplanes start at the same time and fly toward each other from points 1000 km apart at rates of 420 km/h and 330 km/h. When will they meet?

13. *Motorcycle Travel.* A motorcycle breaks down and the rider must walk the rest of the way to work. The motorcycle was being driven at 45 mph, and the rider walks at a speed of 6 mph. The distance from home to work is 25 mi, and the total time for the trip was 2 hr. How far did the motorcycle go before it broke down?

14. *Walking and Jogging.* A student walks and jogs to college each day. She averages 5 km/h walking and 9 km/h jogging. The distance from home to college is 8 km, and she makes the trip in 1 hr. How far does the student jog?

15. DW Discuss the advantages of using a table to organize information when solving a motion problem.

16. DW From the formula $d = rt$, derive two other formulas, one for r and one for t. Discuss the kinds of problems for which each formula might be useful.

SKILL MAINTENANCE

Simplify. [6.1c]

17. $\dfrac{8x^2}{24x}$

18. $\dfrac{5x^8y^4}{10x^3y}$

19. $\dfrac{5a + 15}{10}$

20. $\dfrac{12x - 24}{48}$

21. $\dfrac{2x^2 - 50}{x^2 - 25}$

22. $\dfrac{x^2 - 1}{x^4 - 1}$

23. $\dfrac{x^2 - 3x - 10}{x^2 - 2x - 15}$

24. $\dfrac{6x^2 + 15x - 36}{2x^2 - 5x + 3}$

25. $\dfrac{(x^2 + 6x + 9)(x - 2)}{(x^2 - 4)(x + 3)}$

26. $\dfrac{x^2 + 25}{x^2 - 25}$

27. $\dfrac{6x^2 + 18x + 12}{6x^2 - 6}$

28. $\dfrac{x^3 + 3x^2 + 2x + 6}{2x^3 + 6x^2 + x + 3}$

SYNTHESIS

29. *Lindbergh's Flight.* Charles Lindbergh flew the Spirit of St. Louis in 1927 from New York to Paris at an average speed of 107.4 mph. Eleven years later, Howard Hughes flew the same route, averaged 217.1 mph, and took 16 hr and 57 min less time. Find the length of their route.

30. *Car Travel.* A car travels from one town to another at a speed of 32 mph. If it had gone 4 mph faster, it could have made the trip in $\frac{1}{2}$ hr less time. How far apart are the towns?

31. *River Cruising.* An afternoon sightseeing cruise up river and back down river is scheduled to last 1 hr. The speed of the current is 4 mph, and the speed of the riverboat in still water is 12 mph. How far upstream should the pilot travel before turning around?

CHAPTER

8 Summary and Review

The review that follows is meant to prepare you for a chapter exam. It consists of two parts. The first part is a checklist of some of the Study Tips referred to in this and preceding chapters, as well as a list of important properties and formulas. The second part is the Review Exercises. These provide practice exercises for the exam, together with references to section objectives so you can go back and review. Before beginning, stop and look back over the skills you have obtained. What skills in mathematics do you have now that you did not have before studying this chapter?

STUDY TIPS CHECKLIST

The foundation of all your study skills is TIME!

- ☐ Have you begun to study for the final examination? If not, try to follow the Study Tips in Section 8.2.
- ☐ Are you dealing with your trouble spots in a constructive manner?
- ☐ Are you trying to vary the order in which you study material to prepare for your tests?
- ☐ Did you use the five-step problem-solving strategy when doing the applications in Sections 8.2, 8.4, and 8.5?
- ☐ Did you study the examples in this chapter carefully?

IMPORTANT PROPERTIES AND FORMULAS

Motion Formula: $d = rt$

REVIEW EXERCISES

Determine whether the given ordered pair is a solution of the system of equations. [8.1a]

1. $(6, -1)$; $x - y = 3,$ $2x + 5y = 6$

2. $(2, -3)$; $2x + y = 1,$ $x - y = 5$

3. $(-2, 1)$; $x + 3y = 1,$ $2x - y = -5$

4. $(-4, -1)$; $x - y = 3,$ $x + y = -5$

Solve the system by graphing. [8.1b]

5. $x + y = 4,$ $x - y = 8$

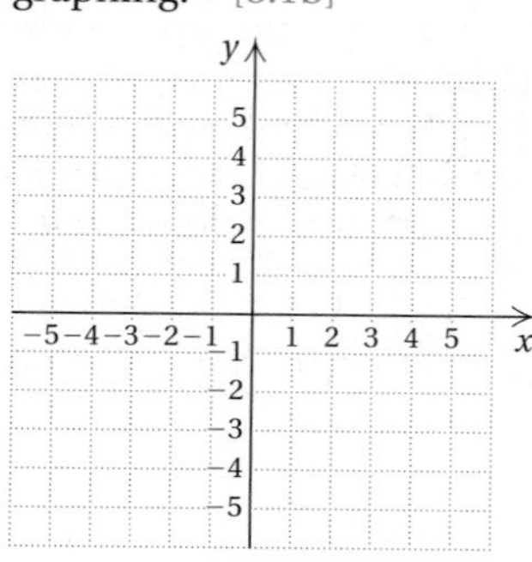

6. $x + 3y = 12,$ $2x - 4y = 4$

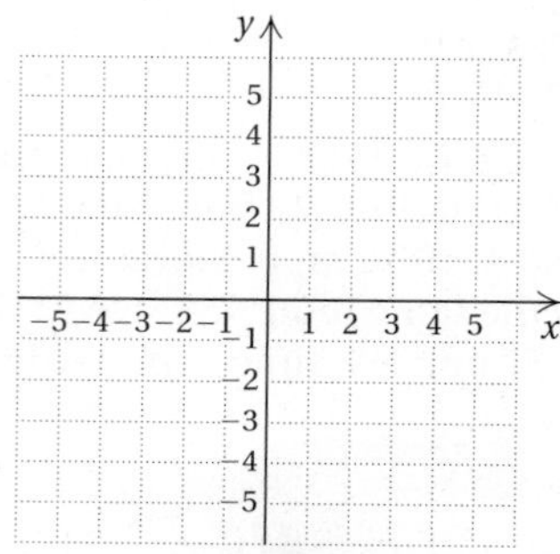

7. $y = 5 - x,$ $3x - 4y = -20$

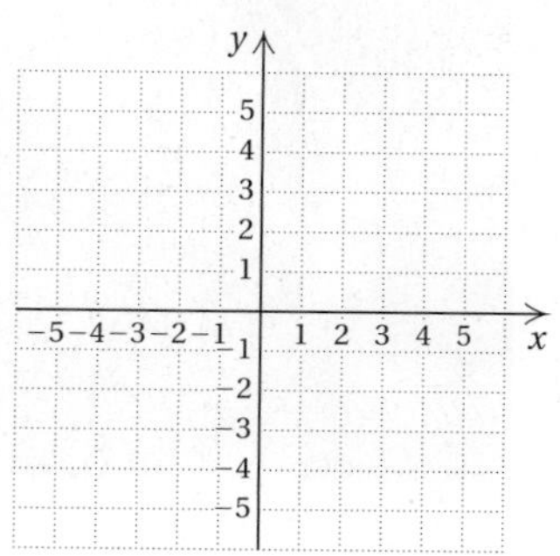

8. $3x - 2y = -4,$
$2y - 3x = -2$

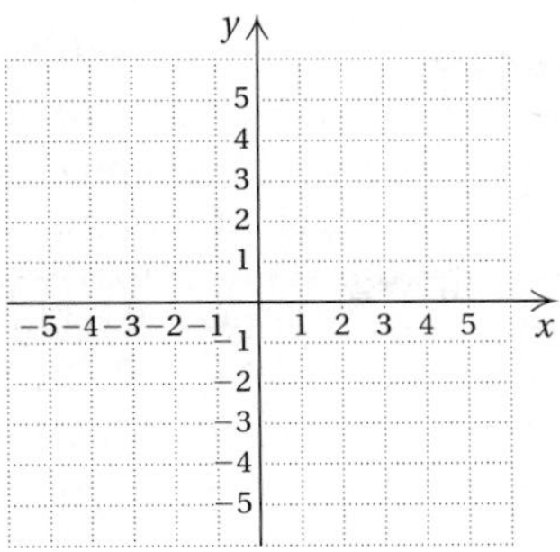

Solve the system using the substitution method. [8.2a]

9. $y = 5 - x,$
$3x - 4y = -20$

10. $x + y = 6,$
$y = 3 - 2x$

11. $x - y = 4,$
$y = 2 - x$

12. $s + t = 5,$
$s = 13 - 3t$

Solve the system using the substitution method. [8.2b]

13. $x + 2y = 6,$
$2x + 3y = 8$

14. $3x + y = 1,$
$x - 2y = 5$

Solve the system using the elimination method. [8.3a]

15. $x + y = 4,$
$2x - y = 5$

16. $x + 2y = 9,$
$3x - 2y = -5$

17. $x - y = 8,$
$2x + y = 7$

Solve the system using the elimination method. [8.3b]

18. $2x + 3y = 8,$
$5x + 2y = -2$

19. $5x - 2y = 2,$
$3x - 7y = 36$

20. $-x - y = -5,$
$2x - y = 4$

21. $6x + 2y = 4,$
$10x + 7y = -8$

22. $-6x - 2y = 5,$
$12x + 4y = -10$

23. $\frac{2}{3}x + y = -\frac{5}{3},$
$x - \frac{1}{3}y = -\frac{13}{3}$

Solve. [8.2c], [8.4a]

24. *Rectangle Dimensions.* The perimeter of a rectangle is 96 cm. The length is 27 cm more than the width. Find the length and the width.

25. *Paid Admissions.* There were 508 people at a rock concert. Orchestra seats cost \$25 each and balcony seats cost \$18 each. The total receipts were \$11,223. Find the number of orchestra seats and the number of balcony seats sold for the concert.

26. *Window Cleaner.* Clear Shine window cleaner is 30% alcohol, whereas Sunstream window cleaner is 60% alcohol. How much of each is needed to make 80 L of a cleaner that is 45% alcohol?

27. *Weights of Elephants.* A zoo has both an Asian and an African elephant. The African elephant weighs 2400 kg more than the Asian elephant. Together, they weigh 12,000 kg. How much does each elephant weigh?

28. *Mixed Nuts.* Sandy's Catering needs to provide 10 lb of mixed nuts for a wedding reception. The wedding couple has allocated \$40 for nuts. Peanuts cost \$2.50 per pound and fancy nuts cost \$7 per pound. How many pounds of each type should be mixed?

29. *Phone Rates.* Recently, Sprint offered one calling plan that charges 25¢ a minute for calling-card calls. Another plan charges 7¢ a minute for calling-card calls, but costs an additional \$4 per month. For what number of minutes will the two plans cost the same?

30. *Octane Ratings.* The octane rating of a gasoline is a measure of the amount of isooctane in the gas. How much 87-octane gas and 95-octane gas should be blended in order to end up with a 10-gal batch of 93-octane gas?
Source: Champlain Electric and Petroleum Equipment

31. *Age.* Jeff is three times as old as his son. In 9 yr, Jeff will be twice as old as his son. How old is each now?

Solve. [8.5a]

32. *Air Travel.* An airplane flew for 4 hr with a 15-km/h tail wind. The return flight against the wind took 5 hr. Find the speed of the airplane in still air.

33. *Car Travel.* One car leaves Phoenix, Arizona, on Interstate highway I-10 traveling at a speed of 55 mph. Two hours later, another car leaves Phoenix on the same highway, but travels at the new speed limit of 75 mph. How far from Phoenix will the second car catch up to the other?

34. *Complementary Angles.* Two angles are complementary. Their difference is 26°. Find the measure of each angle.

35. $^{D}_{W}$ Briefly compare the strengths and weaknesses of the graphical, substitution, and elimination methods. [8.3b]

36. $^{D}_{W}$ Janine can tell by inspection that the system

$$y = 2x - 1,$$
$$y = 2x + 3$$

has no solution. How did she determine this? [8.1b]

SKILL MAINTENANCE

Certain objectives from four particular sections will be retested on the chapter test. The objectives are listed with the practice problems that follow.

Simplify.

37. $t^{-5} \cdot t^{13}$ [4.1d, f]

38. $\frac{t^{-5}}{t^{13}}$ [4.1e, f]

39. Subtract: [6.5a]

$$\frac{x}{x^2 - 9} - \frac{x - 1}{x^2 - 5x + 6}.$$

40. Simplify: [6.1c]

$$\frac{5x^2 - 20}{5x^2 + 40x - 100}.$$

41. Find the intercepts. Then graph the equation. [3.3a]

$$2y - x = 6$$

SYNTHESIS

42. *Value of a Horse.* Stephanie agreed to work as a stablehand for 1 yr. At the end of that time, she was to receive \$2400 and one horse. After 7 months, she quit the job, but still received the horse and \$1000. What was the value of the horse? [8.4a]

43. The solution of the following system is (6, 2). Find C and D. [8.1a]

$$2x - Dy = 6,$$
$$Cx + 4y = 14$$

44. Solve: [8.2a]

$$3(x - y) = 4 + x,$$
$$x = 5y + 2.$$

Each of the following shows the graph of a system of equations. Find the equations. [7.1c], [8.1b]

45. 

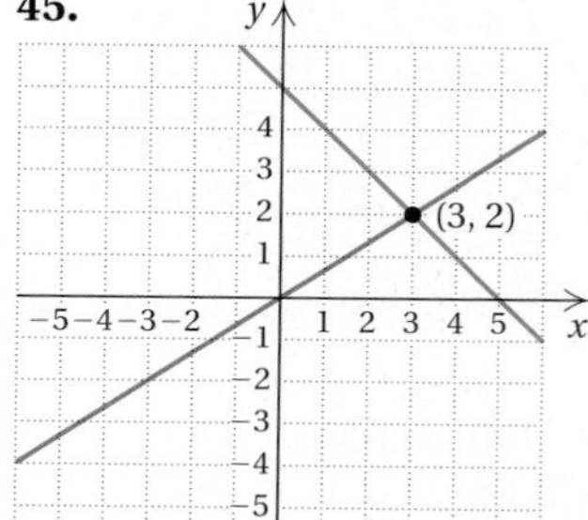

46. 

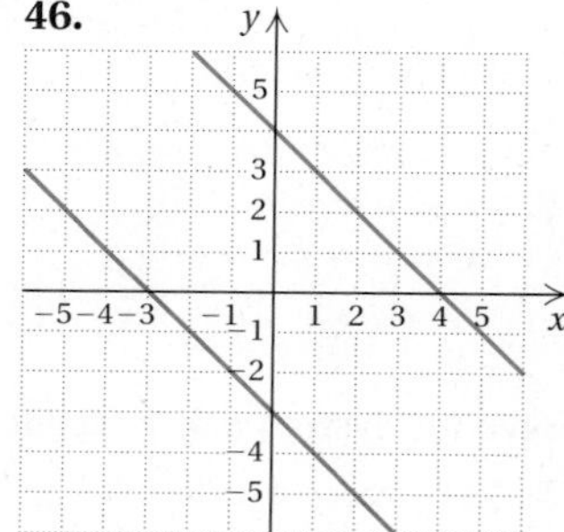

47. *Ancient Chinese Math Problem.* Several ancient Chinese books included problems that can be solved by translating to systems of equations. *Arithmetical Rules in Nine Sections* is a book of 246 problems compiled by a Chinese mathematician, Chang Tsang, who died in 152 B.C. One of the problems is: Suppose there are a number of rabbits and pheasants confined in a cage. In all, there are 35 heads and 94 feet. How many rabbits and how many pheasants are there? Solve the problem. [8.4a]

CHAPTER 8

Chapter Test

1. Determine whether the given ordered pair is a solution of the system of equations.

$$(-2, -1); \quad x = 4 + 2y, \quad 2y - 3x = 4$$

2. Solve this system by graphing. Show your work.

$$x - y = 3,$$
$$x - 2y = 4.$$

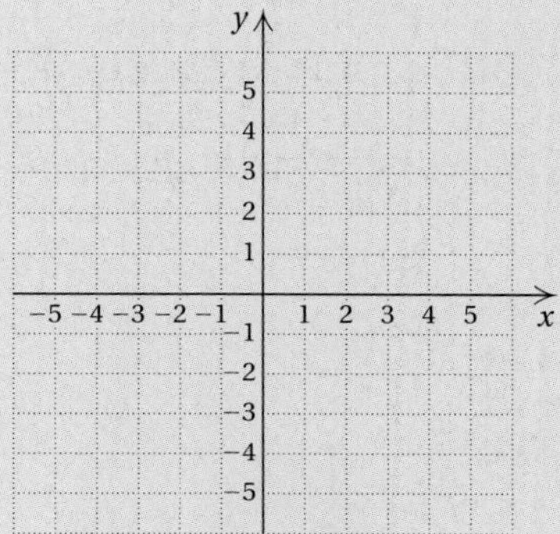

Solve the system using the substitution method.

3. $y = 6 - x,$ $2x - 3y = 22$

4. $x + 2y = 5,$ $x + y = 2$

5. $y = 5x - 2,$ $y - 2 = 5x$

Solve the system using the elimination method.

6. $x - y = 6,$ $3x + y = -2$

7. $\frac{1}{2}x - \frac{1}{3}y = 8,$ $\frac{2}{3}x + \frac{1}{2}y = 5$

8. $4x + 5y = 5,$ $6x + 7y = 7$

9. $2x + 3y = 13,$ $3x - 5y = 10$

Solve.

10. *Rectangle Dimensions.* The perimeter of a rectangular field is 8266 yd. The length is 84 yd more than the width. Find the length and the width.

11. *Mixture of Solutions.* Solution A is 25% acid, and solution B is 40% acid. How much of each is needed to make 60 L of a solution that is 30% acid?

12. *Motorboat Travel.* A motorboat traveled for 2 hr with an 8-km/h current. The return trip against the same current took 3 hr. Find the speed of the motorboat in still water.

13. *Carnival Prices.* A carnival comes to town and makes an income of $4275 one day. Twice as much was made on concessions as on the rides. How much did the concessions bring in? How much did the rides bring in?

14. *Farm Acreage.* The Rolling Velvet Horse Farm allots 650 acres to plant hay and oats. The owners know that their needs are best met if they plant 180 acres more of hay than of oats. How many acres of each should they plant?

15. *Supplementary Angles.* Two angles are supplementary. One angle measures 45° more than twice the measure of the other. Find the measure of each angle.

16. *Octane Ratings.* The octane rating of a gasoline is a measure of the amount of isooctane in the gas. How much 87-octane gas and 93-octane gas should be blended in order to end up with 12 gal of 91-octane gas?
Source: Champlain Electric and Petroleum Equipment

17. *Phone Rates.* One calling plan offered by MCI World Com charges 12.9¢ per minute for daytime long-distance phone calls. A competing plan offered by AT&T charges a monthly fee of $4.95 plus 7¢ per minute for daytime long-distance phone calls. For how many minutes of long-distance calls per month are the costs of the two plans the same?

18. *Ski Trip.* A group of students drive both a car and an SUV on a ski trip. The car left first and traveled at 55 mph. The SUV left 2 hr later and traveled at 65 mph. How long will it take the SUV to catch up to the car?

SKILL MAINTENANCE

19. Subtract: $\frac{1}{x^2 - 16} - \frac{x - 4}{x^2 - 3x - 4}$.

20. Graph: $3x - 4y = -12$.

Simplify.

21. $(2x^{-2}y^7)(5x^6y^{-9})$

22. $\frac{a^4b^2}{a^{-6}b^8}$

23. $\frac{5x^2 + 40x - 100}{10x^2 - 40}$

SYNTHESIS

24. Find the numbers C and D such that $(-2, 3)$ is a solution of the system

$$Cx - 4y = 7,$$
$$3x + Dy = 8.$$

25. *Ticket Line.* You are in line at a ticket window. There are two more people ahead of you than there are behind you. In the entire line, there are three times as many people as there are behind you. How many are ahead of you in line?

Each of the following shows the graph of a system of equations. Find the equations.

26.

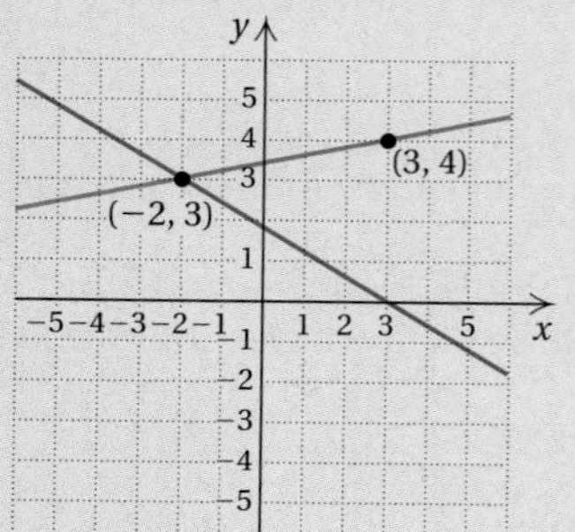

27.

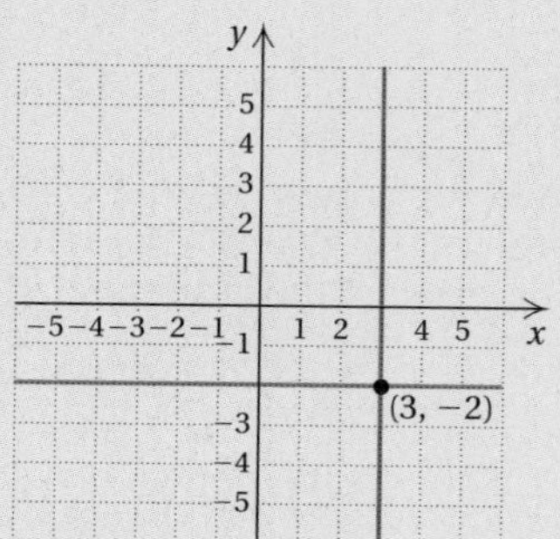

CHAPTERS

1–8 Cumulative Review

Compute and simplify.

1. $-2[1.4 - (-0.8 - 1.2)]$

2. $(1.3 \times 10^8)(2.4 \times 10^{-10})$

3. $\left(-\frac{1}{6}\right) \div \left(\frac{2}{9}\right)$

4. $\frac{2^{12}2^{-7}}{2^8}$

Simplify.

5. $\frac{x^2 - 9}{2x^2 - 7x + 3}$

6. $\frac{t^2 - 16}{(t + 4)^2}$

7. $\frac{x - \frac{x}{x + 2}}{\frac{2}{x} - \frac{1}{x + 2}}$

Perform the indicated operations and simplify.

8. $(1 - 3x^2)(2 - 4x^2)$

9. $(2a^2b - 5ab^2)^2$

10. $(3x^2 + 4y)(3x^2 - 4y)$

11. $-2x^2(x - 2x^2 + 3x^3)$

12. $(1 + 2x)(4x^2 - 2x + 1)$

13. $\left(8 - \frac{1}{3}x\right)\left(8 + \frac{1}{3}x\right)$

14. $(-8y^2 - y + 2) - (y^3 - 6y^2 + y - 5)$

15. $(2x^3 - 3x^2 - x - 1) \div (2x - 1)$

16. $\frac{7}{5x - 25} + \frac{x + 7}{5 - x}$

17. $\frac{2x - 1}{x - 2} - \frac{2x}{2 - x}$

18. $\frac{y^2 + y}{y^2 + y - 2} \cdot \frac{y + 2}{y^2 - 1}$

19. $\frac{7x + 7}{x^2 - 2x} \div \frac{14}{3x - 6}$

Factor completely.

20. $6x^5 - 36x^3 + 9x^2$

21. $16y^4 - 81$

22. $3x^2 + 10x - 8$

23. $4x^4 - 12x^2y + 9y^2$

24. $3m^3 + 6m^2 - 45m$

25. $x^3 + x^2 - x - 1$

Solve.

26. $3x - 4(x + 1) = 5$

27. $x(2x - 5) = 0$

28. $5x + 3 \geq 6(x - 4) + 7$

29. $1.5x - 2.3x = 0.4(x - 0.9)$

30. $2x^2 = 338$

31. $3x^2 + 15 = 14x$

32. $\frac{2}{x} - \frac{3}{x - 2} = \frac{1}{x}$

33. $1 + \frac{3}{x} + \frac{x}{x + 1} = \frac{1}{x^2 + x}$

34. $y = 2x - 9,$
$2x + 3y = -3$

35. $6x + 3y = -6,$
$-2x + 5y = 14$

36. $2x = y - 2,$
$3y - 6x = 6$

Solve.

37. *Desert Dwellers.* The world's population is 6 billion, 96% of whom do not live in the desert. How many do live in the desert?
Source: Kathy Wollard, *How Come Planet Earth?* New York: Workman, 1999.

38. *Elephant Food Consumption.* The average elephant eats 600–700 lb of food per day and drinks 20–40 gal of water per day. The average elephant weighs 6 tons. What percentage of its body weight does an elephant eat and drink each day, at most? (*Hint*: 1 gal of water weighs 8.3453 lb.)
Source: Kathy Wollard, *How Come Planet Earth?* New York: Workman, 1999.

39. *Triangle Dimensions.* The length of one leg of a right triangle is 8 m. The length of the hypotenuse is 4 m longer than the length of the other leg. Find the lengths of the hypotenuse and the other leg.

40. *Fish Population.* To determine the number of fish in a lake, a conservationist catches 85 fish, tags them, and throws them back into the lake. Later, 60 fish are caught, 25 of which are tagged. How many fish are in the lake?

41. *Triangle Dimensions.* The height of a triangle is 3 cm less than the base. The area is 27 cm^2. Find the height and the base.

42. *Heights of Parallelograms.* The height h of a parallelogram of fixed area varies inversely as the base b. Suppose that the height is 24 ft when the base is 15 ft. Find the height when the base is 5 ft. What is the variation constant?

43. *Car Travel.* Two cars leave town at the same time going in the same direction. One travels 50 mph and the other travels 55 mph. In how many hours will they be 50 mi apart?

44. *Mixture of Solutions.* Solution A is 20% alcohol, and solution B is 60% alcohol. How much of each should be used in order to make 10 L of a solution that is 50% alcohol?

45. *Costs of Promotional Buttons.* The vice-president of a sorority has \$100 to spend on promotional buttons. There is a set-up fee of \$18 and a cost of 35¢ per button. How many buttons can she purchase?

46. *Work Time.* It takes David 15 hr to put a roof on a house. It takes Loren 9 hr to put a roof on the same type of house. How long would it take to complete the job if they worked together?

47. Find an equation of variation in which y varies directly as x and $y = 2.4$ when $x = 12$. Then find the value of y when $x = 50$.

48. Find the slope of the line containing the points $(2, 3)$ and $(-1, 3)$.

49. Find the slope and the y-intercept of the line $2x + 3y = 6$.

50. Find an equation of the line that contains the points $(-5, 6)$ and $(2, -4)$.

51. Find an equation of the line containing the point $(0, -3)$ and having the slope $m = 6$.

Graph on a plane.

52. $y = -2$

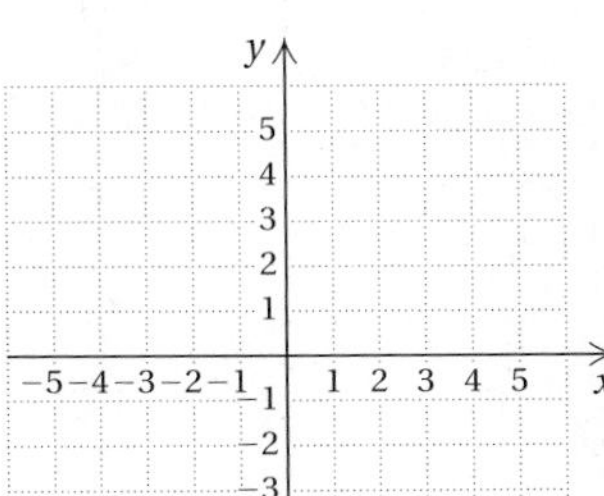

53. $2x + 5y = 10$

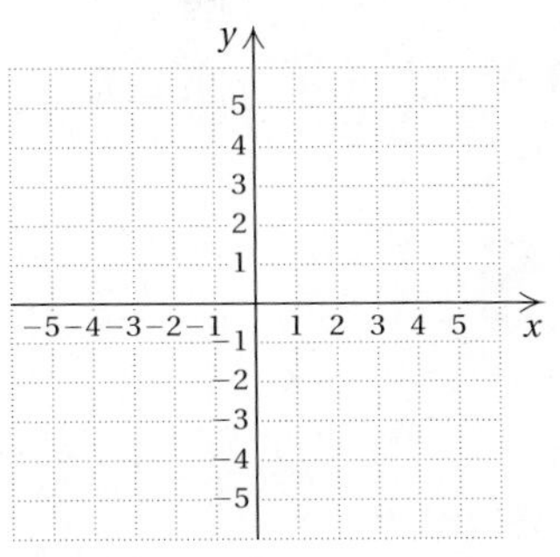

54. $y \leq 5x$

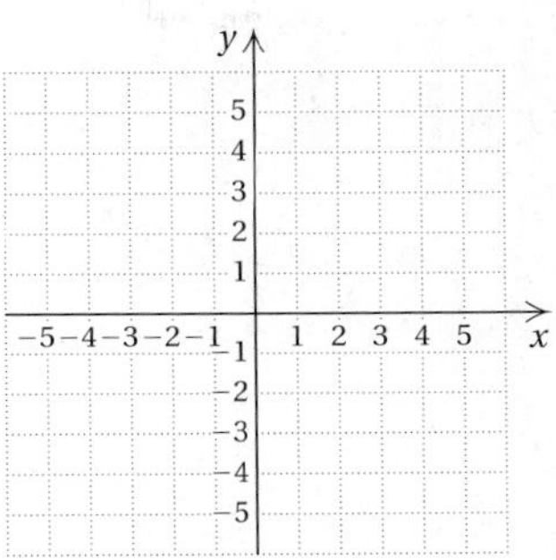

55. $5x - 1 < 24$

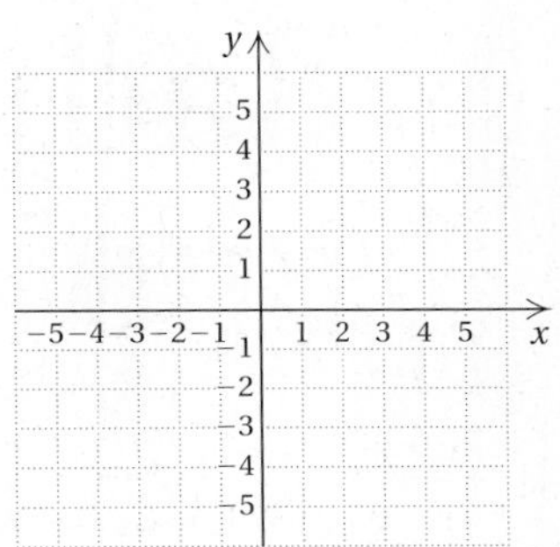

Solve by graphing. Show your work.

56. $x = 5 + y,$
$x - y = 1$

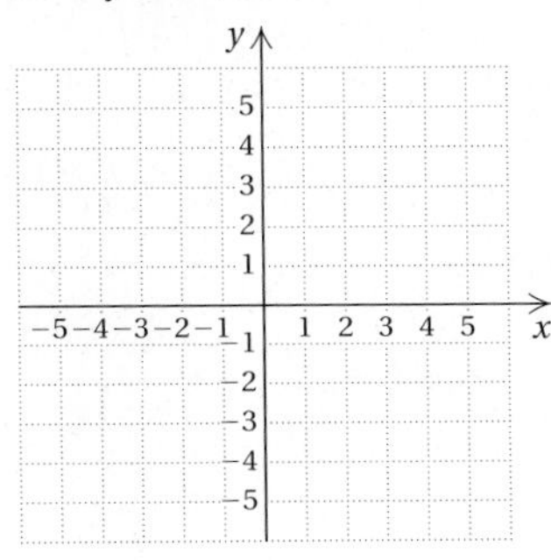

57. $3x - y = 4,$
$x + 3y = -2$

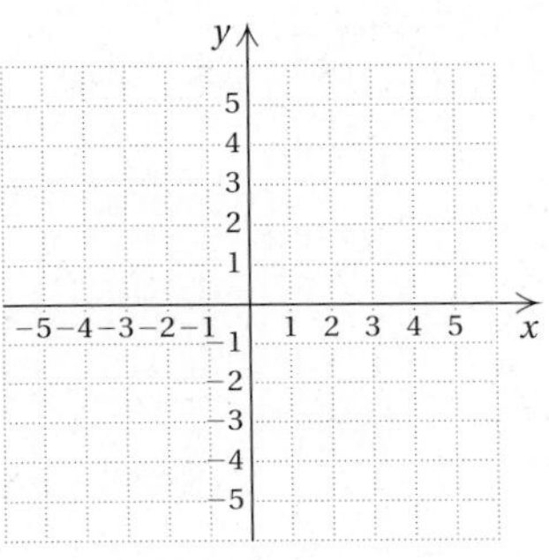

For each of Exercises 58–61, choose the correct answer from the selections given.

58. In math, which of these terms describes the reference line on a graph that is usually labeled "x" or "y"? (This was an \$8000 question on the television quiz show "Who Wants To Be a Millionaire?".)

a) Cosine **b)** Algorithm
c) Axis **d)** Base

59. A cage contains rabbits and ducks. An observer counts 81 heads and 218 feet. Let $x =$ the number of rabbits and $y =$ the number of ducks. Which system represents a translation of the problem?

a) $4x + 2y = 218,$ $x + y = 81$
b) $2x + 4y = 218,$ $x + y = 81$
c) $2x + y = 109,$ $x - y = 81$
d) $x + 2y = 109,$ $x + y = 81$
e) None of these

60. Solve: $\dfrac{x}{x+4} + \dfrac{4x}{x+2} = \dfrac{8}{x^2 + 6x + 8}$.

a) $\dfrac{5x^2 + 18x}{x^2 + 6x + 8}$
b) $\dfrac{8}{x^2 + 6x + 8}$
c) $-4, -2$
d) There are two solutions, one of which is between -1 and 0.
e) There is one solution, and it is between 0 and 1.

61. A tank can be filled in 8 hr by pipe A working alone and in 12 hr by pipe B working alone. How long would it take to fill the tank if both pipes are working together?

a) 10 hr **b)** 20 hr **c)** $4\frac{4}{5}$ hr
d) 5.6 hr **e)** None of these

SYNTHESIS

62. The solution of the following system is $(-5, 2)$. Find A and B.

$$3x - Ay = -7,$$
$$Bx + 4y = 15$$

63. Solve: $x^2 + 2 < 0$.

64. Simplify:

$$\frac{x-5}{x+3} - \frac{x^2 - 6x + 5}{x^2 + x - 2} \div \frac{x^2 + 4x + 3}{x^2 + 3x + 2}.$$

65. Find the value of k such that $y - kx = 4$ and $10x - 3y = -12$ are perpendicular.

Radical Expressions and Equations

Gateway to Chapter 9

We now begin a study of radical expressions. We say that 3 is a square root of 9 because its square is 9; that is, $3^2 = 9$. Similarly, -3 is a square root of 9 because $(-3)^2 = 9$. To express that 3 is the positive square root of 9, we write $\sqrt{9} = 3$. We call $\sqrt{9}$ a radical expression.

In this chapter, we study manipulations with radical expressions such as addition, subtraction, multiplication, division, and simplifying. Finally, we consider another equation-solving principle and apply it to applications and problem solving.

Real-World Application

After an accident, how do police determine the speed at which the car had been traveling? The formula $r = 2\sqrt{5L}$ can be used to approximate the speed r, in miles per hour, of a car that has left a skid mark of length L, in feet. What was the speed of a car that left skid marks of length 30 ft?

This problem appears as Example 7 in Section 9.1.

CHAPTER 9

CHAPTER 9

Pretest

1. Find the square roots of 49. [9.1a]

2. Identify the radicand in $\sqrt{3t}$. [9.1d]

Determine whether the expression represents a real number. Write "yes" or "no." [9.1e]

3. $\sqrt{-47}$

4. $\sqrt{81}$

5. Approximate $\sqrt{47}$ to three decimal places. [9.1b]

6. Solve: $\sqrt{2x+1} = 3$. [9.5a]

Assume henceforth that *all* expressions under radicals represent positive numbers.

Simplify.

7. $\sqrt{4x^2}$ [9.2a]

8. $4\sqrt{18} - 2\sqrt{8} + \sqrt{32}$ [9.4a]

Multiply and simplify.

9. $(2 - \sqrt{3})^2$ [9.4b]

10. $(2 - \sqrt{3})(2 + \sqrt{3})$ [9.4b]

11. $\sqrt{6}\sqrt{10}$ [9.2c]

12. $(2\sqrt{6} - 1)^2$ [9.4b]

Divide and simplify.

13. $\dfrac{\sqrt{15}}{\sqrt{3}}$ [9.3a]

14. $\sqrt{\dfrac{24a^7}{3a^3}}$ [9.3b]

15. In a right triangle, $a = 5$ and $b = 8$. Find c, the length of the hypotenuse. Give an exact answer and an approximation to three decimal places. [9.6a]

16. **Guy Wire.** How long is a guy wire reaching from the top of a 12-m pole to a point 7 m from the base of the pole? Give an exact answer and an approximation to three decimal places. [9.6b]

Rationalize the denominator.

17. $\dfrac{\sqrt{5}}{\sqrt{x}}$ [9.3c]

18. $\dfrac{8}{6 + \sqrt{5}}$ [9.4c]

19. **Sightings to the Horizon.** At a height of h meters, you can see V kilometers to the horizon. These numbers are related with the following formula:

$$V = 3.5\sqrt{h}.$$

How far can a pilot see to the horizon from an altitude of 8400 m? [9.5c]

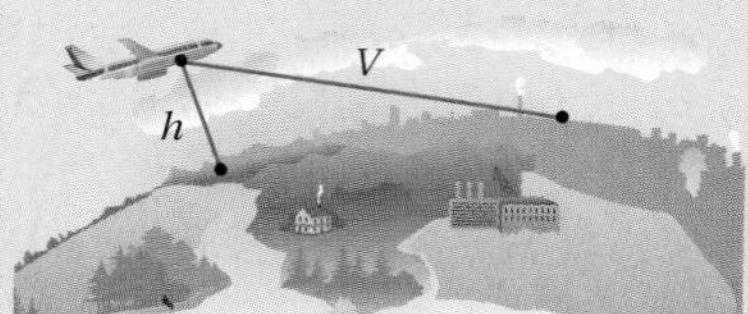

9.1 INTRODUCTION TO RADICAL EXPRESSIONS

Objectives

a Find the principal square roots and their opposites of the whole numbers from 0^2 to 25^2.

b Approximate square roots of real numbers using a calculator.

c Solve applied problems involving square roots.

d Identify radicands of radical expressions.

e Identify whether a radical expression represents a real number.

f Simplify a radical expression with a perfect-square radicand.

a Square Roots

When we raise a number to the second power, we have squared the number. Sometimes we may need to find the number that was squared. We call this process finding a square root of a number.

SQUARE ROOT

The number c is a **square root** of a if $c^2 = a$.

Every positive number has two square roots. For example, the square roots of 25 are 5 and -5 because $5^2 = 25$ and $(-5)^2 = 25$. The positive square root is also called the **principal square root.** The symbol $\sqrt{}$ is called a **radical*** (or **square root**) symbol. The radical symbol represents only the principal square root. Thus, $\sqrt{25} = 5$. To name the negative square root of a number, we use $-\sqrt{}$. The number 0 has only one square root, 0.

EXAMPLE 1 Find the square roots of 81.

The square roots are 9 and -9.

EXAMPLE 2 Find $\sqrt{225}$.

There are two square roots, 15 and -15. We want the principal, or positive, square root since this is what $\sqrt{}$ represents. Thus, $\sqrt{225} = 15$.

EXAMPLE 3 Find $-\sqrt{64}$.

The symbol $\sqrt{64}$ represents the positive square root. Then $-\sqrt{64}$ represents the negative square root. That is, $\sqrt{64} = 8$, so $-\sqrt{64} = -8$.

We can think of the processes of "squaring" and "finding square roots" as inverses of each other. We square a number and get one answer. When we find the square roots of the answer, we get the original number *and* its opposite.

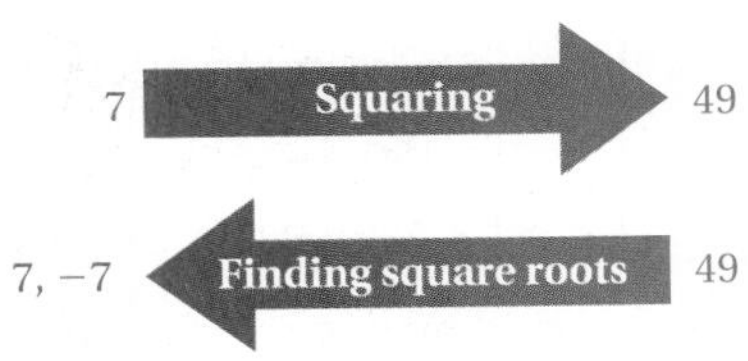

Find the square roots.

1. 36 **2.** 64

3. 121 **4.** 144

Find the following.

5. $\sqrt{16}$ **6.** $\sqrt{49}$

7. $\sqrt{100}$ **8.** $\sqrt{441}$

9. $-\sqrt{49}$ **10.** $-\sqrt{169}$

b Approximating Square Roots

We often need to use rational numbers to *approximate* square roots that are irrational. Such approximations can be found using a calculator with a square-root key $\boxed{\sqrt{}}$.

Use a calculator to approximate each of the following square roots to three decimal places.

11. $\sqrt{15}$ **12.** $\sqrt{30}$

13. $\sqrt{980}$ **14.** $-\sqrt{667.8}$

15. $\sqrt{\dfrac{2}{3}}$ **16.** $-\sqrt{\dfrac{203.4}{67.82}}$

Answers on page A-42

*Radicals can be other than square roots, but we will consider only square-root radicals in Chapter 9. See Appendix B for other types of radicals.

CALCULATOR CORNER

Approximating Square Roots We can use a graphing calculator to approximate square roots. To approximate $\sqrt{18}$, for example, we press [2nd] [√] [1] [8] [)] [ENTER]. (√ is the second operation associated with the [x^2] key.) Although it is not necessary for this example to include the left parenthesis, we do so here in order to close the set of parentheses that are opened when the graphing calculator displays "√(." To approximate $-\sqrt{8.65}$, we press [(−)] [2nd] [√] [8] [.] [6] [5] [)] [ENTER]. We see that $\sqrt{18} \approx 4.243$ and $-\sqrt{8.65} \approx -2.941$.

```
√(18)
          4.242640687
-√(8.65)
         -2.941088234
```

Exercises: Use a graphing calculator to approximate each of the following to three decimal places.

1. $\sqrt{43}$ **2.** $\sqrt{101}$

3. $\sqrt{10{,}467}$ **4.** $\sqrt{\frac{2}{5}}$

5. $-\sqrt{9406}$ **6.** $-\sqrt{\frac{11}{17}}$

17. Speed of a Skidding Car. Referring to Example 7, determine the speed of a car that left skid marks of length **(a)** 40 ft; **(b)** 123 ft.

Answers on page A-42

EXAMPLES Use a calculator to approximate each of the following.

	Number	*Using a calculator with a 10-digit readout*	*Rounded to three decimal places*
4.	$\sqrt{10}$	3.162277660	3.162
5.	$-\sqrt{583.8}$	−24.16195356	−24.162
6.	$\sqrt{\frac{48}{55}}$	0.934198733	0.934

Do Exercises 1–16 on the preceding page.

C Applications of Square Roots

We now consider an application involving a formula with a radical expression.

EXAMPLE 7 *Speed of a Skidding Car.* After an accident, how do police determine the speed at which the car had been traveling? The formula $r = 2\sqrt{5L}$ can be used to approximate the speed r, in miles per hour, of a car that has left a skid mark of length L, in feet. What was the speed of a car that left skid marks of length **(a)** 30 ft? **(b)** 150 ft?

a) We substitute 30 for L and find an approximation:

$$r = 2\sqrt{5L} = 2\sqrt{5 \cdot 30} = 2\sqrt{150} \approx 24.495.$$

The speed of the car was about 24.5 mph.

b) We substitute 150 for L and find an approximation:

$$r = 2\sqrt{5L} = 2\sqrt{5 \cdot 150} \approx 54.772.$$

The speed of the car was about 54.8 mph.

Do Exercise 17.

d Radicands and Radical Expressions

When an expression is written under a radical, we have a **radical expression.** Here are some examples:

$$\sqrt{14}, \qquad \sqrt{x}, \qquad 8\sqrt{x^2+4}, \qquad \sqrt{\frac{x^2-5}{2}}.$$

The expression written under the radical is called the **radicand.**

EXAMPLES Identify the radicand in each expression.

8. $\sqrt{105}$ The radicand is 105.

9. $\sqrt{x}$ The radicand is x.

10. $6\sqrt{y^2-5}$ The radicand is y^2-5.

11. $\sqrt{\dfrac{a-b}{a+b}}$ The radicand is $\dfrac{a-b}{a+b}$.

Do Exercises 18–21.

e Expressions That Are Meaningful as Real Numbers

The square of any nonzero number is always positive. For example, $8^2 = 64$ and $(-11)^2 = 121$. There are no real numbers that when squared yield negative numbers. For example, $\sqrt{-100}$ does not represent a real number because there are no real numbers that when squared yield -100. We can try to square 10 and -10, but we know that $10^2 = 100$ and $(-10)^2 = 100$. Neither square is -100. Thus the following expressions do not represent real numbers (they are meaningless as real numbers):

$$\sqrt{-100}, \qquad \sqrt{-49}, \qquad -\sqrt{-3}.$$

EXCLUDING NEGATIVE RADICANDS

Radical expressions with negative radicands do not represent real numbers.

Later in your study of mathematics, you may encounter a number system called the **complex numbers** in which negative numbers have square roots.

Do Exercises 22–25.

Identify the radicand.

18. $\sqrt{227}$

19. $\sqrt{45+x}$

20. $\sqrt{\dfrac{x}{x+2}}$

21. $8\sqrt{x^2+4}$

Determine whether the expression represents a real number. Write "yes" or "no."

22. $-\sqrt{25}$

23. $\sqrt{-25}$

24. $-\sqrt{-36}$

25. $-\sqrt{36}$

Answers on page A-42

Simplify. Assume that expressions under radicals represent any real number.

26. $\sqrt{(-13)^2}$

27. $\sqrt{(7w)^2}$

28. $\sqrt{(xy)^2}$

29. $\sqrt{x^2y^2}$

30. $\sqrt{(x-11)^2}$

31. $\sqrt{x^2+8x+16}$

Answers on page A-42

f Perfect-Square Radicands

The expression $\sqrt{x^2}$, with a perfect-square radicand, x^2, can be troublesome to simplify. Recall that $\sqrt{}$ denotes the principal square root. That is, the answer is nonnegative (either positive or zero). If x represents a nonnegative number, $\sqrt{x^2}$ simplifies to x. If x represents a negative number, $\sqrt{x^2}$ simplifies to $-x$ (the opposite of x), which is positive.

Suppose that $x = 3$. Then

$$\sqrt{x^2} = \sqrt{3^2} = \sqrt{9} = 3.$$

Suppose that $x = -3$. Then

$$\sqrt{x^2} = \sqrt{(-3)^2} = \sqrt{9} = 3, \quad \text{the } \textit{opposite} \text{ of } -3.$$

Note that 3 is the *absolute value* of both 3 and -3. In general, when replacements for x are considered to be *any* real numbers, it follows that

$$\sqrt{x^2} = |x|,$$

and when $x = 3$ or $x = -3$,

$$\sqrt{x^2} = \sqrt{3^2} = |3| = 3 \quad \text{and} \quad \sqrt{x^2} = \sqrt{(-3)^2} = |-3| = 3.$$

PRINCIPAL SQUARE ROOT OF A^2

For any real number A,

$$\sqrt{A^2} = |A|.$$

(That is, for any real number A, the principal square root of A^2 is the absolute value of A.)

EXAMPLES Simplify. Assume that expressions under radicals represent any real number.

12. $\sqrt{10^2} = |10| = 10$
13. $\sqrt{(-7)^2} = |-7| = 7$
14. $\sqrt{(3x)^2} = |3x|$ Absolute-value notation is necessary.
15. $\sqrt{a^2b^2} = \sqrt{(ab)^2} = |ab|$
16. $\sqrt{x^2+2x+1} = \sqrt{(x+1)^2} = |x+1|$

Do Exercises 26–31.

Fortunately, in most uses of radicals, it can be assumed that expressions under radicals are nonnegative or positive. Indeed, many computers and calculators are programmed to consider only nonnegative radicands. Suppose that $x \geq 0$. Then

$$\sqrt{x^2} = |x| = x,$$

since x is nonnegative.

PRINCIPAL SQUARE ROOT OF A^2

For any nonnegative real number A,

$$\sqrt{A^2} = A.$$

(That is, for any nonnegative real number A, the principal square root of A^2 is A.)

EXAMPLES Simplify. Assume that expressions under radicals represent nonnegative real numbers.

17. $\sqrt{(3x)^2} = 3x$ Since $3x$ is assumed to be nonnegative

18. $\sqrt{a^2b^2} = \sqrt{(ab)^2} = ab$ Since ab is assumed to be nonnegative

19. $\sqrt{x^2 + 2x + 1} = \sqrt{(x + 1)^2} = x + 1$ Since $x + 1$ is assumed to be nonnegative

Do Exercises 32–37.

ASSUMING NONNEGATIVE RADICANDS

Henceforth, in this text we will assume that all expressions under radicals represent nonnegative real numbers.

We make this assumption in order to eliminate some confusion and because it is valid in many applications. As you study further in mathematics, however, you will frequently have to make a determination about expressions under radicals being nonnegative or positive. This will often be necessary in calculus.

Simplify. Assume that expressions under radicals represent nonnegative real numbers.

32. $\sqrt{(xy)^2}$ **33.** $\sqrt{x^2y^2}$

34. $\sqrt{(x - 11)^2}$

35. $\sqrt{x^2 + 8x + 16}$

36. $\sqrt{25y^2}$ **37.** $\sqrt{\frac{1}{4}t^2}$

Answers on page A-42

Study Tips

BEGINNING TO STUDY FOR THE FINAL EXAM (PART 2)

The best scenario for preparing for a final exam is to do so over a period of at least two weeks. Work in a diligent, disciplined manner, doing some final-exam preparation *each* day. Here is a detailed plan that many find useful.

1. **Begin by browsing through each chapter, reviewing the highlighted or boxed information regarding important formulas in both the text and the Summary and Review.** There may be some formulas that you will need to memorize.
2. **Retake each chapter test that you took in class, assuming your instructor has returned it. Otherwise, use the chapter test in the book.** Restudy the objectives in the text that correspond to each question you missed.
3. **Then work the Cumulative Review that covers all chapters up to that point.** Be careful to avoid any questions corresponding to objectives not covered. Again, restudy the objectives in the text that correspond to each question you missed.
4. **If you are still missing questions, use the supplements for extra review.** For example, you might check out the video- or audiotapes, the *Student's Solutions Manual,* the InterAct Math Tutorial Software, or MathXL.
5. **For remaining difficulties, see your instructor, go to a tutoring session, or participate in a study group.**
6. **Check for former final exams that may be on file in the math department or a study center, or with students who have already taken the course.** Use them for practice, being alert to trouble spots.
7. **Take the Final Examination in the text during the last couple of days before the final.** Set aside the same amount of time that you will have for the final. See how much of the final exam you can complete under test-like conditions.

"The door of opportunity won't open unless you do some pushing."

Anonymous

9.1 EXERCISE SET

For Extra Help

Digital Video Tutor CD 8 Videotape 16 | InterAct Math | Math Tutor Center | MathXL | MyMathLab

a Find the square roots.

1. 4

2. 1

3. 9

4. 16

5. 100

6. 121

7. 169

8. 144

9. 256

10. 625

Simplify.

11. $\sqrt{4}$

12. $\sqrt{1}$

13. $-\sqrt{9}$

14. $-\sqrt{25}$

15. $-\sqrt{36}$

16. $-\sqrt{81}$

17. $-\sqrt{225}$

18. $\sqrt{400}$

19. $\sqrt{361}$

20. $\sqrt{441}$

b Use a calculator to approximate the square roots. Round to three decimal places.

21. $\sqrt{5}$

22. $\sqrt{8}$

23. $\sqrt{432}$

24. $\sqrt{8196}$

25. $-\sqrt{347.7}$

26. $-\sqrt{204.788}$

27. $\sqrt{\frac{278}{36}}$

28. $-\sqrt{\frac{567}{788}}$

29. $\sqrt{8 \cdot 9 \cdot 200}$

30. $\sqrt{\frac{47 \cdot 83}{947.03}}$

c *Parking-Lot Arrival Spaces.* The attendants at a parking lot park cars in temporary spaces before the cars are taken to permanent parking stalls. The number N of such spaces needed is approximated by the formula $N = 2.5\sqrt{A}$, where A is the average number of arrivals during peak hours.

31. Find the number of spaces needed when the average number of arrivals is **(a)** 25; **(b)** 89.

32. Find the number of spaces needed when the average number of arrivals is **(a)** 62; **(b)** 100.

Hang Time. An athlete's *hang time* (time airborne for a jump), T, in seconds, is given by $T = 0.144\sqrt{V}$, where V is the athlete's vertical leap, in inches.
Source: Peter Brancazio, "The Mechanics of a Slam Dunk," *Popular Mechanics*, November 1991. Courtesy of Peter Brancazio, Brooklyn College.

33. Kobe Bryant of the Los Angeles Lakers can jump 36 in. vertically. Find his hang time.

34. Brian Grant of the Miami Heat can jump 25 in. vertically. Find his hang time.

d Identify the radicand.

35. $\sqrt{200}$

36. $\sqrt{16z}$

37. $\sqrt{a-4}$

38. $\sqrt{3t+10}$

39. $5\sqrt{t^2+1}$

40. $9\sqrt{x^2+16}$

41. $x^2y\sqrt{\dfrac{3}{x+2}}$

42. $ab^2\sqrt{\dfrac{a}{a+b}}$

e Determine whether the expression represents a real number. Write "yes" or "no."

43. $\sqrt{-16}$

44. $\sqrt{-81}$

45. $-\sqrt{81}$

46. $-\sqrt{64}$

f Simplify. Remember that we have assumed that expressions under radicals represent nonnegative real numbers.

47. $\sqrt{c^2}$

48. $\sqrt{x^2}$

49. $\sqrt{9x^2}$

50. $\sqrt{16y^2}$

51. $\sqrt{(8p)^2}$

52. $\sqrt{(7pq)^2}$

53. $\sqrt{(ab)^2}$

54. $\sqrt{(6y)^2}$

55. $\sqrt{(34d)^2}$

56. $\sqrt{(53b)^2}$

57. $\sqrt{(x + 3)^2}$

58. $\sqrt{(d - 3)^2}$

59. $\sqrt{a^2 - 10a + 25}$

60. $\sqrt{x^2 + 2x + 1}$

61. $\sqrt{4a^2 - 20a + 25}$

62. $\sqrt{9p^2 + 12p + 4}$

63. $^D{}_W$ What is the difference between "**the** square root of 10" and "**a** square root of 10"?

64. $^D{}_W$ Explain the difference between the two descriptions of the principal square root of A^2 given on p. 648.

SKILL MAINTENANCE

65. *Food Expenses.* The amount F that a family spends on food varies directly as its income I. A family making \$39,200 a year will spend \$10,192 on food. At this rate, how much would a family making \$41,000 spend on food? [7.5b]

Divide and simplify. [6.2b]

66. $\dfrac{x - 3}{x + 4} \div \dfrac{x^2 - 9}{x + 4}$

67. $\dfrac{x^2 + 10x - 11}{x^2 - 1} \div \dfrac{x + 11}{x + 1}$

68. $\dfrac{x^4 - 16}{x^4 - 1} \div \dfrac{x^2 + 4}{x^2 + 1}$

SYNTHESIS

69. Use only the graph of $y = \sqrt{x}$, shown below, to approximate $\sqrt{3}$, $\sqrt{5}$, and $\sqrt{7}$. Answers may vary.

70. Between what two consecutive integers is $\sqrt{78}$?

Solve.

71. $\sqrt{x^2} = 16$

72. $\sqrt{y^2} = -7$

73. $t^2 = 49$

74. Suppose that the area of a square is 3. Find the length of a side.

9.2 MULTIPLYING AND SIMPLIFYING WITH RADICAL EXPRESSIONS

Objectives

a Simplify radical expressions.

b Simplify radical expressions where radicands are powers.

c Multiply radical expressions and simplify if possible.

a Simplifying by Factoring

To see how to multiply with radical notation, consider the following.

a) $\sqrt{9} \cdot \sqrt{4} = 3 \cdot 2 = 6$ This is a product of square roots.

b) $\sqrt{9 \cdot 4} = \sqrt{36} = 6$ This is the square root of a product.

Note that

$$\sqrt{9} \cdot \sqrt{4} = \sqrt{9 \cdot 4}.$$

Do Exercise 1.

We can multiply radical expressions by multiplying the radicands.

THE PRODUCT RULE FOR RADICALS

For any nonnegative radicands A and B,

$$\sqrt{A} \cdot \sqrt{B} = \sqrt{A \cdot B}.$$

(The product of square roots is the square root of the product of the radicands.)

EXAMPLES Multiply.

1. $\sqrt{5}\sqrt{7} = \sqrt{5 \cdot 7} = \sqrt{35}$
2. $\sqrt{8}\sqrt{8} = \sqrt{8 \cdot 8} = \sqrt{64} = 8$
3. $\sqrt{\frac{2}{3}}\sqrt{\frac{4}{5}} = \sqrt{\frac{2}{3} \cdot \frac{4}{5}} = \sqrt{\frac{8}{15}}$
4. $\sqrt{2x}\sqrt{3x - 1} = \sqrt{2x(3x - 1)} = \sqrt{6x^2 - 2x}$

Do Exercises 2–5.

To factor radical expressions, we can use the product rule for radicals in reverse.

FACTORING RADICAL EXPRESSIONS

$$\sqrt{AB} = \sqrt{A}\sqrt{B}.$$

In some cases, we can simplify after factoring.

A square-root radical expression is simplified when its radicand has no factors that are perfect squares.

When simplifying a square-root radical expression, we first determine whether a radicand is a perfect square. Then we determine whether it has perfect-square factors. The radicand is then factored and the radical expression simplified using the preceding rule.

1. Simplify.

a) $\sqrt{25} \cdot \sqrt{16}$

b) $\sqrt{25 \cdot 16}$

Multiply.

2. $\sqrt{3}\sqrt{11}$

3. $\sqrt{5}\sqrt{5}$

4. $\sqrt{x}\sqrt{x + 1}$

5. $\sqrt{x + 2}\sqrt{x - 2}$

Answers on page A-42

Simplify by factoring.

6. $\sqrt{32}$

7. $\sqrt{x^2 + 14x + 49}$

8. $\sqrt{25x^2}$

9. $\sqrt{36m^2}$

10. $\sqrt{92}$

11. $\sqrt{x^2 - 20x + 100}$

12. $\sqrt{64t^2}$

13. $\sqrt{100a^2}$

Answers on page A-42

Compare the following:

$$\sqrt{50} = \sqrt{10 \cdot 5} = \sqrt{10}\sqrt{5};$$
$$\sqrt{50} = \sqrt{25 \cdot 2} = \sqrt{25}\sqrt{2} = 5\sqrt{2}.$$

In the second case, the radicand has the perfect-square factor 25. If you do not recognize perfect-square factors, try factoring the radicand into its prime factors. For example,

$$\sqrt{50} = \sqrt{2 \cdot \underbrace{5 \cdot 5}} = 5\sqrt{2}.$$

Perfect square (a pair of the same factors)

Square-root radical expressions in which the radicand has no perfect-square factors, such as $5\sqrt{2}$, are considered to be in simplest form.

EXAMPLES Simplify by factoring.

5. $\sqrt{18} = \sqrt{9 \cdot 2}$ — Identifying a perfect-square factor and factoring the radicand. The factor 9 is a perfect square.
$= \sqrt{9} \cdot \sqrt{2}$ — Factoring into a product of radicals
$= 3\sqrt{2}$ — The radicand has no factors that are perfect squares.

6. $\sqrt{48t} = \sqrt{16 \cdot 3 \cdot t}$ — Identifying a perfect-square factor and factoring the radicand. The factor 16 is a perfect square.
$= \sqrt{16}\sqrt{3t}$ — Factoring into a product of radicals
$= 4\sqrt{3t}$ — Taking a square root

7. $\sqrt{20t^2} = \sqrt{4 \cdot 5 \cdot t^2}$ — Identifying perfect-square factors and factoring the radicand. The factors 4 and t^2 are perfect squares.
$= \sqrt{4}\sqrt{t^2}\sqrt{5}$ — Factoring into a product of several radicals
$= 2t\sqrt{5}$ — Taking square roots. No absolute-value signs are necessary since we have assumed that expressions under radicals are nonnegative.

8. $\sqrt{x^2 - 6x + 9} = \sqrt{(x-3)^2} = x - 3$ — No absolute-value signs are necessary since we have assumed that expressions under radicals are nonnegative.

9. $\sqrt{36x^2} = \sqrt{36}\sqrt{x^2} = 6x$, or $\sqrt{36x^2} = \sqrt{(6x)^2} = 6x$

10. $\sqrt{3x^2 + 6x + 3} = \sqrt{3(x^2 + 2x + 1)}$ — Factoring the radicand
$= \sqrt{3(x+1)^2}$ — Factoring further
$= \sqrt{3}\sqrt{(x+1)^2}$ — Factoring into a product of radicals
$= \sqrt{3}(x+1)$ — Taking the square root

Do Exercises 6–13.

b Simplifying Square Roots of Powers

To take the square root of an even power such as x^{10}, we note that $x^{10} = (x^5)^2$. Then

$$\sqrt{x^{10}} = \sqrt{(x^5)^2} = x^5.$$

We can find the answer by taking half the exponent. That is,

$$\sqrt{x^{10}} = x^5. \leftarrow \tfrac{1}{2}(10) = 5$$

EXAMPLES Simplify.

11. $\sqrt{x^6} = \sqrt{(x^3)^2} = x^3 \leftarrow \tfrac{1}{2}(6) = 3$
12. $\sqrt{x^8} = x^4$
13. $\sqrt{t^{22}} = t^{11}$

Do Exercises 14–16.

If an odd power occurs, we express the power in terms of the largest even power. Then we simplify the even power as in Examples 11–13.

EXAMPLE 14 Simplify by factoring: $\sqrt{x^9}$.

$$\sqrt{x^9} = \sqrt{x^8 \cdot x}$$
$$= \sqrt{x^8}\sqrt{x}$$
$$= x^4\sqrt{x}$$

CAUTION! Note that $\sqrt{x^9} \neq x^3$.

EXAMPLE 15 Simplify by factoring: $\sqrt{32x^{15}}$.

$\sqrt{32x^{15}} = \sqrt{16 \cdot 2 \cdot x^{14} \cdot x}$ We factor the radicand, looking for perfect-square factors. The largest even power is 14.

$= \sqrt{16}\sqrt{x^{14}}\sqrt{2x}$ Factoring into a product of radicals. Perfect-square factors are usually listed first.

$= 4x^7\sqrt{2x}$ Simplifying

Do Exercises 17 and 18.

c Multiplying and Simplifying

Sometimes we can simplify after multiplying. We leave the radicand in factored form and factor further to determine perfect-square factors. Then we simplify the perfect-square factors.

EXAMPLE 16 Multiply and then simplify by factoring: $\sqrt{2}\sqrt{14}$.

$\sqrt{2}\sqrt{14} = \sqrt{2 \cdot 14}$ Multiplying

$= \sqrt{2 \cdot 2 \cdot 7}$ Factoring

$= \sqrt{2 \cdot 2}\sqrt{7}$ Looking for perfect-square factors; pairs of factors

$= 2\sqrt{7}$

Do Exercises 19 and 20.

Simplify.

14. $\sqrt{t^4}$

15. $\sqrt{t^{20}}$

16. $\sqrt{h^{46}}$

Simplify by factoring.

17. $\sqrt{x^7}$

18. $\sqrt{24x^{11}}$

Multiply and simplify.

19. $\sqrt{3}\sqrt{6}$

20. $\sqrt{2}\sqrt{50}$

Answers on page A-42

Multiply and simplify.

21. $\sqrt{2x^3}\sqrt{8x^3y^4}$

22. $\sqrt{10xy^2}\sqrt{5x^2y^3}$

23. $\sqrt{28q^2r} \cdot \sqrt{21q^3r^7}$

Answers on page A-42

CALCULATOR CORNER

Simplifying Radical Expressions

Exercises: Use a table or a graph to determine whether each of the following is true.

1. $\sqrt{x+4} = \sqrt{x} + 2$
2. $\sqrt{3+x} = \sqrt{3} + x$
3. $\sqrt{x-2} = \sqrt{x} - \sqrt{2}$
4. $\sqrt{9x} = 3\sqrt{x}$

EXAMPLE 17 Multiply and then simplify by factoring: $\sqrt{3x^2}\sqrt{9x^3}$.

$$\sqrt{3x^2}\sqrt{9x^3} = \sqrt{3x^2 \cdot 9x^3} \quad \text{Multiplying}$$
$$= \sqrt{3 \cdot x^2 \cdot 9 \cdot x^2 \cdot x} \quad \text{Looking for perfect-square factors or largest even powers}$$

Perfect-square factors are usually listed first.

$$= \sqrt{9}\sqrt{x^2}\sqrt{x^2}\sqrt{3x}$$
$$= 3 \cdot x \cdot x \cdot \sqrt{3x}$$
$$= 3x^2\sqrt{3x}$$

In doing an example like the preceding one, it might be helpful to do more factoring, as follows:

$$\sqrt{3x^2} \cdot \sqrt{9x^3} = \sqrt{3 \cdot \underline{x \cdot x} \cdot \underline{3 \cdot 3} \cdot \underline{x \cdot x} \cdot x}.$$

Then we look for pairs of factors, as shown, and simplify perfect-square factors:

$$= 3 \cdot x \cdot x\sqrt{3x}$$
$$= 3x^2\sqrt{3x}.$$

EXAMPLE 18 Simplify: $\sqrt{20cd^2}\sqrt{35cd^5}$.

$$\sqrt{20cd^2}\sqrt{35cd^5}$$
$$= \sqrt{20cd^2 \cdot 35cd^5} \quad \text{Multiplying}$$
$$= \sqrt{\underline{2 \cdot 2} \cdot 5 \cdot c \cdot \underline{d \cdot d} \cdot 5 \cdot 7 \cdot c \cdot \underline{d \cdot d} \cdot \underline{d \cdot d} \cdot d} \quad \text{Looking for pairs of factors}$$
$$= 2 \cdot 5 \cdot c \cdot d \cdot d \cdot d\sqrt{7d}$$
$$= 10cd^3\sqrt{7d}$$

Do Exercises 21–23.

We know that $\sqrt{AB} = \sqrt{A}\sqrt{B}$. That is, the square root of a product is the product of the square roots. What about the square root of a sum? That is, is the square root of a sum equal to the sum of the square roots? To check, consider $\sqrt{A+B}$ and $\sqrt{A} + \sqrt{B}$ when $A = 16$ and $B = 9$:

$$\sqrt{A+B} = \sqrt{16+9} = \sqrt{25} = 5;$$

and

$$\sqrt{A} + \sqrt{B} = \sqrt{16} + \sqrt{9} = 4 + 3 = 7.$$

Thus we see the following.

CAUTION!

The square root of a sum is not the sum of the square roots.

$$\sqrt{A+B} \neq \sqrt{A} + \sqrt{B}$$

9.2 EXERCISE SET

For Extra Help

Digital Video Tutor CD 8 Videotape 16 | InterAct Math | Math Tutor Center | MathXL | MyMathLab

a Simplify by factoring.

1. $\sqrt{12}$

2. $\sqrt{8}$

3. $\sqrt{75}$

4. $\sqrt{50}$

5. $\sqrt{20}$

6. $\sqrt{45}$

7. $\sqrt{600}$

8. $\sqrt{300}$

9. $\sqrt{486}$

10. $\sqrt{567}$

11. $\sqrt{9x}$

12. $\sqrt{4y}$

13. $\sqrt{48x}$

14. $\sqrt{40m}$

15. $\sqrt{16a}$

16. $\sqrt{49b}$

17. $\sqrt{64y^2}$

18. $\sqrt{9x^2}$

19. $\sqrt{13x^2}$

20. $\sqrt{23s^2}$

21. $\sqrt{8t^2}$

22. $\sqrt{125a^2}$

23. $\sqrt{180}$

24. $\sqrt{320}$

25. $\sqrt{288y}$

26. $\sqrt{363p}$

27. $\sqrt{28x^2}$

28. $\sqrt{20x^2}$

29. $\sqrt{x^2 - 6x + 9}$

30. $\sqrt{t^2 + 22t + 121}$

31. $\sqrt{8x^2 + 8x + 2}$

32. $\sqrt{20x^2 - 20x + 5}$

33. $\sqrt{36y + 12y^2 + y^3}$

34. $\sqrt{x - 2x^2 + x^3}$

b Simplify by factoring.

35. $\sqrt{x^6}$

36. $\sqrt{x^{18}}$

37. $\sqrt{x^{12}}$

38. $\sqrt{x^{16}}$

39. $\sqrt{x^5}$

40. $\sqrt{x^3}$

41. $\sqrt{t^{19}}$

42. $\sqrt{p^{17}}$

43. $\sqrt{(y-2)^8}$

44. $\sqrt{(x+3)^6}$

45. $\sqrt{4(x+5)^{10}}$

46. $\sqrt{16(a-7)^4}$

47. $\sqrt{36m^3}$

48. $\sqrt{250y^3}$

49. $\sqrt{8a^5}$

50. $\sqrt{12b^7}$

51. $\sqrt{104p^{17}}$

52. $\sqrt{284m^{23}}$

53. $\sqrt{448x^6y^3}$

54. $\sqrt{243x^5y^4}$

c Multiply and then, if possible, simplify by factoring.

55. $\sqrt{3}\sqrt{18}$

56. $\sqrt{5}\sqrt{10}$

57. $\sqrt{15}\sqrt{6}$

58. $\sqrt{3}\sqrt{27}$

59. $\sqrt{18}\sqrt{14x}$

60. $\sqrt{12}\sqrt{18x}$

61. $\sqrt{3x}\sqrt{12y}$

62. $\sqrt{7x}\sqrt{21y}$

63. $\sqrt{13}\sqrt{13}$

64. $\sqrt{11}\sqrt{11x}$

65. $\sqrt{5b}\sqrt{15b}$

66. $\sqrt{6a}\sqrt{18a}$

67. $\sqrt{2t}\sqrt{2t}$

68. $\sqrt{7a}\sqrt{7a}$

69. $\sqrt{ab}\sqrt{ac}$

70. $\sqrt{xy}\sqrt{xz}$

71. $\sqrt{2x^2y}\sqrt{4xy^2}$

72. $\sqrt{15mn^2}\sqrt{5m^2n}$

73. $\sqrt{18}\sqrt{18}$

74. $\sqrt{16}\sqrt{16}$

75. $\sqrt{5}\sqrt{2x-1}$

76. $\sqrt{3}\sqrt{4x+2}$

77. $\sqrt{x+2}\sqrt{x+2}$

78. $\sqrt{x-9}\sqrt{x-9}$

79. $\sqrt{18x^2y^3}\sqrt{6xy^4}$

80. $\sqrt{12x^3y^2}\sqrt{8xy}$

81. $\sqrt{50x^4y^6}\sqrt{10xy}$

82. $\sqrt{10xy^2}\sqrt{5x^2y^3}$

83. $\sqrt{99p^4q^3}\sqrt{22p^5q^2}$

84. $\sqrt{75m^8n^9}\sqrt{50m^5n^7}$

85. $\sqrt{24a^2b^3c^4}\sqrt{32a^5b^4c^7}$

86. $\sqrt{18p^5q^2r^{11}}\sqrt{108p^3q^6r^9}$

87. $^{\mathrm{D}}\mathrm{W}$ Are the rules for manipulating expressions with exponents important when simplifying radical expressions? Why or why not?

88. $^{\mathrm{D}}\mathrm{W}$ Explain the error(s) in the following:
$$\sqrt{x^2-25} = \sqrt{x^2} - \sqrt{25} = x - 5.$$

SKILL MAINTENANCE

Solve. [8.3a, b]

89. $x - y = -6,$
$x + y = 2$

90. $3x + 5y = 6,$
$5x + 3y = 4$

91. $3x - 2y = 4,$
$2x + 5y = 9$

92. $4a - 5b = 25,$
$a - b = 7$

Solve. [8.4a]

93. *Storage Area Dimensions.* The perimeter of a rectangular storage area is 84 ft. The length is 18 ft greater than the width. Find the area of the rectangle.

94. *Movie Revenue.* There were 411 people at a movie. Admission was \$7.00 each for adults and \$3.75 each for children, and receipts totaled \$2678.75. How many adults and how many children attended?

95. *Insecticide Mixtures.* A solution containing 30% insecticide is to be mixed with a solution containing 50% insecticide in order to make 200 L of a solution containing 42% insecticide. How much of each solution should be used?

96. *Canoe Travel.* Greg and Beth paddled to a picnic spot downriver in 2 hr. It took them 3 hr to return against the current. If the speed of the current was 2 mph, at what speed were they paddling the canoe? [8.5a]

SYNTHESIS

Factor.

97. $\sqrt{5x - 5}$

98. $\sqrt{x^2 - x - 2}$

99. $\sqrt{x^2 - 36}$

100. $\sqrt{2x^2 - 5x - 12}$

101. $\sqrt{x^3 - 2x^2}$

102. $\sqrt{a^2 - b^2}$

Simplify.

103. $\sqrt{0.25}$

104. $\sqrt{0.01}$

Multiply and then simplify by factoring.

105. $(\sqrt{2y})(\sqrt{3})(\sqrt{8y})$

106. $\sqrt{18(x - 2)}\sqrt{20(x - 2)^3}$

107. $\sqrt{27(x + 1)}\sqrt{12y(x + 1)^2}$

108. $\sqrt{2^{109}}\sqrt{x^{306}}\sqrt{x^{11}}$

109. $\sqrt{x}\sqrt{2x}\sqrt{10x^5}$

110. $\sqrt{a}(\sqrt{a^3} - 5)$

9.3 QUOTIENTS INVOLVING RADICAL EXPRESSIONS

a Dividing Radical Expressions

Consider the expressions

$$\frac{\sqrt{25}}{\sqrt{16}} \quad \text{and} \quad \sqrt{\frac{25}{16}}.$$

Let's evaluate them separately:

a) $\frac{\sqrt{25}}{\sqrt{16}} = \frac{5}{4}$ because $\sqrt{25} = 5$ and $\sqrt{16} = 4$;

b) $\sqrt{\frac{25}{16}} = \frac{5}{4}$ because $\frac{5}{4} \cdot \frac{5}{4} = \frac{25}{16}$.

We see that both expressions represent the same number. This suggests that the quotient of two square roots is the square root of the quotient of the radicands.

THE QUOTIENT RULE FOR RADICALS

For any nonnegative number A and any positive number B,

$$\frac{\sqrt{A}}{\sqrt{B}} = \sqrt{\frac{A}{B}}.$$

(The quotient of two square roots is the square root of the quotient of the radicands.)

EXAMPLES Divide and simplify.

1. $\frac{\sqrt{27}}{\sqrt{3}} = \sqrt{\frac{27}{3}} = \sqrt{9} = 3$

2. $\frac{\sqrt{30a^5}}{\sqrt{6a^2}} = \sqrt{\frac{30a^5}{6a^2}} = \sqrt{5a^3} = \sqrt{5 \cdot a^2 \cdot a} = \sqrt{a^2} \cdot \sqrt{5a} = a\sqrt{5a}$

Do Exercises 1–3.

b Square Roots of Quotients

To find the square root of certain quotients, we can reverse the quotient rule for radicals. We can take the square root of a quotient by taking the square roots of the numerator and the denominator separately.

SQUARE ROOTS OF QUOTIENTS

For any nonnegative number A and any positive number B,

$$\sqrt{\frac{A}{B}} = \frac{\sqrt{A}}{\sqrt{B}}.$$

(We can take the square roots of the numerator and the denominator separately.)

Objectives

a Divide radical expressions.

b Simplify square roots of quotients.

c Rationalize the denominator of a radical expression.

Divide and simplify.

1. $\frac{\sqrt{96}}{\sqrt{6}}$

2. $\frac{\sqrt{75}}{\sqrt{3}}$

3. $\frac{\sqrt{42x^5}}{\sqrt{7x^2}}$

Answers on page A-42

Simplify.

4. $\sqrt{\frac{16}{9}}$

5. $\sqrt{\frac{1}{25}}$

6. $\sqrt{\frac{36}{x^2}}$

Simplify.

7. $\sqrt{\frac{18}{32}}$

8. $\sqrt{\frac{2250}{2560}}$

9. $\sqrt{\frac{98y}{2y^{11}}}$

Answers on page A-42

EXAMPLES Simplify by taking the square roots of the numerator and the denominator separately.

3. $\sqrt{\frac{25}{9}} = \frac{\sqrt{25}}{\sqrt{9}} = \frac{5}{3}$ Taking the square roots of the numerator and the denominator

4. $\sqrt{\frac{1}{16}} = \frac{\sqrt{1}}{\sqrt{16}} = \frac{1}{4}$ Taking the square roots of the numerator and the denominator

5. $\sqrt{\frac{49}{t^2}} = \frac{\sqrt{49}}{\sqrt{t^2}} = \frac{7}{t}$

Do Exercises 4–6.

We are assuming that expressions for numerators are nonnegative and expressions for denominators are positive. Thus we need not be concerned about absolute-value signs or zero denominators.

Sometimes a rational expression can be simplified to one that has a perfect-square numerator and a perfect-square denominator.

EXAMPLES Simplify.

6. $\sqrt{\frac{18}{50}} = \sqrt{\frac{9 \cdot 2}{25 \cdot 2}} = \sqrt{\frac{9}{25} \cdot \frac{2}{2}} = \sqrt{\frac{9}{25} \cdot 1}$

$= \sqrt{\frac{9}{25}} = \frac{\sqrt{9}}{\sqrt{25}} = \frac{3}{5}$

7. $\sqrt{\frac{2560}{2890}} = \sqrt{\frac{256 \cdot 10}{289 \cdot 10}} = \sqrt{\frac{256}{289} \cdot \frac{10}{10}} = \sqrt{\frac{256}{289} \cdot 1}$

$= \sqrt{\frac{256}{289}} = \frac{\sqrt{256}}{\sqrt{289}} = \frac{16}{17}$

8. $\frac{\sqrt{48x^3}}{\sqrt{3x^7}} = \sqrt{\frac{48x^3}{3x^7}} = \sqrt{\frac{16}{x^4}}$ Simplifying the radicand

$= \frac{\sqrt{16}}{\sqrt{x^4}} = \frac{4}{x^2}$

Do Exercises 7–9.

C Rationalizing Denominators

Sometimes in mathematics it is useful to find an equivalent expression without a radical in the denominator. This provides a standard notation for expressing results. The procedure for finding such an expression is called **rationalizing the denominator.** We carry this out by multiplying by 1 in either of two ways.

To rationalize a denominator:

Method 1. Multiply by 1 under the radical to make the radicand in the denominator a perfect square.

Method 2. Multiply by 1 outside the radical to make the radicand in the denominator a perfect square.

EXAMPLE 9 Rationalize the denominator: $\sqrt{\frac{2}{3}}$.

METHOD 1 We multiply by 1, choosing $\frac{3}{3}$ for 1. This makes the denominator a perfect square:

$$\sqrt{\frac{2}{3}} = \sqrt{\frac{2}{3} \cdot \frac{3}{3}} \quad \text{Multiplying by 1}$$

$$= \sqrt{\frac{6}{9}} = \frac{\sqrt{6}}{\sqrt{9}} \quad \text{The radicand in the denominator, 9, is a perfect square.}$$

$$= \frac{\sqrt{6}}{3}.$$

METHOD 2 We can also rationalize by first taking the square roots of the numerator and the denominator. Then we multiply by 1, using $\sqrt{3}/\sqrt{3}$:

$$\sqrt{\frac{2}{3}} = \frac{\sqrt{2}}{\sqrt{3}}$$

$$= \frac{\sqrt{2}}{\sqrt{3}} \cdot \frac{\sqrt{3}}{\sqrt{3}} \quad \text{Multiplying by 1}$$

$$= \frac{\sqrt{2} \cdot \sqrt{3}}{\sqrt{3} \cdot \sqrt{3}} = \frac{\sqrt{6}}{\sqrt{9}} \quad \text{The radicand in the denominator, 9, is a perfect square.}$$

$$= \frac{\sqrt{6}}{3}.$$

Do Exercise 10.

We can always multiply by 1 to make a denominator a perfect square. Then we can take the square root of the denominator.

EXAMPLE 10 Rationalize the denominator: $\sqrt{\frac{5}{18}}$.

The denominator, 18, is not a perfect square. Factoring, we get $18 = 3 \cdot 3 \cdot 2$. If we had another factor of 2, however, we would have a perfect square, 36. Thus we multiply by 1, choosing $\frac{2}{2}$. This makes the denominator a perfect square.

$$\sqrt{\frac{5}{18}} = \sqrt{\frac{5}{3 \cdot 3 \cdot 2}} = \sqrt{\frac{5}{3 \cdot 3 \cdot 2} \cdot \frac{2}{2}} = \sqrt{\frac{10}{36}} = \frac{\sqrt{10}}{\sqrt{36}} = \frac{\sqrt{10}}{6}$$

EXAMPLE 11 Rationalize the denominator: $\frac{8}{\sqrt{7}}$.

This time we obtain an expression without a radical in the denominator by multiplying by 1, choosing $\sqrt{7}/\sqrt{7}$:

$$\frac{8}{\sqrt{7}} = \frac{8}{\sqrt{7}} \cdot \frac{\sqrt{7}}{\sqrt{7}} = \frac{8\sqrt{7}}{\sqrt{49}} = \frac{8\sqrt{7}}{7}.$$

CAUTION! $8\sqrt{7} \neq \sqrt{56}$.

Do Exercises 11 and 12.

10. Rationalize the denominator:

$$\sqrt{\frac{3}{5}}.$$

a) Use method 1.

b) Use method 2.

Rationalize the denominator.

11. $\sqrt{\frac{5}{8}}$

(*Hint*: Multiply the radicand by $\frac{2}{2}$.)

12. $\frac{10}{\sqrt{3}}$

Answers on page A-42

Rationalize the denominator.

13. $\dfrac{\sqrt{3}}{\sqrt{7}}$

14. $\dfrac{\sqrt{5}}{\sqrt{r}}$

15. $\dfrac{\sqrt{64y^2}}{\sqrt{7}}$

Answers on page A-42

EXAMPLE 12 Rationalize the denominator: $\dfrac{\sqrt{3}}{\sqrt{2}}$.

We look at the denominator. It is $\sqrt{2}$. We multiply by 1, choosing $\sqrt{2}/\sqrt{2}$:

$$\frac{\sqrt{3}}{\sqrt{2}} = \frac{\sqrt{3}}{\sqrt{2}} \cdot \frac{\sqrt{2}}{\sqrt{2}} = \frac{\sqrt{3}\cdot\sqrt{2}}{\sqrt{2}\cdot\sqrt{2}} = \frac{\sqrt{6}}{\sqrt{4}} = \frac{\sqrt{6}}{2}, \text{ or } \frac{1}{2}\sqrt{6}.$$

EXAMPLES Rationalize the denominator.

13. $\dfrac{\sqrt{5}}{\sqrt{x}} = \dfrac{\sqrt{5}}{\sqrt{x}} \cdot \dfrac{\sqrt{x}}{\sqrt{x}}$ Multiplying by 1

$= \dfrac{\sqrt{5}\sqrt{x}}{\sqrt{x}\sqrt{x}}$

$= \dfrac{\sqrt{5x}}{x}$ $\sqrt{x}\cdot\sqrt{x} = x$ by the definition of square root

14. $\dfrac{\sqrt{49a^5}}{\sqrt{12}} = \dfrac{\sqrt{49a^5}}{\sqrt{12}} \cdot \dfrac{\sqrt{3}}{\sqrt{3}}$ Multiplying by 1 using $\sqrt{3}/\sqrt{3}$ because $\sqrt{3}\cdot\sqrt{12} = \sqrt{3\cdot 2\cdot 2\cdot 3}$, which gives a perfect-square radicand in $\sqrt{36}$

$= \dfrac{\sqrt{49a^5}\sqrt{3}}{\sqrt{12}\sqrt{3}}$

$= \dfrac{\sqrt{49a^4\cdot 3a}}{\sqrt{36}} = \dfrac{\sqrt{49}\sqrt{a^4}\sqrt{3a}}{\sqrt{36}}$

$= \dfrac{7a^2\sqrt{3a}}{6}$

Do Exercises 13–15.

Study Tips

BEGINNING TO STUDY FOR THE FINAL EXAM (PART 3): THREE DAYS TO TWO WEEKS OF STUDY TIME

1. **Begin by browsing through each chapter, reviewing the highlighted or boxed information regarding important formulas in both the text and the Summary and Review.** There may be some formulas that you will need to memorize.
2. **Retake each chapter test that you took in class, assuming your instructor has returned it. Otherwise, use the chapter test in the book.** Restudy the objectives in the text that correspond to each question you missed.
3. **Work the Cumulative Review/Final Examination during the last couple of days before the final.** Set aside the same amount of time that you will have for the final. See how much of the final exam you can complete under test-like conditions. Be careful to avoid any questions corresponding to objectives not covered. Again, restudy the objectives in the text that correspond to each question you missed.
4. **For remaining difficulties, see your instructor, go to a tutoring session, or participate in a study group.**

"It is a great piece of skill to know how to guide your luck, even while waiting for it."

Baltasar Gracian,
seventeenth-century Spanish philosopher and writer

9.3 EXERCISE SET

For Extra Help

Digital Video Tutor CD 8 Videotape 16 | InterAct Math | Math Tutor Center | MathXL | MyMathLab

a Divide and simplify.

1. $\frac{\sqrt{18}}{\sqrt{2}}$

2. $\frac{\sqrt{20}}{\sqrt{5}}$

3. $\frac{\sqrt{108}}{\sqrt{3}}$

4. $\frac{\sqrt{60}}{\sqrt{15}}$

5. $\frac{\sqrt{65}}{\sqrt{13}}$

6. $\frac{\sqrt{45}}{\sqrt{15}}$

7. $\frac{\sqrt{3}}{\sqrt{75}}$

8. $\frac{\sqrt{3}}{\sqrt{48}}$

9. $\frac{\sqrt{12}}{\sqrt{75}}$

10. $\frac{\sqrt{18}}{\sqrt{32}}$

11. $\frac{\sqrt{8x}}{\sqrt{2x}}$

12. $\frac{\sqrt{18b}}{\sqrt{2b}}$

13. $\frac{\sqrt{63y^3}}{\sqrt{7y}}$

14. $\frac{\sqrt{48x^3}}{\sqrt{3x}}$

b Simplify.

15. $\sqrt{\frac{16}{49}}$

16. $\sqrt{\frac{9}{49}}$

17. $\sqrt{\frac{1}{36}}$

18. $\sqrt{\frac{1}{4}}$

19. $-\sqrt{\frac{16}{81}}$

20. $-\sqrt{\frac{25}{49}}$

21. $\sqrt{\frac{64}{289}}$

22. $\sqrt{\frac{81}{361}}$

23. $\sqrt{\frac{1690}{1960}}$

24. $\sqrt{\frac{1210}{6250}}$

25. $\sqrt{\frac{25}{x^2}}$

26. $\sqrt{\frac{36}{a^2}}$

27. $\sqrt{\frac{9a^2}{625}}$

28. $\sqrt{\frac{x^2y^2}{256}}$

C Rationalize the denominator.

29. $\sqrt{\frac{2}{5}}$

30. $\sqrt{\frac{2}{7}}$

31. $\sqrt{\frac{7}{8}}$

32. $\sqrt{\frac{3}{8}}$

33. $\sqrt{\frac{1}{12}}$

34. $\sqrt{\frac{7}{12}}$

35. $\sqrt{\frac{5}{18}}$

36. $\sqrt{\frac{1}{18}}$

37. $\frac{3}{\sqrt{5}}$

38. $\frac{4}{\sqrt{3}}$

39. $\sqrt{\frac{8}{3}}$

40. $\sqrt{\frac{12}{5}}$

41. $\sqrt{\frac{3}{x}}$

42. $\sqrt{\frac{2}{x}}$

43. $\sqrt{\frac{x}{y}}$

44. $\sqrt{\frac{a}{b}}$

45. $\sqrt{\frac{x^2}{20}}$

46. $\sqrt{\frac{x^2}{18}}$

47. $\frac{\sqrt{7}}{\sqrt{2}}$

48. $\frac{\sqrt{3}}{\sqrt{5}}$

49. $\dfrac{\sqrt{9}}{\sqrt{8}}$

50. $\dfrac{\sqrt{4}}{\sqrt{27}}$

51. $\dfrac{\sqrt{3}}{\sqrt{2}}$

52. $\dfrac{\sqrt{2}}{\sqrt{5}}$

53. $\dfrac{2}{\sqrt{2}}$

54. $\dfrac{3}{\sqrt{3}}$

55. $\dfrac{\sqrt{5}}{\sqrt{11}}$

56. $\dfrac{\sqrt{7}}{\sqrt{27}}$

57. $\dfrac{\sqrt{7}}{\sqrt{12}}$

58. $\dfrac{\sqrt{5}}{\sqrt{18}}$

59. $\dfrac{\sqrt{48}}{\sqrt{32}}$

60. $\dfrac{\sqrt{56}}{\sqrt{40}}$

61. $\dfrac{\sqrt{450}}{\sqrt{18}}$

62. $\dfrac{\sqrt{224}}{\sqrt{14}}$

63. $\dfrac{\sqrt{3}}{\sqrt{x}}$

64. $\dfrac{\sqrt{2}}{\sqrt{y}}$

65. $\dfrac{4y}{\sqrt{5}}$

66. $\dfrac{8x}{\sqrt{3}}$

67. $\dfrac{\sqrt{a^3}}{\sqrt{8}}$

68. $\dfrac{\sqrt{x^3}}{\sqrt{27}}$

69. $\dfrac{\sqrt{56}}{\sqrt{12x}}$

70. $\dfrac{\sqrt{45}}{\sqrt{8a}}$

71. $\dfrac{\sqrt{27c}}{\sqrt{32c^3}}$

72. $\dfrac{\sqrt{7x^3}}{\sqrt{12x}}$

73. $\dfrac{\sqrt{y^5}}{\sqrt{xy^2}}$

74. $\dfrac{\sqrt{x^3}}{\sqrt{xy}}$

75. $\dfrac{\sqrt{45mn^2}}{\sqrt{32m}}$

76. $\dfrac{\sqrt{16a^4b^6}}{\sqrt{128a^6b^6}}$

77. D$_W$ Why is it important to know how to multiply radical expressions before learning how to divide them?

78. D$_W$ Describe a method that could be used to rationalize the *numerator* of a radical expression.

SKILL MAINTENANCE

Solve. [8.3a, b]

79. $x = y + 2,$
$x + y = 6$

80. $4x - y = 10,$
$4x + y = 70$

81. $2x - 3y = 7,$
$2x - 3y = 9$

82. $2x - 3y = 7,$
$-4x + 6y = -14$

83. $x + y = -7,$
$x - y = 2$

84. $2x + 3y = 8,$
$5x - 4y = -2$

Multiply. [4.6b]

85. $(3x - 7)(3x + 7)$

86. $(4a - 5b)(4a + 5b)$

Collect like terms. [1.7e]

87. $9x - 5y + 12x - 4y$

88. $17a + 9b - 3a - 15b$

SYNTHESIS

Periods of Pendulums. The period T of a pendulum is the time it takes the pendulum to move from one side to the other and back. A formula for the period is

$$T = 2\pi\sqrt{\frac{L}{32}},$$

where T is in seconds and L is in feet. Use 3.14 for π.

89. Find the periods of pendulums of lengths 2 ft, 8 ft, 64 ft, and 100 ft.

90. Find the period of a pendulum of length $\frac{2}{3}$ in.

91. The pendulum of a grandfather clock is $(32/\pi^2)$ ft long. How long does it take to swing from one side to the other?

92. The pendulum of a grandfather clock is $(45/\pi^2)$ ft long. How long does it take to swing from one side to the other?

Rationalize the denominator.

93. $\sqrt{\frac{5}{1600}}$

94. $\sqrt{\frac{3}{1000}}$

95. $\sqrt{\frac{1}{5x^3}}$

96. $\sqrt{\frac{3x^2y}{a^2x^5}}$

97. $\sqrt{\frac{3a}{b}}$

98. $\sqrt{\frac{1}{5zw^2}}$

99. $\sqrt{0.009}$

100. $\sqrt{0.012}$

Simplify.

101. $\sqrt{\frac{1}{x^2} - \frac{2}{xy} + \frac{1}{y^2}}$

102. $\sqrt{2 - \frac{4}{z^2} + \frac{2}{z^4}}$

9.4 ADDITION, SUBTRACTION, AND MORE MULTIPLICATION

a Addition and Subtraction

We can add any two real numbers. The sum of 5 and $\sqrt{2}$ can be expressed as

$$5 + \sqrt{2}.$$

We cannot simplify this unless we use rational approximations such as $5 + \sqrt{2} \approx 5 + 1.414 = 6.414$. However, when we have *like radicals*, a sum can be simplified using the distributive laws and collecting like terms. **Like radicals** have the same radicands.

EXAMPLE 1 Add: $3\sqrt{5} + 4\sqrt{5}$.

Suppose we were considering $3x + 4x$. Recall that to add, we use a distributive law as follows:

$$3x + 4x = (3 + 4)x = 7x.$$

The situation is similar in this example, but we let $x = \sqrt{5}$:

$$3\sqrt{5} + 4\sqrt{5} = (3 + 4)\sqrt{5} \quad \text{Using a distributive law to factor out } \sqrt{5}$$
$$= 7\sqrt{5}.$$

If we wish to add or subtract as we did in Example 1, the radicands must be the same. Sometimes after simplifying the radical terms, we discover that we have like radicals.

EXAMPLES Add or subtract. Simplify, if possible, by collecting like radical terms.

2. $5\sqrt{2} - \sqrt{18} = 5\sqrt{2} - \sqrt{9 \cdot 2}$ Factoring 18
$= 5\sqrt{2} - \sqrt{9}\sqrt{2}$
$= 5\sqrt{2} - 3\sqrt{2}$
$= (5 - 3)\sqrt{2}$ Using a distributive law to factor out the common factor, $\sqrt{2}$
$= 2\sqrt{2}$

3. $\sqrt{4x^3} + 7\sqrt{x} = \sqrt{4 \cdot x^2 \cdot x} + 7\sqrt{x}$
$= 2x\sqrt{x} + 7\sqrt{x}$
$= (2x + 7)\sqrt{x}$ Using a distributive law to factor out $\sqrt{x}$

Don't forget the parentheses!

4. $\sqrt{x^3 - x^2} + \sqrt{4x - 4} = \sqrt{x^2(x - 1)} + \sqrt{4(x - 1)}$ Factoring radicands
$= \sqrt{x^2}\sqrt{x - 1} + \sqrt{4}\sqrt{x - 1}$
$= x\sqrt{x - 1} + 2\sqrt{x - 1}$
$= (x + 2)\sqrt{x - 1}$ Using a distributive law to factor out the common factor, $\sqrt{x - 1}$. Don't forget the parentheses!

Objectives

a Add or subtract with radical notation, using the distributive law to simplify.

b Multiply expressions involving radicals, where some of the expressions contain more than one term.

c Rationalize denominators having two terms.

Add or subtract and simplify by collecting like radical terms, if possible.

1. $3\sqrt{2} + 9\sqrt{2}$

2. $8\sqrt{5} - 3\sqrt{5}$

3. $2\sqrt{10} - 7\sqrt{40}$

4. $\sqrt{24} + \sqrt{54}$

5. $\sqrt{9x + 9} - \sqrt{4x + 4}$

Answers on page A-42

Add or subtract.

6. $\sqrt{2} + \sqrt{\frac{1}{2}}$

7. $\sqrt{\frac{5}{3}} + \sqrt{\frac{3}{5}}$

Answers on page A-42

Do Exercises 1–5 on the preceding page.

Sometimes rationalizing denominators enables us to combine like radicals.

EXAMPLE 5 Add: $\sqrt{3} + \sqrt{\frac{1}{3}}$.

$$\sqrt{3} + \sqrt{\frac{1}{3}} = \sqrt{3} + \sqrt{\frac{1}{3} \cdot \frac{3}{3}}$$ Multiplying by 1 in order to rationalize the denominator

$$= \sqrt{3} + \sqrt{\frac{3}{9}}$$

$$= \sqrt{3} + \frac{\sqrt{3}}{\sqrt{9}}$$

$$= \sqrt{3} + \frac{\sqrt{3}}{3}$$

$$= 1 \cdot \sqrt{3} + \frac{1}{3}\sqrt{3}$$

$$= \left(1 + \frac{1}{3}\right)\sqrt{3}$$ Factoring out the common factor, $\sqrt{3}$

$$= \frac{4}{3}\sqrt{3}$$

Do Exercises 6 and 7.

b Multiplication

Now let's multiply where some of the expressions may contain more than one term. To do this, we use procedures already studied in this chapter as well as the distributive laws and special products for multiplying with polynomials.

EXAMPLE 6 Multiply: $\sqrt{2}(\sqrt{3} + \sqrt{7})$.

$$\sqrt{2}(\sqrt{3} + \sqrt{7}) = \sqrt{2}\sqrt{3} + \sqrt{2}\sqrt{7}$$ Multiplying using a distributive law

$$= \sqrt{6} + \sqrt{14}$$ Using the rule for multiplying with radicals

EXAMPLE 7 Multiply: $(2 + \sqrt{3})(5 - 4\sqrt{3})$.

$$(2 + \sqrt{3})(5 - 4\sqrt{3}) = 2 \cdot 5 - 2 \cdot 4\sqrt{3} + \sqrt{3} \cdot 5 - \sqrt{3} \cdot 4\sqrt{3}$$ Using FOIL

$$= 10 - 8\sqrt{3} + 5\sqrt{3} - 4 \cdot 3$$

$$= 10 - 8\sqrt{3} + 5\sqrt{3} - 12$$

$$= -2 - 3\sqrt{3}$$

EXAMPLE 8 Multiply: $(\sqrt{3} - \sqrt{x})(\sqrt{3} + \sqrt{x})$.

$$(\sqrt{3} - \sqrt{x})(\sqrt{3} + \sqrt{x}) = (\sqrt{3})^2 - (\sqrt{x})^2 \quad \text{Using } (A - B)(A + B) = A^2 - B^2$$
$$= 3 - x$$

EXAMPLE 9 Multiply: $(3 - \sqrt{p})^2$.

$$(3 - \sqrt{p})^2 = 3^2 - 2 \cdot 3 \cdot \sqrt{p} + (\sqrt{p})^2 \quad \text{Using } (A - B)^2 = A^2 - 2AB + B^2$$
$$= 9 - 6\sqrt{p} + p$$

EXAMPLE 10 Multiply: $(2 - \sqrt{5})(2 + \sqrt{5})$.

$$(2 - \sqrt{5})(2 + \sqrt{5}) = 2^2 - (\sqrt{5})^2 \quad \text{Using } (A - B)(A + B) = A^2 - B^2$$
$$= 4 - 5$$
$$= -1$$

Do Exercises 8–12.

C More on Rationalizing Denominators

Note in Examples 8 and 10 that the results have no radicals. This will happen whenever we multiply expressions such as $\sqrt{a} - \sqrt{b}$ and $\sqrt{a} + \sqrt{b}$. We see this in the following:

$$(\sqrt{a} + \sqrt{b})(\sqrt{a} - \sqrt{b}) = (\sqrt{a})^2 - (\sqrt{b})^2 = a - b.$$

Expressions such as $\sqrt{3} - \sqrt{x}$ and $\sqrt{3} + \sqrt{x}$ are known as **conjugates**; so too are $2 + \sqrt{5}$ and $2 - \sqrt{5}$. We can use conjugates to rationalize a denominator that involves a sum or difference of two terms, where one or both are radicals. To do so, we multiply by 1 using the conjugate in the numerator and the denominator of the expression for 1.

Do Exercises 13–15.

EXAMPLE 11 Rationalize the denominator: $\dfrac{3}{2 + \sqrt{5}}$.

We multiply by 1 using the conjugate of $2 + \sqrt{5}$, which is $2 - \sqrt{5}$, as the numerator and the denominator of the expression for 1:

$$\frac{3}{2 + \sqrt{5}} = \frac{3}{2 + \sqrt{5}} \cdot \frac{2 - \sqrt{5}}{2 - \sqrt{5}} \quad \text{Multiplying by 1}$$
$$= \frac{3(2 - \sqrt{5})}{(2 + \sqrt{5})(2 - \sqrt{5})} \quad \text{Multiplying}$$
$$= \frac{6 - 3\sqrt{5}}{2^2 - (\sqrt{5})^2} \quad \text{Using } (A + B)(A - B) = A^2 - B^2$$
$$= \frac{6 - 3\sqrt{5}}{4 - 5}$$
$$= \frac{6 - 3\sqrt{5}}{-1}$$
$$= -6 + 3\sqrt{5}, \text{ or } 3\sqrt{5} - 6.$$

Multiply.

8. $\sqrt{3}(\sqrt{5} + \sqrt{2})$

9. $(1 - \sqrt{2})(4 + 3\sqrt{5})$

10. $(\sqrt{2} + \sqrt{a})(\sqrt{2} - \sqrt{a})$

11. $(5 + \sqrt{x})^2$

12. $(3 - \sqrt{7})(3 + \sqrt{7})$

Find the conjugate of the expression.

13. $7 + \sqrt{5}$

14. $\sqrt{5} - \sqrt{2}$

15. $1 - \sqrt{x}$

Answers on page A-42

Rationalize the denominator.

16. $\dfrac{6}{7 + \sqrt{5}}$

17. $\dfrac{\sqrt{5} + \sqrt{2}}{\sqrt{5} - \sqrt{2}}$

18. Rationalize the denominator:

$$\frac{7}{1 - \sqrt{x}}.$$

Answers on page A-43

EXAMPLE 12 Rationalize the denominator: $\dfrac{\sqrt{3} + \sqrt{5}}{\sqrt{3} - \sqrt{5}}$.

We multiply by 1 using the conjugate of $\sqrt{3} - \sqrt{5}$, which is $\sqrt{3} + \sqrt{5}$, as the numerator and the denominator of the expression for 1:

$$\frac{\sqrt{3} + \sqrt{5}}{\sqrt{3} - \sqrt{5}} = \frac{\sqrt{3} + \sqrt{5}}{\sqrt{3} - \sqrt{5}} \cdot \frac{\sqrt{3} + \sqrt{5}}{\sqrt{3} + \sqrt{5}} \qquad \text{Multiplying by 1}$$

$$= \frac{(\sqrt{3} + \sqrt{5})^2}{(\sqrt{3} - \sqrt{5})(\sqrt{3} + \sqrt{5})}$$

$$= \frac{(\sqrt{3})^2 + 2\sqrt{3}\sqrt{5} + (\sqrt{5})^2}{(\sqrt{3})^2 - (\sqrt{5})^2} \qquad \text{Using } (A + B)^2 = A^2 + 2AB + B^2 \text{ and } (A + B)(A - B) = A^2 - B^2$$

$$= \frac{3 + 2\sqrt{15} + 5}{3 - 5}$$

$$= \frac{8 + 2\sqrt{15}}{-2}$$

$$= \frac{2(4 + \sqrt{15})}{2(-1)} \qquad \text{Factoring in order to simplify}$$

$$= \frac{2}{2} \cdot \frac{4 + \sqrt{15}}{-1}$$

$$= \frac{4 + \sqrt{15}}{-1}$$

$$= -4 - \sqrt{15}.$$

Do Exercises 16 and 17.

EXAMPLE 13 Rationalize the denominator: $\dfrac{5}{2 + \sqrt{x}}$.

We multiply by 1 using the conjugate of $2 + \sqrt{x}$, which is $2 - \sqrt{x}$, as the numerator and the denominator of the expression for 1:

$$\frac{5}{2 + \sqrt{x}} = \frac{5}{2 + \sqrt{x}} \cdot \frac{2 - \sqrt{x}}{2 - \sqrt{x}} \qquad \text{Multiplying by 1}$$

$$= \frac{5(2 - \sqrt{x})}{(2 + \sqrt{x})(2 - \sqrt{x})}$$

$$= \frac{5 \cdot 2 - 5 \cdot \sqrt{x}}{2^2 - (\sqrt{x})^2} \qquad \text{Using } (A + B)(A - B) = A^2 - B^2$$

$$= \frac{10 - 5\sqrt{x}}{4 - x}.$$

Do Exercise 18.

9.4 EXERCISE SET

For Extra Help

Digital Video Tutor CD 8 Videotape 17 | InterAct Math | Math Tutor Center | MathXL | MyMathLab

a Add or subtract. Simplify by collecting like radical terms, if possible.

1. $7\sqrt{3} + 9\sqrt{3}$

2. $6\sqrt{2} + 8\sqrt{2}$

3. $7\sqrt{5} - 3\sqrt{5}$

4. $8\sqrt{2} - 5\sqrt{2}$

5. $6\sqrt{x} + 7\sqrt{x}$

6. $9\sqrt{y} + 3\sqrt{y}$

7. $4\sqrt{d} - 13\sqrt{d}$

8. $2\sqrt{a} - 17\sqrt{a}$

9. $5\sqrt{8} + 15\sqrt{2}$

10. $3\sqrt{12} + 2\sqrt{3}$

11. $\sqrt{27} - 2\sqrt{3}$

12. $7\sqrt{50} - 3\sqrt{2}$

13. $\sqrt{45} - \sqrt{20}$

14. $\sqrt{27} - \sqrt{12}$

15. $\sqrt{72} + \sqrt{98}$

16. $\sqrt{45} + \sqrt{80}$

17. $2\sqrt{12} + \sqrt{27} - \sqrt{48}$

18. $9\sqrt{8} - \sqrt{72} + \sqrt{98}$

19. $\sqrt{18} - 3\sqrt{8} + \sqrt{50}$

20. $3\sqrt{18} - 2\sqrt{32} - 5\sqrt{50}$

21. $2\sqrt{27} - 3\sqrt{48} + 3\sqrt{12}$

22. $3\sqrt{48} - 2\sqrt{27} - 3\sqrt{12}$

23. $\sqrt{4x} + \sqrt{81x^3}$

24. $\sqrt{12x^2} + \sqrt{27}$

25. $\sqrt{27} - \sqrt{12x^2}$

26. $\sqrt{81x^3} - \sqrt{4x}$

27. $\sqrt{8x + 8} + \sqrt{2x + 2}$

28. $\sqrt{12x + 12} + \sqrt{3x + 3}$

29. $\sqrt{x^5 - x^2} + \sqrt{9x^3 - 9}$

30. $\sqrt{16x - 16} + \sqrt{25x^3 - 25x^2}$

31. $4a\sqrt{a^2b} + a\sqrt{a^2b^3} - 5\sqrt{b^3}$

32. $3x\sqrt{y^3x} - x\sqrt{yx^3} + y\sqrt{y^3x}$

33. $\sqrt{3} - \sqrt{\frac{1}{3}}$

34. $\sqrt{2} - \sqrt{\frac{1}{2}}$

35. $5\sqrt{2} + 3\sqrt{\frac{1}{2}}$

36. $4\sqrt{3} + 2\sqrt{\frac{1}{3}}$

37. $\sqrt{\frac{2}{3}} - \sqrt{\frac{1}{6}}$

38. $\sqrt{\frac{1}{2}} - \sqrt{\frac{1}{8}}$

b Multiply.

39. $\sqrt{3}(\sqrt{5} - 1)$

40. $\sqrt{2}(\sqrt{2} + \sqrt{3})$

41. $(2 + \sqrt{3})(5 - \sqrt{7})$

42. $(\sqrt{5} + \sqrt{7})(2\sqrt{5} - 3\sqrt{7})$

43. $(2 - \sqrt{5})^2$

44. $(\sqrt{3} + \sqrt{10})^2$

45. $(\sqrt{2} + 8)(\sqrt{2} - 8)$

46. $(1 + \sqrt{7})(1 - \sqrt{7})$

47. $(\sqrt{6} - \sqrt{5})(\sqrt{6} + \sqrt{5})$

48. $(\sqrt{3} + \sqrt{10})(\sqrt{3} - \sqrt{10})$

49. $(3\sqrt{5} - 2)(\sqrt{5} + 1)$

50. $(\sqrt{5} - 2\sqrt{2})(\sqrt{10} - 1)$

51. $(\sqrt{x} - \sqrt{y})^2$

52. $(\sqrt{w} + 11)^2$

C Rationalize the denominator.

53. $\dfrac{2}{\sqrt{3} - \sqrt{5}}$

54. $\dfrac{5}{3 + \sqrt{7}}$

55. $\dfrac{\sqrt{3} - \sqrt{2}}{\sqrt{3} + \sqrt{2}}$

56. $\dfrac{2 - \sqrt{7}}{\sqrt{3} - \sqrt{2}}$

57. $\dfrac{4}{\sqrt{10} + 1}$

58. $\dfrac{6}{\sqrt{11} - 3}$

59. $\dfrac{1 - \sqrt{7}}{3 + \sqrt{7}}$

60. $\dfrac{2 + \sqrt{8}}{1 - \sqrt{5}}$

61. $\dfrac{3}{4 + \sqrt{x}}$

62. $\dfrac{8}{2 - \sqrt{x}}$

63. $\dfrac{3 + \sqrt{2}}{8 - \sqrt{x}}$

64. $\dfrac{4 - \sqrt{3}}{6 + \sqrt{y}}$

65. D_W Explain why it is important for the signs within a pair of conjugates to differ.

66. D_W Describe a method that could be used to rationalize a numerator that contains the sum of two radical expressions.

SKILL MAINTENANCE

Solve.

67. $3x + 5 + 2(x - 3) = 4 - 6x$ [2.3c]

68. $3(x - 4) - 2 = 8(2x + 3)$ [2.3c]

69 $x^2 - 5x = 6$ [5.7b]

70. $x^2 + 10 = 7x$ [5.7b]

Solve.

71. *Juice Mixtures.* Jolly Juice is 3% real fruit juice, and Real Squeeze is 6% real fruit juice. How many liters of each should be combined in order to make an 8-L mixture that is 5.4% real fruit juice? [8.4a]

72. *Travel Time.* The time t that it takes a bus to travel a fixed distance varies inversely as its speed r. At a speed of 40 mph, it takes $\frac{1}{2}$ hr to travel a fixed distance. How long will it take to travel the same distance at 60 mph? [7.5d]

73. The graph of the polynomial equation $y = x^3 - 5x^2 + x - 2$ is shown at right. Use either the graph or the equation to estimate or find the value of the polynomial when $x = -1$, $x = 0$, $x = 1$, $x = 3$, and $x = 4.85$. [4.3a]

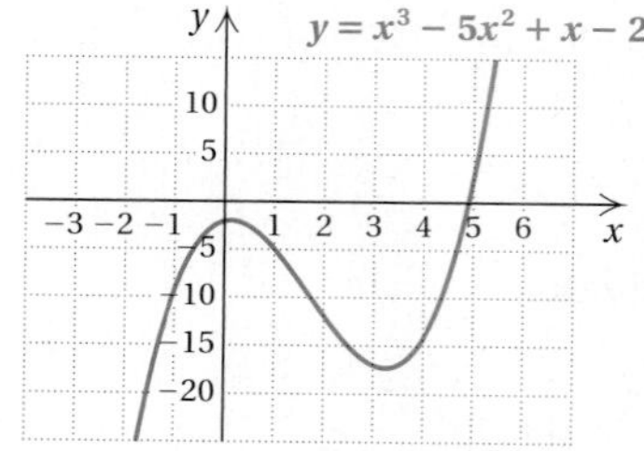

SYNTHESIS

74. Evaluate $\sqrt{a^2 + b^2}$ and $\sqrt{a^2} + \sqrt{b^2}$ when $a = 2$ and $b = 3$.

75. On the basis of Exercise 74, determine whether $\sqrt{a^2 + b^2}$ and $\sqrt{a^2} + \sqrt{b^2}$ are equivalent.

Use the GRAPH and TABLE features to determine whether each of the following is correct.

76. $\sqrt{9x^3} + \sqrt{x} = \sqrt{9x^3 + x}$

77. $\sqrt{x^2 + 4} = \sqrt{x} + 2$

Add or subtract as indicated.

78. $\frac{3}{5}\sqrt{24} + \frac{2}{5}\sqrt{150} - \sqrt{96}$

79. $\frac{1}{3}\sqrt{27} + \sqrt{8} + \sqrt{300} - \sqrt{18} - \sqrt{162}$

80. Three students were asked to simplify $\sqrt{10} + \sqrt{50}$. Their answers were $\sqrt{10}(1 + \sqrt{5})$, $\sqrt{10} + 5\sqrt{2}$, and $\sqrt{2}(5 + \sqrt{5})$. Which, if any, are correct?

Determine whether each of the following is true. Show why or why not.

81. $(3\sqrt{x + 2})^2 = 9(x + 2)$

82. $(\sqrt{x + 2})^2 = x + 2$

9.5 RADICAL EQUATIONS

Objectives

a Solve radical equations with one or two radical terms isolated, using the principle of squaring once.

b Solve radical equations with two radical terms, using the principle of squaring twice.

c Solve applied problems using radical equations.

a Solving Radical Equations

The following are examples of *radical equations*:

$$\sqrt{2x} - 4 = 7, \qquad \sqrt{x+1} = \sqrt{2x-5}.$$

A **radical equation** has variables in one or more radicands. To solve radical equations, we first convert them to equations without radicals. We do this for square-root radical equations by squaring both sides of the equation, using the following principle.

THE PRINCIPLE OF SQUARING

If an equation $a = b$ is true, then the equation $a^2 = b^2$ is true.

To solve square-root radical equations, we first try to get a radical by itself. That is, we try to isolate the radical. Then we use the principle of squaring. This allows us to eliminate one radical.

EXAMPLE 1 Solve: $\sqrt{2x} - 4 = 7$.

$$\begin{aligned} \sqrt{2x} - 4 &= 7 \\ \sqrt{2x} &= 11 && \text{Adding 4 to isolate the radical} \\ (\sqrt{2x})^2 &= 11^2 && \text{Squaring both sides} \\ 2x &= 121 && \sqrt{2x}\cdot\sqrt{2x} = 2x\text{, by the definition of square root} \\ x &= \frac{121}{2} && \text{Dividing by 2} \end{aligned}$$

CHECK: $\sqrt{2x} - 4 = 7$

$$\sqrt{2\cdot\frac{121}{2}} - 4 \; ? \; 7$$
$$\sqrt{121} - 4$$
$$11 - 4$$
$$7 \qquad \text{TRUE}$$

The solution is $\frac{121}{2}$.

Do Exercise 1.

1. Solve: $\sqrt{3x} - 5 = 3$.

Answer on page A-43

EXAMPLE 2 Solve: $2\sqrt{x+2} = \sqrt{x+10}$.

Each radical is isolated. We proceed with the principle of squaring.

$$\begin{aligned} (2\sqrt{x+2})^2 &= (\sqrt{x+10})^2 && \text{Squaring both sides} \\ 2^2(\sqrt{x+2})^2 &= (\sqrt{x+10})^2 && \text{Raising each factor of the product to the second power on the left} \\ 4(x+2) &= x+10 && \text{Simplifying} \\ 4x + 8 &= x + 10 && \text{Removing parentheses} \\ 3x &= 2 && \text{Subtracting } x \text{ and 8} \\ x &= \frac{2}{3} && \text{Dividing by 3} \end{aligned}$$

Solve.

2. $\sqrt{3x+1} = \sqrt{2x+3}$

3. $3\sqrt{x+1} = \sqrt{x+12}$

4. Solve: $x - 1 = \sqrt{x+5}$.

Answers on page A-43

CHECK:
$$\begin{array}{c|c} 2\sqrt{x+2} & \sqrt{x+10} \\ \hline 2\sqrt{\frac{2}{3}+2} \ ? & \sqrt{\frac{2}{3}+10} \\ 2\sqrt{\frac{8}{3}} & \sqrt{\frac{32}{3}} \\ 4\sqrt{\frac{2}{3}} & 4\sqrt{\frac{2}{3}} \end{array} \quad \text{TRUE}$$

The number $\frac{2}{3}$ checks. The solution is $\frac{2}{3}$.

Do Exercises 2 and 3.

It is important to check when using the principle of squaring. This principle may not produce equivalent equations. When we square both sides of an equation, the new equation may have solutions that the first one does not. For example, the equation

$$x = 1 \qquad (1)$$

has just one solution, the number 1. When we square both sides, we get

$$x^2 = 1, \qquad (2)$$

which has two solutions, 1 and -1. The equations $x = 1$ and $x^2 = 1$ do not have the same solutions and thus are not equivalent. Whereas it is true that any solution of equation (1) is a solution of equation (2), it is *not* true that any solution of equation (2) is a solution of equation (1).

CAUTION!

When the principle of squaring is used to solve an equation, all possible solutions *must* be checked in the original equation!

Sometimes we may need to apply the principle of zero products after squaring. (See Section 5.7.)

EXAMPLE 3 Solve: $x - 5 = \sqrt{x+7}$.

$$x - 5 = \sqrt{x+7}$$
$$(x-5)^2 = \left(\sqrt{x+7}\right)^2 \quad \text{Using the principle of squaring}$$
$$x^2 - 10x + 25 = x + 7$$
$$x^2 - 11x + 18 = 0$$
$$(x-9)(x-2) = 0 \quad \text{Factoring}$$
$$x - 9 = 0 \quad or \quad x - 2 = 0 \quad \text{Using the principle of zero products}$$
$$x = 9 \quad or \quad x = 2$$

CHECK: For 9:
$$\begin{array}{c|c} x - 5 & \sqrt{x+7} \\ \hline 9 - 5 \ ? & \sqrt{9+7} \\ 4 & 4 \end{array} \quad \text{TRUE}$$

For 2:
$$\begin{array}{c|c} x - 5 & \sqrt{x+7} \\ \hline 2 - 5 \ ? & \sqrt{2+7} \\ -3 & 3 \end{array} \quad \text{FALSE}$$

The number 9 checks, but 2 does not. Thus the solution is 9.

Do Exercise 4.

ALGEBRAIC–GRAPHICAL CONNECTION

We can visualize or check the solutions of a radical equation graphically. Consider the equation of Example 3:

$$x - 5 = \sqrt{x + 7}.$$

We can examine the solutions by graphing the equations

$$y = x - 5 \quad \text{and} \quad y = \sqrt{x + 7}$$

using the same set of axes. A hand-drawn graph of $y = \sqrt{x + 7}$ would involve approximating square roots on a calculator.

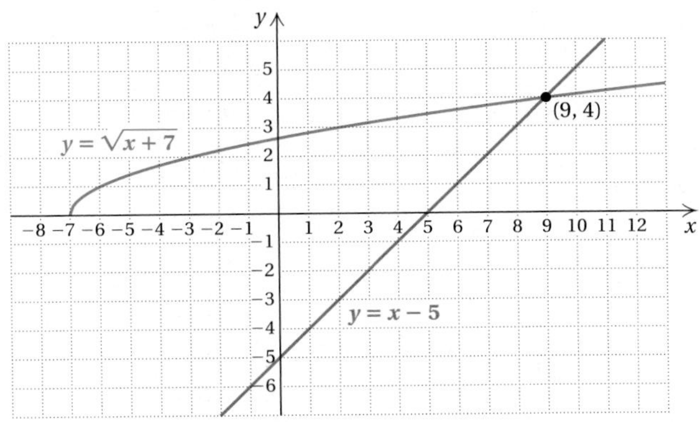

It appears that when $x = 9$, the values of $y = x - 5$ and $y = \sqrt{x + 7}$ are the same, 4. We can check this as we did in Example 3. Note also that the graphs *do not* intersect at $x = 2$.

CALCULATOR CORNER

Solving Radical Equations We can solve radical equations on a graphing calculator. Consider the equation in Example 3, $x - 5 = \sqrt{x + 7}$. We first graph each side of the equation. We enter $y_1 = x - 5$ and $y_2 = \sqrt{x + 7}$ on the equation-editor screen and graph the equations, using the window $[-2, 12, -6, 6]$. Note that there is one point of intersection. Use the INTERSECT feature to find its coordinates. (See the Calculator Corner on p. 595 for the procedure.)

The first coordinate, 9, is the value of x for which $y_1 = y_2$, or $x - 5 = \sqrt{x + 7}$. It is the solution of the equation. Note that the graph shows a single solution whereas the algebraic solution in Example 3 yields two possible solutions, 9 and 2, that must be checked. The check shows that 9 is the only solution.

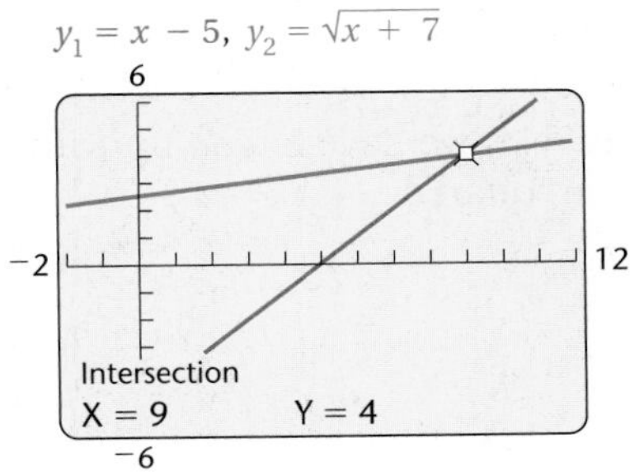

Exercises

1. Solve the equations in Examples 4 and 5 graphically.
2. Solve the equations in Margin Exercises 1–6 graphically.

5. Solve: $1 + \sqrt{1 - x} = x$.

EXAMPLE 4 Solve: $3 + \sqrt{27 - 3x} = x$.

In this case, we must first isolate the radical.

$$3 + \sqrt{27 - 3x} = x$$

$$\sqrt{27 - 3x} = x - 3 \quad \text{Subtracting 3 to isolate the radical}$$

$$(\sqrt{27 - 3x})^2 = (x - 3)^2 \quad \text{Using the principle of squaring}$$

$$27 - 3x = x^2 - 6x + 9$$

$$0 = x^2 - 3x - 18 \quad \text{Adding } 3x \text{ and subtracting 27 to obtain 0 on the left}$$

$$0 = (x - 6)(x + 3) \quad \text{Factoring}$$

$$x - 6 = 0 \quad or \quad x + 3 = 0 \quad \text{Using the principle of zero products}$$

$$x = 6 \quad or \quad x = -3$$

CHECK: For 6:

$$3 + \sqrt{27 - 3x} = x$$

$3 + \sqrt{27 - 3x}$	x
$3 + \sqrt{27 - 3 \cdot 6}$	? 6
$3 + \sqrt{9}$	
$3 + 3$	
6	TRUE

For -3:

$3 + \sqrt{27 - 3x}$	x
$3 + \sqrt{27 - 3 \cdot (-3)}$	? -3
$3 + \sqrt{27 + 9}$	
$3 + \sqrt{36}$	
$3 + 6$	
9	FALSE

The number 6 checks, but -3 does not. The solution is 6.

Do Exercise 5.

Suppose that in Example 4 we do not isolate the radical before squaring. Then we get an expression on the left side of the equation in which we have *not* eliminated the radical:

$$(3 + \sqrt{27 - 3x})^2 = (x)^2$$

$$3^2 + 2 \cdot 3 \cdot \sqrt{27 - 3x} + (\sqrt{27 - 3x})^2 = x^2$$

$$9 + 6\sqrt{27 - 3x} + (27 - 3x) = x^2.$$

In fact, we have ended up with a more complicated expression than the one we squared.

Answer on page A-43

b Using the Principle of Squaring More Than Once

Sometimes when we have two radical terms, we may need to apply the principle of squaring a second time.

EXAMPLE 5 Solve: $\sqrt{x} - 1 = \sqrt{x - 5}$.

$$\sqrt{x} - 1 = \sqrt{x - 5}$$

$$(\sqrt{x} - 1)^2 = (\sqrt{x - 5})^2 \quad \text{Using the principle of squaring}$$

$$(\sqrt{x})^2 - 2 \cdot \sqrt{x} \cdot 1 + 1^2 = x - 5 \quad \text{Using } (A - B)^2 = A^2 - 2AB + B^2 \text{ on the left side}$$

$$x - 2\sqrt{x} + 1 = x - 5 \quad \text{Simplifying. Only one radical term remains.}$$

$$-2\sqrt{x} = -6 \quad \text{Isolating the radical}$$

$$\sqrt{x} = 3$$

$$(\sqrt{x})^2 = 3^2 \quad \text{Using the principle of squaring}$$

$$x = 9$$

The check is left to the student. The number 9 checks and is the solution.

The following is a procedure for solving square-root radical equations.

SOLVING SQUARE-ROOT RADICAL EQUATIONS

To solve square-root radical equations:

1. Isolate one of the radical terms.
2. Use the principle of squaring.
3. If a radical term remains, perform steps (1) and (2) again.
4. Solve the equation and check possible solutions.

Do Exercise 6.

6. Solve: $\sqrt{x} - 1 = \sqrt{x - 3}$.

Answer on page A-43

7. How far to the horizon can you see through an airplane window at a height, or altitude, of 38,000 ft?

8. A sailor climbs 40 ft up the mast of a ship to a crow's nest. How far can he see to the horizon?

9. How far above sea level must a sailor climb on the mast of a ship in order to see 10.2 mi out to an iceberg?

Answers on page A-43

C Applications

Sighting to the Horizon. How far can you see from a given height? The equation

$$D = \sqrt{2h}$$

can be used to approximate the distance D, in miles, that a person can see to the horizon from a height h, in feet.

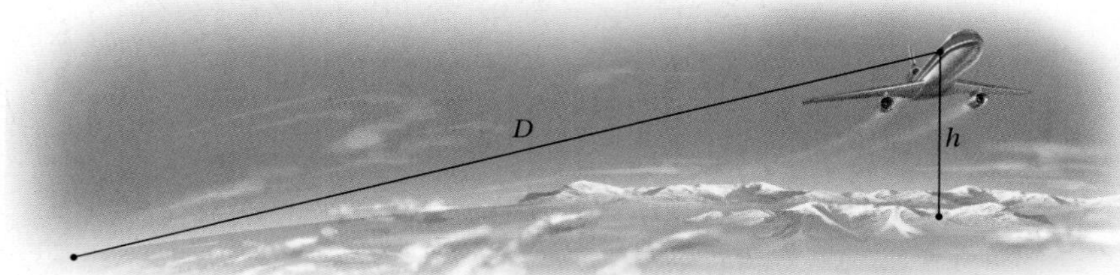

EXAMPLE 6 How far to the horizon can you see through an airplane window at a height, or altitude, of 30,000 ft?

We substitute 30,000 for h in $D = \sqrt{2h}$ and find an approximation using a calculator:

$$D = \sqrt{2 \cdot 30{,}000} \approx 245 \text{ mi.}$$

You can see for about 245 mi to the horizon.

Do Exercises 7 and 8.

EXAMPLE 7 *Height of a Ranger Station.* How high must a ranger station be in order for the ranger to see out to a fire on the horizon 15.4 mi away?

We substitute 15.4 for D in $D = \sqrt{2h}$ and solve:

$$15.4 = \sqrt{2h}$$

$$(15.4)^2 = (\sqrt{2h})^2 \qquad \text{Using the principle of squaring}$$

$$237.16 = 2h$$

$$\frac{237.16}{2} = h$$

$$118.58 = h.$$

The height of the ranger tower must be about 119 ft in order for the ranger to see out to a fire 15.4 mi away.

Do Exercise 9.

For Extra Help

Digital Video Tutor CD 8 Videotape 17 | InterAct Math | Math Tutor Center | MathXL | MyMathLab

a Solve.

1. $\sqrt{x} = 6$

2. $\sqrt{x} = 1$

3. $\sqrt{x} = 4.3$

4. $\sqrt{x} = 6.2$

5. $\sqrt{y + 4} = 13$

6. $\sqrt{y - 5} = 21$

7. $\sqrt{2x + 4} = 25$

8. $\sqrt{2x + 1} = 13$

9. $3 + \sqrt{x - 1} = 5$

10. $4 + \sqrt{y - 3} = 11$

11. $6 - 2\sqrt{3n} = 0$

12. $8 - 4\sqrt{5n} = 0$

13. $\sqrt{5x - 7} = \sqrt{x + 10}$

14. $\sqrt{4x - 5} = \sqrt{x + 9}$

15. $\sqrt{x} = -7$

16. $\sqrt{x} = -5$

17. $\sqrt{2y + 6} = \sqrt{2y - 5}$

18. $2\sqrt{3x - 2} = \sqrt{2x - 3}$

19. $x - 7 = \sqrt{x - 5}$

20. $\sqrt{x + 7} = x - 5$

21. $x - 9 = \sqrt{x - 3}$

22. $\sqrt{x + 18} = x - 2$

23. $2\sqrt{x - 1} = x - 1$

24. $x + 4 = 4\sqrt{x + 1}$

25. $\sqrt{5x + 21} = x + 3$

26. $\sqrt{27 - 3x} = x - 3$

27. $\sqrt{2x - 1} + 2 = x$

28. $x = 1 + 6\sqrt{x - 9}$

29. $\sqrt{x^2 + 6} - x + 3 = 0$

30. $\sqrt{x^2 + 5} - x + 2 = 0$

31. $\sqrt{x^2 - 4} - x = 6$

32. $\sqrt{x^2 - 5x + 7} = x - 3$

33. $\sqrt{(p + 6)(p + 1)} - 2 = p + 1$

34. $\sqrt{(4x + 5)(x + 4)} = 2x + 5$

35. $\sqrt{4x - 10} = \sqrt{2 - x}$

36. $\sqrt{2 - x} = \sqrt{3x - 7}$

b Solve. Use the principle of squaring twice.

37. $\sqrt{x - 5} = 5 - \sqrt{x}$

38. $\sqrt{x + 9} = 1 + \sqrt{x}$

39. $\sqrt{y + 8} - \sqrt{y} = 2$

40. $\sqrt{3x + 1} = 1 - \sqrt{x + 4}$

41. $\sqrt{x - 4} + \sqrt{x + 1} = 5$

42. $1 + \sqrt{x + 7} = \sqrt{3x - 2}$

c Solve.

Use the formula $D = \sqrt{2h}$ for Exercises 43–46.

43. How far to the horizon can you see through an airplane window at a height, or altitude, of 27,000 ft?

44. How far to the horizon can you see through an airplane window at a height, or altitude, of 32,000 ft?

45. How far above sea level must a pilot fly in order to see to a horizon that is 180 mi away?

46. A person can see 220 mi to the horizon through an airplane window. How high above sea level is the airplane?

Speed of a Skidding Car. How do police determine how fast a car had been traveling after an accident has occurred? The formula

$$r = 2\sqrt{5L}$$

can be used to approximate the speed r, in miles per hour, of a car that has left a skid mark of length L, in feet. (See Example 7 in Section 9.1.) Use this formula for Exercises 47 and 48.

47. How far will a car skid at 65 mph? at 75 mph?

48. How far will a car skid at 55 mph? at 90 mph?

49. $^D W$ Explain why possible solutions of radical equations must be checked.

50. $^D W$ Determine whether the statement below is true or false and explain your answer.

The solution of $\sqrt{11 - 2x} = -3$ is 1.

SKILL MAINTENANCE

Divide and simplify. [6.2b]

51. $\dfrac{x^2 - 49}{x + 8} \div \dfrac{x^2 - 14x + 49}{x^2 + 15x + 56}$

52. $\dfrac{x - 2}{x - 3} \div \dfrac{x - 4}{x - 5}$

53. $\dfrac{a^2 - 25}{6} \div \dfrac{a + 5}{3}$

54. $\dfrac{x - 2}{x + 3} \div \dfrac{x^2 - 4x + 4}{x^2 - 9}$

Solve. [8.4a]

55. *Supplementary Angles.* Two angles are supplementary. One angle is 3° less than twice the other. Find the measures of the angles.

56. *Complementary Angles.* Two angles are complementary. The sum of the measure of the first angle and half the measure of the second is 64°. Find the measures of the angles.

Multiply and simplify. [6.1d]

57. $\dfrac{7x^9}{27} \cdot \dfrac{9}{7x^3}$

58. $\dfrac{3}{x^2 - 9} \cdot \dfrac{x^2 - 6x + 9}{12}$

SYNTHESIS

Solve.

59. $\sqrt{5x^2 + 5} = 5$

60. $\sqrt{x} = -x$

61. $4 + \sqrt{19 - x} = 6 + \sqrt{4 - x}$

62. $x = (x - 2)\sqrt{x}$

63. $\sqrt{x + 3} = \dfrac{8}{\sqrt{x - 9}}$

64. $\dfrac{12}{\sqrt{5x + 6}} = \sqrt{2x + 5}$

65.–68. Use a graphing calculator to check your answers to Exercises 11–14.

9.6 APPLICATIONS WITH RIGHT TRIANGLES

Objectives

a Given the lengths of any two sides of a right triangle, find the length of the third side.

b Solve applied problems involving right triangles.

a Right Triangles

A **right triangle** is a triangle with a 90° angle, as shown in the figure below. The small square in the corner indicates the 90° angle.

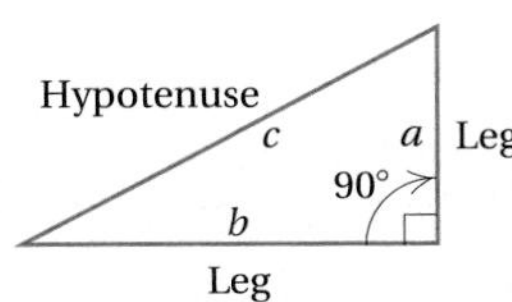

In a right triangle, the longest side is called the **hypotenuse.** It is also the side opposite the right angle. The other two sides are called **legs.** We generally use the letters a and b for the lengths of the legs and c for the length of the hypotenuse. They are related as follows.

THE PYTHAGOREAN THEOREM

In any right triangle, if a and b are the lengths of the legs and c is the length of the hypotenuse, then

$$a^2 + b^2 = c^2.$$

The equation $a^2 + b^2 = c^2$ is called the **Pythagorean equation.**

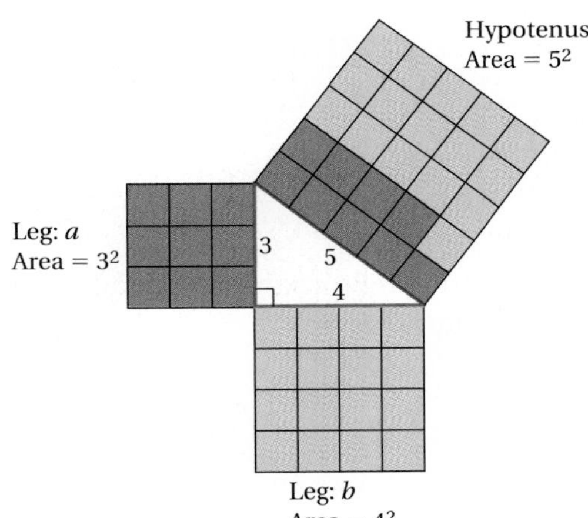

$$a^2 + b^2 = c^2$$
$$3^2 + 4^2 = 5^2$$
$$9 + 16 = 25$$

The Pythagorean theorem is named after the ancient Greek mathematician Pythagoras (569?–500? B.C.). It is uncertain who actually proved this result the first time. The proof can be found in most geometry books.

If we know the lengths of any two sides of a right triangle, we can find the length of the third side.

1. Find the length of the hypotenuse of this right triangle. Give an exact answer and an approximation to three decimal places.

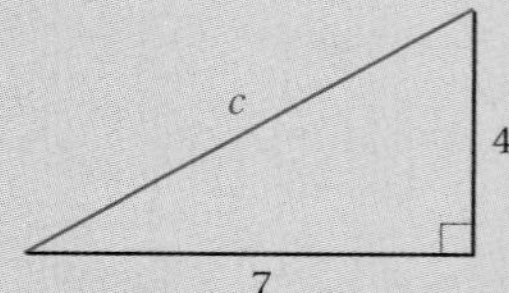

EXAMPLE 1 Find the length of the hypotenuse of this right triangle. Give an exact answer and an approximation to three decimal places.

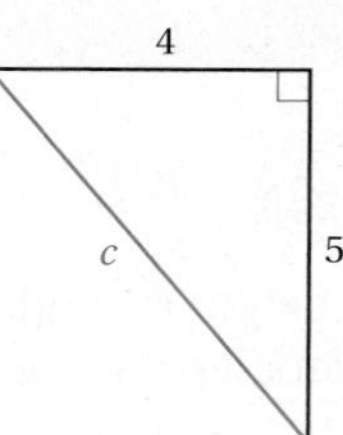

$4^2 + 5^2 = c^2$ Substituting in the Pythagorean equation

$16 + 25 = c^2$

$41 = c^2$

$c = \sqrt{41}$

≈ 6.403 Using a calculator

EXAMPLE 2 Find the length of the leg of this right triangle. Give an exact answer and an approximation to three decimal places.

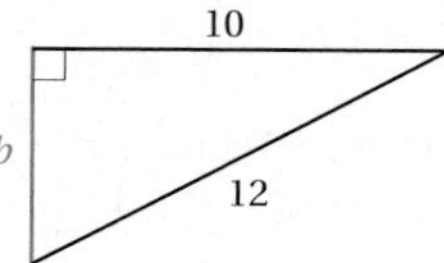

$10^2 + b^2 = 12^2$ Substituting in the Pythagorean equation

$100 + b^2 = 144$

$b^2 = 144 - 100$

$b^2 = 44$

$b = \sqrt{44}$

≈ 6.633 Using a calculator

Do Exercises 1 and 2.

2. Find the length of the leg of this right triangle. Give an exact answer and an approximation to three decimal places.

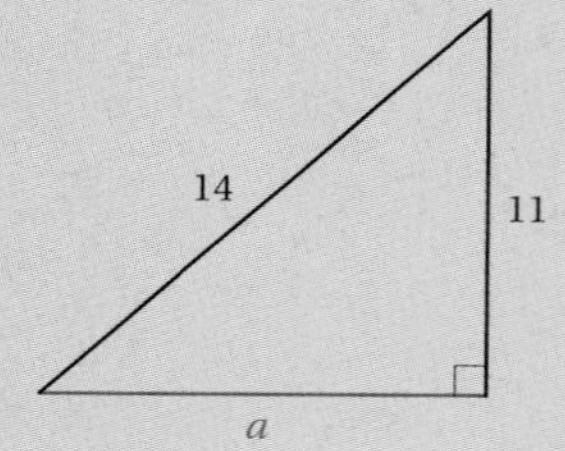

EXAMPLE 3 Find the length of the leg of this right triangle. Give an exact answer and an approximation to three decimal places.

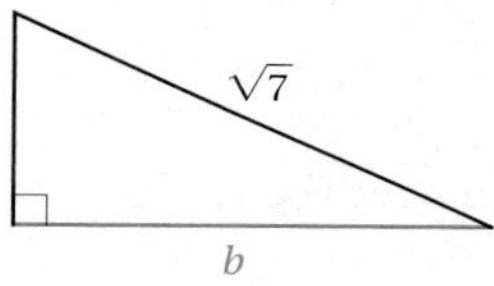

$1^2 + b^2 = (\sqrt{7})^2$ Substituting in the Pythagorean equation

$1 + b^2 = 7$

$b^2 = 7 - 1 = 6$

$b = \sqrt{6}$

≈ 2.449 Using a calculator

EXAMPLE 4 Find the length of the leg of this right triangle. Give an exact answer and an approximation to three decimal places.

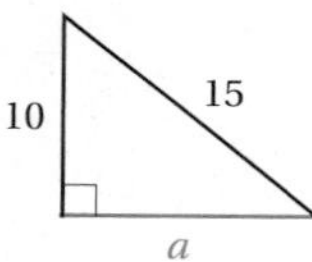

$a^2 + 10^2 = 15^2$

$a^2 + 100 = 225$

$a^2 = 225 - 100$

$a^2 = 125$

$a = \sqrt{125}$

≈ 11.180 Using a calculator

Do Exercises 3 and 4 on the following page.

Answers on page A-43

b Applications

EXAMPLE 5 *Dimensions of a Softball Diamond.* A slow-pitch softball diamond is actually a square 65 ft on a side. How far is it from home plate to second base? (This can be helpful information when lining up the bases.) Give an exact answer and an approximation to three decimal places.

a) We first make a drawing. We note that the first and second base lines, together with a line from home to second, form a right triangle. We label the unknown distance d.

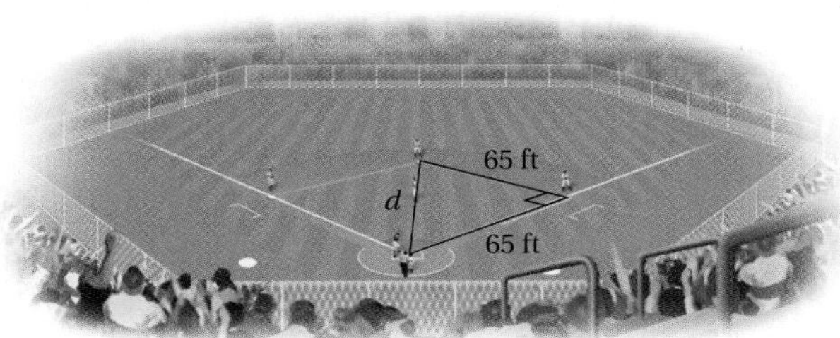

b) We know that $65^2 + 65^2 = d^2$. We solve this equation:

$$4225 + 4225 = d^2$$
$$8450 = d^2.$$

Exact answer: $\sqrt{8450}\text{ ft} = d$

Approximation: $91.924\text{ ft} \approx d$

Do Exercise 5.

Find the length of the leg of the right triangle. Give an exact answer and an approximation to three decimal places.

3.

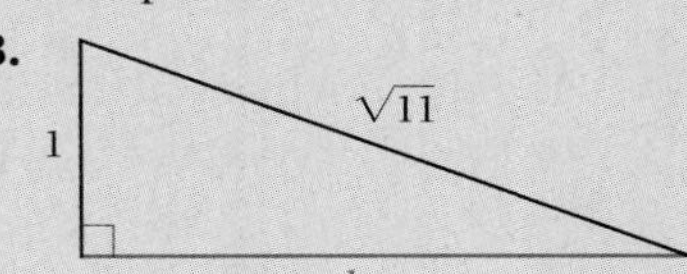

4.

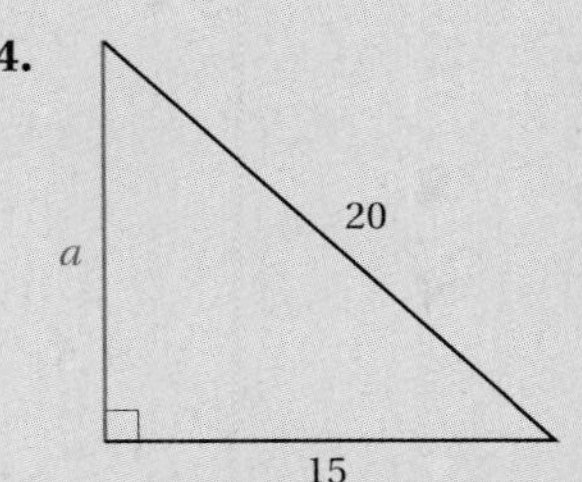

5. Guy Wire. How long is a guy wire reaching from the top of a 15-ft pole to a point on the ground 10 ft from the pole? Give an exact answer and an approximation to three decimal places.

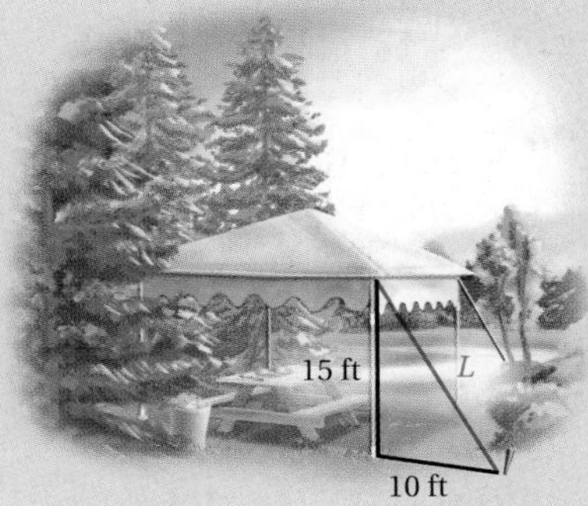

Answers on page A-43

9.6 EXERCISE SET

For Extra Help

Digital Video Tutor CD 8 Videotape 17 | InterAct Math | Math Tutor Center | MathXL | MyMathLab

a Find the length of the third side of the right triangle. Give an exact answer and an approximation to three decimal places.

1.

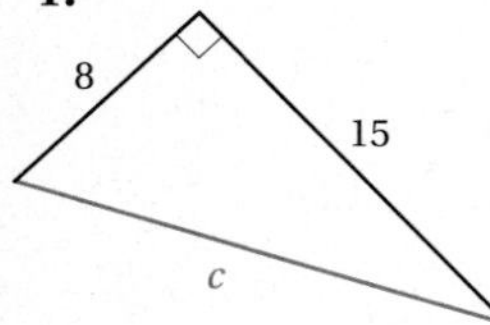

2.

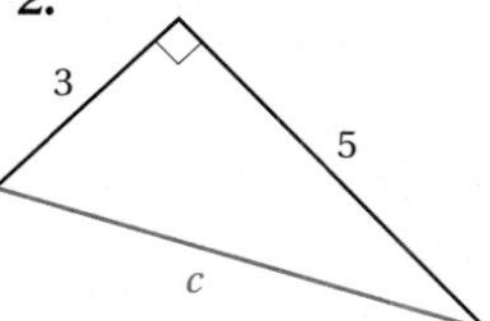

3.

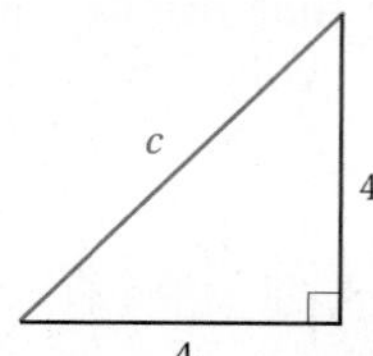

4.

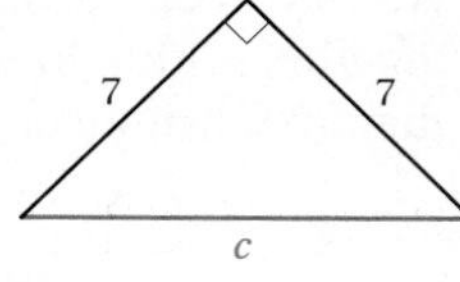

5.

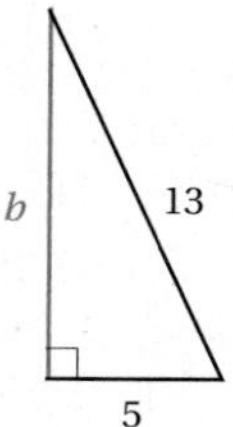

6.

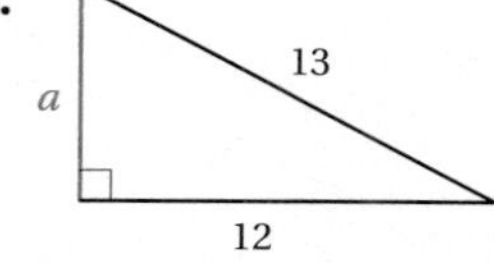

7.

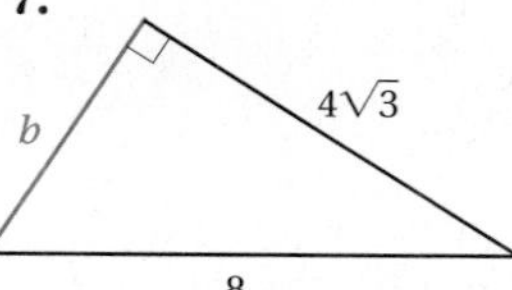

8.

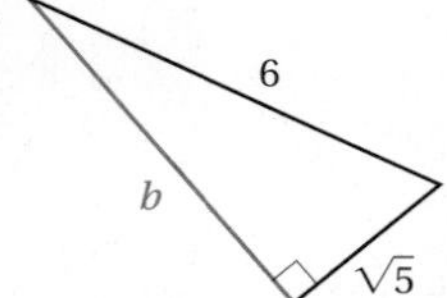

In a right triangle, find the length of the side not given. Give an exact answer and an approximation to three decimal places.

9. $a = 10,\quad b = 24$

10. $a = 5,\quad b = 12$

11. $a = 9,\quad c = 15$

12. $a = 18,\quad c = 30$

13. $b = 1,\quad c = \sqrt{5}$

14. $b = 1,\quad c = \sqrt{2}$

15. $a = 1,\quad c = \sqrt{3}$

16. $a = \sqrt{3},\quad b = \sqrt{5}$

17. $c = 10,\quad b = 5\sqrt{3}$

18. $a = 5,\quad b = 5$

19. $a = \sqrt{2},\quad b = \sqrt{7}$

20. $c = \sqrt{7},\quad a = \sqrt{2}$

b Solve. Don't forget to make a drawing. Give an exact answer and an approximation to three decimal places.

21. *Airport Distance.* An airplane is flying at an altitude of 4100 ft. The slanted distance directly to the airport is 15,100 ft. How far is the airplane horizontally from the airport?

22. *Surveying Distance.* A surveyor had poles located at points P, Q, and R. The distances that the surveyor was able to measure are marked on the drawing. What is the approximate distance from P to R?

23. *Cordless Telephones.* Becky's new cordless telephone has clear reception up to 300 ft from its base. Her phone is located near a window in her apartment, 180 ft above ground level. How far into her backyard can Becky use her phone?

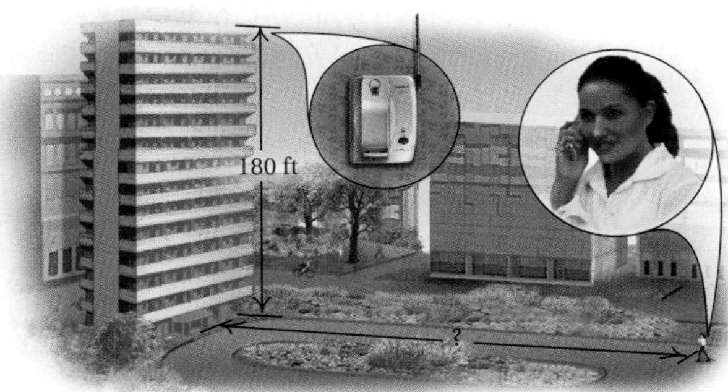

24. *Rope Course.* An outdoor rope course consists of a cable that slopes downward from a height of 37 ft to a resting place 30 ft above the ground. The trees that the cable connects are 24 ft apart. How long is the cable?

25. *Diagonal of a Square.* Find the length of a diagonal of a square whose sides are 3 cm long.

26. *Ladder Height.* A 10-m ladder is leaning against a building. The bottom of the ladder is 5 m from the building. How high is the top of the ladder?

27. *Guy Wire.* How long is a guy wire reaching from the top of a 12-ft pole to a point on the ground 8 ft from the base of the pole?

28. *Diagonal of a Soccer Field.* The largest regulation soccer field is 100 yd wide and 130 yd long. Find the length of a diagonal of such a field.

29. $^D{}_W$ Can a carpenter use a 28-ft ladder to repair clapboard that is 28 ft above ground level? Why or why not?

30. $^D{}_W$ In an **equilateral triangle,** all sides have the same length. Can a right triangle ever be equilateral? Why or why not?

SKILL MAINTENANCE

Solve. [8.3a, b]

31. $5x + 7 = 8y,$
$3x = 8y - 4$

32. $5x + y = 17,$
$-5x + 2y = 10$

33. $3x - 4y = -11,$
$5x + 6y = 12$

34. $x + y = -9,$
$x - y = -11$

35. Find the slope of the line $4 - x = 3y$. [7.1b]

36. Find the slope of the line containing the points $(8, -3)$ and $(0, -8)$. [3.4a]

SYNTHESIS

Find x.

37.

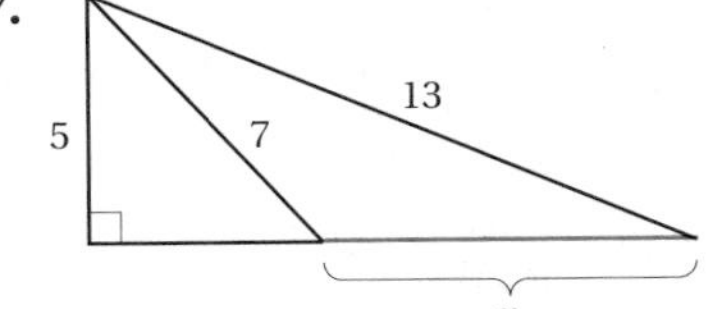

38. *Cordless Telephones.* Virginia's AT&T 9002 cordless phone has a range of 1000 ft. Her apartment is a corner unit, located as shown in the figure below. Will Virginia be able to use the phone at the community pool?
Source: AT&T

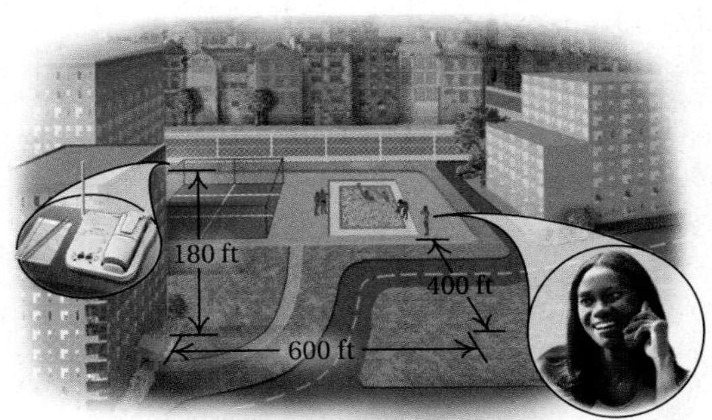

CHAPTER 9

Summary and Review

The review that follows is meant to prepare you for a chapter exam. It consists of two parts. The first part is a checklist of some of the Study Tips referred to in this and preceding chapters, as well as a list of important properties and formulas. The second part is the Review Exercises. These provide practice exercises for the exam, together with references to section objectives so you can go back and review. Before beginning, stop and look back over the skills you have obtained. What skills in mathematics do you have now that you did not have before studying this chapter?

STUDY TIPS CHECKLIST

The foundation of all your study skills is TIME!

- [ ] Have you begun to prepare for the final exam?
- [ ] Have you learned how to use MyMathLab? It can help in preparing for the final exam.
- [ ] Are you practicing the five-step problem-solving strategy?
- [ ] Are you using the tutoring resources on campus?
- [ ] Have you found someone with whom to study for the final exam?

IMPORTANT PROPERTIES AND FORMULAS

Product Rule for Radicals: $\sqrt{A}\,\sqrt{B} = \sqrt{AB}$

Quotient Rule for Radicals: $\dfrac{\sqrt{A}}{\sqrt{B}} = \sqrt{\dfrac{A}{B}}$

Principle of Squaring: If an equation $a = b$ is true, then the equation $a^2 = b^2$ is true.

Pythagorean Equation: $a^2 + b^2 = c^2$, where a and b are the lengths of the legs of a right triangle and c is the length of the hypotenuse.

REVIEW EXERCISES

Find the square roots. [9.1a]

1. 64

2. 400

Simplify. [9.1a]

3. $\sqrt{36}$

4. $-\sqrt{169}$

Use a calculator to approximate each of the following square roots to three decimal places. [9.1b]

5. $\sqrt{3}$

6. $\sqrt{99}$

7. $-\sqrt{320.12}$

8. $\sqrt{\dfrac{11}{20}}$

9. $-\sqrt{\dfrac{47.3}{11.2}}$

10. $18\sqrt{11 \cdot 43.7}$

Identify the radicand. [9.1d]

11. $\sqrt{x^2 + 4}$

12. $\sqrt{5ab^3}$

Determine whether the expression represents a real number. Write "yes" or "no." [9.1e]

13. $\sqrt{-22}$

14. $-\sqrt{49}$

15. $\sqrt{-36}$

16. $\sqrt{-10.2}$

17. $-\sqrt{-4}$

18. $\sqrt{2(-3)}$

Simplify. [9.1f]

19. $\sqrt{m^2}$

20. $\sqrt{(x-4)^2}$

Multiply. [9.2c]

21. $\sqrt{3}\,\sqrt{7}$

22. $\sqrt{x-3}\,\sqrt{x+3}$

Simplify by factoring. [9.2a]

23. $-\sqrt{48}$

24. $\sqrt{32t^2}$

25. $\sqrt{t^2-49}$

26. $\sqrt{x^2+16x+64}$

Simplify by factoring. [9.2b]

27. $\sqrt{x^8}$

28. $\sqrt{m^{15}}$

Multiply and simplify. [9.2c]

29. $\sqrt{6}\,\sqrt{10}$

30. $\sqrt{5x}\,\sqrt{8x}$

31. $\sqrt{5x}\,\sqrt{10xy^2}$

32. $\sqrt{20a^3b}\,\sqrt{5a^2b^2}$

Simplify. [9.3b]

33. $\sqrt{\frac{25}{64}}$

34. $\sqrt{\frac{20}{45}}$

35. $\sqrt{\frac{49}{t^2}}$

Rationalize the denominator. [9.3c]

36. $\sqrt{\frac{1}{2}}$

37. $\sqrt{\frac{1}{8}}$

38. $\sqrt{\frac{5}{y}}$

39. $\frac{2}{\sqrt{3}}$

Divide and simplify. [9.3a, c]

40. $\frac{\sqrt{27}}{\sqrt{45}}$

41. $\frac{\sqrt{45x^2y}}{\sqrt{54y}}$

42. Rationalize the denominator: [9.4c]

$$\frac{4}{2+\sqrt{3}}.$$

Simplify. [9.4a]

43. $10\sqrt{5}+3\sqrt{5}$

44. $\sqrt{80}-\sqrt{45}$

45. $3\sqrt{2}-5\sqrt{\frac{1}{2}}$

Simplify. [9.4b]

46. $(2+\sqrt{3})^2$

47. $(2+\sqrt{3})(2-\sqrt{3})$

Solve. [9.5a]

48. $\sqrt{x-3}=7$

49. $\sqrt{5x+3}=\sqrt{2x-1}$

50. $1+x=\sqrt{1+5x}$

51. Solve: [9.5b]

$$\sqrt{x}=\sqrt{x-5}+1.$$

In a right triangle, find the length of the side not given. Give an exact answer and an approximation to three decimal places. [9.6a]

52. $a=15, \quad c=25$

53. $a=1, \quad b=\sqrt{2}$

Solve. [9.6b]

54. *Airplane Descent.* A pilot is instructed to descend from 30,000 ft to 20,000 ft over a horizontal distance of 50,000 ft. What distance will the plane travel during this descent?

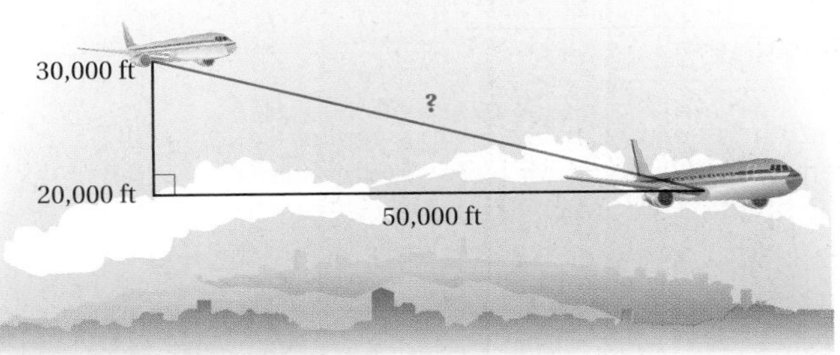

55. *Lookout Tower.* The diagonal braces in a lookout tower are 15 ft long and span a distance of 12 ft. How high does each brace reach vertically?

Solve. [9.1c], [9.5c]

56. *Speed of a Skidding Car.* The formula $r = 2\sqrt{5L}$ can be used to approximate the speed r, in miles per hour, of a car that has left a skid mark of length L, in feet.

a) What was the speed of a car that left skid marks of length 200 ft?

b) How far will a car skid at 90 mph?

57. DW Explain why the following is incorrect: [9.3b]

$$\sqrt{\frac{9 + 100}{25}} = \frac{3 + 10}{5}.$$

58. DW Determine whether each of the following is correct for all real numbers. Explain why or why not. [9.2a]

a) $\sqrt{5x^2} = |x|\sqrt{5}$

b) $\sqrt{b^2 - 4} = b - 2$

c) $\sqrt{x^2 + 16} = x + 4$

SKILL MAINTENANCE

Certain objectives from four particular sections will be retested on the chapter test. The objectives are listed with the practice problems that follow.

59. Solve: [8.3b]

$$2x - 3y = 4,$$
$$3x + 4y = 2.$$

60. Divide and simplify: [6.2b]

$$\frac{x^2 - 10x + 25}{x^2 + 14x + 49} \div \frac{x^2 - 25}{x^2 - 49}.$$

Solve.

61. *Paycheck and Hours Worked.* A person's paycheck varies directly as the number of hours H worked. For 15 hr of work, the pay is $168.75. Find the pay for 40 hr of work. [7.5b]

62. *Tickets Purchased.* There were 14,000 people at an AIDS benefit rock concert. Tickets were $12.00 at the door and $10.00 if purchased in advance. Total receipts were $159,400. How many people bought tickets in advance? [8.4a]

SYNTHESIS

63. *Distance Driven.* Two cars leave a service station at the same time. One car travels east at a speed of 50 mph, and the other travels south at a speed of 60 mph. After one-half hour, how far apart are they? [9.6b]

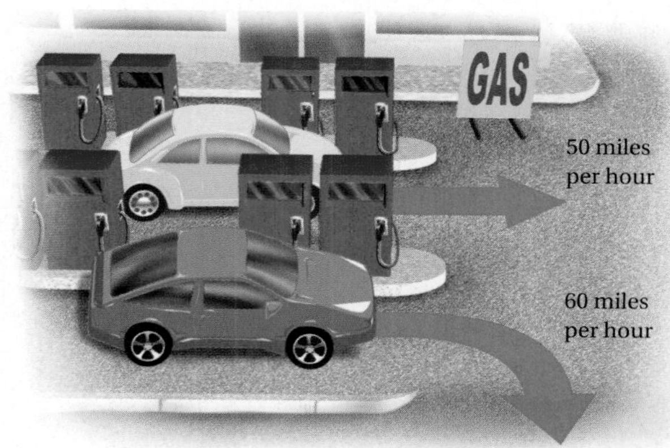

64. Simplify: $\sqrt{\sqrt{\sqrt{256}}}$. [9.2a]

65. Solve $A = \sqrt{a^2 + b^2}$ for b. [9.5a]

66. Find x. [9.6a]

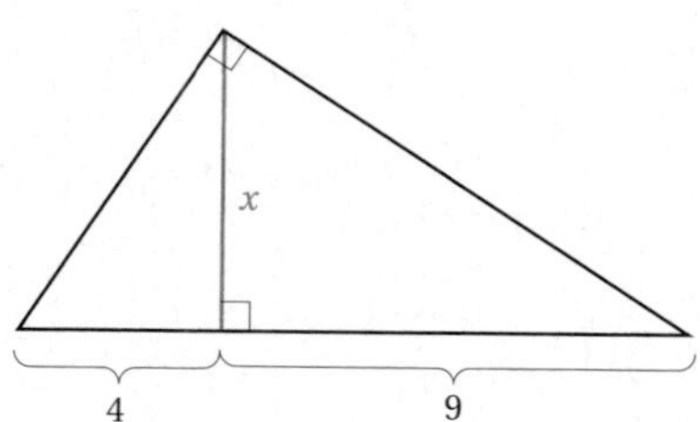

CHAPTER 9

Chapter Test

1. Find the square roots of 81.

Simplify.

2. $\sqrt{64}$

3. $-\sqrt{25}$

Approximate the expression involving square roots to three decimal places.

4. $\sqrt{116}$

5. $-\sqrt{87.4}$

6. $\sqrt{\dfrac{96 \cdot 38}{214.2}}$

7. Identify the radicand in $\sqrt{4 - y^3}$.

Determine whether the expression represents a real number. Write "yes" or "no."

8. $\sqrt{24}$

9. $\sqrt{-23}$

Simplify.

10. $\sqrt{a^2}$

11. $\sqrt{36y^2}$

Multiply.

12. $\sqrt{5}\,\sqrt{6}$

13. $\sqrt{x - 8}\,\sqrt{x + 8}$

Simplify by factoring.

14. $\sqrt{27}$

15. $\sqrt{25x - 25}$

16. $\sqrt{t^5}$

Multiply and simplify.

17. $\sqrt{5}\,\sqrt{10}$

18. $\sqrt{3ab}\,\sqrt{6ab^3}$

Simplify.

19. $\sqrt{\dfrac{27}{12}}$

20. $\sqrt{\dfrac{144}{a^2}}$

Rationalize the denominator.

21. $\sqrt{\dfrac{2}{5}}$

22. $\sqrt{\dfrac{2x}{y}}$

Divide and simplify.

23. $\dfrac{\sqrt{27}}{\sqrt{32}}$

24. $\dfrac{\sqrt{35x}}{\sqrt{80xy^2}}$

Add or subtract.

25. $3\sqrt{18} - 5\sqrt{18}$

26. $\sqrt{5} + \sqrt{\frac{1}{5}}$

Simplify.

27. $(4 - \sqrt{5})^2$

28. $(4 - \sqrt{5})(4 + \sqrt{5})$

29. Rationalize the denominator: $\frac{10}{4 - \sqrt{5}}$.

30. In a right triangle, $a = 8$ and $b = 4$. Find c. Give an exact answer and an approximation to three decimal places.

Solve.

31. $\sqrt{3x} + 2 = 14$

32. $\sqrt{6x + 13} = x + 3$

33. $\sqrt{1 - x} + 1 = \sqrt{6 - x}$

34. *Sighting to the Horizon.* The equation $D = \sqrt{2h}$ can be used to approximate the distance D, in miles, that a person can see to the horizon from a height h, in feet.

a) How far to the horizon can you see through an airplane window at a height of 28,000 ft?

b) Christina can see about 261 mi to the horizon through an airplane window. How high is the airplane?

35. *Lacrosse.* A regulation lacrosse field is 60 yd wide and 110 yd long. Find the length of a diagonal of such a field.

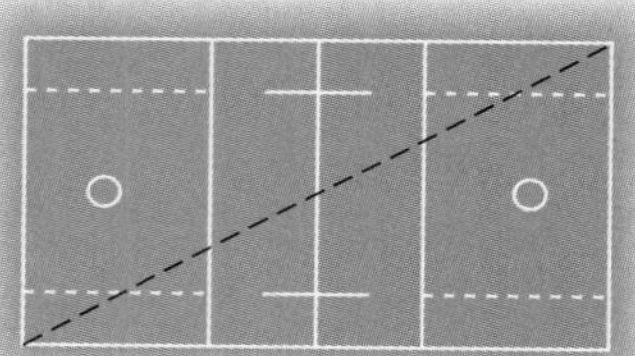

SKILL MAINTENANCE

36. *Rectangle Dimensions.* The perimeter of a rectangle is 118 yd. The width is 18 yd less than the length. Find the area of the rectangle.

37. *Productivity.* The number of switches N that a production line can make varies directly as the time it operates. It can make 7240 switches in 6 hr. How many can it make in 13 hr?

38. Solve:

$$-6x + 5y = 10,$$
$$5x + 6y = 12.$$

39. Divide and simplify:

$$\frac{x^2 - 11x + 30}{x^2 - 12x + 35} \div \frac{x^2 - 36}{x^2 - 14x + 49}.$$

SYNTHESIS

Simplify.

40. $\sqrt{\sqrt{\sqrt{625}}}$

41. $\sqrt{y^{16n}}$

CHAPTERS

1–9 Cumulative Review

1. Evaluate $x^3 - x^2 + x - 1$ when $x = -2$.

2. Collect like terms:

$$2x^3 - 7 + \frac{3}{7}x^2 - 6x^3 - \frac{4}{7}x^2 + 5.$$

3. Find all numbers for which the expression is not defined:

$$\frac{x-6}{2x+1}.$$

4. Determine whether the expression represents a real number. Write "yes" or "no."

$$\sqrt{-24}$$

Simplify.

5. $(2 + \sqrt{3})(2 - \sqrt{3})$

6. $-\sqrt{196}$

7. $\sqrt{3}\,\sqrt{75}$

8. $(1 - \sqrt{2})^2$

9. $\dfrac{\sqrt{162}}{\sqrt{125}}$

10. $2\sqrt{45} + 3\sqrt{20}$

Perform the indicated operations and simplify.

11. $(3x^4 - 2y^5)(3x^4 + 2y^5)$

12. $(x^2 + 4)^2$

13. $\left(2x + \dfrac{1}{4}\right)\left(4x - \dfrac{1}{2}\right)$

14. $\dfrac{x}{2x-1} - \dfrac{3x+2}{1-2x}$

15. $(3x^2 - 2x^3) - (x^3 - 2x^2 + 5) + (3x^2 - 5x + 5)$

16. $\dfrac{2x+2}{3x-9} \cdot \dfrac{x^2 - 8x + 15}{x^2 - 1}$

17. $\dfrac{2x^2 - 2}{2x^2 + 7x + 3} \div \dfrac{4x - 4}{2x^2 - 5x - 3}$

18. $(3x^3 - 2x^2 + x - 5) \div (x - 2)$

Simplify.

19. $\sqrt{2x^2 - 4x + 2}$

20. $x^{-9} \cdot x^{-3}$

21. $\sqrt{\dfrac{50}{2x^8}}$

22. $\dfrac{x - \dfrac{1}{x}}{1 - \dfrac{x-1}{2x}}$

Factor completely.

23. $3 - 12x^8$

24. $12t - 4t^2 - 48t^4$

25. $6x^2 - 28x + 16$

26. $4x^3 + 4x^2 - x - 1$

27. $16x^4 - 56x^2 + 49$

28. $x^2 + 3x - 180$

Solve.

29. $x^2 = -17x$

30. $-3x < 30 + 2x$

31. $\dfrac{1}{x} + \dfrac{2}{3} = \dfrac{1}{4}$

32. $x^2 - 30 = x$

33. $-4(x + 5) \geq 2(x + 5) - 3$

34. $2x^2 = 162$

35. $\sqrt{2x - 1} + 5 = 14$

36. $\sqrt{4x} + 1 = \sqrt{x} + 4$

37. $\dfrac{1}{4}x + \dfrac{2}{3}x = \dfrac{2}{3} - \dfrac{3}{4}x$

38. $\dfrac{x}{x-1} - \dfrac{x}{x+1} = \dfrac{1}{2x-2}$

39. $x = y + 3,$
$3y - 4x = -13$

40. $2x - 3y = 30,$
$5y - 2x = -46$

41. Solve $4A = pr + pq$ for p.

Graph on a plane.

42. $3y - 3x > -6$

43. $x = 5$

44. $2x - 6y = 12$

45. Find an equation of the line containing the points $(1, -2)$ and $(5, 9)$.

46. Find the slope and the y-intercept of the line $5x - 3y = 9$.

47. The graph of the polynomial equation $y = x^3 - 4x - 2$ is shown at right. Use either the graph or the equation to estimate the value of the polynomial when $x = -2$, $x = -1$, $x = 0$, $x = 1$, and $x = 2$.

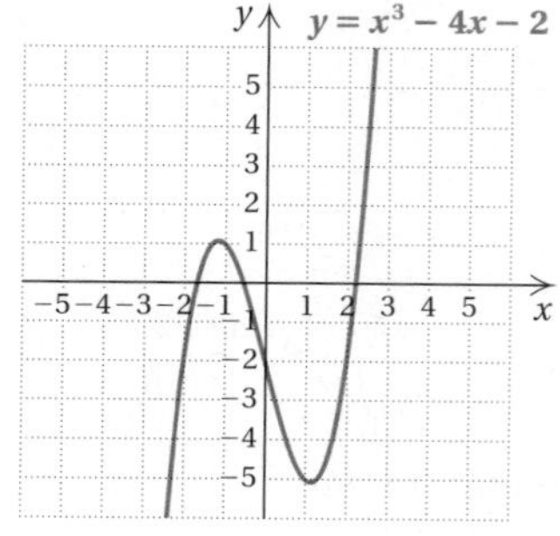

Solve.

48. *Home Run Race.* At the All-Star Break in the 2001 baseball season, Barry Bonds of the San Francisco Giants had hit 39 home runs in 88 games. Assuming he continued to hit home runs at this rate, how many home runs would he hit in the regular season of 162 games?
Source: Major League Baseball

49. *Apparent Size.* The apparent size A of an object varies inversely as the distance d of the object from the eye. You are sitting at a concert 100 ft from the stage. The musicians appear to be 4 ft tall. How tall would they appear to be if you were sitting 1000 ft away in the lawn seats?

50. *Cost Breakdown.* The cost of 6 hamburgers and 4 milkshakes is \$27.70. Three hamburgers and 1 milkshake cost \$11.35. Find the cost of a hamburger and the cost of a milkshake.

51. *Ladder Height.* An 8-m ladder is leaning against a building. The bottom of the ladder is 4 m from the building. How high is the top of the ladder? Give an exact answer and an approximation to three decimal places.

52. *Angles of a Triangle.* The second angle of a triangle is twice as large as the first. The third angle is 48° less than the sum of the other two angles. Find the measures of the angles.

53. *Quality Control.* A sample of 150 resistors contained 12 defective resistors. How many defective resistors would you expect to find in a sample of 250 resistors?

54. *Rectangle Dimensions.* The length of a rectangle is 3 m greater than the width. The area of the rectangle is 180 m^2. Find the length and the width.

55. *Coin Mixture.* A collection of dimes and quarters is worth \$19.00. There are 115 coins in all. How many of each are there?

56. *Amount Invested.* Money is invested in an account at 10.5% simple interest. At the end of 1 yr, there is \$2873 in the account. How much was originally invested?

57. *Car Travel.* A person traveled by car 600 mi in one direction. The return trip took 2 hr longer at a speed that was 10 mph less. Find the speed going.

In each of Exercises 58–61, choose the correct answer from the selections given.

58. Rationalize the denominator: $\frac{\sqrt{27}}{\sqrt{45}}$.

a) $\frac{\sqrt{15}}{5}$ **b)** $\frac{3}{\sqrt{15}}$ **c)** $\frac{3\sqrt{15}}{5}$
d) $\frac{3\sqrt{3}}{5}$ **e)** None of these

59. Solve: $\frac{3}{x-2} + \frac{6}{x^2-4} = \frac{5}{x+2}$.

a) $\frac{3(x+4)}{x^2-4}$ **b)** The solution is an odd integer greater than 10.
c) There is no solution. **d)** The solution is between 0 and 2.
e) None of these

60. Simplify: $\frac{\frac{1}{x}+\frac{1}{y}}{\frac{x}{2}+\frac{y}{2}}$.

a) $\frac{2}{xy}$ **b)** 1 **c)** $\frac{xy}{2}$
d) $\frac{2(x+y)}{xy(y+x)}$ **e)** None of these

61. Multiply: $(4a^2b + 5c)(3a^2b - 2c)$.

a) $12a^4b^2 - 10c^2$ **b)** $12a^4b^2 - 8a^2bc - 10c^2$
c) $12a^4b^2 + 15a^2bc - 10c^2$ **d)** $12a^4b^2 - 7a^2bc - 10c^2$
e) None of these

SYNTHESIS

Write a true sentence using $<$ or $>$.

62. $-4 \square |-3|$

63. $|-4| \square |-3|$

64. *Salt Solutions.* A tank contains 200 L of a 30%-salt solution. How much pure water should be added in order to make a solution that is 12% salt?

65. Solve: $\sqrt{x} + 1 = y,$
$\sqrt{x} + \sqrt{y} = 5.$

66. *Cordless Telephones.* The Panasonic KXTG-2500B cordless phone has a range of one quarter mile. Vance has a corner office in the Empire State Building, 900 ft above street level. Can Vance locate the Panasonic's phone base in his office and use the handset at a restaurant at street level on the opposite corner? Use the figure below and show your work.
Source: Panasonic

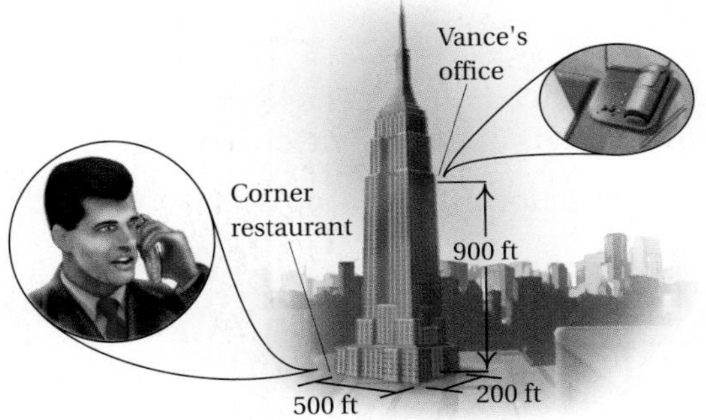

67. Find x.

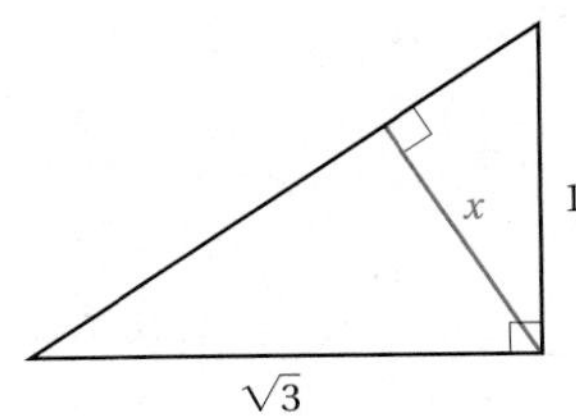

Quadratic Equations

Gateway to Chapter 10

A quadratic equation contains a polynomial of second degree. We begin this chapter by reviewing how to solve quadratic equations by factoring, as we did in Section 5.7. Because certain quadratic equations are difficult to solve by factoring, we also learn to use the quadratic formula to find solutions. Next, we apply these equation-solving skills to applications and problem solving, which extends the problem-solving skills discussed in Sections 5.8 and 6.7. Finally, we graph quadratic equations.

Real-World Application

The area of a rectangular red raspberry patch is 76 ft^2. The length is 7 ft longer than three times the width. Find the dimensions of the raspberry patch.

This problem appears as Example 1 in Section 10.5.

CHAPTER 10

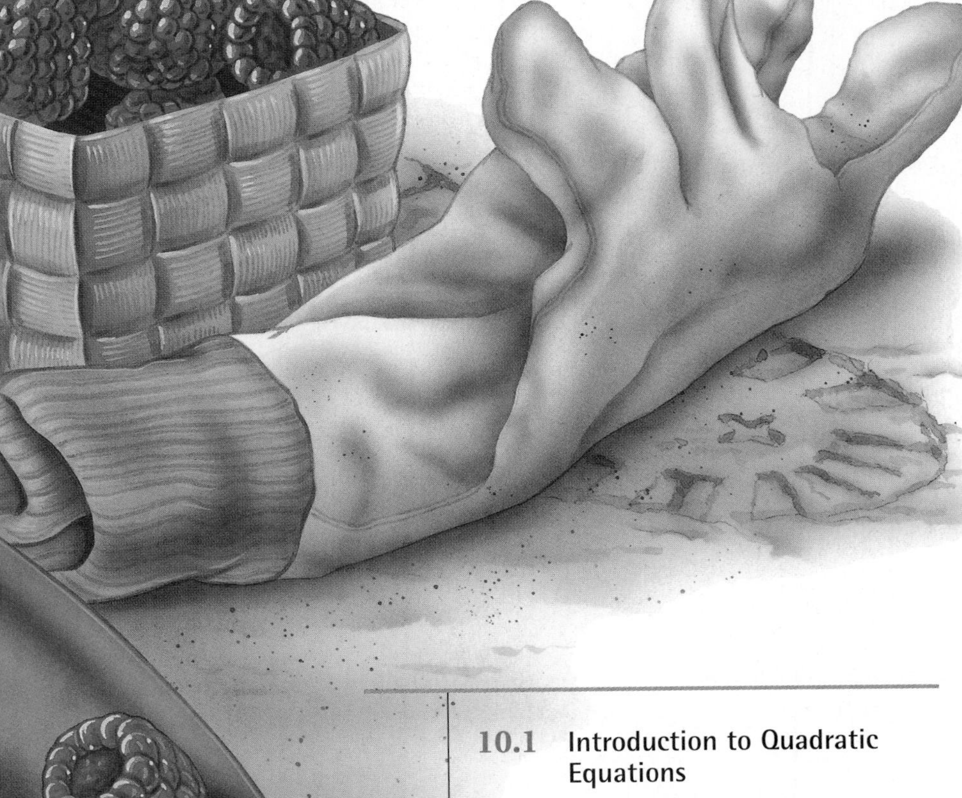

CHAPTER

10 Pretest

Solve.

1. $x^2 + 9 = 6x$ [10.1c]

2. $x^2 - 7 = 0$ [10.2a]

3. $3x^2 + 3x - 1 = 0$ [10.3a]

4. $5y^2 - 3y = 0$ [10.1b]

5. $\frac{3}{3x + 2} - \frac{2}{3x + 4} = 1$ [10.1b]

6. $(x + 4)^2 = 5$ [10.2b]

7. Solve $x^2 - 2x - 5 = 0$ by completing the square. Show your work. [10.2c]

8. Solve $A = n^2 - pn$ for n. [10.4a]

9. **Rectangle Dimensions.** The length of a rectangle is three times the width. The area is 48 cm^2. Find the length and the width. [10.5a]

10. Find the x-intercepts: $y = 2x^2 + x - 4$. [10.6b]

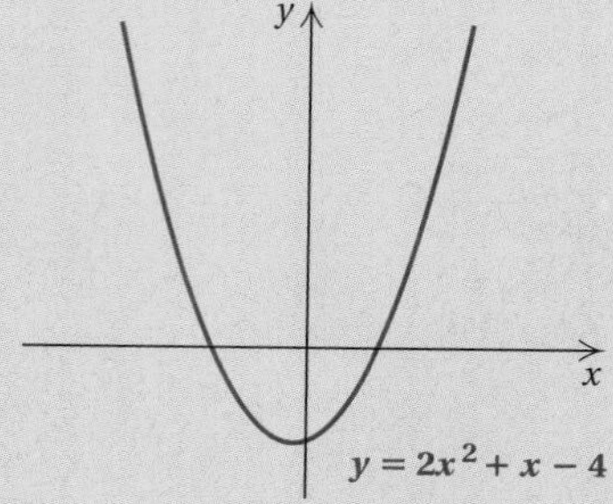

11. **Boat Travel.** The current in a stream moves at a speed of 2 km/h. A boat travels 24 km upstream and 24 km downstream in a total time of 5 hr. What is the speed of the boat in still water? [10.5a]

12. Graph: $y = 4 - x^2$. [10.6a]

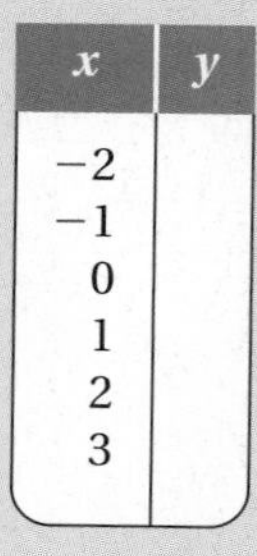

x	y
−2	
−1	
0	
1	
2	
3	

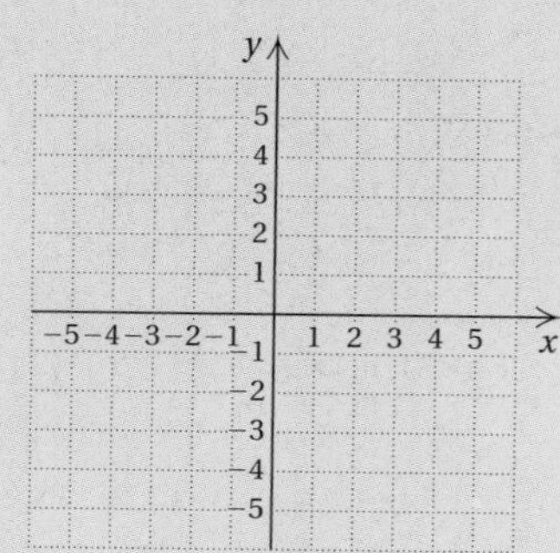

10.1 INTRODUCTION TO QUADRATIC EQUATIONS

Objectives

a Write a quadratic equation in standard form $ax^2 + bx + c = 0$, $a > 0$, and determine the coefficients a, b, and c.

b Solve quadratic equations of the type $ax^2 + bx = 0$, where $b \neq 0$, by factoring.

c Solve quadratic equations of the type $ax^2 + bx + c = 0$, where $b \neq 0$ and $c \neq 0$, by factoring.

d Solve applied problems involving quadratic equations.

ALGEBRAIC–GRAPHICAL CONNECTION

Before we begin this chapter, let's look back at some algebraic–graphical equation-solving concepts and their interrelationships. In Chapter 3, we considered the graph of a *linear equation* $y = mx + b$. For example, the graph of the equation $y = \frac{5}{2}x - 4$ and its x-intercept are shown below.

If $y = 0$, then $x = \frac{8}{5}$. Thus the x-intercept is $\left(\frac{8}{5}, 0\right)$. This point is also the intersection of the graphs of $y = \frac{5}{2}x - 4$ and $y = 0$.

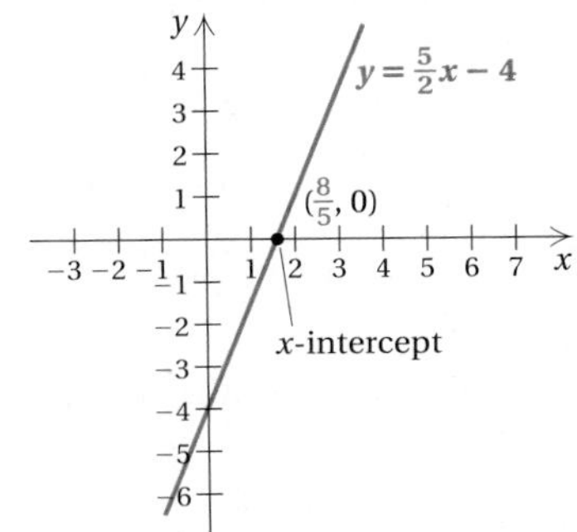

In Chapter 2, we learned how to solve linear equations like $0 = \frac{5}{2}x - 4$ algebraically (using algebra). We proceeded as follows:

$0 = \frac{5}{2}x - 4$

$4 = \frac{5}{2}x$ Adding 4

$8 = 5x$ Multiplying by 2

$\frac{8}{5} = x.$ Dividing by 5

We see that $\frac{8}{5}$, the solution of $0 = \frac{5}{2}x - 4$, is the first coordinate of the x-intercept of the graph of $y = \frac{5}{2}x - 4$.

Do Exercise 1.

In this chapter, we build on these ideas by applying them to quadratic equations. In Section 5.7, we briefly considered the graph of a *quadratic equation*

$$y = ax^2 + bx + c, \quad a \neq 0.$$

For example, the graph of the equation $y = x^2 + 6x + 8$ and its x-intercepts are shown below.

The x-intercepts are $(-4, 0)$ and $(-2, 0)$. We will develop in detail the creation of such graphs in Section 10.6. The points $(-4, 0)$ and $(-2, 0)$ are the intersections of the graphs of $y = x^2 + 6x + 8$ and $y = 0$.

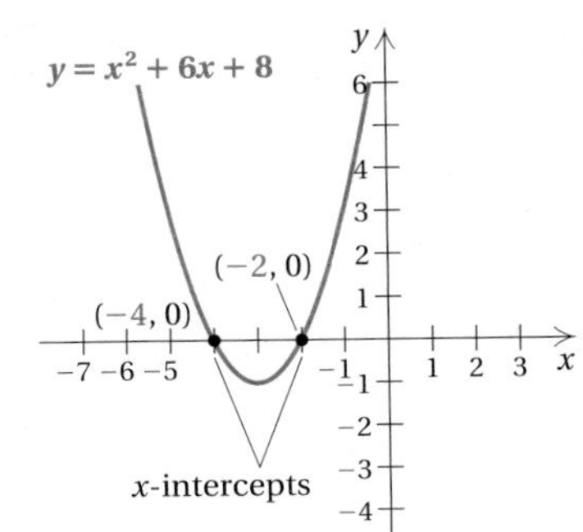

1. a) Consider $y = -\frac{2}{3}x - 3$. Find the intercepts and graph the equation.

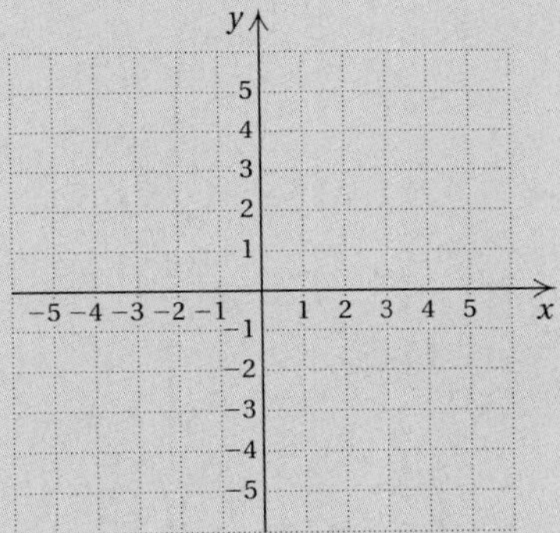

b) Solve the equation

$$0 = -\frac{2}{3}x - 3.$$

c) Complete: The solution of the equation $0 = -\frac{2}{3}x - 3$ is ________. This value is the ________ of the x-intercept, (____, ____), of the graph of $y = -\frac{2}{3}x - 3$.

Answers on page A-44

Write in standard form and determine a, b, and c.

2. $y^2 = 8y$

3. $3 - x^2 = 9x$

4. $3x + 5x^2 = x^2 - 4 + 2x$

5. $5x^2 = 21$

Answers on page A-45

We began studying the solution of quadratic equations like $x^2 + 6x + 8 = 0$ in Section 5.7. There we used factoring for such solutions:

$$x^2 + 6x + 8 = 0$$

$$(x + 4)(x + 2) = 0 \quad \text{Factoring}$$

$$x + 4 = 0 \quad or \quad x + 2 = 0 \quad \text{Using the principle of zero products}$$

$$x = -4 \quad or \quad x = -2.$$

We see that the solutions of $x^2 + 6x + 8 = 0$, -4 and -2, are the first coordinates of the x-intercepts, $(-4, 0)$ and $(-2, 0)$, of the graph of $y = x^2 + 6x + 8$.

We will enhance our ability to solve quadratic equations in Sections 10.1–10.3.

a Standard Form

The following are **quadratic equations.** They contain polynomials of second degree.

$$4x^2 + 7x - 5 = 0,$$

$$3t^2 - \tfrac{1}{2}t = 9,$$

$$5y^2 = -6y,$$

$$5m^2 = 15$$

The quadratic equation $4x^2 + 7x - 5 = 0$ is said to be in **standard form.** Although the quadratic equation $4x^2 = 5 - 7x$ is equivalent to the preceding equation, it is *not* in standard form.

QUADRATIC EQUATION

A **quadratic equation** is an equation equivalent to an equation of the type

$$ax^2 + bx + c = 0, \quad a > 0,$$

where a, b, and c are real-number constants. We say that the preceding is the **standard form of a quadratic equation.**

We define $a > 0$ to ease the proof of the quadratic formula, which we consider later, and to ease solving by factoring, which we review in this section. Suppose we are studying an equation like $-3x^2 + 8x - 2 = 0$. It is not in standard form. We can find an equivalent equation that is in standard form by multiplying both sides by -1:

$$-1(-3x^2 + 8x - 2) = -1(0)$$

$$3x^2 - 8x + 2 = 0.$$

EXAMPLES Write in standard form and determine a, b, and c.

1. $4x^2 + 7x - 5 = 0$ The equation is already in standard form.

$a = 4;\quad b = 7;\quad c = -5$

2. $3x^2 - 0.5x = 9$

$3x^2 - 0.5x - 9 = 0$ Subtracting 9. This is standard form.

$a = 3;\quad b = -0.5;\quad c = -9$

3. $-4y^2 = 5y$

$-4y^2 - 5y = 0$ Subtracting $5y$

Not positive!

$4y^2 + 5y = 0$ Multiplying by -1. This is standard form.

$a = 4;\quad b = 5;\quad c = 0$

Do Exercises 2–5 on the preceding page.

b Solving Quadratic Equations of the Type $ax^2 + bx = 0$

Sometimes we can use factoring and the principle of zero products to solve quadratic equations. We are actually reviewing methods that we introduced in Section 5.7.

When $c = 0$ and $b \neq 0$, we can always factor and use the principle of zero products (see Section 5.7 for a review).

EXAMPLE 4 Solve: $7x^2 + 2x = 0$.

$7x^2 + 2x = 0$

$x(7x + 2) = 0$ Factoring

$x = 0 \quad or \quad 7x + 2 = 0$ Using the principle of zero products

$x = 0 \quad or \quad 7x = -2$

$x = 0 \quad or \quad x = -\frac{2}{7}$

CHECK: For 0:

$7x^2 + 2x = 0$

$7 \cdot 0^2 + 2 \cdot 0 \;?\; 0$

0 TRUE

For $-\frac{2}{7}$:

$7x^2 + 2x = 0$

$7\left(-\frac{2}{7}\right)^2 + 2\left(-\frac{2}{7}\right) \;?\; 0$

$7\left(\frac{4}{49}\right) - \frac{4}{7}$

$\frac{4}{7} - \frac{4}{7}$

0 TRUE

The solutions are 0 and $-\frac{2}{7}$.

CAUTION!

You may be tempted to divide each term in an equation like the one in Example 4 by x. This method would yield the equation

$7x + 2 = 0,$

whose only solution is $-\frac{2}{7}$. In effect, since 0 is also a solution of the original equation, we have divided by 0. The error of such division means the loss of one of the solutions.

Solve.

6. $2x^2 + 8x = 0$

7. $10x^2 - 6x = 0$

Solve.

8. $4x^2 + 5x - 6 = 0$

9. $(x - 1)(x + 1) = 5(x - 1)$

Answers on page A-45

EXAMPLE 5 Solve: $4x^2 - 8x = 0$.

We have

$$4x^2 - 8x = 0$$
$$4x(x - 2) = 0 \qquad \text{Factoring}$$
$$4x = 0 \quad or \quad x - 2 = 0 \qquad \text{Using the principle of zero products}$$
$$x = 0 \quad or \quad x = 2.$$

The solutions are 0 and 2.

ALGEBRAIC–GRAPHICAL CONNECTION

Let's visualize the solutions in Example 5.

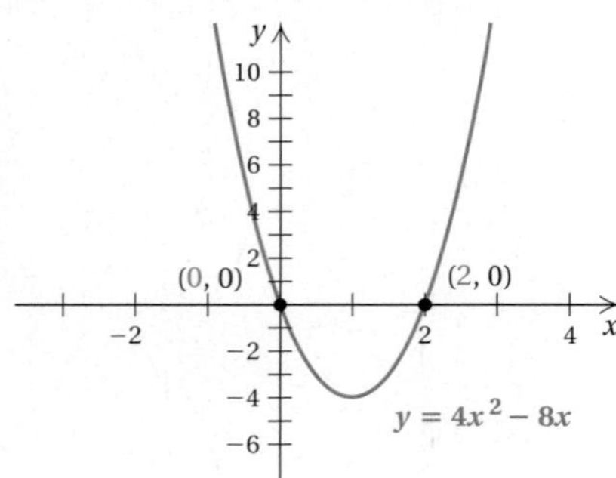

We see that the solutions of $4x^2 - 8x = 0$, 0 and 2, are the first coordinates of the x-intercepts, $(0, 0)$ and $(2, 0)$ of the graph of $y = 4x^2 - 8x$.

A quadratic equation of the type $ax^2 + bx = 0$, where $c = 0$ and $b \neq 0$, will always have 0 as one solution and a nonzero number as the other solution.

Do Exercises 6 and 7 on the preceding page.

C Solving Quadratic Equations of the Type $ax^2 + bx + c = 0$

When neither b nor c is 0, we can sometimes solve by factoring.

EXAMPLE 6 Solve: $2x^2 - x - 21 = 0$.

We have

$$2x^2 - x - 21 = 0$$
$$(2x - 7)(x + 3) = 0 \qquad \text{Factoring}$$
$$2x - 7 = 0 \quad or \quad x + 3 = 0 \qquad \text{Using the principle of zero products}$$
$$2x = 7 \quad or \quad x = -3$$
$$x = \tfrac{7}{2} \quad or \quad x = -3.$$

The solutions are $\frac{7}{2}$ and -3.

ALGEBRAIC–GRAPHICAL CONNECTION

Let's visualize the solutions in Example 6.

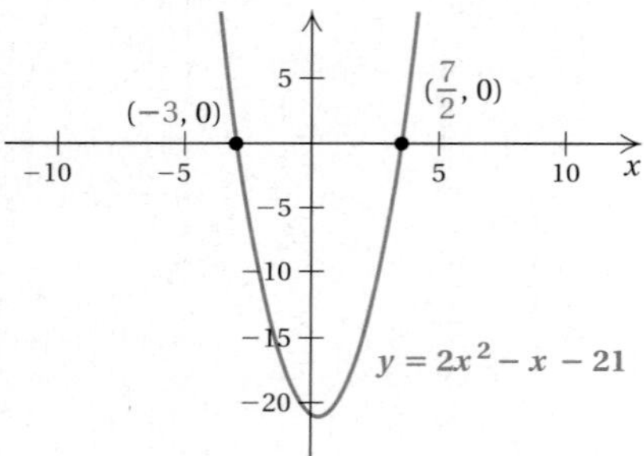

We see that the solutions of $2x^2 - x - 21 = 0$, -3 and $\frac{7}{2}$, are the first coordinates of the x-intercepts, $(-3, 0)$ and $\left(\frac{7}{2}, 0\right)$, of the graph of $y = 2x^2 - x - 21$.

EXAMPLE 7 Solve: $(y - 3)(y - 2) = 6(y - 3)$.

We write the equation in standard form and then try to factor:

$$y^2 - 5y + 6 = 6y - 18 \qquad \text{Multiplying}$$
$$y^2 - 11y + 24 = 0 \qquad \text{Standard form}$$
$$(y - 8)(y - 3) = 0 \qquad \text{Factoring}$$
$$y - 8 = 0 \quad or \quad y - 3 = 0 \qquad \text{Using the principle of zero products}$$
$$y = 8 \quad or \quad y = 3.$$

The solutions are 8 and 3.

Do Exercises 8 and 9 on the preceding page.

Recall that to solve a rational equation, we multiply both sides by the LCM of all the denominators. We may obtain a quadratic equation after a few steps. When that happens, we know how to finish solving, but we must remember to check possible solutions because a replacement may result in division by 0. See Section 6.6.

10. Solve:

$$\frac{20}{x+5} - \frac{1}{x-4} = 1.$$

EXAMPLE 8 Solve: $\dfrac{3}{x-1} + \dfrac{5}{x+1} = 2.$

We multiply by the LCM, which is $(x-1)(x+1)$:

$$(x-1)(x+1)\cdot\left(\frac{3}{x-1} + \frac{5}{x+1}\right) = 2\cdot(x-1)(x+1).$$

We use the distributive law on the left:

$$\begin{aligned}
(x-1)(x+1)\cdot\frac{3}{x-1} + (x-1)(x+1)\cdot\frac{5}{x+1} &= 2(x-1)(x+1)\\
3(x+1) + 5(x-1) &= 2(x-1)(x+1)\\
3x + 3 + 5x - 5 &= 2(x^2 - 1)\\
8x - 2 &= 2x^2 - 2\\
0 &= 2x^2 - 8x\\
0 &= 2x(x-4) \quad \text{Factoring}\\
2x = 0 \quad &or \quad x - 4 = 0\\
x = 0 \quad &or \quad x = 4.
\end{aligned}$$

CHECK: For 0:

$$\begin{array}{c|c}
\multicolumn{2}{c}{\dfrac{3}{x-1} + \dfrac{5}{x+1} = 2}\\
\hline
\dfrac{3}{0-1} + \dfrac{5}{0+1} & ?\ 2\\
\dfrac{3}{-1} + \dfrac{5}{1} & \\
-3 + 5 & \\
2 & \text{TRUE}
\end{array}$$

For 4:

$$\begin{array}{c|c}
\multicolumn{2}{c}{\dfrac{3}{x-1} + \dfrac{5}{x+1} = 2}\\
\hline
\dfrac{3}{4-1} + \dfrac{5}{4+1} & ?\ 2\\
\dfrac{3}{3} + \dfrac{5}{5} & \\
1 + 1 & \\
2 & \text{TRUE}
\end{array}$$

The solutions are 0 and 4.

Do Exercise 10.

d Solving Applied Problems

EXAMPLE 9 *Diagonals of a Polygon.* The number of diagonals d of a polygon of n sides is given by the formula

$$d = \frac{n^2 - 3n}{2}.$$

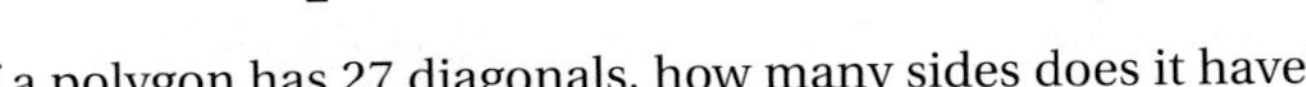

If a polygon has 27 diagonals, how many sides does it have?

1. **Familiarize.** We can make a drawing to familiarize ourselves with the problem. We draw an octagon (8 sides) and count the diagonals and see that there are 20. Let's check this in the formula. We evaluate the formula for $n = 8$:

$$d = \frac{8^2 - 3(8)}{2} = \frac{64 - 24}{2} = \frac{40}{2} = 20.$$

Answer on page A-45

2. **Translate.** We know that the number of diagonals is 27. We substitute 27 for d:

$$27 = \frac{n^2 - 3n}{2}.$$

3. **Solve.** We solve the equation for n, reversing the equation first for convenience:

$$\frac{n^2 - 3n}{2} = 27$$

$$n^2 - 3n = 54 \quad \text{Multiplying by 2 to clear fractions}$$

$$n^2 - 3n - 54 = 0$$

$$(n - 9)(n + 6) = 0$$

$$n - 9 = 0 \quad or \quad n + 6 = 0$$

$$n = 9 \quad or \quad n = -6.$$

4. **Check.** Since the number of sides cannot be negative, -6 cannot be a solution. We leave it to the student to show by substitution that 9 checks.
5. **State.** The polygon has 9 sides (it is a nonagon).

Do Exercise 11 on the following page.

Study Tips

READ FOR SUCCESS

In his article "The Daily Dozen Disciplines for Massive Success in 2001 & Beyond," Jerry Clark comments, "Research has shown that 58% of high school graduates never read another book from cover to cover the rest of their adult lives, that 78% of the population have not been in a bookstore in the last 5 years, and that 97% of the population of the U.S. do not have library cards." Clark then suggests spending at least 15 minutes each day reading an empowering and uplifting book or article. The following books are some suggestions from your author. Their motivating words may empower you in your study of mathematics.

1. *Fish*, by Stephen C. Lundin, Harry Paul, and John Christensen (Hyperion). This was a Wall Street Journal Business Bestseller. Though it has a strange title, it discusses a remarkable way to boost morale and improve results.
2. *True Success*: *A New Philosophy of Excellence*, by Tom Morris (Grosset/Putnam). Morris was a well-loved philosophy professor at Notre Dame. Students, especially athletes, flocked to his classes.
3. *The Road Less Traveled, Abounding Grace,* by M. Scott Peck. Noted psychiatrist and author of many excellent books, Peck has amazing insights and wisdom about life.
4. *The Weight of Glory, The Great Divorce,* by C. S. Lewis. British author and scholar, noted for his philosophical exposition, Lewis also wrote science-fiction fantasies with moral overtones.

"*Far worse than not reading books is not realizing that it matters!*"

Jim Rohn, motivational speaker

CALCULATOR CORNER

Solving Quadratic Equations A quadratic equation written with 0 on one side of the equals sign can be solved using the ZERO feature of a graphing calculator. (See the Calculator Corner on p. 437 for the procedure.)

We can also use the INTERSECT feature to solve a quadratic equation. Consider the equation in Margin Exercise 9,

$$(x - 1)(x + 1) = 5(x - 1).$$

First, we enter $y_1 = (x - 1)(x + 1)$ and $y_2 = 5(x - 1)$ on the equation-editor screen and graph the equations, using the window $[-5, 5, -5, 20]$, Yscl $= 2$. We see that there are two points of intersection, so the equation has two solutions.

Next, we use the INTERSECT feature to find the coordinates of the lefthand point of intersection. (See the Calculator Corner on p. 595 for the procedure.) The first coordinate of this point, 1, is one solution of the equation. We use the INTERSECT feature again to find the other solution, 4.

$y_1 = (x - 1)(x + 1),\ y_2 = 5(x - 1)$

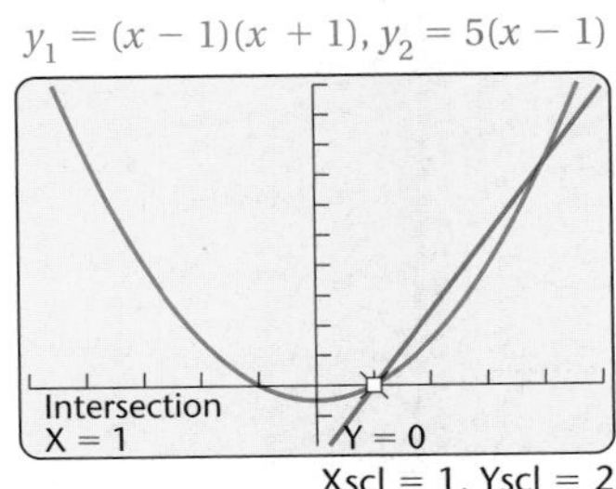

Xscl = 1, Yscl = 2

$y_1 = (x - 1)(x + 1),\ y_2 = 5(x - 1)$

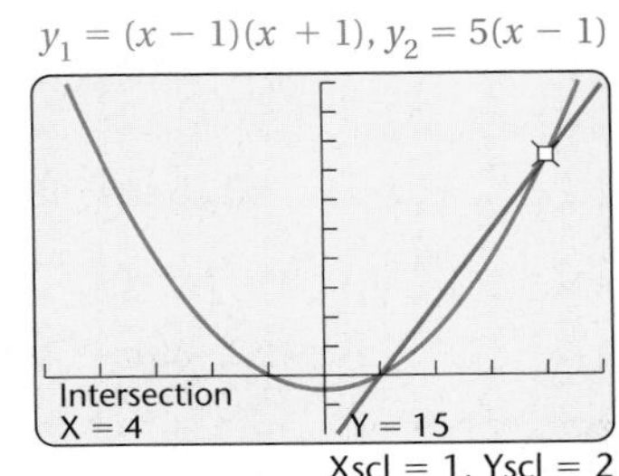

Xscl = 1, Yscl = 2

Note that we could use the ZERO feature to solve this equation if we first write it with 0 on one side, that is, $(x - 1)(x + 1) - 5(x - 1) = 0$.

Exercises: Solve.

1. $5x^2 - 8x + 3 = 0$
2. $2x^2 - 7x - 15 = 0$
3. $6(x - 3) = (x - 3)(x - 2)$
4. $(x + 1)(x - 4) = 3(x - 4)$

11. Consider the following heptagon, that is, a polygon with 7 sides.

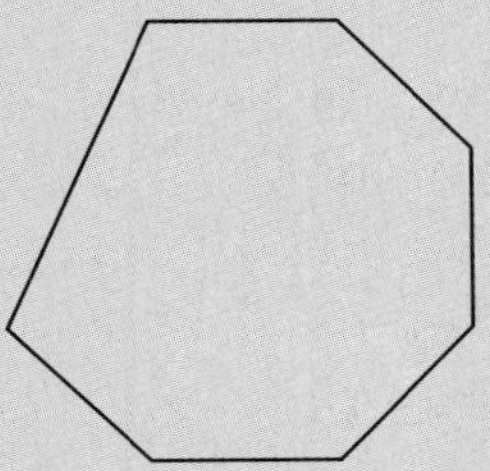

a) Draw all the diagonals and then count them.

b) Use the formula

$$d = \frac{n^2 - 3n}{2}$$

to check your answer to part (a) by evaluating the formula for $n = 7$.

c) A polygon has 44 diagonals. How many sides does it have?

Answers on page A-45

10.1 EXERCISE SET

For Extra Help

Digital Video Tutor CD 9 Videotape 18 | InterAct Math | Math Tutor Center | MathXL | MyMathLab

a Write in standard form and determine a, b, and c.

1. $x^2 - 3x + 2 = 0$

2. $x^2 - 8x - 5 = 0$

3. $7x^2 = 4x - 3$

4. $9x^2 = x + 5$

5. $5 = -2x^2 + 3x$

6. $3x - 1 = 5x^2 + 9$

b Solve.

7. $x^2 + 5x = 0$

8. $x^2 + 7x = 0$

9. $3x^2 + 6x = 0$

10. $4x^2 + 8x = 0$

11. $5x^2 = 2x$

12. $11x = 3x^2$

13. $4x^2 + 4x = 0$

14. $8x^2 - 8x = 0$

15. $0 = 10x^2 - 30x$

16. $0 = 10x^2 - 50x$

17. $11x = 55x^2$

18. $33x^2 = -11x$

19. $14t^2 = 3t$

20. $6m = 19m^2$

21. $5y^2 - 3y^2 = 72y + 9y$

22. $63p - 16p^2 = 17p + 58p^2$

c Solve.

23. $x^2 + 8x - 48 = 0$

24. $x^2 - 16x + 48 = 0$

25. $5 + 6x + x^2 = 0$

26. $x^2 + 10 + 11x = 0$

27. $18 = 7p + p^2$

28. $t^2 + 14t = -24$

29. $-15 = -8y + y^2$

30. $q^2 + 14 = 9q$

31. $x^2 + 10x + 25 = 0$

32. $x^2 + 6x + 9 = 0$

33. $r^2 = 8r - 16$

34. $x^2 + 1 = 2x$

35. $6x^2 + x - 2 = 0$

36. $2x^2 - 11x + 15 = 0$

37. $3a^2 = 10a + 8$

38. $15b - 9b^2 = 4$

39. $6x^2 - 4x = 10$

40. $3x^2 - 7x = 20$

41. $2t^2 + 12t = -10$

42. $12w^2 - 5w = 2$

43. $t(t - 5) = 14$

44. $6z^2 + z - 1 = 0$

45. $t(9 + t) = 4(2t + 5)$

46. $3y^2 + 8y = 12y + 15$

47. $16(p - 1) = p(p + 8)$

48. $(2x - 3)(x + 1) = 4(2x - 3)$

49. $(t - 1)(t + 3) = t - 1$

50. $(x - 2)(x + 2) = x + 2$

Solve.

51. $\frac{24}{x-2}+\frac{24}{x+2}=5$

52. $\frac{8}{x+2}+\frac{8}{x-2}=3$

53. $\frac{1}{x}+\frac{1}{x+6}=\frac{1}{4}$

54. $\frac{1}{x}+\frac{1}{x+9}=\frac{1}{20}$

55. $1+\frac{12}{x^2-4}=\frac{3}{x-2}$

56. $\frac{5}{t-3}-\frac{30}{t^2-9}=1$

57. $\frac{r}{r-1}+\frac{2}{r^2-1}=\frac{8}{r+1}$

58. $\frac{x+2}{x^2-2}=\frac{2}{3-x}$

59. $\frac{x-1}{1-x}=-\frac{x+8}{x-8}$

60. $\frac{4-x}{x-4}+\frac{x+3}{x-3}=0$

61. $\frac{5}{y+4}-\frac{3}{y-2}=4$

62. $\frac{2z+11}{2z+8}=\frac{3z-1}{z-1}$

d Solve.

63. *Diagonals.* A decagon is a figure with 10 sides. How many diagonals does a decagon have?

64. *Diagonals.* A hexagon is a figure with 6 sides. How many diagonals does a hexagon have?

65. *Diagonals.* A polygon has 14 diagonals. How many sides does it have?

66. *Diagonals.* A polygon has 9 diagonals. How many sides does it have?

67. ^{D}W Explain how the graph of $y=(x-2)(x+3)$ is related to the solutions of the equation $(x-2)(x+3)=0$.

68. ^{D}W Explain how you might go about constructing a quadratic equation whose solutions are -5 and 7.

SKILL MAINTENANCE

Simplify. [9.1a], [9.2a]

69. $\sqrt{64}$

70. $-\sqrt{169}$

71. $\sqrt{8}$

72. $\sqrt{12}$

73. $\sqrt{20}$

74. $\sqrt{88}$

75. $\sqrt{405}$

76. $\sqrt{1020}$

Use a calculator to approximate the square roots. Round to three decimal places. [9.1b]

77. $\sqrt{7}$

78. $\sqrt{23}$

79. $\sqrt{\frac{7}{3}}$

80. $\sqrt{524.77}$

SYNTHESIS

Solve.

81. $4m^2-(m+1)^2=0$

82. $x^2+\sqrt{22}x=0$

83. $\sqrt{5}x^2-x=0$

84. $\sqrt{7}x^2+\sqrt{3}x=0$

Use a graphing calculator to solve the equation.

85. $3x^2-7x=20$

86. $x(x-5)=14$

87. $3x^2+8x=12x+15$

88. $(x-2)(x+2)=x+2$

89. $(x-2)^2+3(x-2)=4$

90. $(x+3)^2=4$

91. $16(x-1)=x(x+8)$

92. $x^2+2.5x+1.5625=9.61$

Objectives

a Solve quadratic equations of the type $ax^2 = p$.

b Solve quadratic equations of the type $(x + c)^2 = d$.

c Solve quadratic equations by completing the square.

d Solve certain problems involving quadratic equations of the type $ax^2 = p$.

1. Solve: $x^2 = 10$.

2. Solve: $6x^2 = 0$.

Answers on page A-45

10.2 SOLVING QUADRATIC EQUATIONS BY COMPLETING THE SQUARE

a Solving Quadratic Equations of the Type $ax^2 = p$

For equations of the type $ax^2 = p$, we first solve for x^2 and then apply the *principle of square roots,* which states that a positive number has two square roots. The number 0 has one square root, 0.

THE PRINCIPLE OF SQUARE ROOTS

- The equation $x^2 = d$ has two real solutions when $d > 0$. The solutions are $\sqrt{d}$ and $-\sqrt{d}$.
- The equation $x^2 = d$ has no real-number solution when $d < 0$.
- The equation $x^2 = 0$ has 0 as its only solution.

EXAMPLE 1 Solve: $x^2 = 3$.

$$x^2 = 3$$

$$x = \sqrt{3} \quad or \quad x = -\sqrt{3} \qquad \text{Using the principle of square roots}$$

CHECK: For $\sqrt{3}$:

$$\begin{array}{c|c} x^2 & 3 \\ \hline (\sqrt{3})^2 \ ? & 3 \\ 3 & \end{array} \quad \text{TRUE}$$

For $-\sqrt{3}$:

$$\begin{array}{c|c} x^2 & 3 \\ \hline (-\sqrt{3})^2 \ ? & 3 \\ 3 & \end{array} \quad \text{TRUE}$$

The solutions are $\sqrt{3}$ and $-\sqrt{3}$.

Do Exercise 1.

EXAMPLE 2 Solve: $\frac{1}{8}x^2 = 0$.

$$\frac{1}{8}x^2 = 0$$

$$x^2 = 0 \qquad \text{Multiplying by 8}$$

$$x = 0 \qquad \text{Using the principle of square roots}$$

The solution is 0.

Do Exercise 2.

EXAMPLE 3 Solve: $-3x^2 + 7 = 0$.

$$-3x^2 + 7 = 0$$

$$-3x^2 = -7 \qquad \text{Subtracting 7}$$

$$x^2 = \frac{-7}{-3}, \text{ or } \frac{7}{3} \qquad \text{Dividing by } -3$$

$$x = \sqrt{\frac{7}{3}} \quad or \quad x = -\sqrt{\frac{7}{3}} \qquad \text{Using the principle of square roots}$$

$$x = \sqrt{\frac{7}{3} \cdot \frac{3}{3}} \quad or \quad x = -\sqrt{\frac{7}{3} \cdot \frac{3}{3}} \qquad \text{Rationalizing the denominators}$$

$$x = \frac{\sqrt{21}}{3} \quad or \quad x = -\frac{\sqrt{21}}{3}$$

CHECK: For $\frac{\sqrt{21}}{3}$:

$$\begin{array}{r|l} -3x^2 + 7 = 0 & \\ \hline -3\left(\frac{\sqrt{21}}{3}\right)^2 + 7 & ?\ 0 \\ -3 \cdot \frac{21}{9} + 7 & \\ -7 + 7 & \\ 0 & \quad \text{TRUE} \end{array}$$

For $-\frac{\sqrt{21}}{3}$:

$$\begin{array}{r|l} -3x^2 + 7 = 0 & \\ \hline -3\left(-\frac{\sqrt{21}}{3}\right)^2 + 7 & ?\ 0 \\ -3 \cdot \frac{21}{9} + 7 & \\ -7 + 7 & \\ 0 & \quad \text{TRUE} \end{array}$$

The solutions are $\frac{\sqrt{21}}{3}$ and $-\frac{\sqrt{21}}{3}$.

Do Exercise 3.

3. Solve: $2x^2 - 3 = 0$.

b Solving Quadratic Equations of the Type $(x + c)^2 = d$

In an equation of the type $(x + c)^2 = d$, we have the square of a binomial equal to a constant. We can use the principle of square roots to solve such an equation.

EXAMPLE 4 Solve: $(x - 5)^2 = 9$.

$$(x - 5)^2 = 9$$

$x - 5 = 3$ *or* $x - 5 = -3$ — Using the principle of square roots

$x = 8$ *or* $x = 2$

The solutions are 8 and 2.

EXAMPLE 5 Solve: $(x + 2)^2 = 7$.

$$(x + 2)^2 = 7$$

$x + 2 = \sqrt{7}$ *or* $x + 2 = -\sqrt{7}$ — Using the principle of square roots

$x = -2 + \sqrt{7}$ *or* $x = -2 - \sqrt{7}$

The solutions are $-2 + \sqrt{7}$ and $-2 - \sqrt{7}$, or simply $-2 \pm \sqrt{7}$ (read "-2 plus or minus $\sqrt{7}$").

Do Exercises 4 and 5.

Solve.

4. $(x - 3)^2 = 16$

5. $(x + 4)^2 = 11$

In Examples 4 and 5, the left sides of the equations are squares of binomials. If we can express an equation in such a form, we can proceed as we did in those examples.

EXAMPLE 6 Solve: $x^2 + 8x + 16 = 49$.

$x^2 + 8x + 16 = 49$ — The left side is the square of a binomial.

$(x + 4)^2 = 49$

$x + 4 = 7$ *or* $x + 4 = -7$ — Using the principle of square roots

$x = 3$ *or* $x = -11$

The solutions are 3 and -11.

Do Exercises 6 and 7.

Solve.

6. $x^2 - 6x + 9 = 64$

7. $x^2 - 2x + 1 = 5$

Answers on page A-45

C Completing the Square

We have seen that a quadratic equation like $(x - 5)^2 = 9$ can be solved by using the principle of square roots. We also noted that an equation like $x^2 + 8x + 16 = 49$ can be solved in the same manner because the expression on the left side is the square of a binomial, $(x + 4)^2$. This second procedure is the basis for a method of solving quadratic equations called **completing the square.** *It can be used to solve any quadratic equation.*

Suppose we have the following quadratic equation:

$$x^2 + 10x = 4.$$

If we could add to both sides of the equation a constant that would make the expression on the left the square of a binomial, we could then solve the equation using the principle of square roots.

How can we determine what to add to $x^2 + 10x$ in order to construct the square of a binomial? We want to find a number a such that the following equation is satisfied:

$$x^2 + 10x + a^2 = (x + a)(x + a) = x^2 + 2ax + a^2.$$

Thus, a is such that $2a = 10$. Solving for a, we get $a = 5$; that is, a is half of the coefficient of x in $x^2 + 10x$. Since $a^2 = (\frac{10}{2})^2 = 5^2 = 25$, we add 25 to our original expression:

$$x^2 + 10x + 25 \text{ is the square of } x + 5;$$

that is,

$$x^2 + 10x + 25 = (x + 5)^2.$$

COMPLETING THE SQUARE

To **complete the square** of an expression like $x^2 + bx$, we take half of the coefficient of x and square it. Then we add that number, which is $(b/2)^2$.

Returning to solve our original equation, we first add 25 to *both* sides to complete the square. Then we solve as follows:

$$x^2 + 10x = 4 \qquad \text{Original equation}$$

$$x^2 + 10x + 25 = 4 + 25 \qquad \text{Adding 25: } (\tfrac{10}{2})^2 = 5^2 = 25$$

$$(x + 5)^2 = 29$$

$$x + 5 = \sqrt{29} \quad or \quad x + 5 = -\sqrt{29} \qquad \text{Using the principle of square roots}$$

$$x = -5 + \sqrt{29} \quad or \quad x = -5 - \sqrt{29}.$$

The solutions are $-5 \pm \sqrt{29}$.

We have seen that a quadratic equation $(x + c)^2 = d$ can be solved by using the principle of square roots. Any quadratic equation can be put in this form by completing the square. Then we can solve as before.

EXAMPLE 7 Solve: $x^2 + 6x + 8 = 0$.

We have

$$x^2 + 6x + 8 = 0$$
$$x^2 + 6x = -8. \quad \text{Subtracting 8}$$

We take half of 6, $\frac{6}{2} = 3$, and square it, to get 3^2, or 9. Then we add 9 to *both* sides of the equation. This makes the left side the square of a binomial. We have now completed the square.

$$x^2 + 6x + 9 = -8 + 9 \quad \text{Adding 9}$$
$$(x + 3)^2 = 1$$
$$x + 3 = 1 \quad or \quad x + 3 = -1 \quad \text{Using the principle of square roots}$$
$$x = -2 \quad or \quad x = -4$$

The solutions are -2 and -4.

Do Exercises 8 and 9.

EXAMPLE 8 Solve $x^2 - 4x - 7 = 0$ by completing the square.

We have

$$x^2 - 4x - 7 = 0$$
$$x^2 - 4x = 7 \quad \text{Adding 7}$$
$$x^2 - 4x + 4 = 7 + 4 \quad \text{Adding 4: } \left(\frac{-4}{2}\right)^2 = (-2)^2 = 4$$
$$(x - 2)^2 = 11$$
$$x - 2 = \sqrt{11} \quad or \quad x - 2 = -\sqrt{11} \quad \text{Using the principle of square roots}$$
$$x = 2 + \sqrt{11} \quad or \quad x = 2 - \sqrt{11}.$$

The solutions are $2 \pm \sqrt{11}$.

Do Exercise 10.

Example 7, as well as the following example, can be solved more easily by factoring. We solve them by completing the square only to illustrate that completing the square can be used to solve *any* quadratic equation.

EXAMPLE 9 Solve $x^2 + 3x - 10 = 0$ by completing the square.

We have

$$x^2 + 3x - 10 = 0$$
$$x^2 + 3x = 10$$
$$x^2 + 3x + \frac{9}{4} = 10 + \frac{9}{4} \quad \text{Adding } \tfrac{9}{4}\text{: } \left(\tfrac{3}{2}\right)^2 = \tfrac{9}{4}$$
$$\left(x + \frac{3}{2}\right)^2 = \frac{40}{4} + \frac{9}{4} = \frac{49}{4}$$
$$x + \frac{3}{2} = \frac{7}{2} \quad or \quad x + \frac{3}{2} = -\frac{7}{2} \quad \text{Using the principle of square roots}$$
$$x = \frac{4}{2} \quad or \quad x = -\frac{10}{2}$$
$$x = 2 \quad or \quad x = -5.$$

The solutions are 2 and -5.

Do Exercise 11.

Solve.

8. $x^2 - 6x + 8 = 0$

9. $x^2 + 8x - 20 = 0$

10. Solve: $x^2 - 12x + 23 = 0$.

11. Solve: $x^2 - 3x - 10 = 0$.

Answers on page A-45

12. Solve: $2x^2 + 3x - 3 = 0$.

When the coefficient of x^2 is not 1, we can make it 1, as shown in the following example.

EXAMPLE 10 Solve $2x^2 = 3x + 1$ by completing the square.

We first obtain standard form. Then we multiply both sides by $\frac{1}{2}$ to make the x^2-coefficient 1.

$$2x^2 = 3x + 1$$

$$2x^2 - 3x - 1 = 0 \qquad \text{Finding standard form}$$

$$\frac{1}{2}(2x^2 - 3x - 1) = \frac{1}{2} \cdot 0 \qquad \text{Multiplying by } \frac{1}{2} \text{ to make the } x^2\text{-coefficient 1}$$

$$x^2 - \frac{3}{2}x - \frac{1}{2} = 0$$

$$x^2 - \frac{3}{2}x = \frac{1}{2} \qquad \text{Adding } \frac{1}{2}$$

$$x^2 - \frac{3}{2}x + \frac{9}{16} = \frac{1}{2} + \frac{9}{16} \qquad \text{Adding } \frac{9}{16}: \left[\frac{1}{2}\left(-\frac{3}{2}\right)\right]^2 = \left[-\frac{3}{4}\right]^2 = \frac{9}{16}$$

$$\left(x - \frac{3}{4}\right)^2 = \frac{8}{16} + \frac{9}{16}$$

$$\left(x - \frac{3}{4}\right)^2 = \frac{17}{16}$$

$$x - \frac{3}{4} = \frac{\sqrt{17}}{4} \quad or \quad x - \frac{3}{4} = -\frac{\sqrt{17}}{4} \qquad \text{Using the principle of square roots}$$

$$x = \frac{3}{4} + \frac{\sqrt{17}}{4} \quad or \quad x = \frac{3}{4} - \frac{\sqrt{17}}{4}$$

The solutions are $\frac{3 \pm \sqrt{17}}{4}$.

SOLVING BY COMPLETING THE SQUARE

To solve a quadratic equation $ax^2 + bx + c = 0$ by completing the square:

1. If $a \neq 1$, multiply by $1/a$ so that the x^2-coefficient is 1.
2. If the x^2-coefficient is 1, add so that the equation is in the form

   $$x^2 + bx = -c, \quad \text{or} \quad x^2 + \frac{b}{a}x = -\frac{c}{a} \text{ if step (1) has been applied.}$$

3. Take half of the x-coefficient and square it. Add the result to both sides of the equation.
4. Express the side with the variables as the square of the binomial.
5. Use the principle of square roots and complete the solution.

Completing the square provides a base for the quadratic formula, which we will discuss in Section 10.3. It also has other uses in later mathematics courses.

Do Exercise 12.

Answer on page A-45

d Applications

EXAMPLE 11 *Falling Object.* As of this writing, the CN Tower in Toronto is considered the world's tallest building and free-standing structure. It is about 1815 ft tall. How long would it take an object to fall to the ground from the top?

1. **Familiarize.** If we did not know anything about this problem, we might consider looking up a formula in a mathematics or physics book. A formula that fits this situation is

$$s = 16t^2,$$

where s is the distance, in feet, traveled by a body falling freely from rest in t seconds. This formula is actually an approximation in that it does not account for air resistance. In this problem, we know the distance s to be 1815 ft. We want to determine the time t for the object to reach the ground.

2. **Translate.** We know that the distance is 1815 ft and that we need to solve for t. We substitute 1815 for s: $1815 = 16t^2$. This gives us a translation.

3. **Solve.** We solve the equation:

$$1815 = 16t^2$$

$$\frac{1815}{16} = t^2 \quad \text{Solving for } t^2$$

$$113.4 \approx t^2 \quad \text{Dividing and rounding}$$

$$\sqrt{113.4} = t \quad or \quad -\sqrt{113.4} = t \quad \text{Using the principle of square roots}$$

$$10.6 \approx t \quad or \quad -10.6 \approx t. \quad \text{Using a calculator to find the square root and rounding to the nearest tenth}$$

4. **Check.** The number -10.6 cannot be a solution because time cannot be negative in this situation. We substitute 10.6 in the original equation:

$$s = 16(10.6)^2 = 16(112.36) = 1797.76.$$

This answer is close: $1797.76 \approx 1815$. Remember that we rounded twice to approximate our solution, $t \approx 10.6$. Thus we have a check.

5. **State.** It takes about 10.6 sec for an object to fall to the ground from the top of the CN Tower.

Do Exercise 13.

13. **Falling Object.** The Transco Tower in Houston is 901 ft tall. How long would it take an object to fall to the ground from the top?
Source: *The New York Times Almanac*

Answer on page A-45

10.2 EXERCISE SET

For Extra Help

Digital Video Tutor CD 9 Videotape 18 | InterAct Math | Math Tutor Center | MathXL | MyMathLab

a Solve.

1. $x^2 = 121$

2. $x^2 = 100$

3. $5x^2 = 35$

4. $5x^2 = 45$

5. $5x^2 = 3$

6. $2x^2 = 9$

7. $4x^2 - 25 = 0$

8. $9x^2 - 4 = 0$

9. $3x^2 - 49 = 0$

10. $5x^2 - 16 = 0$

11. $4y^2 - 3 = 9$

12. $36y^2 - 25 = 0$

13. $49y^2 - 64 = 0$

14. $8x^2 - 400 = 0$

b Solve.

15. $(x + 3)^2 = 16$

16. $(x - 4)^2 = 25$

17. $(x + 3)^2 = 21$

18. $(x - 3)^2 = 6$

19. $(x + 13)^2 = 8$

20. $(x - 13)^2 = 64$

21. $(x - 7)^2 = 12$

22. $(x + 1)^2 = 14$

23. $(x + 9)^2 = 34$

24. $(t + 5)^2 = 49$

25. $\left(x + \frac{3}{2}\right)^2 = \frac{7}{2}$

26. $\left(y - \frac{3}{4}\right)^2 = \frac{17}{16}$

27. $x^2 - 6x + 9 = 64$

28. $p^2 - 10p + 25 = 100$

29. $x^2 + 14x + 49 = 64$

30. $t^2 + 8t + 16 = 36$

c Solve by completing the square. Show your work.

31. $x^2 - 6x - 16 = 0$

32. $x^2 + 8x + 15 = 0$

33. $x^2 + 22x + 21 = 0$

34. $x^2 + 14x - 15 = 0$

35. $x^2 - 2x - 5 = 0$

36. $x^2 - 4x - 11 = 0$

37. $x^2 - 22x + 102 = 0$

38. $x^2 - 18x + 74 = 0$

39. $x^2 + 10x - 4 = 0$

40. $x^2 - 10x - 4 = 0$

41. $x^2 - 7x - 2 = 0$

42. $x^2 + 7x - 2 = 0$

43. $x^2 + 3x - 28 = 0$

44. $x^2 - 3x - 28 = 0$

45. $x^2 + \frac{3}{2}x - \frac{1}{2} = 0$

46. $x^2 - \frac{3}{2}x - 2 = 0$

47. $2x^2 + 3x - 17 = 0$

48. $2x^2 - 3x - 1 = 0$

49. $3x^2 + 4x - 1 = 0$

50. $3x^2 - 4x - 3 = 0$

51. $2x^2 = 9x + 5$

52. $2x^2 = 5x + 12$

53. $6x^2 + 11x = 10$

54. $4x^2 + 12x = 7$

d Solve.

55. *Petronas Towers.* At a height of 1483 ft, the Petronas Towers in Kuala Lumpur is one of the tallest buildings in the world. How long would it take an object to fall from the top?
Source: *The New York Times Almanac*

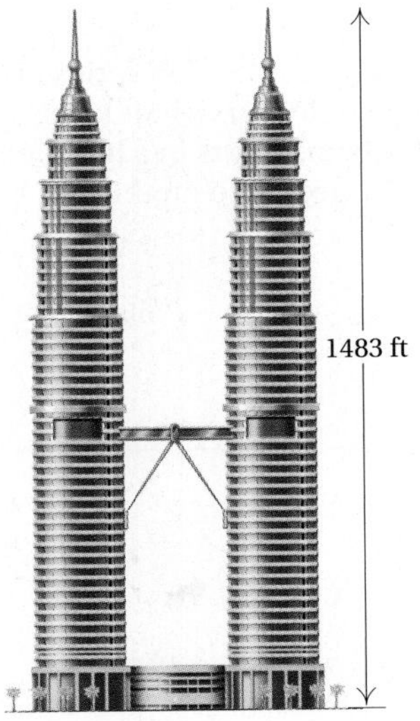

56. *Jin Mao Building.* At a height of 1381 ft, the Jin Mao Building in Shanghai is one of the tallest buildings in the world. How long would it take an object to fall from the top?
Source: *The New York Times Almanac*

57. *Free-Fall Record.* The world record for free-fall to the ground, by a man without a parachute, is 311 ft and is held by Dar Robinson. Approximately how long did the fall take?
Source: *Sports Illustrated*

58. *Free-Fall Record.* The world record for free-fall to the ground, by a woman without a parachute, into a cushioned landing area is 175 ft and is held by Kitty O'Neill. Approximately how long did the fall take?

59. $^{\mathrm{D}}\mathrm{W}$ Corey asserts that the solution of a quadratic equation is $3 \pm \sqrt{14}$ and states that there is only one solution. What mistake is being made?

60. $^{\mathrm{D}}\mathrm{W}$ If a quadratic equation can be solved by factoring, what type of number(s) will generally be solutions?

SKILL MAINTENANCE

61. Find an equation of variation in which y varies inversely as x and $y = 235$ when $x = 0.6$. [7.5c]

62. The time T to do a certain job varies inversely as the number N of people working. It takes 5 hr for 24 people to wash and wax the floors in a building. How long would it take 36 people to do the job? [7.5d]

Multiply and simplify. [9.2c]

63. $\sqrt{3x} \cdot \sqrt{6x}$

64. $\sqrt{8x^2} \cdot \sqrt{24x^3}$

65. $3\sqrt{t} \cdot \sqrt{t}$

66. $\sqrt{x^2} \cdot \sqrt{x^5}$

SYNTHESIS

Find b such that the trinomial is a square.

67. $x^2 + bx + 36$

68. $x^2 + bx + 55$

69. $x^2 + bx + 128$

70. $4x^2 + bx + 16$

71. $x^2 + bx + c$

72. $ax^2 + bx + c$

Solve.

73. $4.82x^2 = 12{,}000$

74. $\frac{x}{2} = \frac{32}{x}$

75. $\frac{x}{9} = \frac{36}{4x}$

76. $\frac{4}{m^2 - 7} = 1$

10.3 THE QUADRATIC FORMULA

Objectives

a Solve quadratic equations using the quadratic formula.

b Find approximate solutions of quadratic equations using a calculator.

We learn to complete the square to prove a general formula that can be used to solve quadratic equations even when they cannot be solved by factoring.

a Solving Using the Quadratic Formula

Each time you solve by completing the square, you perform nearly the same steps. When we repeat the same kind of computation many times, we look for a formula so we can speed up our work. Consider

$$ax^2 + bx + c = 0, \quad a > 0.$$

Let's solve by completing the square. As we carry out the steps, compare them with Example 10 in the preceding section.

$$x^2 + \frac{b}{a}x + \frac{c}{a} = 0 \qquad \text{Multiplying by } \frac{1}{a}$$

$$x^2 + \frac{b}{a}x = -\frac{c}{a} \qquad \text{Adding } -\frac{c}{a}$$

Half of $\frac{b}{a}$ is $\frac{b}{2a}$. The square is $\frac{b^2}{4a^2}$. Thus we add $\frac{b^2}{4a^2}$ to both sides.

$$x^2 + \frac{b}{a}x + \frac{b^2}{4a^2} = -\frac{c}{a} + \frac{b^2}{4a^2} \qquad \text{Adding } \frac{b^2}{4a^2}$$

$$\left(x + \frac{b}{2a}\right)^2 = -\frac{4ac}{4a^2} + \frac{b^2}{4a^2} \qquad \text{Factoring the left side and finding a common denominator on the right}$$

$$\left(x + \frac{b}{2a}\right)^2 = \frac{b^2 - 4ac}{4a^2}$$

$$x + \frac{b}{2a} = \sqrt{\frac{b^2 - 4ac}{4a^2}} \quad \text{or} \quad x + \frac{b}{2a} = -\sqrt{\frac{b^2 - 4ac}{4a^2}} \qquad \text{Using the principle of square roots}$$

Since $a > 0$, $\sqrt{4a^2} = 2a$, so we can simplify as follows:

$$x + \frac{b}{2a} = \frac{\sqrt{b^2 - 4ac}}{2a} \quad \text{or} \quad x + \frac{b}{2a} = -\frac{\sqrt{b^2 - 4ac}}{2a}.$$

Thus,

$$x = -\frac{b}{2a} + \frac{\sqrt{b^2 - 4ac}}{2a} \quad \text{or} \quad x = -\frac{b}{2a} - \frac{\sqrt{b^2 - 4ac}}{2a},$$

so $$x = -\frac{b}{2a} \pm \frac{\sqrt{b^2 - 4ac}}{2a},$$

or $$x = \frac{-b \pm \sqrt{b^2 - 4ac}}{2a}.$$

We now have the following.

THE QUADRATIC FORMULA

The solutions of $ax^2 + bx + c = 0$ are given by

$$x = \frac{-b \pm \sqrt{b^2 - 4ac}}{2a}.$$

1. Solve using the quadratic formula:

$$2x^2 = 4 - 7x.$$

Answer on page A-45

The formula also holds when $a < 0$. A similar proof would show this, but we will not consider it here.

EXAMPLE 1 Solve $5x^2 - 8x = -3$ using the quadratic formula.

We first find standard form and determine a, b, and c:

$$5x^2 - 8x + 3 = 0;$$

$$a = 5, \quad b = -8, \quad c = 3.$$

We then use the quadratic formula:

$$x = \frac{-b \pm \sqrt{b^2 - 4ac}}{2a}$$

$$x = \frac{-(-8) \pm \sqrt{(-8)^2 - 4 \cdot 5 \cdot 3}}{2 \cdot 5} \quad \text{Substituting}$$

CAUTION!

Be sure to write the fraction bar all the way across.

$$x = \frac{8 \pm \sqrt{64 - 60}}{10}$$

$$x = \frac{8 \pm \sqrt{4}}{10}$$

$$x = \frac{8 \pm 2}{10}$$

$$x = \frac{8 + 2}{10} \quad or \quad x = \frac{8 - 2}{10}$$

$$x = \frac{10}{10} \quad or \quad x = \frac{6}{10}$$

$$x = 1 \quad or \quad x = \frac{3}{5}.$$

The solutions are 1 and $\frac{3}{5}$.

Do Exercise 1.

It would have been easier to solve the equation in Example 1 by factoring. We used the quadratic formula only to illustrate that it can be used to solve any quadratic equation. The following is a general procedure for solving a quadratic equation.

SOLVING USING THE QUADRATIC FORMULA

To solve a quadratic equation:

1. Check to see if it is in the form $ax^2 = p$ or $(x + c)^2 = d$. If it is, use the principle of square roots as in Section 10.2.
2. If it is not in the form of (1), write it in standard form, $ax^2 + bx + c = 0$ with a and b nonzero.
3. Then try factoring.
4. If it is not possible to factor or if factoring seems difficult, use the quadratic formula.

The solutions of a quadratic equation can always be found using the quadratic formula. They cannot always be found by factoring. (When $b^2 - 4ac \geq 0$, the equation has real-number solutions. When $b^2 - 4ac < 0$, the equation has no real-number solutions.)

EXAMPLE 2 Solve $x^2 + 3x - 10 = 0$ using the quadratic formula.

The equation is in standard form. So we determine a, b, and c:

$$x^2 + 3x - 10 = 0;$$
$$a = 1, \quad b = 3, \quad c = -10.$$

We then use the quadratic formula:

$$x = \frac{-b \pm \sqrt{b^2 - 4ac}}{2a}$$
$$= \frac{-3 \pm \sqrt{3^2 - 4 \cdot 1 \cdot (-10)}}{2 \cdot 1} \quad \text{Substituting}$$
$$= \frac{-3 \pm \sqrt{9 + 40}}{2}$$
$$= \frac{-3 \pm \sqrt{49}}{2} = \frac{-3 \pm 7}{2}.$$

Thus,

$$x = \frac{-3 + 7}{2} = \frac{4}{2} = 2 \quad or \quad x = \frac{-3 - 7}{2} = \frac{-10}{2} = -5.$$

The solutions are 2 and -5.

Note that the radicand ($b^2 - 4ac = 49$) in the quadratic formula is a perfect square, so we could have used factoring to solve.

Do Exercise 2.

ALGEBRAIC–GRAPHICAL CONNECTION

Let's visualize the solutions in Example 2.

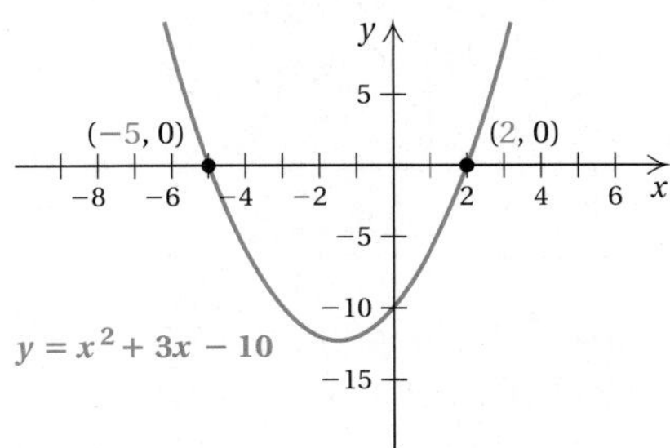

We see that the solutions of $x^2 + 3x - 10 = 0$, -5 and 2, are the first coordinates of the x-intercepts, $(-5, 0)$ and $(2, 0)$, of the graph of $y = x^2 + 3x - 10$.

2. Solve using the quadratic formula:

$$x^2 - 3x - 10 = 0.$$

EXAMPLE 3 Solve $x^2 = 4x + 7$ using the quadratic formula. Compare with Example 8 in Section 10.2.

We first find standard form and determine a, b, and c:

$$x^2 - 4x - 7 = 0;$$
$$a = 1, \quad b = -4, \quad c = -7.$$

We then use the quadratic formula:

$$x = \frac{-b \pm \sqrt{b^2 - 4ac}}{2a} = \frac{-(-4) \pm \sqrt{(-4)^2 - 4 \cdot 1 \cdot (-7)}}{2 \cdot 1} \quad \text{Substituting}$$
$$= \frac{4 \pm \sqrt{16 + 28}}{2} = \frac{4 \pm \sqrt{44}}{2}$$
$$= \frac{4 \pm \sqrt{4 \cdot 11}}{2} = \frac{4 \pm \sqrt{4}\sqrt{11}}{2}$$
$$= \frac{4 \pm 2\sqrt{11}}{2} = \frac{2 \cdot 2 \pm 2\sqrt{11}}{2 \cdot 1}$$
$$= \frac{2(2 \pm \sqrt{11})}{2 \cdot 1} = \frac{2}{2} \cdot \frac{2 \pm \sqrt{11}}{1} \quad \text{Factoring out 2 in the numerator and the denominator}$$
$$= 2 \pm \sqrt{11}.$$

The solutions are $2 + \sqrt{11}$ and $2 - \sqrt{11}$, or $2 \pm \sqrt{11}$.

Do Exercise 3.

3. Solve using the quadratic formula:

$$x^2 + 4x = 7.$$

4. Solve using the quadratic formula:

$$x^2 = x - 1.$$

Answers on page A-45

5. Solve using the quadratic formula:

$$5x^2 - 8x = 3.$$

Answer on page A-45

CALCULATOR CORNER

Visualizing Solutions of Quadratic Equations To see that there are no real-number solutions of the equation in Example 4,

$$x^2 + x = -1,$$

we graph $y_1 = x^2 + x$ and $y_2 = -1$.

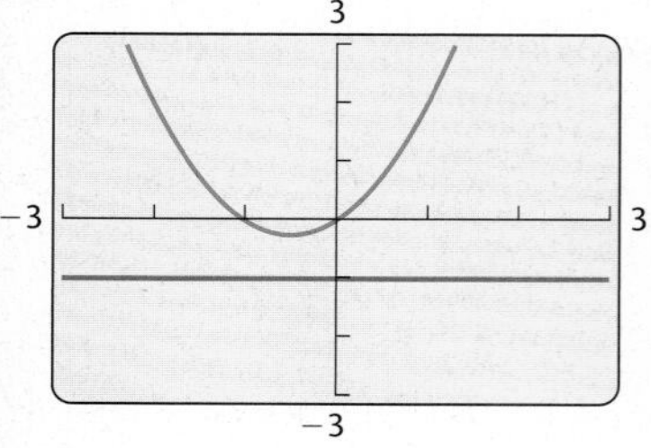

We see that the graphs do not intersect. Thus there is no real number for which $y_1 = y_2$, or $x^2 + x = -1$.

Exercises

1. Explain how the graph of $y = x^2 + x + 1$ shows that the equation in Example 4, $x^2 + x = -1$, has no real-number solutions.
2. Use a graph to determine whether the equation $x^2 + x = 1$ has real-number solutions.
3. Use a graph to determine whether the equation $x^2 = x - 1$ has real-number solutions.

EXAMPLE 4 Solve $x^2 + x = -1$ using the quadratic formula.

We first find standard form and determine a, b, and c:

$$x^2 + x + 1 = 0;$$
$$a = 1, \quad b = 1, \quad c = 1.$$

We then use the quadratic formula:

$$x = \frac{-b \pm \sqrt{b^2 - 4ac}}{2a} = \frac{-1 \pm \sqrt{1^2 - 4 \cdot 1 \cdot 1}}{2 \cdot 1} = \frac{-1 \pm \sqrt{-3}}{2}.$$

Note that the radicand $(b^2 - 4ac = -3)$ in the quadratic formula is negative. Thus there are no real-number solutions because square roots of negative numbers do not exist as real numbers.

Do Exercise 4 on the preceding page.

EXAMPLE 5 Solve $3x^2 = 7 - 2x$ using the quadratic formula.

We first find standard form and determine a, b, and c:

$$3x^2 + 2x - 7 = 0;$$
$$a = 3, \quad b = 2, \quad c = -7.$$

We then use the quadratic formula:

$$x = \frac{-b \pm \sqrt{b^2 - 4ac}}{2a} = \frac{-2 \pm \sqrt{2^2 - 4 \cdot 3 \cdot (-7)}}{2 \cdot 3} = \frac{-2 \pm \sqrt{4 + 84}}{2 \cdot 3}$$
$$= \frac{-2 \pm \sqrt{88}}{6} = \frac{-2 \pm \sqrt{4 \cdot 22}}{6} = \frac{-2 \pm \sqrt{4}\sqrt{22}}{6} = \frac{-2 \pm 2\sqrt{22}}{6}$$
$$= \frac{2(-1 \pm \sqrt{22})}{2 \cdot 3} = \frac{2}{2} \cdot \frac{-1 \pm \sqrt{22}}{3} = \frac{-1 \pm \sqrt{22}}{3}.$$

The solutions are $\frac{-1 + \sqrt{22}}{3}$ and $\frac{-1 - \sqrt{22}}{3}$, or $\frac{-1 \pm \sqrt{22}}{3}$.

Do Exercise 5.

b Approximate Solutions

A calculator can be used to approximate solutions.

EXAMPLE 6 Use a calculator to approximate to the nearest tenth the solutions to the equation in Example 5.

Using a calculator, we have

$$\frac{-1 + \sqrt{22}}{3} \approx 1.230138587 \approx 1.2 \text{ to the nearest tenth, and}$$
$$\frac{-1 - \sqrt{22}}{3} \approx -1.896805253 \approx -1.9 \text{ to the nearest tenth.}$$

The approximate solutions are 1.2 and -1.9.

Do Exercise 6 on the following page.

CALCULATOR CORNER

Approximating Solutions of Quadratic Equations In Example 5, we found that the solutions of the equation $3x^2 = 7 - 2x$ are $\frac{-1 + \sqrt{22}}{3}$ and $\frac{-1 - \sqrt{22}}{3}$. We can use a graphing calculator to approximate these solutions. To approximate $\frac{-1 + \sqrt{22}}{3}$, we press ((−) 1 + 2nd √ 2 2)) ÷ 3 ENTER. To approximate $\frac{-1 - \sqrt{22}}{3}$, we press ((−) 1 − 2nd √ 2 2)) ÷ 3 ENTER. We see that the solutions are approximately 1.2 and −1.9.

```
(−1+√(22))/3
                1.230138587
(−1−√(22))/3
               −1.896805253
```

Exercises: Use a graphing calculator to approximate the solutions to each of the following to the nearest tenth.

1. Example 3
2. Margin Exercise 3
3. Margin Exercise 5

COMPARING METHODS OF SOLVING QUADRATIC EQUATIONS

In Sections 10.1–10.3, we have studied three different methods of solving quadratic equations. Each of these methods has advantages and disadvantages, as outlined in the table below. Note that although the quadratic formula can be used to solve *any* quadratic equation, the other methods are sometimes faster and easier to use.

METHOD	ADVANTAGES	DISADVANTAGES
The quadratic formula	Can be used to solve *any* quadratic equation.	Can be slower than factoring or the principle of square roots.
The principle of square roots	Fastest way to solve equations of the form $ax^2 = p$, or $(x + k)^2 = p$. Can be used to solve *any* quadratic equation.	Can be slow when completing the square is required.
Factoring	Can be very fast.	Can be used only on certain equations. Many equations are difficult or impossible to solve by factoring.

6. Approximate the solutions to the equation in Margin Exercise 5. Round to the nearest tenth.

Answer on page A-45

Study Tips

BEGINNING TO STUDY FOR THE FINAL EXAM (PART 4): ONE OR TWO DAYS OF STUDY TIME

1. **Begin by browsing through each chapter, reviewing the highlighted or boxed information regarding important formulas in both the text and the Summary and Review.** There may be some formulas that you will need to memorize.
2. **Take the Final Examination in the text during the last couple of days before the final.** Set aside the same amount of time that you will have for the final. See how much of the final exam you can complete under test-like conditions. Be careful to avoid any questions corresponding to objectives not covered. Restudy the objectives in the text that correspond to each question you missed.
3. **Attend a final-exam review session if one is available.**

"Great is the art of beginning, but greater is the art of ending."

Henry Wadsworth Longfellow, nineteenth-century American poet

10.3 EXERCISE SET

For Extra Help

Digital Video Tutor CD 9 Videotape 18 | InterAct Math | Math Tutor Center | MathXL | MyMathLab

a Solve. Try factoring first. If factoring is not possible or is difficult, use the quadratic formula.

1. $x^2 - 4x = 21$

2. $x^2 + 8x = 9$

3. $x^2 = 6x - 9$

4. $x^2 = 24x - 144$

5. $3y^2 - 2y - 8 = 0$

6. $3y^2 - 7y + 4 = 0$

7. $4x^2 + 4x = 15$

8. $4x^2 + 12x = 7$

9. $x^2 - 9 = 0$

10. $x^2 - 16 = 0$

11. $x^2 - 2x - 2 = 0$

12. $x^2 - 2x - 11 = 0$

13. $y^2 - 10y + 22 = 0$

14. $y^2 + 6y - 1 = 0$

15. $x^2 + 4x + 4 = 7$

16. $x^2 - 2x + 1 = 5$

17. $3x^2 + 8x + 2 = 0$

18. $3x^2 - 4x - 2 = 0$

19. $2x^2 - 5x = 1$

20. $4x^2 + 4x = 5$

21. $2y^2 - 2y - 1 = 0$

22. $4y^2 + 4y - 1 = 0$

23. $2t^2 + 6t + 5 = 0$

24. $4y^2 + 3y + 2 = 0$

25. $3x^2 = 5x + 4$

26. $2x^2 + 3x = 1$

27. $2y^2 - 6y = 10$

28. $5m^2 = 3 + 11m$

29. $\frac{x^2}{x+3} - \frac{5}{x+3} = 0$

30. $\frac{x^2}{x-4} - \frac{7}{x-4} = 0$

31. $x + 2 = \frac{3}{x+2}$

32. $x - 3 = \frac{5}{x-3}$

33. $\frac{1}{x} + \frac{1}{x+1} = \frac{1}{3}$

34. $\frac{1}{x} + \frac{1}{x+6} = \frac{1}{5}$

b Solve using the quadratic formula. Use a calculator to approximate the solutions to the nearest tenth.

35. $x^2 - 4x - 7 = 0$

36. $x^2 + 2x - 2 = 0$

37. $y^2 - 6y - 1 = 0$

38. $y^2 + 10y + 22 = 0$

39. $4x^2 + 4x = 1$

40. $4x^2 = 4x + 1$

41. $3x^2 - 8x + 2 = 0$

42. $3x^2 + 4x - 2 = 0$

43. DW List a quadratic equation with no real-number solutions. How can you use the equation to find an equation in the form $y = ax^2 + bx + c$ that does not cross the x-axis?

44. DW Under what condition(s) would using the quadratic formula *not* be the easiest way to solve a quadratic equation?

SKILL MAINTENANCE

Add or subtract. [9.4a]

45. $\sqrt{40} - 2\sqrt{10} + \sqrt{90}$

46. $\sqrt{54} - \sqrt{24}$

47. $\sqrt{18} + \sqrt{50} - 3\sqrt{8}$

48. $\sqrt{81x^3} - \sqrt{4x}$

49. Simplify: $\sqrt{80}$. [9.2a]

50. Multiply and simplify: $\sqrt{3x^2}\sqrt{9x^3}$. [9.2c]

51. Simplify: $\sqrt{9000x^{10}}$. [9.2b]

52. Rationalize the denominator: $\sqrt{\frac{7}{3}}$. [9.3c]

SYNTHESIS

Solve.

53. $5x + x(x - 7) = 0$

54. $x(3x + 7) - 3x = 0$

55. $3 - x(x - 3) = 4$

56. $x(5x - 7) = 1$

57. $(y + 4)(y + 3) = 15$

58. $(y + 5)(y - 1) = 27$

59. $x^2 + (x + 2)^2 = 7$

60. $x^2 + (x + 1)^2 = 5$

61.–68. Use a graphing calculator to approximate the solutions of the equations in Exercises 35–42. Compare your answers with those found using the quadratic formula.

Objective

a Solve a formula for a given letter.

1. a) Solve for I: $E = \frac{9R}{I}$.

b) Solve for R: $E = \frac{9R}{I}$.

2. Solve for x: $y = ax - bx + 5$.

Answers on page A-45

10.4 FORMULAS

a Solving Formulas

Formulas arise frequently in the natural and social sciences, business, engineering, and health care. In Section 2.4, we saw that the same steps that are used to solve linear equations can be used to solve a formula that appears in this form. Similarly, the steps that are used to solve a rational, radical, or quadratic equation can also be used to solve a formula that appears in one of these forms.

EXAMPLE 1 *Intelligence Quotient.* The formula $Q = \frac{100m}{c}$ is used to determine the intelligence quotient, Q, of a person of mental age m and chronological age c. Solve for c.

$$Q = \frac{100m}{c}$$

$$c \cdot Q = c \cdot \frac{100m}{c} \qquad \text{Multiplying both sides by } c$$

$$cQ = 100m \qquad \text{Simplifying}$$

$$c = \frac{100m}{Q} \qquad \text{Dividing both sides by } Q$$

This formula can be used to determine a person's chronological, or actual, age from his or her mental age and intelligence quotient.

Do Exercise 1.

EXAMPLE 2 Solve for x: $y = ax + bx - 4$.

$$y = ax + bx - 4 \qquad \text{We want this letter alone on one side.}$$

$$y + 4 = ax + bx \qquad \text{Adding 4}$$

$$y + 4 = (a + b)x \qquad \text{Collecting like terms}$$

$$\frac{y + 4}{(a + b)} = \frac{(a + b)x}{(a + b)} \qquad \text{Dividing both sides by } a + b$$

$$\frac{y + 4}{a + b} = x \qquad \text{Simplifying. The answer can also be written as } x = \frac{y + 4}{a + b}.$$

Do Exercise 2.

CAUTION!

Had we performed the following steps in Example 2, we would *not* have solved for x:

$$y = ax + bx - 4$$

$$y - ax + 4 = bx \quad \text{Subtracting } ax \text{ and adding 4}$$

Two occurrences of x

$$\frac{y - ax + 4}{b} = x. \quad \text{Dividing by } b$$

The mathematics of each step is correct, but since x occurs on both sides of the formula, *we have not solved the formula for x*. Remember that the letter being solved for should be **alone** on one side of the equation, with no occurrence of that letter on the other side!

EXAMPLE 3 Solve the following work formula for t:

$$\frac{t}{a} + \frac{t}{b} = 1.$$

We multiply by the LCM, which is ab:

$$ab \cdot \left(\frac{t}{a} + \frac{t}{b}\right) = ab \cdot 1 \quad \text{Multiplying by } ab$$

$$ab \cdot \frac{t}{a} + ab \cdot \frac{t}{b} = ab \quad \text{Using a distributive law to remove parentheses}$$

$$bt + at = ab \quad \text{Simplifying}$$

$$(b + a)t = ab \quad \text{Factoring out } t$$

$$t = \frac{ab}{b + a}. \quad \text{Dividing by } b + a$$

Do Exercise 3.

EXAMPLE 4 *Distance to the Horizon.* Solve for h: $D = \sqrt{2h}$. (See Exercises 43–46 in Exercise Set 9.5.)

This is a radical equation. Recall that we first isolate the radical. Then we use the principle of squaring.

$$D = \sqrt{2h}$$

$$D^2 = (\sqrt{2h})^2 \quad \text{Using the principle of squaring (Section 9.5)}$$

$$D^2 = 2h \quad \text{Simplifying}$$

$$\frac{D^2}{2} = h \quad \text{Dividing by 2}$$

EXAMPLE 5 Solve for g: $T = 2\pi\sqrt{\frac{L}{g}}$ (the period of a pendulum).

$$\frac{T}{2\pi} = \sqrt{\frac{L}{g}} \quad \text{Dividing by } 2\pi \text{ to isolate the radical}$$

$$\left(\frac{T}{2\pi}\right)^2 = \left(\sqrt{\frac{L}{g}}\right)^2 \quad \text{Using the principle of squaring}$$

3. Optics Formula. Solve for f:

$$\frac{1}{p} + \frac{1}{q} = \frac{1}{f}.$$

Answer on page A-45

4. Solve for L: $r = 2\sqrt{5L}$ (the speed of a skidding car).

5. Solve for L: $T = 2\pi\sqrt{\dfrac{L}{g}}$.

6. Solve for m: $c = \sqrt{\dfrac{E}{m}}$.

7. Solve for r: $A = \pi r^2$ (the area of a circle).

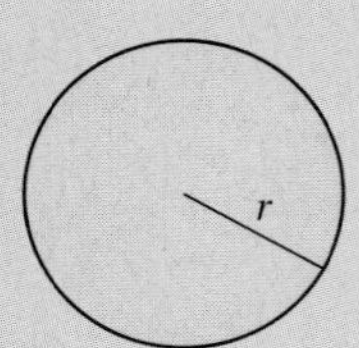

8. Solve for n: $N = n^2 - n$.

Answers on page A-45

Then

$$\frac{T^2}{4\pi^2} = \frac{L}{g}$$

$$gT^2 = 4\pi^2 L \qquad \text{Multiplying by } 4\pi^2 g \text{ to clear fractions}$$

$$g = \frac{4\pi^2 L}{T^2}. \qquad \text{Dividing by } T^2 \text{ to get } g \text{ alone}$$

Do Exercises 4–6.

In most formulas, the letters represent nonnegative numbers, so we need not use absolute values when taking square roots.

EXAMPLE 6 *Torricelli's Theorem.* The speed v of a liquid leaving a bucket from an opening is related to the height h of the top of the liquid above the opening by the formula

$$h = \frac{v^2}{2g}.$$

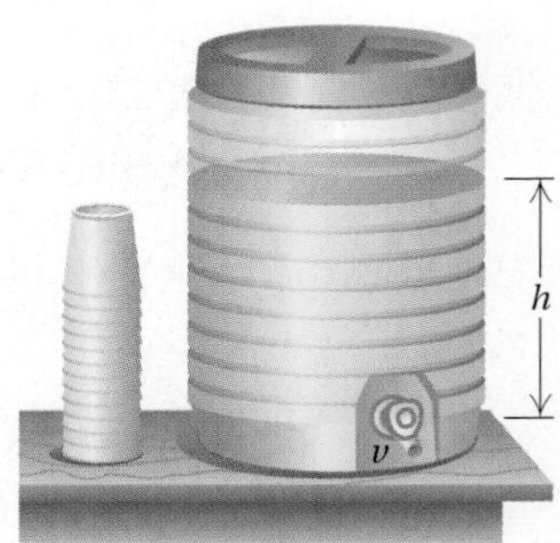

Solve for v.

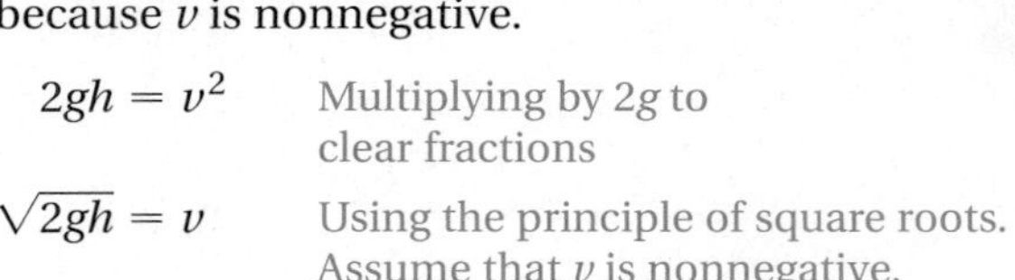

Since v^2 appears by itself and there is no expression involving v, we first solve for v^2. Then we use the principle of square roots, taking only the nonnegative square root because v is nonnegative.

$$2gh = v^2 \qquad \text{Multiplying by } 2g \text{ to clear fractions}$$

$$\sqrt{2gh} = v \qquad \text{Using the principle of square roots. Assume that } v \text{ is nonnegative.}$$

Do Exercise 7.

EXAMPLE 7 Solve for n: $d = \dfrac{n^2 - 3n}{2}$, where d is the number of diagonals of an n-sided polygon. (See Example 9 of Section 10.1.)

In this case, there is a term involving n as well as an n^2-term. Thus we must use the quadratic formula.

$$d = \frac{n^2 - 3n}{2}$$

$$n^2 - 3n = 2d \qquad \text{Multiplying by 2 to clear fractions}$$

$$n^2 - 3n - 2d = 0 \qquad \text{Finding standard form}$$

$$a = 1, \quad b = -3, \quad c = -2d \qquad \text{The variable is } n\text{; } d \text{ represents a constant.}$$

$$n = \frac{-b \pm \sqrt{b^2 - 4ac}}{2a} \qquad \text{Quadratic formula}$$

$$= \frac{-(-3) \pm \sqrt{(-3)^2 - 4 \cdot 1 \cdot (-2d)}}{2 \cdot 1} \qquad \text{Substituting into the quadratic formula}$$

$$= \frac{3 + \sqrt{9 + 8d}}{2} \qquad \text{Using the positive root}$$

Do Exercise 8.

10.4 EXERCISE SET

For Extra Help

Digital Video Tutor CD 9 Videotape 18 | InterAct Math | Math Tutor Center | MathXL | MyMathLab

a Solve for the indicated letter.

1. $q = \dfrac{VQ}{I}$, for I
(An engineering formula)

2. $y = \dfrac{4A}{a}$, for a

3. $S = \dfrac{kmM}{d^2}$, for m

4. $S = \dfrac{kmM}{d^2}$, for M

5. $S = \dfrac{kmM}{d^2}$, for d^2

6. $T = \dfrac{10t}{W^2}$, for W^2

7. $T = \dfrac{10t}{W^2}$, for W

8. $S = \dfrac{kmM}{d^2}$, for d

9. $A = at + bt$, for t

10. $S = rx + sx$, for x

11. $y = ax + bx + c$, for x

12. $y = ax - bx - c$, for x

13. $\dfrac{t}{a} + \dfrac{t}{b} = 1$, for a
(A work formula)

14. $\dfrac{t}{a} + \dfrac{t}{b} = 1$, for b
(A work formula)

15. $\dfrac{1}{p} + \dfrac{1}{q} = \dfrac{1}{f}$, for p
(An optics formula)

16. $\dfrac{1}{p} + \dfrac{1}{q} = \dfrac{1}{f}$, for q
(An optics formula)

17. $A = \dfrac{1}{2}bh$, for b
(The area of a triangle)

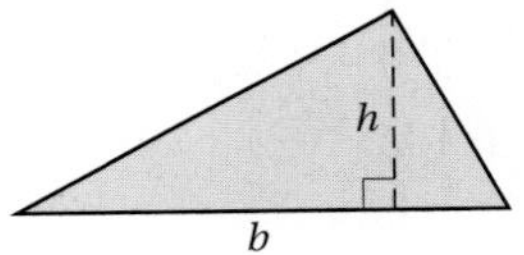

18. $s = \dfrac{1}{2}gt^2$, for g

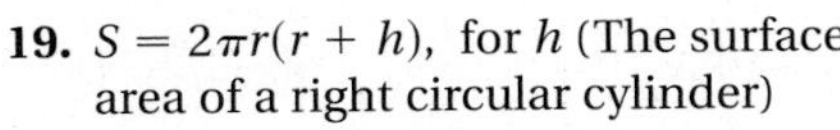

19. $S = 2\pi r(r + h)$, for h (The surface area of a right circular cylinder)

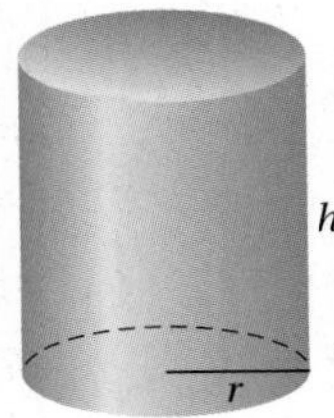

20. $S = 2\pi(r + h)$, for r

21. $\dfrac{1}{R} = \dfrac{1}{r_1} + \dfrac{1}{r_2}$, for R
(An electricity formula)

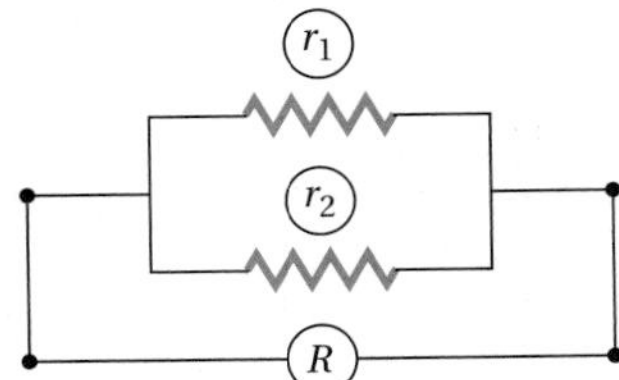

22. $\dfrac{1}{R} = \dfrac{1}{r_1} + \dfrac{1}{r_2}$, for r_1

23. $P = 17\sqrt{Q}$, for Q

24. $A = 1.4\sqrt{t}$, for t

25. $v = \sqrt{\dfrac{2gE}{m}}$, for E

26. $Q = \sqrt{\dfrac{aT}{c}}$, for T

27. $S = 4\pi r^2$, for r

28. $E = mc^2$, for c

29. $P = kA^2 + mA$, for A

30. $Q = ad^2 - cd$, for d

31. $c^2 = a^2 + b^2$, for a

32. $c = \sqrt{a^2 + b^2}$, for b

33. $s = 16t^2$, for t

34. $V = \pi r^2 h$, for r

35. $A = \pi r^2 + 2\pi rh$, for r

36. $A = 2\pi r^2 + 2\pi rh$, for r

37. $F = \dfrac{Av^2}{400}$, for v

38. $A = \dfrac{\pi r^2 S}{360}$, for r

39. $c = \sqrt{a^2 + b^2}$, for a

40. $c^2 = a^2 + b^2$, for b

41. $h = \dfrac{a}{2}\sqrt{3}$, for a
(The height of an equilateral triangle with sides of length a)

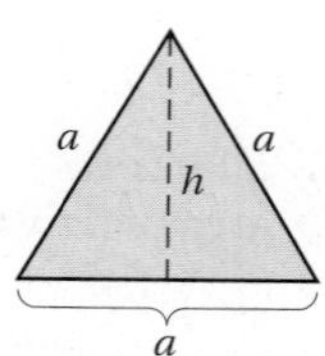

42. $d = s\sqrt{2}$, for s
(The hypotenuse of an isosceles right triangle with s the length of the legs)

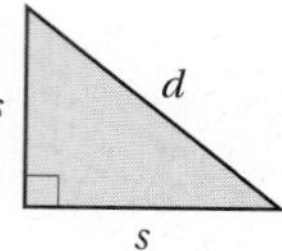

43. $n = aT^2 - 4T + m$, for T

44. $y = ax^2 + bx + c$, for x

45. $v = 2\sqrt{\dfrac{2kT}{\pi m}}$, for T

46. $E = \dfrac{1}{2}mv^2 + mgy$, for v

47. $3x^2 = d^2$, for x

48. $c = \sqrt{\dfrac{E}{m}}$, for E

49. $N = \dfrac{n^2 - n}{2}$, for n

50. $M = \dfrac{m}{\sqrt{1 - \left(\dfrac{v}{c}\right)^2}}$, for c

51. $S = \dfrac{a + b}{3b}$, for b

52. $Q = \dfrac{a - b}{2b}$, for b

53. $\dfrac{A - B}{AB} = Q$, for B

54. $L = \dfrac{Mt + g}{t}$, for t

55. $S = 180(n - 2)$, for n

56. $S = \frac{n}{2}(a + 1)$, for a

57. $A = P(1 + rt)$, for t
(An interest formula)

58. $A = P(1 + rt)$, for r
(An interest formula)

59. $\frac{A}{B} = \frac{C}{D}$, for D

60. $\frac{A}{B} = \frac{C}{D}$, for B

61. $C = \frac{Ka - b}{a}$, for a

62. $Q = \frac{Pt - h}{t}$, for t

63. D$_W$ Describe a situation in which the result of Example 3,

$$t = \frac{ab}{a + b},$$

would be especially useful.

64. D$_W$ Explain how you would solve the equation $0 = ax^2 + bx + c$ for x.

SKILL MAINTENANCE

In a right triangle, find the length of the side not given. Give an exact answer and an approximation to three decimal places. [9.6a]

65. $a = 4, b = 7$

66. $b = 11, c = 14$

67. $a = 4, b = 5$

68. $a = 10, c = 12$

69. $c = 8\sqrt{17}, a = 2$

70. $a = \sqrt{2}, b = \sqrt{3}$

Solve. [9.6b]

71. *Guy Wire.* How long is a guy wire reaching from the top of an 18-ft pole to a point on the ground 10 ft from the pole? Give an exact answer and an approximation to three decimal places.

72. *Soccer Fields.* The smallest regulation soccer field is 50 yd wide and 100 yd long. Find the length of a diagonal of such a field.

SYNTHESIS

73. The circumference C of a circle is given by $C = 2\pi r$.

a) Solve $C = 2\pi r$ for r.

b) The area is given by $A = \pi r^2$. Express the area in terms of the circumference C.

74. Referring to Exercise 73, express the circumference C in terms of the area A.

75. Solve $3ax^2 - x - 3ax + 1 = 0$ for x.

76. Solve $h = 16t^2 + vt + s$ for t.

Objective

a Solve applied problems using quadratic equations.

1. **Pool Dimensions.** The area of a rectangular swimming pool is 68 yd^2. The length is 1 yd longer than three times the width. Find the dimensions of the rectangular swimming pool. Round to the nearest tenth.

Answer on page A-46

10.5 APPLICATIONS AND PROBLEM SOLVING

a Using Quadratic Equations to Solve Applied Problems

EXAMPLE 1 *Red Raspberry Patch.* The area of a rectangular red raspberry patch is 76 ft^2. The length is 7 ft longer than three times the width. Find the dimensions of the raspberry patch.

1. **Familiarize.** We first make a drawing and label it with both known and unknown information. We let $w =$ the width of the rectangle. The length of the rectangle is 7 ft longer than three times the width. Thus the length is $3w + 7$.

2. **Translate.** Recall that area is length $\times$ width. Thus we have two expressions for the area of the rectangle: $(3w + 7)(w)$ and 76. This gives us a translation:

$$(3w + 7)(w) = 76.$$

3. **Solve.** We solve the equation:

$$3w^2 + 7w = 76$$
$$3w^2 + 7w - 76 = 0$$
$$(3w + 19)(w - 4) = 0 \qquad \text{Factoring (the quadratic formula could also be used)}$$
$$3w + 19 = 0 \quad or \quad w - 4 = 0 \qquad \text{Using the principle of zero products}$$
$$3w = -19 \quad or \quad w = 4$$
$$w = -\tfrac{19}{3} \quad or \quad w = 4.$$

4. **Check.** We check in the original problem. We know that $-\frac{19}{3}$ is not a solution because width cannot be negative. When $w = 4$, $3w + 7 = 19$, and the area is $4(19)$, or 76. This checks.

5. **State.** The width of the rectangular raspberry patch is 4 ft, and the length is 19 ft.

Do Exercise 1.

EXAMPLE 2 *Staircase.* A mason builds a staircase in such a way that the portion underneath the stairs forms a right triangle. The hypotenuse is 6 m long. The leg across the ground is 1 m longer than the leg next to the wall at the back. Find the lengths of the legs. Round to the nearest tenth.

1. **Familiarize.** We first make a drawing, letting s = the length of one leg. Then $s + 1$ = the length of the other leg.
2. **Translate.** To translate, we use the Pythagorean equation:

$$s^2 + (s + 1)^2 = 6^2.$$

3. **Solve.** We solve the equation:

$$s^2 + (s + 1)^2 = 6^2$$
$$s^2 + s^2 + 2s + 1 = 36$$
$$2s^2 + 2s - 35 = 0.$$

Since we cannot factor, we use the quadratic formula:

$$a = 2, \quad b = 2, \quad c = -35$$

$$s = \frac{-b \pm \sqrt{b^2 - 4ac}}{2a} = \frac{-2 \pm \sqrt{2^2 - 4 \cdot 2(-35)}}{2 \cdot 2}$$
$$= \frac{-2 \pm \sqrt{4 + 280}}{4} = \frac{-2 \pm \sqrt{284}}{4}$$
$$= \frac{-2 \pm \sqrt{4 \cdot 71}}{4} = \frac{-2 \pm 2 \cdot \sqrt{71}}{2 \cdot 2}$$
$$= \frac{2(-1 \pm \sqrt{71})}{2 \cdot 2} = \frac{2}{2} \cdot \frac{-1 \pm \sqrt{71}}{2} = \frac{-1 \pm \sqrt{71}}{2}.$$

Using a calculator, we get approximations:

$$\frac{-1 + \sqrt{71}}{2} \approx 3.7 \quad \text{or} \quad \frac{-1 - \sqrt{71}}{2} \approx -4.7.$$

4. **Check.** Since the length of a leg cannot be negative, -4.7 does not check. But 3.7 does check. If the smaller leg s is 3.7, the other leg is $s + 1$, or 4.7. Then

$$(3.7)^2 + (4.7)^2 = 13.69 + 22.09 = 35.78.$$

Using a calculator, we get $\sqrt{35.78} \approx 5.98 \approx 6$. Note that our check is not exact because we are using an approximation for $\sqrt{71}$.

5. **State.** One leg is about 3.7 m long, and the other is about 4.7 m long.

2. Animal Pen. The hypotenuse of a right triangular animal pen at the zoo is 4 yd long. One leg is 1 yd longer than the other. Find the lengths of the legs. Round to the nearest tenth.

Do Exercise 2.

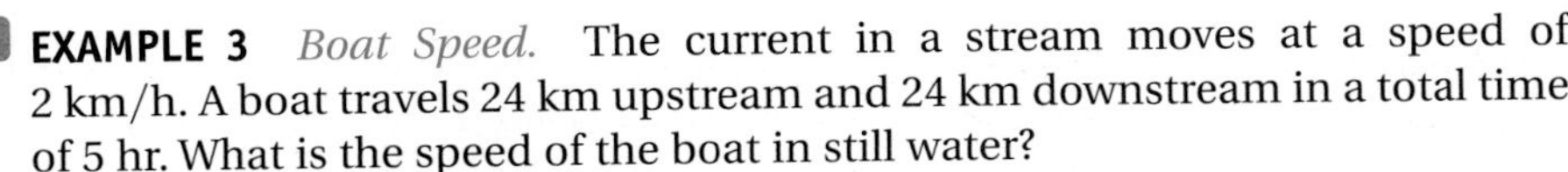

EXAMPLE 3 *Boat Speed.* The current in a stream moves at a speed of 2 km/h. A boat travels 24 km upstream and 24 km downstream in a total time of 5 hr. What is the speed of the boat in still water?

1. **Familiarize.** We first make a drawing. The distances are the same. We let r = the speed of the boat in still water. Then when the boat is traveling upstream, its speed is $r - 2$. When it is traveling downstream, its speed is $r + 2$. We let t_1 represent the time it takes the boat to go upstream and t_2 the time it takes to go downstream. We summarize in a table.

Answer on page A-46

3. Speed of a Stream. The speed of a boat in still water is 12 km/h. The boat travels 45 km upstream and 45 km downstream in a total time of 8 hr. What is the speed of the stream? (*Hint*: Let s = the speed of the stream. Then $12 - s$ is the speed upstream and $12 + s$ is the speed downstream. Note also that $12 - s$ cannot be negative, because the boat must be going faster than the current if it is moving forward.)

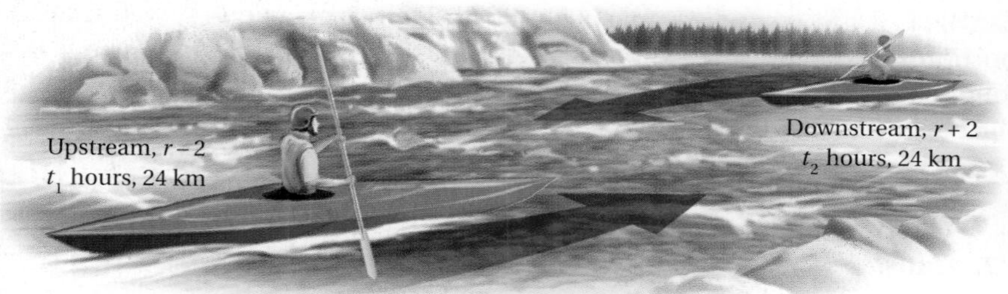

	d	r	t
Upstream	24	$r - 2$	t_1
Downstream	24	$r + 2$	t_2
Total time			5

$\rightarrow t_1 = \dfrac{24}{r-2}$

$\rightarrow t_2 = \dfrac{24}{r+2}$

2. Translate. Recall the basic formula for motion: $d = rt$. From it we can obtain an equation for time: $t = d/r$. Total time consists of the time to go upstream, t_1, plus the time to go downstream, t_2. Using $t = d/r$ and the rows of the table, we have

$$t_1 = \frac{24}{r-2} \quad \text{and} \quad t_2 = \frac{24}{r+2}.$$

Since the total time is 5 hr, $t_1 + t_2 = 5$, and we have

$$\frac{24}{r-2} + \frac{24}{r+2} = 5.$$

3. Solve. We solve the equation. We multiply both sides by the LCM, which is $(r-2)(r+2)$:

$$(r-2)(r+2)\cdot\left[\frac{24}{r-2} + \frac{24}{r+2}\right] = (r-2)(r+2)5$$

$$(r-2)(r+2)\cdot\frac{24}{r-2} + (r-2)(r+2)\cdot\frac{24}{r+2} = (r^2-4)5$$

$$24(r+2) + 24(r-2) = 5r^2 - 20$$

$$24r + 48 + 24r - 48 = 5r^2 - 20$$

$$-5r^2 + 48r + 20 = 0$$

$$5r^2 - 48r - 20 = 0 \quad \text{Multiplying by } -1$$

$$(5r+2)(r-10) = 0 \quad \text{Factoring}$$

$$5r + 2 = 0 \quad or \quad r - 10 = 0 \quad \text{Using the principle of zero products}$$

$$5r = -2 \quad or \quad r = 10$$

$$r = -\tfrac{2}{5} \quad or \quad r = 10.$$

4. Check. Since speed cannot be negative, $-\frac{2}{5}$ cannot be a solution. But suppose the speed of the boat in still water is 10 km/h. The speed upstream is then $10 - 2$, or 8 km/h. The speed downstream is $10 + 2$, or 12 km/h. The time upstream, using $t = d/r$, is 24/8, or 3 hr. The time downstream is 24/12, or 2 hr. The total time is 5 hr. This checks.

5. State. The speed of the boat in still water is 10 km/h.

Answer on page A-46

Do Exercise 3.

For Extra Help

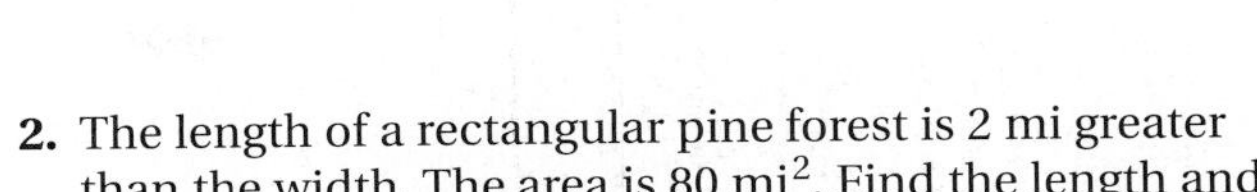

a Solve.

1. *Standard-Sized Television.* When we say that a television is 30 in., we mean that the diagonal is 30 in. For a standard-sized 30-in. television, the width is 6 in. more than the height. Find the dimensions of a standard-sized 30-in. television.

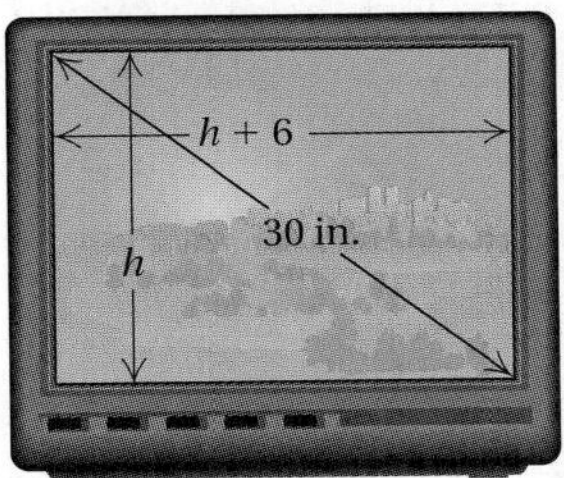

2. The length of a rectangular pine forest is 2 mi greater than the width. The area is 80 mi^2. Find the length and the width.

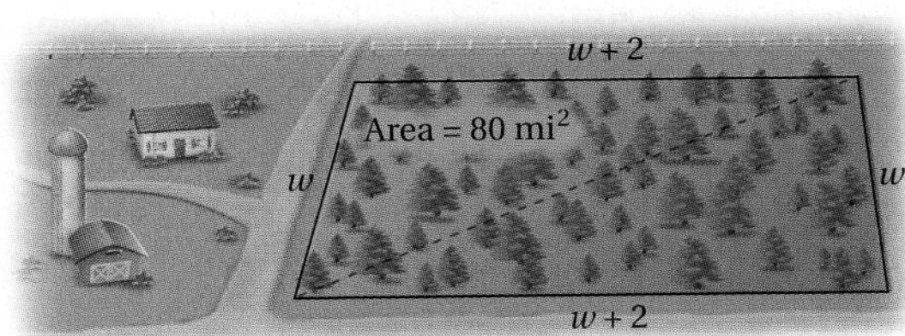

3. The length of a rectangular area rug is 3 ft greater than the width. The area is 70 ft^2. Find the length and the width.

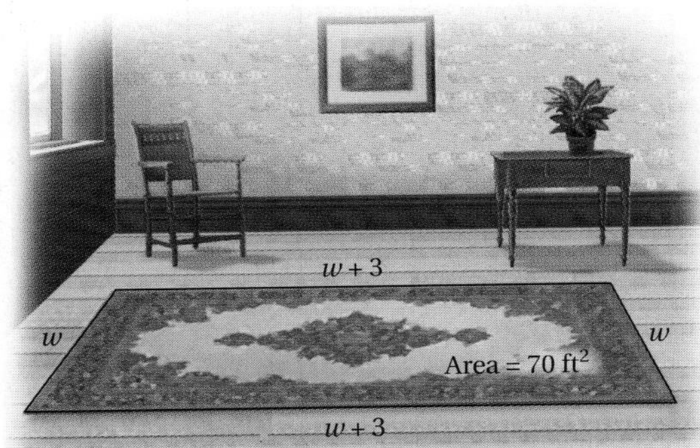

4. *HDTV Dimensions.* A high-definition television (HDTV) features a larger screen and greater clarity than a standard television. An HDTV might have a 70-in. diagonal screen with the width 27 in. greater than the height. Find the width and the height of a 70-in. HDTV screen.

5. *Rectangle Dimensions.* The length of a rectangle is twice the width. The area is 50 m^2. Find the length and the width.

6. *Carpenter's Square.* A *square* is a carpenter's tool in the shape of a right triangle. One side, or leg, of a square is 8 in. longer than the other. The length of the hypotenuse is $8\sqrt{13}$ in. Find the lengths of the legs of the square.

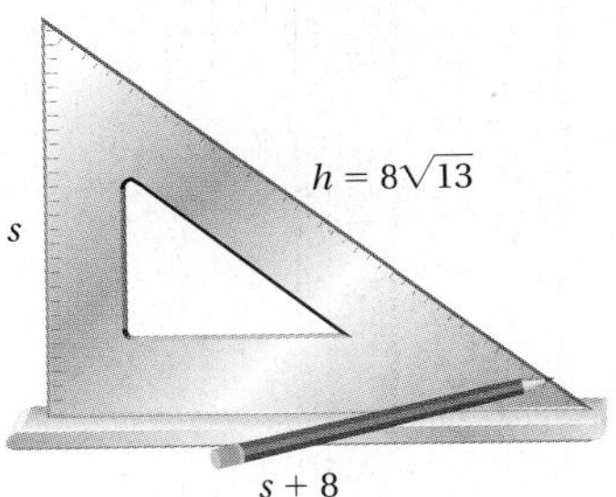

7. *Rectangle Dimensions.* The width of a rectangle is 4 cm less than the length. The area is 320 cm^2. Find the length and the width.

8. *Rectangle Dimensions.* The width of a rectangle is 3 cm less than the length. The area is 340 cm^2. Find the length and the width.

Find the approximate answers for Exercises 9–14. Round to the nearest tenth.

9. *Right-Triangle Dimensions.* The hypotenuse of a right triangle is 8 m long. One leg is 2 m longer than the other. Find the lengths of the legs.

10. *Right-Triangle Dimensions.* The hypotenuse of a right triangle is 5 cm long. One leg is 2 cm longer than the other. Find the lengths of the legs.

11. *Rectangle Dimensions.* The length of a rectangle is 2 in. greater than the width. The area is 20 in^2. Find the length and the width.

12. *Rectangle Dimensions.* The length of a rectangle is 3 ft greater than the width. The area is 15 ft^2. Find the length and the width.

13. *Rectangle Dimensions.* The length of a rectangle is twice the width. The area is 20 cm^2. Find the length and the width.

14. *Rectangle Dimensions.* The length of a rectangle is twice the width. The area is 10 m^2. Find the length and the width.

15. *Picture Frame.* A picture frame measures 25 cm by 20 cm. There is 266 cm^2 of picture showing. The frame is of uniform thickness. Find the thickness of the frame.

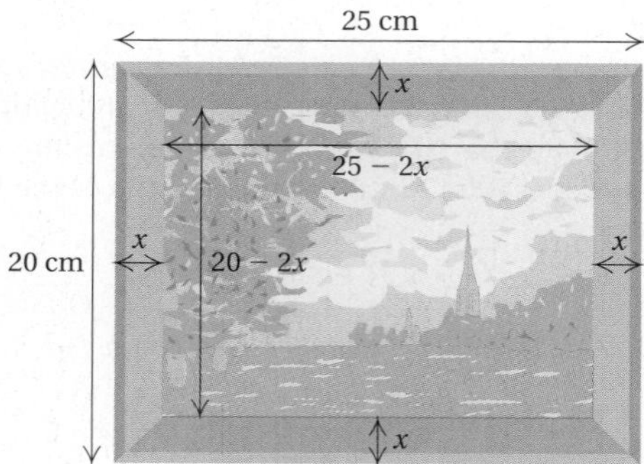

16. *Tablecloth.* A tablecloth measures 96 in. by 72 in. It is laid on a tabletop with an area of 5040 in^2, and hangs over the edge by the same amount on all sides. By how many inches does the cloth hang over the edge?

For Exercises 17–22, complete the table to help with the familiarization.

17. *Boat Speed.* The current in a stream moves at a speed of 3 km/h. A boat travels 40 km upstream and 40 km downstream in a total time of 14 hr. What is the speed of the boat in still water? Complete the following table to help with the familiarization.

	d	r	t
Upstream		$r - 3$	t_1
Downstream	40		t_2
Total Time			

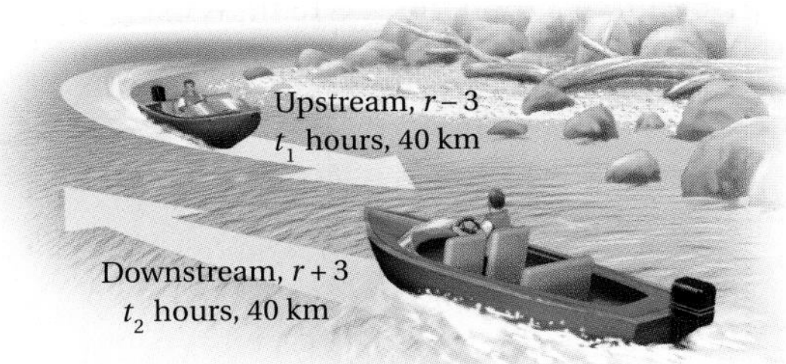

18. *Boat Speed.* The current in a stream moves at a speed of 3 km/h. A boat travels 45 km upstream and 45 km downstream in a total time of 8 hr. What is the speed of the boat in still water?

	d	r	t
Upstream	45		
Downstream		$r + 3$	
Total Time			

19. *Boat Speed.* The current in a stream moves at a speed of 4 mph. A boat travels 4 mi upstream and 12 mi downstream in a total time of 2 hr. What is the speed of the boat in still water?

	d	r	t
Upstream		$r - 4$	
Downstream	12		
Total Time			

20. *Boat Speed.* The current in a stream moves at a speed of 4 mph. A boat travels 5 mi upstream and 13 mi downstream in a total time of 2 hr. What is the speed of the boat in still water?

	d	r	t
Upstream			
Downstream			
Total Time			

21. *Speed of a Stream.* The speed of a boat in still water is 10 km/h. The boat travels 12 km upstream and 28 km downstream in a total time of 4 hr. What is the speed of the stream?

	d	r	t
Upstream			
Downstream			
Total Time			

22. *Speed of a Stream.* The speed of a boat in still water is 8 km/h. The boat travels 60 km upstream and 60 km downstream in a total time of 16 hr. What is the speed of the stream?

	d	r	t
Upstream			
Downstream			
Total Time			

23. *Wind Speed.* An airplane flies 738 mi against the wind and 1062 mi with the wind in a total time of 9 hr. The speed of the airplane in still air is 200 mph. What is the speed of the wind?

24. *Wind Speed.* An airplane flies 520 km against the wind and 680 km with the wind in a total time of 4 hr. The speed of the airplane in still air is 300 km/h. What is the speed of the wind?

25. *Speed of a Stream.* The speed of a boat in still water is 9 km/h. The boat travels 80 km upstream and 80 km downstream in a total time of 18 hr. What is the speed of the stream?

26. *Speed of a Stream.* The speed of a boat in still water is 10 km/h. The boat travels 48 km upstream and 48 km downstream in a total time of 10 hr. What is the speed of the stream?

D_W Find and explain the error(s) in each of the following solutions of a quadratic equation.

27. $(x + 6)^2 = 16$
$x + 6 = \sqrt{16}$
$x + 6 = 4$
$x = -2$

28. $x^2 + 2x - 8 = 0$
$(x + 4)(x - 2) = 0$
$x = 4$ *or* $x = -2$

SKILL MAINTENANCE

Add or subtract. [9.4a]

29. $5\sqrt{2} + \sqrt{18}$

30. $7\sqrt{40} - 2\sqrt{10}$

31. $\sqrt{4x^3} - 7\sqrt{x}$

32. $\sqrt{24} - \sqrt{54}$

33. $\sqrt{2} + \sqrt{\frac{1}{2}}$

34. $\sqrt{3} - \sqrt{\frac{1}{3}}$

35. $\sqrt{24} + \sqrt{54} - \sqrt{48}$

36. $\sqrt{4x} + \sqrt{81x^3}$

SYNTHESIS

37. *Pizza.* What should the diameter d of a pizza be so that it has the same area as two 12-in. pizzas? Do you get more to eat with a 16-in. pizza or with two 12-in. pizzas?

38. *Golden Rectangle.* The so-called *golden rectangle* is said to be extremely pleasing visually and was used often by ancient Greek and Roman architects. The length of a golden rectangle is approximately 1.6 times the width. Find the dimensions of a golden rectangle if its area is 9000 m^2.

10.6 GRAPHS OF QUADRATIC EQUATIONS

Objectives

a Graph quadratic equations.

b Find the x-intercepts of a quadratic equation.

In this section, we will graph equations of the form

$$y = ax^2 + bx + c, \quad a \neq 0.$$

The polynomial on the right side of the equation is of second degree, or **quadratic.** Examples of the types of equations we are going to graph are

$$y = x^2, \qquad y = x^2 + 2x - 3, \qquad y = -2x^2 + 3.$$

a Graphing Quadratic Equations of the Type $y = ax^2 + bx + c$

Graphs of quadratic equations of the type $y = ax^2 + bx + c$ (where $a \neq 0$) are always cup-shaped. They have a **line of symmetry** like the dashed lines shown in the figures below. If we fold on this line, the two halves will match exactly. The curve goes on forever. The top or bottom point where the curve changes is called the **vertex.** The second coordinate is either the largest value of y or the smallest value of y. The vertex is also thought of as a turning point. Graphs of quadratic equations are called **parabolas.**

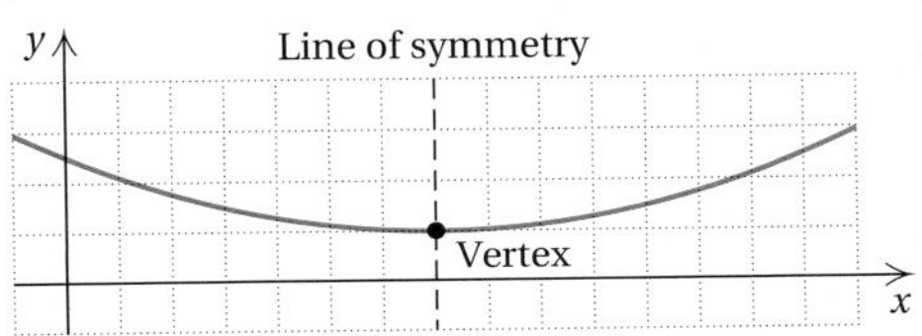

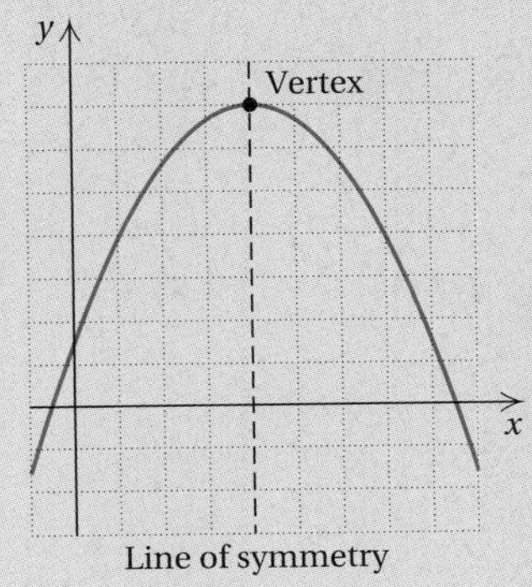

To graph a quadratic equation, we begin by choosing some numbers for x and computing the corresponding values of y.

EXAMPLE 1 Graph: $y = x^2$.

We choose numbers for x and find the corresponding values for y. Then we plot the ordered pairs (x, y) resulting from the computations and connect them with a smooth curve.

For $x = -3$, $y = x^2 = (-3)^2 = 9$.
For $x = -2$, $y = x^2 = (-2)^2 = 4$.
For $x = -1$, $y = x^2 = (-1)^2 = 1$.
For $x = 0$, $y = x^2 = (0)^2 = 0$.
For $x = 1$, $y = x^2 = (1)^2 = 1$.
For $x = 2$, $y = x^2 = (2)^2 = 4$.
For $x = 3$, $y = x^2 = (3)^2 = 9$.

x	y	(x, y)
-3	9	$(-3, 9)$
-2	4	$(-2, 4)$
-1	1	$(-1, 1)$
0	0	$(0, 0)$
1	1	$(1, 1)$
2	4	$(2, 4)$
3	9	$(3, 9)$

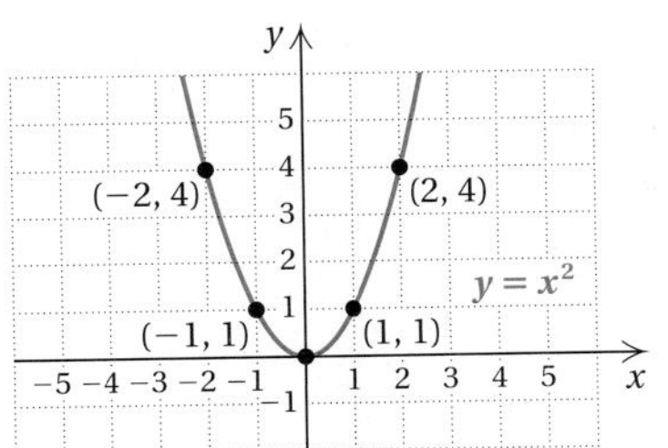

In Example 1, the vertex is the point $(0, 0)$. The second coordinate of the vertex, 0, is the smallest y-value. The y-axis is the line of symmetry. Parabolas whose equations are $y = ax^2$ always have the origin $(0, 0)$ as the vertex and the y-axis as the line of symmetry.

How do we graph a general equation? There are many methods, some of which you will study in your next mathematics course. Our goal here is to give you a basic graphing technique that is fairly easy to apply. A key in the graphing is knowing the vertex. By graphing it and then choosing x-values on both sides of the vertex, we can compute more points and complete the graph.

FINDING THE VERTEX

For a parabola given by the quadratic equation $y = ax^2 + bx + c$:

1. The x-coordinate of the vertex is $-\dfrac{b}{2a}$.
2. The second coordinate of the vertex is found by substituting the x-coordinate into the equation and computing y.

The proof that the vertex can be found in this way can be shown by completing the square in a manner similar to the proof of the quadratic formula, but it will not be considered here.

CALCULATOR CORNER

Graphing Quadratic Equations Use a graphing calculator to make a table of values for $y = x^2$. (See the Calculator Corner on p. 250 for the procedure.)

EXAMPLE 2 Graph: $y = -2x^2 + 3$.

We first find the vertex. The x-coordinate of the vertex is

$$-\frac{b}{2a} = -\frac{0}{2(-2)} = 0.$$

We substitute 0 for x into the equation to find the second coordinate of the vertex:

$$y = -2x^2 + 3 = -2(0)^2 + 3 = 3.$$

The vertex is $(0, 3)$. The line of symmetry is $x = 0$, which is the y-axis. We choose some x-values on both sides of the vertex and graph the parabola.

For $x = 1$, $y = -2x^2 + 3 = -2(1)^2 + 3 = -2 + 3 = 1$.
For $x = -1$, $y = -2x^2 + 3 = -2(-1)^2 + 3 = -2 + 3 = 1$.
For $x = 2$, $y = -2x^2 + 3 = -2(2)^2 + 3 = -8 + 3 = -5$.
For $x = -2$, $y = -2x^2 + 3 = -2(-2)^2 + 3 = -8 + 3 = -5$.

x	y	
0	3	← This is the vertex.
1	1	
−1	1	
2	−5	
−2	−5	

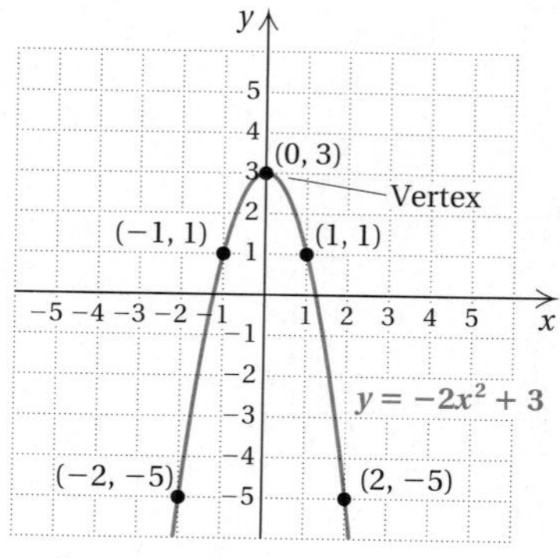

There are two other tips you might use when graphing quadratic equations. The first involves the coefficient of x^2. Note that a in $y = ax^2 + bx + c$ tells us whether the graph opens up or down. When a is positive, as in Example 1, the graph opens up; when a is negative, as in Example 2, the graph opens down. It is also helpful to plot the y-intercept. It occurs when $x = 0$.

TIPS FOR GRAPHING QUADRATIC EQUATIONS

1. Graphs of quadratic equations $y = ax^2 + bx + c$ are all parabolas. They are *smooth* cup-shaped symmetric curves, with no sharp points or kinks in them.
2. The graph of $y = ax^2 + bx + c$ opens up if $a > 0$. It opens down if $a < 0$.
3. Find the y-intercept. It occurs when $x = 0$, and it is easy to compute.

EXAMPLE 3 Graph: $y = x^2 + 2x - 3$.

We first find the vertex. The x-coordinate of the vertex is

$$-\frac{b}{2a} = -\frac{2}{2(1)} = -1.$$

We substitute -1 for x into the equation to find the second coordinate of the vertex:

$$\begin{aligned} y &= x^2 + 2x - 3 \\ &= (-1)^2 + 2(-1) - 3 \\ &= 1 - 2 - 3 \\ &= -4. \end{aligned}$$

The vertex is $(-1, -4)$. The line of symmetry is $x = -1$.

We choose some x-values on both sides of $x = -1$—say, -2, -3, -4 and 0, 1, 2—and graph the parabola. Since the coefficient of x^2 is 1, which is positive, we know that the graph opens up. Be sure to find y when $x = 0$. This gives the y-intercept.

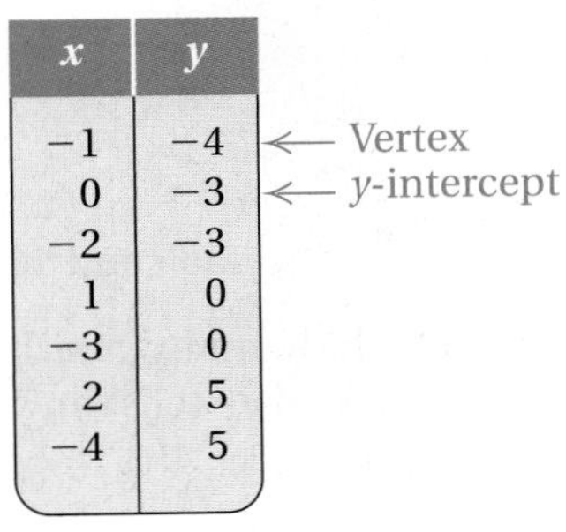

x	y	
-1	-4	← Vertex
0	-3	← y-intercept
-2	-3	
1	0	
-3	0	
2	5	
-4	5	

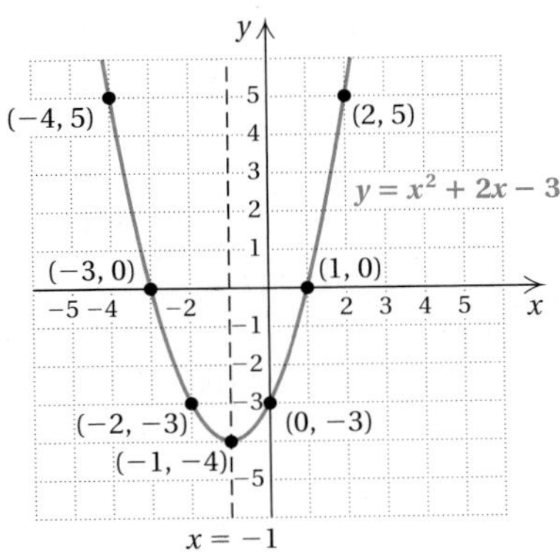

Do Exercises 1–3.

Graph. List the ordered pair for the vertex.

1. $y = x^2 - 3$

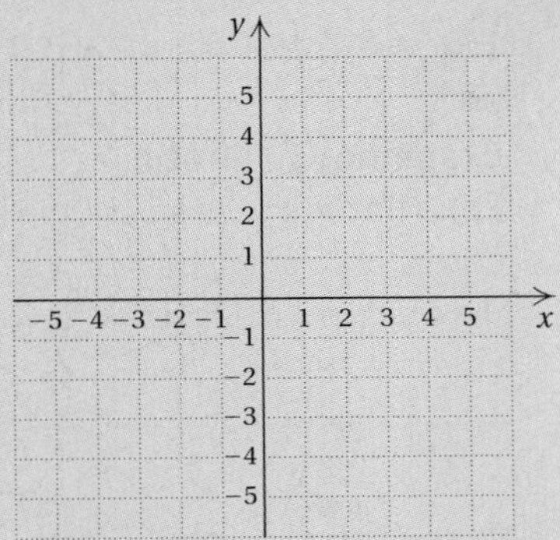

2. $y = -3x^2 + 6x$

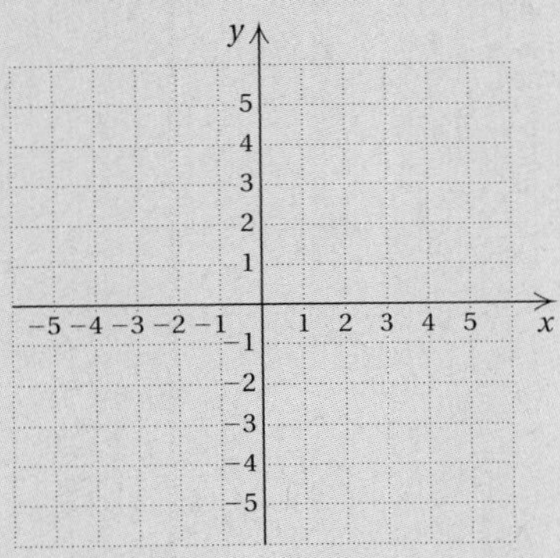

3. $y = x^2 - 4x + 4$

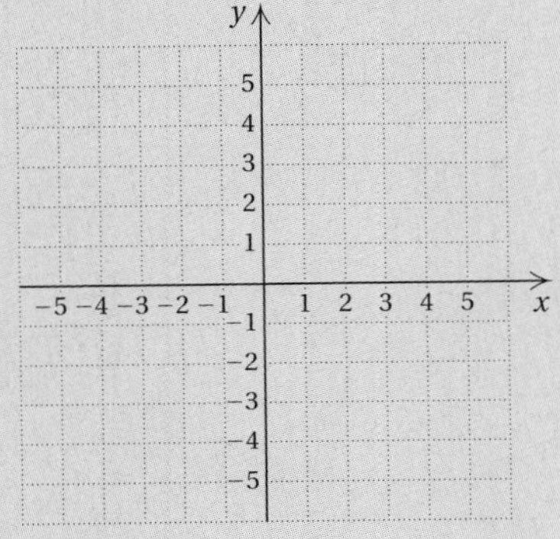

Answers on page A-46

CALCULATOR CORNER

Graphing Quadratic Equations Use a graphing calculator to graph each of the equations in Margin Exercises 1–3.

Find the x-intercepts.

4. $y = x^2 - 3$

5. $y = x^2 + 6x + 8$

6. $y = -2x^2 - 4x + 1$

7. $y = x^2 + 3$

Answers on page A-46

b Finding the x-Intercepts of a Quadratic Equation

The x-intercepts of $y = ax^2 + bx + c$ occur at those values of x for which $y = 0$. Thus the first coordinates of the x-intercepts are solutions of the equation

$$0 = ax^2 + bx + c.$$

We have been studying how to find such numbers in Sections 10.1–10.3.

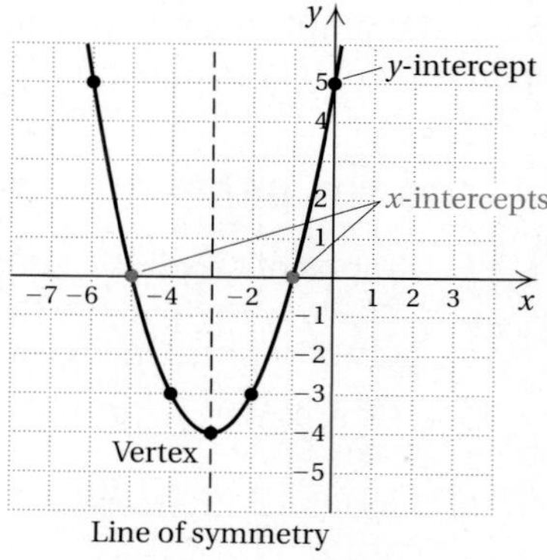

EXAMPLE 4 Find the x-intercepts of $y = x^2 - 4x + 1$.

We solve the equation

$$x^2 - 4x + 1 = 0.$$

Factoring is not convenient, so we use the quadratic formula:

$$a = 1, \quad b = -4, \quad c = 1$$

$$x = \frac{-b \pm \sqrt{b^2 - 4ac}}{2a}$$

$$= \frac{-(-4) \pm \sqrt{(-4)^2 - 4(1)(1)}}{2(1)}$$

$$= \frac{4 \pm \sqrt{16 - 4}}{2}$$

$$= \frac{4 \pm \sqrt{12}}{2} = \frac{4 \pm \sqrt{4 \cdot 3}}{2}$$

$$= \frac{4 \pm 2\sqrt{3}}{2} = \frac{2 \cdot 2 \pm 2\sqrt{3}}{2 \cdot 1}$$

$$= \frac{2}{2} \cdot \frac{2 \pm \sqrt{3}}{1} = 2 \pm \sqrt{3}.$$

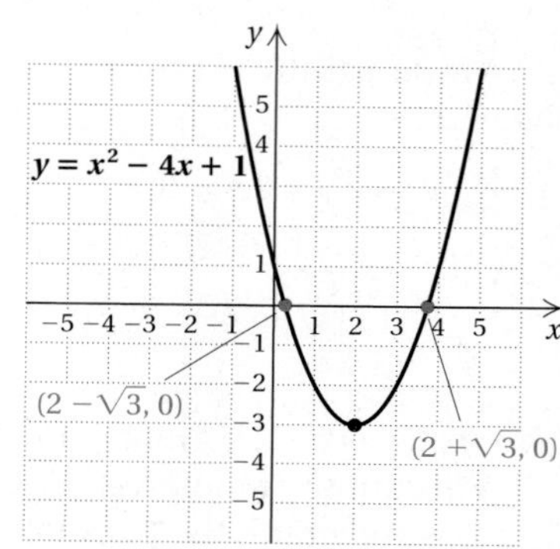

The x-intercepts are $(2 - \sqrt{3}, 0)$ and $(2 + \sqrt{3}, 0)$.

In the quadratic formula $x = \dfrac{-b \pm \sqrt{b^2 - 4ac}}{2a}$, the radicand $b^2 - 4ac$ is called the **discriminant.** The discriminant tells how many real-number solutions the equation $0 = ax^2 + bx + c$ has, so it also tells how many x-intercepts there are.

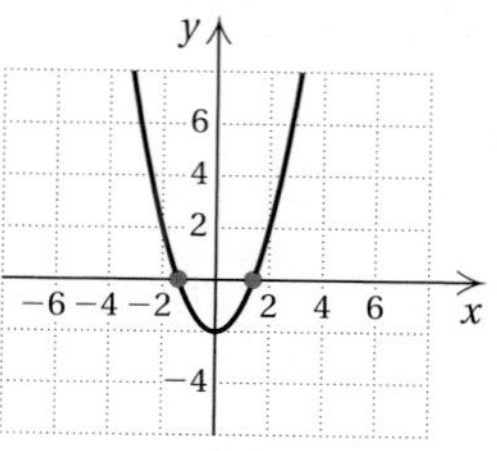

$y = x^2 - 2$
$b^2 - 4ac = 8 > 0$
Two real solutions
Two x-intercepts

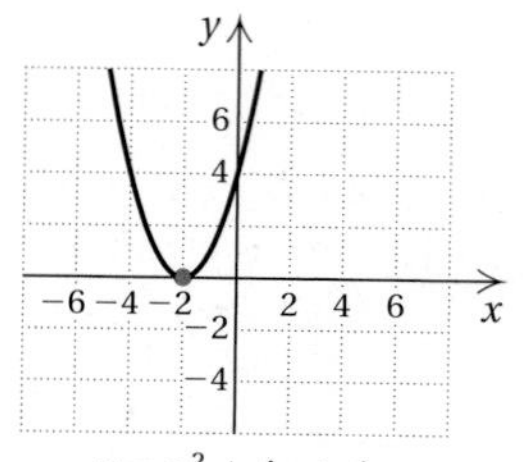

$y = x^2 + 4x + 4$
$b^2 - 4ac = 0$
One real solution
One x-intercept

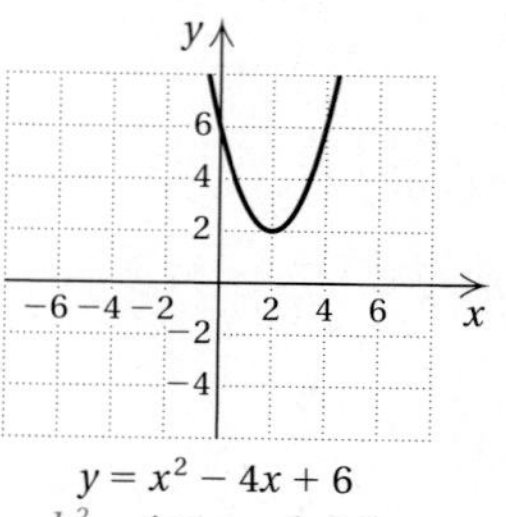

$y = x^2 - 4x + 6$
$b^2 - 4ac = -8 < 0$
No real solutions
No x-intercepts

Do Exercises 4–7.

10.6 EXERCISE SET

For Extra Help

Digital Video Tutor CD 9 Videotape 19 | InterAct Math | Math Tutor Center | MathXL | MyMathLab

a Graph the quadratic equation. In Exercises 1–8, label the ordered pairs for the vertex and the y-intercept.

1. $y = x^2 + 1$

x	y
-2	
-1	
0	
1	
2	
3	

2. $y = 2x^2$

x	y
-2	
-1	
0	
1	
2	
3	

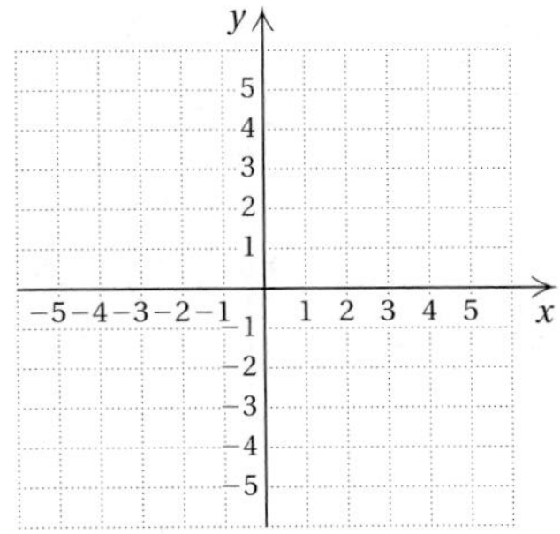

3. $y = -1 \cdot x^2$

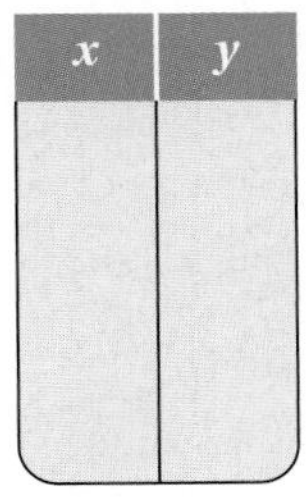

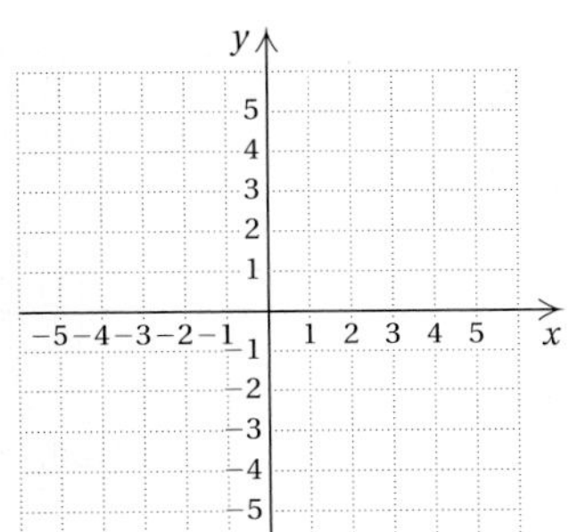

4. $y = x^2 - 1$

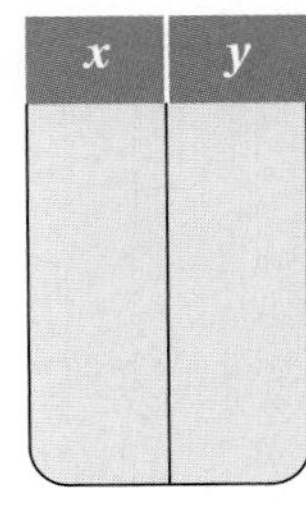

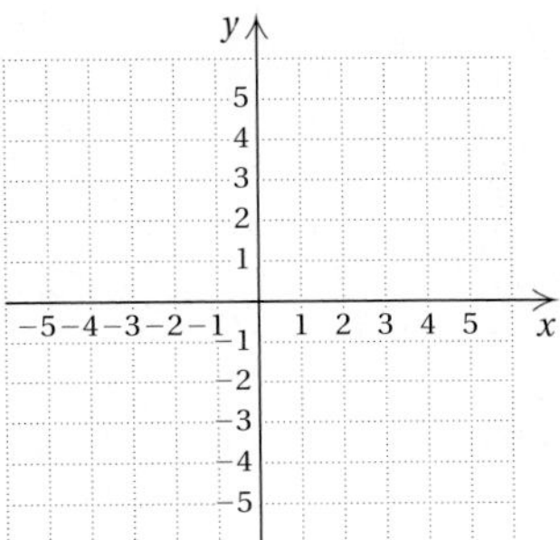

5. $y = -x^2 + 2x$

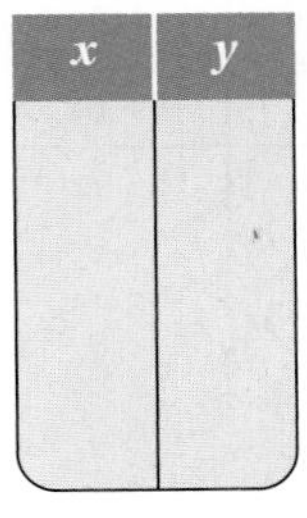

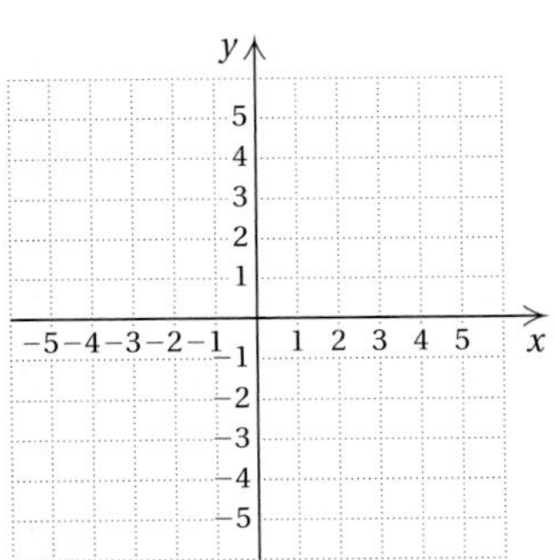

6. $y = x^2 + x - 2$

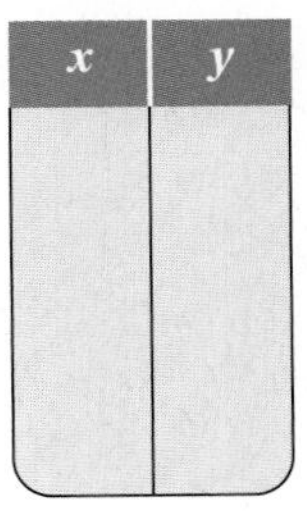

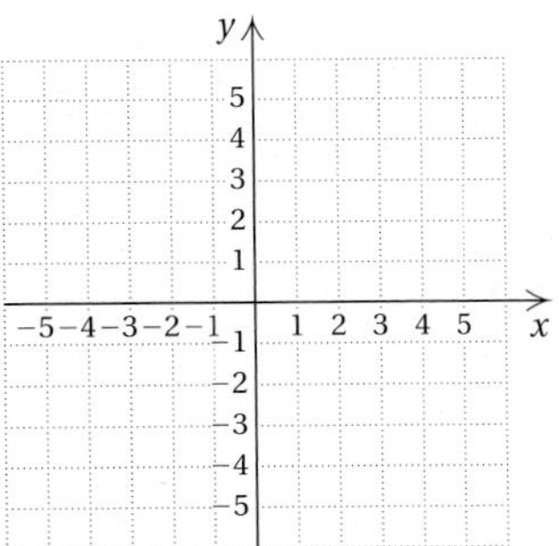

7. $y = 5 - x - x^2$

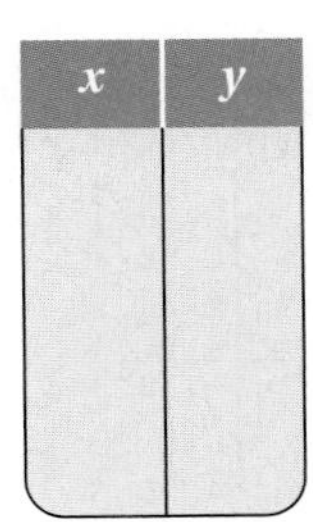

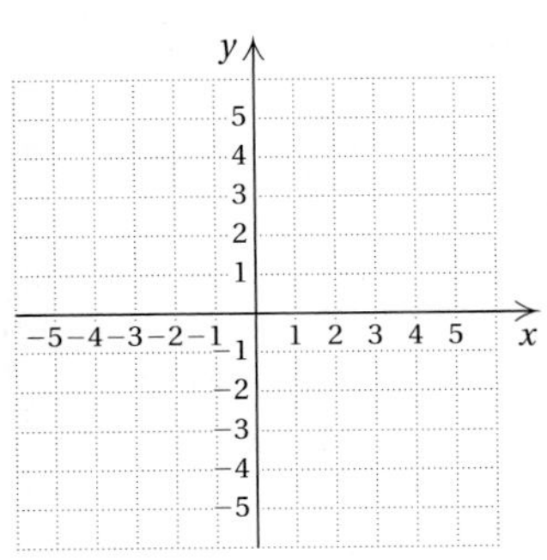

8. $y = x^2 + 2x + 1$

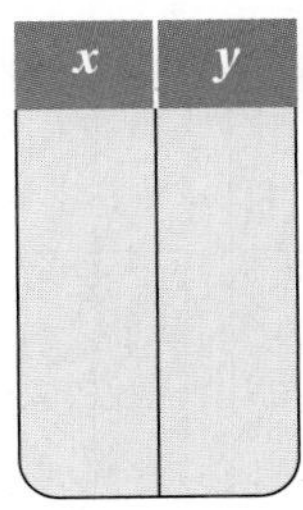

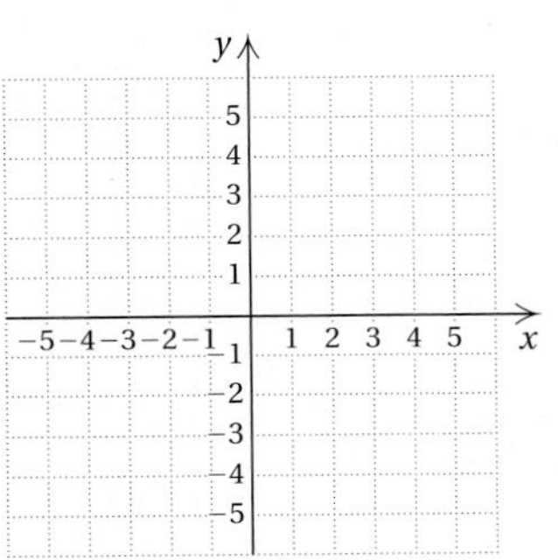

9. $y = x^2 - 2x + 1$

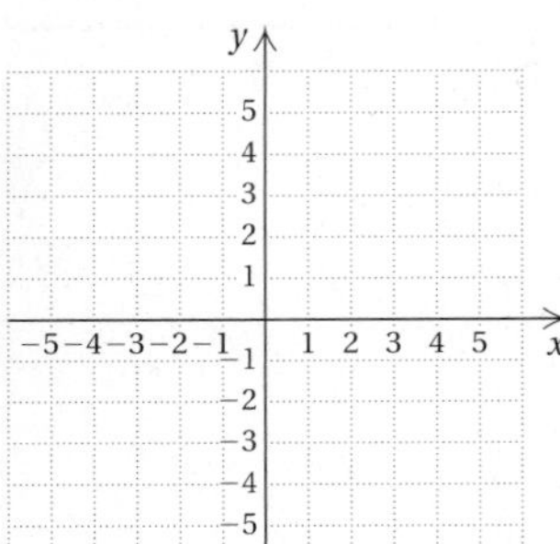

10. $y = -\frac{1}{2}x^2$

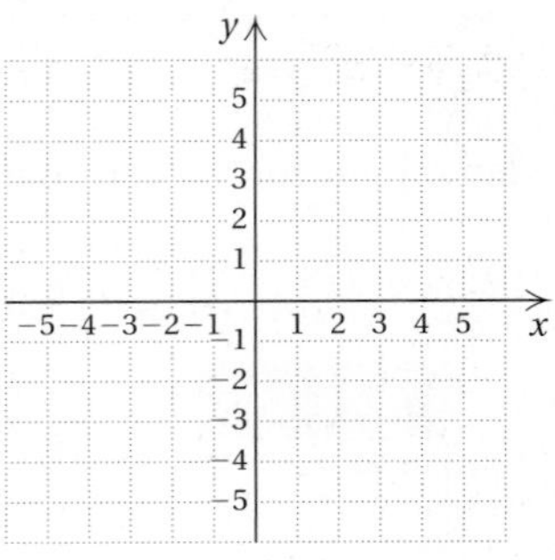

11. $y = -x^2 + 2x + 3$

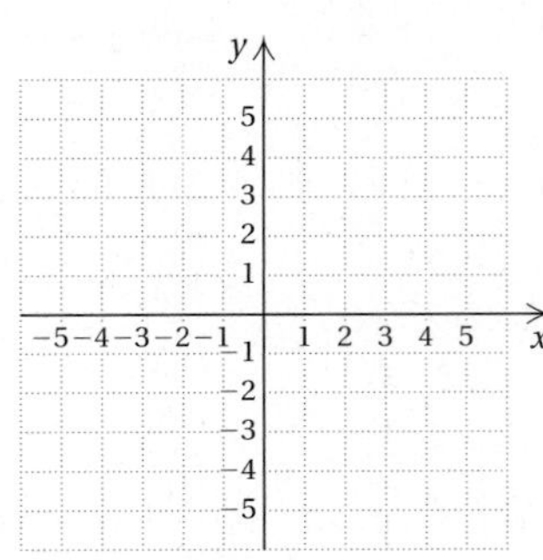

12. $y = -x^2 - 2x + 3$

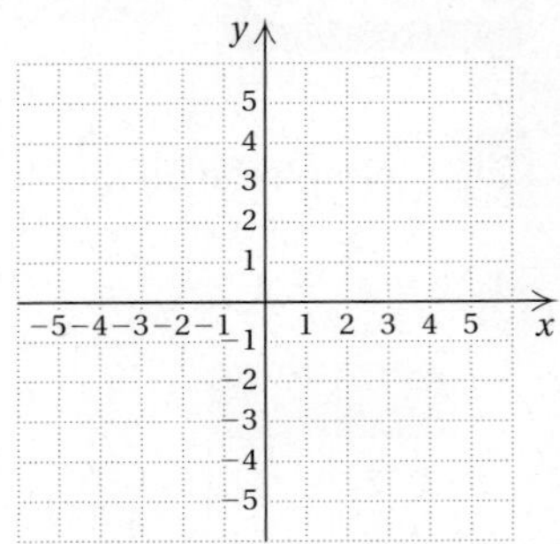

13. $y = -2x^2 - 4x + 1$

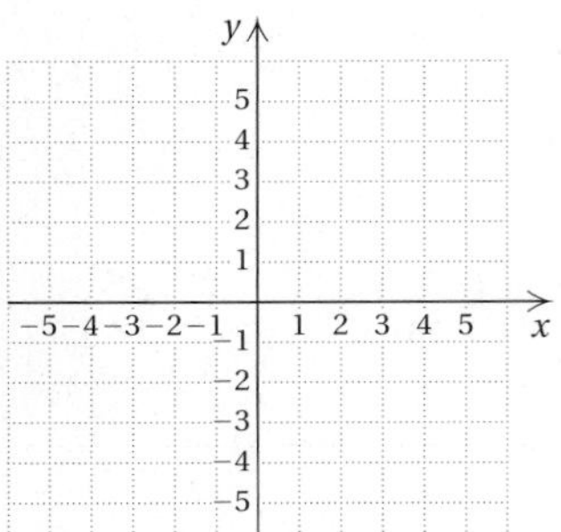

14. $y = 2x^2 + 4x - 1$

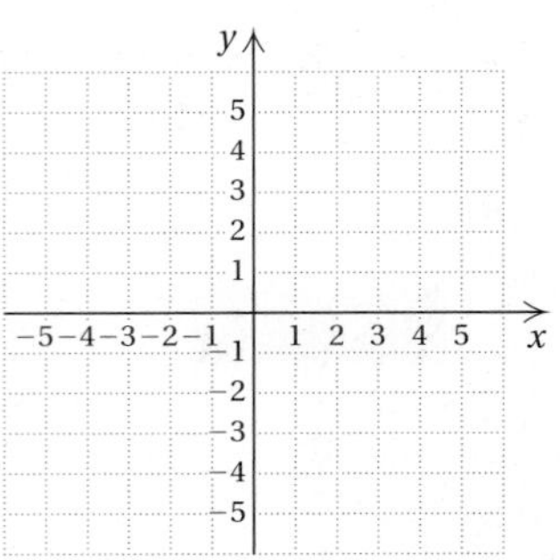

15. $y = 5 - x^2$

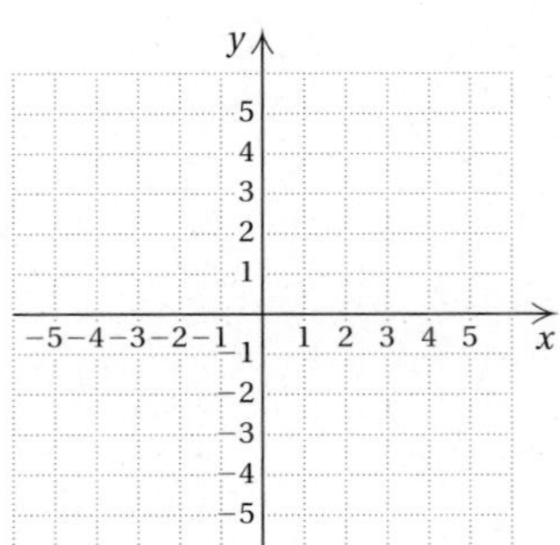

16. $y = 4 - x^2$

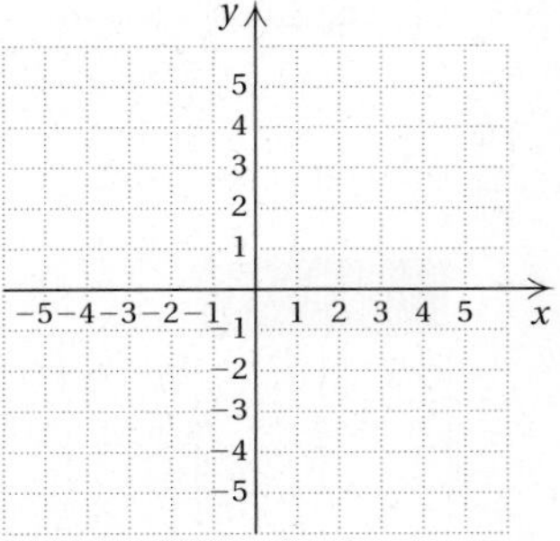

17. $y = \frac{1}{4}x^2$

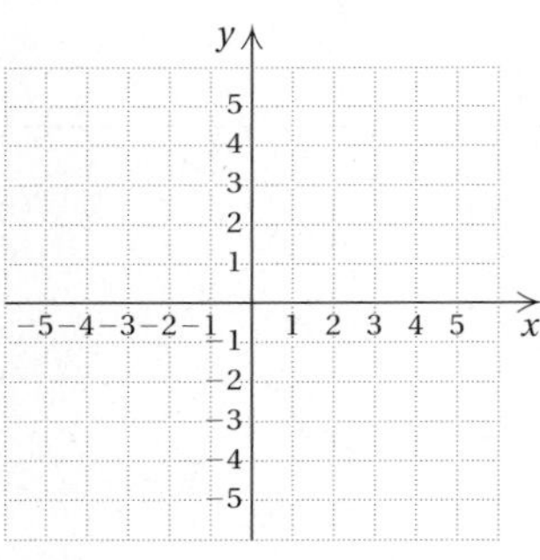

18. $y = -0.1x^2$

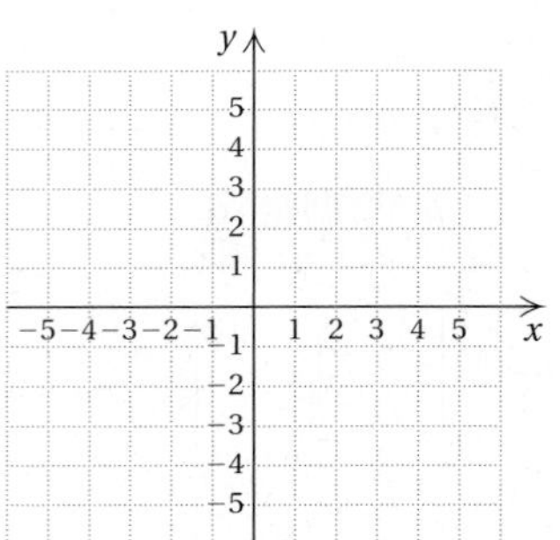

19. $y = -x^2 + x - 1$

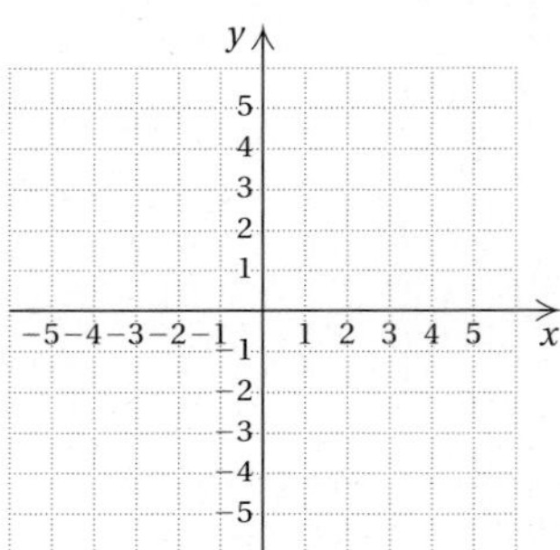

20. $y = x^2 + 2x$

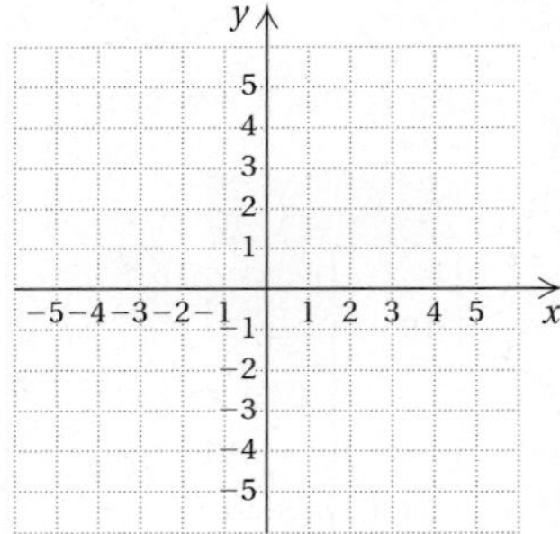

21. $y = -2x^2$

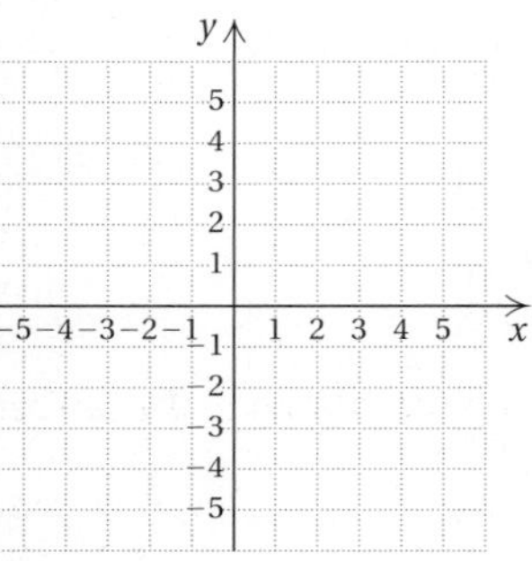

22. $y = -x^2 - 1$

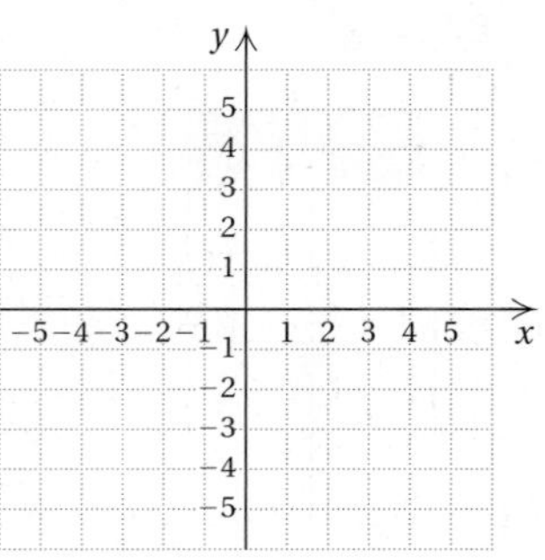

23. $y = x^2 - x - 6$

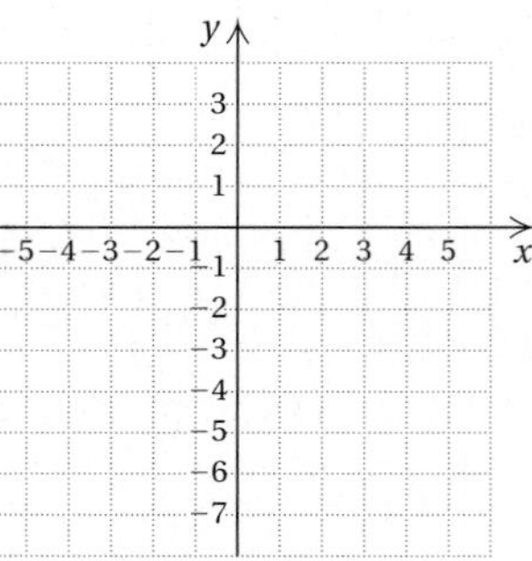

24. $y = 6 + x - x^2$

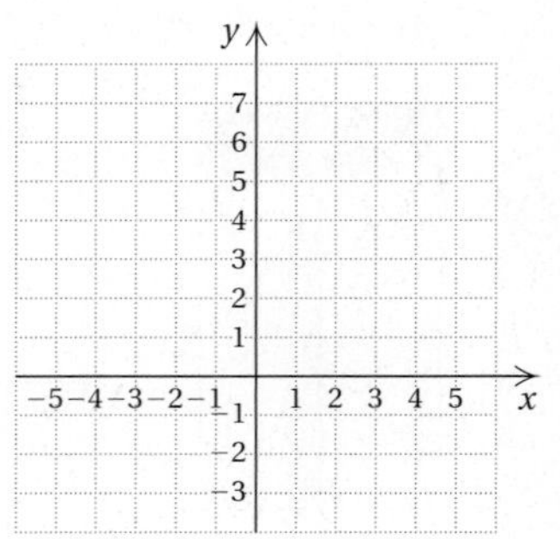

b Find the x-intercepts.

25. $y = x^2 - 2$

26. $y = x^2 - 7$

27. $y = x^2 + 5x$

28. $y = x^2 - 4x$

29. $y = 8 - x - x^2$

30. $y = 8 + x - x^2$

31. $y = x^2 - 6x + 9$

32. $y = x^2 + 10x + 25$

33. $y = -x^2 - 4x + 1$

34. $y = x^2 + 4x - 1$

35. $y = x^2 + 9$

36. $y = x^2 + 1$

37. D_W Suppose that the x-intercepts of a parabola are $(a_1, 0)$ and $(a_2, 0)$. What is the easiest way to find an equation for the line of symmetry? the coordinates of the vertex?

38. D_W Discuss the effect of the sign of a on the graph of $y = ax^2 + bx + c$.

SKILL MAINTENANCE

39. Add: $\sqrt{8} + \sqrt{50} + \sqrt{98} + \sqrt{128}$. [9.4a]

40. Multiply and simplify: $\sqrt{5y^4}\sqrt{125y}$. [9.2c]

41. Find an equation of variation in which y varies inversely as x and $y = 12.4$ when $x = 2.4$. [7.5c]

42. Evaluate $3x^4 + 3x - 7$ when $x = -2$. [4.3a]

SYNTHESIS

43. *Height of a Projectile.* The height H, in feet, of a projectile with an initial velocity of 96 ft/sec is given by the equation

$$H = -16t^2 + 96t,$$

where t is the time, in seconds. Use the graph of this equation, shown here, or any equation-solving technique to answer the following questions.

a) How many seconds after launch is the projectile 128 ft above ground?

b) When does the projectile reach its maximum height?

c) How many seconds after launch does the projectile return to the ground?

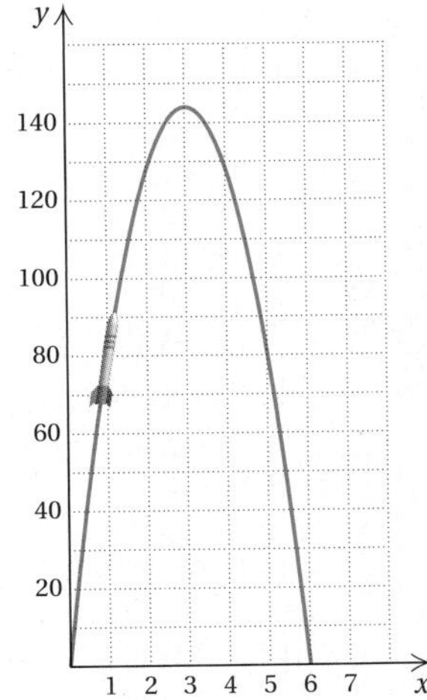

For each equation in Exercises 44–47, evaluate the discriminant $b^2 - 4ac$. Then use the answer to state how many real-number solutions exist for the equation.

44. $y = x^2 + 8x + 16$

45. $y = x^2 + 2x - 3$

46. $y = -2x^2 + 4x - 3$

47. $y = -0.02x^2 + 4.7x - 2300$

Objectives

a Determine whether a correspondence is a function.

b Given a function described by an equation, find function values (outputs) for specified values (inputs).

c Draw a graph of a function.

d Determine whether a graph is that of a function.

e Solve applied problems involving functions and their graphs.

10.7 FUNCTIONS

a Identifying Functions

We now develop one of the most important concepts in mathematics, **functions.** We have actually been studying functions all through this text; we just haven't identified them as such. Ordered pairs form a correspondence between first and second coordinates. A function is a special correspondence from one set of numbers to another. For example:

To each student in a college, there corresponds his or her student ID.

To each item in a store, there corresponds its price.

To each real number, there corresponds the cube of that number.

In each case, the first set is called the **domain** and the second set is called the **range.** Given a member of the domain, there is *just one* member of the range to which it corresponds. This kind of correspondence is called a **function.**

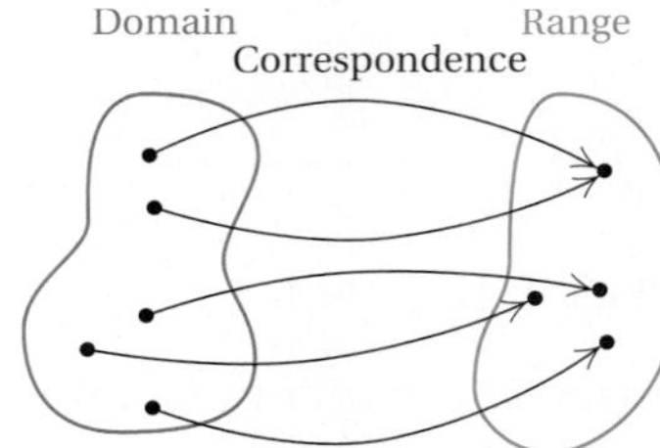

EXAMPLE 1 Determine whether the correspondence is a function.

f:

Domain	*Range*
1 →	$107.4
2 →	$ 34.1
3 →	$ 29.6
4 →	$ 19.6

g:

Domain	*Range*
3 →	5
4 →	9
5 →	−7
6 →	−7

h:

Domain	*Range*
Chicago →	Cubs
Chicago →	White Sox
Baltimore →	Orioles
San Diego →	Padres

p:

Domain	*Range*
Cubs →	Chicago
White Sox →	Chicago
Orioles →	Baltimore
Padres →	San Diego

The correspondence f *is* a function because each member of the domain is matched to only one member of the range.

The correspondence g *is* also a function because each member of the domain is matched to only one member of the range.

The correspondence h *is not* a function because one member of the domain, Chicago, is matched to more than one member of the range.

The correspondence p *is* a function because each member of the domain is paired with only one member of the range.

FUNCTION, DOMAIN, AND RANGE

A **function** is a correspondence between a first set, called the **domain,** and a second set, called the **range,** such that each member of the domain corresponds to *exactly one* member of the range.

Do Exercises 1–4.

EXAMPLE 2 Determine whether the correspondence is a function.

Domain	*Correspondence*	*Range*
a) A family	Each person's weight	A set of positive numbers
b) The natural numbers	Each number's square	A set of natural numbers
c) The set of all states	Each state's members of the U.S. Senate	A set of U.S. Senators

a) The correspondence *is* a function because each person has *only one* weight.

b) The correspondence *is* a function because each natural number has *only one* square.

c) The correspondence *is not* a function because each state has two U.S. Senators.

Do Exercises 5 and 6.

When a correspondence between two sets is not a function, it may still be an example of a *relation.*

RELATION

A **relation** is a correspondence between a first set, called the **domain,** and a second set, called the **range,** such that each member of the domain corresponds to *at least one* member of the range.

Thus, although the correspondences of Examples 1 and 2 are not all functions, they *are* all relations. A function is a special type of relation—one in which each member of the domain is paired with *exactly one* member of the range.

b Finding Function Values

Most functions considered in mathematics are described by equations. A linear equation like $y = 2x + 3$, studied in Chapters 3 and 7, is called a **linear function.** A quadratic equation like $y = 4 - x^2$, studied in Chapter 10, is called a **quadratic function.**

Determine whether the correspondence is a function.

1. *Domain* → *Range*

Cheetah → 70 mph
Human → 28 mph
Lion → 50 mph
Chicken → 9 mph

2. *Domain* *Range*

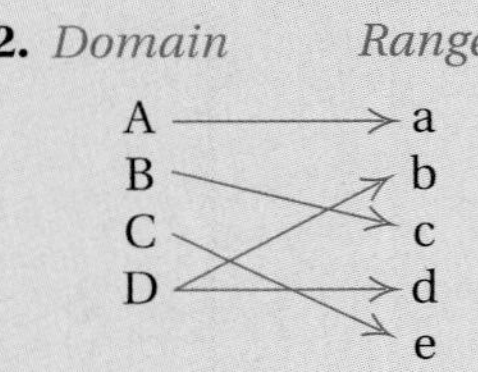

3. *Domain* *Range*

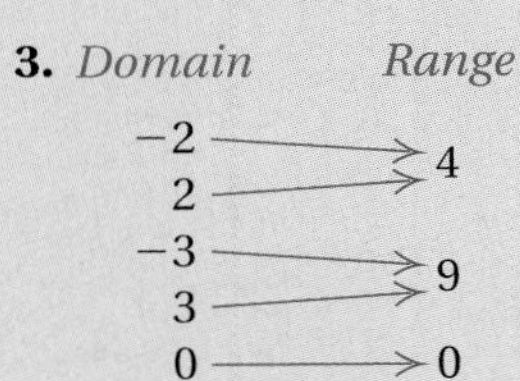

4. *Domain* *Range*

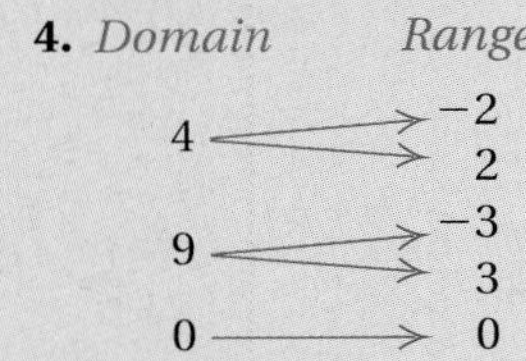

Determine whether each of the following is a function.

5. *Domain*
A set of numbers

Correspondence
Square each number and subtract 10.

Range
A set of numbers

6. *Domain*
A set of polygons

Correspondence
Find the perimeter of each polygon.

Range
A set of numbers

Answers on page A-47

Find the function values.

7. $f(x) = 5x - 3$

a) $f(-6)$

b) $f(0)$

c) $f(1)$

d) $f(20)$

e) $f(-1.2)$

8. $g(x) = x^2 - 4x + 9$

a) $g(-2)$

b) $g(0)$

c) $g(5)$

d) $g(10)$

Answers on page A-47

Recall that when graphing $y = 2x + 3$, we chose x-values and then found corresponding y-values. For example, when $x = 4$,

$$y = 2x + 3 = 2 \cdot 4 + 3 = 11.$$

When thinking of functions, we call the number 4 an **input** and the number 11 an **output.**

It helps to think of a function as a machine; that is, think of putting a member of the domain (an input) into the machine. The machine knows the correspondence and gives out a member of the range (the output).

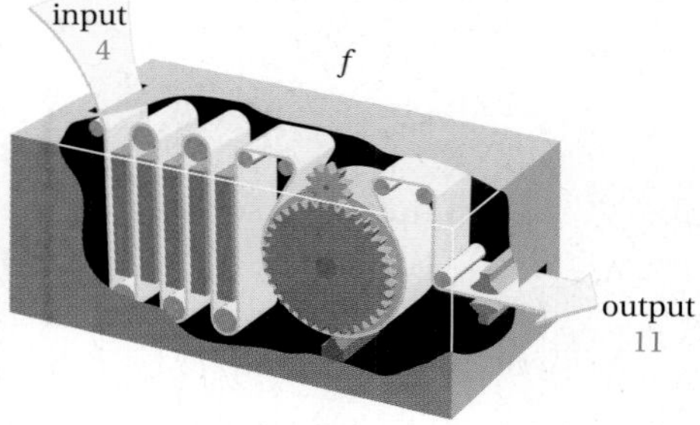

The function $y = 2x + 3$ can be named f and described by the equation $f(x) = 2x + 3$. We call the input x and the output $f(x)$. This is read "f of x," or "f at x," or "the value of f at x."

CAUTION!

The notation $f(x)$ *does not mean* "f times x" and should not be read that way.

The equation $f(x) = 2x + 3$ describes the function that takes an input x, multiplies it by 2, and then adds 3.

Input

$$f(x) = 2x + 3$$

Multiply by 2 **Add 3**

To find the output $f(4)$, we take the input 4, double it, and add 3 to get 11. That is, we substitute 4 into the formula for $f(x)$:

$$f(4) = 2 \cdot 4 + 3 = 11.$$

Outputs of functions are also called **function values.** For $f(x) = 2x + 3$, we know that $f(4) = 11$. We can say that "the function value at 4 is 11."

EXAMPLE 3 Find the indicated function value.

a) $f(5)$, for $f(x) = 3x + 2$

b) $g(3)$, for $g(z) = 5z^2 - 4$

c) $A(-2)$, for $A(r) = 3r^2 - 2r$

d) $f(-5)$, for $f(x) = x^2 + 3x - 4$

a) $f(5) = 3 \cdot 5 + 2 = 17$

b) $g(3) = 5(3)^2 - 4 = 41$

c) $A(-2) = 3(-2)^2 + 2(-2) = 8$

d) $f(-5) = (-5)^2 + 3(-5) - 4 = 25 - 15 - 4 = 6$

Do Exercises 7 and 8.

CALCULATOR CORNER

Finding Function Values We can find function values on a graphing calculator. One method is to substitute inputs directly into the formula. Consider the function in Example 3(d), $f(x) = x^2 + 3x - 4$. To find $f(-5)$, we press [(] [(−)] [5] [)] [x^2] [+] [3] [(] [(−)] [5] [)] [−] [4] [ENTER]. We find that $f(-5) = 6$.

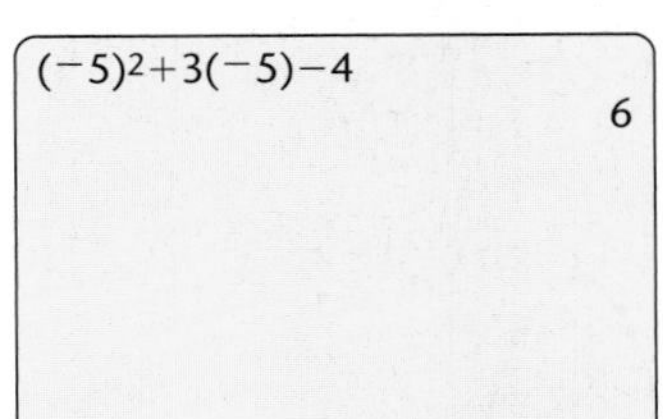

X	Y1	
−5	6	

X =

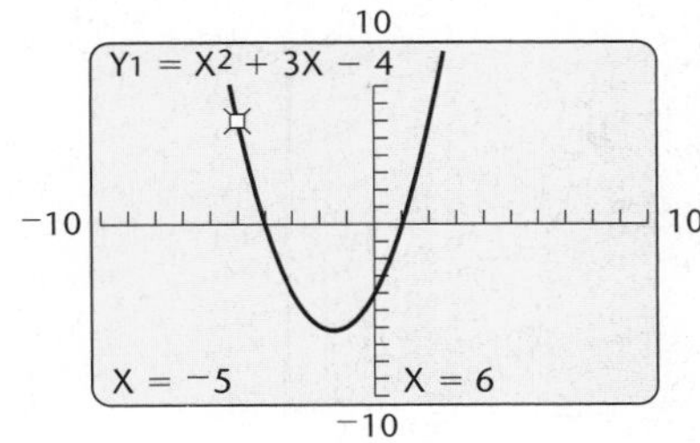

After we have entered the function as $y_1 = x^2 + 3x - 4$ on the equation-editor screen, there are several other methods that we can use to find function values. We can use a table set in ASK mode and enter $x = -5$. We see that the function value, y_1, is 6. We can also use the VALUE feature to evaluate the function. To do this, we first graph the function in a window that includes $x = -5$ and then press [2nd] [CALC] [1] to access the VALUE feature. Next, we supply the desired x-value by pressing [(−)] [5]. Finally, we press [ENTER] to see $x = -5$, $y = 6$ at the bottom of the screen. Again we see that the function value is 6.

There are other ways to find function values, but we will not discuss them here.

Exercises: Find the function values.

1. $f(-3.4)$, for $f(x) = 2x - 6$
2. $f(4)$, for $f(x) = -2.3x$
3. $f(-1)$, for $f(x) = x^2 - 3$
4. $f(3)$, for $f(x) = 2x^2 - x + 5$

C Graphs of Functions

To graph a function, we find ordered pairs (x, y) or $(x, f(x))$, plot them, and connect the points. Note that y and $f(x)$ are used interchangeably when working with functions and their graphs.

EXAMPLE 4 Graph: $f(x) = x + 2$.

A list of some function values is shown in this table. We plot the points and connect them. The graph is a straight line.

x	$f(x)$
−4	−2
−3	−1
−2	0
−1	1
0	2
1	3
2	4
3	5
4	6

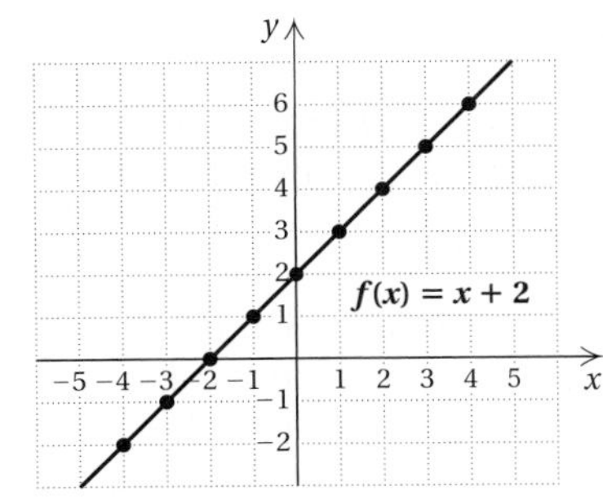

EXAMPLE 5 Graph: $g(x) = 4 - x^2$.

Recall from Section 10.6 that the graph is a parabola. We calculate some function values and draw the curve.

$$g(0) = 4 - 0^2 = 4 - 0 = 4,$$
$$g(-1) = 4 - (-1)^2 = 4 - 1 = 3,$$
$$g(2) = 4 - (2)^2 = 4 - 4 = 0,$$
$$g(-3) = 4 - (-3)^2 = 4 - 9 = -5$$

x	$g(x)$
−3	−5
−2	0
−1	3
0	4
1	3
2	0
3	−5

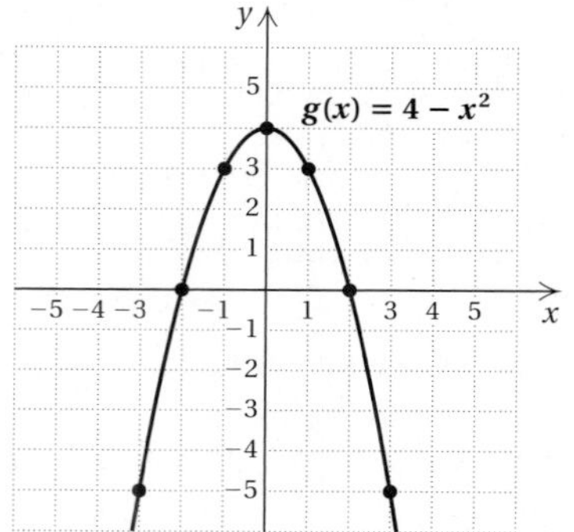

EXAMPLE 6 Graph: $h(x) = |x|$.

A list of some function values is shown in the following table. We plot the points and connect them. The graph is a V-shaped "curve" that rises on either side of the vertical axis.

x	$h(x)$
−3	3
−2	2
−1	1
0	0
1	1
2	2
3	3

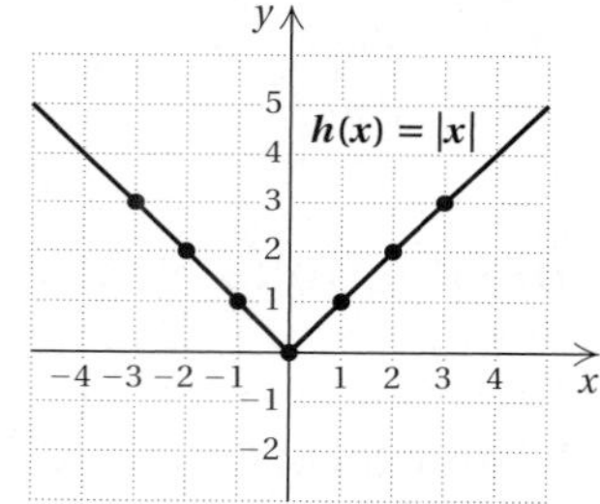

Do Exercises 9–11 on the following page.

d The Vertical-Line Test

Consider the function f described by $f(x) = x^2 - 5$. Its graph is shown at right. It is also the graph of the equation $y = x^2 - 5$.

To find a function value, like $f(3)$, from a graph, we locate the input on the horizontal axis, move vertically to the graph of the function, and then horizontally to find the output on the vertical axis, where members of the range can be found.

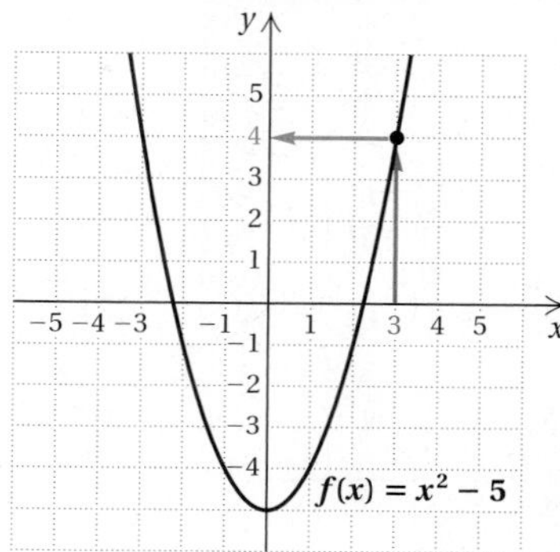

Study Tips

FINAL STUDY TIP

You are arriving at the end of your course in Introductory Algebra. If you have not begun to prepare for the final examination, be sure to read the comments in the Study Tips on pp. 598, 649, 664, and 725.

"We make a living by what we get, but we make a life by what we give."

Winston Churchill

Recall that when one member of the domain is paired with two or more different members of the range, the correspondence is *not* a function. Thus, when a graph contains two or more different points with the same first coordinate, the graph cannot represent a function. Points sharing a common first coordinate are vertically above or below each other (see the following graph).

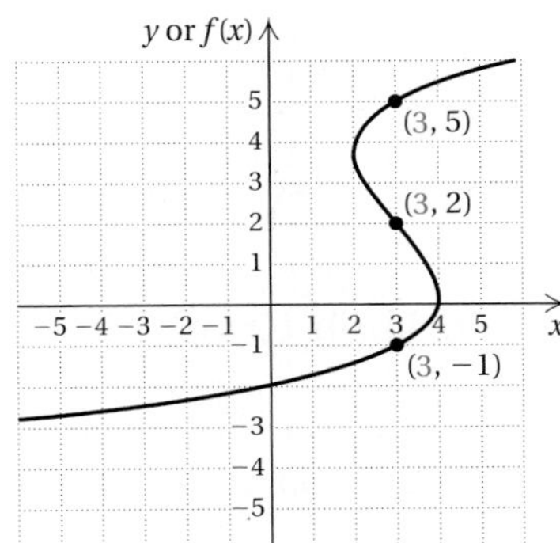

Since 3 is paired with more than one member of the range, the graph does not represent a function.

This observation leads to the *vertical-line test.*

THE VERTICAL-LINE TEST

A graph represents a function if it is impossible to draw a vertical line that intersects the graph more than once.

EXAMPLE 7 Determine whether each of the following is the graph of a function.

a)

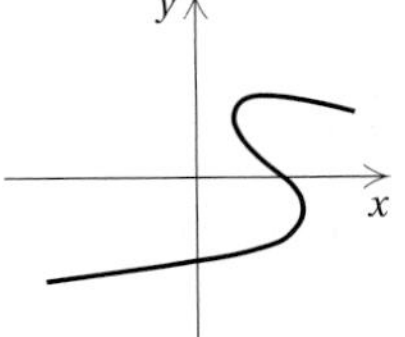

b)

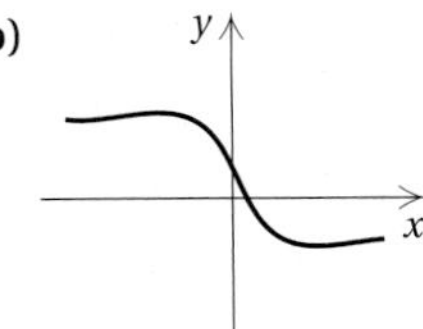

c)

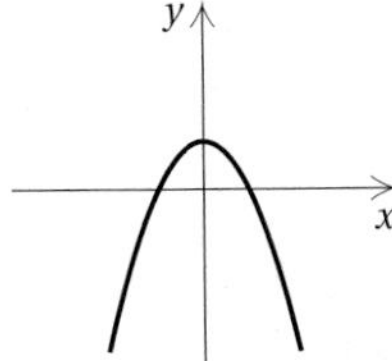

d)

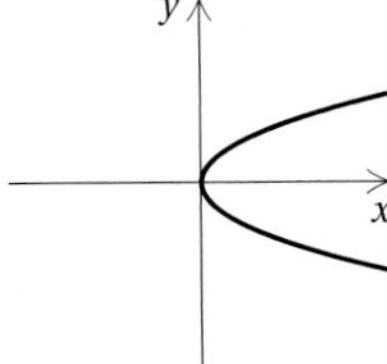

a) The graph *is not* that of a function because a vertical line crosses the graph at more than one point.

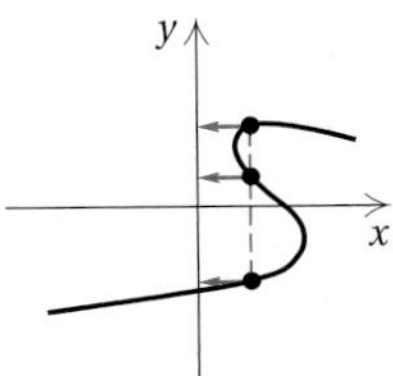

b) The graph *is* that of a function because no vertical line can cross the graph at more than one point. This can be confirmed with a ruler or straightedge.

Graph.

9. $f(x) = x - 4$

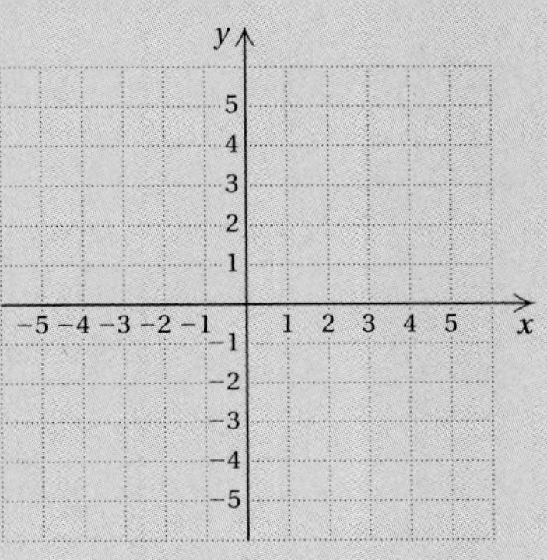

10. $g(x) = 5 - x^2$

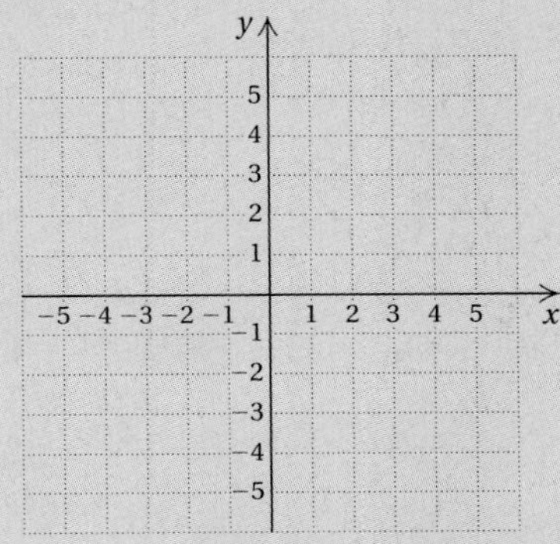

11. $t(x) = 3 - |x|$

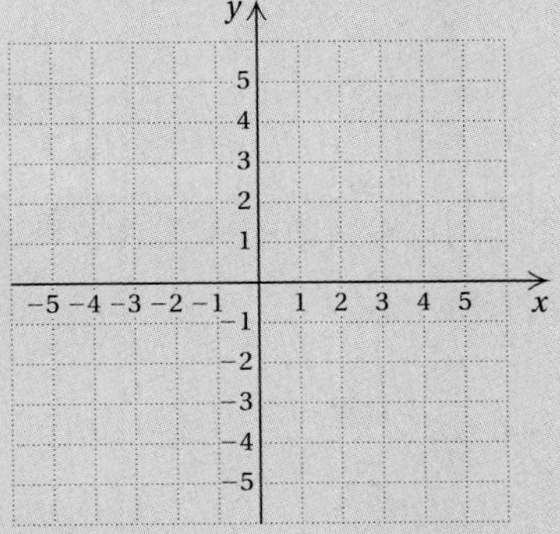

Answers on page A-47

Determine whether each of the following is the graph of a function.

12.

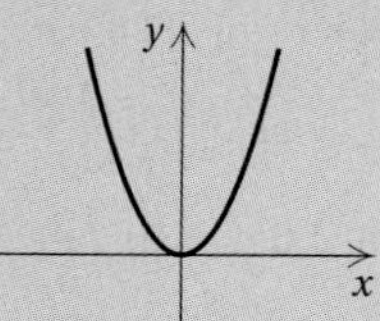

13.

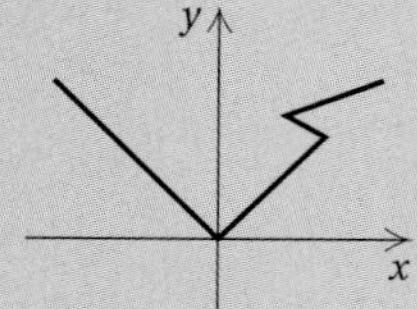

14.

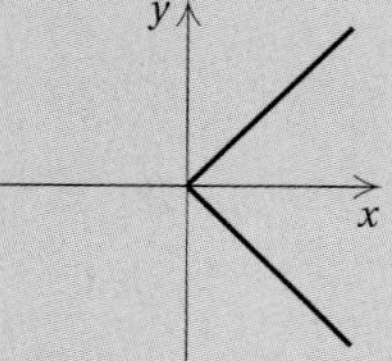

15.

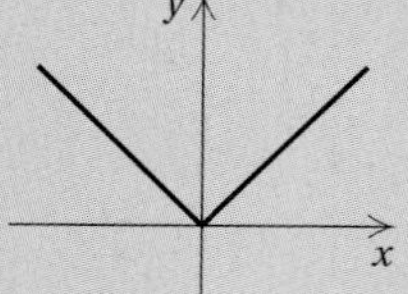

Referring to the graph in Example 8:

16. What was the movie revenue for week 2?

17. What was the movie revenue for week 6?

Answers on page A-47

c) The graph *is* that of a function.

d) The graph *is not* that of a function. There is a vertical line that crosses the graph more than once.

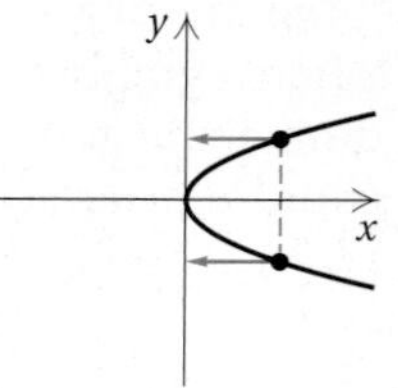

Do Exercises 12–15.

e Applications of Functions and Their Graphs

Functions are often described by graphs, whether or not an equation is given. To use a graph in an application, we note that each point on the graph represents a pair of values.

EXAMPLE 8 *Movie Revenue.* The following graph approximates the weekly revenue, in millions of dollars, from the movie *Jurassic Park*. The revenue is a function of the week, and no equation is given for the function.

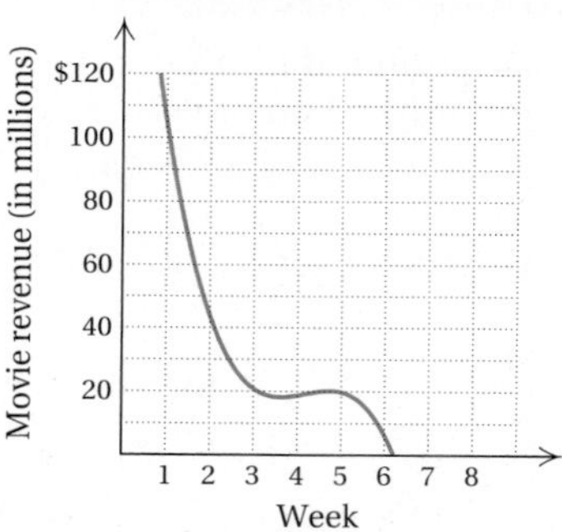

Source: Exhibitor Relations Co., Inc.

Use the graph to answer the following.

a) What was the movie revenue for week 1?

b) What was the movie revenue for week 5?

a) To estimate the revenue for week 1, we locate 1 on the horizontal axis and move directly up until we reach the graph. Then we move across to the vertical axis. We estimate that value to be about $105 million.

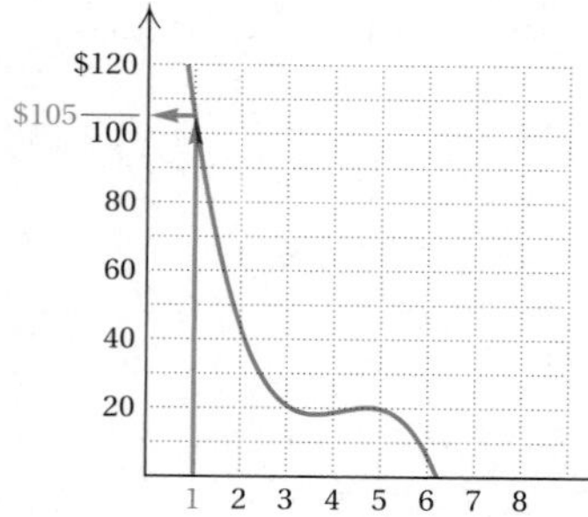

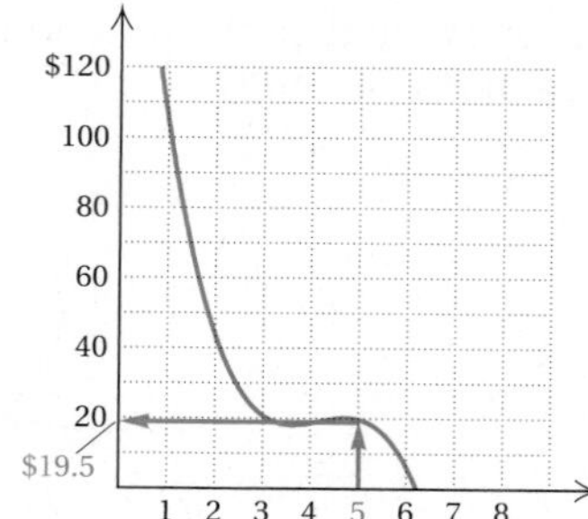

b) To estimate the revenue for week 5, we locate 5 on the horizontal axis and move directly up until we reach the graph. Then we move across to the vertical axis. We estimate that value to be about $19.5 million.

Do Exercises 16 and 17.

10.7 EXERCISE SET

For Extra Help

Digital Video Tutor CD 9 Videotape 19

InterAct Math

Math Tutor Center

MathXL

MyMathLab

a Determine whether the correspondence is a function.

1. *Domain* → *Range*

2 → 9
5 → 8
19 → 8

2. *Domain* → *Range*

5 → 3
−3 → 3
7 → 7
−7 → 7

3.

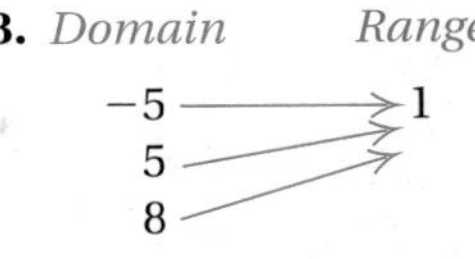

4.

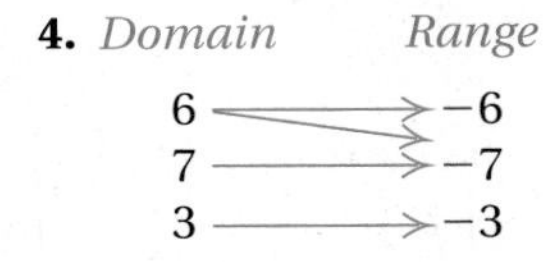

5.

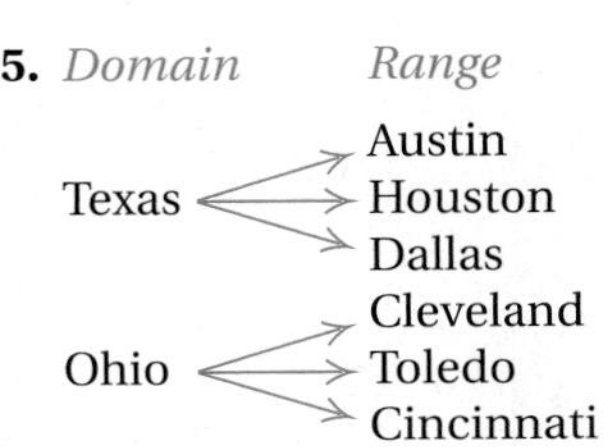

6. *Domain* → *Range*

Austin → Texas
Houston → Texas
Dallas → Texas
Cleveland → Ohio
Toledo → Ohio
Cincinnati → Ohio

7.

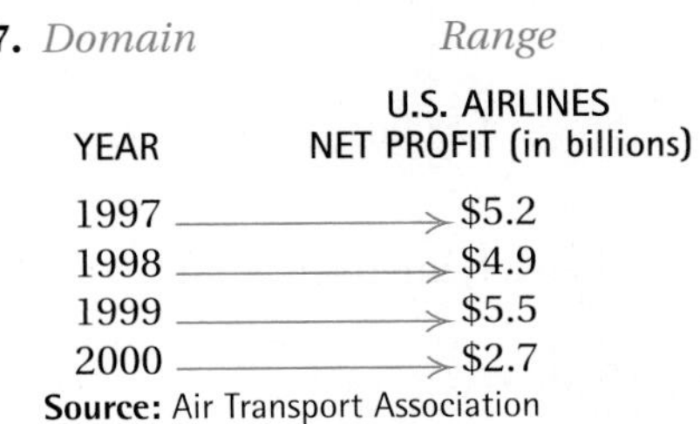

Domain YEAR	*Range* U.S. AIRLINES NET PROFIT (in billions)
1997	\$5.2
1998	\$4.9
1999	\$5.5
2000	\$2.7

Source: Air Transport Association

8.

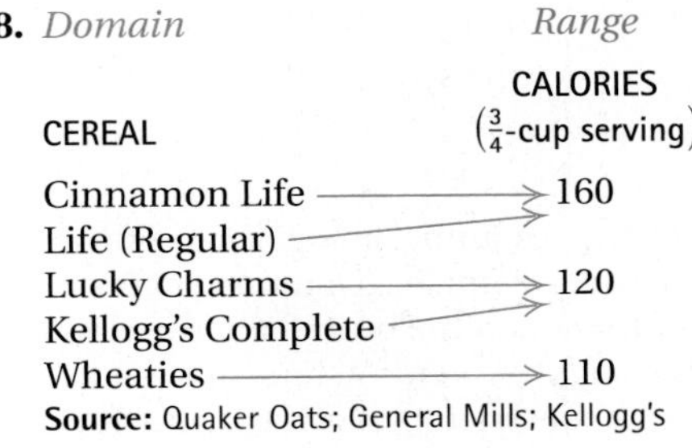

Domain CEREAL	*Range* CALORIES ($\frac{3}{4}$-cup serving)
Cinnamon Life	160
Life (Regular)	160
Lucky Charms	120
Kellogg's Complete	120
Wheaties	110

Source: Quaker Oats; General Mills; Kellogg's

Determine whether each of the following is a function. Identify any relations that are not functions.

	Domain	*Correspondence*	*Range*
9.	A math class	Each person's seat number	A set of numbers
10.	A set of numbers	Square each number and then add 4.	A set of numbers
11.	A set of shapes	Find the area of each shape.	A set of numbers
12.	A family	Each person's eye color	A set of colors
13.	The people in a town	Each person's aunt	A set of females
14.	A set of avenues	Find an intersecting road.	A set of cross streets

b Find the function values.

15. $f(x) = x + 5$

a) $f(4)$ **b)** $f(7)$
c) $f(-3)$ **d)** $f(0)$
e) $f(2.4)$ **f)** $f\left(\frac{2}{3}\right)$

16. $g(t) = t - 6$

a) $g(0)$ **b)** $g(6)$
c) $g(13)$ **d)** $g(-1)$
e) $g(-1.08)$ **f)** $g\left(\frac{7}{8}\right)$

17. $h(p) = 3p$

a) $h(-7)$ **b)** $h(5)$
c) $h(14)$ **d)** $h(0)$
e) $h\left(\frac{2}{3}\right)$ **f)** $h(-54.2)$

18. $f(x) = -4x$

a) $f(6)$
b) $f\left(-\frac{1}{2}\right)$
c) $f(20)$
d) $f(11.8)$
e) $f(0)$
f) $f(-1)$

19. $g(s) = 3s + 4$

a) $g(1)$
b) $g(-7)$
c) $g(6.7)$
d) $g(0)$
e) $g(-10)$
f) $g\left(\frac{2}{3}\right)$

20. $h(x) = 19$, a constant function

a) $h(4)$
b) $h(-6)$
c) $h(12.5)$
d) $h(0)$
e) $h\left(\frac{2}{3}\right)$
f) $h(1234)$

21. $f(x) = 2x^2 - 3x$

a) $f(0)$
b) $f(-1)$
c) $f(2)$
d) $f(10)$
e) $f(-5)$
f) $f(-10)$

22. $f(x) = 3x^2 - 2x + 1$

a) $f(0)$
b) $f(1)$
c) $f(-1)$
d) $f(10)$
e) $f(2)$
f) $f(-3)$

23. $f(x) = |x| + 1$

a) $f(0)$
b) $f(-2)$
c) $f(2)$
d) $f(-3)$
e) $f(-10)$
f) $f(22)$

24. $g(t) = \sqrt{t}$

a) $g(4)$
b) $g(25)$
c) $g(16)$
d) $g(100)$
e) $g(50)$
f) $g(84)$

25. $f(x) = x^3$

a) $f(0)$
b) $f(-1)$
c) $f(2)$
d) $f(10)$
e) $f(-5)$
f) $f(-10)$

26. $f(x) = x^4 - 3$

a) $f(1)$
b) $f(-1)$
c) $f(0)$
d) $f(2)$
e) $f(-2)$
f) $f(10)$

27. *Estimating Heights.* An anthropologist can estimate the height of a male or a female, given the lengths of certain bones. A *humerus* is the bone from the elbow to the shoulder. The height, in centimeters, of a female with a humerus of x centimeters is given by the function

$$F(x) = 2.75x + 71.48.$$

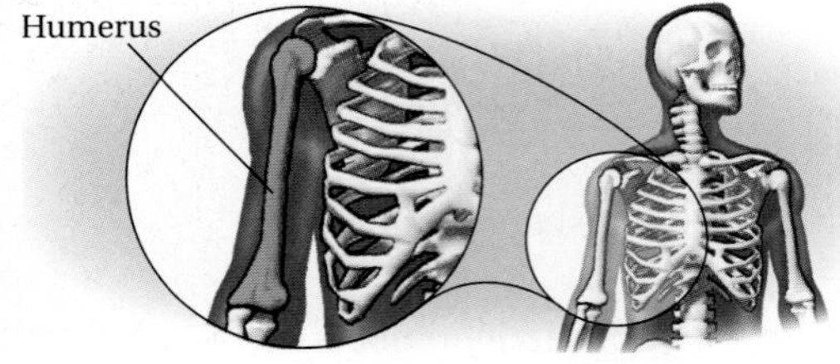

If a humerus is known to be from a female, how tall was she if the bone is **(a)** 32 cm long? **(b)** 30 cm long?

28. Refer to Exercise 27. When a humerus is from a male, the function

$$M(x) = 2.89x + 70.64$$

can be used to find the male's height, in centimeters. If a humerus is known to be from a male, how tall was he if the bone is **(a)** 30 cm long? **(b)** 35 cm long?

29. *Pressure at Sea Depth.* The function $P(d) = 1 + (d/33)$ gives the pressure, in *atmospheres* (atm), at a depth of d feet in the sea. Note that $P(0) = 1$ atm, $P(33) = 2$ atm, and so on. Find the pressure at 20 ft, 30 ft, and 100 ft.

30. *Temperature as a Function of Depth.* The function $T(d) = 10d + 20$ gives the temperature, in degrees Celsius, inside the earth as a function of the depth d, in kilometers. Find the temperature at 5 km, 20 km, and 1000 km.

31. *Melting Snow.* The function $W(d) = 0.112d$ approximates the amount, in centimeters, of water that results from d centimeters of snow melting. Find the amount of water that results from snow melting from depths of 16 cm, 25 cm, and 100 cm.

32. *Temperature Conversions.* The function $C(F) = \frac{5}{9}(F - 32)$ determines the Celsius temperature that corresponds to F degrees Fahrenheit. Find the Celsius temperature that corresponds to 62°F, 77°F, and 23°F.

C Graph the function.

33. $f(x) = 3x - 1$

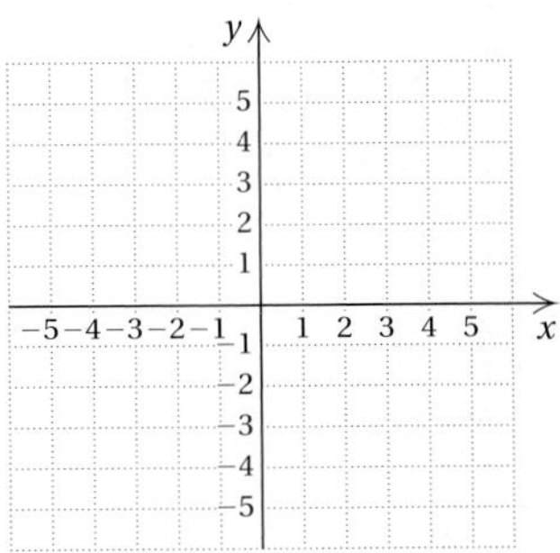

34. $g(x) = 2x + 5$

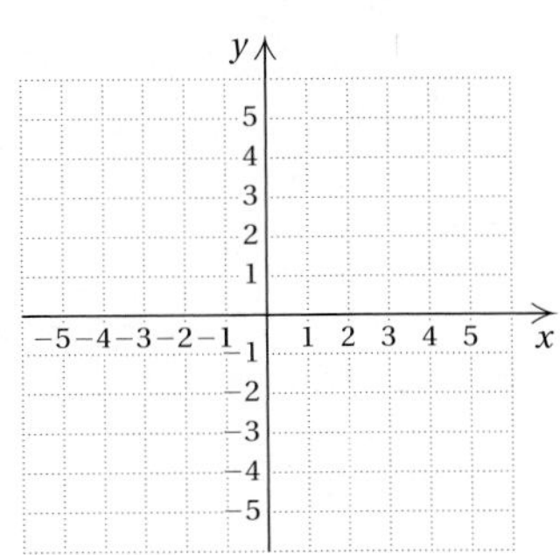

35. $g(x) = -2x + 3$

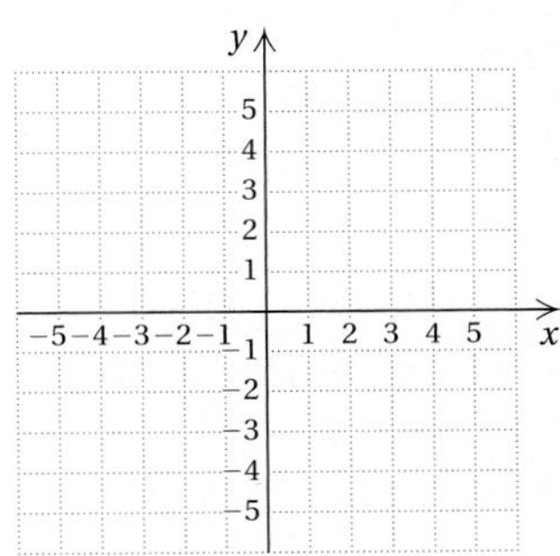

36. $f(x) = -\frac{1}{2}x + 2$

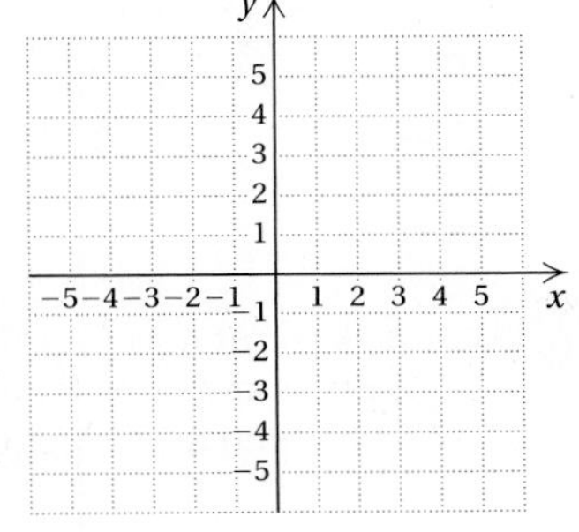

37. $f(x) = \frac{1}{2}x + 1$

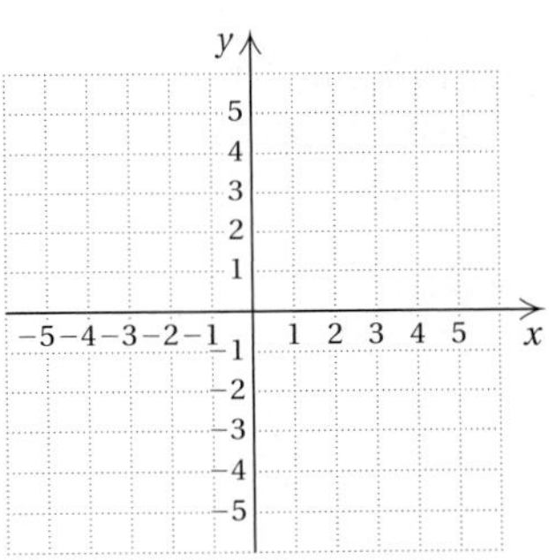

38. $f(x) = -\frac{3}{4}x - 2$

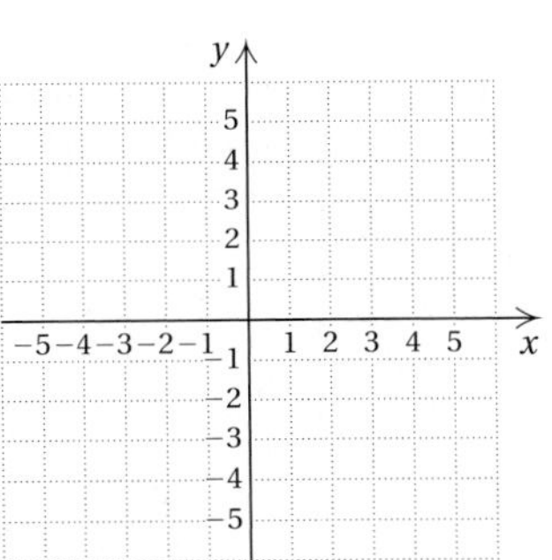

39. $f(x) = 2 - |x|$

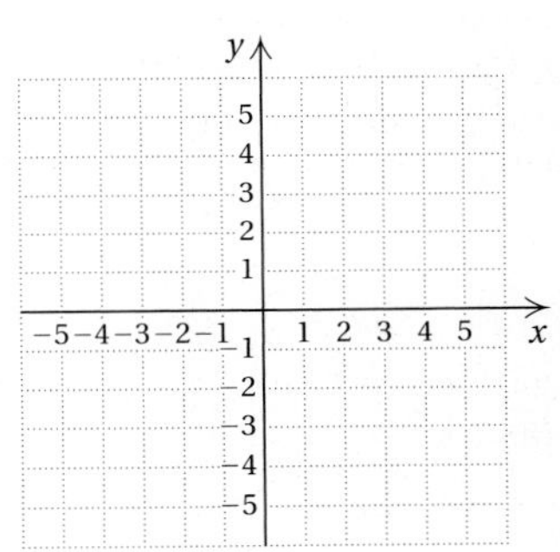

40. $f(x) = |x| - 4$

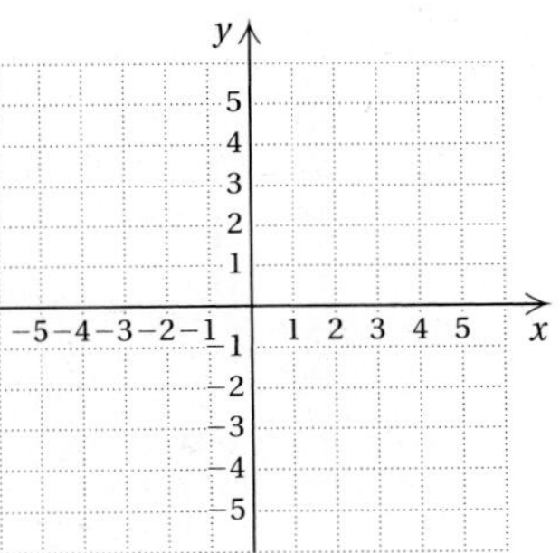

41. $f(x) = x^2$

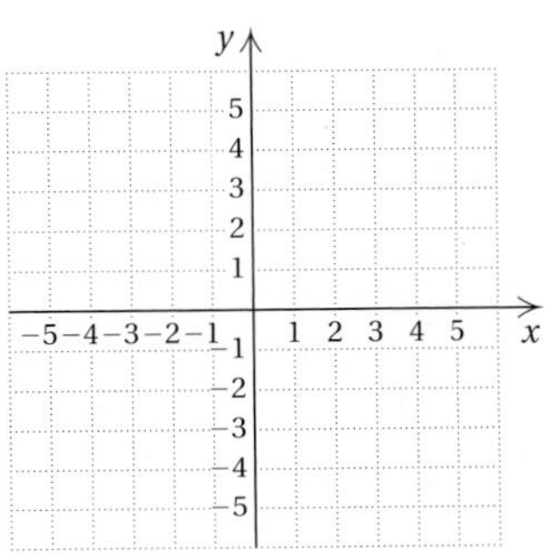

42. $f(x) = x^2 - 1$

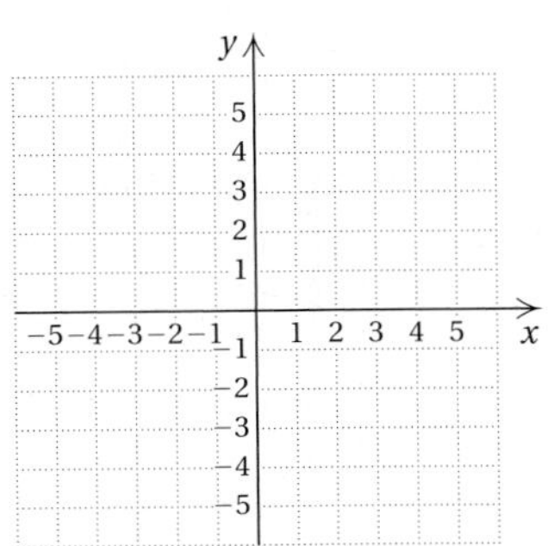

43. $f(x) = x^2 - x - 2$

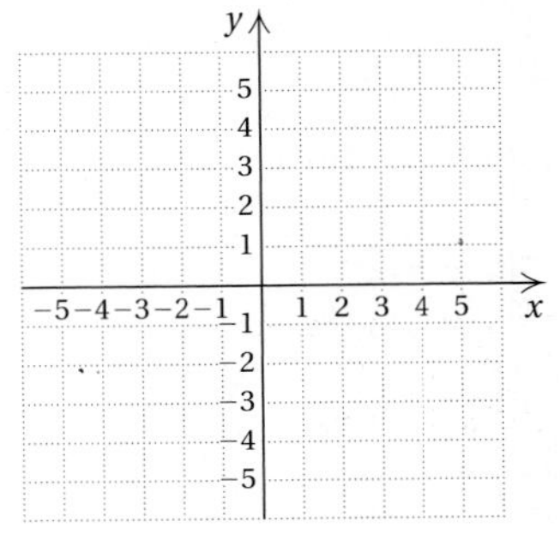

44. $f(x) = x^2 + 6x + 5$

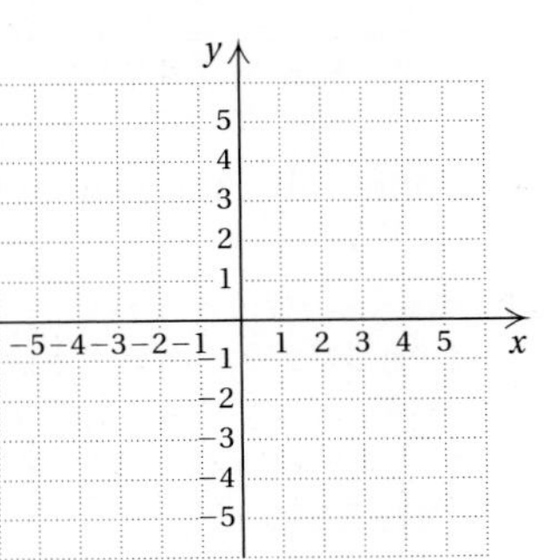

d Determine whether each of the following is the graph of a function.

45.

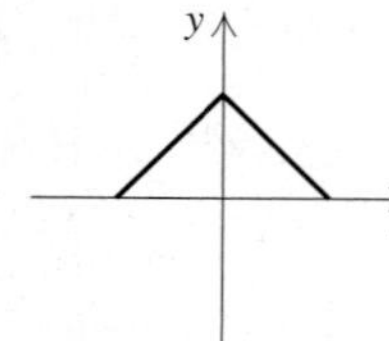

46.

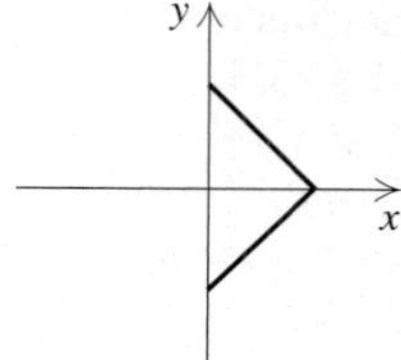

47.

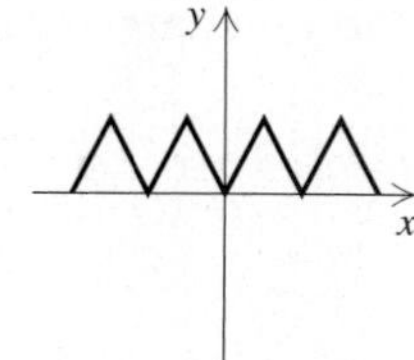

48.

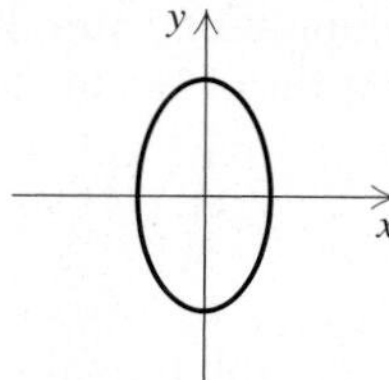

49.

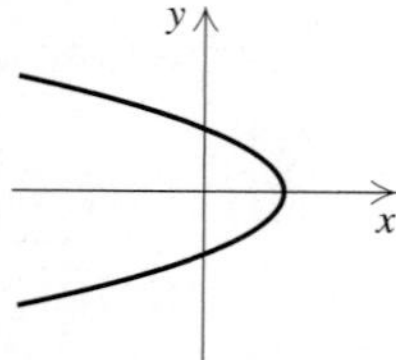

50.

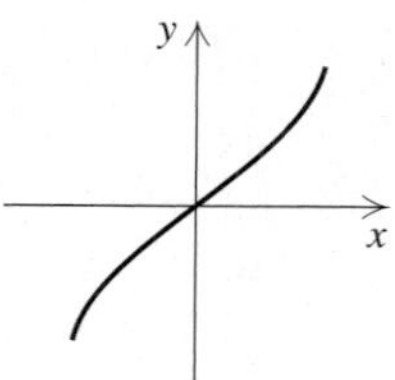

51.

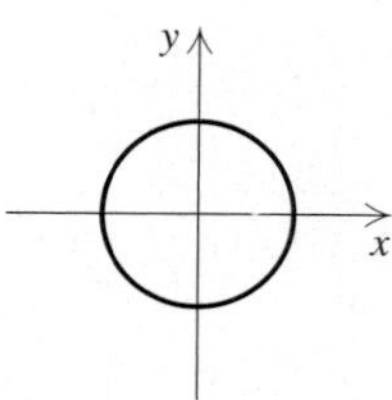

52.

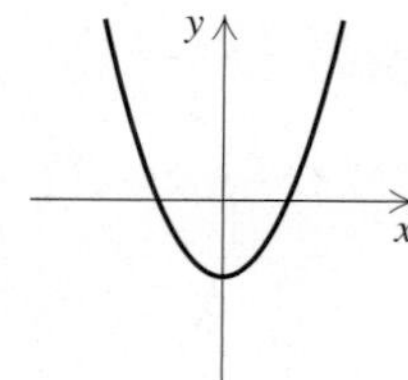

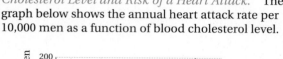

e *Cholesterol Level and Risk of a Heart Attack.* The graph below shows the annual heart attack rate per 10,000 men as a function of blood cholesterol level.

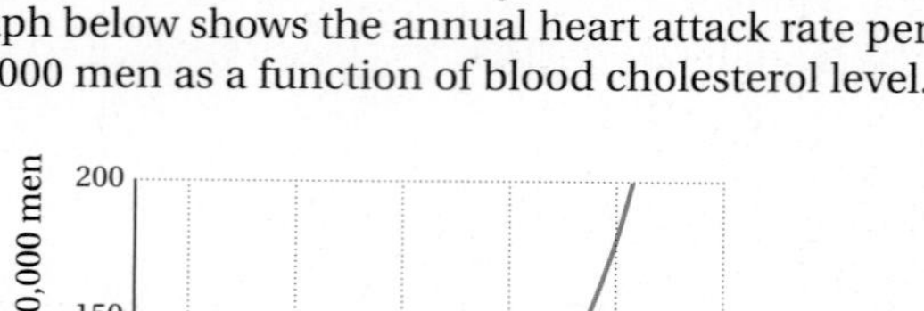

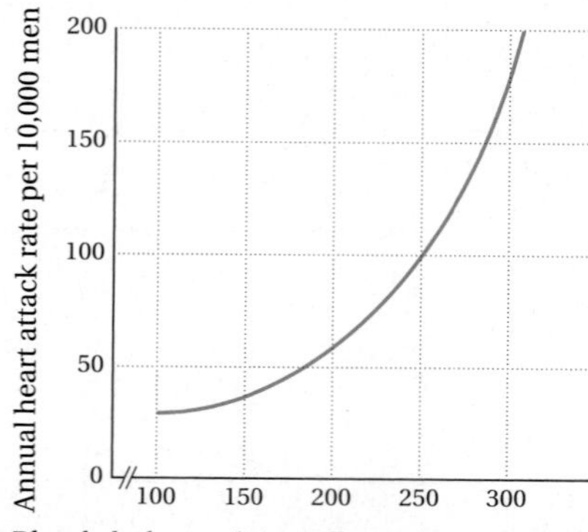

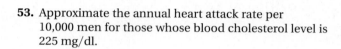

Blood cholesterol (in milligrams per deciliter)

Source: Copyright 1989, CSPI. Adapted from *Nutrition Action Healthletter* (1875 Connecticut Avenue, N.W., Suite 300, Washington, DC 20009-5728)

53. Approximate the annual heart attack rate per 10,000 men for those whose blood cholesterol level is 225 mg/dl.

54. Approximate the annual heart attack rate per 10,000 men for those whose blood cholesterol level is 275 mg/dl.

55. D_W Is it possible for a function to have more numbers as outputs than as inputs? Why or why not?

56. D_W Look up the word "function" in a dictionary. Explain how that definition might be related to the mathematical one given in this section.

SKILL MAINTENANCE

Determine whether the pair of equations represents parallel lines. [7.3a]

57. $y = \frac{3}{4}x - 7,$
$3x + 4y = 7$

58. $y = \frac{3}{5},$
$y = -\frac{5}{3}$

Solve the system using the substitution method. [8.2b]

59. $2x - y = 6,$
$4x - 2y = 5$

60. $x - 3y = 2,$
$3x - 9y = 6$

SYNTHESIS

Graph.

61. $g(x) = x^3$

62. $f(x) = 2 + \sqrt{x}$

63. $f(x) = |x| + x$

64. $g(x) = |x| - x$

CHAPTER 10

Summary and Review

The review that follows is meant to prepare you for a chapter exam. It consists of two parts. The first part is a checklist of some of the Study Tips referred to in this and preceding chapters, as well as a list of important properties and formulas. The second part is the Review Exercises. These provide practice exercises for the exam, together with references to section objectives so you can go back and review. Before beginning, stop and look back over the skills you have obtained. What skills in mathematics do you have now that you did not have before studying this chapter?

STUDY TIPS CHECKLIST

The foundation of all your study skills is TIME!

- ☐ Have you begun to prepare for the final exam?
- ☐ Have you learned how to use MyMathLab? It can help in preparing for the final exam.
- ☐ Are you practicing the five-step problem-solving strategy?
- ☐ Are you using the tutoring resources on campus?
- ☐ Have you found someone with whom to study for the final exam?

IMPORTANT PROPERTIES AND FORMULAS

Standard Form: $ax^2 + bx + c = 0, a > 0$

Principle of Square Roots: The equation $x^2 = d$, where $d > 0$, has two solutions, $\sqrt{d}$ and $-\sqrt{d}$. The solution of $x^2 = 0$ is 0.

Quadratic Formula: $x = \dfrac{-b \pm \sqrt{b^2 - 4ac}}{2a}$

Discriminant: $b^2 - 4ac$

The x-coordinate of the vertex of a parabola $= -\dfrac{b}{2a}$.

REVIEW EXERCISES

Solve.

1. $8x^2 = 24$ [10.2a]

2. $40 = 5y^2$ [10.2a]

3. $5x^2 - 8x + 3 = 0$ [10.1c]

4. $3y^2 + 5y = 2$ [10.1c]

5. $(x + 8)^2 = 13$ [10.2b]

6. $9x^2 = 0$ [10.2a]

7. $5t^2 - 7t = 0$ [10.1b]

Solve. [10.3a]

8. $x^2 - 2x - 10 = 0$

9. $9x^2 - 6x - 9 = 0$

10. $x^2 + 6x = 9$

11. $1 + 4x^2 = 8x$

12. $6 + 3y = y^2$

13. $3m = 4 + 5m^2$

14. $3x^2 = 4x$

Solve. [10.1c]

15. $\frac{15}{x} - \frac{15}{x + 2} = 2$

16. $x + \frac{1}{x} = 2$

Solve by completing the square. Show your work. [10.2c]

17. $x^2 - 5x + 2 = 0$

18. $3x^2 - 2x - 5 = 0$

Approximate the solutions to the nearest tenth. [10.3b]

19. $x^2 - 5x + 2 = 0$

20. $4y^2 + 8y + 1 = 0$

21. Solve for T: $V = \frac{1}{2}\sqrt{1 + \frac{T}{L}}$. [10.4a]

Graph the quadratic equation. Label the ordered pairs for the vertex and the y-intercept. [10.6a]

22. $y = 2 - x^2$

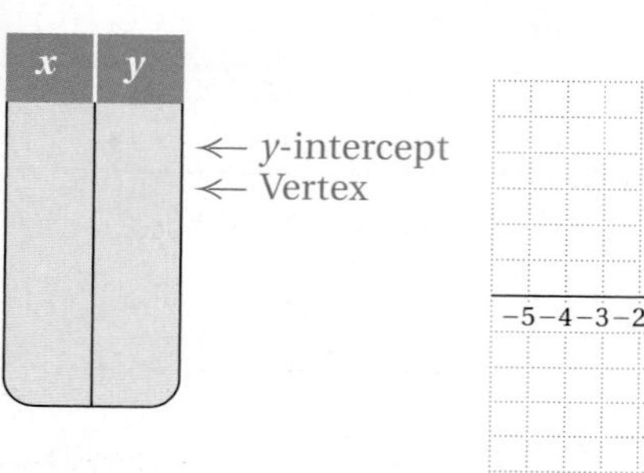

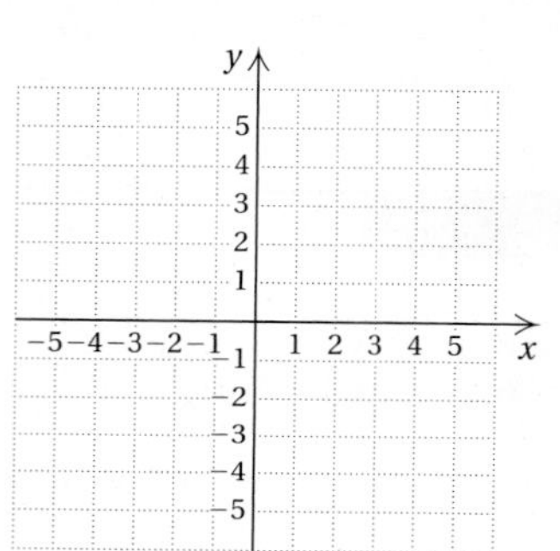

23. $y = x^2 - 4x - 2$

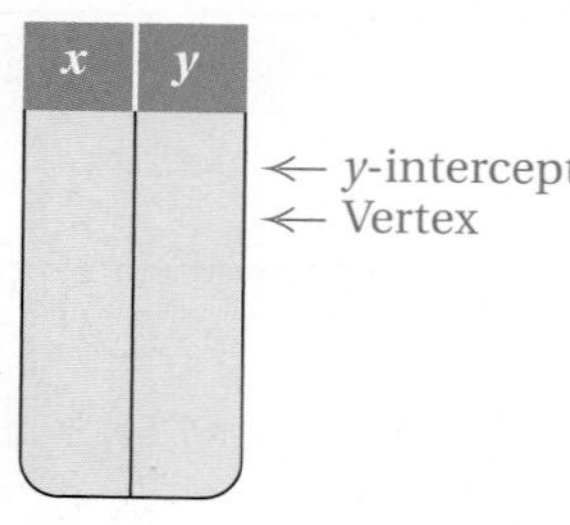

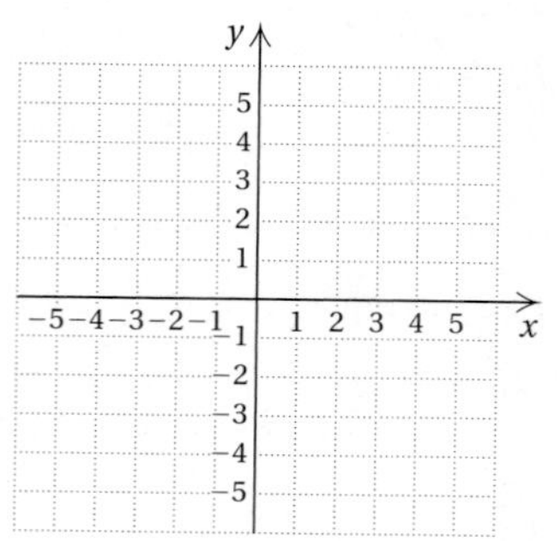

Find the x-intercepts. [10.6b]

24. $y = 2 - x^2$

25. $y = x^2 - 4x - 2$

Solve.

26. *Right-Triangle Dimensions.* The hypotenuse of a right triangle is 5 cm long. One leg is 3 cm longer than the other. Find the lengths of the legs. Round to the nearest tenth. [10.5a]

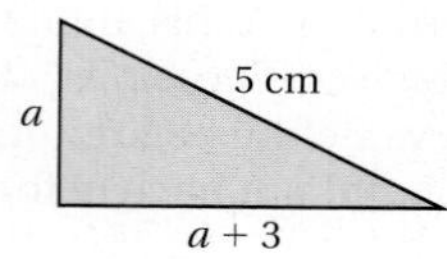

27. *Freight Ramp.* The hypotenuse of a right triangular freight ramp is 26 yd long. One leg is 14 yd longer than the other. Find the lengths of the legs. [10.5a]

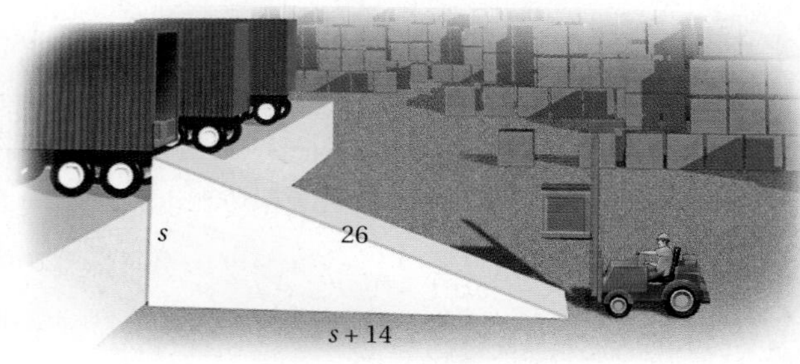

28. *Lake Point Towers.* The height of Lake Point Towers in Chicago is 645 ft. How long would it take an object to fall to the ground from the top? [10.2d]

Find the function values. [10.7b]

29. If $f(x) = 2x - 5$, find $f(2)$, $f(-1)$, and $f(3.5)$.

30. If $g(x) = |x| - 1$, find $g(1)$, $g(-1)$, and $g(-20)$.

31. *Caloric Needs.* If you are moderately active, you need to consume each day about 15 calories per pound of body weight. The function $C(p) = 15p$ approximates the number of calories C that are needed to maintain body weight p, in pounds. How many calories are needed to maintain a body weight of 180 lb? [10.7e]

Graph the function. [10.7c]

32. $g(x) = 4 - x$

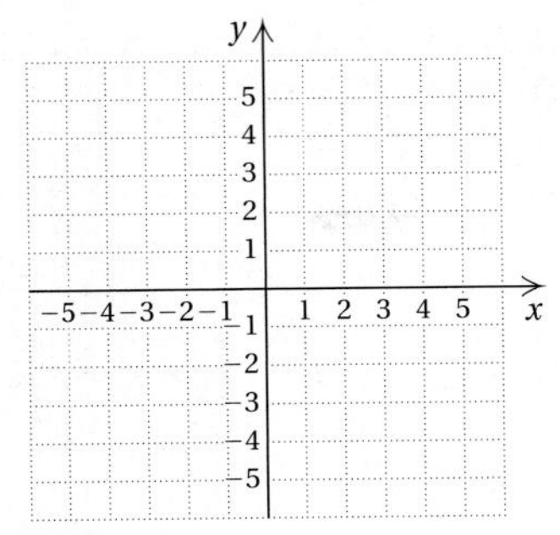

33. $f(x) = x^2 - 3$

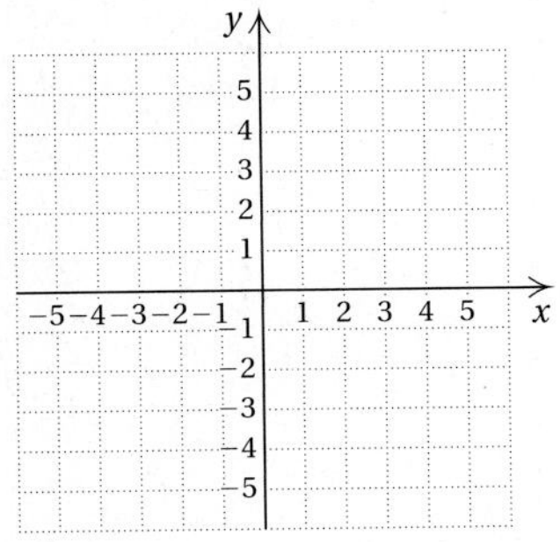

34. $h(x) = |x| - 5$

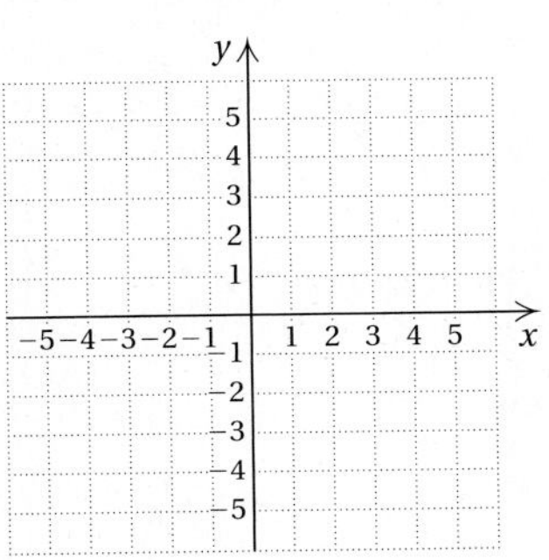

Determine whether each of the following is the graph of a function. [10.7d]

35.

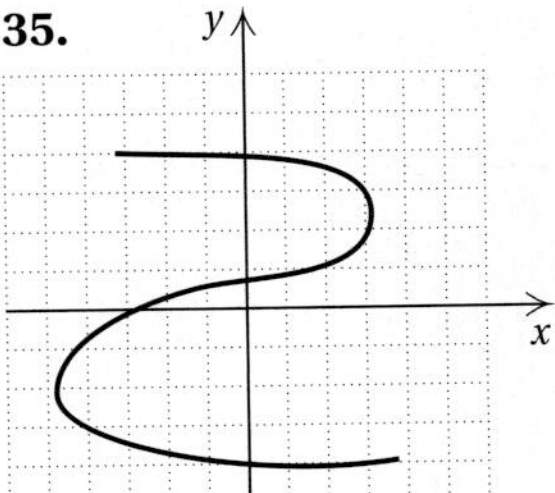

36.

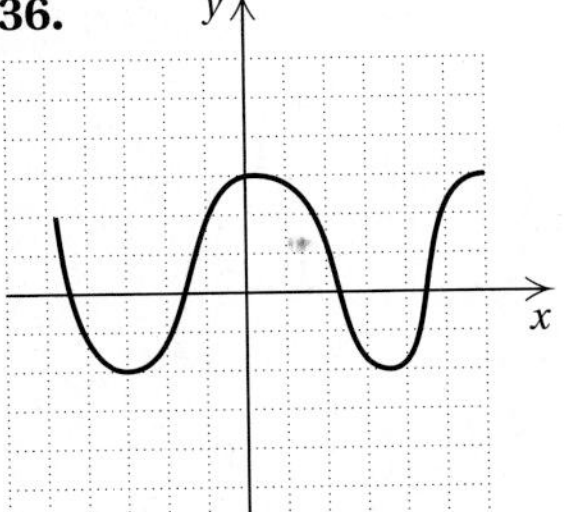

37. ^{D}W List the names and give an example of as many types of equation as you can that you have learned to solve in this text. [2.1a], [6.6a], [8.1a], [9.5a], [10.1a]

38. ^{D}W Find the errors in each of the following solutions of equations. [10.1b]

a)
$$x^2 + 20x = 0$$
$$x(x + 20) = 0$$
$$x + 20 = 0$$
$$x = 20$$

b)
$$x^2 + x = 6$$
$$x(x + 1) = 6$$
$$x = 6 \quad \text{or} \quad x + 1 = 6$$
$$x = 6 \quad \text{or} \quad x = 5$$

SKILL MAINTENANCE

Certain objectives from four particular sections will be retested on the chapter test. The objectives are listed with the practice problems that follow.

Multiply and simplify. [9.2c]

39. $\sqrt{18a}\,\sqrt{2}$

40. $\sqrt{12xy^2}\,\sqrt{5xy}$

41. Find an equation of variation in which y varies inversely as x and $y = 10$ when $x = 0.0625$. Then find the value of y when $x = 200$. [7.5c]

42. The sides of a rectangle are of lengths 1 and $\sqrt{2}$. Find the length of a diagonal. [9.6b]

Add or subtract. [9.4a]

43. $5\sqrt{11} + 7\sqrt{11}$

44. $2\sqrt{90} - \sqrt{40}$

SYNTHESIS

45. Two consecutive integers have squares that differ by 63. Find the integers. [10.5a]

46. A square with sides of length s has the same area as a circle with a radius of 5 in. Find s. [10.5a]

47. Solve: $x - 4\sqrt{x} - 5 = 0$. [10.1c]

Use the graph of

$$y = (x + 3)^2$$

to solve each equation. [10.6b]

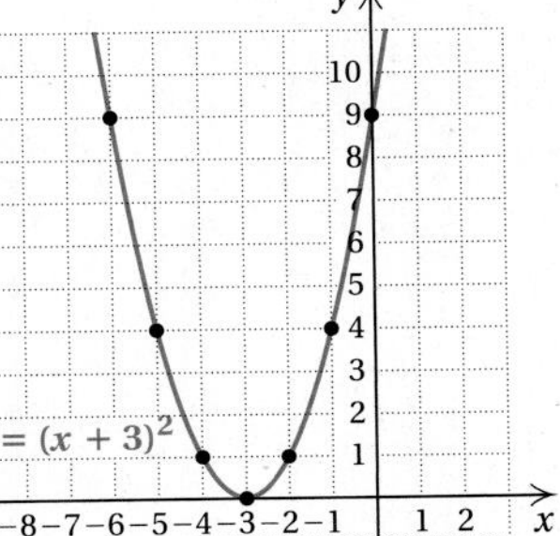

48. $(x + 3)^2 = 1$

49. $(x + 3)^2 = 4$

50. $(x + 3)^2 = 9$

51. $(x + 3)^2 = 0$

Chapter Test

Solve.

1. $7x^2 = 35$

2. $7x^2 + 8x = 0$

3. $48 = t^2 + 2t$

4. $3y^2 - 5y = 2$

5. $(x - 8)^2 = 13$

6. $x^2 = x + 3$

7. $m^2 - 3m = 7$

8. $10 = 4x + x^2$

9. $3x^2 - 7x + 1 = 0$

10. $x - \frac{2}{x} = 1$

11. $\frac{4}{x} - \frac{4}{x + 2} = 1$

12. Solve $x^2 - 4x - 10 = 0$ by completing the square. Show your work.

13. Approximate the solutions to $x^2 - 4x - 10 = 0$ to the nearest tenth.

14. Solve for n: $d = an^2 + bn$.

15. Find the x-intercepts: $y = -x^2 + x + 5$.

Graph. Label the ordered pairs for the vertex and the y-intercept.

16. $y = 4 - x^2$

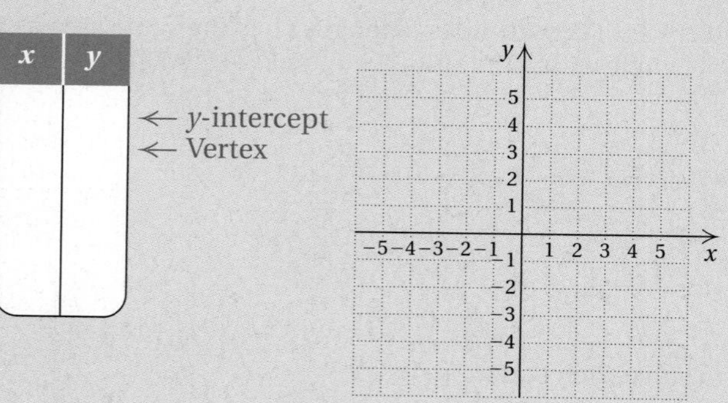

17. $y = -x^2 + x + 5$

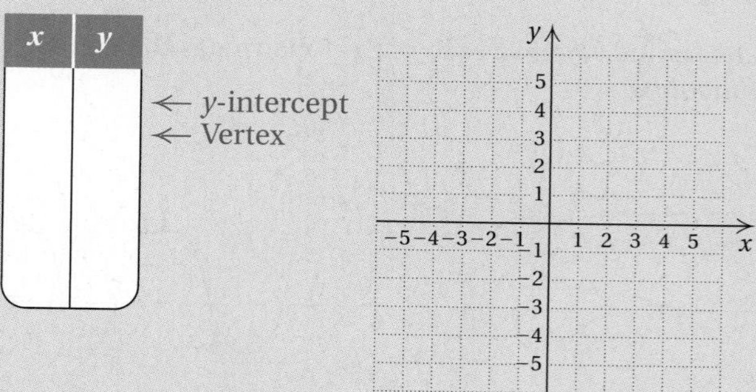

18. If $f(x) = \frac{1}{2}x + 1$, find $f(0)$, $f(1)$, and $f(2)$.

19. If $g(t) = -2|t| + 3$, find $g(-1)$, $g(0)$, and $g(3)$.

Solve.

20. *Rug Dimensions.* The width of a rectangular area rug is 4 m less than the length. The area is 16.25 m^2. Find the length and the width.

21. *Boat Speed.* The current in a stream moves at a speed of 2 km/h. A boat travels 44 km upstream and 52 km downstream in a total of 4 hr. What is the speed of the boat in still water?

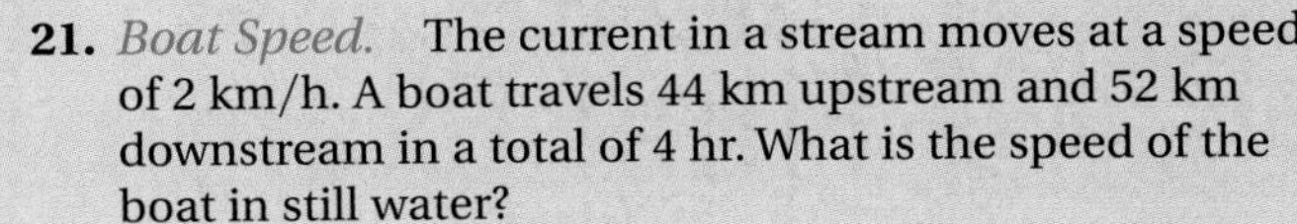

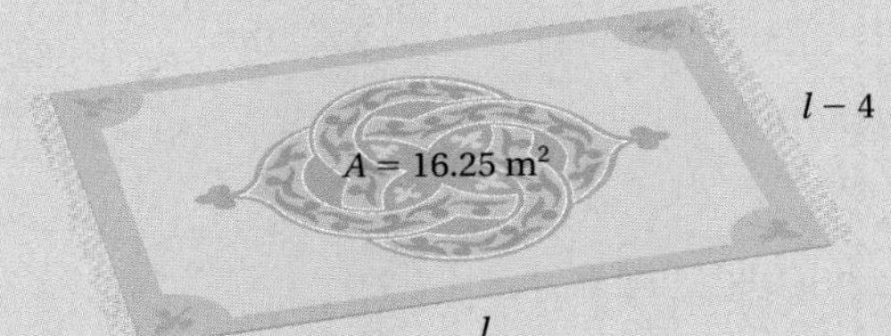

22. *World Record for 10,000-m Run.* The world record for the 10,000-m run has been decreasing steadily since 1940. The record is approximately 30.18 min minus 0.06 times the number of years since 1940. The function $R(t) = 30.18 - 0.06t$ estimates the record R, in minutes, as a function of t, the time in years since 1940. Predict what the record will be in 2010.

Graph.

23. $h(x) = x - 4$

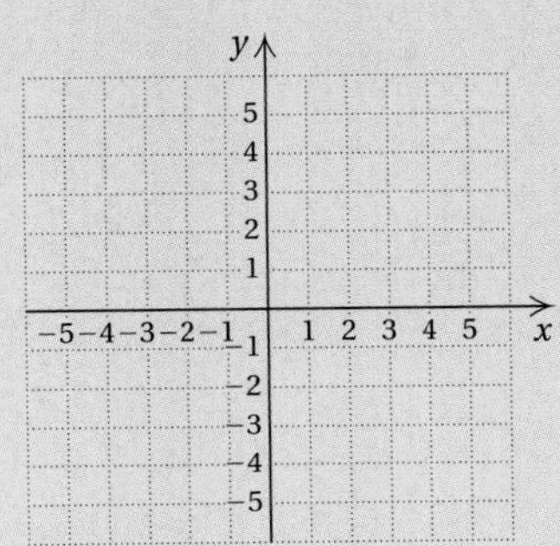

24. $g(x) = x^2 - 4$

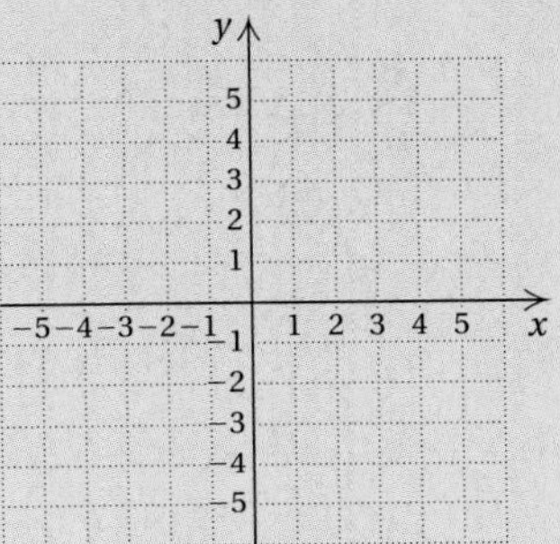

Determine whether each of the following is the graph of a function.

25.

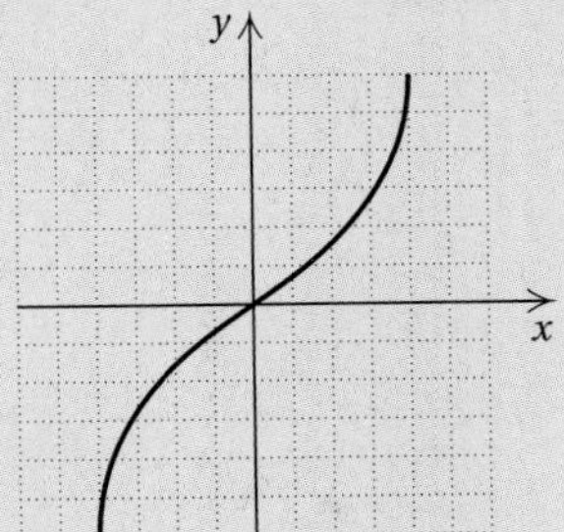

26.

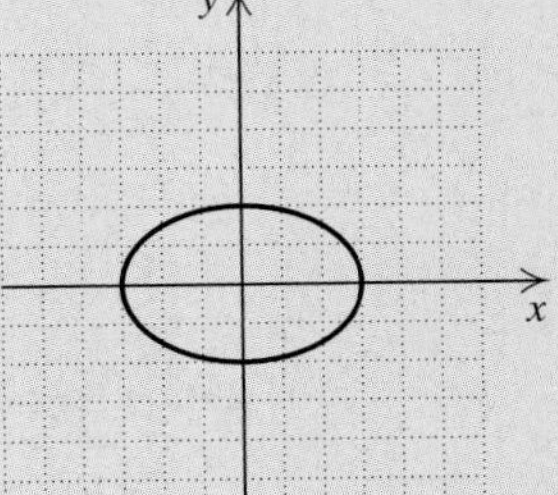

SKILL MAINTENANCE

27. Subtract: $\sqrt{240} - \sqrt{60}$.

28. Multiply and simplify: $\sqrt{7xy}\,\sqrt{14x^2y}$.

29. Find an equation of variation in which y varies inversely as x and $y = 32$ when $x = 0.125$. Find the value of y when $x = 16$.

30. The sides of a rectangle are of lengths $\sqrt{2}$ and $\sqrt{3}$. Find the length of a diagonal.

SYNTHESIS

31. Find the side of a square whose diagonal is 5 ft longer than a side.

32. Solve this system for x. Use the substitution method.

$$x - y = 2,$$
$$xy = 4$$

CHAPTERS 1–10

Cumulative Review/ Final Examination

1. What is the meaning of x^3?

2. Evaluate $(x - 3)^2 + 5$ when $x = 10$.

3. Find decimal notation: $-\frac{3}{11}$.

4. Find the LCM of 15 and 48.

5. Find the absolute value: $|-7|$.

Compute and simplify.

6. $-6 + 12 + (-4) + 7$

7. $2.8 - (-12.2)$

8. $-\frac{3}{8} \div \frac{5}{2}$

9. $13 \cdot 6 \div 3 \cdot 2 \div 13$

10. Remove parentheses and simplify: $4m + 9 - (6m + 13)$.

Solve.

11. $3x = -24$

12. $3x + 7 = 2x - 5$

13. $3(y - 1) - 2(y + 2) = 0$

14. $x^2 - 8x + 15 = 0$

15. $y - x = 1,$
$y = 3 - x$

16. $x + y = 17,$
$x - y = 7$

17. $4x - 3y = 3,$
$3x - 2y = 4$

18. $x^2 - x - 6 = 0$

19. $x^2 + 3x = 5$

20. $3 - x = \sqrt{x^2 - 3}$

21. $5 - 9x \leq 19 + 5x$

22. $-\frac{7}{8}x + 7 = \frac{3}{8}x - 3$

23. $0.6x - 1.8 = 1.2x$

24. $-3x > 24$

25. $23 - 19y - 3y \geq -12$

26. $3y^2 = 30$

27. $(x - 3)^2 = 6$

28. $\frac{6x - 2}{2x - 1} = \frac{9x}{3x + 1}$

29. $\frac{2x}{x + 1} = 2 - \frac{5}{2x}$

30. $\frac{2x}{x + 3} + \frac{6}{x} + 7 = \frac{18}{x^2 + 3x}$

31. $\sqrt{x + 9} = \sqrt{2x - 3}$

Solve the formula for the given letter.

32. $A = \frac{4b}{t}$, for b

33. $\frac{1}{t} = \frac{1}{m} - \frac{1}{n}$, for m

34. $r = \sqrt{\frac{A}{\pi}}$, for A

35. $y = ax^2 - bx$, for x

Simplify.

36. $x^{-6} \cdot x^2$

37. $\dfrac{y^3}{y^{-4}}$

38. $(2y^6)^2$

39. Collect like terms and arrange in descending order: $2x - 3 + 5x^3 - 2x^3 + 7x^3 + x$.

Compute and simplify.

40. $(4x^3 + 3x^2 - 5) + (3x^3 - 5x^2 + 4x - 12)$

41. $(6x^2 - 4x + 1) - (-2x^2 + 7)$

42. $-2y^2(4y^2 - 3y + 1)$

43. $(2t - 3)(3t^2 - 4t + 2)$

44. $\left(t - \dfrac{1}{4}\right)\left(t + \dfrac{1}{4}\right)$

45. $(3m - 2)^2$

46. $(15x^2y^3 + 10xy^2 + 5) - (5xy^2 - x^2y^2 - 2)$

47. $(x^2 - 0.2y)(x^2 + 0.2y)$

48. $(3p + 4q^2)^2$

49. $\dfrac{4}{2x - 6} \cdot \dfrac{x - 3}{x + 3}$

50. $\dfrac{3a^4}{a^2 - 1} \div \dfrac{2a^3}{a^2 - 2a + 1}$

51. $\dfrac{3}{3x - 1} + \dfrac{4}{5x}$

52. $\dfrac{2}{x^2 - 16} - \dfrac{x - 3}{x^2 - 9x + 20}$

Factor.

53. $8x^2 - 4x$

54. $25x^2 - 4$

55. $6y^2 - 5y - 6$

56. $m^2 - 8m + 16$

57. $x^3 - 8x^2 - 5x + 40$

58. $3a^4 + 6a^2 - 72$

59. $16x^4 - 1$

60. $49a^2b^2 - 4$

61. $9x^2 + 30xy + 25y^2$

62. $2ac - 6ab - 3db + dc$

63. $15x^2 + 14xy - 8y^2$

Simplify.

64. $\dfrac{\dfrac{3}{x} + \dfrac{1}{2x}}{\dfrac{1}{3x} - \dfrac{3}{4x}}$

65. $\sqrt{49}$

66. $-\sqrt{625}$

67. $\sqrt{64x^2}$

68. Multiply: $\sqrt{a+b}\,\sqrt{a-b}$.

69. Multiply and simplify: $\sqrt{32ab}\,\sqrt{6a^4b^2}$.

Simplify.

70. $\sqrt{150}$

71. $\sqrt{243x^3y^2}$

72. $\sqrt{\dfrac{100}{81}}$

73. $\sqrt{\dfrac{64}{x^2}}$

74. $4\sqrt{12} + 2\sqrt{12}$

75. Divide and simplify: $\dfrac{\sqrt{72}}{\sqrt{45}}$.

76. In a right triangle, $a = 9$ and $c = 41$. Find b.

Graph.

77. $y = \dfrac{1}{3}x - 2$

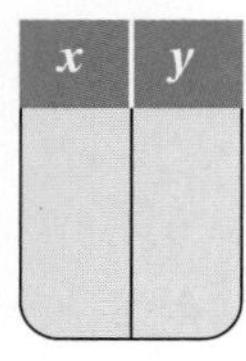

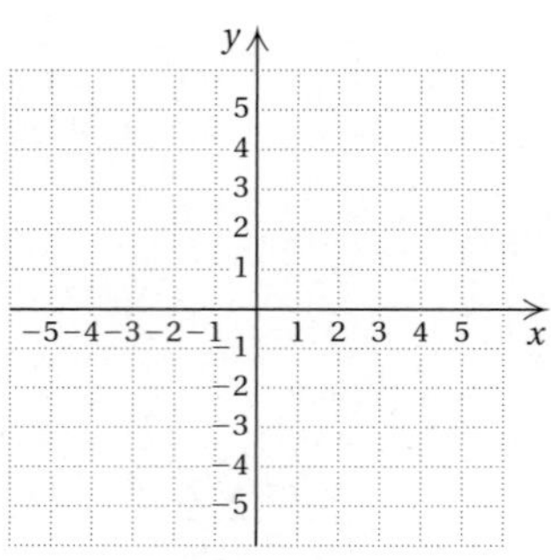

78. $2x + 3y = -6$

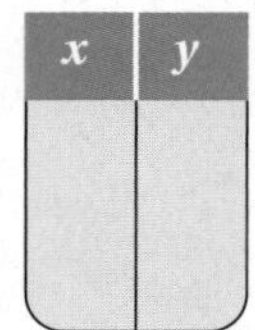

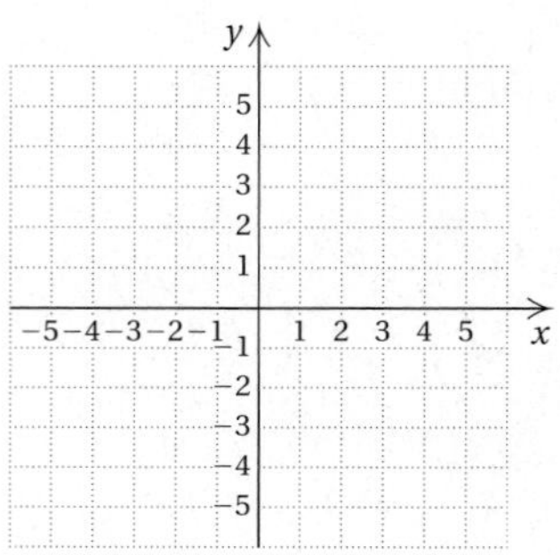

79. $y = -3$

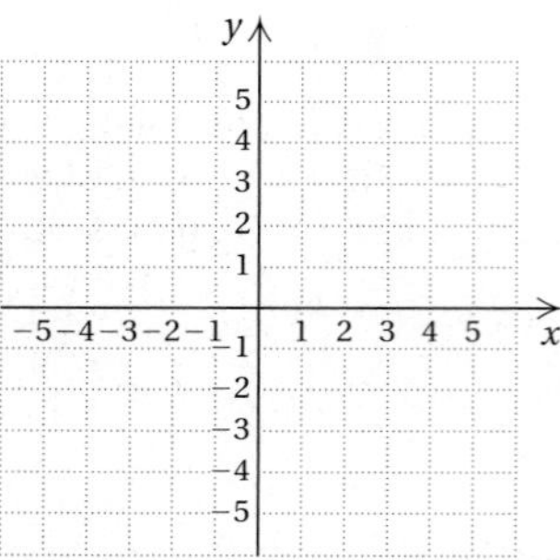

80. $x \geq -3$

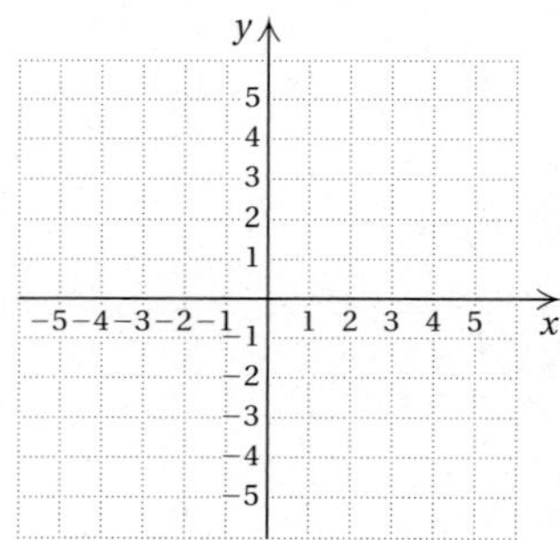

81. $4x - 3y > 12$

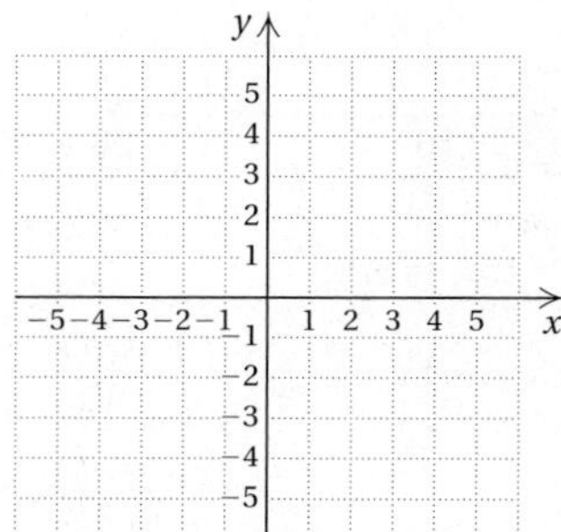

82. Graph: $y = x^2 + 2x + 1$. Label the vertex and the y-intercept.

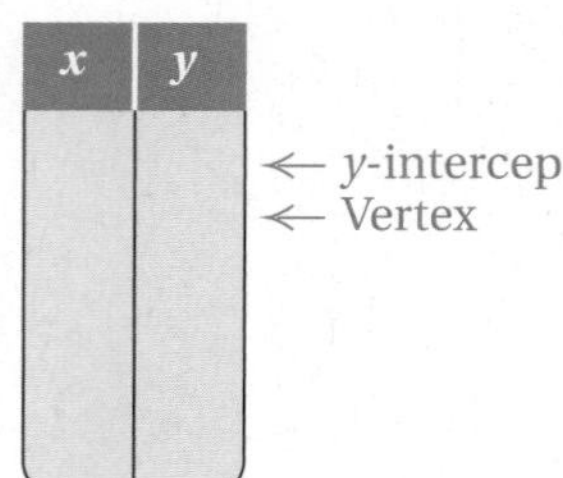

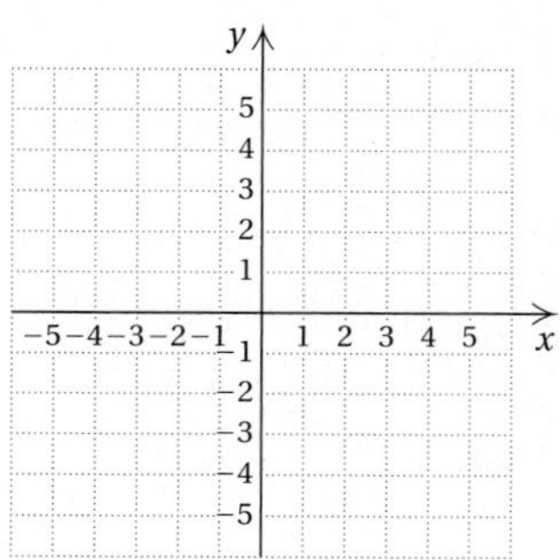

83. Solve $9x^2 - 12x - 2 = 0$ by completing the square. Show your work.

84. Approximate the solutions of $4x^2 = 4x + 1$ to the nearest tenth.

Solve.

85. What percent of 52 is 13?

86. 12 is 20% of what?

87. *Work Time.* In checking records, a contractor finds that crew A can resurface a tennis court in 8 hr. Crew B can do the same job in 10 hr. How long would they take if they worked together?

88. *Movie Screen.* The area of a rectangular movie screen is 96 ft^2. The length is 4 ft longer than the width. Find the length and the width of the movie screen.

89. *Speed of a Stream.* The speed of a boat in still water is 8 km/h. It travels 60 km upstream and 60 km downstream in a total time of 16 hr. What is the speed of the stream?

90. *Garage Length.* The length of a rectangular garage floor is 7 m more than the width. The length of a diagonal is 13 m. Find the length of the garage floor.

91. *Consecutive Odd Integers.* The sum of the squares of two consecutive odd integers is 74. Find the integers.

92. *Alcohol Solutions.* Solution A is 75% alcohol and solution B is 50% alcohol. How much of each is needed in order to make 60 L of a solution that is $66\frac{2}{3}$% alcohol?

93. *Eiffel Tower.* The Eiffel Tower in Paris is 984 ft tall. How long would it take an object to fall to the ground from the top?
Source: *The New York Times Almanac*

94. *Paycheck and Hours Worked.* A student's paycheck varies directly as the number of hours worked. The pay was \$242.52 for 43 hr of work. What would the pay be for 80 hr of work? Explain the meaning of the variation constant.

95. *Parking Spaces.* Three-fifths of the automobiles entering the city each morning will be parked in city parking lots. There are 3654 such parking spaces filled each morning. How many cars enter the city each morning?

96. *Candy Mixture.* A candy shop wants to mix nuts worth \$3.30 per pound with another variety worth \$2.40 per pound in order to make 42 lb of a mixture worth \$2.70 per pound. How many pounds of each kind of nuts should be used?

97. *Air Travel.* An airplane flew for 6 hr with a 10-km/h tail wind. The return flight against the same wind took 8 hr. Find the speed of the plane in still air.

98. Use *only* the graph below to solve $x^2 + x - 6 = 0$.

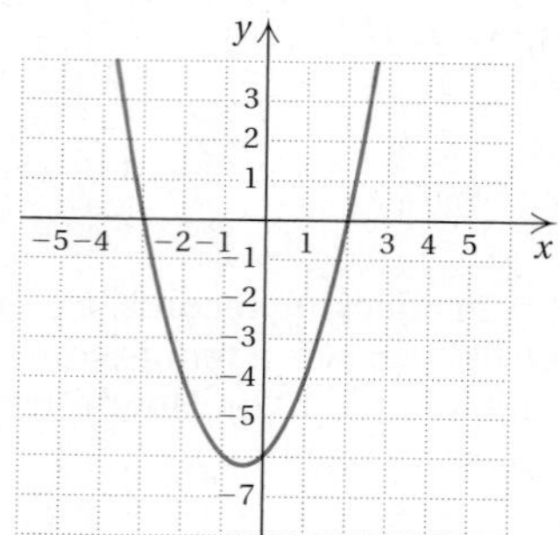

99. Find the x-intercepts of $y = x^2 + 4x + 1$.

100. Find the slope and the y-intercept:

$$-6x + 3y = -24.$$

101. Determine whether the graphs of the following equations are parallel, perpendicular, or neither.

$$y - x = 4,$$
$$3y + x = 8$$

102. Find the slope of the line containing the points $(-5, -6)$ and $(-4, 9)$.

103. Find an equation of variation in which y varies directly as x and $y = 100$ when $x = 10$. Then find the value of y when $x = 64$.

104. Find an equation of variation in which y varies inversely as x and $y = 100$ when $x = 10$. Then find the value of y when $x = 125$.

Determine whether each of the following is the graph of a function.

105.

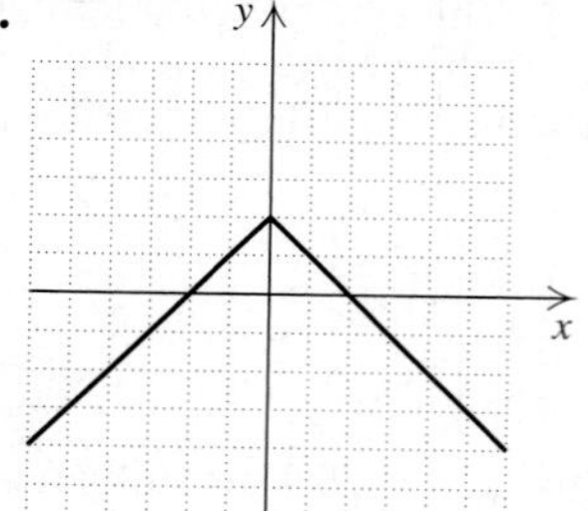

106.

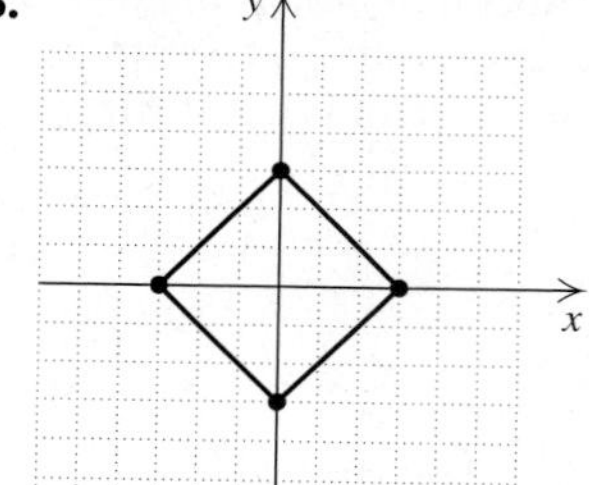

Graph the function.

107. $f(x) = x^2 + x - 2$

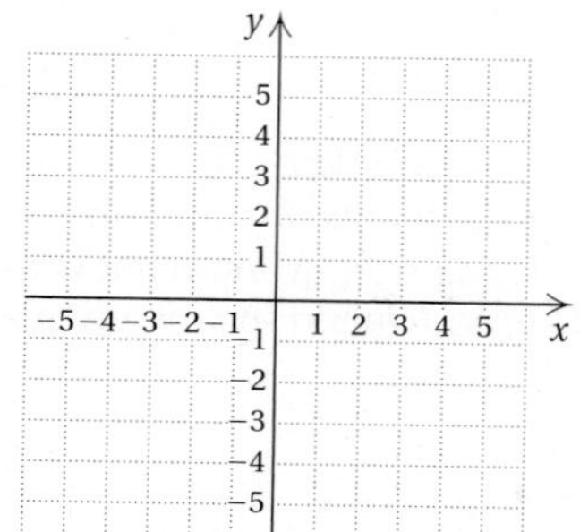

108. $g(x) = |x + 2|$

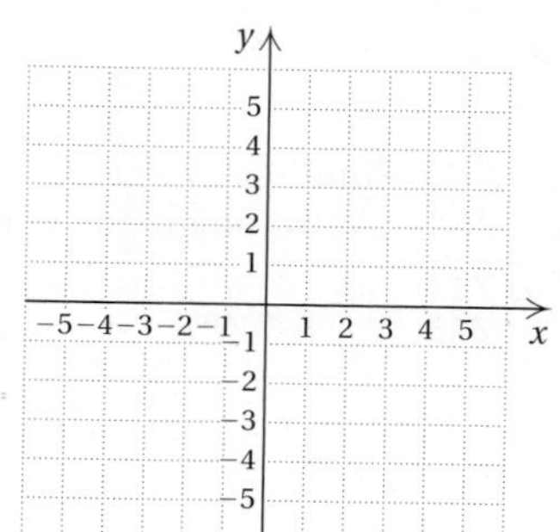

109. For the function f described by $f(x) = 2x^2 + 7x - 4$, find $f(0)$, $f(-4)$, and $f(\frac{1}{2})$.

For each of Questions 110–113, choose the correct answer from the selections given.

110. An airplane flies 408 mi against the wind and 492 mi with the wind in a total time of 3 hr. The speed of the airplane in still air is 300 mph. The speed of the wind is between:

a) 8 and 15 mph. **b)** 15 and 22 mph.
c) 22 and 29 mph. **d)** 29 and 36 mph.
e) None of these

111. Solve: $2x^2 + 6x + 5 = 4$.

a) $-3 \pm \sqrt{7}$ **b)** $-3 \pm 2\sqrt{7}$
c) No real solutions **d)** $\dfrac{-3 \pm \sqrt{7}}{2}$
e) None of these

112. Solve for b: $S = \dfrac{a + b}{3b}$.

a) $b = 3bS - a$ **b)** $b = \dfrac{a + b}{3S}$ **c)** $a = b(3S - 1)$ **d)** $b = \dfrac{a}{3S - 1}$ **e)** None of these

113. Graph: $3x - 4y = 12$.

a)

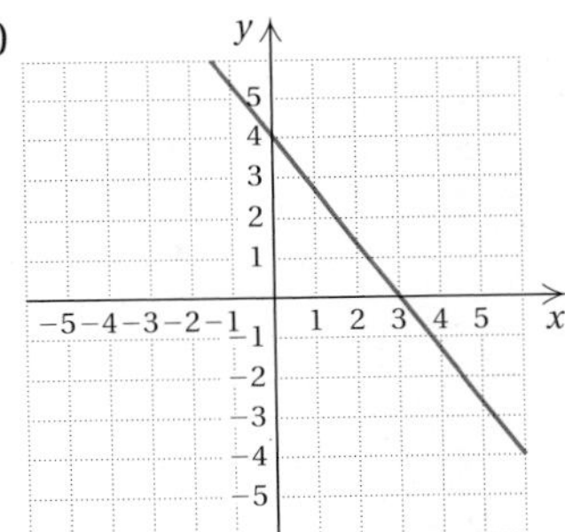

b)

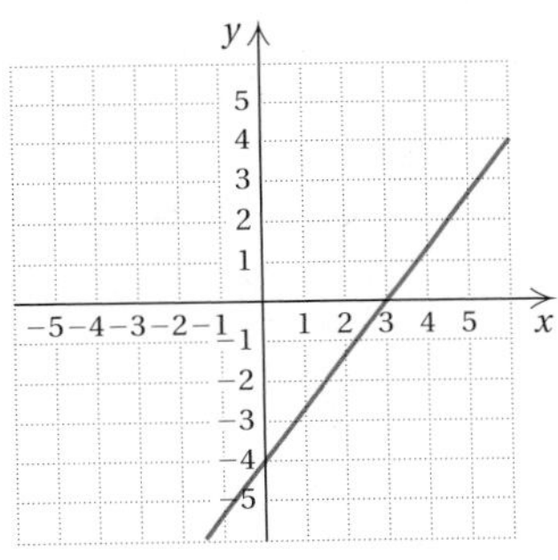

c)

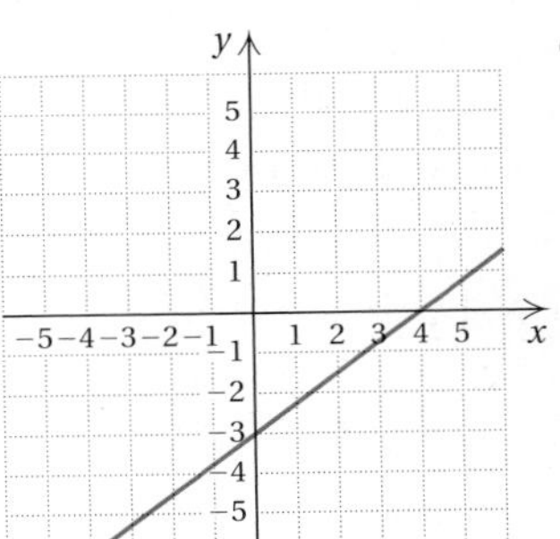

d)

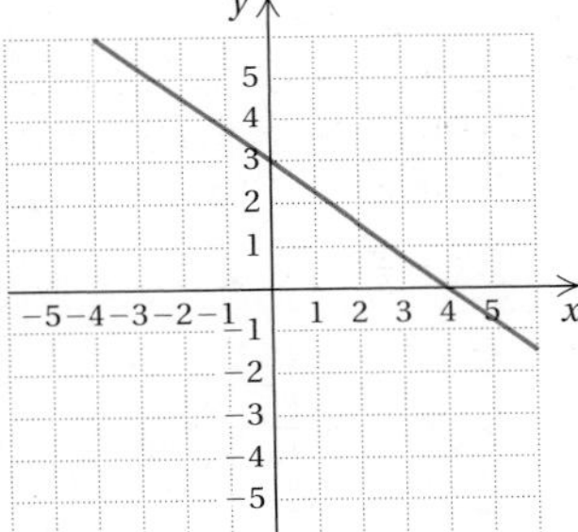

e) None of these

SYNTHESIS

114. Solve: $|x| = 12$.

115. Simplify: $\sqrt{\sqrt{\sqrt{81}}}$.

116. Find b such that the trinomial $x^2 - bx + 225$ is a square.

117. Find x.

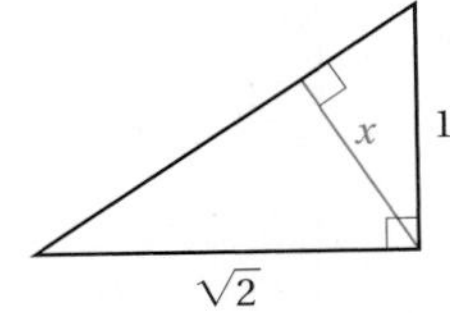

Determine whether the pair of expressions is equivalent.

118. $x^2 - 9$, $(x - 3)(x + 3)$

119. $\dfrac{x + 3}{3}$, x

120. $(x + 5)^2$, $x^2 + 25$

121. $\sqrt{x^2 + 16}$, $x + 4$

122. $\sqrt{x^2}$, $|x|$

Appendixes

A FACTORING SUMS OR DIFFERENCES OF CUBES

B HIGHER ROOTS

C SETS

D MEAN, MEDIAN, AND MODE

A FACTORING SUMS OR DIFFERENCES OF CUBES

Objective

a Factor sums and differences of two cubes.

N	N^3
0.2	0.008
0.1	0.001
0	0
1	1
2	8
3	27
4	64
5	125
6	216
7	343
8	512
9	729
10	1000

Factor.

1. $x^3 - 27$

2. $64 - y^3$

Factor.

3. $y^3 + 8$

4. $125 + t^3$

Answers on page A-51

a Factoring Sums or Differences of Cubes

We can factor the sum or the difference of two expressions that are cubes.
Consider the following products:

$$\begin{aligned}(A + B)(A^2 - AB + B^2) &= A(A^2 - AB + B^2) + B(A^2 - AB + B^2)\\ &= A^3 - A^2B + AB^2 + A^2B - AB^2 + B^3\\ &= A^3 + B^3\end{aligned}$$

and

$$\begin{aligned}(A - B)(A^2 + AB + B^2) &= A(A^2 + AB + B^2) - B(A^2 + AB + B^2)\\ &= A^3 + A^2B + AB^2 - A^2B - AB^2 - B^3\\ &= A^3 - B^3.\end{aligned}$$

The above equations (reversed) show how we can factor a sum or a difference of two cubes.

FACTORING SUMS OR DIFFERENCES OF CUBES

$A^3 + B^3 = (A + B)(A^2 - AB + B^2)$,
$A^3 - B^3 = (A - B)(A^2 + AB + B^2)$

Note that what we are considering here is a sum or a difference of cubes. We are not cubing a binomial. For example, $(A + B)^3$ is *not* the same as $A^3 + B^3$. The table of cubes in the margin is helpful.

EXAMPLE 1 Factor: $x^3 - 8$.

We have

$$x^3 - 8 = x^3 - 2^3 = (x - 2)(x^2 + x \cdot 2 + 2^2).$$
$$A^3 - B^3 = (A - B)(A^2 + A\ B + B^2)$$

This tells us that $x^3 - 8 = (x - 2)(x^2 + 2x + 4)$. Note that we cannot factor $x^2 + 2x + 4$. (It is not a trinomial square nor can it be factored by trial and error or the *ac*-method.) The check is left to the student.

Do Exercises 1 and 2.

EXAMPLE 2 Factor: $x^3 + 125$.

We have

$$x^3 + 125 = x^3 + 5^3 = (x + 5)(x^2 - x \cdot 5 + 5^2).$$
$$A^3 + B^3 = (A + B)(A^2 - A\ B + B^2)$$

Thus, $x^3 + 125 = (x + 5)(x^2 - 5x + 25)$. The check is left to the student.

Do Exercises 3 and 4.

EXAMPLE 3 Factor: $x^3 - 27t^3$.

We have

$$x^3 - 27t^3 = x^3 - (3t)^3 = (x - 3t)(x^2 + x \cdot 3t + (3t)^2)$$

$$A^3 - B^3 = (A - B)(A^2 + A\ B + B^2)$$

$$= (x - 3t)(x^2 + 3xt + 9t^2)$$

Do Exercises 5 and 6.

EXAMPLE 4 Factor: $128y^7 - 250x^6y$.

We first look for a common factor:

$$128y^7 - 250x^6y = 2y(64y^6 - 125x^6) = 2y[(4y^2)^3 - (5x^2)^3]$$
$$= 2y(4y^2 - 5x^2)(16y^4 + 20x^2y^2 + 25x^4).$$

EXAMPLE 5 Factor: $a^6 - b^6$.

We can express this polynomial as a difference of squares:

$$(a^3)^2 - (b^3)^2.$$

We factor as follows:

$$a^6 - b^6 = (a^3 + b^3)(a^3 - b^3).$$

One factor is a sum of two cubes, and the other factor is a difference of two cubes. We factor them:

$$(a + b)(a^2 - ab + b^2)(a - b)(a^2 + ab + b^2).$$

We have now factored completely.

In Example 5, had we thought of factoring first as a difference of two cubes, we would have had

$$(a^2)^3 - (b^2)^3 = (a^2 - b^2)(a^4 + a^2b^2 + b^4)$$
$$= (a + b)(a - b)(a^4 + a^2b^2 + b^4).$$

In this case, we might have missed some factors; $a^4 + a^2b^2 + b^4$ can be factored as $(a^2 - ab + b^2)(a^2 + ab + b^2)$, but we probably would not have known to do such factoring.

EXAMPLE 6 Factor: $64a^6 - 729b^6$.

$$64a^6 - 729b^6 = (8a^3 - 27b^3)(8a^3 + 27b^3)$$ Factoring a difference of squares

$$= [(2a)^3 - (3b)^3][(2a)^3 + (3b)^3].$$

Each factor is a sum or a difference of cubes. We factor each:

$$= (2a - 3b)(4a^2 + 6ab + 9b^2)(2a + 3b)(4a^2 - 6ab + 9b^2)$$

Sum of cubes:	$A^3 + B^3 = (A + B)(A^2 - AB + B^2)$;
Difference of cubes:	$A^3 - B^3 = (A - B)(A^2 + AB + B^2)$;
Difference of squares:	$A^2 - B^2 = (A + B)(A - B)$;
Sum of squares:	$A^2 + B^2$ cannot be factored using real numbers if the largest common factor has been removed.

Do Exercises 7–10.

Factor.

5. $27x^3 - y^3$

6. $8y^3 + z^3$

Factor.

7. $m^6 - n^6$

8. $16x^7y + 54xy^7$

9. $729x^6 - 64y^6$

10. $x^3 - 0.027$

Answers on page A-51

A EXERCISE SET

a Factor.

1. $z^3 + 27$

2. $a^3 + 8$

3. $x^3 - 1$

4. $c^3 - 64$

5. $y^3 + 125$

6. $x^3 + 1$

7. $8a^3 + 1$

8. $27x^3 + 1$

9. $y^3 - 8$

10. $p^3 - 27$

11. $8 - 27b^3$

12. $64 - 125x^3$

13. $64y^3 + 1$

14. $125x^3 + 1$

15. $8x^3 + 27$

16. $27y^3 + 64$

17. $a^3 - b^3$

18. $x^3 - y^3$

19. $a^3 + \frac{1}{8}$

20. $b^3 + \frac{1}{27}$

21. $2y^3 - 128$

22. $3z^3 - 3$

23. $24a^3 + 3$

24. $54x^3 + 2$

25. $rs^3 + 64r$

26. $ab^3 + 125a$

27. $5x^3 - 40z^3$

28. $2y^3 - 54z^3$

29. $x^3 + 0.001$

30. $y^3 + 0.125$

31. $64x^6 - 8t^6$

32. $125c^6 - 8d^6$

33. $2y^4 - 128y$

34. $3z^5 - 3z^2$

35. $z^6 - 1$

36. $t^6 + 1$

37. $t^6 + 64y^6$

38. $p^6 - q^6$

SYNTHESIS

Consider these polynomials:

$$(a + b)^3; \quad a^3 + b^3; \quad (a + b)(a^2 - ab + b^2);$$
$$(a + b)(a^2 + ab + b^2); \quad (a + b)(a + b)(a + b).$$

39. Evaluate each polynomial when $a = -2$ and $b = 3$.

40. Evaluate each polynomial when $a = 4$ and $b = -1$.

Factor. Assume that variables in exponents represent natural numbers.

41. $x^{6a} + y^{3b}$

42. $a^3x^3 - b^3y^3$

43. $3x^{3a} + 24y^{3b}$

44. $\frac{8}{27}x^3 + \frac{1}{64}y^3$

45. $\frac{1}{24}x^3y^3 + \frac{1}{3}z^3$

46. $7x^3 - \frac{7}{8}$

47. $(x + y)^3 - x^3$

48. $(1 - x)^3 + (x - 1)^6$

49. $(a + 2)^3 - (a - 2)^3$

50. $y^4 - 8y^3 - y + 8$

B HIGHER ROOTS

Objectives

a Find higher roots of real numbers.

b Simplify radical expressions using the product and quotient rules.

In this appendix, we study *higher* roots, such as cube roots, or fourth roots.

a Higher Roots

Recall that c is a square root of a if $c^2 = a$. A similar definition can be made for *cube roots.*

CUBE ROOT

The number c is the **cube root** of a if $c^3 = a$.

Every real number has exactly *one* real-number cube root. The symbolism $\sqrt[3]{a}$ is used to represent the cube root of a. In the radical $\sqrt[3]{a}$, the number 3 is called the **index** and a is called the **radicand**.

EXAMPLE 1 Find $\sqrt[3]{8}$.

The cube root of 8 is the number whose cube is 8. Since $2^3 = 2 \cdot 2 \cdot 2 = 8$, the cube root of 8 is 2, so $\sqrt[3]{8} = 2$.

EXAMPLE 2 Find $\sqrt[3]{-125}$.

The cube root of -125 is the number whose cube is -125. Since $(-5)^3 = (-5)(-5)(-5) = -125$, the cube root of -125 is -5, so $\sqrt[3]{-125} = -5$.

Do Exercises 1–3.

Positive real numbers always have *two* nth roots (one positive and one negative) when n is even, but we refer to the *positive nth root* of a positive number a as the *nth root* and denote it $\sqrt[n]{a}$. For example, although both -3 and 3 are fourth roots of 81, since $(-3)^4 = 81$ and $3^4 = 81$, 3 is considered to be *the* fourth root of 81. In symbols, $\sqrt[4]{81} = 3$.

***n*TH ROOT**

The number c is the ***n*th root** of a if $c^n = a$.

If n is odd, then there is exactly one real-number nth root of a and $\sqrt[n]{a}$ represents that root.

If n is even and a is positive, then $\sqrt[n]{a}$ represents the nonnegative nth root.

Even roots of negative numbers are not real numbers.

EXAMPLES Find the root of each of the following.

3. $\sqrt[4]{16} = 2$ Since $2^4 = 2 \cdot 2 \cdot 2 \cdot 2 = 16$

4. $\sqrt[4]{-16}$ is not a real number, because it is an even root of a negative number.

---

1. Find $\sqrt[3]{27}$.

2. Find $\sqrt[3]{-8}$.

3. Find $\sqrt[3]{216}$.

Find the root, if it exists, of each of the following.

4. $\sqrt[5]{1}$

5. $\sqrt[5]{-1}$

6. $\sqrt[4]{-81}$

7. $\sqrt[4]{81}$

8. $\sqrt[3]{-216}$

9. $-\sqrt[3]{216}$

Answers on page A-51

5. $\sqrt[5]{32} = 2$ Since $2^5 = 2 \cdot 2 \cdot 2 \cdot 2 \cdot 2 = 32$

6. $\sqrt[5]{-32} = -2$ Since $(-2)^5 = (-2)(-2)(-2)(-2)(-2) = -32$

7. $-\sqrt[3]{64} = -(\sqrt[3]{64})$ This is the opposite of $\sqrt[3]{64}$.
$= -4$ Since $4^3 = 4 \cdot 4 \cdot 4 = 64$

Do Exercises 4–9 on the preceding page.

Some roots occur so frequently that you may want to memorize them.

SQUARE ROOTS		CUBE ROOTS	FOURTH ROOTS	FIFTH ROOTS
$\sqrt{1} = 1$	$\sqrt{4} = 2$	$\sqrt[3]{1} = 1$	$\sqrt[4]{1} = 1$	$\sqrt[5]{1} = 1$
$\sqrt{9} = 3$	$\sqrt{16} = 4$	$\sqrt[3]{8} = 2$	$\sqrt[4]{16} = 2$	$\sqrt[5]{32} = 2$
$\sqrt{25} = 5$	$\sqrt{36} = 6$	$\sqrt[3]{27} = 3$	$\sqrt[4]{81} = 3$	$\sqrt[5]{243} = 3$
$\sqrt{49} = 7$	$\sqrt{64} = 8$	$\sqrt[3]{64} = 4$	$\sqrt[4]{256} = 4$	
$\sqrt{81} = 9$	$\sqrt{100} = 10$	$\sqrt[3]{125} = 5$	$\sqrt[4]{625} = 5$	
$\sqrt{121} = 11$	$\sqrt{144} = 12$	$\sqrt[3]{216} = 6$		

b Products and Quotients Involving Higher Roots

The rules for working with products and quotients of square roots can be extended to products and quotients of nth roots.

THE PRODUCT AND QUOTIENT RULES FOR RADICALS

For any nonnegative real numbers a and b and any index n, $n \geq 2$,

$$\sqrt[n]{AB} = \sqrt[n]{A} \cdot \sqrt[n]{B} \quad \text{and} \quad \sqrt[n]{\frac{A}{B}} = \frac{\sqrt[n]{A}}{\sqrt[n]{B}}.$$

EXAMPLES Simplify.

8. $\sqrt[3]{40} = \sqrt[3]{8 \cdot 5}$ Factoring the radicand. 8 is a perfect cube.
$= \sqrt[3]{8} \cdot \sqrt[3]{5}$ Using the product rule
$= 2\sqrt[3]{5}$

9. $\sqrt[3]{\frac{125}{27}} = \frac{\sqrt[3]{125}}{\sqrt[3]{27}}$ Using the quotient rule
$= \frac{5}{3}$ Simplifying. 125 and 27 are perfect cubes.

10. $\sqrt[4]{1250} = \sqrt[4]{2 \cdot 625}$ Factoring the radicand. 625 is a perfect fourth power.
$= \sqrt[4]{2 \cdot 5 \cdot 5 \cdot 5 \cdot 5}$
$= 5\sqrt[4]{2}$ Simplifying

11. $\sqrt[5]{\frac{2}{243}} = \frac{\sqrt[5]{2}}{\sqrt[5]{243}}$ Using the quotient rule
$= \frac{\sqrt[5]{2}}{3}$ Simplifying. 243 is a perfect fifth power.

Do Exercises 10–13.

Simplify.

10. $\sqrt[3]{24}$

11. $\sqrt[4]{\frac{81}{256}}$

12. $\sqrt[5]{96}$

13. $\sqrt[3]{\frac{4}{125}}$

Answers on page A-51

B EXERCISE SET

a Simplify. If an expression does not represent a real number, state this.

1. $\sqrt[3]{125}$
2. $\sqrt[3]{-27}$
3. $\sqrt[3]{-1000}$
4. $\sqrt[3]{8}$
5. $\sqrt[4]{1}$

6. $-\sqrt[5]{32}$
7. $\sqrt[4]{-256}$
8. $\sqrt[6]{-1}$
9. $-\sqrt[3]{-216}$
10. $\sqrt[3]{-125}$

11. $\sqrt[4]{256}$
12. $-\sqrt[3]{-8}$
13. $\sqrt[4]{10{,}000}$
14. $\sqrt[3]{-64}$
15. $-\sqrt[4]{81}$

16. $-\sqrt[3]{1}$
17. $-\sqrt[4]{-16}$
18. $\sqrt[6]{64}$
19. $-\sqrt[3]{125}$
20. $\sqrt[3]{1000}$

21. $\sqrt[5]{t^5}$
22. $\sqrt[7]{y^7}$
23. $-\sqrt[3]{x^3}$
24. $-\sqrt[9]{a^9}$
25. $\sqrt[3]{64}$

26. $-\sqrt[3]{216}$
27. $\sqrt[3]{-343}$
28. $\sqrt[5]{-243}$
29. $\sqrt[5]{-3125}$
30. $\sqrt[4]{625}$

31. $\sqrt[6]{1{,}000{,}000}$
32. $\sqrt[5]{243}$
33. $-\sqrt[5]{-100{,}000}$
34. $-\sqrt[4]{-10{,}000}$
35. $-\sqrt[3]{343}$

36. $\sqrt[3]{512}$
37. $\sqrt[8]{-1}$
38. $\sqrt[6]{-64}$
39. $\sqrt[5]{3125}$
40. $\sqrt[4]{-625}$

b Simplify.

41. $\sqrt[3]{54}$
42. $\sqrt[5]{64}$
43. $\sqrt[4]{324}$
44. $\sqrt[3]{81}$

45. $\sqrt[3]{\frac{27}{64}}$
46. $\sqrt[3]{\frac{125}{64}}$
47. $\sqrt[4]{512}$
48. $\sqrt[3]{24}$

49. $\sqrt[5]{128}$
50. $\sqrt[4]{112}$
51. $\sqrt[4]{\frac{256}{625}}$
52. $\sqrt[5]{\frac{243}{32}}$

53. $\sqrt[3]{\frac{17}{8}}$
54. $\sqrt[5]{\frac{11}{32}}$
55. $\sqrt[3]{250}$
56. $\sqrt[5]{96}$

57. $\sqrt[5]{486}$
58. $\sqrt[3]{128}$
59. $\sqrt[4]{\frac{13}{81}}$
60. $\sqrt[3]{\frac{10}{27}}$

61. $\sqrt[4]{\frac{7}{16}}$
62. $\sqrt[4]{\frac{27}{256}}$
63. $\sqrt[4]{\frac{16}{625}}$
64. $\sqrt[3]{\frac{216}{27}}$

SYNTHESIS

Simplify. If an expression does not represent a real number, state this.

65. $\sqrt[3]{\sqrt{64}}$
66. $\sqrt{\sqrt[3]{-64}}$
67. $\sqrt[3]{\sqrt[3]{1{,}000{,}000{,}000}}$
68. $\sqrt{-\sqrt[3]{-1}}$

C SETS

Objectives

a Name sets using the roster method.

b Classify statements regarding set membership and subsets as true or false.

c Find the intersection and the union of sets.

a Naming Sets

To name the set of whole numbers less than 6, we can use the **roster method,** as follows: $\{0, 1, 2, 3, 4, 5\}$.

The set of real numbers x such that x is less than 6 cannot be named by listing all its members because there are infinitely many. We name such a set using **set-builder notation,** as follows: $\{x \mid x < 6\}$. This is read "The set of all x such that x is less than 6." See Section 2.7 for more on this notation.

Do Exercises 1 and 2.

Name the set using the roster method.

1. The set of whole numbers 0 through 7

2. $\{x \mid \text{the square of } x \text{ is } 25\}$

b Set Membership and Subsets

The symbol $\in$ means **is a member of** or **belongs to,** or **is an element of.** Thus, $x \in A$ means x is a member of A or x belongs to A or x is an element of A.

EXAMPLE 1 Classify each of the following as true or false.

a) $1 \in \{1, 2, 3\}$

b) $1 \in \{2, 3\}$

c) $4 \in \{x \mid x \text{ is an even whole number}\}$

d) $5 \in \{x \mid x \text{ is an even whole number}\}$

a) Since 1 *is* listed as a member of the set, $1 \in \{1, 2, 3\}$ is true.

b) Since 1 is *not* a member of $\{2, 3\}$, the statement $1 \in \{2, 3\}$ is false.

c) Since 4 *is* an even whole number, $4 \in \{x \mid x \text{ is an even whole number}\}$ is a true statement.

d) Since 5 is *not* even, $5 \in \{x \mid x \text{ is an even whole number}\}$ is false.

Determine whether each of the following is true or false.

3. $8 \in \{x \mid x \text{ is an even whole number}\}$

4. $2 \in \{x \mid x \text{ is a prime number}\}$

Set membership can be illustrated with a diagram, as shown here.

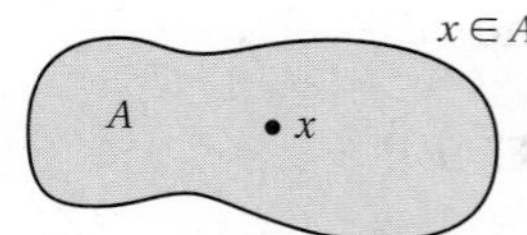

Do Exercises 3 and 4.

If every element of A is an element of B, then A is a **subset** of B. This is denoted $A \subseteq B$. The set of whole numbers is a subset of the set of integers. The set of rational numbers is a subset of the set of real numbers.

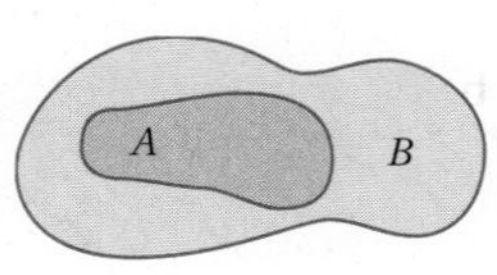

EXAMPLE 2 Classify each of the following as true or false.

a) $\{1, 2\} \subseteq \{1, 2, 3, 4\}$

b) $\{p, q, r, w\} \subseteq \{a, p, r, z\}$

c) $\{x \mid x < 6\} \subseteq \{x \mid x \le 11\}$

a) Since every element of $\{1, 2\}$ is in the set $\{1, 2, 3, 4\}$, the statement $\{1, 2\} \subseteq \{1, 2, 3, 4\}$ is true.

Answers on page A-51

b) Since $q \in \{p, q, r, w\}$, but $q \notin \{a, p, r, z\}$, the statement $\{p, q, r, w\} \subseteq \{a, p, r, z\}$ is false.

c) Since every number that is less than 6 is also less than 11, the statement $\{x|x < 6\} \subseteq \{x|x \leq 11\}$ is true.

Do Exercises 5–7.

Determine whether each of the following is true or false.

5. $\{-2, -3, 4\} \subseteq \{-5, -4, -2, 7, -3, 5, 4\}$

6. $\{a, e, i, o, u\} \subseteq$ The set of all consonants

7. $\{x|x \leq -8\} \subseteq \{x|x \leq -7\}$

C Intersections and Unions

The **intersection** of sets A and B, denoted $A \cap B$, is the set of members that are common to both sets.

EXAMPLE 3 Find the intersection.

a) $\{0, 1, 3, 5, 25\} \cap \{2, 3, 4, 5, 6, 7, 9\}$

b) $\{a, p, q, w\} \cap \{p, q, t\}$

a) $\{0, 1, 3, 5, 25\} \cap \{2, 3, 4, 5, 6, 7, 9\} = \{3, 5\}$

b) $\{a, p, q, w\} \cap \{p, q, t\} = \{p, q\}$

Set intersection can be illustrated with a diagram, as shown here.

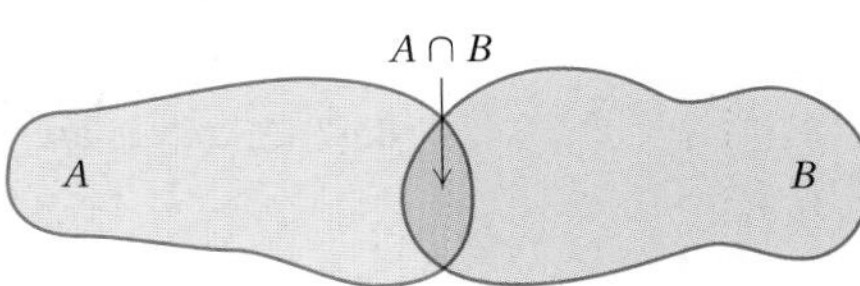

The set without members is known as the **empty set,** and is often named $\varnothing$, and sometimes { }. Each of the following is a description of the empty set:

$\{2, 3\} \cap \{5, 6, 7\}$;

$\{x|x$ is an even natural number$\} \cap \{x|x$ is an odd natural number$\}$.

Do Exercises 8–10.

Find the intersection.

8. $\{-2, -3, 4, -4, 8\} \cap \{-5, -4, -2, 7, -3, 5, 4\}$

9. $\{a, e, i, o, u\} \cap \{m, a, r, v, i, n\}$

10. $\{a, e, i, o, u\} \cap$ The set of all consonants

Two sets A and B can be combined to form a set that contains the members of A as well as those of B. The new set is called the **union** of A and B, denoted $A \cup B$.

EXAMPLE 4 Find the union.

a) $\{0, 5, 7, 13, 27\} \cup \{0, 2, 3, 4, 5\}$

b) $\{a, c, e, g\} \cup \{b, d, f\}$

a) $\{0, 5, 7, 13, 27\} \cup \{0, 2, 3, 4, 5\} = \{0, 2, 3, 4, 5, 7, 13, 27\}$

Note that the 0 and the 5 are *not* listed twice in the solution.

b) $\{a, c, e, g\} \cup \{b, d, f\} = \{a, b, c, d, e, f, g\}$

Set union can be illustrated with a diagram, as shown here.

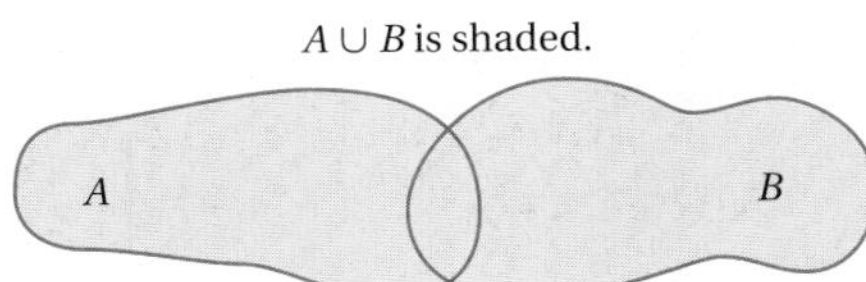

The solution set of the equation $(x - 3)(x + 2) = 0$ is $\{3, -2\}$. This set is the union of the solution sets of $x - 3 = 0$ and $x + 2 = 0$, which are $\{3\}$ and $\{-2\}$.

Do Exercises 11–13.

Find the union.

11. $\{-2, -3, 4, -4, 8\} \cup \{-5, -4, -2, 7, -3, 5, 4\}$

12. $\{a, e, i, o, u\} \cup \{m, a, r, v, i, n\}$

13. $\{a, e, i, o, u\} \cup$ The set of all consonants

Answers on page A-51

C EXERCISE SET

a Name the set using the roster method.

1. The set of whole numbers 3 through 8

2. The set of whole numbers 101 through 107

3. The set of odd numbers between 40 and 50

4. The set of multiples of 5 between 11 and 39

5. $\{x \mid \text{the square of } x \text{ is } 9\}$

6. $\{x \mid x \text{ is the cube of } 0.2\}$

b Classify the statement as true or false.

7. $2 \in \{x \mid x \text{ is an odd number}\}$

8. $7 \in \{x \mid x \text{ is an odd number}\}$

9. Jeff Gordon $\in$ The set of all NASCAR drivers

10. Apple $\in$ The set of all fruit

11. $-3 \in \{-4, -3, 0, 1\}$

12. $0 \in \{-4, -3, 0, 1\}$

13. $\frac{2}{3} \in \{x \mid x \text{ is a rational number}\}$

14. Heads $\in$ The set of outcomes of flipping a penny

15. $\{4, 5, 8,\} \subseteq \{1, 3, 4, 5, 6, 7, 8, 9\}$

16. The set of vowels $\subseteq$ The set of consonants

17. $\{-1, -2, -3, -4, -5\} \subseteq \{-1, 2, 3, 4, 5\}$

18. The set of integers $\subseteq$ The set of rational numbers

c Find the intersection.

19. $\{a, b, c, d, e\} \cap \{c, d, e, f, g\}$

20. $\{a, e, i, o, u\} \cap \{q, u, i, c, k\}$

21. $\{1, 2, 5, 10\} \cap \{0, 1, 7, 10\}$

22. $\{0, 1, 7, 10\} \cap \{0, 1, 2, 5\}$

23. $\{1, 2, 5, 10\} \cap \{3, 4, 7, 8\}$

24. $\{a, e, i, o, u\} \cap \{m, n, f, g, h\}$

d Find the union.

25. $\{a, e, i, o, u\} \cup \{q, u, i, c, k\}$

26. $\{a, b, c, d, e\} \cup \{c, d, e, f, g\}$

27. $\{0, 1, 7, 10\} \cup \{0, 1, 2, 5\}$

28. $\{1, 2, 5, 10\} \cup \{0, 1, 7, 10\}$

29. $\{a, e, i, o, u\} \cup \{m, n, f, g, h\}$

30. $\{1, 2, 5, 10\} \cup \{a, b\}$

SYNTHESIS

31. Find the union of the set of integers and the set of whole numbers.

32. Find the intersection of the set of odd integers and the set of even integers.

33. Find the union of the set of rational numbers and the set of irrational numbers.

34. Find the intersection of the set of even integers and the set of positive rational numbers.

35. Find the intersection of the set of rational numbers and the set of irrational numbers.

36. Find the union of the set of negative integers, the set of positive integers, and the set containing 0.

37. For a set A, find each of the following.
   a) $A \cup \varnothing$
   b) $A \cup A$
   c) $A \cap A$
   d) $A \cap \varnothing$

38. A set is *closed* under an operation if, when the operation is performed on its members, the result is in the set. For example, the set of real numbers is closed under the operation of addition since the sum of any two real numbers is a real number.
   a) Is the set of even numbers closed under addition?
   b) Is the set of odd numbers closed under addition?
   c) Is the set $\{0, 1\}$ closed under addition?
   d) Is the set $\{0, 1\}$ closed under multiplication?
   e) Is the set of real numbers closed under multiplication?
   f) Is the set of integers closed under division?

39. Experiment with sets of various types and determine whether the following distributive law for sets is true:

$$A \cap (B \cup C) = (A \cap B) \cup (A \cap C).$$

D MEAN, MEDIAN, AND MODE

Objective

a Find the mean (average), the median, and the mode of a set of data and solve related applied problems.

Find the mean. Round to the nearest tenth.

1. 28, 103, 39

2. 85, 46, 105.7, 22.1

3. A student scored the following on five tests:

78, 95, 84, 100, 82.

What was the average score?

Answers on page A-51

a Mean, Median, and Mode

One way to analyze data is to look for a single representative number, called a **center point** or **measure of central tendency.** Those most often used are the **mean** (or **average**), the **median,** and the **mode.**

MEAN

Let's first consider the *mean,* or *average.*

MEAN, OR AVERAGE

The **mean,** or **average,** of a set of numbers is the sum of the numbers divided by the number of addends.

EXAMPLE 1 Consider the following data on revenue, in billions of dollars, at McDonald's restaurants in five recent years:

$12.5, $13.2, $14.2, $14.9, $15.9.

What is the mean of the numbers?

Source: McDonalds Corporation

First, we add the numbers:

$$12.5 + 13.2 + 14.2 + 14.9 + 15.9 = 70.7.$$

Then we divide by the number of addends, 5:

$$\frac{(12.5 + 13.2 + 14.2 + 14.9 + 15.9)}{5} = \frac{70.7}{5} = 14.14.$$

The mean, or average, revenue of McDonald's for those five years is $14.14 billion.

Note that $14.14 + 14.14 + 14.14 + 14.14 + 14.14 = 70.7$. If we use this center point, 14.14, repeatedly as the addend, we get the same sum that we do when adding individual data numbers.

Do Exercises 1–3.

MEDIAN

The *median* is useful when we wish to de-emphasize extreme scores. For example, suppose five workers in a technology company manufactured the following number of computers during one day's work:

Sarah:	88	Jen:	94
Matt:	92	Mark:	91
Pat:	66		

Let's first list the scores in order from smallest to largest:

66 88 91 92 94.

↑ Middle number

The middle number—in this case, 91—is the **median.**

MEDIAN

Once a set of data has been arranged from smallest to largest, the **median** of the set of data is the middle number if there is an odd number of data numbers. If there is an even number of data numbers, then there are two middle numbers and the median is the *average* of the two middle numbers.

EXAMPLE 2 What is the median of the following set of yearly salaries?

\$76,000, \$58,000, \$87,000, \$32,500, \$64,800, \$62,500

We first rearrange the numbers in order from smallest to largest.

\$32,500, \$58,000, \$62,500, ↑ \$64,800, \$76,000, \$87,000

Median

There is an even number of numbers. We look for the middle two, which are \$62,500 and \$64,800. In this case, the median is the average of \$62,500 and \$64,800:

$$\frac{\$62{,}500 + \$64{,}800}{2} = \$63{,}650.$$

Do Exercises 4–6.

MODE

The last center point we consider is called the *mode*. A number that occurs most often in a set of data can be considered a representative number or center point.

MODE

The **mode** of a set of data is the number or numbers that occur most often. If each number occurs the same number of times, there is *no* mode.

EXAMPLE 3 Find the mode of the following data:

23, 24, 27, 18, 19, 27

The number that occurs most often is 27. Thus the mode is 27.

EXAMPLE 4 Find the mode of the following data:

83, 84, 84, 84, 85, 86, 87, 87, 87, 88, 89, 90.

There are two numbers that occur most often, 84 and 87. Thus the modes are 84 and 87.

EXAMPLE 5 Find the mode of the following data:

115, 117, 211, 213, 219.

Each number occurs the same number of times. The set of data has *no* mode.

Do Exercises 7–10.

Find the median.

4. 17, 13, 18, 14, 19

5. 17, 18, 16, 19, 13, 14

6. 122, 102, 103, 91, 83, 81, 78, 119, 88

Find any modes that exist.

7. 33, 55, 55, 88, 55

8. 90, 54, 88, 87, 87, 54

9. 23.7, 27.5, 54.9, 17.2, 20.1

10. In conducting laboratory tests, Carole discovers bacteria in different lab dishes grew to the following areas, in square millimeters:

25, 19, 29, 24, 28.

a) What is the mean?

b) What is the median?

c) What is the mode?

Answers on page A-51

D EXERCISE SET

a For each set of numbers, find the mean (average), the median, and any modes that exist.

1. 17, 19, 29, 18, 14, 29

2. 72, 83, 85, 88, 92

3. 5, 37, 20, 20, 35, 5, 25

4. 13, 32, 25, 27, 13

5. 4.3, 7.4, 1.2, 5.7, 7.4

6. 13.4, 13.4, 12.6, 42.9

7. 234, 228, 234, 229, 234, 278

8. \$29.95, \$28.79, \$30.95, \$29.95

9. *Atlantic Storms and Hurricanes.* The following bar graph shows the number of Atlantic storms or hurricanes that formed in various months from 1980 to 2000. What is the average number for the 9 months given? the median? the mode?

Source: Colorado State University

10. *Cheddar Cheese Prices.* The following prices per pound of sharp cheddar cheese were found at five supermarkets:

\$5.99, \$6.79, \$5.99, \$6.99, \$6.79.

What was the average price per pound? the median price? the mode?

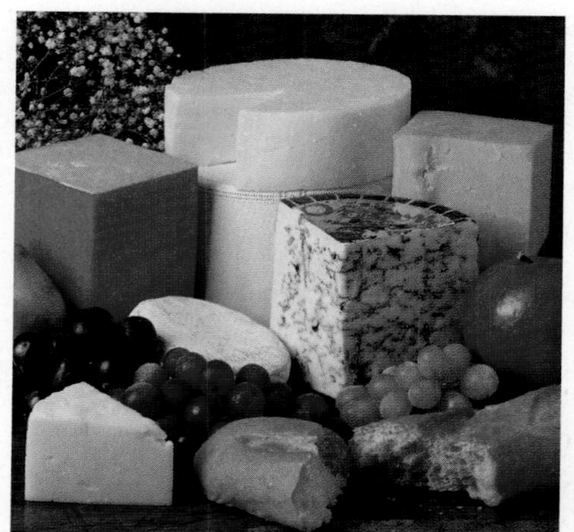

11. *Coffee Consumption.* The following lists the annual coffee consumption, in cups per person, for various countries. Find the mean, the median, and the mode.

Germany	1113
United States	610
Switzerland	1215
France	798
Italy	750

Source: Beverage Marketing Corporation

12. *NBA Tall Men.* The following is a list of the heights, in inches, of the tallest men in the NBA in a recent year. Find the mean, the median, and the mode.

Shaquille O'Neal	85
Gheorghe Muresan	91
Shawn Bradley	90
Priest Lauderdale	88
Rik Smits	88
David Robinson	85
Arvydas Sabonis	87

Source: National Basketball Association

13. *Salmon Prices.* The following prices per pound of Atlantic salmon were found at five fish markets:

$6.99, $8.49, $8.99, $6.99, $9.49.

What was the average price per pound? the median price? the mode?

14. *PBA Scores.* Chris Barnes rolled scores of 224, 224, 254, and 187 in a recent tournament of the Professional Bowlers Association. What was his average? his median? his mode?
Source: Professional Bowlers Association

SYNTHESIS

Grade Point Average. The tables in Exercises 15 and 16 show the grades of a student for one semester. In each case, find the grade point average. Assume that the grade point values are 4.0 for an A, 3.0 for a B, and so on. Round to the nearest tenth.

15.

GRADE	NUMBER OF CREDIT HOURS IN COURSE
B	4
A	5
D	3
C	4

16.

GRADE	NUMBER OF CREDIT HOURS IN COURSE
A	5
C	4
F	3
B	5

17. *Hank Aaron.* Hank Aaron averaged $34\frac{7}{22}$ home runs per year over a 22-yr career. After 21 yr, Aaron had averaged $35\frac{10}{21}$ home runs per year. How many home runs did Aaron hit in his final year?

18. The ordered set of data 18, 21, 24, a, 36, 37, b has a median of 30 and an average of 32. Find a and b.

19. *Length of Pregnancy.* Marta was pregnant 270 days, 259 days, and 272 days for her first three pregnancies. In order for Marta's average length of pregnancy to equal the worldwide average of 266 days, how long must her fourth pregnancy last?
Source: David Crystal (ed.), *The Cambridge Factfinder.* Cambridge CB2 1RP: Cambridge University Press, 1993, p. 84.

20. *Male Height.* Jason's brothers are 174 cm, 180 cm, 179 cm, and 172 cm tall. The average male is 176.5 cm tall. How tall is Jason if he and his brothers have an average height of 176.5 cm?

Answers

Chapter R

Pretest: Chapter R, p. 2

1. [R.1a] $2 \cdot 2 \cdot 2 \cdot 31$ **2.** [R.1b] 168 **3.** [R.2a] $\frac{44}{48}$
4. [R.2b] $\frac{23}{64}$ **5.** [R.2b] $\frac{2}{3}$ **6.** [R.2c] $\frac{11}{10}$ **7.** [R.2c] $\frac{2}{21}$
8. [R.2c] $\frac{1}{2}$ **9.** [R.2c] $\frac{5}{21}$ **10.** [R.3b] 134.0362
11. [R.3b] 212.05 **12.** [R.3b] 350.5824 **13.** [R.3b] 12.4
14. [R.3a] $\frac{3217}{100}$ **15.** [R.3a] 0.0789 **16.** [R.3b] $1.\overline{4}$
17. [R.3c] 345.84 **18.** [R.3c] 345.8 **19.** [R.4a] 0.116
20. [R.4b] $\frac{87}{100}$ **21.** [R.4d] 87.5% **22.** [R.5a] 5^4
23. [R.5b] 8 **24.** [R.5b] 1.21 **25.** [R.5c] 18
26. [R.6b] 100 ft^2 **27.** [R.6a] 131 mm
28. [R.6d] 160 cm^3 **29.** [R.6c] 9.6 m
30. [R.6c] 72.3456 m^2 **31.** [R.6b] 22 cm^2
32. [R.6b] 4 m^2

Margin Exercises, Section R.1, pp. 3–6

1. 1, 3, 9 **2.** 1, 2, 4, 8, 16 **3.** 1, 2, 3, 4, 6, 8, 12, 24
4. 1, 2, 3, 4, 5, 6, 9, 10, 12, 15, 18, 20, 30, 36, 45, 60, 90, 180
5. 13 **6.** $2 \cdot 2 \cdot 2 \cdot 2 \cdot 3$ **7.** $2 \cdot 5 \cdot 5$ **8.** $2 \cdot 5 \cdot 7 \cdot 11$
9. 15, 30, 45, 60, . . . **10.** 45, 90, 135, 180, . . . **11.** 40
12. 54 **13.** 360 **14.** 18 **15.** 24 **16.** 36 **17.** 210

Exercise Set R.1, p. 8

1. 1, 2, 4, 5, 10, 20 **3.** 1, 2, 3, 4, 6, 8, 9, 12, 18, 24, 36, 72
5. $3 \cdot 5$ **7.** $2 \cdot 11$ **9.** $3 \cdot 3$ **11.** $7 \cdot 7$ **13.** $2 \cdot 3 \cdot 3$
15. $2 \cdot 2 \cdot 2 \cdot 5$ **17.** $2 \cdot 3 \cdot 3 \cdot 5$ **19.** $2 \cdot 3 \cdot 5 \cdot 7$
21. $7 \cdot 13$ **23.** $7 \cdot 17$ **25.** $2 \cdot 2$; 5; 20
27. $2 \cdot 2 \cdot 2 \cdot 3$; $2 \cdot 2 \cdot 3 \cdot 3$; 72 **29.** 3; $3 \cdot 5$; 15
31. $2 \cdot 3 \cdot 5$; $2 \cdot 2 \cdot 2 \cdot 5$; 120 **33.** 13; 23; 299
35. $2 \cdot 3 \cdot 3$; $2 \cdot 3 \cdot 5$; 90 **37.** $2 \cdot 3 \cdot 5$; $2 \cdot 2 \cdot 3 \cdot 3$; 180
39. $2 \cdot 2 \cdot 2 \cdot 3$; $2 \cdot 3 \cdot 5$; 120 **41.** 17; 29; 493
43. $2 \cdot 2 \cdot 3$; $2 \cdot 2 \cdot 7$; 84 **45.** 2; 3; 5; 30
47. $2 \cdot 2 \cdot 2 \cdot 3$; $2 \cdot 2 \cdot 3 \cdot 3$; $2 \cdot 2 \cdot 3$; 72
49. 5; $2 \cdot 2 \cdot 3$; $3 \cdot 5$; 60 **51.** $2 \cdot 3$; $2 \cdot 2 \cdot 3$; $2 \cdot 3 \cdot 3$; 36
53. Every 60 yr **55.** Every 420 yr **57.** DW
59. **(a)** No; not a multiple of 8; **(b)** no; not a multiple of 8; **(c)** no; not a multiple of 8 or 12; **(d)** yes; it is a multiple of both 8 and 12 and is the smallest such multiple.
61. 70,200

Margin Exercises, Section R.2, pp. 11–17

1. $\frac{8}{12}$ **2.** $\frac{21}{28}$ **3.** $\frac{14}{16}, \frac{21}{24}, \frac{28}{32}$; answers may vary **4.** $\frac{2}{5}$
5. $\frac{19}{9}$ **6.** $\frac{8}{3}$ **7.** $\frac{1}{2}$ **8.** 4 **9.** $\frac{5}{2}$ **10.** $\frac{35}{16}$ **11.** $\frac{7}{5}$
12. 2 **13.** $\frac{23}{15}$ **14.** $\frac{3}{4}$ **15.** $\frac{19}{40}$ **16.** $\frac{7}{36}$ **17.** $\frac{11}{4}$
18. $\frac{7}{15}$ **19.** $\frac{1}{5}$ **20.** 3 **21.** $\frac{21}{20}$ **22.** $\frac{8}{21}$ **23.** $\frac{5}{6}$
24. $\frac{8}{15}$ **25.** $\frac{1}{64}$ **26.** 81

Calculator Corner, p. 18

1. $\frac{7}{12}$ **2.** $\frac{41}{24}$ **3.** $\frac{11}{10}$ **4.** $\frac{153}{112}$ **5.** $\frac{35}{16}$ **6.** $\frac{4}{3}$ **7.** $\frac{3}{10}$
8. $\frac{35}{12}$

Exercise Set R.2, p. 19

1. $\frac{9}{12}$ **3.** $\frac{60}{100}$ **5.** $\frac{104}{160}$ **7.** $\frac{21}{24}$ **9.** $\frac{20}{16}$ **11.** $\frac{2}{3}$ **13.** 4
15. $\frac{1}{7}$ **17.** 8 **19.** $\frac{1}{4}$ **21.** 5 **23.** $\frac{17}{21}$ **25.** $\frac{13}{7}$
27. $\frac{4}{3}$ **29.** $\frac{1}{12}$ **31.** $\frac{45}{16}$ **33.** $\frac{2}{3}$ **35.** $\frac{7}{6}$ **37.** $\frac{5}{6}$
39. $\frac{1}{2}$ **41.** $\frac{13}{24}$ **43.** $\frac{31}{60}$ **45.** $\frac{35}{18}$ **47.** $\frac{10}{3}$ **49.** $\frac{1}{2}$
51. $\frac{5}{36}$ **53.** 500 **55.** $\frac{3}{40}$ **57.** DW **59.** $2 \cdot 2 \cdot 7$

60. $2 \cdot 2 \cdot 2 \cdot 7$ 61. $2 \cdot 2 \cdot 2 \cdot 5 \cdot 5 \cdot 5$
62. $2 \cdot 2 \cdot 2 \cdot 2 \cdot 2 \cdot 2 \cdot 3$ 63. $3 \cdot 23 \cdot 29$ 64. 126
65. 48 66. 392 67. 192 68. 150 69. $\frac{3}{4}$ 71. 4
73. 1

Margin Exercises, Section R.3, pp. 21–26

1. $\frac{568}{1000}$ 2. $\frac{23}{10}$ 3. $\frac{8904}{100}$ 4. 4.131 5. 0.4131
6. 5.73 7. 284.455 8. 268.63 9. 27.676
10. 64.683 11. 99.59 12. 239.883 13. 5.868
14. 0.5868 15. 51.53808 16. 48.9 17. 15.82
18. 1.28 19. 17.95 20. 856 21. 0.85 22. 0.625
23. $0.\overline{6}$ 24. $7.\overline{63}$ 25. 2.8 26. 13.9 27. 7.0
28. 7.83 29. 34.68 30. 0.03 31. 0.943 32. 8.004
33. 43.112 34. 37.401 35. 7459.355 36. 7459.35
37. 7459.4 38. 7459 39. 7460

Calculator Corner, p. 27

1. 40.42 2. 2.6 3. 3.33 4. 0.69324 5. 7.5
6. 2.38

Exercise Set R.3, p. 28

1. $\frac{53}{10}$ 3. $\frac{67}{100}$ 5. $\frac{20,007}{10,000}$ 7. $\frac{78,898}{10}$ 9. 0.1
11. 0.0001 13. 9.999 15. 0.4578 17. 444.94
19. 390.617 21. 155.724 23. 63.79 25. 32.234
27. 26.835 29. 47.91 31. 1.9193 33. 13.212
35. 0.7998 37. 179.5 39. 1.40756 41. 3.60558
43. 2.3 45. 5.2 47. 0.023 49. 18.75 51. 660
53. 0.68 55. 0.34375 57. $1.\overline{18}$ 59. $0.\overline{5}$ 61. $2.\overline{1}$
63. 745.07; 745.1; 745; 750; 700 65. 6780.51; 6780.5; 6781; 6780; 6800 67. \$17.99; \$18 69. \$346.08; \$346
71. \$17 73. \$190 75. 12.3457; 12.346; 12.35; 12.3; 12
77. 0.5897; 0.590; 0.59; 0.6; 1 79. D_W 81. $\frac{33}{32}$ 82. $\frac{1}{48}$
83. $\frac{55}{64}$ 84. $\frac{5}{4}$ 85. $5 \cdot 17$ 86. $2 \cdot 43$ 87. $3 \cdot 29$
88. $2 \cdot 2 \cdot 2 \cdot 11$ 89. $2 \cdot 2 \cdot 2 \cdot 2 \cdot 13$
90. $2 \cdot 2 \cdot 2 \cdot 2 \cdot 2 \cdot 2 \cdot 2$ 91. $2 \cdot 5 \cdot 5 \cdot 5 \cdot 5$
92. $2 \cdot 2 \cdot 2 \cdot 2 \cdot 2 \cdot 2 \cdot 2 \cdot 2 \cdot 2 \cdot 5$

Margin Exercises, Section R.4, pp. 30–32

1. 0.134 2. 1 3. 0.6667 4. $\frac{90}{100}$ 5. $\frac{53}{100}$ 6. $\frac{459}{1000}$
7. $\frac{23}{10,000}$ 8. 10.6% 9. 677% 10. 99.44%
11. $33.\overline{3}\%$, or $33\frac{1}{3}\%$ 12. 25% 13. 87.5%

Exercise Set R.4, p. 33

1. 0.55 3. 0.0523 5. 0.63 7. 0.941 9. 0.01
11. 0.0061 13. 2.4 15. 0.0325 17. $\frac{7}{100}$ 19. $\frac{77}{100}$
21. $\frac{60}{100}$ 23. $\frac{289}{1000}$ 25. $\frac{110}{100}$ 27. $\frac{42}{100,000}$ 29. $\frac{250}{100}$
31. $\frac{347}{10,000}$ 33. 64% 35. 100% 37. 99.6%
39. 0.47% 41. 7.2% 43. 920% 45. 0.68%
47. $16.\overline{6}\%$, or $16\frac{2}{3}\%$ 49. 65% 51. 29% 53. 80%
55. 60% 57. $66.\overline{6}\%$, or $66\frac{2}{3}\%$ 59. 175% 61. 75%
63. 0.1186% 65. 0.19; $\frac{19}{100}$ 67. $\frac{23}{100}$; 23%
69. 0.3; 30% 71. 0.21; $\frac{21}{100}$ 73. D_W 75. 2.25
76. 1.375 77. $1.41\overline{6}$ 78. $0.\overline{8}$ 79. $0.\overline{90}$ 80. $1.\overline{54}$
81. 164.90974 82. 56.43 83. 896.559 84. 722.579
85. 32% 87. 70% 89. 2700% 91. 345% 93. 2.5%

Margin Exercises, Section R.5, pp. 36–39

1. 4^3 2. 6^5 3. $(1.08)^2$ 4. 10,000 5. 512 6. 1.331
7. 5 8. 14 9. 13 10. 1000 11. 250 12. 178
13. 2 14. 125 15. 48 16. $\frac{11}{2}$

Calculator Corner, p. 37

1. 1024 2. 40,353,607 3. 361 4. 32.695524
5. 10.4976 6. 12,812.904 7. 0.0423150361
8. 0.0260122949

Calculator Corner, p. 39

1. 38 2. 81 3. 72 4. 5932 5. 25.011 6. 743.027
7. 450 8. 14,321,949.1 9. 4 10. 2 11. 783
12. 228,112.96 13. 40; the calculator adds 39 to 141 ÷ 47 to get 42 and then subtracts 2 to get 40.

Exercise Set R.5, p. 40

1. 5^4 3. 10^3 5. 10^6 7. 49 9. 59,049 11. 100
13. 1 15. 5.29 17. 0.008 19. 416.16 21. $\frac{9}{64}$
23. 125 25. 1061.208 27. 25 29. 114 31. 33
33. 5 35. 12 37. 324 39. 100 41. 1000 43. 22
45. 1 47. 4 49. 102 51. 96 53. 24 55. 90
57. 8 59. 1 61. 50,000 63. $\frac{22}{45}$ 65. $\frac{19}{66}$ 67. 9
69. D_W 71. 31.25% 72. $91.\overline{6}\%$, or $91\frac{2}{3}\%$
73. 53.125% 74. $183.\overline{3}\%$, or $183\frac{1}{3}\%$ 75. $\frac{5}{13}$ 76. $\frac{3}{667}$
77. $\frac{2}{3}$ 78. $\frac{401}{728}$ 79. $2 \cdot 2 \cdot 2 \cdot 2 \cdot 3$ 80. 168
81. 10^2 83. 5^6 85. $3 = \frac{5+5}{5} + \frac{5}{5}$; $4 = \frac{5+5+5+5}{5}$; $5 = \frac{5(5+5)}{5} - 5$; $6 = \frac{5}{5} + \frac{5 \cdot 5}{5}$;

$7 = \frac{5}{5} + \frac{5}{5} + 5$; $8 = 5 + \frac{5+5+5}{5}$; $9 = \frac{5 \cdot 5 - 5}{5} + 5$;
$10 = \frac{5 \cdot 5 + 5 \cdot 5}{5}$

Margin Exercises, Section R.6, pp. 42–48

1. 26 cm **2.** 46 in. **3.** 12 cm **4.** 17.5 yd **5.** 40 km **6.** 21 yd **7.** 8 cm^2 **8.** 56 km^2 **9.** 16.96 yd^2 **10.** 118.81 m^2 **11.** $\frac{4}{9}$ yd^2 **12.** 43.8 cm^2 **13.** 12.375 km^2 **14.** 96 m^2 **15.** 18.7 cm^2 **16.** 9 in. **17.** 7 ft **18.** 62.8 m **19.** 88 m **20.** $78\frac{4}{7}$ km^2 **21.** 339.62 cm^2 **22.** 128 ft^3

Calculator Corner, p. 47

1. Varies by calculator; 3.141592654 gives 9 decimal places. **2.** 1417.98926 in.; 160,005.9081 in^2 **3.** 1705.539236 in^2 **4.** 125,663.7061 ft^2

Exercise Set R.6, p. 49

1. 17 mm **3.** 15.25 in. **5.** 30 ft **7.** 79.14 cm **9.** 88 ft **11.** 182 mm **13.** **(a)** 228 ft; **(b)** \$1046.52 **15.** 15 km^2 **17.** 1.4 in^2 **19.** $\frac{4}{9}$ yd^2 **21.** 8100 ft^2 **23.** 50 ft^2 **25.** 169.883 cm^2 **27.** $\frac{5}{9}$ in^2 **29.** 484 ft^2 **31.** 3237.61 km^2 **33.** $\frac{9}{64}$ yd^2 **35.** 630.36 m^2 **37.** 32 cm^2 **39.** 60 in^2 **41.** 8.05 cm^2 **43.** 7 m^2 **45.** 14 cm; 44 cm; 154 cm^2 **47.** $1\frac{1}{2}$ in.; $4\frac{5}{7}$ in.; $1\frac{43}{56}$ in^2 **49.** 16 ft; 100.48 ft; 803.84 ft^2 **51.** 0.7 cm; 4.396 cm; 1.5386 cm^2 **53.** 154 ft^2 **55.** 768 cm^3 **57.** 45 in^3 **59.** 75 m^3 **61.** 357.5 yd^3 **63.** D_W **65.** 87.5% **66.** 58% **67.** $66.\overline{6}\%$, or $66\frac{2}{3}\%$ **68.** 43.61% **69.** 37.5% **70.** 62.5% **71.** $66.\overline{6}\%$, or $66\frac{2}{3}\%$ **72.** 20% **73.** 803.84 in^3 **75.** 353.25 cm^3 **77.** 41,580,000 yd^3

Summary and Review: Chapter R, p. 55

1. $2 \cdot 2 \cdot 23$ **2.** $2 \cdot 2 \cdot 2 \cdot 5 \cdot 5 \cdot 7$ **3.** 416 **4.** 90 **5.** $\frac{12}{30}$ **6.** $\frac{96}{184}$ **7.** $\frac{40}{64}$ **8.** $\frac{91}{84}$ **9.** $\frac{5}{12}$ **10.** $\frac{51}{91}$ **11.** $\frac{31}{36}$ **12.** $\frac{1}{4}$ **13.** $\frac{3}{5}$ **14.** $\frac{72}{25}$ **15.** $\frac{1797}{100}$ **16.** 0.2337 **17.** 2442.905 **18.** 86.0298 **19.** 9.342 **20.** 133.264 **21.** 430.8 **22.** 110.483 **23.** 55.6 **24.** 0.45 **25.** $1.58\overline{3}$ **26.** 34.1 **27.** 0.18 **28.** 8.2% **29.** $\frac{22}{100}$ **30.** 0.1141% **31.** 62.5% **32.** 116% **33.** 6^3 **34.** 1.1236 **35.** 119 **36.** 29 **37.** 7 **38.** $\frac{103}{17}$ **39.** 23 m **40.** 4.4 m **41.** 228 ft; 2808 ft^2 **42.** 36 ft; 81 ft^2 **43.** 17.6 cm; 12.6 cm^2 **44.** 60 cm^2 **45.** 22.5 m^2 **46.** 27.5 cm^2 **47.** 126 in^2 **48.** 840 ft^2 **49.** 8 m **50.** $\frac{14}{11}$ in. **51.** 14 ft **52.** 20 cm **53.** 50.24 m **54.** 8 in. **55.** 200.96 m^2 **56.** $5\frac{1}{11}$ in^2 **57.** 93.6 m^3 **58.** 193.2 cm^3 **59.** See the formulas for area listed at the beginning of the Summary and Review for Chapter R. **60.** When expressed in simplified fraction notation, the numerator is 11 and the denominator is 16. **61.** 0.077104 m^2

Test: Chapter R, p. 57

1. [R.1a] $2 \cdot 2 \cdot 3 \cdot 5 \cdot 5$ **2.** [R.1b] 120 **3.** [R.2a] $\frac{21}{49}$ **4.** [R.2a] $\frac{33}{48}$ **5.** [R.2b] $\frac{2}{3}$ **6.** [R.2b] $\frac{37}{61}$ **7.** [R.2c] $\frac{5}{36}$ **8.** [R.2c] $\frac{11}{40}$ **9.** [R.3a] $\frac{678}{100}$ **10.** [R.3a] 1.895 **11.** [R.3b] 99.0187 **12.** [R.3b] 1796.58 **13.** [R.3b] 435.072 **14.** [R.3b] 1.6 **15.** [R.3b] $2.\overline{09}$ **16.** [R.3c] 234.7 **17.** [R.3c] 234.728 **18.** [R.4a] 0.007 **19.** [R.4b] $\frac{91}{100}$ **20.** [R.4d] 44% **21.** [R.5b] 625 **22.** [R.5b] 1.44 **23.** [R.5c] 242 **24.** [R.5c] 20,000 **25.** [R.4d] 0.1014% **26.** [R.4a] 0.0471 **27.** [R.6a, b] 32.82 cm; 65.894 cm^2 **28.** [R.6a, b] 100 m; 625 m^2 **29.** [R.6b] 25 cm^2 **30.** [R.6b] 12 m^2 **31.** [R.6c] $\frac{1}{4}$ in.; $\frac{11}{14}$ in.; $\frac{11}{224}$ in^2 **32.** [R.6c] 9 cm; 56.52 cm; 254.34 cm^2 **33.** [R.6d] 84 cm^3 **34.** [R.6b, c] 26.28 ft^2

Chapter 1

Pretest: Chapter 1, p. 62

1. $\frac{5}{16}$ **2.** $78\%x$, or $0.78x$ **3.** 360 ft^2 **4.** 12 **5.** $>$ **6.** $>$ **7.** $>$ **8.** $<$ **9.** 12 **10.** 2.3 **11.** 0 **12.** -5.4 **13.** $\frac{2}{3}$ **14.** $\frac{1}{10}$ **15.** $-\frac{3}{2}$ **16.** -17 **17.** 38.6 **18.** $-\frac{17}{15}$ **19.** -5 **20.** 63 **21.** $-\frac{5}{12}$ **22.** -98 **23.** 8 **24.** 24 **25.** 26 **26.** $9z - 18$ **27.** $-4a - 2b + 10c$ **28.** $4(x - 3)$ **29.** $3(2y - 3z - 6)$ **30.** $-y - 13$ **31.** $y + 18$ **32.** $12 < x$ **33.** 50°C higher

Margin Exercises, Section 1.1, pp. 63–66

1. $8 + x = 21$; 13 **2.** 64 **3.** 28 **4.** 60 **5.** 192 ft^2 **6.** 25 **7.** 16 **8.** 12 hr **9.** $x - 8$ **10.** $y + 8$, or $8 + y$ **11.** $m - 4$ **12.** $\frac{1}{2}p$ **13.** $6 + 8x$, or $8x + 6$ **14.** $a - b$ **15.** $59\%x$, or $0.59x$ **16.** $xy - 200$ **17.** $p + q$

Calculator Corner, p. 65

1. 56 **2.** 11.9 **3.** 1.8 **4.** 34,427.16 **5.** 20.1 **6.** 29.9

Exercise Set 1.1, p. 68

1. \$20,400; \$46,800; \$150,000 **3.** 1935 m^2 **5.** 260 mi **7.** 24 ft^2 **9.** 56 **11.** 8 **13.** 1 **15.** 6 **17.** 2 **19.** $b + 7$, or $7 + b$ **21.** $c - 12$ **23.** $4 + q$, or $q + 4$ **25.** $a + b$, or $b + a$ **27.** $x \div y$, or $\frac{x}{y}$, or x/y, or $x \cdot \frac{1}{y}$ **29.** $x + w$, or $w + x$ **31.** $n - m$ **33.** $x + y$, or $y + x$ **35.** $2z$ **37.** $3m$ **39.** 89%s, or $0.89s$, where s is the salary **41.** $65t$ miles **43.** \$50 $- x$ **45.** DW **47.** $2 \cdot 3 \cdot 3 \cdot 3$ **48.** $2 \cdot 2 \cdot 2 \cdot 2 \cdot 2$ **49.** $2 \cdot 2 \cdot 3 \cdot 3 \cdot 3$ **50.** $2 \cdot 2 \cdot 2 \cdot 2 \cdot 2 \cdot 2 \cdot 3$ **51.** $3 \cdot 23 \cdot 29$ **52.** 18 **53.** 96 **54.** 60 **55.** 96 **56.** 396 **57.** $x + 3y$ **59.** $2x - 3$

Margin Exercises, Section 1.2, pp. 73–78

1. 8; −5 **2.** 950,000,000 **3.** −6 **4.** −10; 156 **5.** −120; 50; −80 **6.**

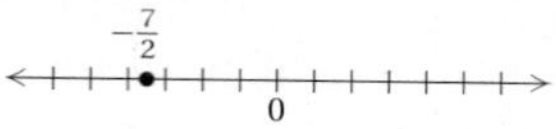

7.

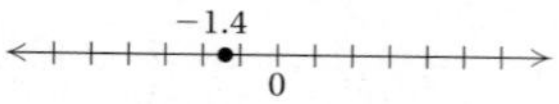

8.

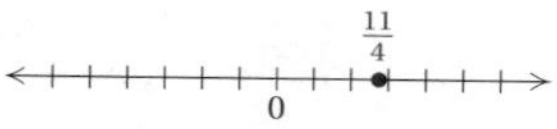

9. −0.375 **10.** $-0.\overline{54}$ **11.** $1.\overline{3}$ **12.** < **13.** < **14.** > **15.** > **16.** > **17.** < **18.** < **19.** > **20.** $7 > -5$ **21.** $4 < x$ **22.** False **23.** True **24.** True **25.** 8 **26.** 9 **27.** $\frac{2}{3}$ **28.** 5.6

Calculator Corner, p. 74

1. −0.75 **2.** −0.45 **3.** −0.125 **4.** −1.8 **5.** −0.675 **6.** −0.6875 **7.** −3.5 **8.** −0.76

Calculator Corner, p. 75

1. 8.717797887 **2.** 17.80449381 **3.** 67.08203932 **4.** 35.4807407 **5.** 3.141592654 **6.** 91.10618695 **7.** 530.9291585 **8.** 138.8663978

Calculator Corner, p. 78

1. 5 **2.** 17 **3.** 0 **4.** 6.48 **5.** 12.7 **6.** 0.9 **7.** $\frac{5}{7}$ **8.** $\frac{4}{3}$

Exercise Set 1.2, p. 79

1. −1286; 14,410 **3.** 24; −2 **5.** −5,600,000,000,000 **7.** Alley Cats: −34; Strikers: 34 **9.**

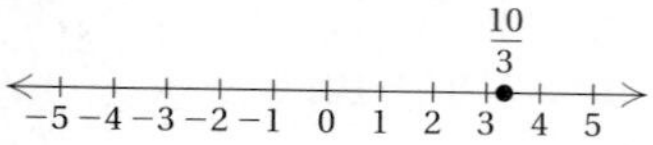

11.

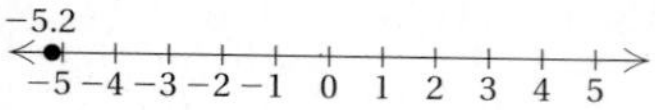

13. −0.875

15. $0.8\overline{3}$ **17.** $-1.1\overline{6}$ **19.** $0.\overline{6}$ **21.** −0.5 **23.** 0.1 **25.** > **27.** < **29.** < **31.** < **33.** > **35.** < **37.** > **39.** < **41.** < **43.** < **45.** True **47.** False **49.** $x < -6$ **51.** $y \geq -10$ **53.** 3 **55.** 10 **57.** 0 **59.** 24 **61.** $\frac{2}{3}$ **63.** 0 **65.** $3\frac{5}{8}$ **67.** DW **69.** 0.63 **70.** 0.083 **71.** 1.1 **72.** 0.2276 **73.** 75% **74.** 62.5%, or $62\frac{1}{2}$% **75.** $83.\overline{3}$%, or $83\frac{1}{3}$% **76.** 59.375%, or $59\frac{3}{8}$% **77.** $-\frac{5}{6}, -\frac{3}{4}, -\frac{2}{3}, \frac{1}{6}, \frac{3}{8}, \frac{1}{2}$ **79.** $\frac{1}{9}$ **81.** $5\frac{5}{9}$, or $\frac{50}{9}$

Margin Exercises, Section 1.3, pp. 82–86

1. −3 **2.** −3 **3.** −5 **4.** 4 **5.** 0 **6.** −2 **7.** −11 **8.** −12 **9.** 2 **10.** −4 **11.** −2 **12.** 0 **13.** −22 **14.** 3 **15.** 0.53 **16.** 2.3 **17.** −7.7 **18.** −6.2 **19.** $-\frac{2}{9}$ **20.** $-\frac{19}{20}$ **21.** −58 **22.** −56 **23.** −14 **24.** −12 **25.** 4 **26.** −8.7 **27.** 7.74 **28.** $\frac{8}{9}$ **29.** 0 **30.** −12 **31.** −14; 14 **32.** −1; 1 **33.** 19; −19 **34.** 1.6; −1.6 **35.** $-\frac{2}{3}; \frac{2}{3}$ **36.** $\frac{9}{8}; -\frac{9}{8}$ **37.** 4 **38.** 13.4 **39.** 0 **40.** $-\frac{1}{4}$ **41.** 24 students

Exercise Set 1.3, p. 87

1. −7 **3.** −6 **5.** 0 **7.** −8 **9.** −7 **11.** −27 **13.** 0 **15.** −42 **17.** 0 **19.** 0 **21.** 3 **23.** −9 **25.** 7 **27.** 0 **29.** 35 **31.** −3.8 **33.** −8.1 **35.** $-\frac{1}{5}$ **37.** $-\frac{7}{9}$ **39.** $-\frac{3}{8}$ **41.** $-\frac{19}{24}$ **43.** $\frac{1}{24}$ **45.** 37 **47.** 50 **49.** −1409 **51.** −24 **53.** 26.9 **55.** −8 **57.** $\frac{13}{8}$ **59.** −43 **61.** $\frac{4}{3}$ **63.** 24 **65.** $\frac{3}{8}$ **67.** 13,796 ft **69.** −3°F **71.** −\$20,300 **73.** −\$85 **75.** DW **77.** 0.57 **78.** 0.713 **79.** 0.238 **80.** 0.92875 **81.** 125% **82.** 12.5% **83.** 52% **84.** 40.625% **85.** All positive **87.** (b)

Margin Exercises, Section 1.4, pp. 90–92

1. −10 **2.** 3 **3.** −5 **4.** −1 **5.** 2 **6.** −4 **7.** −2 **8.** −11 **9.** 4 **10.** −2 **11.** −6 **12.** −16 **13.** 7.1 **14.** 3 **15.** 0 **16.** $\frac{3}{2}$ **17.** −8 **18.** 7 **19.** −3 **20.** −23.3 **21.** 0 **22.** −9 **23.** 17 **24.** 12.7 **25.** 214°F higher

Exercise Set 1.4, p. 93

1. −7 **3.** −4 **5.** −6 **7.** 0 **9.** −4 **11.** −7 **13.** −6 **15.** 0 **17.** 0 **19.** 14 **21.** 11 **23.** −14

25. 5 **27.** -7 **29.** -1 **31.** 18 **33.** -10 **35.** -3 **37.** -21 **39.** 5 **41.** -8 **43.** 12 **45.** -23 **47.** -68 **49.** -73 **51.** 116 **53.** 0 **55.** -1 **57.** $\frac{1}{12}$ **59.** $-\frac{17}{12}$ **61.** $\frac{1}{8}$ **63.** 19.9 **65.** -8.6 **67.** -0.01 **69.** -193 **71.** 500 **73.** -2.8 **75.** -3.53 **77.** $-\frac{1}{2}$ **79.** $\frac{6}{7}$ **81.** $-\frac{41}{30}$ **83.** $-\frac{2}{15}$ **85.** 37 **87.** -62 **89.** -139 **91.** 6 **93.** 107 **95.** 219 **97.** 2385 m **99.** $347.94 **101.** **(a)** 77; **(b)** -41 **103.** 383 ft **105.** DW **107.** 125 **108.** 243 **109.** 6561 **110.** 10,000 **111.** $2 \cdot 2 \cdot 2 \cdot 2 \cdot 2 \cdot 3 \cdot 3 \cdot 3$ **112.** $5 \cdot 7 \cdot 11 \cdot 11$ **113.** 100.5 **114.** 226 **115.** 0.583 **116.** $\frac{41}{64}$ **117.** $-309{,}882$ **119.** False; $3 - 0 \neq 0 - 3$ **121.** True **123.** True **125.** **(a)** -2; **(b)** yes

Margin Exercises, Section 1.5, pp. 98–101

1. 20; 10; 0; -10; -20; -30 **2.** -18 **3.** -100 **4.** -80 **5.** $-\frac{5}{9}$ **6.** -30.033 **7.** $-\frac{7}{10}$ **8.** -10; 0; 10; 20; 30 **9.** 27 **10.** 32 **11.** 35 **12.** $\frac{20}{63}$ **13.** $\frac{2}{3}$ **14.** 13.455 **15.** -30 **16.** 30 **17.** 0 **18.** $-\frac{8}{3}$ **19.** 0 **20.** 0 **21.** -30 **22.** -30.75 **23.** $-\frac{5}{3}$ **24.** 120 **25.** -120 **26.** 6 **27.** 4; -4 **28.** 9; -9 **29.** 48; 48 **30.** 55°C

Exercise Set 1.5, p. 102

1. -8 **3.** -48 **5.** -24 **7.** -72 **9.** 16 **11.** 42 **13.** -120 **15.** -238 **17.** 1200 **19.** 98 **21.** -72 **23.** -12.4 **25.** 30 **27.** 21.7 **29.** $-\frac{2}{5}$ **31.** $\frac{1}{12}$ **33.** -17.01 **35.** $-\frac{5}{12}$ **37.** 420 **39.** $\frac{2}{7}$ **41.** -60 **43.** 150 **45.** $-\frac{2}{45}$ **47.** 1911 **49.** 50.4 **51.** $\frac{10}{189}$ **53.** -960 **55.** 17.64 **57.** $-\frac{5}{784}$ **59.** 0 **61.** -720 **63.** $-30{,}240$ **65.** 441; -147 **67.** 20; 20 **69.** -20 lb **71.** -54°C **73.** $12.71 **75.** -32 m **77.** DW **79.** 180 **80.** $2 \cdot 2 \cdot 2 \cdot 2 \cdot 2 \cdot 2 \cdot 2 \cdot 2 \cdot 2 \cdot 3 \cdot 3$ **81.** $\frac{2}{3}$ **82.** $\frac{8}{9}$ **83.** $\frac{6}{11}$ **84.** $\frac{41}{265}$ **85.** $\frac{11}{32}$ **86.** $\frac{37}{67}$ **87.** $\frac{1}{24}$ **88.** 6 **89.** (a)

91.

$x-2y$, $-y$, $-x$, $x-y$, 0, x, y, $2x$, $x+y$, $3x$, $2y$

Margin Exercises, Section 1.6, pp. 105–110

1. -2 **2.** 5 **3.** -3 **4.** 8 **5.** -6 **6.** $-\frac{30}{7}$ **7.** Not defined **8.** 0 **9.** $\frac{3}{2}$ **10.** $-\frac{4}{5}$ **11.** $-\frac{1}{3}$ **12.** -5 **13.** $\frac{1}{1.6}$ **14.** $\frac{2}{3}$

15.

Number	Opposite	Reciprocal
$\frac{2}{3}$	$-\frac{2}{3}$	$\frac{3}{2}$
$-\frac{5}{4}$	$\frac{5}{4}$	$-\frac{4}{5}$
0	0	Not defined
1	-1	1
-8	8	$-\frac{1}{8}$
-4.5	4.5	$-\frac{1}{4.5}$

16. $\frac{4}{7} \cdot \left(-\frac{5}{3}\right)$ **17.** $5 \cdot \left(-\frac{1}{8}\right)$ **18.** $(a - b) \cdot \left(\frac{1}{7}\right)$ **19.** $-23 \cdot a$ **20.** $-5 \cdot \left(\frac{1}{7}\right)$ **21.** $-\frac{20}{21}$ **22.** $-\frac{12}{5}$ **23.** $\frac{16}{7}$ **24.** -7 **25.** $\frac{5}{-6}, -\frac{5}{6}$ **26.** $\frac{-8}{7}, \frac{8}{-7}$ **27.** $\frac{-10}{3}, -\frac{10}{3}$ **28.** -3.4°F per minute

Calculator Corner, p. 110

1. -4 **2.** -0.3 **3.** -12 **4.** -9.5 **5.** -12 **6.** 2.7 **7.** -2 **8.** -5.7 **9.** -32 **10.** -1.8 **11.** 35 **12.** 14.44 **13.** -2 **14.** -0.8 **15.** 1.4 **16.** 4

Exercise Set 1.6, p. 111

1. -8 **3.** -14 **5.** -3 **7.** 3 **9.** -8 **11.** 2 **13.** -12 **15.** -8 **17.** Not defined **19.** $\frac{23}{2}$ **21.** $\frac{7}{15}$ **23.** $-\frac{13}{47}$ **25.** $\frac{1}{13}$ **27.** $\frac{1}{4.3}$ **29.** -7.1 **31.** $\frac{q}{p}$ **33.** $4y$ **35.** $\frac{3b}{2a}$ **37.** $4 \cdot \left(\frac{1}{17}\right)$ **39.** $8 \cdot \left(-\frac{1}{13}\right)$ **41.** $13.9 \cdot \left(-\frac{1}{1.5}\right)$ **43.** $x \cdot y$ **45.** $(3x + 4)\left(\frac{1}{5}\right)$

47. $(5a - b)\left(\frac{1}{5a+b}\right)$ **49.** $-\frac{9}{8}$ **51.** $\frac{5}{3}$ **53.** $\frac{9}{14}$
55. $\frac{9}{64}$ **57.** -2 **59.** $\frac{11}{13}$ **61.** -16.2 **63.** Not defined
65. -7.4% **67.** 42.4% **69.** DW **71.** 33 **72.** 129
73. 1 **74.** 1296 **75.** $\frac{22}{39}$ **76.** 0.477 **77.** 87.5%
78. $\frac{2}{3}$ **79.** $\frac{9}{8}$ **80.** $\frac{128}{625}$ **81.** $\frac{1}{-10.5}$; -10.5, the reciprocal of the reciprocal is the original number
83. Negative **85.** Positive **87.** Negative

Margin Exercises, Section 1.7, pp. 114–122

1.

Value	$x + x$	$2x$
$x = 3$	6	6
$x = -6$	-12	-12
$x = 4.8$	9.6	9.6

2.

Value	$x + 3x$	$5x$
$x = 2$	8	10
$x = -6$	-24	-30
$x = 4.8$	19.2	24

3. $\frac{6}{8}$ **4.** $\frac{3t}{4t}$ **5.** $\frac{3}{4}$ **6.** $-\frac{4}{3}$ **7.** 1; 1 **8.** -10; -10
9. $9 + x$ **10.** qp **11.** $t + xy$, or $yx + t$, or $t + yx$
12. 19; 19 **13.** 150; 150 **14.** $(r + s) + 7$ **15.** $(9a)b$
16. $(4t)u$, $(tu)4$, $t(4u)$; answers may vary
17. $(2 + r) + s$, $(r + s) + 2$, $s + (r + 2)$; answers may vary
18. (a) 63; (b) 63 **19.** (a) 80; (b) 80 **20.** (a) 28; (b) 28
21. (a) 8; (b) 8 **22.** (a) -4; (b) -4 **23.** (a) -25; (b) -25
24. $5x$, $-8y$, 3 **25.** $-4y$, $-2x$, $3z$ **26.** $3x - 15$
27. $5x + 5$ **28.** $\frac{3}{5}p + \frac{3}{5}q - \frac{3}{5}t$ **29.** $-2x + 6$
30. $5x - 10y + 20z$ **31.** $-5x + 10y - 20z$ **32.** $6(x - 2)$
33. $3(x - 2y + 3)$ **34.** $b(x + y - z)$
35. $2(8a - 18b + 21)$ **36.** $\frac{1}{8}(3x - 5y + 7)$
37. $-4(3x - 8y + 4z)$ **38.** $3x$ **39.** $6x$ **40.** $-8x$
41. $0.59x$ **42.** $3x + 3y$ **43.** $-4x - 5y - 7$
44. $-\frac{2}{3} + \frac{1}{10}x + \frac{7}{9}y$

Exercise Set 1.7, p. 123

1. $\frac{3y}{5y}$ **3.** $\frac{10x}{15x}$ **5.** $-\frac{3}{2}$ **7.** $-\frac{7}{6}$ **9.** $8 + y$ **11.** nm
13. $xy + 9$, or $9 + yx$ **15.** $c + ab$, or $ba + c$
17. $(a + b) + 2$ **19.** $8(xy)$ **21.** $a + (b + 3)$
23. $(3a)b$
25. $2 + (b + a)$, $(2 + a) + b$, $(b + 2) + a$; answers may vary **27.** $(5 + w) + v$; $(v + 5) + w$; $(w + v) + 5$; answers may vary
29. $(3x)y$, $y(x \cdot 3)$, $3(yx)$; answers may vary
31. $a(7b)$, $b(7a)$, $(7b)a$; answers may vary **33.** $2b + 10$
35. $7 + 7t$ **37.** $30x + 12$ **39.** $7x + 28 + 42y$
41. $7x - 21$ **43.** $-3x + 21$ **45.** $\frac{2}{3}b - 4$
47. $7.3x - 14.6$ **49.** $-\frac{3}{5}x + \frac{3}{5}y - 6$
51. $45x + 54y - 72$ **53.** $-4x + 12y + 8z$
55. $-3.72x + 9.92y - 3.41$ **57.** $4x$, $3z$ **59.** $7x$, $8y$, $-9z$
61. $2(x + 2)$ **63.** $5(6 + y)$ **65.** $7(2x + 3y)$
67. $5(x + 2 + 3y)$ **69.** $8(x - 3)$ **71.** $4(8 - y)$
73. $2(4x + 5y - 11)$ **75.** $a(x - 1)$ **77.** $a(x - y - z)$
79. $6(3x - 2y + 1)$ **81.** $\frac{1}{3}(2x - 5y + 1)$ **83.** $19a$
85. $9a$ **87.** $8x + 9z$ **89.** $7x + 15y^2$ **91.** $-19a + 88$
93. $4t + 6y - 4$ **95.** b **97.** $\frac{13}{4}y$ **99.** $8x$ **101.** $5n$
103. $-16y$ **105.** $17a - 12b - 1$ **107.** $4x + 2y$
109. $7x + y$ **111.** $0.8x + 0.5y$ **113.** $\frac{35}{6}a + \frac{3}{2}b - 42$
115. DW **117.** 144 **118.** 72 **119.** 144 **120.** 60
121. 32 **122.** 72 **123.** 90 **124.** 108 **125.** $\frac{89}{48}$
126. $\frac{5}{24}$ **127.** $-\frac{5}{24}$ **128.** 30% **129.** Not equivalent; $3 \cdot 2 + 5 \neq 3 \cdot 5 + 2$ **131.** Equivalent; commutative law of addition **133.** $q(1 + r + rs + rst)$

Margin Exercises, Section 1.8, pp. 127–132

1. $-x - 2$ **2.** $-5x - 2y - 8$ **3.** $-6 + t$ **4.** $-x + y$
5. $4a - 3t + 10$ **6.** $-18 + m + 2n - 4z$ **7.** $2x - 9$
8. $3y + 2$ **9.** $2x - 7$ **10.** $3y + 3$ **11.** $-2a + 8b - 3c$
12. $-9x - 8y$ **13.** $-16a + 18$ **14.** $-26a + 41b - 48c$
15. $3x - 7$ **16.** 2 **17.** 18 **18.** 6 **19.** 17
20. $5x - y - 8$ **21.** -1237 **22.** 8 **23.** 4 **24.** 381
25. -12

Calculator Corner, p. 131

1. -11 **2.** 9 **3.** 114 **4.** 117,649 **5.** $-1{,}419{,}857$
6. $-1{,}124{,}864$ **7.** $-117{,}649$ **8.** $-1{,}419{,}857$
9. $-1{,}124{,}864$ **10.** -4 **11.** -2 **12.** 787

Exercise Set 1.8, p. 133

1. $-2x - 7$ **3.** $-8 + x$ **5.** $-4a + 3b - 7c$
7. $-6x + 8y - 5$ **9.** $-3x + 5y + 6$ **11.** $8x + 6y + 43$
13. $5x - 3$ **15.** $-3a + 9$ **17.** $5x - 6$ **19.** $-19x + 2y$
21. $9y - 25z$ **23.** $-7x + 10y$ **25.** $37a - 23b + 35c$
27. 7 **29.** -40 **31.** 19 **33.** $12x + 30$ **35.** $3x + 30$
37. $9x - 18$ **39.** $-4x - 64$ **41.** -7 **43.** -7
45. -16 **47.** -334 **49.** 14 **51.** 1880 **53.** 12
55. 8 **57.** -86 **59.** 37 **61.** -1 **63.** -10
65. -67 **67.** -7988 **69.** -3000 **71.** 60 **73.** 1

75. 10 **77.** $-\frac{13}{45}$ **79.** $-\frac{23}{18}$ **81.** -118 **83.** $^{\mathbf{D}}\mathbf{W}$
85. $2 \cdot 2 \cdot 59$ **86.** 252 **87.** $\frac{8}{5}$ **88.** $\frac{5}{18}$ **89.** $\frac{13}{12}$
90. $\frac{1}{4}$ **91.** 81 **92.** 1000 **93.** 100 **94.** 225
95. $6y - (-2x + 3a - c)$ **97.** $6m - (-3n + 5m - 4b)$
99. $-2x - f$ **101.** **(a)** 52; 52; 28.130169; **(b)** -24; -24; -108.307025 **103.** -6

Summary and Review: Chapter 1, p. 137

1. 4 **2.** $19\%x$, or $0.19x$ **3.** -45, 72 **4.** 38
5. 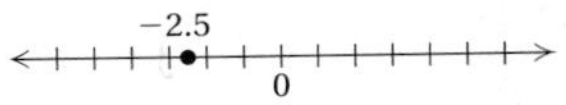

6. 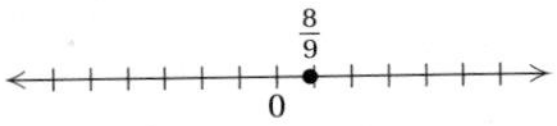

7. < **8.** > **9.** > **10.** < **11.** -3.8 **12.** $\frac{3}{4}$
13. $\frac{8}{3}$ **14.** $-\frac{1}{7}$ **15.** 34 **16.** 5 **17.** -3 **18.** -4
19. -5 **20.** 4 **21.** $-\frac{7}{5}$ **22.** -7.9 **23.** 54
24. -9.18 **25.** $-\frac{2}{7}$ **26.** -210 **27.** -7 **28.** -3
29. $\frac{3}{4}$ **30.** 40.4 **31.** -2 **32.** 2 **33.** -180
34. 8-yd gain **35.** $-\$130$ **36.** \$4.64 **37.** \$18.95
38. $15x - 35$ **39.** $-8x + 10$ **40.** $4x + 15$
41. $-24 + 48x$ **42.** $2(x - 7)$ **43.** $6(x - 1)$
44. $5(x + 2)$ **45.** $3(4 - x)$ **46.** $7a - 3b$
47. $-2x + 5y$ **48.** $5x - y$ **49.** $-a + 8b$ **50.** $-3a + 9$
51. $-2b + 21$ **52.** 6 **53.** $12y - 34$ **54.** $5x + 24$
55. $-15x + 25$ **56.** True **57.** False **58.** $x > -3$
59. $^{\mathbf{D}}\mathbf{W}$ If the sum of two numbers is 0, they are opposites, or additive inverses of each other. For every real number a, the opposite of a can be named $-a$, and $a + (-a) = (-a) + a = 0$. **60.** $^{\mathbf{D}}\mathbf{W}$ No; $|0| = 0$, and 0 is not positive. **61.** $\frac{55}{42}$ **62.** $\frac{109}{18}$
63. $2 \cdot 2 \cdot 2 \cdot 3 \cdot 3 \cdot 3 \cdot 3$ **64.** 62.5% **65.** 0.0567
66. 270 **67.** $-\frac{5}{8}$ **68.** -2.1 **69.** 1000 **70.** $4a + 2b$

Test: Chapter 1, p. 140

1. [1.1a] 6 **2.** [1.1b] $x - 9$ **3.** [1.1a] 240 ft^2
4. [1.2d] < **5.** [1.2d] > **6.** [1.2d] > **7.** [1.2d] <
8. [1.2e] 7 **9.** [1.2e] $\frac{9}{4}$ **10.** [1.2e] 2.7 **11.** [1.3b] $-\frac{2}{3}$
12. [1.3b] 1.4 **13.** [1.3b] 8 **14.** [1.6b] $-\frac{1}{2}$
15. [1.6b] $\frac{7}{4}$ **16.** [1.4a] 7.8 **17.** [1.3a] -8
18. [1.3a] $\frac{7}{40}$ **19.** [1.4a] 10 **20.** [1.4a] -2.5
21. [1.4a] $\frac{7}{8}$ **22.** [1.5a] -48 **23.** [1.5a] $\frac{3}{16}$
24. [1.6a] -9 **25.** [1.6c] $\frac{3}{4}$ **26.** [1.6c] -9.728
27. [1.8d] -173 **28.** [1.8d] -5 **29.** [1.4b] 14°F
30. [1.3c], [1.4b] Up 15 points **31.** [1.5b] 16,080
32. [1.6d] $\frac{33}{35}$°C per minute **33.** [1.7c] $18 - 3x$
34. [1.7c] $-5y + 5$ **35.** [1.7d] $2(6 - 11x)$
36. [1.7d] $7(x + 3 + 2y)$ **37.** [1.4a] 12
38. [1.8b] $2x + 7$ **39.** [1.8b] $9a - 12b - 7$
40. [1.8c] $68y - 8$ **41.** [1.8d] -4 **42.** [1.8d] 448
43. [1.2d] $-2 \geq x$ **44.** [R.5b] 1.728 **45.** [R.4d] 12.5%
46. [R.1a] $2 \cdot 2 \cdot 2 \cdot 5 \cdot 7$ **47.** [R.1b] 240
48. [1.2e], [1.8d] 15 **49.** [1.8c] $4a$
50. [R.6a], [1.7e] $4x + 4y$

Chapter 2

Pretest: Chapter 2, p. 144

1. 8 **2.** -7 **3.** 2 **4.** -1 **5.** -5 **6.** $\frac{135}{32}$ **7.** 1
8. $\{y \mid y > -4\}$ **9.** $\{x \mid x \geq -6\}$ **10.** $\{a \mid a > -1\}$
11. $\{x \mid x \geq 3\}$ **12.** $\left\{y \mid y < -\frac{9}{4}\right\}$ **13.** No solution
14. $x = \frac{y}{A}$ **15.** $a = \frac{A + b}{3}$
16. Width: 34 in.; length: 39 in. **17.** \$460 **18.** 81, 82, 83
19. $\{l \mid l \geq 174 \text{ yd}\}$
20. 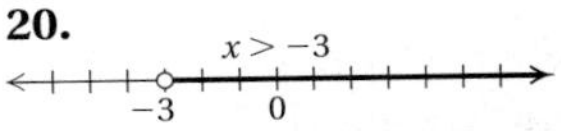

21. 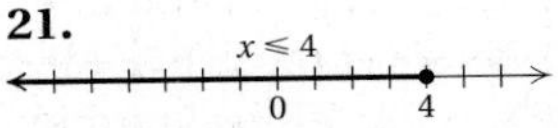

22. 20.4 **23.** 54 **24.** 20% **25.** About 26.5%

Margin Exercises, Section 2.1, pp. 145–148

1. False **2.** True **3.** Neither **4.** Yes **5.** No
6. No **7.** 9 **8.** -13 **9.** 22 **10.** 13.2 **11.** -6.5
12. -2 **13.** $\frac{31}{8}$

Exercise Set 2.1, p. 149

1. Yes **3.** No **5.** No **7.** Yes **9.** No **11.** No
13. 4 **15.** -20 **17.** -14 **19.** -18 **21.** 15
23. -14 **25.** 2 **27.** 20 **29.** -6 **31.** $6\frac{1}{2}$ **33.** 19.9
35. $\frac{7}{3}$ **37.** $-\frac{7}{4}$ **39.** $\frac{41}{24}$ **41.** $-\frac{1}{20}$ **43.** 5.1 **45.** 12.4
47. -5 **49.** $1\frac{5}{6}$ **51.** $-\frac{10}{21}$ **53.** $^{\mathbf{D}}\mathbf{W}$ **55.** -11 **56.** 5
57. $-\frac{5}{12}$ **58.** $\frac{1}{3}$ **59.** $-\frac{3}{2}$ **60.** -5.2 **61.** $-\frac{1}{24}$
62. 172.72 **63.** $\$83 - x$ **64.** $65t$ miles **65.** 342.246
67. $-\frac{26}{15}$ **69.** -10 **71.** All real numbers **73.** $-\frac{5}{17}$
75. 13, -13

Margin Exercises, Section 2.2, pp. 151–154

1. 15 **2.** $-\frac{7}{4}$ **3.** -18 **4.** 10 **5.** 10 **6.** $-\frac{4}{5}$
7. 7800 **8.** -3 **9.** 28

Exercise Set 2.2, p. 155

1. 6 **3.** 9 **5.** 12 **7.** -40 **9.** 1 **11.** -7 **13.** -6
15. 6 **17.** -63 **19.** 36 **21.** -21 **23.** $-\frac{3}{5}$ **25.** $-\frac{3}{2}$

27. $\frac{9}{2}$ 29. 7 31. -7 33. 8 35. 15.9 37. -50
39. -14 41. $\mathbf{D_W}$ 43. $7x$ 44. $-x + 5$ 45. $8x + 11$
46. $-32y$ 47. $x - 4$ 48. $-5x - 23$ 49. $-10y - 42$
50. $-22a + 4$ 51. $8r$ miles 52. $\frac{1}{2}b \cdot 10\text{ m}^2$, or $5b\text{ m}^2$
53. -8655 55. No solution 57. No solution
59. $\frac{b}{3a}$ 61. $\frac{4b}{a}$

Margin Exercises, Section 2.3, pp. 157–163

1. 5 2. 4 3. 4 4. 39 5. $-\frac{3}{2}$ 6. -4.3 7. -3
8. 800 9. 1 10. 2 11. 2 12. $\frac{17}{2}$ 13. $\frac{8}{3}$
14. $-\frac{43}{10}$, -4.3 15. 2 16. 3 17. -2 18. $-\frac{1}{2}$
19. Yes 20. Yes 21. Yes 22. Yes 23. No
24. No 25. No 26. No 27. All real numbers
28. No solution

Calculator Corner, p. 164

1. Left to the student 2. Left to the student

Exercise Set 2.3, p. 165

1. 5 3. 8 5. 10 7. 14 9. -8 11. -8 13. -7
15. 15 17. 6 19. 4 21. 6 23. -3 25. 1
27. 6 29. -20 31. 7 33. 2 35. 5 37. 2
39. 10 41. 4 43. 0 45. -1 47. $-\frac{4}{3}$ 49. $\frac{2}{5}$
51. -2 53. -4 55. $\frac{4}{5}$ 57. $-\frac{28}{27}$ 59. 6 61. 2
63. No solution 65. All real numbers 67. 6 69. 8
71. 1 73. All real numbers 75. No solution
77. 17 79. $-\frac{5}{3}$ 81. -3 83. 2 85. $\frac{4}{7}$
87. No solution 89. All real numbers 91. $-\frac{51}{31}$
93. $\mathbf{D_W}$ 95. -6.5 96. -75.14 97. $7(x - 3 - 2y)$
98. $8(y - 11x + 1)$ 99. 4.4233464 101. $-\frac{5}{32}$

Margin Exercises, Section 2.4, pp. 169–172

1. 2.8 mi 2. 341 mi 3. $q = 3B$ 4. $r = \frac{d}{t}$ 5. $I = \frac{E}{R}$
6. $x = y - 5$ 7. $x = y + 7$ 8. $x = y + b$
9. $y = \frac{5x}{9}$, or $\frac{5}{9}x$ 10. $p = \frac{bq}{a}$ 11. $x = \frac{y - b}{m}$
12. $Q = \frac{a + p}{t}$ 13. $D = \frac{C}{\pi}$ 14. $c = 4A - a - b - d$

Exercise Set 2.4, p. 173

1. (a) 57,000 Btu's; (b) $a = \frac{B}{30}$ 3. (a) $1\frac{3}{5}$ mi; (b) $t = 5M$
5. (a) 1423; (b) $n = 15f$ 7. 10.5 calories per ounce
9. 42 games 11. $x = \frac{y}{5}$ 13. $c = \frac{a}{b}$ 15. $x = y - 13$
17. $x = y - b$ 19. $x = 5 - y$ 21. $x = a - y$
23. $y = \frac{5x}{8}$, or $\frac{5}{8}x$ 25. $x = \frac{By}{A}$ 27. $t = \frac{W - b}{m}$
29. $x = \frac{y - c}{b}$ 31. $b = 3A - a - c$ 33. $t = \frac{A - b}{a}$
35. $h = \frac{A}{b}$ 37. $w = \frac{P - 2l}{2}$, or $\frac{1}{2}P - l$ 39. $a = 2A - b$
41. $a = \frac{F}{m}$ 43. $c^2 = \frac{E}{m}$ 45. $x = \frac{c - By}{A}$ 47. $t = \frac{3k}{v}$
49. $\mathbf{D_W}$ 51. 0.92 52. -90 53. -9.325 54. 44
55. -13.2 56. $-21a + 12b$ 57. 0.031 58. 0.671
59. $\frac{1}{6}$ 60. $-\frac{3}{2}$
61. (a) 1901 calories;
(b) $a = \frac{917 + 6w + 6h - K}{6}$;
$h = \frac{K - 917 - 6w + 6a}{6}$;
$w = \frac{K - 917 - 6h + 6a}{6}$
63. $b = \frac{2A - ah}{h}$; $h = \frac{2A}{a + b}$
65. A quadruples. 67. A increases by $2h$ units.

Margin Exercises, Section 2.5, pp. 177–180

1. $13\% \cdot 80 = a$ 2. $a = 60\% \cdot 70$ 3. $43 = 20\% \cdot b$
4. $110\% \cdot b = 30$ 5. $16 = n \cdot 80$ 6. $n \cdot 94 = 10.5$
7. 1.92 8. 115 9. 36% 10. 111,416 mi^2
11. About 3.9 million 12. About 58%

Exercise Set 2.5, p. 181

1. 20% 3. 150 5. 546 7. 24% 9. 2.5 11. 5%
13. 25% 15. 84 17. 24% 19. 16% 21. $46\frac{2}{3}$
23. 0.8 25. 5 27. 40 29. \$198 31. \$1584
33. \$528 35. U.S.: 68.4%; Asia: 25.9%; Europe: 5.7%
37. About 603 at-bats 39. \$280 41. (a) 16%; (b) \$29
43. (a) \$3.75; (b) \$28.75 45. (a) \$28.80; (b) \$33.12
47. 200 women 49. About 31.5 lb 51. \$305; 71%
53. \$1560; \$780 55. \$1310; 80% 57. $\mathbf{D_W}$
59. 181.52 60. 0.4538 61. 12.0879 62. 844.1407
63. $a + c$ 64. $7x - 9y$ 65. -3.9 66. $-6\frac{1}{8}$
67. 6 ft 7 in.

Margin Exercises, Section 2.6, pp. 186–194

1. $62\frac{2}{3}$ mi 2. Top: 24 ft; middle: 72 ft; bottom: 144 ft
3. 313 and 314 4. 60,417 copies
5. Length: 84 ft; width: 50 ft 6. First: 30°; second: 90°; third: 60° 7. \$8400 8. \$658

Exercise Set 2.6, p. 195

1. 180 in.; 60 in. 3. \$3.67 5. \$6.3 billion 7. $699\frac{1}{3}$ mi
9. 273, 274 11. 41, 42, 43 13. 61, 63, 65
15. Length: 48 ft; width: 14 ft 17. \$75 19. \$85
21. 11 visits 23. 28°, 84°, 68° 25. 33°, 38°, 109°
27. \$350 29. \$852.94 31. 12 mi 33. \$36 35. $\mathbf{D_W}$
37. $-\frac{47}{40}$ 38. $-\frac{17}{40}$ 39. $-\frac{3}{10}$ 40. $-\frac{32}{15}$ 41. -10
42. 1.6 43. 409.6 44. -9.6 45. -41.6 46. 0.1
47. 120 apples 49. About 0.65 in. 51. \$9.17, not \$9.10

Margin Exercises, Section 2.7, pp. 200–207

1. **(a)** No; **(b)** no; **(c)** no; **(d)** yes; **(e)** no; **(f)** no **2.** **(a)** Yes; **(b)** yes; **(c)** yes; **(d)** no; **(e)** yes; **(f)** yes

3. $x \le 4$ (0, 4)

4. $x > -2$ (−2, 0)

5. $-2 < x \le 4$ (−2, 0, 4)

6. $\{x | x > 2\}$; (0, 2)

7. $\{x | x \le 3\}$; (0, 3)

8. $\{x | x < -3\}$; (−3, 0)

9. $\left\{x \middle| x \ge \frac{2}{15}\right\}$ **10.** $\{y | y \le -3\}$

11. $\{x | x < 8\}$; (0, 8)

12. $\{y | y \ge 32\}$; (0 10, 32)

13. $\{x | x \ge -6\}$ **14.** $\left\{y \middle| y < -\frac{13}{5}\right\}$ **15.** $\left\{x \middle| x > -\frac{1}{4}\right\}$ **16.** $\left\{y \middle| y \ge \frac{19}{9}\right\}$ **17.** $\left\{y \middle| y \ge \frac{19}{9}\right\}$ **18.** $\{x | x \ge -2\}$ **19.** $\{x | x \ge -4\}$ **20.** $\left\{x \middle| x > \frac{8}{3}\right\}$

Exercise Set 2.7, p. 208

1. **(a)** Yes; **(b)** yes; **(c)** no; **(d)** yes; **(e)** yes **3.** **(a)** No; **(b)** no; **(c)** no; **(d)** yes; **(e)** no

5. $x > 4$ (0, 4)

7. $t < -3$ (−3, 0)

9. $m \ge -1$ (−1 0)

11. $-3 < x \le 4$ (−3, 0, 4)

13. $0 < x < 3$ (0, 3)

15. $\{x | x > -5\}$; (−5, 0)

17. $\{x | x \le -18\}$; (−18; −20, 0)

19. $\{y | y > -5\}$

21. $\{x | x > 2\}$ **23.** $\{x | x \le -3\}$ **25.** $\{x | x < 4\}$ **27.** $\{t | t > 14\}$ **29.** $\left\{y \middle| y \le \frac{1}{4}\right\}$ **31.** $\left\{x \middle| x > \frac{7}{12}\right\}$

33. $\{x | x < 7\}$; (0, 7)

35. $\{x | x < 3\}$; (0, 3)

37. $\left\{y \middle| y \ge -\frac{2}{5}\right\}$ **39.** $\{x | x \ge -6\}$ **41.** $\{y | y \le 4\}$ **43.** $\left\{x \middle| x > \frac{17}{3}\right\}$ **45.** $\left\{y \middle| y < -\frac{1}{14}\right\}$ **47.** $\left\{x \middle| x \le \frac{3}{10}\right\}$ **49.** $\{x | x < 8\}$ **51.** $\{x | x \le 6\}$ **53.** $\{x | x < -3\}$ **55.** $\{x | x > -3\}$ **57.** $\{x | x \le 7\}$ **59.** $\{x | x > -10\}$ **61.** $\{y | y < 2\}$ **63.** $\{y | y \ge 3\}$ **65.** $\{y | y > -2\}$ **67.** $\{x | x > -4\}$ **69.** $\{x | x \le 9\}$ **71.** $\{y | y \le -3\}$ **73.** $\{y | y < 6\}$ **75.** $\{m | m \ge 6\}$ **77.** $\left\{t \middle| t < -\frac{5}{3}\right\}$ **79.** $\{r | r > -3\}$ **81.** $\left\{x \middle| x \ge -\frac{57}{34}\right\}$ **83.** $\{x | x > -2\}$ **85.** D_W **87.** −74 **88.** 4.8 **89.** $-\frac{5}{8}$ **90.** −1.11 **91.** −38 **92.** $-\frac{7}{8}$ **93.** −9.4 **94.** 1.11 **95.** 140 **96.** 41 **97.** $-2x - 23$ **98.** $37x - 1$ **99.** **(a)** Yes; **(b)** yes; **(c)** no; **(d)** no; **(e)** no; **(f)** yes; **(g)** yes **101.** All real numbers

Margin Exercises, Section 2.8, pp. 212–214

1. $m \ge 92$ **2.** $c \ge 4000$ **3.** $p \le 21{,}900$ **4.** $45 < t < 55$ **5.** $d > 15$ **6.** $w < 110$ **7.** $n > -2$ **8.** $c \le 12{,}500$ **9.** $d \le 11.4\%$ **10.** $s \ge 23$ **11.** $\frac{9}{5}C + 32 < 88$; $\left\{C \middle| C < 31\frac{1}{9}^{\circ}\right\}$ **12.** $\dfrac{91 + 86 + 89 + s}{4} \ge 90$; $\{s | s \ge 94\}$

Exercise Set 2.8, p. 215

1. $n \ge 7$ **3.** $w > 2$ kg **5.** 90 mph $< s <$ 110 mph **7.** $a \le 1{,}200{,}000$ **9.** $c \ge \$1.50$ **11.** $x > 8$ **13.** $y \le -4$ **15.** $n \ge 1300$ **17.** $A \le 500$ L **19.** $2 + 3x < 13$ **21.** $\{x | x \ge 84\}$ **23.** $\{C | C < 1063^{\circ}\}$ **25.** $\{Y | Y \ge 1935\}$ **27.** $\{L | L \ge 5 \text{ in.}\}$ **29.** 15 or fewer copies **31.** 5 min or more **33.** 2 courses **35.** 4 servings or more **37.** Lengths greater than or equal to 92 ft; lengths less than or equal to 92 ft **39.** Lengths less than 21.5 cm **41.** The blue-book value is greater than or equal to $10,625. **43.** It has at least 16 g of fat. **45.** Dates at least 6 weeks after July 1 **47.** Heights greater than or equal to 4 ft **49.** 21 calls or more **51.** D_W **53.** −160 **54.** $-17x + 18$ **55.** $91x - 242$ **56.** 0.25 **57.** Temperatures between −15°C and $-9\frac{4}{9}$°C **59.** They contain at least 7.5 g of fat per serving.

Summary and Review: Chapter 2, p. 220

1. −22 **2.** 1 **3.** 25 **4.** 9.99 **5.** $\frac{1}{4}$ **6.** 7 **7.** −192 **8.** $-\frac{7}{3}$ **9.** $-\frac{15}{64}$ **10.** −8 **11.** 4 **12.** −5 **13.** $-\frac{1}{3}$ **14.** 3 **15.** 4 **16.** 16 **17.** All real numbers **18.** 6 **19.** −3 **20.** 28 **21.** 4 **22.** No solution **23.** Yes **24.** No **25.** Yes **26.** $\left\{y \middle| y \ge -\frac{1}{2}\right\}$ **27.** $\{x | x \ge 7\}$ **28.** $\{y | y > 2\}$ **29.** $\{y | y \le -4\}$ **30.** $\{x | x < -11\}$ **31.** $\{y | y > -7\}$ **32.** $\left\{x \middle| x > -\frac{9}{11}\right\}$ **33.** $\left\{x \middle| x \ge -\frac{1}{12}\right\}$

34. $x < 3$ (0, 3)

35. $-2 < x \le 5$ (−2, 0, 5)

36. $y > 0$ (0)

37. $d = \dfrac{C}{\pi}$ **38.** $B = \dfrac{3V}{h}$

39. $a = 2A - b$ **40.** $x = \dfrac{y - b}{m}$ **41.** Length: 365 mi; width: 275 mi **42.** 345, 346 **43.** $2117 **44.** 27 **45.** 35°, 85°, 60° **46.** 15 **47.** 18.75% **48.** 600 **49.** About 26% **50.** $220 **51.** $53,400 **52.** $138.95 **53.** 86 **54.** $\{w | w > 17 \text{ cm}\}$ **55.** D_W The end result is the same either way. If *s* is the original salary, the new salary after a 5% raise followed by an 8% raise is 1.08(1.05*s*). If the raises occur the other way around, the new salary is 1.05(1.08*s*). By the commutative and associative laws of multiplication, we see that these are equal. However, it would be better to receive the 8% raise first, because this increase yields a higher salary initially than a 5% raise.

56. $\mathbf{D_W}$ The inequalities are equivalent by the multiplication principle for inequalities. If we multiply both sides of one inequality by -1, the other inequality results. **57.** $\frac{41}{4}$ **58.** $58t$ **59.** -45 **60.** $-43x + 8y$
61. $23, -23$ **62.** $20, -20$ **63.** $a = \dfrac{y - 3}{2 - b}$

Test: Chapter 2, p. 223

1. [2.1b] 8 **2.** [2.1b] 26 **3.** [2.2a] -6 **4.** [2.2a] 49
5. [2.3b] -12 **6.** [2.3a] 2 **7.** [2.3a] -8 **8.** [2.1b] $-\frac{7}{20}$
9. [2.3d] 7 **10.** [2.3d] $\frac{5}{3}$ **11.** [2.3b] $\frac{5}{2}$
12. [2.3c] No solution **13.** [2.3c] All real numbers
14. [2.7c] $\{x \mid x \le -4\}$ **15.** [2.7c] $\{x \mid x > -13\}$
16. [2.7d] $\{x \mid x \le 5\}$ **17.** [2.7d] $\{y \mid y \le -13\}$
18. [2.7d] $\{y \mid y \ge 8\}$ **19.** [2.7d] $\{x \mid x \le -\frac{1}{20}\}$
20. [2.7e] $\{x \mid x < -6\}$ **21.** [2.7e] $\{x \mid x \le -1\}$
22. [2.7b] $y \le 9$ (0, 4, 9) **23.** [2.7b, e] $x < 1$ (0, 1)
24. [2.7b] $-2 \le x \le 2$ (-2, 0, 2) **25.** [2.5a] 18
26. [2.5a] 16.5% **27.** [2.5a] 40,000
28. [2.5a] About 30.3% **29.** [2.6a] Width: 7 cm; length: 11 cm **30.** [2.5a] About \$141.5 billion
31. [2.6a] 2509, 2510, 2511 **32.** [2.6a] \$880
33. [2.6a] 3 m, 5 m **34.** [2.8b] $\{l \mid l \ge 174 \text{ yd}\}$
35. [2.8b] $\{b \mid b \le \$105\}$ **36.** [2.8a] $\{c \mid c \le 143{,}750\}$
37. [2.4b] $r = \dfrac{A}{2\pi h}$ **38.** [2.4b] $x = \dfrac{y - b}{8}$ **39.** [1.3a] $-\frac{2}{9}$
40. [1.1a] $\frac{8}{3}$ **41.** [1.1b] 73% p, or $0.73p$
42. [1.8b] $-18x + 37y$ **43.** [2.4b] $d = \dfrac{1 - ca}{-c}$, or $\dfrac{ca - 1}{c}$
44. [1.2e], [2.3a] $15, -15$ **45.** [2.6a] 60 tickets

Cumulative Review: Chapters 1–2, p. 225

1. [1.1a] $\frac{3}{2}$ **2.** [1.1a] $\frac{15}{4}$ **3.** [1.1a] 0 **4.** [1.1b] $2w - 4$
5. [1.2d] $>$ **6.** [1.2d] $>$ **7.** [1.2d] $<$
8. [1.3b], [1.6b] $-\frac{2}{5}, \frac{5}{2}$ **9.** [1.2e] 3 **10.** [1.2e] $\frac{3}{4}$
11. [1.2e] 0 **12.** [1.3a] -4.4
13. [1.4a] $-\frac{5}{2}$ **14.** [1.5a] $\frac{5}{6}$ **15.** [1.5a] -105
16. [1.6a] -9 **17.** [1.6c] -3 **18.** [1.6c] $\frac{32}{125}$
19. [1.7c] $15x + 25y + 10z$ **20.** [1.7c] $-12x - 8$
21. [1.7c] $-12y + 24x$ **22.** [1.7d] $2(32 + 9x + 12y)$
23. [1.7d] $8(2y - 7)$ **24.** [1.7d] $5(a - 3b + 5)$
25. [1.7e] $15b + 22y$ **26.** [1.7e] $4 + 9y + 6z$
27. [1.7e] $1 - 3a - 9d$ **28.** [1.7e] $-2.6x - 5.2y$
29. [1.8b] $3x - 1$ **30.** [1.8b] $-2x - y$
31. [1.8b] $-7x + 6$ **32.** [1.8b] $8x$ **33.** [1.8c] $5x - 13$
34. [2.1b] 4.5 **35.** [2.2a] $\frac{4}{25}$ **36.** [2.1b] 10.9
37. [2.1b] $3\frac{5}{6}$ **38.** [2.2a] -48 **39.** [2.2a] 12
40. [2.2a] -6.2 **41.** [2.3a] -3 **42.** [2.3b] $-\frac{12}{5}$
43. [2.3b] 8 **44.** [2.3d] 7 **45.** [2.3b] $-\frac{4}{5}$
46. [2.3b] $-\frac{10}{3}$ **47.** [2.3c] All real numbers
48. [2.3c] No solution **49.** [2.3c] All real numbers
50. [2.7c] $\{x \mid x < 2\}$ **51.** [2.7e] $\{y \mid y \ge 4\}$
52. [2.7e] $\{y \mid y < -3\}$ **53.** [2.4b] $m = 65 - H$
54. [2.4b] $P = \dfrac{I}{rt}$ **55.** [2.5a] 25.2 **56.** [2.5a] 45%
57. [2.5a] $363.\overline{5}$ **58.** [2.5a] 168.6 million
59. [2.8a] $\{s \mid s \ge 84\}$ **60.** [2.6a] \$45 **61.** [2.5a] \$1050
62. [2.6a] 50 m, 53 m, 40 m **63.** [2.8a] $\{d \mid d \le 128\frac{1}{3} \text{ mi}\}$
64. [2.6a] \$24.60 **65.** [1.8d] (c) **66.** [1.8c] (d)
67. [2.4b] (b) **68.** [2.5a] \$45,200 **69.** [2.5a] 30%
70. [1.2e], [2.3a] $4, -4$ **71.** [2.3b] 3
72. [2.4a] $Q = \dfrac{2 - pm}{p}$

Chapter 3

Pretest: Chapter 3, p. 230

1.

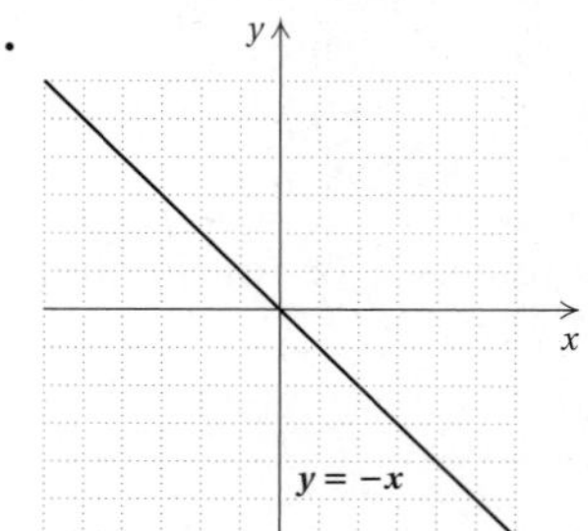

2.

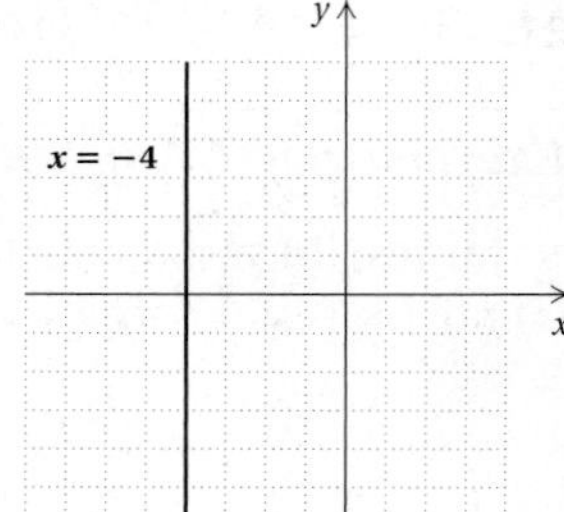

3.

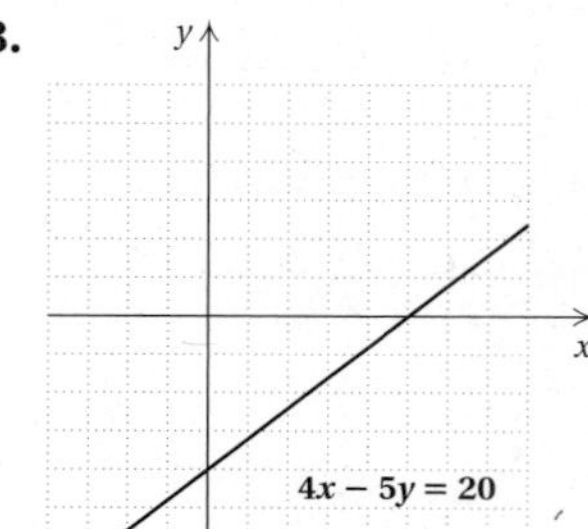

4.

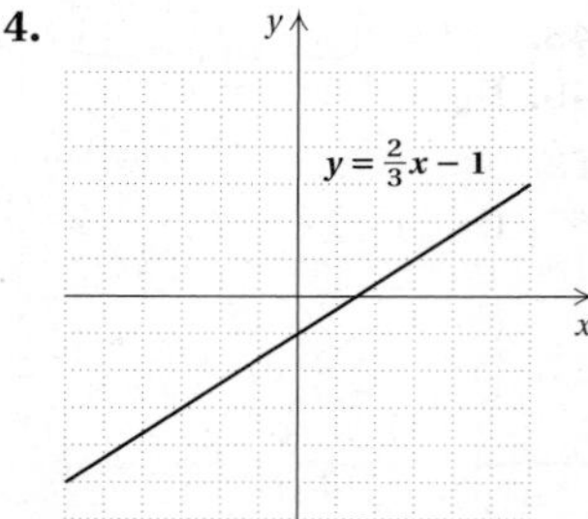

5. III **6.** No **7.** y-intercept: $(0, -4)$; x-intercept: $(5, 0)$
8. $(0, -8)$ **9.** Graph of $P = \frac{7}{2}n + 20$; Price (in cents) vs. Number of pages. 320¢, or \$3.20
10. $-\frac{2}{3}$ **11.** $2\frac{1}{2}$ cars per hour
12. 15 calories per minute **13.** $\frac{3}{5}$

Margin Exercises, Section 3.1, pp. 231–235

1. 270 travelers **2.** **(a)** 3 A.M.–6 A.M.; **(b)** midnight–3 A.M.; 3 A.M.–6 A.M.; 6 A.M.–9 A.M.; 9 A.M.–noon **3.** **(a)** 2 months; **(b)** 60 beats per minute

4.–11.

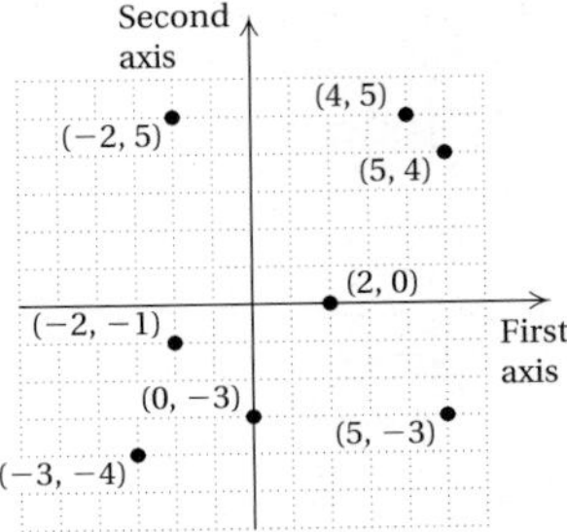

12. Both are negative numbers. **13.** First, positive; second, negative **14.** I **15.** III **16.** IV **17.** II **18.** Not in any quadrant **19.** A: $(-5, 1)$; B: $(-3, 2)$; C: $(0, 4)$; D: $(3, 3)$; E: $(1, 0)$; F: $(0, -3)$; G: $(-5, -4)$

Exercise Set 3.1, p. 236

1. 47.6% **3.** 44.6% **5.** 49,500 **7.** 18,900 **9.** 6 drinks **11.** The weight is greater than 200 lb. **13.** The weight is greater than 120 lb. **15.** 17,000 **17.** 1998 **19.** About 1000 **21.**

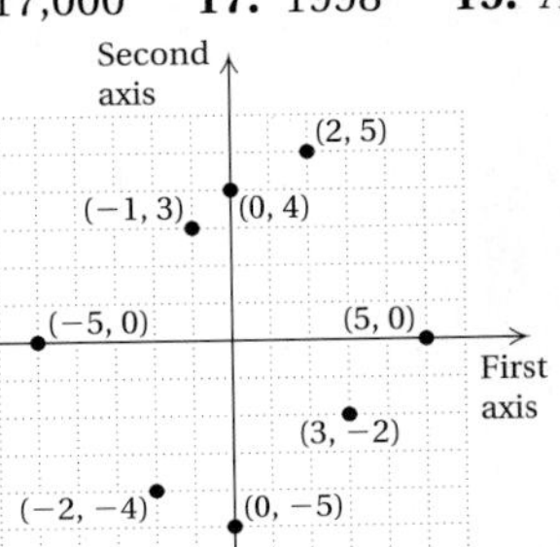

23. II **25.** IV **27.** III **29.** I **31.** II **33.** IV **35.** I **37.** Positive **39.** I, IV **41.** I, III **43.** A: $(3, 3)$; B: $(0, -4)$; C: $(-5, 0)$; D: $(-1, -1)$; E: $(2, 0)$ **45.** 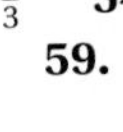 **47.** 12 **48.** 4.89 **49.** 0 **50.** $\frac{4}{5}$ **51.** 3.4 **52.** $\sqrt{2}$ **53.** $\frac{2}{3}$ **54.** $\frac{7}{8}$ **55.** \$28.32 **56.** \$18.40 **57.** $(-1, -5)$ **59.**

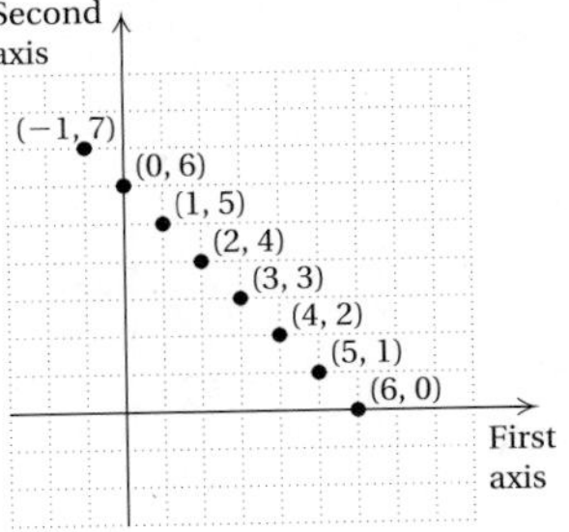

61. 26

Margin Exercises, Section 3.2, pp. 240–248

1. No **2.** Yes **3.** $(-2, -3)$, $(1, 3)$; answers may vary

4.

x	y	(x, y)
−3	6	(−3, 6)
−1	2	(−1, 2)
0	0	(0, 0)
1	−2	(1, −2)
3	−6	(3, −6)

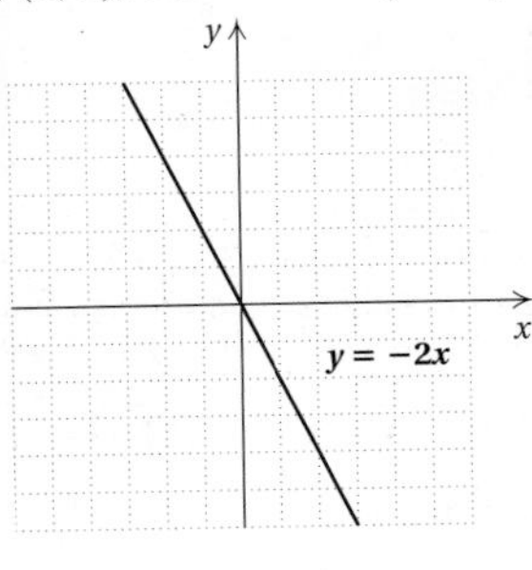

5.

x	y	(x, y)
4	2	(4, 2)
2	1	(2, 1)
0	0	(0, 0)
−2	−1	(−2, −1)
−4	−2	(−4, −2)
−1	$-\frac{1}{2}$	$(-1, -\frac{1}{2})$

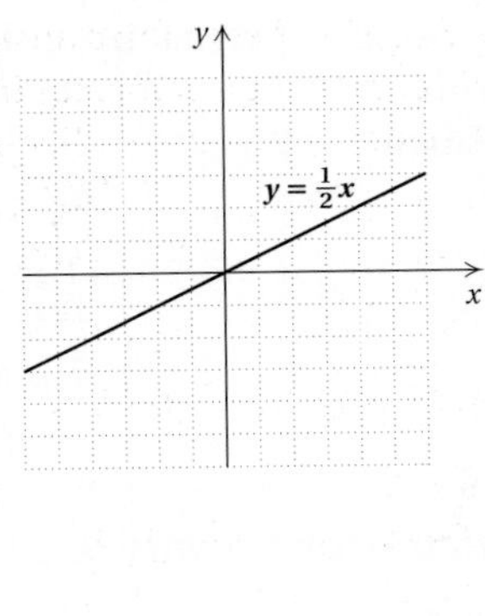

6.

$y = 2x + 3$

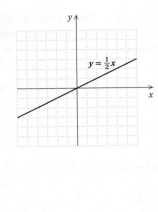

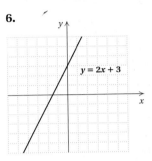

7.

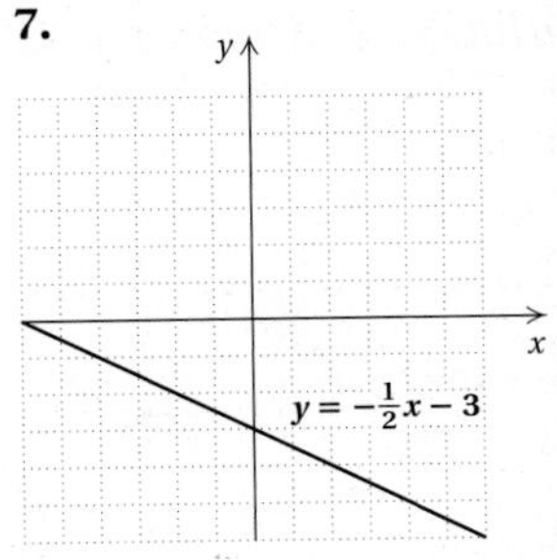

8.

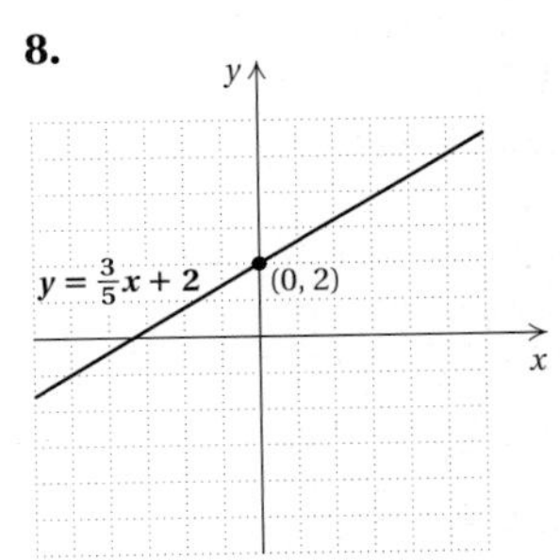

9.

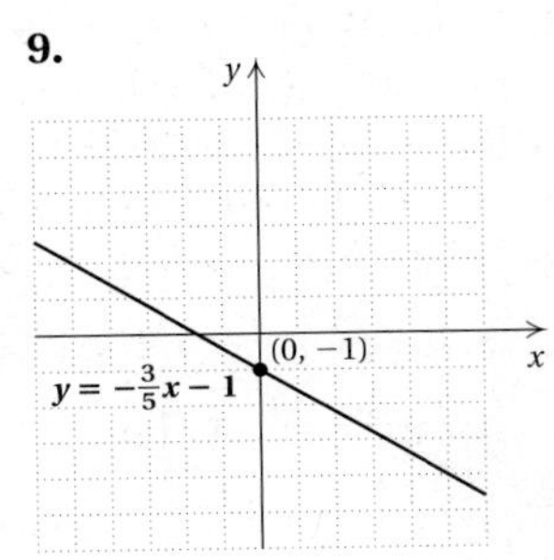

10.

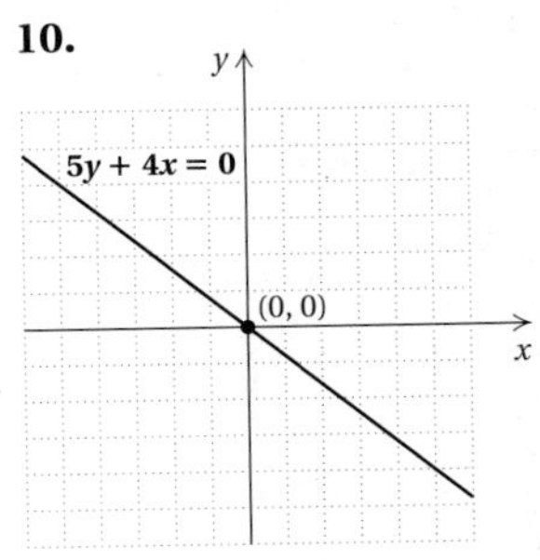

11.

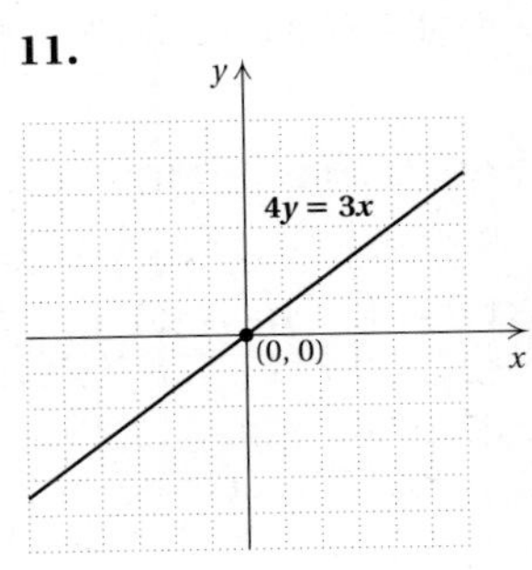

12.

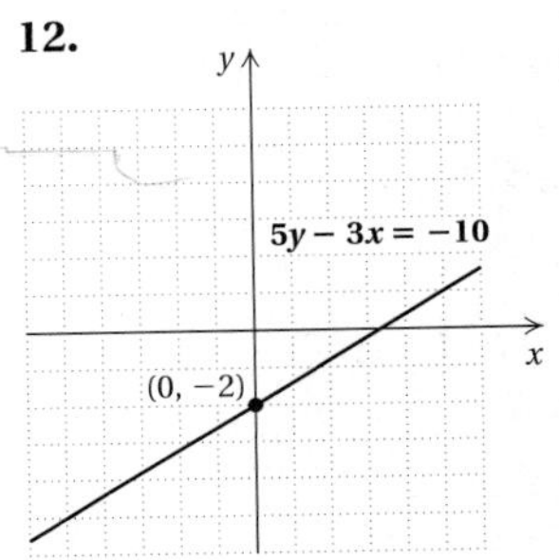

13.

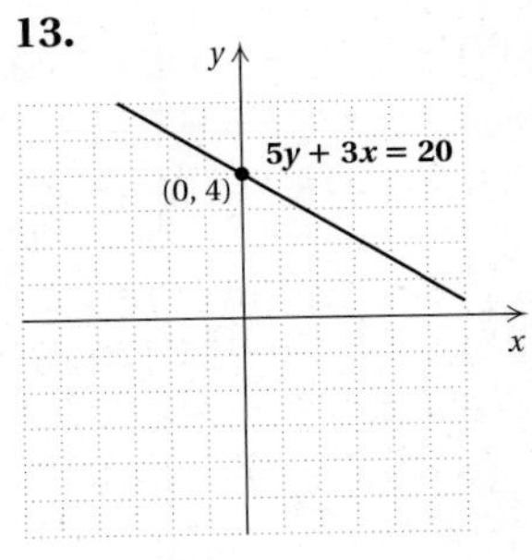

14. **(a)** \$2720; \$2040; \$680; \$0; **(b)** about \$1700; **(c)** about 2.8 yr

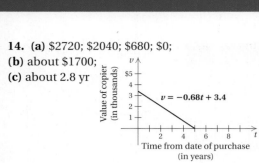

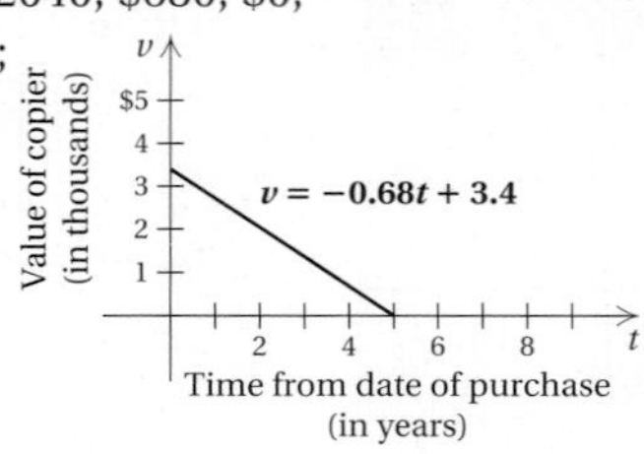

Calculator Corner, p. 244

1. Left to the student

Calculator Corner, p. 250

1. $y = 2x + 1$

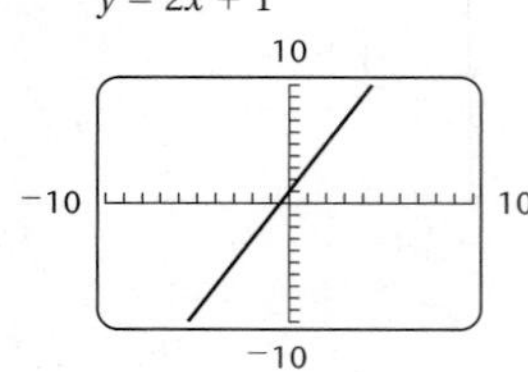

2. $y = -3x + 1$

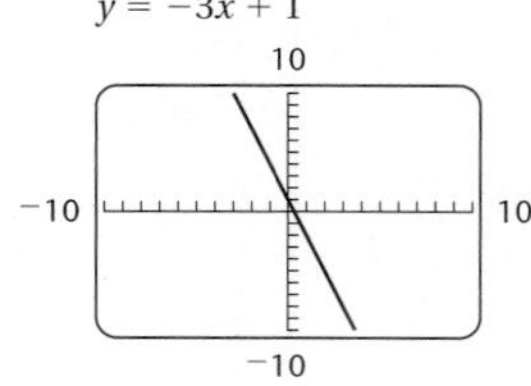

3. $y = -5x + 3$

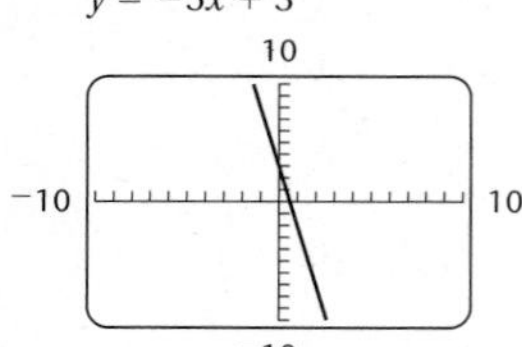

4. $y = 4x - 5$

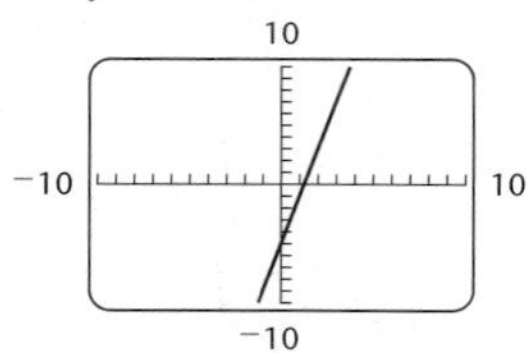

5. $y = \frac{4}{5}x + 2$

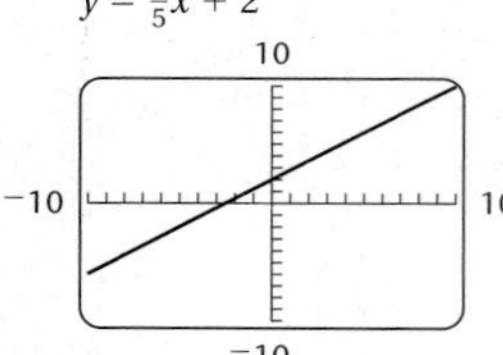

6. $y = -\frac{3}{5}x - 1$

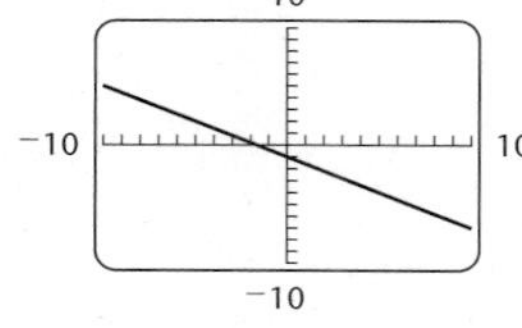

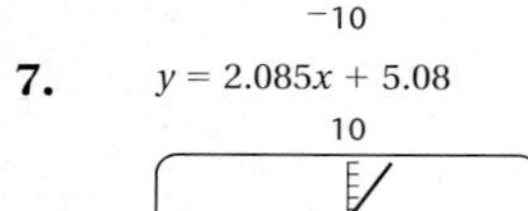

7. $y = 2.085x + 5.08$

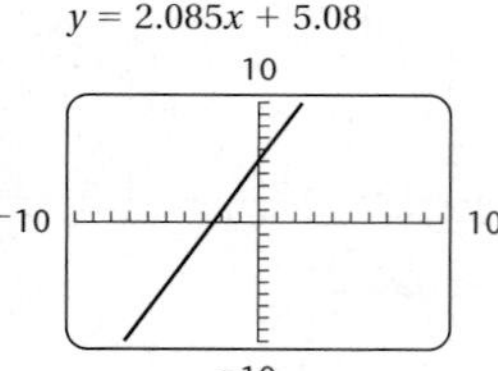

8. $y = -3.45x - 1.68$

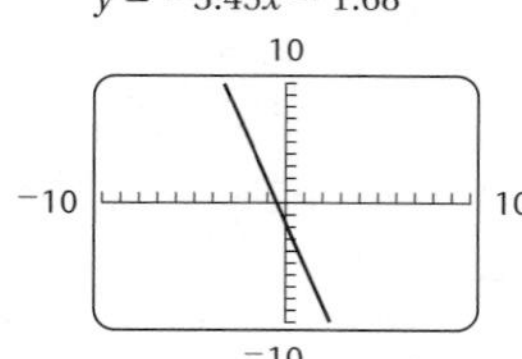

Exercise Set 3.2, p. 251

1. No **3.** No **5.** Yes

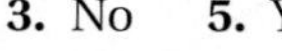

7.

$$\begin{array}{c|l} y & = x - 5 \\ \hline -1 & ?\ 4 - 5 \\ & -1 \end{array} \quad \text{TRUE}$$

$$\begin{array}{c|l} y & = x - 5 \\ \hline -4 & ?\ 1 - 5 \\ & -4 \end{array} \quad \text{TRUE}$$

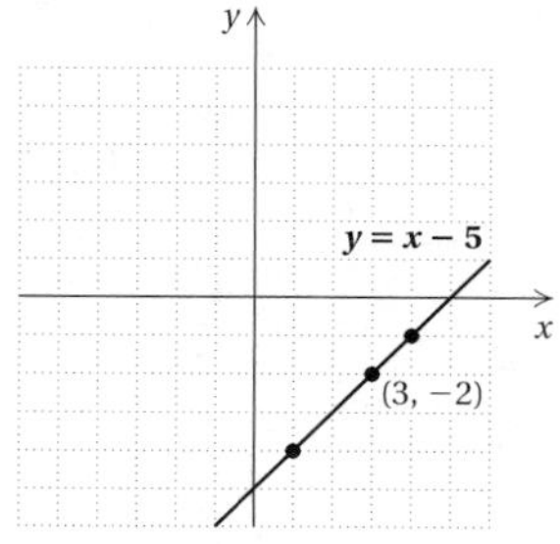

9.

$$\begin{array}{c|l} y & = \frac{1}{2}x + 3 \\ \hline 5 & ?\ \frac{1}{2}\cdot 4 + 3 \\ & 2 + 3 \\ & 5 \end{array} \quad \text{TRUE}$$

$$\begin{array}{c|l} y & = \frac{1}{2}x + 3 \\ \hline 2 & ?\ \frac{1}{2}(-2) + 3 \\ & -1 + 3 \\ & 2 \end{array} \quad \text{TRUE}$$

y
$y = \frac{1}{2}x + 3$
(−4, 1)
x

11.

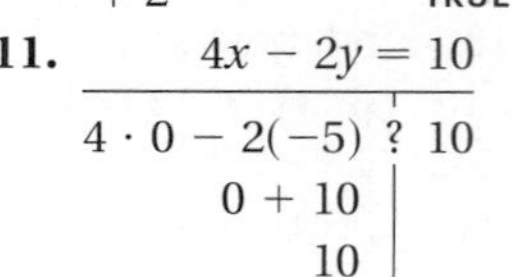

$$\begin{array}{l|c} 4x - 2y & = 10 \\ \hline 4\cdot 0 - 2(-5)\ ? & 10 \\ 0 + 10 & \\ 10 & \end{array} \quad \text{TRUE}$$

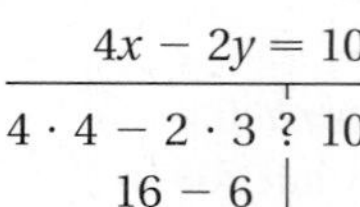

$$\begin{array}{l|c} 4x - 2y & = 10 \\ \hline 4\cdot 4 - 2\cdot 3\ ? & 10 \\ 16 - 6 & \\ 10 & \end{array} \quad \text{TRUE}$$

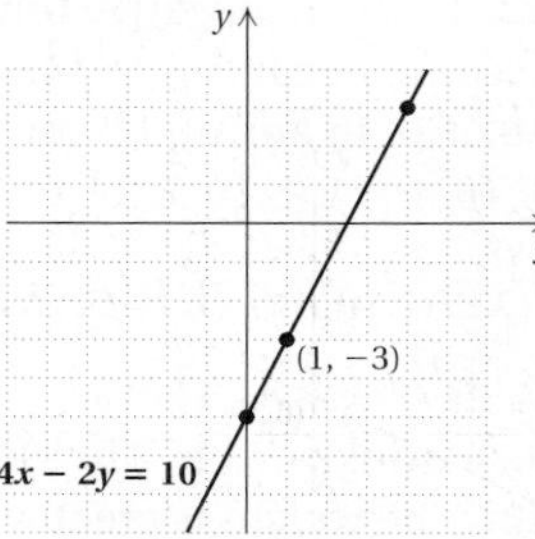

13.

x	y
−2	−1
−1	0
0	1
1	2
2	3
3	4

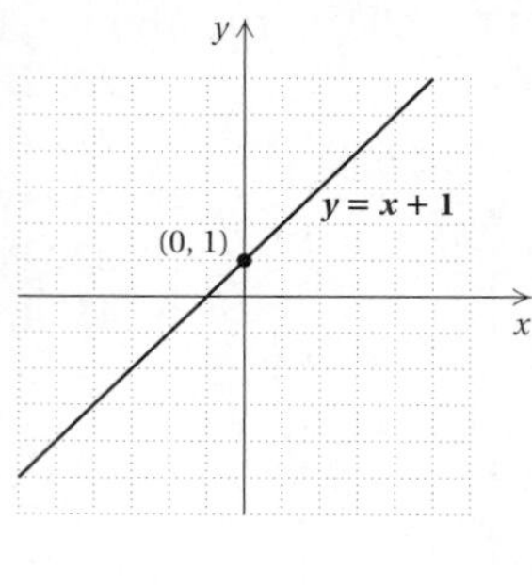

15.

x	y
−2	−2
−1	−1
0	0
1	1
2	2
3	3

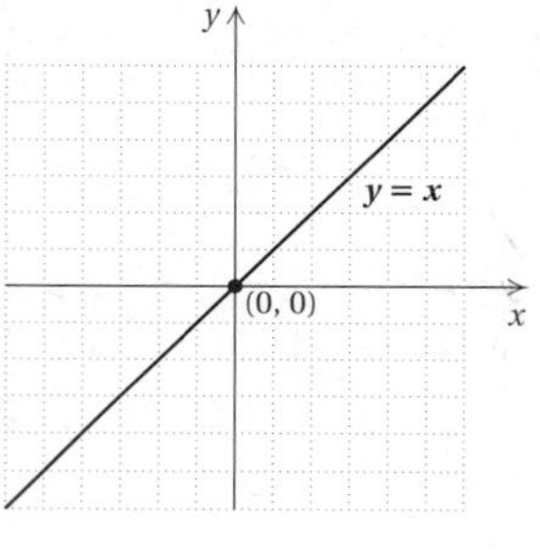

17.

x	y
−2	−1
0	0
4	2

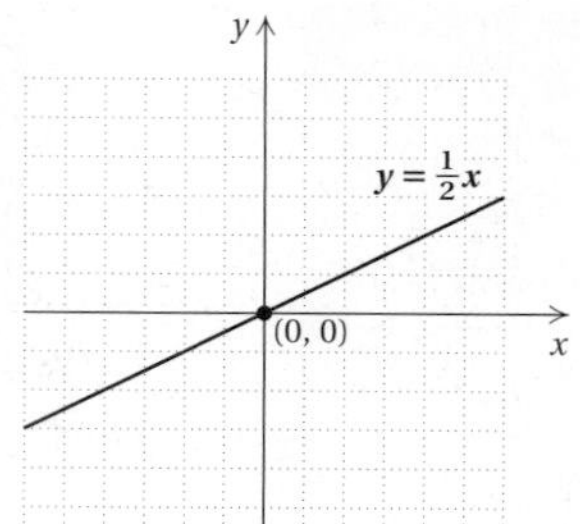

19.

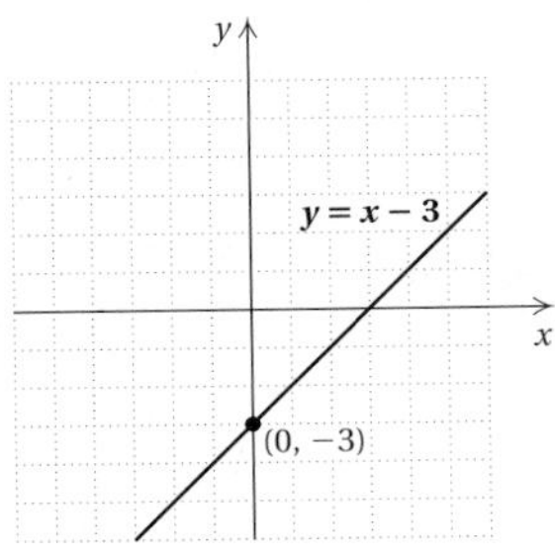

21.

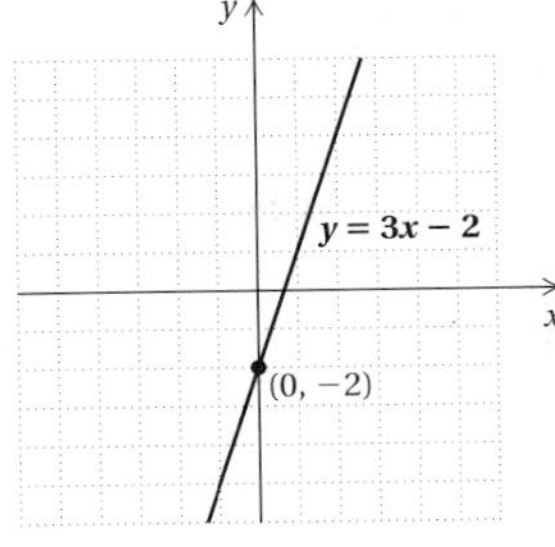

23.

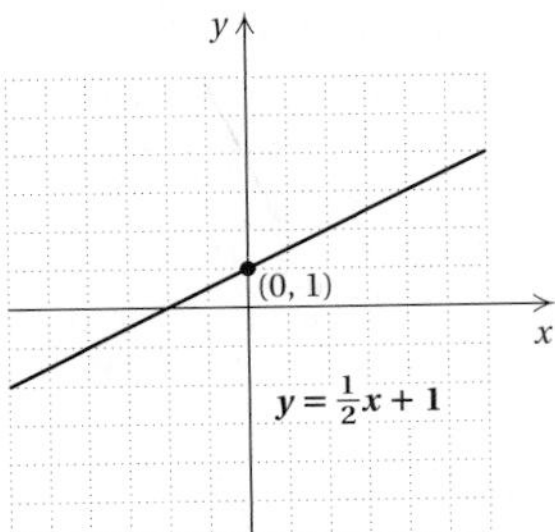

25.

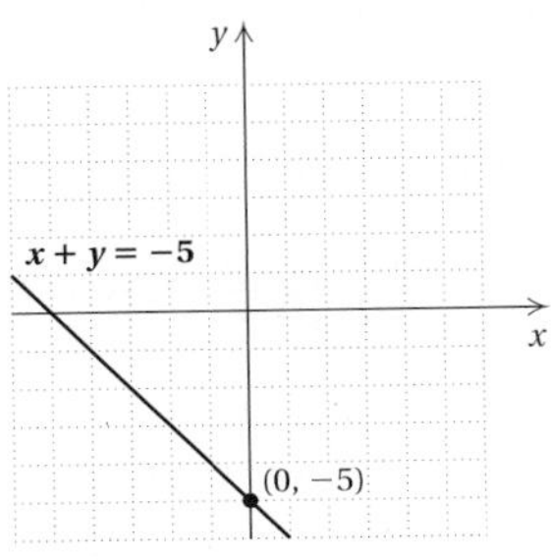

27.

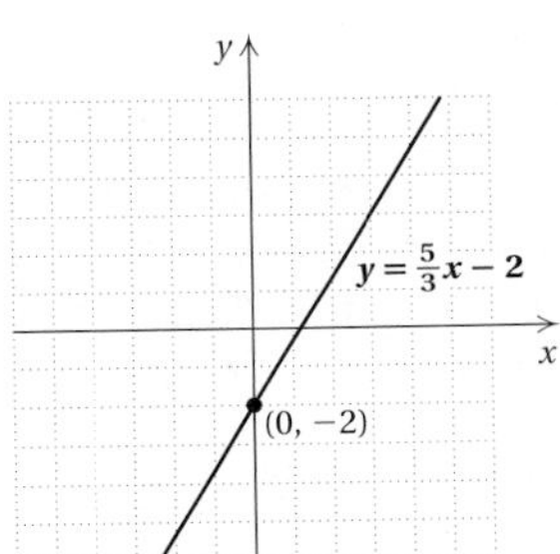

29.

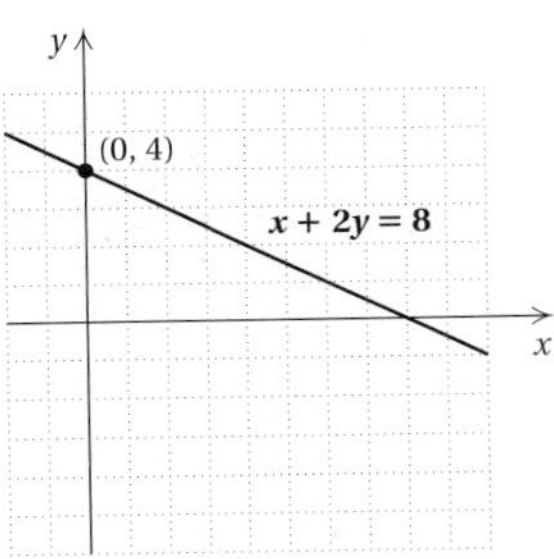

31.

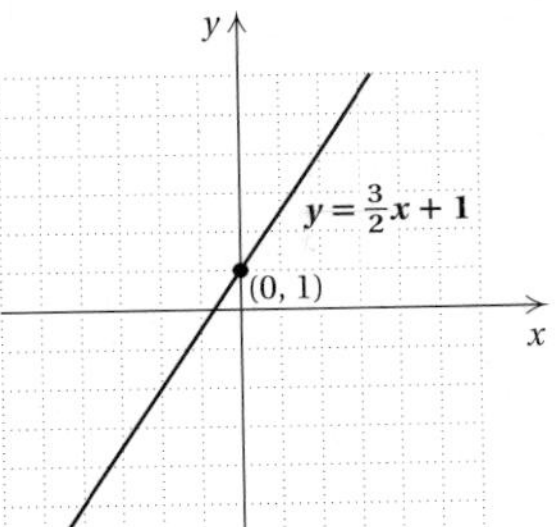

33.

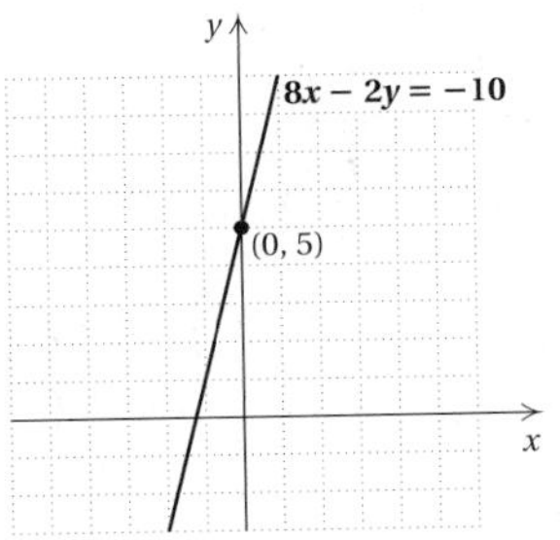

35.

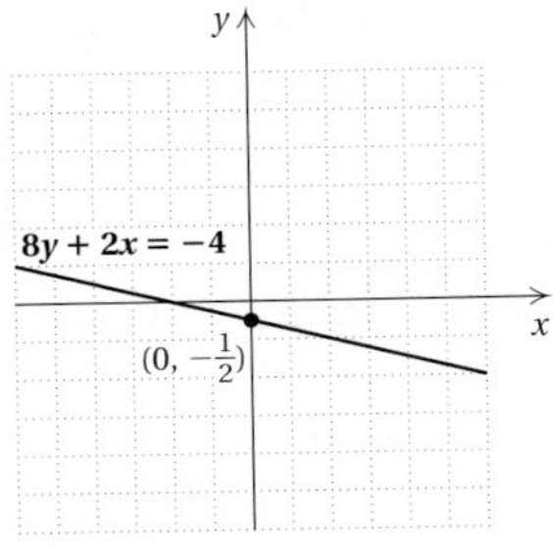

37. **(a)** \$300, \$100, \$0; **(b)**

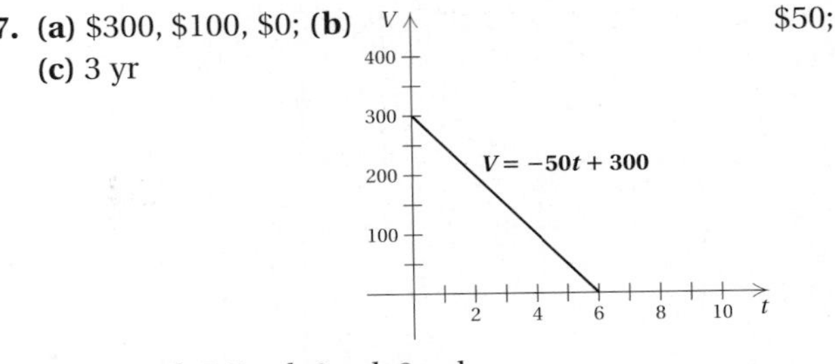

\$50; **(c)** 3 yr

39. **(a)** 7.1 gal, 7.5 gal, 8 gal, 9 gal; **(b)**

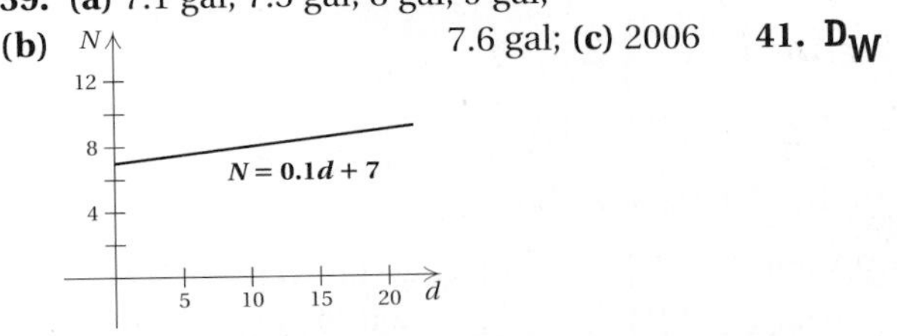

7.6 gal; **(c)** 2006 **41.** D_W

43. 3000 **44.** 125,000 **45.** 0 **46.** 6,078,000 **47.** 3000 **48.** 0.71875 **49.** −0.875 **50.** −2.25 **51.** 1.828125 **52.** −1.0625 **53.** $y = -x + 5$ **55.** $y = x + 2$

Margin Exercises, Section 3.3, pp. 257–261

1. **(a)** $(0, 3)$; **(b)** $(4, 0)$

2.

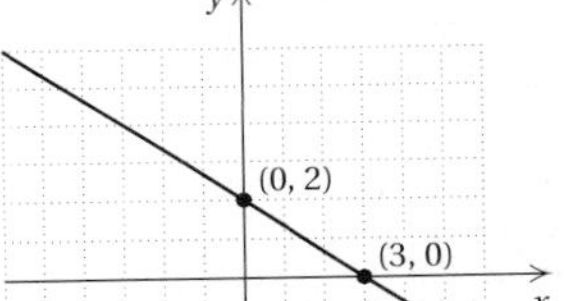

3.

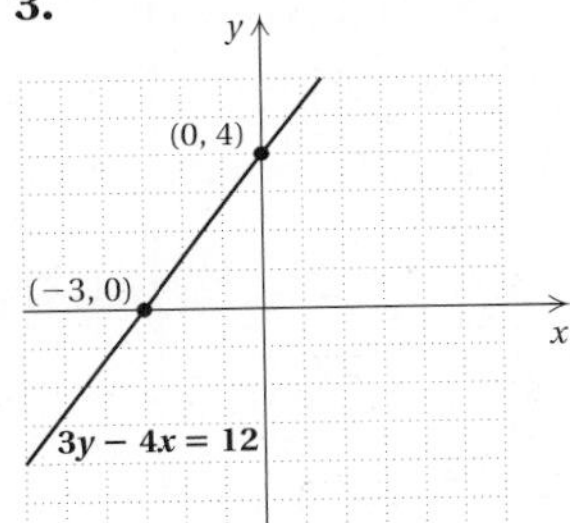

4.

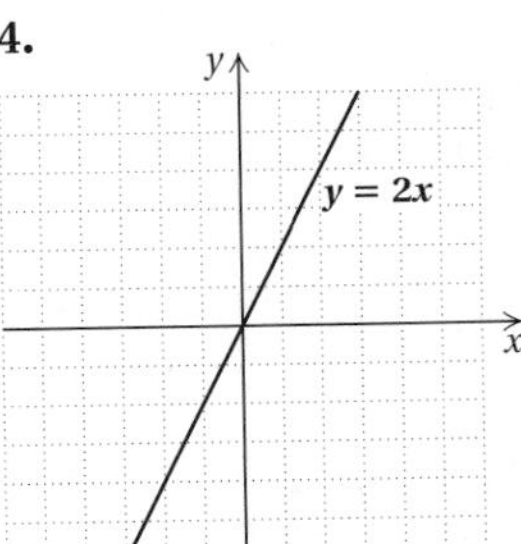

5.

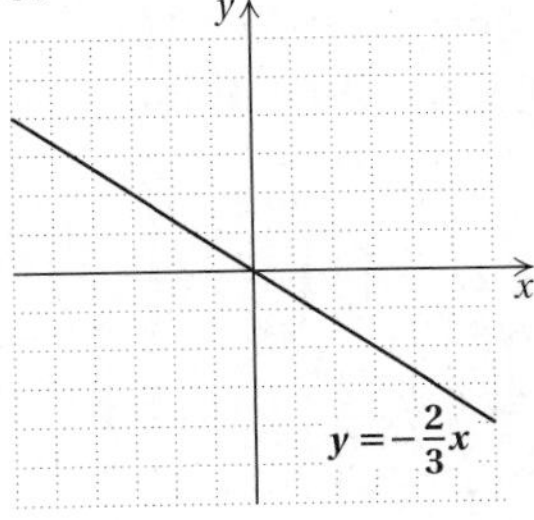

6.

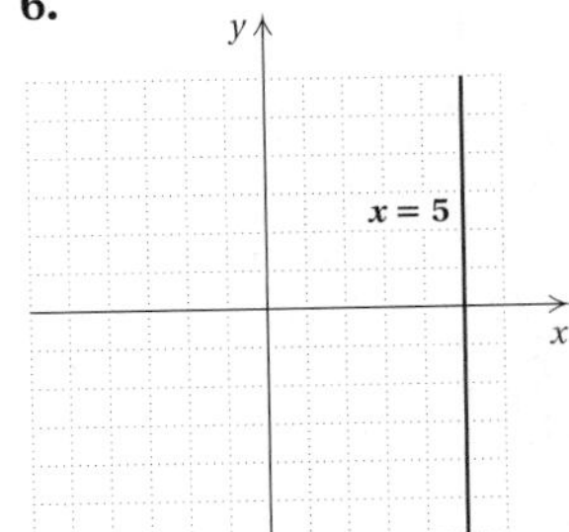

7.

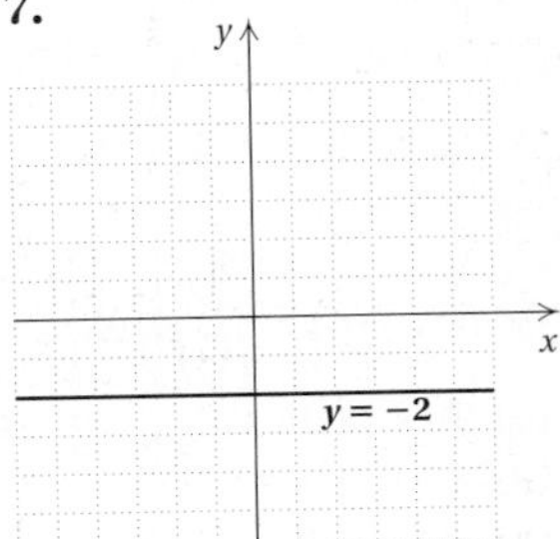

8.

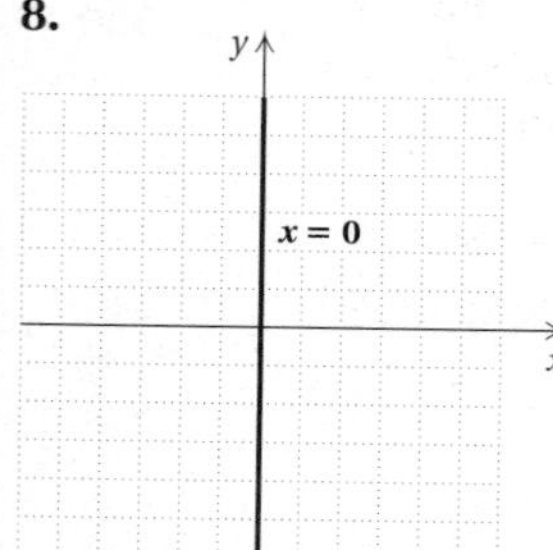

9.

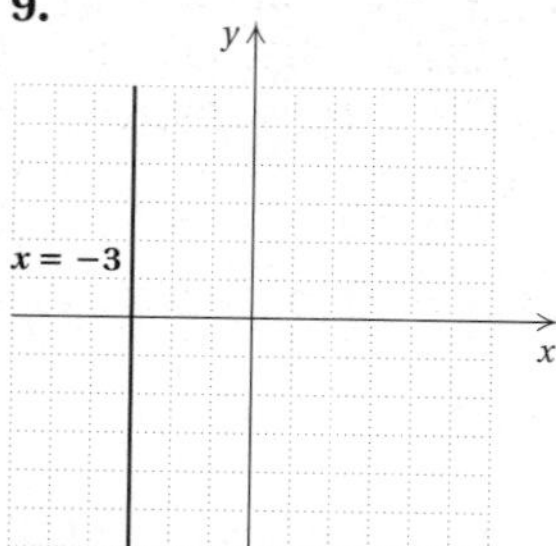

Calculator Corner, p. 259

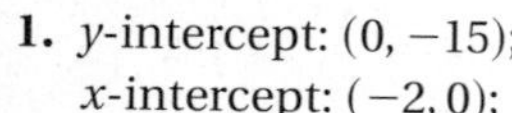

1. y-intercept: $(0, -15)$; x-intercept: $(-2, 0)$;

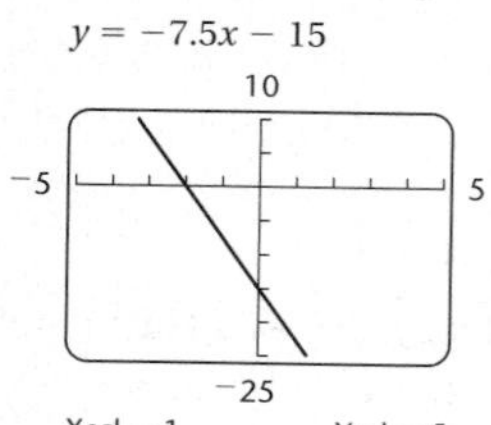

2. y-intercept: $(0, 43)$; x-intercept: $(-20, 0)$;

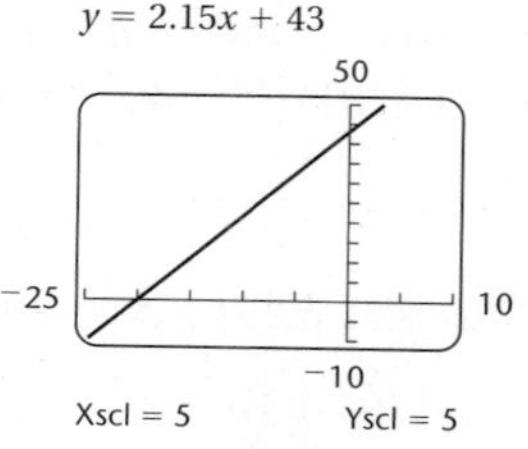

3. y-intercept: $(0, -30)$; x-intercept: $(25, 0)$;

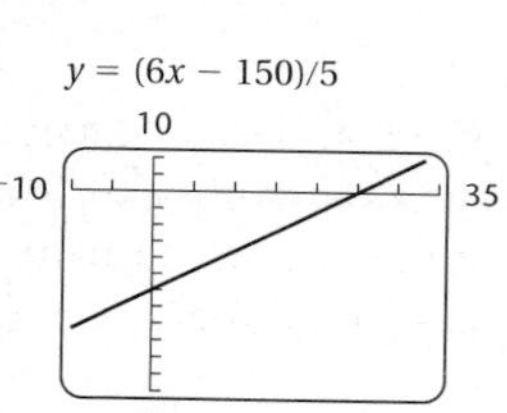

4. y-intercept: $(0, -4)$; x-intercept: $(20, 0)$;

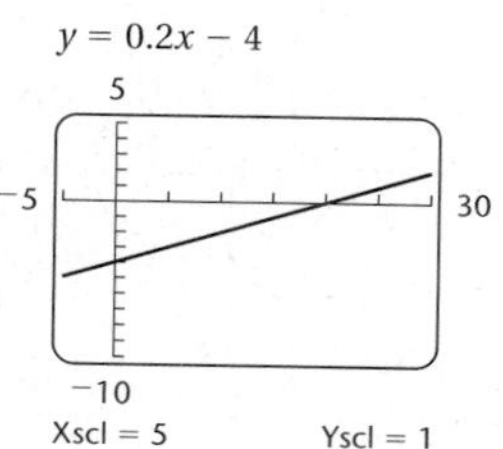

5. y-intercept: $(0, -15)$; x-intercept: $(10, 0)$;

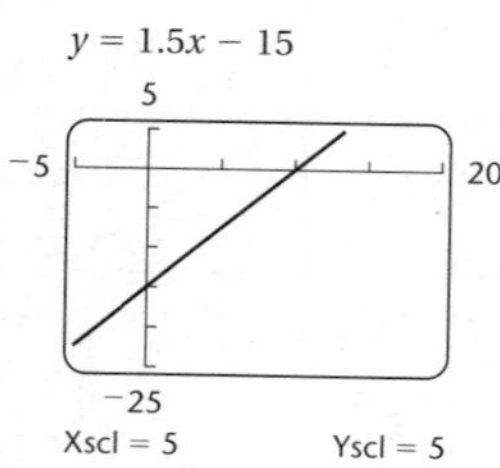

6. y-intercept: $\left(0, -\frac{1}{2}\right)$; x-intercept: $\left(\frac{2}{5}, 0\right)$;

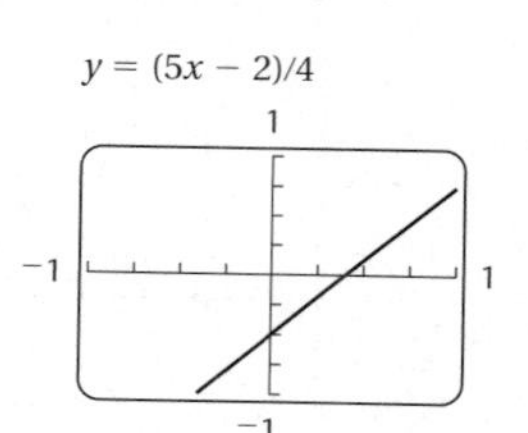

Exercise Set 3.3, p. 262

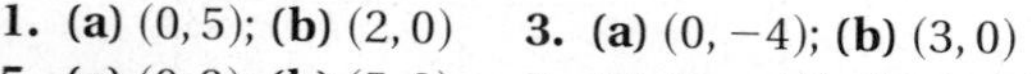

1. **(a)** $(0, 5)$; **(b)** $(2, 0)$ **3.** **(a)** $(0, -4)$; **(b)** $(3, 0)$
5. **(a)** $(0, 3)$; **(b)** $(5, 0)$ **7.** **(a)** $(0, -14)$; **(b)** $(4, 0)$
9. **(a)** $\left(0, \frac{10}{3}\right)$; **(b)** $\left(-\frac{5}{2}, 0\right)$ **11.** **(a)** $\left(0, -\frac{1}{3}\right)$; **(b)** $\left(\frac{1}{2}, 0\right)$

13.

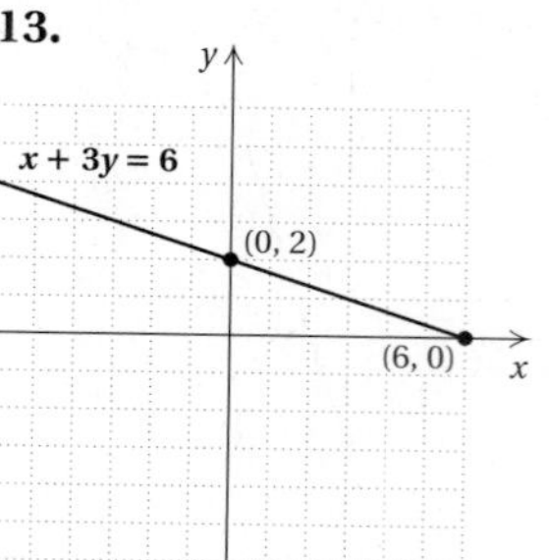

15.

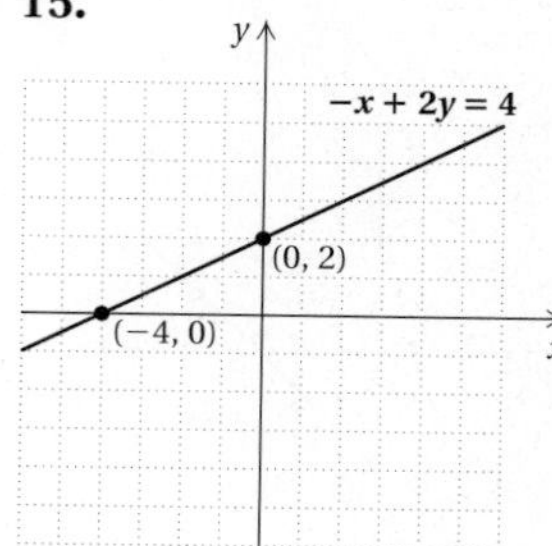

17.

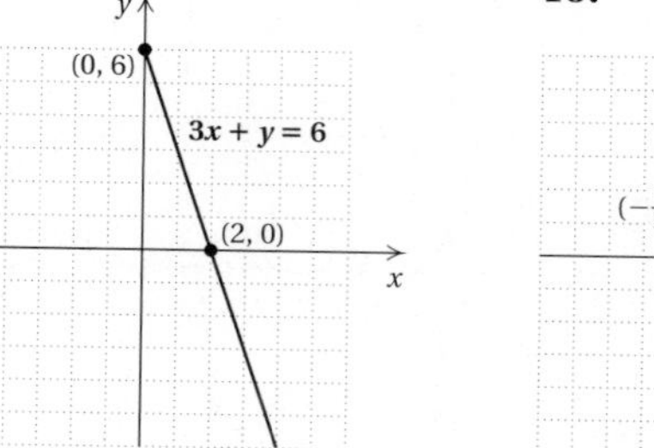

19.

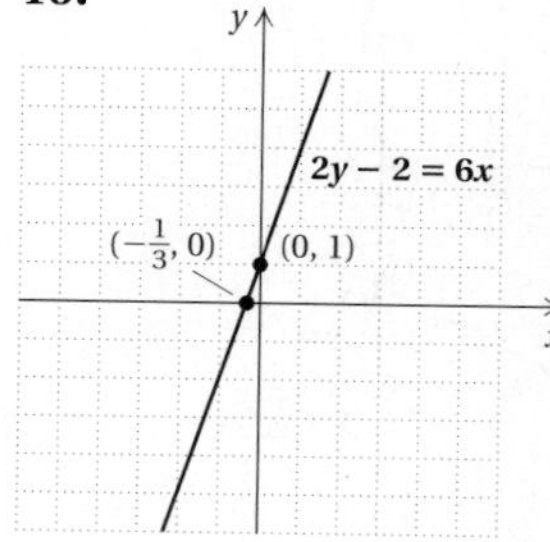

21.

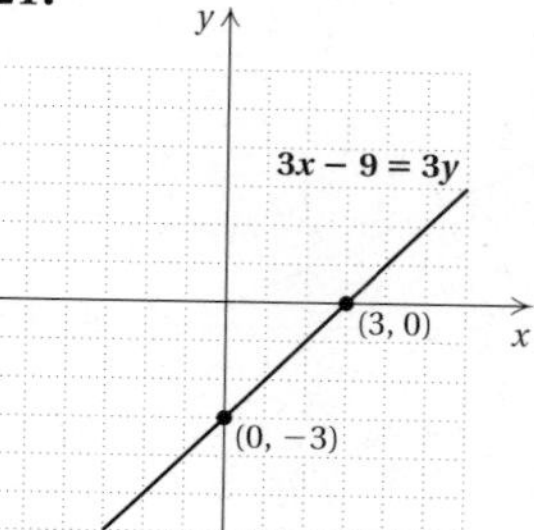

23.

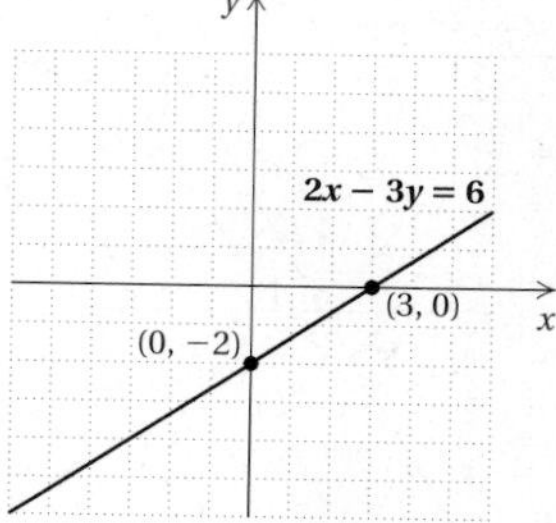

25.

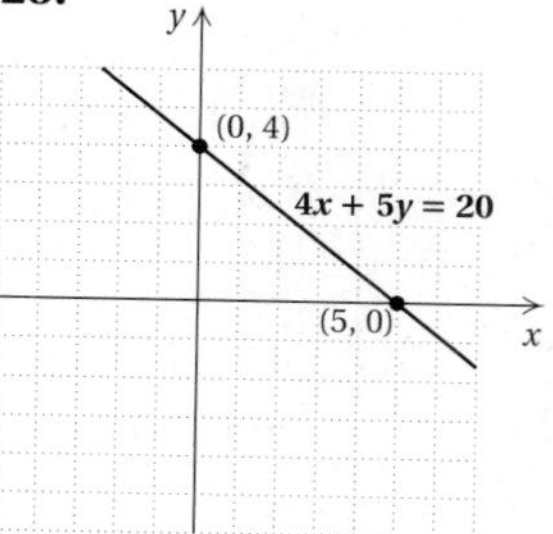

27.

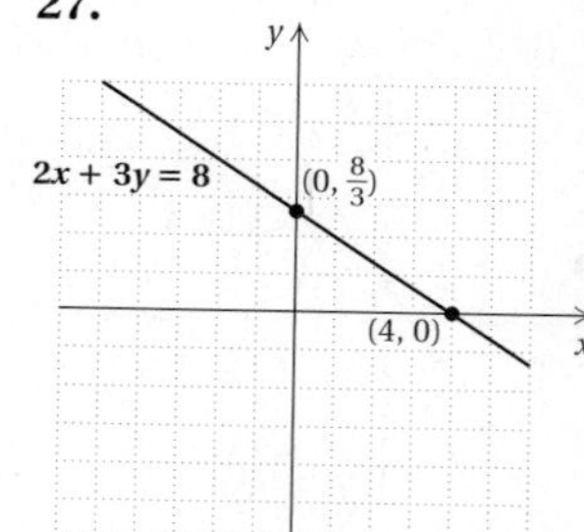

29.

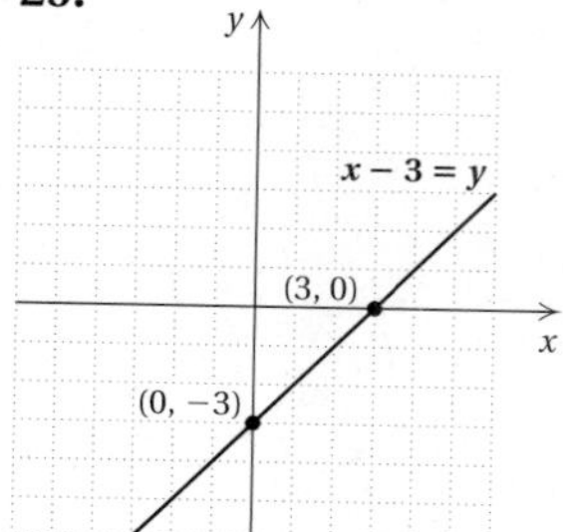

31.

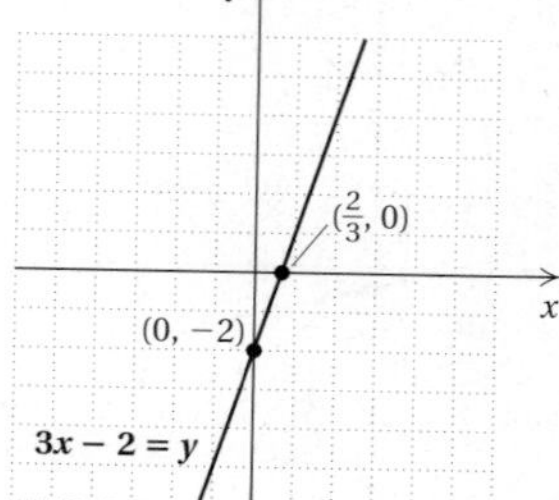

33.

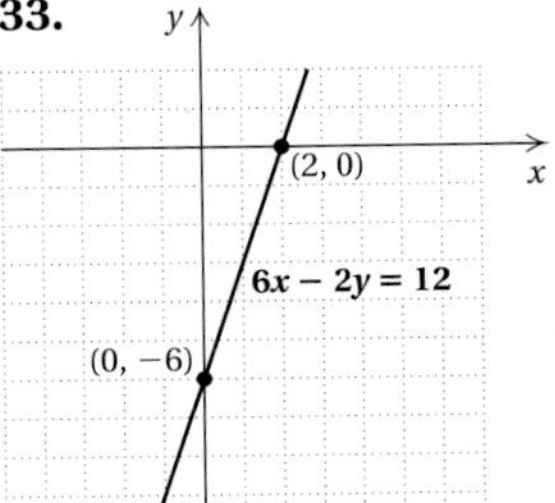

35.

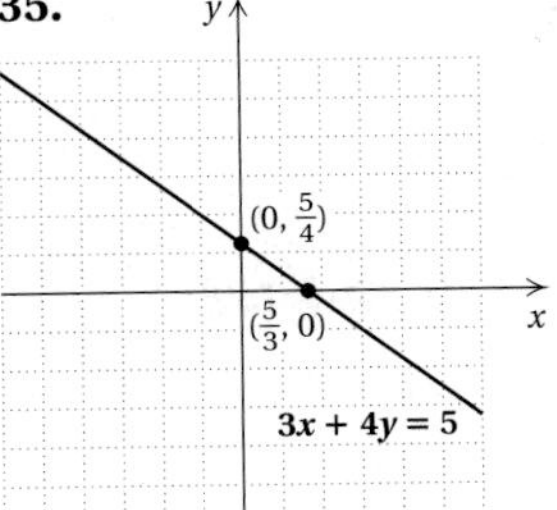

37.

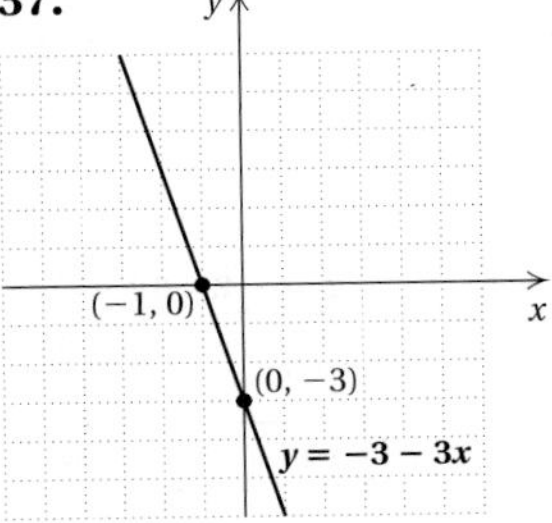

39.

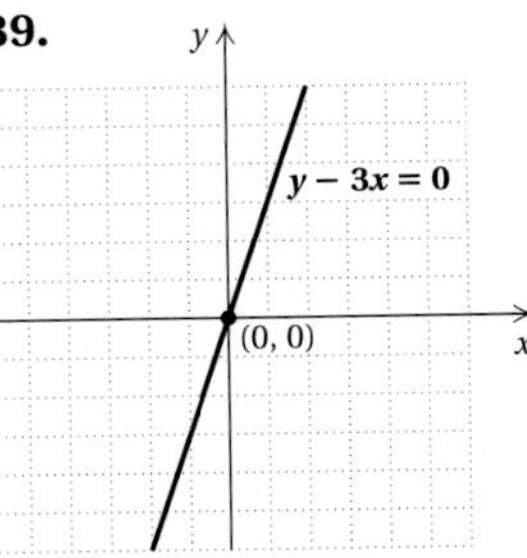

41.

y

x = −2

x

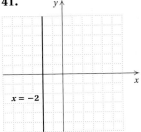

43.

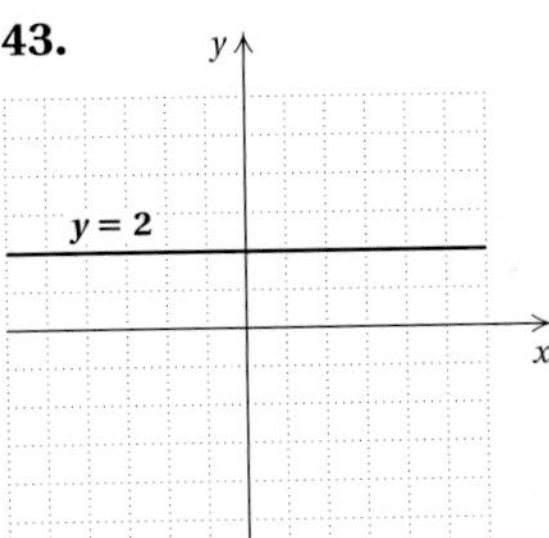

45.

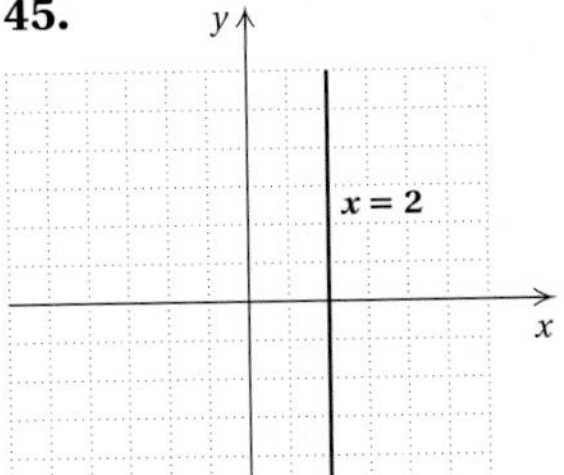

47.

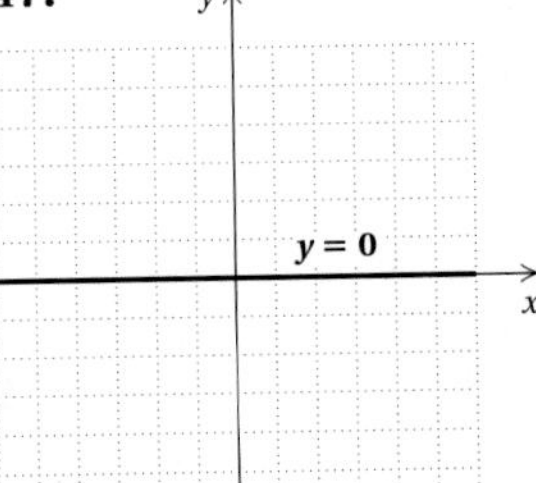

49.

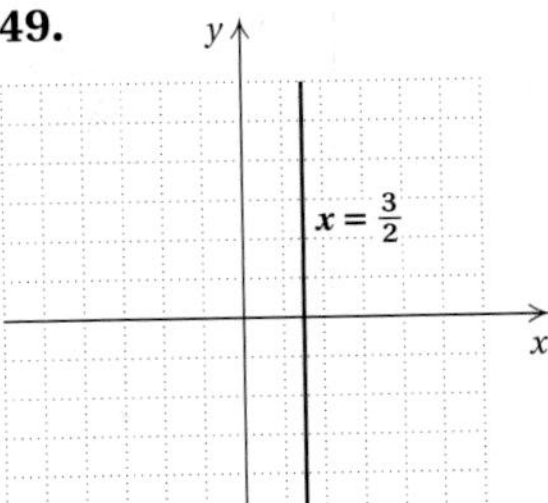

51.

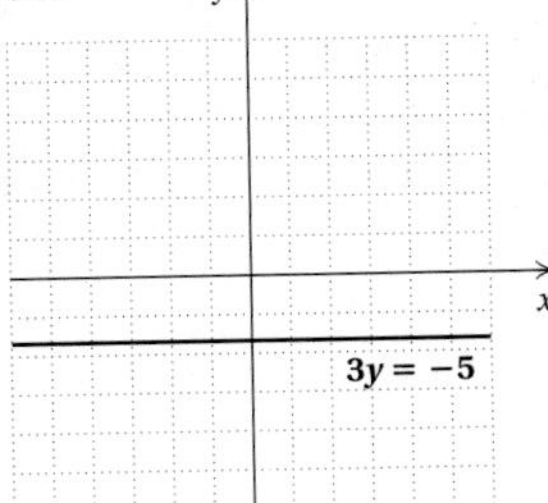

53.

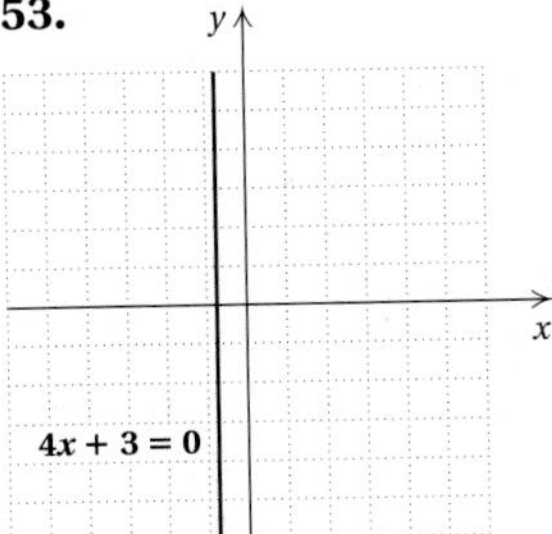

55.

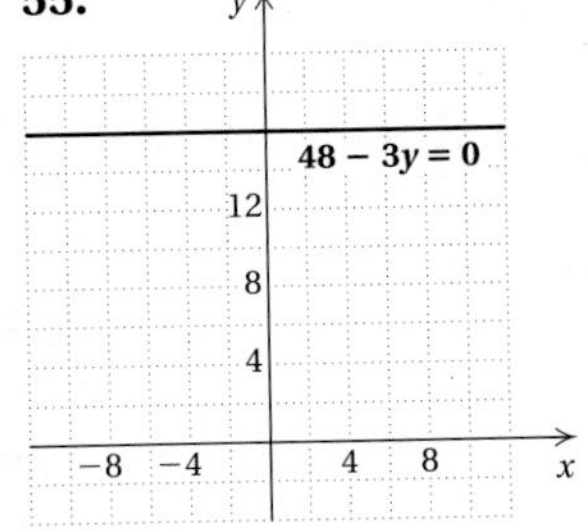

57. $y = -1$ **59.** $x = 4$ **61.** $\mathbf{D_W}$ **63.** 16% **64.** \$32.50 **65.** $\{x \mid x > -40\}$ **66.** $\{x \mid x \leq -7\}$ **67.** $\{x \mid x < 1\}$ **68.** $\{x \mid x \geq 2\}$ **69.** $y = -4$ **71.** $k = 12$

Margin Exercises, Section 3.4, pp. 269–274

1. $\frac{2}{5}$

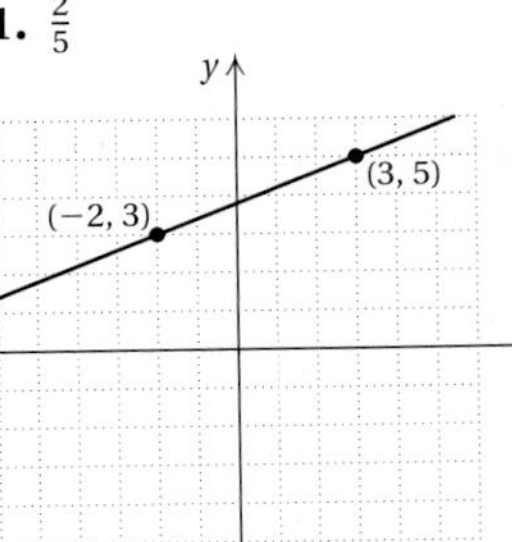

2. $-\frac{5}{3}$

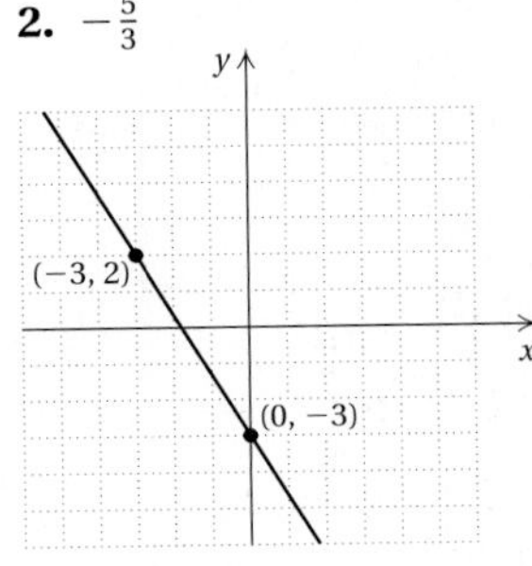

3. $63\frac{7}{11}\%$, or $63.\overline{63}\%$ **4.** 7 cents per minute **5.** −0.4% per year **6.** 4 **7.** −17 **8.** −1 **9.** $\frac{2}{3}$ **10.** −1 **11.** $\frac{5}{4}$ **12.** Not defined **13.** 0

Calculator Corner, p. 273

1. This line will pass through the origin and slant up from left to right. This line will be steeper than $y = 10x$.
2. This line will pass through the origin and slant up from left to right. This line will be less steep than $y = \frac{5}{32}x$.

Calculator Corner, p. 274

1. This line will pass through the origin and slant down from left to right. This line will be steeper than $y = -10x$.
2. This line will pass through the origin and slant down from left to right. This line will be less steep than $y = -\frac{5}{32}x$.

Exercise Set 3.4, p. 275

1. $-\frac{3}{7}$ **3.** $\frac{2}{3}$ **5.** $\frac{3}{4}$ **7.** 0

9. $-\frac{4}{5}$;

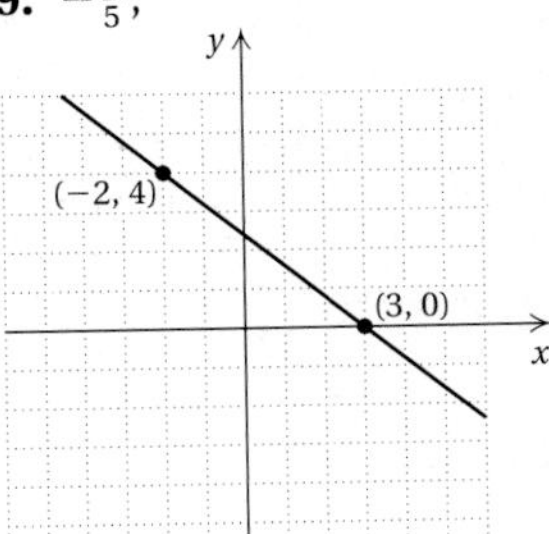

11. 3;

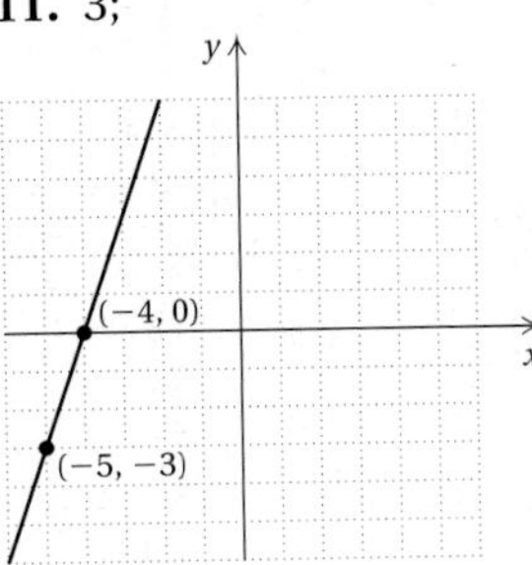

13. $-\frac{5}{6}$;

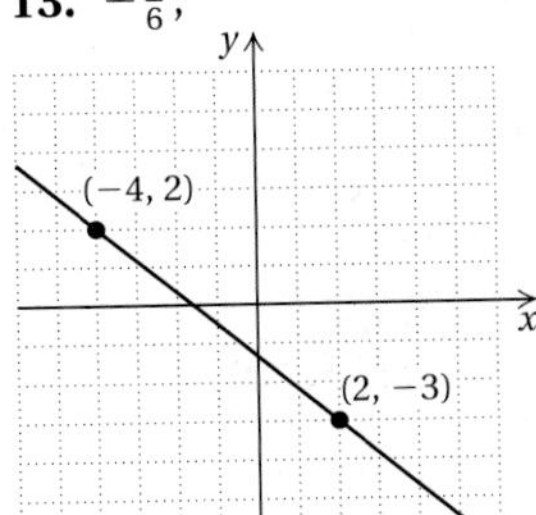

15. $\frac{7}{8}$;

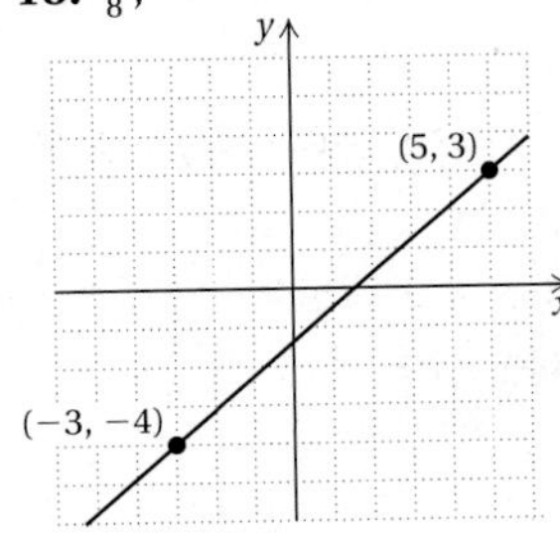

17. $\frac{2}{3}$ **19.** Not defined **21.** $\frac{12}{41}$ **23.** $\frac{28}{129}$ **25.** About 29.4% **27.** 25 miles per gallon **29.** $-\$500$ per year **31.** About 7689 people per year **33.** -10 **35.** 3.78 **37.** 3 **39.** $-\frac{1}{5}$ **41.** $-\frac{3}{2}$ **43.** $\frac{5}{7}$ **45.** -2.74 **47.** 3 **49.** $\frac{5}{4}$ **51.** 0 **53.** $\mathbf{D_W}$ **55.** $\frac{4}{25}$ **56.** $\frac{1}{3}$ **57.** $\frac{3}{8}$ **58.** $\frac{3}{4}$ **59.** \$3.57 **60.** \$48.60 **61.** 20% **62.** \$18 **63.** \$45.15 **64.** \$55

65. $y = 0.35x - 7$

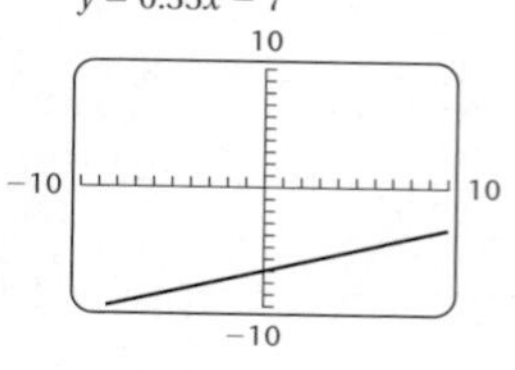

X	Y1
−10	−10.5
−9.9	−10.47
−9.8	−10.43
−9.7	−10.4
−9.6	−10.36
−9.5	−10.33
−9.4	−10.29

X = −10

67. $y = x^3 - 5$

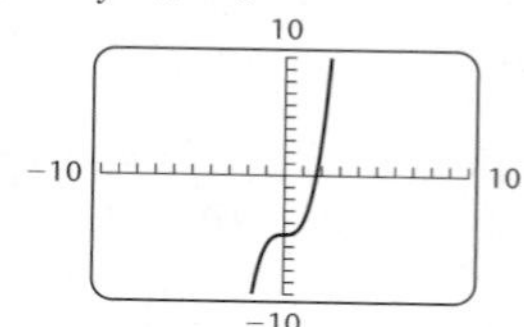

X	Y1
−10	−1005
−9.9	−975.3
−9.8	−946.2
−9.7	−917.7
−9.6	−889.7
−9.5	−862.4
−9.4	−835.6

X = −10

Summary and Review: Chapter 3, p. 280

1. \$775.50; \$634.50 **2.** 47 lb **3.** 80 lb **4.** 33 lb **5.** 1993 **6.** 1990–1995 **7.** One shower **8.** One toilet flush **9.** One shave, wash dishes, one shower **10.** One toilet flush **11.** $(-5, -1)$ **12.** $(-2, 5)$ **13.** $(3, 0)$

14.–16.

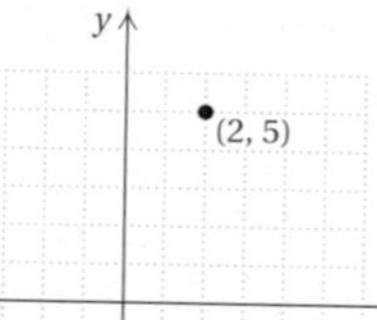
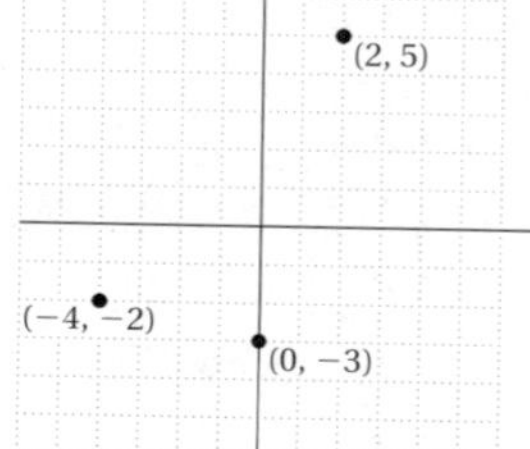

17. IV **18.** III **19.** I **20.** No **21.** Yes

22.

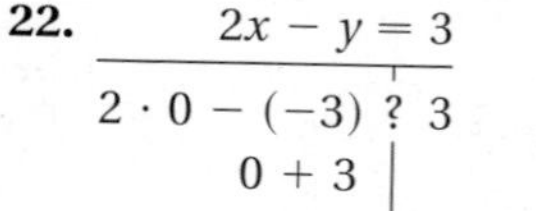

$2x - y = 3$	
$2 \cdot 0 - (-3)$	? 3
$0 + 3$	
3	TRUE

$2x - y = 3$	
$2 \cdot 2 - 1$	? 3
$4 - 1$	
3	TRUE

y
(3, 3)
x
$2x - y = 3$

23.

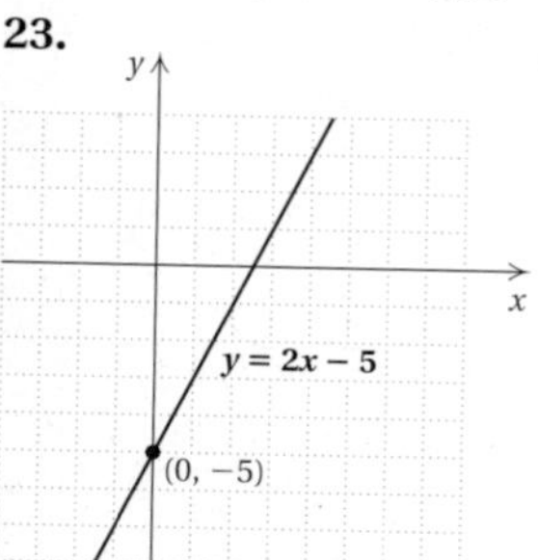

24.

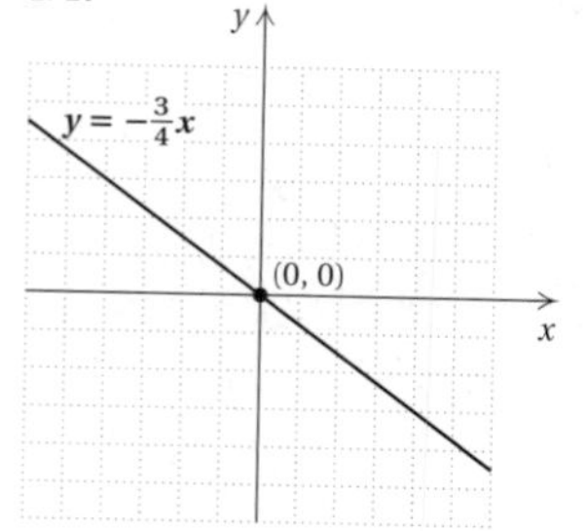

25.

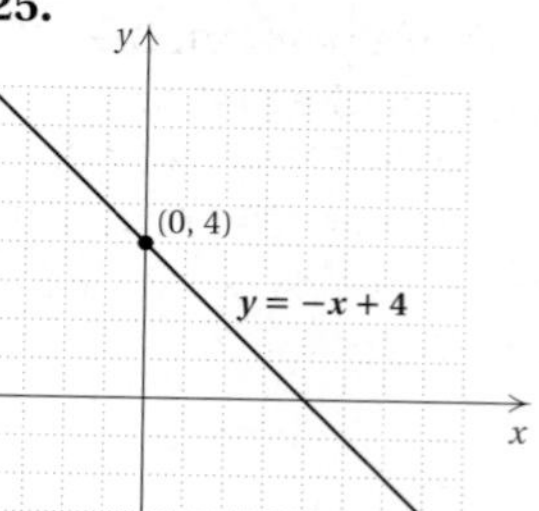

26.

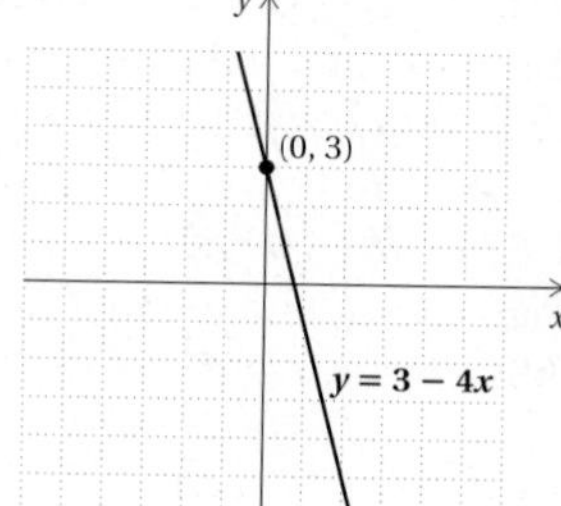

27.

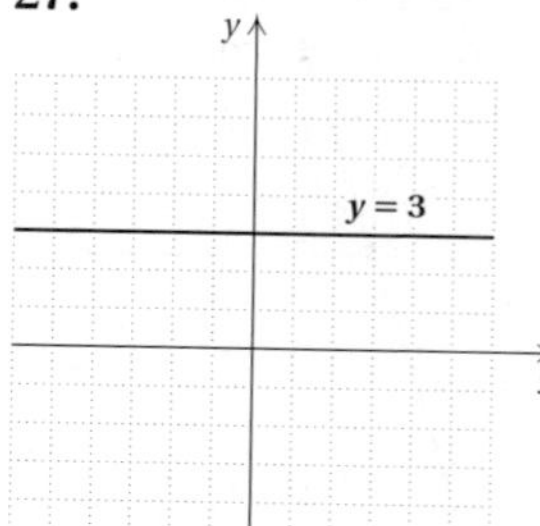

28.

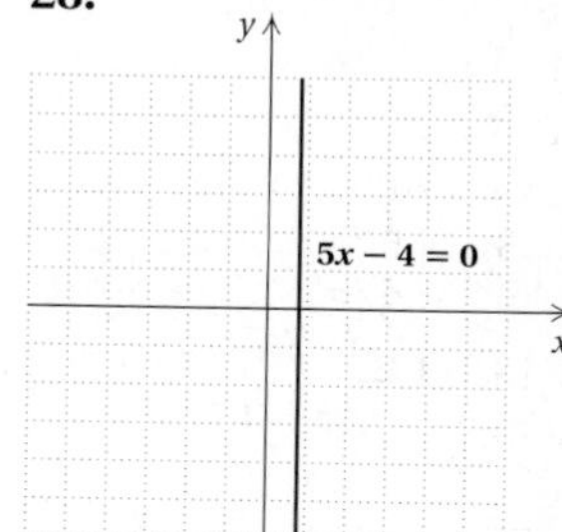

29.

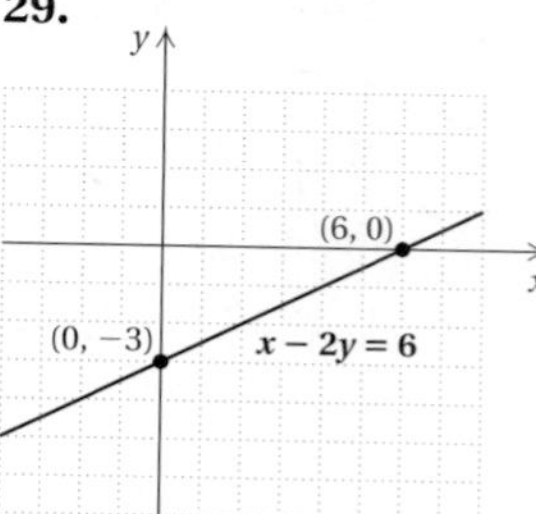

30.

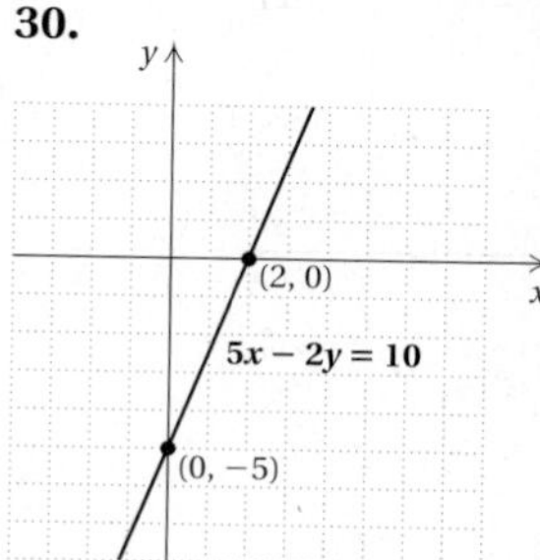

31. (a) $14\frac{1}{2}$ ft^3, 16 ft^3, $20\frac{1}{2}$ ft^3, 28 ft^3; **(b)**

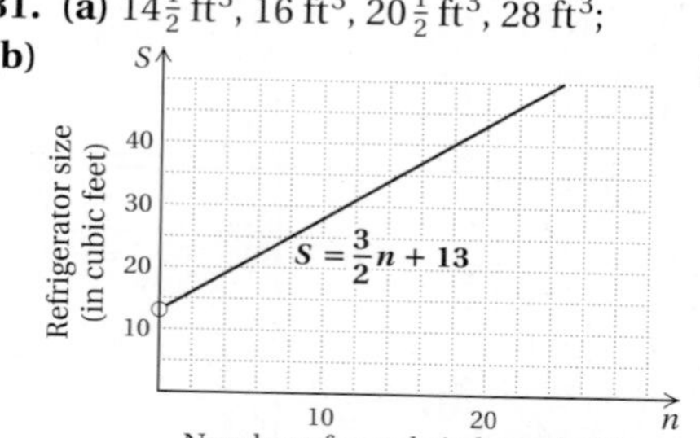

$17\frac{1}{2}$ ft^3; **(c)** 6

32. (a) 2.4 driveways per hour; **(b)** 25 minutes per driveway **33.** 4 manicures per hour **34.** $\frac{1}{3}$ **35.** $-\frac{1}{3}$ **36.** $\frac{3}{5}$;

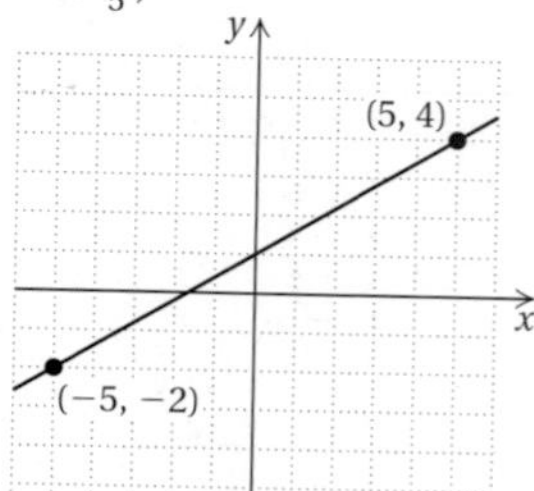

37. -1;

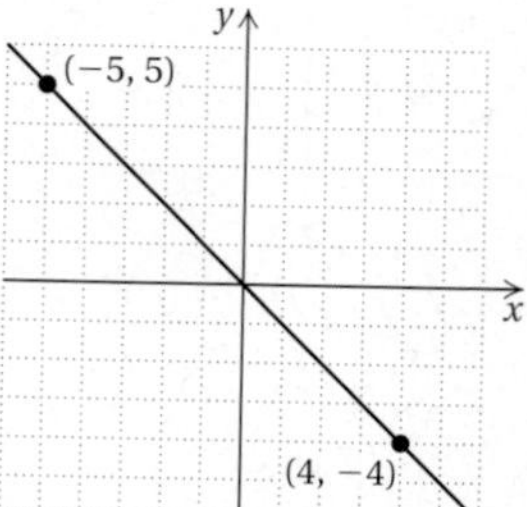

38. 7% **39.** $-\frac{5}{8}$ **40.** $\frac{1}{2}$ **41.** Not defined **42.** 0 **43.** $\mathbf{D_W}$ A small business might use a graph to look up prices quickly (as in the FedEx mailing costs example) or to plot change in sales over a period of time. Many other applications exist. **44.** $\mathbf{D_W}$ The *y*-intercept is the point at which the graph crosses the y-axis. Since a point on the y-axis is neither left nor right of the origin, the first or x-coordinate of the point is 0. **45.** -0.34375 **46.** $0.\overline{8}$ **47.** 3.2 **48.** $\frac{17}{19}$ **49.** 42.71 **50.** 112.53 **51.** \$9755.09 **52.** \$79.95 **53.** $m = -1$ **54.** 45 square units; 28 linear units **55.** **(a)** 3.709 feet per minute; **(b)** about 0.2696 minute per foot

Test: Chapter 3, p. 285

1. [3.1a] \$495,000,000 **2.** [3.1a] Crest and Colgate **3.** [3.1a] Crest **4.** [3.1a] Arm & Hammer **5.** [3.1a] June **6.** [3.1a] January **7.** [3.1a] March, April, May, June, July **8.** [3.1a] August **9.** [3.1a] 2001 **10.** [3.1a] 1996 **11.** [3.1a] 5002 **12.** [3.1a] 2000 and 2001 **13.** [3.1a] 3481 **14.** [3.1a] 2309 **15.** [3.1c] II **16.** [3.1c] III **17.** [3.1d] (3, 4) **18.** [3.1d] (0, −4)
19. [3.2a]

$$\begin{array}{c|l} y - 2x = 5 & \\ \hline -3 - 2(-4) \ ? \ 5 & \\ -3 + 8 & \\ 5 & \quad \text{TRUE} \end{array}$$

$$\begin{array}{c|l} y - 2x = 5 & \\ \hline 3 - 2(-1) \ ? \ 5 & \\ 3 + 2 & \\ 5 & \quad \text{TRUE} \end{array}$$

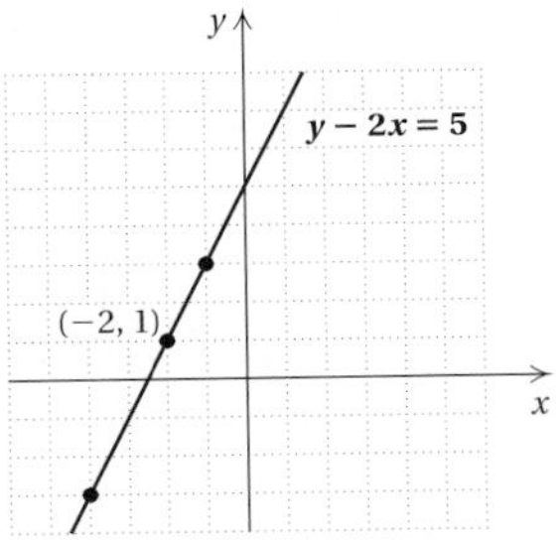

20. [3.2b]

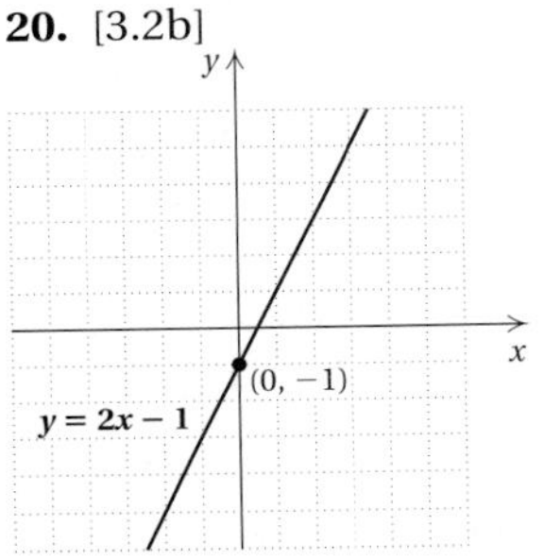

21. [3.2b]

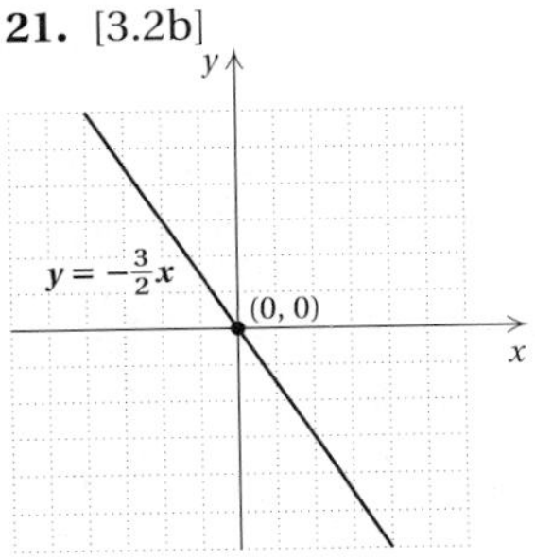

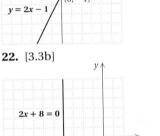

22. [3.3b]

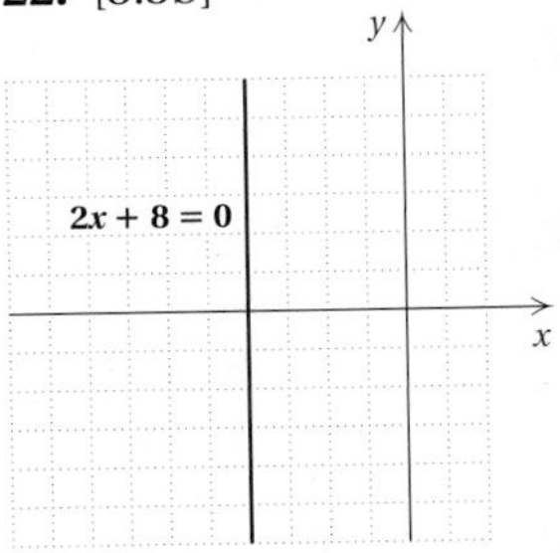

23. [3.3b]

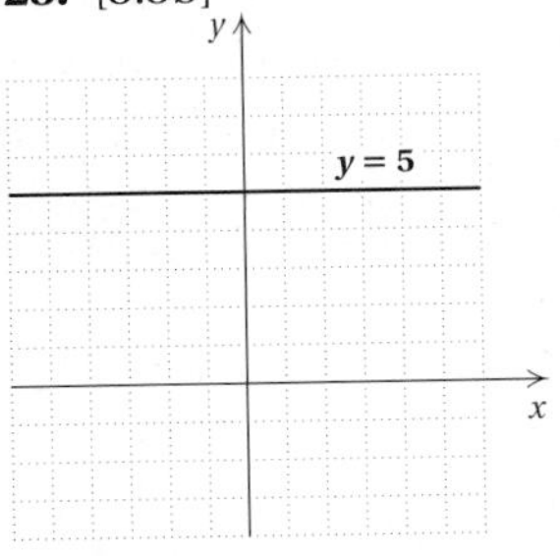

24. [3.3a]

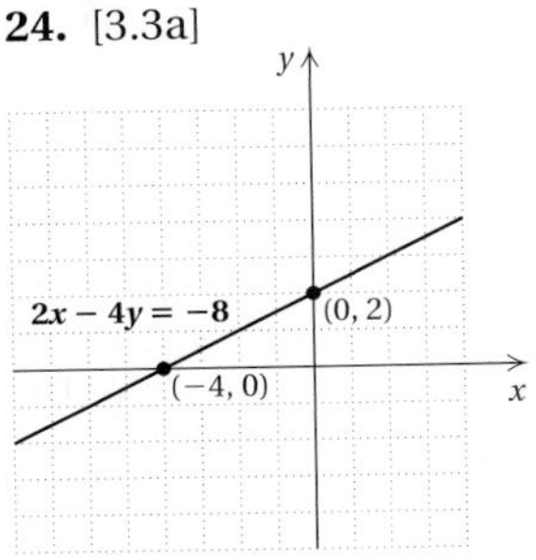

25. [3.3a]

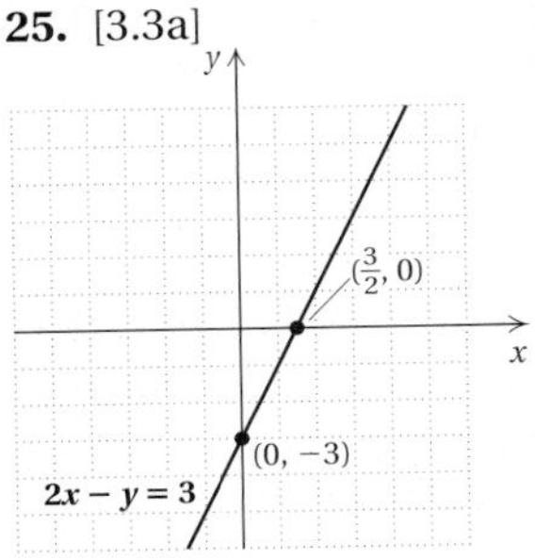

26. [3.2c] **(a)** \$17,000; \$19,400; \$22,600; \$24,200; \$27,500; **(b)**

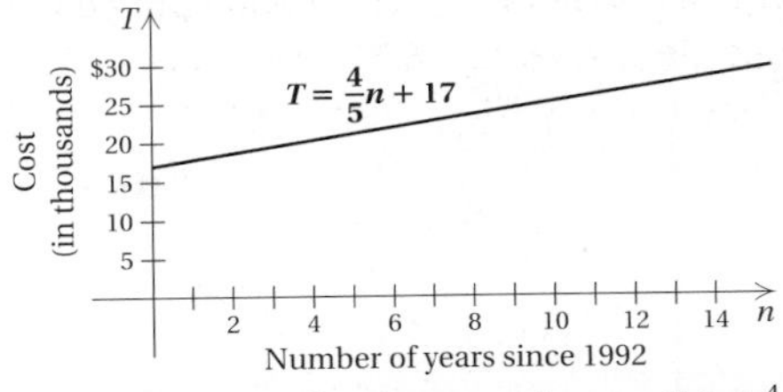

(c) 2002

27. [3.4b] **(a)** 14.5 floors per minute; **(b)** $4\frac{4}{29}$ seconds per floor **28.** [3.4b] 87.5 miles per hour **29.** [3.4a] −2 **30.** [3.4a] $\frac{3}{8}$;

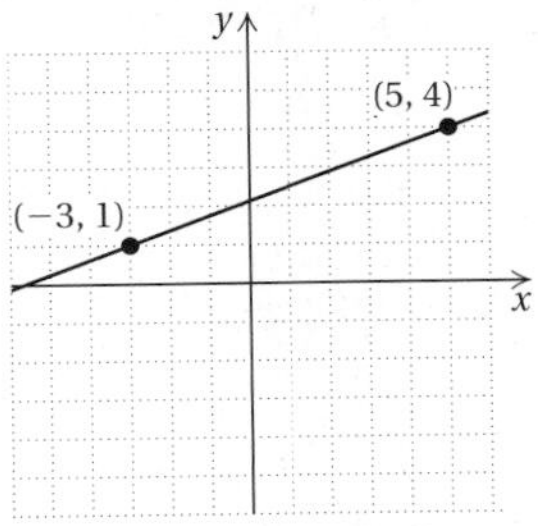

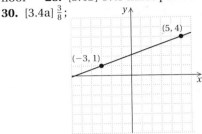

31. [3.4b] $-\frac{1}{20}$ **32.** [3.4c] **(a)** $\frac{2}{5}$; **(b)** not defined **33.** [1.2c] 0.975 **34.** [1.2c] $-1.08\overline{3}$ **35.** [1.2e] 71.2 **36.** [1.2e] $\frac{13}{47}$ **37.** [R.3c] 42.705 **38.** [R.3c] 112.527 **39.** [2.5a] \$84.50 **40.** [2.5a] \$36,400 **41.** [3.1b] 25 square units, 20 linear units **42.** [3.3b] $y = 3$

Cumulative Review: Chapters 1–3, p. 289

1. [1.1a] $\frac{5}{2}$ **2.** [1.7c] $12x - 15y + 21$ **3.** [1.7d] $3(5x - 3y + 1)$ **4.** [R.1a] $2 \cdot 3 \cdot 7$ **5.** [1.2c] 0.45 **6.** [1.2e] 4 **7.** [1.3b] 3.08 **8.** [1.6b] $-\frac{7}{8}$ **9.** [1.7e] $-x - y$ **10.** [R.4a] 0.785 **11.** [R.2c] $\frac{1}{3}$ **12.** [1.3a] 2.6 **13.** [1.5a] 7.28 **14.** [1.6c] $-\frac{5}{12}$ **15.** [1.8d] −2 **16.** [1.8d] 27 **17.** [1.8b] $-2y - 7$ **18.** [1.8c] $5x + 11$ **19.** [2.1b] −1.2 **20.** [2.2a] −21 **21.** [2.3a] 9 **22.** [2.2a] $\frac{4}{25}$ **23.** [2.3b] 2 **24.** [2.1b] $\frac{13}{8}$ **25.** [2.3c] $-\frac{17}{21}$ **26.** [2.3b] −17 **27.** [2.3b] 2 **28.** [2.7e] $\{x \mid x < 16\}$ **29.** [2.7e] $\{x \mid x \le -\frac{11}{8}\}$ **30.** [2.3c] All real numbers **31.** [2.3c] No solution **32.** [2.3c] All real numbers **33.** [2.4b] $h = \frac{2A}{b + c}$ **34.** [3.1c] IV **35.** [2.7b]

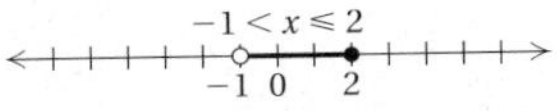

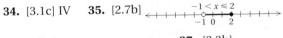

36. [3.3a]

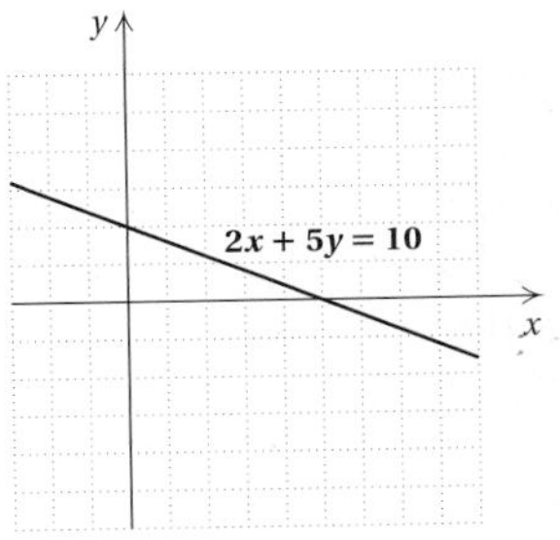

37. [3.3b]

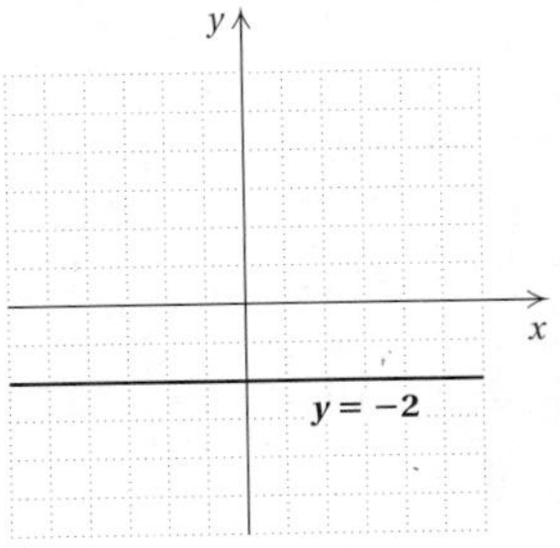

38. [3.2b]

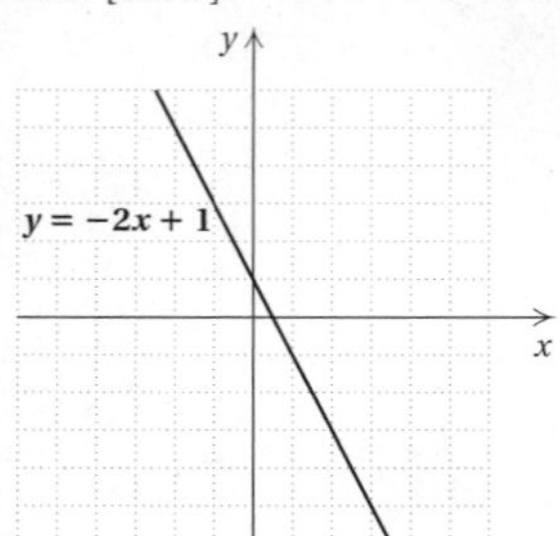

39. [3.2b]

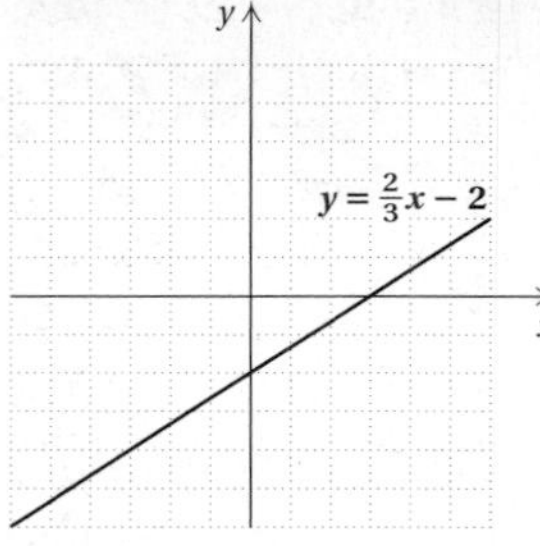

40. [3.2b]

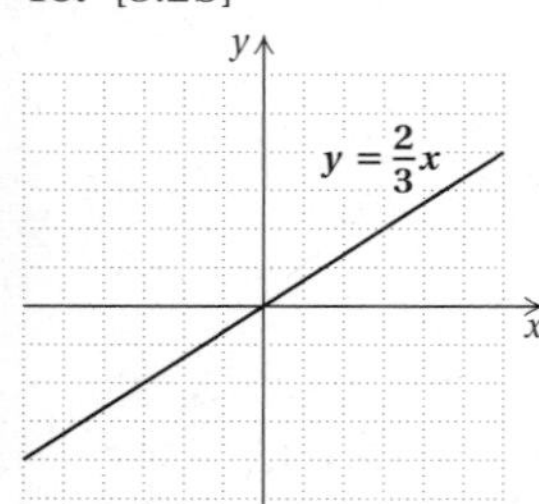

41. [3.2b]

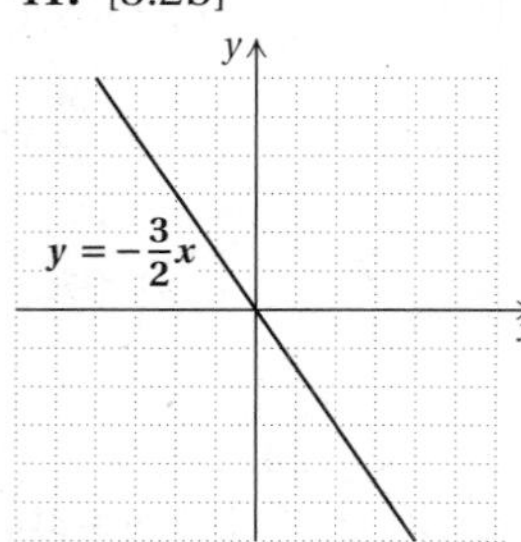

42. [3.3a] y-intercept: $(0, -3)$; x-intercept: $(10.5, 0)$
43. [3.3a] y-intercept: $(0, 5)$; x-intercept: $\left(-\frac{5}{4}, 0\right)$
44. [2.5a] 160 million **45.** [2.6a] 15.6 million
46. [2.5a] \$120 **47.** [2.6a] First: 50 m; second: 53 m; third: 40 m **48.** [2.8b] $\{x \mid x \leq 8\}$
49. [3.2c] **(a)** \$375, \$450, \$525, \$825; \$1050; **(c)** 196 months
(b)

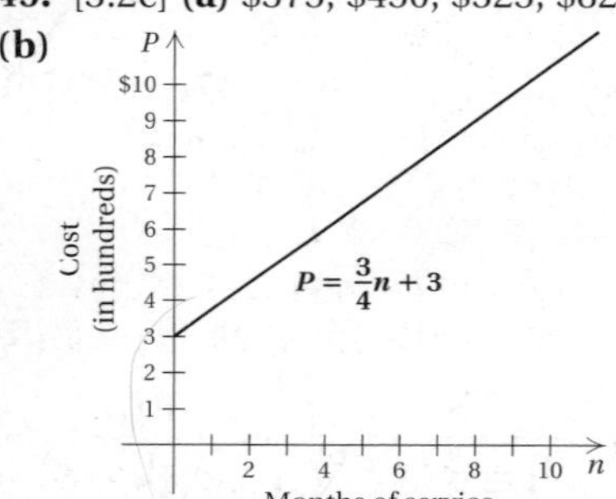

50. [3.4b] $\frac{1}{12}$ gal per mile **51.** [3.4a] $-\frac{1}{4}$ **52.** [3.3a] (b)
53. [1.8d] (e) **54.** [1.8b] (d) **55.** [3.4a] (d)
56. [2.3a], [1.2e] -4, 4 **57.** [2.3b] 3
58. [2.4b] $Q = \frac{2 - pm}{p}$, or $\frac{2}{p} - m$

Chapter 4

Pretest: Chapter 4, p. 294

1. x^2 **2.** $\frac{1}{x^7}$ **3.** $\frac{16x^4}{y^6}$ **4.** $\frac{1}{p^3}$ **5.** 3.47×10^{-4}
6. 3,400,000 **7.** 1.395×10^3 **8.** 8×10^{-2}
9. 3, 2, 1, 0; 3 **10.** $-3a^3b - 2a^2b^2 + ab^3 + 12b^3 + 9$
11. $11x^2 + 4x - 11$ **12.** $-x^2 - 18x + 27$
13. $15x^4 - 20x^3 + 5x^2$ **14.** $x^2 + 10x + 25$ **15.** $x^2 - 25$
16. $4x^6 + 19x^3 - 30$ **17.** $4x^2 - 12xy + 9y^2$
18. $x^2 + x + 3$, R 8; or $x^2 + x + 3 + \frac{8}{x - 2}$

19. **(a)** $4w + 16$; **(b)** $w^2 + 8w$

Margin Exercises, Section 4.1, pp. 295–300

1. $5 \cdot 5 \cdot 5 \cdot 5$ **2.** $x \cdot x \cdot x \cdot x \cdot x$ **3.** $3t \cdot 3t$ **4.** $3 \cdot t \cdot t$
5. $(-x) \cdot (-x) \cdot (-x) \cdot (-x)$ **6.** 6 **7.** 1 **8.** 8.4 **9.** 1
10. 125 **11.** 3215.36 cm^2 **12.** 119 **13.** 3; -3
14. **(a)** 144; **(b)** 36; **(c)** no **15.** 3^{10} **16.** x^{10} **17.** p^{24}
18. x^5 **19.** a^9b^8 **20.** 4^3 **21.** y^4 **22.** p^9 **23.** a^4b^2
24. $\frac{1}{4^3} = \frac{1}{64}$ **25.** $\frac{1}{5^2} = \frac{1}{25}$ **26.** $\frac{1}{2^4} = \frac{1}{16}$
27. $\frac{1}{(-2)^3} = -\frac{1}{8}$ **28.** $\frac{4}{p^3}$ **29.** x^2 **30.** 5^2 **31.** $\frac{1}{x^7}$
32. $\frac{1}{7^5}$ **33.** b **34.** t^6

Exercise Set 4.1, p. 301

1. $3 \cdot 3 \cdot 3 \cdot 3$ **3.** $(-1.1)(-1.1)(-1.1)(-1.1)(-1.1)$
5. $\left(\frac{2}{3}\right)\left(\frac{2}{3}\right)\left(\frac{2}{3}\right)\left(\frac{2}{3}\right)$ **7.** $(7p)(7p)$ **9.** $8 \cdot k \cdot k \cdot k$ **11.** 1
13. b **15.** 1 **17.** 1 **19.** ab **21.** ab **23.** 27
25. 19 **27.** 256 **29.** 93 **31.** 10; 4 **33.** 3629.84 ft^2
35. $\frac{1}{3^2} = \frac{1}{9}$ **37.** $\frac{1}{10^3} = \frac{1}{1000}$ **39.** $\frac{1}{7^3} = \frac{1}{343}$ **41.** $\frac{1}{a^3}$
43. $8^2 = 64$ **45.** y^4 **47.** z^n **49.** 4^{-3} **51.** x^{-3}
53. a^{-5} **55.** 2^7 **57.** 8^{14} **59.** x^7 **61.** 9^{38} **63.** $(3y)^{12}$
65. $(7y)^{17}$ **67.** 3^3 **69.** $\frac{1}{x}$ **71.** x^{17} **73.** $\frac{1}{x^{13}}$
75. $\frac{1}{a^{10}}$ **77.** 1 **79.** 7^3 **81.** 8^6 **83.** y^4 **85.** $\frac{1}{16^6}$
87. $\frac{1}{m^6}$ **89.** $\frac{1}{(8x)^4}$ **91.** 1 **93.** x^2 **95.** x^9 **97.** $\frac{1}{z^4}$
99. x^3 **101.** 1 **103.** $5^2 = 25$; $5^{-2} = \frac{1}{25}$; $\left(\frac{1}{5}\right)^2 = \frac{1}{25}$; $\left(\frac{1}{5}\right)^{-2} = 25$; $-5^2 = -25$; $(-5)^2 = 25$; $-\left(-\frac{1}{5}\right)^2 = -\frac{1}{25}$; $\left(-\frac{1}{5}\right)^{-2} = 25$ **105.** $\mathbf{D_W}$ **107.** $64\%t$, or $0.64t$ **108.** 1
109. 64 **110.** 1579.5 **111.** $\frac{4}{3}$ **112.** $8(x - 7)$
113. 8 in., 4 in. **114.** 228, 229 **115.** No **117.** No
119. y^{5x} **121.** a^{4t} **123.** 1 **125.** $>$ **127.** $<$
129. Let $x = 2$; then $3x^2 = 12$, but $(3x)^2 = 36$.

Margin Exercises, Section 4.2, pp. 305–312

1. 3^{20} **2.** $\frac{1}{x^{12}}$ **3.** y^{15} **4.** $\frac{1}{x^{32}}$ **5.** $\frac{16x^{20}}{y^{12}}$ **6.** $\frac{25x^{10}}{y^{12}z^6}$
7. x^{74} **8.** $\frac{27z^{24}}{y^6x^{15}}$ **9.** $\frac{x^{12}}{25}$ **10.** $\frac{8t^{15}}{w^{12}}$ **11.** $\frac{9}{x^8}$
12. 5.17×10^{-4} **13.** 5.23×10^8 **14.** 689,300,000,000
15. 0.0000567 **16.** 5.6×10^{-15} **17.** 7.462×10^{-13}
18. 2.0×10^3 **19.** 5.5×10^2 **20.** 1.884672×10^{11} L
21. The mass of Saturn is 9.5×10 times the mass of Earth.

Calculator Corner, p. 310

1. 1.3545×10^{-4} **2.** 9.044×10^5 **3.** 3.2×10^5
4. 3.6×10^{12} **5.** 3×10^{-6} **6.** 4×10^5 **7.** 8×10^{-26}
8. 3×10^{13}

Exercise Set 4.2, p. 313

1. 2^6 **3.** $\frac{1}{5^6}$ **5.** x^{12} **7.** $\frac{1}{a^{18}}$ **9.** t^{18} **11.** $\frac{1}{t^{12}}$
13. x^8 **15.** a^3b^3 **17.** $\frac{1}{a^3b^3}$ **19.** $\frac{1}{m^3n^6}$ **21.** $16x^6$
23. $\frac{9}{x^8}$ **25.** $\frac{1}{x^{12}y^{15}}$ **27.** $x^{24}y^8$ **29.** $\frac{a^{10}}{b^{35}}$ **31.** $\frac{25t^6}{r^8}$
33. $\frac{b^{21}}{a^{15}c^6}$ **35.** $\frac{9x^6}{y^{16}z^6}$ **37.** $\frac{y^6}{4}$ **39.** $\frac{a^8}{b^{12}}$ **41.** $\frac{8}{y^6}$
43. $49x^6$ **45.** $\frac{x^6y^3}{z^3}$ **47.** $\frac{c^2d^6}{a^4b^2}$ **49.** 2.8×10^{10}
51. 9.07×10^{17} **53.** 3.04×10^{-6} **55.** 1.8×10^{-8}
57. 10^{11} **59.** 2.81×10^{8} **61.** 10^{-7} **63.** 87,400,000
65. 0.00000005704 **67.** 10,000,000 **69.** 0.00001
71. 6×10^9 **73.** 3.38×10^4 **75.** 8.1477×10^{-13}
77. 2.5×10^{13} **79.** 5.0×10^{-4} **81.** 3.0×10^{-21}
83. Approximately 1.325×10^{14} ft^3 **85.** The mass of Jupiter is 3.18×10^2 times the mass of Earth. **87.** 1×10^{22}
89. The mass of the sun is 3.33×10^5 times the mass of Earth.
91. 4.375×10^2 days **93.** DW **95.** $9(x - 4)$
96. $2(2x - y + 8)$ **97.** $3(s + t + 8)$ **98.** $-7(x + 2)$
99. $\frac{7}{4}$ **100.** 2 **101.** $-\frac{12}{7}$ **102.** $-\frac{11}{2}$

103.

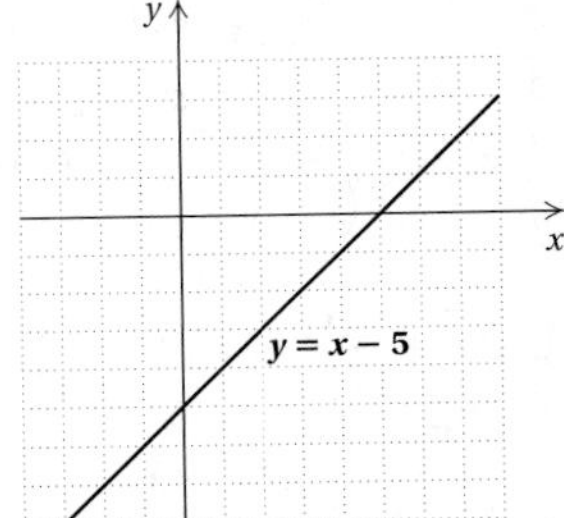

104.

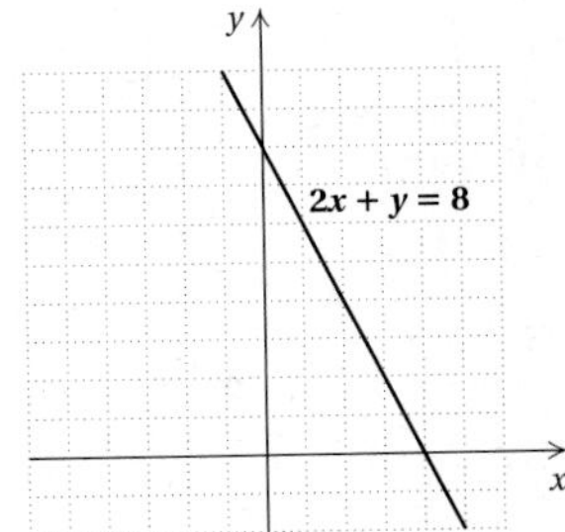

105. 2.478125×10^{-1} **107.** $\frac{1}{5}$ **109.** 3^{11} **111.** 7
113. $\frac{1}{0.4}$, or 2.5 **115.** False **117.** False

Margin Exercises, Section 4.3, pp. 318–325

1. $4x^2 - 3x + \frac{5}{4}$; $15y^3$; $-7x^3 + 1.1$; answers may vary
2. -19 **3.** -104 **4.** -18 **5.** 21 **6.** 6; -4
7. 132 games **8.** 360 ft **9. (a)** 7.55 parts per million; **(b)** When $t = 3$, $C \approx 7.5$; so the value found in part (a) appears to be correct. **10.** 20 parts per million
11. $-9x^3 + (-4x^5)$ **12.** $-2y^3 + 3y^7 + (-7y)$
13. $3x^2, 6x, \frac{1}{2}$ **14.** $-4y^5, 7y^2, -3y, -2$ **15.** $4x^3$ and $-x^3$
16. $4t^4$ and $-7t^4$; $-9t^3$ and $10t^3$ **17.** $5x^2$ and $7x^2$; $3x$ and $-8x$; -10 and 11 **18.** 2, -7, -8.5, 10, -4 **19.** $8x^2$
20. $2x^3 + 7$ **21.** $-\frac{1}{4}x^5 + 2x^2$ **22.** $-4x^3$ **23.** $5x^3$
24. $25 - 3x^5$ **25.** $6x$ **26.** $4x^3 + 4$
27. $-\frac{1}{4}x^3 + 4x^2 + 7$ **28.** $3x^2 + x^3 + 9$
29. $6x^7 + 3x^5 - 2x^4 + 4x^3 + 5x^2 + x$
30. $7x^5 - 5x^4 + 2x^3 + 4x^2 - 3$
31. $14t^7 - 10t^5 + 7t^2 - 14$ **32.** $-2x^2 - 3x + 2$
33. $10x^4 - 8x - \frac{1}{2}$ **34.** 4, 2, 1, 0; 4 **35.** x **36.** x^3, x^2, x, x^0 **37.** x^2, x **38.** x^3
39. $2x^3 + 4x^2 + 0x - 2$; $2x^3 + 4x^2 \quad - 2$
40. $a^4 + 0a^3 + 0a^2 + 0a + 10$; $a^4 \quad + 10$
41. Monomial **42.** None of these **43.** Binomial
44. Trinomial

Calculator Corner, p. 321

1. 3; 2.25; -27 **2.** 44; 0; 9.28 **3.** 13; -3.32; 7
4. -1; -7; -40.6

Exercise Set 4.3, p. 326

1. -18; 7 **3.** 19; 14 **5.** -12; -7 **7.** -1; 5 **9.** 9; 1
11. 56; -2 **13.** 1112 ft **15. (a)** 3.93 million gigawatt hours, 4.12 million gigawatt hours, 4.5 million gigawatt hours, 4.88 million gigawatt hours, 5.45 million gigawatt hours, 5.83 million gigawatt hours; **(b)** left to the student
17. \$18,750; \$24,000 **19.** -4, 4, 5, 2.75, 1
21. 1,820,000; 3,660,000 **23.** 9 words **25.** 6 **27.** 15
29. 2, $-3x$, x^2 **31.** $6x^2$ and $-3x^2$
33. $2x^4$ and $-3x^4$; $5x$ and $-7x$ **35.** $3x^5$ and $14x^5$; $-7x$ and $-2x$; 8 and -9 **37.** -3, 6 **39.** 5, 3, 3
41. -5, 6, -3, 8, -2 **43.** $-3x$ **45.** $-8x$ **47.** $11x^3 + 4$
49. $x^3 - x$ **51.** $4b^5$ **53.** $\frac{3}{4}x^5 - 2x - 42$ **55.** x^4
57. $\frac{15}{16}x^3 - \frac{7}{6}x^2$ **59.** $x^5 + 6x^3 + 2x^2 + x + 1$
61. $15y^9 + 7y^8 + 5y^3 - y^2 + y$ **63.** $x^6 + x^4$
65. $13x^3 - 9x + 8$ **67.** $-5x^2 + 9x$ **69.** $12x^4 - 2x + \frac{1}{4}$
71. 1, 0; 1 **73.** 2, 1, 0; 2 **75.** 3, 2, 1, 0; 3 **77.** 2, 1, 6, 4; 6
79.

Term	Coefficient	Degree of the Term	Degree of the Polynomial
$-7x^4$	-7	4	
$6x^3$	6	3	
$-3x^2$	-3	2	4
$8x$	8	1	
-2	-2	0	

81. x^2, x **83.** x^3, x^2, x^0 **85.** None missing
87. $x^3 + 0x^2 + 0x - 27$; $x^3 \quad - 27$
89. $x^4 + 0x^3 + 0x^2 - x + 0x^0$; $x^4 \quad - x$
91. None missing **93.** Trinomial **95.** None of these
97. Binomial **99.** Monomial **101.** DW
103. 27 apples **104.** -19 **105.** $-\frac{17}{24}$ **106.** $\frac{5}{8}$
107. -2.6 **108.** $\frac{15}{2}$ **109.** $b = \frac{C + r}{a}$ **110.** 45%; 37.5%; 17.5% **111.** $3(x - 5y + 21)$ **113.** $3x^6$ **115.** 10
117. -4, 4, 5, 2.75, 1 **119.** 1,820,000; 3,660,000

Margin Exercises, Section 4.4, pp. 332–335

1. $x^2 + 7x + 3$ **2.** $-4x^5 + 7x^4 + x^3 + 2x^2 + 4$
3. $24x^4 + 5x^3 + x^2 + 1$ **4.** $2x^3 + \frac{10}{3}$ **5.** $2x^2 - 3x - 1$
6. $8x^3 - 2x^2 - 8x + \frac{5}{2}$ **7.** $-8x^4 + 4x^3 + 12x^2 + 5x - 8$
8. $-x^3 + x^2 + 3x + 3$ **9.** $-4x^3 + 6x - 3$
10. $-5x^4 - 3x^2 - 7x + 5$
11. $-14x^{10} + \frac{1}{2}x^5 - 5x^3 + x^2 - 3x$ **12.** $2x^3 + 2x + 8$
13. $x^2 - 6x - 2$ **14.** $-8x^4 - 5x^3 + 8x^2 - 1$
15. $x^3 - x^2 - \frac{4}{3}x - 0.9$ **16.** $2x^3 + 5x^2 - 2x - 5$
17. $-x^5 - 2x^3 + 3x^2 - 2x + 2$ **18.** Sum of perimeters: $13x$; sum of areas: $\frac{7}{2}x^2$ **19.** $x^2 - 64$ ft^2

Calculator Corner, p. 335

1. Yes **2.** Yes **3.** No **4.** Yes **5.** No **6.** Yes

Exercise Set 4.4, p. 336

1. $-x + 5$ **3.** $x^2 - 5x - 1$ **5.** $2x^2$ **7.** $5x^2 + 3x - 30$
9. $-2.2x^3 - 0.2x^2 - 3.8x + 23$ **11.** $6 + 12x^2$
13. $-\frac{1}{2}x^4 + \frac{2}{3}x^3 + x^2$
15. $0.01x^5 + x^4 - 0.2x^3 + 0.2x + 0.06$
17. $9x^8 + 8x^7 - 6x^4 + 8x^2 + 4$
19. $1.05x^4 + 0.36x^3 + 14.22x^2 + x + 0.97$ **21.** $5x$
23. $x^2 - 10x + 2$ **25.** $-12x^4 + 3x^3 - 3$ **27.** $-3x + 7$
29. $-4x^2 + 3x - 2$ **31.** $4x^4 - 6x^2 - \frac{3}{4}x + 8$
33. $7x - 1$ **35.** $-x^2 - 7x + 5$ **37.** -18
39. $6x^4 + 3x^3 - 4x^2 + 3x - 4$
41. $4.6x^3 + 9.2x^2 - 3.8x - 23$ **43.** $\frac{3}{4}x^3 - \frac{1}{2}x$
45. $0.06x^3 - 0.05x^2 + 0.01x + 1$ **47.** $3x + 6$
49. $11x^4 + 12x^3 - 9x^2 - 8x - 9$ **51.** $x^4 - x^3 + x^2 - x$
53. $5x^2 + 4x$ **55.** $\frac{23}{2}a + 12$ **57.** $(r + 11)(r + 9)$; $9r + 99 + 11r + r^2$, or $r^2 + 20r + 99$
59. $(x + 3)(x + 3)$, or $(x + 3)^2$; $x^2 + 3x + 9 + 3x$, or $x^2 + 6x + 9$ **61.** $\pi r^2 - 25\pi$ **63.** $18z - 64$ **65.** DW
67. 6 **68.** -19 **69.** $-\frac{7}{22}$ **70.** 5 **71.** 5 **72.** 1
73. $\frac{39}{2}$ **74.** $\frac{37}{2}$ **75.** $\{x|x \geq -10\}$ **76.** $\{x|x < 0\}$
77. $20w + 42$ **79.** $2x^2 + 20x$ **81.** $y^2 - 4y + 4$
83. $12y^2 - 23y + 21$ **85.** $-3y^4 - y^3 + 5y - 2$

Margin Exercises, Section 4.5, pp. 340–343

1. $-15x$ **2.** $-x^2$ **3.** x^2 **4.** $-x^5$ **5.** $12x^7$ **6.** $-8y^{11}$
7. $7y^5$ **8.** 0 **9.** $8x^2 + 16x$ **10.** $-15t^3 + 6t^2$
11. $-5x^6 - 25x^5 + 30x^4 - 40x^3$
12. (a)
$$\begin{aligned}(y + 2)(y + 7) &= y \cdot (y + 7) + 2(y + 7)\\ &= y \cdot y + y \cdot 7 + 2 \cdot y + 2 \cdot 7\\ &= y^2 + 7y + 2y + 14\\ &= y^2 + 9y + 14;\end{aligned}$$
(b) $y^2 + 2y + 7y + 14$
13. $x^2 + 13x + 40$ **14.** $x^2 + x - 20$
15. $5x^2 - 17x - 12$ **16.** $6x^2 - 19x + 15$
17. $x^4 + 3x^3 + x^2 + 15x - 20$
18. $6y^5 - 20y^3 + 15y^2 + 14y - 35$
19. $3x^3 + 13x^2 - 6x + 20$
20. $20x^4 - 16x^3 + 32x^2 - 32x - 16$
21. $6x^4 - x^3 - 18x^2 - x + 10$

Calculator Corner, p. 343

1. Correct **2.** Correct **3.** Not correct **4.** Not correct

Exercise Set 4.5, p. 344

1. $40x^2$ **3.** x^3 **5.** $32x^8$ **7.** $0.03x^{11}$ **9.** $\frac{1}{15}x^4$ **11.** 0
13. $-24x^{11}$ **15.** $-2x^2 + 10x$ **17.** $-5x^2 + 5x$
19. $x^5 + x^2$ **21.** $6x^3 - 18x^2 + 3x$ **23.** $-6x^4 - 6x^3$
25. $18y^6 + 24y^5$ **27.** $x^2 + 9x + 18$ **29.** $x^2 + 3x - 10$
31. $x^2 - 7x + 12$ **33.** $x^2 - 9$ **35.** $25 - 15x + 2x^2$
37. $4x^2 + 20x + 25$ **39.** $x^2 - \frac{21}{10}x - 1$
41. $x^2 + 2.4x - 10.81$ **43.**

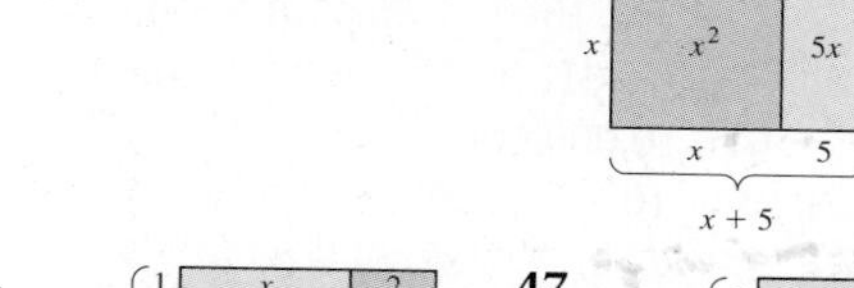

45.

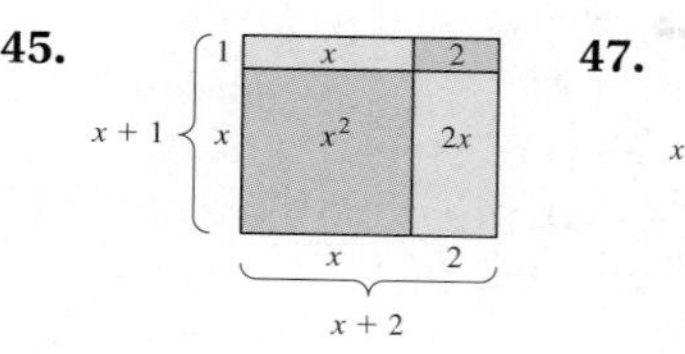

47.

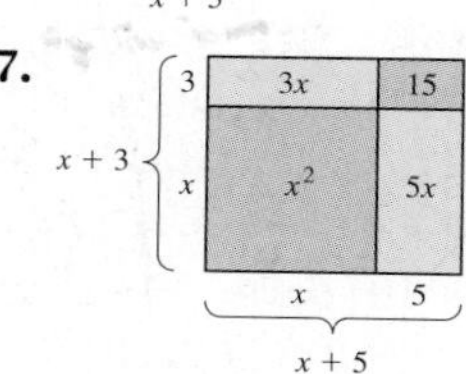

49.

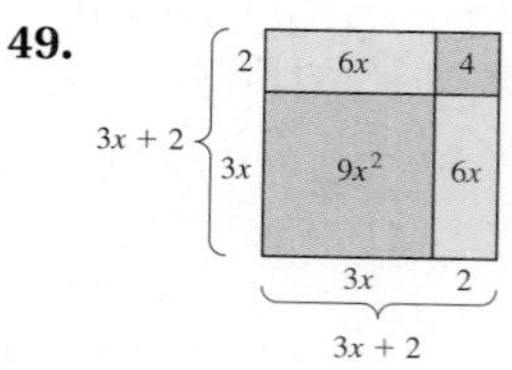

51. $x^3 - 1$

53. $4x^3 + 14x^2 + 8x + 1$ **55.** $3y^4 - 6y^3 - 7y^2 + 18y - 6$
57. $x^6 + 2x^5 - x^3$ **59.** $-10x^5 - 9x^4 + 7x^3 + 2x^2 - x$
61. $-1 - 2x - x^2 + x^4$ **63.** $6t^4 + t^3 - 16t^2 - 7t + 4$
65. $x^9 - x^5 + 2x^3 - x$ **67.** $x^4 - 1$
69. $x^4 + 8x^3 + 12x^2 + 9x + 4$
71. $2x^4 - 5x^3 + 5x^2 - \frac{19}{10}x + \frac{1}{5}$ **73.** DW **75.** $-\frac{3}{4}$
76. 6.4 **77.** 96 **78.** 32 **79.** $3(5x - 6y + 4)$
80. $4(4x - 6y + 9)$ **81.** $-3(3x + 15y - 5)$
82. $100(x - y + 10a)$ **83.**

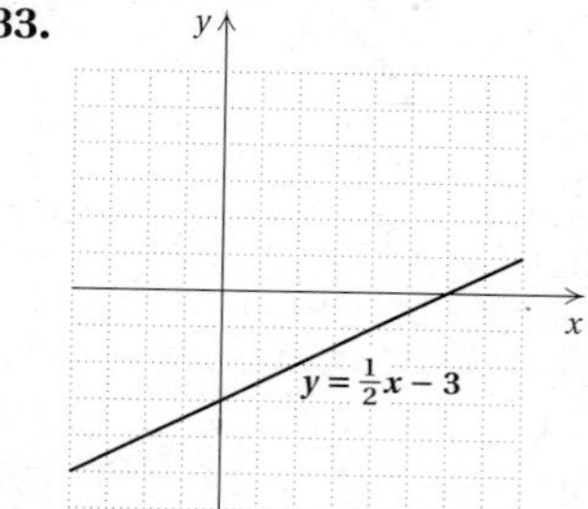

84. $\frac{23}{19}$ **85.** $75y^2 - 45y$ **87.** $V = 4x^3 - 48x^2 + 144x$ in^3; $S = -4x^2 + 144$ in^2 **89.** 5 **91.** $x^3 + 2x^2 - 210$ m^3
93. 0 **95.** 0

Margin Exercises, Section 4.6, pp. 348–352

1. $x^2 + 7x + 12$ **2.** $x^2 - 2x - 15$ **3.** $2x^2 - 9x + 4$
4. $2x^3 - 4x^2 - 3x + 6$ **5.** $12x^5 + 10x^3 + 6x^2 + 5$
6. $y^6 - 49$ **7.** $t^2 + 8t + 15$ **8.** $-2x^7 + x^5 + x^3$
9. $x^2 - \frac{16}{25}$ **10.** $x^5 + 0.5x^3 - 0.5x^2 - 0.25$
11. $8 + 2x^2 - 15x^4$ **12.** $30x^5 - 27x^4 + 6x^3$

13. $x^2 - 25$ **14.** $4x^2 - 9$ **15.** $x^2 - 4$ **16.** $x^2 - 49$
17. $36 - 16y^2$ **18.** $4x^6 - 1$ **19.** $x^2 - \frac{4}{25}$
20. $x^2 + 16x + 64$ **21.** $x^2 - 10x + 25$ **22.** $x^2 + 4x + 4$
23. $a^2 - 8a + 16$ **24.** $4x^2 + 20x + 25$
25. $16x^4 - 24x^3 + 9x^2$ **26.** $60.84 + 18.72y + 1.44y^2$
27. $9x^4 - 30x^2 + 25$ **28.** $x^2 + 11x + 30$ **29.** $t^2 - 16$
30. $-8x^5 + 20x^4 + 40x^2$ **31.** $81x^4 + 18x^2 + 1$
32. $4a^2 + 6a - 40$ **33.** $25x^2 + 5x + \frac{1}{4}$ **34.** $4x^2 - 2x + \frac{1}{4}$
35. $x^3 - 3x^2 + 6x - 8$

Exercise Set 4.6, p. 353

1. $x^3 + x^2 + 3x + 3$ **3.** $x^4 + x^3 + 2x + 2$
5. $y^2 - y - 6$ **7.** $9x^2 + 12x + 4$ **9.** $5x^2 + 4x - 12$
11. $9t^2 - 1$ **13.** $4x^2 - 6x + 2$ **15.** $p^2 - \frac{1}{16}$
17. $x^2 - 0.01$ **19.** $2x^3 + 2x^2 + 6x + 6$
21. $-2x^2 - 11x + 6$ **23.** $a^2 + 14a + 49$
25. $1 - x - 6x^2$ **27.** $x^5 + 3x^3 - x^2 - 3$
29. $3x^6 - 2x^4 - 6x^2 + 4$ **31.** $13.16x^2 + 18.99x - 13.95$
33. $6x^7 + 18x^5 + 4x^2 + 12$ **35.** $8x^6 + 65x^3 + 8$
37. $4x^3 - 12x^2 + 3x - 9$ **39.** $4y^6 + 4y^5 + y^4 + y^3$
41. $x^2 - 16$ **43.** $4x^2 - 1$ **45.** $25m^2 - 4$ **47.** $4x^4 - 9$
49. $9x^8 - 16$ **51.** $x^{12} - x^4$ **53.** $x^8 - 9x^2$ **55.** $x^{24} - 9$
57. $4y^{16} - 9$ **59.** $\frac{25}{64}x^2 - 18.49$ **61.** $x^2 + 4x + 4$
63. $9x^4 + 6x^2 + 1$ **65.** $a^2 - a + \frac{1}{4}$ **67.** $9 + 6x + x^2$
69. $x^4 + 2x^2 + 1$ **71.** $4 - 12x^4 + 9x^8$
73. $25 + 60t^2 + 36t^4$ **75.** $x^2 - \frac{5}{4}x + \frac{25}{64}$
77. $9 - 12x^3 + 4x^6$ **79.** $4x^3 + 24x^2 - 12x$
81. $4x^4 - 2x^2 + \frac{1}{4}$ **83.** $9p^2 - 1$ **85.** $15t^5 - 3t^4 + 3t^3$
87. $36x^8 + 48x^4 + 16$ **89.** $12x^3 + 8x^2 + 15x + 10$
91. $64 - 96x^4 + 36x^8$ **93.** $t^3 - 1$ **95.** 25; 49
97. 56; 16 **99.** $a^2 + 2a + 1$ **101.** $t^2 + 10t + 24$
103. D_W **105.** Lamps: 500 watts; air conditioner: 2000 watts; television: 50 watts **106.** $\frac{28}{27}$ **107.** $-\frac{41}{7}$
108. $\frac{27}{4}$ **109.** $y = \dfrac{3x - 12}{2}$, or $y = \dfrac{3}{2}x - 6$
110. $a = \dfrac{5d + 4}{3}$, or $a = \dfrac{5}{3}d + \dfrac{4}{3}$
111. $30x^3 + 35x^2 - 15x$ **113.** $a^4 - 50a^2 + 625$
115. $81t^{16} - 72t^8 + 16$ **117.** -7 **119.** First row: 90, -432, -63; second row: 7, -18, -36, -14, 12, -6, -21, -11; third row: 9, -2, -2, 10, -8, -8, -8, -10, 21; fourth row: -19, -6 **121.** Yes **123.** No

Margin Exercises, Section 4.7, pp. 357–360

1. -7940 **2.** -176 **3.** 1889 calories **4.** $-3, 3, -2, 1, 2$
5. 3, 7, 1, 1, 0; 7 **6.** $2x^2y + 3xy$ **7.** $5pq - 8$
8. $-4x^3 + 2x^2 - 4y + 2$ **9.** $14x^3y + 7x^2y - 3xy - 2y$
10. $-5p^2q^4 + 2p^2q^2 + 3p^2q + 6pq^2 + 3q + 5$
11. $-8s^4t + 6s^3t^2 + 2s^2t^3 - s^2t^2$
12. $-9p^4q + 9p^3q^2 - 4p^2q^3 - 9q^4 + 5$
13. $x^5y^5 + 2x^4y^2 + 3x^3y^3 + 6x^2$
14. $p^5q - 4p^3q^3 + 3pq^3 + 6q^4$
15. $3x^3y + 6x^2y^3 + 2x^3 + 4x^2y^2$

16. $2x^2 - 11xy + 15y^2$ **17.** $16x^2 + 40xy + 25y^2$
18. $9x^4 - 12x^3y^2 + 4x^2y^4$ **19.** $4x^2y^4 - 9x^2$
20. $16y^2 - 9x^2y^4$ **21.** $9y^2 + 24y + 16 - 9x^2$
22. $4a^2 - 25b^2 - 10bc - c^2$

Exercise Set 4.7, p. 361

1. -1 **3.** -15 **5.** 240 **7.** -145 **9.** 3.715 liters
11. 110.4 m **13.** 44.46 in^2 **15.** 63.78125 in^2
17. Coefficients: 1, -2, 3, -5; degrees: 4, 2, 2, 0; 4
19. Coefficients: 17, -3, -7; degrees: 5, 5, 0; 5
21. $-a - 2b$ **23.** $3x^2y - 2xy^2 + x^2$ **25.** $20au + 10av$
27. $8u^2v - 5uv^2$ **29.** $x^2 - 4xy + 3y^2$ **31.** $3r + 7$
33. $-a^3b^2 - 3a^2b^3 + 5ab + 3$ **35.** $ab^2 - a^2b$
37. $2ab - 2$ **39.** $-2a + 10b - 5c + 8d$
41. $6z^2 + 7zu - 3u^2$ **43.** $a^4b^2 - 7a^2b + 10$
45. $a^6 - b^2c^2$ **47.** $y^6x + y^4x + y^4 + 2y^2 + 1$
49. $12x^2y^2 + 2xy - 2$ **51.** $12 - c^2d^2 - c^4d^4$
53. $m^3 + m^2n - mn^2 - n^3$
55. $x^9y^9 - x^6y^6 + x^5y^5 - x^2y^2$ **57.** $x^2 + 2xh + h^2$
59. $r^6t^4 - 8r^3t^2 + 16$ **61.** $p^8 + 2m^2n^2p^4 + m^4n^4$
63. $4a^6 - 2a^3b^3 + \frac{1}{4}b^6$ **65.** $3a^3 - 12a^2b + 12ab^2$
67. $4a^2 - b^2$ **69.** $c^4 - d^2$ **71.** $a^2b^2 - c^2d^4$
73. $x^2 + 2xy + y^2 - 9$ **75.** $x^2 - y^2 - 2yz - z^2$
77. $a^2 - b^2 - 2bc - c^2$ **79.** D_W **81.** IV **82.** III
83. I **84.** II **85.**

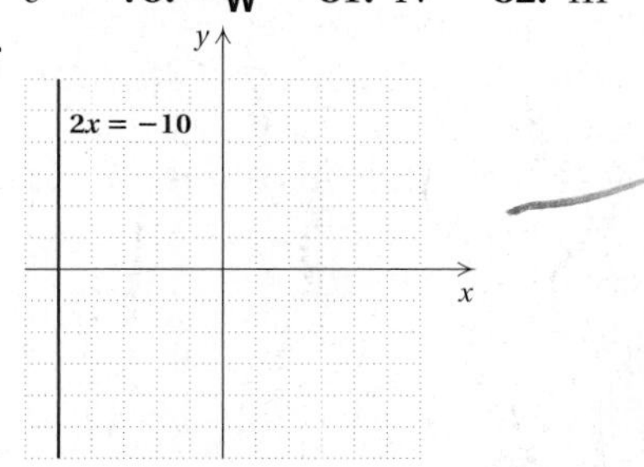

86.

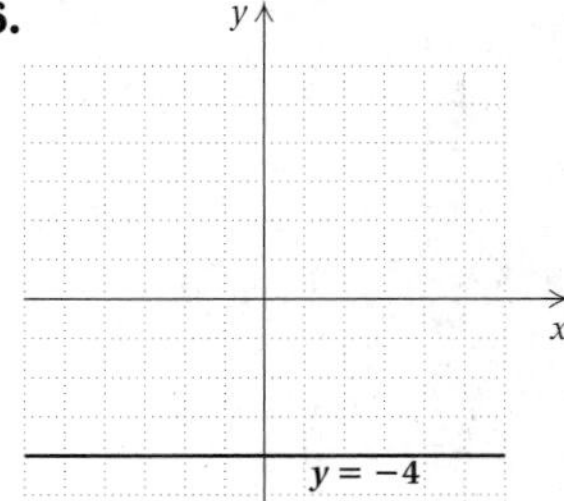

87.

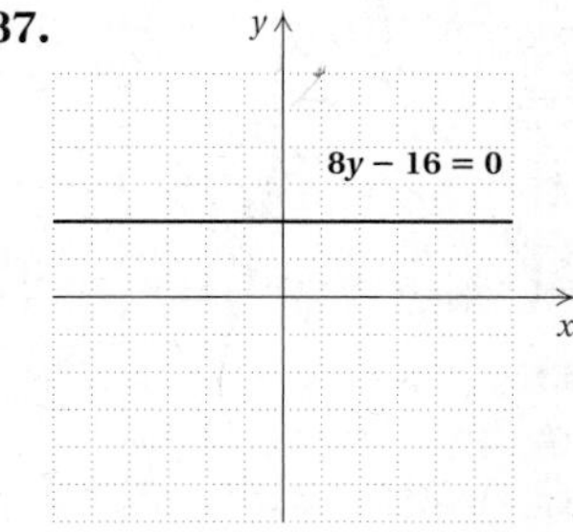

88.

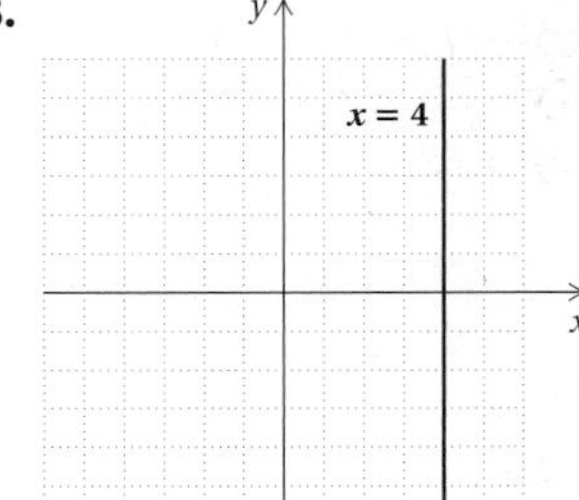

89. $4xy - 4y^2$

91. $2xy + \pi x^2$ **93.** $2\pi nh + 2\pi mh + 2\pi n^2 - 2\pi m^2$
95. 16 gal **97.** \$15,638.03

Margin Exercises, Section 4.8, pp. 366–369

1. $4x^2$ **2.** $-7x^{11}$ **3.** $-28p^3q$ **4.** $\frac{1}{4}x^4$ **5.** $7x^4 + 8x^2$ **6.** $x^2 + 3x + 2$ **7.** $2x^2 + x - \frac{2}{3}$ **8.** $4x^2 - \frac{3}{2}x + \frac{1}{2}$ **9.** $2x^2y^4 - 3xy^2 + 5y$ **10.** $x - 2$ **11.** $x + 4$ **12.** $x + 4$, R -2, or $x + 4 + \dfrac{-2}{x + 3}$ **13.** $x^2 + x + 1$

Exercise Set 4.8, p. 370

1. $3x^4$ **3.** $5x$ **5.** $18x^3$ **7.** $4a^3b$ **9.** $3x^4 - \frac{1}{2}x^3 + \frac{1}{8}x^2 - 2$ **11.** $1 - 2u - u^4$ **13.** $5t^2 + 8t - 2$ **15.** $-4x^4 + 4x^2 + 1$ **17.** $6x^2 - 10x + \frac{3}{2}$ **19.** $9x^2 - \frac{5}{2}x + 1$ **21.** $6x^2 + 13x + 4$ **23.** $3rs + r - 2s$ **25.** $x + 2$ **27.** $x - 5 + \dfrac{-50}{x - 5}$ **29.** $x - 2 + \dfrac{-2}{x + 6}$ **31.** $x - 3$ **33.** $x^4 - x^3 + x^2 - x + 1$ **35.** $2x^2 - 7x + 4$ **37.** $x^3 - 6$ **39.** $x^3 + 2x^2 + 4x + 8$ **41.** $t^2 + 1$ **43.** D_W **45.** -28 **46.** -59 **47.** 6.8 **48.** $-\frac{11}{8}$ **49.** 25, 543.75 ft² **50.** 51°, 27°, 102° **51.** $\frac{23}{14}$ **52.** $\frac{11}{10}$ **53.** $4(x - 3 + 6y)$ **54.** $2(128 - a - 2b)$ **55.** $x^2 + 5$ **57.** $a + 3 + \dfrac{5}{5a^2 - 7a - 2}$ **59.** $2x^2 + x - 3$ **61.** $a^5 + a^4b + a^3b^2 + a^2b^3 + ab^4 + b^5$ **63.** -5 **65.** 1

Summary and Review: Chapter 4, p. 372

1. $\dfrac{1}{7^2}$ **2.** y^{11} **3.** $(3x)^{14}$ **4.** t^8 **5.** 4^3 **6.** $\dfrac{1}{a^3}$ **7.** 1 **8.** $9t^8$ **9.** $36x^8$ **10.** $\dfrac{y^3}{8x^3}$ **11.** t^{-5} **12.** $\dfrac{1}{y^4}$ **13.** 3.28×10^{-5} **14.** 8,300,000 **15.** 2.09×10^4 **16.** 5.12×10^{-5} **17.** 4.4676×10^9 gal **18.** 10 **19.** $-4y^5, 7y^2, -3y, -2$ **20.** x^2, x^0 **21.** 3, 2, 1, 0; 3 **22.** Binomial **23.** None of these **24.** Monomial **25.** $-2x^2 - 3x + 2$ **26.** $10x^4 - 7x^2 - x - \frac{1}{2}$ **27.** $x^5 - 2x^4 + 6x^3 + 3x^2 - 9$ **28.** $-2x^5 - 6x^4 - 2x^3 - 2x^2 + 2$ **29.** $2x^2 - 4x$ **30.** $x^5 - 3x^3 - x^2 + 8$ **31.** Perimeter: $4w + 6$; area: $w^2 + 3w$ **32.** $(t + 3)(t + 4)$, $t^2 + 7t + 12$ **33.** $x^2 + \frac{7}{6}x + \frac{1}{3}$ **34.** $49x^2 + 14x + 1$ **35.** $12x^3 - 23x^2 + 13x - 2$ **36.** $9x^4 - 16$ **37.** $15x^7 - 40x^6 + 50x^5 + 10x^4$ **38.** $x^2 - 3x - 28$ **39.** $9y^4 - 12y^3 + 4y^2$ **40.** $2t^4 - 11t^2 - 21$ **41.** 49 **42.** Coefficients: 1, −7, 9, −8; degrees: 6, 2, 2, 0; 6 **43.** $-y + 9w - 5$ **44.** $m^6 - 2m^2n + 2m^2n^2 + 8n^2m - 6m^3$ **45.** $-9xy - 2y^2$ **46.** $11x^3y^2 - 8x^2y - 6x^2 - 6x + 6$ **47.** $p^3 - q^3$ **48.** $9a^8 - 2a^4b^3 + \frac{1}{9}b^6$ **49.** $5x^2 - \frac{1}{2}x + 3$ **50.** $3x^2 - 7x + 4 + \dfrac{1}{2x + 3}$ **51.** 0, 3.75, −3.75, 0, 2.25 **52.** D_W 578.6×10^{-7} is not in scientific notation because 578.6 is larger than 10. **53.** D_W A monomial is an expression of the type ax^n, where n is a whole number and a is a real number. A binomial is a sum of two monomials and has two terms. A trinomial is a sum of three monomials and has three terms. A general polynomial is a monomial or a sum of monomials and has one or more terms. **54.** $25(t - 2 + 4m)$ **55.** $\frac{9}{4}$ **56.** -12 **57.** -11.2 **58.** Width: 125.5 m; length: 144.5 m **59.** $\frac{1}{2}x^2 - \frac{1}{2}y^2$ **60.** $400 - 4a^2$ **61.** $-28x^8$ **62.** $\frac{94}{13}$ **63.** $x^4 + x^3 + x^2 + x + 1$ **64.** 16 ft by 8 ft

Test: Chapter 4, p. 375

1. [4.1d, f] $\dfrac{1}{6^5}$ **2.** [4.1d] x^9 **3.** [4.1d] $(4a)^{11}$ **4.** [4.1e] 3^3 **5.** [4.1e, f] $\dfrac{1}{x^5}$ **6.** [4.1b, e] 1 **7.** [4.2a] x^6 **8.** [4.2a, b] $-27y^6$ **9.** [4.2a, b] $16a^{12}b^4$ **10.** [4.2b] $\dfrac{a^3b^3}{c^3}$ **11.** [4.1d], [4.2a, b] $-216x^{21}$ **12.** [4.1d], [4.2a, b] $-24x^{21}$ **13.** [4.1d], [4.2a, b] $162x^{10}$ **14.** [4.1d], [4.2a, b] $324x^{10}$ **15.** [4.1f] $\dfrac{1}{5^3}$ **16.** [4.1f] y^{-8} **17.** [4.2c] 3.9×10^9 **18.** [4.2c] 0.00000005 **19.** [4.2d] 1.75×10^{17} **20.** [4.2d] 1.296×10^{22} **21.** [4.2e] 1.5×10^4 **22.** [4.3a] -43 **23.** [4.3d] $\frac{1}{3}$, −1, 7 **24.** [4.3g] 3, 0, 1, 6; 6 **25.** [4.3i] Binomial **26.** [4.3e] $5a^2 - 6$ **27.** [4.3e] $\frac{7}{4}y^2 - 4y$ **28.** [4.3f] $x^5 + 2x^3 + 4x^2 - 8x + 3$ **29.** [4.4a] $4x^5 + x^4 + 2x^3 - 8x^2 + 2x - 7$ **30.** [4.4a] $5x^4 + 5x^2 + x + 5$ **31.** [4.4c] $-4x^4 + x^3 - 8x - 3$ **32.** [4.4c] $-x^5 + 0.7x^3 - 0.8x^2 - 21$ **33.** [4.5b] $-12x^4 + 9x^3 + 15x^2$ **34.** [4.6c] $x^2 - \frac{2}{3}x + \frac{1}{9}$ **35.** [4.6b] $9x^2 - 100$ **36.** [4.6a] $3b^2 - 4b - 15$ **37.** [4.6a] $x^{14} - 4x^8 + 4x^6 - 16$ **38.** [4.6a] $48 + 34y - 5y^2$ **39.** [4.5d] $6x^3 - 7x^2 - 11x - 3$ **40.** [4.6c] $25t^2 + 20t + 4$ **41.** [4.7c] $-5x^3y - y^3 + xy^3 - x^2y^2 + 19$ **42.** [4.7e] $8a^2b^2 + 6ab - 4b^3 + 6ab^2 + ab^3$ **43.** [4.7f] $9x^{10} - 16y^{10}$ **44.** [4.8a] $4x^2 + 3x - 5$ **45.** [4.8b] $2x^2 - 4x - 2 + \dfrac{17}{3x + 2}$ **46.** [4.3a] 3, 1.5, −3.5, −5, −5.25 **47.** [4.4d] $28a + 90$ **48.** [4.4d] $(5t + 2)(5t + 2)$, $25t^2 + 20t + 4$ **49.** [2.3b] 13 **50.** [2.3c] -3 **51.** [1.7d] $16(4t - 2m + 1)$ **52.** [1.4a] $\frac{23}{20}$ **53.** [2.6a] 100°, 25°, 55° **54.** [4.5b], [4.6a] $V = l^3 - 3l^2 + 2l$ **55.** [2.3b], [4.6b, c] $-\frac{61}{12}$

Cumulative Review: Chapters 1–4, p. 377

1. [4.3a] 66.6 ft, 86.6 ft, 66.6 ft, 41.6 ft **2.** [1.1a] $\frac{5}{2}$ **3.** [4.3a] -4 **4.** [4.7a] -14 **5.** [1.2e] 4 **6.** [1.6b] $\frac{1}{5}$ **7.** [1.3a] $-\frac{11}{60}$ **8.** [1.4a] 4.2 **9.** [1.5a] 7.28 **10.** [1.6c] $-\frac{5}{12}$ **11.** [4.2d] 2.2×10^{22} **12.** [4.2d] 4×10^{-5} **13.** [1.7a] -3 **14.** [1.8b] $-2y - 7$ **15.** [1.8c] $5x + 11$ **16.** [1.8d] -2 **17.** [4.4a] $2x^5 - 2x^4 + 3x^3 + 2$ **18.** [4.7d] $3x^2 + xy - 2y^2$ **19.** [4.4c] $x^3 + 5x^2 - x - 7$ **20.** [4.4c] $-\frac{1}{3}x^2 - \frac{3}{4}x$ **21.** [1.7c] $12x - 15y + 21$ **22.** [4.5a] $6x^8$ **23.** [4.5b] $2x^5 - 4x^4 + 8x^3 - 10x^2$

24. [4.5d] $3y^4 + 5y^3 - 10y - 12$
25. [4.5d] $2p^4 + 3p^3q + 2p^2q^2 - 2p^4q - p^3q^2 - p^2q^3 + pq^3$
26. [4.6a] $6x^2 + 13x + 6$ 27. [4.6c] $9x^4 + 6x^2 + 1$
28. [4.6b] $t^2 - \frac{1}{4}$ 29. [4.6b] $4y^4 - 25$
30. [4.6a] $4x^6 + 6x^4 - 6x^2 - 9$ 31. [4.6c] $t^2 - 4t^3 + 4t^4$
32. [4.7f] $15p^2 - pq - 2q^2$ 33. [4.8a] $6x^2 + 2x - 3$
34. [4.8b] $3x^2 - 2x - 7$ 35. [2.1b] -1.2 36. [2.2a] -21
37. [2.3a] 9 38. [2.2a] $-\frac{20}{3}$ 39. [2.3b] 2 40. [2.1b] $\frac{13}{8}$
41. [2.3c] $-\frac{17}{21}$ 42. [2.3b] -17 43. [2.3b] 2
44. [2.7e] $\{x|x < 16\}$ 45. [2.7e] $\{x|x \leq -\frac{11}{8}\}$
46. [2.4b] $x = \dfrac{A - P}{Q}$ 47. [4.4d] $\pi r^2 - 18\text{ ft}^2$
48. [2.6a] 18 and 19 49. [2.6a] 20 ft, 24 ft 50. [2.6a] 10°
51. [2.5a] \$3.50 52. [4.2e] 6.1612×10^9 gal
53. [4.1d, f] y^4 54. [4.1e, f] $\dfrac{1}{x}$ 55. [4.2a, b] $-\dfrac{27x^9}{y^6}$
56. [4.1d, e, f] x^3 57. [4.3d] $\frac{2}{3}, 4, -6$ 58. [4.3g] 4, 2, 1, 0; 4
59. [4.3i] Binomial 60. [4.3i] Trinomial
61. [3.3a] y-intercept: $(0, -4)$; x-intercept: $(5, 0)$
62. [3.3a]

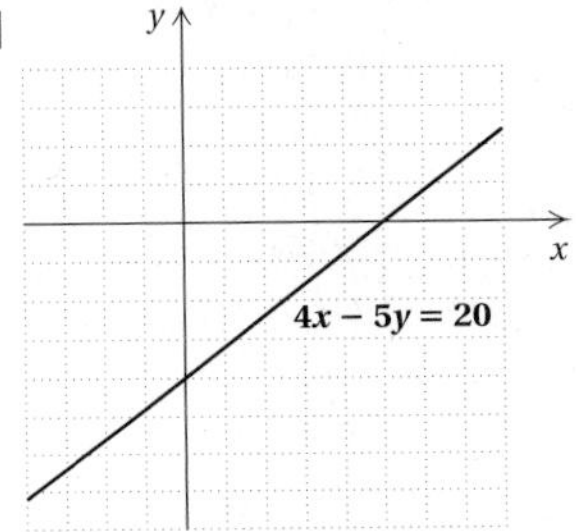

63. [4.4d] $(3x + 4)(3x + 4)$, $9x^2 + 24x + 16$
64. [4.1a, f] $3^2 = 9$, $3^{-2} = \frac{1}{9}$, $(\frac{1}{3})^2 = \frac{1}{9}$, $(\frac{1}{3})^{-2} = 9$, $-3^2 = -9$, $(-3)^2 = 9$, $(-\frac{1}{3})^2 = \frac{1}{9}$, $(-\frac{1}{3})^{-2} = 9$ 65. [4.4d] $4x - 4\text{ in}^2$
66. [4.1d], [4.2a, b], [4.4a] $12x^5 - 15x^4 - 27x^3 + 4x^2$
67. [4.4a], [4.6c] $5x^2 - 2x + 10$
68. [4.4a], [4.8b] $4x^2 - 2x + 7$ 69. [2.3b], [4.6a, c] $\frac{11}{7}$
70. [2.3b], [4.8b] 1 71. [1.2e], [2.3a] $-5, 5$
72. [2.3b], [4.6a], [4.8b] All real numbers except 5
73. [4.5c, d] $V = x^3 + 6x^2 + 12x + 8\text{ cm}^3$

Chapter 5

Pretest: Chapter 5, p. 382

1. $4(-5x^6)$, $(-2x^3)(10x^3)$, $x^2(-20x^4)$; answers may vary
2. $2(x + 1)^2$ 3. $(x + 4)(x + 2)$ 4. $4a(2a^4 + a^2 - 5)$
5. $(5x + 2)(x - 3)$ 6. $(9 + z^2)(3 + z)(3 - z)$
7. $(y^3 - 2)^2$ 8. $(x^2 + 4)(3x + 2)$ 9. $(p - 6)(p + 5)$
10. $(x^2y + 8)(x^2y - 8)$ 11. $(2p - q)(p + 4q)$ 12. 0, 5
13. $4, \frac{3}{5}$ 14. $\frac{2}{3}, -4$ 15. $-3, 3$ 16. Base: 8 cm; height: 11 cm 17. Length: 18 in.; width: 9 in. 18. 5 ft

Margin Exercises, Section 5.1, pp. 383–386

1. **(a)** $12x^2$; **(b)** $(3x)(4x)$, $(2x)(6x)$; answers may vary
2. **(a)** $-16x^3$; **(b)** $(2x)(-8x^2)$, $(-4x)(4x^2)$; answers may vary
3. $(8x)(x^3)$, $(4x^2)(2x^2)$, $(2x^3)(4x)$; answers may vary
4. $(-7x)(3x)$, $(7x)(-3x)$, $(-21x)(x)$; answers may vary
5. $(6x^4)(x)$, $(-2x^3)(-3x^2)$, $(3x^3)(2x^2)$; answers may vary
6. **(a)** $3x + 6$; **(b)** $3(x + 2)$
7. **(a)** $2x^3 + 10x^2 + 8x$; **(b)** $2x(x^2 + 5x + 4)$ 8. $x(x + 3)$
9. $y^2(3y^4 - 5y + 2)$ 10. $3x^2(3x^2 - 5x + 1)$
11. $\frac{1}{4}(3t^3 + 5t^2 + 7t + 1)$ 12. $7x^3(5x^4 - 7x^3 + 2x^2 - 9)$
13. $2.8(3x^2 - 2x + 1)$ 14. $(x^2 + 3)(x + 7)$
15. $(x^2 + 2)(a + b)$ 16. $(x^2 + 3)(x + 7)$
17. $(2t^2 + 3)(4t + 1)$ 18. $(3m^3 + 2)(m^2 - 5)$
19. $(3x^2 - 1)(x - 2)$ 20. $(2x^2 - 3)(2x - 3)$
21. Not factorable using factoring by grouping

Exercise Set 5.1, p. 387

1. $(4x^2)(2x)$, $(-8)(-x^3)$, $(2x^2)(4x)$; answers may vary
3. $(-5a^5)(2a)$, $(10a^3)(-a^3)$, $(-2a^2)(5a^4)$; answers may vary
5. $(8x^2)(3x^2)$, $(-8x^2)(-3x^2)$, $(4x^3)(6x)$; answers may vary
7. $x(x - 6)$ 9. $2x(x + 3)$ 11. $x^2(x + 6)$
13. $8x^2(x^2 - 3)$ 15. $2(x^2 + x - 4)$
17. $17xy(x^4y^2 + 2x^2y + 3)$ 19. $x^2(6x^2 - 10x + 3)$
21. $x^2y^2(x^3y^3 + x^2y + xy - 1)$
23. $2x^3(x^4 - x^3 - 32x^2 + 2)$
25. $0.8x(2x^3 - 3x^2 + 4x + 8)$
27. $\frac{1}{3}x^3(5x^3 + 4x^2 + x + 1)$ 29. $(x^2 + 2)(x + 3)$
31. $(5a^3 - 1)(2a - 7)$ 33. $(x^2 + 2)(x + 3)$
35. $(2x^2 + 1)(x + 3)$ 37. $(4x^2 + 3)(2x - 3)$
39. $(4x^2 + 1)(3x - 4)$ 41. $(5x^2 - 1)(x - 1)$
43. $(x^2 - 3)(x + 8)$ 45. $(2x^2 - 9)(x - 4)$ 47. D_W
49. $\{x|x > -24\}$ 50. $\{x|x \leq \frac{14}{5}\}$ 51. 27
52. $p = 2A - q$ 53. $y^2 + 12y + 35$ 54. $y^2 + 14y + 49$
55. $y^2 - 49$ 56. $y^2 - 14y + 49$

57.

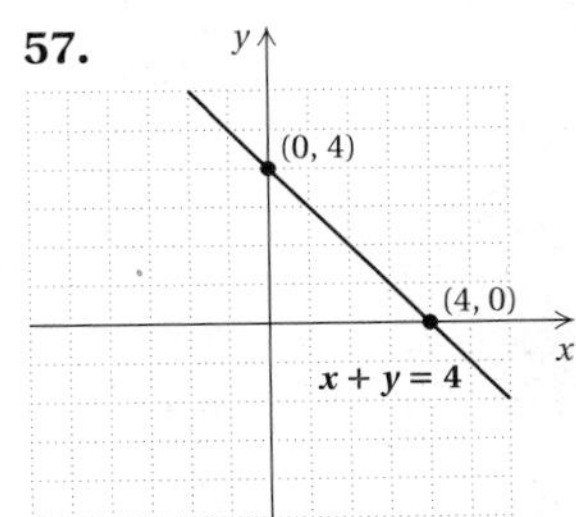

58.

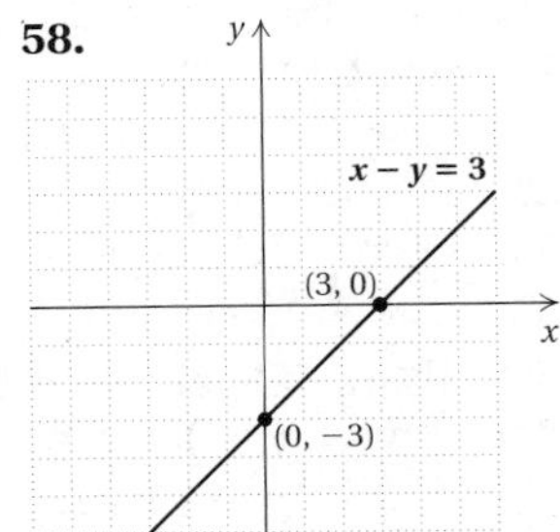

59.

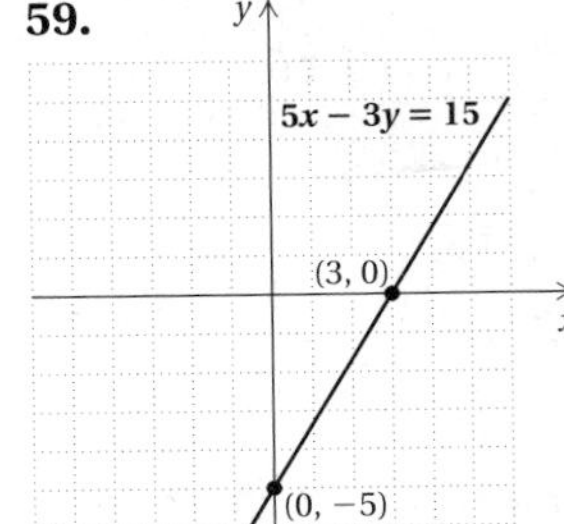

60.

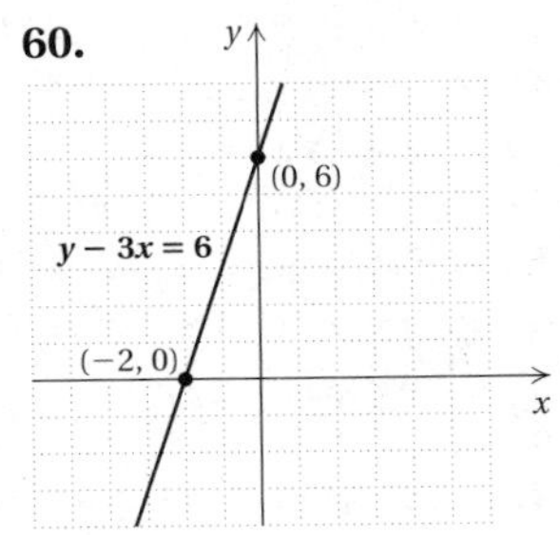

61. $(2x^3 + 3)(2x^2 + 3)$ 63. $(x^7 + 1)(x^5 + 1)$
65. Not factorable by grouping

Margin Exercises, Section 5.2, pp. 389–394

1. (a) $-13, 8, -8, 7, -7$; **(b)** $13, 8, 7$; both 7 and 12 are positive; **(c)** $(x + 3)(x + 4)$ **2.** $(x + 9)(x + 4)$ **3.** The coefficient of the middle term, -8, is negative. **4.** $(x - 5)(x - 3)$ **5.** $(t - 5)(t - 4)$ **6. (a)** $23, 10, 5, 2$; the positive factor has the larger absolute value; **(b)** $-23, -10, -5, -2$; the negative factor has the larger absolute value; **(c)** $(x + 3)(x - 8)$ **7. (a)** $-23, -10, -5, -2$; the negative factor has the larger absolute value; **(b)** $23, 10, 5, 2$; the positive factor has the larger absolute value; **(c)** $(x - 2)(x + 12)$ **8.** $(a - 2)(a + 12)$ **9.** $(t + 2)(t - 12)$ **10.** $(y - 6)(y + 2)$ **11.** $(t^2 + 7)(t^2 - 2)$ **12.** Prime **13.** $x(x + 6)(x - 2)$ **14.** $p(p - q - 3q^2)$ **15.** $3x(x + 4)^2$ **16.** $-1(x + 2)(x - 7)$, or $(-x - 2)(x - 7)$, or $(x + 2)(-x + 7)$ **17.** $-1(x + 3)(x - 6)$, or $(-x - 3)(x - 6)$, or $(x + 3)(-x + 6)$

Exercise Set 5.2, p. 395

1.

Pairs of Factors	Sums of Factors
1, 15	16
−1, −15	−16
3, 5	8
−3, −5	−8

$(x + 3)(x + 5)$

3.

Pairs of Factors	Sums of Factors
1, 12	13
−1, −12	−13
2, 6	8
−2, −6	−8
3, 4	7
−3, −4	−7

$(x + 3)(x + 4)$

5.

Pairs of Factors	Sums of Factors
1, 9	10
−1, −9	−10
3, 3	6
−3, −3	−6

$(x - 3)^2$

7.

Pairs of Factors	Sums of Factors
−1, 14	13
1, −14	−13
−2, 7	5
2, −7	−5

$(x + 2)(x - 7)$

9.

Pairs of Factors	Sums of Factors
1, 4	5
−1, −4	−5
2, 2	4
−2, −2	−4

$(b + 1)(b + 4)$

11.

Pairs of Factors	Sums of Factors
$\frac{1}{3}, \frac{1}{3}$	$\frac{2}{3}$
$-\frac{1}{3}, -\frac{1}{3}$	$-\frac{2}{3}$

$(x + \frac{1}{3})^2$

13. $(d - 2)(d - 5)$ **15.** $(y - 1)(y - 10)$ **17.** Prime **19.** $(x - 9)(x + 2)$ **21.** $x(x - 8)(x + 2)$ **23.** $y(y - 9)(y + 5)$ **25.** $(x - 11)(x + 9)$ **27.** $(c^2 + 8)(c^2 - 7)$ **29.** $(a^2 + 7)(a^2 - 5)$ **31.** $(x - 6)(x + 7)$ **33.** Prime **35.** $(x + 10)^2$ **37.** $-1(x - 10)(x + 3)$, or $(-x + 10)(x + 3)$, or $(x - 10)(-x - 3)$ **39.** $-1(a - 2)(a + 12)$, or $(-a + 2)(a + 12)$, or $(a - 2)(-a - 12)$ **41.** $x^2(x - 25)(x + 4)$ **43.** $(x - 24)(x + 3)$ **45.** $(x - 9)(x - 16)$ **47.** $(a + 12)(a - 11)$ **49.** $(x - 15)(x - 8)$ **51.** $-1(x + 12)(x - 9)$, or $(-x - 12)(x - 9)$, or $(x + 12)(-x + 9)$ **53.** $(y - 0.4)(y + 0.2)$ **55.** $(p + 5q)(p - 2q)$ **57.** $-1(t + 14)(t - 6)$, or $(-t - 14)(t - 6)$, or $(t + 14)(-t + 6)$ **59.** $(m + 4n)(m + n)$ **61.** $(s + 3t)(s - 5t)$ **63.** $6a^8(a + 2)(a - 7)$ **65.** D_W **67.** D_W **69.** $16x^3 - 48x^2 + 8x$ **70.** $28w^2 - 53w - 66$ **71.** $49w^2 + 84w + 36$ **72.** $16w^2 - 88w + 121$ **73.** $16w^2 - 121$ **74.** $27x^{12}$ **75.** $\frac{8}{3}$ **76.** $-\frac{7}{2}$ **77.** 29,555 **78.** 100°, 25°, 55° **79.** $15, -15, 27, -27, 51, -51$ **81.** $(x + \frac{1}{4})(x - \frac{3}{4})$ **83.** $(x + 5)(x - \frac{5}{7})$ **85.** $(b^n + 5)(b^n + 2)$ **87.** $2x^2(4 - \pi)$

89. First consider all the factorizations of 36 that contain three factors. We also find the sum of the factors in each factorization.

Factorization	Sum of Factors
$1 \cdot 1 \cdot 36$	38
$1 \cdot 2 \cdot 18$	21
$1 \cdot 3 \cdot 12$	16
$1 \cdot 4 \cdot 9$	14
$1 \cdot 6 \cdot 6$	13
$2 \cdot 2 \cdot 9$	13
$2 \cdot 3 \cdot 6$	11
$3 \cdot 3 \cdot 4$	10

We can conclude that the number on the house next door is 13, because two sums are 13. This is what causes the census taker to be puzzled. She cannot determine which trio of factors gives the children's ages. When the mother supplies the additional information that there is an oldest child, the census taker knows that the ages of the children cannot be 1, 6, and 6 because there is not an oldest child in this group. Therefore, the children's ages must be 2, 2, and 9.

Margin Exercises, Section 5.3, pp. 400–403

1. $(2x + 5)(x - 3)$ **2.** $(4x + 1)(3x - 5)$ **3.** $(3x - 4)(x - 5)$ **4.** $2(5x - 4)(2x - 3)$ **5.** $(2x + 1)(3x + 2)$ **6.** $-1(2x - 1)(3x + 2)$, or $(2x - 1)(-3x - 2)$, or $(-2x + 1)(3x + 2)$ **7.** $-2(3x - 4)(x + 1)$, or $2(-3x + 4)(x + 1)$, or $2(3x - 4)(-x - 1)$ **8.** $(2a - b)(3a - b)$ **9.** $3(2x + 3y)(x + y)$

Calculator Corner, p. 404

1. Correct **2.** Correct **3.** Not correct **4.** Not correct **5.** Not correct **6.** Correct **7.** Not correct **8.** Correct

Exercise Set 5.3, p. 405

1. $(2x + 1)(x - 4)$ **3.** $(5x + 9)(x - 2)$ **5.** $(3x + 1)(2x + 7)$ **7.** $(3x + 1)(x + 1)$ **9.** $(2x - 3)(2x + 5)$ **11.** $(2x + 1)(x - 1)$ **13.** $(3x - 2)(3x + 8)$ **15.** $(3x + 1)(x - 2)$ **17.** $(3x + 4)(4x + 5)$ **19.** $(7x - 1)(2x + 3)$ **21.** $(3x + 2)(3x + 4)$ **23.** $(3x - 7)^2$ **25.** $(24x - 1)(x + 2)$ **27.** $(5x - 11)(7x + 4)$ **29.** $-2(x - 5)(x + 2)$, or $2(-x + 5)(x + 2)$, or $2(x - 5)(-x - 2)$ **31.** $4(3x - 2)(x + 3)$ **33.** $6(5x - 9)(x + 1)$ **35.** $2(3y + 5)(y - 1)$ **37.** $(3x - 1)(x - 1)$ **39.** $4(3x + 2)(x - 3)$ **41.** $(2x + 1)(x - 1)$ **43.** $(3x + 2)(3x - 8)$ **45.** $5(3x + 1)(x - 2)$ **47.** $p(3p + 4)(4p + 5)$ **49.** $-1(3x + 2)(3x - 8)$, or $(-3x - 2)(3x - 8)$, or $(3x + 2)(-3x + 8)$ **51.** $-1(5x - 3)(3x - 2)$, or $(-5x + 3)(3x - 2)$, or $(5x - 3)(-3x + 2)$ **53.** $x^2(7x - 1)(2x + 3)$ **55.** $3x(8x - 1)(7x - 1)$ **57.** $(5x^2 - 3)(3x^2 - 2)$ **59.** $(5t + 8)^2$ **61.** $2x(3x + 5)(x - 1)$ **63.** Prime **65.** Prime **67.** $(4m + 5n)(3m - 4n)$ **69.** $(2a + 3b)(3a - 5b)$ **71.** $(3a + 2b)(3a + 4b)$ **73.** $(5p + 2q)(7p + 4q)$ **75.** $6(3x - 4y)(x + y)$ **77.** DW **79.** $q = \dfrac{A + 7}{p}$ **80.** $x = \dfrac{y - b}{m}$ **81.** $y = \dfrac{6 - 3x}{2}$ **82.** $q = p + r - 2$ **83.** $\{x|x > 4\}$ **84.** $\left\{x|x \leq \frac{8}{11}\right\}$ **85.**

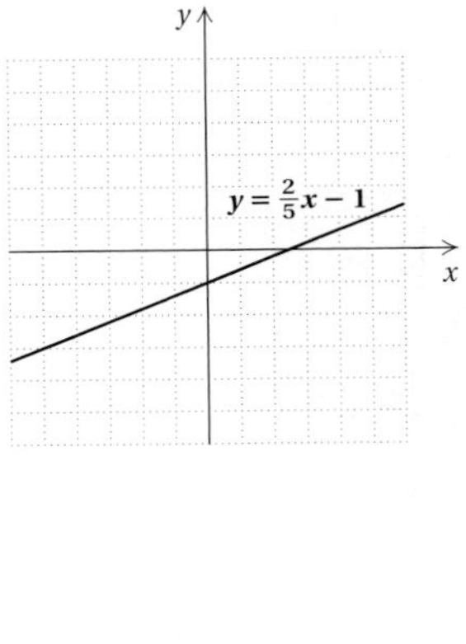

86. y^8 **87.** $9x^2 - 25$ **88.** $16a^2 - 24a + 9$ **89.** $(2x^n + 1)(10x^n + 3)$ **91.** $(x^{3a} - 1)(3x^{3a} + 1)$ **93.–101.** Left to the student

Margin Exercises, Section 5.4, p. 409

1. $(2x + 1)(3x + 2)$ **2.** $(4x + 1)(3x - 5)$ **3.** $3(2x + 3)(x + 1)$ **4.** $2(5x - 4)(2x - 3)$

Exercise Set 5.4, p. 410

1. $(x + 7)(x + 2)$ **3.** $(x - 1)(x - 4)$ **5.** $(2x + 3)(3x + 2)$ **7.** $(x - 4)(3x - 4)$ **9.** $(5x + 3)(7x - 8)$ **11.** $(2x - 3)(2x + 3)$ **13.** $(2x^2 + 5)(x^2 + 3)$ **15.** $(2x - 1)(x + 4)$ **17.** $(3x + 5)(x - 3)$ **19.** $(2x + 7)(3x + 1)$ **21.** $(3x - 1)(x - 1)$ **23.** $(2x + 3)(2x - 5)$ **25.** $(2x - 1)(x + 1)$ **27.** $(3x + 2)(3x - 8)$ **29.** $(3x - 1)(x + 2)$ **31.** $(3x - 4)(4x - 5)$ **33.** $(7x + 1)(2x - 3)$ **35.** $(3x + 2)(3x + 4)$ **37.** $(3x - 7)^2$ **39.** $(24x + 1)(x - 2)$ **41.** $-1(3a - 1)(3a + 5)$, or $(-3a + 1)(3a + 5)$, or $(3a - 1)(-3a - 5)$ **43.** $-2(x - 5)(x + 2)$, or $2(-x + 5)(x + 2)$, or $2(x - 5)(-x - 2)$ **45.** $4(3x - 2)(x + 3)$ **47.** $6(5x - 9)(x + 1)$ **49.** $2(3y + 5)(y - 1)$ **51.** $(3x - 1)(x - 1)$ **53.** $4(3x + 2)(x - 3)$ **55.** $(2x + 1)(x - 1)$ **57.** $(3x - 2)(3x + 8)$ **59.** $5(3x + 1)(x - 2)$ **61.** $p(3p + 4)(4p + 5)$ **63.** $-1(5x - 4)(x + 1)$, or $(-5x + 4)(x + 1)$, or $(5x - 4)(-x - 1)$ **65.** $-3(2t - 1)(t - 5)$, or $3(-2t + 1)(t - 5)$, or $3(2t - 1)(-t + 5)$ **67.** $x^2(7x - 1)(2x + 3)$ **69.** $3x(8x - 1)(7x - 1)$ **71.** $(5x^2 - 3)(3x^2 - 2)$ **73.** $(5t + 8)^2$ **75.** $2x(3x + 5)(x - 1)$ **77.** Prime **79.** Prime **81.** $(4m + 5n)(3m - 4n)$ **83.** $(2a + 3b)(3a - 5b)$ **85.** $(3a - 2b)(3a - 4b)$ **87.** $(5p + 2q)(7p + 4q)$ **89.** $6(3x - 4y)(x + y)$ **91.** $-6x(x - 5)(x + 2)$, or $6x(-x + 5)(x + 2)$, or $6x(x - 5)(-x - 2)$ **93.** $x^3(5x - 11)(7x + 4)$ **95.** DW **97.** $\{x|x < -100\}$ **98.** $\{x|x \geq 217\}$ **99.** $\{x|x \leq 8\}$ **100.** $\{x|x < 2\}$ **101.** $\left\{x|x \geq \frac{20}{3}\right\}$ **102.** $\{x|x > 17\}$ **103.** $\left\{x|x > \frac{26}{7}\right\}$ **104.** $\left\{x|x \geq \frac{77}{17}\right\}$ **105.** About 6369 km, or 3949 mi **106.** 40° **107.** $(3x^5 - 2)^2$ **109.** $(4x^5 + 1)^2$ **111.–119.** Left to the student

Margin Exercises, Section 5.5, pp. 415–419

1. Yes **2.** No **3.** No **4.** Yes **5.** No **6.** Yes **7.** No **8.** Yes **9.** $(x + 1)^2$ **10.** $(x - 1)^2$ **11.** $(t + 2)^2$ **12.** $(5x - 7)^2$ **13.** $(7 - 4y)^2$ **14.** $3(4m + 5)^2$ **15.** $(p^2 + 9)^2$ **16.** $z^3(2z - 5)^2$ **17.** $(3a + 5b)^2$ **18.** Yes **19.** No **20.** No **21.** No **22.** Yes **23.** Yes **24.** Yes **25.** $(x + 3)(x - 3)$ **26.** $4(4 + t)(4 - t)$ **27.** $(a + 5b)(a - 5b)$ **28.** $x^4(8 + 5x)(8 - 5x)$ **29.** $5(1 + 2t^3)(1 - 2t^3)$

30. $(9x^2 + 1)(3x + 1)(3x - 1)$
31. $(7p^2 + 5q^3)(7p^2 - 5q^3)$

Exercise Set 5.5, p. 420

1. Yes **3.** No **5.** No **7.** No **9.** $(x - 7)^2$
11. $(x + 8)^2$ **13.** $(x - 1)^2$ **15.** $(x + 2)^2$ **17.** $(q^2 - 3)^2$
19. $(4y + 7)^2$ **21.** $2(x - 1)^2$ **23.** $x(x - 9)^2$
25. $3(2q - 3)^2$ **27.** $(7 - 3x)^2$ **29.** $5(y^2 + 1)^2$
31. $(1 + 2x^2)^2$ **33.** $(2p + 3q)^2$ **35.** $(a - 3b)^2$
37. $(9a - b)^2$ **39.** $4(3a + 4b)^2$ **41.** Yes **43.** No
45. No **47.** Yes **49.** $(y + 2)(y - 2)$
51. $(p + 3)(p - 3)$ **53.** $(t + 7)(t - 7)$
55. $(a + b)(a - b)$ **57.** $(5t + m)(5t - m)$
59. $(10 + k)(10 - k)$ **61.** $(4a + 3)(4a - 3)$
63. $(2x + 5y)(2x - 5y)$ **65.** $2(2x + 7)(2x - 7)$
67. $x(6 + 7x)(6 - 7x)$ **69.** $(7a^2 + 9)(7a^2 - 9)$
71. $(a^2 + 4)(a + 2)(a - 2)$ **73.** $5(x^2 + 9)(x + 3)(x - 3)$
75. $(1 + y^4)(1 + y^2)(1 + y)(1 - y)$
77. $(x^6 + 4)(x^3 + 2)(x^3 - 2)$ **79.** $\left(y + \frac{1}{4}\right)\left(y - \frac{1}{4}\right)$
81. $\left(5 + \frac{1}{7}x\right)\left(5 - \frac{1}{7}x\right)$ **83.** $(4m^2 + t^2)(2m + t)(2m - t)$
85. D_W **87.** -11 **88.** 400 **89.** $-\frac{5}{6}$ **90.** -0.9
91. 2 **92.** -160 **93.** $x^2 - 4xy + 4y^2$ **94.** $\frac{1}{2}\pi x^2 + 2xy$
95. y^{12} **96.** $25a^4b^6$ **97.**

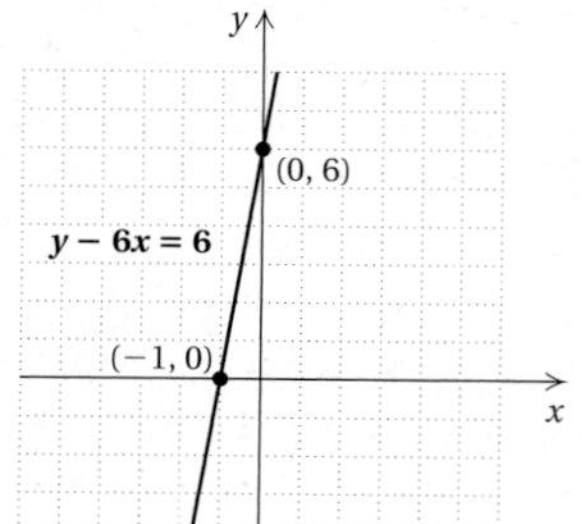

98.

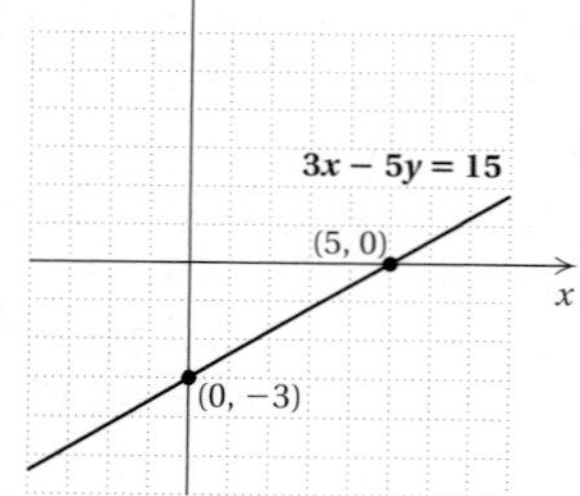

99. Prime **101.** $(x + 11)^2$
103. $2x(3x + 1)^2$ **105.** $(x^4 + 2^4)(x^2 + 2^2)(x + 2)(x - 2)$
107. $3x^3(x + 2)(x - 2)$ **109.** $2x\left(3x + \frac{2}{5}\right)\left(3x - \frac{2}{5}\right)$
111. $p(0.7 + p)(0.7 - p)$ **113.** $(0.8x + 1.1)(0.8x - 1.1)$
115. $x(x + 6)$ **117.** $\left(x + \dfrac{1}{x}\right)\left(x - \dfrac{1}{x}\right)$
119. $(9 + b^{2k})(3 - b^k)(3 + b^k)$ **121.** $(3b^n + 2)^2$
123. $(y + 4)^2$ **125.** 9 **127.** Not correct
129. Not correct

Margin Exercises, Section 5.6, pp. 425–427

1. $3(m^2 + 1)(m + 1)(m - 1)$ **2.** $(x^3 + 4)^2$
3. $2x^2(x + 1)(x + 3)$ **4.** $(3x^2 - 2)(x + 4)$
5. $8x(x - 5)(x + 5)$ **6.** $x^2y(x^2y + 2x + 3)$
7. $2p^4q^2(5p^2 + 2pq + q^2)$ **8.** $(a - b)(2x + 5 + y^2)$
9. $(a + b)(x^2 + y)$ **10.** $(x^2 + y^2)^2$ **11.** $(xy + 1)(xy + 4)$
12. $(p^2 + 9q^2)(p + 3q)(p - 3q)$

Exercise Set 5.6, p. 428

1. $3(x + 8)(x - 8)$ **3.** $(a - 5)^2$ **5.** $(2x - 3)(x - 4)$
7. $x(x + 12)^2$ **9.** $(x + 2)(x - 2)(x + 3)$
11. $3(4x + 1)(4x - 1)$ **13.** $3x(3x - 5)(x + 3)$
15. Prime **17.** $x(x - 3)(x^2 + 7)$ **19.** $x^3(x - 7)^2$
21. $-2(x - 2)(x + 5)$, or $2(-x + 2)(x + 5)$, or $2(x - 2)(-x - 5)$ **23.** Prime
25. $4(x^2 + 4)(x + 2)(x - 2)$
27. $(1 + y^4)(1 + y^2)(1 + y)(1 - y)$ **29.** $x^3(x - 3)(x - 1)$
31. $\frac{1}{9}\left(\frac{1}{3}x^3 - 4\right)^2$ **33.** $m(x^2 + y^2)$ **35.** $9xy(xy - 4)$
37. $2\pi r(h + r)$ **39.** $(a + b)(2x + 1)$
41. $(x + 1)(x - 1 - y)$ **43.** $(n + p)(n + 2)$
45. $(3q + p)(2q - 1)$ **47.** $(2b - a)^2$, or $(a - 2b)^2$
49. $(4x + 3y)^2$ **51.** $(7m^2 - 8n)^2$ **53.** $(y^2 + 5z^2)^2$
55. $\left(\frac{1}{2}a + \frac{1}{3}b\right)^2$ **57.** $(a + b)(a - 2b)$
59. $(m + 20n)(m - 18n)$ **61.** $(mn - 8)(mn + 4)$
63. $r^3(rs - 2)(rs - 8)$ **65.** $a^3(a - b)(a + 5b)$
67. $\left(a + \frac{1}{5}b\right)\left(a - \frac{1}{5}b\right)$ **69.** $(x + y)(x - y)$
71. $(4 + p^2q^2)(2 + pq)(2 - pq)$
73. $(1 + 4x^6y^6)(1 + 2x^3y^3)(1 - 2x^3y^3)$
75. $(q + 1)(q - 1)(q + 8)$ **77.** $(7x + 8y)^2$ **79.** D_W
81. 1999 **82.** 1990 **83.** 1996 **84.** 847 million
85. 10.9% **86.** 2.6% **87.** $-\frac{14}{11}$ **88.** $25x^2 - 10xt + t^2$
89. $X = \dfrac{A + 7}{a + b}$ **90.** $\{x|x < 32\}$ **91.** $(a + 1)^2(a - 1)^2$
93. $(3.5x - 1)^2$ **95.** $(5x + 4)(x + 1.8)$
97. $(y + 3)(y - 3)(y - 2)$ **99.** $(a^2 + 1)(a + 4)$
101. $(x + 2)(x - 2)(x - 1)$ **103.** $(y - 1)^3$
105. $(y + 4 + x)^2$

Margin Exercises, Section 5.7, pp. 433–436

1. $3, -4$ **2.** $7, 3$ **3.** $-\frac{1}{4}, \frac{2}{3}$ **4.** $0, \frac{17}{3}$ **5.** $-2, 3$
6. $-4, 7$ **7.** 3 **8.** $0, 4$ **9.** $-\frac{4}{3}, \frac{4}{3}$ **10.** $3, \frac{7}{2}$ **11.** $-5, 2$
12. $-3, 3$ **13.** $(-5, 0), (1, 0)$ **14.** $0, 3$

Calculator Corner, p. 437

1. Left to the student

Exercise Set 5.7, p. 438

1. $-4, -9$ **3.** $-3, 8$ **5.** $-12, 11$ **7.** $0, -3$ **9.** $0, -18$
11. $-\frac{5}{2}, -4$ **13.** $-\frac{1}{5}, 3$ **15.** $4, \frac{1}{4}$ **17.** $0, \frac{2}{3}$ **19.** $-\frac{1}{10}, \frac{1}{27}$
21. $\frac{1}{3}, -20$ **23.** $0, \frac{2}{3}, \frac{1}{2}$ **25.** $-5, -1$ **27.** $-9, 2$
29. $3, 5$ **31.** $0, 8$ **33.** $0, -18$ **35.** $-4, 4$ **37.** $-\frac{2}{3}, \frac{2}{3}$
39. -3 **41.** 4 **43.** $0, \frac{6}{5}$ **45.** $-1, \frac{5}{3}$ **47.** $-\frac{1}{4}, \frac{2}{3}$
49. $-1, \frac{2}{3}$ **51.** $-\frac{7}{10}, \frac{7}{10}$ **53.** $-2, 9$ **55.** $\frac{4}{5}, \frac{3}{2}$
57. $(-4, 0), (1, 0)$ **59.** $\left(-\frac{5}{2}, 0\right), (2, 0)$ **61.** $(-3, 0), (5, 0)$
63. $-1, 4$ **65.** $-1, 3$ **67.** D_W **69.** $(a + b)^2$
70. $a^2 + b^2$ **71.** -16 **72.** -4.5 **73.** $-\frac{10}{3}$ **74.** $\frac{3}{10}$
75. $-5, 4$ **77.** $-3, 9$ **79.** $-\frac{1}{8}, \frac{1}{8}$ **81.** $-4, 4$

83. Answers may vary. **(a)** $x^2 - x - 12 = 0$; **(b)** $x^2 + 7x + 12 = 0$; **(c)** $4x^2 - 4x + 1 = 0$; **(d)** $x^2 - 25 = 0$; **(e)** $40x^3 - 14x^2 + x = 0$ **85.** 2.33, 6.77 **87.** −9.15, −4.59 **89.** 0, 2.74

Margin Exercises, Section 5.8, pp. 441–446

1. Length: 24 in.; width: 12 in. **2.** Height: 25 ft; width: 10 ft **3. (a)** 342 games; **(b)** 9 teams **4.** 22 and 23 **5.** 24 ft **6.** 3 m, 4 m

Exercise Set 5.8, p. 447

1. Length: 12 ft; width: 2 ft **3.** Length: 18 cm; width: 8 cm **5.** Height: 4 cm; base: 14 cm **7.** Base: 8 m; height: 16 m **9.** 182 games **11.** 12 teams **13.** 4950 handshakes **15.** 25 people **17.** 20 people **19.** 14 and 15 **21.** 12 and 14; −12 and −14 **23.** 15 and 17; −15 and −17 **25.** Hypotenuse: 17 ft; leg: 15 ft **27.** 32 ft **29.** 25 ft **31.** Dining room: 12 ft by 12 ft; kitchen: 12 ft by 10 ft **33.** 4 sec **35.** 5 and 7 **37.** DW **39.** $9x^2 - 25y^2$ **40.** $9x^2 - 30xy + 25y^2$ **41.** $9x^2 + 30xy + 25y^2$ **42.** $6x^2 + 11xy - 35y^2$ **43.** y-intercept: $(0, -4)$; x-intercept: $(16, 0)$ **44.** y-intercept: $(0, 4)$; x-intercept: $(16, 0)$ **45.** y-intercept: $(0, -5)$; x-intercept: $(6.5, 0)$ **46.** y-intercept: $\left(0, \frac{2}{3}\right)$; x-intercept: $\left(\frac{5}{8}, 0\right)$ **47.** y-intercept: $(0, 4)$; x-intercept: $\left(\frac{4}{5}, 0\right)$ **48.** y-intercept: $(0, -5)$; x-intercept: $\left(\frac{5}{2}, 0\right)$ **49.** 35 ft **51.** 5 ft **53.** 30 cm by 15 cm **55.** 39 cm

Summary and Review, Chapter 5, p. 453

1. $(-10x)(x)$; $(-5x)(2x)$; $(5x)(-2x)$; answers may vary **2.** $(6x)(6x^4)$; $(4x^2)(9x^3)$; $(-2x^4)(-18x)$; answers may vary **3.** $5(1 + 2x^3)(1 - 2x^3)$ **4.** $x(x - 3)$ **5.** $(3x + 2)(3x - 2)$ **6.** $(x + 6)(x - 2)$ **7.** $(x + 7)^2$ **8.** $3x(2x^2 + 4x + 1)$ **9.** $(x^2 + 3)(x + 1)$ **10.** $(3x - 1)(2x - 1)$ **11.** $(x^2 + 9)(x + 3)(x - 3)$ **12.** $3x(3x - 5)(x + 3)$ **13.** $2(x + 5)(x - 5)$ **14.** $(x^3 - 2)(x + 4)$ **15.** $(4x^2 + 1)(2x + 1)(2x - 1)$ **16.** $4x^4(2x^2 - 8x + 1)$ **17.** $3(2x + 5)^2$ **18.** Prime **19.** $x(x - 6)(x + 5)$ **20.** $(2x + 5)(2x - 5)$ **21.** $(3x - 5)^2$ **22.** $2(3x + 4)(x - 6)$ **23.** $(x - 3)^2$ **24.** $(2x + 1)(x - 4)$ **25.** $2(3x - 1)^2$ **26.** $3(x + 3)(x - 3)$ **27.** $(x - 5)(x - 3)$ **28.** $(5x - 2)^2$ **29.** $(7b^5 - 2a^4)^2$ **30.** $(xy + 4)(xy - 3)$ **31.** $3(2a + 7b)^2$ **32.** $(m + t)(m + 5)$ **33.** $32(x^2 - 2y^2z^2)(x^2 + 2y^2z^2)$ **34.** 1, −3 **35.** −7, 5 **36.** −4, 3 **37.** $\frac{2}{3}$, 1 **38.** $-4, \frac{3}{2}$ **39.** −2, 8 **40.** Height: 6 cm; base: 5 cm **41.** −18 and −16; 16 and 18 **42.** −19 and −17; 17 and 19 **43.** 3 ft **44.** 6 km **45.** $(-5, 0)$, $(-4, 0)$ **46.** $\left(-\frac{3}{2}, 0\right)$, $(5, 0)$ **47.** DW Answers may vary. Because Sheri did not first factor out the largest common factor, 4, her factorization will not be "complete" until she removes a common factor of 2 from each binomial. Awarding 5 to 7 points seems reasonable. **48.** DW The equations solved in this chapter have an x^2-term (are quadratic), whereas those solved previously have no x^2-term (are linear). The principle of zero products is used to solve quadratic equations; it is not used to solve linear equations. **49.** $\frac{8}{35}$ **50.** $\left\{x \middle| x \leq \frac{4}{3}\right\}$ **51.** $4a^2 - 9$ **52.**

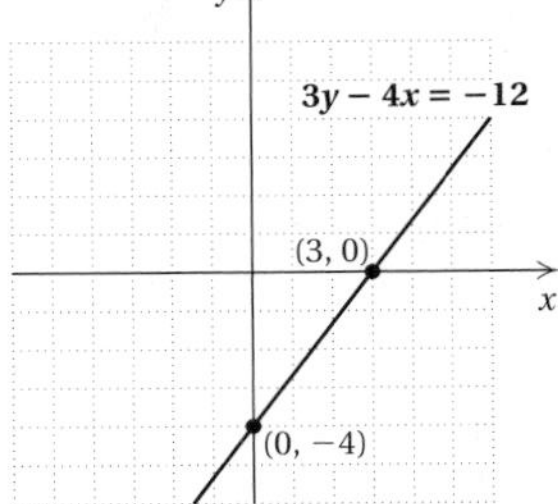

53. 2.5 cm **54.** 0, 2 **55.** Length: 12; width: 6 **56.** No solution **57.** $2, -3, \frac{5}{2}$ **58.** $-2, \frac{5}{4}, 3$ **59.** $(\pi - 2)x^2$

Test: Chapter 5, p. 456

1. [5.1a] $(4x)(x^2)$; $(2x^2)(2x)$; $(-2x)(-2x^2)$; answers may vary **2.** [5.2a] $(x - 5)(x - 2)$ **3.** [5.5b] $(x - 5)^2$ **4.** [5.1b] $2y^2(2y^2 - 4y + 3)$ **5.** [5.1c] $(x^2 + 2)(x + 1)$ **6.** [5.1b] $x(x - 5)$ **7.** [5.2a] $x(x + 3)(x - 1)$ **8.** [5.3a], [5.4a] $2(5x - 6)(x + 4)$ **9.** [5.5d] $(2x + 3)(2x - 3)$ **10.** [5.2a] $(x - 4)(x + 3)$ **11.** [5.3a], [5.4a] $3m(2m + 1)(m + 1)$ **12.** [5.5d] $3(w + 5)(w - 5)$ **13.** [5.5b] $5(3x + 2)^2$ **14.** [5.5d] $3(x^2 + 4)(x + 2)(x - 2)$ **15.** [5.5b] $(7x - 6)^2$ **16.** [5.3a], [5.4a] $(5x - 1)(x - 5)$ **17.** [5.1c] $(x^3 - 3)(x + 2)$ **18.** [5.5d] $5(4 + x^2)(2 + x)(2 - x)$ **19.** [5.3a], [5.4a] $(2x + 3)(2x - 5)$ **20.** [5.3a], [5.4a] $3t(2t + 5)(t - 1)$ **21.** [5.2a] $3(m + 2n)(m - 5n)$ **22.** [5.7b] −4, 5 **23.** [5.7b] $-5, \frac{3}{2}$ **24.** [5.7b] −4, 7 **25.** [5.8a] Length: 8 m; width: 6 m **26.** [5.8a] Height: 4 cm; base: 14 cm **27.** [5.8a] 5 ft **28.** [5.7b] $(-5, 0)$, $(7, 0)$ **29.** [5.7b] $\left(\frac{2}{3}, 0\right)$, $(1, 0)$ **30.** [1.6c] $-\frac{10}{11}$ **31.** [2.7e] $\left\{x \middle| x < \frac{19}{3}\right\}$ **32.** [3.3a]

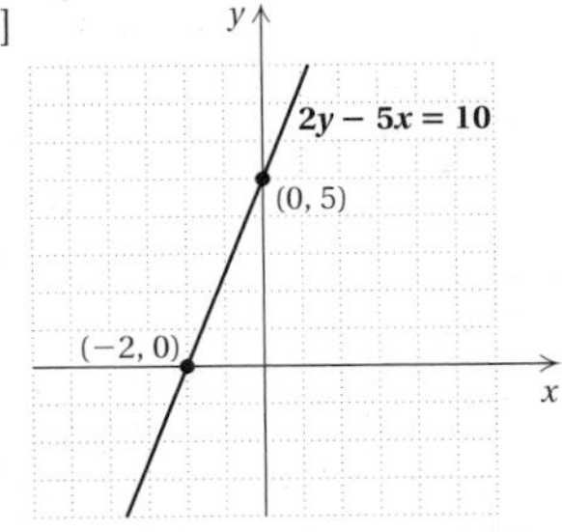

33. [4.6c] $25x^4 - 70x^2 + 49$ **34.** [5.8a] Length: 15; width: 3 **35.** [5.2a] $(a - 4)(a + 8)$ **36.** [5.7b] $-\frac{8}{3}, 0, \frac{2}{5}$ **37.** [4.6b], [5.5d] (d)

Cumulative Review: Chapters 1–5, p. 458

1. [4.4d] $20 + 5(m - 4) + 4(m - 5) + (m - 5)(m - 4)$; m^2 **2.** [4.4d] $36 + 4t + t^2 + 9t$; $(9 + t)(4 + t)$ **3.** [4.4d] $12x + 16 + 12x + 9x^2$; $(3x + 4)^2$ **4.** [1.2d] < **5.** [1.2d] > **6.** [1.4a] 0.35 **7.** [1.6c] −1.57

8. [1.5a] $-\frac{1}{14}$ 9. [1.6c] $-\frac{6}{5}$ 10. [1.8c] $4x + 1$
11. [1.8d] -8 12. [4.2a, b] $\frac{8x^6}{y^3}$
13. [4.1d, e, f] $-\frac{1}{6x^3}$
14. [4.4a] $x^4 - 3x^3 - 3x^2 - 4$
15. [4.7e] $2x^2y^2 - x^2y - xy$
16. [4.8b] $x^2 + 3x + 2 + \frac{3}{x-1}$
17. [4.6c] $4t^2 - 12t + 9$
18. [4.6b] $x^4 - 9$ 19. [4.6a] $6x^2 + 4x - 16$
20. [4.5b] $2x^4 + 6x^3 + 8x^2$
21. [4.5d] $4y^3 + 4y^2 + 5y - 4$
22. [4.6b] $x^2 - \frac{4}{9}$ 23. [5.2a] $(x + 4)(x - 2)$
24. [5.5d] $(2x + 5)(2x - 5)$
25. [5.1c] $(3x - 4)(x^2 + 1)$
26. [5.5b] $(x - 13)^2$ 27. [5.5d] $3(5x + 6y)(5x - 6y)$
28. [5.3a], [5.4a] $(3x + 7)(2x - 9)$
29. [5.2a] $(x^2 - 3)(x^2 + 1)$
30. [5.6a] $2(y - 1)(y + 1)(2y - 3)$
31. [5.3a], [5.4a] $(3p - q)(2p + q)$
32. [5.3a], [5.4a] $2x(5x + 1)(x + 5)$
33. [5.5b] $x(7x - 3)^2$
34. [5.3a], [5.4a] Prime 35. [5.1b] $3x(25x^2 + 9)$
36. [5.5d] $3(x^4 + 4y^4)(x^2 + 2y^2)(x^2 - 2y^2)$
37. [5.2a] $14(x + 2)(x + 1)$
38. [5.6a] $(x^3 + 1)(x + 1)(x - 1)$ 39. [2.3b] 15
40. [2.7e] $\{y|y < 6\}$ 41. [5.7a] $15, -\frac{1}{4}$
42. [5.7a] $0, -37$ 43. [5.7b] $-5, -1, 5$ 44. [5.7b] $-6, 6$
45. [5.7b] $\frac{1}{3}$ 46. [5.7b] $-10, -7$ 47. [5.7b] $0, \frac{3}{2}$
48. [2.3a] 0.2 49. [5.7b] $-4, 5$ 50. [2.7e] $\{x|x \leq 20\}$
51. [2.3c] All real numbers 52. [2.4b] $m = \frac{y-b}{x}$
53. [2.6a] 50, 52 54. [5.8a] -20 and -18; 18 and 20
55. [5.8a] 6 ft, 3 ft 56. [2.6a] 150 m by 350 m
57. [2.6a] 30 m, 60 m, 10 m 58. [5.8a] 17 m
59. [2.5a] \$6500 60. [2.5a] \$29 61. [5.8a] 18 cm, 16 cm
62. [3.3a]

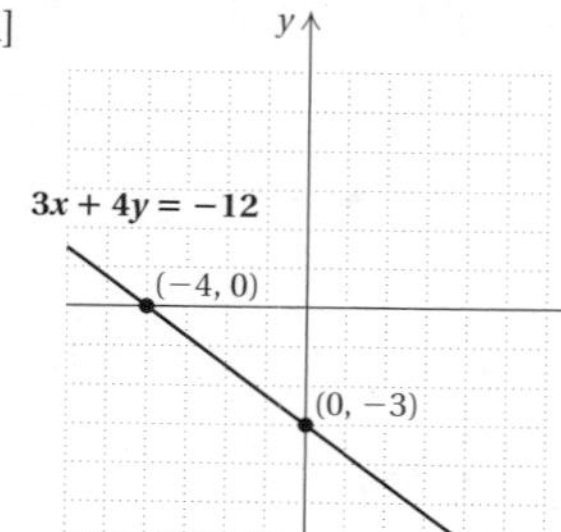

63. [5.7b] $(-2, 0), (7, 0)$ 64. [4.1c] $\left(\frac{1}{8}\right)^3 = \frac{1}{512}$; $\left(\frac{1}{8}\right)^{-3} = 512$; $8^{-3} = \frac{1}{512}$; $8^3 = 512$; $-8^3 = -512$; $(-8)^3 = -512$; $\left(-\frac{1}{8}\right)^{-3} = -512$; $\left(-\frac{1}{8}\right)^3 = -\frac{1}{512}$
65. [3.3a] (e) 66. [1.8d] (b) 67. [5.7b] (b)
68. [3.4a] (d) 69. [2.7e], [4.6a] $\left\{x|x \geq -\frac{13}{3}\right\}$
70. [2.3b] 22 71. [5.7b] $-6, 4$
72. [5.6a] $(x - 2)(x + 1)(x - 3)$
73. [5.6a] $(2a + 3b + 3)(2a - 3b - 5)$
74. [5.5a] 25 75. [5.8a] 2 cm

Chapter 6

Pretest: Chapter 6, p. 464

1. $(x + 2)(x + 3)^2$ 2. $\frac{-b-1}{b^2-4}$, or $\frac{b+1}{4-b^2}$ 3. $\frac{1}{y-2}$
4. $\frac{7a+6}{a(a+2)}$ 5. $\frac{2x}{x+1}$ 6. $\frac{2(x-3)}{x-2}$ 7. $\frac{x-3}{x+3}$
8. $\frac{y+x}{y-x}$ 9. -5 10. 0 11. About 55.6 gal
12. 10.5 hr 13. $2\frac{8}{11}$ hr 14. 60 mph, 80 mph

Margin Exercises, Section 6.1, pp. 465–470

1. 3 2. $-8, 3$ 3. None 4. $\frac{(2x+1)x}{(3x-2)x}$
5. $\frac{(x+1)(x+2)}{(x-2)(x+2)}$ 6. $\frac{(x-8)(-1)}{(x-y)(-1)}$ 7. 5 8. $\frac{x}{4}$
9. $\frac{2x+1}{3x+2}$ 10. $\frac{x+1}{2x+1}$ 11. $x + 2$ 12. $\frac{y+2}{4}$
13. -1 14. -1 15. -1 16. $\frac{a-2}{a-3}$ 17. $\frac{x-5}{2}$

Calculator Corner, p. 471

1. Correct 2. Correct 3. Not correct 4. Not correct
5. Not correct 6. Not correct 7. Correct 8. Correct

Exercise Set 6.1, p. 472

1. 0 3. 8 5. $-\frac{5}{2}$ 7. $-4, 7$ 9. $-5, 5$ 11. None
13. $\frac{(4x)(3x^2)}{(4x)(5y)}$ 15. $\frac{2x(x-1)}{2x(x+4)}$ 17. $\frac{-1(3-x)}{-1(4-x)}$
19. $\frac{(y+6)(y-7)}{(y+6)(y+2)}$ 21. $\frac{x^2}{4}$ 23. $\frac{8p^2q}{3}$ 25. $\frac{x-3}{x}$
27. $\frac{m+1}{2m+3}$ 29. $\frac{a-3}{a+2}$ 31. $\frac{a-3}{a-4}$ 33. $\frac{x+5}{x-5}$
35. $a + 1$ 37. $\frac{x^2+1}{x+1}$ 39. $\frac{3}{2}$ 41. $\frac{6}{t-3}$
43. $\frac{t+2}{2(t-4)}$ 45. $\frac{t-2}{t+2}$ 47. -1 49. -1 51. -6
53. $-x - 1$ 55. $\frac{56x}{3}$ 57. $\frac{2}{dc^2}$ 59. $\frac{x+2}{x-2}$
61. $\frac{(a+3)(a-3)}{a(a+4)}$ 63. $\frac{2a}{a-2}$ 65. $\frac{(t+2)(t-2)}{(t+1)(t-1)}$
67. $\frac{x+4}{x+2}$ 69. $\frac{5(a+6)}{a-1}$ 71. D_W 73. 18 and 20; -18 and -20 74. 3.125 L 75. $(x - 8)(x + 7)$
76. $(a - 8)^2$ 77. $x^3(x - 7)(x + 5)$
78. $(2y^2 + 1)(y - 5)$ 79. $(2 - t)(2 + t)(4 + t^2)$
80. $10(x + 7)(x + 1)$ 81. $(x - 7)(x - 2)$ 82. Prime
83. $(4x - 5y)^2$ 84. $(a - 7b)(a - 2b)$ 85. $x + 2y$
87. $\frac{(t-9)^2(t-1)}{(t^2+9)(t+1)}$ 89. $\frac{x-y}{x-5y}$

91. $\dfrac{5(2x+5)-25}{10} = \dfrac{10x+25-25}{10}$

$= \dfrac{10x}{10}$

$= x$

You get the same number you selected. To do a number trick, ask someone to select a number and then perform these operations. The person will probably be surprised that the result is the original number.

Margin Exercises, Section 6.2, pp. 476–478

1. $\frac{2}{7}$ **2.** $\frac{2x^3-1}{x^2+5}$ **3.** $\frac{1}{x-5}$ **4.** x^2-3 **5.** $\frac{6}{7}$
6. $\frac{5}{8}$ **7.** $\frac{(x-3)(x-2)}{(x+5)(x+5)}$ **8.** $\frac{x-3}{x+2}$ **9.** $\frac{(x-3)(x-2)}{x+2}$
10. $\frac{y+1}{y-1}$

Exercise Set 6.2, p. 479

1. $\frac{x}{4}$ **3.** $\frac{1}{x^2-y^2}$ **5.** $a+b$ **7.** $\frac{x^2-4x+7}{x^2+2x-5}$ **9.** $\frac{3}{10}$
11. $\frac{1}{4}$ **13.** $\frac{b}{a}$ **15.** $\frac{(a+2)(a+3)}{(a-3)(a-1)}$ **17.** $\frac{(x-1)^2}{x}$
19. $\frac{1}{2}$ **21.** $\frac{15}{8}$ **23.** $\frac{15}{4}$ **25.** $\frac{a-5}{3(a-1)}$ **27.** $\frac{(x+2)^2}{x}$
29. $\frac{3}{2}$ **31.** $\frac{c+1}{c-1}$ **33.** $\frac{y-3}{2y-1}$ **35.** $\frac{x+1}{x-1}$ **37.** D_W
39. $\{x \mid x \geq 77\}$ **40.** Height: 4 cm; base: 14 cm
41. $8x^3-11x^2-3x+12$ **42.** $-2p^2+4pq-4q^2$
43. $\frac{4y^8}{x^6}$ **44.** $\frac{125x^{18}}{y^{12}}$ **45.** $\frac{4x^6}{y^{10}}$ **46.** $\frac{1}{a^{15}b^{20}}$ **47.** $-\frac{1}{b^2}$
49. $\frac{(x-7)^2}{x+y}$

Margin Exercises, Section 6.3, pp. 481–482

1. 144 **2.** 12 **3.** 10 **4.** 120 **5.** $\frac{35}{144}$ **6.** $\frac{1}{4}$ **7.** $\frac{11}{10}$
8. $\frac{9}{40}$ **9.** $60x^3y^2$ **10.** $(y+1)^2(y+4)$
11. $7(t^2+16)(t-2)$ **12.** $3x(x+1)^2(x-1)$

Exercise Set 6.3, p. 483

1. 108 **3.** 72 **5.** 126 **7.** 360 **9.** 500 **11.** $\frac{65}{72}$
13. $\frac{29}{120}$ **15.** $\frac{23}{180}$ **17.** $12x^3$ **19.** $18x^2y^2$
21. $6(y-3)$ **23.** $t(t+2)(t-2)$
25. $(x+2)(x-2)(x+3)$ **27.** $t(t+2)^2(t-4)$
29. $(a+1)(a-1)^2$ **31.** $(m-3)(m-2)^2$
33. $(2+3x)(2-3x)$ **35.** $10v(v+4)(v+3)$
37. $18x^3(x-2)^2(x+1)$ **39.** $6x^3(x+2)^2(x-2)$
41. D_W **43.** $(x-3)^2$ **44.** $2x(3x+2)$
45. $(x+3)(x-3)$ **46.** $(x+7)(x-3)$ **47.** $(x+3)^2$
48. $(x-7)(x+3)$ **49.** 54% **50.** 64% **51.** 74%
52. 98%; this seems unreasonably high. **53.** 1965
54. 1999 **55.** 24 min

Margin Exercises, Section 6.4, pp. 485–488

1. $\frac{7}{9}$ **2.** $\frac{3+x}{x-2}$ **3.** $\frac{6x+4}{x-1}$ **4.** $\frac{10x^2+9x}{48}$ **5.** $\frac{9x+10}{48x^2}$
6. $\frac{4x^2-x+3}{x(x-1)(x+1)^2}$ **7.** $\frac{2x^2+16x+5}{(x+3)(x+8)}$
8. $\frac{8x+88}{(x+16)(x+1)(x+8)}$ **9.** $\frac{x-5}{4}$ **10.** $\frac{x-1}{x-3}$
11. $\frac{-2x-11}{3(x+4)(x-4)}$

Exercise Set 6.4, p. 489

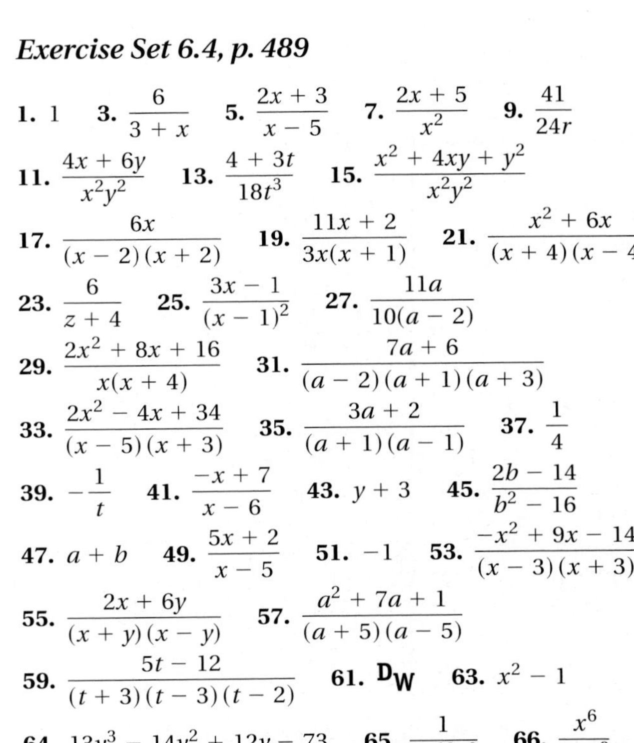

1. 1 **3.** $\frac{6}{3+x}$ **5.** $\frac{2x+3}{x-5}$ **7.** $\frac{2x+5}{x^2}$ **9.** $\frac{41}{24r}$
11. $\frac{4x+6y}{x^2y^2}$ **13.** $\frac{4+3t}{18t^3}$ **15.** $\frac{x^2+4xy+y^2}{x^2y^2}$
17. $\frac{6x}{(x-2)(x+2)}$ **19.** $\frac{11x+2}{3x(x+1)}$ **21.** $\frac{x^2+6x}{(x+4)(x-4)}$
23. $\frac{6}{z+4}$ **25.** $\frac{3x-1}{(x-1)^2}$ **27.** $\frac{11a}{10(a-2)}$
29. $\frac{2x^2+8x+16}{x(x+4)}$ **31.** $\frac{7a+6}{(a-2)(a+1)(a+3)}$
33. $\frac{2x^2-4x+34}{(x-5)(x+3)}$ **35.** $\frac{3a+2}{(a+1)(a-1)}$ **37.** $\frac{1}{4}$
39. $-\frac{1}{t}$ **41.** $\frac{-x+7}{x-6}$ **43.** $y+3$ **45.** $\frac{2b-14}{b^2-16}$
47. $a+b$ **49.** $\frac{5x+2}{x-5}$ **51.** -1 **53.** $\frac{-x^2+9x-14}{(x-3)(x+3)}$
55. $\frac{2x+6y}{(x+y)(x-y)}$ **57.** $\frac{a^2+7a+1}{(a+5)(a-5)}$

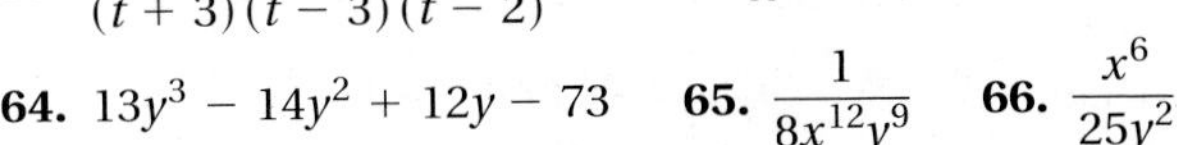

59. $\frac{5t-12}{(t+3)(t-3)(t-2)}$ **61.** D_W **63.** x^2-1
64. $13y^3-14y^2+12y-73$ **65.** $\frac{1}{8x^{12}y^9}$ **66.** $\frac{x^6}{25y^2}$
67. $\frac{1}{x^{12}y^{21}}$ **68.** $\frac{25}{x^4y^6}$ **69.**

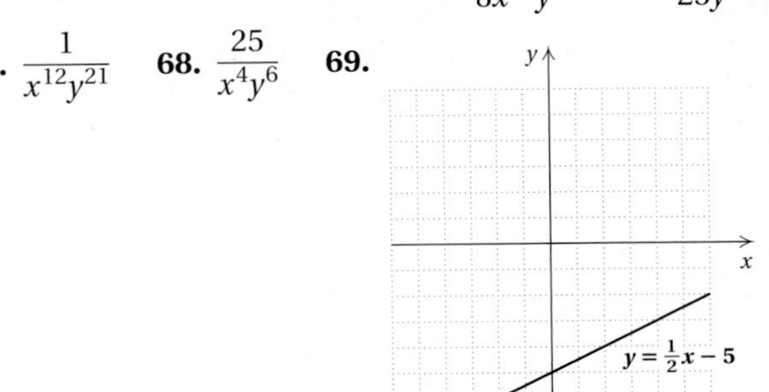

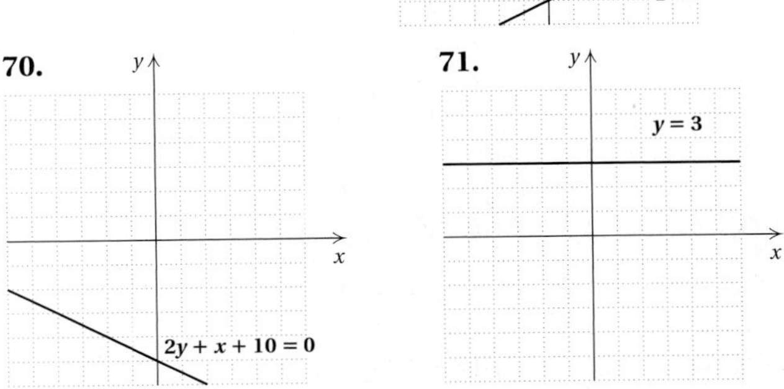

72. 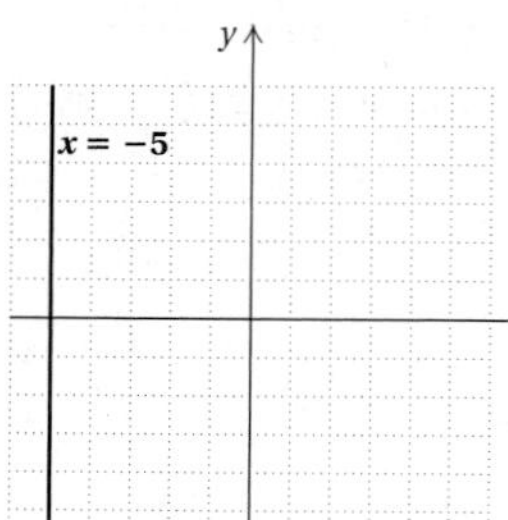

73. −8 74. $\frac{5}{6}$
75. 3, 5 76. −2, 9
77. Perimeter: $\frac{16y + 28}{15}$; area: $\frac{y^2 + 2y - 8}{15}$
79. $\frac{(z + 6)(2z - 3)}{(z + 2)(z - 2)}$
81. $\frac{11z^4 - 22z^2 + 6}{(z^2 + 2)(z^2 - 2)(2z^2 - 3)}$
83.–85. Left to the student

Margin Exercises, Section 6.5, pp. 493–496

1. $\frac{4}{11}$ 2. $\frac{5}{y}$ 3. $\frac{x^2 + 2x + 1}{2x + 1}$ 4. $\frac{-x - 7}{15x}$
5. $\frac{x^2 - 48}{(x + 7)(x + 8)(x + 6)}$ 6. $\frac{3x - 1}{3}$ 7. $\frac{4x - 3}{x - 2}$
8. $\frac{-8y - 28}{(y + 4)(y - 4)}$ 9. $\frac{x - 13}{(x + 3)(x - 3)}$ 10. $\frac{6x^2 - 2x - 2}{3x(x + 1)}$

Exercise Set 6.5, p. 497

1. $\frac{4}{x}$ 3. 1 5. $\frac{1}{x - 1}$ 7. $\frac{-a - 4}{10}$ 9. $\frac{7z - 12}{12z}$
11. $\frac{4x^2 - 13xt + 9t^2}{3x^2t^2}$ 13. $\frac{2x - 40}{(x + 5)(x - 5)}$ 15. $\frac{3 - 5t}{2t(t - 1)}$
17. $\frac{2s - st - s^2}{(t + s)(t - s)}$ 19. $\frac{y - 19}{4y}$ 21. $\frac{-2a^2}{(x + a)(x - a)}$
23. $\frac{8}{3}$ 25. $\frac{13}{a}$ 27. $\frac{8}{y - 1}$ 29. $\frac{x - 2}{x - 7}$ 31. $\frac{4}{a^2 - 25}$
33. $\frac{2x - 4}{x - 9}$ 35. $\frac{9x + 12}{(x + 3)(x - 3)}$ 37. $\frac{1}{2}$
39. $\frac{x - 3}{(x + 3)(x + 1)}$ 41. $\frac{18x + 5}{x - 1}$ 43. 0 45. $\frac{-9}{2x - 3}$
47. $\frac{20}{2y - 1}$ 49. $\frac{2a - 3}{2 - a}$ 51. $\frac{z - 3}{2z - 1}$ 53. $\frac{2}{x + y}$
55. $\mathbf{D_W}$ 57. x^5 58. $30x^{12}$ 59. $\frac{b^{20}}{a^8}$ 60. $18x^3$
61. $\frac{6}{x^3}$ 62. $\frac{10}{x^3}$ 63. $x^2 - 9x + 18$ 64. $(4 - \pi)r^2$
65. $\frac{30}{(x - 3)(x + 4)}$ 67. $\frac{x^2 + xy - x^3 + x^2y - xy^2 + y^3}{(x^2 + y^2)(x + y)^2(x - y)}$
69. Missing side: $\frac{-2a - 15}{a - 6}$; area: $\frac{-2a^3 - 15a^2 + 12a + 90}{2(a - 6)^2}$
71.–73. Left to the student

Margin Exercises, Section 6.6, pp. 501–505

1. $\frac{33}{2}$ 2. $\frac{3}{2}$ 3. 3 4. $-\frac{1}{8}$ 5. 1 6. 2 7. 4

Calculator Corner, p. 504

1.–8. Left to the student

Study Tip, p. 506

1. Rational expression 2. Solutions 3. Rational expression 4. Rational expression 5. Rational expression 6. Solutions 7. Rational expression 8. Solutions 9. Solutions 10. Solutions 11. Rational expression 12. Solutions 13. Rational expression

Exercise Set 6.6, p. 507

1. $\frac{6}{5}$ 3. $\frac{40}{29}$ 5. $\frac{47}{2}$ 7. −6 9. $\frac{24}{7}$ 11. −4, −1
13. −4, 4 15. 3 17. $\frac{14}{3}$ 19. 5 21. 5 23. $\frac{5}{2}$
25. −2 27. $-\frac{13}{2}$ 29. $\frac{17}{2}$ 31. No solution 33. −5
35. $\frac{5}{3}$ 37. $\frac{1}{2}$ 39. No solution 41. No solution
43. 4 45. $\mathbf{D_W}$ 47. $\frac{1}{a^6b^{15}}$ 48. x^8y^{12} 49. $\frac{16x^4}{t^8}$
50. $\frac{w^4}{y^6}$ 51. $32x^6$ 52. $\frac{64x^{10}}{y^8}$

53. 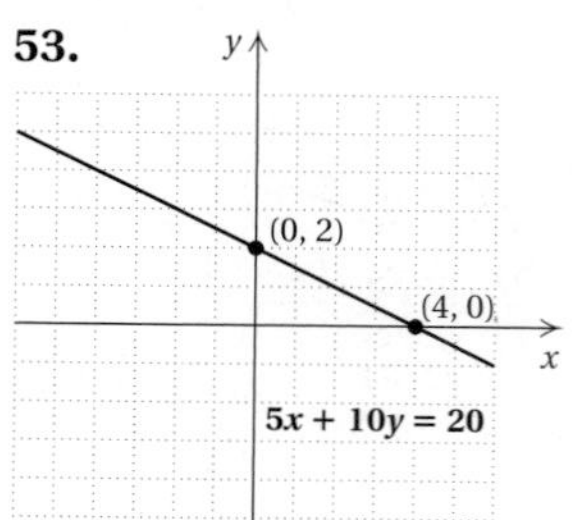

54. 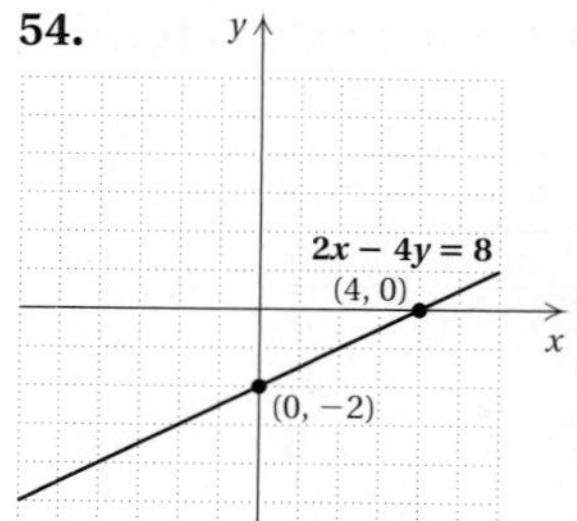

55. 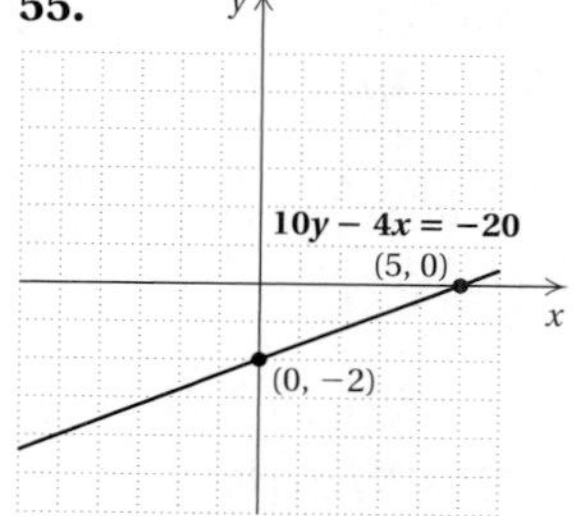

56. 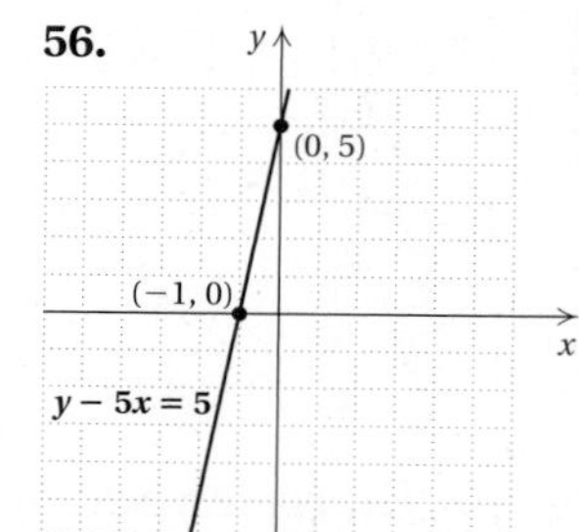

57. 7 59. No solution 61. −2, 2 63. 4
65. Left to the student

Margin Exercises, Section 6.7, pp. 513–518

1. $3\frac{3}{7}$ hr 2. Greg: 40 mph; Nancy: 60 mph 3. 58 km/L
4. 0.280 5. 124 km/h 6. 2.4 fish/yd^2 7. About 34.6 gal 8. 90 whales 9. No 10. 24.75 ft
11. About 34.9 ft

Exercise Set 6.7, p. 519

1. $2\frac{2}{9}$ hr 3. $25\frac{5}{7}$ min 5. $3\frac{15}{16}$ hr 7. $22\frac{2}{9}$ min
9. $7\frac{1}{2}$ min 11. Sarah: 30 km/h; Rick: 70 km/h
13. Passenger: 80 mph; freight: 66 mph 15. 20 mph
17. Hank: 14 km/h; Kelly: 19 km/h 19. Ralph: 5 km/h; Bonnie: 8 km/h 21. 3 hr 23. $\frac{5}{9}$ divorce/marriage

25. 2.3 km/h **27.** 66 g **29.** 1.92 g **31.** 1.75 lb **33.** $1\frac{11}{39}$ kg **35.** (a) 0.361; (b) 268 hits; (c) 202 hits **37.** 22 in.; 55.8 cm **39.** $7\frac{1}{4}$; 57.9 cm **41.** $7\frac{1}{2}$; $23\frac{3}{5}$ in. **43.** 10,000 blue whales **45.** (a) 4.8 tons; (b) 48 lb **47.** 200 duds **49.** $\frac{21}{2}$ **51.** $\frac{8}{3}$ **53.** $\frac{35}{3}$ **55.** 15 ft

57.  **59.** x^{11} **60.** x **61.** $\frac{1}{x^{11}}$ **62.** $\frac{1}{x}$

63.

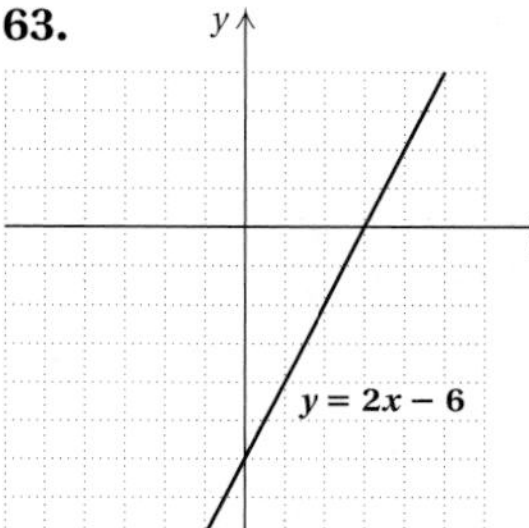

64.

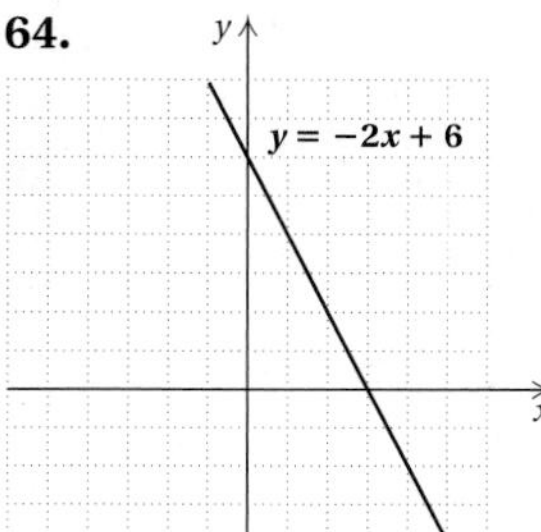

65.

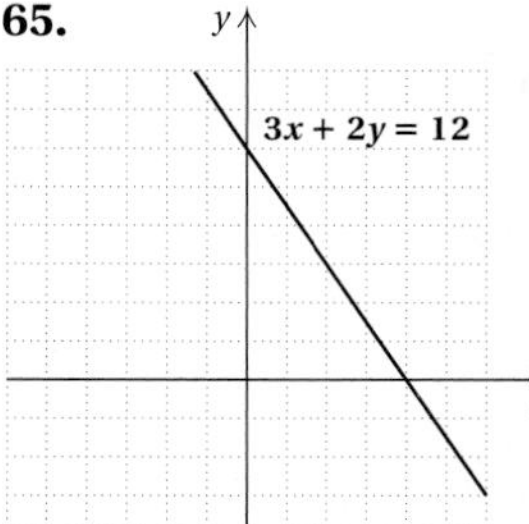

66.

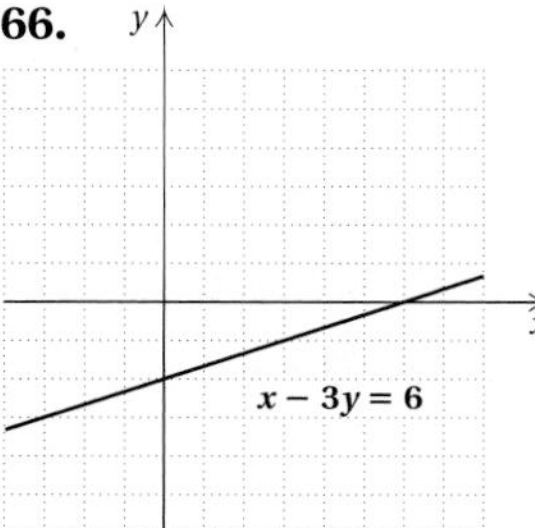

67.

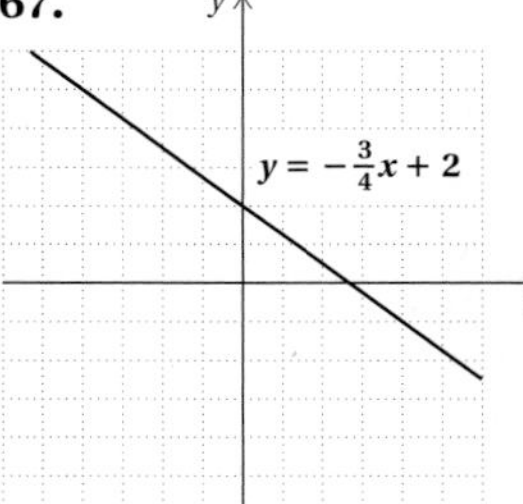

68.

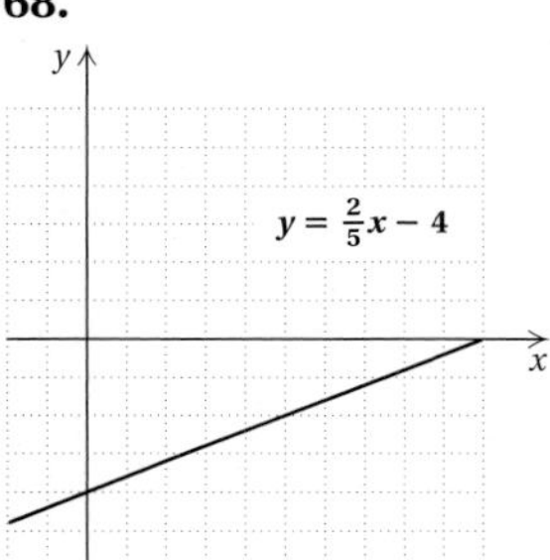

69. Ann: 6 hr; Betty: 12 hr **71.** $27\frac{3}{11}$ min **73.** $t = \frac{ab}{b + a}$

Margin Exercises, Section 6.8, pp. 527–529

1. $\frac{136}{5}$ **2.** $\frac{7x^2}{3(2 - x^2)}$ **3.** $\frac{x}{x - 1}$ **4.** $\frac{136}{5}$ **5.** $\frac{7x^2}{3(2 - x^2)}$ **6.** $\frac{x}{x - 1}$

Exercise Set 6.8, p. 530

1. $\frac{25}{4}$ **3.** $\frac{1}{3}$ **5.** -6 **7.** $\frac{1 + 3x}{1 - 5x}$ **9.** $\frac{2x + 1}{x}$ **11.** 8 **13.** $x - 8$ **15.** $\frac{y}{y - 1}$ **17.** $-\frac{1}{a}$ **19.** $\frac{ab}{b - a}$ **21.** $\frac{p^2 + q^2}{q + p}$ **23.** $\frac{2a^2 + 4a}{5 - 3a^2}$ **25.** $\frac{60 - 15a^3}{126a^2 + 28a^3}$ **27.** $\frac{ac}{bd}$ **29.** 1 **31.** $\frac{4x + 1}{5x + 3}$ **33.** DW

35. $4x^4 + 3x^3 + 2x - 7$ **36.** 0 **37.** $(p - 5)^2$ **38.** $(p + 5)^2$ **39.** $50(p^2 - 2)$ **40.** $5(p + 2)(p - 10)$ **41.** 14 yd **42.** 12 ft, 5 ft **43.** $\frac{(x - 1)(3x - 2)}{5x - 3}$ **45.** $\frac{5x + 3}{3x + 2}$

Summary and Review: Chapter 6, p. 532

1. 0 **2.** 6 **3.** $-6, 6$ **4.** $-6, 5$ **5.** -2 **6.** 0, 3, 5 **7.** $\frac{x - 2}{x + 1}$ **8.** $\frac{7x + 3}{x - 3}$ **9.** $\frac{y - 5}{y + 5}$ **10.** $\frac{a - 6}{5}$ **11.** $\frac{6}{2t - 1}$ **12.** $-20t$ **13.** $\frac{2x^2 - 2x}{x + 1}$ **14.** $30x^2y^2$ **15.** $4(a - 2)$ **16.** $(y - 2)(y + 2)(y + 1)$ **17.** $\frac{-3x + 18}{x + 7}$ **18.** -1 **19.** $\frac{2a}{a - 1}$ **20.** $d + c$ **21.** $\frac{4}{x - 4}$ **22.** $\frac{x + 5}{2x}$ **23.** $\frac{2x + 3}{x - 2}$ **24.** $\frac{-x^2 + x + 26}{(x - 5)(x + 5)(x + 1)}$ **25.** $\frac{2(x - 2)}{x + 2}$ **26.** $\frac{z}{1 - z}$ **27.** $c - d$ **28.** 8 **29.** $-5, 3$ **30.** $5\frac{1}{7}$ hr **31.** 240 km/h, 280 km/h **32.** 95 mph, 175 mph **33.** 160 defective calculators **34.** (a) $\frac{12}{13}$ c; (b) $4\frac{1}{5}$ c; (c) $9\frac{1}{3}$ c **35.** 7200 frogs **36.** 6 **37.** DW $\frac{5x + 6}{(x + 2)(x - 2)}$; used to find an equivalent expression for each rational expression with the LCM as the least common denominator **38.** DW $\frac{3x + 10}{(x - 2)(x + 2)}$; used to find an equivalent expression for each rational expression with the LCM as the least common denominator **39.** DW 4; used to clear fractions **40.** DW $\frac{4(x - 2)}{x(x + 4)}$; method 1: used to multiply by 1 using LCM/LCM; method 2: LCM of the denominators in the numerator used to subtract in the numerator and LCM of the denominators in the denominator used to add in the denominator **41.** $(5x^2 - 3)(x + 4)$ **42.** $\frac{1}{125x^9y^6}$ **43.** $-2x^3 + 3x^2 + 12x - 18$ **44.** Length: 5 cm; width: 3 cm; perimeter: 16 cm **45.** $\frac{5(a + 3)^2}{a}$ **46.** $\frac{10a}{(a - b)(b - c)}$ **47.** They are equivalent proportions.

Test: Chapter 6, p. 535

1. [6.1a] 0 **2.** [6.1a] -8 **3.** [6.1a] $-7, 7$ **4.** [6.1a] 1, 2 **5.** [6.1a] 1 **6.** [6.1a] $-5, -3, 0$ **7.** [6.1c] $\frac{3x + 7}{x + 3}$ **8.** [6.1d] $\frac{a + 5}{2}$ **9.** [6.2b] $\frac{(5x + 1)(x + 1)}{3x(x + 2)}$

10. [6.3c] $(y-3)(y+3)(y+7)$ **11.** [6.4a] $\frac{23-3x}{x^3}$

12. [6.5a] $\frac{8-2t}{t^2+1}$ **13.** [6.4a] $\frac{-3}{x-3}$ **14.** [6.5a] $\frac{2x-5}{x-3}$

15. [6.4a] $\frac{8t-3}{t(t-1)}$ **16.** [6.5a] $\frac{-x^2-7x-15}{(x+4)(x-4)(x+1)}$

17. [6.5b] $\frac{x^2+2x-7}{(x-1)^2(x+1)}$ **18.** [6.8a] $\frac{3y+1}{y}$

19. [6.6a] 12 **20.** [6.6a] $-3, 5$ **21.** [6.7b] 16 defective spark plugs **22.** [6.7b] 50 zebras **23.** [6.7a] 12 min **24.** [6.7a] Craig: 65 km/h; Marilyn: 45 km/h **25.** [6.7b] 15

26. [5.6a] $(4a+7)(4a-7)$ **27.** [4.2a, b] $\frac{y^{12}}{81x^8}$

28. [4.4c] $13x^2-29x+76$ **29.** [5.8a] 21 and 22; -22 and -21 **30.** [6.7a] Rema: 4 hr; Reggie: 10 hr

31. [6.8a] $\frac{3a+2}{2a+1}$

Cumulative Review: Chapters 1–6, p. 537

1. [2.5a] About \$316,623 **2.** [6.7b] About 844 mi **3.** [6.7a] 12 km/h, 10 km/h **4.** [6.7a] $2\frac{8}{11}$ hr **5.** [2.8b] At most 225 sheets **6.** [5.8a] 14 ft **7.** [2.6a] -278 and -276 **8.** [5.8a] 9 and 11 **9.** [1.1a] $-\frac{9}{5}$ **10.** [4.3a] 12

11. [1.8c] $2x+6$ **12.** [4.1d], [4.2a, b] $\frac{27x^7}{4}$

13. [4.1e] $\frac{4x^{10}}{3}$ **14.** [6.1c] $\frac{2(t-3)}{2t-1}$ **15.** [6.8a] $\frac{(x+2)^2}{x^2}$

16. [6.1c] $\frac{a+4}{a-4}$ **17.** [1.3a] $\frac{17}{42}$ **18.** [6.4a] $\frac{x^2+4xy+y^2}{x^2y^2}$

19. [6.4a] $\frac{3z-2}{z^2-1}$ **20.** [4.4a] $2x^4+8x^3-2x+9$

21. [1.4a] 2.33 **22.** [4.7e] $-xy$ **23.** [6.5a] $\frac{1}{x-3}$

24. [6.5a] $\frac{2x^2-14x-16}{(x+4)(x-5)^2}$ **25.** [1.5a] -1.3

26. [4.5b] $6x^4+12x^3-15x^2$ **27.** [4.6b] $9t^2-\frac{1}{4}$

28. [4.6c] $4p^2-4pq+q^2$ **29.** [4.6a] $3x^2-7x-20$

30. [4.6b] $4x^4-1$ **31.** [6.1d] $\frac{2(t-1)}{t}$

32. [6.1d] $-\frac{2(a+1)}{a}$ **33.** [4.8b] $3x^2-4x+5$

34. [1.6c] $-\frac{9}{20}$ **35.** [6.2b] $\frac{x-2}{2x(x-3)}$ **36.** [6.2b] $-\frac{12}{x}$

37. [5.1c], [5.5d] $(2x+3)(2x-3)(x+3)$ **38.** [5.2a] $(x+8)(x-1)$ **39.** [5.3a], [5.4a] $(3x+1)(x-5)$ **40.** [5.5b] $(4y+5x)^2$ **41.** [5.1b], [5.2a] $3x(x+5)(x+3)$ **42.** [5.1b], [5.5d] $2(x+1)(x-1)$ **43.** [5.5b] $(x-14)^2$ **44.** [5.1b, c] $2(y^2+3)(2y+5)$ **45.** [2.3c] -7 **46.** [5.7a] $0, -\frac{4}{3}$ **47.** [5.7b] 0, 8 **48.** [5.7b] 4 **49.** [2.7c] $\{x \mid x \geq -9\}$ **50.** [5.7b] $-3, 3$ **51.** [2.3b] $-\frac{11}{7}$ **52.** [6.6a] $-3, 3$ **53.** [6.6a] $-\frac{1}{11}$ **54.** [5.7a] $0, \frac{1}{10}$ **55.** [6.6a] No solution **56.** [6.6a] $\frac{5}{7}$

57. [3.3a]

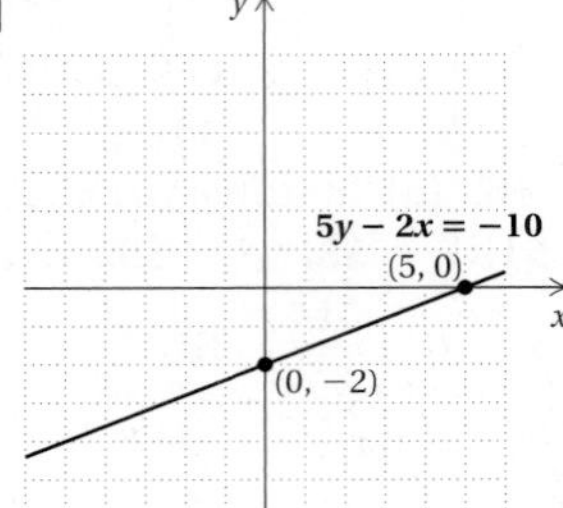

58. [5.7b] $-\frac{2}{3}, 4$

59. [5.7b], [6.6a] $-1, 2$ **60.** [5.7b], [6.6a], [6.8a] $-4, 1$

61. [4.1d, e], [5.7b] $-5, 5$ **62.** [6.2a], [6.4a] $\frac{-x^2-x+6}{2x^2+x+5}$

63. [4.2d] 5×10^7

Chapter 7

Pretest: Chapter 7, p. 542

1. Slope: $\frac{1}{3}$; y-intercept: $\left(0, -\frac{7}{3}\right)$ **2.** $y=-4.7x+8$ **3.** $y=x-4$ **4.** $y=4x+7$ **5.** $y=\frac{5}{2}x$; 125

6. 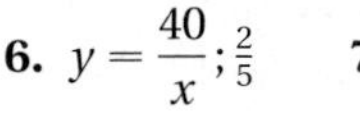 $y=\frac{40}{x}$; $\frac{2}{5}$ **7.**

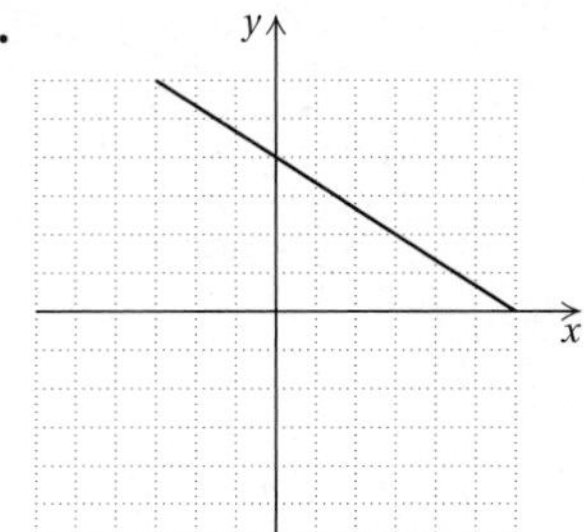

8.

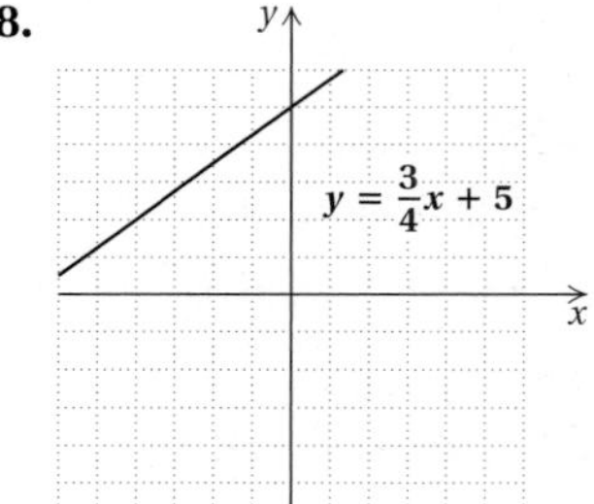

9. **(a)** $S=\frac{70}{9}W$; **(b)** $93\frac{1}{3}$ servings

10. **(a)**  $N=\frac{80{,}000{,}000}{S}$; **(b)** 640 files

11.

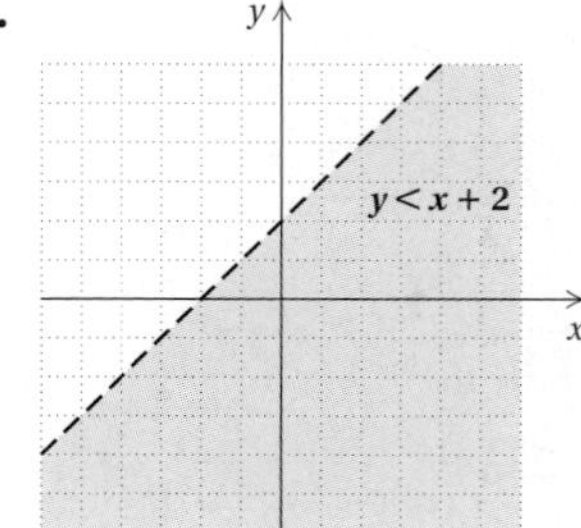

12.

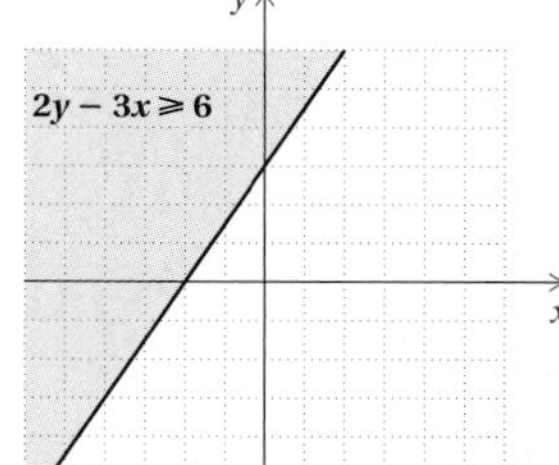

13. Parallel

14. Perpendicular **15.** Perpendicular **16.** Yes
17. **(a)** $y = 0.47x + 0.49$; **(b)** \$0.47 billion per year; **(c)** \$6.6 billion

Margin Exercises, Section 7.1, pp. 543–547

1. **(a)** 15; **(b)** 15 calories per minute; **(c)** slope: 15; y-intercept: (0, 0) **2.** Slope: 5; y-intercept: (0, 0)
3. Slope: $-\frac{3}{2}$; y-intercept: $(0, -6)$
4. Slope: $-\frac{3}{4}$; y-intercept: $\left(0, \frac{15}{4}\right)$
5. Slope: 2; y-intercept: $\left(0, -\frac{17}{2}\right)$
6. Slope: $-\frac{7}{5}$; y-intercept: $\left(0, -\frac{22}{5}\right)$
7. $y = 3.5x - 23$ **8.** $y = 13$ **9.** $y = -7.29x$
10. $y = 5x - 18$ **11.** $y = -3x - 5$ **12.** $y = 6x - 13$
13. $y = -\frac{2}{3}x + \frac{14}{3}$ **14.** $y = x + 2$ **15.** $y = 2x + 4$

Exercise Set 7.1, p. 548

1. Slope: -4; y-intercept: $(0, -9)$
3. Slope: 1.8; y-intercept: (0, 0)
5. Slope: $-\frac{8}{7}$; y-intercept: $(0, -3)$
7. Slope: $\frac{4}{9}$; y-intercept: $\left(0, -\frac{7}{9}\right)$
9. Slope: $-\frac{3}{2}$; y-intercept: $\left(0, -\frac{1}{2}\right)$
11. Slope: 0; y-intercept: $(0, -17)$ **13.** $y = -7x - 13$
15. $y = 1.01x - 2.6$ **17.** $y = -2x - 6$
19. $y = \frac{3}{4}x + \frac{5}{2}$ **21.** $y = x - 8$ **23.** $y = -3x + 3$
25. $y = x + 4$ **27.** $y = -\frac{1}{2}x + 4$ **29.** $y = -\frac{3}{2}x + \frac{13}{2}$
31. $y = -4x - 11$ **33.** **(a)** $T = -0.75a + 165$; **(b)** -0.75 beat per minute per year; **(c)** 127.5 beats per minute **35.** $\mathbf{D_W}$ **37.** $0, -3$ **38.** $-7, 7$
39. $-2, 3$ **40.** $-5, 1$ **41.** $-7, \frac{3}{2}$ **42.** $-\frac{6}{5}, 4$
43. $-7, 2$ **44.** $-2, \frac{2}{3}$ **45.** $\frac{53}{7}$ **46.** $\frac{3}{8}$ **47.** $\frac{24}{19}$
48. $\frac{125}{7}$ **49.** $y = 3x - 9$ **51.** $y = \frac{3}{2}x - 2$

Margin Exercises, Section 7.2, pp. 550–552

1.

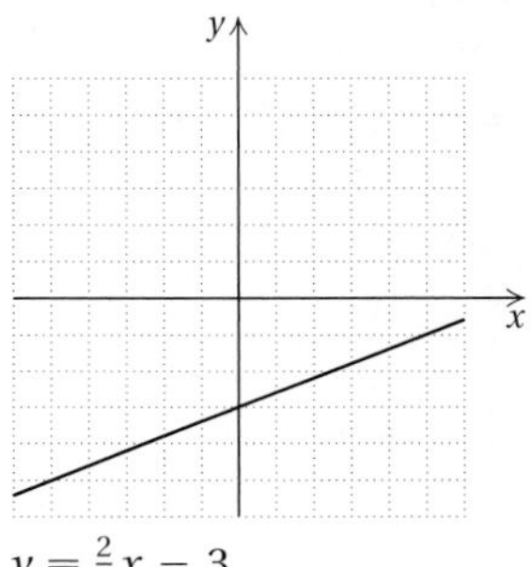

$y = \frac{2}{5}x - 3$

2.

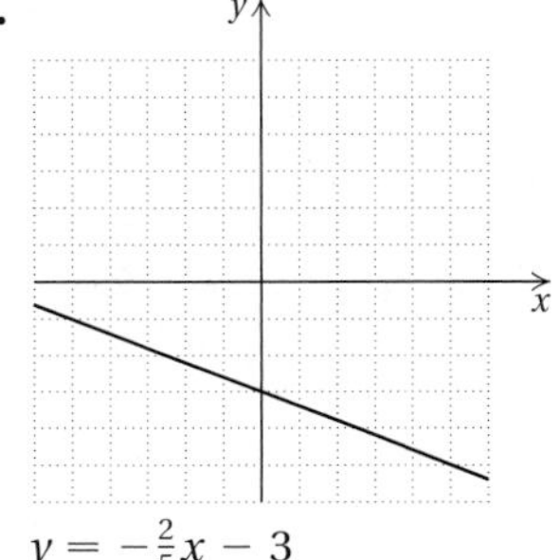

$y = -\frac{2}{5}x - 3$

3.

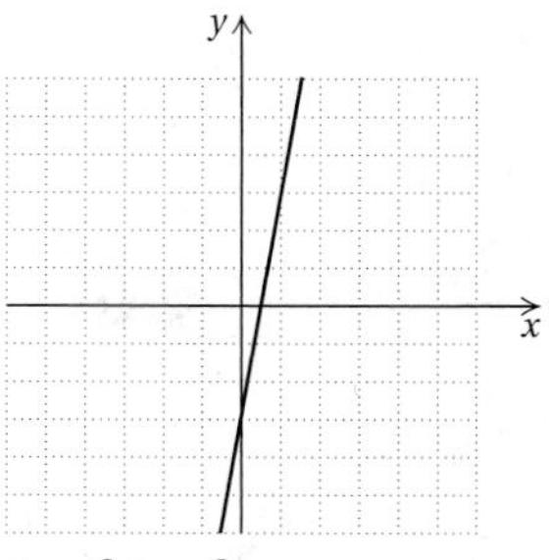

$y = 6x - 3$

4.

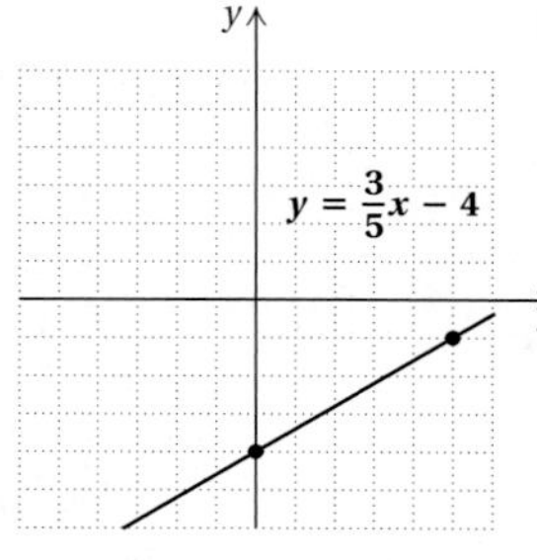

5.

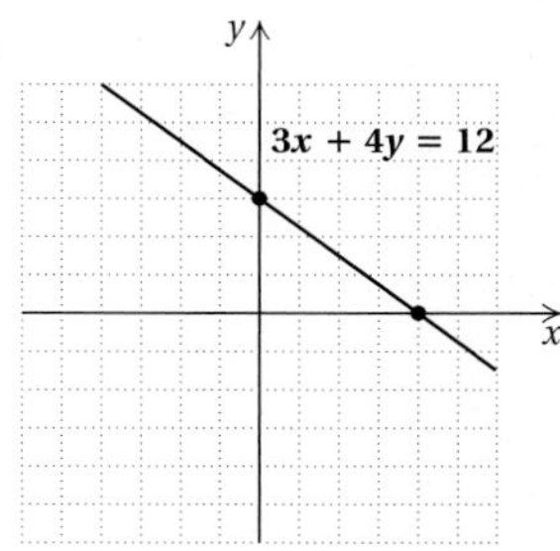

Exercise Set 7.2, p. 553

1.

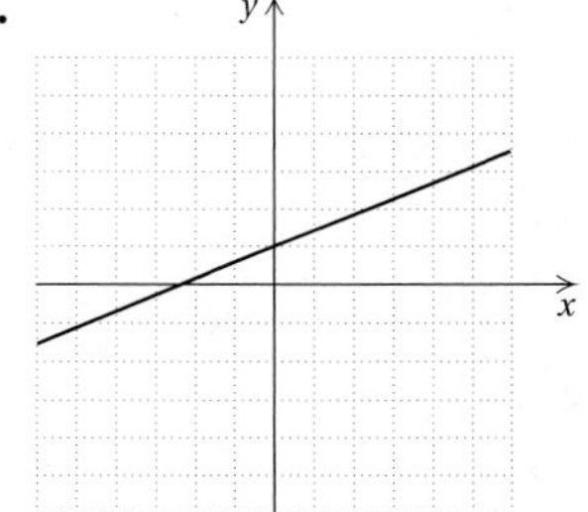

3.

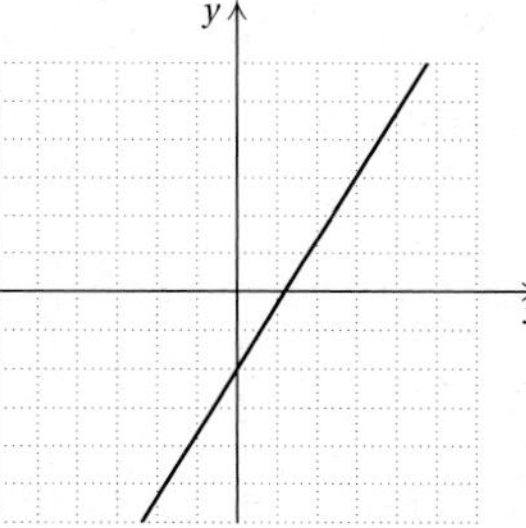

5.

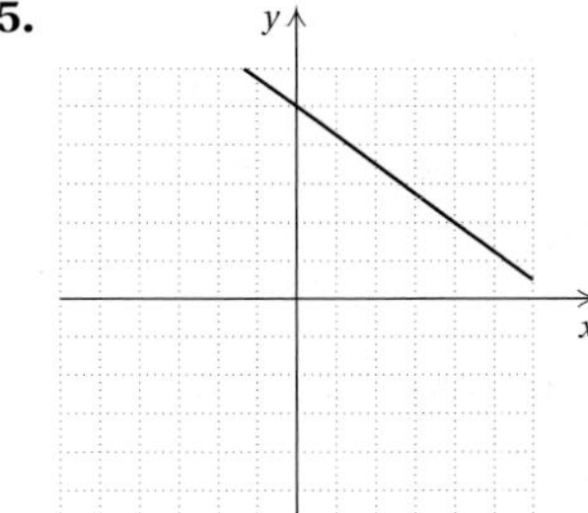

7.

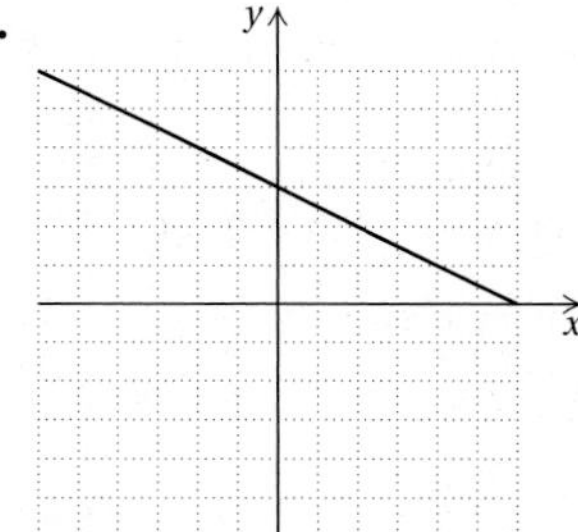

9.

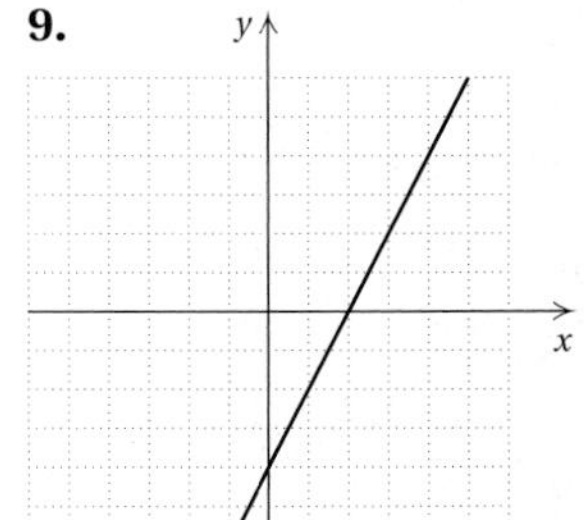

11.

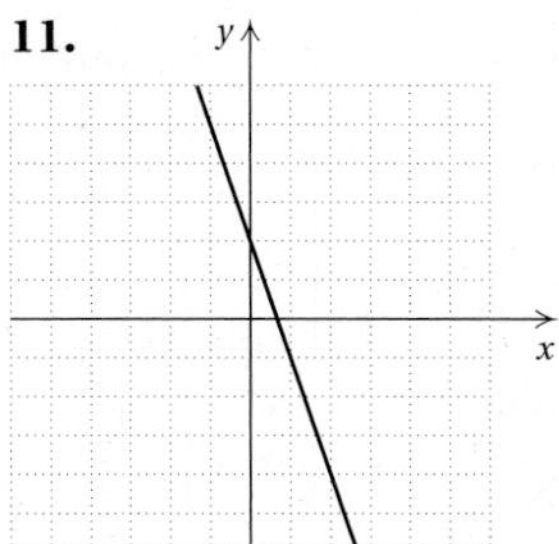

13.

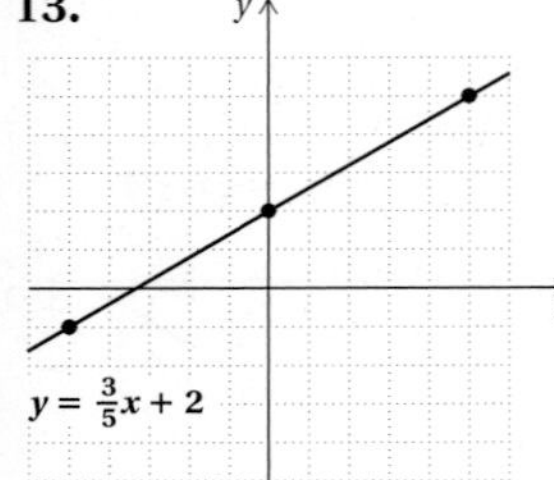

15.

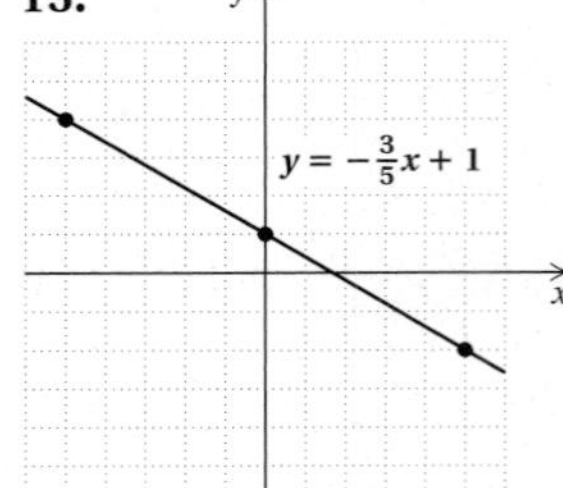

17.

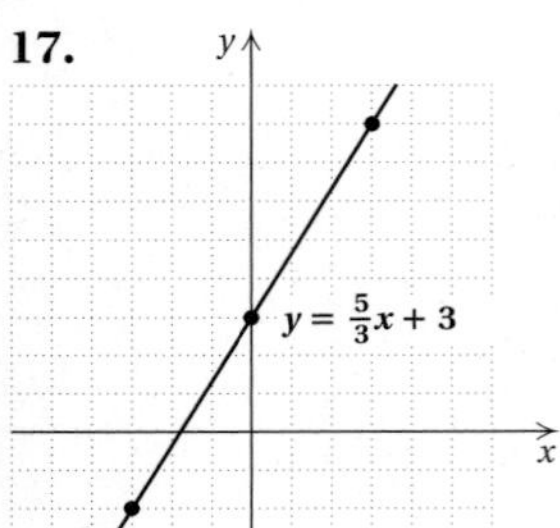

19.

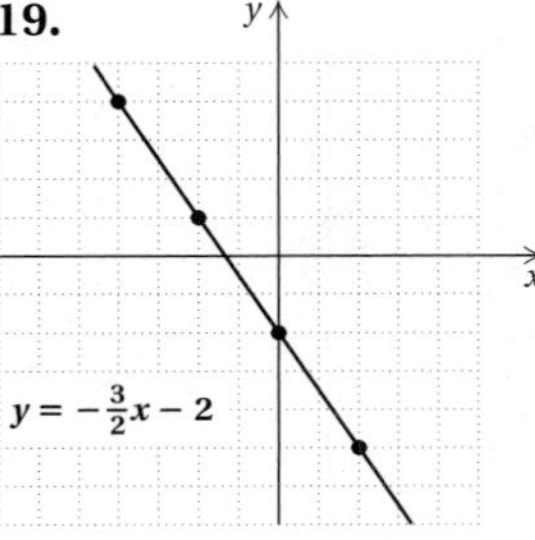

21.

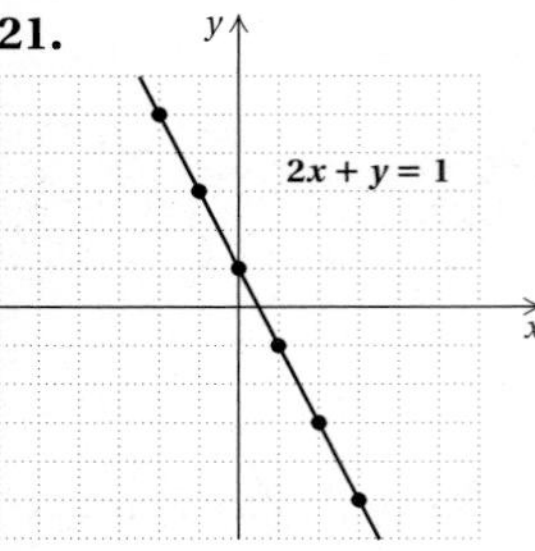

23.

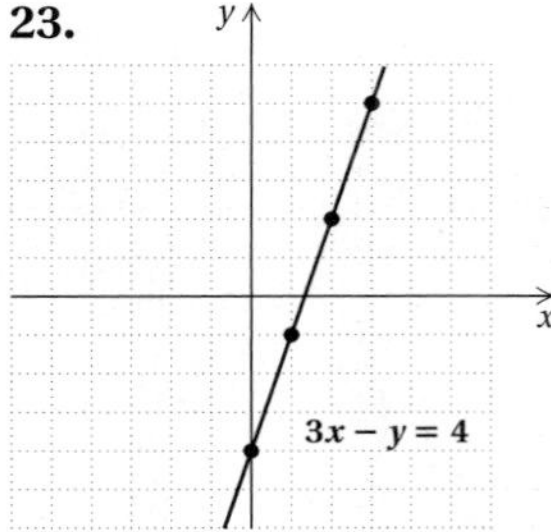

25.

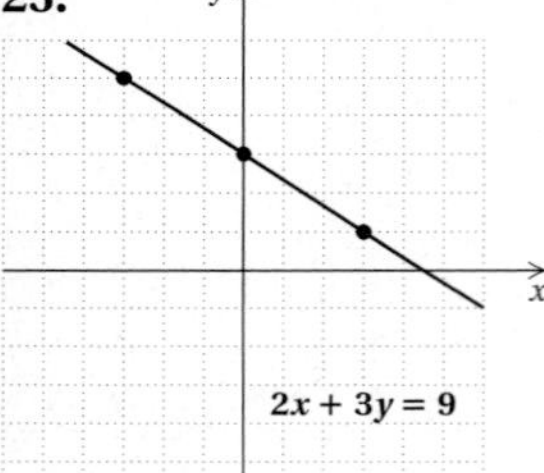

27.

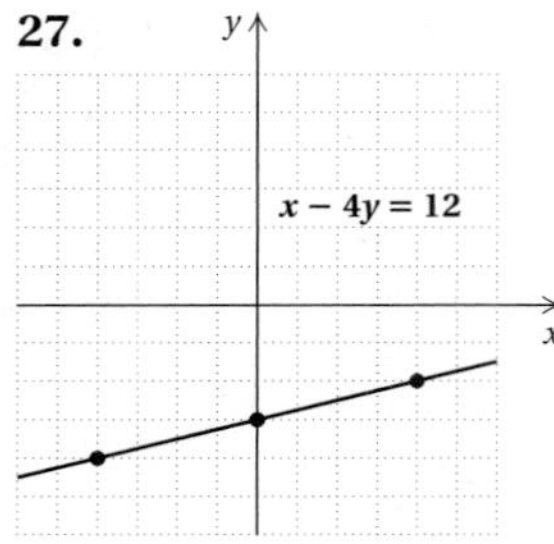

29.

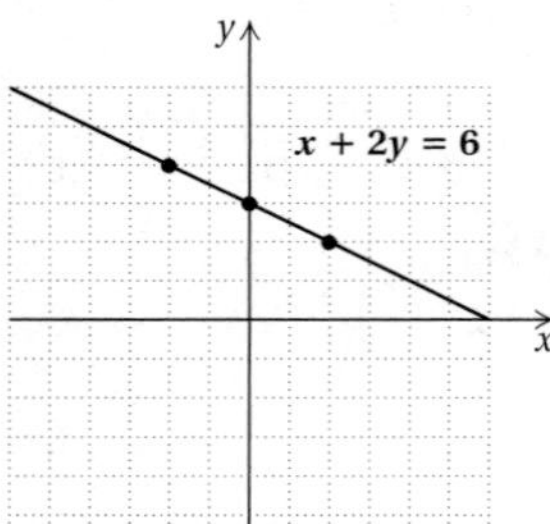

31. D_W **33.** $\frac{13}{10}$

34. $-\frac{1}{6}$ **35.** $\frac{69}{100}$, or 0.69 **36.** $-\frac{3}{5}$, or -0.6 **37.** 0
38. Not defined **39.** Not defined **40.** 0
41. 3.6 million vehicles per year; 3.6 **42.** 1.04 million cars per year; 1.04 **43.** $y = 1.5x + 16$

45.

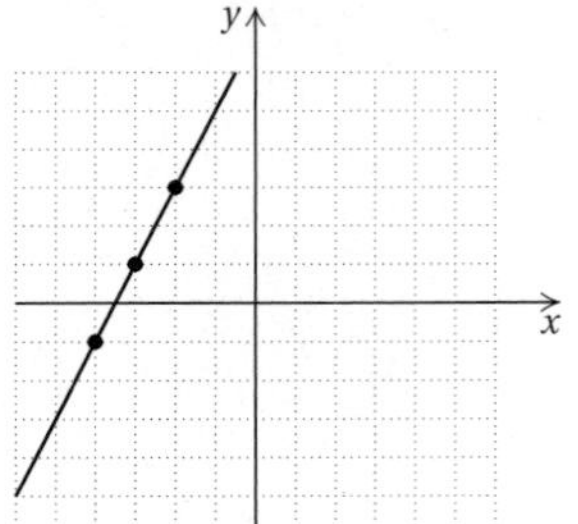

Margin Exercises, Section 7.3, pp. 558–559

1. No **2.** Yes **3.** Yes **4.** No

Exercise Set 7.3, p. 560

1. Yes **3.** No **5.** No **7.** No **9.** Yes **11.** Yes
13. No **15.** Yes **17.** Yes **19.** Yes **21.** No
23. Yes **25.** D_W **27.** In 7 hr **28.** 130 km/h; 140 km/h **29.** 4 **30.** -6 **31.** $\frac{30}{13}$ **32.** -11
33. $\frac{36}{11}$ **34.** 1 **35.–45.** Left to the student
47. $y = 3x + 6$ **49.** $y = -3x + 2$ **51.** $y = \frac{1}{2}x + 1$
53. 16 **55.** A: $y = \frac{4}{3}x - \frac{7}{3}$; B: $y = -\frac{3}{4}x - \frac{1}{4}$

Margin Exercises, Section 7.4, pp. 562–565

1. No **2.** No. **3.**

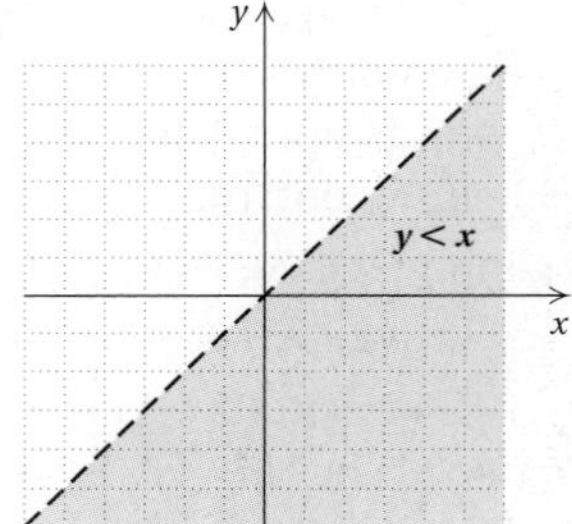

4.

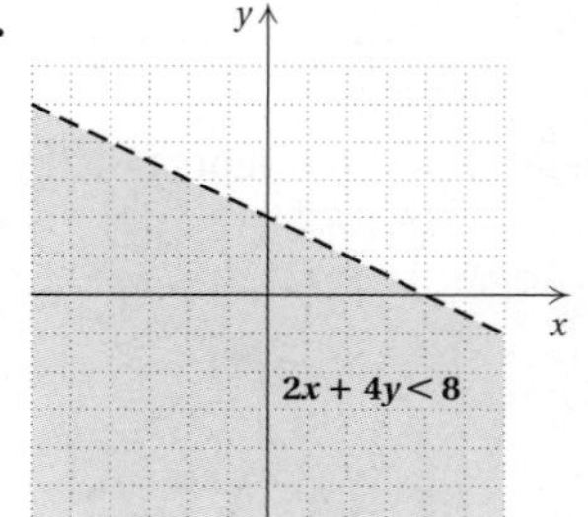

5.

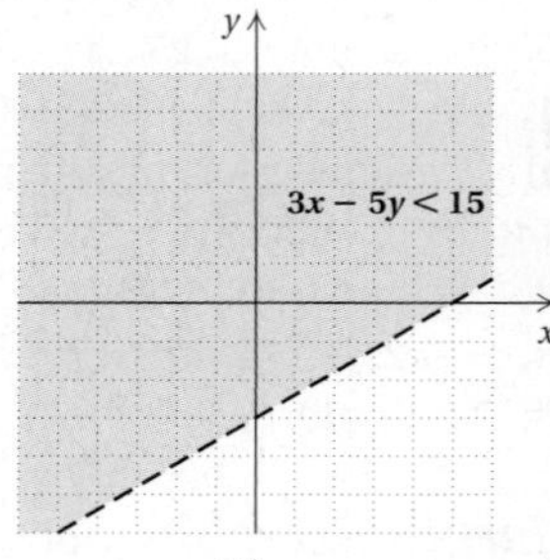

6.

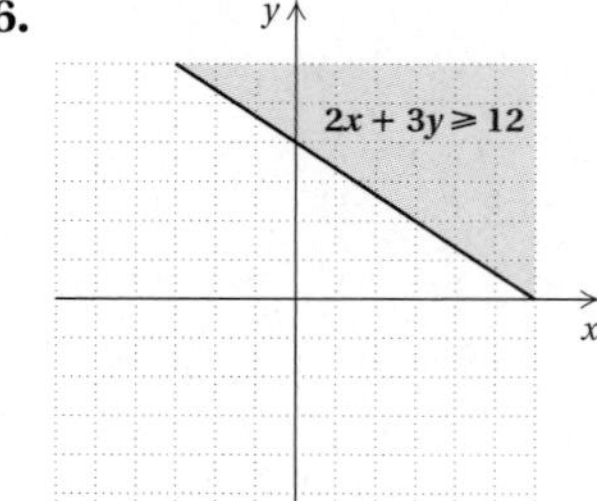

7.

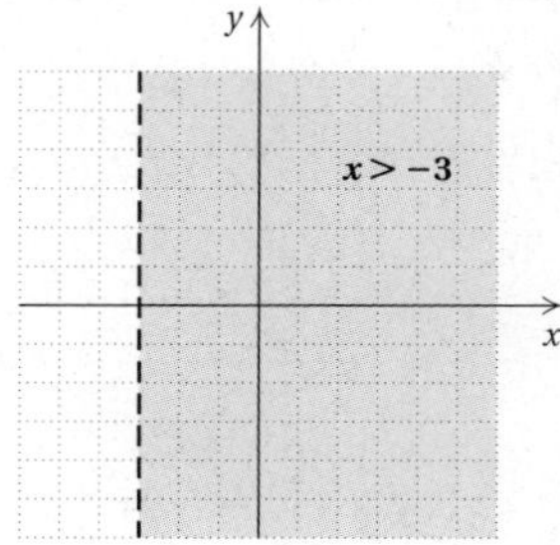

8.

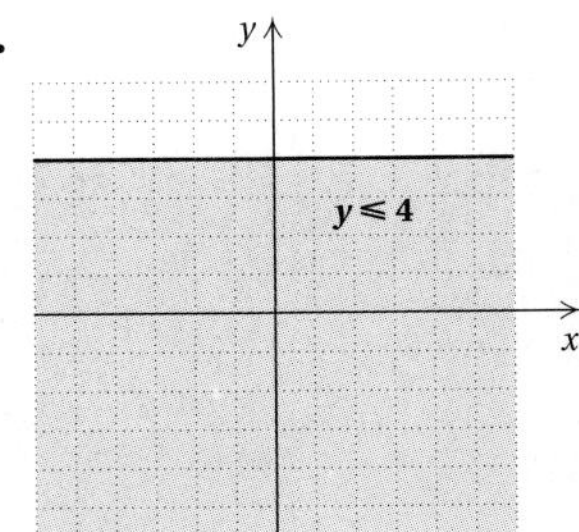

Calculator Corner, p. 566

1. Left to the student **2.** Left to the student

Exercise Set 7.4, p. 567

1. No **3.** Yes **5.**

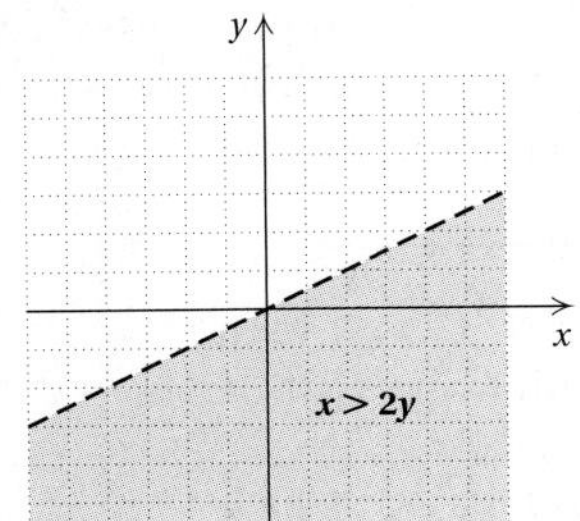

7.

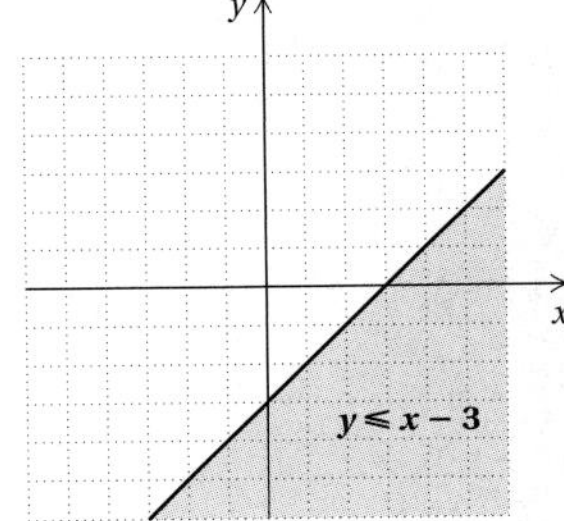

9.

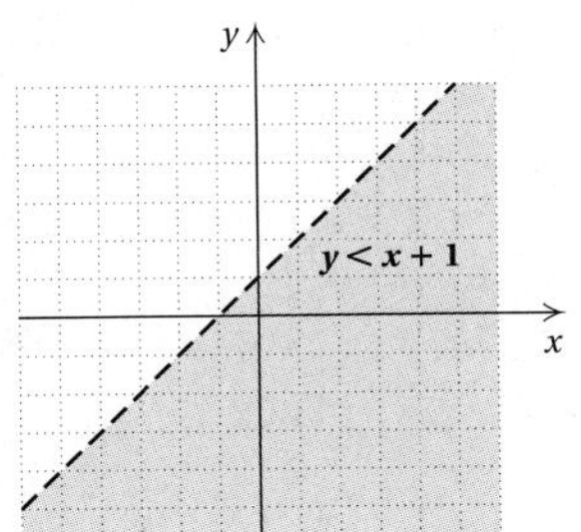

11.

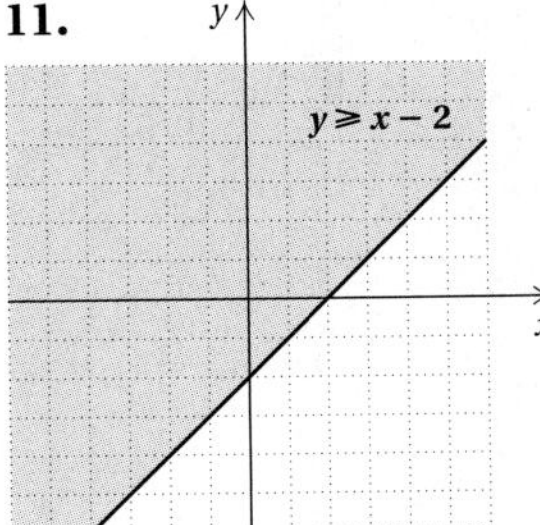

13.

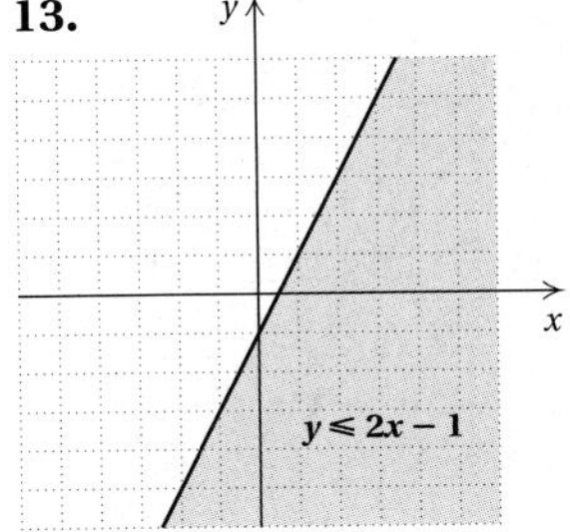

15.

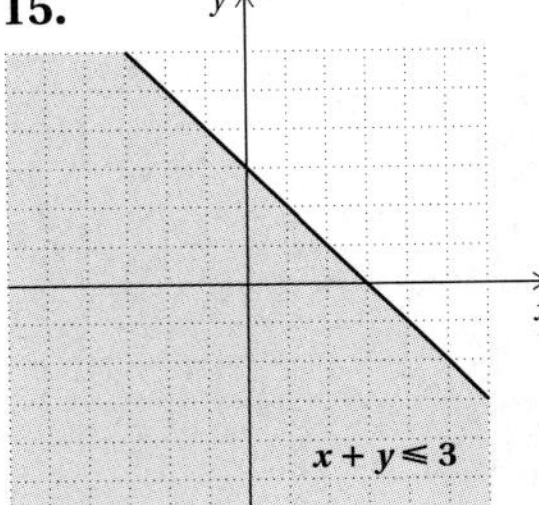

17.

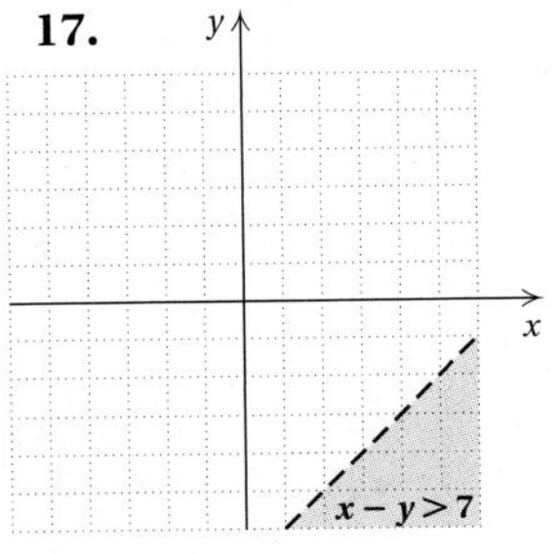

19.

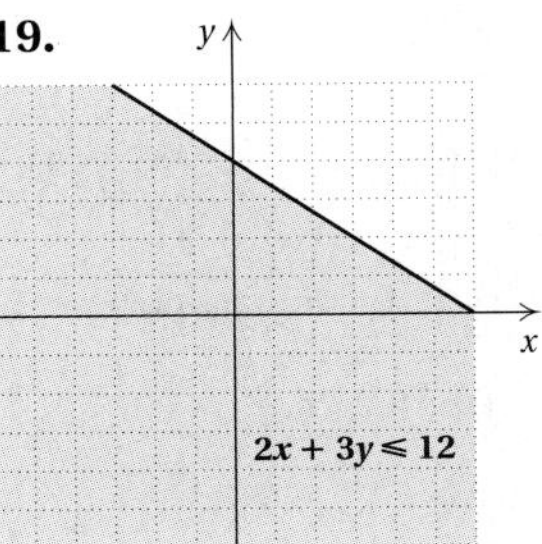

21.

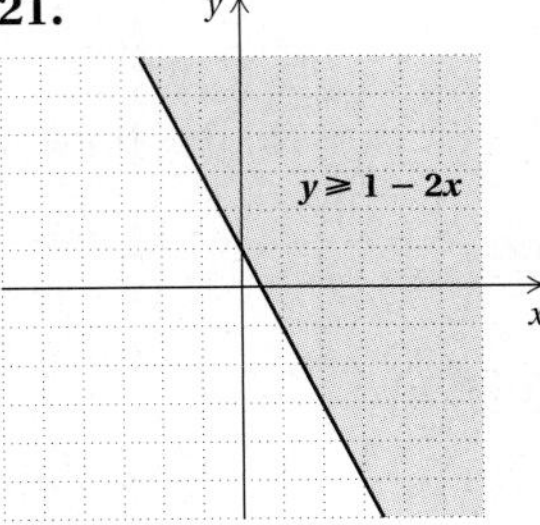

23.

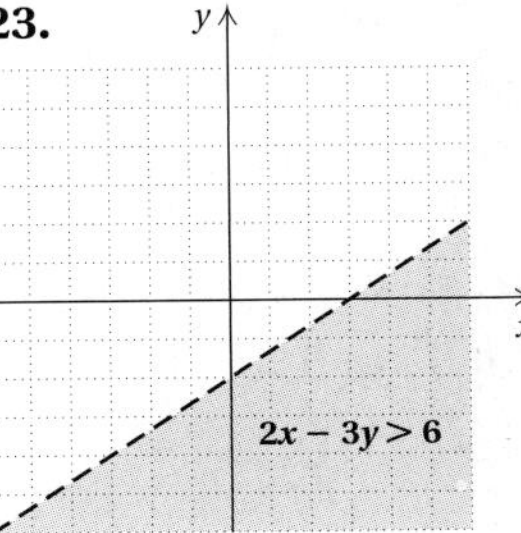

25.

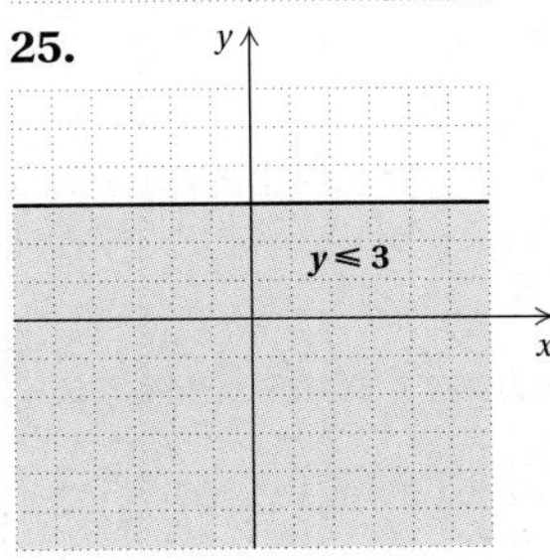

27.

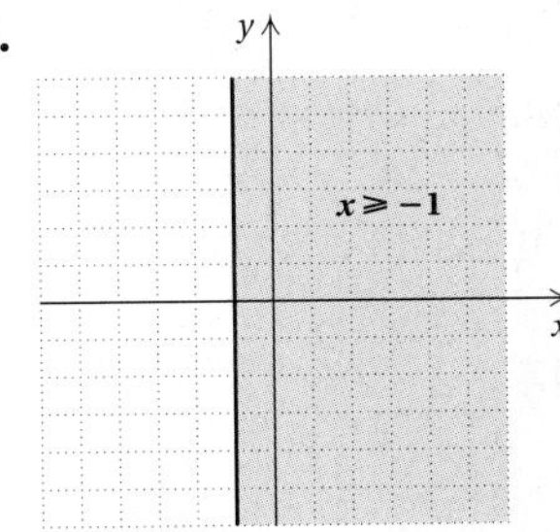

29. DW 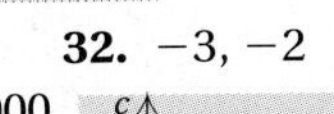 **31.** 3 **32.** $-3, -2$ **33.** 4 **34.** $-\frac{5}{3}, \frac{1}{4}$
35. $35c + 75a > 1000$

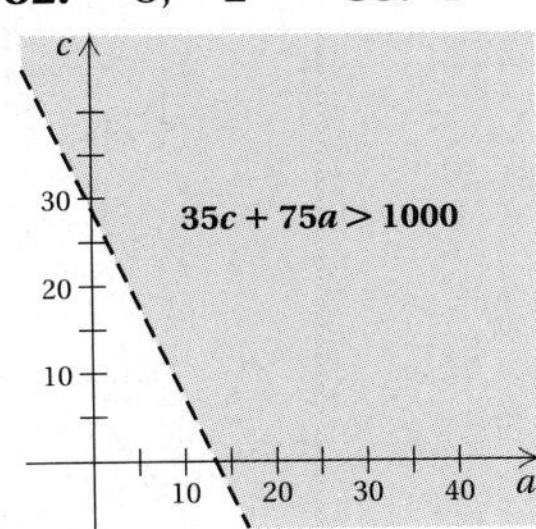

Margin Exercises, Section 7.5, pp. 570–574

1. $y = 7x$; 287 **2.** $y = \frac{5}{8}x$; $\frac{25}{2}$ **3. (a)** $C = \frac{7}{360}n$; **(b)** \$0.4667; \$0.0194 **4. (a)** $V = 0.88E$; **(b)** 174.24 lb **5.** $y = \frac{63}{x}$; 3.15 **6.** $y = \frac{900}{x}$; 562.5 **7.** 8 hr **8. (a)** $t = \frac{300}{r}$; **(b)** 7.5 hr

Exercise Set 7.5, p. 575

1. $y = 4x$; 80 **3.** $y = \frac{8}{5}x$; 32 **5.** $y = 3.6x$; 72 **7.** $y = \frac{25}{3}x$; $\frac{500}{3}$ **9. (a)** $P = 5.6H$; **(b)** \$196 **11. (a)** $C = 12.5S$; **(b)** \$100 **13. (a)** $M = \frac{1}{6}E$; **(b)** $18.\overline{3}$ lb; **(c)** 30 lb **15. (a)** $N = 80{,}000S$; **(b)** 16,000,000 instructions/sec **17.** $93\frac{1}{3}$ servings **19.** $y = \frac{75}{x}$; $\frac{15}{2}$, or 7.5 **21.** $y = \frac{80}{x}$; 8 **23.** $y = \frac{1}{x}$; $\frac{1}{10}$

25. $y = \frac{2100}{x}$; 210 27. $y = \frac{0.06}{x}$; 0.006
29. (a) Direct; (b) $69\frac{3}{8}$ players 31. (a) Inverse; (b) $4\frac{1}{2}$ hr
33. (a) $N = \frac{280}{P}$; (b) 10 gal 35. (a) $I = \frac{1920}{R}$;
(b) 32 amperes 37. (a) $m = \frac{40}{n}$; (b) 10 questions
39. 8.25 ft 41. D_W 43. D_W 45. $\frac{8}{5}$ 46. 11
47. 9, 16 48. $-12, -9$ 49. $\frac{2}{5}, \frac{4}{7}$ 50. $-\frac{1}{7}, \frac{3}{2}$
51. $\frac{1}{3}$ 52. $\frac{47}{20}$ 53. -1 54. 49
55. The y-values become larger. 57. $P^2 = kt$
59. $P = kV^3$

Summary and Review: Chapter 7, p. 580

1. Slope: -9; y-intercept: $(0, 46)$
2. Slope: -1; y-intercept: $(0, 9)$
3. Slope: $\frac{3}{5}$; y-intercept: $\left(0, -\frac{4}{5}\right)$
4. $y = -2.8x + 19$ 5. $y = \frac{5}{8}x - \frac{7}{8}$ 6. $y = 3x - 1$
7. $y = \frac{2}{3}x - \frac{11}{3}$ 8. $y = -2x - 4$ 9. $y = x + 2$
10. $y = \frac{1}{2}x - 1$ 11. (a) $A = 0.233x + 5.87$;
(b) 0.233 year per year (c) 8.2 yr

12.
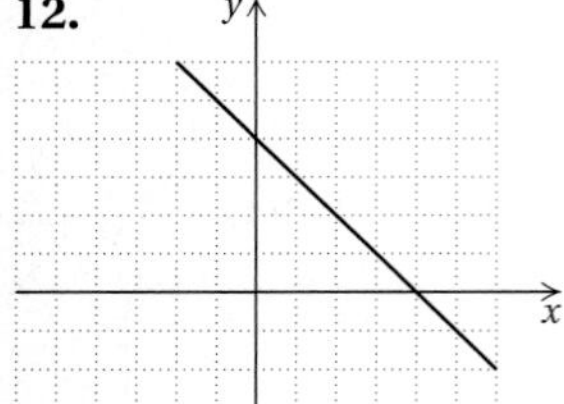

13.
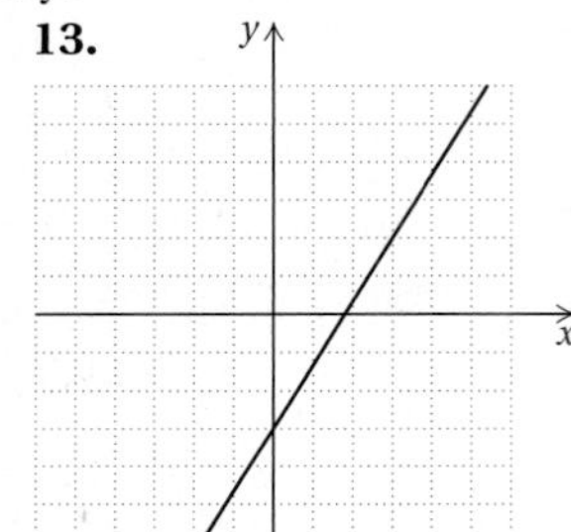

14.
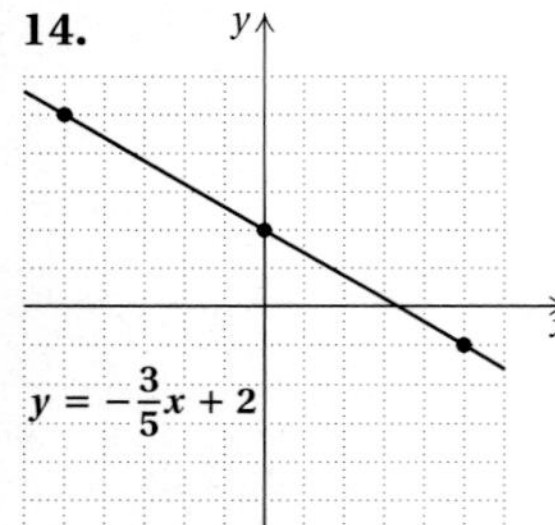

15.
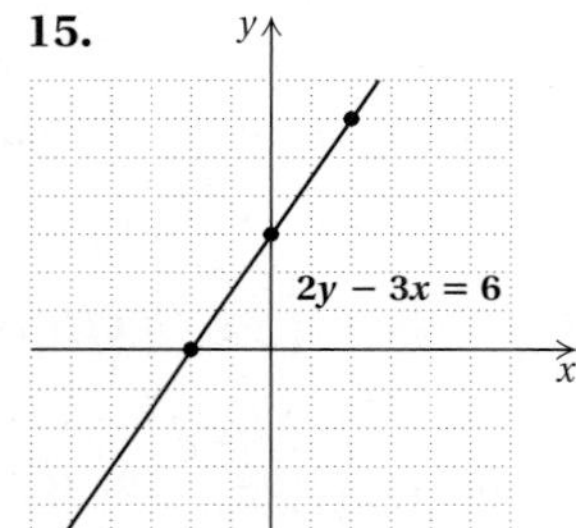

16. Parallel 17. Perpendicular 18. Parallel
19. Neither 20. No 21. No 22. Yes

23.
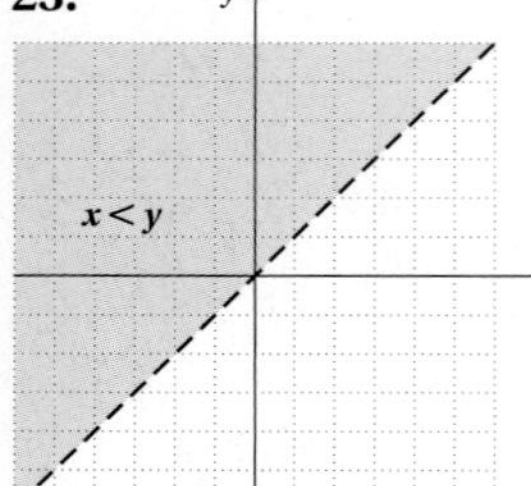

24.
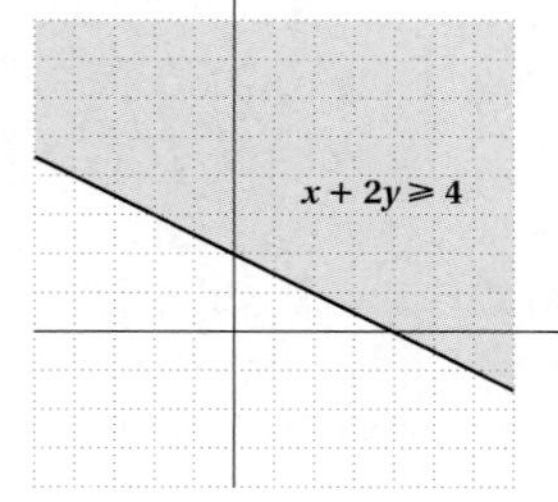

25.
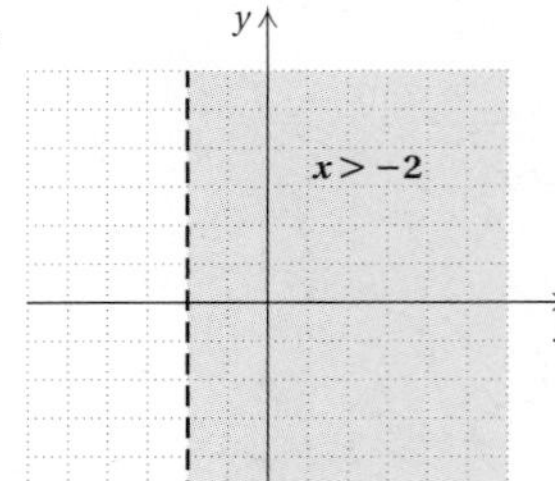

26. $y = 3x$; 60
27. $y = \frac{1}{2}x$; 10
28. $y = \frac{4}{5}x$; 16
29. $y = \frac{30}{x}$; 6

30. $y = \frac{1}{x}$; $\frac{1}{5}$ 31. $y = \frac{0.65}{x}$; 0.13 32. \$288.75
33. 1 hr
34. D_W First plot the y-intercept, $(0, 2458)$. Then, thinking of the slope as $\frac{37}{100}$, plot a second point on the line by moving up 37 units and to the right 100 units from the y-intercept. Next, thinking of the slope as $\frac{-37}{-100}$, start at the y-intercept and plot a third point by moving down 37 units and to the left 100 units. Finally, draw a line through the three points.

35. D_W
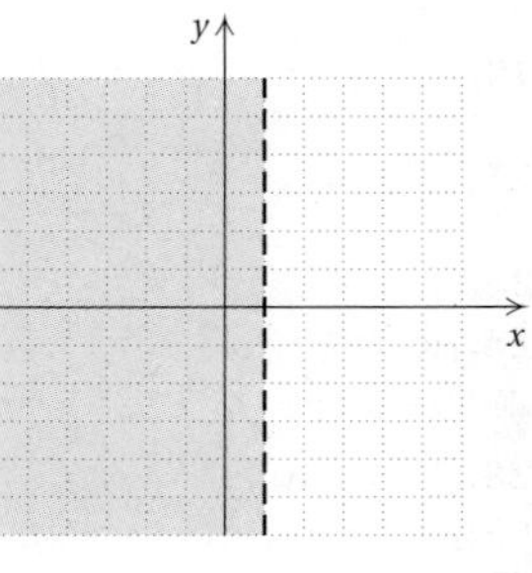

The graph of $x < 1$ on a number line consists of the points in the set $\{x | x < 1\}$. The graph of $x < 1$ on a plane consists of the points, or ordered pairs, in the set $\{(x, y) | x + 0 \cdot y < 1\}$. This is the set of ordered pairs with first coordinate less than 1.

36. $3\frac{1}{3}$ hr 37. 52 38. -4 39. $-11, 5$
40. $-\frac{1}{2}, \frac{1}{2}, -2, 2$

Test: Chapter 7, p. 583

1. [7.2a]
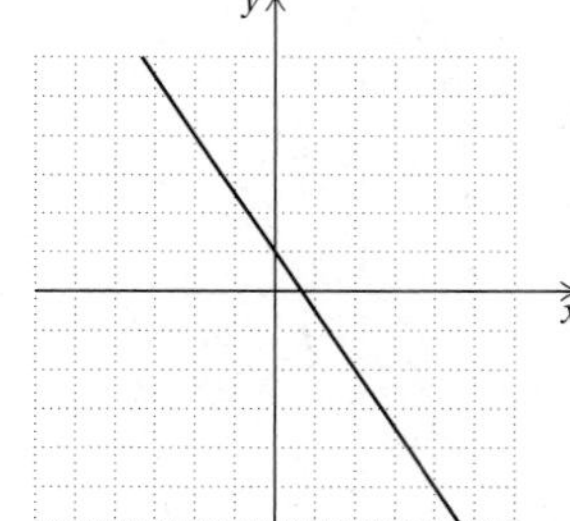

2. [7.2a]
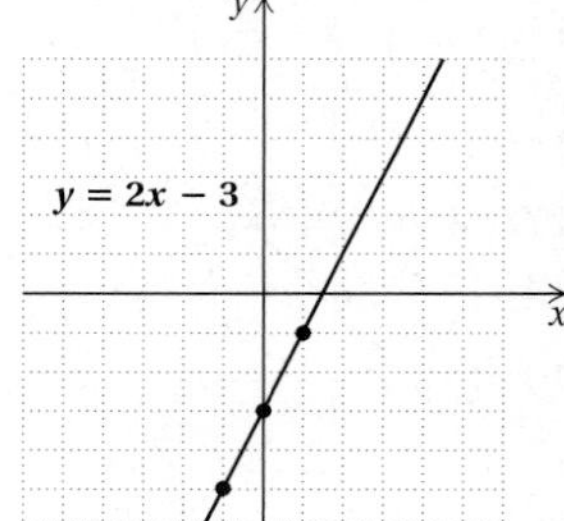

3. [7.1a] Slope: 2; y-intercept: $\left(0, -\frac{1}{4}\right)$
4. [7.1a] Slope: $\frac{4}{3}$; y-intercept: $(0, -2)$
5. [7.1a] $y = 1.8x - 7$ 6. [7.1a] $y = -\frac{3}{8}x - \frac{1}{8}$
7. [7.1b] $y = x + 2$ 8. [7.1b] $y = -3x - 6$
9. [7.1c] $y = -3x + 4$ 10. [7.1c] $y = \frac{1}{4}x - 2$
11. [7.1c] (a) $M = 102x + 2313$; (b) \$102 million per year; (c) \$3129 million 12. [7.3a, b] Parallel

13. [7.3a, b] Neither **14.** [7.3a, b] Perpendicular
15. [7.4a] No **16.** [7.4a] Yes
17. [7.4b]

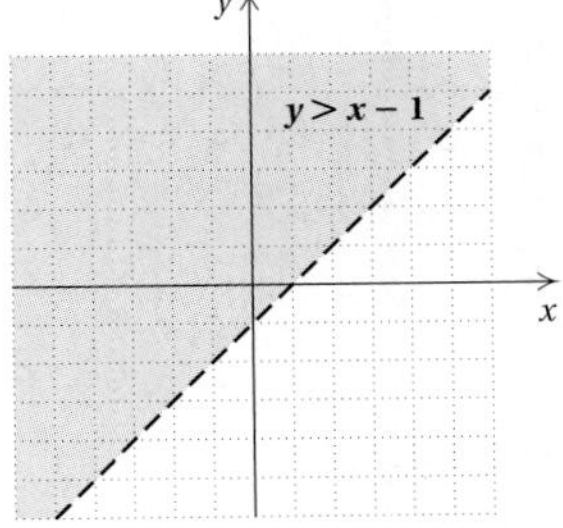

18. [7.4b]

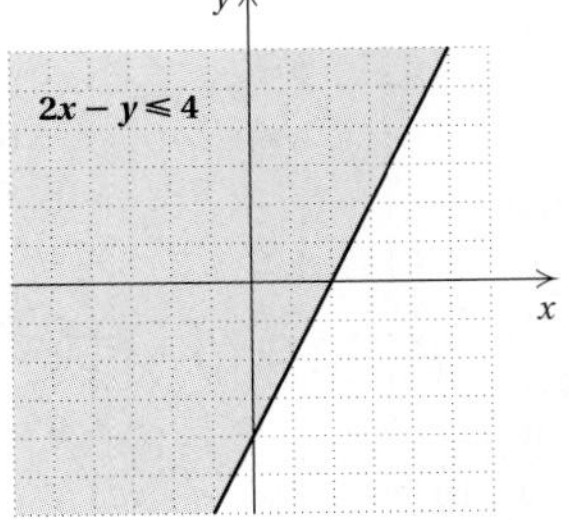

19. [7.5a] $y = 2x$; 50 **20.** [7.5a] $y = 0.5x$; 12.5
21. [7.5c] $y = \frac{18}{x}$; $\frac{9}{50}$ **22.** [7.5c] $y = \frac{22}{x}$; $\frac{11}{50}$
23. [7.5b] 240 km **24.** [7.5d] $1\frac{1}{5}$ hr
25. [6.7a] Freight: 90 mph; passenger: 105 mph
26. [1.8d] 1 **27.** [6.6a] 10 **28.** [5.7b] −7, 4
29. [7.3b] 3 **30.** [7.1b] $y = \frac{2}{3}x + \frac{11}{3}$

Cumulative Review: Chapters 1–7, p. 585

1. [1.2e] 3.5 **2.** [4.3d] 1, −2, 1, −1
3. [4.3g] 3, 2, 1, 0; 3 **4.** [4.3i] None of these
5. [2.5a] $16.74 **6.** [6.7b] 27 lb **7.** [6.7a] 30 min
8. [7.5b] **(a)** Let M = muscle weight, in pounds, and B = body weight, in pounds; $M = 0.4B$; **(b)** 76.8 lb
9. [2.5a] $2500 **10.** [6.7a] 35 mph, 25 mph
11. [5.8a] 14 ft **12.** [2.6a] 34 and 35
13. [4.3e] $2x^3 - 3x^2 - 2$ **14.** [1.8c] $\frac{3}{8}x + 1$
15. [4.1e], [4.2a, b] $\frac{9}{4x^8}$ **16.** [6.8a] $\frac{8x - 12}{17x}$
17. [4.7e] $-2xy^2 - 4x^2y^2 + xy^3$
18. [4.4a] $2x^5 + 6x^4 + 2x^3 - 10x^2 + 3x - 9$
19. [6.1d] $\frac{2}{3(y + 2)}$ **20.** [6.2b] 2 **21.** [6.4a] $x + 4$
22. [6.5a] $\frac{2x - 6}{(x + 2)(x - 2)}$ **23.** [4.6a] $a^2 - 9$
24. [4.6c] $36x^2 - 60x + 25$ **25.** [4.6b] $4x^6 - 1$
26. [5.3a], [5.4a] $(9a - 2)(a + 6)$ **27.** [5.5b] $(3x - 5y)^2$
28. [5.5d] $(7x - 1)(7x + 1)$ **29.** [2.3c] 3
30. [5.7b] $-4, \frac{1}{2}$ **31.** [5.7b] −5, 4
32. [2.7e] $\{x | x \geq -26\}$ **33.** [5.7a] 0, 4
34. [5.7b] 0, 10 **35.** [5.7b] −20, 20
36. [2.4b] $a = \frac{t}{x + y}$ **37.** [6.6a] 2
38. [6.6a] No solution

39. [3.2b]

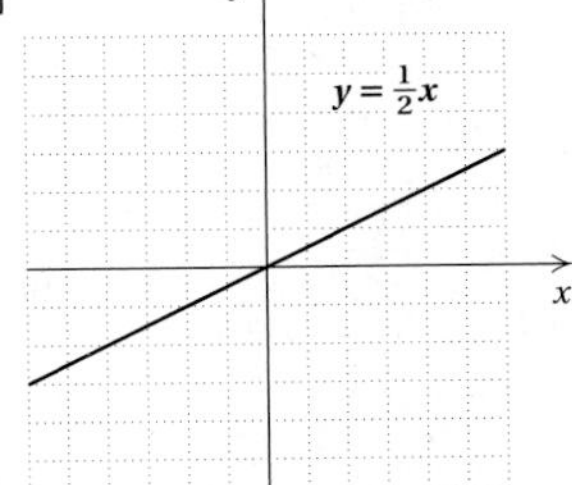

40. [3.3a]

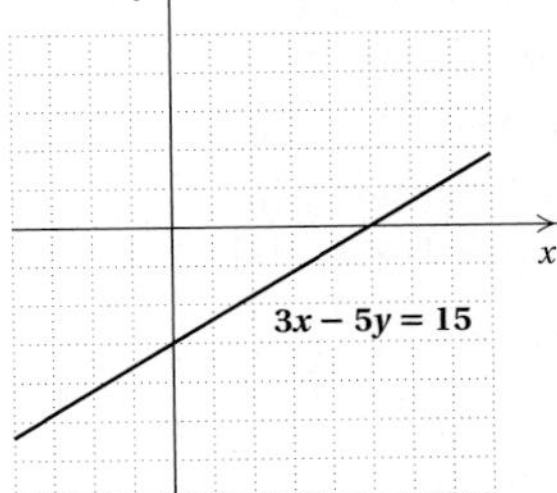

41. [3.3b]

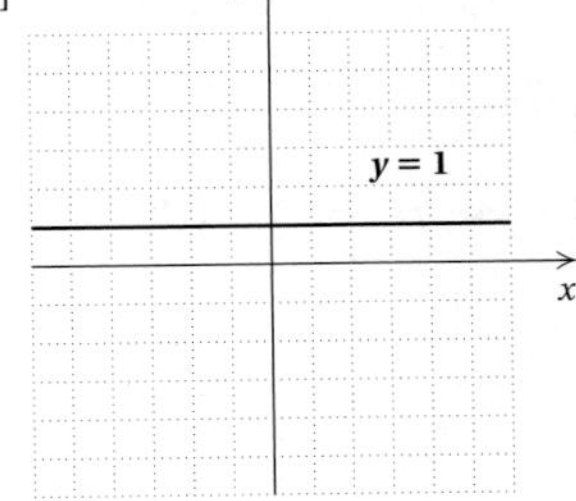

42. [3.3b]

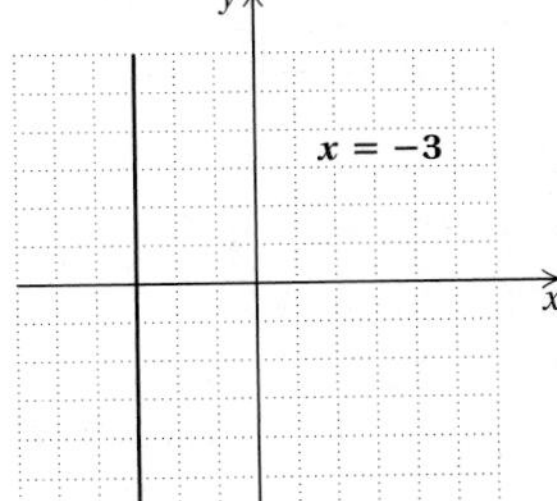

43. [7.4b]

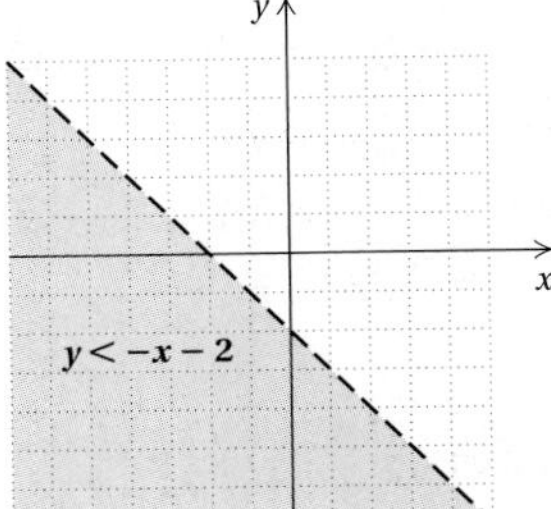

44. [7.4b]

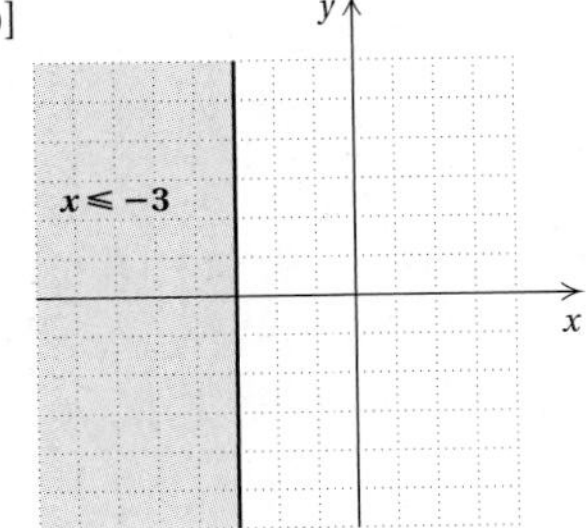

45. [7.5a] $y = \frac{2}{3}x$; $\frac{250}{3}$ **46.** [7.5c] $y = \frac{10}{x}$; $\frac{2}{25}$

47. [3.4a] Not defined **48.** [3.4a] $-\frac{3}{7}$

49. [7.1a] Slope: $\frac{4}{3}$; y-intercept: $(0, -2)$

50. [7.1b] $y = -4x + 5$ **51.** [7.1c] $y = \frac{1}{6}x - \frac{17}{6}$

52. [7.3a, b] Neither **53.** [7.3a, b] Parallel

54. [3.4a] 733 **55.** [3.3a] (0, 17,634)

56. [3.4b] \$733 per year **57.** [7.1c] $C = 733x + 17{,}634$

58. [7.1c] \$28,629

59. [2.3a] 44.2 yr after 1995, or in 2040

60. [R.6c] (b) **61.** [4.4c], [4.6a] 12

62. [4.6b, c] $16y^6 - y^4 + 6y^2 - 9$

63. [5.5d] $2(a^{16} + 81b^{20})(a^8 + 9b^{10})(a^4 + 3b^5)(a^4 - 3b^5)$

64. [5.7a] 4, −7, 12 **65.** [7.1b], [7.3a] $y = \frac{2}{3}x$

66. [6.1a], [6.8a] 0, 3, $\frac{5}{2}$

Chapter 8

Pretest: Chapter 8, p. 590

1. Yes **2.**

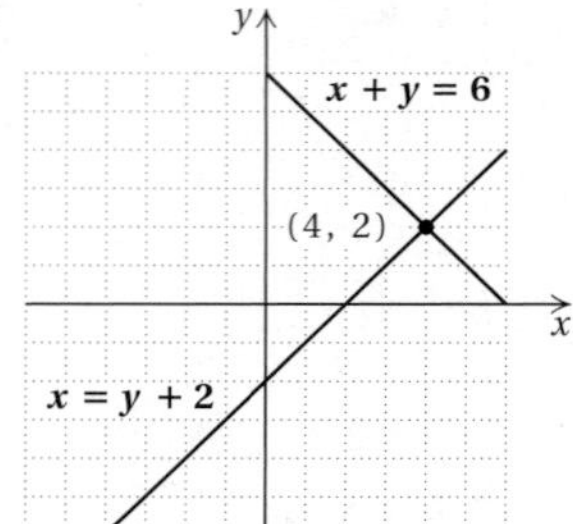

3. (5, 2)

4. (2, −1) **5.** $\left(\frac{3}{4}, \frac{1}{2}\right)$ **6.** (−5, −2) **7.** (10, 8)

8. Adults: 320; children: 627

9. Foul shots: 28; field goals: 36

10. 8 hr after the second train has left **11.** 25°, 65°

Margin Exercises, Section 8.1, p. 592–595

1. Yes **2.** No **3.** (2, −3) **4.** (−4, 3) **5.** No solution

6. Infinite number of solutions **7.** **(a)** 3; **(b)** 3; **(c)** same

8. **(a)** 3; **(b)** same

Calculator Corner, p. 595

1. (−1, 3) **2.** (0, 5) **3.** (2, −3) **4.** (−4, 1)

Exercise Set 8.1, p. 596

1. Yes **3.** No **5.** Yes **7.** Yes **9.** Yes **11.** (4, 2)

13. (4, 3) **15.** (−3, −3) **17.** No solution **19.** (2, 2)

21. $\left(\frac{1}{2}, 1\right)$ **23.** Infinite number of solutions

25. (5, −3) **27.** D_W **29.** $\frac{2x^2 - 1}{x^2(x + 1)}$ **30.** $\frac{-4}{x - 2}$

31. $\frac{9x + 12}{(x - 4)(x + 4)}$ **32.** $\frac{2x + 5}{x + 3}$ **33.** Trinomial

34. Binomial **35.** Monomial **36.** None of these

37. $A = 2, B = 2$ **39.** $x + 2y = 2, x - y = 8$

41.–47. Left to the student

Margin Exercises, Section 8.2, pp. 599–601

1. (3, 2) **2.** (3, −1) **3.** $\left(\frac{24}{5}, -\frac{8}{5}\right)$

4. Length: 25 m; width: 21 m

Exercise Set 8.2, p. 602

1. (1, 9) **3.** (2, −4) **5.** (4, 3) **7.** (−2, 1) **9.** (2, −4)

11. $\left(\frac{17}{3}, \frac{16}{3}\right)$ **13.** $\left(\frac{25}{8}, -\frac{11}{4}\right)$ **15.** (−3, 0) **17.** (6, 3)

19. Length: 94 ft; width: 50 ft

21. Length: $3\frac{1}{2}$ in.; width: $1\frac{3}{4}$ in.

23. Length: 365 mi; width: 275 mi

25. Length: 40 ft; width: 20 ft

27. Length: 110 yd; width: 60 yd **29.** 16 and 21

31. 12 and 40 **33.** 20 and 8 **35.** D_W

37.

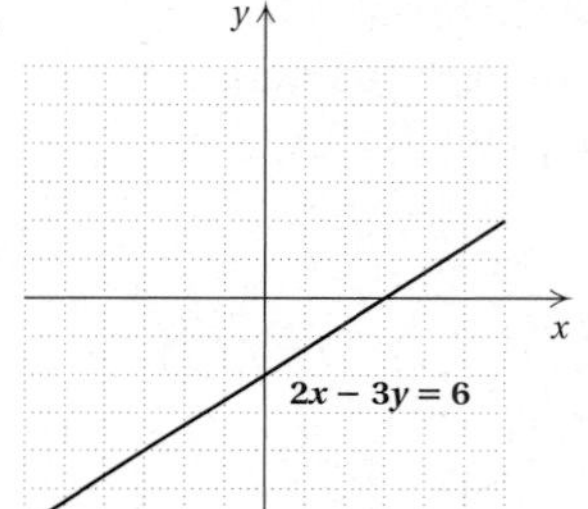

38.

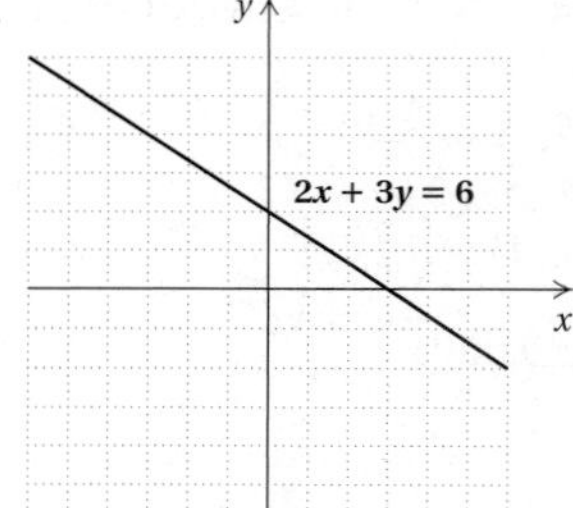

39.

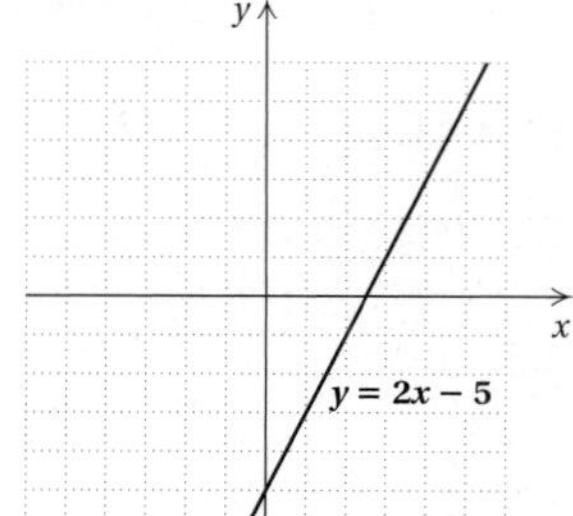

40.

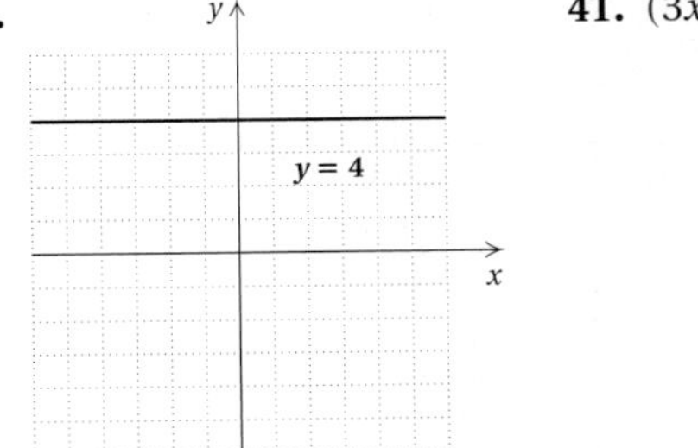

41. $(3x - 2)(2x - 3)$

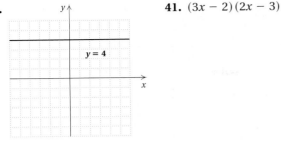

42. $(4p + 3)(p - 1)$ 43. Not factorable
44. $(3a - 5)(3a + 5)$ 45. $(5.\overline{6}, 0.\overline{6})$ 47. $(4.38, 4.33)$
49. Baseball: 30 yd; softball: 20 yd

Margin Exercises, Section 8.3, pp. 606–611

1. $(3, 2)$ 2. $(1, -2)$ 3. $(1, 4)$ 4. $(-8, 3)$ 5. $(1, 1)$
6. $(-2, -1)$ 7. $\left(\frac{17}{13}, -\frac{7}{13}\right)$ 8. No solution
9. Infinite number of solutions 10. $(1, -1)$
11. $(2, -2)$

Calculator Corner, p. 610

1. We get equations of two lines with the same slope but different y-intercepts. This indicates that the lines are parallel, so the system of equations has no solution.
2. We get equivalent equations. This indicates that the lines are the same, so the system of equations has an infinite number of solutions.

Exercise Set 8.3, p. 612

1. $(6, -1)$ 3. $(3, 5)$ 5. $(2, 5)$ 7. $\left(-\frac{1}{2}, 3\right)$
9. $\left(-1, \frac{1}{5}\right)$ 11. No solution 13. $(-1, -6)$ 15. $(3, 1)$
17. $(8, 3)$ 19. $(4, 3)$ 21. $(1, -1)$ 23. $(-3, -1)$
25. $(3, 2)$ 27. $(50, 18)$ 29. Infinite number of solutions
31. $(2, -1)$ 33. $\left(\frac{231}{202}, \frac{117}{202}\right)$ 35. $(-38, -22)$ 37. DW
39. $\frac{1}{x^7}$ 40. x^3 41. $\frac{1}{x^3}$ 42. x^7 43. x^3 44. x^7
45. $\frac{a^7}{b^9}$ 46. $\frac{b^3}{a^3}$ 47. $\frac{x - 3}{x + 2}$ 48. $\frac{x + 5}{x - 5}$
49. $\frac{-x^2 - 7x + 23}{(x + 3)(x - 4)}$ 50. $\frac{-2x + 10}{(x + 1)(x - 1)}$
51.–69. Left to the student 71. $(5, 2)$ 73. $(0, -1)$
75. $(0, 3)$ 77. $x = \frac{c - b}{a - 1}, y = \frac{ac - b}{a - 1}$

Margin Exercises, Section 8.4, pp. 615–620

1. Hamburger: \$1.99; chicken: \$1.70 2. Adults: 125; children: 41 3. 50% alcohol: 22.5 L; 70% alcohol: 7.5 L
4. Seed A: 30 lb; seed B: 20 lb 5. Quarters: 7; dimes: 13

Exercise Set 8.4, p. 622

1. Two-pointers: 33; three-pointers: 10
3. 24-exposure: 13; 36-exposure: 6 5. Hay: 10 lb; grain: 5 lb 7. One soda: \$1.45; one slice of pizza: \$2.25
9. Cardholders: 128; non-cardholders: 75
11. Upper Box: 17; Lower Reserved: 12
13. Solution A: 40 L; solution B: 60 L 15. Dimes: 70; quarters: 33 17. Brazilian: 200 lb; Turkish: 100 lb
19. 28% fungicide: 100 L; 40% fungicide: 200 L
21. Large type: $6\frac{17}{25}$ pages; small type: $5\frac{8}{25}$ pages
23. Seed A: 39 lb; seed B: 36 lb
25. Type A: 12; type B: 4; 180 27. Kuyatts': 32 yr; Marconis': 16 yr 29. Randy: 24; Mandy: 6 31. 50°, 130°
33. 28°, 62° 35. 87-octane: 12 gal; 93-octane: 6 gal
37. Kinney's: $33\frac{1}{3}$ fl oz; Coppertone: $16\frac{2}{3}$ fl oz 39. DW
41. $(5x + 9)(5x - 9)$ 42. $(6 + a)(6 - a)$
43. $4(x^2 + 25)$ 44. $4(x + 5)(x - 5)$

45.

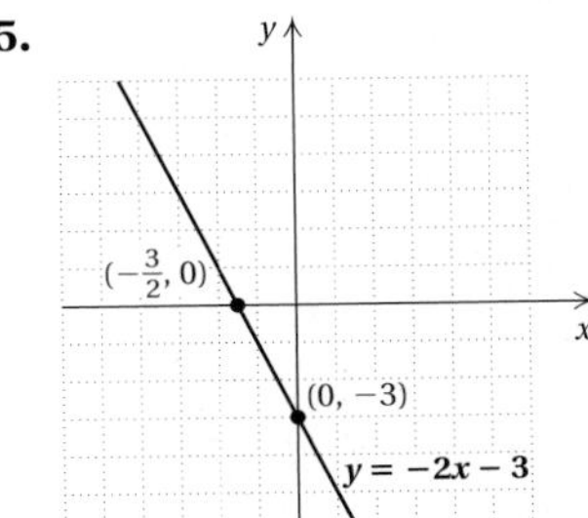

46.

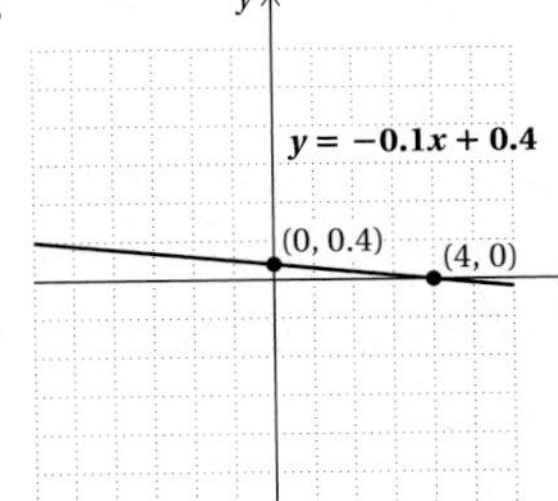

47.

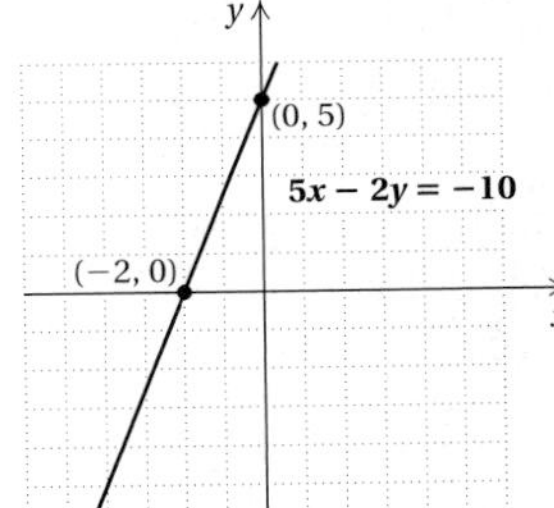

48.

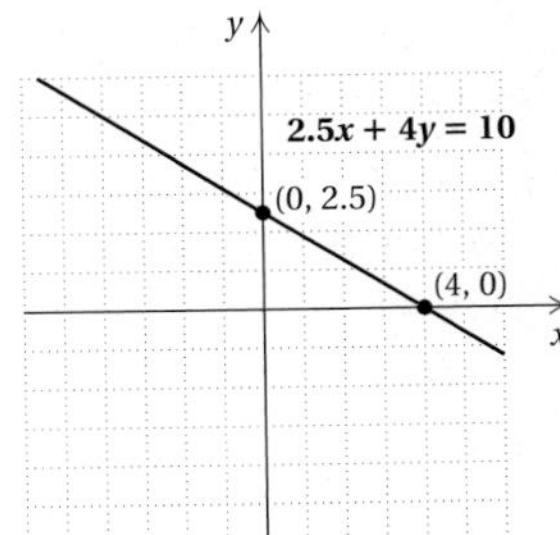

49. 43.75 L

51. $4\frac{4}{7}$ L 53. Flavoring: 0.8 oz; sugar: 19.2 oz
55. \$1100 at 9%; \$1800 at $10\frac{1}{2}$%

Margin Exercises, Section 8.5, pp. 629–631

1. 324 mi 2. 3 hr 3. 168 km 4. 275 km/h

Exercise Set 8.5, p. 632

1.

Speed	Time
30	t
46	t

4.5 hr

3.

Speed	Time	
72	$t + 3$	$\rightarrow d = 72(t + 3)$
120	t	$\rightarrow d = 120t$

$4\frac{1}{2}$ hr

5.

Speed	Time	
$r + 6$	4	$\rightarrow d = (r + 6)4$
$r - 6$	10	$\rightarrow d = (r - 6)10$

14 km/h

7. 384 km **9.** **(a)** 24 mph; **(b)** 90 mi **11.** $1\frac{23}{43}$ min after the toddler starts running, or $\frac{23}{43}$ min after the mother starts running **13.** 15 mi **15.** $\mathbf{D_W}$ **17.** $\frac{x}{3}$ **18.** $\frac{x^5y^3}{2}$ **19.** $\frac{a + 3}{2}$ **20.** $\frac{x - 2}{4}$ **21.** 2 **22.** $\frac{1}{x^2 + 1}$ **23.** $\frac{x + 2}{x + 3}$ **24.** $\frac{3(x + 4)}{x - 1}$ **25.** $\frac{x + 3}{x + 2}$ **26.** $\frac{x^2 + 25}{x^2 - 25}$ **27.** $\frac{x + 2}{x - 1}$ **28.** $\frac{x^2 + 2}{2x^2 + 1}$ **29.** Approximately 3603 mi **31.** $5\frac{1}{3}$ mi

Summary and Review: Chapter 8, p. 634

1. No **2.** Yes **3.** Yes **4.** No **5.** $(6, -2)$ **6.** $(6, 2)$ **7.** $(0, 5)$ **8.** No solution **9.** $(0, 5)$ **10.** $(-3, 9)$ **11.** $(3, -1)$ **12.** $(1, 4)$ **13.** $(-2, 4)$ **14.** $(1, -2)$ **15.** $(3, 1)$ **16.** $(1, 4)$ **17.** $(5, -3)$ **18.** $(-2, 4)$ **19.** $(-2, -6)$ **20.** $(3, 2)$ **21.** $(2, -4)$ **22.** Infinite number of solutions **23.** $(-4, 1)$ **24.** Length: 37.5 cm; width: 10.5 cm **25.** Orchestra: 297; balcony: 211 **26.** 40 L of each **27.** Asian: 4800 kg; African: 7200 kg **28.** Peanuts: $6\frac{2}{3}$ lb; fancy nuts: $3\frac{1}{3}$ lb **29.** $22.\overline{2}$ min **30.** 87-octane: 2.5 gal; 95-octane: 7.5 gal **31.** Jeff: 27; son: 9 **32.** 135 km/h **33.** 412.5 mi **34.** 32°, 58° **35.** $\mathbf{D_W}$ The strengths and weaknesses of these methods are summarized in the table in Section 8.3. **36.** $\mathbf{D_W}$ The equations have the same slope but different y-intercepts, so they represent parallel lines. Thus the system of equations has no solution. **37.** t^8 **38.** $\frac{1}{t^{18}}$ **39.** $\frac{-4x + 3}{(x - 2)(x - 3)(x + 3)}$ **40.** $\frac{x + 2}{x + 10}$

41.

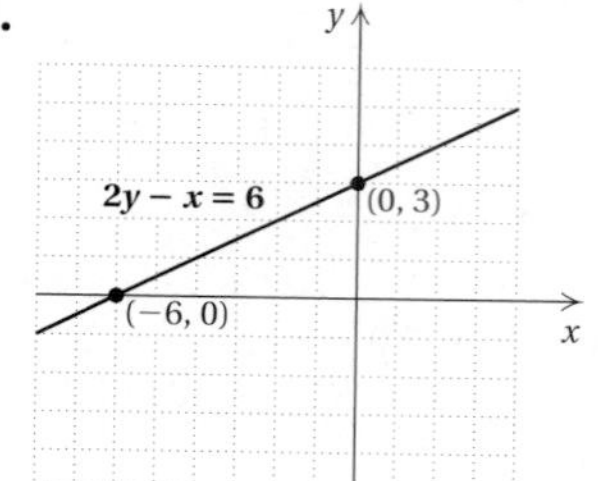

42. $960

43. $C = 1, D = 3$ **44.** $(2, 0)$ **45.** $y = -x + 5, y = \frac{2}{3}x$ **46.** $x + y = 4, x + y = -3$ **47.** Rabbits: 12; pheasants: 23

Test: Chapter 8, p. 637

1. [8.1a] No
2. [8.1b]

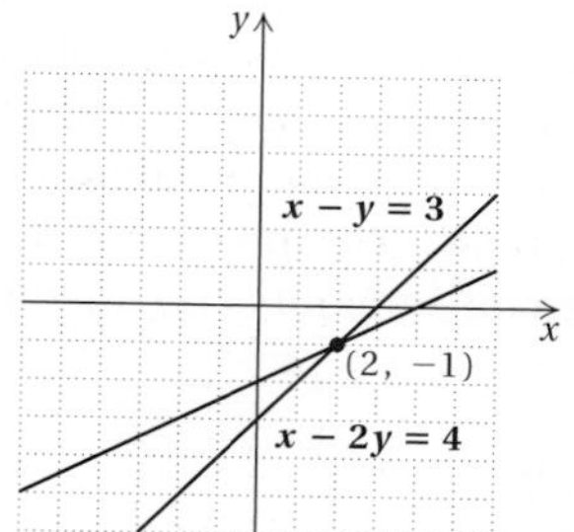

3. [8.2a] $(8, -2)$ **4.** [8.2b] $(-1, 3)$ **5.** [8.2a] No solution **6.** [8.3a] $(1, -5)$ **7.** [8.3b] $(12, -6)$ **8.** [8.3b] $(0, 1)$ **9.** [8.3b] $(5, 1)$ **10.** [8.2c] Length: 2108.5 yd; width: 2024.5 yd **11.** [8.4a] Solution A: 40 L; solution B: 20 L **12.** [8.5a] 40 km/h **13.** [8.2c] Concessions: $2850; rides: $1425 **14.** [8.2c] Hay: 415 acres; oats: 235 acres **15.** [8.2c] 45°, 135° **16.** [8.4a] 87-octane: 4 gal; 93-octane: 8 gal **17.** [8.4a] About 84 min **18.** [8.5a] 11 hr **19.** [6.5a] $\frac{-x^2 + x + 17}{(x - 4)(x + 4)(x + 1)}$

20. [3.3a]

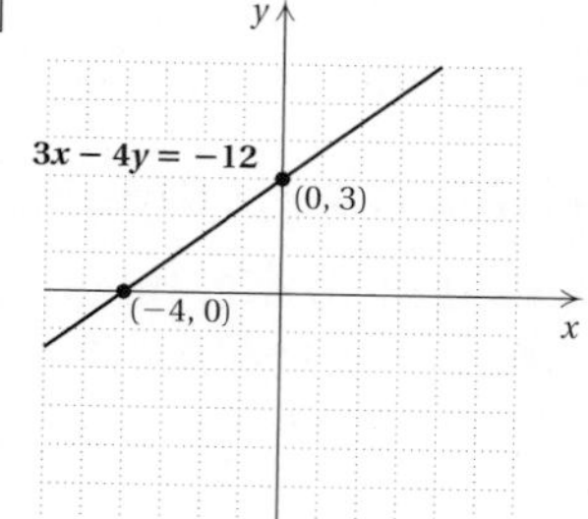

21. [4.1d, f] $\frac{10x^4}{y^2}$ **22.** [4.1e, f] $\frac{a^{10}}{b^6}$ **23.** [6.1c] $\frac{x+10}{2(x+2)}$

24. [8.1a] $C = -\frac{19}{2}$; $D = \frac{14}{3}$ **25.** [8.4a] 5

26. [7.1c], [8.1b] $y = \frac{1}{5}x + \frac{17}{5}$, $y = -\frac{3}{5}x + \frac{9}{5}$

27. [7.1c], [8.1b] $x = 3$, $y = -2$

Cumulative Review: Chapters 1–8, p. 639

1. [1.8d] -6.8 **2.** [4.2d] 3.12×10^{-2} **3.** [1.6c] $-\frac{3}{4}$

4. [4.1d, e] $\frac{1}{8}$ **5.** [6.1c] $\frac{x+3}{2x-1}$ **6.** [6.1c] $\frac{t-4}{t+4}$

7. [6.8a] $\frac{x^2(x+1)}{x+4}$ **8.** [4.6a] $2 - 10x^2 + 12x^4$

9. [4.6c] $4a^4b^2 - 20a^3b^3 + 25a^2b^4$ **10.** [4.6b] $9x^4 - 16y^2$

11. [4.5b] $-2x^3 + 4x^4 - 6x^5$ **12.** [4.5d] $8x^3 + 1$

13. [4.6b] $64 - \frac{1}{9}x^2$ **14.** [4.4c] $-y^3 - 2y^2 - 2y + 7$

15. [4.8b] $x^2 - x - 1 + \frac{-2}{2x-1}$ **16.** [6.4a] $\frac{-5x-28}{5x-25}$

17. [6.5a] $\frac{4x-1}{x-2}$ **18.** [6.1d] $\frac{y}{(y-1)^2}$

19. [6.2b] $\frac{3(x+1)}{2x}$ **20.** [5.1b] $3x^2(2x^3 - 12x + 3)$

21. [5.5d] $(4y^2 + 9)(2y + 3)(2y - 3)$

22. [5.3a], [5.4a] $(3x - 2)(x + 4)$ **23.** [5.5b] $(2x^2 - 3y)^2$

24. [5.1b], [5.2a] $3m(m + 5)(m - 3)$

25. [5.6a] $(x + 1)^2(x - 1)$ **26.** [2.3c] -9

27. [5.7a] $0, \frac{5}{2}$ **28.** [2.7e] $\{x \mid x \le 20\}$ **29.** [2.3c] 0.3

30. [5.7b] $-13, 13$ **31.** [5.7b] $\frac{5}{3}, 3$ **32.** [6.6a] -1

33. [6.6a] No solution **34.** [8.2a] $(3, -3)$

35. [8.3b] $(-2, 2)$ **36.** [8.3b] Infinite number of solutions

37. [2.5a] 0.24 billion people **38.** [2.5a] About 8.6%

39. [5.8a] 6 m, 10 m **40.** [6.7b] 204 fish

41. [5.8a] 6 cm, 9 cm **42.** [7.5d] 72 ft; 360

43. [8.5a] 10 hr

44. [8.4a] Solution A: 2.5 L; solution B: 7.5 L

45. [2.6a] 234 buttons **46.** [6.7a] $5\frac{5}{8}$ hr

47. [7.5a] $y = 0.2x$; 10 **48.** [3.4a] 0

49. [7.1a] Slope: $-\frac{2}{3}$; y-intercept: $(0, 2)$

50. [7.1c] $y = -\frac{10}{7}x - \frac{8}{7}$ **51.** [7.1a] $y = 6x - 3$

52. [3.3b]

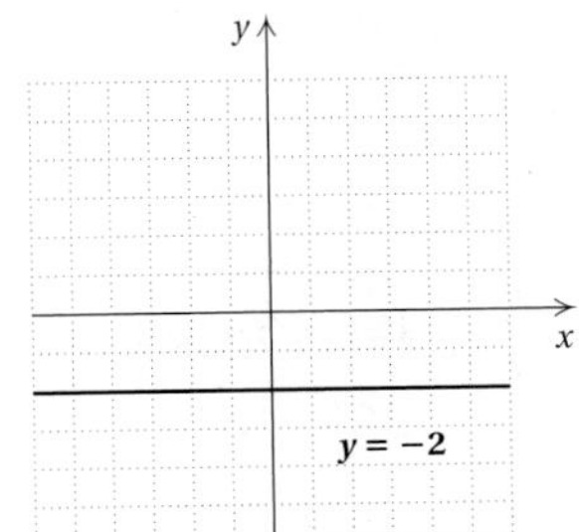

53. [3.3a]

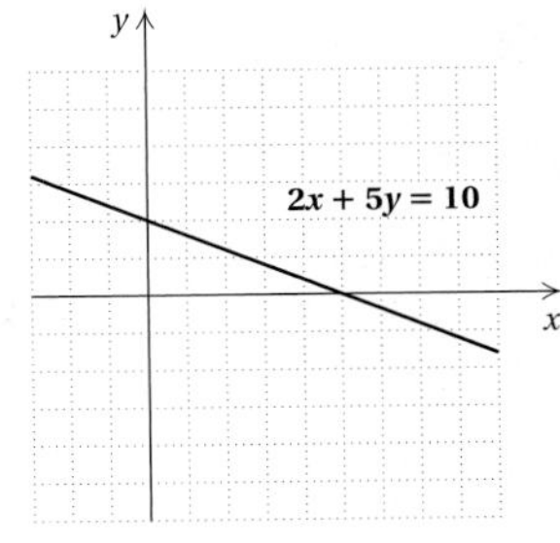

54. [7.4b]

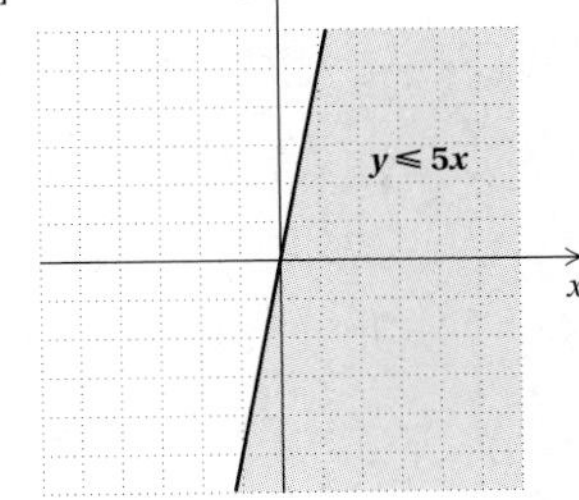

55. [7.4b]

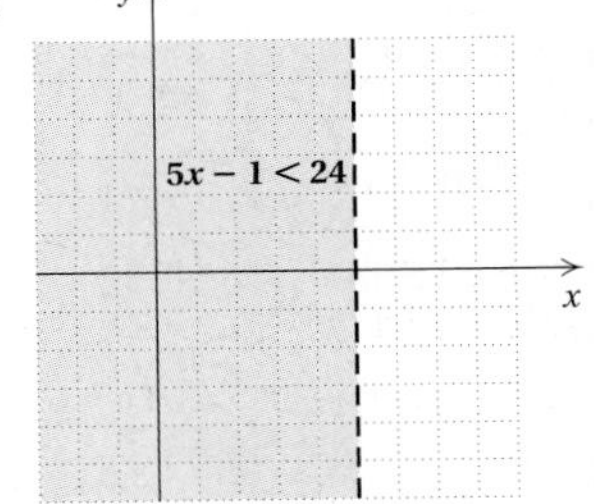

56. [8.1b]

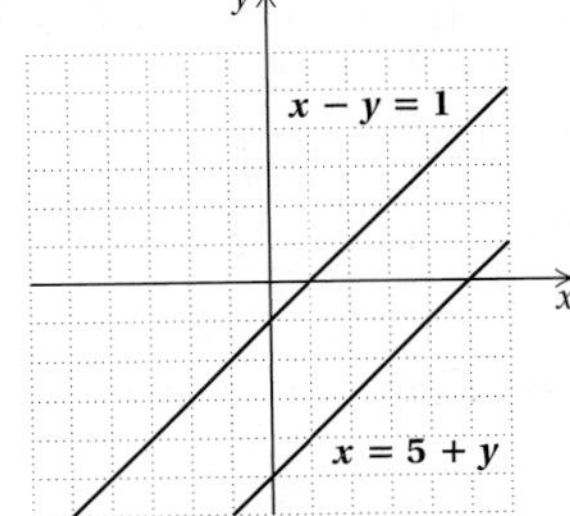

No solution

57. [8.1b]

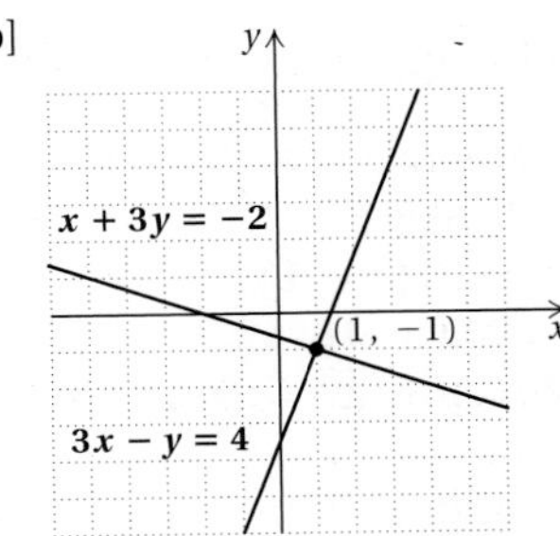

58. [3.1b] (c)

59. [8.4a] (a) **60.** [6.6a] (e) **61.** [6.7a] (c)

62. [8.1a] $A = -4$, $B = -\frac{7}{5}$ **63.** [2.7e] No solution

64. [1.8d], [6.2b], [6.5a] 0 **65.** [7.3b] $-\frac{3}{10}$

Chapter 9

Pretest: Chapter 9, p. 644

1. 7, −7 **2.** $3t$ **3.** No **4.** Yes **5.** 6.856 **6.** 4 **7.** $2x$ **8.** $12\sqrt{2}$ **9.** $7 - 4\sqrt{3}$ **10.** 1 **11.** $2\sqrt{15}$ **12.** $25 - 4\sqrt{6}$ **13.** $\sqrt{5}$ **14.** $2a^2\sqrt{2}$ **15.** $\sqrt{89} \approx 9.434$ **16.** $\sqrt{193}$ m ≈ 13.892 m **17.** $\frac{\sqrt{5x}}{x}$ **18.** $\frac{48 - 8\sqrt{5}}{31}$ **19.** 320.780 km

Margin Exercises, Section 9.1, pp. 645–649

1. 6, −6 **2.** 8, −8 **3.** 11, −11 **4.** 12, −12 **5.** 4 **6.** 7 **7.** 10 **8.** 21 **9.** −7 **10.** −13 **11.** 3.873 **12.** 5.477 **13.** 31.305 **14.** −25.842 **15.** 0.816 **16.** −1.732 **17.** **(a)** About 28.3 mph; **(b)** about 49.6 mph **18.** 227 **19.** $45 + x$ **20.** $\frac{x}{x+2}$ **21.** $x^2 + 4$ **22.** Yes **23.** No **24.** No **25.** Yes **26.** 13 **27.** $|7w|$ **28.** $|xy|$ **29.** $|xy|$ **30.** $|x - 11|$ **31.** $|x + 4|$ **32.** xy **33.** xy **34.** $x - 11$ **35.** $x + 4$ **36.** $5y$ **37.** $\frac{1}{2}t$

Calculator Corner, p. 646

1. 6.557 **2.** 10.050 **3.** 102.308 **4.** 0.632 **5.** −96.985 **6.** −0.804

Exercise Set 9.1, p. 650

1. 2, −2 **3.** 3, −3 **5.** 10, −10 **7.** 13, −13 **9.** 16, −16 **11.** 2 **13.** −3 **15.** −6 **17.** −15 **19.** 19 **21.** 2.236 **23.** 20.785 **25.** −18.647 **27.** 2.779 **29.** 120 **31.** **(a)** 13; **(b)** 24 **33.** 0.864 sec **35.** 200 **37.** $a - 4$ **39.** $t^2 + 1$ **41.** $\frac{3}{x+2}$ **43.** No **45.** Yes **47.** c **49.** $3x$ **51.** $8p$ **53.** ab **55.** $34d$ **57.** $x + 3$ **59.** $a - 5$ **61.** $2a - 5$ **63.** DW **65.** $10,660 **66.** $\frac{1}{x+3}$ **67.** 1 **68.** $\frac{(x+2)(x-2)}{(x+1)(x-1)}$ **69.** 1.7, 2.2, 2.6 **71.** 16, −16 **73.** 7, −7

Margin Exercises, Section 9.2, pp. 653–656

1. **(a)** 20; **(b)** 20 **2.** $\sqrt{33}$ **3.** 5 **4.** $\sqrt{x^2 + x}$ **5.** $\sqrt{x^2 - 4}$ **6.** $4\sqrt{2}$ **7.** $x + 7$ **8.** $5x$ **9.** $6m$ **10.** $2\sqrt{23}$ **11.** $x - 10$ **12.** $8t$ **13.** $10a$ **14.** t^2 **15.** t^{10} **16.** h^{23} **17.** $x^3\sqrt{x}$ **18.** $2x^5\sqrt{6x}$ **19.** $3\sqrt{2}$ **20.** 10 **21.** $4x^3y^2$ **22.** $5xy^2\sqrt{2xy}$ **23.** $14q^2r^4\sqrt{3q}$

Calculator Corner, p. 656

1. False **2.** False **3.** False **4.** True

Exercise Set 9.2, p. 657

1. $2\sqrt{3}$ **3.** $5\sqrt{3}$ **5.** $2\sqrt{5}$ **7.** $10\sqrt{6}$ **9.** $9\sqrt{6}$ **11.** $3\sqrt{x}$ **13.** $4\sqrt{3x}$ **15.** $4\sqrt{a}$ **17.** $8y$ **19.** $x\sqrt{13}$ **21.** $2t\sqrt{2}$ **23.** $6\sqrt{5}$ **25.** $12\sqrt{2y}$ **27.** $2x\sqrt{7}$ **29.** $x - 3$ **31.** $\sqrt{2}(2x + 1)$ **33.** $\sqrt{y}(6 + y)$ **35.** x^3 **37.** x^6 **39.** $x^2\sqrt{x}$ **41.** $t^9\sqrt{t}$ **43.** $(y - 2)^4$ **45.** $2(x + 5)^5$ **47.** $6m\sqrt{m}$ **49.** $2a^2\sqrt{2a}$ **51.** $2p^8\sqrt{26p}$ **53.** $8x^3y\sqrt{7y}$ **55.** $3\sqrt{6}$ **57.** $3\sqrt{10}$ **59.** $6\sqrt{7x}$ **61.** $6\sqrt{xy}$ **63.** 13 **65.** $5b\sqrt{3}$ **67.** $2t$ **69.** $a\sqrt{bc}$ **71.** $2xy\sqrt{2xy}$ **73.** 18 **75.** $\sqrt{10x - 5}$ **77.** $x + 2$ **79.** $6xy^3\sqrt{3xy}$ **81.** $10x^2y^3\sqrt{5xy}$ **83.** $33p^4q^2\sqrt{2pq}$ **85.** $16a^3b^3c^5\sqrt{3abc}$ **87.** DW **89.** $(-2, 4)$ **90.** $\left(\frac{1}{8}, \frac{9}{8}\right)$ **91.** $(2, 1)$ **92.** $(10, 3)$ **93.** 360 ft^2 **94.** Adults: 350; children: 61 **95.** 80 L of 30%; 120 L of 50% **96.** 10 mph **97.** $\sqrt{5}\sqrt{x - 1}$ **99.** $\sqrt{x + 6}\sqrt{x - 6}$ **101.** $x\sqrt{x - 2}$ **103.** 0.5 **105.** $4y\sqrt{3}$ **107.** $18(x + 1)\sqrt{y(x + 1)}$ **109.** $2x^3\sqrt{5x}$

Margin Exercises, Section 9.3, pp. 661–664

1. 4 **2.** 5 **3.** $x\sqrt{6x}$ **4.** $\frac{4}{3}$ **5.** $\frac{1}{5}$ **6.** $\frac{6}{x}$ **7.** $\frac{3}{4}$ **8.** $\frac{15}{16}$ **9.** $\frac{7}{y^5}$ **10.** **(a)** $\frac{\sqrt{15}}{5}$; **(b)** $\frac{\sqrt{15}}{5}$ **11.** $\frac{\sqrt{10}}{4}$ **12.** $\frac{10\sqrt{3}}{3}$ **13.** $\frac{\sqrt{21}}{7}$ **14.** $\frac{\sqrt{5r}}{r}$ **15.** $\frac{8y\sqrt{7}}{7}$

Exercise Set 9.3, p. 665

1. 3 **3.** 6 **5.** $\sqrt{5}$ **7.** $\frac{1}{5}$ **9.** $\frac{2}{5}$ **11.** 2 **13.** $3y$ **15.** $\frac{4}{7}$ **17.** $\frac{1}{6}$ **19.** $-\frac{4}{9}$ **21.** $\frac{8}{17}$ **23.** $\frac{13}{14}$ **25.** $\frac{5}{x}$ **27.** $\frac{3a}{25}$ **29.** $\frac{\sqrt{10}}{5}$ **31.** $\frac{\sqrt{14}}{4}$ **33.** $\frac{\sqrt{3}}{6}$ **35.** $\frac{\sqrt{10}}{6}$ **37.** $\frac{3\sqrt{5}}{5}$ **39.** $\frac{2\sqrt{6}}{3}$ **41.** $\frac{\sqrt{3x}}{x}$ **43.** $\frac{\sqrt{xy}}{y}$ **45.** $\frac{x\sqrt{5}}{10}$ **47.** $\frac{\sqrt{14}}{2}$ **49.** $\frac{3\sqrt{2}}{4}$ **51.** $\frac{\sqrt{6}}{2}$ **53.** $\sqrt{2}$ **55.** $\frac{\sqrt{55}}{11}$ **57.** $\frac{\sqrt{21}}{6}$ **59.** $\frac{\sqrt{6}}{2}$ **61.** 5 **63.** $\frac{\sqrt{3x}}{x}$ **65.** $\frac{4y\sqrt{5}}{5}$ **67.** $\frac{a\sqrt{2a}}{4}$ **69.** $\frac{\sqrt{42x}}{3x}$ **71.** $\frac{3\sqrt{6}}{8c}$ **73.** $\frac{y\sqrt{xy}}{x}$ **75.** $\frac{3n\sqrt{10}}{8}$ **77.** DW **79.** $(4, 2)$ **80.** $(10, 30)$ **81.** No solution **82.** Infinite number of solutions **83.** $\left(-\frac{5}{2}, -\frac{9}{2}\right)$ **84.** $\left(\frac{26}{23}, \frac{44}{23}\right)$ **85.** $9x^2 - 49$ **86.** $16a^2 - 25b^2$ **87.** $21x - 9y$ **88.** $14a - 6b$ **89.** 1.57 sec; 3.14 sec; 8.88 sec; 11.10 sec **91.** 1 sec **93.** $\frac{\sqrt{5}}{40}$ **95.** $\frac{\sqrt{5x}}{5x^2}$ **97.** $\frac{\sqrt{3ab}}{b}$ **99.** $\frac{3\sqrt{10}}{100}$ **101.** $\frac{y - x}{xy}$

Margin Exercises, Section 9.4, pp. 669–672

1. $12\sqrt{2}$ **2.** $5\sqrt{5}$ **3.** $-12\sqrt{10}$ **4.** $5\sqrt{6}$ **5.** $\sqrt{x + 1}$ **6.** $\frac{3}{2}\sqrt{2}$ **7.** $\frac{8\sqrt{15}}{15}$ **8.** $\sqrt{15} + \sqrt{6}$ **9.** $4 + 3\sqrt{5} - 4\sqrt{2} - 3\sqrt{10}$ **10.** $2 - a$ **11.** $25 + 10\sqrt{x} + x$ **12.** 2 **13.** $7 - \sqrt{5}$

14. $\sqrt{5} + \sqrt{2}$ **15.** $1 + \sqrt{x}$ **16.** $\frac{21 - 3\sqrt{5}}{22}$

17. $\frac{7 + 2\sqrt{10}}{3}$ **18.** $\frac{7 + 7\sqrt{x}}{1 - x}$

Exercise Set 9.4, p. 673

1. $16\sqrt{3}$ **3.** $4\sqrt{5}$ **5.** $13\sqrt{x}$ **7.** $-9\sqrt{d}$ **9.** $25\sqrt{2}$ **11.** $\sqrt{3}$ **13.** $\sqrt{5}$ **15.** $13\sqrt{2}$ **17.** $3\sqrt{3}$ **19.** $2\sqrt{2}$ **21.** 0 **23.** $(2 + 9x)\sqrt{x}$ **25.** $(3 - 2x)\sqrt{3}$ **27.** $3\sqrt{2x + 2}$ **29.** $(x + 3)\sqrt{x^3 - 1}$ **31.** $(4a^2 + a^2b - 5b)\sqrt{b}$ **33.** $\frac{2\sqrt{3}}{3}$ **35.** $\frac{13\sqrt{2}}{2}$ **37.** $\frac{\sqrt{6}}{6}$ **39.** $\sqrt{15} - \sqrt{3}$ **41.** $10 + 5\sqrt{3} - 2\sqrt{7} - \sqrt{21}$ **43.** $9 - 4\sqrt{5}$ **45.** -62 **47.** 1 **49.** $13 + \sqrt{5}$ **51.** $x - 2\sqrt{xy} + y$ **53.** $-\sqrt{3} - \sqrt{5}$ **55.** $5 - 2\sqrt{6}$ **57.** $\frac{4\sqrt{10} - 4}{9}$ **59.** $5 - 2\sqrt{7}$ **61.** $\frac{12 - 3\sqrt{x}}{16 - x}$ **63.** $\frac{24 + 3\sqrt{x} + 8\sqrt{2} + \sqrt{2x}}{64 - x}$ **65.** $\mathbf{D_W}$ **67.** $\frac{5}{11}$ **68.** $-\frac{38}{13}$ **69.** $-1, 6$ **70.** 2, 5 **71.** Jolly Juice: 1.6 L; Real Squeeze: 6.4 L **72.** $\frac{1}{3}$ hr **73.** $-9, -2, -5, -17, -0.678375$ **75.** Not equivalent **77.** Not correct **79.** $11\sqrt{3} - 10\sqrt{2}$ **81.** True; $(3\sqrt{x + 2})^2 = (3\sqrt{x + 2})(3\sqrt{x + 2}) = (3 \cdot 3)(\sqrt{x + 2} \cdot \sqrt{x + 2}) = 9(x + 2)$

Margin Exercises, Section 9.5, pp. 677–682

1. $\frac{64}{3}$ **2.** 2 **3.** $\frac{3}{8}$ **4.** 4 **5.** 1 **6.** 4 **7.** About 276 mi **8.** About 9 mi **9.** About 52 ft

Calculator Corner, p. 679

1. Left to the student **2.** Left to the student

Exercise Set 9.5, p. 683

1. 36 **3.** 18.49 **5.** 165 **7.** $\frac{621}{2}$ **9.** 5 **11.** 3 **13.** $\frac{17}{4}$ **15.** No solution **17.** No solution **19.** 9 **21.** 12 **23.** 1, 5 **25.** 3 **27.** 5 **29.** No solution **31.** $-\frac{10}{3}$ **33.** 3 **35.** No solution **37.** 9 **39.** 1 **41.** 8 **43.** About 232 mi **45.** 16,200 ft **47.** 211.25 ft; 281.25 ft **49.** $\mathbf{D_W}$ **51.** $\frac{(x + 7)^2}{x - 7}$ **52.** $\frac{(x - 2)(x - 5)}{(x - 3)(x - 4)}$ **53.** $\frac{a - 5}{2}$ **54.** $\frac{x - 3}{x - 2}$ **55.** 61°, 119° **56.** 38°, 52° **57.** $\frac{x^6}{3}$ **58.** $\frac{x - 3}{4(x + 3)}$ **59.** $-2, 2$ **61.** $-\frac{57}{16}$ **63.** 13 **65.** Left to the student **67.** Left to the student

Margin Exercises, Section 9.6, pp. 688–689

1. $\sqrt{65} \approx 8.062$ **2.** $\sqrt{75} \approx 8.660$ **3.** $\sqrt{10} \approx 3.162$ **4.** $\sqrt{175} \approx 13.229$ **5.** $\sqrt{325}$ ft ≈ 18.028 ft

Exercise Set 9.6, p. 690

1. 17 **3.** $\sqrt{32} \approx 5.657$ **5.** 12 **7.** 4 **9.** 26 **11.** 12 **13.** 2 **15.** $\sqrt{2} \approx 1.414$ **17.** 5 **19.** 3 **21.** $\sqrt{211{,}200{,}000}$ ft ≈ 14.533 ft **23.** 240 ft **25.** $\sqrt{18}$ cm ≈ 4.243 cm **27.** $\sqrt{208}$ ft ≈ 14.422 ft **29.** $\mathbf{D_W}$ **31.** $\left(-\frac{3}{2}, -\frac{1}{16}\right)$ **32.** $\left(\frac{8}{5}, 9\right)$ **33.** $\left(-\frac{9}{19}, \frac{91}{38}\right)$ **34.** $(-10, 1)$ **35.** $-\frac{1}{3}$ **36.** $\frac{5}{8}$ **37.** $12 - 2\sqrt{6} \approx 7.101$

Summary and Review: Chapter 9, p. 692

1. 8, -8 **2.** 20, -20 **3.** 6 **4.** -13 **5.** 1.732 **6.** 9.950 **7.** -17.892 **8.** 0.742 **9.** -2.055 **10.** 394.648 **11.** $x^2 + 4$ **12.** $5ab^3$ **13.** No **14.** Yes **15.** No **16.** No **17.** No **18.** No **19.** m **20.** $x - 4$ **21.** $\sqrt{21}$ **22.** $\sqrt{x^2 - 9}$ **23.** $-4\sqrt{3}$ **24.** $4t\sqrt{2}$ **25.** $\sqrt{t - 7}\sqrt{t + 7}$ **26.** $x + 8$ **27.** x^4 **28.** $m^7\sqrt{m}$ **29.** $2\sqrt{15}$ **30.** $2x\sqrt{10}$ **31.** $5xy\sqrt{2}$ **32.** $10a^2b\sqrt{ab}$ **33.** $\frac{5}{8}$ **34.** $\frac{2}{3}$ **35.** $\frac{7}{t}$ **36.** $\frac{\sqrt{2}}{2}$ **37.** $\frac{\sqrt{2}}{4}$ **38.** $\frac{\sqrt{5y}}{y}$ **39.** $\frac{2\sqrt{3}}{3}$ **40.** $\frac{\sqrt{15}}{5}$ **41.** $\frac{x\sqrt{30}}{6}$ **42.** $8 - 4\sqrt{3}$ **43.** $13\sqrt{5}$ **44.** $\sqrt{5}$ **45.** $\frac{\sqrt{2}}{2}$ **46.** $7 + 4\sqrt{3}$ **47.** 1 **48.** 52 **49.** No solution **50.** 0, 3 **51.** 9 **52.** 20 **53.** $\sqrt{3} \approx 1.732$ **54.** About 50,990 ft **55.** 9 ft **56.** **(a)** About 63 mph; **(b)** 405 ft **57.** $\mathbf{D_W}$ It is incorrect to take the square roots of the terms in the numerator individually—that is, $\sqrt{a + b}$ and $\sqrt{a} + \sqrt{b}$ are not equivalent. The following is correct:

$$\sqrt{\frac{9 + 100}{25}} = \frac{\sqrt{9 + 100}}{\sqrt{25}} = \frac{\sqrt{109}}{5}.$$

58. **(a)** $\sqrt{5x^2} = \sqrt{5}\sqrt{x^2} = \sqrt{5} \cdot |x| = |x|\sqrt{5}$. The given statement is correct.
(b) Let $b = 3$. Then $\sqrt{b^2 - 4} = \sqrt{3^2 - 4} = \sqrt{9 - 4} = \sqrt{5}$, but $b - 2 = 3 - 2 = 1$. The given statement is false.
(c) Let $x = 3$. Then $\sqrt{x^2 + 16} = \sqrt{3^2 + 16} = \sqrt{9 + 16} = \sqrt{25} = 5$, but $x + 4 = 3 + 4 = 7$. The given statement is false.
59. $\left(\frac{22}{17}, -\frac{8}{17}\right)$ **60.** $\frac{(x - 5)(x - 7)}{(x + 7)(x + 5)}$ **61.** \$450 **62.** 4300 **63.** $\sqrt{1525}$ mi ≈ 39.051 mi **64.** 2 **65.** $b = \sqrt{A^2 - a^2}$ **66.** 6

Test: Chapter 9, p. 695

1. [9.1a] 9, -9 **2.** [9.1a] 8 **3.** [9.1a] -5 **4.** [9.1b] 10.770 **5.** [9.1b] -9.349 **6.** [9.1b] 4.127 **7.** [9.1d] $4 - y^3$ **8.** [9.1e] Yes **9.** [9.1e] No **10.** [9.1f] a **11.** [9.1f] $6y$ **12.** [9.2c] $\sqrt{30}$ **13.** [9.2c] $\sqrt{x^2 - 64}$ **14.** [9.2a] $3\sqrt{3}$ **15.** [9.2a] $5\sqrt{x - 1}$ **16.** [9.2b] $t^2\sqrt{t}$ **17.** [9.2c] $5\sqrt{2}$ **18.** [9.2c] $3ab^2\sqrt{2}$ **19.** [9.3b] $\frac{3}{2}$ **20.** [9.3b] $\frac{12}{a}$ **21.** [9.3c] $\frac{\sqrt{10}}{5}$ **22.** [9.3c] $\frac{\sqrt{2xy}}{y}$ **23.** [9.3a, c] $\frac{3\sqrt{6}}{8}$ **24.** [9.3a] $\frac{\sqrt{7}}{4y}$ **25.** [9.4a] $-6\sqrt{2}$

26. [9.4a] $\frac{6\sqrt{5}}{5}$ **27.** [9.4b] $21 - 8\sqrt{5}$ **28.** [9.4b] 11
29. [9.4c] $\frac{40 + 10\sqrt{5}}{11}$ **30.** [9.6a] $\sqrt{80} \approx 8.944$
31. [9.5a] 48 **32.** [9.5a] $-2, 2$ **33.** [9.5b] -3
34. [9.5c] **(a)** About 237 mi; **(b)** 34,060.5 ft **35.** [9.6b] $\sqrt{15{,}700}$ yd ≈ 125.300 yd **36.** [8.4a] 789.25 yd^2
37. [7.5b] $15{,}686\frac{2}{3}$ switches **38.** [8.3b] $(0, 2)$
39. [6.2b] $\frac{x-7}{x+6}$ **40.** [9.1a] $\sqrt{5}$ **41.** [9.2b] y^{8n}

Cumulative Review: Chapters 1–9, p. 697

1. [4.3a] -15 **2.** [4.3e] $-4x^3 - \frac{1}{7}x^2 - 2$ **3.** [6.1a] $-\frac{1}{2}$
4. [9.1e] No **5.** [9.4b] 1 **6.** [9.1a] -14 **7.** [9.2c] 15
8. [9.4b] $3 - 2\sqrt{2}$ **9.** [9.3a, c] $\frac{9\sqrt{10}}{25}$ **10.** [9.4a] $12\sqrt{5}$
11. [4.7f] $9x^8 - 4y^{10}$ **12.** [4.6c] $x^4 + 8x^2 + 16$
13. [4.6a] $8x^2 - \frac{1}{8}$ **14.** [6.5a] $\frac{4x+2}{2x-1}$
15. [4.4a, c] $-3x^3 + 8x^2 - 5x$ **16.** [6.1d] $\frac{2(x-5)}{3(x-1)}$
17. [6.2b] $\frac{(x+1)(x-3)}{2(x+3)}$
18. [4.8b] $3x^2 + 4x + 9 + \frac{13}{x-2}$ **19.** [9.2a] $\sqrt{2}(x-1)$
20. [4.1d, f] $\frac{1}{x^{12}}$ **21.** [9.3b] $\frac{5}{x^4}$ **22.** [6.8a] $2(x-1)$
23. [5.5d] $3(1 + 2x^4)(1 - 2x^4)$
24. [5.1b] $4t(3 - t - 12t^3)$ **25.** [5.3a], [5.4a] $2(3x-2)(x-4)$ **26.** [5.6a] $(2x+1)(2x-1)(x+1)$
27. [5.5b] $(4x^2 - 7)^2$ **28.** [5.2a] $(x+15)(x-12)$
29. [5.7b] $-17, 0$ **30.** [2.7e] $\{x \mid x > -6\}$ **31.** [6.6a] $-\frac{12}{5}$
32. [5.7b] $-5, 6$ **33.** [2.7e] $\{x \mid x \le -\frac{9}{2}\}$ **34.** [5.7b] $-9, 9$
35. [9.5a] 41 **36.** [9.5b] 9 **37.** [2.3b] $\frac{2}{5}$ **38.** [6.6a] $\frac{1}{3}$
39. [8.2a] $(4, 1)$ **40.** [8.3a] $(3, -8)$ **41.** [2.4b] $p = \frac{4A}{r+q}$

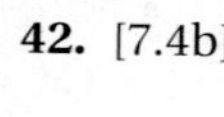
42. [7.4b]

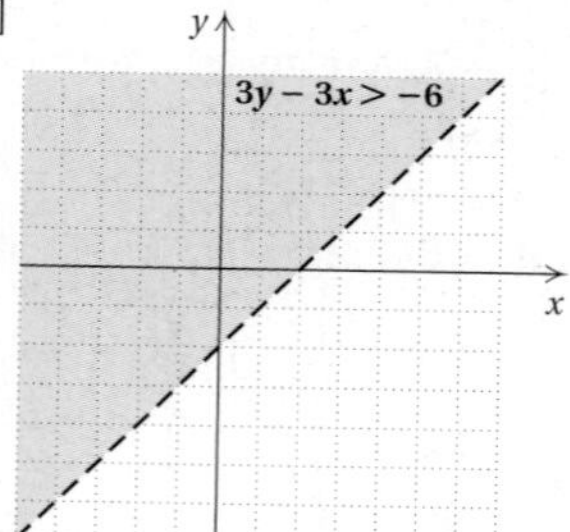

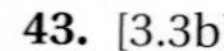
43. [3.3b]

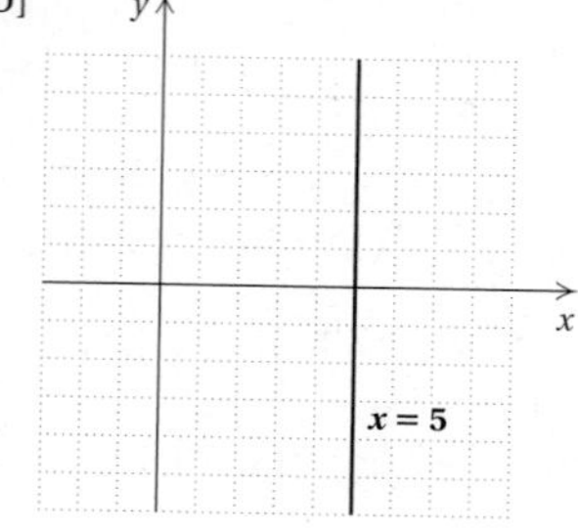

44. [3.3a]

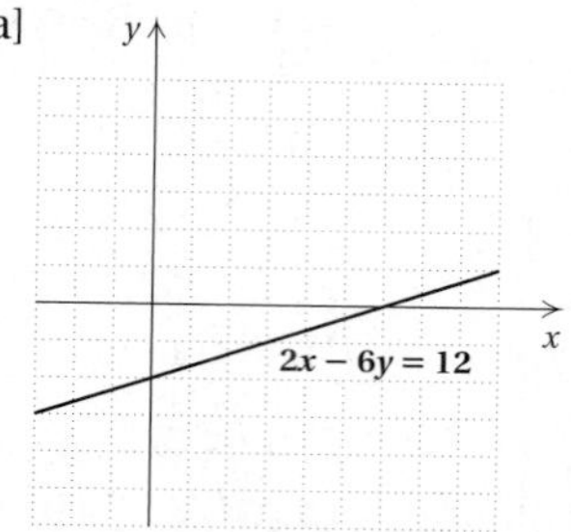

45. [7.1c] $y = \frac{11}{4}x - \frac{19}{4}$ **46.** [7.1a] Slope: $\frac{5}{3}$; y-intercept: $(0, -3)$ **47.** [4.3a] -2; 1; -2; -5; -2 **48.** [6.7b] About 72 home runs **49.** [7.5d] 0.4 ft **50.** [8.4a] Hamburger: \$2.95; milkshake: \$2.50 **51.** [9.6b] $\sqrt{48}$ m ≈ 6.9 m
52. [2.6a] 38°, 76°, 66° **53.** [6.7b] 20 defective resistors
54. [5.8a] Length: 15 m; width: 12 m **55.** [8.4a] Dimes: 65; quarters: 50 **56.** [2.5a] \$2600 **57.** [6.7a] 60 mph
58. [9.3c] (a) **59.** [6.6a] (b) **60.** [6.8a] (a)
61. [4.7f] (e) **62.** [1.2d] $<$ **63.** [1.2d, e] $>$
64. [8.4a] 300 L **65.** [8.3b], [9.5a] $(9, 4)$ **66.** [9.6b] Yes
67. [9.6a] $\frac{\sqrt{3}}{2} \approx 0.866$

Chapter 10

Pretest: Chapter 10, p. 702

1. 3 **2.** $\sqrt{7}, -\sqrt{7}$ **3.** $\frac{-3 \pm \sqrt{21}}{6}$ **4.** $0, \frac{3}{5}$
5. $-\frac{5}{3}, 0$ **6.** $-4 \pm \sqrt{5}$ **7.** $1 \pm \sqrt{6}$
8. $n = \frac{p \pm \sqrt{p^2 + 4A}}{2}$ **9.** Width: 4 cm; length: 12 cm
10. $\left(\frac{-1 - \sqrt{33}}{4}, 0\right), \left(\frac{-1 + \sqrt{33}}{4}, 0\right)$
11. 10 km/h **12.**

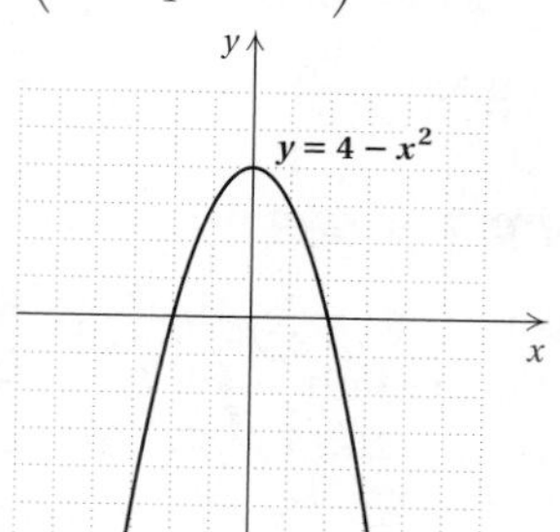

Margin Exercises, Section 10.1, pp. 703–709

1. **(a)** y-intercept: $(0, -3)$; x-intercept: $\left(-\frac{9}{2}, 0\right)$;

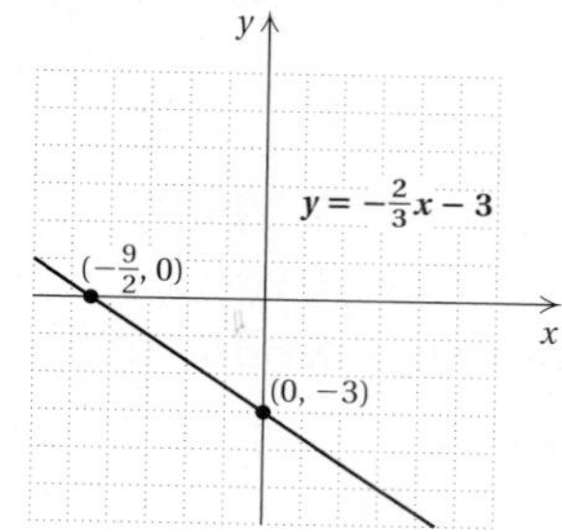

(b) $-\frac{9}{2}$; **(c)** $-\frac{9}{2}$; first coordinate; $\left(-\frac{9}{2}, 0\right)$
2. $y^2 - 8y = 0$; $a = 1, b = -8, c = 0$
3. $x^2 + 9x - 3 = 0$; $a = 1, b = 9, c = -3$
4. $4x^2 + x + 4 = 0$; $a = 4, b = 1, c = 4$
5. $5x^2 - 21 = 0$; $a = 5, b = 0, c = -21$ **6.** $0, -4$
7. $0, \frac{3}{5}$ **8.** $-2, \frac{3}{4}$ **9.** $1, 4$ **10.** $5, 13$
11. (a) 14 diagonals;

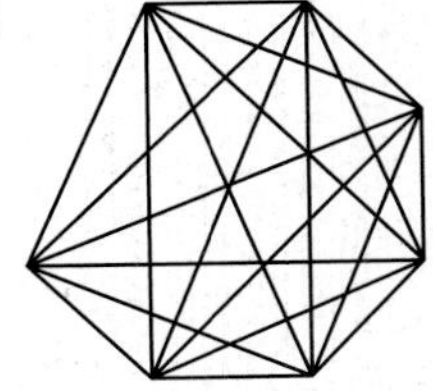

(b) 14 diagonals; **(c)** 11 sides

Calculator Corner, p. 709

1. $0.6, 1$ **2.** $-1.5, 5$ **3.** $3, 8$ **4.** $2, 4$

Exercise Set 10.1, p. 710

1. $x^2 - 3x + 2 = 0$; $a = 1, b = -3, c = 2$
3. $7x^2 - 4x + 3 = 0$; $a = 7, b = -4, c = 3$
5. $2x^2 - 3x + 5 = 0$; $a = 2, b = -3, c = 5$
7. $0, -5$ **9.** $0, -2$ **11.** $0, \frac{2}{5}$ **13.** $0, -1$
15. $0, 3$ **17.** $0, \frac{1}{5}$ **19.** $0, \frac{3}{14}$ **21.** $0, \frac{81}{2}$ **23.** $-12, 4$
25. $-5, -1$ **27.** $-9, 2$ **29.** $3, 5$ **31.** -5 **33.** 4
35. $-\frac{2}{3}, \frac{1}{2}$ **37.** $-\frac{2}{3}, 4$ **39.** $-1, \frac{5}{3}$ **41.** $-5, -1$
43. $-2, 7$ **45.** $-5, 4$ **47.** 4 **49.** $-2, 1$
51. $-\frac{2}{5}, 10$ **53.** $-4, 6$ **55.** 1 **57.** $2, 5$
59. No solution **61.** $-\frac{5}{2}, 1$ **63.** 35 diagonals
65. 7 sides **67.** $\mathbf{D_W}$ **69.** 8 **70.** -13 **71.** $2\sqrt{2}$
72. $2\sqrt{3}$ **73.** $2\sqrt{5}$ **74.** $2\sqrt{22}$ **75.** $9\sqrt{5}$
76. $2\sqrt{255}$ **77.** 2.646 **78.** 4.796 **79.** 1.528
80. 22.908 **81.** $-\frac{1}{3}, 1$ **83.** $0, \frac{\sqrt{5}}{5}$ **85.** $-1.7, 4$
87. $-1.7, 3$ **89.** $-2, 3$ **91.** 4

Margin Exercises, Section 10.2, pp. 712–717

1. $\sqrt{10}, -\sqrt{10}$ **2.** 0 **3.** $\frac{\sqrt{6}}{2}, -\frac{\sqrt{6}}{2}$ **4.** $7, -1$
5. $-4 \pm \sqrt{11}$ **6.** $-5, 11$ **7.** $1 \pm \sqrt{5}$ **8.** $2, 4$
9. $-10, 2$ **10.** $6 \pm \sqrt{13}$ **11.** $-2, 5$ **12.** $\frac{-3 \pm \sqrt{33}}{4}$
13. About 7.5 sec

Exercise Set 10.2, p. 718

1. $11, -11$ **3.** $\sqrt{7}, -\sqrt{7}$ **5.** $\frac{\sqrt{15}}{5}, -\frac{\sqrt{15}}{5}$ **7.** $\frac{5}{2}, -\frac{5}{2}$
9. $\frac{7\sqrt{3}}{3}, -\frac{7\sqrt{3}}{3}$ **11.** $\sqrt{3}, -\sqrt{3}$ **13.** $\frac{8}{7}, -\frac{8}{7}$
15. $-7, 1$ **17.** $-3 \pm \sqrt{21}$ **19.** $-13 \pm 2\sqrt{2}$
21. $7 \pm 2\sqrt{3}$ **23.** $-9 \pm \sqrt{34}$ **25.** $\frac{-3 \pm \sqrt{14}}{2}$
27. $-5, 11$ **29.** $-15, 1$ **31.** $-2, 8$ **33.** $-21, -1$
35. $1 \pm \sqrt{6}$ **37.** $11 \pm \sqrt{19}$ **39.** $-5 \pm \sqrt{29}$
41. $\frac{7 \pm \sqrt{57}}{2}$ **43.** $-7, 4$ **45.** $\frac{-3 \pm \sqrt{17}}{4}$
47. $\frac{-3 \pm \sqrt{145}}{4}$ **49.** $\frac{-2 \pm \sqrt{7}}{3}$ **51.** $-\frac{1}{2}, 5$
53. $-\frac{5}{2}, \frac{2}{3}$ **55.** About 9.6 sec **57.** About 4.4 sec
59. $\mathbf{D_W}$ **61.** $y = \frac{141}{x}$ **62.** $3\frac{1}{3}$ hr **63.** $3x\sqrt{2}$
64. $8x^2\sqrt{3x}$ **65.** $3t$ **66.** $x^3\sqrt{x}$ **67.** $-12, 12$
69. $-16\sqrt{2}, 16\sqrt{2}$ **71.** $-2\sqrt{c}, 2\sqrt{c}$
73. $49.896, -49.896$ **75.** $-9, 9$

Margin Exercises, Section 10.3, pp. 722–725

1. $-4, \frac{1}{2}$ **2.** $-2, 5$ **3.** $-2 \pm \sqrt{11}$
4. No real-number solutions **5.** $\frac{4 \pm \sqrt{31}}{5}$
6. $-0.3, 1.9$

Calculator Corner, p. 724

1. The equations $x^2 + x = -1$ and $x^2 + x + 1 = 0$ are equivalent. The graph of $y = x^2 + x + 1$ has no x-intercepts, so the equation $x^2 + x = -1$ has no real-number solutions.
2. Yes **3.** No

Calculator Corner, p. 725

1. $-1.3, 5.3$ **2.** $-5.3, 1.3$ **3.** $-0.3, 1.9$

Exercise Set 10.3, p. 726

1. $-3, 7$ **3.** 3 **5.** $-\frac{4}{3}, 2$ **7.** $-\frac{5}{2}, \frac{3}{2}$ **9.** $-3, 3$
11. $1 \pm \sqrt{3}$ **13.** $5 \pm \sqrt{3}$ **15.** $-2 \pm \sqrt{7}$
17. $\frac{-4 \pm \sqrt{10}}{3}$ **19.** $\frac{5 \pm \sqrt{33}}{4}$ **21.** $\frac{1 \pm \sqrt{3}}{2}$
23. No real-number solutions **25.** $\frac{5 \pm \sqrt{73}}{6}$
27. $\frac{3 \pm \sqrt{29}}{2}$ **29.** $-\sqrt{5}, \sqrt{5}$ **31.** $-2 \pm \sqrt{3}$
33. $\frac{5 \pm \sqrt{37}}{2}$ **35.** $-1.3, 5.3$ **37.** $-0.2, 6.2$
39. $-1.2, 0.2$ **41.** $0.3, 2.4$ **43.** $\mathbf{D_W}$ **45.** $3\sqrt{10}$
46. $\sqrt{6}$ **47.** $2\sqrt{2}$ **48.** $(9x - 2)\sqrt{x}$ **49.** $4\sqrt{5}$
50. $3x^2\sqrt{3x}$ **51.** $30x^5\sqrt{10}$ **52.** $\frac{\sqrt{21}}{3}$ **53.** $0, 2$
55. $\frac{3 \pm \sqrt{5}}{2}$ **57.** $\frac{-7 \pm \sqrt{61}}{2}$ **59.** $\frac{-2 \pm \sqrt{10}}{2}$
61.–67. Left to the student

Margin Exercises, Section 10.4, pp. 728–730

1. (a) $I = \frac{9R}{E}$; **(b)** $R = \frac{EI}{9}$ **2.** $x = \frac{y - 5}{a - b}$ **3.** $f = \frac{pq}{q + p}$
4. $L = \frac{r^2}{20}$ **5.** $L = \frac{T^2 g}{4\pi^2}$ **6.** $m = \frac{E}{c^2}$ **7.** $r = \sqrt{\frac{A}{\pi}}$
8. $n = \frac{1 + \sqrt{1 + 4N}}{2}$

Exercise Set 10.4, p. 731

1. $I = \dfrac{VQ}{q}$ **3.** $m = \dfrac{Sd^2}{kM}$ **5.** $d^2 = \dfrac{kmM}{S}$
7. $W = \sqrt{\dfrac{10t}{T}}$ **9.** $t = \dfrac{A}{a + b}$ **11.** $x = \dfrac{y - c}{a + b}$
13. $a = \dfrac{bt}{b - t}$ **15.** $p = \dfrac{qf}{q - f}$ **17.** $b = \dfrac{2A}{h}$
19. $h = \dfrac{S - 2\pi r^2}{2\pi r}$, or $h = \dfrac{S}{2\pi r} - r$ **21.** $R = \dfrac{r_1 r_2}{r_2 + r_1}$
23. $Q = \dfrac{P^2}{289}$ **25.** $E = \dfrac{mv^2}{2g}$ **27.** $r = \dfrac{1}{2}\sqrt{\dfrac{S}{\pi}}$
29. $A = \dfrac{-m + \sqrt{m^2 + 4kP}}{2k}$ **31.** $a = \sqrt{c^2 - b^2}$
33. $t = \dfrac{\sqrt{s}}{4}$ **35.** $r = \dfrac{-\pi h + \sqrt{\pi^2 h^2 + \pi A}}{\pi}$
37. $v = 20\sqrt{\dfrac{F}{A}}$ **39.** $a = \sqrt{c^2 - b^2}$ **41.** $a = \dfrac{2h\sqrt{3}}{3}$
43. $T = \dfrac{2 + \sqrt{4 - a(m - n)}}{a}$ **45.** $T = \dfrac{v^2\pi m}{8k}$
47. $x = \dfrac{d\sqrt{3}}{3}$ **49.** $n = \dfrac{1 + \sqrt{1 + 8N}}{2}$ **51.** $b = \dfrac{a}{3S - 1}$
53. $B = \dfrac{A}{QA + 1}$ **55.** $n = \dfrac{S + 360}{180}$, or $n = \dfrac{S}{180} + 2$
57. $t = \dfrac{A - P}{Pr}$ **59.** $D = \dfrac{BC}{A}$ **61.** $a = \dfrac{-b}{C - K}$, or $a = \dfrac{b}{K - C}$ **63.** D_W **65.** $\sqrt{65} \approx 8.062$
66. $\sqrt{75} \approx 8.660$ **67.** $\sqrt{41} \approx 6.403$ **68.** $\sqrt{44} \approx 6.633$
69. $\sqrt{1084} \approx 32.924$ **70.** $\sqrt{5} \approx 2.236$
71. $\sqrt{424}$ ft ≈ 20.591 ft **72.** $\sqrt{12{,}500}$ yd ≈ 111.803 yd
73. (a) $r = \dfrac{C}{2\pi}$; (b) $A = \dfrac{C^2}{4\pi}$ **75.** $\dfrac{1}{3a}, 1$

Margin Exercises, Section 10.5, pp. 734–736

1. Length: 14.8 yd; width: 4.6 yd **2.** 2.3 yd, 3.3 yd
3. 3 km/h

Exercise Set 10.5, p. 737

1. Width: 24 in.; height: 18 in. **3.** Length: 10 ft; width: 7 ft
5. Length: 10 m; width: 5 m
7. Length: 20 cm; width: 16 cm
9. 4.6 m; 6.6 m **11.** Length: 5.6 in.; width: 3.6 in.
13. Length: 6.4 cm; width: 3.2 cm **15.** 3 cm
17. 7 km/h **19.** 8 mph **21.** 0 km/h (stream is still) or 4 km/h **23.** 0 mph (no wind) or 36 mph **25.** 1 km/h
27. D_W **29.** $8\sqrt{2}$ **30.** $12\sqrt{10}$ **31.** $(2x - 7)\sqrt{x}$
32. $-\sqrt{6}$ **33.** $\dfrac{3\sqrt{2}}{2}$ **34.** $\dfrac{2\sqrt{3}}{3}$ **35.** $5\sqrt{6} - 4\sqrt{3}$
36. $(9x + 2)\sqrt{x}$
37. $12\sqrt{2}$ in. ≈ 16.97 in.; two 12-in. pizzas

Margin Exercises, Section 10.6, pp. 743–744

1. $(0, -3)$

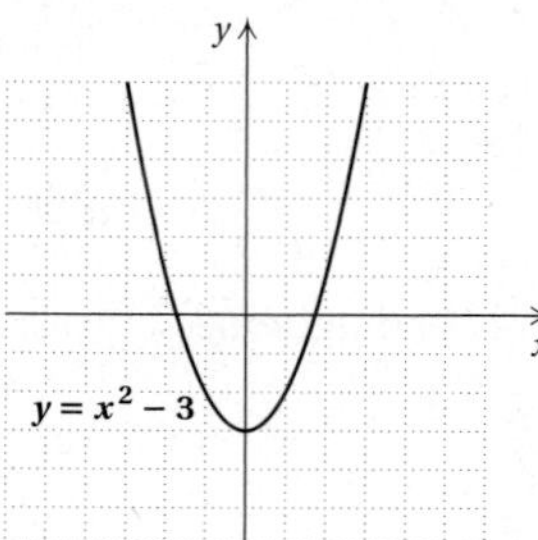

2. $(1, 3)$

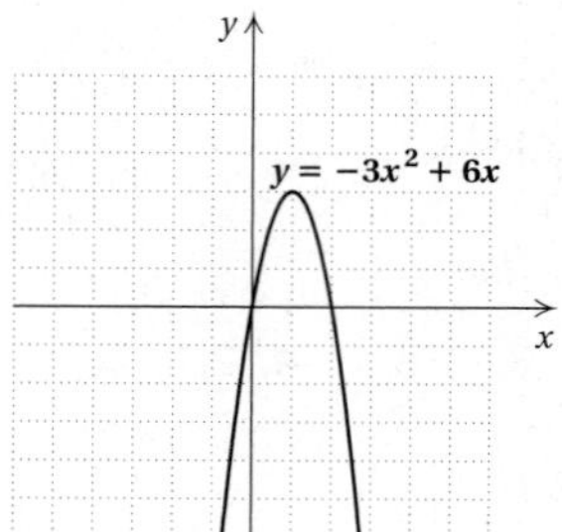

3. $(2, 0)$

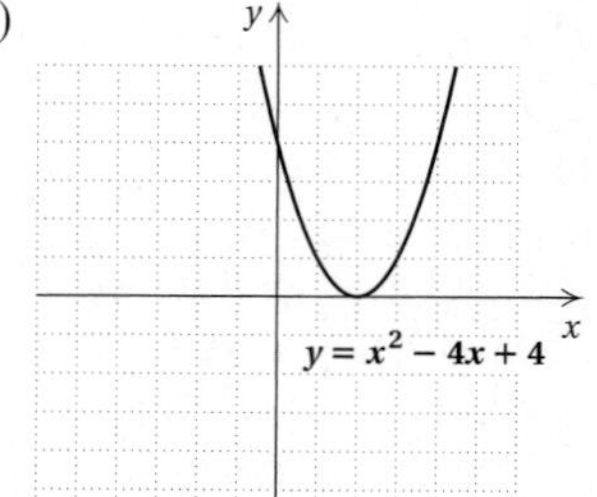

4. $(-\sqrt{3}, 0); (\sqrt{3}, 0)$

5. $(-4, 0); (-2, 0)$ **6.** $\left(\dfrac{-2 - \sqrt{6}}{2}, 0\right); \left(\dfrac{-2 + \sqrt{6}}{2}, 0\right)$
7. None

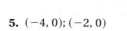

Calculator Corner, p. 742

Left to the student

Calculator Corner, p. 744

Left to the student

Exercise Set 10.6, p. 745

1.

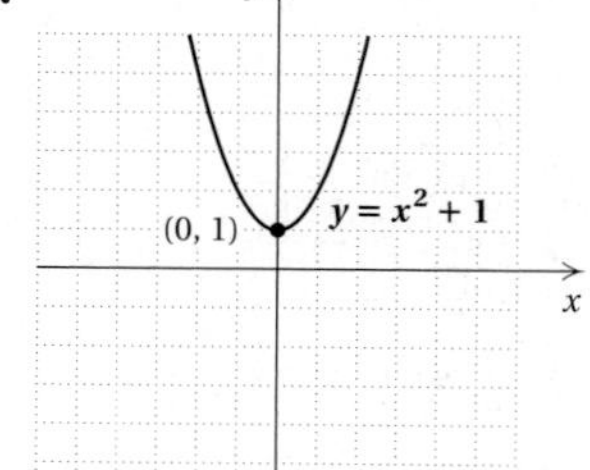

3.

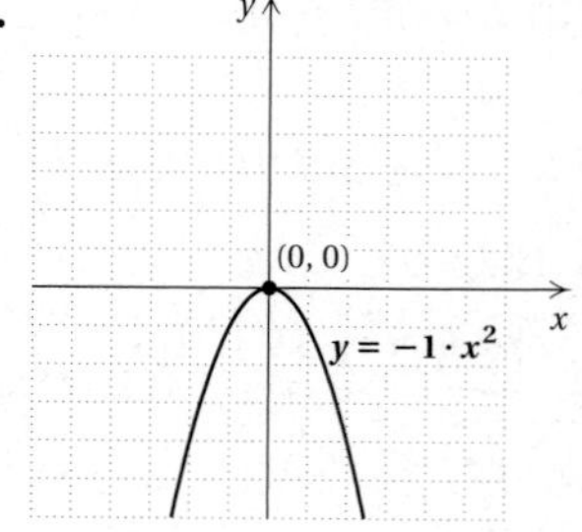

5.
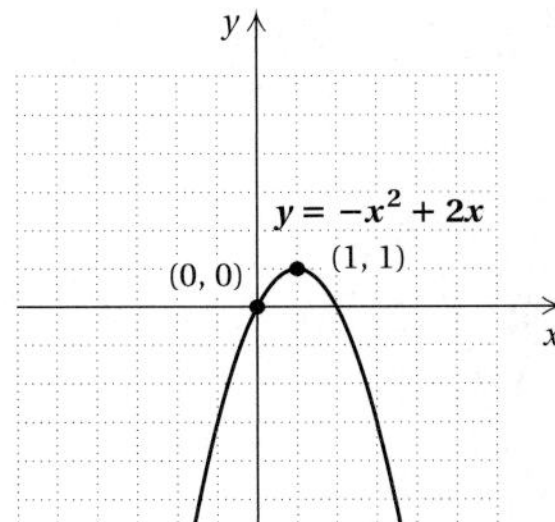

7.
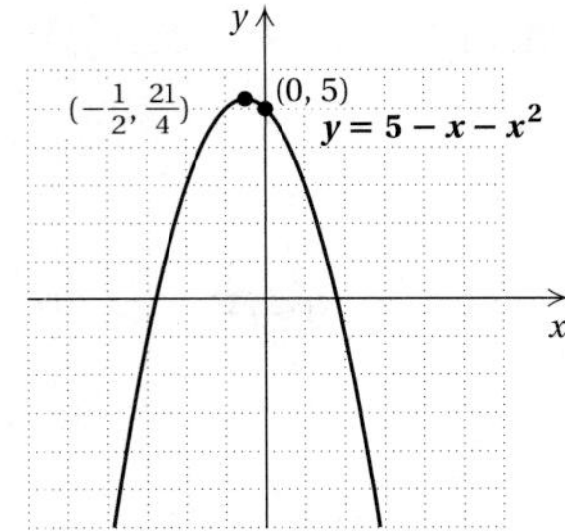

9.
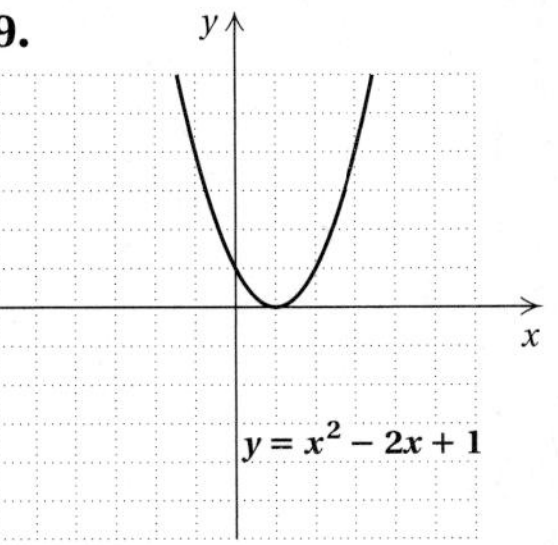

11.
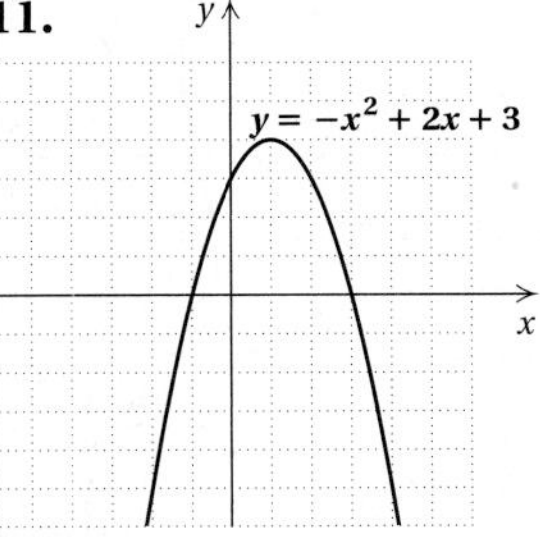

13.
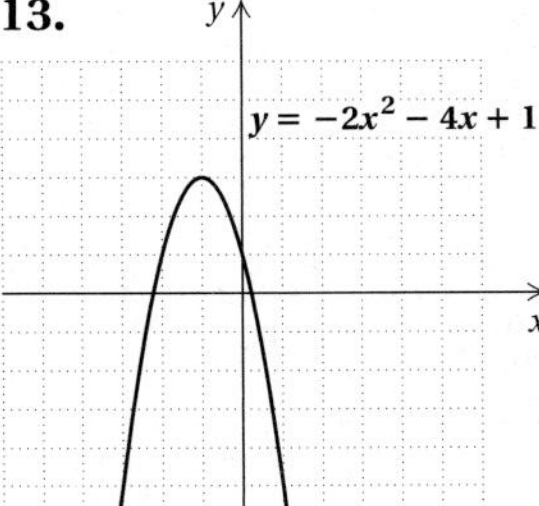

15.
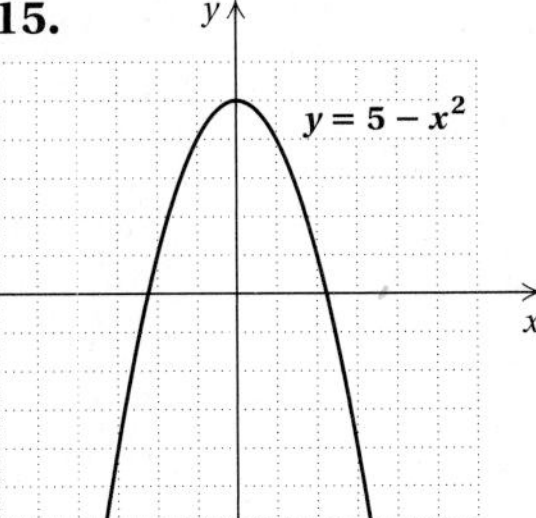

17.
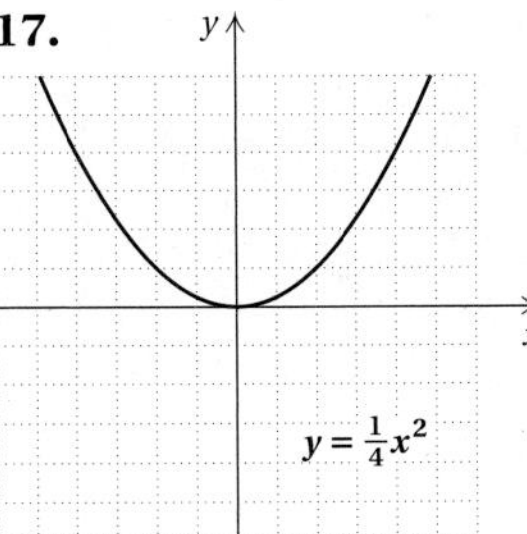

19.
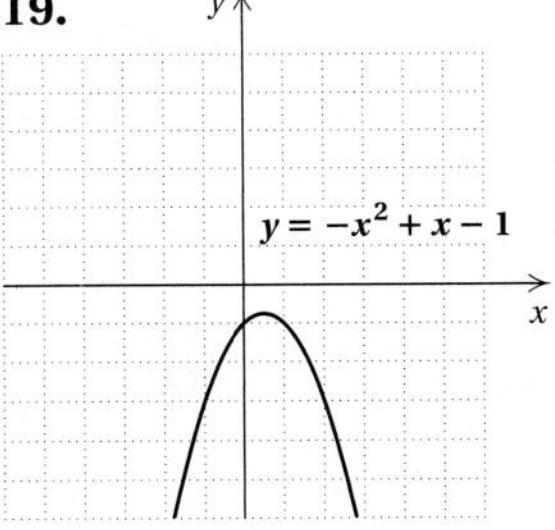

21.
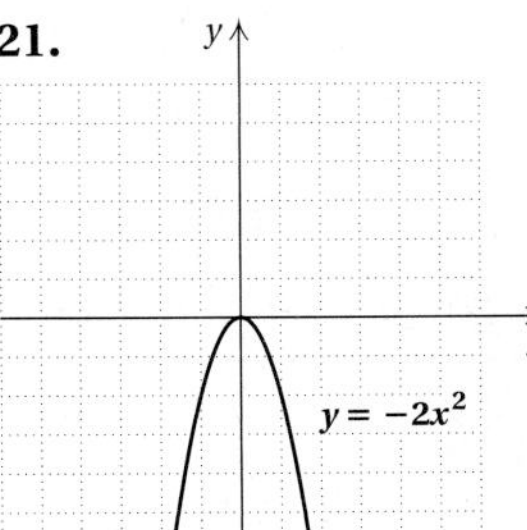

23.
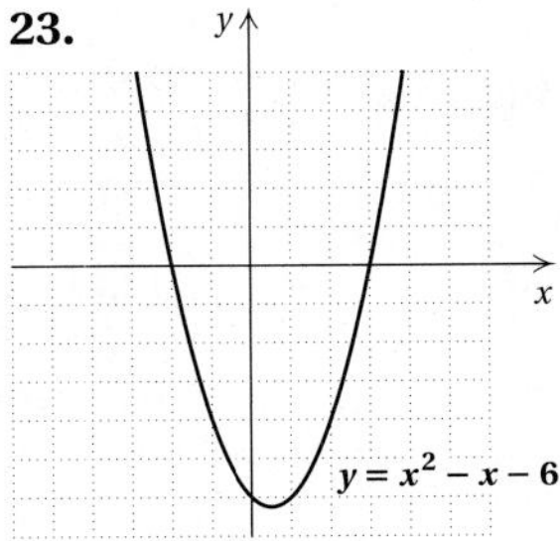

25. $(-\sqrt{2}, 0); (\sqrt{2}, 0)$ **27.** $(-5, 0); (0, 0)$
29. $\left(\dfrac{-1-\sqrt{33}}{2}, 0\right); \left(\dfrac{-1+\sqrt{33}}{2}, 0\right)$ **31.** $(3, 0)$
33. $(-2-\sqrt{5}, 0); (-2+\sqrt{5}, 0)$ **35.** None
37. DW **39.** $22\sqrt{2}$
40. $25y^2\sqrt{y}$ **41.** $y = \dfrac{29.76}{x}$

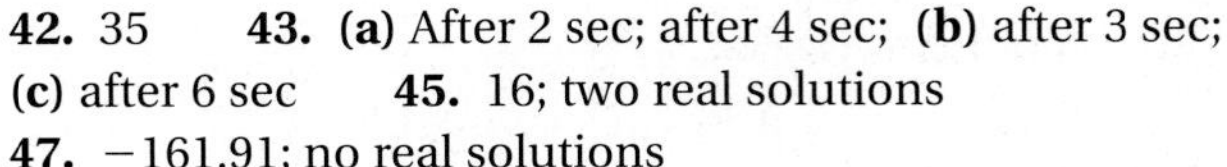

42. 35 **43.** **(a)** After 2 sec; after 4 sec; **(b)** after 3 sec; **(c)** after 6 sec **45.** 16; two real solutions
47. -161.91; no real solutions

Margin Exercises, Section 10.7, pp. 749–754

1. Yes **2.** No **3.** Yes **4.** No **5.** Yes **6.** Yes
7. **(a)** -33; **(b)** -3; **(c)** 2; **(d)** 97; **(e)** -9
8. **(a)** 21; **(b)** 9; **(c)** 14; **(d)** 69
9.
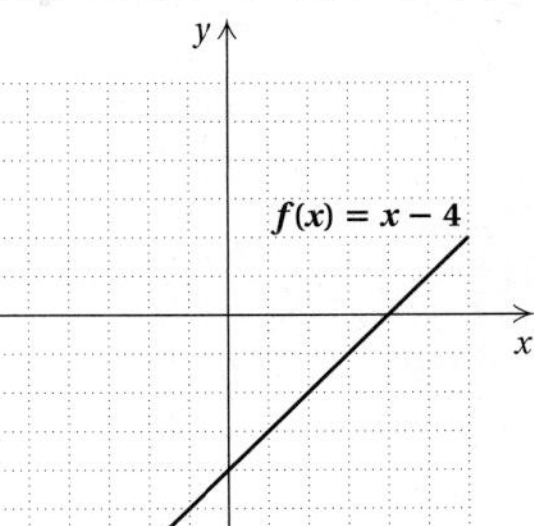

10.
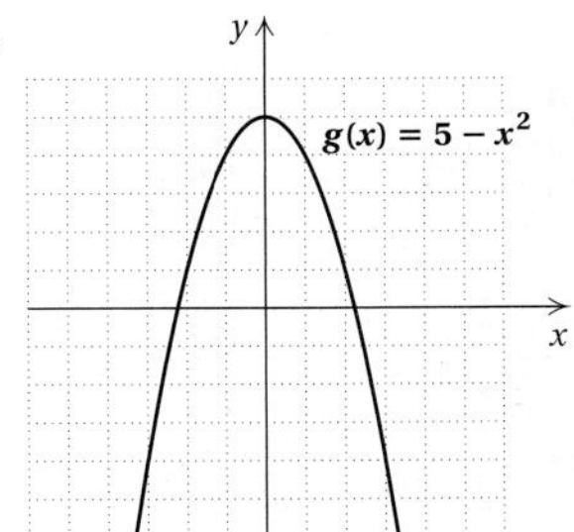

11.
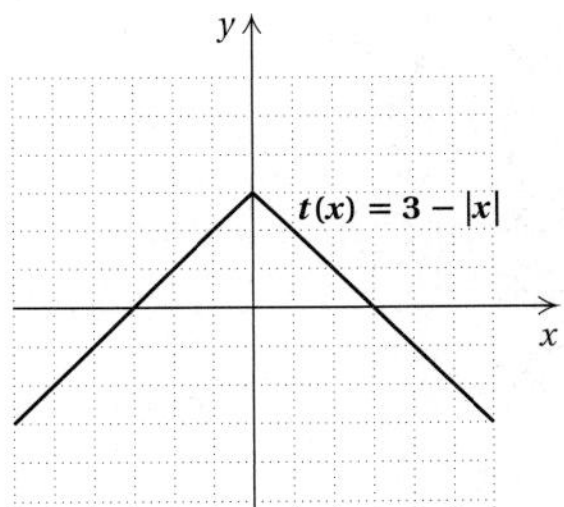

12. Yes **13.** No
14. No **15.** Yes **16.** About \$43 million
17. About \$6 million

Calculator Corner, p. 751

1. -12.8 **2.** -9.2 **3.** -2 **4.** 20

Exercise Set 10.7, p. 755

1. Yes **3.** Yes **5.** No **7.** Yes **9.** Yes **11.** Yes
13. A relation but not a function **15.** **(a)** 9; **(b)** 12; **(c)** 2; **(d)** 5; **(e)** 7.4; **(f)** $5\frac{2}{3}$ **17.** **(a)** -21; **(b)** 15; **(c)** 42; **(d)** 0; **(e)** 2; **(f)** -162.6 **19.** **(a)** 7; **(b)** -17; **(c)** 24.1; **(d)** 4; **(e)** -26; **(f)** 6 **21.** **(a)** 0; **(b)** 5; **(c)** 2; **(d)** 170; **(e)** 65; **(f)** 230 **23.** **(a)** 1; **(b)** 3; **(c)** 3; **(d)** 4; **(e)** 11; **(f)** 23 **25.** **(a)** 0; **(b)** -1; **(c)** 8; **(d)** 1000; **(e)** -125; **(f)** -1000 **27.** **(a)** 159.48 cm; **(b)** 153.98 cm
29. $1\frac{20}{33}$ atm; $1\frac{10}{11}$ atm; $4\frac{1}{33}$ atm **31.** 1.792 cm; 2.8 cm; 11.2 cm

33.

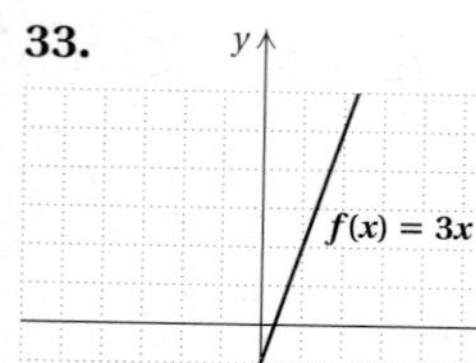

35.

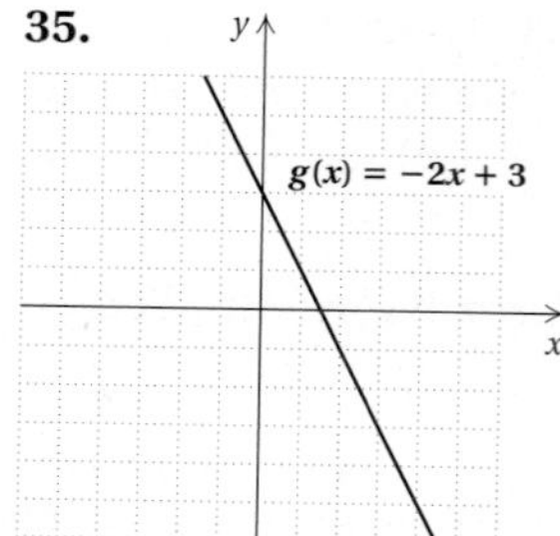

37.

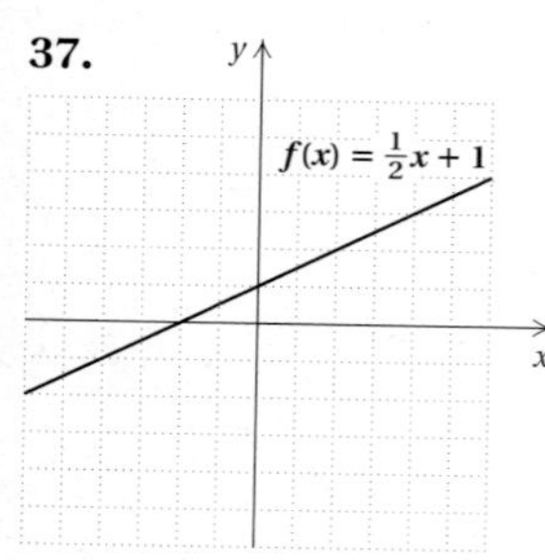

39.

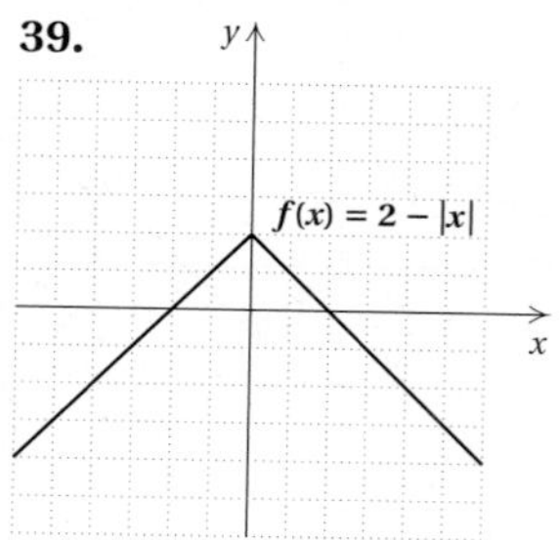

41.

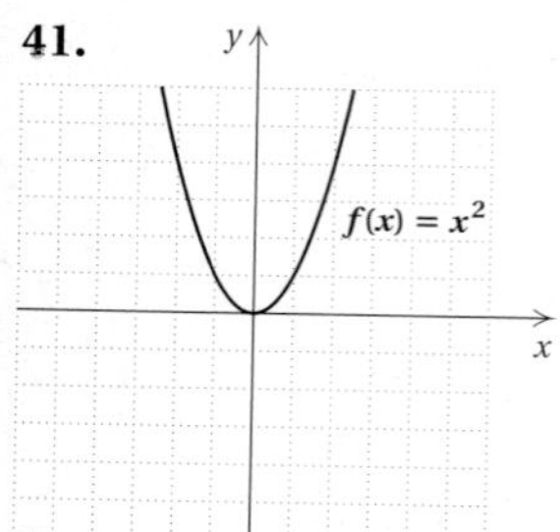

43.

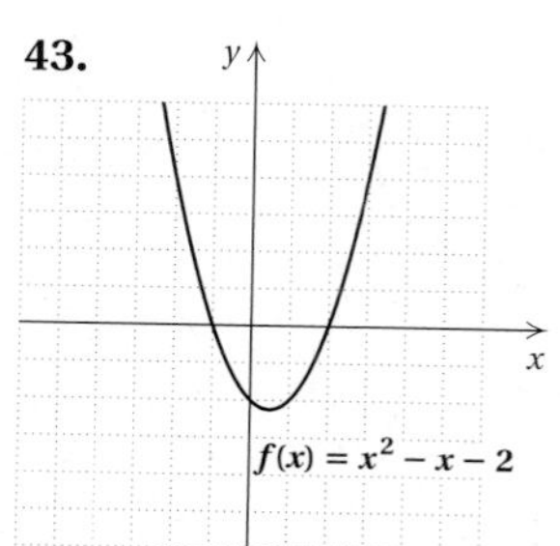

45. Yes **47.** Yes **49.** No **51.** No **53.** About 75 per 10,000 men **55.** DW **57.** No **58.** Yes **59.** No solution **60.** Infinite number of solutions

61.

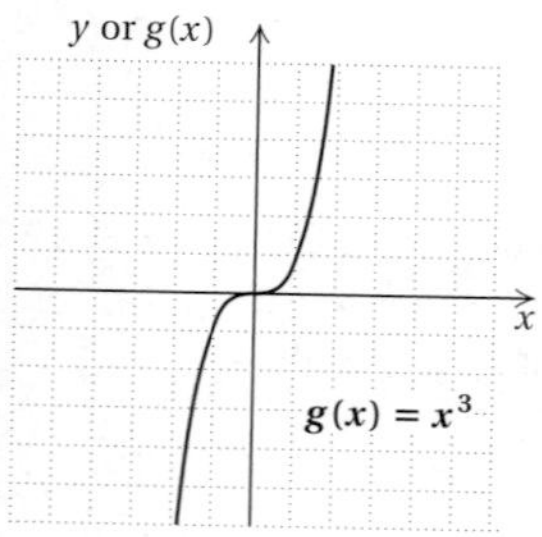

63.

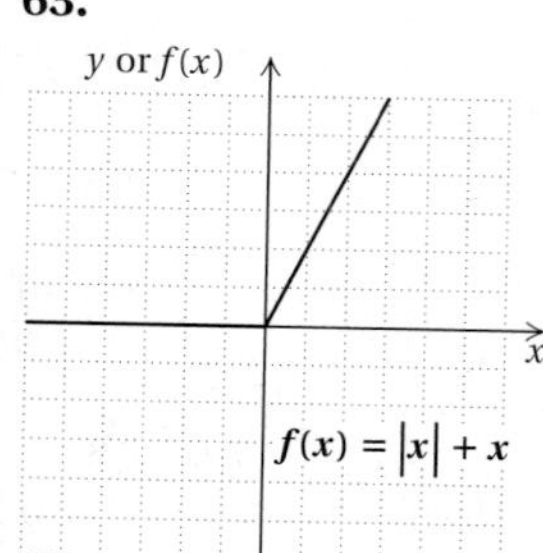

Summary and Review: Chapter 10, p. 759

1. $-\sqrt{3}, \sqrt{3}$ **2.** $-2\sqrt{2}, 2\sqrt{2}$ **3.** $\frac{3}{5}, 1$ **4.** $-2, \frac{1}{3}$ **5.** $-8 \pm \sqrt{13}$ **6.** 0 **7.** $0, \frac{7}{5}$ **8.** $1 \pm \sqrt{11}$ **9.** $\frac{1 \pm \sqrt{10}}{3}$ **10.** $-3 \pm 3\sqrt{2}$ **11.** $\frac{2 \pm \sqrt{3}}{2}$ **12.** $\frac{3 \pm \sqrt{33}}{2}$ **13.** No real-number solutions **14.** $0, \frac{4}{3}$ **15.** $-5, 3$ **16.** 1 **17.** $\frac{5 \pm \sqrt{17}}{2}$ **18.** $-1, \frac{5}{3}$ **19.** 0.4, 4.6 **20.** $-1.9, -0.1$

21. $T = L(4V^2 - 1)$ **22.**

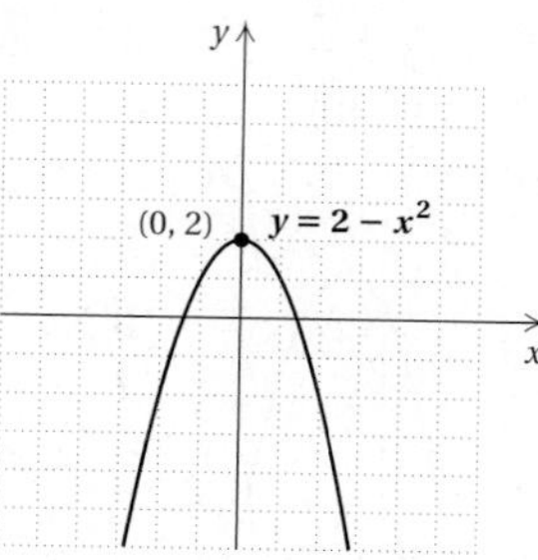

23.

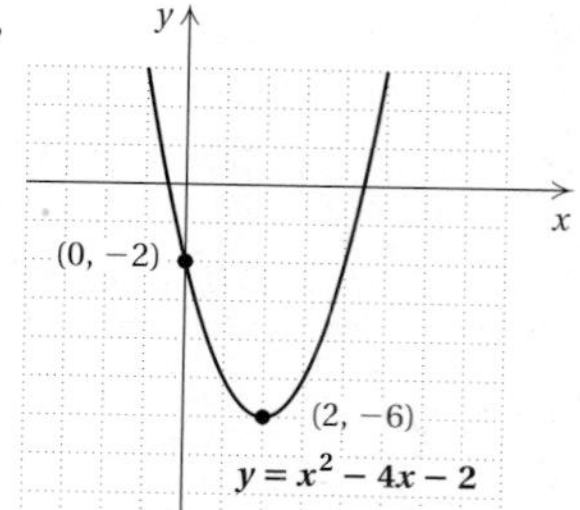

24. $(-\sqrt{2}, 0); (\sqrt{2}, 0)$

25. $(2 - \sqrt{6}, 0); (2 + \sqrt{6}, 0)$ **26.** 4.7 cm, 1.7 cm **27.** 10 yd, 24 yd **28.** About 6.3 sec **29.** $-1, -7, 2$ **30.** 0, 0, 19 **31.** 2700 calories

32.

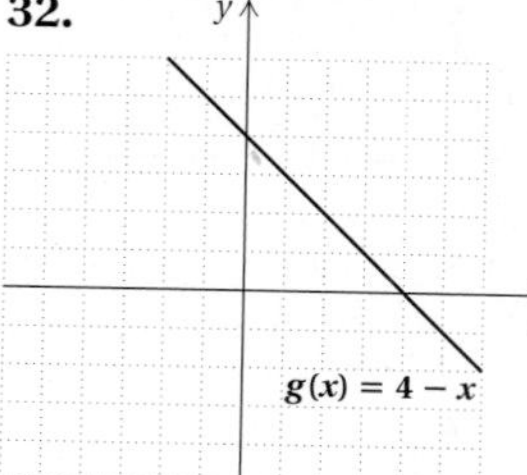

33.

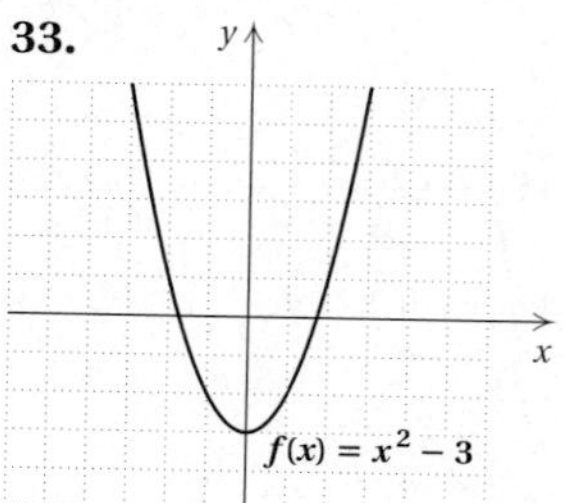

34.

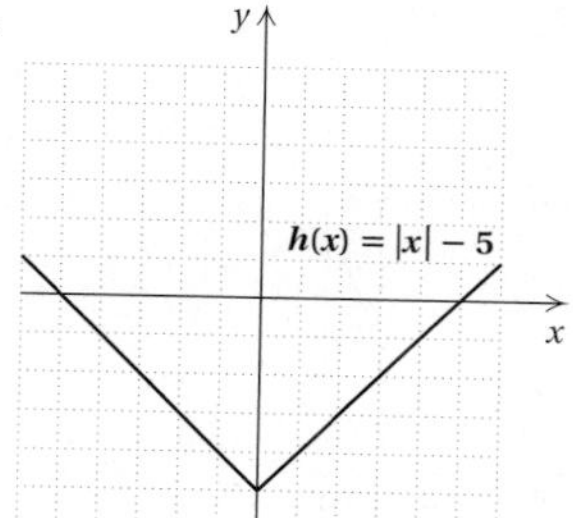

35. No **36.** Yes

37. DW

Equation	Form	Example
Linear	Reducible to $x = a$	$3x - 5 = 8$
Quadratic	$ax^2 + bx + c = 0$	$2x^2 - 3x + 1 = 0$
Rational	Contains one or more rational expressions	$\frac{x}{3} + \frac{4}{x-1} = 1$
Radical	Contains one or more radical expressions	$\sqrt{3x - 1} = x - 7$
Systems of equations	$Ax + By = C$, $Dx + Ey = F$	$4x - 5y = 3$, $3x + 2y = 1$

38. DW **(a)** The third line should be $x = 0$ *or* $x + 20 = 0$; the solution 0 gets lost in the given procedure. Also, the last line should be $x = -20$. **(b)** The addition principle should be used at the outset to get 0 on one side of the equation. Since this was not done in the given procedure, the principle of zero products was not applied correctly.

39. $6\sqrt{a}$ **40.** $2xy\sqrt{15y}$ **41.** $y = \frac{0.625}{x}$; 0.003125
42. $\sqrt{3}$ **43.** $12\sqrt{11}$ **44.** $4\sqrt{10}$
45. 31 and 32; -32 and -31 **46.** $5\sqrt{\pi}$ in., or about 8.9 in.
47. 25 **48.** $-4, -2$ **49.** $-5, -1$ **50.** $-6, 0$
51. -3

Test: Chapter 10, p. 762

1. [10.2a] $-\sqrt{5}, \sqrt{5}$ **2.** [10.1b] $-\frac{8}{7}, 0$
3. [10.1c] $-8, 6$ **4.** [10.1c] $-\frac{1}{3}, 2$ **5.** [10.2b] $8 \pm \sqrt{13}$
6. [10.3a] $\frac{1 \pm \sqrt{13}}{2}$ **7.** [10.3a] $\frac{3 \pm \sqrt{37}}{2}$
8. [10.3a] $-2 \pm \sqrt{14}$ **9.** [10.3a] $\frac{7 \pm \sqrt{37}}{6}$
10. [10.1c] $-1, 2$ **11.** [10.1c] $-4, 2$
12. [10.2c] $2 \pm \sqrt{14}$ **13.** [10.3b] $-1.7, 5.7$
14. [10.4a] $n = \frac{-b + \sqrt{b^2 + 4ad}}{2a}$
15. [10.6b] $\left(\frac{1 - \sqrt{21}}{2}, 0\right), \left(\frac{1 + \sqrt{21}}{2}, 0\right)$
16. [10.6a]

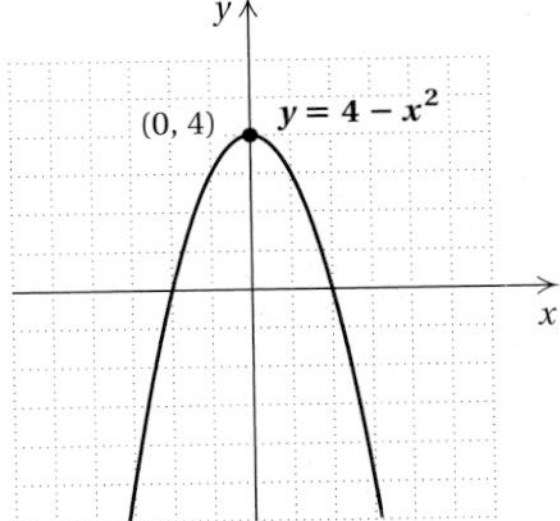

17. [10.6a]

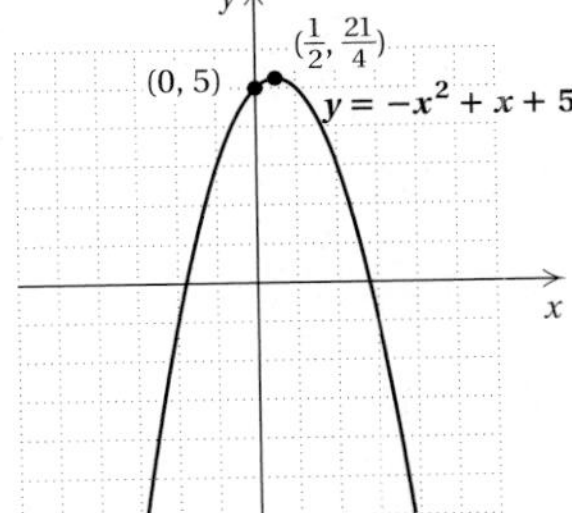

18. [10.7b] 1; $1\frac{1}{2}$; 2 **19.** [10.7b] 1; 3; -3
20. [10.5a] Length: 6.5 m; width: 2.5 m
21. [10.5a] 24 km/h **22.** [10.7e] 25.98 min
23. [10.7c]

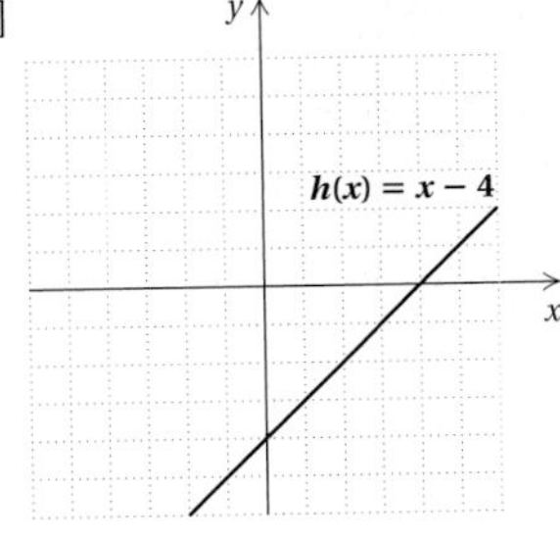

24. [10.7c]

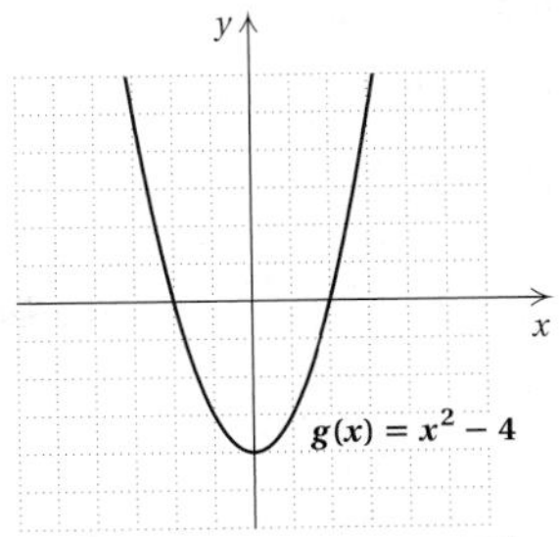

25. [10.7d] Yes
26. [10.7d] No **27.** [9.4a] $2\sqrt{15}$ **28.** [9.2c] $7xy\sqrt{2x}$
29. [7.5c] $y = \frac{4}{x}$; 0.25 **30.** [9.6b] $\sqrt{5}$
31. [10.5a] $5 + 5\sqrt{2}$ **32.** [8.2b], [10.3a] $1 \pm \sqrt{5}$

Cumulative Review/Final Examination: Chapters 1–10, p. 764

1. [4.1a] $x \cdot x \cdot x$ **2.** [4.1c] 54 **3.** [1.2c] $-0.\overline{27}$
4. [6.3a] 240 **5.** [1.2e] 7 **6.** [1.3a] 9 **7.** [1.4a] 15
8. [1.6c] $-\frac{3}{20}$ **9.** [1.8d] 4 **10.** [1.8b] $-2m - 4$
11. [2.2a] -8 **12.** [2.3b] -12 **13.** [2.3c] 7
14. [5.7b] 3, 5 **15.** [8.2a] (1, 2) **16.** [8.3a] (12, 5)
17. [8.3b] (6, 7) **18.** [5.7b] $-2, 3$
19. [10.3a] $\frac{-3 \pm \sqrt{29}}{2}$ **20.** [9.5a] 2
21. [2.7e] $\{x \mid x \geq -1\}$ **22.** [2.3b] 8 **23.** [2.3b] -3
24. [2.7d] $\{x \mid x < -8\}$ **25.** [2.7e] $\{y \mid y \leq \frac{35}{22}\}$
26. [10.2a] $-\sqrt{10}, \sqrt{10}$ **27.** [10.2b] $3 \pm \sqrt{6}$
28. [6.6a] $\frac{2}{9}$ **29.** [6.6a] -5
30. [6.6a], [10.1b] No solution **31.** [9.5a] 12
32. [2.4b] $b = \frac{At}{4}$ **33.** [10.4a] $m = \frac{tn}{t + n}$
34. [10.4a] $A = \pi r^2$ **35.** [10.4a] $x = \frac{b + \sqrt{b^2 + 4ay}}{2a}$
36. [4.1d, f] $\frac{1}{x^4}$ **37.** [4.1e, f] y^7 **38.** [4.2a, b] $4y^{12}$
39. [4.3f] $10x^3 + 3x - 3$ **40.** [4.4a] $7x^3 - 2x^2 + 4x - 17$
41. [4.4c] $8x^2 - 4x - 6$ **42.** [4.5b] $-8y^4 + 6y^3 - 2y^2$
43. [4.5d] $6t^3 - 17t^2 + 16t - 6$ **44.** [4.6b] $t^2 - \frac{1}{16}$
45. [4.6c] $9m^2 - 12m + 4$
46. [4.7e] $15x^2y^3 + x^2y^2 + 5xy^2 + 7$
47. [4.7f] $x^4 - 0.04y^2$ **48.** [4.7f] $9p^2 + 24pq^2 + 16q^4$
49. [6.1d] $\frac{2}{x + 3}$ **50.** [6.2b] $\frac{3a(a - 1)}{2(a + 1)}$
51. [6.4a] $\frac{27x - 4}{5x(3x - 1)}$ **52.** [6.5a] $\frac{-x^2 + x + 2}{(x + 4)(x - 4)(x - 5)}$
53. [5.1b] $4x(2x - 1)$ **54.** [5.5d] $(5x - 2)(5x + 2)$
55. [5.3a], [5.4a] $(3y + 2)(2y - 3)$ **56.** [5.5b] $(m - 4)^2$
57. [5.1c] $(x^2 - 5)(x - 8)$
58. [5.6a] $3(a^2 + 6)(a + 2)(a - 2)$
59. [5.5d] $(4x^2 + 1)(2x + 1)(2x - 1)$
60. [5.5d] $(7ab + 2)(7ab - 2)$ **61.** [5.5b] $(3x + 5y)^2$
62. [5.1c] $(2a + d)(c - 3b)$
63. [5.3a], [5.4a] $(5x - 2y)(3x + 4y)$
64. [6.8a] $-\frac{42}{5}$ **65.** [9.1a] 7 **66.** [9.1a] -25

67. [9.1f] $8x$ **68.** [9.2c] $\sqrt{a^2 - b^2}$
69. [9.2c] $8a^2b\sqrt{3ab}$ **70.** [9.2a] $5\sqrt{6}$
71. [9.2b] $9xy\sqrt{3x}$ **72.** [9.3b] $\frac{10}{9}$ **73.** [9.3b] $\frac{8}{x}$
74. [9.4a] $12\sqrt{3}$ **75.** [9.3a, c] $\frac{2\sqrt{10}}{5}$ **76.** [9.6a] 40
77. [3.2b]

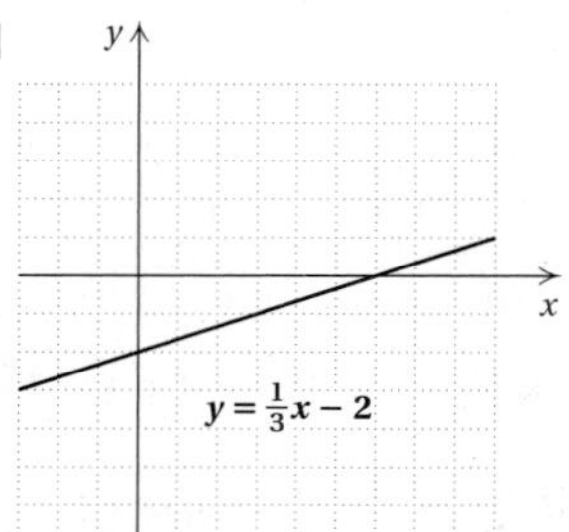

78. [3.3a]

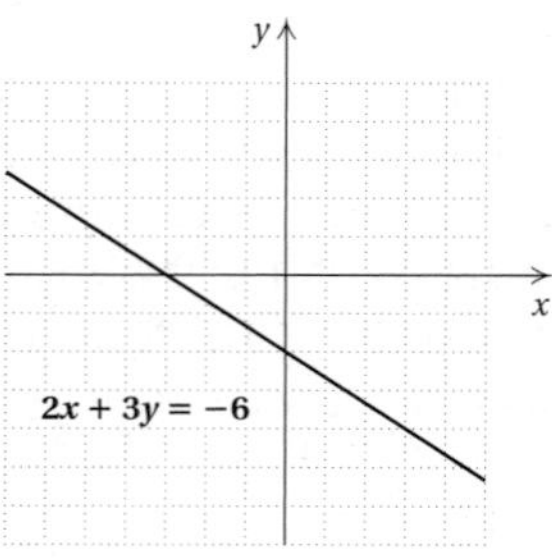

79. [3.3b]

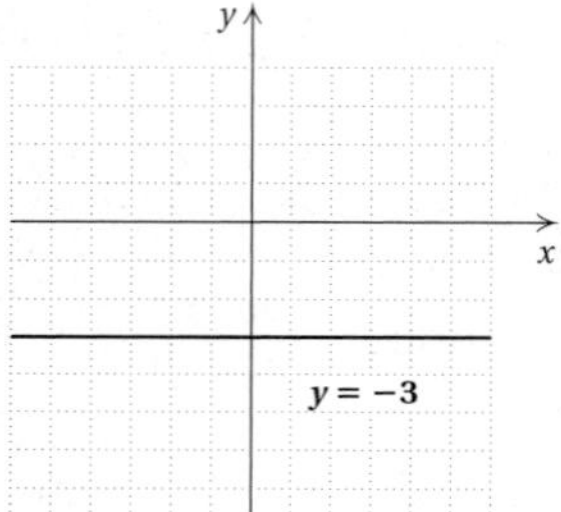

80. [7.4b]

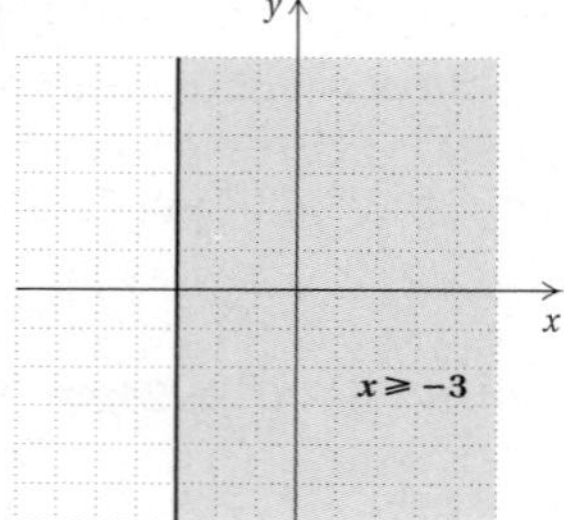

81. [7.4b]

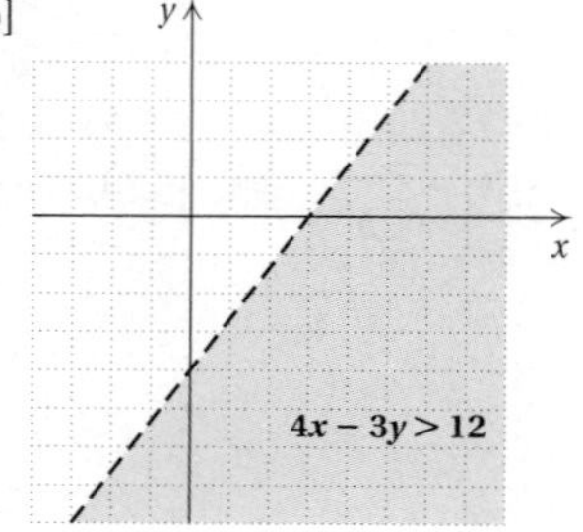

82. [10.6a]

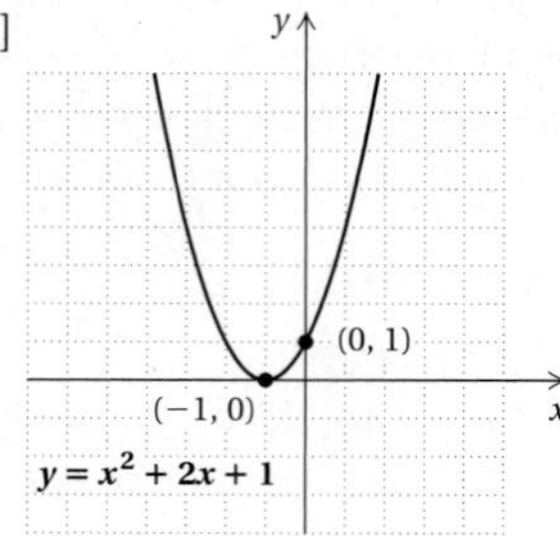

83. [10.2c] $\frac{2 \pm \sqrt{6}}{3}$

84. [10.3b] −0.2, 1.2 **85.** [2.5a] 25% **86.** [2.5a] 60
87. [6.7a] $4\frac{4}{9}$ hr **88.** [10.5a] Length: 12 ft; width: 8 ft
89. [10.5a] 2 km/h **90.** [5.8a] 12 m
91. [10.5a] 5 and 7; −7 and −5
92. [8.4a] 40 L of A; 20 L of B **93.** [10.2d] About 7.8 sec
94. [7.5b] $451.20; the variation constant is the amount earned per hour **95.** [2.6a] 6090 cars
96. [8.4a] $3.30 per pound: 14 lb; $2.40 per pound: 28 lb
97. [8.5a] 70 km/h **98.** [10.6b] −3, 2
99. [10.6b] $(-2 - \sqrt{3}, 0), (-2 + \sqrt{3}, 0)$
100. [7.1a] Slope: 2; y-intercept: (0, −8)
101. [7.3a, b] Neither **102.** [3.4a] 15
103. [7.5a] $y = 10x$; 640 **104.** [7.5c] $y = \frac{1000}{x}$; 8
105. [10.7d] Yes **106.** [10.7d] No
107. [10.7c]

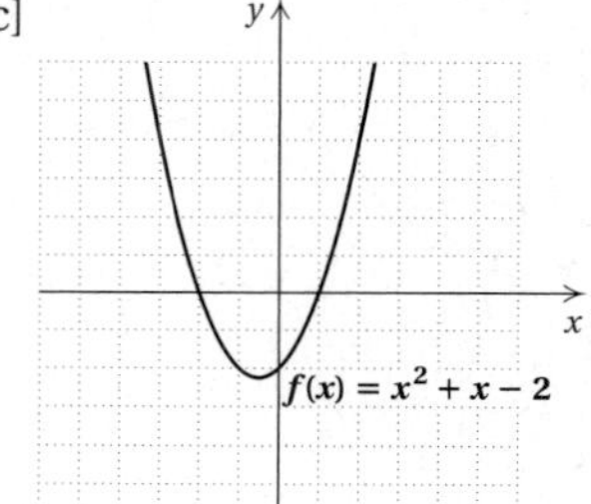

108. [10.7c]

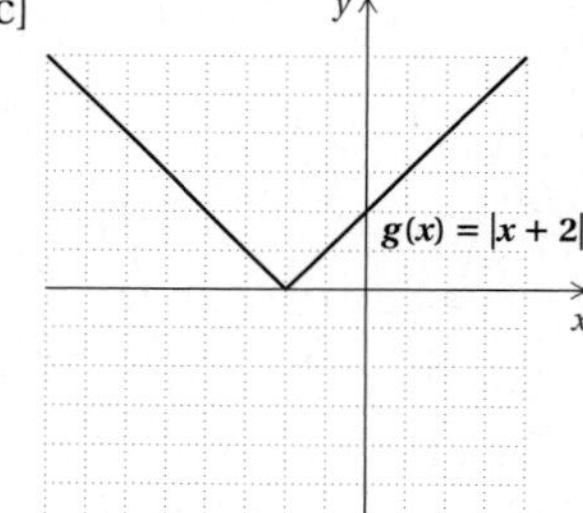

109. [10.7b] −4; 0; 0 **110.** [6.7a], [10.5a] (c)
111. [10.3a] (d) **112.** [10.4a] (d) **113.** [3.3a] (c)
114. [1.2e] −12, 12 **115.** [9.1a] $\sqrt{3}$
116. [10.2c] −30, 30 **117.** [9.6a] $\frac{\sqrt{6}}{3}$ **118.** [5.5d] Yes
119. [6.1c] No **120.** [4.6c] No **121.** [5.5a], [9.2a] No
122. [9.1f] Yes

Appendixes

Margin Exercises, Appendix A, pp. 772–773

1. $(x-3)(x^2+3x+9)$ **2.** $(4-y)(16+4y+y^2)$ **3.** $(y+2)(y^2-2y+4)$ **4.** $(5+t)(25-5t+t^2)$ **5.** $(3x-y)(9x^2+3xy+y^2)$ **6.** $(2y+z)(4y^2-2yz+z^2)$ **7.** $(m+n)(m^2-mn+n^2)(m-n)(m^2+mn+n^2)$ **8.** $2xy(2x^2+3y^2)(4x^4-6x^2y^2+9y^4)$ **9.** $(3x+2y)(9x^2-6xy+4y^2)(3x-2y)(9x^2+6xy+4y^2)$ **10.** $(x-0.3)(x^2+0.3x+0.09)$

Exercise Set A, p. 774

1. $(z+3)(z^2-3z+9)$ **3.** $(x-1)(x^2+x+1)$ **5.** $(y+5)(y^2-5y+25)$ **7.** $(2a+1)(4a^2-2a+1)$ **9.** $(y-2)(y^2+2y+4)$ **11.** $(2-3b)(4+6b+9b^2)$ **13.** $(4y+1)(16y^2-4y+1)$ **15.** $(2x+3)(4x^2-6x+9)$ **17.** $(a-b)(a^2+ab+b^2)$ **19.** $\left(a+\frac{1}{2}\right)\left(a^2-\frac{1}{2}a+\frac{1}{4}\right)$ **21.** $2(y-4)(y^2+4y+16)$ **23.** $3(2a+1)(4a^2-2a+1)$ **25.** $r(s+4)(s^2-4s+16)$ **27.** $5(x-2z)(x^2+2xz+4z^2)$ **29.** $(x+0.1)(x^2-0.1x+0.01)$ **31.** $8(2x^2-t^2)(4x^4+2x^2t^2+t^4)$ **33.** $2y(y-4)(y^2+4y+16)$ **35.** $(z-1)(z^2+z+1)(z+1)(z^2-z+1)$ **37.** $(t^2+4y^2)(t^4-4t^2y^2+16y^4)$ **39.** 1; 19; 19; 7; 1 **41.** $(x^{2a}+y^b)(x^{4a}-x^{2a}y^b+y^{2b})$ **43.** $3(x^a+2y^b)(x^{2a}-2x^ay^b+4y^{2b})$ **45.** $\frac{1}{3}\left(\frac{1}{2}xy+z\right)\left(\frac{1}{4}x^2y^2-\frac{1}{2}xyz+z^2\right)$ **47.** $y(3x^2+3xy+y^2)$ **49.** $4(3a^2+4)$

Margin Exercises, Appendix B, pp. 776–777

1. 3 **2.** -2 **3.** 6 **4.** 1 **5.** -1 **6.** Not a real number **7.** 3 **8.** -6 **9.** -6 **10.** $2\sqrt[3]{3}$ **11.** $\frac{3}{4}$ **12.** $2\sqrt[5]{3}$ **13.** $\frac{\sqrt[3]{4}}{5}$

Exercise Set B, p. 778

1. 5 **3.** -10 **5.** 1 **7.** Not a real number **9.** 6 **11.** 4 **13.** 10 **15.** -3 **17.** Not a real number **19.** -5 **21.** t **23.** $-x$ **25.** 4 **27.** -7 **29.** -5 **31.** 10 **33.** 10 **35.** -7 **37.** Not a real number **39.** 5 **41.** $3\sqrt[3]{2}$ **43.** $3\sqrt[4]{4}$ **45.** $\frac{3}{4}$ **47.** $4\sqrt[4]{2}$ **49.** $2\sqrt[5]{4}$ **51.** $\frac{4}{5}$ **53.** $\frac{\sqrt[3]{17}}{2}$ **55.** $5\sqrt[3]{2}$ **57.** $3\sqrt[5]{2}$ **59.** $\frac{\sqrt[4]{13}}{3}$ **61.** $\frac{\sqrt[4]{7}}{2}$ **63.** $\frac{2}{5}$ **65.** 2 **67.** 10

Margin Exercises, Appendix C, pp. 780–781

1. {0, 1, 2, 3, 4, 5, 6, 7} **2.** {−5, 5} **3.** True **4.** True **5.** True **6.** False **7.** True **8.** {−2, −3, 4, −4} **9.** {a, i} **10.** { }, or ∅ **11.** {−2, −3, 4, −4, 8, −5, 7, 5} **12.** {a, e, i, o, u, m, r, v, n} **13.** {a, b, c, d, e, f, g, h, i, j, k, l, m, n, o, p, q, r, s, t, u, v, w, x, y, z}

Exercise Set C, p. 782

1. {3, 4, 5, 6, 7, 8} **3.** {41, 43, 45, 47, 49} **5.** {−3, 3} **7.** False **9.** True **11.** True **13.** True **15.** True **17.** False **19.** {c, d, e} **21.** {1, 10} **23.** { }, or ∅ **25.** {a, e, i, o, u, q, c, k} **27.** {0, 1, 7, 10, 2, 5} **29.** {a, e, i, o, u, m, n, f, g, h} **31.** $\{x \mid x$ is an integer$\}$ **33.** $\{x \mid x$ is a real number$\}$ **35.** { }, or ∅ **37.** **(a)** A; **(b)** A; **(c)** A; **(d)** { }, or ∅ **39.** True

Margin Exercises, Appendix D, pp. 784–785

1. 56.7 **2.** 64.7 **3.** 87.8 **4.** 17 **5.** 16.5 **6.** 91 **7.** 55 **8.** 54, 87 **9.** No mode exists. **10.** **(a)** 25 mm^2; **(b)** 25 mm^2; **(c)** No mode exists.

Exercise Set D, p. 786

1. Mean: 21; median: 18.5; mode: 29 **3.** Mean: 21; median: 20; modes: 5, 20 **5.** Mean: 5.2; median: 5.7; mode: 7.4 **7.** Mean: 239.5; median: 234; mode: 234 **9.** Mean: $23.\bar{8}$; median: 15; mode: 1 **11.** Mean: 897.2; median: 798; no mode exists **13.** Mean: \$8.19; median: \$8.49; mode: \$6.99 **15.** 2.7 **17.** 10 home runs **19.** 263 days

Index

Photo Credits

31, © Paul Barton/Corbisstockmarket **32,** © ML Sinibaldi/corbisstockmarket **33 (top),** © Antonio N. Rosario/The Image Bank **33 (bottom),** Harry How/Allsport **56 (top),** © Chuck Savage/Corbis Stock Market **56 (bottom),** © Jim Cummins/Corbis Stock Market **65,** © Dale O'Dell/Corbis Stock Market **79 (left),** © Michel Tcherevkoff Ltd/The Image Bank **79 (right),** David Cannon/Allsport **88,** © Ruth Dixon/Stock Boston **96,** AP/Wide World Photos **97,** John W. Banagan/The Image Bank **138,** © Comstock Images **144,** Jack Hollingsworth, PhotoDisc **179,** Sharp Electronics Corporation **180 (left),** FPG International **180 (right),** Compaq **183 (top),** Doug Meneuz, PhotoDisc **183 (bottom),** Brian Spurlock **187,** courtesy of SUBWAY® **191,** AP/Wide World Photos **193,** Ryan McVay, PhotoDisc **195,** EyeWire Collection **196,** © Kevin Horan, Tony Stone **198,** EyeWire Collection **213,** Christine D. Tuff **216,** Tom Tracy/FPG International **222,** © Lawrence Migdale, Tony Stone **226,** © ML Sinibaldi/The Stock Market **231,** EyeWire Collection **271,** The Image Bank **277 (left),** Mary Clay/Tom Stack and Associates **277 (right),** Terje Rakke/The Image Bank **283,** Federal Highway Administration and Washington Infrastructure Services, Inc. **284,** Copyright © by Klev Schoening **307 (left),** © Dave G. Houser/CORBIS **307 (middle),** FPG International **307 (right),** EyeWire Collection **314,** Jamie Squire/Allsport **315 (right),** Hoby Finn, PhotoDisc **316,** Bob Daemmrich/The Image Works **327,** Ron Chapple, FPG International **357,** Brian Spurlock **377,** AP/Wide World Photos **382,** Karen Bittinger **413,** FPG International **441,** Karen Bittinger **443,** Chris Salvo, FPG International **446,** Judy Gelles/Stock Boston **448 (top),** Duncan Smith/PhotoDisc **448 (bottom), 511,** EyeWire Collection **515,** American Honda **516 (left),** © John Warden/Index Stock Image **516 (right),** © 1990 Glenn Randall **521,** PhotoLink/PhotoDisc **522 (left),** Geostock/PhotoDisc **522 (right),** Tom Turpin, Purdue University **524,** Joseph Sohm, ChromoSohn Inc./CORBIS **533,** Hisham E. Ibrahim, Photodisc **534 (top),** Steve Mason, PhotoDisc **534 (bottom),** PhotoLink/PhotoDisc **535 (left),** Sergio Carmona/CORBIS **535 (right),** Robert Holmes/CORBIS **574,** Mark E. Gibson at CLM/CORBIS OUTLINE **576 (left),** © W. Cody/CORBIS **576 (right),** © Norbert Schafer/Corbis Stock Market **585 (left),** Joseph Sohm, ChromoSohn, Inc./CORBIS **585 (right),** AFP Photo/Jeff Haynes, CORBIS **590,** Dave G. Houser/CORBIS **615,** © Bob Krist/CORBIS **620,** Keith Brofsky/PhotoDisc **622 (top),** © Archive Photos **622 (right),** © Wildlife Conservation Society headquartered at the Bronx Zoo **624 (top),** Greg Kuchik/PhotoDisc **624 (bottom),** Digital Vision/PictureQuest **633,** Bettmann/CORBIS **635,** © Ghislain and Marie David de Lossy/The Image Bank **640 (left),** Victoria Pearson, FPG International **640 (right),** Jeremy Woodhouse, PhotoDisc **650,** Andre Lichenberg/FPG International **651,** © AFB Photo, John G. Mabanglo, AFP/CORBIS **660,** Doug Meneuz/PhotoDisc **685,** © Galen Rowell/CORBIS **737,** Robert Severi/Gamma Liaison **740,** AP/Wide World Photos **754,** Doug Meneuz/PhotoDisc **760,** © Joseph Sohm, ChromoSohn Inc./CORBIS **784,** AP/Wide World Photos **786,** © Corbis Images **787,** PBA

Index of Study Tips

Statistics and Demographics

Transportation

Geometry

Physical Science

Sports, Leisure, and Entertainment

Index of Applications

Biology, Health, and Life Science

Business and Economics